The ELECTRIC POWER ENGINEERING Series
series editor Leo Grigsby

Electromechanical Systems, Electric Machines, AND Applied Mechatronics

Sergey E. Lyshevski

CRC Press
Boca Raton London New York Washington, D.C.

Library of Congress Cataloging-in-Publication Data

Lyshevski, Sergey Edward.
Electromechanical systems, electric machines, and applied mechatronics / Sergey Edward Lyshevski.
p. cm. -- (Electric power engineering series)
Includes bibliographical references.
ISBN 0-8493-2275-8 (alk. paper)
1. Power electronics. 2. Mechatronics. 3. Microelectromechanical systems. I. Title. II. Series.
TK7881.15.L97 1999
621.3 —dc21 99-15197
CIP

Visit the CRC Press Web site at www.crcpress.com

International Standard Book Number 0-8493-2275.8
Library of Congress Card Number 99-15197
Printed in the United States of America 3 4 5 6 7 8 9 0
Printed on acid-free paper

Dedication

To my family

Preface

The author invites you, the reader, to embark on an exciting journey in analysis, design, and implementation of advanced state-of-the-art electromechanical systems. Electric machines (AC and DC motors and generators), power electronic devices and power converters, transducers and sensors, microprocessors, and PC-DSP hardware will be thoroughly covered. Recent trends in engineering have increased the emphasis on integrated analysis, design, and control of advanced electromechanical systems. The scope of electromechanical systems has continued to expand and, in addition to actuators, includes power electronics and sensors, microprocessors, and digital signal processors (DSP), etc. Although, the basic foundations have been developed, some urgent areas have been downgraded, less emphasized and covered. To guarantee an eventual consensus and ensure descriptive features, extend the results to the modern software and hardware developments of recent years, acquire engineering background and include a wide variety of interdisciplinary areas, link and place the integrated perspectives of electromechanical systems and power electronics in the engineering curriculum in favor of a common structure requested, it is essential to introduce the *mechatronic* concept. The author feels quite strongly that the study of electromechanical systems should be considered as the unified cornerstone of the engineering curriculum. The unified study of actuators and power electronics, microprocessors and DSP hardware, advanced software applied in motion control and analysis, transducers, and sensors have barely been introduced in the engineering curriculum. Integrated features are coming quickly, and *mechatronics* takes place. *Mechatronics* deals with emerging engineering technologies and challenging problems that have not been considered previously. This area reflects obvious curriculum changes in response to engineering enterprise and entreaties of steady evolutionary industrial demands to fulfill the needs. A great majority of engineers and scholars encounter *mechatronics* in daily activity; and in view of the enormous impact that electromechanical and computer-software engineering have had in recent years, this book is written. A simple answer does not exist to the question, "What is *mechatronics*?" *Mechatronics* can be viewed as an integrated comprehensive study of electromechanical motion devices and mechanisms or processes actuated, motion control through the use of advanced microprocessors and DSP hardware, power electronics and sensors, fully featured design and virtual prototyping of electromechanical systems. *Mechatronics,* as a breakthrough concept in the design and analysis of electromechanical systems, was introduced to attack, integrate, and solve a great variety of emerging problems in engineering. The great power and versatility of *mechatronics,* not to mention the prime importance in all areas of engineering of the results it approaches, make it worthwhile for all engineers to be acquainted with the basic theoretical standpoints and practical applications. There is no end to the application of *mechatronics* (which can be called the endless frontier) and to the further contribution to this interdisciplinary concept. It seems that we have just skimmed the surface of *mechatronics;* new trends will be accomplished, discovered, and researched in the near future. It is gratifying to see that this program is being enthusiastically explored by a variety of institutions, ranging from educationally oriented schools to research-oriented universities.

This book is a unique combination of traditional engineering topics and the latest technologies and developments to integrate and stimulate new advances in analysis and design of advanced state-of-the-art electromechanical systems. The major goal of this text is to ensure a deep understanding of the engineering underpinnings of integrated technologies and develop the modern picture of electromechanical engineering using the bedrock fundamentals. It is evident why this book was written. The demand for an educational textbook in electromechanical systems and *mechatronics* far exceeds what was previously anticipated by the academia, industry, and the engineering community. While excellent textbooks in electric machinery (Fitzgerald, Kingsley, and Umans, 1990; Krause, Wasynczuk, and Sudhoff, 1995; Leonhard, 1996; Novotny and Lipo, 1996; Ong, 1998; Sarma, 1996; Sen, 1989; Slemon, 1992), power electronics (Hart, 1997; Kassakian, Schlecht, and Verghese, 1991; Mohan, Undeland, and Robbins, 1995; Sen, 1989), microelectronic circuits (Sedra and Smith, 1997), and sensors (Kovacs, 1998) have been published, and educational examples in analysis and design of linear electromechanical systems are available from control books (Dorf and Bishop, 1995; Franklin, Powell, and Emami-Naeini, 1994; Kuo, 1995; Ogata, 1995 and 1997; Phillips and Harbor, 1996), the need for a comprehensive treatment of nonlinear electromechanical systems using the *mechatronic* concept is evident. However, as far as this author knows, there does not exist a textbook in electromechanical systems and *mechatronics* that is comprehensive in its coverage and global in its perspective. It seems that the time has come to target new frontiers using the developed engineering enterprise, emerging technologies, advanced software and hardware, and advanced motion control. Therefore, the opportunity to write an entirely new book is a challenge because the target is high and a goal might be viewed as quite ambiguous. However, by mapping engineering principles, which are available in books and papers devoted to the studied subjects, this pioneering book comprehensively covers the general spectrum of emerging engineering areas, and the well-known gap has been overcome.

The author is disturbed by the recent increase in the number of students whose excellent programming skills and solid theoretical background are matched by their complete inability to solve the simplest engineering problems. The fundamental goal of *Electromechanical Systems, Electric Machines, and Applied Mechatronics* is to demonstrate the application of theoretical results in analysis, design, and deployment of electromechanical systems, to cover emerging software and hardware developments, to develop the rigorous theory of electromechanical systems, and to help the reader develop strong problem-solving skills. Engineers and students come from exceedingly diverse backgrounds: many have extensive electrical engineering experience, while others have expertise in aerospace, computer, or mechanical engineering. For this reason, to avoid excessive and unnecessary superficiality, this book offers an in-depth presentation of numerical and analytical methods, as well as develops and promotes the reader's problem-solving skills using the developed concepts. The text is intended to be readable, interesting, and accessible to students and engineers. This book develops a thorough understanding of integrated perspectives in analysis and design of electromechanical systems, covers the basic concepts in *mechatronics*, and by means of practical worked-out examples, prepares the reader to use the results in engineering practice. Electromechanical motion devices, power electronics and sensors, and advanced software and hardware of engineering importance are comprehensively covered and studied in detail. Engineers and students who master this book will know what they are doing, why they are doing it, and how to do it. It is shown that the mathematical complexity of mathematical models and design methods do not tend to obscure analysis, control, implementation, and deployment.

While students are familiar with calculus and physics, it is assumed that the reader does not have a sufficient background in nonlinear dynamics and control, electric machinery and mechanics, power electronics and sensors, microprocessors, and DSP. To avoid the possible obstacles, the material is presented in sufficient detail, and the basic theory needed to fully understand, appreciate, and apply the knowledge is given for use by those whose background and expertise are deficient to some extent. A wide range of worked-out examples and qualitative illustrations, which are treated in-depth, bridge the gap between theory, practical problems, and solutions. Step-by-step, *Electromechanical Systems, Electric Machines, and Applied Mechatronics* guides the reader through all aspects in engineering areas: from a rigorous theoretical foundation to advanced applications and deployment of results. In addition to achieving a good balance between theory and application, this book covers state-of-the-art software and hardware, as well as highlights analysis, control, and modeling. Throughout the text, the metric (*Systeme International*) system and SI units are used.

It is evident that simulation is a critical and urgently important aspect for developing and prototyping advanced systems. Competition has prompted cost and product cycle reductions. To speed up analysis and design, facilitate enormous gains in productivity and creativity, integrate control laws using advanced microprocessors and DSP, accelerate prototyping features, generate realtime C code, and visualize the results, the MATLAB® environment (SIMULINK®, Real-Time Workshop™, Control System, Nonlinear Control Design, Optimization, Robust Control, Signal Processing, Symbolic Math, System Identification, and other application-specific toolboxes) can be efficiently used. As a flexible high-performance modeling and design environment, MATLAB has become a standard cost-effective tool within the engineering community. The book demonstrates the MATLAB capabilities and helps the reader master this viable enterprisewide environment, focuses on programming techniques and attacks important practical examples, and helps increase designer productivity by showing how to use the advanced software. The MATLAB environment offers a rich set of capabilities to efficiently solve a variety of complex analysis and control problems and promotes enormous gains in productivity and creativity. The reader can easily modify the studied application-specific problems and run MATLAB files and SIMULINK models for solving and prototyping particular systems. The examples modeled in this book consist of a wide spectrum of practical electromechanical systems and motion devices actuated by AC and DC motors controlled by power converters. A variety of most complex and challenging augmented nonlinear models of electromechanical systems are studied, and block diagrams and simulation results are reported. Users can easily apply these results and develop new MATLAB files and SIMULINK block diagrams using the treated enterprisewide practical examples that cover all known rotating electric machines and basic power converters topologies. Most efficient and straightforward analysis and design methods to approach and solve the motion control problem and to approach virtual prototyping are presented and demonstrated. The developed scripts and models are easily assessed, and the documented results are thoroughly demonstrated in analysis, control, and modeling of advanced state-of-the-art electromechanical systems.

In line with the specifically related topics philosophy and flexibility of course content, this text is adaptable to a wide variety of course outlines and engineering curricula, and chapters are essentially independent of one another. *Electromechanical Systems, Electric Machines, and Applied Mechatronics* forms a versatile and modular basis for courses in energy conversion, electric machinery, electromechanical motion devices, electric drives, power electronics, and mechatronics, which are offered at Electrical and Mechanical Engineering departments at most universities. One can find that there is too much material for a one-semester course. The author agrees that the book contains more material than can be covered in a three-credit-hour course. This book has been tested for a two-semester sequence of junior/senior undergraduate and dual-level undergraduate/graduate courses (Energy Conversion (Electromechanical Motion Devices) and Electromechanical Systems, Electric Machines, and Applied Mechatronics (Electromechanical Systems)). It is worth mentioning that this book is balanced and works nicely for short one-semester courses because the material is arranged to guarantee great flexibility in the choice of topics; as well, different course outlines are possible. In this case, sections can be omitted without interfering with the study of other chapters. The book was written such that the studied topics are concentrated in specific chapters and sections, and not covered in different places. Some sections, which are challenging or might be found outside the major course mainstream, are optional. These sections contain advanced topics that are usually covered in comprehensive undergraduate and graduate programs. Therefore, these topics can be omitted without loss of continuity. Instructors can select the chapters and sections that fit the program needs, course objectives, inclination of the instructor, and time available. However, these uncovered sections (if any) are very useful because they provide an indispensable source for anyone who needs complete coverage of theoretical foundation with engineering implementation. Furthermore, the author has successfully tested the advanced topics in the graduate course Electromechanical Systems and Applied Mechatronics.

To guarantee the comprehensive balanced coverage, while keeping in mind the various scenarios of courses, program objectives, and goals, this book is organized as follows. An introduction to electromechanical systems and mechatronics, basic concepts of electromagnetics and mechanics, as well as the introduction to the MATLAB environment, are given in Chapter one. Chapter two provides a case study of electromechanical system dynamics using Newtonian mechanics, Lagrange equations of motion, and Kirchhoff's laws. Energy conversion and electromechanical analogies are thoroughly covered in Chapter two to guarantee a complete understanding of electromechanical systems. Introduction to motion control of electromechanical systems, given in Chapter three, affords a perfect opportunity to demonstrate that the growing interrelationship between electromechanical systems and their control is the key to understanding Mechatronics, as applied in many areas of science and technology. Chapter four covers operational amplifiers, and a great variety of power electronic devices and converters commonly used are thoroughly studied. Direct current electric machines (motors and generators), their steady-state and dynamic characteristics, power converters used to control DC machines, nonlinear analysis and modeling, are studied in detail in Chapter five. Chapter six is devoted to performing a thorough analysis of induction motors, and background foundations are illustrated through nonlinear analysis, simulation, and design. Modeling, analysis, and control of synchronous machines (synchronous, permanent-magnet synchronous, and synchronous reluctance motors and generators) are discussed in Chapter seven. Permanent-magnet and variable-reluctance stepper motors, which belong to the class of synchronous machines, are also covered in Chapter seven. Mechatronic systems are studied in Chapter eight and illustrative examples are treated in detail to illustrate the *mechatronics* concept in practice.

There exists a great number of solutions in course planning. As the instructor might appreciate, the author prefers the following course modules: electromechanical motion devices, energy conversion, electric machines (direct-current, induction, and synchronous motors and generators) should be covered in courses Electromechanical Motion Devices, or Electric Machines, or Energy Conversion (Chapters one, two, five to seven). Electromechanical motion devices, digital control and power electronics, analysis and motion control, mechatronic systems should be covered in Electromechanical Systems and Applied Mechatronics (chapters one to four, the corresponding topics related to the electromechanical motion devices covered in chapters five to seven, as well as chapter eight).

Sincere words of thanks go to those who are deeply and permanently linked to this book. I express my sincere appreciation to many students and faculty who have helped shape my view and vision of how electromechanical motion devices and electromechanical systems, power electronics, and *mechatronics* should be presented in the classroom by showing me what works and what does not. Over the years, students and colleges have offered much valuable criticism and advice. To all of these people go my deepest gratitude and thanks. I am deeply indebted to Professor Oleg Wasynczuk for helpful comments and suggestions. It is always a privilege to express my gratitude to my wife Marina and our children Lydia and Alexander for their love and sustenance.

The revision of this book has depended on students, engineers, scholars, and faculty who have been kind enough to send their suggestions. This valuable communication, especially concerning deficiencies and errors, is greatly appreciated and deeply treasured. The author welcomes any comments and corrections to ensure the highest degree of quality and accuracy. It is my hope that through class testing and implementation, students, engineers, scholars, and faculty will become the author's partners in crafting an exciting emerging concept called *mechatronics.* Your comments and criticism, praise, and suggestions are very welcome. This feedback will help shape the book and crystallize my vision of what the book should be. Thank you for your interest in *Electromechanical Systems, Electric Machines, and Applied Mechatronics.* Your support and indulgence are truly appreciated. I have thoroughly enjoyed writing this book. I hope that you, the reader, will enjoy reading the text. I am delighted to have you as a reader and look forward to hearing from you.

Sergey Edward Lyshevski
Department of Electrical Engineering
Purdue University at Indianapolis
Indianapolis, Indiana 46202
E-mail: lyshevs@engr.iupui.edu

How to study
Electromechanical Systems, Electric Machines, and Applied Mechatronics

The major objectives of this readable and user-friendly book are to establish students' and engineers' confidence in their ability to apply advanced, fully integrated theoretical foundations, enhance learning, as well as to provide a gradual progression from versatile theoretical to practical topics. It is the author's belief that the student is a partner, not an audience to be lectured to from on top a platform. This book is written for engineers and students interested in electromechanical systems and *mechatronics*. Usually, students and engineers are not primarily interested in theory and encyclopedic treatments. They are frustrated by excessive and unnecessary superficial theory, which frequently appears as a misguided desire to shelter an inadequate engineering vision of professional theoreticians from the rigor of mathematics. Here, I would like to mention remarkably clear textbooks in applied mathematics (Wylie and Barrett, 1995), control (Dorf and Bishop, 1995; Franklin, Powell, and Emami-Naeini, 1994; Kuo, 1995; Ogata, 1995; 1997; Phillips and Harbor, 1996), electric machinery (Fitzgerald, Kingsley, and Umans, 1990; Krause, Wasynczuk, and Sudhoff, 1995; Leonhard, 1996; Novotny and Lipo, 1996; Ong, 1998; Sarma, 1996; Sen, 1989; Slemon, 1992), electromagnetics (Hayt, 1989; Paul, Whites, and Nasar, 1998), electromechanical systems and motion devices (Krause and Wasynczuk, 1989), mechatronics (Bradley, Dawson, Burd, and Loader, 1996; Fraser and Milne, 1994; Shetty and Kolk, 1997), power electronics and microelectronics (Hart, 1997; Kassakian, Schlecht, and Verghese, 1991; Mohan, Undeland, and Robbins, 1995; Sedra and Smith, 1997; Sen, 1989), and sensors (Kovacs, 1998). These books have been widely used by generations of students and engineers all over the world. For many years, these excellent books (or their previous editions) have inspired and guided the author; and as a result, this project is successfully finished. Engineers and students have told me that they need straightforward and practical instructions on how to approach and solve specific practical challenging problems. This book stresses the practical aspects to help one understand and apply the advanced developments in analysis and design of electromechanical systems and power electronics, as well as utilize the full potential of microprocessor-DSP technology. In analysis and control of electromechanical systems, advanced theory, innovative methods, leading-edge software, and state-of-the-art hardware are of great importance. This book presents a well-defined theoretical base with step-by-step instructions on how to apply it by stressing systematic philosophy. Numerous practical worked-out examples prepare the reader to use the methods covered. The theory has been successfully transferred into straightforward practical methods that one can understand, visualize, grasp, use, and deploy.

Usually, students efficiently learn by reading and reviewing text and notes, understanding the basic concept and solving the real-world examples at a steady pace over many hours in a disturbance-free environment. An important component of this course is lectures. Listening and notetaking are essential ingredients. The goal in taking notes in class is to capture the instructor's key ideas, not words. To write good notes, one must be a good listener. Section-by-section reading, taking notes, absorbing the basic features, and working-out the problems will help in mastering this quite difficult discipline. There exist a great number of study techniques and principles, and each of us has a different learning style and a preferred strategy. One might use the following five-step-system to succeed in the course:

- *Overview:* glance at the headings and formulas to understand and determine a general picture and basic ideas.
- *Read* and *record:* read section by section to answer the questions, summarize the central idea of each section, understand concepts qualitatively, and apply formulas in specific applications; during your reading, make brief notes and underline key words, phrases, and formulas.
- *Recite:* recite and visualize the main ideas in each section.
- *Review:* covering up your notes and recalling the basic ideas, one should establish a fundamental general picture of the entire section and chapter in a systematic and logical manner.
- *Reflect:* consider, redesign, and attack the studied topics, problems, and quantitative and numerical examples from different emphases and standpoints.

By reorganizing notes, one is advised to make master information sheets. It seems that all of these learning activities will help one understand the basic principles, master the text, and competently apply a balanced knowledge and skills in attacking and solving practical problems. I hope that the suggestions above will benefit your study of *Electromechanical Systems, Electric Machines, and Applied Mechatronics*. Good luck!

References

Applied Mathematics

Wylie, C. R. and Barrett, L. C., *Advanced Engineering Mathematics*, McGraw-Hill, New York, 1995.

Control

Dorf, R. C. and Bishop, R. H., Modern Control Systems, Addison-Wesley, Reading, MA, 1995.

Franklin, J. F., Powell, J. D., and Emami-Naeini, A., *Feedback Control of Dynamic Systems*, Addison-Wesley, Reading, MA, 1994.

Kuo, B. C., *Automatic Control Systems*, Prentice Hall, Englewood Cliffs, NJ, 1995.

Ogata, K., *Discrete-Time Control Systems*, Prentice-Hall, Upper Saddle River, NJ, 1995.

Ogata, K., *Modern Control Engineering*, Prentice-Hall, Upper Saddle River, NJ, 1997.

Phillips, C. L. and Harbor, R. D., *Feedback Control Systems*, Prentice-Hall, Englewood Cliffs, NJ, 1996.

Electric Machinery

Chapman, S. J., *Electric Machinery Fundamentals*, McGraw-Hill, New York, 1999.

Fitzgerald, A. E., Kingsley, C., and Umans, S. D., *Electric Machinery*, McGraw-Hill, New York, 1990.

Krause, P. C., Wasynczuk, O., and Sudhoff, S. D., *Analysis of Electric Machinery*, IEEE Press, New York, 1995.

Leonhard, W., *Control of Electrical Drives*, Springer-Verlag, Berlin, 1996.

Ong, C.-M., *Dynamic Simulation of Electric Machines*, Prentice-Hall, Upper Saddle River, NJ, 1998.

Novotny, D. W. and Lipo, T. A., *Vector Control and Dynamics of AC Drives*, Clarendon Press, Oxford, 1996.

Sarma, M. S., *Electric Machines: Steady-State Theory and Dynamic Performance*, PWS Press, New York, 1996.

Sen, P. C., *Principles of Electric Machines and Power Electronics*, John Wiley & Sons, New York, 1989.

Slemon, G. R., *Electric Machines and Drives*, Addison-Wesley, Reading, MA, 1992.

Electromagnetics

Hayt, W. H., *Engineering Electromagnetics*, McGraw-Hill, New York, 1989.

Paul, C. R., Whites, K. W., and Nasar, S. A., *Introduction to Electromagnetic Fields*, McGraw-Hill, New York, 1998.

Electromechanical Motion Devices

Krause, P. C. and Wasynczuk, O., *Electromechanical Motion Devices*, McGraw-Hill, New York, 1989.

White, D. C. and Woodson, H. H., *Electromechanical Energy Conversion*, John Wiley & Sons, New York, 1959.

Mechatronics

Bradley, D. A., Dawson, D., Burd, N. C., and Loader, A. J., *Mechatronics*, Chapman and Hall, New York, 1996.

Fraser, C. and Milne, J., *Electro-Mechanical Engineering*, IEEE Press, New York, 1994.

Shetty, D. and Kolk, R. A., *Mechatronics System Design*, PWS Publishing, New York, 1997.

Power Electronics and Microelectronics

Hart, D. W., *Introduction to Power Electronics*, Prentice-Hall, Upper Saddle River, NJ, 1997.

Kassakian, J. G., Schlecht, M. F., and Verghese, G. C., *Principles of Power Electronics*, Addison-Wesley, Reading, MA, 1991.

Mohan, N., Undeland, T. M., and Robbins, W. P., *Power Electronics: Converters, Applications, and Design*, John Wiley & Sons, New York, 1995.

Sedra, A. S. and Smith, K. C., *Microelectronic Circuits*, Oxford University Press, New York, 1997.

Sensors

Kovacs, G. T. A., *Micromachined Transducers Sourcebook*, McGraw-Hill, New York, 1998.

Contents

chapter one

Introduction

1.1 Mechatronics and emerging trends in engineering

To enhance and optimize the curriculum and to increase the content of the engineering program, this book is devoted to performing comprehensive analysis, design, optimization and control of state-of-the-art electromechanical systems. Electromechanical systems integrate electromechanical motion devices (actuators and sensors), transducers (analog and digital), power electronics and converters, controllers (analog or digital), data acquisition systems, etc. Industrial, manufacturing, and technological processes must be controlled, and Figure 1.1 represents a functional block diagram of the process that is controlled (operated) using the following key elements: electromechanical actuators, power electronics, sensors, controllers, data acquisition systems, and decision and analysis systems.

The next step is to introduce the key elements to perform actuation, measurement (sensing), analysis, decision, and control. The majority of actuators and sensors (transducers) are electromechanical motion devices that must be studied using electromagnetics and mechanics. Electromechanical energy-conversion devices allow one to convert electrical energy to mechanical energy and mechanical energy to electrical energy. It is evident that electromechanical energy conversion is highly coupled and nonlinear, and energy-conversion processes attract attention and will be thoroughly studied herein. Actuators are controlled by power converters (e.g., to regulate the angular velocity of electric motors, the applied voltages or currents are regulated by voltage or current converters), and control algorithms should be designed and deployed to control electromechanical systems. These control laws can be implemented using analog and digital controllers, and PC-DSP hardware and microcontrollers have been widely used to control and optimize energy management, accomplish filtering and data acquisition, etc. It has been recognized that the multi-disciplinary functionality of electromechanical systems does not allow one to successfully approach the challenging problems through design-by-discipline philosophy, and a new framework, called *mechatronics*, is found and pursued by practicing engineers and scholars. Electromechanical systems have become increasingly dependent on PC-DSP and microprocessor hardware used to attain the desired degree of adaptation and functionality, flexibility and reliability, failure tolerance and survivability, as well as to achieve the specified performance demanded by users and industrial standards.

Mechatronics is a multi-disciplinary conceptual framework, based on a concurrent integrated concept, and used to perform the comprehensive analysis and design, optimization and control, implementation and deployment of advanced state-of-the-art electromechanical systems. *Mechatronics* integrates electrical, mechanical, computer, and software

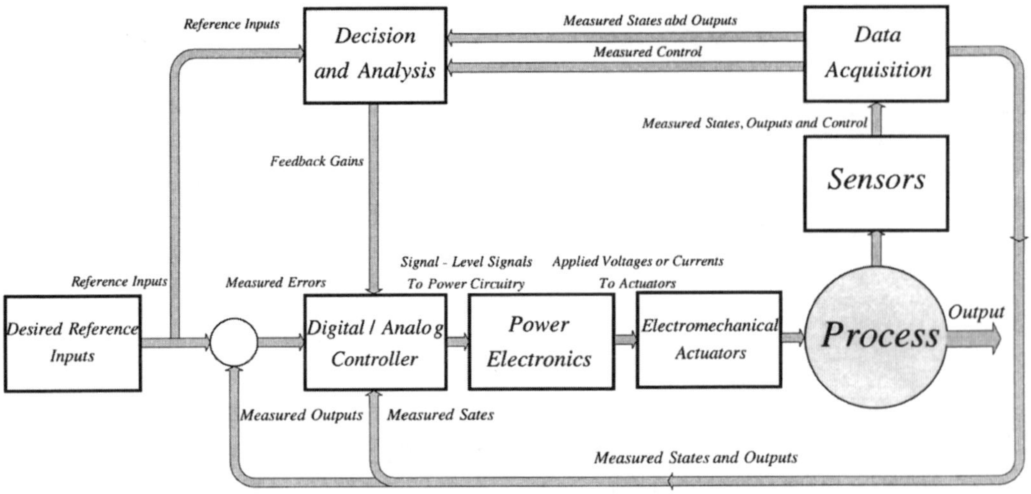

Figure 1.1 High-level functional block diagram of the controlled process.

engineering to approach and solve a broad spectrum of extremely challenging problems (see Figure 1.2).

As an example of the application of the mechatronic concept, consider the motion control problem as applied to advanced aircraft. The aircraft is controlled by flight control surfaces (aileron, canard, elevator, flap, rudder, stabilizer), and direct-drive servo-systems are used to actuate (displace) these control surfaces. A fly-by-wire control surface *servo* with stepper motor is illustrated in Figure 1.3. The desired angular displacement of the control surface (reference input) is assigned either by the pilot or aircraft flight computer. Using this reference signal (the specified angular displacement of the control surface), as

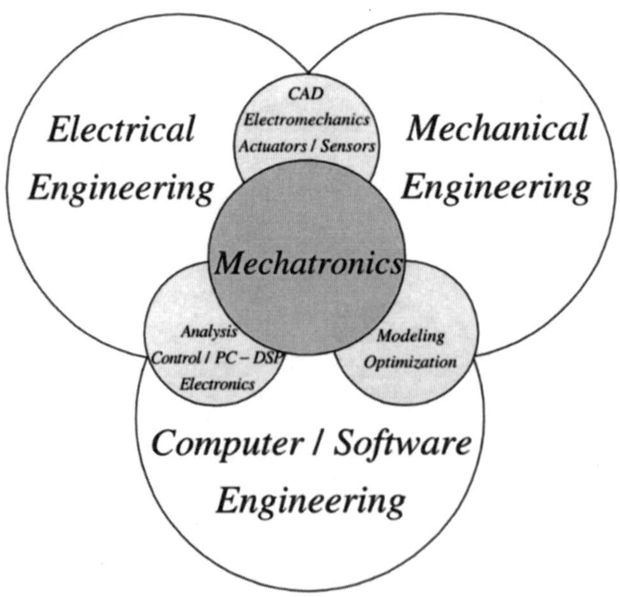

Figure 1.2 The view of mechatronics as a multi-disciplinary concept.

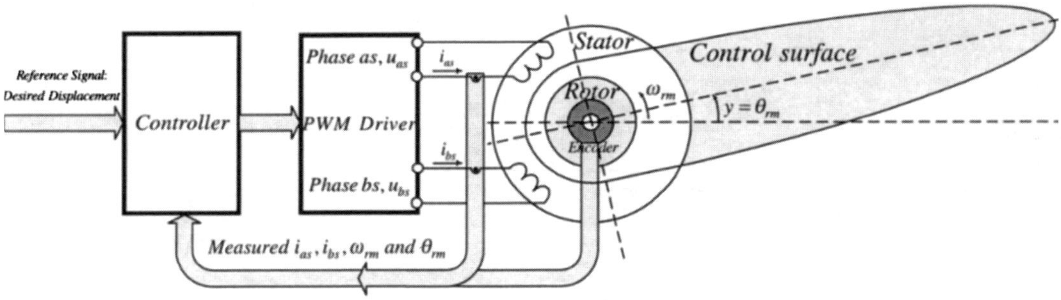

Figure 1.3 Fly-by-wire flight servo-system actuated by stepper motor.

well as the measured by sensors currents in the *ab* phase windings i_{as} and i_{bs}, mechanical angular velocity ω_{rm} and actual mechanical angular displacement of the control surface θ_{rm}, the controller develops signal-level signals that drive high-frequency switches, and the magnitude and frequency of the applied voltages to the *ab* phase windings u_{as} and u_{bs} of the permanent-magnet stepper motor are controlled by the PWM driver (amplifier) (see Figure 1.3).

This electromechanical servomechanism (flight servo) integrates electromechanical motion devices (the stepper motor is the actuator, and the optical encoder, which measures the angular velocity and displacement, is the sensor), PWM amplifier, transducers, and controller. It should be emphasized that microcontrollers and DSP are applied to deploy the control algorithms synthesized, and analog–digital (A/D) and digital–analog (D/A) converters are used.

Advances in computers and digital signal processing, state-of-the-art software development tools and hardware platforms, and computer-aided design have stimulated the application of the *mechatronic* concept in avionics. A spectrum of extremely important problems in nonlinear analysis, identification, optimization, and control of current and next-generation advanced aircraft must be solved. The conventional design-by-discipline concept does not allow one to attain the desired flying and handling requirements (controllability, maneuverability, agility, and other quantities) in the full operating envelope. The *mechatronic* concept is applied to guarantee the required performance specifications, as tailored as necessary to perform the specified mission and other objectives for a fully functional and damaged/crippled/failed aircraft, by solving the motion control problem that integrates nonlinear analysis and modeling, real-time identification, control redesign, and controller reconfiguration, as well as control algorithms deployment using advanced avionics platforms. The high-level closed-loop flight control system is shown in Figure 1.4. Here, the state variables used to map the longitudinal and lateral dynamics are: v is the forward velocity; α is the angle of attack; q is the pitch rate; θ is the pitch angle; β is the sideslip angle; p is the roll rate; r is the yaw rate; ϕ is the roll angle; and ψ is the yaw angle. The reference inputs r_θ, r_θ, and r_ψ are the desired (assigned by the pilot or flight computer) Euler angles. The Euler angles θ, ϕ, and ψ are the aircraft outputs.

Multi-degree-of-freedom manipulators and robotic systems are considered as an important subclass of electromechanical (mechatronic) systems. Figure 1.5 illustrates *j*-th single-joint for direct drive robotic manipulators. In particular, the arm is actuated by *j*-th brushless limited angle torque motor with permanent-magnet rotor.

Advanced actuators and power converters (called PWM amplifiers), state-of-the-art sensors, and microprocessor-DSP hardware are used to attain the required objectives and specified characteristics imposed on industrial robots. The use of the latest direct-drive

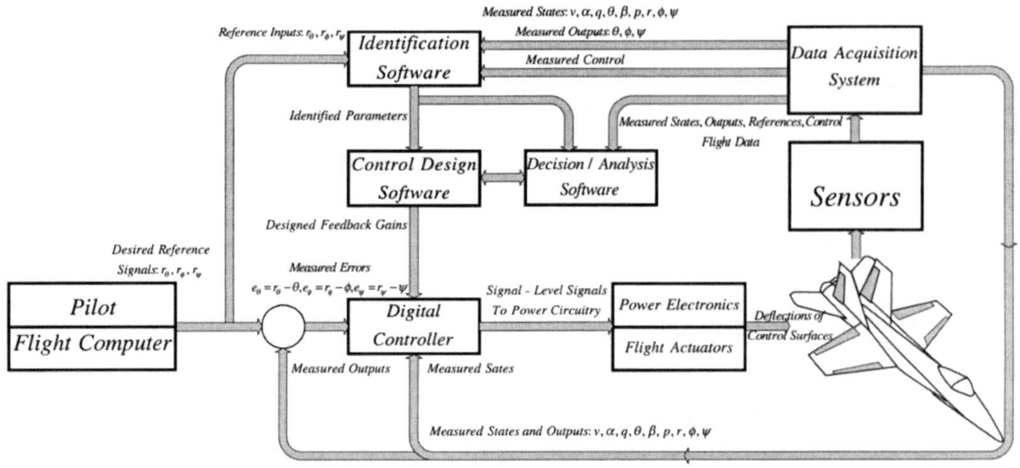

Figure 1.4 The mechatronic concept (synergetic combination of electrical, mechanical, computer, and software engineering) is applied to approach and solve flight dynamics and control problems.

actuator technology in robots is motivated by their high efficiency and superior performance capabilities, accuracy and compactness, simplicity and ruggedness, durability and reliability, low weight-to-torque and volume-to-torque ratios, etc. To optimize the robot performance and to attain the specified accuracy, to guarantee tracking and robustness, as well as to enhance stability margins and ensure disturbance attenuation, there is a critical need to use complete nonlinear electromechanical system models (without simplifications and assumptions) to rigorously approach virtual prototyping, solve the motion control problem, and implement the advanced software and hardware. The need for an innovative nonlinear model-design dimension and application of emerging technologies is real. As a result, these trends have been recognized and pursued in recent years by academia and industry. *Mechatronics* emerges as a most general concept that provides a systematic and unified conceptual framework in analysis and optimization, prototyping and design, and control and deployment of robotic manipulators.

The basic configuration of electromechanical systems was depicted in Figure 1.1. Modern electromechanical systems rely increasingly on digital controllers (microprocessors and DSP). Digital control systems are widely used, and the block diagram is shown in Figure 1.6.

Electromechanical systems are controlled using the difference between the desired reference input $r(t)$ and the system output $y(t)$ (which are assumed to be analog for the example studied). In order to use microprocessors or DSPs to control industrial and technological systems and processes, this continuous-time error signal $e(t) = r(t) - y(t)$ must be converted into digital form. The sample-and-hold circuit (S-H circuit) receives the continuous-time (analog) signal and holds this signal at a constant value for the specified period of time (the sampling period). The analog-to-digital converter (A/D converter) converts the piecewise continuous-time signal to digital format (binary numbers). The transformation of continuous-time signals to discrete-time signals is called sampling or discretization. Hence, the input signal to the digital controller is the sampled version of the continuous-time error signal $e(t) = r(t) - y(t)$. In each sampling time, the controller "reads" the discretized value of the error signal e_k. That is, the binary numbers are fed to the digital controller. The controller generates the control signal in binary number form, which then must be converted to analog form to be fed to the power converter. The digital-to-analog conversion (decoding) is performed by the digital-to-analog converter

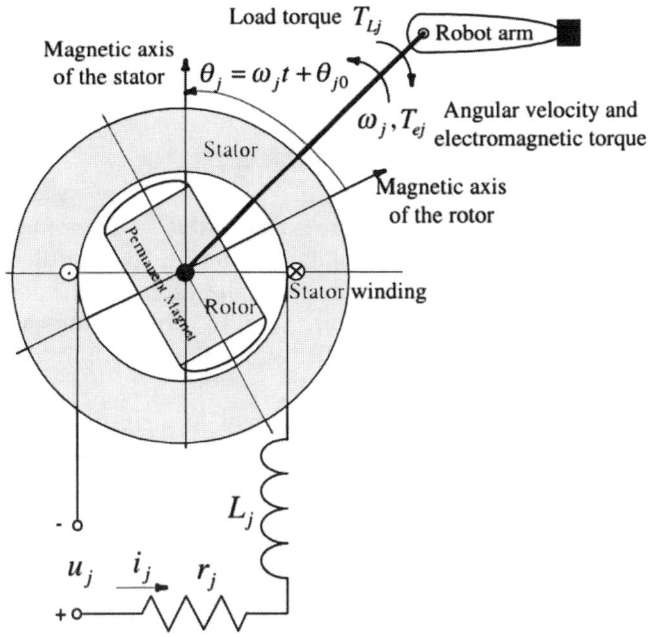

Figure 1.5 j-th link of a direct drive robot: the arm is actuated by j-th brushless limited angle torque motor with permanent-magnet rotor.

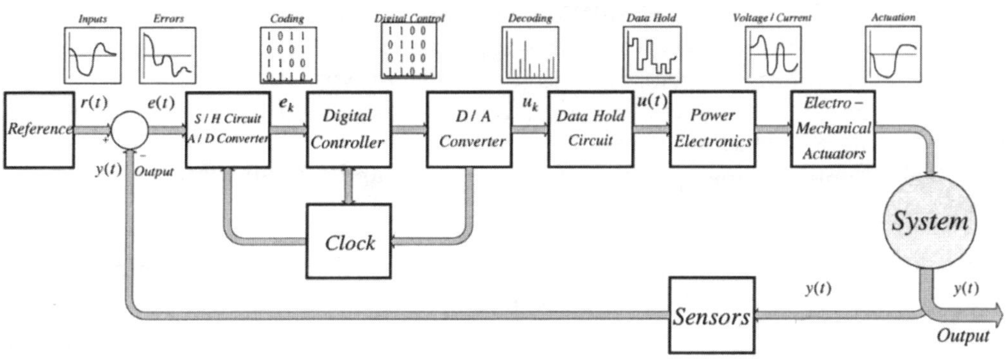

Figure 1.6 Block diagram of a typical digitally controlled system.

(D/A converter) and the data hold circuit. The A/D and D/A converters are built with finite word length (if the operating voltage of the n-bit A/D converter is $[-V_{A/D} V_{A/D}]$, the quantization step cannot be smaller than $2V_{A/D}/2^n$). As the binary numbers are converted into the piecewice continuous-time signals, they are fed to the power circuitry. The clock synchronizes coding and decoding.

The signal conversion involves multiplexing and demultiplexing, sample and hold, and analog-to-digital (quantizing and encoding) and digital-to-analog (decoding) conversion.

The *multiplexer* is a multiple switch that sequentially switches among analog input channels in a particular sequence to read the channel information in a sequential order. At a given instant, only one switch is *on* for a specified period of time, and the sample-and-hold circuit samples this analog signal and holds its value. The operation of the 2^p-input m-bit *multiplexer* is represented by m sum-of-product. The Boolean equations is

$$z_j = (x_{0,j}a_0 + x_{1,j}a_1 + \ldots + x_{2^p-1,j}a_{2^p-1})e \qquad j = 0,\ 1,\ \ldots,\ m-1$$

and using the single word-based equation one obtains

$$Z = (X_0 a_0 + X_1 a_1 + \ldots + X_{2^p-1} a_{2^p-1})e$$

Here, a_i is the selection of the input data bus x_i. The analysis shows that the data from the m-bit input data bus $X_i = (x_{i,0}, x_{i,1}, \ldots, x_{i,m-1})$ is transferred to the output Z bus.

The *decoder* decodes the signals from n or more than n lines (the so-called tree configuration). For example, a $1/2^n$ decoder is a circuit with n input lines and 2^n output lines. Addressing and routing the data is accomplished by the decoder, which is also called a *demultiplexer*.

The A/D converter transforms an analog signal to a digital word, while the D/A converter transforms a binary number (digital input) to an electrical signal (analog output) of corresponding magnitude. Various A/D and D/A converter configurations are available, and converters and interfacing are thoroughly studied and covered in the undergraduate and graduate classes in relation to microprocessors and interfacing.

The application of analog and digital controllers allows one to solve a spectrum of problems. In general, analysis, optimization, design, and deployment of electromechanical systems can be divided into the requirements-specifications phase (application specific), the software-hardware phase (virtual prototyping, technology transfer, and validation), and the deployment phase. The basic ingredients are:

- Analysis and modeling, simulation and optimization, and visualization and validation of electromechanical systems applying computer-aided design (using databases and libraries developed) with experimental verification
- Development and deployment of advanced software and state-of-the-art hardware using emerging technologies
- Real-time interfacing, digital signal processing, and motion control
- Deployment and testing of electromechanical systems

The design specifications imposed on electromechanical systems are given in the *systems domain*. The commonly used performance criteria to be attained are:

- Stability with the desired stability margins in the full operating envelope
- Robustness to parameter variations and structural changes
- Tracking and disturbance attenuation
- Dynamic and steady-state accuracy
- Specifications imposed on the states and outputs transient responses; for example:
  - The settling time: the settling time is the time needed for the system output $y(t)$ to reach and stay within the step reference input $r(t) = const$ assigned (the absolute allowable difference between $y(t)$ and $r(t)$ is used to find the settling time, and usually this difference is assigned in the range of 2 to 5%).
  - The maximum overshoot: the maximum overshoot is the difference between the maximum peak value of the systems output $y(t)$ and the steady-state value $y_{steady\ state}$ divided by the $y_{steady\ state}$, the maximum overshoot is

$$\Delta y = \frac{y_{\max} - y_{steady\ state}}{y_{steady\ state}} \quad \text{or} \quad \Delta y = \frac{y_{\max} - y_{steady\ state}}{y_{steady\ state}} \times 100\%$$

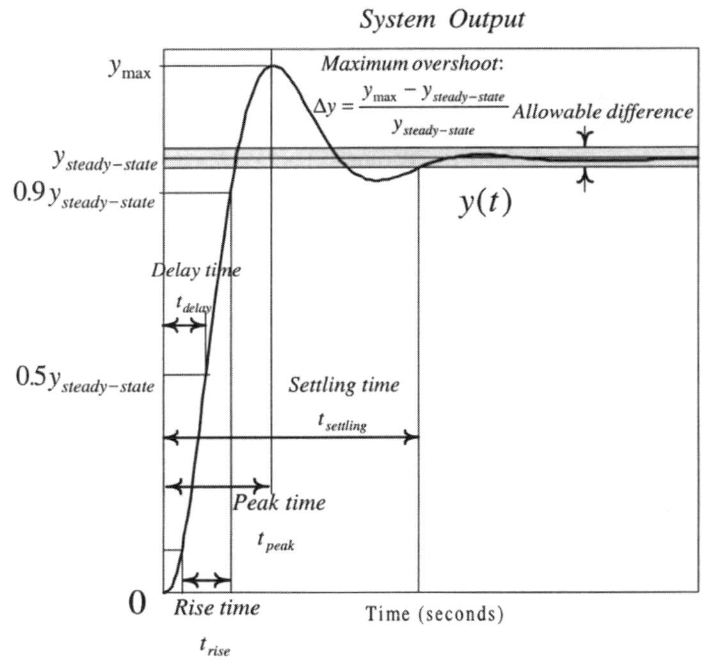

Figure 1.7 Electromechanical system output.

- The rise time: the rise time is the time needed for the system output $y(t)$ to increase from 10 to 90% of the steady-state value $y_{steady\ state}$ if $r(t) = const.$
- The delay time: the delay time is the time needed for the system output $y(t)$ to reach 50% of $y_{steady\ state}$ if $r(t) = const.$
- The peak time: the peak time is the time required for the system output to reach the first peak of the overshoot.

The typical output transient response is shown in Figure 1.7.

The major goal of this book is to introduce and educate students, engineers, and scholars in how to solve challenging problems and perform systematic integrated research using the emerging *mechatronic* concept in order to design high-performance electromechanical systems. In particular, this book introduces and conducts fundamental (analytical and numerical) and experimental studies in analysis and design of electromechanical systems, and illustrates the application of analysis and design methods through numerous illustrative examples.

1.2 Basic foundations

Mathematics and mechanics were the first disciplines to be thoroughly researched, and these studies can be traced back to ancient civilizations. Historically, mathematics was developed first. However, the author must apologize, but due to obvious limitations, the basic principles of classical mathematics are not covered in this book. The undergraduate curriculum covers the major concepts in mathematics needed, and one should concentrate on the fundamental electromechanical features. Electromechanical systems must be studied using the fundamental laws and basic principles of mechanics and electromagnetics. One can identify and study these key concepts to illustrate the use of cornerstone principles.

1.2.1 Mechanics

The study of the motion of systems with the corresponding analysis of forces that cause motion is of interest. The foundation of classical Newtonian mechanics was established in the 17th century by Sir Isaac Newton. According to Newton's second law of motion, the net force (the vector sum of all forces) acting on an object of mass m is related to its acceleration vector $\vec{a}$. In particular,

$$\sum \vec{F} = m\vec{a}$$

In the Cartesian (XYZ) coordinate system, Newton's second law is expressed as

$$\sum F_x = ma_x, \quad \sum F_y = ma_y, \quad \text{and} \quad \sum F_z = ma_z$$

It is worth noting that $m\vec{a}$ represents the magnitude and direction of the applied net force acting on the object. Hence, $m\vec{a}$ is not a force. A body is at equilibrium, that is, the object is at rest or is moving with constant speed, if

$$\sum \vec{F} = 0$$

Newton's second law in terms of the linear momentum, which is found as $\vec{p} = m\vec{v}$, is given by

$$\sum \vec{F} = \frac{d\vec{p}}{dt} = \frac{d(m\vec{v})}{dt}$$

where $\vec{v}$ is the vector of the object velocity.

For rotational motion, the net torque and angular acceleration must be used. In particular, the torque vector $\vec{T}$ is found as

$$\vec{T} = \vec{r} \times \vec{F}$$

where $\vec{r}$ is the position vector with respect to the origin.

The rotational analog of Newton's second law for a rigid body is

$$\sum \vec{T} = J\vec{\alpha}$$

where $\sum \vec{T}$ is the applied net torque

J is the moment of inertia (*rotational inertia*);

$\vec{\alpha}$ is the angular acceleration vector $\vec{\alpha} = \frac{d}{dt}\frac{d\vec{\theta}}{dt} = \frac{d^2\vec{\theta}}{dt^2} = \frac{d\vec{\omega}}{dt}$ because $\frac{d\vec{\theta}}{dt} = \vec{\omega}$;

$\vec{\theta}$ is the angular displacement

ω is the angular velocity.

The angular momentum of the system $\vec{L}_M$ is expressed as

$$\vec{L}_M = \vec{r} \times \vec{p} = \vec{r} \times m\vec{v}$$

and

$$\sum \vec{T} = \frac{d\vec{L}_M}{dt} = \vec{r} \times \vec{F}$$

For the rigid body rotating around the axis of symmetry, one obtains $\vec{L}_M = J\vec{\omega}$.

Example 1.1

An electric motor has the equivalent moment of inertia $J = 0.06$ kg-m^2. The motor starts from stall, and the angular velocity of the rotor is $\omega_r = 100t^{1/3}$. Find the angular momentum and the developed electromagnetic torque as functions of time. The load and friction torques are zero.

Solution

The angular momentum is found as

$$L_M = J\omega_r = 0.06 \times 100t^{1/3} = 6t^{1/3}$$

The developed electromagnetic torque is

$$T_e = \frac{dL_M}{dt} = 2t^{-2/3}$$ □

From Newtonian mechanics, one concludes that the applied net force plays a central role in quantitatively describing the motion. An alternative analysis of motion can be performed in terms of the energy or momentum quantities, which are conserved. The principle of conservation of energy states that energy can be only converted from one form to another. Kinetic energy is associated with motion, while potential energy is associated with position. The sum of the kinetic (Γ), potential (Π), and dissipated (D) energies is called the total energy of the system (Σ_T), which is conserved, and the total amount of energy remains constant; that is,

$$\Sigma_T = \Gamma + \Pi + D = const$$

For example, consider the translational motion of a body that is attached to an ideal spring that obeys Hooke's law. Neglecting friction, one obtains the following expression for the total energy

$$\Sigma_T = \Gamma + \Pi = \frac{1}{2}(mv^2 + k_s x^2) = const$$

Here, the translational kinetic energy is $\Gamma = \frac{1}{2}mv^2$; the elastic potential energy is $\Pi = \frac{1}{2}k_s x^2$; k_s is the force constant of the spring; x is the displacement.

For a rotating spring, one obtains

$$\Sigma_T = \Gamma + \Pi = \frac{1}{2}(J\omega^2 + k_s\theta^2) = const$$

where the rotational kinetic energy is $\Gamma = \frac{1}{2}J\omega^2$, and the elastic potential energy is obtained as $\Pi = \frac{1}{2}k_s\theta^2$.

The kinetic energy of a rigid body having translational and rotational components of motion is found to be

$$\Gamma = \frac{1}{2}(mv^2 + J\omega^2)$$

That is, the motion of the rigid body is represented as a combination of translational motion of the center of mass and rotational motion about the axis through the center of mass. The moment of inertia depends on how the mass is distributed with respect to the axis, and J is different for different axes of rotation. If the body is uniform in density, J can be easily calculated for regularly shaped bodies in terms of their dimensions. For example, a rigid cylinder with mass m (which is uniformly distributed), radius R, and length l has the following horizontal and vertical moments of inertia: $J_{horizontal} = \frac{1}{2}mR^2$ and $J_{vertical} = \frac{1}{4}mR^2 + \frac{1}{12}ml^2$. The *radius of gyration* can be found for irregularly shaped objects, and the moment of inertia can be easily obtained. Table 1.1 gives some formulas for a number of solid objects that rotate about the axes specified.

The center of mass (x_c, y_c, z_c) of a solid region S with density $\sigma(x, y, z)$ can be found using the so-called first moments of region S about the yz-, xz-, and xy-planes, which are given by $\iiint_S x\sigma(x,y,x)dV$, $\iiint_S y\sigma(x,y,z)dV$, and $\iiint_S z\sigma(x,y,z)dV$, respectively. The following expressions are used to obtain the center of mass:

$$x_c = \frac{1}{m}\iiint_S x\sigma(x,y,z)dV$$

$$y_c = \frac{1}{m}\iiint_S y\sigma(x,y,z)dV$$

$$z_c = \frac{1}{m}\iiint_S z\sigma(x,y,z)dV$$

where $m = \iiint_S \sigma(x,y,z)dV$ is the mass.

Table 1.1 Moments of Inertia

Object	Location of axis	Moment of inertia
Long uniform rod of length l	Through center 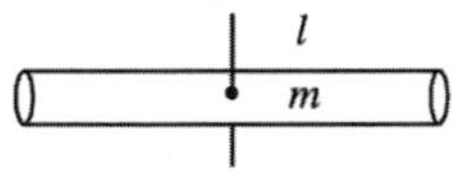	$J = \frac{1}{12}ml^2$
	Through end 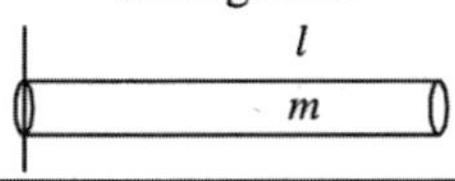	$J = \frac{1}{3}ml^2$
Rectangular plate	Through center 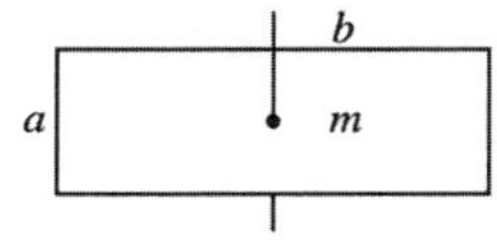	$J = \frac{1}{12}m(a^2 + b^2)$
	Along edge 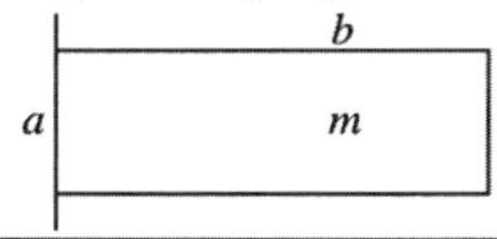	$J = \frac{1}{3}mb^2$
Uniform solid cylinder of radius R	Through center 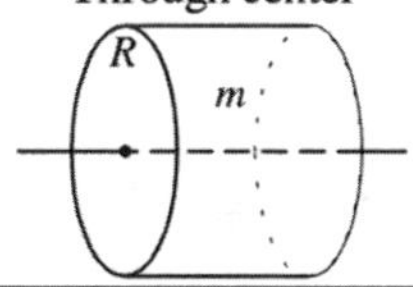	$J = \frac{1}{2}mR^2$
Hollow cylinder	Through center 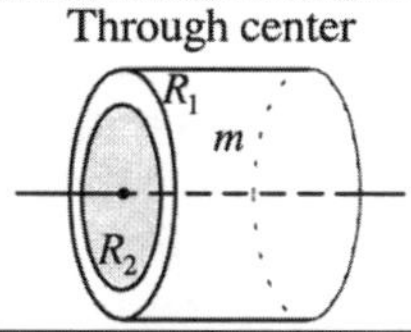	$J = \frac{1}{2}m(R_1^2 + R_2^2)$
Uniform solid sphere of radius R	Through center 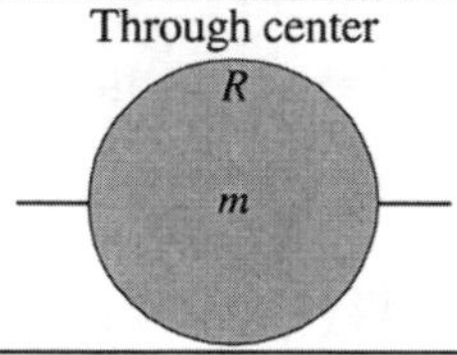	$J = \frac{2}{5}mR^2$

To find the triple (volume) integral, it is assumed that the density $\sigma(x, y, z)$ is continuous; and by using six possible permutations of dx, dy, and dz, the volume integral $\iiint_S \sigma(x,y,z)dV$ can be found. For example, one has $\int_{f_1}^{f_2}\int_{f_3(z)}^{f_4(z)}\int_{f_5(z,y)}^{f_6(z,y)} \sigma(x,y,z)dx\,dy\,dz.$

Example 1.2

Find the center of mass (x_c, y_c, z_c) of the unit cube $0 \le x \le 1$, $0 \le y \le 1$, $0 \le z \le 1$ if the density at any point is equal to the square of its distance from the origin (0, 0, 0); that is, $\sigma(x, y, z) = x^2 + y^2 + z^2$.

Solution

The mass is found as

$$m = \iiint_S \sigma(x,y,z)dV = \int_0^1\int_0^1\int_0^1 (x^2+y^2+z^2)dz\,dy\,dx = \int_0^1\int_0^1 \left(x^2z + y^2z + \frac{1}{3}z^3\right)\Big|_0^1 dy\,dx$$

$$= \int_0^1\int_0^1 \left(x^2+y^2+\frac{1}{3}\right)dy\,dx = \int_0^1 \left(x^2y + \frac{1}{3}y^3 + \frac{1}{3}y\right)\Big|_0^1 dx = \int_0^1 \left(x^2 + \frac{2}{3}\right)dx = \left(\frac{1}{3}x^3 + \frac{2}{3}x\right)\Big|_0^1 = 1$$

To find x_c, the first moment about the yz-plane, as given by $\iiint_S x\sigma(x,y,z)dV$, should be derived. We have

$$\iiint_S \sigma(x,y,z)dV = \int_0^1\int_0^1\int_0^1 x(x^2+y^2+z^2)dz\,dy\,dx = \int_0^1 x\left(\int_0^1\int_0^1 x(x^2+y^2+z^2)dz\,dy\right)dx = \frac{7}{12}$$

Hence, one obtains

$$x_c = \frac{1}{m}\iiint_S x\sigma(x,y,z)dV = \frac{7}{12}$$

Due to the symmetry, $x_c = y_c = z_c$. That is, the center of mass of the unit cube is located at $\left(\frac{7}{12}, \frac{7}{12}, \frac{7}{12}\right)$. □

To find the moments of inertia of a solid region S, one must use the so-called second moments of region S about the x, y, and z axes, which are found using the density $\sigma(x, y, z)$. In particular,

$$J_x = \iiint_S (y^2+z^2)\sigma(x,y,z)dV,\quad J_y = \iiint_S (x^2+z^2)\sigma(x,y,z)dV,\quad J_z = \iiint_S (x^2+y^2)\sigma(x,y,z)dV$$

It should be emphasized that

$$J_x = J_{xy} + J_{xz},\quad J_y = J_{xy} + J_{yz},\quad \text{and}\quad J_z = J_{xz} + J_{yz}$$

where
$$J_{xy} = \iiint_S z^2\sigma(x,y,z)dV$$
$$J_{xz} = \iiint_S y^2\sigma(x,y,z)dV$$
$$J_{yz} = \iiint_S x^2\sigma(x,y,z)dV.$$

Example 1.3

Find the moment of inertia about the x-axis of the solid region bounded by the hemisphere $z=\sqrt{4-x^2-y^2}$; that is, the circular base within the xy-plane is $x^2 + y^2 = 4$, as well as $-2 \le x \le -2$, $-\sqrt{4-x^2} \le y \le \sqrt{4-x^2}$, $0 \le z \le \sqrt{4-x^2-y^2}$. The density at any point is proportional to the distance between (x, y, z) and xy-plane; hence, $\sigma(x, y, z) = \sigma_z z$.

Solution

The moment of inertia about the x-axis is found as

$$J_x = \iiint_S (y^2+z^2)\sigma(x,y,z)dV = \sigma_z \int_{-2}^{2}\int_{-\sqrt{4-x^2}}^{\sqrt{4-x^2}}\int_{0}^{\sqrt{4-x^2-y^2}} (y^2+z^2)z dz\, dy\, dx$$

$$= \sigma_z \int_{-2}^{2}\int_{-\sqrt{4-x^2}}^{\sqrt{4-x^2}} \left(\frac{1}{2}y^2z^2+\frac{1}{4}z^4\right)\Bigg|_0^{\sqrt{4-x^2-y^2}} dy\, dx$$

$$= \frac{1}{4}\sigma_z \int_{-2}^{2}\int_{-\sqrt{4-x^2}}^{\sqrt{4-x^2}} ((4-x^2)^2-y^4)dy\, dx = \frac{1}{4}\sigma_z \int_{-2}^{2}\left((4-x^2)^2 y-\frac{1}{5}y^5\right)\Bigg|_{-\sqrt{4-x^2}}^{\sqrt{4-x^2}} dx = \frac{4}{5}\sigma_z \int_{0}^{2}(4-x^2)^{\frac{5}{2}}dx$$

$$= \frac{4}{5}\sigma_z \int_{0}^{\pi/2} 64\cos^6\rho d\rho = \frac{256}{5}\sigma_z \frac{5}{32}\pi = 8\sigma_z\pi$$

It should be emphasized that to find the integral $\int_0^2 (4-x^2)^{\frac{5}{2}}dx$, let $x = 2\sin\rho$; hence, $dx = 2(\cos\rho)d\rho$.

Due to the symmetry, we have $J_x = J_y = 8\sigma_z\pi$. □

In electromechanical motion devices, the force and torque are thoroughly studied. Assuming that the body is rigid and the moment of inertia is constant, one has

$$\vec{T}d\vec{\theta} = J\vec{\alpha}d\vec{\theta} = J\frac{d\vec{\omega}}{dt}d\vec{\theta} = J\frac{d\vec{\theta}}{dt}d\vec{\omega} = J\vec{\omega}d\vec{\omega}$$

The total work, as given by

$$W = \int_{\theta_0}^{\theta_f}\vec{T}d\vec{\theta} = \int_{\omega_0}^{\omega_f} J\vec{\omega}d\vec{\omega} = \frac{1}{2}(J\omega_f^2 - J\omega_0^2)$$

represents the change of the kinetic energy. Furthermore,

$$\frac{dW}{dt} = \vec{T}\frac{d\vec{\theta}}{dt} = \vec{T}\times\vec{\omega}$$

and the power is defined by

$$P = \vec{T} \times \vec{\omega}$$

This equation is an analog of $P = \vec{F} \times \vec{v}$, which is applied for translational motion.

Example 1.4

Consider a positioning table actuated by an electric motor. How much work is required to accelerate a 10 kg payload (m = 10 kg) from v_0 = 0 m-s^{-1} to v_f = 5 m-s^{-1}?

Solution

The work needed is calculated as $W = \frac{1}{2}(mv_f^2 - mv_0^2) = \frac{1}{2}10 \times 5^2 = 125$ J. □

Example 1.5

The rated power and angular velocity of a permanent-magnet DC motor are 100 W and 200 rad-s^{-1}, respectively. Calculate the rated electromagnetic torque.

Solution

The result is found to be $T_e = \frac{P}{\omega_r} = \frac{100}{200} = 0.5$ N-m. □

Example 1.6

An electric motor drives a load, and the electromagnetic torque developed by the motor is 0.2 N-m (T_e = 0.2 N-m). The motor starts from stall and has the equivalent moment of inertia 0.01 kg-m^2 (J = 0.01 kg-m^2). Calculate the work and kinetic energy at t = 15 s.

Solution

The angular acceleration is found using $T_e = J\alpha$. By applying this equation, one obtains $\alpha = \frac{T_e}{J} = 20$ rad-s^{-2}. Hence, the initial and final angular velocities of the motor are ω_0 = 0 rad-s^{-1} and ω_f = 300 rad-s^{-1} because $\omega = \alpha t$.

The initial kinetic energy is zero, while the final value can be easily found as

$$\Gamma = \frac{1}{2}J\omega_f^2 = \frac{1}{2}0.01 \times 300^2 = 450 \text{ J}.$$

The total angular displacement can be obtained using the angular acceleration or angular velocity. In particular,

$$\theta = \frac{1}{2}\alpha t^2 = \frac{1}{2}20 \times 15^2 = 2250 \text{ rad}$$

or

$$\theta = \omega t = 300 \times 15 = 2250 \text{ rad}$$

The total work is found as

$$W = T_e\theta = 0.2 \times 2250 = 450 \text{ J}$$

Of course, $W = \Gamma$ because the system is lossless. □

1.2.2 *Electromagnetics*

The basic concepts of electromagnetic field theory constitute the foundation of electromechanical motion devices (motors, generators, and transducers), as well as electromechanical systems analysis and design. Electric machines have been of great importance since the beginning of the industrial revolution, and their role has increased significantly. Modern industrial trends require new innovative mainstreams; and while the theory of electromechanical motion devices has been essentially developed and completed, advanced nonlinear analysis and control methods should be applied. In electromechanical systems, electric machines convert energy from one form to another, and electromagnetic principles play a key role in analysis and design of AC and DC motors and generators. Therefore, it is appropriate to discuss the basic theoretical foundations.

Certain facts regarding electromagnetism that may be unknown to many engineers are worth mentioning. The attraction of magnetized iron ore (magnetic phenomena) was first discovered in Magnesia (western Turkey), and a compass needle, which was used as a navigation tool in the 10th century, orients itself along the earth *magnetic field lines.* The Greek philosophers Democritus and Leucippus in 400 B.C. suggested that matter is made by not(*a*)-divided(*tomos*) particles. However, in 1897, J. Thomson discovered the electron. Electric force holds atoms and molecules together. Electrostatic interaction plays a central role in molecular biology. For example, two DNA (deoxyribonucleic acid) chains wrap about one another in the shape of a double helix. These two strands are held together by electrostatic forces. Electric force is responsible for energy-transforming processes in all living organisms (metabolism). Electromagnetism is used to study protein synthesis and structure, nervous system and nerve conditions, etc.

The word *electricity* comes from the Greek word *electron,* which means amber. The ancient Greeks observed that when they rubbed amber with wool, the amber attracted objects. Electrostatic interaction was investigated by Charles Coulomb. For charges q_1 and q_2, separated by a distance x in free space, the magnitude of the electric force is obtained as

$$F = \frac{|q_1 q_2|}{4\pi\varepsilon_0 x^2}$$

where ε_0 is the permittivity of free space, $\varepsilon_0 = 8.85 \times 10^{-12}$ F-m^{-1} or C^2-N^{-1}-m^{-2}, $\frac{1}{4\pi\varepsilon_0} = 9 \times 10^9$ N-m^2-C^{-1}. The unit for the force is the newton N, while the charges are given in coulombs, C.

In vector form, we have

$$\vec{F} = \frac{q_1 q_2}{4\pi\varepsilon_0 x^2}\vec{a}_x$$

where $\vec{a}_x$ is the unit vector directed along the line joining these two charges.

The capacity, elegance, and uniformity of electromagnetics arise from a sequence of fundamental laws linked one to another and needed to study the field quantities. Using the Gauss law and denoting the vector of electric flux density as $\vec{D}$ [F-m^{-1}] and the vector of electric field intensity as $\vec{E}$ [V-m^{-1} or N-C^{-1}], the total electric flux Φ[C] through a closed surface is found to be equal to the total force charge enclosed by the surface; that is,

$$\Phi = \oint_s \vec{D} \cdot d\vec{s} = Q_s, \quad \vec{D} = \varepsilon \vec{E}$$

where $d\vec{s}$ is the vector surface area, $d\vec{s} = ds\vec{a}_n$; $\vec{a}_n$ is the unit vector that is normal to the surface; ε is the permittivity of the medium; and Q_s is the total charge enclosed by the surface.

Ohm's law relates the volume charge density $\vec{J}$ and electric field intensity $\vec{E}$; in particular,

$$\vec{J} = \sigma \vec{E}$$

where σ is the conductivity [A-V^{-1}m^{-1}], for copper, $\sigma = 5.8 \times 10^7$, and for aluminum, $\sigma = 3.5 \times 10^7$.

The current i is proportional to the potential difference, and the resistivity ρ of the conductor is the ratio between the electric field $\vec{E}$ and current density $\vec{J}$; that is,

$$\rho = \frac{\vec{E}}{\vec{J}}$$

The resistance r of the conductor is related to the resistivity and conductivity by

$$r = \frac{\rho l}{A} \text{ and } r = \frac{l}{\sigma A}$$

where l is the length and A is the cross-sectional area.

It is important to emphasize that parameters of electromechanical motion devices vary. In fact, it is evident that the resistances of electric machine windings vary due to heating. The resistivity depends on temperature T [°C], and

$$\rho(T) = \rho_0[1 + \alpha_{\rho 1}(T - T_0) + \alpha_{\rho 2}(T - T_0)^2 + \ldots]$$

where α_{r1} and $\alpha_{\rho 2}$ are the coefficients. As an example, over the small temperature range (up to 160°C) for copper at $T_0 = 20$°C, one has

$$\rho(T) = 1.7 \times 10^{-8}[1 + 0.0039(T - 20)]$$

To study electromechanical motion devices, the basic principles of electromagnetic theory should be briefly reviewed. The total magnetic flux through the surface is given by

$$\Phi = \int \vec{B} \cdot d\vec{s}$$

where $\vec{B}$ is the magnetic flux density.

The Ampere circuital law is

$$\int_l \vec{B} \cdot d\vec{l} = \mu_o \int_s \vec{J} \cdot d\vec{s}$$

where μ_o is the permeability of free space, $\mu_o = 4\pi \times 10^{-7}$ H-m^{-1} or T-m-A^{-1}.

For the filamentary current, Ampere's law connects the magnetic flux with the algebraic sum of the enclosed (linked) currents (*net current*) i_n, and

$$\oint_l \vec{B} \cdot d\vec{l} = \mu_o i_n$$

The time-varying magnetic field produces the electromotive force (*emf*),denoted as $\mathscr{E}$, which induces the current in the closed circuit. Faraday's law relates the *emf*, which is merely the induced voltage due to conductor motion in the magnetic field, to the rate of change of the magnetic flux Φ penetrating in the loop. In approaching the analysis of electromechanical energy transformation, Lenz's law should be mentioned. Lenz's law allows one to find the direction of *emf* and the current induced. In particular, the *emf* is in such a direction as to produce a current whose flux, if added to the original flux, would reduce the magnitude of the *emf*. In particular, according to Faraday's law, the induced *emf* in a closed-loop circuit is defined in terms of the rate of change of the magnetic flux Φ as

$$\mathscr{E} = \oint_l \vec{E}(t) \cdot d\vec{l} = -\frac{d}{dt} \int_s \vec{B}(t) \cdot d\vec{s} = -N\frac{d\Phi}{dt} = -\frac{d\psi}{dt}$$

where N is the number of turns and ψ denotes the flux linkages. This formula represents the Faraday law of induction, and in electromechanical motion devices, the induced *emf* (*induced voltage*), as given by

$$\mathscr{E} = -\frac{d\psi}{dt} = -N\frac{d\Phi}{dt}$$

is of particular interest.

The current flows in an opposite direction to the flux linkages. The electromotive force (energy-per-unit-charge quantity) represents the magnitude of the potential difference V in a circuit carrying a current. One obtains

$$V = -ir + \mathscr{E} = -ir - \frac{d\psi}{dt}$$

The unit for the *emf* is volts [V].

The Kirchhoff voltage law states that around a closed path in an electric circuit, the algebraic sum of the *emf* is equal to the algebraic sum of the voltage drop across the resistance. Another formulation is: the algebraic sum of the voltages around any closed path in a circuit is zero. The Kirchhoff current law states that the algebraic sum of the currents at any node in a circuit is zero.

The magnetomotive force (*mmf*) is the line integral of the time-varying magnetic field intensity $\vec{H}(t)$; that is,

$$mmf = \oint_l \vec{H}(t) \cdot d\vec{l}$$

One concludes that the induced *mmf* is the sum of the induced current and the rate of change of the flux penetrating the surface bounded by the contour. To show that, apply Stoke's theorem to find the integral form of Ampere's law (second Maxwell's equation), as given by

$$\oint_l \vec{H}(t) \cdot d\vec{l} = \int_s \vec{J}(t) \cdot d\vec{s} + \int_s \frac{d\vec{D}(t)}{dt} d\vec{s}$$

where $\vec{J}(t)$ is the time-varying current density vector. The unit for the magnetomotive force is amperes or ampere-turns.

The duality of the *emf* and *mmf* can be observed using

$$\mathscr{E} = \oint_l \vec{E}(t) \cdot d\vec{l} \quad and \quad mmf = \oint_l \vec{H}(t) \cdot d\vec{l}$$

The inductance (the ratio of the total flux linkages to the current which they link, $L = \frac{N\Phi}{i}$) and reluctance (the ratio of the *mmf* to the total flux, $\Re = \frac{mmf}{\Phi}$) are used to find *emf* and *mmf*.

Using the following equation for self-inductance $L = \frac{\psi}{i}$, one has

$$\mathscr{E} = -\frac{d\psi}{dt} = -\frac{d(Li)}{dt} = -L\frac{di}{dt} - i\frac{dL}{dt}$$

If $L = const$, one obtains

$$\mathscr{E} = -L\frac{di}{dt}$$

That is, the self-inductance is the magnitude of the self-induced *emf* per unit rate of change of current.

Example 1.7

Find the self-inductances of a long solenoid with air-core and iron-core ($\mu = 2000\ \mu_o$). The solenoid has 100 turns ($N = 100$), the length is 0.2 m ($l = 0.2$ m), and the uniform circular cross-sectional area is 0.005 m² ($A = 0.005$ m²).

Solution

The magnetic field inside a solenoid is given by $B = \frac{\mu_0 N i}{l}$. Using $\mathscr{E} = -N\frac{d\Phi}{dt} = -L\frac{di}{dt}$ and applying $\Phi = BA = \frac{\mu_0 N i A}{l}$, one obtains

$$L = \frac{\mu_0 N^2 A}{l}$$

Then, $L = 3.14 \times 10^{-6}$ H. If solenoid is filled with a magnetic material, then

$$L = \frac{\mu N^2 A}{l} \text{ and } L = 6.28 \times 10^{-3}\ \text{H}$$ □

Example 1.8

Derive a formula for the self-inductance of a torroidal solenoid that has rectangular cross section ($2a \times b$) and mean radius r.

Solution

The magnetic flux through a cross section is found as

$$\Phi = \int_{r-a}^{r+a} B b dr = \int_{r-a}^{r+a} \frac{\mu N i}{2\pi r} b dr = \frac{\mu N i b}{2\pi}\int_{r-a}^{r+a} \frac{1}{r} dr = \frac{\mu N i b}{2\pi} \ln\left(\frac{r+a}{r-a}\right)$$

Hence,

$$L = \frac{N\Phi}{i} = \frac{\mu N^2 b}{2\pi} \ln\left(\frac{r+a}{r-a}\right)$$ □

By studying the electromagnetic torque $\vec{T}$ [N-m] in a current loop, one obtains the following equation

$$\vec{T} = \vec{M} \times \vec{B}$$

where $\vec{M}$ denotes the magnetic moment.

In examining the torque-energy relations in electromechanical motion devices, the goal is to study the magnetic field energy. It is known that the energy stored in the capacitor is $\frac{1}{2}CV^2$, while energy stored in the inductor is $\frac{1}{2}Li^2$. Observe that the energy in the capacitor is stored in the electric field between plates, while the energy in the inductor is stored in the magnetic field within the coils.

Let us find the expressions for energies stored in electrostatic and magnetic fields in terms of field quantities. The total potential energy stored in the electrostatic field is found using the potential difference *V*, and one obtains

$$W_e = \frac{1}{2}\int_v \rho_v V dv \text{ [J]}$$

where ρ_v is the volume charge density [C-m^{-3}], $\rho_v = \vec{\nabla}\cdot\vec{D}$, $\vec{\nabla}$ is the curl operator.

This expression for W_e is interpreted in the following way. The potential energy should be found using the amount of work that is required to position the charge in the electrostatic field. In particular, the work is found as the product of the charge and the potential. Considering the region with a continuous charge distribution ($\rho_v = const$), each charge is replaced by $\rho_v dv$, and hence the equation $W_e = \frac{1}{2}\int_v \rho_v V dv$ results.

In the Gauss form, using $\rho_v = \vec{\nabla}\cdot\vec{D}$ and making use $\vec{E} = -\vec{\nabla}V$, one obtains the following expression for the energy stored in the electrostatic field:

$$W_e = \frac{1}{2}\int_v \vec{D}\cdot\vec{E} dv$$

and the electrostatic volume energy density is $\frac{1}{2}\vec{D}\cdot\vec{E}$ [J - m$^{\pm 3}$].

For a linear isotropic medium

$$W_e = \frac{1}{2}\int_v \varepsilon\left|\vec{E}\right|^2 dv = \frac{1}{2}\int_v \frac{1}{\varepsilon}\left|\vec{D}\right|^2 d\imath$$

The electric field $\vec{E}(x, y, z)$ is found using the scalar electrostatic potential function $V(x, y, z)$ as

$$\vec{E}(x,y,z) = -\vec{\nabla}V(x,y,z)$$

In the cylindrical and spherical coordinate systems, we have

$$\vec{E}(r,\phi,z) = -\vec{\nabla}V(r,\phi,z) \text{ and } \vec{E}(r,\theta,\phi) = -\vec{\nabla}V(r,\theta,\phi)$$

Using $W_e = \frac{1}{2}\int_v \rho_v V dv$, the potential energy that is stored in the electric field between two surfaces (e.g., in capacitor) is found to be

$$W_e = \frac{1}{2}QV = \frac{1}{2}CV^2$$

Using the principle of virtual work, for the lossless conservative system, the differential change of the electrostatic energy dW_e is equal to the differential change of mechanical energy dW_{mec}; that is,

$$dW_e = dW_{mec}$$

For translational motion,

$$dW_{mec} = \vec{F}_e \cdot d\vec{l}$$

where $d\vec{l}$ is the differential displacement.

One obtains $dW_e = \vec{\nabla} W_e \cdot d\vec{l}$.

Hence, the force is the gradient of the stored electrostatic energy, $\vec{F}_e = \vec{\nabla} W_e$.

In the Cartesian coordinates, we have

$$F_{ex} = \frac{\partial W_e}{\partial x},\quad F_{ey} = \frac{\partial W_e}{\partial y},\quad F_{ez} = \frac{\partial W_e}{dz}$$

Example 1.9

Consider the capacitor (the plates have area A and they are separated by x), that is charged to a voltage V. The permittivity of the dielectric is ε. Find the stored electrostatic energy and the force F_{ex} in the x direction.

Solution

Neglecting the fringing effect at the edges, one concludes that the electric field is uniform, and $E = \frac{V}{x}$. Therefore, we have

$$W_e = \frac{1}{2}\int_v \varepsilon \left|\vec{E}\right|^2 dv = \frac{1}{2}\int_v \varepsilon \left(\frac{V}{x}\right)^2 dv = \frac{1}{2}\varepsilon \frac{V^2}{x^2} Ax = \frac{1}{2}\varepsilon \frac{A}{x} V^2 = \frac{1}{2} C(x) V^2$$

The force is

$$F_{ex} = \frac{\partial W_e}{\partial x} = \frac{\partial\left(\frac{1}{2} C(x) V^2\right)}{\partial x} = \frac{1}{2} V^2 \frac{\partial C(x)}{\partial x}$$ □

To find the stored energy in the magnetostatic field in terms of field quantities, the following formula is used:

$$W_m = \frac{1}{2}\int_v \vec{B} \cdot \vec{H} dv$$

The magnetic volume energy density is $\frac{1}{2}\vec{B} \cdot \vec{H}$ [J-m^{-3}].

Using $\vec{B} = \mu\vec{H}$, one obtains two alternative formulas

$$W_m = \frac{1}{2}\int_v \mu\left|\vec{H}\right|^2 dv = \frac{1}{2}\int_v \frac{\left|\vec{B}\right|^2}{\mu} dv$$

To show how the energy concept studied is applied to electromechanical devices, one finds the energy stored in inductors. To approach this problem, substitute $\vec{B} = \vec{\nabla}\times\vec{A}$, and using the following vector identity $\vec{H}\cdot\vec{\nabla}\times\vec{A} = \vec{\nabla}\cdot(\vec{A}\times\vec{H}) + \vec{A}\cdot\vec{\nabla}\times\vec{H}$, one obtains

$$W_m = \frac{1}{2}\int_v \vec{B}\cdot\vec{H}dv = \frac{1}{2}\int_v \vec{\nabla}\cdot(\vec{A}\times\vec{H})dv + \frac{1}{2}\int_v \vec{A}\cdot\vec{\nabla}\times\vec{H}dv = \frac{1}{2}\int_s (\vec{A}\times\vec{H})\cdot d\vec{s} +$$
$$\frac{1}{2}\int_v \vec{A}\cdot\vec{J}dv = \frac{1}{2}\int_v \vec{A}\cdot\vec{J}dv$$

Using the general expression for the vector magnetic potential $\vec{A}(\vec{r})$ [Wb-m^{-1}], as given by

$$\vec{A}(\vec{r}) = \frac{\mu_0}{4\pi}\int_{v_A} \frac{\vec{J}(\vec{r}_A)}{x} dv_J, \quad \vec{\nabla}\cdot\vec{A} = 0$$

we have

$$W_m = \frac{\mu}{8\pi}\int_v\int_{v_J} \frac{\vec{J}(\vec{r}_A)\cdot\vec{J}(\vec{r})}{x} dv_J dv$$

where v_J is the volume of the medium where $\vec{J}$ exists.

The general formula for the self-inductance $i = j$ and the mutual inductance $i \neq j$ of loops i and j is

$$L_{ij} = \frac{N_i\Phi_{ij}}{i_j} = \frac{\psi_{ij}}{i_j}$$

where ψ_{ij} is the flux linkage through i-th coil due to the current in j-th coil; i_j is the current in j-th coil.

The so-called *Neumann* formula is applied to find the mutual inductance. In particular,

$$L_{ij} = L_{ji} = \frac{\mu}{4\pi}\oint_{l_i}\oint_{l_j} \frac{d\vec{l}_j\cdot d\vec{l}_i}{x_{ij}}, \quad i \neq j$$

Then, using $W_m = \frac{\mu}{8\pi}\int_v\int_{v_J}\frac{\vec{J}(\vec{r}_A)\cdot\vec{J}(\vec{r})}{x}dv_J dv$ one immediately obtains

$$W_m = \frac{\mu}{8\pi}\int_{l_i}\int_{l_j}\frac{i_j d\vec{l}_j \cdot i_i d\vec{l}_i}{x_{ij}}$$

Hence, the energy stored in the magnetic field is found to be

$$W_m = \frac{1}{2}i_i L_{ij} i_j$$

As an example, the energy, stored in the inductor is

$$W_m = \frac{1}{2}Li^2$$

The differential change in the stored magnetic energy should be found. Using

$$\frac{dW_m}{dt} = \frac{1}{2}\left(L_{ij}i_j\frac{di_i}{dt} + L_{ij}i_i\frac{di_j}{dt} + i_i i_j\frac{dL_{ij}}{dt}\right)$$

we have $dW_m = \frac{1}{2}(L_{ij}i_j di_i + L_{ij}i_i di_j + i_i i_j dL_{ij})$.

For translational motion, the differential change in the mechanical energy is

$$dW_{mec} = \vec{F}_m \cdot d\vec{l}$$

Assuming that the system is conservative (for lossless systems $dW_{mec} = dW_m$) in the rectangular coordinate system, we obtain the following equation:

$$dW_m = \frac{\partial W_m}{\partial x}dx + \frac{\partial W_m}{\partial y}dy + \frac{\partial W_m}{\partial z}dz = \vec{\nabla}W_m \cdot d\vec{l}$$

Hence, the force is the gradient of the stored magnetic energy, and

$$\vec{F}_m = \vec{\nabla}W_m$$

In the XYZ coordinate system for the translational motion, we have

$$F_{mx} = \frac{\partial W_m}{\partial x},\ F_{my} = \frac{\partial W_m}{\partial y},\ \text{and } F_{mz} = \frac{\partial W_m}{\partial z}$$

For the rotational motion, the torque should be used. Using the differential change in the mechanical energy as a function of the angular displacement θ, the following formula results if the rigid body is constrained to rotate about the z-axis:

$$dW_{mec} = T_e d\theta$$

where T_e is the z-component of the electromagnetic torque.

Assuming that the system is lossless, one obtains the following expression for the electromagnetic torque

$$T_e = \frac{\partial W_m}{\partial \theta}$$

Example 1.10

Calculate the magnetic energy of the torroidal solenoid if its self-inductance is 6.28×10^{-3} H ($L = 6.28 \times 10^{-3}$ H) when the current is 10 A ($i = 10$ A).

Solution

The stored field energy is $W_m = \frac{1}{2}Li^2$; therefore,

$$W_m = \frac{1}{2}6.28 \times 10^{-3} \times 10^2 = 0.314 \text{ J} \qquad \square$$

In electromechanical motion devices, the coupling (magnetic interaction) between windings that are carrying currents is represented by their mutual inductances. In fact, the current in each winding causes the magnetic field in other windings. The mutually induced *emf* is characterized by the mutual inductance, which is a function of the rotor position x or the angular displacement θ. By applying the expression for the coenergy $W_c[i, L(x)]$ or $W_c[i, L(\theta)]$, the developed electromagnetic torque can be easily found; in particular,

$$T_e(i,x) = \frac{\partial W_c[i,L(x)]}{\partial x} \text{ and } T_e(i,\theta) = \frac{\partial W_c[i,L(\theta)]}{\partial \theta}$$

Example 1.11

Consider the electromagnet that has N turns (see Figure 1.8). The distance between the stationary and movable members is denoted as $x(t)$. The mean lengths of the stationary and movable members are l_1 and l_2, respectively, and the cross-sectional area is A. Neglecting the leakage flux, find the force exerted on the movable member if the time-varying current $i_a(t)$ is supplied. The permeabilities of stationary and movable members are μ_1 and μ_2, respectively.

Solution

The magnetostatic force to be found is

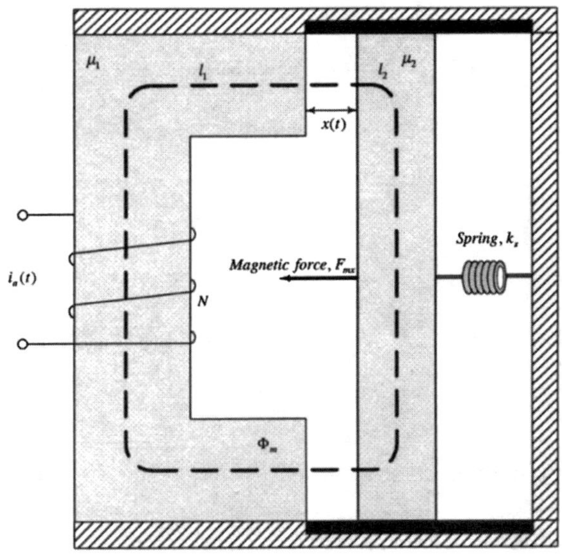

Figure 1.8 Schematic of an electromagnet.

$$F_{mx} = \frac{\partial W_m}{\partial x}$$

where $W_m = \frac{1}{2} L i_a^2(t)$.

The magnetizing inductance should be calculated, and we have

$$L = \frac{N\Phi}{i_a(t)} = \frac{\psi}{i_a(t)}$$

where the magnetic flux is $\Phi = \dfrac{N i_a(t)}{\Re_1 + \Re_x + \Re_x + \Re_2}$.

The reluctances of the ferromagnetic materials of stationary and movable members $\Re_1$ and $\Re_2$, as well as the reluctance of the air gap $\Re_x$, are found as

$$\Re_1 = \frac{l_1}{\mu_0 \mu_1 A}, \quad \Re_2 = \frac{l_2}{\mu_0 \mu_2 A}, \quad \Re_x = \frac{x(t)}{\mu_0 A}$$

and the circuit analog with the reluctances of the various paths is illustrated in Figure 1.9.

Making use of the reluctances in the movable and stationary members and air gap, one obtains the following formula for the flux linkages

$$\psi = N\Phi = \frac{N^2 i_a(t)}{\dfrac{l_1}{\mu_0 \mu_1 A} + \dfrac{2x(t)}{\mu_0 A} + \dfrac{l_2}{\mu_0 \mu_2 A}}$$

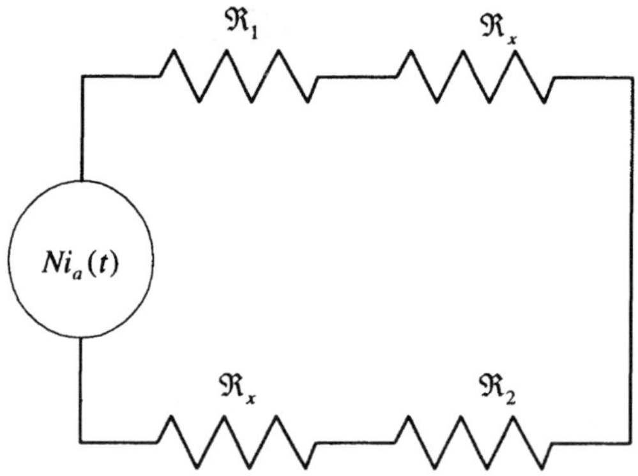

Figure 1.9 Circuit analog.

and the magnetizing inductance is a nonlinear function of the displacement; in particular, we have

$$L(x)=\frac{N^2}{\dfrac{l_1}{\mu_0\mu_1 A}+\dfrac{2x(t)}{\mu_0 A}+\dfrac{l_2}{\mu_0\mu_2 A}}=\frac{N^2\mu_0\mu_1\mu_2 A}{\mu_2 l_1+\mu_1\mu_2 2x(t)+\mu_1 l_2}$$

Using $F_{mx}=\dfrac{\partial W_m}{\partial x}=\dfrac{1}{2}\dfrac{\partial(L(x(t))i_a^2(t))}{\partial x}$, the force in the x direction is found to be

$$F_{mx}=-\frac{N^2\mu_0\mu_1^2\mu_2^2 A i_a^2}{(\mu_2 l_1+\mu_1\mu_2 2x(t)+\mu_1 l_2)^2}$$

It should be emphasized that as differential equations must be developed to model the electromagnet studied, using Newton's second law (see Section 2.1.1 for details), one obtains

$$\frac{dx}{dt}=v$$

$$\frac{dv}{dt}=\frac{1}{m}\left(-\frac{N^2\mu_0\mu_1^2\mu_2^2 A i_a^2}{(\mu_2 l_1+\mu_1\mu_2 2x(t)+\mu_1 l_2)^2}-k_s x^2\right)$$

□

Example 1.12

Two coils have mutual inductance 0.1 H (L_{12} = 0.1 H). The current in the first coil is $i_1=\sqrt{\sin 4t}$. Find the induced *emf* in the second coil.

Solution

The induced *emf* is given as $\mathscr{E}_2=L_{12}\dfrac{di_1}{dt}$. Using the power rule for the time-varying current in the first coil $i_1=\sqrt{\sin 4t}$, we have

$$\frac{di_1}{dt} = \frac{2\cos 4t}{\sqrt{\sin 4t}}$$

Hence,

$$\mathscr{E}_2 = \frac{0.2\cos 4t}{\sqrt{\sin 4t}} \qquad \square$$

Example 1.13

Consider the nonlinear magnetic circuit. Ampere's law states that the line integral of $\vec{B}$ around the closed path is proportional to the *net current* through the enclosed area. However, the value of permeability of the ferromagnetic material depends on the external field, and μ is not constant. This effect is observed due to the *saturation magnetization* and, in general, one should use the nonlinear magnetization curves. That is, the equation $L = \frac{N\Phi}{i} = \frac{\psi}{i} = const$ is valid only if the magnetic system is linear.

As an illustrative example, consider a set of data for the flux linkage–current relation for an electric motor. In particular, we use

$$i = [0 \quad 0.2 \quad 0.4 \quad 0.6 \quad 0.8 \quad 1 \quad 1.2 \quad 1.4 \quad 1.6 \quad 1.8 \quad 2]$$

and

$$\Phi = [0 \quad 0.022 \quad 0.044 \quad 0.065 \quad 0.084 \quad 0.1 \quad 0.113 \quad 0.123 \quad 0.131 \quad 0.136 \quad 0.138]$$

Using MATLAB, the curve fitting, interpolation, and approximation can be performed; see the script.

MATLAB script

```
% NUMERICAL ANALYSIS OF THE CURRENT - FLUX DATA
    % The current-flux data points
I=linspace(0,2,11);
F=[0 0.022 0.044 0.065 0.084 0.1 0.113 0.123 0.131 0.136 0.138];
subplot(3,2,1)
plot(I,F,'o'); axis([0 2 0 0.14]);
xlabel('Current'),ylabel('Flux')
title('The plot of the Current - Flux data')
pause;
    % Differentiation of the current-flux data
dI=I(1:length(I)-1);
dF=diff(F)./diff(I);
subplot(3,2,2)
plot(dI,dF); axis([0 2 0 0.14]);
xlabel('Current'),ylabel('dFlux / dCurrent')
title('Derivative dFlux /dCurrent as a function of Current')
pause;
    % Interpolation of the current-flux curve
I_interval=linspace(0,2,1001);
```

```
F_interpol=interp1(I,F,I_interval,'spline');
subplot(3,2,3)
plot(I,F,'o',I_interval,F_interpol,'-'); axis([0 2 0 0.14]);
xlabel('Current'),ylabel('Flux')
title('Smoothing of the Current - Flux curve using cubic spline
 interpolation')
pause;
      % Differentiation of the interpolated current-flux curve
dI_interval=I_interval(1:length(I_interval)-1);
dF_interpol=diff(F_interpol)'./diff(I_interval);
subplot(3,2,4)
plot(dI_interval,dF_interpol); axis([0 2 0 0.14]);
xlabel('Current'),ylabel('dFlux / dCurrent')
title('Derivative dFlux /dCurrent as a function of Current')
pause;
    % Interpolation of the current-flux curve by N-order polynomials
N=3
polynomial=polyfit(I,F,N)
F_interpolated=polyval(polynomial,I_interval);
subplot(3,2,5)
plot(I,F,'o',I_interval,F_interpolated,'-'); axis([0 2 0 0.14]);
xlabel('Current'),ylabel('Flux')
title('Interpolation of the Current - Flux curve by N-order polynomials')
pause;
      % Differentiation of the interpolated current-flux curve
polynomial_derivatives=polyder(polynomial)
dF_inter=polyval(polynomial_derivatives,I_interval);
subplot(3,2,6)
plot(I_interval,dF_inter); axis([0 2 0 0.14]);
xlabel('Current'),ylabel('dFlux / dCurrent')
title('Derivative dFlux /dCurrent as a function of Current')
clear; disp('End');
```

Figure 1.10 plots the data, and the variation of the inductance is evident. Observe that approximating $\Phi = f(i)$, the third-order polynomial is found as

$$\Phi = -0.0048i^3 - 0.016i^2 + 0.122i - 0.00075$$

1.3 *Engineering computations using* MATLAB

To integrate computation, visualization, and programming in flexible and user-friendly environments, there exist computationally efficient tools, such as MATLAB, MATEMATICA, MATCAD, MATRIX, etc. The MATLAB environment is found to be one of the most useful tools available to efficiently model complex electromechanical systems, design advanced control algorithms, optimize system performance, accomplish data analysis and visualization, perform hardware-in-the-loop simulation, and deploy developed control laws through generation of C-code.

A family of application-specific MATLAB toolboxes with a specialized collection of m-files for solving a particular class of problems provides comprehensiveness in advanced analysis and design, system optimization, controller synthesis, debug, and deployment. The professional version of MATLAB with standard toolboxes is usually available in computing environments of academia and industry. Three toolboxes (Control Systems Toolbox,

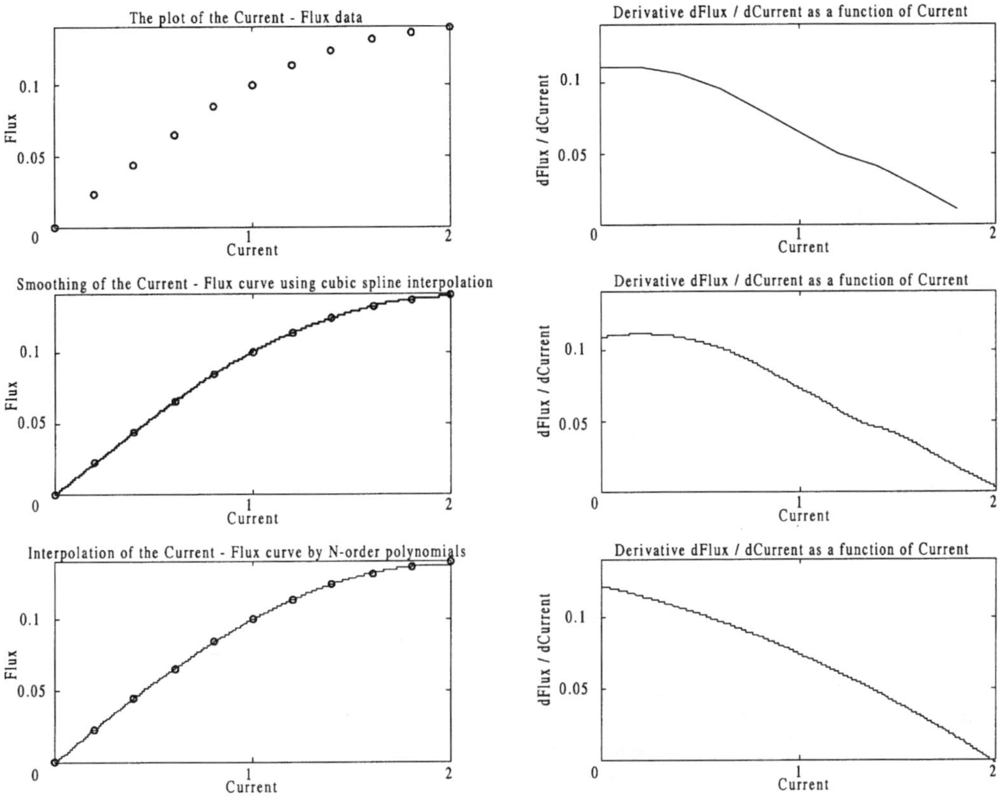

Figure 1.10 Application of numerical analysis in curve fitting, interpolation, and approximation.

Signal Processing Toolbox, and Symbolic Math Toolbox) are included in *The Student Edition of* Matlab*.* This section, as well as Section 1.4, are intended to help students and practicing engineers to use the MATLAB environment efficiently. The topics frequently encountered in analysis and design of electromechanical systems will be covered. A great number of books on MATLAB and SIMULINK are available. Important topics are covered in *The Student Edition of MATLAB* and *The Student Edition of SIMULINK*, as well as in *The MATLAB User's Guide* supplemented by the MathWorks, Inc. The MathWorks educational Web site can be reached at http://education.mathworks.com and http://www.mathworks.com.

This section is not intended to be a substitute for the MATLAB books available. Due to obvious limitations, only the most basic features will be demonstrated to introduce the MATLAB environment and show how it can be applied to a wide variety of problems. It is the author's goal to encourage students and engineers to master MATLAB, even those who have little or no experience. The major aim is to build problem-solving skills and to provide the background and capabilities needed to efficiently proceed in analysis and design, optimization and control, data acquisition, and filtering. It is the author's hope that through this book, the reader will find that advanced software complements theory.

To start MATLAB, one clicks on the appropriate icon or uses the menu option. After each MATLAB command, the Enter (Return) key must be pressed. The Command window is the primary place where one interacts with MATLAB. The MATLAB prompt » is displayed in the Command window, and a blinking cursor appears to the right of the prompt when the Command window is active.

The use of the help and demo commands is the simplest way to find the information and help needed. For example, typing

```
» help sin
```

and pressing the Enter key, the following information is displayed

```
SIN   Sine
SIN(X) is the sine of the elements of X
```

MATLAB offers the following help topics:

- Data analysis (help datafun)
- Demonstration (help demo);
- Differential equations solvers (help funfun);
- General purpose command (help general);
- Matrices and linear algebra (help elmat and help matfun);
- Mathematical functions (help elfun and help specfun);
- Operators and special characters (help ops);
- Polynomials (help polyfun);
- Programming language (help lang);
- Two- and three-dimensional graphics (help graph2d and help graph3d).

At the beginning, to illustrate the basic arithmetic operations (addition, subtraction, multiplication, division, and exponentiation), one can calculate $\frac{2+3-1}{4\times 5-6^2}$.

One types in the MATLAB Command window

```
» (2+3-1)/(4*5-6^2)
```

Next, one presses the Enter key, and the following answer appears.

```
ans =
   -0.2500
```

As an alternative, one can use variables; in particular, for the example considered, type

```
» a=2;b=3;c=1;d=4;e=5;f=6;(a+b-c)/(d*e-f^2)
```

then press the Enter key, and the answer is displayed.

```
ans =
   -0.2500
```

To display the value of variables used, one should enter the variable name at the prompt and press the Enter key; for example,

```
» a
a =
     2
```

A list of the variables used can be obtained using the MATLAB command `who`. In particular,

```
» who
Your variables are:
a          b          d          f
ans        c          e
```

The basic complex and coordinate transformations, exponential, remainder and rounding, and specialized mathematic and trigonometric functions are supported by MATLAB. For example, to find and plot the following function $f(t) = \cos(t) + \sin(2t + 1)e^{-0.5t}$ for $0 \le t \le 10$ s, one has to assign the time interval of interest, calculate $f(t)$, and plot this function using the `plot` command. We have

```
» t=linspace(0,10,1000); f=cos(t)+sin(2*t+1).*exp(-0.5*t); plot(t,f)
```

The plot of the function $f(t)$ is given in Figure 1.11.

Polynomial analysis, curve fitting and interpolation, are easily performed. For example, consider two polynomials

$$p_1(x) = 2x^6 + 4x^4 + 5x^3 + 7x^2 - 8x - 10 \text{ and } p_2(x) = -3x^5 - 8x^4 - 7x^3 - 4x^2 - 9x - 1$$

The goal is to multiply these polynomials and find the roots of the resulting polynomial. To solve this problem, enter $p_1(x)$ and $p_2(x)$, and then multiply them using the `conv` command.

```
» p1=[2 0 4 5 7 -8 -10];p2=[-3 -8 -7 -4 -9 -1];p=conv(p1,p2)
p =
  -6 -16 -26 -55 -107 -85 -11   59   34 105   98   10
```

The roots of the resulting polynomial

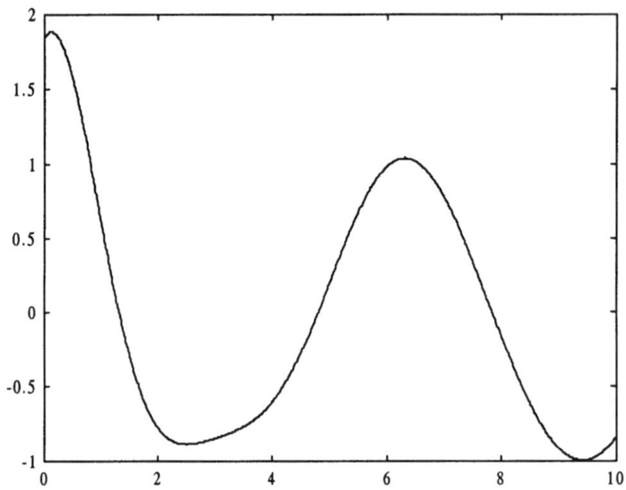

Figure 1.11 Plot of function $f(t) = \cos(t) + \sin(2t + 1)e^{-0.5t}$.

$$p(x) = -6x^{11} - 16x^{10} - 26x^9 - 55x^8 - 107x^7 - 85x^6 - 11x^5 + 59x^4 + 34x^3 + 105x^2 + 98x + 10$$

are found using the `roots` command. We have

```
» roots(p)
ans =
   0.7405 + 1.7253i
   0.7405 - 1.7253i
  -1.5840 + 0.7370i
  -1.5840 - 0.7370i
  -0.8518 + 1.0485i
  -0.8518 - 1.0485i
   1.0000
   0.3087 + 0.9197i
   0.3087 - 0.9197i
  -0.7773
  -0.1160
```

Four different formats are supported by MATLAB. In `format long` and `format long e`, the value is displayed in fifteen-digit-fixed and fifteen-digit-floating point formats, respectively. Using `format short` and `format short e`, only five digits are displayed.

MATLAB is a very efficient tool when one works with arrays, matrices, and linear algebra. For example, create the matrix

$$A = \begin{bmatrix} -6 & -5 & -4 & -3 & -2 \\ 3 & 6 & 9 & 12 & 15 \\ 5 & 10 & 15 & 20 & 25 \\ 5 & 4 & 3 & 2 & 1 \\ 6 & 7 & 8 & 9 & 10 \end{bmatrix}$$

and find $B = 2A - 3A^2 - A^{-1}$. The matrix A is assigned as

```
» A=[-6:1:-2;3:3:15;5:5:25;5 4 3 2 1;6 7 8 9 10]
A =
   -6    -5    -4    -3    -2
    3     6     9    12    15
    5    10    15    20    25
    5     4     3     2     1
    6     7     8     9    10
```

To calculate $B = 2A - 3A^2 - A^{-1}$, we have

```
» B=2*A-3*A^2-inv(A)

Warning: Matrix is close to singular or badly scaled.
     Results may be inaccurate. RCOND = 4.271668e-018.

B =
  1.0e+015 *
     0.6769   -0.3753    0.2305    0.7506    0.0469
    -2.1580    0.7506   -0.0751   -1.5012   -1.2197
     1.6084   -1.1259   -0.0965    0.0000    2.2518
     0.5495    1.5012   -0.5040    1.5012   -1.0321
    -0.6769   -0.7506    0.4450   -0.7506   -0.0469
```

The warning message appears, and the eigenvalues should be studied. In the `format short`, the eigenvalues of matrix A are found using the `eig` command; that is,

```
» eig(A)
ans =
  35.1391
  -8.1391
   0.0000
   0.0000 + 0.0000i
   0.0000 - 0.0000i
```

while in the `format short e`, we have

```
» format short e; eig(A)
ans =
   3.5139e+001
  -8.1391e+000
   6.1065e-015
  -1.3856e-015 +4.9312e-016i
  -1.3856e-015 -4.9312e-016i
```

The determinant of matrix can be easily obtained. For example, for matrix

$$C=\begin{bmatrix}1 & 0 & 0\\ 2 & 3 & 4\\ 5 & 0 & 6\end{bmatrix}$$

one finds the determinant using the `det` command. We have

```
» C=[1 0 0;2 3 4;5 0 6];det(C)
ans =
   18
```

To solve the linear algebraic equation

$$Cz = D$$

one finds the solution as

$$z = C^{-1} D$$

Using MATLAB, and assigning $C=\begin{bmatrix}1 & 0 & 0\\ 2 & 3 & 4\\ 5 & 0 & 6\end{bmatrix}$ and $D=\begin{bmatrix}7\\ 8\\ 9\end{bmatrix}$, we can find z. In particular,

```
» C=[1 0 0;2 3 4;5 0 6];D=[7; 8; 9];z=inv(C)*D
z =
    7.0000e+000
    3.7778e+000
   -4.3333e+000
```

That is, the solution is $z=\begin{bmatrix}7\\ 3.7778\\ -4.3333\end{bmatrix}$.

1.4 Analysis and modeling of dynamic system using MATLAB: An introduction

1.4.1 Continuous-time systems

Linear and nonlinear differential equations should be solved to perform the analysis of electromechanical systems. In the state-space vector form, n states $x \in \mathbb{R}^n$ and m controls $u \in \mathbb{R}^m$ are used. In particular, mathematical models of electromechanical systems are found as n first-order differential equations. The transient dynamics of linear continuous-time electromechanical systems are mapped by linear differential equations

$$\frac{dx_1}{dt} = a_{11}x_1 + a_{12}x_2 + \ldots + a_{1n-1}x_{n-1} + a_{1n}x_n + b_{11}u_1 + b_{12}u_2 + \ldots + b_{1m-1}u_{m-1} + b_{1m}u_m, \ x_1(t_0) = x_{10}$$

$$\frac{dx_2}{dt} = a_{21}x_1 + a_{22}x_2 + \ldots + a_{2n-1}x_{n-1} + a_{2n}x_n + b_{21}u_1 + b_{22}u_2 + \ldots + b_{2m-1}u_{m-1} + b_{2m}u_m, \ x_2(t_0) = x_{20}$$

$$\vdots$$

$$\frac{dx_{n-1}}{dt} = a_{n-11}x_1 + a_{n-12}x_2 + \ldots + a_{n-1\ n-1}x_{n-1} + a_{n-1\ n}x_n + b_{n-11}u_1 + b_{n-12}u_2 + \ldots + b_{n-1\ m-1}u_{m-1} +$$

$$b_{n-1\ m}u_m, \ x_{n-1}(t_0) = x_{n-10}$$

$$\frac{dx_n}{dt} = a_{n1}x_1 + a_{n2}x_2 + \ldots + a_{n\,n-1}x_{n-1} + a_{n\ n}x_x + b_{n1}u_1 + b_{n2}u_2 + \ldots + b_{n\ m-1}u_{m-1} + b_{n\ m}u_m, \ x_n(t_0) = x_{n0}$$

which can be represented in matrix form as

$$\frac{dx}{dt} = \begin{bmatrix} \frac{dx_1}{dt} \\ \frac{dx_2}{dt} \\ \vdots \\ \frac{dx_{n-1}}{dt} \\ \frac{dx_n}{dt} \end{bmatrix} = \begin{bmatrix} a_{11} & a_{12} & \cdots & a_{1n-1} & a_{1n} \\ a_{21} & a_{22} & \cdots & a_{2n-1} & a_{2n} \\ \vdots & \vdots & \ddots & \vdots & \vdots \\ a_{n-11} & a_{n-12} & \cdots & a_{n-1\ n-1} & a_{n-1\ n} \\ a_{n1} & a_{n2} & \cdots & a_{n\ n-1} & a_{n\ n} \end{bmatrix} \begin{bmatrix} x_1 \\ x_2 \\ \vdots \\ x_{n-1} \\ x_n \end{bmatrix} +$$

$$\begin{bmatrix} b_{11} & b_{12} & \cdots & b_{1m-1} & b_{1m} \\ b_{21} & b_{22} & \cdots & b_{2m-1} & b_{2m} \\ \vdots & \vdots & \ddots & \vdots & \vdots \\ b_{n-11} & b_{n-12} & \cdots & b_{n-1\ m-1} & b_{n-1\ m} \\ b_{n1} & b_{n2} & \cdots & b_{n\ m-1} & b_{n\ m} \end{bmatrix} \begin{bmatrix} u_1 \\ u_2 \\ \vdots \\ u_{m-1} \\ u_m \end{bmatrix} = Ax + Bu$$

$$x(t_0) = x_0$$

where x is the state vector with initial conditions $x(t_0) = x_0$; u is the control vector; and $A \in \mathbb{R}^{n \times n}$ and $B \in \mathbb{R}^{n \times m}$ are the matrices of constant coefficients.

Warning: Using the common notations, x denotes the states (currents and voltages, angular and linear velocities, angular and linear displacements, and other physical quantities) in differential equations. One must distinguish the states and the displacement (position), also denoted by x. □

In contrast, nonlinear electromechanical systems, mapped by a set of n first-order nonlinear differential equations, are modeled as

$$\frac{dx_1}{dt} = F_1(t, \ x_1, \ x_2, \ \ldots, \ x_{n-1}, \ x_n, \ u_1, \ u_2, \ \ldots, \ u_{m-1}, \ u_m), \ x_1(t_0) = x_{10}$$

$$\frac{dx_2}{dt} = F_2(t, \ x_1, \ x_2, \ \ldots, \ x_{n-1}, \ x_n, \ u_1, \ u_2, \ \ldots, \ u_{m-1}, \ u_m), \ x_2(t_0) = x_{20}$$

$$\vdots$$

$$\frac{dx_{n-1}}{dt} = F_{n-1}(t,\ x_1,\ x_2,\ \ldots,\ x_{n-1},\ x_n,\ u_1,\ u_2,\ \ldots,\ u_{m-1},\ u_m),\ x_{n-1}(t_0) = x_{n-10}$$

$$\frac{dx_n}{dt} = F_n(t,\ x_1,\ x_2,\ \ldots,\ x_{n-1},\ x_n,\ u_1,\ u_2,\ \ldots,\ u_{m-1},\ u_m),\ x_n(t_0) = x_{n0}$$

and matrix form is given by

$$\frac{dx}{dt} = F(t, x, u),\ \ x(t_0) = x_0$$

where t is the time; $F(t, x, u)$ is the nonlinear map.

Hence, for n-degree-of-freedom electromechanical systems, we use n-dimensional state variables (current, voltage, angular velocity, displacement, and acceleration)

$$x = \begin{bmatrix} x_1 \\ x_2 \\ \vdots \\ x_{n-1} \\ x_n \end{bmatrix}$$

m-dimensional control inputs (current and voltage applied to control electric machines and power converters; for example, the voltages applied to the *abc* stator and rotor windings of induction motors),

$$u = \begin{bmatrix} u_1 \\ u_2 \\ \vdots \\ u_{m-1} \\ u_m \end{bmatrix}$$

In addition to the states and controls, the disturbance vector, as denoted by d, should be considered. The l-dimensional disturbance vector (the load and disturbances torques and forces) is

$$d = \begin{bmatrix} d_1 \\ d_2 \\ \vdots \\ d_{l-1} \\ d_l \end{bmatrix}$$

Five solvers `ode113`, `ode15s`, `ode23`, `ode23s`, and `ode45` are available and can be used to solve linear and nonlinear differential equations given in Cauchy's form. The commonly applied `ode45` solver uses the fourth-fifth Runge-Kutta method to find the solution of differential equations. Nonlinear differential equations are difficult to solve analytically. To show the use of the `ode45` solver, let us demonstrate the application of the MATLAB environment.

Example 1.14

Solve the following system of highly nonlinear differential equations

$$\frac{dx_1(t)}{dt} = -2x_1 + |x_2|, \ \ x_1(t_0) = x_{10}$$

$$\frac{dx_2(t)}{dt} = -x_1x_2 - |x_2| + 10(-x_2 + x_3), \ \ x_2(t_0) = x_{20}$$

$$\frac{dx_3(t)}{dt} = -10x_1x_2 + 30x_2 - x_3 + u, \ \ x_3(t_0) = x_{30}$$

The modeling of these differential equations was performed by assigning the initial conditions to be

$$x_0 = \begin{bmatrix} x_{10} \\ x_{20} \\ x_{30} \end{bmatrix} = \begin{bmatrix} 10 \\ 20 \\ -20 \end{bmatrix}$$

It is important to emphasize that control laws significantly influence the dynamic performance. To illustrate this, we model the dynamics by applying two control inputs as given by

$$u = 0$$

and

$$u = -100\tanh(0.06x_1 + 0.06x_2 + x_3)$$

Two MATLAB m-files are developed. Comments and remarks, that are not executed (ignored), appear after the % symbol. These comments explain particular steps in MATLAB scripts. MATLAB and SIMULINK `m`- and `mdl`-files are macros that are stored as ordinary files with the extension `m` or `mdl` (`filename.m` or `filename.mdl`). An `m`-file can be either a function with input and output variables or a list of commands. These `m`- and `mdl`-files must be stored either in the working directory or in a directory that is specified in the MATLAB path list. If the user stored the files in the directory files, then to access to `m`- and `mdl`-files, one should go to the directory and type `c:\Matlab\files` in the Command window. Alternatively, the directory can be assigned in the path.

MATLAB `ode45` and `plot3` script (`ch_1_01.m`)

```
echo on
clc
    %  Differential equations to be solved are
    %  dy1/dt=-2*y1+|y2|
```

```
echo on
   %  dy2/dt=-y1*y2-|y2|+10*(-y2-y3)
   %  dy3/dt=-10*y1*y2+30*y2-y3+u
tspan=[0 20];        % The assigned initial and final time
y0=[10 20 -20]';     % The assigned initial conditions
[t,y]=ode45('ch_1_02',tspan,y0);
% Plot of the time history found solving the third-order differential
% equation
plot(t,y(:,1),'--',t,y(:,2),'-',t,y(:,3),':');
xlabel('Time (seconds)');
title('Differential equation time history');
pause
% 3-D plot
plot3(y(:,1),y(:,2),y(:,3))
xlabel('x1'), ylabel('x2'), zlabel('x3')
text(10,20,-20,'o')
text(11,25,-20,'Initial')
text(0,0,0,'o')
text(1,5,-10,'Origin')
v=axis
pause
clc
disp('END')
```

The second file to be used is

MATLAB ode45 and plot3 script (ch_1_02.m)

```
function yprime = difer(t,y);
%  Solution of differential equations
%  Differential equations are:
%  dy1/dt=-2*y1+|y2|
%  dy2/dt=-y1*y2-|y2|+10*(-y2+y3)
%  dy3/dt=-10*y1*y2+30*y2-y3+u
%u=-100*tanh(0.06*y(1,:)+0.06*y(2,:)+y(3,:));
u=0;
a11=-2; a22=10; a31=-10; a32=30;
yprime=[a11*y(1,:)+abs(y(2,:));...
-y(1,:)*y(2,:)-abs(y(2,:))+a22*(-y(2,:)+y(3,:));...
a31*y(1,:)*y(2,:)+a32*y(2,:)-y(3,:)+u];
```

Using these m-files, the transient dynamics are found, and the state evolution is demonstrated in Figure 1.12.

Two- and three-dimensional graphics illustrate the MATLAB visualization capabilities and allow one to plot two or three variables. These two- and three-dimensional plots are found to be very illustrative in addition to the transient analysis. For the example studied, a three-dimensional plot is obtained using x_1, x_2, and x_3 as the variables to be analyzed.

With the controller, as given by $u = -100\tanh(0.06x_1 + 0.06x_2 + x_3)$, the dynamic behavior of the closed-loop system and the corresponding three-dimensional plot are documented in Figure 1.13.

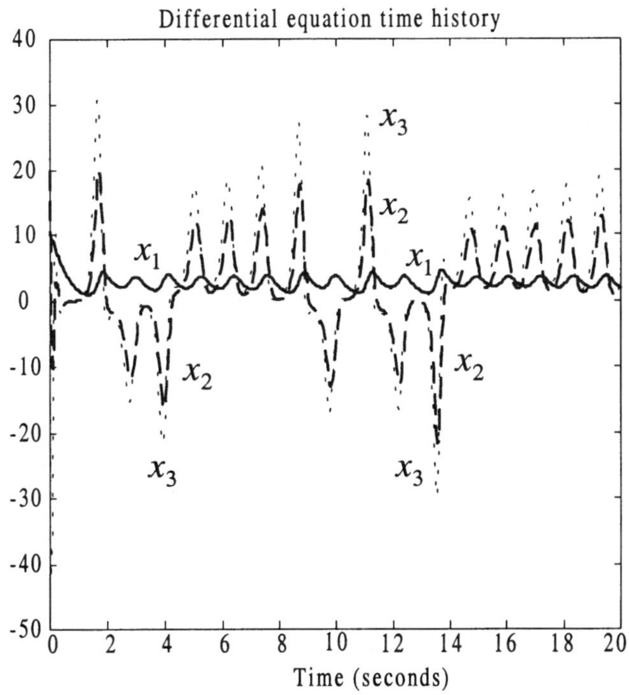

Figure 1.12 Transient dynamics if $u = 0$.

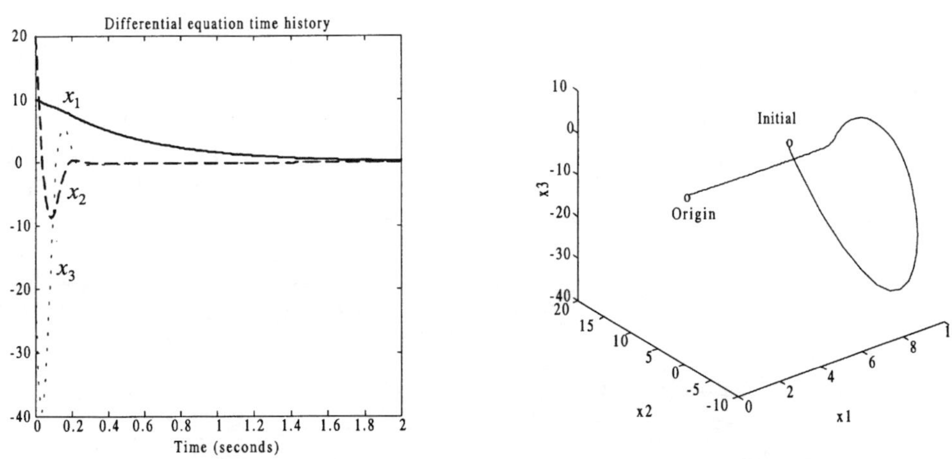

Figure 1.13 System dynamics and three-dimensional plot if $u = -100\tanh(0.06x_1 + 0.06x_2 + x_3)$.

□

Multi-input/multi-output electromechanical systems are also studied. For example, consider the motion control problem for the aircraft studied in Section 1.1 (see Figures 1.3 and 1.4). The aircraft outputs are the Euler angles, and the flight vehicle is controlled by deflecting the control surfaces. To actuate ailerons, canards, fins, flaps, rudders, stabilizers, and other control surfaces, permanent-magnet stepper motors can be used. The number of actuators is equal to the number of control surfaces needed to be driven. Furthermore, voltages u_{as} and u_{bs} are applied to the *as* and *bs* stator windings of stepper motors. That is, the multi-input (two voltages are applied to the stepper motors to actuate the control surfaces) -multi-output (three Euler angles θ, ϕ, and ψ to be controlled) nature is obvious. The pilot assigns the desired Euler angles r_θ, r_ϕ, and r_ψ (reference commands). Using the errors between the reference vector $r = \begin{bmatrix} r_\theta \\ r_\phi \\ r_\psi \end{bmatrix}$ and output vector $y = \begin{bmatrix} \theta \\ \phi \\ \psi \end{bmatrix}$, as defined by $e = r - y = \begin{bmatrix} r_\theta \\ r_\phi \\ r_\psi \end{bmatrix} - \begin{bmatrix} \theta \\ \phi \\ \psi \end{bmatrix}$, the flight tracking control is achieved by applying the stator voltages to stepper motors to deflect the control surfaces.

It is evident that the aircraft outputs (θ, ϕ, and ψ) can be obtained using the state variables (v, α, q, θ, β, p, r, ϕ, and ψ). The output equation is given by

$$y = \begin{bmatrix} \theta \\ \phi \\ \psi \end{bmatrix} = \begin{bmatrix} 0 & 0 & 0 & 1 & 0 & 0 & 0 & 0 & 0 \\ 0 & 0 & 0 & 0 & 0 & 0 & 0 & 1 & 0 \\ 0 & 0 & 0 & 0 & 0 & 0 & 0 & 0 & 1 \end{bmatrix} \begin{bmatrix} v \\ \alpha \\ q \\ \theta \\ \beta \\ p \\ r \\ \phi \\ \psi \end{bmatrix} = Hx$$

Figure 1.14 illustrates the block diagram representation of the multi-variable fighter with nine states (v, α, q, θ, β, p, r, ϕ, and ψ), $x \in \mathbb{R}^9$, eight control surfaces (right and left horizontal stabilizers, right and left leading and trailing edge laps, right and left rudders), three outputs (θ, ϕ, and ψ), $y \in \mathbb{R}^3$, and three reference inputs (r_θ, r_ϕ, and r_ψ), $r \in \mathbb{R}^3$.

To actuate eight control surfaces, eight permanent-magnet stepper motors are used. For each motor, the ap ages to the *ab* windings must be regulated. For *j*-th flight actuator, we have, $u_j = \begin{bmatrix} u_{jas} \\ u_{jbs} \end{bmatrix}$. That is, we have 16 control inputs, $u \in \mathbb{R}^{16}$.

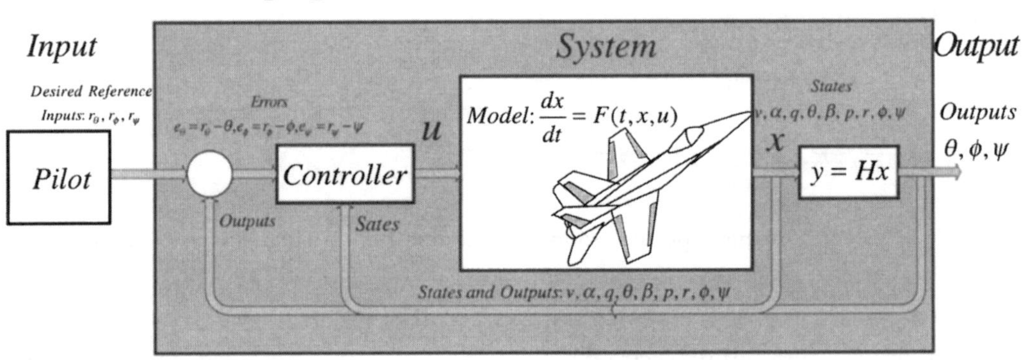

Figure 1.14 Block diagram representation of a multi-input/multi-output aircraft.

In electric drives, the output is the angular velocity, which is proportional to the rotor angular velocity ω_{motor}, and $y = k_{gear}\omega_{motor}$. In servo-systems, the output is either the linear position or the angular displacement, which is proportional to the angular rotor displacement θ_{rotor}. Hence, the output equation is $y = k_{gear}\theta_{rotor}$.

In general, for multi-input/multi-output electromechanical systems, one has b reference inputs $r \in \mathbb{R}^b$ and b outputs $y \in \mathbb{R}^b$, which are obtained in terms of n states $x \in \mathbb{R}^n$. It must be emphasized that m control inputs $u \in \mathbb{R}^m$ and l disturbances $d \in \mathbb{R}^1$ can be also viewed as the input signals. The output equation is given as

$$\begin{aligned}
y_1 &= h_{11}x_1 + h_{12}x_2 + \ldots + h_{1n-1}x_{n-1} + h_{1n}x_n \\
y_2 &= h_{21}x_1 + h_{22}x_2 + \ldots + h_{2n-1}x_{n-1} + h_{2n}x_n \\
&\vdots \\
y_{b-1} &= h_{b-11}x_1 + h_{b-12}x_2 + \ldots + h_{b-1\ n-1}x_{n-1} + h_{b-1\ n}x_n \\
y_b &= h_{b1}x_1 + h_{b2}x_2 + \ldots + h_{b\ n-1}x_{n-1} + h_{b\ n}x_n
\end{aligned}$$

and in matrix form

$$y = \begin{bmatrix} y_1 \\ y_2 \\ \vdots \\ y_{b-1} \\ y_b \end{bmatrix} = \begin{bmatrix} h_{11} & h_{12} & \cdots & h_{1n-1} & h_{1n} \\ h_{21} & h_{22} & \cdots & h_{2n-1} & h_{2n} \\ \vdots & \vdots & \ddots & \vdots & \vdots \\ h_{b-11} & h_{b-12} & \cdots & h_{b-1\ n-1} & h_{b-1\ n} \\ h_{b1} & h_{b2} & \cdots & h_{b\ n-1} & h_{b\ n} \end{bmatrix} \begin{bmatrix} x_1 \\ x_2 \\ \vdots \\ x_{n-1} \\ x_n \end{bmatrix} = Hx$$

where $H \in \mathbb{R}^{b\times n}$ is the matrices of coefficients.

The functional block diagram of nonlinear electromechanical systems (n states, m controls, b reference inputs, and b outputs), which are described by the state-space equation

$$\frac{dx}{dt} = F(t, x, u), \quad x(t_0) = x_0$$

with the output equation

$$y = Hx$$

is illustrated in Figure 1.15.

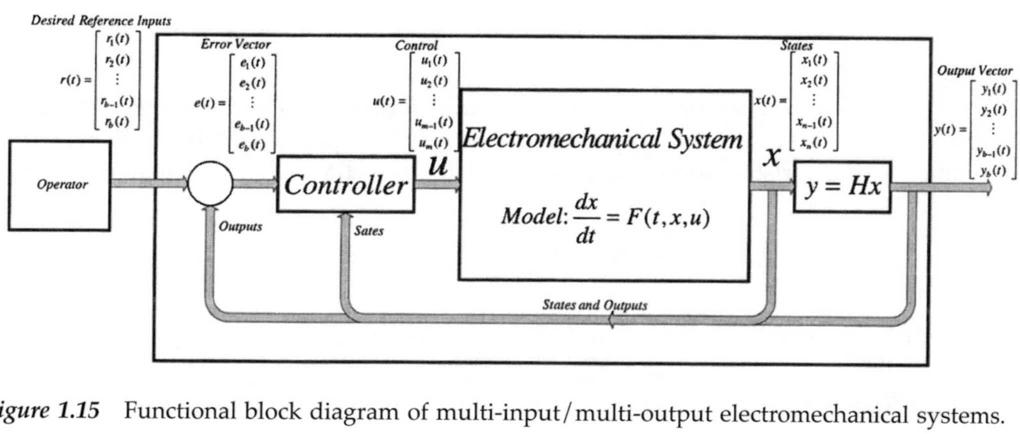

Figure 1.15 Functional block diagram of multi-input/multi-output electromechanical systems.

The transfer function concept is thoroughly covered in the undergraduate engineering curriculum. The transfer function of a linear time-invariant (real constant coefficients) n-order differential equation

$$a_n \frac{d^n x(t)}{dt^n} + a_{n-1} \frac{d^{n-1} x(t)}{dt^{n-1}} + \ldots + a_1 \frac{dx(t)}{dt} + a_0 = b_m \frac{d^m u(t)}{dt^m} + b_{m-1} \frac{d^{m-1} u(t)}{dt^{m-1}} + \ldots + b_1 \frac{du(t)}{dt} + b_0,$$

$$n \geq m$$

$$\sum_{k=0}^{n} a_k \frac{d^k x(t)}{dt^k} = \sum_{k=0}^{m} b_k \frac{d^k u(t)}{dt^k}, \quad n \geq m$$

is the ratio of the Laplace transform of the system output to the Laplace transform of the system input with initial conditions set to zero.

Assuming that initial conditions are zero, apply the Laplace transform to both sides of equation

$$\sum_{k=0}^{n} a_k \frac{d^k x(t)}{dt^k} = \sum_{k=0}^{m} b_m \frac{d^k u(t)}{dt^k}$$

Using the linearity and differentiation properties, we have

$$\left(\sum_{k=0}^{n} a_k s^k \right) X(s) = \left(\sum_{k=0}^{m} b_m s^k \right) U(s)$$

Hence, the transfer function is the ratio of polynomials, and

$$G(s) = \frac{X(s)}{U(s)} = \frac{b_m s^m + b_{m-1} s^{m-1} + \ldots + b_1 s + b_0}{a_n s^n + a_{n-1} s^{n-1} + \ldots + a_1 s + a_0}$$

In MATLAB, the numerator and denominator of this transfer function are assigned as

$$num = [b_m b_{m-1} \ldots b_1 b_0] \text{ and } den = [a_n a_{n-1} \ldots a_1 a_0]$$

By setting the denominator polynomial of the transfer function to zero, one obtains the characteristic equation. The stability of linear systems is guaranteed if all characteristic eigenvalues (characteristic roots) obtained by solving the characteristic equations

$$a_n s^n + a_{n-1} s^{n-1} + \ldots + a_1 s + a_0 = 0$$

have negative real parts.

Example 1.15

Consider the following transfer function

$$G(s) = \frac{X(s)}{U(s)} = \frac{s^3 + 2s^2 + 3s + 4}{s^7 + 2s^6 + 3s^5 + 4s^4 + 5s^3 + 6s^2 + 7s + 8}$$

Find the eigenvalues and analyze the stability of the system.

Solution

The characteristic equation is

$$s^7 + 2s^6 + 3s^5 + 4s^4 + 5s^3 + 6s^2 + 7s + 8 = 0$$

One can find the eigenvalues (characteristic roots) using the `roots` command. In particular, typing

```
» den=[1 2 3 4 5 6 7 8];eigenvalues=roots(den)
```

and pressing the Enter key, the following roots are displayed.

```
eigenvalues =
  -1.3995
  -1.0164 + 0.9455i
  -1.0164 - 0.9455i
  -0.0946 + 1.3479i
  -0.0946 - 1.3479i
   0.8107 + 0.9836i
   0.8107 - 0.9836i
```

One conclues that the system is unstable. □

To find the solution using the inverse Laplace transform, the partial-fraction expansion of the transfer function must be found, and the `residue` MATLAB command is applied.

Example 1.16

Find the inverse Laplace transform of

$$G(s) = \frac{X(s)}{U(s)} = \frac{10s^2 + 50s + 30}{(s^2 + 2s)(s^2 + 9s + 20)}$$

and express the solution $x(t)$.

Solution

To solve the problems, type

```
» num=[10 50 30]; den1=[1 2 0]; den2=[1 9 20]; den=conv(den1,den2);
» [residues,eigenvalues,dc_component]=residue(num,den)
```

The following eigenvalues, residues, and dc component are displayed on the screen as one presses the Enter key.

```
residues =
   -2.0000
   -1.2500
    2.5000
    0.7500

eigenvalues =
   -5.0000
   -4.0000
   -2.0000
    0

dc_component =
    []
```

Using the residues, eigenvalues, and dc component obtained, we have the following formula for $x(t)$

$$x(t) = (0.75 + 2.5e^{-2t} - 1.25e^{-4t} - 2e^{-5t})1(t) \qquad \square$$

In addition to possibilities to find analytical solutions, the system dynamics can be obtained numerically. For example, one can use the `step` command to find the unit response.

Example 1.17

Using MATLAB, find the unit step response for the following transfer function

$$G(s) = \frac{X(s)}{U(s)} = \frac{10s + 30}{0.2s^4 + s^3 + 2s^2 + 3s + 4}$$

Solution

The unit response is found using the `step` command and plotting the resulting transient response. It must be emphasized that one assigns the numerator and denominator in the Command window, and then the `step` command and plotting statement are used. We have

```
» num=[10 30]; den=[0.2 1 2 3 4];t=0:0.01:10;
» y=step(num,den,t);
» plot(t,y); title('Output dynamics'); xlabel('Time (seconds)');
```

The system output on the unit step input is shown in Figure 1.16.

It is important to analyze the frequency response of dynamic systems to perform the analysis and design of electromechanical systems and filters. The frequency response is obtained by letting $s = j\omega$, where ω is the angular frequency in rad-s^{-1}. We plot two separate

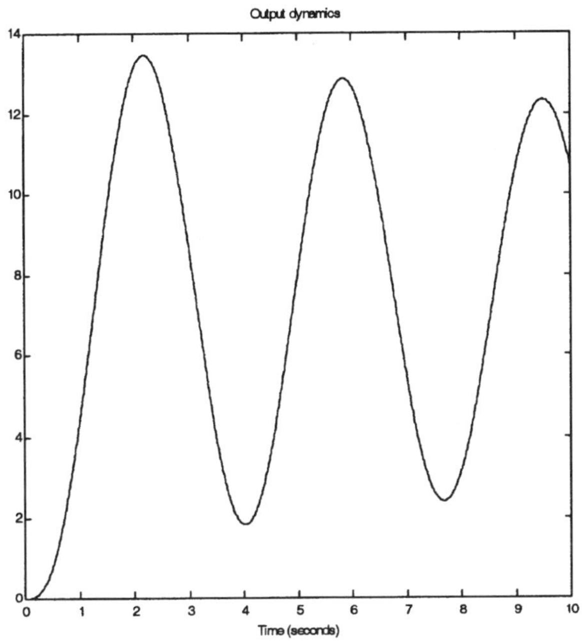

Figure 1.16 The unit step response. □

plots. The first shows how the amplitude of $G(j\omega)$ varies with the angular frequency (the log scale is used for ω), and the second plot illustrates how the phase angle of $G(j\omega)$ varies with ω. The magnitude of $G(j\omega)$ in decibels is denoted as $|G(j\omega)|_{dB}$ and,

$$|G(j\omega)|_{dB} = 20\log_{10}|G(j\omega)|$$

Using the bode MATLAB command, the `Bode` plot can be easily found.

Example 1.18

Plot the Bode plots for

$$G(s) = \frac{X(s)}{U(s)} = \frac{10s + 30}{0.2s^4 + s^3 + 2s^2 + 3s + 4}$$

Solution

As one assigns the numerator and denominator, the `bode` function is used as given below

```
» num=[10 30]; den=[0.2 1 2 3 4]; bode(num,den)
```

The Bode plots for $|G(j\omega)|_{dB}$ and the phase angle of $G(j\omega)$, which are displayed on the screen, are illustrated in Figure 1.17.

The transfer function can be found using the state-space equations. Consider the linear time-invariant systems as described by

$$\frac{dx}{dt} = Ax + Bu, \quad y = Hx$$

The goal is to find the transfer function

$$G(s) = \frac{Y(s)}{U(s)}$$

Taking the Laplace transform of the state-space and output equations, we obtain

$$sX(s) - x(t_0) = AX(s) + BU(s), \quad Y(s) = HX(s)$$

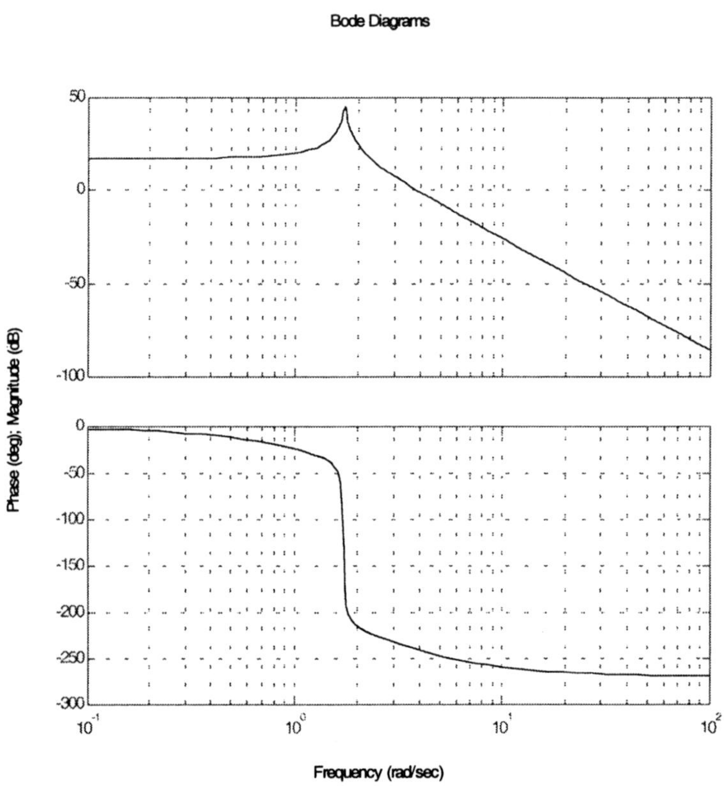

Figure 1.17 Bode plots for $G(s) = \dfrac{X(s)}{U(s)} = \dfrac{10s + 30}{0.2s^4 + s^3 + 2s^2 + 3s + 4}$. □

Assuming that $x(t_0) = 0$, we have

$$X(s) = (sI - A)^{-1} BU(s)$$

Hence,

$$Y(s) = H(sI - A)^{-1}BU(s)$$

Therefore, the transfer function is found to be

$$G(s) = \frac{Y(s)}{U(s)} = H(sI - A)^{-1}B$$

One can transform the linear differential equations $\frac{dx}{dt} = Ax + Bu,\ \ y = Hx$ to the transfer function

$$G(s) = \frac{Y(s)}{U(s)}$$

using the `ss2tf` command, and the transfer function can be converted to the differential equations using the `tf2ss` function.

Example 1.19

Consider the following transfer function

$$G(s) = \frac{Y(s)}{U(s)} = \frac{10s + 30}{0.2s^4 + s^3 + 2s^2 + 3s + 4}$$

Obtain the state-space model.

Solution

In the Command window, one types

```
» num=[10 30]; den=[0.2 1 2 3 4]; [A,B,H,D]=tf2ss(num,den)
```

and presses the Enter key. The resulting matrices are found to be

```
A =
    -5   -10   -15   -20
     1     0     0     0
     0     1     0     0
     0     0     1     0

B =
     1
     0
     0
     0
```

```
H =
      0     0    50   150
D =
      0
```

That is, we have

$$\frac{dx}{dt}=\begin{bmatrix}\frac{dx_1}{dt}\\ \frac{dx_2}{dt}\\ \frac{dx_3}{dt}\\ \frac{dx_4}{dt}\end{bmatrix}=Ax+Bu=\begin{bmatrix}-5 & -10 & -15 & -20\\ 1 & 0 & 0 & 0\\ 0 & 1 & 0 & 0\\ 0 & 0 & 1 & 0\end{bmatrix}\begin{bmatrix}x_1\\ x_2\\ x_3\\ x_4\end{bmatrix}+\begin{bmatrix}1\\ 0\\ 0\\ 0\end{bmatrix}u$$

□

$$y=Hx+Du=[0 \quad 0 \quad 50 \quad 150]\begin{bmatrix}x_1\\ x_2\\ x_3\\ x_4\end{bmatrix}+[0]u$$

Remark. Mathematically, the system output is supposed to be a function of states and control. Therefore, the output equation is expressed as $y = Hx + Du$, $D \in \mathbb{R}^{b\times m}$. In electromechanical systems, the output is not a function of the control inputs, and therefore,

$$D=\begin{bmatrix}0 & 0 & \dots & 0 & 0\\ 0 & 0 & \dots & 0 & 0\\ \vdots & \vdots & \ddots & \vdots & \vdots\\ 0 & 0 & \dots & 0 & 0\\ 0 & 0 & \dots & 0 & 0\end{bmatrix}\in \mathbb{R}^{b\times m}$$

However, using the MATLAB environment, the general form of the output equation in $y = Hx + Du$ is used.

□

Example 1.20

Consider the system as described by

$$\frac{dx}{dt}=\begin{bmatrix}\frac{dx_1}{dt}\\ \frac{dx_2}{dt}\\ \frac{dx_3}{dt}\\ \frac{dx_4}{dt}\end{bmatrix}=Ax+Bu=\begin{bmatrix}1 & 2 & 3 & 4\\ 5 & 6 & 0 & 0\\ 0 & 7 & 8 & 0\\ 0 & 0 & 9 & 10\end{bmatrix}\begin{bmatrix}x_1\\ x_2\\ x_3\\ x_4\end{bmatrix}+\begin{bmatrix}2\\ 3\\ 4\\ 5\end{bmatrix}u$$

$$y = Hx + Du = [3 \quad 4 \quad 5 \quad 6]\begin{bmatrix} x_1 \\ x_2 \\ x_3 \\ x_4 \end{bmatrix} + [0]u$$

Find the transfer function.

Solution

We use the `ss2tf` command. In particular, entering the matrices *A*, *B*, *C*, and *D*, the following is written in the Command window:

```
» A=[1 2 3 4;5 6 0 0;0 7 8 0;0 0 9 10];B=[2; 3; 4; 5];
» H=[3 4 5 6]; D=[0];
» [num,den]=ss2tf(A,B,H,D)
```

The corresponding numerator and denominator are displayed on the screen.

```
num =
  1.0e+003  *
         0    0.0680    -0.6870      2.8020      2.3480
den =
   1.0000 -25.0000  202.0000  -593.0000  -530.0000
```

That is, we have

$$G(s) = \frac{Y(s)}{U(s)} = \frac{68s^3 - 687s^2 + 2802s + 2348}{1000s^4 - 25000s^3 + 202000s^2 - 593000s - 530000} \qquad \square$$

Statistical analysis and filtering can be easily performed. In particular, to design the Butterworth and Chebyshev filters, one uses `butter` and `cheby1` or `cheby2` commands.

Example 1.21

Find the transfer function of the third-order Butterworth filter with bandwidth 20 Hz.

Solution

Typing in the Command window

```
»N=3;Wn=1;W0=2*pi*10;[num,den]=butter(N,W0,'s');bode(num,den);num,den
```

the following numerator and denominator are found.

```
num =
            0             0             0  2.4805e+005
den =
  1.0000e+000  1.2566e+002  7.8957e+003  2.4805e+005
```

That is, the transfer function is

$$G(s)=\frac{2.4805\times10^{5}}{s^{3}+1.2566\times10^{2}s^{2}+7.8957\times10^{3}s+2.4805\times10^{5}}$$

The Bode plots for the $G(j\omega)$ found are plotted in Figure 1.18.

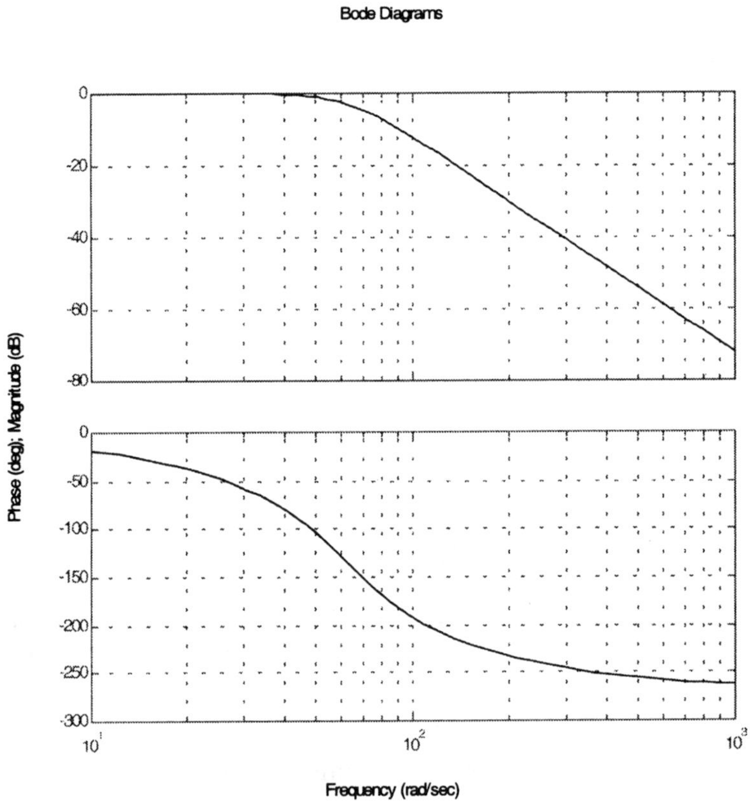

Figure 1.18 Bode plots for $G(s)=\frac{2.4805\times10^{5}}{s^{3}+1.2566\times10^{2}s^{2}+7.8957\times10^{3}s+2.4805\times10^{5}}$.

□

The Symbolic Math Toolbox allows one to perform the symbolic analysis and arithmetic using symbolic expressions. Differentiation, integration, simplification, and other capabilities are available. Furthermore, differential and integral equations can be solved.

Example 1.22

Solve the second-order differential equation

$$\frac{d^2x}{dt^2} + 4\frac{dx}{dt} + 5x = d$$

Solution

Using dsolve command, and typing

```
» x=dsolve('D2x+4*Dx+5*x=d')
```

one obtains

```
x =
1/5*d+C1*exp(-2*t)*sin(t)+C2*exp(-2*t)*cos(t)
```

Hence, the solution of the equation $\frac{d^2x}{dt^2} + 4\frac{dx}{dt} + 5x = d$ is found to be

$$x(t) = \frac{1}{5}d + c_1e^{-2t}\sin(t) + c_2e^{-2t}\cos(t)$$

The initial conditions can be incorporated to find the constants C1 and C2. For example, solving

$$\frac{d^2x}{dt^2} + 4\frac{dx}{dt} + 5x = d \text{ with } \left(\frac{dx}{dt}\right)_0 = 1,\ x_0 = 2$$

we have

```
» x=dsolve('D2x+4*Dx+5*x=d','Dx(0)=1','x(0)=2')
x =
1/5*d+(-2/5*d+5)*exp(-2*t)*sin(t)+(-1/5*d+2)*exp(-2*t)*cos(t)
```

that is; $c_1 = -\frac{2}{5}d + 5$ and $c_2 = -\frac{1}{5}d + 2$.

Assigning the forcing function $d(t)$ to be a function of time, one has

$$\frac{d^2x}{dt^2} + 4\frac{dx}{dt} + 5x = d(t)$$

The general solution can be found by using the Symbolic Toolbox. The result is

```
» x=dsolve('D2x+4*Dx+5*x=d(t)')
x =
int(cos(t)*d(t)*exp(2*t),t)*exp(-2*t)*sin(t)
-int(sin(t)*d(t)*exp(2*t),t)*exp(-2*t)*cos(t)
+C1*exp(-2*t)*sin(t)+C2*exp(-2*t)*cos(t)
```

□

To plot the functions, the `ezplot` plotter is used. For example, typing

```
» ezplot('tanh(x)')
```

one obtains the plot of the hyperbolic tangent (see Figure 1.19).

Four decision-making (control flow) the corresponding structures are available; in particular, For Loops, White Loops, If-Else-End construction, and Switch-Case construction.

To build continuous- and discrete-time models of dynamic systems, the block diagram concept is used, and SIMULINK extends the MATLAB environment. SIMULINK offers a large variety of ready-to-use building block-diagram blocks. By clicking the SIMULINK icon (or typing `simulink` in the Command window at the MATLAB prompt), two new windows (SIMULINK block library window and `untitled` empty model window, where the new model should be built) appear on the screen (see Figure 1.20).

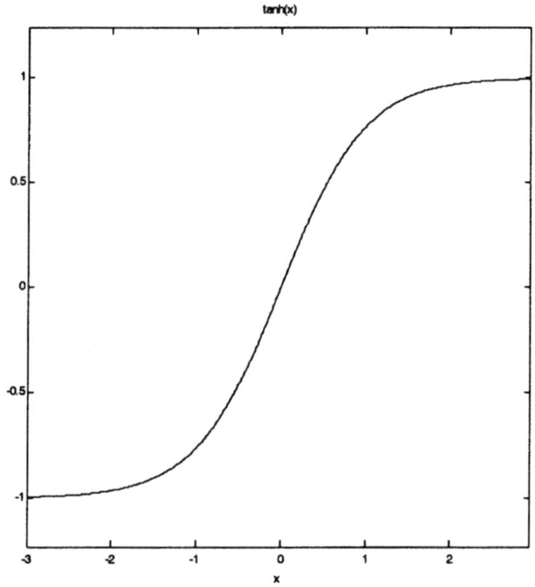

Figure 1.19 Plot of tanh(x).

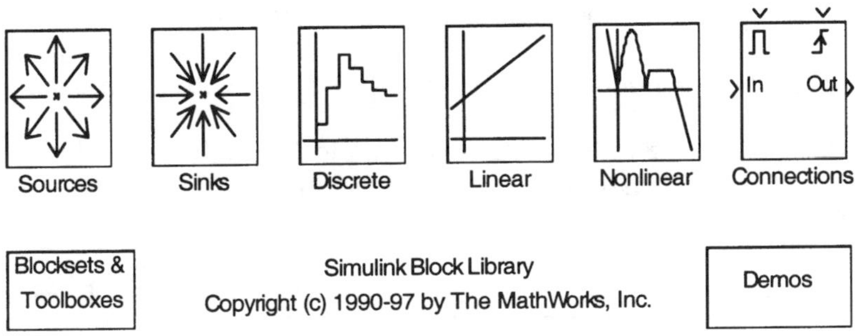

Figure 1.20 SIMULINK block library window.

One can open the `Sources, Sinks, Discrete, Linear, Nonlinear, Connections, Blocksets,` and `Toolboxes` block libraries, as well as the `Demos,` by double-clicking the corresponding icon. The ready-to-use building blocks, commonly used in the comprehensive analysis and design, are available. It should be emphasized that the `Demos` is available, and one who does not have enough experience to use SIMULINK will find a great deal of help using the `Demos`. A complete list of blocks available is given in the Block Browers.

To demonstrate how to use the SIMULINK block libraries, consider Example 1.23.

Example 1.23

Model the Van der Pol oscillator, as given by the following differential equation

$$\frac{d^2x}{dt^2} - k(1-x^2)\frac{dx}{dt} + x = d(t)$$

where $d(t)$ is the disturbance.

Solution

This second-order differential equation can be rewritten as a system of coupled first-order differential equations; in particular,

$$\frac{dx_1(t)}{dt} = x_2, \quad x_1(t_0) = x_{10}$$

$$\frac{dx_2(t)}{dt} = -x_1 + kx_2 - kx_1^2x_2 + d(t), \quad x_2(t_0) = x_{20}$$

The SIMULINK diagram was built using the following blocks: `Gain, Integrator,` and `Sum` (form the `Linear` blocks library); `Scope` (from the `Sinks` blocks library); and `Signal Generator` (from the `Sources` blocks library). These blocks were dragged from the above-mentioned block libraries to the `untitled` window, positioned (placed), and connected using the signal lines as shown in Figure 1.21, which illustrates the SIMULINK diagram built to model the Van der Pol oscillator `(file ch_1_03.m)`.

Modeling of the dynamics was performed by assigning $k = 5$ (in the Command window, one types `k=5`, and presses the Enter key), and $d(t) = 2\sin(2t)$ (by double-clicking on the `Signal Generator` block, one selects sinusoidal function and assigns the corresponding magnitude and frequency), and $x_0 = \begin{bmatrix} x_{10} \\ x_{20} \end{bmatrix} = \begin{bmatrix} 1 \\ -1 \end{bmatrix}$ (these initial conditions are assigned by double-clicking of two `integrator` blocks used). It should be emphasized that the coefficient k can be assigned by double-clicking the `Gain` block and entering the value needed.

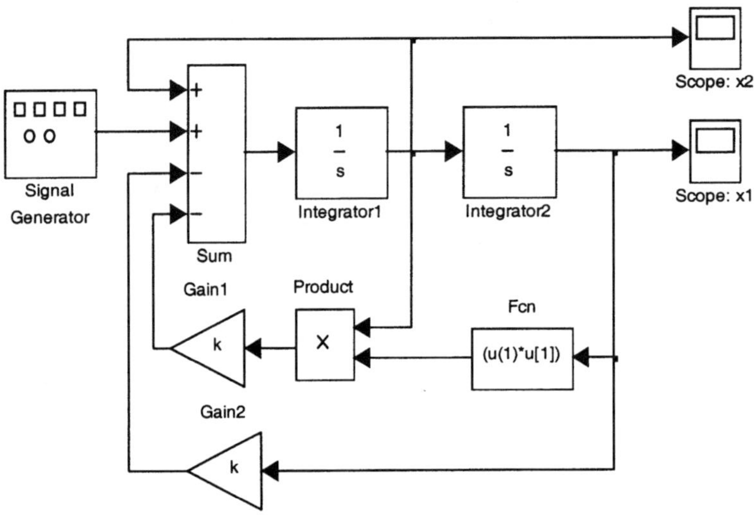

Figure 1.21 SIMULINK diagram.

The transient dynamic waveforms, which are displayed by double-clicking the `Scope` blocks, are shown in Figure 1.22.

□

1.4.2 *Discrete-time systems*

The discrete-time counterpart of casual ($n \ge m$) differential equation is the n-order linear constant-coefficient difference equation as given by

$$\sum_{k=0}^{n} a_k x_{n-k} = \sum_{k=0}^{m} b_k u_{n-k}, \quad n \ge m$$

Using the z-transform and assuming that initial conditions are zero, one obtains

$$\left(\sum_{k=0}^{n} a_k z^k \right) X(z) = \left(\sum_{k=0}^{m} b_m z^k \right) U(z)$$

and the transfer function is

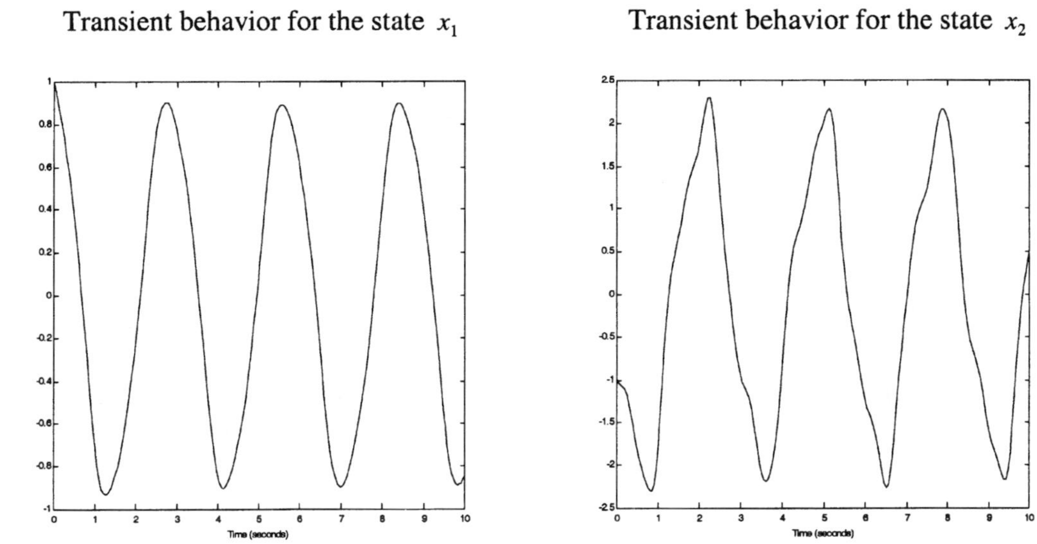

Figure 1.22 Dynamics of the Van der Pol oscillator.

$$G(z)=\frac{X(z)}{U(z)}=\frac{b_m z^m+b_{m-1}z^{m-1}+\ldots+b_1 z+b_0}{a_n z^n+a_{n-1}z^{n-1}+\ldots+a_1 z+a_0}$$

If $n = m$, we have

$$G(z)=\frac{X(z)}{U(z)}=\frac{b_n z^n+b_{n-1}z^{n-1}+\ldots+b_1 z+b_0}{a_n z^n+a_{n-1}z^{n-1}+\ldots+a_1 z+a_0} \quad \text{or}$$

$$G(z)=\frac{X(z)}{U(z)}=\frac{b_n+b_{n-1}z^{-1}+\ldots+b_1 z^{-n+1}+b_0 z^{-n}}{a_n+a_{n-1}z^{-1}+\ldots+a_1 z^{-n+1}+a_0 z^{-n}}$$

Example 1.24

Consider the first-order linear constant-coefficient differential equation (the transient dynamics of linear time-invariant electromechanical systems are mapped by differential equations)

$$\frac{dx}{dt}=-ax(t)+bu(t)$$

where a and b are the constants. Find the discrete-time model.

Solution

We have

$$\left.\frac{dx}{dt}\right|_{t=kT_s}=-ax(kT_s)+bu(kT_s)$$

For small sampling period T_s, one uses the *Euler approximation* (the forward rectangular rule) as given by

$$\frac{dx}{dt} \approx \frac{x(t+T_s)-x(t)}{T_s}$$

and

$$\left.\frac{dx}{dt}\right|_{t=kT_s} = \frac{x(kT_s+T_s)-x(kT_s)}{T_s}$$

Therefore, using the forward difference, we have

$$\frac{x(kT_s+T_s)-x(kT_s)}{T_s} = -ax(kT_s)+bu(kT_s)$$

Denoting values of $x(t)$ and $u(t)$ at t_k and t_{k+1} as

$$x_k = x(t)\big|_{t=kT_s},\; x_{k+1} = x(t)\big|_{t=(k+1)T_s},\; \text{and}\; u_k = u(t)\big|_{t=kT_s}$$

we have

$$\frac{x_{k+1}-x_k}{T_s} = -ax_k + bu_k$$

where $x_{k+1} = x[(k+1)T_s]$, $x_k = x(kT_s)$, and $u_k = u(kT_s)$.

Finally, the following difference equation is found

$$x_{k+1} = (1-aT_s)x_k + bT_s u_k$$

or

$$x_k = (1-aT_s)x_{k-1} + bT_s u_{k-1}$$

In the z-domain, one obtains the transfer function

$$G(z) = \frac{X(z)}{U(z)} = \frac{bT_s z^{-1}}{1-(1-aT_s)z^{-1}} = \frac{bT_s}{z-(1-aT_s)}$$ □

Example 1.25

Consider the following constant-coefficient difference equation

$$x_{k-4} + 2x_{k-3} + 3x_{k-1} + 4x_k = 5u_{k-2} + 6u_k$$

Find the transfer function and numerically calculate the dynamics due to step input.

Solution

Using the z-transform, the transfer function is found as

$$G(z)=\frac{X(z)}{U(z)}=\frac{6+5z^{-2}}{4+3z^{-1}+2z^{-3}+z^{-4}}=\frac{6z^4+5z^2}{4z^4+3z^3+2z+1}$$

The `dstep` and `dimpulse` commands are used to find the step response, and the `stem` command is applied to plot the dynamics.

□

For example, find and plot the transient behavior of discrete-time systems with the transfer function

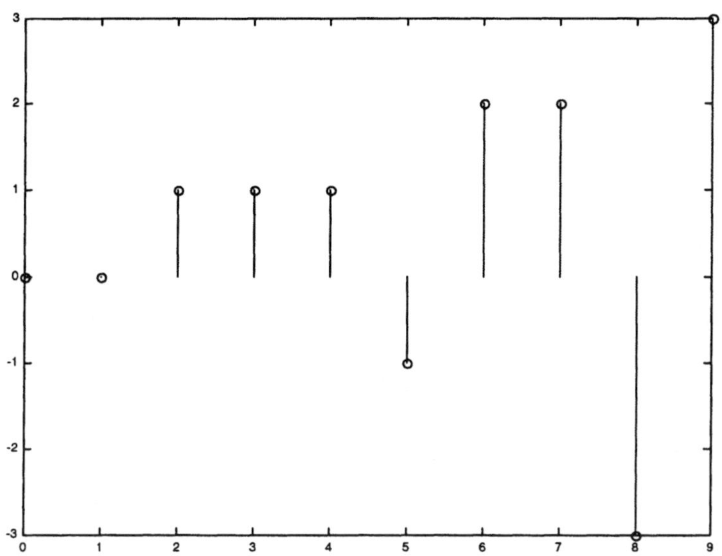

Figure 1.23 Transient dynamics for x_k if the unit-step input is applied, $u_k = 1$, $k \geq 0$.

$$G(z)=\frac{X(z)}{U(z)}=\frac{z^2+2z+3}{z^4+2z^3+3z^2+2z+1}$$

for ten instants $k = 0, 1, \ldots, 8, 9$ if the unit step input is applied. We type in the Command window

```
»k=0:1:9;num=[1 2 3];den=[1 2 3 2 1];x=dstep(num,den,10);stem(k,x)
```

As one presses the Enter key, the resulting plot is displayed. The transient behavior is illustrated in Figure 1.23.

The `filter` command can be used to simulate the discrete-time systems.

Example 1.27

Find the dynamics of the systems with the following transfer function

$$G(z)=\frac{X(z)}{U(z)}=\frac{3z^2+2z+1}{z^2-0.5z+1}$$

if $u_k = 1$ for $k = 0, 1, 2, 3, 4$ and $u_k = 0$ for $k = 5, 6, 7, 8, 9$. Perform analysis of stability based on numerical and analytical results.

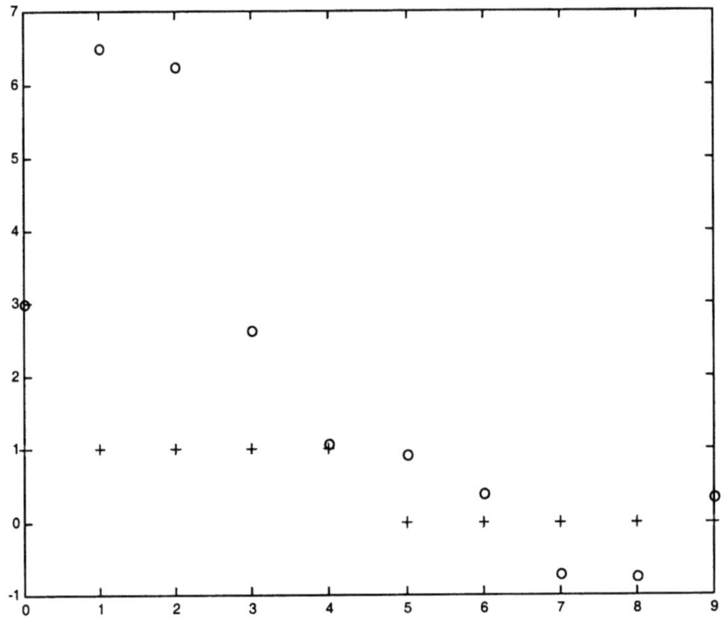

Figure 1.24 Transient dynamics of the discrete-time output x_k.

Solution

Typing in the Command window

```
» k=0:1:9;u=[ones(1,5) zeros(1,5)];num=[3 2 1];den=[1 -0.5 1];
» x=filter(num,den,u);plot(k,x,'o',k,u,'+')
```

the output dynamics is obtained. We use the '+' symbol to plot the input signal and 'o' illustrates the system output (see Figure 1.24).

The analysis shows that the single-input/single-output discrete system studied is unstable. This result is easily justified if one finds the characteristic eigenvalues, which for stable systems lie in the unit circle. Using the `roots` command, we find that this criterion for stability is not guaranteed, and

```
» format short; roots(den)
ans =
  -0.5000 + 0.8660i
  -0.5000 - 0.8660i
  -0.5000 + 0.8660i
  -0.5000 - 0.8660i
```

□

Consider n-degree-of-freedom electromechanical systems that are modeled using difference equations. As for continuous-time systems, to perform analysis and design, the dynamics of electromechanical motion devices and power electronics should be mapped. We use n-dimensional state, m-dimensional control, and b-dimensional output vectors; that is,

$$x_k = \begin{bmatrix} x_{k1} \\ x_{k2} \\ \vdots \\ x_{k\,n-1} \\ x_{k\,n} \end{bmatrix}, \quad u_k = \begin{bmatrix} u_{k1} \\ u_{k2} \\ \vdots \\ u_{k\,m-1} \\ u_{k\,m} \end{bmatrix}, \quad y_k = \begin{bmatrix} y_{k1} \\ y_{k2} \\ \vdots \\ y_{k\,b-1} \\ y_{k\,b} \end{bmatrix}$$

In matrix form, the state-space equations are give for linear discrete-time electromechanical systems in the following form

$$x_{k+1} = \begin{bmatrix} x_{k+1,1} \\ x_{k+1,2} \\ \vdots \\ x_{k+1,n-1} \\ x_{k+1,n} \end{bmatrix} = \begin{bmatrix} a_{k11} & a_{k12} & \cdots & a_{k1\,n-1} & a_{k1\,n} \\ a_{k21} & a_{k22} & \cdots & a_{k2\,n-1} & a_{k2\,n} \\ \vdots & \vdots & \ddots & \vdots & \vdots \\ a_{kn-11} & a_{kn-12} & \cdots & a_{kn-1\,n-1} & a_{kn-1n} \\ a_{kn\,1} & a_{kn\,2} & \cdots & a_{kn\,n-1} & a_{kn\,n} \end{bmatrix} \begin{bmatrix} x_{k1} \\ x_{k2} \\ \vdots \\ x_{k\,n-1} \\ x_{k\,n} \end{bmatrix} +$$

$$\begin{bmatrix} b_{k11} & b_{k12} & \cdots & b_{k1\,m-1} & b_{k1m} \\ b_{k21} & b_{k22} & \cdots & b_{k2\,m-1} & b_{k2m} \\ \vdots & \vdots & \ddots & \vdots & \vdots \\ b_{kn-11} & b_{kn-12} & \cdots & b_{kn-1\,m-1} & b_{kn-1m} \\ b_{kn1} & b_{kn2} & \cdots & b_{kn\,m-1} & b_{kn\,m} \end{bmatrix} \begin{bmatrix} u_{k1} \\ u_{k2} \\ \vdots \\ u_{k\,m-1} \\ u_{k\,m} \end{bmatrix}$$

$$= A_k x_k + B_k u_k$$

$$x_{k=k_0} = x_{k0}$$

while for nonlinear systems we have

$$x_{k+1} = F_k(k, x_k, u_k), \quad x_{k=k_0} = x_{k0}$$

where $A_k \in \mathbb{R}^{n\times n}$ and $B_k \in \mathbb{R}^{n\times m}$ are the matrices of constant coefficients, and $F_k(k, x_k, u_k)$ is the nonlinear map.

The output equation is

$$y_k = H_k x_k$$

where $H_k \in \mathbb{R}^{b\times n}$ is the matrix of constant coefficients.

The state-space and output equations for continuous-time systems were obtained as

$$\frac{dx}{dt} = Ax + Bu, \quad y = Hx$$

while for discrete-time systems, we have

$$x_{k+1} = A_k x_k + B_k u_k, \quad y_k = H_k x_k$$

Approaching and solving a great variety of problems in analysis and design of discrete-time electromechanical systems, the digital controller must be synthesized, and one can discretize the continuous-time system.

From $\frac{dx(t)}{dt} = Ax(t) + Bu(t)$, we have $\frac{dx(t)}{dt} - Ax(t) = Bu(t)$.
Premultiplying both sides of this equation by the inverse matrix exponential e^{-At} and rearranging terms, we have

$$e^{-At}\left(\frac{dx(t)}{dt} - Ax(t)\right) = \frac{d(e^{-At}x(t))}{dt} = e^{-At}Bu(t)$$

where the matrix exponential is given as

$$e^{At} = \sum_{k=0}^{\infty} \frac{A^k t^k}{k!}$$

$$e^{At}e^{-At} = I$$

and

$$\frac{de^{At}}{dt} = Ae^{At} = e^{At}A$$

Then, the integration results in

$$e^{-At}x(t) = x(0) + \int_0^t e^{-A\tau}Bu(\tau)d\tau$$

and multiplying both sides by e^{At}, one obtains the solution of the differential equation in the following form

$$x(t) = e^{At}x(0) + \int_0^t e^{A(t-\tau)}Bu(\tau)d\tau, \ \ t \geq 0$$

Let us sample the input function $u(t)$ and feed it to the zero-order data hold. Hence,

$$u(t) = u_k \ \text{ if } \ kT_s \leq t < (k+1)T_s$$

Using

$$x(t) = e^{At}x(0) + \int_0^t e^{A(t-\tau)}B(\tau)u(\tau)$$

one obtains two equations

$$x_{k+1} = x((k+1)T_s) = e^{A(k+1)T_s}x(0) + e^{A(k+1)T_s}\int_0^{(k+1)T_s} e^{-A\tau}Bu(\tau)d\tau$$

$$x_k = x(kT_s) = e^{AkT_s}x(0) + e^{AkT_s}\int_0^{kT_s} e^{-A\tau}Bu(\tau)d\tau$$

Subtracting these two equation, we have

$$x_{k+1} = x((k+1)T_s) = e^{AT_s}x(kT_s) + \int_0^{T_s} e^{A(T_s-t)}Bu(kT_s)d(T_s-t) = e^{AT_s}x_k + \int_0^{T_s} e^{A(T_s-t)}Bu_k d(T_s-t)$$

Therefore,

$$A_k = e^{AT_s} \ \text{ and } \ B_k = \left(\int_0^{T_s} e^{A(T_s-t)}d(T_s-t)\right)B$$

and if matrix A is nonsingular,

$$B_k = \left(\int_0^{T_s} e^{A(T_s-t)}d(T_s-t)\right)B = A^{-1}(e^{AT_s} - I)B = (e^{AT_s} - I)A^{-1}B$$

One concludes that A_k and B_k depend on the sampling time T_s.

The output equations are

$$y_k = H_k x_k,\ y(kT_s) = H_k x(kT_s)$$

and

$$y(t) = H_x(t)$$

That is, $H_k = H$.

Example 1.27

The continuous-time system is given as

$$\frac{dx}{dt} = \begin{bmatrix} \frac{dx_1}{dt} \\ \frac{dx_2}{dt} \\ \frac{dx_3}{dt} \\ \frac{dx_4}{dt} \end{bmatrix} = Ax + Bu = \begin{bmatrix} 1 & 2 & 3 & 4 \\ 5 & 6 & 0 & 0 \\ 0 & 7 & 8 & 0 \\ 0 & 0 & 9 & 10 \end{bmatrix} \begin{bmatrix} x_1 \\ x_2 \\ x_3 \\ x_4 \end{bmatrix} + \begin{bmatrix} 2 \\ 3 \\ 4 \\ 5 \end{bmatrix} u$$

Find the discrete-time model in $x_{k+1} = A_k x_k + B_k u_k$ if the sampling period is 0.01 s.

Solution

To solve the problem, the MATLAB command c2d is used. We type

```
» A=[1 2 3 4;5 6 0 0;0 7 8 0;0 0 9 10];B=[2; 3; 4; 5];Ts=0.01;
» [Ak,Bk]=c2d(A,B,Ts)
```

and the following matrices of the discrete-time system result

```
Ak =
    1.0106   0.0219   0.0333   0.0423
    0.0518   1.0624   0.0008   0.0011
    0.0018   0.0751   1.0833   0.0000
    0.0001   0.0034   0.0985   1.1052

Bk =
    0.0221
    0.0315
    0.0428
    0.0545
```

Hence, one has

$$x_{k+1} = \begin{bmatrix} x_{k+11} \\ x_{k+12} \\ x_{k+13} \\ x_{k+14} \end{bmatrix} = A_k x_k + B_k u_k = \begin{bmatrix} 1.01 & 0.022 & 0.033 & 0.042 \\ 0.052 & 1.06 & 0.0008 & 0.001 \\ 0.002 & 0.075 & 1.08 & 0 \\ 0 & 0.0034 & 0.099 & 1.11 \end{bmatrix} \begin{bmatrix} x_{k1} \\ x_{k2} \\ x_{k3} \\ x_{k4} \end{bmatrix} + \begin{bmatrix} 0.022 \\ 0.032 \\ 0.043 \\ 0.055 \end{bmatrix} u_k$$

□

Example 1.28

Note the rotating platform, actuated by an actuator that develops electromagnetic torque T_e. The angular displacement θ_r is the output, and the *torsional-mechanical* dynamics are given by

$$\frac{d\theta_r}{dt} = \omega_r$$

$$\frac{d\omega_r}{dt} = \frac{1}{J}(-B_m \omega_r + T_e)$$

Find the state-space form and derive the analytic expressions for matrices A_k and B_k as functions of the sampling time T_s.

Solution

Denoting $x = \begin{bmatrix} x_1 \\ x_2 \end{bmatrix} = \begin{bmatrix} \theta_r \\ \omega_r \end{bmatrix}$ and $u = T_e$, we have

$$\frac{dx(t)}{dt} = Ax + Bu = \begin{bmatrix} 0 & 1 \\ 0 & -\frac{B_m}{J} \end{bmatrix} \begin{bmatrix} x_1 \\ x_2 \end{bmatrix} + \begin{bmatrix} 0 \\ \frac{1}{J} \end{bmatrix} u$$

To obtain the matrix exponential, we find

$$(sI - A)^{-1} = \begin{bmatrix} s & -1 \\ 0 & s + \frac{B_m}{J} \end{bmatrix}^{-1} = \frac{1}{s\left(s + \frac{B_m}{J}\right)} \begin{bmatrix} s + \frac{B_m}{J} & 1 \\ 0 & s \end{bmatrix}$$

Using the Laplace transforms, we have

$$e^{At} = \begin{bmatrix} 1 & \frac{B_m}{J}\left(1 - e^{-\frac{B_m}{J}t}\right) \\ 0 & e^{-\frac{B_m}{J}t} \end{bmatrix}$$

Analytic expressions for matrices A_k and B_k are found to be

$$A_k = e^{AT_s} = \begin{bmatrix} 1 & \frac{B_m}{J}\left(1 - e^{-\frac{B_m}{J}T_s}\right) \\ 0 & e^{-\frac{B_m}{J}T_s} \end{bmatrix}$$

and

$$B_k = \left(\int_0^{T_s} e^{A(T_s-t)} d(T_s - t)\right)B = \left(\int_0^{T_s} e^{A(T_s-t)} d(T_s - 1)\right)\begin{bmatrix} 0 \\ \frac{1}{J} \end{bmatrix} = \begin{bmatrix} \frac{T_s}{B_m} - \frac{J}{B_m^2}\left(1 - e^{-\frac{B_m}{J}T_s}\right) \\ \frac{1}{B_m}\left(1 - e^{-\frac{B_m}{J}T_s}\right) \end{bmatrix} \qquad \square$$

References

1. Fraser, C. and Milne, J., *Electro-Mechanical Engineering,* IEEE Press, New York, 1994.
2. Hanselman, D. and Littlefield, B., *The Student Edition of MATLAB,* Prentice-Hall, Upper Saddle River, NJ, 1997.
3. Hanselman, D. and Littlefield, B., *Mastering MATLAB 5,* Prentice-Hall, Upper Saddle River, NJ, 1998.
4. Hanselman, D. and Littlefield, B., *Mastering SIMULINK 2,* Prentice-Hall, Upper Saddle River, NJ, 1998.
5. Krause, P. C. and Wasynczuk, O., *Electromechanical Motion Devices,* McGraw-Hill, New York, 1989.
6. Kuo, B. C., *Automatic Control Systems,* Prentice-Hall, Englewood Cliffs, NJ, 1995.
7. Ogata, K., *Discrete-Time Control Systems,* Prentice-Hall, Upper Saddle River, NJ, 1995.
8. Ogata, K., *Modern Control Engineering,* Prentice-Hall, Upper Saddle River, NJ, 1997.
9. Paul, C. R., Whites, K. W., and Nasar, S. A., *Introduction to Electromagnetic Fields,* McGraw-Hill, New York, 1998.
10. White, D. C. and Woodson, H. H., *Electromechanical Energy Conversion,* John Wiley & Sons, New York, 1959.
11. Wylie, C. R. and Barrett, L. C., *Advanced Engineering Mathematics,* McGraw-Hill, New York, 1995.

chapter two

Electromechanical system dynamics, energy conversion, and electromechanical analogies

2.1 Mathematical models and system dynamics

2.1.1 Newtonian mechanics

Newtonian mechanics: translational motion

The equations of motion of mechanical systems can be found using Newton's second law of motion, which is given by

$$\sum \vec{F} = m\vec{a} \tag{2.1}$$

where $\vec{F}$ is the vector sum of all forces applied to the body; $\vec{a}$ is the vector of acceleration of the body with respect to an inertial reference frame (the so-called inertial acceleration); and m is the mass of the body.

To apply Newton's law, the free-body diagram in the coordinate system used should be studied.

Example 2.1

Consider a body of mass m in the XY coordinate system. The force $\vec{F}_a$ is applied in the x direction. Neglecting *Coulomb* and static friction, and assuming that the viscous friction force is $F_{fr} = B_v \frac{dx}{dt}$, find the equations of motion. Here, B_v is the viscous friction coefficient.

Solution

The free-body diagram developed is illustrated in Figure 2.1.

The sum of the forces, acting in the y direction, is expressed as

$$\sum \vec{F}_Y = \vec{F}_N - \vec{F}_g,$$

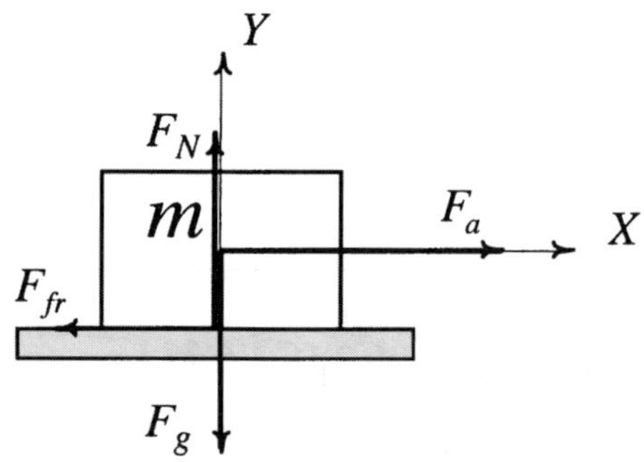

Figure 2.1 Free-body diagram.

where $\vec{F}_g = mg$ is the gravitational force acting on the mass m; and $\vec{F}_N$ is the normal force which is equal and opposite to the gravitational force.

From (2.1), the equation of motion in the y direction is expressed as

$$\vec{F}_N - \vec{F}_g = ma_y = m\frac{d^2y}{dt^2}$$

where a_y is the acceleration in the y direction, $a_y = \frac{d^2y}{dt^2}$.

Making use $\vec{F}_N = \vec{F}_g$, we have

$$\frac{d^2y}{dt^2} = 0$$

The sum of the forces acting in the x direction is found using the applied force $\vec{F}_a$ and the friction force $\vec{F}_{fr}$; in particular, we have

$$\sum \vec{F}_X = \vec{F}_a - \vec{F}_{fr}$$

The applied force can be time-invariant $\vec{F}_a = const$, and time-varying $\vec{F}_a(t) = f(t, x, y, z)$. For example, $\vec{F}_a(t) = x\sin(6t-4)e^{-0.5t} + \frac{dy}{dt}t^2 + z^3\cos\left(\frac{dx}{dt}t - y^2t^4\right)$.

Using (2.1), the equation motion in the x direction is found to be

$$\vec{F}_a - \vec{F}_{fr} = ma_x = m\frac{d^2x}{dt^2}$$

where a_x is the acceleration in the x direction, $a_x = \frac{d^2x}{dt^2}$, and the velocity in the x direction is $v = \frac{dx}{dt}$.

Assuming that *Coulomb* and static friction can be neglected, the friction force, as a function of the viscous friction coefficient B_v and velocity $v = \frac{dx}{dt}$. is given by $F_{fr} = B_v \frac{dx}{dt}$.

Hence, one obtains the second-order differential equation to map the body dynamics in the x direction

$$\frac{d^2x}{dt^2} = \frac{1}{m}\left(F_a - B_v \frac{dx}{dt}\right)$$

A set of two first-order differential equations results, and

$$\frac{dx}{dt} = v$$

$$\frac{dv}{dt} = \frac{1}{m}(F_a - B_v v)$$

□

Friction models in electromechanical systems

A thorough consideration of friction is essential for understanding the operation of electromechanical systems. Friction is a very complex nonlinear phenomenon, and it is difficult to model friction. The classical *Coulomb* friction is a retarding frictional force (for translational motion) or torque (for rotational motion) that changes its sign with the reversal of the direction of motion, and the amplitude of the frictional force or torque are constant. For translational and rotational motions, the *Coulomb* friction is modeled as

$$\text{force: } F_{Coulomb} = F_{F_c}\,\text{sgn}(v) = k_{F_c}\,\text{sgn}\left(\frac{dx}{dt}\right)$$

$$\text{torque: } T_{Coulomb} = k_{T_c}\,\text{sgn}(\omega) = k_{T_c}\,\text{sgn}\left(\frac{d\theta}{dt}\right)$$

where k_{F_c} and k_{T_c} are the Coulomb friction coefficients.

Figure 2.2.a illustrates the *Coulomb* friction.

Viscous friction is a retarding force or torque that is a linear function of linear or angular velocity. The viscous friction force and torque versus linear and angular velocities are shown in Figure 2.2b. The following expressions are commonly used to model the viscous friction in electromechanical motion devices

$$F_{viscous} = B_v v = B_v \frac{dx}{dt} \text{ for translational motion}$$

and

$$T_{viscous} = B_m \omega = B_m \frac{d\theta}{dt} \text{ for rotational motion}$$

where B_v and B_m are the viscous friction coefficients.

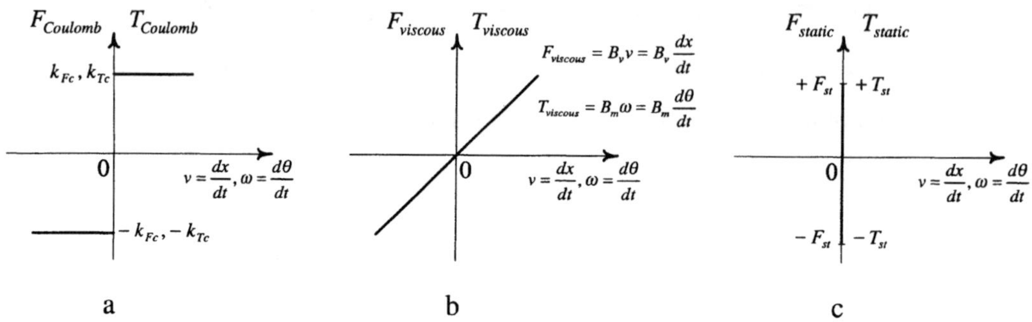

Figure 2.2 Functional representations of: (a) Coulomb friction; (b) viscous friction; and (c) static friction.

The static friction exists only when the body is stationary and vanishes as motion begins. The static friction is a force F_{static} or torque T_{static}, and we have the following expressions

$$F_{static} = \pm F_{st}\Big|_{v=\frac{dx}{dt}=0}$$

and

$$T_{static} = \pm T_{st}\Big|_{\omega=\frac{d\theta}{dt}=0}$$

One concludes that the static friction is a retarding force or torque that tends to prevent the initial translational or rotational motion at beginning (see Figure 2.2c).

In general, the friction force and torque are nonlinear functions that must be modeled using frictional memory, presliding conditions, etc. The empirical formulas, commonly used to express F_{static} and T_{static}, are

$$F_{fr} = (k_{fr1} - k_{fr2}e^{-k|v|} + k_{fr3}|v|)\,\text{sgn}(v_t) = \left(k_{fr1} - k_{fr2}e^{-k\left|\frac{dx}{dt}\right|} + k_{fr3}\left|\frac{dx}{dt}\right|\right)\text{sgn}\left(\frac{dx}{dt}\right)$$

and

$$T_{fr} = (k_{fr1} - k_{fr2}e^{-k|\omega|} + k_{fr3}|\omega|)\,\text{sgn}(\omega) = \left(k_{fr1} - k_{fr2}e^{-k\left|\frac{d\theta}{dt}\right|} + k_{fr3}\left|\frac{d\theta}{dt}\right|\right)\text{sgn}\left(\frac{d\theta}{dt}\right)$$

These F_{fr} and T_{fr} are shown in Figure 2.3.

□

Example 2.2

Consider a two-mass wheel suspension system assuming that the motion is only within y direction (the vertical position of the wheel is denoted as y_1, while the vertical position

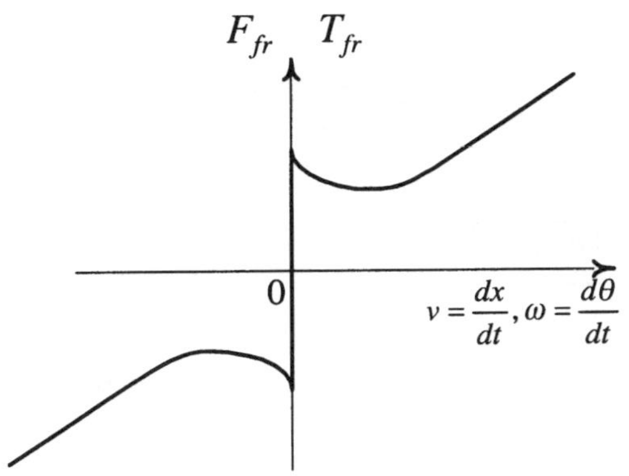

Figure 2.3 Friction force and torque as nonlinear functions of linear and angular velocities.

of the automobile frame relative to the equilibrium position is y_2) (see Figure 2.4). The mass m_1 is the equivalent mass of the wheel, and the mass m_2 is the equivalent mass of the automobile frame. The constants k_{s1} and k_{s2} are the stiffness of the automobile suspension spring and the tire (spring constants), respectively. The shock absorber has the damping constant B_v (viscous friction coefficient). The goal is to find equations that map the system dynamics.

Solution

The automobile suspension system is studied to illustrate model development using Newtonian mechanics. The suspension system can be represented by two free-body diagrams, as shown in Figure 2.5.

Using these free-body diagrams, one can find Newton's force equations. First consider the free-body diagram for the mass m_1. The sum of the forces acting in the y direction is found to be

$$\sum \vec{F}_{Y_1} = k_{s1}(y_2 - y_1) + B_v\left(\frac{dy_2}{dt} - \frac{dy_1}{dt}\right) - k_{s2}y_1$$

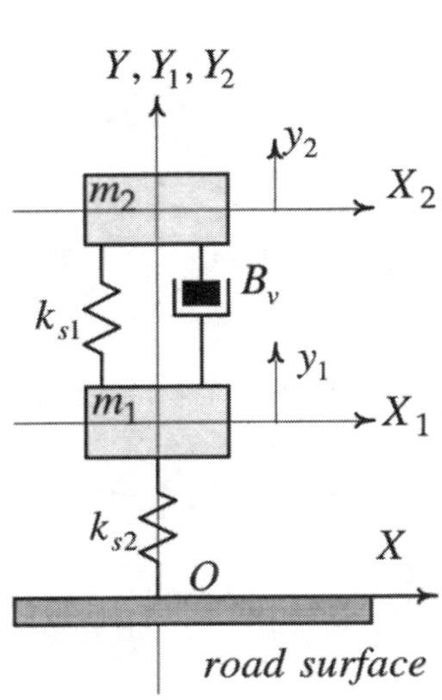

Figure 2.4 Suspension system with two masses.

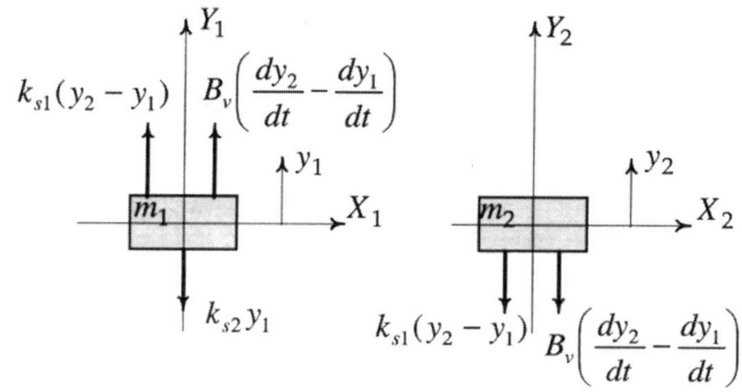

Figure 2.5 Free-body diagrams of the suspension system.

where y_1 and y_2 are the displacement of the masses m_1 and m_2 along the Y-axis, respectively; and $\frac{dy_1}{dt}$ and $\frac{dy_2}{dt}$ are the velocities of the masses m_1 and m_2 in the Y direction.

Hence, using (2.1), the equation of motion for the mass m_1 in the y direction is

$$m_1 \frac{d^2 y_1}{dt^2} = k_{s1}(y_2 - y_1) + B_v\left(\frac{dy_2}{dt} - \frac{dy_1}{dt}\right) - k_{s2} y_1$$

which leads to

$$\frac{d^2 y_1}{dt^2} = \frac{1}{m_1}\left(k_{s1}(y_2 - y_1) + B_v\left(\frac{dy_2}{dt} - \frac{dy_1}{dt}\right) - k_{s2} y_1\right)$$

where $\frac{d^2 y_1}{dt^2}$ is the acceleration of the mass m_1 in the Y direction.

Consider the free-body diagram for the mass m_2 (see Figure 2.5). The sum of the forces acting in the Y direction is

$$\sum \vec{F}_{Y_2} = -k_{s1}(y_2 - y_1) - B_v\left(\frac{dy_2}{dt} - \frac{dy_1}{dt}\right)$$

The equation of motion for the mass m_2 in the Y direction, as given by

$$m_2 \frac{d^2 y_2}{dt^2} = -k_{s1}(y_2 - y_1) - B_v\left(\frac{dy_2}{dt} - \frac{dy_1}{dt}\right)$$

yields the following second-order differential equation

$$\frac{d^2 y_2}{dt^2} = \frac{1}{m_2}\left(-k_{s1}(y_2 - y_1) - B_v\left(\frac{dy_2}{dt} - \frac{dy_1}{dt}\right)\right)$$

The resulting system of differential equations is found as

$$\frac{d^2y_1}{dt^2}=\frac{1}{m_1}\left(k_{s1}(y_2-y_1)+B_v\left(\frac{dy_2}{dt}-\frac{dy_1}{dt}\right)-k_{s2}y_1\right)$$

$$\frac{d^2y_2}{dt^2}=\frac{1}{m_2}\left(-k_{s1}(y_2-y_1)-B_v\left(\frac{dy_2}{dt}-\frac{dy_1}{dt}\right)\right)$$

As an alternative, the suspension system model can be represented using the first-order differential equations, and the state variables are applied. For example, one can assign the states as

$$x_1=y_1,\ x_2=\frac{dx_1}{dt}=\frac{dy_1}{dt},\ x_3=y_2,\ \text{and}\ x_4=\frac{dx_3}{dt}=\frac{dy_2}{dt}$$

In this analysis, the gravitational forces m_1g and m_2g are equal and opposite to the forces from the springs. Therefore, m_1g and m_2g have been omitted. It should be emphasized that the gravitational forces can be omitted from vertical-spring mass systems if the position coordinates are defined using the equilibrium position. □

Example 2.3 Transducer model development and modeling of dynamics

Figure 2.6 shows a simple electromechanical device (actuator) with a stationary member and movable plunger. Using Newton's second law, find the differential equations and perform the modeling using MATLAB.

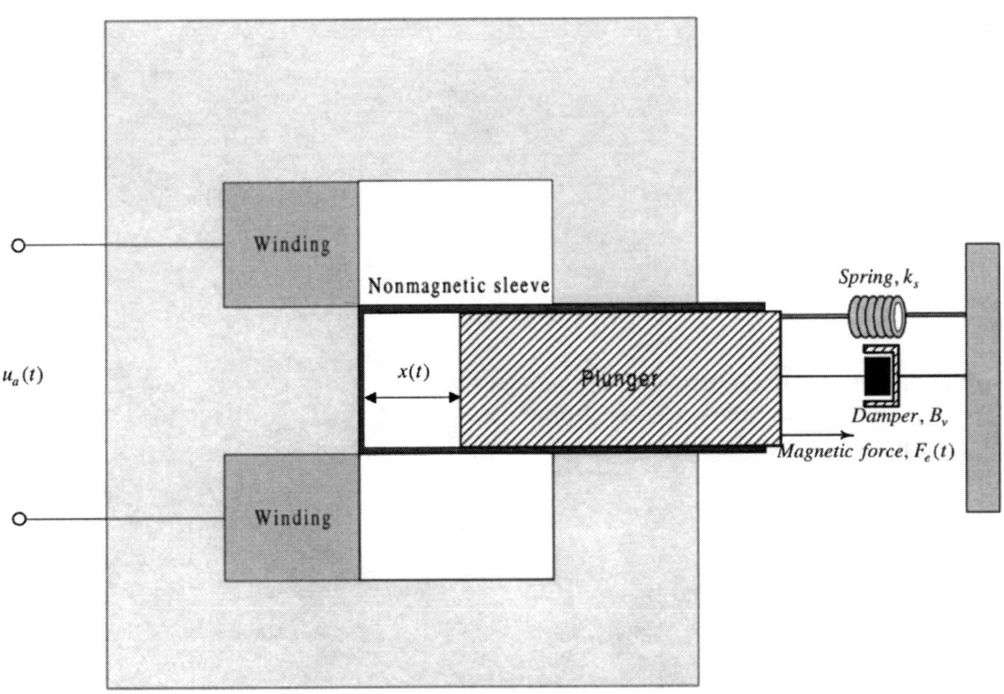

Figure 2.6 Schematic of a transducer.

Solution

Let us apply Newton's second law of motion to find the equations of motion and study the dynamics. Newton's law states that the acceleration of an object is proportional to the net force. The vector sum of all forces acting on the object can be found using a free-body diagram. In particular, for the studied translational mechanical system, one obtains

$$F(t) = m\frac{d^2x}{dt^2} + B_v\frac{dx}{dt} + (k_{s1}x + k_{s2}x^2) + F_e(t)$$

where x denotes the displacement of a plunger; m is the mass of a movable member; B_v is the viscous friction coefficient; k_{s1} and k_{s2} are the spring constants; and $F_e(t)$ is the magnetic force,

$$F_e(i,x) = \frac{\partial W_c(i,x)}{\partial x}$$

It should be emphasized that Hooke's law is valid only for sufficiently small displacements. The stretch and restoring forces are not directly proportional to the displacement, and these forces are different on either side of the equilibrium position. The restoring/stretching force exerted by the spring is expressed by $(k_{s1}x + k_{s2}x^2)$.

Assuming that the magnetic system is linear, the coenergy is expressed as

$$W_c(i,x) = \frac{1}{2}L(x)i^2$$

then,

$$F_e(i,x) = \frac{1}{2}i^2\frac{dL(x)}{dx}$$

The inductance is found using the following formula

$$L(x) = \frac{N^2}{\Re_f + \Re_g} = \frac{N^2\mu_f\mu_0 A_f A_g}{A_g l_f + 2A_f\mu_f(x+2d)}$$

where $\Re_f$ and $\Re_g$ are the reluctances of the ferromagnetic material and air gap, respectively; A_f and A_g are the associated cross-sectional areas; and l_f and $(x + 2d)$ are the lengths of the magnetic material and the air gap, respectively. Hence,

$$\frac{dL}{dx} = -\frac{2N^2\mu_f^2\mu_0 A_f^2 A_g}{[A_g l_f + 2A_f\mu_f(x+2d)]^2}$$

Using Kirchhoff's law, the voltage equation for the electric circuit is given as

$$u_a = ri + \frac{d\psi}{dt}$$

where the flux linkage ψ is expressed as $\psi = L(x)i$. One obtains

$$u_a = ri + L(x)\frac{di}{dt} + i\frac{dL(x)}{dx}\frac{dx}{dt}$$

and

$$\frac{di}{dt} = -\frac{r}{L(x)}i + \frac{2N^2\mu_f^2\mu_0 A_f^2 A_g}{L(x)[A_g l_f + 2A_f\mu_f(x+2d)]^2}iv + \frac{1}{L(x)}u_a$$

Augmenting this equation with the differential equation for mechanical systems

$$F(t) = m\frac{d^2x}{dt^2} + B_v\frac{dx}{dt} + (k_{s1}x + k_{s2}x^2) + F_e(t)$$

three nonlinear differential equations for the considered transducer are found as

$$\frac{di}{dt} = -\frac{r[A_g l_f + 2A_f\mu_f(x+2d)]}{N^2\mu_f\mu_0 A_f A_g}i + \frac{2\mu_f A_f}{A_g l_f + 2A_f\mu_f(x+2d)}iv + \frac{A_g l_f + 2A_f\mu_f(x+2d)}{N^2\mu_f\mu_0 A_f A_g}u_a$$

$$\frac{dx}{dt} = v$$

$$\frac{dv}{dt} = -\frac{N^2\mu_f^2\mu_0 A_f^2 A_g}{m[A_g l_f + 2A_f\mu_f(x+2d)]}i^2 - \frac{1}{m}(k_{s1} + k_{s2}x^2) - \frac{B_v}{m}v$$

The nonlinear mathematical model is found. Using the differential equations obtained, the dynamics can be studied. The amplitude and form of the applied voltage, initial conditions, and disturbances play a key role in analysis of the transient dynamics. Think of a transducer with these parameters: $r = 6$ ohm, $N = 25000$, $\mu_f = 4000\mu_0$, $A_f = 0.0015$ m^2, $A_g = 0.0003$ m^2, $l_f = 0.25$ m, $d = 0.0005$ m, $m = 0.4$ kg, $k_{s1} = 3$ N-m^{-1}, $k_{s2} = 5$ N-m^{-2}, and $B_v = 1.5$ N-s-m^{-1}. The transient responses for $u_a = 50$ V are shown in Figure 2.7, and the corresponding MATLAB scripts are presented.

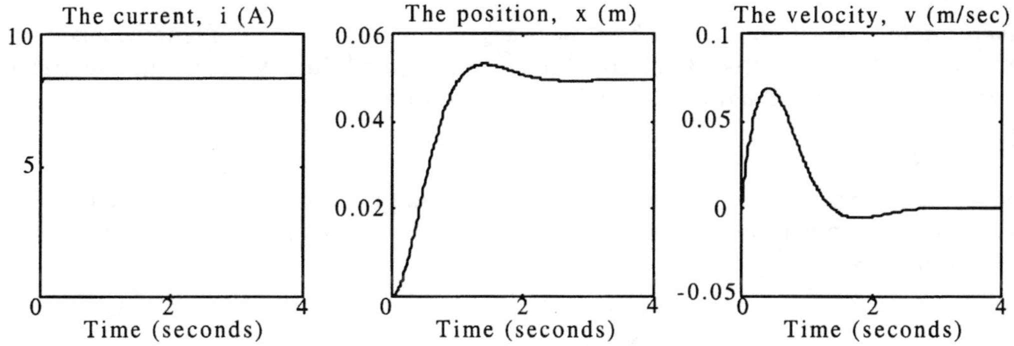

Figure 2.7 Transducer dynamics.

MATLAB script

```
t0=0; tfinal=4; tol=1e-4; trace=1e-4; y0=[0 0 0]';
[t,y]=ode45('ms01_002',t0,tfinal,y0,tol,trace);
subplot(2,3,1); plot(t,y(:,1),'-');
xlabel('Time (seconds)'); title('The current'); pause;
subplot(2,3,2); plot(t,y(:,2),'-');
xlabel('Time (seconds)'); title('The position'); pause;
subplot(2,3,3); plot(t,y(:,3),'-');
xlabel('Time (seconds)'); title('The velocity'); pause;
clear; disp('END');
```

MATLAB script

```
%   Analysis of an electromechanical motion device
function yprime=difer(t,y);
r=6; N=25000; mf=0.005; m0=0.00000126; Af=0.0015; Ag=0.0003;
l=0.25; d=0.0005; m=0.4; ks1=3; ks2=5; Bv=1.5; u=50;
L=N*N*mf*m0*Af*Ag/(Ag*l+2*Af*mf*(y(2,:)+2*d));
dL=-2*mf*Af*L/(Ag*l+2*Af*mf*(y(2,:)+2*d));
yprime=[(-y(1,:)*r-y(1,:)*y(3,:)*dL+u)/L; y(3,:);...
(-y(1,:)*y(1,:)*0.5*dL-y(2,:)*(ks1+y(2,:)*ks2)-y(3,:)*Bv)/m];
```

□

Newtonian mechanics: rotational motion

For one-dimensional rotational systems, Newton's second law of motion is expressed as

$$M = J\alpha \tag{2.2}$$

where M is the sum of all moments about the center of mass of a body, (N-m); J is the moment of inertia about its center of mass, (kg-m^2); and α is the angular acceleration of the body, (rad-s^{-2}).

Example 2.4

Given a point mass m suspended by a massless unstretchable string of length l (see Figure 2.8), derive the equations of motion for a simple pendulum with negligible friction. Develop the SIMULINK diagram to model the pendulum.

Solution

The restoring force, which is proportional to sinθ and given by $-mg\sin\theta$, is the tangential component of the net force. Therefore, the sum of the moments about the pivot point O is found as

$$\sum M = -mgl\sin\theta + T_a$$

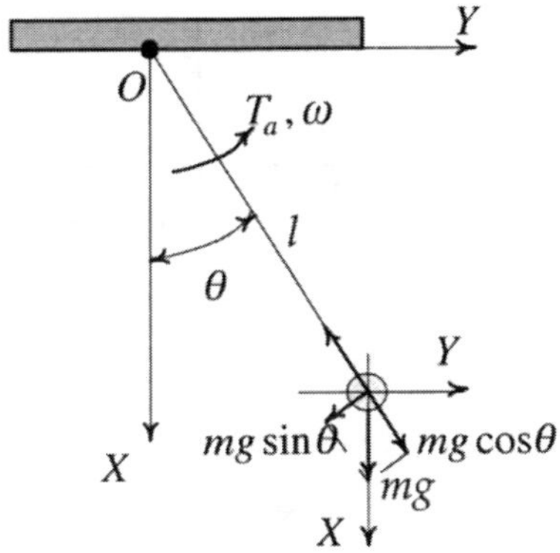

Figure 2.8 A simple pendulum.

where T_a is the applied torque and l is the length of the pendulum measured from the point of rotation.

Using (2.2), one obtains the equation of motion as

$$J\alpha = J\frac{d^2\theta}{dt^2} = -mgl\sin\theta + T_a$$

where J is the moment of inertia of the mass about the point O.

Hence, the second-order differential equation is found to be

$$\frac{d^2\theta}{dt^2} = \frac{1}{J}(-mgl\sin\theta + T_a)$$

Using the following differential equation for the angular displacement

$$\frac{d\theta}{dt} = \omega$$

one obtains the following set of two first-order differential equations

$$\frac{d\omega}{dt} = \frac{1}{J}(-mgl\sin\theta + T_a)$$

$$\frac{d\theta}{dt} = \omega$$

The moment of inertia is expressed by $J = ml^2$. Hence, the following differential equations can be used in modeling:

$$\frac{d\omega}{dt} = -\frac{g}{l}\sin\theta + \frac{1}{ml^2}T_a$$

$$\frac{d\theta}{dt} = \omega$$

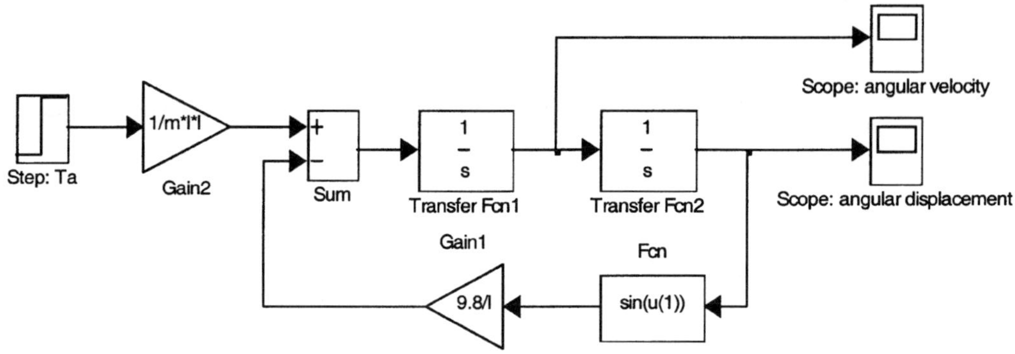

Figure 2.9 SIMULINK diagram.

In the MATLAB environment, the nonlinear dynamics can be easily modeled. Applying the torque T_a = 1N-m, and letting m = 0.5 kg and l = 1 m (these parameter values must be downloaded entering `m=0.5` and `l=1` in the Command window, or by using the m-file), the SIMULINK diagram is developed (see Figure 2.9).

The simulation results are illustrated in Figure 2.10. The data is plotted by setting the variable names `velocity` and `displacement` in the corresponding scopes, and plotting the dynamics simulated using the plot command. In particular, in the Command window we type `plot((velocity(:,1)),velocity(:,2))` to plot the angular velocity and `plot((displacement(:,1)),displacement(:,2))` to plot the angular displacement. Online help capabilities are available in the MATLAB environment. For example, typing

```
» help plot
```

the following is displayed.

```
PLOT Linear plot.
PLOT(X,Y) plots vector Y versus vector X. If X or Y is a matrix,
then the vector is plotted versus the rows or columns of the matrix,
whichever line up.
```

□

2.1.2 *The Lagrange equations of motion*

Although Newton's laws of motion form the fundamental foundation for the study of mechanical systems, they cannot be straightforwardly used to derive the dynamics of electromechanical motion devices because electromagnetic and circuitry transient behavior must be considered. That is, the circuitry dynamics must be incorporated to find augmented models. This can be performed by integrating *torsional-mechanical* dynamics and actuator/sensor circuitry equations, which can be derived using Kirchhoff's laws. The goal is to show that the Lagrange concept allows one to integrate the dynamics of mechanical and electrical components. It will be shown that in modeling, the Lagrange

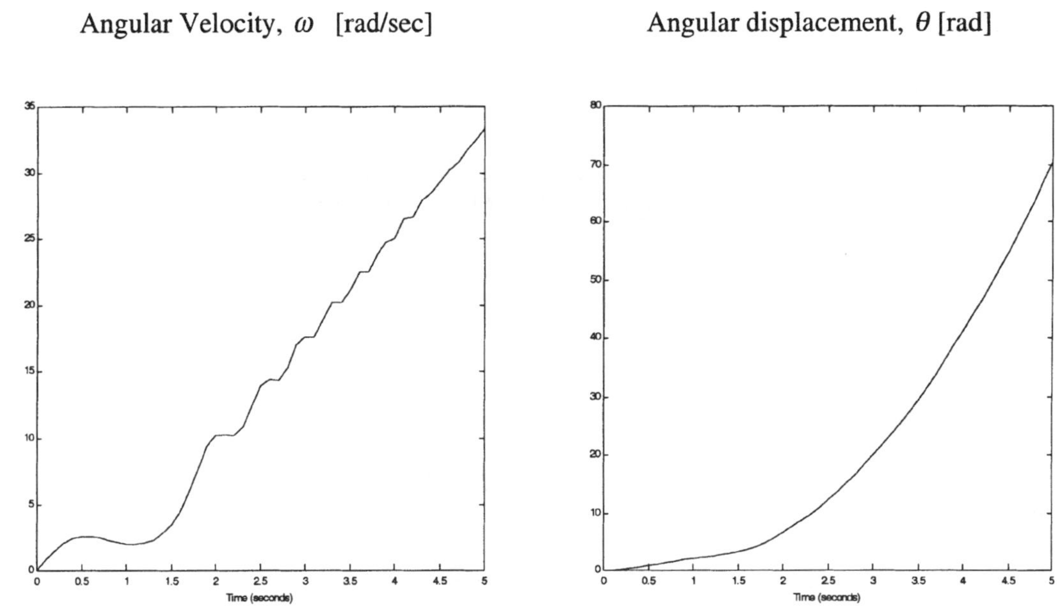

Figure 2.10 Plotted transient dynamics.

equations of motion allow one to develop a unified inroad to derive augmented dynamics. The Lagrange equations of motion are given as

$$\frac{d}{dt}\left(\frac{\partial \Gamma}{\partial \dot{q}_i}\right) - \frac{\partial \Gamma}{\partial q_i} + \frac{\partial D}{\partial \dot{q}_i} + \frac{\partial \Pi}{\partial q_i} = Q_i$$

Here, Γ, D, and Π are the total kinetic, dissipation, and potential energies of the system, respectively; q_i and Q_i are the generalized coordinates and the generalized applied forces, respectively.

Example 2.5

Derive the mathematical model for a simple pendulum as illustrated in Figure 2.8 using the Lagrange equations of motion.

Solution

Derivation of the mathematical model for the simple pendulum, shown in Figure 2.8, was performed in Example 2.4 using Newtonian mechanics. For the studied conservative (lossless) system $D = 0$. Therefore, the Lagrange equations of motion are

$$\frac{d}{dt}\left(\frac{\partial \Gamma}{\partial \dot{q}_i}\right) - \frac{\partial \Gamma}{\partial q_i} + \frac{\partial \Pi}{\partial q_i} = Q_i$$

The kinetic energy of the pendulum bob is

$$\Gamma = \frac{1}{2} m(l\dot{\theta})^2$$

and the potential energy is found as

$$\Pi = mgl(1-\cos\theta)$$

As the generalized coordinate, the angular displacement is used, $\theta_i = \theta$. The generalized force is the torque applied, $Q_i = T_a$. One obtains

$$\frac{\partial \Gamma}{\partial \dot{q}_i} = \frac{\partial \Gamma}{\partial \dot{\theta}} = ml^2\dot{\theta}$$

$$\frac{\partial \Gamma}{\partial q_i} = \frac{\partial \Gamma}{\partial \theta} = 0$$

$$\frac{\partial \Pi}{\partial q_i} = \frac{\partial \Pi}{\partial \theta} = mgl\sin\theta$$

The first term of the Lagrange equation is found to be

$$\frac{d}{dt}\left(\frac{\partial \Gamma}{\partial \dot{\theta}}\right) = ml^2\frac{d^2\theta}{dt^2} + 2ml\frac{dl}{dt}\frac{d\theta}{dt}$$

Assuming that the string is unstretchable, we have $\frac{dl}{dt} = 0.$ Hence,

$$ml^2\frac{d^2\theta}{dt^2} + mgl\sin\theta = T_a$$

and one obtains

$$\frac{d^2\theta}{dt^2} = \frac{1}{ml^2}(-mgl\sin\theta + T_a)$$

Recalling the equation of motion derived using Newtonian mechanics

$$\frac{d^2\theta}{dt^2} = \frac{1}{J}(-mgl\sin\theta + T_a) \text{ where } J = ml^2$$

one concludes that the results are the same. Two first-order differential equations are

$$\frac{d\omega}{dt} = -\frac{g}{l}\sin\theta + \frac{1}{ml^2}T_a$$

$$\frac{d\theta}{dt} = \omega$$

□

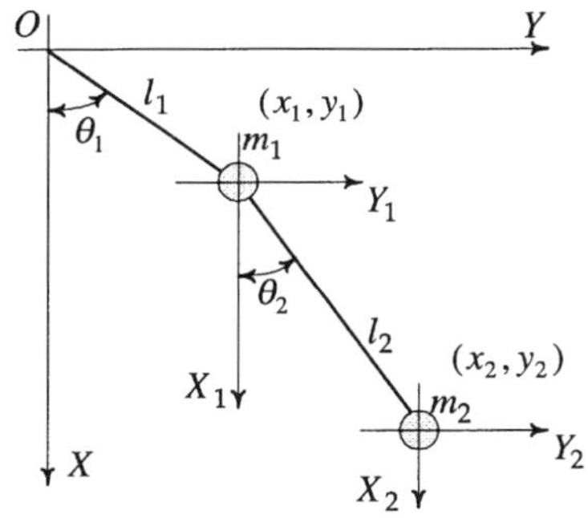

Figure 2.11 Double pendulum.

Example 2.6

Consider a double pendulum of two degrees of freedom with no external forces applied to the system (see Figure 2.11). Using the Lagrange equations of motion, derive the differential equations.

Solution

The angular displacements θ_1 and θ_2 are chosen as the independent generalized coordinates. In the XY plane studied, let (x_1, y_1) and (x_2, y_2) be the rectangular coordinates of m_1 and m_2. Then, one obtains

$$x_1 = l_1 \cos\theta_1, \ x_2 = l_1 \cos\theta_1 + l_2 \cos\theta_2, \ y_1 = l_1 \sin\theta_1, \ y_2 = l_1 \sin\theta_1 + l_2 \sin\theta_2$$

The total kinetic energy Γ is found to be

$$\Gamma = \frac{1}{2} m_1(\dot{x}_1^2 + \dot{y}_1^2) + \frac{1}{2} m_2(\dot{x}_2^2 + \dot{y}_2^2) = \frac{1}{2}(m_1 + m_2) l_1^2 \dot{\theta}_1^2 + m_2 l_1 l_2 \dot{\theta}_1 \dot{\theta}_2 \cos(\theta_2 - \theta_1) + \frac{1}{2} m_2 l_2^2 \dot{\theta}_2^2$$

Then, one obtains

$$\frac{\partial \Gamma}{\partial \theta_1} = m_2 l_1 l_2 \sin(\theta_2 - \theta_1)\dot{\theta}_1 \dot{\theta}_2, \ \frac{\partial \Gamma}{\partial \dot{\theta}_1} = (m_1 + m_2) l_1^2 \dot{\theta}_1 + m_2 l_1 l_2 \cos(\theta_2 - \theta_1)\dot{\theta}_2$$

$$\frac{\partial \Gamma}{\partial \theta_2} = -m_2 l_1 l_2 \sin(\theta_1 - \theta_2)\dot{\theta}_1 \dot{\theta}_2, \ \frac{\partial \Gamma}{\partial \dot{\theta}_2} = m_2 l_1 l_2 \cos(\theta_2 - \theta_1)\dot{\theta}_1 + m_2 l_1^2 \dot{\theta}_2$$

The total potential energy is given by

$$\Pi = m_1 g y_1 + m_2 g y_2 = -(m_1 + m_2) g l_1 \cos\theta_1 - m_2 g l_2 \cos\theta_2$$

and hence,

$$\frac{\partial \Gamma}{\partial \theta_1} = (m_1 + m_2) g l_1 \sin\theta_1 \ \text{ and } \ \frac{\partial \Pi}{\partial \theta_2} = m_2 g l_2 \sin\theta_2$$

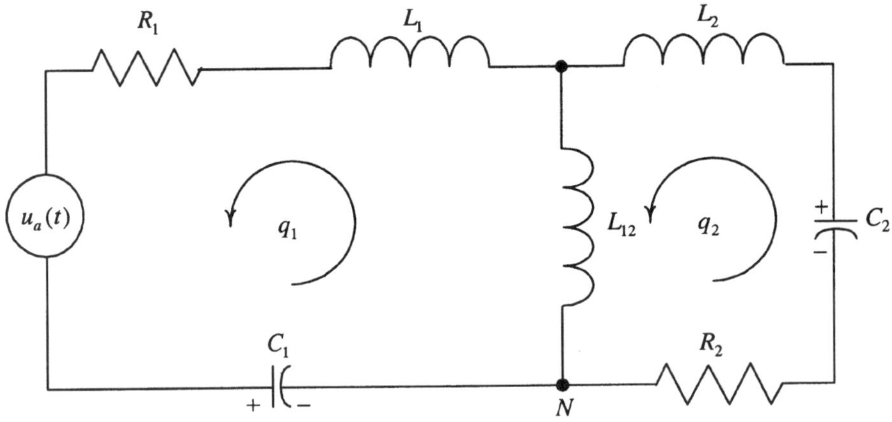

Figure 2.12 Two-mesh circuit network.

The Lagrange equations of motion in terms of the independent coordinates used are

$$\frac{d}{dt}\left(\frac{\partial \Gamma}{\partial \dot{\theta}_1}\right) - \frac{\partial \Gamma}{\partial \theta_1} + \frac{\partial \Pi}{\partial \theta_1} = 0$$

$$\frac{d}{dt}\left(\frac{\partial \Gamma}{\partial \dot{\theta}_2}\right) - \frac{\partial \Gamma}{\partial \theta_2} + \frac{\partial \Pi}{\partial \theta_2} = 0$$

Hence, the dynamic equations of the system are

$$(m_1 + m_2)l_1\ddot{\theta}_1 + m_2 l_2 \cos(\theta_2 - \theta_1)\ddot{\theta}_2 - m_2 l_2 \sin(\theta_2 - \theta_1)\dot{\theta}_2^2 + (m_1 + m_2) g \sin\theta_1 = 0$$

$$l_2\ddot{\theta}_2 + l_1 \cos(\theta_2 - \theta_1)\ddot{\theta}_1 + l_1 \sin(\theta_2 - \theta_1)\dot{\theta}_1^2 + g \sin\theta_2 = 0$$

It should be emphasized that if the torques T_1 and T_2 are applied to the first and second joints, the following equations of motion result

$$(m_1 + m_2)l_1\ddot{\theta}_1 + m_2 l_2 \cos(\theta_2 - \theta_1)\ddot{\theta}_2 - m_2 l_2 \sin(\theta_2 - \theta_1)\dot{\theta}_2^2 + (m_1 + m_2) g \sin\theta_1 = T_1$$

$$l_2\ddot{\theta}_2 + l_1 \cos(\theta_2 - \theta_1)\ddot{\theta}_1 + l_1 \sin(\theta_2 - \theta_1)\dot{\theta}_1^2 + g \sin\theta_2 = T_2$$

□

Example 2.7

Consider a two-mesh electric circuit, as shown in Figure 2.12. Find the circuitry dynamics.

Solution

Let us use q_1 and q_2 as the independent generalized coordinates, where q_1 is the electric charge in the first loop and q_2 represents the electric charge in the second loop. The generalized force applied to the system is denoted as Q_1.

It is important that the generalized coordinates used can be viewed as the circuitry variables. In particular, the currents i_1 and i_2 are found in terms of charges, $i_1 = \dot{q}_1$ and $i_2 = \dot{q}_2$. That is,

$$q_1 = \frac{i_1}{s} \text{ and } q_2 = \frac{i_2}{s}.$$

The generalized force is the applied voltage, $u_a(t) = Q_1$.

The total magnetic energy (kinetic energy) is expressed by

$$\Gamma = \frac{1}{2}L_1\dot{q}_1^2 + \frac{1}{2}L_{12}(\dot{q}_1 - \dot{q}_2)^2 + \frac{1}{2}L_2\dot{q}_2^2$$

Using this equation for Γ, one obtains

$$\frac{\partial \Gamma}{\partial q_1} = 0, \quad \frac{\partial \Gamma}{\partial \dot{q}_1} = (L_1 + L_{12})\dot{q}_1 - L_{12}\dot{q}_2$$

$$\frac{\partial \Gamma}{\partial q_2} = 0, \quad \frac{\partial \Gamma}{\partial \dot{q}_2} = -L_{12}\dot{q}_1 + (L_2 + L_{12})\dot{q}_2$$

Using the equation for the total electric energy (potential energy)

$$\Pi = \frac{1}{2}\frac{q_1^2}{C_1} + \frac{1}{2}\frac{q_2^2}{C_2}$$

one finds

$$\frac{\partial \Pi}{\partial q_1} = \frac{q_1}{C_1} \text{ and } \frac{\partial \Pi}{\partial q_2} = \frac{q_2}{C_2}$$

The total heat energy dissipated is

$$D = \frac{1}{2}R_1\dot{q}_1^2 + \frac{1}{2}R_2\dot{q}_2^2$$

Hence,

$$\frac{\partial D}{\partial \dot{q}_1} = R_1\dot{q}_1 \text{ and } \frac{\partial D}{\partial \dot{q}_2} = R_2\dot{q}_2$$

The Lagrange equations of motion are expressed for the independent coordinates used. In particular, one obtains

$$\frac{d}{dt}\left(\frac{\partial \Gamma}{\partial \dot{q}_1}\right) - \frac{\partial \Gamma}{\partial q_1} + \frac{\partial D}{\partial \dot{q}_1} + \frac{\partial \Pi}{\partial q_1} = Q_1$$

$$\frac{d}{dt}\left(\frac{\partial \Gamma}{\partial \dot{q}_2}\right) - \frac{\partial \Gamma}{\partial q_2} + \frac{\partial D}{\partial \dot{q}_2} + \frac{\partial \Pi}{\partial q_2} = 0$$

Hence, the differential equations for the circuit studied are

$$(L_1 + L_{12})\ddot{q}_1 - L_{12}\ddot{q}_2 + R_1\dot{q}_1 + \frac{q_1}{C_1} = u_a$$

$$-L_{12}\ddot{q}_1 + (L_2 + L_{12})\ddot{q}_2 + R_2\dot{q}_2 + \frac{q_2}{C_2} = 0$$

The SIMULINK model can be built using these derived nonlinear differential equations. In particular,

$$\ddot{q}_1 = \frac{1}{(L_1 + L_{12})}\left(-\frac{q_1}{C_1} - R_1\dot{q}_1 + L_{12}\ddot{q}_2 + u_a\right) \ \textit{and} \ \ddot{q}_2 = \frac{1}{(L_2 + L_{12})}\left(L_{12}\ddot{q}_1 - \frac{q_2}{C_2} - R_2\dot{q}_2\right)$$

The corresponding SIMULINK diagram is shown in Figure 2.13. The circuitry parameters are assigned to be: $L_1 = 0.01$ H, $L_2 = 0.005$ H, $L_{12} = 0.0025$ H, $C_1 = 0.02$ F, $C_2 = 0.1$ F, $R_1 = 10$ ohm, $R_1 = 5$ ohm, and $u_a = 100\sin(200t)$ V.

It should be emphasized that the currents i_1 and i_2 are expressed in terms of charges as $i_1 = \dot{q}_1$ and $i_2 = \dot{q}_2$. That is,

$$q_1 = \frac{i_1}{s} \text{ and } q_2 = \frac{i_2}{s}$$

Simulation results, that is, the time history of $q_1(t)$, $q_2(t)$, $i_1(t)$, and $i_2(t)$, are documented in Figure 2.14.

Example 2.8

Using the Lagrange equations of motion, develop the mathematical models for the circuit shown in the Figure 2.15. Prove that the model derived using the Lagrange equations of motion are equivalent to the model developed using Kirchhoff's law.

Solution

Using q_1 and q_2 as the independent generalized coordinates, where q_1 is the electric charge in the first loop $i_a = \dot{q}_1$, and q_2, is the electric charge in the second loop, $i_L = \dot{q}_2$. The generalized force applied to the system is Q_1, and $u_a(t) = Q_1$.

The total kinetic energy is

$$\Gamma = \frac{1}{2}L\dot{q}_2^2$$

We have

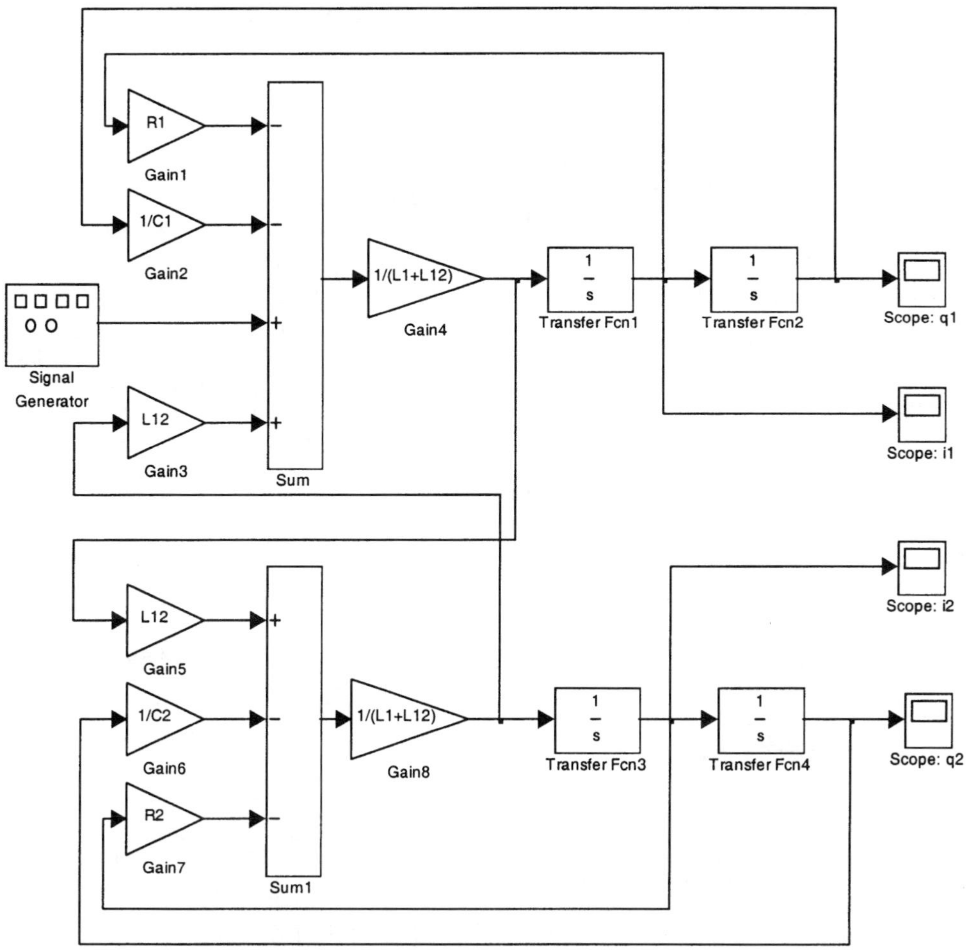

Figure 2.13 SIMULINK diagram.

$$\frac{\partial \Gamma}{\partial q_1} = 0, \quad \frac{\partial \Gamma}{\partial \dot{q}_1} = 0, \quad \frac{d}{dt}\left(\frac{\partial \Gamma}{\partial \dot{q}_1}\right) = 0$$

$$\frac{\partial \Gamma}{\partial q_2} = 0, \quad \frac{\partial \Gamma}{\partial \dot{q}_2} = L\dot{q}_2, \quad \frac{d}{dt}\left(\frac{\partial \Gamma}{\partial \dot{q}_2}\right) = L\ddot{q}_2$$

The total potential energy is expressed as

$$\Pi = \frac{1}{2}\frac{(q_1 - q_2)^2}{C}$$

and hence,

$$\frac{\partial \Pi}{\partial q_1} = \frac{q_1 - q_2}{C} \quad \text{and} \quad \frac{\partial \Pi}{\partial q_2} = \frac{-q_1 + q_2}{C}$$

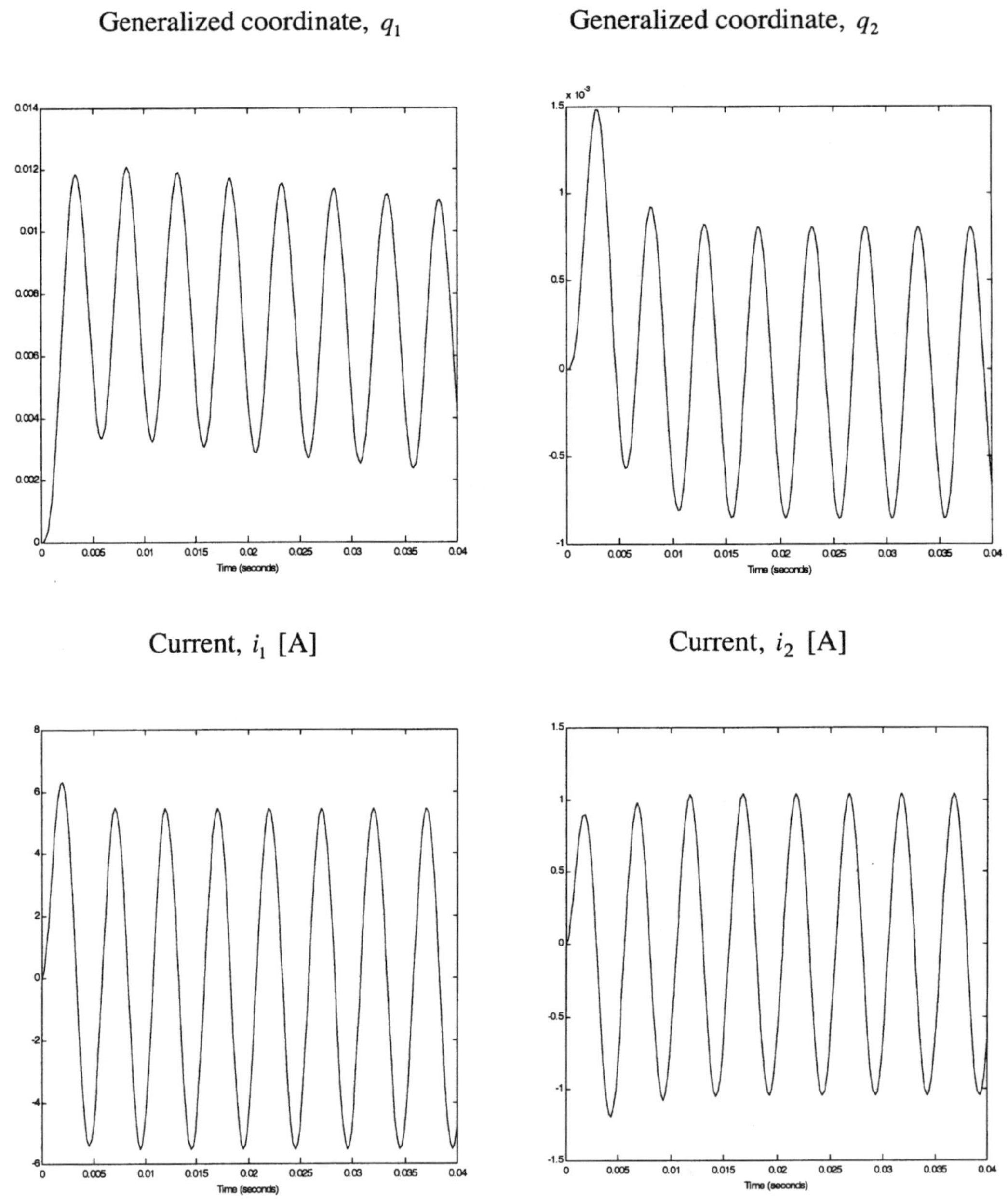

Figure 2.14 Transient dynamics of the generalized coordinates and currents. □

The total dissipated energy is

$$D = \frac{1}{2} R \dot{q}_1^2 + \frac{1}{2} R_L \dot{q}_2^2$$

Therefore,

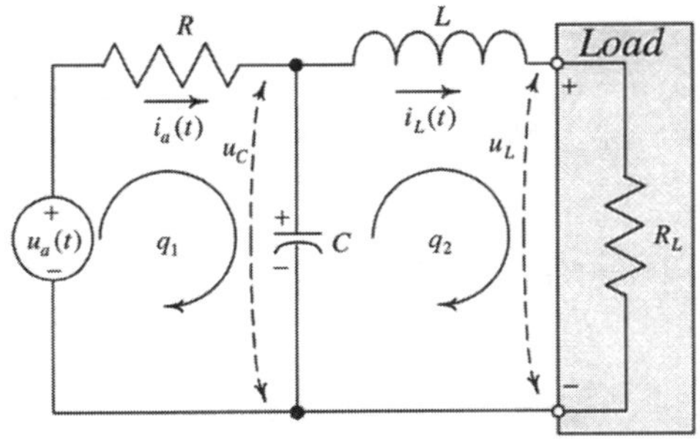

Figure 2.15 Electric circuit.

$$\frac{\partial D}{\partial \dot{q}_1} = R\dot{q}_1 \text{ and } \frac{\partial D}{\partial \dot{q}_2} = R_L\dot{q}_2$$

The Lagrange equations of motion

$$\frac{d}{dt}\left(\frac{\partial \Gamma}{\partial \dot{q}_1}\right) - \frac{\partial \Gamma}{\partial q_1} + \frac{\partial D}{\partial \dot{q}_1} + \frac{\partial \Pi}{\partial q_1} = Q_1, \quad \frac{d}{dt}\left(\frac{\partial \Gamma}{\partial \dot{q}_2}\right) - \frac{\partial \Gamma}{\partial q_2} + \frac{\partial D}{\partial \dot{q}_2} + \frac{\partial \Pi}{\partial q_2} = 0$$

lead one to the following two differential equations.

$$R\dot{q}_1 + \frac{q_1 - q_2}{C} = u_a, \quad L\ddot{q}_2 + R_L\dot{q}_2 + \frac{-q_1 + q_2}{C} = 0$$

Hence, we have found a set of two differential equations. In particular,

$$\dot{q}_1 = \frac{1}{R}\left(\frac{-q_1 + q_2}{C} + u_a\right)$$

$$\ddot{q}_2 = \frac{1}{L}\left(-R_L\dot{q}_2 + \frac{q_1 - q_2}{C}\right)$$

Using Kirchhoff's law, two differential equations result

$$\frac{du_C}{dt} = \frac{1}{C}\left(-\frac{u_C}{R} - i_L + \frac{u_a(t)}{R}\right)$$

$$\frac{di_L}{dt} = \frac{1}{L}(u_C - R_L i_L)$$

Taking note of $i_a = \dot{q}_1$ and $i_L = \dot{q}_2$, and making use $C\frac{du_C}{dt} = i_a - i_L$, one obtains $u_C = \frac{q_1 - q_2}{C}$. The equivalence of the differential equations derived using the Lagrange equations of motion and Kirchhoff's law is proven. □

Example 2.9

As an important example in analysis and design of directly driven servo-systems, consider a servomechanism actuated by a motor with two independently excited stator and rotor

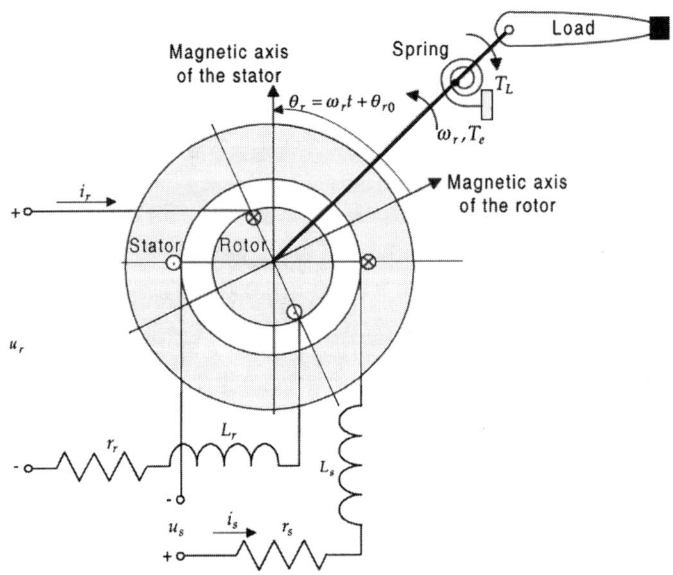

Figure 2.16 A directly driven servo-system.

windings. Figure 2.16 shows a studied configuration of a directly driven servo-system. Derive the differential equations.

The following notations are used: i_s and i_r are the currents in the stator and rotor windings; u_s and u_r are the applied voltages to the stator and rotor windings; ω_r and θ_r are the angular velocity and displacement of the rotor; T_e and T_L are the electromagnetic and load torques; r_s and r_r are the resistances of the stator and rotor windings; L_s and L_r are the self-inductances of the stator and rotor windings; L_{sr} is the mutual inductance of the stator and rotor windings; $\Re_m$ is the reluctance of the magnetizing path; N_s and N_r are the number of turns in the stator and rotor windings; J is the moment of inertia of the rotor and attached load; B_m is the viscous friction coefficient; and k_s is the spring constant.

Solution

The magnetic fluxes that cross an air gap produce a force of attraction, and the developed electromagnetic torque T_e is countered by the tortional spring that causes a *counterclockwise* rotation. The load torque T_L should be considered. The goal is to find a nonlinear mathematical model. In fact, the ability to formulate the modeling problem and find the resulting equations that describe a motion device constitute the most important issues. Using the Lagrange concept, the independent generalized coordinates must be chosen. In particular, let us use q_1, q_2, and q_3, where q_1 and q_2 denote the electric charges in the stator and rotor windings; and q_3 represents the angular displacement of the rotor.

Denote the generalized forces applied to an electromechanical system as Q_1, Q_2, and Q_3, where Q_1 and Q_2 are the applied voltages to the stator and rotor windings, and Q_3 is the load torque.

The first derivative of the generalized coordinates $\dot{q}_1$ and $\dot{q}_2$ represent the stator and rotor currents i_s and i_r, while $\dot{q}_3$ is the angular velocity of the rotor ω_r.

We have,

$$q_1 = \frac{i_s}{s},\ q_2 = \frac{i_r}{s},\ q_3 = \theta_r,\ \dot{q}_1 = i_s,\ \dot{q}_2 = i_r,\ \dot{q}_3 = \omega_r,\ Q_1 = u_s,\ Q_2 = u_r,\ \text{and}\ Q_3 = -T_L$$

The Lagrange equations are expressed in terms of each independent coordinate, and we have

$$\frac{d}{dt}\left(\frac{\partial \Gamma}{\partial \dot{q}_1}\right) - \frac{\partial \Gamma}{\partial q_1} + \frac{\partial D}{\partial \dot{q}_1} + \frac{\partial \Pi}{\partial q_1} = Q_1$$

$$\frac{d}{dt}\left(\frac{\partial \Gamma}{\partial \dot{q}_2}\right) - \frac{\partial \Gamma}{\partial q_2} + \frac{\partial D}{\partial \dot{q}_2} + \frac{\partial \Pi}{\partial q_2} = Q_2$$

$$\frac{d}{dt}\left(\frac{\partial \Gamma}{\partial \dot{q}_3}\right) - \frac{\partial \Gamma}{\partial q_3} + \frac{\partial D}{\partial \dot{q}_3} + \frac{\partial \Pi}{\partial q_3} = Q_3$$

The total kinetic energy of electrical and mechanical systems is found as a sum of the total magnetic (electrical) Γ_E and mechanical Γ_M energies. The total kinetic energy of the stator and rotor circuitry is given as

$$\Gamma_E = \frac{1}{2}L_s\dot{q}_1^2 + L_{sr}\dot{q}_1\dot{q}_2 + \frac{1}{2}L_r\dot{q}_2^2$$

The total kinetic energy of the mechanical system, which is a function of the equivalent moment of inertia of the rotor and the payload attached, is expressed by

$$\Gamma_M = \frac{1}{2}J\dot{q}_3^2$$

Then, we have

$$\Gamma = \Gamma_E + \Gamma_M = \frac{1}{2}L_s\dot{q}_1^2 + L_{sr}\dot{q}_1\dot{q}_2 + \frac{1}{2}L_r\dot{q}_2^2 + \frac{1}{2}J\dot{q}_3^2$$

The mutual inductance is a periodic function of the angular rotor displacement and

$$L_{sr}(\theta_r) = \frac{N_s N_r}{\Re_m(\theta_r)}$$

The magnetizing reluctance is maximum if the stator and rotor windings are not displaced, and $\Re_m(\theta_r)$ is minimum if the coils are displaced by 90°. Then,

$$L_{sr\ \min} \le L_{sr}(\theta_r) \le L_{sr\ \max}$$

Where $L_{sr\ \max} = \dfrac{N_s N_r}{\Re_m(90°)}$ *and* $L_{sr\ \min} = \dfrac{N_s N_r}{\Re_m(0°)}$.

The mutual inductance can be approximated as a cosine function of the rotor angular displacement, and the amplitude of the mutual inductance between the stator and rotor windings is found as

$$L_M = L_{sr\ \max} = \frac{N_s N_r}{\Re_m(90°)}$$

Then,

$$L_{sr}(\theta_r) = L_M \cos\theta_r = L_M \cos q_3$$

One obtains an explicit expression for the total kinetic energy as

$$\Gamma = \frac{1}{2}L_s\dot{q}_1^2 + L_M\dot{q}_1\dot{q}_2\cos q_3 + \frac{1}{2}L_r\dot{q}_2^2 + \frac{1}{2}J\dot{q}_3^2$$

The following partial derivatives result

$$\frac{\partial\Gamma}{\partial q_1} = 0,\ \frac{\partial\Gamma}{\partial\dot{q}_1} = L_s\dot{q}_1 + L_M\dot{q}_2\cos q_3$$

$$\frac{\partial\Gamma}{\partial q_2} = 0,\ \frac{\partial\Gamma}{\partial\dot{q}_2} = L_M\dot{q}_1\cos q_3 + L_r\dot{q}_2$$

$$\frac{\partial\Gamma}{\partial q_3} = -L_M\dot{q}_1\dot{q}_2\sin q_3,\ \frac{\partial\Gamma}{\partial\dot{q}_3} = J\dot{q}_3$$

The potential energy of the spring with constant k_s is

$$\Pi = \frac{1}{2}k_s q_3^2$$

Therefore,

$$\frac{\partial\Pi}{\partial q_1} = 0,\ \frac{\partial\Pi}{\partial q_2} = 0,\ \text{and}\ \frac{\partial\Pi}{\partial q_3} = k_s q_3$$

The total heat energy dissipated is expressed as

$$D = D_E + D_M$$

where D_E is the heat energy dissipated in the stator and rotor windings, $D_E = \frac{1}{2} r_s \dot{q}_1^2 + \frac{1}{2} r_r \dot{q}_2^2$; D_M is the heat energy dissipated by mechanical system, $D_M = \frac{1}{2} B_m \dot{q}_3^2$. Hence,

$$D = \frac{1}{2} r_s \dot{q}_1^2 + \frac{1}{2} r_r \dot{q}_2^2 + \frac{1}{2} B_m \dot{q}_3^2$$

One obtains

$$\frac{\partial D}{\partial \dot{q}_1} = r_s \dot{q}_1, \quad \frac{\partial D}{\partial \dot{q}_2} = r_r \dot{q}_2, \text{ and } \frac{\partial D}{\partial \dot{q}_3} = B_m \dot{q}_3$$

Using $q_1 = \frac{i_s}{s}$, $q_2 = \frac{i_r}{s}$, $q_3 = \theta_r$, $\dot{q}_1 = i_s$, $\dot{q}_2 = i_r$, $\dot{q}_3 = \omega_r$, $Q_1 = u_s$, $Q_2 = u_r$, and $Q_3 = -T_L$, we have three differential equations for a servo-system. In particular,

$$L_s \frac{di_s}{dt} + L_M \cos\theta_r \frac{di_r}{dt} - L_M i_r \sin\theta_r \frac{d\theta_r}{dt} + r_s i_s = u_s$$

$$L_r \frac{di_r}{dt} + L_M \cos\theta_r \frac{di_s}{dt} - L_M i_s \sin\theta_r \frac{d\theta_r}{dt} + r_r i_r = u_r$$

$$J \frac{d^2\theta_r}{dt^2} + L_M i_s i_r \sin\theta_r + B_m \frac{d\theta_r}{dt} + k_s \theta_r = -T_L$$

The last equation should be rewritten making use of the rotor angular velocity; that is, $\frac{d\theta_r}{dt} = \omega_r$. Finally, using the stator and rotor currents, angular velocity, and position as the state variables, the nonlinear differential equations in Cauchy's form are found as

$$\frac{di_s}{dt} = \frac{1}{L_s L_r - L_M^2 \cos^2\theta_r}\left(-r_s L_r i_s - \frac{1}{2} L_M^2 i_s \omega_r \sin 2\theta_r + r_r L_M i_r \cos\theta_r + L_r L_M i_r \omega_r \sin\theta_r + L_r u_s - L_M \cos\theta_r u_r\right)$$

$$\frac{di_r}{dt} = \frac{1}{L_s L_r - L_M^2 \cos^2\theta_r}\left(r_s L_M i_s \cos\theta_r + L_s L_M i_s \omega_r \sin\theta_r - r_r L_s i_r - \frac{1}{2} L_M^2 i_r \omega_r \sin 2\theta_r - L_M \cos\theta_r u_s + L_s u_r\right)$$

$$\frac{d\omega_r}{dt} = \frac{1}{J}(-L_M i_s i_r \sin\theta_r - B_m \omega_r - k_s \theta_r - T_L)$$

$$\frac{d\theta_r}{dt} = \omega_r$$

The developed nonlinear mathematical model in the form of highly coupled nonlinear differential equations cannot be linearized, and one must model the doubly excited transducer studied using the nonlinear differential equations derived. □

2.2 Energy conversion

Energy conversion takes place in electromechanical motion devices and power systems, micro-electromechanical systems, and power electronic devices. This book studies electromagnetic and electromechanical motion devices that are used for conversion of electrical energy to mechanical energy and vice versa (conversion of mechanical energy to electrical energy). Hydro-electric power production is an example of direct conversion of mechanical energy into electrical energy. There are various forms of energy, and multi-channel energy conversion is needed, for example, in nuclear power plants. Consider a steam-driven locomotive. It is not possible to convert the chemical energy of coal directly into electrical energy; chemical energy is converted first into mechanical energy. Coal is burned to heat water and produce steam. This steam is supplied to the cylinders to actuate the pistons that rotate wheels, as well as to drive the generator (mechanical energy results). The voltage is induced by the generator; that is, electrical energy results. To convert electrical energy to chemical energy, one feeds the current to batteries that store the energy in the form of chemical energy. It is evident that the losses of energy should be thoroughly treated, and analysis of efficiency of energy conversion is a very important problem that must be considered.

This section focuses on the comprehensive study of energy conversion in electromechanical systems, which include electromagnetic and electromechanical motion devices. Analysis and design of high-performance systems require a complete understanding of electromagnetic and electromechanical motion devices characteristics and associated interactions with power electronics. Analysis and optimization of steady-state and dynamic characteristics depend critically on an intimate knowledge of energy conversion, and comprehensive nonlinear analytical and numerical studies must be performed using energy conversion principles. Thorough analytical and advanced numerical analysis speed up the design and allow one to optimize systems approaching performance limits that can be achieved. Special attention should be placed on the analysis, modeling, and design of electromechanical motion devices controlled by power converters. The major emphasis of this section focuses on fundamental concepts of energy conversion and thorough analysis using cornerstone methods in the context of the topics covered.

Fundamental principles of energy conversion, applicable to electromagnetic and electromechanical motion devices, must be studied first to provide analytical foundations. Energy conversion fundamentals were already applied, and this section, as well as the following chapters, devoted to electromechanical motion devices and power electronics, will further emphasize the application of this concept. The physics of energy conversion must be researched and used to study the dynamics of electromechanical motion devices. Advanced analytical and numerical methods should be applied to solve a spectrum of complex problems in energy conversion. The computer-based exercises and problems given in this book help one to appreciate fully the theoretical foundations and master the software needed. Rapid advances in computer technology and software have made it possible to apply parallel processing techniques to large system simulation. It is evident that analysis and design of advanced state-of-the-art multi-input/multi-output electromechanical systems continue to challenge engineers and enforce the thorough study of energy conversion. Therefore, the designer should use energy-conversion fundamentals to approach analytical studies as well as to perform numerical analysis.

Using the principle of conservation of energy one can formulate:

> For a lossless electromechanical system (in the conservative system no energy is lost through friction, heat, or other irreversible energy

conversion), the sum of the instantaneous kinetic and potential energies of the system remains constant.

The energy conversion is represented in Figure 2.17.

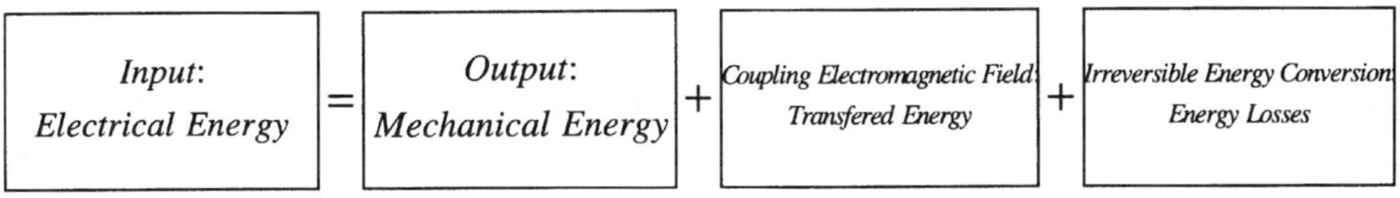

Figure 2.17 Energy transfer in electromechanical systems.

In Section 1.1, it was shown that to find the total energy stored in the magnetic field, the following formula should be applied

$$W_m = \frac{1}{2}\int_v \vec{B}\cdot\vec{H}dv$$

where $\vec{B}$ and $\vec{H}$ are related using the permeability μ, $\vec{B}=\mu\vec{H}$.

The material becomes magnetized in response to the external field $\vec{H}$, and the dimensionless magnetic susceptibility χ_m or relative permeability μ_r are used. We have,

$$\vec{B}=\mu\vec{H}=\mu_0(1+\chi_m)\vec{H}=\mu_0\mu_r\vec{H}=\mu\vec{H}$$

Based on the value of the magnetic susceptibility χ_m, the materials are classified as

- Diamagnetic: $\chi_m \approx -1\times10^{-5}$ (for example, $\chi_m = -9.5\times10^{-6}$ for copper, $\chi_m = -3.2\times10^{-5}$ for gold, and $\chi_m = -2.6\times10^{-5}$ for silver).
- Paramagnetic: $\chi_m \approx 1\times10^{-4}$ (for example, $\chi_m = 1.4\times10^{-3}$ for Fe_2O_3 and $\chi_m = 1.7\times10^{-3}$ for Cr_2O_3).
- Ferromagnetic: $|\chi_m| >> 1$ (iron, nickel, and cobalt, neodymium, iron, boron, and samarium cobalt permanent magnets).

The magnetization behavior of the ferromagnetic materials is mapped by the magnetization curve, where H is the externally applied magnetic field and B is the total magnetic flux density in the medium. Typical B-H curves for hard and soft ferromagnetic materials are given in Figure 2.18, respectively.

The B vs. H curve allows one to establish the energy analysis. Assume that initially $B_0 = 0$ and $H_0 = 0$. Let H increase from $H_0 = 0$ to H_{max}. Then, B increases from $B_0 = 0$ until the maximum value of B, denoted as B_{max}, is reached. If H then decreases to H_{min}, B decreases to B_{min} through the remanent value B_r (the so-called residual magnetic flux density) along a different curve (see Figure 2.18). For variations of H, $H \in [H_{min}H_{max}]$, B changes within the *hysteresis loop*, and $B \in [B_{min}B_{max}]$.

In per-unit volume, the applied field energy is

$$W_F = \oint_B HdB$$

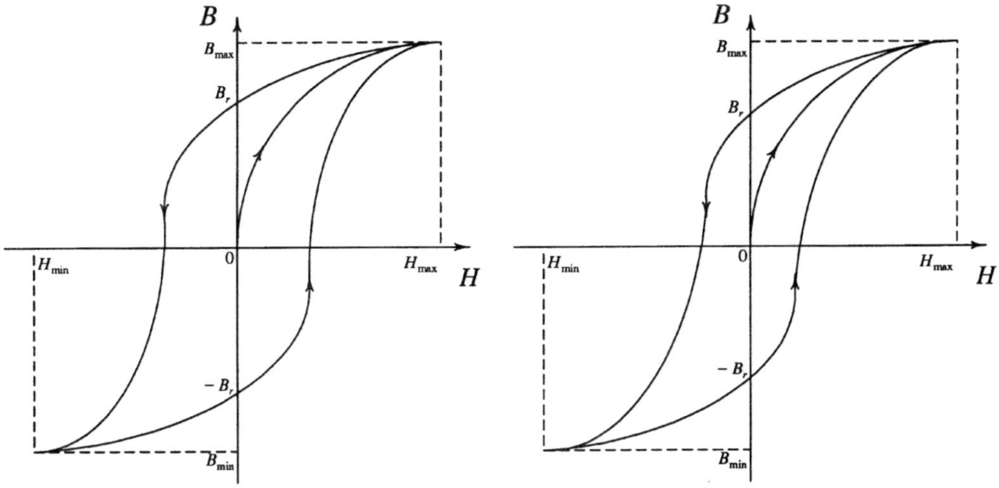

Figure 2.18 *B-H* curves for hard and soft ferromagnetic materials.

while the stored energy is expressed as

$$W_e = \oint_H B dH$$

In the volume v, one has the following expressions for the field and stored energies

$$W_F = v\oint_B H dB \text{ and } W_c = v\oint_H B dH$$

A complete B vs. H loop should be considered, and the equations for field and stored energy represent the areas enclosed by the corresponding curve. It should be emphasized that each point on the B vs. H curve represents the total energy.

In ferromagnetic materials, time-varying magnetic flux produces core losses that consist of hysteresis losses (due to the hysteresis loop of the *B-H* curve) and eddy-current losses, which are proportional to the current frequency and lamination thickness. The area of the hysteresis loop is related to hysteresis losses. Soft ferromagnetic materials have narrow hysteresis loops and are easily magnetized and demagnetized; therefore, lower hysteresis losses, compared with hard ferromagnetic materials, result.

For electromechanical motion devices, the flux linkages are plotted versus the current because the current and flux linkages are used rather than the flux intensity and flux density. In electric machines, almost all energy is stored in the air gap. Using the fact that the air is a conservative medium, one concludes that the coupling filed is lossless. Figure 2.19. illustrates the nonlinear magnetizing characteristic. The energy stored in the magnetic field is

$$W_F = \oint_\psi i d\psi \text{ (the shaded area above the magnetization curve in Figure 2.19)}$$

while the coenergy is expressed as

$$W_c = \oint_i \psi di \text{ (the filled area under the magnetization curve in Figure 2.19)}$$

It should be emphasized that the total energy is

$$W_F + W_c = \oint_\psi i d\psi + \oint_i \psi di = \psi i$$

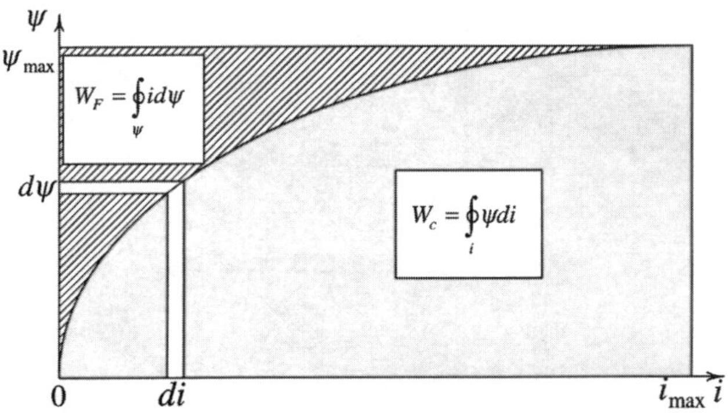

Figure 2.19 Magnetization curve and graphical representation of stored energy and coenergy.

The flux linkages are the function of current i and position x (for translational motion) or angular displacement θ (for rotational motion); that is, $\psi = f(i, x)$ or $\psi = f(i, \theta)$. The current can be found as the nonlinear function of the flux linkages and position or angular displacement. Hence, one obtains

$$d\psi = \frac{\partial\psi(i,x)}{\partial i}di + \frac{\partial\psi(i,x)}{\partial x}dx,\quad d\psi = \frac{\partial\psi(i,\theta)}{\partial i}di + \frac{\partial\psi(i,\theta)}{\partial\theta}d\theta$$

and

$$di = \frac{\partial i(\psi,x)}{\partial\psi}d\psi + \frac{\partial i(\psi,x)}{\partial x}dx,\quad di = \frac{\partial i(\psi,\theta)}{\partial\psi}d\psi + \frac{\partial i(\psi,\theta)}{\partial\theta}d\theta$$

Therefore,

$$W_F = \oint_\psi i d\psi = \oint_i i\frac{\partial\psi(i,x)}{\partial i}di + \oint_x i\frac{\partial\psi(i,x)}{\partial x}dx,\quad W_F = \oint_\psi i d\psi = \oint_i i\frac{\partial\psi(i,\theta)}{\partial i}di + \oint_\theta i\frac{\partial\psi(i,\theta)}{\partial\theta}d\theta$$

and

$$W_c = \oint_i \psi di = \oint_\psi \psi\frac{\partial i(\psi,x)}{\partial\psi}d\psi + \oint_x \psi\frac{\partial i(\psi,x)}{\partial x}dx,\quad W_c = \oint_i \psi di = \oint_\psi \psi\frac{\partial i(\psi,\theta)}{\partial\psi}d\psi + \oint_x \psi\frac{\partial i(\psi,\theta)}{\partial\theta}d\theta$$

Assuming that the coupling field is lossless, the differential change in the mechanical energy (found in Section 1.1 using the differential displacement $d\vec{l}$ as $dW_{mec} = \vec{F}_m \cdot d\vec{l}$) is related to the differential change of the coenergy. In particular, for displacement at constant current, one obtains $dW_{mec} = dW_c$, and hence, the electromagnetic force is found to be

$$F_e(i,x) = \frac{\partial W_c(i,x)}{\partial x}$$

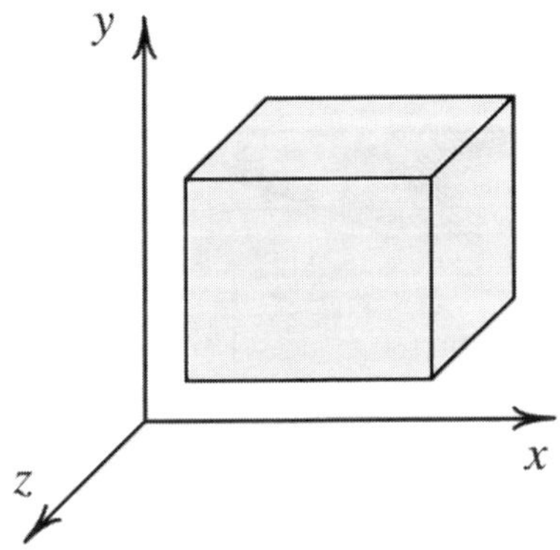

Figure 2.20 Element in the *xyz* coordinate system.

For rotational motion, the electromagnetic torque, as a function of current and angular displacement, is

$$T_e(i,\theta) = \frac{\partial W_c(i,\theta)}{\partial \theta}$$

In micro-electromechanical systems, smart materials are used. Micro- and meso-scale structures, as well as thin magnetic films, exhibit anisotropy. Consider the anisotropic ferromagnetic element in the Cartesian (rectangular) coordinate system shown in Figure 2.20.

The permeability $\mu(x, y, z)$ must be given as a matrix

$$\mu(x,y,z) = \begin{bmatrix} \mu_{xx} & \mu_{xy} & \mu_{xz} \\ \mu_{yx} & \mu_{yy} & \mu_{yz} \\ \mu_{zx} & \mu_{zy} & \mu_{zz} \end{bmatrix}$$

and the following matrix equation results

$$\begin{bmatrix} B_x \\ B_y \\ B_z \end{bmatrix} = \begin{bmatrix} \mu_{xx} & \mu_{xy} & \mu_{xz} \\ \mu_{yx} & \mu_{yy} & \mu_{yz} \\ \mu_{zx} & \mu_{zy} & \mu_{zz} \end{bmatrix} \begin{bmatrix} H_x \\ H_y \\ H_z \end{bmatrix}$$

In expanded form, one obtains

$$B_x = \mu_{xx} H_x + \mu_{xy} H_y + \mu_{xz} H_z$$

$$B_y = \mu_{yx} H_x + \mu_{yy} H_y + \mu_{yz} H_z$$

$$B_z = \mu_{zx} H_x + \mu_{zy} H_y + \mu_{zz} H_z$$

It is evident that the analysis of anisotropic ferromagnetic devices can be performed in a similar way; however, the use of matrices complicates the study. Furthermore, the majority of electromechanical motion devices can be studied assuming that the medium is linear, homogeneous, and isotropic. This assumption is not valid for advanced analysis of micro-electromechanical systems.

2. 3 Electromechanical analogies

The differential equations result as one applies the fundamental laws to electrical and mechanical systems. It has been shown that the transient dynamics of electrical and mechanical systems are mapped by nonlinear differential equations. A thorough analysis of electromechanical analogies, which will be performed in this section, gives striking examples of the role that mathematical models play in unifying the transient behavior of different phenomena in electrical and mechanical systems. Specifically, one finds that the dynamics of different electrical and mechanical subsystems are described by the same form of differential equations. Using this mathematical identity and the common quantities (variables) obtained, electrical systems can be modeled using mechanical systems, and mechanical systems can be analyzed using electrical systems. The transient behavior of electromechanical systems have been mapped by differential equations that use states and controls, outputs and references, errors between the desired inputs and measured outputs, as well as parametric coefficients (resistance, inductance, capacitance, moment of inertia, mass, friction coefficient, and other constants). The corresponding quantities (variables) and parameters of electrical and mechanical elements, mapped by differential equations of the same form, are called analogs. This section examines the basic elements of mechanical and electrical systems. To perform the objectives, to illustrate the similarity, and to visualize the results, the equations of motion for some electromechanical system elements are shown in Table 2.1.

The similarity of equations of motion are evident as one compares the derived dynamics, which are mapped by the corresponding differential equations. Comparing the differential equations of the mechanical and electrical elements, one obtains the mechanical and electrical analogs shown in Table 2.2.

Table 2.1 illustrates and visualizes the results for the simplest elements of electromechanical systems described by the first-order differential equations. One can now perform the comprehensive studies for more complex systems. Consider the translational and rotational (torsional) mechanical systems as shown in Figure 2.21.

From Newton's law or using Lagrange equations of motions, the second-order differential equations of translational-dynamics and torsional-dynamics are found to be

Translational dynamics: $$m\frac{d^2x}{dt^2} + B_v\frac{dx}{dt} + k_s x = F_a(t)$$

Torisional dynamics: $$J\frac{d^2\theta}{dt^2} + B_m\frac{d\theta}{dt} + k_s\theta = T_a(t)$$

Table 2.1 Basic Elements of Electromechanical Systems

Electromechanical system elements	Variables Used	Equation
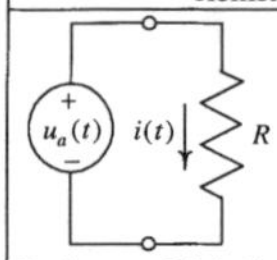 Resistance, R [ohm]	Applied voltage $u_a(t)$ [V] Current $i(t)$ [A]	$u_a(t) = Ri(t)$ $i(t) = \frac{1}{R}u_a(t)$
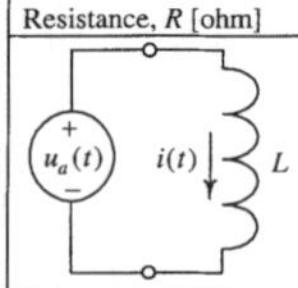 Inductance, L [H]	Applied voltage $u_a(t)$ [V] Current $i(t)$ [A]	$u_a(t) = L\frac{di(t)}{dt}$ $i(t) = \frac{1}{L}\int_{t_0}^{t} u_a(\tau)d\tau$
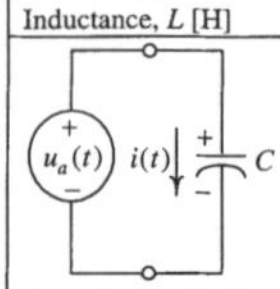 Capacitance, C [F]	Applied voltage $u_a(t)$ [V] Current $i(t)$ [A]	$u_a(t) = \frac{1}{C}\int_{t_0}^{t} i(\tau)d\tau$ $i(t) = C\frac{du_a(t)}{dt}$
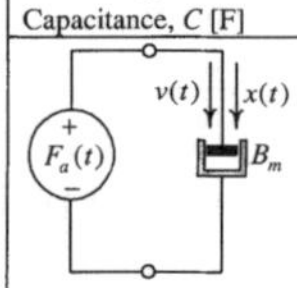 Translational damper, B_v [N-sec]	Applied force $F_a(t)$ [N] Linear velocity $v(t)$ [m/sec] Linear position $x(t)$ [m]	$F_a(t) = B_m v(t)$ $v(t) = \frac{1}{B_m}F_a(t)$ and $F_a(t) = B_m v(t) = B_m\frac{dx(t)}{dt}$ $x(t) = \frac{1}{B_v}\int_{t_0}^{t} F_a(\tau)d\tau$
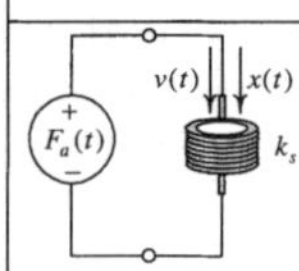 Translational spring, k_s [N]	Applied force $F_a(t)$ [N] Linear velocity $v(t)$ [m/sec] Linear position $x(t)$ [m]	$F_a(t) = k_s x(t)$ $x(t) = \frac{1}{k_s}F_a(t)$ and $v(t) = \frac{dx(t)}{dt} = \frac{1}{k_s}\frac{dF_a(t)}{dt}$ $F_a(t) = k_s\int_{t_0}^{t} v(\tau)d\tau$
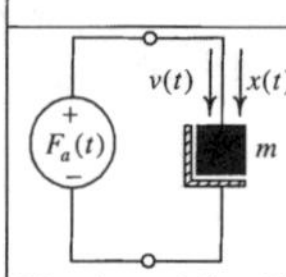 Mass (grounded), m [kg]	Applied force $F_a(t)$ [N] Linear velocity $v(t)$ [m/sec] Linear position $x(t)$ [m]	$F_a(t) = m\frac{dv}{dt} = m\frac{d^2x(t)}{dt^2}$ $v(t) = \frac{1}{m}\int_{t_0}^{t} F_a(\tau)d\tau$
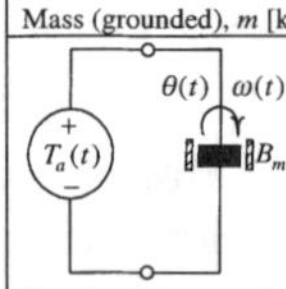 Rotational damper, B_m [N-m-sec/rad]	Applied torque $T_a(t)$ [N-m] Angular velocity $\omega(t)$ [rad/sec] Angular displacement $\theta(t)$ [rad]	$T_a(t) = B_m\omega(t)$ $\omega(t) = \frac{1}{B_m}T_a(t)$ and $T_a(t) = B_m\omega(t) = B_m\frac{d\theta(t)}{dt}$ $\theta(t) = \frac{1}{B_m}\int_{t_0}^{t} T_a(\tau)d\tau$
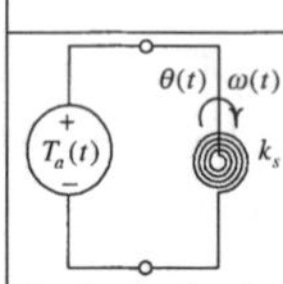 Rotational spring, k_s [N-m-sec/rad]	Applied torque $T_a(t)$ [N-m] Angular velocity $\omega(t)$ [rad/sec] Angular displacement $\theta(t)$ [rad]	$T_a(t) = k_s\theta(t)$ $\theta(t) = \frac{1}{k_s}T_a(t)$ and $\omega(t) = \frac{d\theta(t)}{dt} = \frac{1}{k_s}\frac{dT_a(t)}{dt}$ $T_a(t) = k_s\int_{t_0}^{t}\omega(\tau)d\tau$
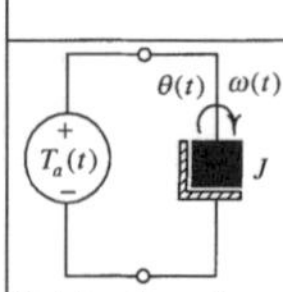 Rotational mass (grounded), J [kg-m^2]	Applied torque $T_a(t)$ [N-m] Angular velocity $\omega(t)$ [rad/sec] Angular displacement $\theta(t)$ [rad]	$T_a(t) = J\frac{d\omega}{dt} = J\frac{d^2\theta(t)}{dt^2}$ $\omega(t) = \frac{1}{J}\int_{t_0}^{t} T_a(\tau)d\tau$

Table 2.2 Physical Analogies

Mechanical Elements			Electrical Elements		
Symbol	Quantity (Variable)	Energy	Symbol	Quantity (Variable)	Energy
$F_a(t)$ $T_a(t)$	Applied force Applied torque		$u_a(t)$	Applied voltage	
$x(t)$ $v(t)=\frac{dx(t)}{dt}$ $\theta(t)$ $\omega(t)=\frac{d\theta(t)}{dt}$	Linear position Linear velocity Angular position Angular velocity		$i(t)$ $i(t)$	Current Current	
B_v B_m	Viscous friction	$\frac{1}{2}B_v v^2$ $\frac{1}{2}B_m\omega^2$ Dissipated energy	R	Resistance	$\frac{1}{2}Ri^2$ Dissipated energy
k_s	Spring constant	$\frac{1}{2}k_s x^2$ $\frac{1}{2}k_s\theta^2$ Potential (elastic energy)	L	Inductance	$\frac{1}{2}Li^2$ Potential energy
m J	Mass Moment of inertia	$\frac{1}{2}mv^2$ $\frac{1}{2}J\omega^2$ Kinetic energy	C	Capacitance	$\frac{1}{2}Cu_a^2$ Kinetic energy

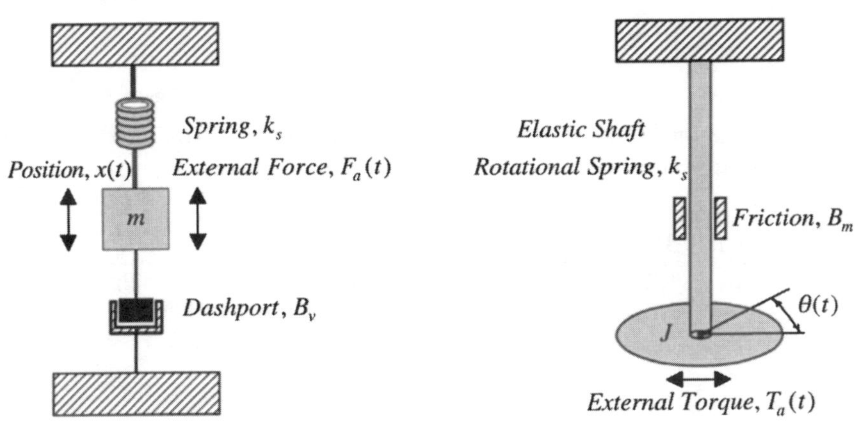

Figure 2.21 Translational and torsional mechanical systems.

where $F_a(t)$ and $T_a(t)$ are the time-varying applied force and torque; B_v and B_m are the viscous friction coefficients; and k_s is the translational and rotational (elasticity) spring coefficient.

Turning one's attention to electric circuits, consider two basic *RLC* circuits assuming that energy was initially stored in the inductor and the capacitor (see Figure 2.22).

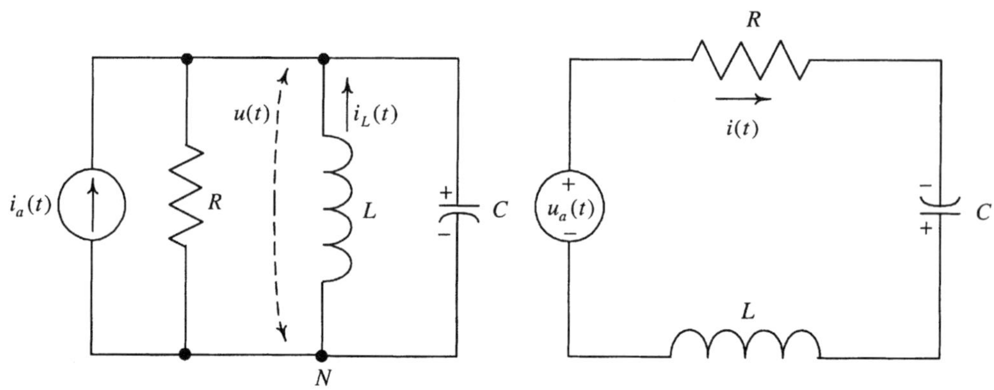

Figure 2.22 Parallel and series *RLC* circuits.

The integro-differential equation (an integral as well as a derivative of the dependent variables appear) for the parallel circuit is obtained by summing the currents away from the tope node

$$\frac{u}{R}+\frac{1}{L}\int_{t_0}^{t} u(\tau)d\tau - i_L(t_0) + C\frac{du}{dt} = i_a(t)$$

The integro-differential equation for the series circuit is found by summing the voltages around the closed path; in particular,

$$Ri+\frac{1}{C}\int_{t_0}^{t} i(\tau)d\tau - v_C(t_0) + L\frac{di}{dt} = u_a(t)$$

By differentiating these equations with respect to time and using the fact that $i_L(t_0)$ and $v_C(t_0)$ are constants, one obtains

$$C\frac{d^2u}{dt^2}+\frac{1}{R}\frac{du}{dt}+\frac{1}{L}u=\frac{di_a}{dt} \quad \text{or} \quad \frac{d^2u}{dt^2}+\frac{1}{RC}\frac{du}{dt}+\frac{1}{LC}u=\frac{1}{C}\frac{di_a}{dt}$$

and

$$L\frac{d^2i}{dt^2}+R\frac{di}{dt}+\frac{1}{C}i=\frac{du_a}{dt} \quad \text{or} \quad \frac{d^2i}{dt^2}+\frac{R}{L}\frac{di}{dt}+\frac{1}{LC}i=\frac{1}{L}\frac{du_a}{dt}$$

Parallel and series *RLC* circuits lead to the second-order differential equations, and mathematical models are dual of one another.

It was shown that the mechanical systems and the *RLC* circuits considered are modeled by second-order differential equations. The analytical solution of linear differential equations with constant coefficients can be easily derived. The general solution of the second-order linear differential equations is found using the roots of the characteristic equation. In particular, using the damping ξ coefficient and the resonant frequency ω_0, as given by $\xi = \frac{1}{2RC}$, $\omega_0 = \frac{1}{\sqrt{LC}}$ and $\xi = \frac{R}{2L}$, $\omega_0 = \frac{1}{\sqrt{LC}}$ for the *RLC* circuits studied, one can write the differential equation as

$$\frac{d^2x}{dt^2} + 2\xi\frac{dx}{dt} + \omega_0^2 x = d$$

to find three possible solutions. The characteristic equation is

$$s^2 + 2\xi s + \omega_0^2 = (s - s_1)(s - s_2) = 0, \quad s_{1,2} = -\xi \pm \sqrt{\xi^2 - \omega_0^2}$$

If $\xi^2 > \omega_0^2$, the real distinct characteristic roots s_1 and s_2 result, and the general solution is found to be

$$x(t) = ae^{s_1 t} + be^{s_2 t} + c_f$$

Coefficients a and b are found using the initial conditions, and c_f is the solution due to the *forcing* function, which are $i_a(t)$ and $u_a(t)$.

For $\xi^2 = \omega_0^2$, the characteristic roots are real and identical, $s_1 = s_2 = -\xi$. The solution is expressed as

$$x(t) = (a + b)e^{-\xi t} + c_f$$

If $\xi^2 < \omega_0^2$, the complex distinct characteristic roots are found as

$$s_{1,2} = -\xi \pm j\sqrt{\omega_0^2 - \xi^2}, \xi^2 < \omega^2.$$

and the general solution for the third possible scenario is

$$x(t) = e^{-\xi t}\left[a\cos\left(\sqrt{\omega_0^2 - \xi^2}\,t\right) + b\sin\left(\sqrt{\omega_0^2 - \xi^2}\,t\right)\right] + c_f$$

$$= e^{-\xi t}\sqrt{a^2 + b^2}\cos\left[\left(\sqrt{\omega_0^2 - \xi^2}\,t\right) + \tan^{-1}\left(\frac{-b}{a}\right)\right] + c_f$$

Example 2.10

For the series *RLC* circuit, find the characteristic equation and define the analytic relationships between the characteristic roots and circuitry parameters.

Solution

For the series *RLC* circuit, the following differential equation was found:

$$\frac{d^2i}{dt^2}+\frac{R}{L}\frac{di}{dt}+\frac{1}{LC}i=\frac{1}{L}\frac{du_a}{dt}$$

The characteristic equation is

$$s^2+\frac{R}{L}s+\frac{1}{LC}=0$$

Hence, the characteristic roots are

$$s_1=-\frac{R}{2L}-\sqrt{\left(\frac{R}{2L}\right)^2-\frac{1}{LC}}\text{ and } s_2=-\frac{R}{2L}+\sqrt{\left(\frac{R}{2L}\right)^2-\frac{1}{LC}}$$

If $\left(\frac{R}{2L}\right)^2>\frac{1}{LC}$, then the characteristic roots are real and distinct.

If $\left(\frac{R}{2L}\right)^2=\frac{1}{LC}$, then the characteristic roots are real and identical.

If $\left(\frac{R}{2L}\right)^2<\frac{1}{LC}$, then the characteristic roots are complex. □

Newton's law, the Lagrange equations of motion, and Kichhoff's laws have been used to find the corresponding differential equations to perform the analysis of mechanical systems and electric circuits. In the differential equations obtained, the following variables have been used: linear position and velocity, angular displacement and velocity, acceleration, current, and voltage. A state-variable description is found to be very efficient in engineering practice. In particular, n-order linear and nonlinear differential equations with time-invariant and time-varying parameters can be represented using state, control, output, reference, and disturbance vectors as a set of n first-order differential equations.

Example 2.11

Consider series *RLC* circuit as given in Figure 2.23. Derive the differential equations that map the circuitry dynamics. Find the analytic solution using MATLAB.

Solution

Using the voltage across the capacitor and the current through the inductor as the state variables, and the supplied voltage $u_a(t)$ as the control, one obtains the following set of first-order differential equations:

$$C\frac{du_C}{dt}=i$$

$$L\frac{di}{dt}=-u_C-Ri+u_a(t)$$

Hence,

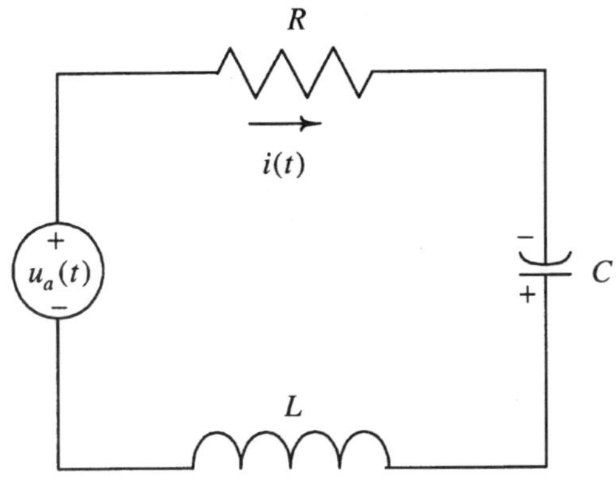

Figure 2.23 Series *RLC* circuit.

$$\frac{du_C}{dt} = \frac{1}{C} i$$

$$\frac{di}{dt} = \frac{1}{L}(-u_C - Ri + u_a(t))$$

The analytic solution can be found using the Symbolic Math Toolbox. In particular, for time-varying $u_a(t)$, one obtains

```
» [V,I]=dsolve('DV=I/C','DI=(-V-R*I+Va(t))/L')

V = (-int(diff(Va(t),t)*exp(1/2*(C*R+(C^2*R^2-
4*L*C)^(1/2))*t/L/C),t)*C*exp(-1/2*(C*R+(C^2*R^2-
4*L*C)^(1/2))*t/L/C)+int(diff(Va(t),t)*exp(1/2*(C*R-(C^2*R^2-
4*L*C)^(1/2))*t/L/C),t)*C*exp(-1/2*(C*R-(C^2*R^2-
4*L*C)^(1/2))*t/L/C)+(C^2*R^2-4*L*C)^(1/2)*C1*exp(-1/2*(C*R+(C^2*R^2-
4*L*C)^(1/2))*t/L/C)+(C^2*R^2-4*L*C)^(1/2)*C2*exp(-1/2*(C*R-(C^2*R^2-
4*L*C)^(1/2))*t/L/C))/(C^2*R^2-4*L*C)^(1/2)

I = (1/2*int(diff(Va(t),t)*exp(1/2*(C*R+(C^2*R^2-
4*L*C)^(1/2))*t/L/C),t)*C*R*exp(-1/2*(C*R+(C^2*R^2-4*L*C)^(1/2))*t/L/C)-
1/2*int(diff(Va(t),t)*exp(1/2*(C*R-(C^2*R^2-
4*L*C)^(1/2))*t/L/C),t)*C*R*exp(-1/2*(C*R-(C^2*R^2-4*L*C)^(1/2))*t/L/C)-
1/2*int(diff(Va(t),t)*exp(1/2*(C*R-(C^2*R^2-
4*L*C)^(1/2))*t/L/C),t)*(C^2*R^2-4*L*C)^(1/2)*exp(-1/2*(C*R-(C^2*R^2-
4*L*C)^(1/2))*t/L/C)-1/2*int(diff(Va(t),t)*exp(1/2*(C*R+(C^2*R^2-
4*L*C)^(1/2))*t/L/C),t)*(C^2*R^2-4*L*C)^(1/2)*exp(-1/2*(C*R+(C^2*R^2-
4*L*C)^(1/2))*t/L/C)+Va(t)*(C^2*R^2-4*L*C)^(1/2)-1/2*(C^2*R^2-
4*L*C)^(1/2)*C1*R*exp(-1/2*(C*R+(C^2*R^2-
4*L*C)^(1/2))*t/L/C)+1/2*C1*C*R^2*exp(-1/2*(C*R+(C^2*R^2-
4*L*C)^(1/2))*t/L/C)-2*C1*L*exp(-1/2*(C*R+(C^2*R^2-4*L*C)^(1/2))*t/L/C)-
1/2*(C^2*R^2-4*L*C)^(1/2)*C2*R*exp(-1/2*(C*R-(C^2*R^2-
4*L*C)^(1/2))*t/L/C)-1/2*C2*C*R^2*exp(-1/2*(C*R-(C^2*R^2-
4*L*C)^(1/2))*t/L/C)+2*C2*L*exp(-1/2*(C*R-(C^2*R^2-
4*L*C)^(1/2))*t/L/C))/(C^2*R^2-4*L*C)^(1/2)
```

If the applied voltage (forcing function) is constant $u_a(t) = const$, then

```
» [V,I]=dsolve('DV=I/C','DI=(-V-R*I+Va)/L')

V = (1/2*C1*exp(-1/2*(C*R-(C^2*R^2-4*L*C)^(1/2))*t/L/C)*(C^2*R^2-
4*L*C)^(1/2)+1/2*C1*C*R*exp(-1/2*(C*R+(C^2*R^2-4*L*C)^(1/2))*t/L/C)-
1/2*C1*C*R*exp(-1/2*(C*R-(C^2*R^2-4*L*C)^(1/2))*t/L/C)+1/2*C1*exp(-
1/2*(C*R+(C^2*R^2-4*L*C)^(1/2))*t/L/C)*(C^2*R^2-4*L*C)^(1/2)+C*C2*exp(-
1/2*(C*R+(C^2*R^2-4*L*C)^(1/2))*t/L/C)-C*C2*exp(-1/2*(C*R-(C^2*R^2-
4*L*C)^(1/2))*t/L/C))/(C^2*R^2-4*L*C)^(1/2)

I = (-L*C1*exp(-1/2*(C*R+(C^2*R^2-4*L*C)^(1/2))*t/L/C)+L*C1*exp(-
1/2*(C*R-(C^2*R^2-4*L*C)^(1/2))*t/L/C)+1/2*C2*exp(-1/2*(C*R-(C^2*R^2-
4*L*C)^(1/2))*t/L/C)*(C^2*R^2-4*L*C)^(1/2)-1/2*C2*C*R*exp(-
1/2*(C*R+(C^2*R^2-4*L*C)^(1/2))*t/L/C)+1/2*C2*C*R*exp(-1/2*(C*R-
(C^2*R^2-4*L*C)^(1/2))*t/L/C)+1/2*C2*exp(-1/2*(C*R+(C^2*R^2-
4*L*C)^(1/2))*t/L/C)*(C^2*R^2-4*L*C)^(1/2))/(C^2*R^2-4*L*C)^(1/2)-
4*L*Va*(C^2*R^2-4*L*C)/(-R^2*C+4*L)/(C*R-(C^2*R^2-
4*L*C)^(1/2))/(C*R+(C^2*R^2-4*L*C)^(1/2))
```

Letting the values of the resistance, inductance, and capacitance to be $R = 2$ ohm $L = 0.1$ H and $C = 0.5$ F, one obtains

```
» [V,I]=dsolve('DV=I/0.5','DI=(-V-2*I+Va)/0.1')

V=1/2*C1*exp(2*(-5+2*5^(1/2))*t)+1/4*C1*5^(1/2)*exp(-2*(5+2*5^(1/2))*t)-
1/4*C1*5^(1/2)*exp(2*(-5+2*5^(1/2))*t)+1/2*C1*exp(-
2*(5+2*5^(1/2))*t)+1/4*C2*5^(1/2)*exp(-2*(5+2*5^(1/2))*t)-
1/4*C2*5^(1/2)*exp(2*(-5+2*5^(1/2))*t)

I=-1/20*C1*5^(1/2)*exp(-2*(5+2*5^(1/2))*t)+1/20*C1*5^(1/2)*exp(2*(-
5+2*5^(1/2))*t)+1/2*C2*exp(2*(-5+2*5^(1/2))*t)-1/4*C2*5^(1/2)*exp(-
2*(5+2*5^(1/2))*t)+1/4*C2*5^(1/2)*exp(2*(-5+2*5^(1/2))*t)+1/2*C2*exp(-
2*(5+2*5^(1/2))*t)-5*Va/(-5+2*5^(1/2))/(5+2*5^(1/2))
```

where the constants *C1* and *C2* must be found by using the initial conditions. These initial conditions can be easily incorporated. Assigning $u(t_0) = u_0 = 15$ V and $i(t_0) = i_0 = 5$ A, we have

```
» [V,I]=dsolve('DV=I/0.5','DI=(-V-2*I+Va)/0.1','V(0)=15,I(0)=5')

V=5/2*exp(2*(-5+2*5^(1/2))*t)+5*5^(1/2)*exp(-2*(5+2*5^(1/2))*t)-
5*5^(1/2)*exp(2*(-5+2*5^(1/2))*t)+5/2*exp(-2*(5+2*5^(1/2))*t)-
5/4*5^(1/2)*Va*(-exp(-2*(5+2*5^(1/2))*t)+exp(2*(-5+2*5^(1/2))*t))/(-
5+2*5^(1/2))/(5+2*5^(1/2))

I=-4*5^(1/2)*exp(-2*(5+2*5^(1/2))*t)+4*5^(1/2)*exp(2*(-
5+2*5^(1/2))*t)+15/2*exp(2*(-5+2*5^(1/2))*t)+15/2*exp(-
2*(5+2*5^(1/2))*t)+5/4*Va*(-4+2*exp(2*(-5+2*5^(1/2))*t)-5^(1/2)*exp(-
2*(5+2*5^(1/2))*t)+5^(1/2)*exp(2*(-5+2*5^(1/2))*t)+2*exp(-
2*(5+2*5^(1/2))*t))/(-5+2*5^(1/2))/(5+2*5^(1/2))
```

which can be simplified using the `simplify` command; that is, `simplify(V)` and `simplify(I)` should be used. □

Recall the state-space concept that is widely used in analysis, design, and optimization of electromechanical systems. The general notations used were introduced in Section 1.2. In particular, a set of n first-order linear ordinary differential equations with n states $x \in \mathbb{R}^n$ and m controls $u \in \mathbb{R}^m$ is found as

$$\frac{dx_1}{dt} = a_{11}x_1 + a_{12}x_2 + \ldots + a_{1n-1}x_{n-1} + a_{1n}x_n + b_{11}u_1 + b_{12}u_2 + \ldots + b_{1m-1}u_{m-1} + b_{1m}u_m$$

$$\frac{dx_2}{dt} = a_{21}x_1 + a_{22}x_2 + \ldots + a_{2n-1}x_{n-1} + a_{2n}x_n + b_{21}u_1 + b_{22}u_2 + \ldots + b_{2m-1}u_{m-1} + b_{2m}u_m$$

$$\vdots$$

$$\frac{dx_{n-1}}{dt} = a_{n-11}x_1 + a_{n-12}x_2 + \ldots + a_{n-1\;n-1}x_{n-1} + a_{n-1\;n}x_n + b_{n-11}u_1 + b_{n-12}u_2 + \ldots + b_{n-1\;m-1}u_{m-1} + b_{m-1\;m}u_m$$

$$\frac{dx_n}{dt} = a_{n1}x_1 + a_{n2}x_2 + \ldots + a_{n\;n-1}x_{n-1} + a_{n\;n}x_n + b_{n1}u_1 + b_{n2}u_2 + \ldots + b_{n\;m-1}u_{m-1} + b_{n\;m}u_m$$

These differential equations can be rewritten in matrix form as

$$\frac{dx}{dt} = \begin{bmatrix} \frac{dx_1}{dt} \\ \frac{dx_2}{dt} \\ \vdots \\ \frac{dx_{n-1}}{dt} \\ \frac{dx_n}{dt} \end{bmatrix} = \begin{bmatrix} a_{11} & a_{12} & \cdots & a_{1n-1} & a_{1n} \\ a_{21} & a_{22} & \cdots & a_{2n-1} & a_{2n} \\ \vdots & \vdots & \ddots & \vdots & \vdots \\ a_{n-11} & a_{n-12} & \cdots & a_{n-1\;n-1} & a_{n-1n} \\ a_{n1} & a_{n2} & \cdots & a_{n\;n-1} & a_{n\;n} \end{bmatrix} \begin{bmatrix} x_1 \\ x_2 \\ \vdots \\ x_{n-1} \\ x_n \end{bmatrix} +$$

$$\begin{bmatrix} b_{11} & b_{12} & \cdots & b_{1m-1} & b_{1m} \\ b_{21} & b_{22} & \cdots & b_{2m-1} & b_{2m} \\ \vdots & \vdots & \ddots & \vdots & \vdots \\ b_{n-11} & b_{n-12} & \cdots & b_{n-1\;m-1} & b_{n-1m} \\ b_{n1} & b_{n2} & \cdots & b_{n\;m-1} & b_{n\;m} \end{bmatrix} \begin{bmatrix} u_1 \\ u_2 \\ \vdots \\ u_{m-1} \\ u_m \end{bmatrix} = Ax + Bu$$

where $A \in \mathbb{R}^{n\times n}$ and $B \in \mathbb{R}^{n\times m}$ are the matrices of coefficients.

In multi-input/multi-output electromechanical systems (e.g., robotic manipulators, multi-degree-of-freedom servomechanisms, and other motion devices), b outputs $y \in \mathbb{R}^b$ are obtained in terms of n states $x \in \mathbb{R}^n$ and m controls $u \in \mathbb{R}^m$. For example, in the drive application, the output is the angular velocity of the output shaft, and the output angular velocity is proportional to the angular velocity of the electric motor (the gear ratio should be used as a coefficient) $y = k_{gear}\omega_{motor}$, where ω_{motor} is the system state. In servo-systems, the output linear position or angular displacement is proportional to the angular rotor displacement, and the output equation is expressed as $y = k_{gear}\theta_{rotor}$.

Hence, in general, the output equation is given as

$$
\begin{aligned}
y_1 &= h_{11}x_1 + h_{12}x_2 + \ldots + h_{1n-1}x_{n-1} + h_{1n}x_n \\
y_2 &= h_{21}x_1 + h_{22}x_2 + \ldots + h_{2n-1}x_{n-1} + h_{2n}x_n \\
&\vdots \\
y_{b-1} &= h_{b-11}x_1 + h_{b-12}x_2 + \ldots + h_{b-1\ n-1}x_{n-1} + h_{b-1n}x_n \\
y_b &= h_{b1}x_1 + h_{b2}x_2 + \ldots + h_{bn-1}x_{n-1} + h_{bn}x_n
\end{aligned}
$$

The following matrix form results

$$
y = \begin{bmatrix} y_1 \\ y_2 \\ \vdots \\ y_{b-1} \\ y_b \end{bmatrix} = \begin{bmatrix} h_{11} & h_{12} & \cdots & h_{1n-1} & h_{1n} \\ h_{21} & h_{22} & \cdots & h_{2n-1} & h_{2n} \\ \vdots & \vdots & \ddots & \vdots & \vdots \\ h_{b-11} & h_{b-12} & \cdots & h_{b-1\ n-1} & h_{b-1n} \\ h_{b1} & h_{b2} & \cdots & h_{b\ n-1} & h_{b\ n} \end{bmatrix} \begin{bmatrix} x_1 \\ x_2 \\ \vdots \\ x_{n-1} \\ x_n \end{bmatrix} = Hx
$$

where $H \in \mathbb{R}^{b \times n}$ is the matrices of coefficients.

Remark. In MATLAB, the output equation has the following form: $y = Hx + Du$, $D \in \mathbb{R}^{b \times m}$. Hence, the system output is a function of states and control. In electromechanical systems, the output is not a function of the control inputs and, therefore, matrix D must be assigned to be zero,

$$
y = \begin{bmatrix} y_1 \\ y_2 \\ \vdots \\ y_{b-1} \\ y_b \end{bmatrix} = \begin{bmatrix} h_{11} & h_{12} & \cdots & h_{1n-1} & h_{1n} \\ h_{21} & h_{22} & \cdots & h_{2n-1} & h_{2n} \\ \vdots & \vdots & \ddots & \vdots & \vdots \\ h_{b-11} & h_{b-12} & \cdots & h_{b-1\ n-1} & h_{b-1n} \\ h_{b1} & h_{b2} & \cdots & h_{b\ n-1} & h_{b\ n} \end{bmatrix} \begin{bmatrix} x_1 \\ x_2 \\ \vdots \\ x_{n-1} \\ x_n \end{bmatrix} = Hx.
$$

However, one must keep in mind that the matrix D must be assigned if one uses the MATLAB environment to solve a spectrum of analysis, design, and optimization problems. □

Example 2.12

Using the state-space concept, find the state-space model and analyze the transient dynamics of the series *RLC* circuit considered in Example 2.11.

Solution

A set of two first-order differential equations to map the circuitry dynamics was found as

$$
\frac{du_C}{dt} = \frac{1}{C}i
$$

$$
\frac{di}{dt} = \frac{1}{L}(-u_C - Ri + u_a(t))
$$

Denoting the states as

$$x_1(t) = u_C(t) \text{ and } x_2(t) = i(t)$$

and the control

$$u(t) = u_a(t)$$

one obtains the state equation in matrix form as given by

$$\frac{dx}{dt} = \begin{bmatrix} \frac{dx_1}{dt} \\ \frac{dx_2}{dt} \end{bmatrix} = \begin{bmatrix} \frac{du_C}{dt} \\ \frac{di}{dt} \end{bmatrix} = \begin{bmatrix} 0 & \frac{1}{C} \\ -\frac{1}{L} & -\frac{R}{L} \end{bmatrix} \begin{bmatrix} u_C \\ i \end{bmatrix} + \begin{bmatrix} 0 \\ \frac{1}{L} \end{bmatrix} u_a = Ax + Bu$$

Here, the matrices of coefficients are

$$A = \begin{bmatrix} 0 & \frac{1}{C} \\ -\frac{1}{L} & -\frac{R}{L} \end{bmatrix} \text{ and } B = \begin{bmatrix} 0 \\ \frac{1}{L} \end{bmatrix}$$

The numerical solution can be found using the `ode` solvers, and the application of the `ode45` solver was demonstrated in Section 1.2. In contrast, the state-space solver `lsim` is used. Typing `help lsim`, the following text is displayed

```
» help lsim

LSIM Simulation of the time response of LTI systems to arbitrary
inputs.
LSIM(SYS,U,T) plots the time response of the LTI system SYS to the
input signal described by U and T. The time vector T consists of
regularly spaced time samples and U is a matrix with as many columns
as inputs and whose i-th row specifies the input value at time T(i).
For instance
            t = 0:0.01:5; u = sin(t); lsim(sys, u, t)
simulates the response of SYS to u(t) = sin(t) during 5 seconds.
In discrete time, U should be sampled at the same rate and the system
T is then redundant and can be omitted or set to the empty matrix).
In continuous time, the sampling period T(2)-T(1) should be chosen
small enough to capture the details of the input signal. The
time vector T is resampled when intersample oscillations may occur.
LSIM(SYS,U,T,X0) specifies an additional nonzero initial state X0
(for state-space systems only).
LSIM(SYS1,SYS2,…,U,T,X0)simulates the response of multiple LTI
systems SYS1,SYS2,… on a single plot. The initial condition X0 is
optional. You can also specify a color, line style, and marker for
each system, as in lsim(sys1,'r',sys2,'y--',sys3,'gx',u,t).
```

```
» help lsim

When invoked with left hand arguments, [Y,T] = LSIM(SYS,U,…)
returns the output history Y and time vector T used for simulation.
No plot is drawn on the screen. The matrix Y has LENGTH(T) rows
and as many columns as outputs in SYS.
For state-space systems, [Y,T,X}= LSIM(SYS,U,…) also returns the
state trajectory X, a matrix with LENGTH(T) rows and as many columns
as states
```

For R = 2 ohm, L = 0.1 H, and C = 0.5 F, one finds the following matrices of coefficients

$$A = \begin{bmatrix} 0 & 2 \\ -10 & -20 \end{bmatrix} \text{ and } B = \begin{bmatrix} 0 \\ 10 \end{bmatrix}$$

As in Example 2.11, the initial conditions are assigned to be $u_C(t_0) = u_{C0}$ = 15 V and $i(t_0) = i_0$ = 5 A. That is,

$$x_0 = \begin{bmatrix} x_{10} \\ x_{20} \end{bmatrix} = \begin{bmatrix} 15 \\ 5 \end{bmatrix}$$

Let the voltage across the capacitor be the output; that is,

$$y(t) = u_C(t)$$

Hence, the output equation is found to be

$$y = \begin{bmatrix} 1 & 0 \end{bmatrix} \begin{bmatrix} u_C \\ i \end{bmatrix} = Hx, \text{ where } H = \begin{bmatrix} 1 & 0 \end{bmatrix}$$

However, to model the dynamic systems, recall that the expanded output equation in $y = Hx + Du$ should be used. In particular,

$$y = \begin{bmatrix} 1 & 0 \end{bmatrix} \begin{bmatrix} u_C \\ i \end{bmatrix} + [0]u_a = Hx + Du, \; H = \begin{bmatrix} 1 & 0 \end{bmatrix}, \text{ and } D = [0]$$

The circuit responses if $u_a(t) = 5\sin(t + 2)$ are found using the following m-file.

MATLAB script (ch_2_01.m)

```
R=2; L=0.1; C=0.5;                          % Circuitry parameters
A=[0 1/C;-1/L -R/L]; B=[0; 1/L];            % Matrices A and B
H=[1 0]; D=[0];                             % Matrices H and D
x0=[15; 5];                                 % Initial conditions
t=0:0.01:20;                   % Start and stop time with increment 0.01
Va=5*sin(t+2);                              % Applied excitation voltage
[y,x]=lsim(A,B,H,D,Va,t,x0);                % Call lsim
```

```
R=2; L=0.1; C=0.5;                       % Circuitry parameters
plot(t,x) % Plotting statement
title('Numerical solution of the series RLC circuit, State dynamics')
xlabel('Time (seconds)'); pause
plot(t,y)                                % Plotting statement
title('Numerical solution of the series RLC circuit, Output
dynamics')
xlabel('Time (seconds)'); pause
disp('End')
```

The transient dynamics for two states and output are plotted in Figure 2.24.

It should be emphasized that the length and size of the input vector (in the example considered, $u_a(t)$ is the circuit input) must have the length and size equal to time. For example, if the applied voltage is $u_a(t) = 1$ V (unit step is applied), one must assign this input voltage `Va` as `Va=1*ones(1,length(t))` or `Va=1*ones(size(t))`. The length and size can be easily checked using `length` or `size`; as an example, typing `length(Va)` or `size(Va)`, the answer is displayed.

The SIMULINK diagram (`ch_2_02.mdl`) is shown in Figure 2.25.

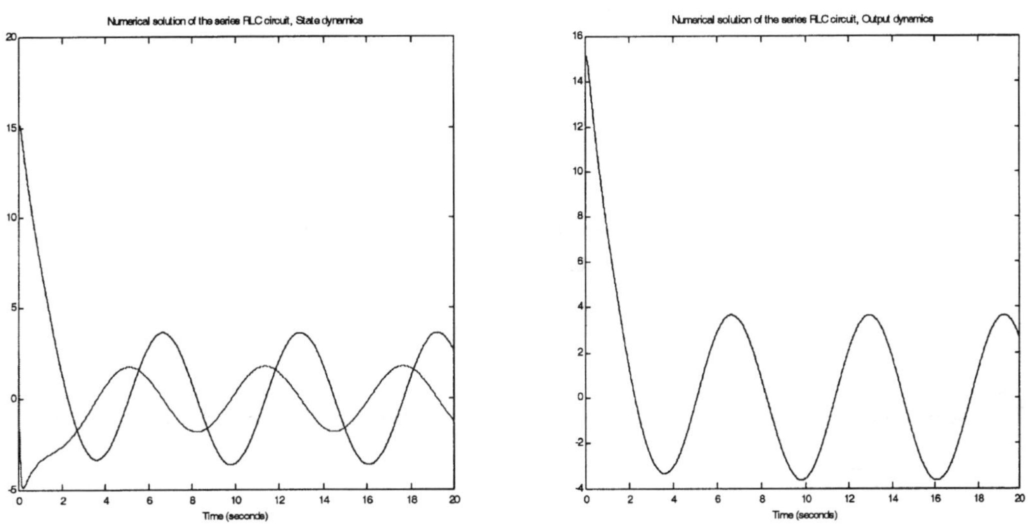

Figure 2.24 Dynamics of the series RLC circuit if $u_a(t) = 5\sin(t+2)$ and $x_0 = \begin{bmatrix} x_{10} \\ x_{20} \end{bmatrix} = \begin{bmatrix} 15 \\ 5 \end{bmatrix}$.

□

Example 2.13

As another illustrative example, consider the network shown in Figure 2.26. Find the state-space model and simulate the circuit studied. Use the voltage at the load terminal $u_L(t)$ as the circuit output.

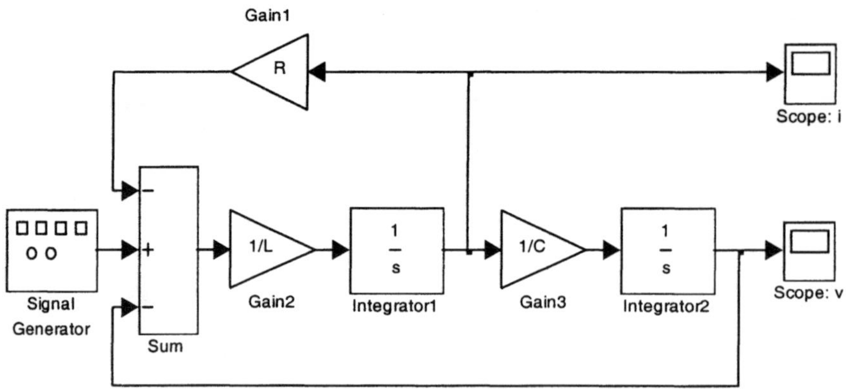

Figure 2.25 SIMULINK diagram to model the series *RLC* circuit.

Solution

Using Kirchhoff's law, the following two differential equations result

$$C\frac{du_C}{dt} = -\frac{u_C}{R} - i_L + \frac{u_a(t)}{R}$$

$$L\frac{di_L}{dt} = u_C - R_L i_L$$

Therefore,

$$\frac{du_C}{dt} = \frac{1}{C}\left(-\frac{u_C}{R} - i_L + \frac{u_a(t)}{R}\right)$$

$$\frac{di_L}{dt} = \frac{1}{L}(u_C - R_L i_L)$$

The output is the voltage at the load. Hence, the output equation is found to be

$$u_L = R_L i_L$$

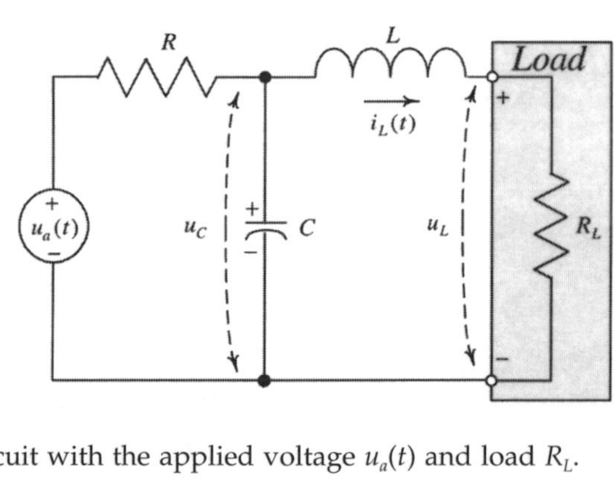

Figure 2.26 Electric circuit with the applied voltage $u_a(t)$ and load R_L.

Using the state-space concept, denoting the states as $x_1(t) = u_C(t)$ and $x_2(t) = i_L(t)$, the control input as $u(t) = u_a(t)$, and the output as $y(t) = R_L i_L(t)$, one obtains the state and output equations in matrix form as

$$\frac{dx}{dt} = \begin{bmatrix} \frac{dx_1}{dt} \\ \frac{dx_2}{dt} \end{bmatrix} = \begin{bmatrix} \frac{du_C}{dt} \\ \frac{di_L}{dt} \end{bmatrix} = \begin{bmatrix} -\frac{1}{RC} & -\frac{1}{C} \\ \frac{1}{L} & -\frac{R_L}{L} \end{bmatrix} \begin{bmatrix} u_C \\ i_L \end{bmatrix} + \begin{bmatrix} \frac{1}{RC} \\ 0 \end{bmatrix} u_a = Ax + Bu$$

$$y = \begin{bmatrix} 0 & R_L \end{bmatrix} \begin{bmatrix} u_C \\ i_L \end{bmatrix} = Hx$$

where the matrices of coefficients are.

$$A = \begin{bmatrix} -\frac{1}{RC} & -\frac{1}{C} \\ \frac{1}{L} & -\frac{R_L}{L} \end{bmatrix}, \quad B = \begin{bmatrix} \frac{1}{RC} \\ 0 \end{bmatrix}, \quad H = \begin{bmatrix} 0 & R_L \end{bmatrix}$$

The MATLAB `lsim` function is used to perform the modeling, and the output equation used in the analysis of dynamics is

$$y = \begin{bmatrix} 0 & R_L \end{bmatrix} \begin{bmatrix} u_C \\ i_L \end{bmatrix} + [0]u_a = Hx + Du$$

where $D = [0]$.

The circuitry parameters are $R = 0.1$ ohm, $R_L = 2$ ohm, $L = 0.05$ H, and $C = 0.25$ F. The applied voltage is assigned to be $u_a(t) = const = 10$ V. The m-file script developed is given below.

MATLAB script (ch_2_03.m)

```
R=0.1; RL=2; L=0.05; C=0.25;                    % Circuitry parameters
A=[-1/R*C -1/C;1/L -RL/L]; B=[1/R*C; 0];       % Matrices A and B
H=[0 RL]; D=[0];                                % Matrices H and D
x0=[10; -10];                                   % Initial conditions
t=0:0.001:1;              % Start and stop time with increment 0.001
Va=10*ones(1,length(t));        % Applied excitation voltage
[y,x]=lsim(A,B,H,D,Va,t,x0);    % Call lsim
plot(t,x)                       % Plotting statement for x1 and x2
title('Numerical solution of the circuit dynamics, states x1 and x2')
xlabel('Time (seconds)'); pause
plot(t,x(:,1))                  % Plotting statement for x1
title('Numerical solution of the circuit dynamics, state x1')
xlabel('Time (seconds)'); pause
plot(t,x(:,2))                  % Plotting statement for x2
title('Numerical solution of the circuit dynamics, state x2')
xlabel('Time (seconds)'); pause
plot(t,y)                       % Plotting statement for y
title('Numerical solution of the circuit dynamics, output y')
```

```
R=0.1; RL=2; L=0.05; C=0.25;                % Circuitry parameters
xlabel('Time (seconds)'); pause
disp('End')
```

Transient behavior for the states x_1 and x_2

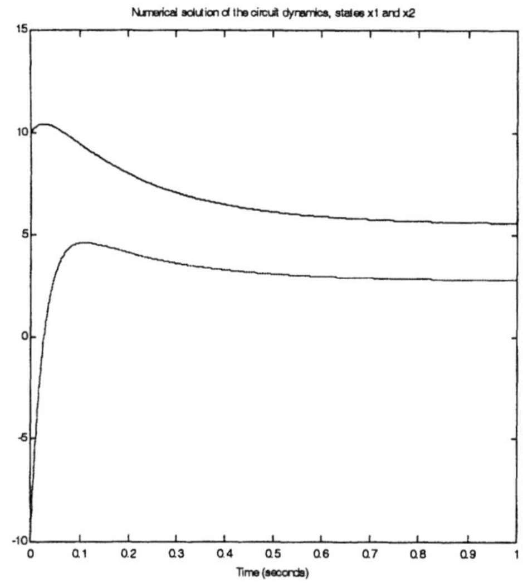

Transient behavior for the state variable x_1

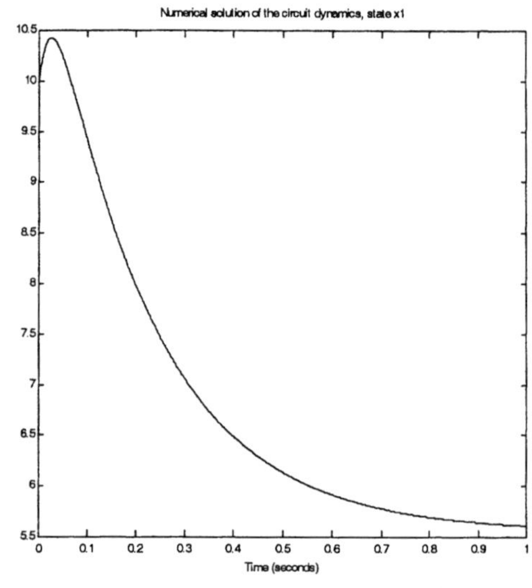

Transient behavior for the state variable x_2

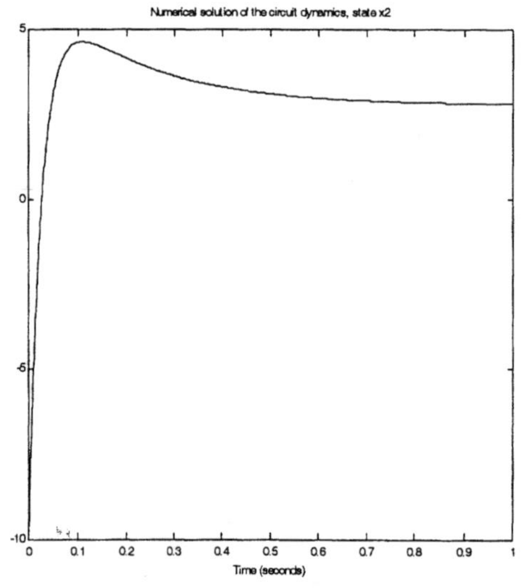

Transient behavior for the output y

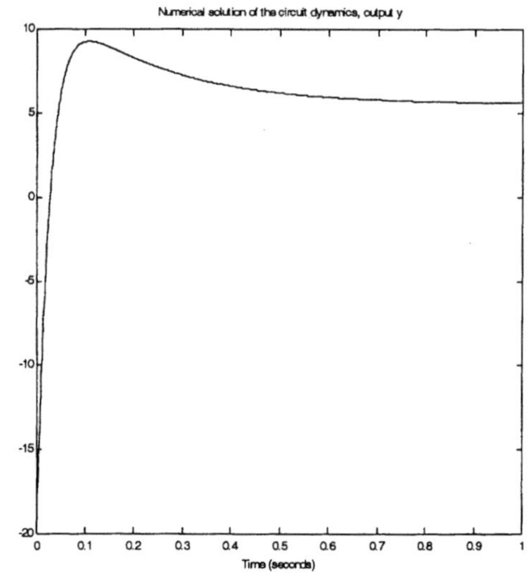

Figure 2.27 Dynamics of the circuit if $u_a(t) = 10$ and $x_0 = \begin{bmatrix} x_{10} \\ x_{20} \end{bmatrix} = \begin{bmatrix} 10 \\ -10 \end{bmatrix}$.

The transient responses in the circuit due to the voltage applied and initial conditions $x_0 = \begin{bmatrix} x_{10} \\ x_{20} \end{bmatrix} = \begin{bmatrix} 10 \\ -10 \end{bmatrix}$ are plotted in Figure 2.27.

References

1. Friedland, B., *Advanced Control System Design*, Prentice-Hall, Enlewood Cliffs, NJ, 1996.
2. Krause, P. C. and Wasynczuk, O., *Electromechanical Motion Devices*, McGraw-Hill, New York, 1989.
3. Kuo, B. C., *Automatic Control Systems*, Prentice-Hall, Englewood Cliffs, NJ, 1995.
4. Paul, C. R., Whites, K. W., and Nasar, S. A., *Introduction to Electromagnetic Fields*, McGraw-Hill, New York, 1998.
5. White, D. C. and Woodson, H. H., *Electromechanical Energy Conversion*, John Wiley & Sons, New York, 1959.
6. Wylie, C. R. and Barrett, L. C., *Advanced Engineering Mathematics*, McGraw-Hill, New York, 1995.

chapter three

Introduction to feedback control of electromechanical systems

This chapter develops the basic relationships to solve the motion control problem using analog and digital proportional-integral-derivative (PID) controllers. Proportional, proportional-integral, and proportional-integral-derivative control algorithms are widely used in electromechanical systems, and power amplifiers integrate PID controllers to attain tracking control and disturbance attenuation. In fact, electromechanical systems, which are used in a wide variety of industrial systems and technological processes, are controlled. Hence, the problems in design and deployment of controllers must be addressed and thoroughly studied. The major goal of this chapter is to solve practical problems using PID-type control algorithms that are found to be straightforward and well-understood, reliable, and efficient for solving the motion control problems for nonlinear electromechanical systems. This chapter provides an introduction to motion control and feedback tracking. Single-input/single-output electromechanical systems and mapping of the electromechanical system dynamics in the s- and z-domains using transfer functions $G_{sys}(s)$ and $G_{sys}(z)$ are studied. Transfer functions of analog and digital PID controllers are employed to find mathematical models of closed-loop systems, to analyze and study closed-loop systems.

This chapter presents the introductory material needed to understand the basic foundations in control of electric machines and power converters, which are thoroughly covered in advanced control books using the Hamilton-Jacobi theory, Lyapunov's concept, or other methods. It should be emphasized that analysis and modeling of nonlinear continuous-time and discrete-time electromechanical systems can be easily accomplished with PID-type controllers as mathematical models of electromechanical motion devices (AC and DC electric machines and power converters) being available. The corresponding differential equations to map electric machines and power converter dynamics will be developed and thoroughly treated in Chapters 4 through 8. As mathematical models are found in the form of differential or difference equations, analysis can be easily performed to determine stability and stability margins, time response, and accuracy, using the PID-type controllers studied in this chapter. The general question approached in this chapter is how to design PID-type controllers that ensure that closed-loop systems satisfy the specifications imposed. The synthesis of analog and digital control laws involve the changing of controller structures, as well as adjusting feedback coefficients to attain certain desired criteria and characteristics. These performance specifications relate stability and

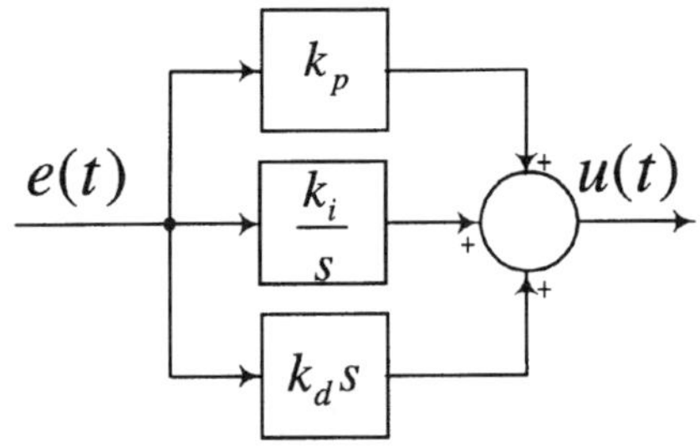

Figure 3.1 PID controller.

robustness, dynamics and accuracy, tracking and disturbance attenuation, and other criteria needed to be achieved through the use of control algorithms. The trade-off between stability and accuracy, as well as between robustness and system response, is well known. The design procedures are reported to find the structures and feedback coefficients of PID-type controllers.

3.1 Continuous-time electromechanical systems and analog PID controllers

The majority of electromechanical motion devices should be considered as continuous-time systems. Assume that we are dealing with linear systems and denote the transfer function of the electromechanical system to be controlled as $G_{sys}(s)$. Whether involved in analysis and design of electric motors and generators, power electronics, or power converters, the synthesis of PID-type controllers is a critical aspect of engineering practice. Furthermore, to attain the performance limits of electromechanical systems, one must remain up-to-date on design methods and control algorithms. The simplest control algorithms available, which the designer should study first, are the PID-type controllers. The linear analog PID control law is given as

$$u(t) = k_p e(t) + k_i \frac{e(t)}{s} + k_d s e(t) = k_p e(t) + k_i \int e(t)dt + k_d \frac{de(t)}{dt}, \quad s = \frac{d}{dt} \tag{3.1}$$

where $e(t) = r(t) - y(t)$ is the error between the reference signal and the system output; and k_p, k_i, and, k_d are the proportional, integral, and derivative feedback gains, respectively.

The block diagram of the PID controller is shown in Figure 3.1. Setting k_d equal to zero, the proportional-integral (PI) controller results, and one obtains

$$u(t) = k_p e(t) + k_i \frac{e(t)}{s} = k_p e(t) + k_i \int e(t)dt$$

Assigning the integral feedback coefficient k_i to be zero, one obtains the following proportional-derivative (PD) controller

$$u(t) = k_p e(t) + k_d s e(t) = k_p e(t) + k_d \frac{de(t)}{dt}$$

If k_i and k_d are set to be equal to zero, the proportional (P) control law results, and

$$u(t) = k_p e(t)$$

Using the Laplace transform for the most general case, when all three components of the PID algorithm (3.1) are in place, one obtains

$$U(s) = \left(k_p + \frac{k_i}{s} + k_d s\right)E(s)$$

One finds, at once, the transfer function for PID controllers to be

$$G_{PID}(s) = \frac{U(s)}{E(s)} = \frac{k_d s^2 + k_p s + k_i}{s}$$

The closed-loop system in the time- and s-domain with the negative unit loop gain are represented in Figure 3.2. This feedback control configuration is commonly used to control electromechanical systems.

If the system output $y(t)$ converges to the reference signal $r(t)$ as time approaches infinity, one concludes that tracking of the reference input is accomplished, and the error vector is zero; that is, tracking is achieved if the following is guaranteed:

$$e(t) = r(t) - y(t) = 0 \ \text{ as } \ t \to \infty$$

At a summing point, two signals $r(t)$ and $y(t)$ are superimposed linearly, and "plus" and "minus" are used to designate the algebraic sign. In Figure 3.2, the operation taking place at the summing point is subtraction, and $e(t) = r(t) - y(t)$. It is evident that

$$E(s) = R(s) - Y(s)$$

is the Laplace transform of the error signal.

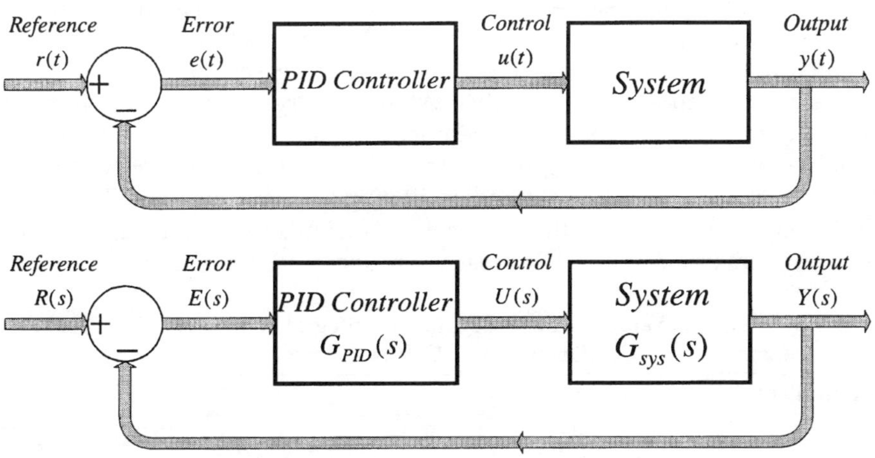

Figure 3.2 Time- and s-domain block diagrams of closed-loop systems.

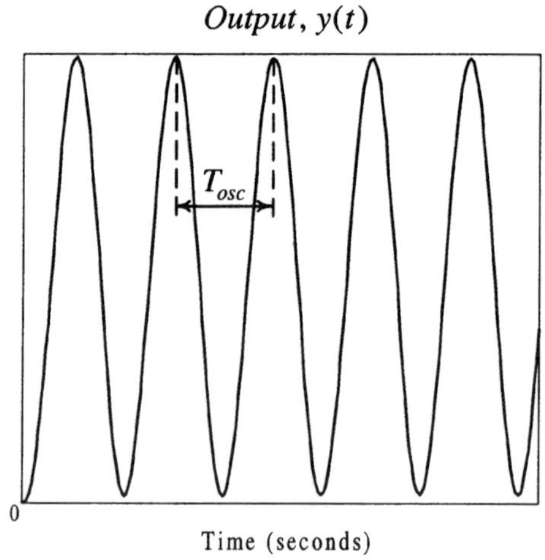

Figure 3.3 The output $y(t)$ for a marginally or critically stable system with critical value of the proportional feedback coefficient $k_{p\ \max}$.

The goal is to find the Laplace transform of the output $y(t)$ for the closed-loop system studied. The analysis is based on the following equation:

$$Y(s) = G_{sys}(s)U(s) = G_{sys}(s)G_{PID}(s)E(s) = G_{sys}(s)G_{PID}(s)(R(s) - Y(s))$$

Hence, the following transfer function of the closed-loop system is obtained:

$$G(s) = \frac{Y(s)}{R(s)} = \frac{G_{sys}(s)G_{PID}(s)}{1 + G_{sys}(s)G_{PID}(s)}$$

The objective is to design the controller from the PID-type algorithms and find the proportional, integral, and derivative feedback coefficients to ensure that the tracking error $e(t)$ converges to zero as time approaches infinity.

For electromechanical systems, k_p, k_i, and k_d feedback gains can be found in the following way. For the specified reference input $r(t) = const$, set the integral and derivative feedback equal to zero; that is, $k_i = 0$ and $k_d = 0$. One increases the proportional feedback gain k_p to the critical value $k_{p\ \max}$ when the system is marginally stable [the output $y(t)$ is bounded, but the system exhibits the sustained oscillates (the so-called limit cycles) or critically stable (the output $y(t)$ is lightly bounded; see Figure 3.3).

Using the period of the self-sustained oscillations T_{osc} and the value $k_{p\ \max}$, one obtains the typically suggested values of the feedback coefficients that can be applied to tune the PID-type controllers (see Table 3.1). It must be emphasized that the proportional feedback

Table 3.1 Typical Values of the Proportional, Integral, and Derivative Feedback Coefficients for PID-Type Controllers

Controller	k_p	k_i	k_d
PID	$k_p \in [0.1 \quad 0.5]k_{p\ \max}$	$k_i \in [0.1 \quad 10]k_{p\ \max}T_{osc}$	$k_d \in [0.05 \quad 1]k_{p\ \max}T_{osc}$
PD	$k_p \in [0.1 \quad 0.5]k_{p\ \max}$	0	$k_d \in [0.05 \quad 1]k_{p\ \max}T_{osc}$
PI	$k_p \in [0.05 \quad 0.5]k_{p\ \max}$	$k_i \in [0.01 \quad 1]k_{p\ \max}T_{osc}$	0
P	$k_p \in [0.05 \quad 0.5]k_{p\ \max}$	0	0

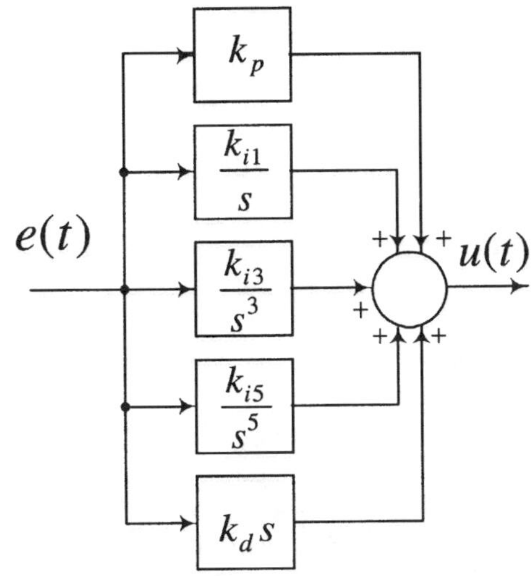

Figure 3.4 PII³I⁵D controller.

coefficient must be adjusted first to provide acceptable performance, and the derivative feedback (which is not implemented in many electromechanical systems due to the noise imposed) improves stability and decreases overshoot and oscillations. The integral feedback allows one to decrease or eliminate the steady-state error because an additional pole (which is at the origin) is added; however, the stability and transient responses are degraded.

To attain the desired steady-state and dynamic performance, ensure accuracy and stability, as well as expand the operating envelope, control algorithms should be implemented. The most general structure (configuration) of PID-type controllers is given as

$$u(t) = \sum_{j=1}^{2N_p-1} k_{p(2j-1)} e(t) + \sum_{j=1}^{2N_i-1} k_{i(2j-1)} \frac{e(t)}{s^{2j-1}} + \sum_{j=1}^{2N_d-1} k_{d(2j-1)} \frac{d^{2j-1}e(t)}{dt^{2j-1}}$$

where N_p, N_i, and N_d are the positive integers assigned by the designer; and $k_{p(2j-1)}$, $k_{i(2j-1)}$, and $k_{d(2j-1)}$ are the proportional, integral, and derivative feedback coefficients, respectively.

For example, letting $N_p = 1$, $N_i = 1$, and $N_d = 1$, the PID control law can be given by (3.1). If $N_p = 1$, $N_i = 2$, and $N_d = 2$, then the PII³I⁵DD³D⁵ control law as given by

$$u(t) = k_{po}e(t) + k_{i1}\frac{e(t)}{s} + k_{i3}\frac{e(t)}{s^3} + k_{i5}\frac{e(t)}{s^5} + k_{d1}\frac{de(t)}{dt} + k_{d3}\frac{d^3e(t)}{dt^3} + k_{d5}\frac{d^5e(t)}{dt^5}$$

Assigning $N_p = 1$, $N_i = 3$, and $N_d = 1$, the following control algorithm results:

$$u(t) = k_{po}e(t) + k_{i1}\frac{e(t)}{s} + k_{i3}\frac{e(t)}{s^3} + k_{i5}\frac{e(t)}{s^5} + k_{i7}\frac{e(t)}{s^7} + k_{i9}\frac{e(t)}{s^9} + k_d\frac{de(t)}{dt}$$

The block diagram of the PII³I⁵D control law is illustrated in Figure 3.4.

In addition to PID-type controllers, nonlinear control laws with nonlinear error maps can be used to improve electromechanical systems performance. The family of PID-type nonlinear controllers is obtained by applying the following general formula

$$u(t)=\sum_{j=0}^{\varsigma} k_{pj} e^{\frac{2j+1}{2\beta+1}}+\sum_{j=0}^{\sigma} k_{ij} \frac{1}{s} e^{\frac{2j+1}{2\mu+1}}+\sum_{j=0}^{\eta} k_{dj} s e^{\frac{2j+1}{2\gamma+1}} \tag{3.2}$$

where ς, β, σ, μ, η, and γ are non-negative integers.

In (3.2), the nonlinear proportional $\sum_{j=0}^{\varsigma} k_{pj} e^{\frac{2j+1}{2\beta+1}}$, integral $\sum_{j=0}^{\sigma} k_{ij} \frac{1}{s} e^{\frac{2j+1}{2\mu+1}}$, and derivative $\sum_{j=0}^{\eta} k_{dj} s e(t)^{\frac{2j+1}{2\gamma+1}}$ feedback are given in terms of the error vector. Non-negative integers ς, β, σ, μ, η, and γ are assigned by the designer to attain the specified criteria in stability and tracking, accuracy and disturbance attenuation, etc.

Letting $\varsigma = \beta = \sigma = \mu = \eta = \gamma = 0$, the linear PID controller in (3.1) results.

If $\varsigma = \beta = 1$, $\sigma = \mu = 2$, and $\eta = \gamma = 3$, the following nonlinear control algorithm results:

$$u(t)=k_{p0} e^{\frac{1}{3}}+k_{p1} e+k_{i0} \frac{1}{s} e^{\frac{1}{5}}+k_{i1} \frac{1}{s} e^{\frac{3}{5}}+k_{2} \frac{1}{s} e+k_{d0} s e^{\frac{1}{7}}+k_{d1} s e^{\frac{3}{7}}+k_{d2} s e^{\frac{5}{7}}+k_{d3} s e$$

As a final example, if the nonlinear proportional controller is needed and the designer assigns $\varsigma = 3$ and $\beta = 0$, then using (3.2), one obtains

$$u(t)=k_{p0} e+k_{p1} e^{3}+k_{p2} e^{5}+k_{p3} e^{7}$$

Control bounds in electromechanical systems

In electromechanical systems, the control inputs (e.g., the armature voltages applied to the motor windings or the signal-level control voltages supplied to the operational amplifiers), as well as the system variables (voltages and currents, angular and linear velocities and displacements), are bounded. These rated (maximum allowed) voltages and currents, as well as the maximum mechanical angular and linear velocities and displacements, are specified. That is, the applied voltage to the motor windings cannot exceed the rated voltage, and the currents in the stator and rotor windings cannot be greater than the maximum allowed peak and continuous currents. One concludes that due to the limits imposed, the allowed control is bounded, and the system variables must be within the maximum admissible set. Furthermore, electromechanical motion devices (actuators) are matched with the power converters (amplifiers), and the current limit is installed. Even if one uses power amplifiers that allow one to avoid the amplifier saturation limits, voltage and current bounds associated with electric machines are in place. Hence, the control authority is constrained, and the typical saturation characteristic is shown in Figure 3.5.

The saturated control, as a function of the error vector, is bounded as $u_{\min} \le u \le u_{\max}$, and the bounded PID control algorithm is

$$u(t)=\operatorname{sat}_{u_{\max}}^{u_{\max}}\left(k_{p} e(t)+k_{i} \int e(t) dt+k_{d} \frac{de(t)}{dt}\right), \quad u_{\min} \le u \le u_{\max} \tag{3.3}$$

Compared with the linear unbounded PID controller of (3.1), the applied voltage u found using (3.3) is bounded by the minimum and maximum values $u_{\min} < 0$ and $u_{\max} > 0$, $u_{\min} \le u \le u_{\max}$. These constraints are assigned using the rated voltages or currents specified.

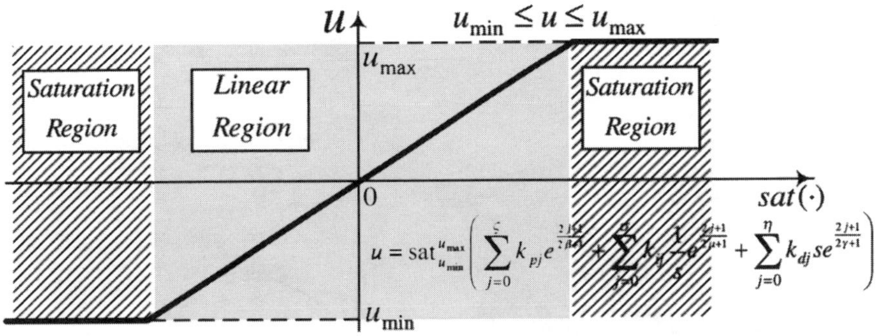

Figure 3.5 Saturation characteristic of the bounded control.

Using nonlinear error maps, one obtains the bounded controller from (3.2) as

$$u(t) = \text{sat}_{u_{\min}}^{u_{\max}}\left(\sum_{j=0}^{\varsigma} k_{pj} e^{\frac{2j+1}{2\beta+1}} + \sum_{j=0}^{\sigma} k_{ij} \frac{1}{s} e^{\frac{2j+1}{2\mu+1}} + \sum_{j=0}^{\eta} k_{dj} s e(t)^{\frac{2j+1}{2\gamma+1}}\right), \quad u_{\min} \le u \le u_{\max}$$

Consider the control algorithm in (3.3). In the linear region, the control varies between the maximum and minimum values $u_{\min}$ and $u_{\max}$; that is, $u_{\min} \le u \le u_{\max}$, and we have

$$u(t) = k_p e(t) + k_i \frac{1}{s} e(t) + k_d s e(t) = k_p e(t) + k_i \int e(t)dt + k_d \frac{de(t)}{dt}.$$

If $k_p e(t) + k_i \int e(t)dt + k_d \frac{de(t)}{dt} > u_{\max}$, the control is saturated, and $u(t) = u_{\max}$. For $k_p e(t) + k_i \int e(t)dt + k_d \frac{de(t)}{dt} < u_{\min}$, we have $u(t) = u_{\min}$. For example, if the maximum applied voltage to the DC motor (the so-called rated voltage) is ± 30 V, we have $u_{\max}$ = 30 V and $u_{\min}$ = –30 V; that is, $-30 \le u \le 30$ V.

3.2 Analog control of permanent-magnet direct current motor

Permanent-magnet DC motors are widely used in servo-systems. Consider the pointing systems (rotating table) actuated by the permanent-magnet DC motor (see Figure 3.6). The motor is attached to the rotating table using a gear with a coefficient k_{gear}. The goal is to guarantee the desired (assigned) angular displacement of the platform, and the reference signal is denoted as $r(t)$, while the angular displacement of the positioning table is $y(t) = k_{gear}\theta_r$. It should be emphasized that $y(t) = k_{gear}\theta_r$ is the output equation. To control the motor angular velocity ω_r, as well as the angular displacements of the motor θ_r and rotating table $k_{gear}\theta_r$, one regulates the armature voltage applied to the motor winding u_a. To guarantee the stability, to attain the desired accuracy, as well as to ensure tracking and disturbance attenuation of the servo-system studied, one should design the control algorithm, and the coefficients of the PID controller must be found.

First, find the transfer functions to develop the block diagrams of the open- and closed-loop systems. These transfer functions are obtained using the differential equations that

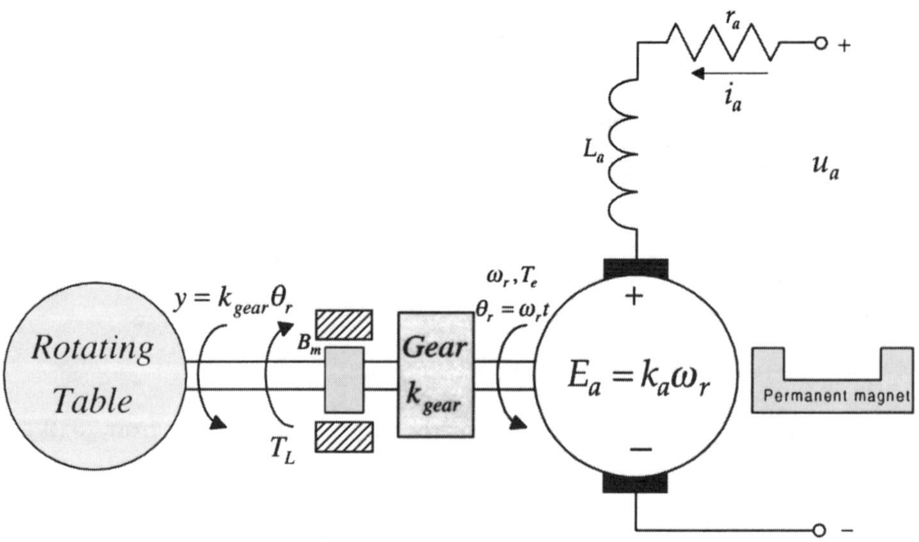

Figure 3.6 Schematic diagram of a rotating table actuated by permanent-magnet motor.

describe the system dynamics. Kirchhoff's voltage law maps the armature circuitry dynamics of the motor. For permanent-magnet DC motors, the *induced emf* E_a is a product of the so-called motor *back emf* constant k_a and the angular velocity; that is, $E_a = k_a\omega_r$. Denoting the armature resistance and inductance as r_a and L_a, respectively, one obtains

$$\frac{di_a}{dt} = -\frac{r_a}{L_a}i_a - \frac{k_a}{L_a}\omega_r + \frac{1}{L_a}u_a$$

Newtonian mechanics is applied to find the differential equations for mechanical systems. Using Newton's second law, the following differential equation for the angular acceleration was introduced in Section 1.2

$$\sum \vec{T} = J\vec{\alpha} = J\frac{d\vec{\omega}}{dt}$$

where J is the equivalent moment of inertia, which in the example studied, is found as a sum of the moments of inertia of rotor, coupling, gear, and rotating platform.

The electromagnetic torque developed by permanent-magnet DC motors is found as

$$T_e = k_a i_a$$

where k_a is the *torque* constant (the *torque* constant [N-m-A^{-1}] is equal to the *back emf* constant [V-sec-rad^{-1}], and therefore both are denoted as k_a).

Denoting the viscous friction coefficient as B_m, the viscous friction torque is

$$T_{viscous} = B_m\omega_r$$

The load torque is denoted as T_L. Then, making use of Newton's second law, we have

$$\frac{d\omega_r}{dt} = \frac{1}{J}(T_e - T_{viscous} - T_L) = \frac{1}{J}(k_a i_a - B_m \omega_r - T_L)$$

The dynamics of the rotor angular displacement is

$$\frac{d\theta_r}{dt} = \omega_r$$

To find the transfer functions, the derived three first-order differential equations

$$\frac{di_a}{dt} = -\frac{r_a}{L_a} i_a - \frac{k_a}{L_a}\omega_r + \frac{1}{L_a}u_a$$

$$\frac{d\omega_r}{dt} = \frac{1}{J}(k_a i_a - B_m \omega_r - T_L)$$

and

$$\frac{d\theta_r}{dt} = \omega_r$$

must be rewritten in the *s*-domain.

Using the Laplace operator $s = \frac{d}{dt}$, one immediately obtains

$$\left(s + \frac{r_a}{L_a}\right) i_a(s) = -\frac{k_a}{L_a}\omega_r(s) + \frac{1}{L_a}u_a(s)$$

$$\left(s + \frac{B_m}{J}\right)\omega_r(s) = \frac{1}{J}k_a i_a(s) - \frac{1}{J}T_L(s)$$

$$s\theta_r(s) = \omega_r(s)$$

The block diagram of the permanent-magnet DC motor, as a single-input (applied armature voltage)/single-output (rotor angular displacement) system, is illustrated by Figure 3.7.

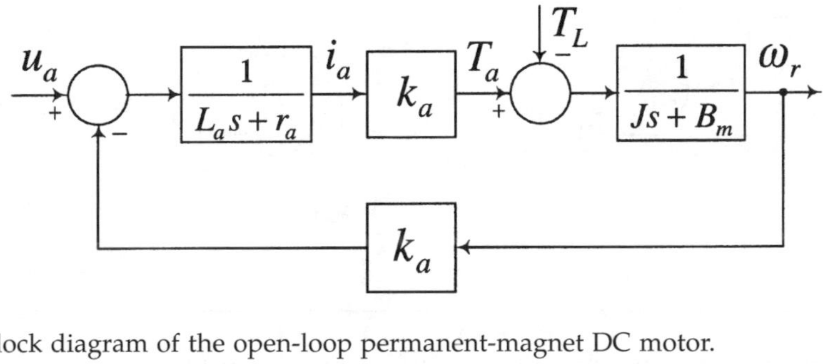

Figure 3.7 Block diagram of the open-loop permanent-magnet DC motor.

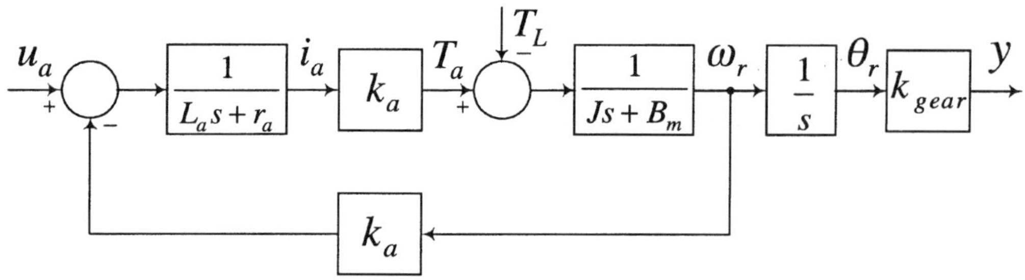

Figure 3.8 Block diagram of the open-loop servo actuated by permanent-magnet motor.

The controller should be designed, and the output equation, which relates the system output (angular displacement of the rotating table) and the angular displacement of the motor, is $y(t) = k_{gear}\theta_r$. Using this output equation, as well as $s\theta_r(s) = \omega_r(s)$, the block diagram of the open-loop servomechanism is drawn in Figure 3.8.

Using the linear PID controller, as given by

$$u_a(t) = k_p e(t) + k_i \frac{e(t)}{s} + k_d s e(t)$$

as well as the block diagram developed and documented in Figure 3.8, one obtains the closed-loop system illustrated in Figure 3.9.

The parameters of the permanent-magnet JDH-2250 Clifton Precision motor are: r_a = 2.7 ohm, L_a = 0.004 H, B_m = 0.0000093 N-m-s-rad^{-1}, k_a = 0.105 V-s-rad^{-1} (the *back emf* constant) and k_a = 0.105 N-m-A^{-1} (the *torque* constant), and J = 0.0001 kg-m^2.

The goal is to find the proportional k_p, integral k_i, and derivative k_d coefficients. Assigning the reference input $r(t)$ = 0.2 rad, and setting $k_i = 0$ and $k_d = 0$, the critical value of the proportional coefficient is found, and $k_{p\,\max} = 1400$. Figure 3.10 illustrates the output dynamics if $k_{p\,\max} = 1400$, $k_i = 0$, and $k_d = 0$.

Using the period of the self-sustained oscillations T_{osc} and $k_{p\,\max}$ (T_{osc} = 0.037 s and $k_{p\,\max}$ =1400), we have $k_p = 700$, $k_i = 30$, and $k_d = 7$. Making use of the closed-loop block diagram as given by Figure 3.9 and the controller found

$$u_a(t) = 700e(t) + 30\int e(t)dt + 7\frac{de(t)}{dt}$$

the SIMULINK block diagram is developed (the mdl file is named as `ch_3_01.mdl`; see Figure 3.11).

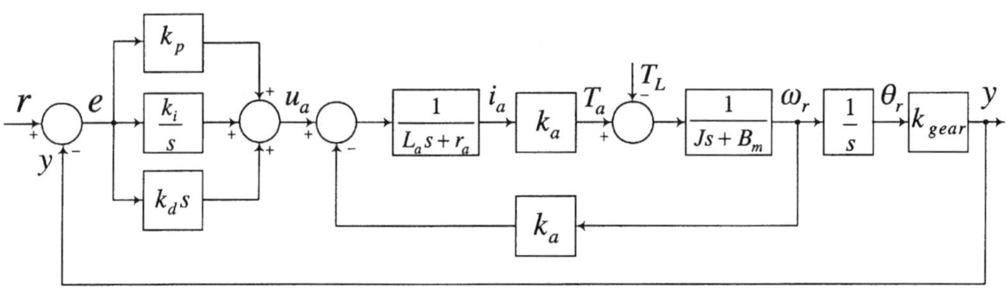

Figure 3.9 Block diagram of the closed-loop servo with the linear PID controller.

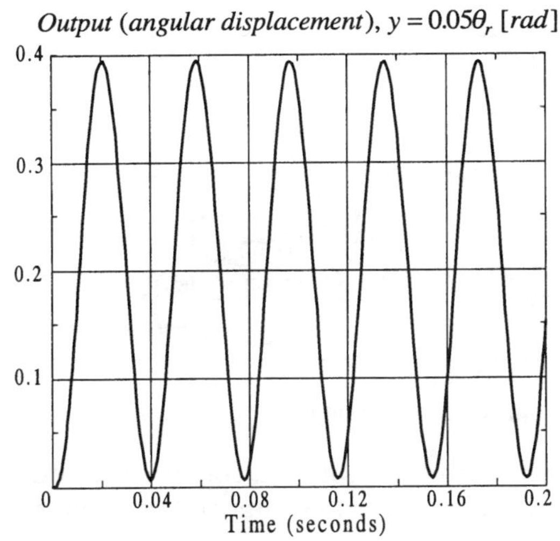

Figure 3.10 SIMULINK diagram of closed-loop servo-system with PID controller.

It should be emphasized that the `PID Controller` building blocks are available in the SIMULINK environment, and one has to set the feedback coefficients designed (see the two PID blocks illustrated at the bottom of Figure 3.11).

To illustrate and visualize the linear PID control law

$$u_a(t) = 700e(t) + 30\int e(t)dt + 7\frac{de(t)}{dt}$$

the detailed and explicit representation of this controller was shown in Figure 3.11 using conventional building blocks.

It is convenient to assign the parameters of electromechanical systems in the additional m-file or in the Command window. These parameters must be used in the form of the symbolic variables employed in the SIMULINK diagrams developed. The SIMULINK diagram of the closed-loop servomechanism is demonstrated in Figure 3.12 using the symbolic variables of the motor parameters r_a, L_a, B_m, k_a, and J, the gear coefficient k_{gear}, as well as

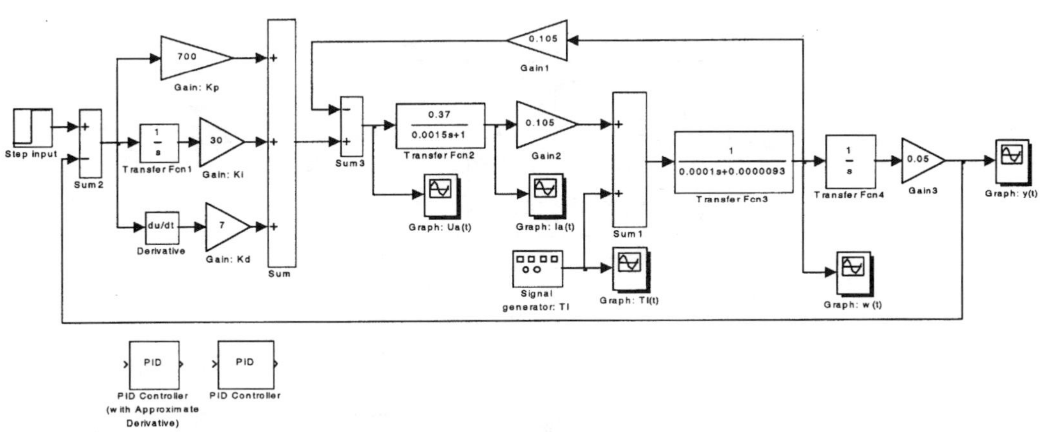

Figure 3.11 SIMULINK diagram of the closed-loop servo-system with PID controller.

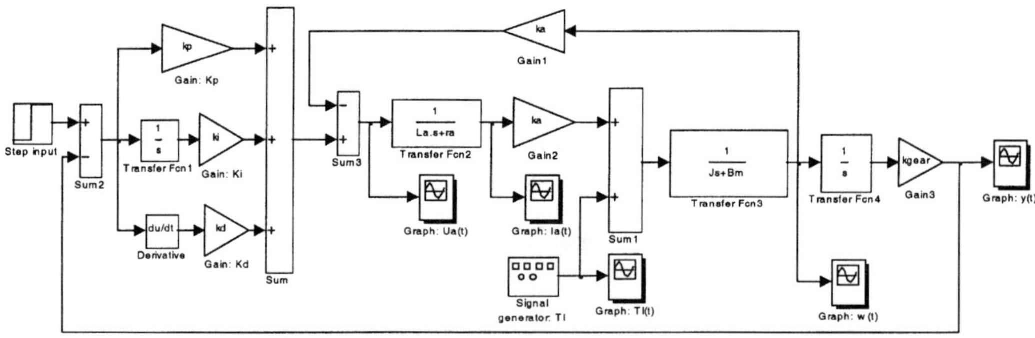

Figure 3.12 SIMULINK diagram with symbolic parameter variables (`ch_3_02.mdl`).

the proportional k_p, integral k_i, and derivative k_d coefficients. These servo parameters and feedback gains are assigned in the MATLAB Command window as

```
» ra=2.7;La=0.004;Bm=0.0000093;ka=0.105;kgear=0.05;J=0.0001;
» kp=700;ki=30;kd=7;
```

The SIMULINK diagrams, as documented in Figures 3.11 and 3.12, are equivalent.

Assigning the desired reference displacement of the rotating table to be $r(t) = 0.2$ rad, the modeling was performed, and Figure 3.13 illustrates the output dynamics. It is evident that the output $y(t)$ converges to the desired reference, and the tracking error is zero.

It was emphasized that the design specifications imposed on high-performance electromechanical systems are given in the *system domain*. The commonly used performance criteria to be attained are:

- Stability with the desired stability margins in the full operating envelope
- Robustness to parameter variations and structural changes

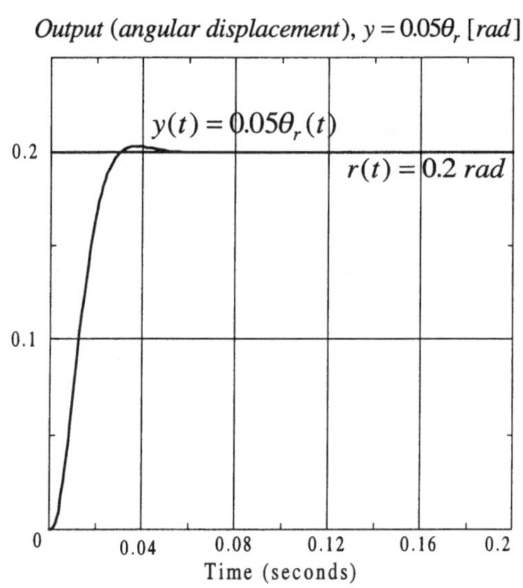

Figure 3.13 Output of the servo-system with PID controller.

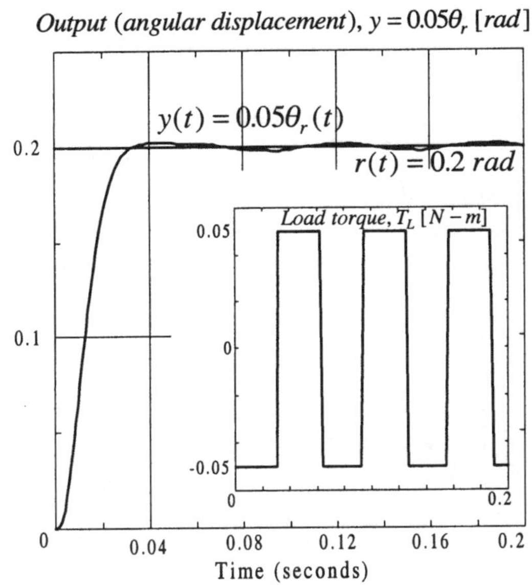

Figure 3.14 Output of the servo-system if the load torque is applied.

- Tracking and disturbance attenuation
- Dynamic and steady-state accuracy
- Dynamic performance specifications imposed on the states and outputs transient responses (i.e., the settling time, the maximum overshoot, the rise time, the delay time, as well as the peak time)

For the servo-system studied and designed, the settling time is 0.03 s, the maximum overshoot is 2%, the rise time is 0.025 s, and the delay time is 0.01 s.

The disturbance attenuation features are studied. Using the `Signal Generator` building block, we assign the square waveform for the load torque with the magnitude ±0.05 N-m and frequency 100 rad-s^{-1}. The modeling is performed, and the output dynamics as well as the load torque are illustrated in Figure 3.14.

The rated applied voltage and armature current for the motor studied are ±30 V and ±11 A, respectively. With these bounds imposed, the modeling was performed. The corresponding SIMULINK block diagram with the `Saturation` block to constrain the supplied

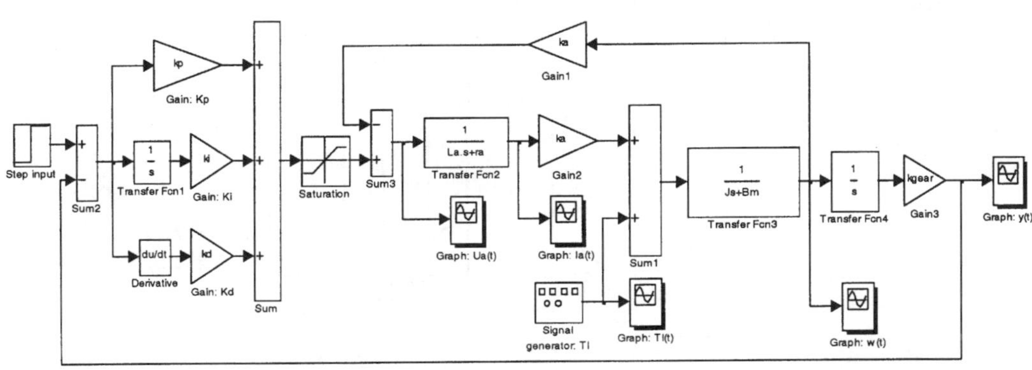

Figure 3.15 SIMULINK diagram of the *servo* with the saturated controller.

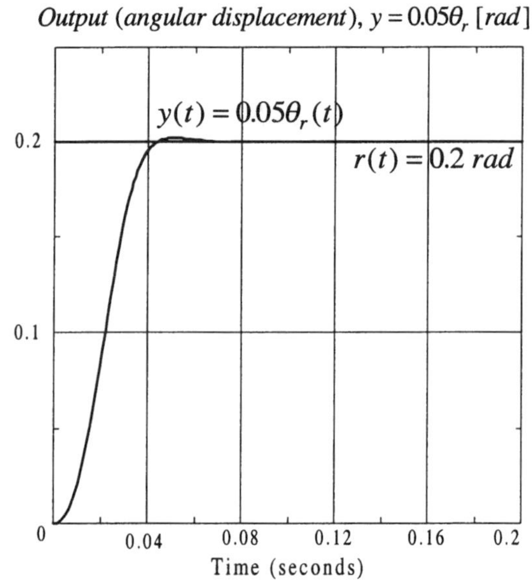

Figure 3.16 Output of the servo-system with saturated control.

voltage is shown in Figure 3.15 (`ch_3_03.mdl`). For the desired reference displacement $r(t) = 0.2$ rad, the output $y(t)$ transient responses are plotted in Figures 3.16 and 3.17 if $T_L = 0$ N-m and if the square load torque is applied with the magnitude ±0.05 N-m and the frequency 100 rad-s^{-1}.

One concludes that the dynamics is degraded due to the saturation effect. In particular, the settling time is 0.05 s, the rise time is 0.04 s, and the delay time is 0.03 s. However, the ability of the PID controller to cope with the control bounds and to guarantee the acceptable tracking performance and accuracy, stability, and disturbance attenuation is evident.

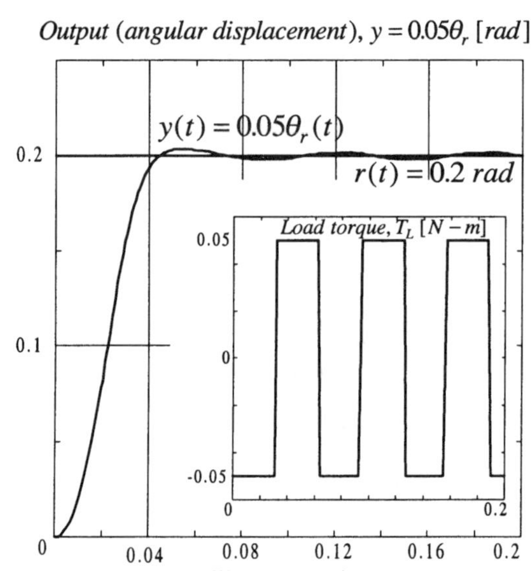

Figure 3.17 Servo-system output if the load is applied.

3.3 Electromechanical systems with digital PID controllers

Digital controllers are widely used to control electromechanical systems. The application of microprocessors, microcontrollers, and DSPs to deploy the control algorithms, perform diagnostics and filtering, and accomplish data acquisition, places the critical needs and facilitates study of discrete-data systems.

Using $t = kT_s$, one can discretize linear and nonlinear differential equations. Here, k denotes an integer and T_s is the sampling period.

Consider the continuous-time signal $e(t)$. The sampled values of the signal are $t = kT_s$; that is, with the sampling period T_s, we have $e(0)$, $e(T_s)$, $e(2T_s)$, … .

Let us sample the continuous-time signal $e(t)$; the impulse-sampled signal is given by

$$e_{sampled}(t) = \sum_{k=0}^{\infty} e(kT_s)\delta(t - kT_s)$$

where δ is the Dirac delta function. The ideal sampler can be considered as a modulator that multiplies two signals: $e(t)$ and a train of unit impulses (see Figure 3.18).

For $t \in [0, +\infty)$, it is immediately apparent that $k = 1, 2, 3, \ldots$, and the two-side z-transform is given as

$$E(z) = Z[e(t)] = Z[e(kT_s)] = \sum_{k=0}^{\infty} e(kT_s)e^{-kT_s s} = \sum_{k=0}^{\infty} e(kT_s)z^{-k}$$

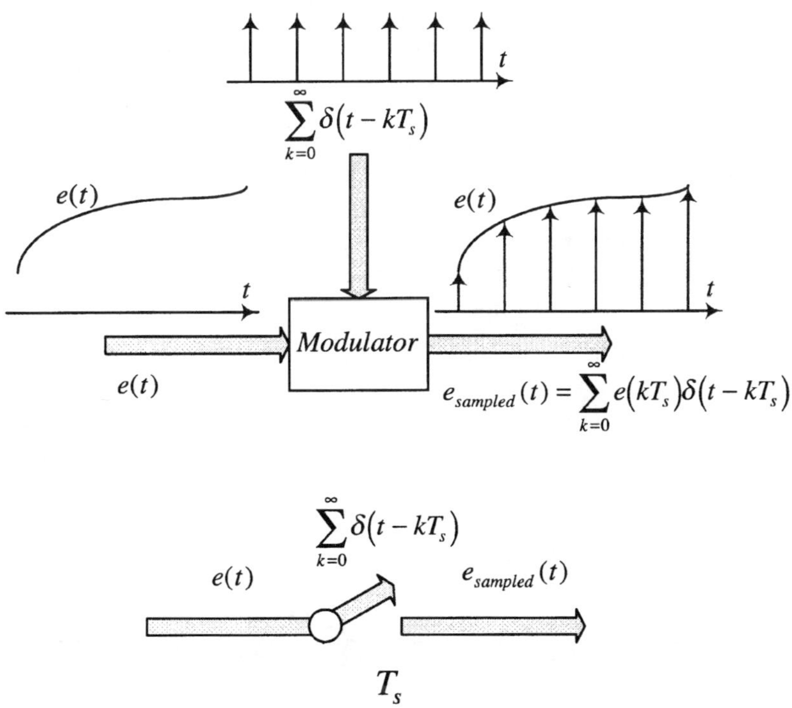

Figure 3.18 Sampling of the continuous-time signal.

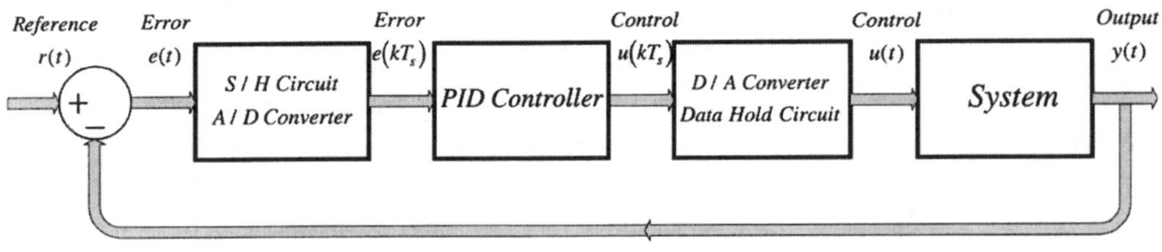

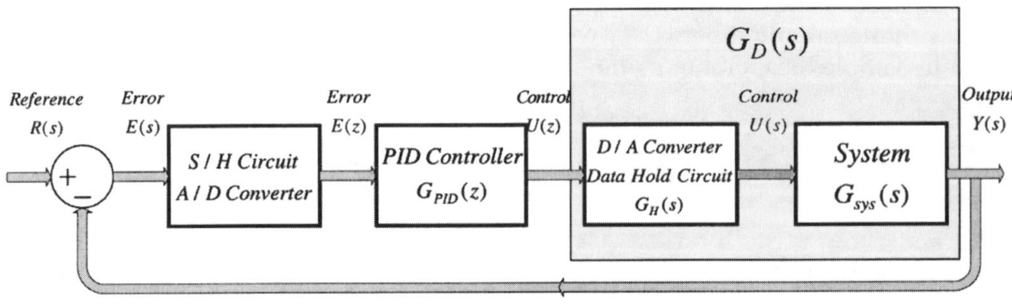

Figure 3.19 Time-, *s*-, and *z*-domain block diagrams of digital systems with PID-type controllers.

The block diagram of the closed-loop hybrid electromechanical system, which integrates continuous- and discrete-time systems, is shown in Figure 3.19a. Mathematical models of electromechanical motion devices with power electronics, as well as data hold circuits, should be integrated into the analysis and design. Assuming that the electromechanical system dynamics are mapped by linear constant-coefficient differential equations, the digitally controlled electromechanical system is drawn in Figure 3.19b using transfer functions for the systems (actuators with amplifiers) $G_{sys}(s)$, data hold circuit $G_H(s)$, and digital PID controller $G_{PID}(z)$.

To convert the discrete-time signals into piecewise continuous signals to be fed to power converters, data hold circuits are used. Mathematically, the data hold circuit can be viewed as a w-order polynomial extrapolator. Zero- and first-order data hold circuits are typically used to avoid the complexity and time delay of higher-order data hold circuits. Consider the sampler and the zero-order data hold circuit shown in Figure 3.20.

If the input continuous-time signal $e(t)$ is causal, the piecewise continuous data hold output is found to be

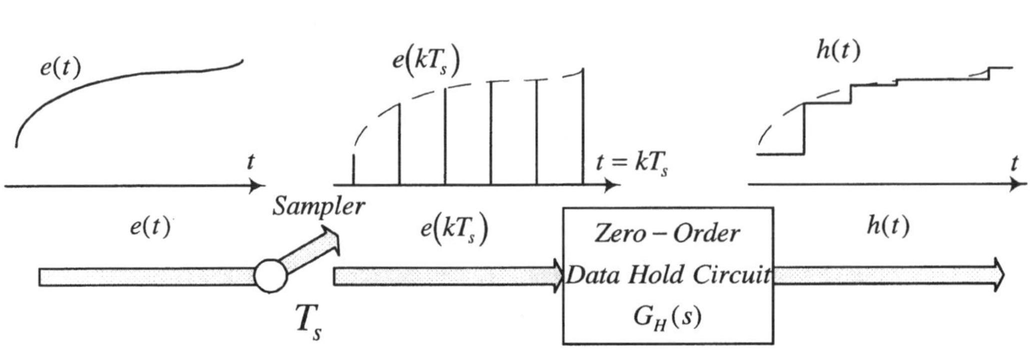

Figure 3.20 Sampler and zero-order data hold circuit.

$$h(t)=\sum_{k=0}^{\infty} e(kT_s)[1(t-kT_s)-1(t-(k+1)T_s)]$$

The output of the zero-order data hold is the signal that is equal to the last sampled value (equal to the value of the continuous-time signal at the instant $t = kT_s$) until the next sampled value is available; that is,

$$h(kT_s+1)=h(kT_s)=e(t)\Big|_{t=kT_s} \quad \text{for} \quad 0 \le t < T_s$$

Using the following Laplace transforms

$$L[1(t)]=\frac{1}{s} \quad \textit{and} \quad L[1(t-kT_s)]=\frac{e^{-kT_s s}}{s}$$

one finds that

$$L[h(t)]=\sum_{k=0}^{\infty} e(kT_s)\frac{e^{-kT_s s}-e^{-(k+1)T_s s}}{s}=\frac{1-e^{-T_s s}}{s}\sum_{k=0}^{\infty} e(kT_s)e^{-kT_s s}$$

Please distinguish between the exponential function e and the error signal, which is denoted as $e(kT_s)$.

The Laplace transform of the data hold input is

$$E_{sampled}(s)=\sum_{k=0}^{\infty} e(kT_s)e^{-kT_s s}$$

Hence, using

$$L[h(t)]=\frac{1-e^{-T_s s}}{s}\sum_{k=0}^{\infty} e(kT_s)e^{-kT_s s}$$

the transfer function of the zero-order data hold is found to be

$$G_H(s)=\frac{1-e^{-T_s s}}{s}$$

The direct linear extrapolation is performed by the first-order data hold, and we have the following expression for the first-order data hold output

$$h(t)=1(t)+\frac{t}{T_s}1(t)-\frac{t-T_s}{T_s}1(t-T_s)-1(t-T_s)$$

and the transfer function is given by

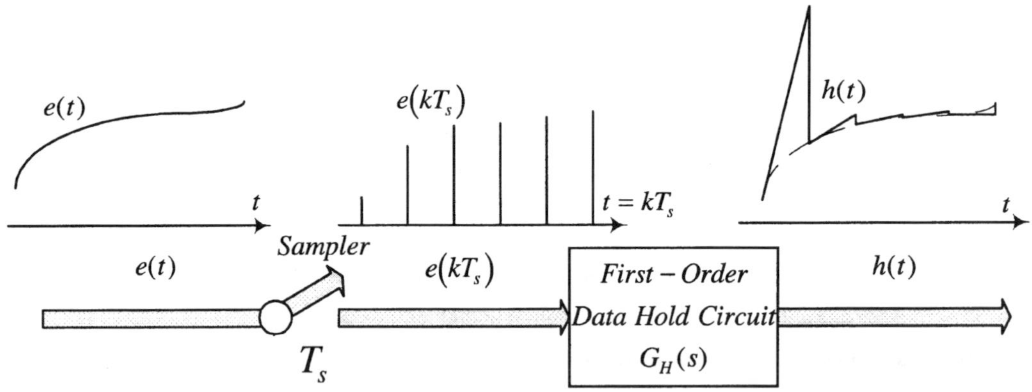

Figure 3.21 Sampler cascaded with first-order data hold.

$$G_H(s) = \frac{1}{s} + \frac{1}{T_s s^2} - \frac{1}{T_s s^2} e^{-T_s s} - \frac{1}{s} e^{-T_s s} = (1 - e^{-T_s s}) \frac{T_s s + 1}{T_s s^2}$$

The sampler and the first-order data hold circuit are illustrated in Figure 3.21.

The augmented transfer function of the electromechanical system and the data hold circuit to be used in analysis, modeling, and design is

$$G_D(s) = G_H(s) G_{sys}(s)$$

and the augmented transfer function $G_D(s)$ obtained is illustrated in Figure 3.19.

Let us find the digital PID controller using the analog PID controller, as expressed by

$$u(t) = k_p e(t) + k_i \frac{e(t)}{s} + k_d s e(t)$$

with transfer function $G_{PID} = \dfrac{U(s)}{E(s)} = \dfrac{k_d s^2 + k_p s + k_i}{s}$.

For the proportional analog controller, we have $u_p(t) = k_p e(t)$ with $G_P(s) = \dfrac{U_p(s)}{E(s)} = k_p$. It is evident that the proportional digital control law is

$$u_p(kT_s) = k_p e(kT_s)$$

and

$$G_P(z) = \frac{U_p(z)}{E(z)} = k_p$$

The analog integral $u_i(t) = k_i \dfrac{e(t)}{s}$ and derivative $u_d(t) = k_d s e(t)$ terms of the PID controller (3.1), with $G_I(s) = \dfrac{U_I(s)}{E(s)} = \dfrac{k_i}{s}$ *and* $G_D(s) = \dfrac{U_D(s)}{E(s)} = k_d s$, can be represented in the discrete-time domain. The corresponding transfer functions in the z-domain can be found using the trapezoidal, forward-rectangular, or backward-rectangular approximations. The transfer functions of the analog PID-type controllers are given in the s-domain, and the $G_{PID}(s)$ are available.

For the linear PID controller in (3.1), it was shown that $G_{PID}(s)=\dfrac{U(s)}{E(s)}=\dfrac{k_d s^2+k_p s+k_i}{s}$.

If the PII³D controller $u(t)=k_p e(t)+k_{i1}\dfrac{e(t)}{s}+k_{i3}\dfrac{e(t)}{s^3}+k_d s e(t)$ is employed, we have

$$G_{PID}(s)=\frac{U(s)}{E(s)}=\frac{k_d s^4+k_p s^3+k_{i1}s^2+k_{i3}}{s^3}$$

Table 3.2 provides the z-transforms to be used.

For the integral part, using the Euler approximation, the difference equation and the corresponding transfer function in z-domain can be easily obtained; and using Table 3.2, we have

$$G_i(z)=\frac{U_i(z)}{E(z)}=\frac{T_s}{1-z^{-1}}=\frac{T_s z}{z-1}$$

To find the derivative term, use the trapezoidal approximation and the first difference results, and one obtains

$$G_d(z)=\frac{U_d(z)}{E(z)}=\frac{1-z^{-1}}{T_s}=\frac{z-1}{T_s z}$$

Hence, through numerical integration and differentiation we have the z-domain representation of the control signal $U(z)$ and the transfer functions of linear digital PID controllers result. In particular, in the *error* form (the *error* signal is used to calculate the control needed), we have

$$U(z)=\left(k_{dp}+\frac{k_{di}}{1-z^{-1}}+k_{dd}(1-z^{-1})\right)E(z)$$

and

$$G_{PID}(z)=\frac{U(z)}{E(z)}=k_{dp}+\frac{k_{di}}{1-z^{-1}}+k_{dd}(1-z^{-1})$$

That is,

$$G_{PID}(z)=\frac{(k_{dp}+k_{di}+k_{dd})z^2-(k_{dp}+2k_{dd})z+k_{dd}}{z^2-z}$$

The *reference-output* form (reference input and system output are used to find the control) of the linear digital PID control law is

Table 3.2 z-Transform Table

Laplace Transform $X(s)$	Time-Domain Signal $x(t)$	Time-Domain Signal $x(kT_s)$	z-Transform $X(z)$
$\frac{1}{s}$	Unit step $1(t)$	$1(kT_s)$	$\frac{1}{1-z^{-1}}=\frac{1}{z-1}$
$\frac{1}{s^2}$	$t1(t)$	$kT_s1(kT_s)$	$\frac{T_sz^{-1}}{(1-z^{-1})^2}=\frac{T_sz}{(z-1)^2}$
$\frac{2}{s^3}$	$t^21(t)$	$(kT_s)^21(kT_s)$	$\frac{T_s^2z^{-1}(1+z^{-1})}{(1-z^{-1})^3}$
$\frac{6}{s^4}$	$t^31(t)$	$(kT_s)^31(kT_s)$	$\frac{T_s^3z^{-1}(1+4z^{-1}+z^{-2})}{(1-z^{-1})^4}$
$\frac{24}{s^5}$	$t^41(t)$	$(kT_s)^41(kT_s)$	$\frac{T_s^4z^{-1}(1+11z^{-1}+11z^{-2}+z^{-3})}{(1-z^{-1})^5}$
$\frac{1}{s+a}$	$e^{-at}1(t)$	$e^{-akT_s}1(kT_s)$	$\frac{1}{1-e^{-aT_s}z^{-1}}$
$\frac{a}{s(s+a)}$	$(1-e^{-at})1(t)$	$(1-e^{-akT_s})1(kT_s)$	$\frac{(1-e^{-aT_s})z^{-1}}{(1-z^{-1})(1-e^{-aT_s}z^{-1})}$
$\frac{b-a}{(s+a)(s+b)}$	$(e^{-at}-e^{-bt})1(t)$	$(e^{-akT_s}-e^{-bkT_s})1(kT_s)$	$\frac{(e^{-aT_s}-e^{-bT_s})z^{-1}}{(1-e^{-aT_s}z^{-1})(1-e^{-bT_s}z^{-1})}$
$\frac{1}{(s+a)^2}$	$te^{-at}1(t)$	$kT_se^{-akT_s}1(kT_s)$	$\frac{T_se^{-aT_s}z^{-1}}{(1-e^{-aT_s}z^{-1})^2}$
$\frac{s}{(s+a)^2}$	$(1-at)e^{-at}1(t)$	$(1-akT_s)e^{-akT_s}1(kT_s)$	$\frac{1-(1+aT_s)e^{-aT_s}z^{-1}}{(1-e^{-aT_s}z^{-1})^2}$
$\frac{\omega_0}{s^2+\omega_0^2}$	$\sin(\omega_0t)1(t)$	$\sin(\omega_0kT_s)1(kT_s)$	$\frac{z^{-1}\sin(\omega_0T_s)}{1-2z^{-1}\cos(\omega_0T_s)+z^{-2}}$
$\frac{s}{s^2+\omega_0^2}$	$\cos(\omega_0t)1(t)$	$\cos(\omega_0kT_s)1(kT_s)$	$\frac{1-z^{-1}\cos(\omega_0T_s)}{1-2z^{-1}\cos(\omega_0T_s)+z^{-2}}$
$\frac{\omega_0}{(s+a)^2+\omega_0^2}$	$e^{-at}\sin(\omega_0t)1(t)$	$e^{-akT_s}\sin(\omega_0kT_s)1(kT_s)$	$\frac{e^{-akT_s}z^{-1}\sin(\omega_0T_s)}{1-2e^{-aT_s}z^{-1}\cos(\omega_0T_s)+e^{-2aT_s}z^{-2}}$
$\frac{s+a}{(s+a)^2+\omega_0^2}$	$e^{-at}\cos(\omega_0t)1(t)$	$e^{-akT_s}\cos(\omega_0kT_s)1(kT_s)$	$\frac{1-e^{-aT_s}z^{-1}\cos(\omega_0T_s)}{1-2e^{-aT_s}z^{-1}\cos(\omega_0T_s)+e^{-2aT_s}z^{-2}}$

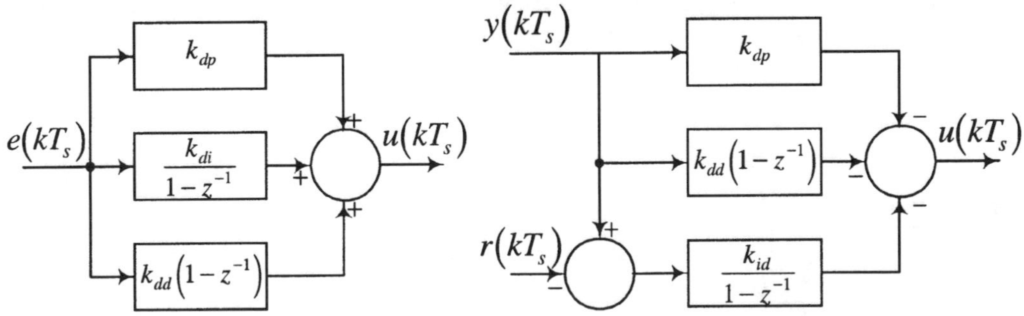

Figure 3.22 Error and reference-output forms of linear digital PID controllers.

$$U(z) = -k_{dp}Y(z) - k_{di}\frac{Y(z)-R(z)}{1-z^{-1}} - k_{dd}(1-z^{-1})Y(z)$$

The block diagrams of the digital PID controllers are shown in Figure 3.22.

The feedback gains k_{dp}, k_{di}, and k_{dd} of the digital control law are found using the proportional, integral, and derivative coefficients of the analog PID controller k_p, k_i, and k_d. In particular, the following formulas to calculate the feedback gains

$$k_{dp} = k_p - \frac{1}{2}k_{di},\ k_{di} = T_s k_i,\ k_{dd} = \frac{k_d}{T_s}$$

must be applied.

The transfer function of the closed-loop systems, which is given in Figure 3.19, is

$$G(z) = \frac{Y(z)}{R(z)} = \frac{G_{PID}(z)G_D(z)}{1+G_{PID}(z)G_D(z)}$$

It should be emphasized that to implement the digital PID controller, one has

$$u(kT_s) = k_p e(kT_s) + \frac{1}{2}k_i T_s \sum_{i=1}^{k}[e((i-1)T_s) + e(iT_s)] + \frac{k_d}{T_s}[e(kT_s) - e((k-1)T_s)]$$

The Tustin substitution method is used to find difference equations that approximate differential equations. Using $z = e^{sT_s}$, we have

$$s = \frac{1}{T_s}\ln(z) \approx \frac{2}{T_s}\frac{z-1}{z+1} = \frac{2}{T_s}\frac{1-z^{-1}}{1+z^{-1}}$$

This approximation is obtained by truncating the series expansion of ln(z), which is given as

$$\ln(z) = 2\left[\frac{z-1}{z+1} + \frac{1}{3}\left(\frac{z-1}{z+1}\right)^3 + \frac{1}{5}\left(\frac{z-1}{z+1}\right)^5 + \dots\right],\ z>0$$

The Tustin approximation is used to transform the s-domain to the z-domain. It was shown that the transfer function of the linear analog PID controller is given by

$$G_{PID}(s)=\frac{U(s)}{E(s)}=\frac{k_d s^2+k_p s+k_i}{s}$$

Then, we have

$$G_{PID}(z)=\frac{U(z)}{E(z)}$$

$$=\frac{k_d\left(\frac{2}{T_s}\frac{1-z^{-1}}{1+z^{-1}}\right)^2+k_p\frac{2}{T_s}\frac{1-z^{-1}}{1+z^{-1}}+k_i}{\frac{2}{T_s}\frac{1-z^{-1}}{1+z^{-1}}}$$

$$=\frac{(2k_pT_s+k_iT_s^2+4k_d)+(2k_iT_s^2-8k_d)z^{-1}+(-2k_pT_s+k_iT_s^2+4k_d)z^{-2}}{2T_s(1-z^{-2})}$$

Hence,

$$U(z)-U(z)z^{-2}=k_{e0}E(z)+k_{e1}E(z)z^{-1}+k_{e2}E(z)z^{-2}$$

where

$$k_{e0}=k_p+\frac{1}{2}k_iT_s+2\frac{k_d}{T_s},\quad k_{e1}=k_iT_s-4\frac{k_d}{T_s},\quad \text{and}\quad k_{e2}=-k_p+\frac{1}{2}k_iT_s+2\frac{k_d}{T_s}$$

That is, taking the z-transform, one obtains the following control input needed to be implemented to accomplish the linear PID control

$$u(k)=u(k-2)+k_{e0}e(k)+k_{e1}e(k-1)+k_{e2}e(k-2)$$

The Tustin approximation used can be accomplished using the `c2dm` command (see Example 3.2).

Example 3.1

Using

$$G_{PID}(s)=\frac{U(s)}{E(s)}=\frac{k_d s^2+k_p s+k_i}{s}$$

and applying the following approximation

$$\ln(z)\approx 2\left(\frac{z-1}{z+1}+\frac{1}{3}\left(\frac{z-1}{z+1}\right)^3\right)$$

find the transfer function $G_{PID}(z)=\dfrac{U(z)}{E(z)}$ and the corresponding expression for $u(k)$.

Solution

Accurate approximation of the logarithm, as given by

$$\ln(z)\approx 2\left(\frac{z-1}{z+1}+\frac{1}{3}\left(\frac{z-1}{z+1}\right)^3\right)$$

leads one to

$$s=\frac{1}{T_s}\ln(z)\approx\frac{2}{T_s}\left(\frac{z-1}{z+1}+\frac{1}{3}\left(\frac{z-1}{z+1}\right)^3\right)$$

The transfer function for the digital PID control law $G_{PID}(z)$ can be found, and the equation for $u(k)$ to implement the controller needs to be derived. Using the Symbolic Math Toolbox, we have

```
» syms s z Ts Kd Kp Ki
» s=(2*(((z-1)/(z+1))+(((z-1)^3)/(3*(z+1)^3))))/Ts;
» g=(Kd*s^2+Kp*s+Ki)/s; g1=simplify(g); pretty(g1)

                6           3                     6              5              4
1/24 (64 Kd z  - 128 Kd z  + 64 Kd + 24 Kp Ts z  + 72 Kp Ts z  + 72 Kp Ts z

                  2                                          2  6             2  5
  - 72 Kp Ts z  - 72 Kp Ts z  - 24 Kp Ts + 9 Ki Ts  z  + 54 Ki Ts  z

                2  4             2  3              2  2             2            2
  + 135 Ki Ts  z  + 180 Ki Ts  z  + 135 Ki Ts  z  + 54 Ki Ts  z + 9 Ki Ts  )

  /          2                                3
  /  (Ts (z  + z + 1)  (z - 1)  (z + 1)  )
  /
» expand(Ts*(z^2+z+1)*(z-1)*(z+1)^3)
ans =
Ts*z^6+3*Ts*z^5+3*Ts*z^4-3*Ts*z^2-3*Ts*z-Ts
```

Hence, the following transfer function results

$$G_{PID}(z)=\frac{U(z)}{E(z)}$$

$$=\frac{(24k_pT_s+9k_iT_s^2+64k_d)z^6+(72k_pT_s+54k_iT_s^2)z^5+(72k_pT_s+135k_iT_s^2)z^4+(180k_iT_s^2-128k_d)z^3}{24T_s(z^6+3z^5+3z^4-3z^2-3z-1)}$$

$$+\frac{(-72k_pT_s+135k_iT_s^2)z^2+(-72k_pT_s+54k_iT_s^2)z+(-24k_pT_s+9k_iT_s^2+64k_d)}{24T_s(z^6+3z^5+3z^4-3z^2-3z-1)}$$

Denoting

$$k_{en0}=24k_pT_s+9k_iT_s^2+64k_d,\ k_{en1}=72k_pT_s+54k_iT_s^2,\ k_{en2}=72k_pT_s+135k_iT_s^2,$$

$$k_{en3}=180k_iT_s^2-128k_d$$

$$k_{en4}=-72k_pT_s+135k_iT_s^2,\ k_{en5}=-72k_pT_s+54k_iT_s^2,\ \textit{and}\ k_{en6}=-24k_pT_s+9k_iT_s^2+64k_d$$

we have

$$G_{PID}(z)=\frac{U(z)}{E(z)}=\frac{k_{en0}+k_{en1}z^{-1}+k_{en2}z^{-2}+k_{en3}z^{-3}+k_{en4}z^{-4}+k_{en5}z^{-5}+k_{en6}z^{-6}}{24T_s(1+3z^{-1}+3z^{-2}-3z^{-4}-3z^{-5}-z^{-6})}$$

Thus, one obtains an expression to implements the PID control algorithm. In particular,

$$u(k)=-3u(k-1)-3u(k-2)+3u(k-4)+3u(k-5)+u(k-6)$$

$$+\frac{1}{24T_s}(k_{en0}e(k)+k_{en1}e(k-1)+k_{en2}e(k-2)+k_{en3}e(k-3)+k_{en4}e(k-4)+k_{en5}e(k-5)+k_{en6}e(k-6))$$

□

Example 3.2

Using the z-transform for $\frac{6}{s^4}$, find the z-transform for $X(s)=\frac{24}{s^5}$.

Solution

From Table 3.2,

$$Z\left[\frac{6}{s^4}\right]=\frac{T_s^3z^{-1}(1+4z^{-1}+z^{-2})}{(1-z^{-1})^4}$$

Using the properties of the z-transform, the following results

$$Z\left[\frac{24}{s^5}\right]=-zT_s^4\frac{d}{dz}\left(\frac{z^{-1}(1+4z^{-1}+z^{-2})}{(1-z^{-1})^4}\right)$$

Letting $a = z^{-1}(1 + 4z^{-1} + z^{-2})$ and $b = (1 - z^{-1})^4$, and applying the quotient rule

$$\frac{d}{dz}\left(\frac{a}{b}\right)=\frac{b\frac{da}{dz}-a\frac{db}{dz}}{b^2}$$

we have

$$\frac{d}{dz}\left(\frac{z^{-1}(1+4z^{-1}+z^{-2})}{(1-z^{-1})^4}\right)=\frac{(1-z^{-1})^4(-z^{-2}-8z^{-3}-3z^{-4})-z^{-1}(1+4z^{-1}+z^{-2})4z^{-2}(1-z^{-1})^3}{(1-z^{-1})^8}$$

$$=-\frac{z^{-2}(1+11z^{-1}+11z^{-2}+z^{-3})}{(1-z^{-1})^5}$$

Hence, one obtains

$$Z\left(\frac{24}{s^5}\right)=\frac{T_s^4z^{-1}(1+11z^{-1}+11z^{-2}+z^{-3})}{(1-z^{-1})^5}$$ □

3.4 Control of permanent-magnet direct-current motor using digital controller

Consider the servo-system studied in Section 3.2. The rotating positioning table is actuated by the permanent-magnet DC motor, and differential equations and transfer functions for the servomechanism considered were obtained. The goal is to design the digital PID controller, and simulate and analyze the closed-loop system. Permanent-magnet DC motors, as well as other electric machines used in electromechanical systems, are continuous-time systems. The transfer function of the open-loop servo-systems is found using the block diagram illustrated in Figure 3.8. We have

$$G_{sys}(s)=\frac{Y(s)}{U(s)}=\frac{k_{gear}k_a}{s(L_aJs^2+(r_aJ+L_aB_m)s+r_aB_m+k_a^2)}$$

Augmenting this $G_{sys}(s)$ with the transfer function of the zero-order data hold $G_H(s)=\frac{1-e^{-T_ss}}{s}$, one obtains

$$G_D(s)=G_H(s)G_{sys}(s)=\frac{1-e^{-T_ss}}{s}\frac{k_{gear}k_a}{s(L_aJs^2+(r_aJ+L_aB_m)s+r_aB_m+k_a^2)}$$

To find $G_D(z)$, we use the $G_D(s)$ obtained, and the `c2dm` command is applied. The `filter` should be used to model the dynamics. Using the `help` command we have

```
» help c2dm
C2DM Conversion of continuous LTI systems to discrete-time.
[Ad,Bd,Cd,Dd]=C2DM(A,B,C,D,Ts,'method') converts the continuous-
time state-space system (A,B,C,D) to discrete time using 'method':
    'zoh'        Convert to discrete time assuming a zero order
                 hold on the inputs.
    'foh'        Convert to discrete time assuming a first order
                 hold on the inputs.
    'tustin'     Convert to discrete time using the bilinear
                 (Tustin) approximation to the derivative.
    'prewarp'    Convert to discrete time using the bilinear
                 (Tustin) approximation with frequency prewarping.
```

```
                    Specify the critical frequency with an additional
                    argument, i.e. C2DM(A,B,C,D,Ts,'prewarp',Wc)
     'matched'      Convert the SISO system to discrete time using the
                    matched pole-zero method.
[NUMd,DENd]=C2DM(NUM,DEN,Ts,'method') converts the continuous-time
polynomial transfer function G(s) = NUM(s)/DEN(s) to discrete-time,
G(z)=NUMd(z)/DENd(z), using 'method'.
See also: C2D, and D2CM.
```

```
» help filter
FILTER One-dimensional digital filter.
Y=FILTER(B,A,X) filters the data in vector X with the filter described
by vectors A and B to create the filtered data Y.
The filter is a "Direct Form II Transposed" implementation of the
standard difference equation:
    a(1)*y(n) = b(1)*x(n) + b(2)*x(n-1) + ... + b(nb+1)*x(n-nb)
                  - a(2)*y(n-1) - ... - a(na+1)*y(n-na)
If a(1) is not equal to 1, FILTER normalizes the filter coefficients
by a(1).
When X is a matrix, FILTER operates on the columns of X.
When X is an N-D array, FILTER operates along the first non-singleton
dimension.
[Y,Zf] = FILTER(B,A,X,Zi) gives access to initial and final
conditions, Zi and Zf, of the delays. Zi is a vector of length
MAX(LENGTH(A),LENGTH(B))-1 or an array of such vectors, one for each
column of X.
FILTER(B,A,X,[],DIM) or FILTER(B,A,X,Zi,DIM) operates along the
dimension DIM.
See also FILTER2, FILTFILT (in the Signal Processing Toolbox).
```

The following MATLAB m-file was written (ch_3_04.m).

```
clc; format short e;
% parameters of the servomechanism
ra=2.7; La=0.004; Bm=0.0000093; ka=0.105; kgear=0.05; J=0.0001;
% numerator and denominator of the s-domain transfer function
% of open-loop servo
num_s=[ka*kgear];
den_s=[ra*J (ra*Bm+ka^2) 0]; % armature inductance is neglected
den_s=[La*J (ra*J+La*Bm) (ra*Bm+ka^2) 0]; % armature inductance is
not neglected
num_s, den_s
pause;
% sampling time Ts
Ts=0.0005;
% numerator and denominator of the transfer function with zero-order
% hold
hold GD(z)
[num_dz,den_dz]=c2dm(num_s,den_s,Ts,'zoh');
num_dz, den_dz
```

```
pause;
% feedback coefficients of the analog PID controller
kp=700;ki=30;kd=7;
% feedback coefficients of the digital PID controller
kdi=Ts*ki; kdp=kp-kdi/2; kdd=kd/Ts;
% numerator and denominator of the z-domain transfer function of
% the PID
num_pidz=[(kdp+kdi+kdd) -(kdp+2*kdd) kdd];den_pidz=[1 -1 0];
num_pidz, den_pidz
pause;
% numerator and denominator of the closed-loop transfer function G(z)
num_z=conv(num_pidz,num_dz);
den_z=conv(den_pidz,den_dz)+conv(num_pidz,num_dz);
num_z, den_z
pause;
% samples, t=k*Ts
k_final=100; k=0:1:k_final;
% reference input r(t)=0.2 rad
r=0.2*ones(1,k_final+1);
% modeling of the servo-system output y(t)
y=filter(num_z,den_z,r);
%plotting statement
plot(k,y,'o',k,y,'--');
title('Output (angular displacement), y(kTs)=0.05Q(kTs) [rad]');
xlabel('k, t=kTs (seconds)');
disp('End')
```

Having found the numerator and denominator of $G_{sys}(s)$ to be

```
»
num_s =
  5.2500e-003
den_s =
  4.0000e-007   2.7004e-004   1.1050e-002   0
```

we have

$$G_{sys}(s) = \frac{Y(s)}{U(s)} = \frac{0.00525}{4\times 10^{-7}s^3 + 2.7\times 10^{-4}s^2 + 0.011s}$$

Using the sampling time $T_s = 0.0005$ s, the transfer function $G_D(z)$ is found, and from

```
num_dz =
  0   2.5175e-007   9.2720e-007   2.1267e-007
den_dz =
  1.0000e+000   -2.7077e+000   2.4212e+000   -7.1352e-001
```

we obtain

$$G_D(z) = \frac{2.52 \times 10^{-7} z^2 + 9.27 \times 10^{-7} z + 2.13 \times 10^{-7}}{z^3 + 2.71z^2 + 2.42z - 0.71}$$

The transfer function of the digital PID controller is

$$G_{PID}(z) = \frac{(k_{dp} + k_{di} + k_{dd})z^2 - (k_{dp} + 2k_{dd})z + k_{dp}}{z^2 - z}, \ k_{dp} = k_p - \frac{1}{2}k_{di}, \ k_{di} = T_s k_i, \text{ and } k_{dd} = \frac{k_d}{T_s}$$

Having calculated

```
num_pidz =
  1.4700e+004    -2.8700e+004    1.4000e+004
den_pidz =
  1    -1    0
```

one obtains

$$G_{PID}(z) = \frac{1.47 \times 10^4 z^2 - 2.87 \times 10^4 z + 1.4 \times 10^4}{z^2 - z}$$

Hence, for the closed loop systems, as given by

$$G(z) = \frac{Y(z)}{U(z)} = \frac{G_{PID}(z)G_D(z)}{1 + G_{PID}(z)G_D(z)}$$

using the derived numerator and denominator

```
num_z =
0            3.7008e-003   6.4046e-003   -1.9960e-002   6.8773e-003
2.9774e-003
den_z =
1.0000e+000  -3.7040e+000  5.1352e+000   -3.1547e+000   7.2040e-001
2.9774e-003
```

we have

$$G(z) = \frac{3.7 \times 10^{-3} z^4 + 6.4 \times 10^{-3} z^3 - 2 \times 10^{-2} z^2 + 6.88 \times 10^{-3} z + 0.00298}{z^5 - 3.7z^4 + 5.14z^3 - 3.15z^2 + 0.72z + 0.00298}$$

The output dynamics for the reference input $r(kT_s) = 0.2$ rad, $k \ge 0$ is shown in Figure 3.23.

It must be observed that the sampling time significantly influences system dynamics. To illustrate the role of the sampling time, we assign $T_s = 0.005$ s. Using the `ch_3_04.m`

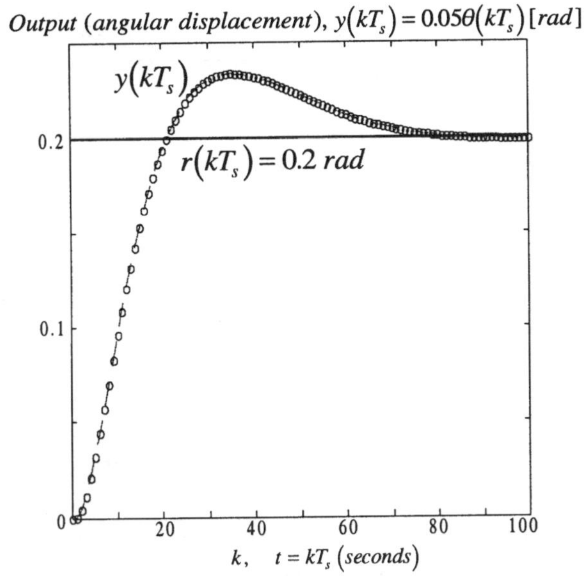

Figure 3.23 Output of the servo-system with digital PID controller; T_s = 0.0005 s.

file developed, one obtains the following transfer functions for the open-loop servo with the zero-order data hold and the digital controller

$$G_D(z) = \frac{1.37 \times 10^{-4} z^2 + 2.83 \times 10^{-4} z + 2.68 \times 10^{-5}}{z^3 - 1.85z^2 + 0.88z - 0.034}$$

and

$$G_{PID}(z) = \frac{2.1 \times 10^3 z^2 - 3.5 \times 10^3 z + 1.4 \times 10^3}{z^2 - z}$$

The transfer function for the closed-loop system is

$$G(z) = \frac{0.287z^4 + 0.117z^3 - 0.745z^2 + 0.303z + 0.0375}{z^5 - 2.56z^4 + 2.84z^3 - 1.66z^2 + 0.337z + 0.0375}$$

Assigning the reference input as $r(kT_s)$ = 0.2 rad, $k \geq 0$, the output $y(kT_s)$ is documented in Figure 3.24.

One can analytically find the transfer functions in z-domain using $G_{sys}(s)$. To accomplish this, the z-transform (Table 3.2) is used. In particular, to perform the design, we augment $G_{sys}(s)$ and $G_H(s)$ in order to find $G_D(z)$. Let us demonstrate how the analytical solution can be found.

It was shown that

$$G_D(s) = G_H(s) G_{sys}(s) = \frac{1 - e^{-T_s s}}{s} \frac{k_{gear} k_a}{s(L_a J s^2 + (r_a J + L_a B_m)s + r_a B_m + k_a^2)}$$

If the characteristic eigenvalues are real and distinct, we have

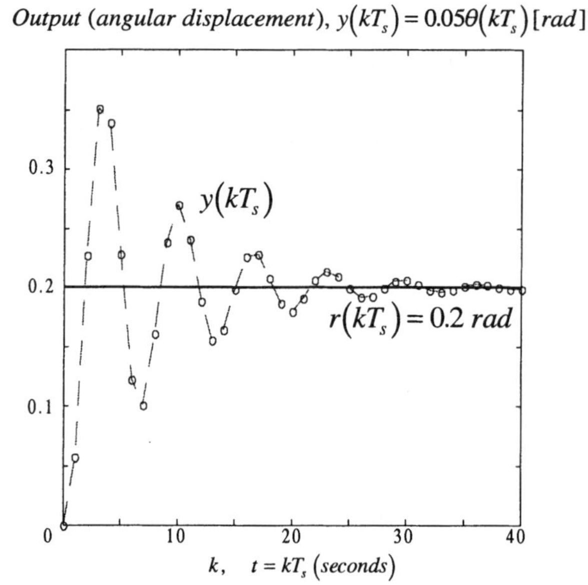

Figure 3.24 Output of the servo-system with digital PID controller; $T_s = 0.005$ s.

$$G_D(s) = G_H(s)G_{sys}(s) = \frac{1-e^{-T_s s}}{s}\frac{k}{s(T_1 s+1)(T_2 s+1)}$$

Let us find $G_D(z)$:

$$G_D(z) = Z[G_H(s)G_{sys}(s)] = \frac{z-1}{z}Z\left[\frac{G_{sys}(s)}{s}\right] = \frac{z-1}{z}Z\left[\frac{k}{s^2(T_1 s+1)(T_2 s+1)}\right]$$

The partial fraction expansion is found assuming that $T_1 \neq T_2$; that is, the eigenvalues $-\frac{1}{T_1}$ and $-\frac{1}{T_2}$ are distinct (nonrepeated). Using Heaviside's expansion formula, we have

$$\frac{k}{s^2(T_1 s+1)(T_2 s+1)} = \frac{c_1}{s} + \frac{c_2}{s^2} + \frac{c_3}{T_1 s+1} + \frac{c_4}{T_2 s+1}$$

where c_1, c_2, c_3, and c_4 are the unknown coefficients, which can be found using Heaviside's theorem.

We have to use the differentiation technique to find c_1 and

$$c_1 = \frac{d}{ds}\left(\frac{ks^2}{s^2(T_1 s+1)(T_2 s+1)}\right)\Bigg|_{s=0} = \frac{-k(2T_1T_2 s + T_1 + T_2)}{(T_1T_2 s^2 + (T_1+T_2)s+1)}\Bigg|_{s=0} = -k(T_1+T_2)$$

By letting $s = 0$, $s = -\frac{1}{T_1}$, and $s = -\frac{1}{T_2}$, we obtain the unknown coefficients c_2, c_3, and c_4. In particular,

$$c_2 = \left.\frac{k}{(T_1 s+1)(T_2 s+1)}\right|_{s=0} = k,\ c_3 = \left.\frac{k}{s^2(T_2 s+1)}\right|_{s=-\frac{1}{T_1}} = \frac{kT_1^3}{T_1 - T_2},$$

$$\text{and } c_4 = \left.\frac{k}{s^2(T_1 s+1)}\right|_{s=-\frac{1}{T_2}} = \frac{kT_2^3}{T_2 - T_1}$$

Hence, we have

$$G_D(z) = \frac{z-1}{z} Z\left[\frac{k}{s^2(T_1 s+1)(T_2 s+1)}\right] = k\frac{z-1}{z} Z\left[-\frac{T_1+T_2}{s} + \frac{1}{s^2} + \frac{T_1^3}{T_1 - T_2}\left(\frac{1}{T_1 s+1}\right) + \frac{T_2^3}{T_2 - T_1}\frac{1}{(T_2 s+1)}\right]$$

$$= k\frac{z-1}{z}\left(-(T_1+T_2)\frac{z}{z-1} + \frac{T_s z}{(z-1)^2} + \frac{T_1^2}{T_1 - T_2}\frac{1}{(1-e^{-T_s/T_1}z^{-1})} + \frac{T_2^2}{T_2 - T_1}\frac{1}{(1-e^{-T_s/T_2}z^{-1})}\right)$$

One obtains

$$G_D(z) = k\left(-T_1 - T_2 + \frac{T_s}{z-1}\frac{T_1^2}{T_1 - T_2}\frac{z-1}{z-e^{-T_s/T_1}} + \frac{T_2^2}{T_2 - T_1}\frac{z-1}{z-e^{-T_s/T_2}}\right)$$

and the following closed-loop transfer function results

$$G(z) = \frac{Y(z)}{R(z)} = \frac{G_{PID}(z)G_D(z)}{1 + G_{PID}(z)G_D(z)}$$

where $G_{PID}(z) = \dfrac{(k_{dp} + k_{di} + k_{dd})z^2 - (k_{dp} + 2k_{dd})z + k_{dd}}{z^2 - z}$.

Neglecting the armature inductance, the transfer function of the open-loop servo-system is simplified to

$$G_{sys}(s) = \frac{k_a k_{gear}}{s(r_a Js + r_a B_m + k_a^2)} = \frac{k}{s(Ts+1)}$$

For the zero-order data hold, we have

$$G_H(s) = \frac{1 - e^{-T_s s}}{s}$$

and

$$G_D(z) = Z[G_D(s)] = Z[G_H(s)G_{sys}(s)] = \frac{z-1}{z} Z\left(\frac{G_{sys}(s)}{s}\right) = k\frac{z-1}{z} Z\left(-\frac{T}{s} + \frac{1}{s^2} + \frac{1}{s + \frac{1}{T}}\right)$$

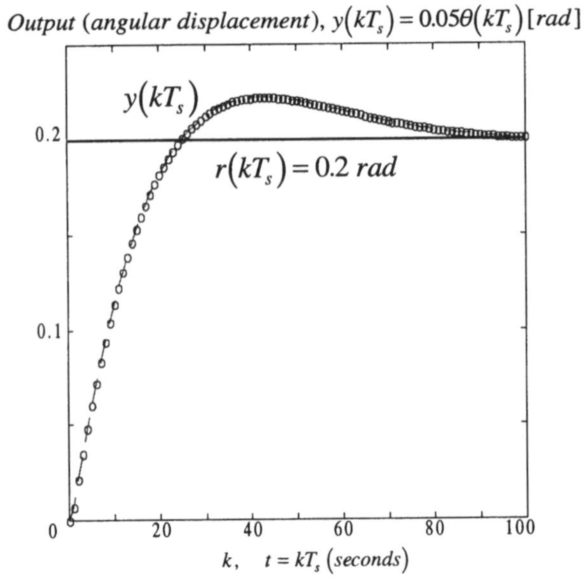

Figure 3.25 Output of the servo-system; T_s = 0.0005 s and L_a = 0.

Using the z-transform table, we have

$$G_D(z) = k\frac{z-1}{z}\left(-\frac{Tz}{z-1}+\frac{T_s z}{(z-1)^2}+\frac{Tz}{z-e^{-T_s/T}}\right)=\frac{kT_s\left(z-e^{-T_s/T}-\frac{T}{T_s}(z-1)(1-e^{-T_s/T})\right)}{(z-1)(z-e^{T_s/T})}$$

For T_s = 0.001 s, one obtains the following values of the proportional, integral, and derivative feedback gains:

$$k_{dp} = 180,\ k_{di} = 0.0067,\ \text{and}\ k_{dd} = 20000$$

The transfer function of the closed-loop system is

$$G(z) = \frac{Y(z)}{R(z)} = \frac{G_{PID}(z)G_D(z)}{1+G_{PID}(z)G_D(z)} = \frac{0.25z^{-1}-0.09z^{-2}-0.15z^{-3}+0.041z^{-4}}{1-2.4z^{-1}+1.8z^{-2}-0.59z^{-3}+0.041z^{-4}}$$

MATLAB is used to model the transient dynamics shown in Figure 3.25. The analysis of transient dynamics for the third- and second-order servo-system models indicates that the armature inductance can be neglected because the sampling period is small (see Figures 3.23 and 3.25). This result is not valid if one increases the sampling time; and even for T_s = 0.005 s, the dynamics are different. Using the `ch_3_04.m` file given, the reader is encouraged to model the electromechanical servomechanism with the different sampling periods assigned.

References

Krause, P. C. and Wasynczuk, O., *Electromechanical Motion Devices*, McGraw-Hill, New York, 1989.
Kuo, B. C., *Automatic Control Systems*, Prentice-Hall, Englewood Cliffs, NJ, 1995.
Ogata, K., *Discrete-Time Control Systems*, Prentice-Hall, Upper Saddle River, NJ, 1995.
Ogata, K., *Modern Control Engineering*, Prentice-Hall, Upper Saddle River, NJ, 1997.

chapter four

Introduction to microelectronic circuits, power electronic devices, and power converters

4.1 Operational amplifiers

Signal processing is accomplished by microelectronic circuits and devices. Transducers convert time-varying physical quantities (displacement and velocity, force and torque, pressure, and temperature) into electrical signals (voltage or current). The signal-level transducer output must be amplified, and operational amplifiers are used to perform amplification. The operational amplifier is a two-port network that has two input terminals (1 and 2) and one output terminal (3), and two DC voltage sources are needed (terminal 4 is connected to a positive voltage u_+, and a negative voltage u_- is supplied to the terminal 5) (see Figure 4.1). Operational amplifiers are monolithic integrated circuits fabricated using complementary metal oxide semiconductor (CMOS) and bipolar-CMOS fabrication processes, and the commonly used materials are silicon, germanium, and gallium arsenic.

The amplifier output is the difference between two input voltages applied to the inverting input terminal 1 $u_1(t)$ and the *noninverting* input terminal 2 $u_2(t)$, multiplied by the differential open-loop coefficient k_{og}. That is, the resulting output voltage is

$$u_0(t) = k_{og}(u_2(t) - u_1(t))$$

The differential open-loop coefficient is positive, and the value of k_{og} is very large, $k_{og} \in [1 \times 10^5 \quad 1 \times 10^7]$. General-purpose operational amplifiers have input and output resistances $R_{input} \in [1 \times 10^5 \quad 1 \times 10^{12}]$ and $R_{output} \in [10 \quad 1000]$ ohm, and the temperature range from –40°C to +105°C. Operational amplifier packages are different, and the commonly used eight-pin package is shown in Figure 4.2.

The *inverting* and *noninverting* input terminals 1 and 2 are distinguished using "–" and "+" signs. Supplying the signal-level input voltage $u_1(t)$ to the *inverting* input terminal using external resistor R_1, and grounding the *noninverting* input terminal, one can find the differential closed-loop coefficient k_{cg} if negative feedback is used (the output terminal is connected to the *inverting* input terminal, and the resistor R_2 is inserted) (see Figure 4.3).

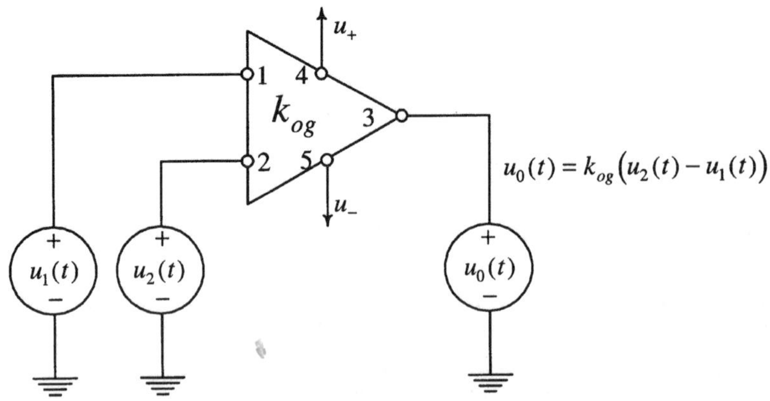

Figure 4.1 Differential-input, single-ended-output operational amplifier.

To find the differential closed-loop coefficient k_{cg}, one has to obtain the ratio between the output and input voltages $u_0(t)$ and $u_1(t)$. The voltage between two input terminals is $\frac{u_0(t)}{k_{og}}$, and the voltage at the *inverting* input terminal is $-\frac{u_0(t)}{k_{og}}$ because the *noninverting* input terminal is grounded. Therefore, we have

$$i_1(t) = \frac{u_1(t) + \frac{u_0(t)}{k_{og}}}{R_1}$$

and the output voltage is

$$u_0(t) = -\frac{u_0(t)}{k_{og}} - \frac{u_1(t) + \frac{u_0(t)}{k_{og}}}{R_1} R_2$$

Hence, the differential closed-loop coefficient is found to be

$$k_{cg} = \frac{u_0(t)}{u_1(t)} = -\frac{\frac{R_2}{R_1}}{1 + \frac{1}{k_{og}} + \frac{R_2}{k_{og} R_1}}$$

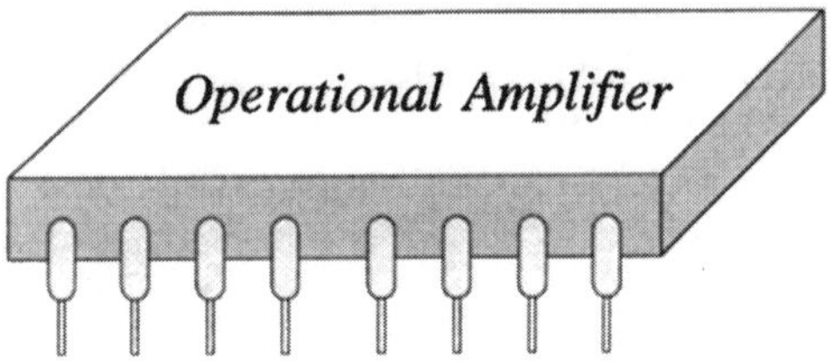

Figure 4.2 Eight-pin operational amplifier.

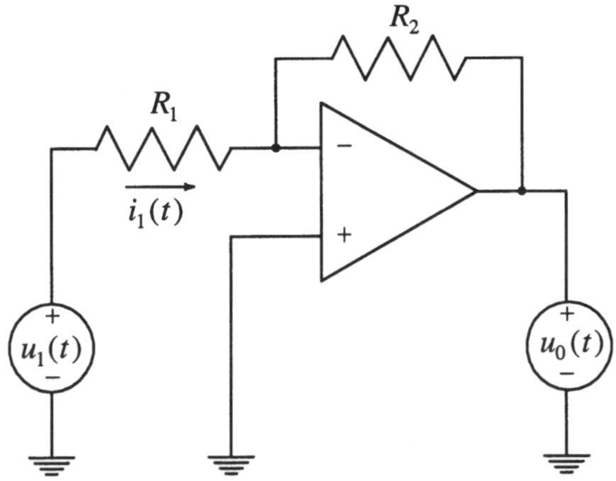

Figure 4.3 Inverting configuration of the operational amplifier.

The analysis performed shows that the configuration studied allows one to invert the signal-level input signal, and using the fact that the differential open-loop coefficient k_{og} is very large (hundred thousands), we have

$$k_{cg} = \frac{u_0(t)}{u_1(t)} \approx -\frac{R_2}{R_1}$$

The inverting summing amplifier (the so-called weighted summer) is shown in Figure 4.4.

We have m input signals; in particular, the applied voltages are $u_{1,1}(t)$, $u_{1,2}(t)$, ..., $u_{1,m-1}(t)$, and $u_{1,m}(t)$. The currents $i_{1,1}(t)$, $i_{1,2}(t)$, ..., $i_{1,m-1}(t)$, and $i_{1,m}(t)$ are found to be

$$i_{1,1}(t) = \frac{u_{1,1}(t)}{R_{1,1}},\ i_{1,2}(t) = \frac{u_{1,2}(t)}{R_{1,2}},\ i_{1,m-1}(t) = \frac{u_{1,m-1}(t)}{R_{1,m-1}},\ \text{and}\ i_{1,m}(t) = \frac{u_{1,m}(t)}{R_{1,m}}$$

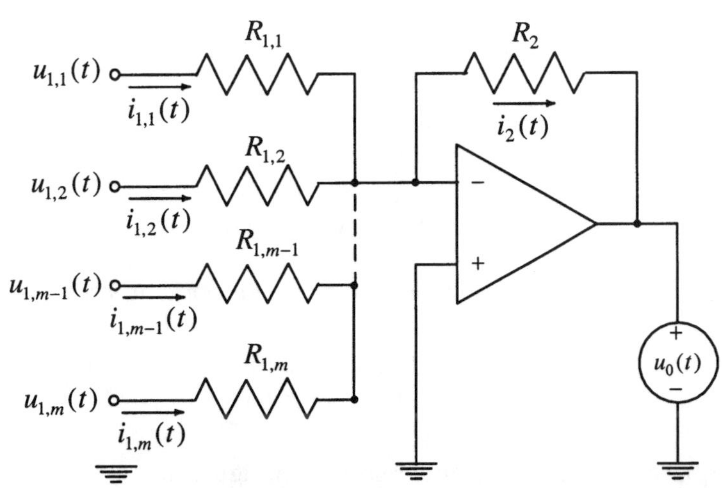

Figure 4.4 The weighted summing amplifier.

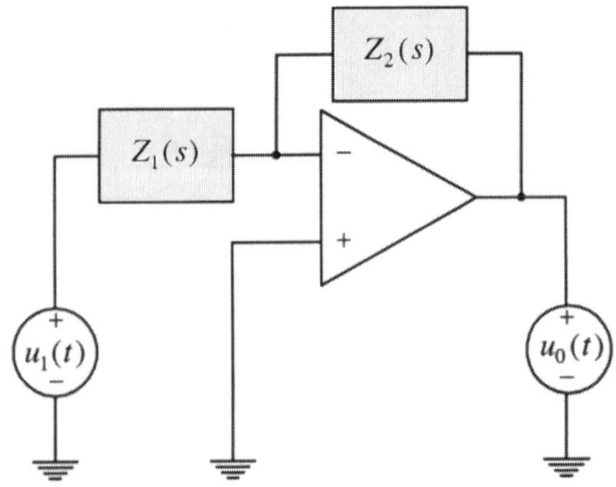

Figure 4.5 Inverting configuration of the operational amplifier with general impedances $Z_1(s)$ and $Z_2(s)$.

The current in the feedback path is given by

$$i_2(t) = i_{1,1}(t) + i_{1,2}(t) + \ldots + i_{1,m-1}(t) + i_{1,m}(t)$$

and hence, the amplifier output is

$$u_0(t) = -\left(\frac{R_2}{R_{1,1}} u_{1,1}(t) + \frac{R_2}{R_{1,2}} u_{1,2}(t) + \ldots + \frac{R_2}{R_{1,m-1}} u_{1,m-1}(t) + \frac{R_2}{R_{1,m}} u_{1,m}(t)\right)$$

To generalize the results derived by studying the inverting configuration, one can use the input impedance $Z_1(s)$ and the feedback path impedance $Z_2(s)$, as illustrated in Figure 4.5. The impedance is the ratio of the phasor voltage to the phasor current at the two terminals. In particular, the impedances of the resistor, capacitor, and inductance are

$$Z_R(s) = Z_R(j\omega) = R$$

$$Z_C(s) = \frac{1}{sC}, \; Z_C(j\omega) = \frac{1}{j\omega C}$$

and

$$Z_L(s) = sL, \; Z_L(j\omega) = j\omega L$$

It is evident that the transfer function of the closed-loop amplifier configuration is

$$G(s) = \frac{U_0(s)}{U_1(s)} = -\frac{Z_2(s)}{Z_1(s)}$$

As an example, consider the amplifier shown in Figure 4.6.

The transfer function is found using the following expressions for impedances

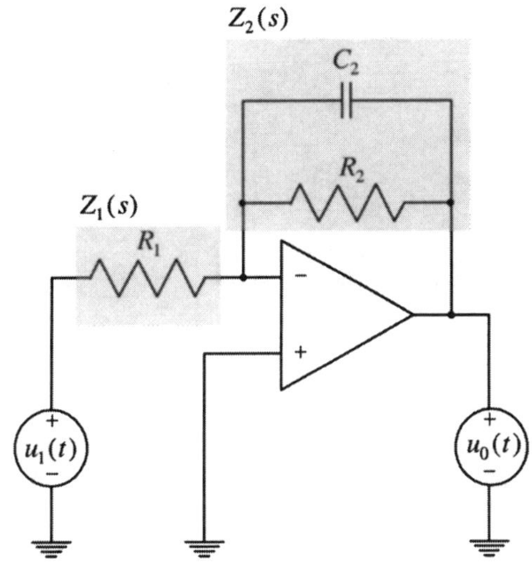

Figure 4.6 Inverting operational amplifier configuration.

$$Z_1(s) = R_1 \text{ and } Z_2(s) = \frac{R_2}{R_2C_2s+1}$$

Therefore,

$$G(s) = \frac{U_0(s)}{U_1(s)} = -\frac{Z_2(s)}{Z_1(s)} = -\frac{\frac{R_2}{R_1}}{R_2C_2s+1}$$

Hence, the closed-loop gain coefficient for the inverting operational amplifier studied is $k_{ca} = -\frac{R_2}{R_1}$, and the time constant is R_2C_2.

In the frequency domain, substituting $s = j\omega$, we have

$$G(j\omega) = \frac{U_0(j\omega)}{U_1(j\omega)} = -\frac{Z_2(j\omega)}{Z_1(j\omega)} = -\frac{\frac{R_2}{R_1}}{R_2C_2j\omega+1}$$

and the Bode plots of the resulting low-pass filter can be easily found and plotted.

Filters are used in electromechanical systems to ensure noise immunity. Sensors are connected to operational amplifiers, and due to electromagnetic interference (*emi*), sensor signals are contaminated by noise. This high-frequency noise is attenuated by filters. A filter configuration is demonstrated in Figure 4.7.

The transfer function is found using the input and feedback impedances. In particular,

$$G(s) = \frac{U_0(s)}{U_1(s)} = -\frac{Z_2(s)}{Z_1(s)} = -\frac{\frac{R_2}{R_2C_2s+1}}{\frac{R_1}{R_1C_1s+1}} = -\frac{\frac{R_2}{R_1}(R_1C_1s+1)}{R_2C_2s+1}$$

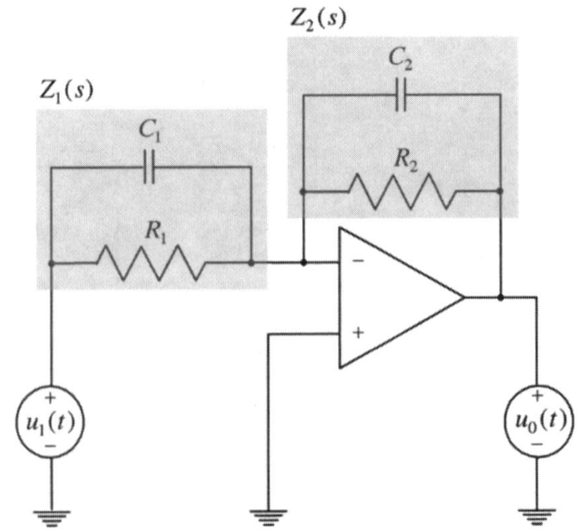

Figure 4.7 Inverting operational amplifier with $Z_1(s)=\dfrac{R_1}{R_1C_1s+1}$ and $Z_2(s)=\dfrac{R_2}{R_2C_2s+1}$.

A large number of transfer functions can be implemented using operational amplifiers and passive elements in the input and feedback paths. Operational amplifiers are widely used in electromechanical systems to implement PID-type control algorithms needed. An inverting integrator is obtained by placing a capacitor C_2 in the feedback path (see Figure 4.8). This circuit performs the integration, and the corresponding transfer function is

$$G(s)=\frac{U_0(s)}{U_1(s)}=-\frac{Z_2(s)}{Z_1(s)}=-\frac{1}{R_1C_2s}$$

Denoting the initial value of the capacitor voltage as $u_C(t_0)$, the amplifier output voltage is found as

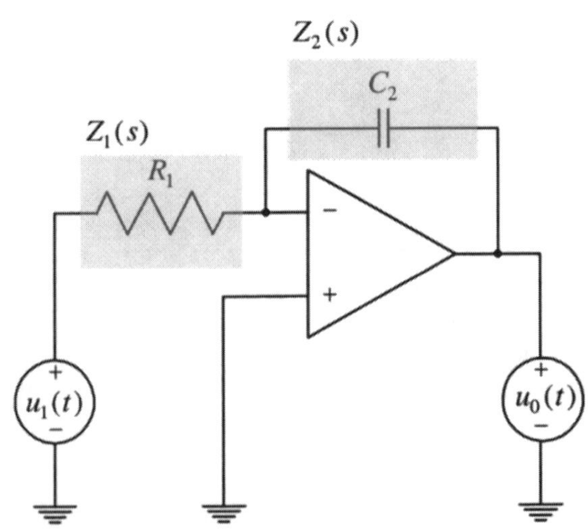

Figure 4.8 Inverting integrator.

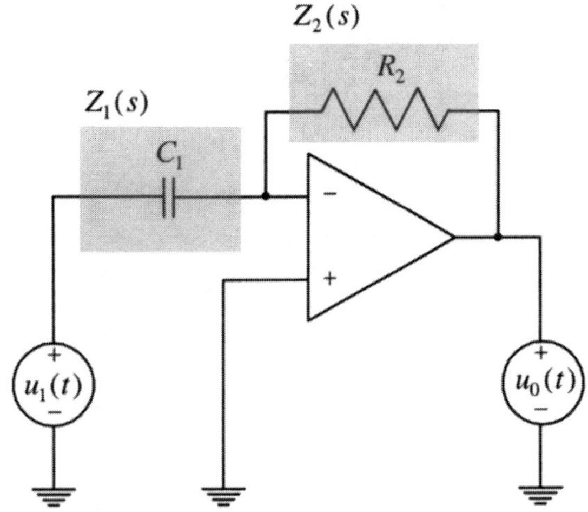

Figure 4.9 Inverting differentiator.

$$u_0(t) = -u_C(t_0) - \frac{1}{R_1C_2}\int_{t_0}^{t_f} u_1(\tau)d\tau$$

The operational differentiator performs the differentiation of the input signal. The current through the input capacitor is $C_1\frac{du_1(t)}{dt}$ (see Figure 4.9). That is, the output voltage is proportional to the derivative of the input voltage with respect to time, and $u_0(t) = -R_2C_1\frac{du_1(t)}{dt}$.

The transfer function is found as

$$G(s) = \frac{U_0(s)}{U_1(s)} = -\frac{Z_2(s)}{Z_1(s)} = -R_2C_1s$$

Table 4.1 gives the transfer functions of the inverting operational amplifier with different passive elements in the input and feedback paths.

The linear PID controller can be implemented using the configuration shown in Figure 4.10.

The transfer function is found as

$$G(s) = \frac{U_0(s)}{U_1(s)} = -\frac{(R_1C_1s+1)(R_2C_2s+1)}{R_1C_2s} = -\frac{R_2C_1s^2 + \frac{R_1C_1+R_2C_2}{R_1C_2}s + \frac{1}{R_1C_2}}{s}$$

The transfer function of the PID controller is

$$G_{PID}(s) = \frac{k_ds^2 + k_ps + k_i}{s}$$

Table 4.1 Transfer Functions of the Inverting Amplifier

Input circuit with impedance $Z_1(s)$	Feedback circuit with impedance $Z_2(s)$	Transfer function
$Z_1(s)$ 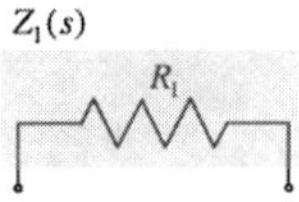 $Z_1(s)=R_1$	$Z_2(s)$ 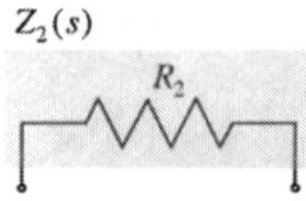 $Z_2(s)=R_2$	$G(s)=\frac{U_0(s)}{U_1(s)}=-\frac{R_2}{R_1}$
$Z_1(s)$ 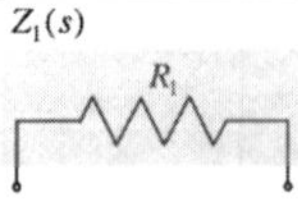 $Z_1(s)=R_1$	$Z_2(s)$ 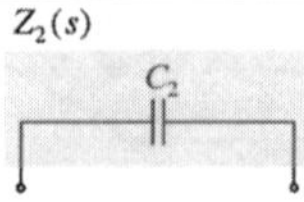 $Z_2(s)=\frac{1}{C_2s}$	$G(s)=\frac{U_0(s)}{U_1(s)}=-\frac{1}{R_1C_2s}$
$Z_1(s)$ 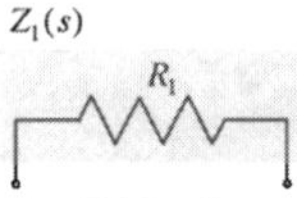 $Z_1(s)=R_1$	$Z_2(s)$ 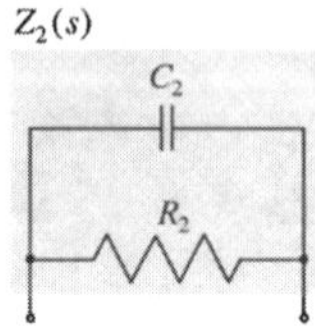 $Z_2(s)=\frac{R_2}{R_2C_2s+1}$	$G(s)=\frac{U_0(s)}{U_1(s)}=-\frac{\frac{R_2}{R_1}}{R_2C_2s+1}$
$Z_1(s)$ 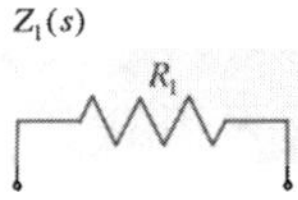 $Z_1(s)=R_1$	$Z_2(s)$ 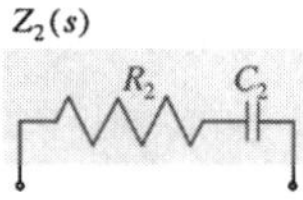 $Z_2(s)=\frac{R_2C_2s+1}{C_2s}$	$G(s)=\frac{U_0(s)}{U_1(s)}=-\frac{R_2C_2s+1}{R_1C_2s}$
$Z_1(s)$ 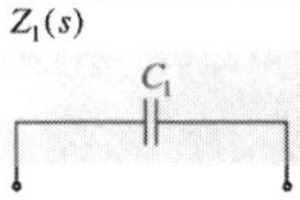 $Z_1(s)=\frac{1}{C_1s}$	$Z_2(s)$ 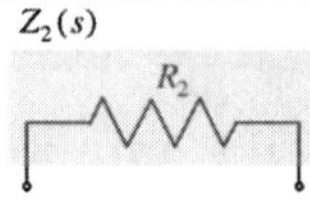 $Z_2(s)=R_2$	$G(s)=\frac{U_0(s)}{U_1(s)}=-R_2C_1s$
$Z_1(s)$ 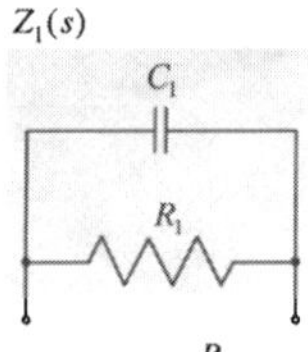 $Z_1(s)=\frac{R_1}{R_1C_1s+1}$	$Z_2(s)$ 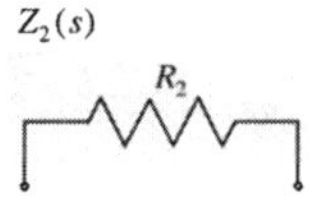 $Z_2(s)=R_2$	$G(s)=\frac{U_0(s)}{U_1(s)}=-\frac{R_1R_2C_1s+R_2}{R_1}$
$Z_1(s)$ 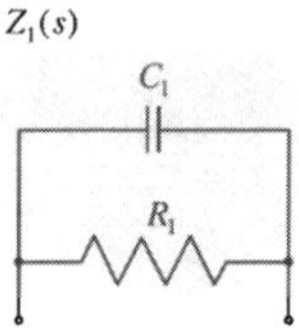 $Z_1(s)=\frac{R_1}{R_1C_1s+1}$	$Z_2(s)$ 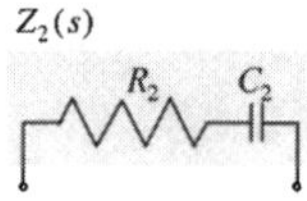 $Z_2(s)=\frac{R_2C_2s+1}{C_2s}$	$G(s)=\frac{U_0(s)}{U_1(s)}$ $=-\frac{(R_1C_1s+1)(R_2C_2s+1)}{R_1C_2s}$
$Z_1(s)$ 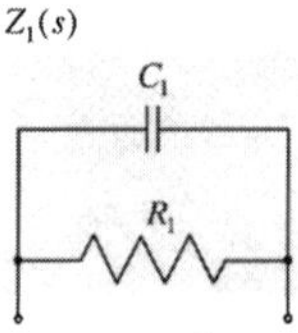 $Z_1(s)=\frac{R_1}{R_1C_1s+1}$	$Z_2(s)$ 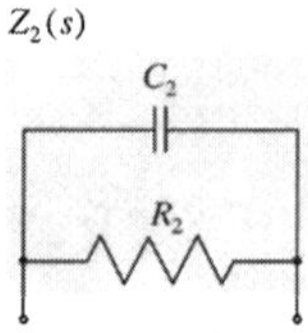 $Z_2(s)=\frac{R_2}{R_2C_2s+1}$	$G(s)=\frac{U_0(s)}{U_1(s)}=-\frac{\frac{R_2}{R_1}(R_1C_1s+1)}{R_2C_2s+1}$

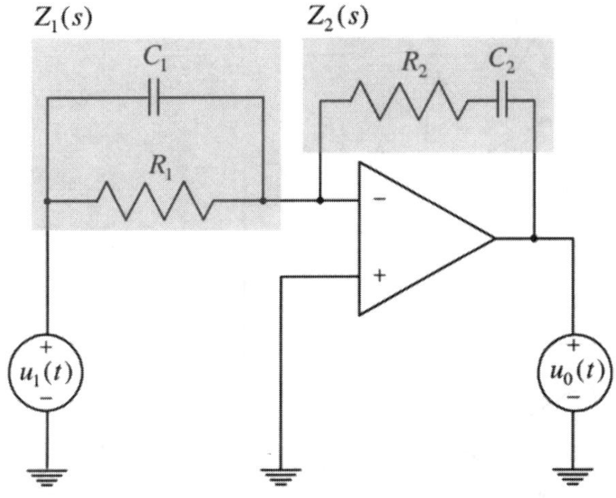

Figure 4.10 Implementation of the PID control law using an inverting operational amplifier.

Hence, the proportional, integral, and derivative feedback result, and k_p, k_i, and k_d gains are implemented by assigning the values for resistances and capacitances of the input and feedback paths. In particular, we have

$$k_p = -\frac{R_1C_1 + R_2C_2}{R_1C_2}$$

$$k_i = -\frac{1}{R_1C_2}$$

$$k_d = -R_2C_1$$

Due to the minus signs, the proportional, integral, and derivative feedback should be inverted. Hence, an additional inverting amplifier, is needed.

In practice, the configuration illustrated in Figure 4.11 is commonly used to implement the PID controller.

It is evident that the transfer function is expressed as

$$G(s) = \frac{U_0(s)}{U_1(s)} = \frac{R_{2p}}{R_{1p}} + \frac{1}{R_{1i}C_{2i}s} + R_{2d}C_{1d}s$$

That is, the proportional, integral, and derivative feedback coefficients are given as

$$k_p = \frac{R_{2p}}{R_{1p}}, \quad k_i = \frac{1}{R_{1i}C_{2i}}, \quad \text{and} \quad k_d = R_{2d}C_{1d}$$

The data, schematic diagram, operating characteristics, frequency response, and transient dynamics for low-power, single-supply, dual and quad operational amplifiers MC33171, MC33172, and MC33174, manufactured by Motorola, are documented in Appendix 4.1. The possible amplifier configurations (noninverting, inverting, notch, bandpass, and other filters) should be examined and studied.

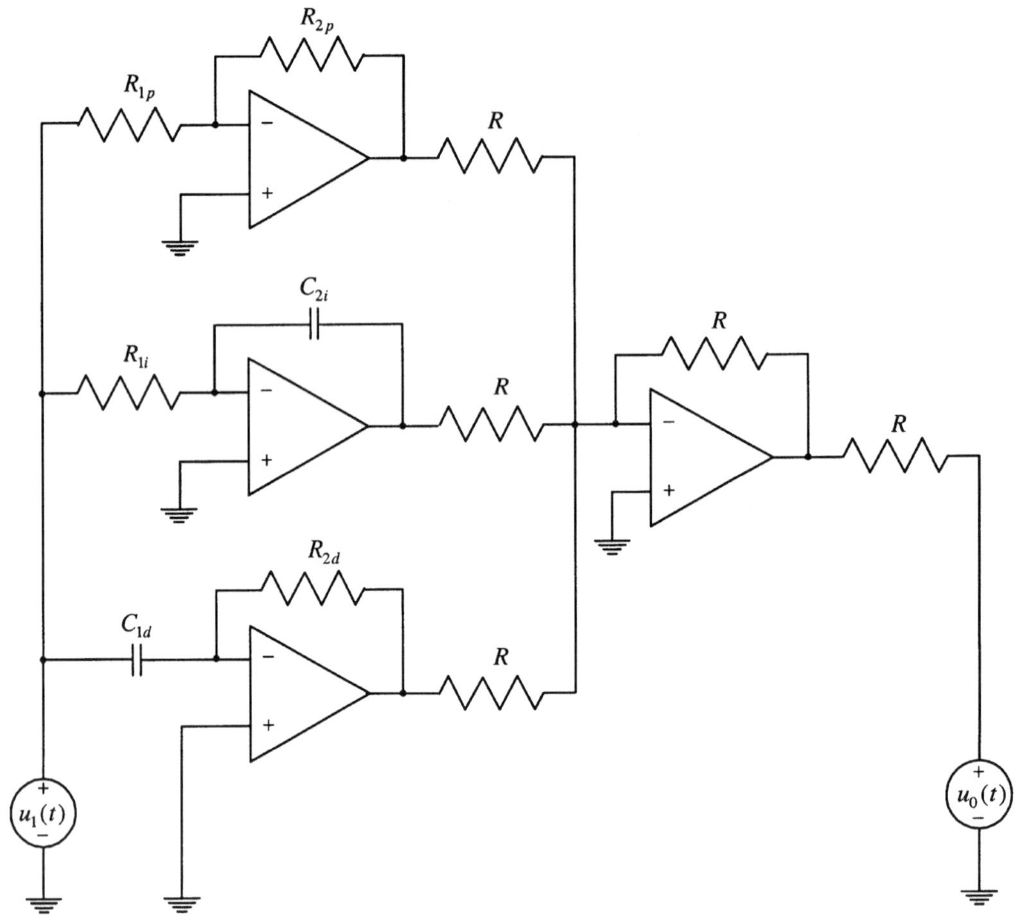

Figure 4.11 Implementation of the analog PID control law.

High slew rate, wide bandwidth, JFET input operational amplifiers MC34080 to MC34085 are illustrated in Appendix 4.2. The basic characteristics are documented, and the steady-state, frequency, and dynamic characteristics in the operating envelope are given.

The dual operational amplifier and comparator MC3405, manufactured by Motorola, are presented in Appendix 4.3. The basic data for amplifier and comparator operations is given. In addition, the transient dynamics, frequency responses, and operating characteristics are documented, and the application of the MC3405 different operational amplifier and comparator is illustrated to attain pulse-width modulation.

4.2 Circuit elements

4.2.1 Diodes

The basic circuit elements (diodes, transistors, and thyristors) should be considered when studying power converters used in electromechanical and micro-electromechanical systems. The simplest fundamental circuit element is the diode, which is a two-terminal device with positive and negative terminals, called anode and cathode, respectively. If a negative voltage is applied to the diode, negligibly small current flows, and the diode

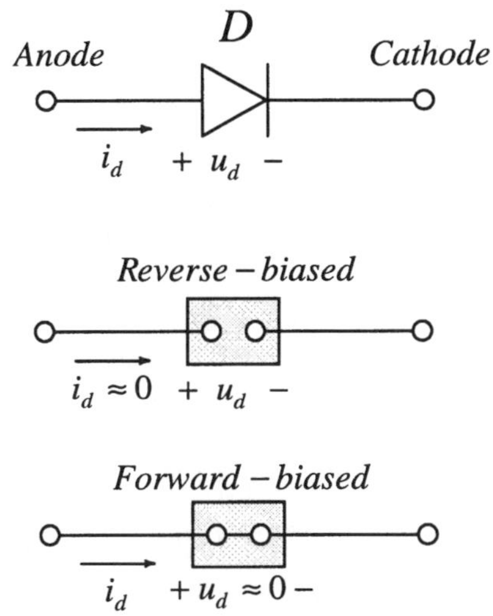

Figure 4.12 Diode circuit and the equivalent circuits for ideal reverse- and forward-biased diodes.

behaves as an open circuit (the diode is reverse-biased). If a positive current is applied to the diode, a negligibly small voltage drop appears across the diode; that is, the forward-biased diode behaves as a conducting circuit. Figure 4.12 illustrates the diode, as well as the equivalent circuits for the ideal diode.

The scaled $i_d - u_d$ piecewise-continuous characteristic of a silicon junction diode is plotted in Figure 4.13, where the forward-biased, reverse-biased, and breakdown regions are denoted as *FB*, *RB*, and *BR*, respectively.

In the forward-biased region (FB), the diode voltage is positive, and the slope is denoted as s_{fb}. It should be emphasized that in the so-called small-signal model of diode,

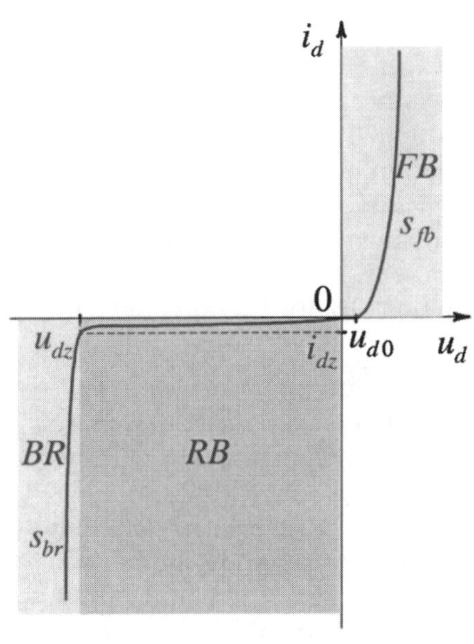

Figure 4.13 The $i_d - u_d$ diode characteristic.

the slope is found as $\frac{1}{r_d}$, and it is assumed that $\frac{1}{r_d} = \frac{\partial i_d}{\partial u_d}$ is constant. However, in nonlinear analysis, the $i_d - u_d$ diode characteristic is approximated as

$$i_d = I_S\left(e^{\frac{u_d}{n_d V_t}} - 1\right), \quad u_d = n_d V_t \ln\frac{i_d + I_s}{I_s}$$

where I_S is the saturation current; n_d is a constant that depends on the material used and diode structure, and for the standard integrated fabrication $n_d = 1$; V_t is the thermal voltage, $V_t = \frac{k_B T_K}{q_e}$, and if $T_K = 293°K$, we have $V_t = 25.8 \times 10^{-3}$ V; k_B is the Boltzmann constant, $k_B = 8.62 \times 10^{-5}$ eV-K^{-1} or $k_B = 1.38 \times 10^{-23}$ J-K^{-1}; T_K is the absolute temperature in Kelvin; q_e is the magnitude of the electron charge, $q_e = 1.6 \times 10^{-19}$ C.

The saturation current is found using the following formula

$$I_S = A_j q_e n_i^2\left(\frac{D_h}{k_s N_D} + \frac{D_e}{l_d N_A}\right)$$

where A_j is the cross-sectional area of the junction; n_i is the concentration of free electrons or holes in intrinsic silicon at a given temperature, $n_i^2 = k_m T_K^3 e_e^{-\frac{E_{bg}}{k_B T_K}}$, $n_i = 1.5 \times 10^{16}$ carriers/m^3 if $T_K = 293°K$; k_m is the material-dependent constant, and for silicon $k_m = 5.4 \times 10^{31}$; E_{bg} is the bandgap energy (the energy required to break a covalent bond and hence generate an electron-hole pair), and for silicon E_{bg} electron volts (eV); D_h is the diffusion constant (the diffusivity of holes); D_e is the diffusivity of electrons; N_D and N_A are the concentrations of the donor and acceptor atoms; k_s is the steepness constant that affects the exponential decay; and l_d is the diffusion length of electrons in the p region.

The saturation current I_S is proportional to n_i^2, which is a nonlinear function of the temperature. One concludes that I_S increases as the temperature increases (for the temperature T ≈ 20°C, the saturation current doubles for every 5°C rise in temperature). It should be emphasized that at the constant current i_d, the voltage drop across the diode decreases as the temperature increases. This temperature dependence is different for various manufactured diodes.

In the reverse-biased region (RB), the diode voltage is negative. If u_d is larger than V_t, the term $e^{\frac{u_d}{n_d V_t}}$ is negligibly small, and one concludes that $i_d \approx I_S$. This is the reason why I_S is called the saturation current. However, the diode exhibits small reverse current due to leakage. For the temperature T ≈ 20°C, this reverse current doubles for every 10°C rise in temperature.

In the breakdown region (BR), the reverse diode voltage is greater than the threshold value, denoted as u_{dz}. The subscript z stands for the zener diode. The reverse current increases rapidly with a small increase of the voltage drop, and in the breakdown region, the $i_d - u_d$ characteristic is almost a vertical line.

The diode transient dynamics can be studied and differential equations can be found. In particular, the depletion-layer C_{dl} and the diffusion C_{dd} capacitances are applied to

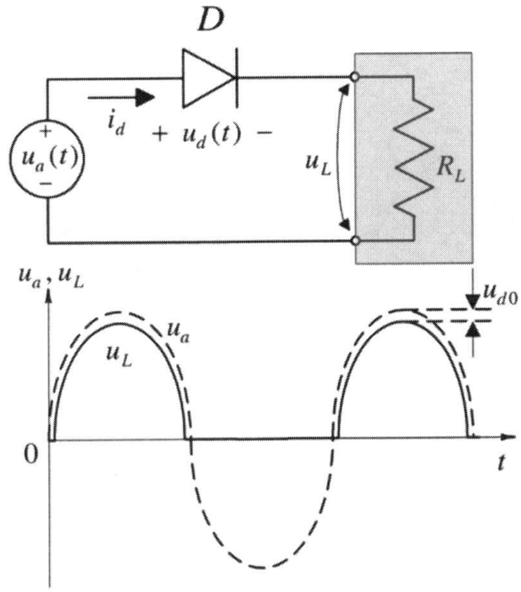

Figure 4.14 Half-wave rectifier circuit, and the waveforms of the applied and output voltages.

develop the mathematical model. However, this problem is beyond the topic of the book because the diode transient dynamics are very fast compared with actuator transient behavior and transient responses of converter and filter topologies needed to implement power converters, filters, etc.

Diodes are widely used in power electronic devices. The half-wave rectifier is illustrated in Figure 4.14. The sinusoidal voltage $u_a(t)$ is applied to the circuit, which consists of the series connection of the diode D and the resistor R_L. During the positive half-cycles of the applied voltage, the current flows through the diode in the forward direction (the diode is forward-biased), and the voltage drop is very small. The voltage $u_L(t)$ is slightly delayed, and the voltage drop is shown. It should be emphasized that the value of u_{d0} ranges from 0.5 to 0.9 V for general-purpose diodes. During the negative half-cycles of $u_a(t)$, the diode is not conducting; hence, the voltage at the load terminal $u_L(t)$ is zero. It is evident that the average value of the applied voltage $u_a(t)$ is zero, while the dc component of the $u_L(t)$ can be easily derived. One concludes that the circuit rectifies the applied voltage.

The full-wave rectifier, which is illustrated in Figure 4.15, provides a unipolar output voltage and inverts the negative half-cycles of the applied voltage $u_a(t)$ to the positive half-cycles of the output voltage $u_L(t)$. The transformer secondary winding is center-tapped to provide two equal applied voltages $u_a(t)$ across two halves of the secondary windings. During the positive half-cycles of $u_a(t)$, the diode $D1$ is forward-biased and diode $D2$ is reverse-biased. The current flows through $D1$ and R_L, and back to the center tap. The circuit behaves as a half-wave rectifier. During the negative half-cycles of $u_a(t)$, diode $D1$ is not conducting, while $D2$ is forward-biased, and hence $D2$ is conducting. In particular, the current flows through $D2$ and R_L, and back to the center tap. The waveforms of the applied and output voltages are documented in Figure 4.15.

In power supplies, the pulsations of the output voltage must be reduced, and the filter capacitor C_F is placed across the load. Figure 4.16 illustrates the full-wave rectifier with the filter capacitor that charges to the maximum value of the voltage $u_L(t)$, and discharges through the load R_L as the capacitor voltage is greater than $u_L(t)$. The capacitor discharges until the time when the applied voltage becomes greater than the capacitor voltage.

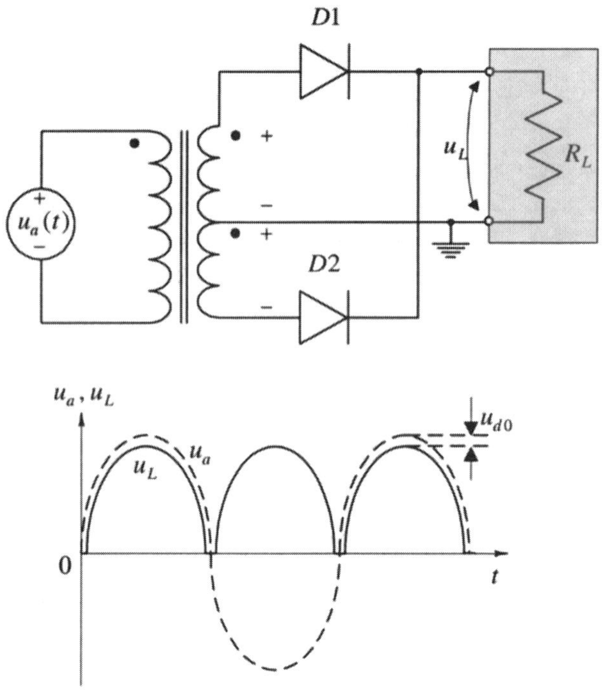

Figure 4.15 Full-wave rectifier circuit, and the waveforms of the applied and output voltages.

4.2.2 Transistors

Transistors are three- or four-terminal semiconductor devices. In power electronic circuits, bipolar junction (BJT), metal-oxide semiconductor field-effect (MOSFET), and insulated-gate bipolar (IGBT) transistors are used. Figure 4.17 illustrates the circuit symbols, voltage polarities, current directions, and cross sections of the *npn* and *pnp* BJTs with three terminals called the base (B), collector (C), and emitter (E).

The voltage between two terminals is used to control the current through the third terminal. An *npn* BJT whose emitter–base junction is forward-biased operates in the active mode if the collector potential is higher than the base potential. In the active mode, the collector current is found using the saturation current as

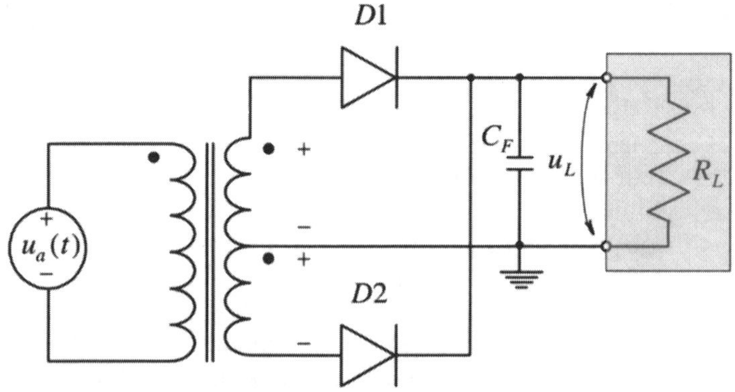

Figure 4.16 Full-wave rectifier circuit with the filter capacitor.

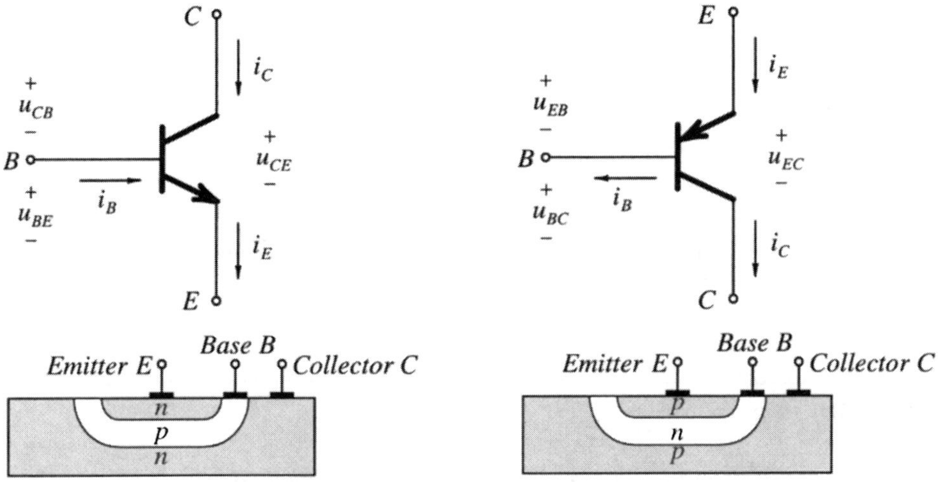

Figure 4.17 Circuit symbols, voltage polarities, and current flows in the BJTs biased in the active mode. Cross sections of the *npn* and *pnp* BJTs.

$$i_C = I_S e^{\frac{u_{BE}}{V_t}}, \quad I_S = A_{BE} q_e n_i^2 \frac{D_{Be}}{l_B N_A}$$

where A_{BE} is the cross-sectional area of the base-emitter junction; D_{Be} is the electron diffusivity in the base; and l_B is the effective width of the base.

The base and emitter currents are

$$i_B = \frac{i_C}{\beta_T} \text{ and } i_E = \frac{i_C}{\alpha_T}$$

where β_T and α_T are the common-base and common-emitter current constants, which are temperature dependent: $\alpha_T = \frac{\beta_T}{\beta_T + 1}$ and $\beta_T \in [100 \quad 1000]$.

It should be mentioned that β_T is the ratio of the collector current to the base current when BJTs operate in the active mode, and β_T is a nonlinear function of the collector current and temperature.

In contrast, the *pnp* BJT operates in the active mode if the collector potential is lower than the base potential. We have the following expressions for currents

$$i_C = I_S e^{\frac{u_{BE}}{V_t}}, \quad i_B = \frac{i_C}{\beta_T}, \text{ and } i_E = \frac{i_C}{\alpha_T}$$

For an *npn* BJT, the $i_C - u_{BE}$ and $i_C - u_{CE}$ characteristics are plotted in Figure 4.18. In particular, the $i_C - u_{BE}$ and $i_C - u_{CE}$ characteristics can be found: $i_C = I_S e^{\frac{u_{BE}}{V_t}}$. The $i_C - u_{CE}$ characteristics have great importance; and using the so-called Early voltage u_{CEe}, which is usually in the range of –50 to –100 V, the following equation results

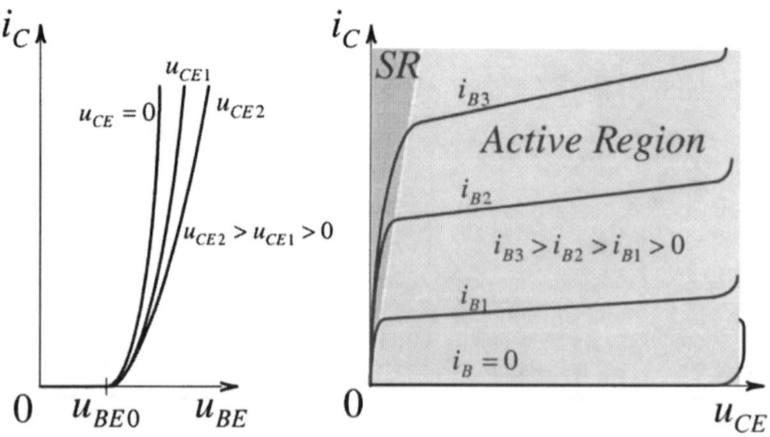

Figure 4.18 Static $i_C - u_{BE}$ and $i_C - u_{CE}$ characteristic curves for *npn* BJTs.

$$i_C = I_S e^{\frac{u_{BE}}{V_t}}\left(1+\frac{u_{CE}}{u_{CEe}}\right)$$

The families of the common-emitter $i_C - u_{CE}$ characteristics at the different constant values of the base current i_B are plotted in Figure 4.18 and 4.19, and the saturation (SR) and active regions are documented.

In electromechanical systems (electric drives and servos), *npn* silicon transistors are widely used, and the data and operating characteristics of the MJ16010, MJW16010, MJ16012, and MJW16012 transistors, manufactured by Motorola, are given in Appendix 4.4.

In BJT, the current-control mechanism is based on the voltage between two terminals. In contrast, the current-control mechanism of the field-effect transistors is based on an electrostatic field established by the voltage applied to the control terminal. The majority of very-large-scale integrated (VLSI) circuits are fabricated using metal-oxide semiconduc

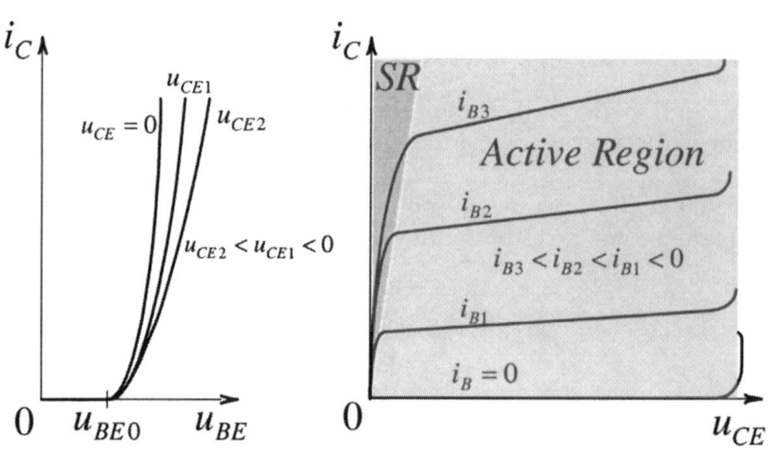

Figure 4.19 Static $i_C - u_{BE}$ and $i_C - u_{CE}$ characteristic curves for *pnp* BJTs.

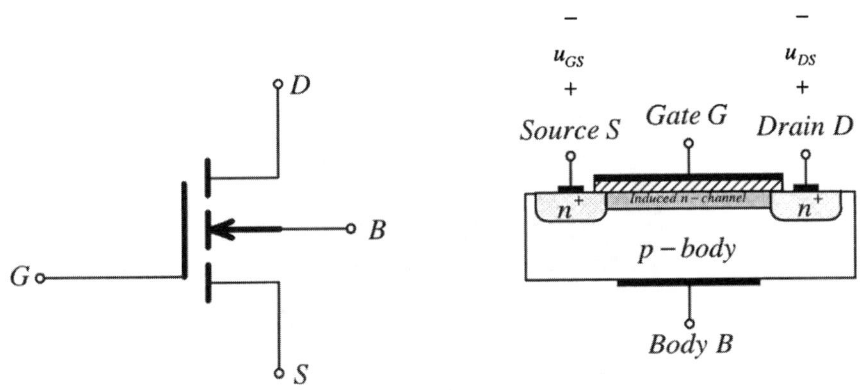

Figure 4.20 Circuit symbol and physical structure of the n-channel enhancement-type MOSFET.

tor technology. Figure 4.20 shows an n-channel enhancement-type MOSFET with four terminals, and a p-channel MOSFET is shown in Figure 4.21. In Figures 4.20 and 4.21, the gate, source, drain, and base (substrate) terminals are denoted as G, S, D, and B, respectively.

Consider the n-channel enhancement-type MOSFET. A positive voltage u_{GS}, applied to the gate, causes free positively charged holes in the n-channel region. These holes are pushed downward into the p-base (substrate). The applied voltage u_{GS} attracts electrons from the source and drain regions into the channel region. The voltage is applied between the drain and source, and the current flows through the induced n-channel region. The gate and body form a parallel-plate capacitor with the oxide layer. The positive gate voltage u_{GS} causes the positive charge at the top plate, and the negative charge at the bottom is formed by the electrons in the induced n-channel region. Hence, the electric field is formed in the vertical direction. As a positive voltage u_{DS} is applied, the drain current i_D flows in the induced n-channel region from source to drain, and the magnitude of i_D is proportional to the effective voltage, which is given as $u_{GS} - u_{GSt}$. If u_{GS} is greater than the threshold value u_{GSt}, $u_{GS} > u_{GSt}$, the induced n-channel is enhanced. If one increases u_{DS}, the n-channel resistance is increased, the drain current i_D saturates, and the saturation region occurs as one increases u_{DS} to the u_{DSsat} value. The $i_D - u_{DS}$ characteristics at constant values of u_{GS}, as well as three distinct regions (triode, saturation, and cut-off regions) are illustrated in Figure 4.22. A sufficient number of mobile electrons accumulate in the channel region to form a conducting channel if u_{GS} is greater than the threshold voltage u_{GSt}, which usually

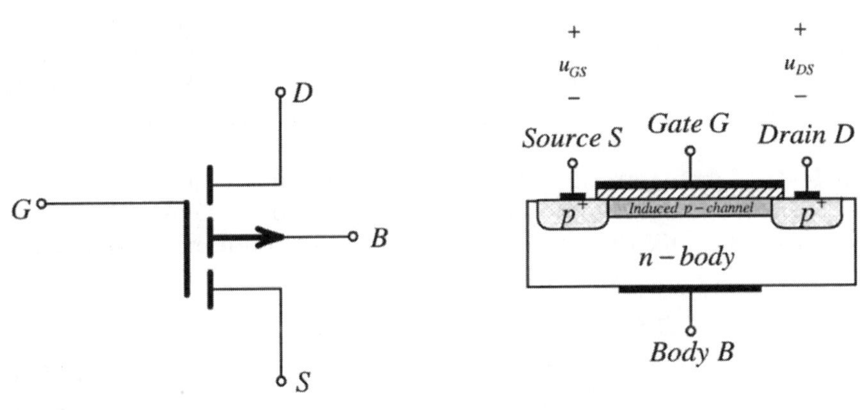

Figure 4.21 Circuit symbol and physical structure of the p-channel enhancement-type MOSFET.

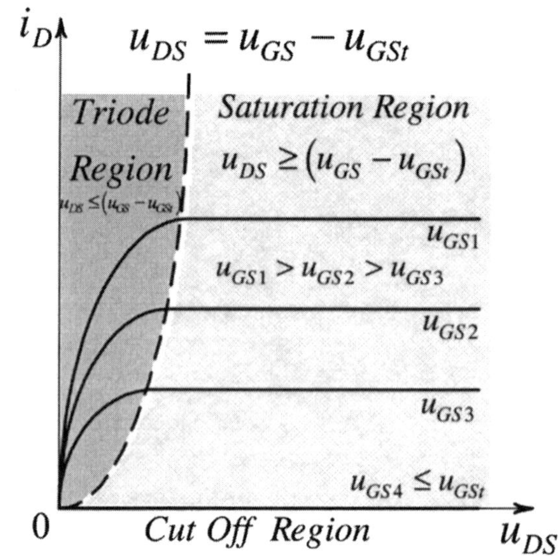

Figure 4.22 Static $i_B - u_{DS}$ characteristic curves for an *n*-channel enhancement-type MOSFET with constant values of u_{GS}.

ranges from 1 to 3 V. It should be emphasized that in the saturation region, the MOSFET operates as an amplifier, while in the triode region and in the cut-off region, the MOSFET can be used as a switch.

It should be mentioned that *p*-channel enhancement-type MOSFETs are fabricated on an *n*-type substrate (body) with p^+ regions for the drain and source (see Figure 4.21). The voltages applied u_{GS} and u_{DS}, and the threshold voltage u_{GSt}, are negative.

High-frequency MOSFET drivers are needed, and the basic configurations and topologies are designed and available. As an example, consider the MC33152 and MC34152 Motorola dual *noninverting* high-speed drivers developed to interface low-current digital circuitry with MOSFETs (see the representative diagram in Appendix 4.5). Using Schottky clamped bipolar analog technology, the operating envelope is enhanced and performance capabilities are extended. To increase robustness to noise and transients, the logic inputs A and B have 170 mV of hysteresis with the input threshold 1.67 V. Output stages A and B have 100 kohm pull-down resistors to keep the MOSFET gate low when the voltage V_{CC} is less than 1.4 V. Load and parasitic inductance, which are in series with the load, cause the driver outputs to be greater or lower than V_{CC} during the turn-on and turn-off transients. The MC33152 and MC34152 are immune to these destructive features. The output stages A and B have internal diodes, and the negative output transients are clamped by the *npn* pull-up transistors. Appendix 4.5 documents the data for MC33152 and MC34152, as well as operating characteristics and transient behavior. Different configurations of the MOSFET driver and application information are given.

A great deal of data for analog and digital integrated circuits and devices can be found in manufacturer catalogs; Appendix 4.6 provides the Motorola catalog data for the single IGBT gate driver MC 33153, which is widely used in electromechanical systems.

4.3 *Power amplifiers and power converters*

Power amplifiers allow one to perform amplification, and various classes of amplifiers are available. The most important part of power amplifiers is the output stage, which deals with medium and large voltages and currents; the power dissipated in the output stage power transistors should be minimized to guarantee power conversion efficiency. Depending on output stage topologies and operating concepts, output stages are classified as A, B, AB, C, and D classes of power amplifiers. In electromechanical systems, D-class output stages are commonly used due to high efficiency, simplicity, reliability, low harmonic distortion, etc. A possible circuit configuration with a standard push-pull D-class output stage to feed the permanent-magnet DC motor regulated by the PID controller is illustrated in Figure 4.23.

If the signal-level voltage u_{PID} of the output inverting operational amplifier of the PID controller is positive, transistor *S*1 is *off* because the voltage between the base and emitter is zero. Transistor *S*2 is *on*, and the negative voltage is applied to the motor armature winding. If the voltage u_{PID} is negative, *S*1 is *on* and *S*2 is *off*, and the voltage u_a supplied is positive. The diodes *D*1 and *D*2 prevent damage to the transistors by the *back emf* E_a from the motor. It should be emphasized that this simplest D-class power stage can be modified to ensure the pulse-width-modulation concept to control the angular velocity. The tachogenerator is used as a sensor to measure the rotor angular velocity to be fed to the input operational amplifier that compares the desired angular velocity with the actual angular velocity of the motor.

4.3.1 *Switching converters*

The basic dc-dc power converters used in electromechanical systems are switching converters. In particular, buck, boost, buck-boost, and Cuk converters are commonly applied. These converters should be considered and studied.

Buck converters

Using a pulse-width-modulation (PWM) switching concept, the voltage at the load terminal is regulated, and a high-frequency *buck* (*step-down*) switching converter is shown in Figure 4.24.

The operation of dc-dc converters, elements selection, as well as the analysis of the operating characteristics and transient dynamics are documented in Appendix 4.7. The *step-down* switching converter is documented in Figure 4.24, with the switch *S* open and closed. The switching frequency is expressed as $\frac{1}{t_{on}+t_{off}}$. Assuming that the switch is lossless, the voltage u_{dN} is equal to the supplied voltage V_d when the switch is closed, and the output voltage is zero if the switch is open (see Figure 4.25).

One concludes that the voltage u_{dN}, as well as the voltage applied to the load u_a, can be regulated by controlling the switching *on* and *off* durations (t_{on} and t_{off}), and the average voltage, applied to the load, depends on t_{on} and t_{off}. In particular, in the steady-state

$$u_{dN} = \frac{t_{on}}{t_{on}+t_{off}} V_d = d_D V_d$$

where d_D is the duty ratio, which is a function of the switching frequency and the time during which the switch is *on*, $d_D = \frac{t_{on}}{t_{on}+t_{off}}$; that is,

$$d_D \in [0 \quad 1], \text{ and } d_D = 0 \text{ if } t_{on} = 0;\ d_D = 1 \text{ if } t_{off} = 0$$

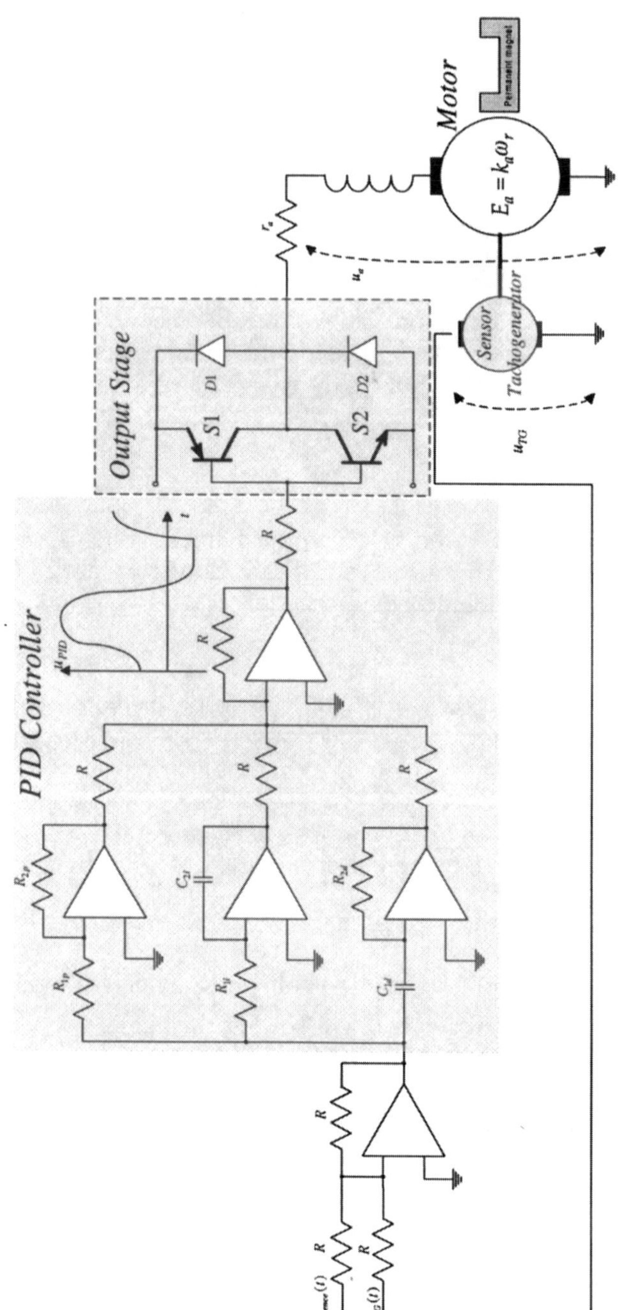

Figure 4.23 Application of the D-class power amplifier to drive the permanent-magnet DC motor: closed-loop configuration with PID controller and sensor.

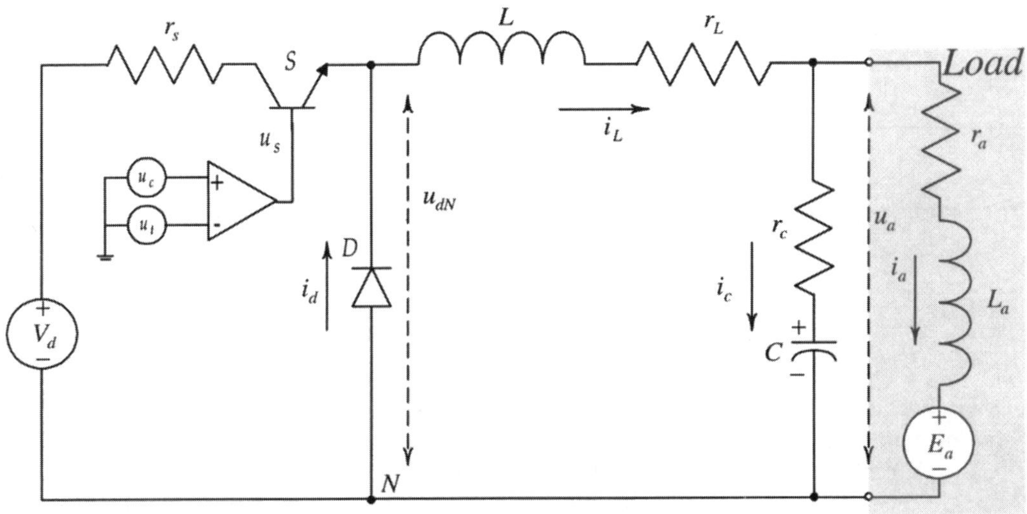

Figure 4.24 High-frequency step-down switching converter.

By changing the duty ratio d_D of the switch, which is bounded by [0 1], the average voltage supplied to the load u_a is regulated. The obvious way of establishing PWM switching is to use a control-triangle concept. The switching signal u_s, which drives the switch, is generated by comparing a signal-level control voltage u_c with a repetitive triangular signal u_t; a comparator is shown in Figure 4.26. The duration of the output pulses u_s represents the *weighted* value between the triangular voltage u_t with the assigned switching frequency and the control signal u_c. The output voltage of the comparator u_s drives the switch, and *on* and *off* durations are obtained by comparing u_c and u_t. Figure 4.27 illustrates the voltage waveforms.

High-speed PWM controllers MC33025 and MC34025, basic characteristics and configurations, as well as transient responses are documented in Appendix 4.8. Appendix 4.9 provides information regarding the PWM control circuits MC33063A and MC34063A used in dc-dc power converters.

A low-pass first-order filter with inductance L and capacitance C is used to ensure the specified voltage ripple, and

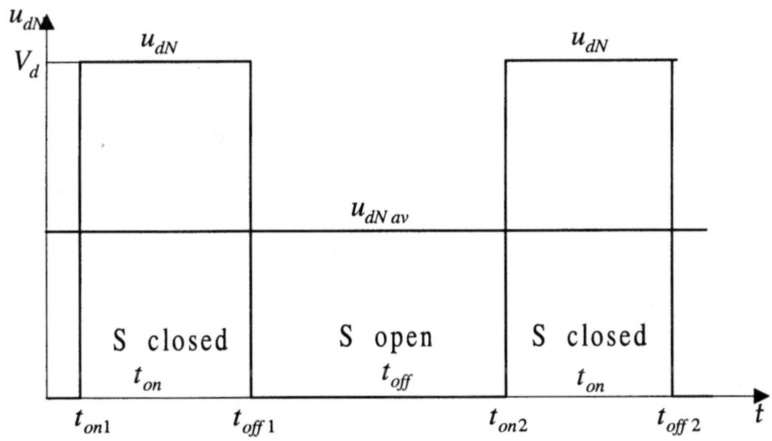

Figure 4.25 Voltage waveforms.

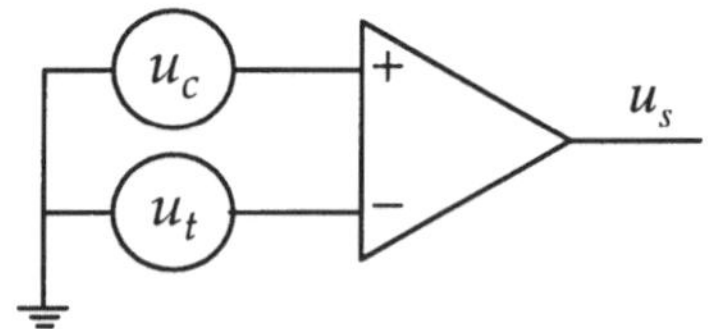

Figure 4.26 The role of a comparator in the pulse-width-modulation concept.

$$\frac{\Delta u_a}{u_a} = \frac{1-d_D}{8LCf^2}, \quad L_{\min} = \frac{(1-d_D)r_a}{2f}$$

Switches, inductors, and capacitors have resistances, denoted as r_s, r_L, and r_c, respectively. Equivalent circuits when the switch is closed and open are illustrated in Figure 4.28.

Using Kirchhoff's law, let us find the differential equations to map the dc-dc converter dynamics. If the switch is closed, the diode is reverse-biased, and the equivalent circuit is shown in Figure 4.28a. We have

$$\frac{du_C}{dt} = \frac{1}{C}(i_L - i_a)$$

$$\frac{di_L}{dt} = \frac{1}{L}(-u_C - (r_L + r_c)i_L + r_c i_a - r_c i_a - r_s i_L + V_d)$$

$$\frac{di_a}{dt} = \frac{1}{L_a}(u_C + r_c i_L - (r_a + r_c)i_a - E_a)$$

Note: One must distinguish between the state variable u_C and comparator signal-level input u_c, which is the control input.

If the switch is open, the diode becomes forward-biased, and carries the inductor current i_L (see Figure 4.28b). One obtains

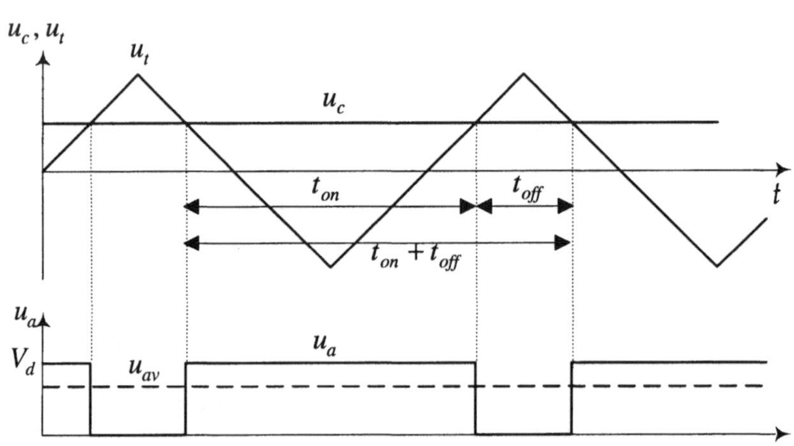

Figure 4.27 Voltage waveforms.

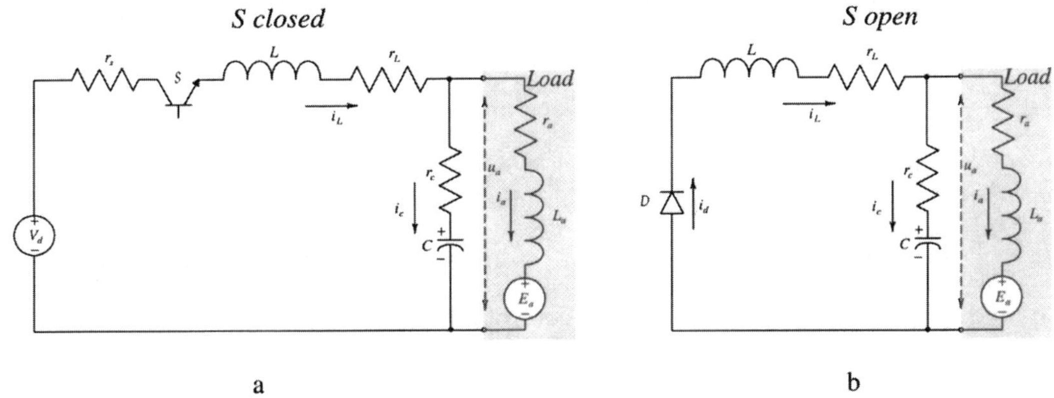

Figure 4.28 Circuits of the *buck* dc-dc converter when the switch is closed (a) and open (b).

$$\frac{du_C}{dt} = \frac{1}{C}(i_L - i_a)$$

$$\frac{di_L}{dt} = \frac{1}{L}(-u_C - (r_L + r_c)i_L + r_c i_a)$$

$$\frac{di_a}{dt} = \frac{1}{L_a}(u_C + r_c i_L - (r_a + r_c)i_a - E_a)$$

When the switch is closed, the duty ratio is one ($d_D = 1$); while if the switch is open, the duty ratio is zero ($d_D = 0$). Using the *averaging* concept and the differential equations derived, one obtains the following nonlinear differential equations for the *buck* switching converter:

$$\frac{du_C}{dt} = \frac{1}{C}(i_L - i_a)$$

$$\frac{di_L}{dt} = \frac{1}{L}(-u_C - (r_L + r_c)i_L + r_c i_a - r_s i_L d_D + V_d d_D)$$

$$\frac{di_a}{dt} = \frac{1}{L_a}(u_C + r_c i_L - (r_a + r_c)i_a - E_a)$$

The duty ratio is regulated by the signal-level control voltage u_c, and u_c is bounded. In particular,

$$d_D = \frac{u_c}{u_{t\max}} \in [0 \quad 1],\ u_c = [0 \quad u_{c\max}],\ u_{c\max} = u_{t\max}$$

Neglecting negligibly small resistances of the switch, inductor, and capacitor, the analysis of the steady-state performance leads one to the following expression

$$\frac{u_a}{V_d} = d_D$$

The signal-level control voltage u_c should be considered as a control, and the converter output is the applied terminal voltage u_a. We have

$$u_a = u_C + r_c i_L - r_c i_a$$

One concludes that the resulting buck converter model is mapped by a set of nonlinear differential equations. In fact, having derived that the duty ratio is expressed by $d_D = \dfrac{u_c}{u_{t\max}}$, a nonlinear term $\dfrac{r_s}{L} i_L d_D = \dfrac{r_s}{L u_{t\max}} i_L u_c$ gives the multiplication of the state variable i_L and control signal u_c. The hard control limit is imposed because $u_c \in [0 \quad u_{c\max}]$.

Example 4.1

Using the MATLAB environment, model the *step-down* converter, as shown in Figure 4.24, if the parameters are $r_s = 0.035$ ohm, $r_L = 0.02$ ohm, $r_c = 0.15$ ohm, $r_a = 3.2$ ohm, $C = 0.003$ F, $L = 0.0007$ H, and $L_a = 0.004$ H. The duty ratio is assigned as 0.5. Plot the converter output voltage u_a. The supplied *dc* voltage is $V_d = 30$ V and $E_a = 6$ V.

Solution

Using the differential equations derived, the following m-files are written to perform the modeling.

MATLAB script (ch_4_01.m)

```
t0=0; tfinal=0.03; tol=1e-6; trace=1e-6; y0=[0 0 0]';
[t,y]=ode45('ch_4_02',t0,tfinal,y0,tol,trace);
subplot(2,2,1); plot(t,y);
xlabel('Time (seconds)');
title('Transient dynamics of the state variables');
subplot(2,2,2); plot(t,y(:,1),'-');
xlabel('Time (seconds)'); title('Voltage uc, [V]');
subplot(2,2,3); plot(t,y(:,2),'-');
xlabel('Time (seconds)'); title('Current il, [A]');
subplot(2,2,4); plot(t,y(:,3),'-');
xlabel('Time (seconds)'); title('Current ia, [A]');
disp('End');
```

MATLAB script (ch_4_02.m)

```
% Dynamics of the buck converter
function yprime=difer(t,y);
% parameters
Vd=30; Ea=6; rs=0.035; rl=0.02; rc=0.15; ra=3.2;
C=0.003; L=0.0007; La=0.004; D=0.5;
% differential equations for buck converters
yprime=[(y(2,:)-y(3,:))/C;...
(-y(1,:)-(rl+rc)*y(2,:)+rc*y(3,:)-rs*y(2,:)*D+Vd*D)/L;...
(y(1,:)+rc*y(2,:)-(rc+ra)*y(3,:)-Ea)/La];
```

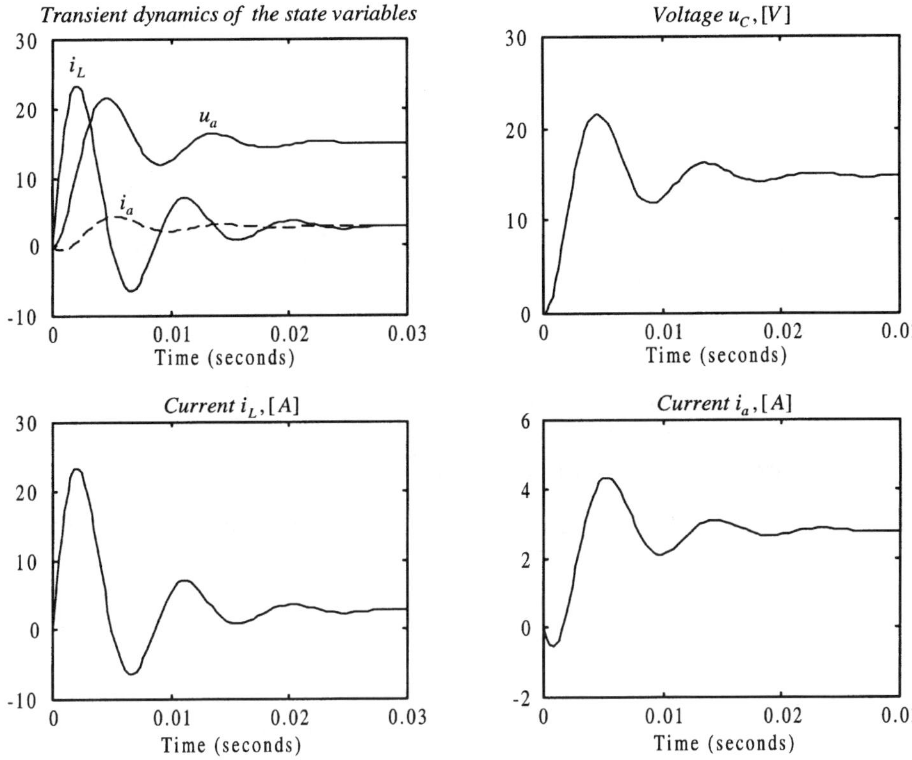

Figure 4.29 Transient dynamics of the *buck* converter; $V_d = 30$ V and $d_D = 0.5$.

The transient dynamics for the state variables $u_C(t)$, $i_L(t)$, and $i_a(t)$ are illustrated in Figure 4.29; the settling time is 0.022 s. The steady-state value of the output voltage is 15 V because the applied voltage is 30 V, and the duty ratio is assigned as $d_D = 0.5$.

Making note of $u_a = u_C + r_c i_L - r_c i_a$, the converter output voltage u_a can be plotted. In particular, one types in the Command window

```
plot(t,y(:,1)+rc*y(:,2)-rc*y(:,3),'-');
```

and the plot for $u_a(t)$ results. □

Boost converters

A typical configuration of a one-quadrant *boost* (*step-up*) dc-dc switching converter is demonstrated in Figure 4.30. When the switch is closed, the diode is reverse-biased, and the following differential equations result

$$\frac{du_C}{dt} = -\frac{1}{C} i_a$$

$$\frac{di_L}{dt} = \frac{1}{L}(-(r_L + r_s) i_L + V_d)$$

$$\frac{di_a}{dt} = \frac{1}{L_a}(u_C - (r_a + r_c) i_a - E_a)$$

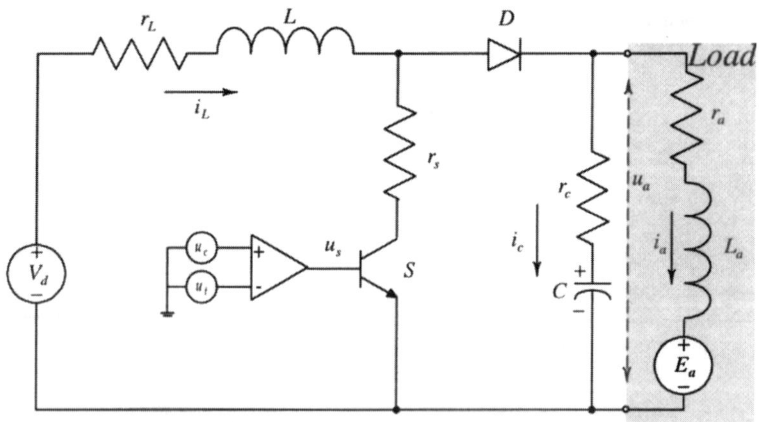

Figure 4.30 *Boost* high-frequency switching converter.

If the switch is open, the diode is forward-biased because the direction of the inductor current i_L does not change instantly. We have

$$\frac{du_C}{dt} = \frac{1}{C}(i_L - i_a)$$

$$\frac{di_L}{dt} = \frac{1}{L}(-u_C - (r_L + r_c)i_L + r_c i_a + V_d)$$

$$\frac{di_a}{dt} = \frac{1}{L_a}(u_C + r_c i_L - (r_a + r_c)i_a - E_a)$$

Using the duty ratio, one finds

$$\frac{du_C}{dt} = \frac{1}{C}(i_L - i_a - i_L d_D)$$

$$\frac{di_L}{dt} = \frac{1}{L}(-u_C - (r_L + r_c)i_L + r_c i_a + u_C d_D + (r_c - r_s)i_L d_D - r_c i_a d_D + V_d)$$

$$\frac{di_a}{dt} = \frac{1}{L_a}(u_C + r_c i_L - (r_a + r_c)i_a - r_c i_L d_D - E_a)$$

For the steady-state, we have

$$\frac{u_a}{V_d} = \frac{1}{1 - d_D}$$

Example 4.2

Perform the simulations of the boost converter, as shown in Figure 4.30, if the parameters are: r_s = 0.035 ohm, r_L = 0.02 ohm, r_c = 0.15 ohm, r_a = 3.2 ohm, C = 0.003 F, L = 0.0007 H, and L_a = 0.004 H. Let the duty ratio be 0.5, V_d = 30 V, and E_a = 6 V.

Solution

Taking note of the differential equations found, two m-files are written. In particular,

MATLAB script (ch_4_03.m)

```
t0=0; tfinal=0.04; tol=1e-7; trace=1e-7; y0=[0 0 0]';
[t,y]=ode45('ch_4_04',t0,tfinal,y0,tol,trace);
subplot(2,2,1); plot(t,y);
xlabel('Time (seconds)');
title('Transient dynamics of the state variables');
subplot(2,2,2); plot(t,y(:,1),'-');
xlabel('Time (seconds)'); title('Voltage uc, [V]');
subplot(2,2,3); plot(t,y(:,2),'-');
xlabel('Time (seconds)'); title('Current il, [A]');
subplot(2,2,4); plot(t,y(:,3),'-');
xlabel('Time (seconds)'); title('Current ia, [A]');
disp('End');
```

MATLAB script (ch_4_04.m)

```
% Dynamics of the boost converter
function yprime=difer(t,y);
% parameters
Vd=30; Ea=6; rs=0.035; rl=0.02; rc=0.15; ra=3.2;
C=0.003; L=0.0007; La=0.004; D=0.5;
% differential equations for boost converters
yprime=[(y(2,:)-y(3,:)-y(2,:)*D)/C;...
(-y(1,:)-(rl+rc)*y(2,:)+rc*y(3,:)+y(1,:)*D+(rc-rs)*y(2,:)*
D-rc*y(3,:)*D+Vd)/L;...
(y(1,:)+rc*y(2,:)-(ra+rc)*y(3,:)-rc*y(2,:)*D-Ea)/La];
```

The transient behaviors for $u_C(t)$, $i_L(t)$, and $i_a(t)$ are plotted in Figure 4.31; the settling time is 0.035 s. One concludes that the transient dynamics of boost converters are fast, and might not be integrated into the differential equations that map the electromechanical systems dynamics. However, the analysis performed is extremely important to demonstrate that the applied voltage to the motor windings cannot be considered as the control input, and $u_a(t)$ is the converter output.

□

Model development using the Lagrange concept: boost dc-dc converters

The Lagrange equations of motion

$$\frac{d}{dt}\left(\frac{\partial \Gamma}{\partial \dot{q}_1}\right) - \frac{\partial \Gamma}{\partial q_1} + \frac{\partial D}{\partial \dot{q}_1} + \frac{\partial \Pi}{\partial q_1} = Q_1$$

$$\frac{d}{dt}\left(\frac{\partial \Gamma}{\partial \dot{q}_2}\right) - \frac{\partial \Gamma}{\partial q_2} + \frac{\partial D}{\partial \dot{q}_2} + \frac{\partial \Pi}{\partial q_2} = Q_2$$

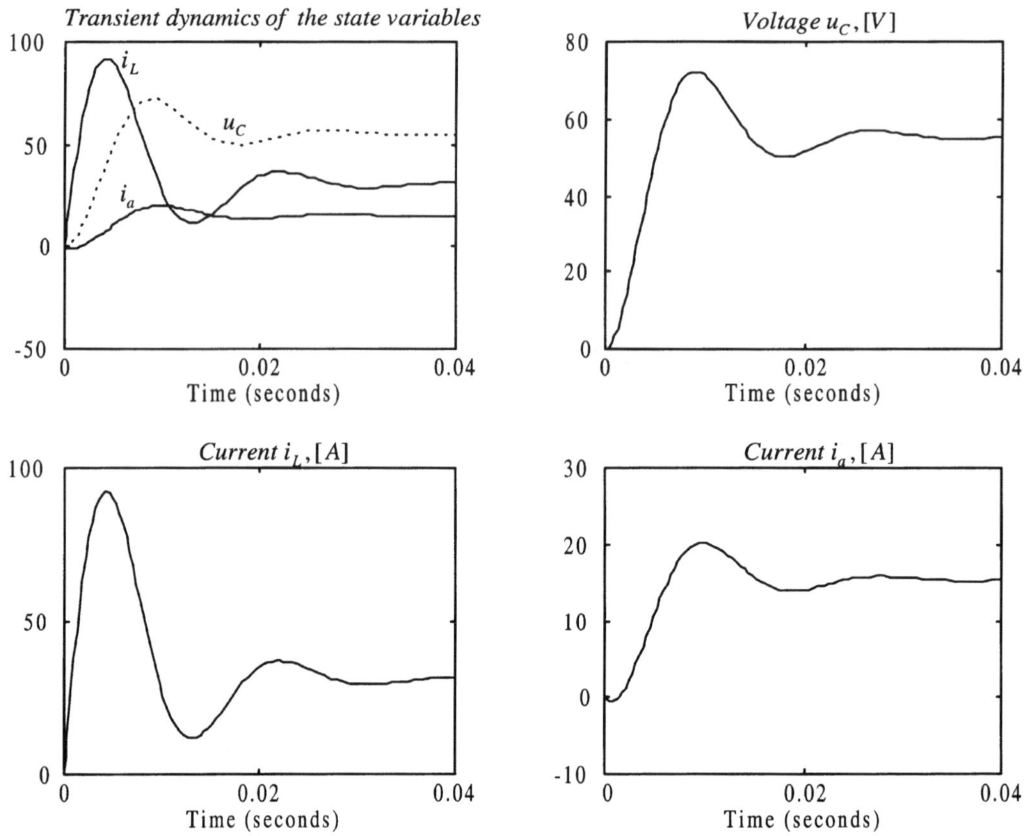

Figure 4.31 Transient dynamics of the *boost* converter; V_d = 30 V and d_D = 0.5.

can be used, and the electric charges in the first and the second loops are denoted as q_1 and q_2. That is, we have $i_L = \dot{q}_1$ and $i_a = \dot{q}_2$.

The generalized forces are Q_1 and Q_2. In particular, $Q_1 = V_d$ and $Q_2 = -E_a$.

When the switch is closed, the total kinetic, potential, and dissipated energies, respectively, are found to be

$$\Gamma = \frac{1}{2}(L\dot{q}_1^2 + L_a\dot{q}_2^2)$$

$$\Pi = \frac{1}{2}\frac{q_2^2}{C}$$

and

$$D = \frac{1}{2}((r_L + r_s)\dot{q}_1^2 + (r_c + r_a)\dot{q}_2^2)$$

Hence, assuming that the resistances, inductances, and capacitance are not varying (time-constant), we have

$$\frac{\partial \Gamma}{\partial q_1}=0,\ \frac{\partial \Gamma}{\partial q_2}=0,\ \frac{\partial \Gamma}{\partial \dot q_1}=L\dot q_1,\ \frac{\partial \Gamma}{\partial \dot q_2}=L_a\dot q_2,\ \frac{d}{dt}\left(\frac{\partial \Gamma}{\partial \dot q_1}\right)=L\ddot q_1,\ \text{and}\ \frac{d}{dt}\left(\frac{\partial \Gamma}{\partial \dot q_2}\right)=L_a\ddot q_2$$

$$\frac{\partial \Pi}{\partial q_1}=0\ \text{and}\ \frac{\partial \Pi}{\partial q_2}=\frac{q_2}{C}$$

$$\frac{\partial D}{\partial \dot q_1}=(r_L+r_s)\dot q_1\ \text{and}\ \frac{\partial D}{\partial \dot q_2}=(r_c+r_a)\dot q_a$$

The application of the Lagrange equations of motion gives

$$L\ddot q_1+(r_L+r_s)\dot q_1=Q_1$$

$$L_a\ddot q_2+(r_c+r_a)\dot q_2+\frac{1}{C}q_2=Q_2$$

and one obtains immediately

$$\ddot q_1=\frac{1}{L}(-(r_L+r_s)\dot q_1+Q_1)$$

$$\ddot q_2=\frac{1}{L_a}\left(-(r_c+r_a)\dot q_2-\frac{1}{C}q_2+Q_2\right)$$

If the switch is open, we have

$$\Gamma=\frac{1}{2}(L\dot q_1^2+L_a\dot q_2^2)$$

$$\Pi=\frac{1}{2}\frac{(q_1-q_2)^2}{C}$$

and

$$D=\frac{1}{2}(r_L\dot q_1^2+r_c(\dot q_1-\dot q_2)^2+r_a\dot q_2^2)$$

Hence,

$$\frac{\partial \Gamma}{\partial q_1}=0,\ \frac{\partial \Gamma}{\partial q_2}=0,\ \frac{\partial \Gamma}{\partial \dot q_1}=L\dot q_1,\ \frac{\partial \Gamma}{\partial \dot q_2}=L_a\dot q_2,\ \frac{d}{dt}\left(\frac{\partial \Gamma}{\partial \dot q_1}\right)=L\ddot q_1,\ \text{and}\ \frac{d}{dt}\left(\frac{\partial \Gamma}{\partial \dot q_2}\right)=L_a\ddot q_2$$

$$\frac{\partial \Pi}{\partial q_1}=\frac{q_1-q_2}{C}\ \text{and}\ \frac{\partial \Pi}{\partial q_2}=-\frac{q_1-q_2}{C}$$

$$\frac{\partial D}{\partial \dot q_1}=(r_L+r_c)\dot q_1-r_c\dot q_2\ \text{and}\ \frac{\partial D}{\partial \dot q_2}=-r_c\dot q_1+(r_c+r_a)\dot q_2$$

Therefore,

$$L\ddot{q}_1 + (r_L + r_c)\dot{q}_1 - r_c\dot{q}_2 + \frac{q_1 - q_2}{C} = Q_1$$

$$L_a\ddot{q}_2 - r_c\dot{q}_1 + (r_c + r_a)\dot{q}_2 - \frac{q_1 - q_2}{C} = Q_2$$

and hence, we have

$$\ddot{q}_1 = \frac{1}{L}\left(-(r_L + r_c)\dot{q}_1 + r_c\dot{q}_2 - \frac{q_1 - q_2}{C} + Q_1\right)$$

$$\ddot{q}_2 = \frac{1}{L_a}\left(r_c\dot{q}_1 - (r_c + r_a)\dot{q}_2 + \frac{q_1 - q_2}{C} + Q_2\right)$$

Taking note of the differential equations derived when the switch is closed and open, Cauchy's form of differential equations can be found using $i_L = \dot{q}_1$ and $i_a = \dot{q}_2$. That is, $\frac{dq_1}{dt} = i_L$ and $\frac{dq_2}{dt} = i_a$. It is obvious that the voltage across the capacitor u_C can be expressed using the charges; that is, when the switch is closed, $u_C = -\frac{q_2}{C}$; and if the switch is open, one obtains $u_C = \frac{q_1 - q_2}{C}$. The analysis of the differential equations derived using Kirchhoff's voltage law and the Lagrange equations of motion leads one to the conclusion that the mathematical model of the boost converter can be found using different state variables. In particular, u_C, i_L, i_a, and q_1, i_L, q_2, i_a are used. However, the resulting differential equations that map the converter dynamics are related, and the solutions are the same.

Buck-Boost converters

The *buck-boost* switching converter is illustrated in Figure 4.32.

If the switch is closed, the diode is reverse-biased; and when the switch is open, the diode is forward-biased. One obtains a set of differential equations using Kirchhoff's law, and the steady-state ratio between the supplied and terminal voltage is $\frac{u_a}{V_d} = \frac{-d_D}{1-d_D}$. That is, the output voltage has opposite polarity compared with the supplied voltage, and depending on the duty ratio, the voltage u_a is less or greater than V_d.

Cuk converters

The Cuk converter is based on a capacitive energy transfer, while the *buck*, *buck-boost*, *boost*, and flyback converter topologies are based on inductive energy transfer. If the switch is *on* or *off*, the currents in the input and output inductors L_1 and L are continuous. The output voltage applied to the load can be either smaller or greater than the supplied voltage V_d. When the switch is turned *off*, the diode is forward-biased. The voltage V_d is supplied, and the capacitor C_1 is charged through the inductor L_1 (see Figure 4.33). To study how the converter operates, assume that the switch is turned *on*. The current through

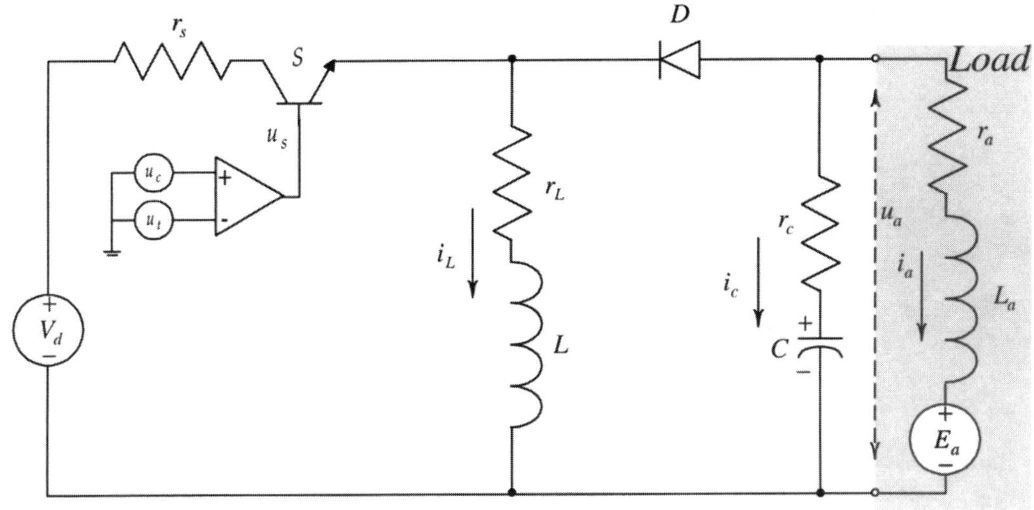

Figure 4.32 High-frequency *buck-boost* switching converter.

the inductor L_1 rises, and at the same time, the voltage of capacitor C_1 reverse-biases the diode, and turns the diode *off*. The capacitor C_1 discharges the stored energy through the circuit formed by capacitors C_1, C, the load $r_a - L_a - E_a$, and the inductor L. Consider the situation if the switch is turned *off*. The voltage V_d is applied, and the capacitor C_1 charges. The energy, stored in the inductor L, transfers to the load. The diode and switch provide a synchronous switching action, and the capacitor C_1 is an element for transferring energy from the source to the load.

Using Kirchhoff's laws and augmenting the differential equations that map the transient dynamics when the switch is open and closed, one finds that the following mathematical model needs to be used

$$\frac{du_{C1}}{dt} = \frac{1}{C_1}(i_{L1} - i_{L1}d_D + i_L d_D)$$

$$\frac{du_C}{dt} = \frac{1}{C}(i_L - i_a)$$

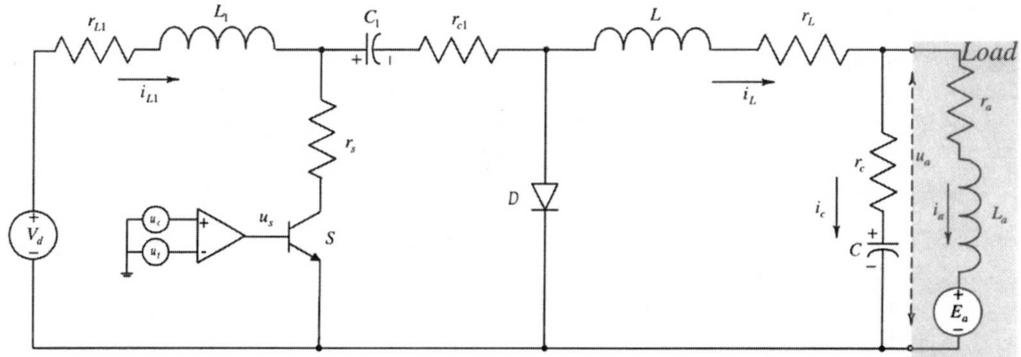

Figure 4.33 Cuk high-frequency switching converter.

$$\frac{di_{L1}}{dt} = \frac{1}{L_1}(-u_{C1} - (r_{L1} + r_{c1})i_{L1} + u_{C1}d_D + (r_{c1} - r_s)i_{L1}d_D + r_s i_L d_D + V_d)$$

$$\frac{di_L}{dt} = \frac{1}{L}(-u_C - (r_L + r_c)i_L + r_c i_a - u_{C1}d_D + r_s i_{L1}d_D - (r_{c1} + r_s)i_L d_D)$$

$$\frac{di_a}{dt} = \frac{1}{L_a}(u_C + r_c i_L - (r_c + r_a)i_a - E_a)$$

From the differential equations derived, and neglecting the resistances of the switch, inductors, and capacitors, one finds the following steady-state equation

$$\frac{u_a}{V_d} = \frac{-d_D}{1 - d_D}$$

Example 4.3

Perform nonlinear modeling of the Cuk converter if $V_d = 30$ V, $E_a = 6$ V, $r_{L1} = 0.035$ ohm, $r_L = 0.02$ ohm, $r_c = 0.15$ ohm, $r_s = 0.03$ ohm, $r_{c1} = 0.018$ ohm, $r_a = 3.2$ ohm, $C_1 = 0.005$ F, $C = 0.003$ F, $L_1 = 0.0004$ H, $L = 0.0007$ H, and $L_a = 0.004$ H. The duty ratio is assigned as 0.25.

Solution

Taking note of the differential equations found, two m-files are developed. In particular,

MATLAB script (ch_4_05.m)

```
tspan=[0 0.06]; y0=[0 0 0 0 0]';
[t,y]=ode45('ch_4_06',tspan,y0);
subplot(3,2,1); plot(t,y);
xlabel('Time (seconds)'); title('Transient dynamics of the state
variables');
subplot(3,2,2); plot(t,y(:,1),'-');
xlabel('Time (seconds)'); title('Voltage uc1, [V]');
subplot(3,2,3); plot(t,y(:,2),'-');
xlabel('Time (seconds)'); title('Voltage uc, [V]');
subplot(3,2,4); plot(t,y(:,3),'-');
xlabel('Time (seconds)'); title('Current il1, [A]');
subplot(3,2,5); plot(t,y(:,4),'-');
xlabel('Time (seconds)'); title('Current il, [A]');
subplot(3,2,6); plot(t,y(:,5),'-');
xlabel('Time (seconds)'); title('Current ia, [A]');
disp('End')
```

MATLAB script (ch_4_06.m)

```
% Dynamics of the Cuk converter
function yprime=difer(t,y);
Vd=30; Ea=6; rl1=0.035; rl=0.02; rc=0.15; rs=0.03; rc1=0.018; ra=3.2;
L1=0.0004; L=0.0007; La=0.004; C1=0.005; C=0.003;
```

```
D=0.25;
yprime=[(y(3,:)-y(3,:)*D+y(4,:)*D)/C1;...
(y(4,:)-y(5,:))/C;...
(-y(1,:)-(rl1+rc1)*y(3,:)+y(1,:)*D+(rc1-
 rs)*y(3,:)*D+rs*y(4,:)*D+Vd)/L1;...
(-y(2,:)-(rl+rc)*y(4,:)+rc*y(5,:)-y(1,:)*D+rs*y(3,:)*D-
 (rc1+rs)*y(4,:)*D)/L;...
(y(2,:)+rc*y(4,:)-(rc+ra)*y(5,:)-ra*y(5,:)-Ea)/La];
```

The transient dynamics for voltages and currents $u_{C1}(t)$, $u_C(t)$, $i_{L1}(t)$, $i_L(t)$, and $i_a(t)$, which are considered as the converter state variables, are illustrated in Figure 4.34, and the settling time is 0.055 s. The steady-state value of the output voltage is 15 V because the applied voltage is 30 V, and the duty ratio is $d_D = 0.25$.

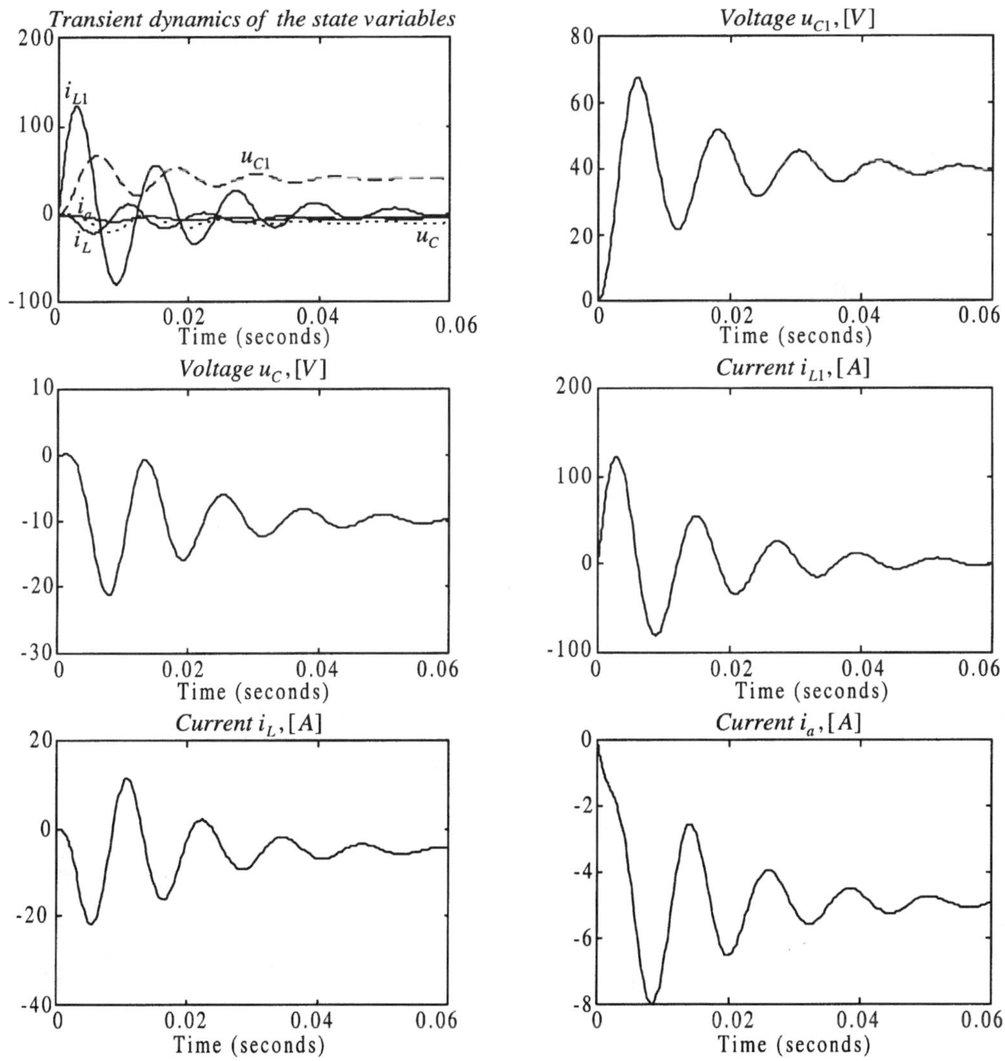

Figure 4.34 Transient dynamics of the Cuk converter; $V_d = 30$ V and $d_D = 0.25$.

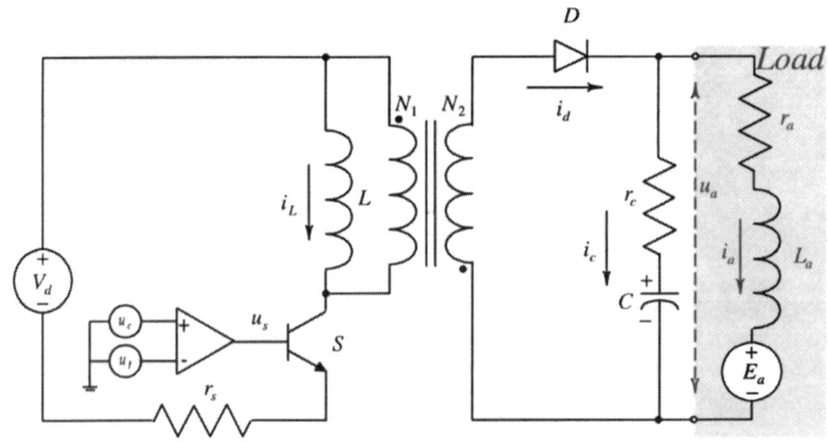

Figure 4.35 Flyback dc-dc converter.

Flyback and forward converters

To avoid the electrical connection between the input and output, which is a serious disadvantage, *flyback* and *forward* converters are used to magnetically isolate the input and output using transformers in the switching scheme. The application of transformers increases the size and cost; however, flyback and forward converters are commonly used to guarantee isolation between input and output. The energy is stored in the inductor when the switch is closed, and the stored energy is transformed to the load when the switch is open (see the flyback and forward magnetically coupled dc-dc converters illustrated in Figures 4.35 and 4.36).

When the switch is closed, the diode is reverse-biased; and if the switch is open, the diode is forward-biased. The switch is closed for time $\frac{d_D}{f}$ and open for $\frac{1-d_D}{f}$. Taking note that for the switch open $i_d = \frac{N_1}{N_2} i_L$, and using the duty ratio d_D, the differential equations can be found.

Using the differential equations derived if the switch is closed

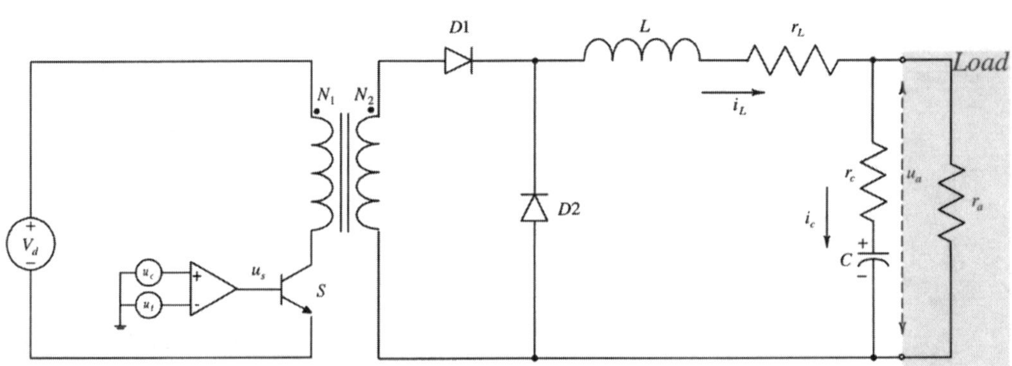

Figure 4.36 Forward dc-dc converter.

$$\frac{du_C}{dt} = \frac{1}{C(r_c + r_a)}(-u_C + r_a i_L)$$

$$\frac{di_L}{dt} = \frac{1}{L}\left(-\frac{r_a}{r_c + r_a}u_C + \left(\frac{r_a^2}{r_c + r_a} - r_L - r_a\right)i_L + \frac{N_2}{N_1}V_d\right)$$

and if the switch is open

$$\frac{du_C}{dt} = \frac{1}{C(r_c + r_a)}(-u_C + r_a i_L)$$

$$\frac{di_L}{dt} = \frac{1}{L}\left(-\frac{r_a}{r_c + r_a}u_C + \left(\frac{r_a^2}{r_c + r_a} - r_L - r_a\right)i_L\right)$$

one finds

$$\frac{du_C}{dt} = \frac{1}{C(r_c + r_a)}(-u_C + r_a i_L)$$

$$\frac{di_L}{dt} = \frac{1}{L}\left(-\frac{r_a}{r_c + r_a}u_C + \left(\frac{r_a^2}{r_c + r_a} - r_L - r_a\right)i_L + \frac{N_2}{N_1}V_d d_D\right)$$

Example 4.4

Using a set of differential equations, simulate the *forward* converter if the applied voltage is $V_d = 30$ V and the duty ratio is $d_D = 0.5$. The parameters are: $r_L = 0.02$ ohm, $r_c = 0.01$ ohm, $r_a = 3.2$ ohm, $L = 0.000005$ H, $C = 0.002$ F, and $\frac{N_2}{N_1} = 1$.

Solution

The following m-files are developed based on the differential equations derived.

MATLAB script (ch_4_07.m)

```
t0=0; tfinal=0.002; tol=1e-6; trace=1e-6; y0=[0 0]';
[t,y]=ode45('ch_4_08',t0,tfinal,y0,tol,trace);
subplot(2,2,1); plot(t,y);
xlabel('Time (seconds)');
title('Transient dynamics of the state variables');
subplot(2,2,2); plot(t,y(:,1),'-');
xlabel('Time (seconds)'); title('Voltage uc, [V]');
subplot(2,2,3); plot(t,y(:,2),'-');
xlabel('Time (seconds)'); title('Current il, [A]');
disp('End');
```

MATLAB script (ch_4_08.m)

```
% Dynamics of the forward converter
function yprime=difer(t,y);
% parameters
Vd=30; rl=0.02; rc=0.01; ra=3.2; C=0.002; L=0.000005; D=0.5;
% differential equations for the forward converter
yprime=[(-y(1,:)+ra*y(2,:))/(C*(rc+ra));...
(-(ra/(rc+ra))*y(1,:)+(ra*ra/(rc+ra)-rl-ra)*y(2,:)+Vd*D)/L];
```

The transient dynamics for the state variables $u_C(t)$ and $i_L(t)$ are documented in Figure 4.37. The settling time is 0.0015 s, and the steady-state voltage applied to the load terminal is 15 V.

4.3.2 Resonant converters

The current trends in development and deployment of advanced switching converters have facilitated the unified activities in topology design, nonlinear analysis, optimization, and control. To attain high efficiency and power density, new topologies were developed. It is recognized that nonlinear analysis and design must be performed to guarantee a spectrum of specifications and requirements imposed on the converter dynamics. The

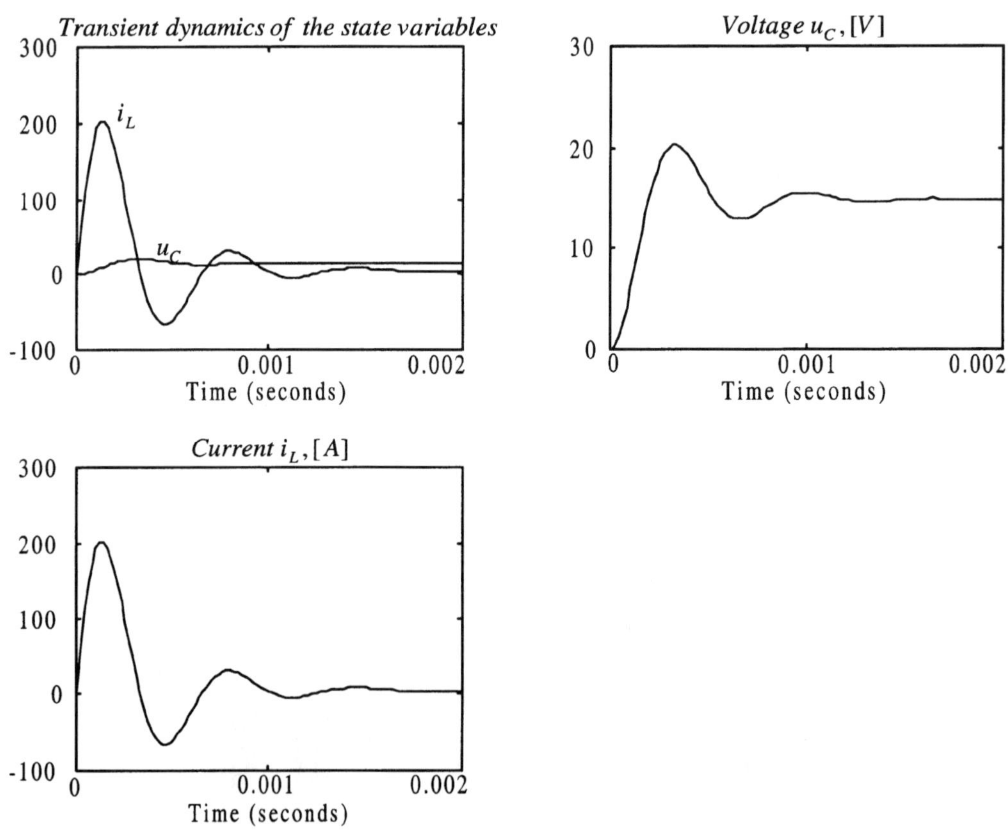

Figure 4.37 Transient dynamics of the forward converter; $V_d = 30$ V and $d_D = 0.5$.

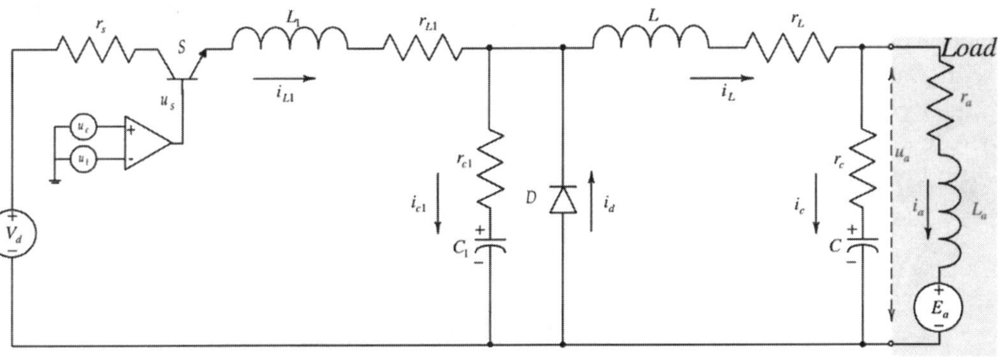

Figure 4.38 Resonant converter with zero-current switching.

output voltage of converters is regulated by changing the duty ratio, which is constrained by lower and upper limits. To approach the design trade-offs and analyze converter performance (settling time, overshoot, stability margins, and other quantities), the constraints and nonlinearities should be thoroughly examined. Resonant converters have been widely used in high-performance electromechanical systems. In fact, zero-voltage zero-current switching has become the enabling technology for the majority of converters to improve power density, efficiency, reliability, and other performance characteristics. Recent innovations include development of advanced converter topologies and control algorithms to maximize efficiency, minimize losses, improve performance, increase power density, etc. The problem of controlling high-frequency switching converters is a very important one in many applications. Nonlinear analysis and control are central issues to be solved to improve the steady-state and dynamic characteristics. A great variety of resonant converter topologies and filters have been developed, and it has been demonstrated that the resulting nonlinear dynamics cannot be linearized and hard bounds imposed cannot be neglected. Specific requirements are assigned, and the *absolute* limit of converter performance must be placed in the scope of practical design. It is recognized that the steady-state and dynamic characteristics of switching converters can be improved through nonlinear design, using complete nonlinear models without simplifications and assumptions. As resonant converter topology and filter circuitry are developed, the corresponding nonlinear differential equations can be found using the Kirchhoff laws to map the converter and filter dynamics. Based on mathematical models derived, nonlinear analysis and design can be performed.

Switching converters

Mathematical models in the form of nonlinear differential equations for power converters can be developed using the Kirchhoff laws or Lagrange equations of motion. There exists a great number of converter topologies and filter configurations, and the model development is straightforward. Consider the resonant converter shown in Figure 4.38.

The output voltage at the load terminal (resistance r_a and inductance L_a) is regulated by controlling the switching *on* and *off* durations t_{on} and t_{off}. The switch S is open and closed, and the switching frequency is $\frac{1}{t_{on}+t_{off}}$. When the switch is open, the diode D is forward-biased to carry the output inductor current i_L, and the voltage across the capacitor C_1 is zero. When the switch is closed, the diode remains forward-biased while $i_{L1} < i_L$; and as i_{L1} reaches i_L, the diode turns off. Hence, the switch turns off and on at zero-current (i_{L1}

is zero). A control-triangle concept is used to establish the PWM switching. The switching signal u_s, which drives the switch, is generated by comparing a signal-level control voltage u_c with a repetitive triangular signal u_t. It should be emphasized that in resonant converters, the frequency usually is controlled to regulate the output voltage.

A set of highly coupled differential equations to map the resonant converter dynamics is found as

$$\frac{du_{C1}}{dt} = \frac{1}{C_1}(i_{L1} - i_L)d_D$$

$$\frac{du_C}{dt} = \frac{1}{C}(i_L - i_a)$$

$$\frac{di_{L1}}{dt} = \frac{1}{L_1}(-u_{C1} - (r_s + r_{L1} + r_{c1})i_{L1} + r_{c1}i_L + V_d)d_D$$

$$\frac{di_L}{dt} = \frac{1}{L}(u_{C1} - u_C + r_{c1}i_{L1} - (r_{c1} + r_L + r_c)i_L + r_c i_a)d_D$$

$$\frac{di_a}{dt} = \frac{1}{L_a}(u_C + r_c i_L - (r_c + r_a)i_a - E_a)$$

Hence, a nonlinear mathematical model results due to the multiplication of the state variables u_{C1}, u_C, i_{L1}, i_L, i_a and duty ratio d_D.

From the differential equations obtained, we have the following nonlinear state-space model

$$\begin{bmatrix} \frac{du_{C1}}{dt} \\ \frac{du_C}{dt} \\ \frac{di_{L1}}{dt} \\ \frac{di_L}{dt} \\ \frac{di_a}{dt} \end{bmatrix} = \begin{bmatrix} 0 & 0 & 0 & 0 & 0 \\ 0 & 0 & 0 & \frac{1}{C} & -\frac{1}{C} \\ 0 & 0 & 0 & 0 & 0 \\ 0 & 0 & 0 & 0 & 0 \\ 0 & \frac{1}{L_a} & 0 & \frac{r_c}{L_a} & -\frac{r_c + r_a}{L_a} \end{bmatrix} \begin{bmatrix} u_{C1} \\ u_C \\ i_{L1} \\ i_L \\ i_a \end{bmatrix} +$$

$$\begin{bmatrix} \frac{1}{C_1}(i_{L1} - i_L) \\ 0 \\ \frac{1}{L_1}(-u_{C1} - (r_s + r_{L1} + r_{c1})i_{L1} + r_{c1}i_L + V_d) \\ \frac{1}{L}(u_{C1} - u_C + r_{c1}i_{L1} - (r_{c1} + r_L + r_c)i_L + r_c i_a) \\ 0 \end{bmatrix} d_D - \begin{bmatrix} 0 \\ 0 \\ 0 \\ 0 \\ E_a \end{bmatrix}$$

Example 4.5

Perform modeling of the resonant converter that is illustrated in Figure 4.38 if V_d = 30 V and E_a = 6 V. The parameters are: r_s = 0.025 ohm, r_{L1} = 0.01 ohm, r_{c1} = 0.04 ohm,

r_L = 0.02 ohm, r_c = 0.15 ohm, L_1 = 0.000005 H, L = 0.0007 H, C_1 = 0.000003 F, and C = 0.003 F. The duty ratio is d_D = 0.5. The resistive-inductive load is formed by r_a and L_a; r_a = 3.2 ohm and L_a = 0.004 H.

Solution

Using differential equations, two m-files are developed.

MATLAB script (ch_4_09.m)

```
y0=[0 0 0 0 0]; tspan=[0 0.04];
[t,y]=ode45('ch_4_10',tspan,y0);
subplot(3,2,1);plot(t,y,'-');
xlabel('Time (seconds)'); title('Transient dynamics of the state
 variables');
subplot(3,2,2); plot(t,y(:,1),'-');
xlabel('Time (seconds)'); title('Voltage uc1, [V]');
subplot(3,2,3); plot(t,y(:,2),'-');
xlabel('Time (seconds)'); title('Voltage uc, [V]');
subplot(3,2,4); plot(t,y(:,3),'-');
xlabel('Time (seconds)'); title('Current il1, [A]');
subplot(3,2,5); plot(t,y(:,4),'-');
xlabel('Time (seconds)'); title('Current il, [A]');
subplot(3,2,6); plot(t,y(:,5),'-');
xlabel('Time (seconds)'); title('Current ia, [A]');
disp('End')
```

MATLAB script (ch_4_10.m)

```
function yprime=difer(t,y);
Vd=30; Ea=6; rs=0.025; rl1=0.01; rc1=0.04; rl=0.02; rc=0.15; ra=3.2;
L1=0.000005; L=0.0007; La=0.004; C1=0.000003; C=0.003;
D=0.5;
yprime=[D*(y(3,:)-y(4,:))/C1;...
(y(4,:)-y(5,:))/C;...
(Vd-(rs+rl1+rc1)*y(3,:)+rc1*y(4,:)-y(1,:))*D/L1;...
(y(1,:)-y(2,:)+rc1*y(3,:)-(rc1+rl+rc1)*y(4,:)-rc*y(5,:))*D/L;...
(y(2,:)+rc*y(4,:)-(rc+ra)*y(5,:)-Ea)/La];
```

Transient dynamics for the resonant converter with zero-current switching is studied, and the transient responses are plotted in Figure 4.39.

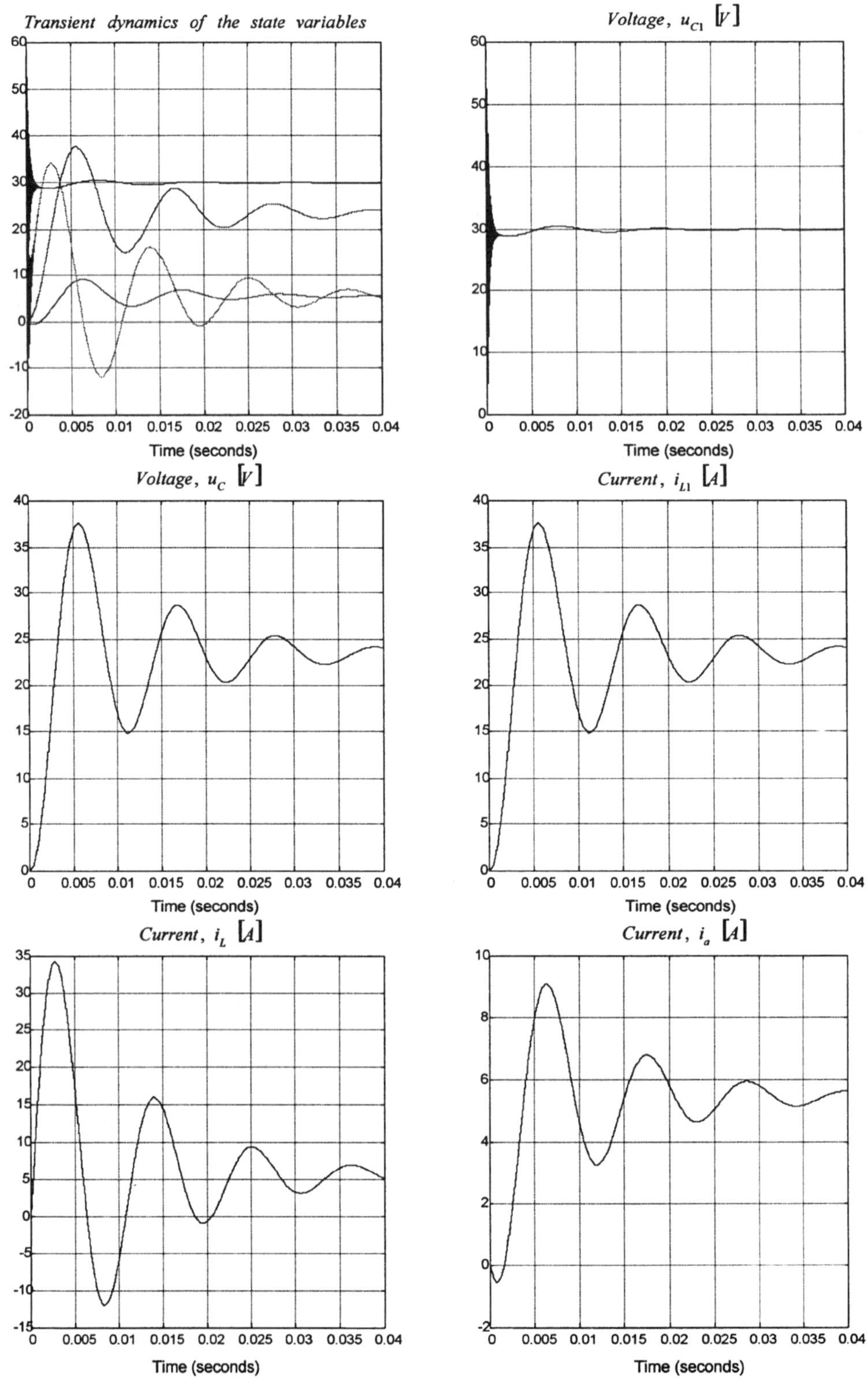

Figure 4.39 Transient dynamics of the resonant converter; V_d = 30 V and d_D = 0.5.

References

Hart, D. W., *Introduction to Power Electronics*, Prentice-Hall, Upper Saddle River, NJ, 1997.

Kassakian, J. G., Schlecht, M. F., and Verghese, G. C., *Principles of Power Electronics*, Addison-Wesley, Reading, MA, 1991.

Mohan, N., Undeland, T. M., and Robbins, W. P., *Power Electronics: Converters, Applications, and Design*, John Wiley & Sons, New York, 1995.

Sedra, A. S. and Smith, K. C., *Microelectronic Circuits*, Oxford University Press, New York, 1997.

Appendix 4.1-1

MOTOROLA

MC33171
MC33172
MC33174

Low Power, Single Supply Operational Amplifiers

Quality bipolar fabrication with innovative design concepts are employed for the MC33171/72/74 series of monolithic operational amplifiers. These devices operate at 180 μA per amplifier and offer 1.8 MHz of gain bandwidth product and 2.1 V/μs slew rate without the use of JFET device technology. Although this series can be operated from split supplies, it is particularly suited for single supply operation, since the common mode input voltage includes ground potential (V_{EE}). With a Darlington input stage, these devices exhibit high input resistance, low input offset voltage and high gain. The all NPN output stage, characterized by no deadband crossover distortion and large output voltage swing, provides high capacitance drive capability, excellent phase and gain margins, low open loop high frequency output impedance and symmetrical source/sink AC frequency response.

The MC33171/72/74 are specified over the industrial/ automotive temperature ranges. The complete series of single, dual and quad operational amplifiers are available in plastic as well as the surface mount packages.

- Low Supply Current: 180 μA (Per Amplifier)
- Wide Supply Operating Range: 3.0 V to 44 V or ±1.5 V to ±22 V
- Wide Input Common Mode Range, Including Ground (V_{EE})
- Wide Bandwidth: 1.8 MHz
- High Slew Rate: 2.1 V/μs
- Low Input Offset Voltage: 2.0 mV
- Large Output Voltage Swing: –14.2 V to +14.2 V (with ±15 V Supplies)
- Large Capacitance Drive Capability: 0 pF to 500 pF
- Low Total Harmonic Distortion: 0.03%
- Excellent Phase Margin: 60°C
- Excellent Gain Margin: 15 dB
- Output Short Circuit Protection
- ESD Diodes Provide Input Protection for Dual and Quad

DUAL

P SUFFIX
PLASTIC PACKAGE
CASE 626

D SUFFIX
PLASTIC PACKAGE
CASE 751
(SO–8)

PIN CONNECTIONS

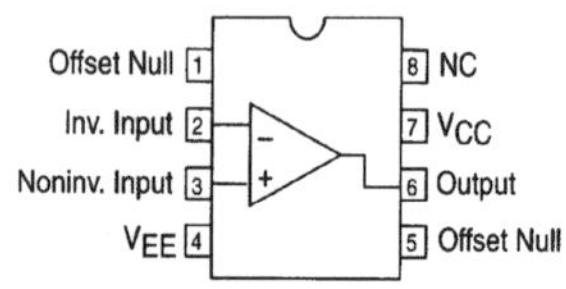

(Single, Top View)

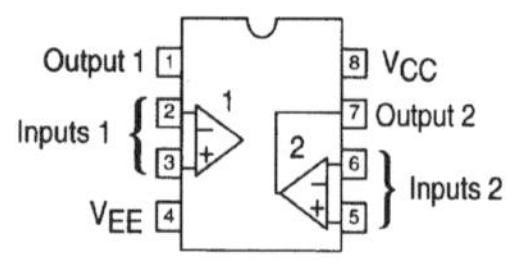

(Top View)

QUAD

P SUFFIX
PLASTIC PACKAGE
CASE 646

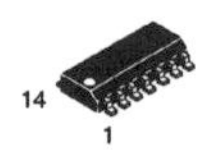

D SUFFIX
PLASTIC PACKAGE
CASE 751A
(SO–14)

PIN CONNECTIONS

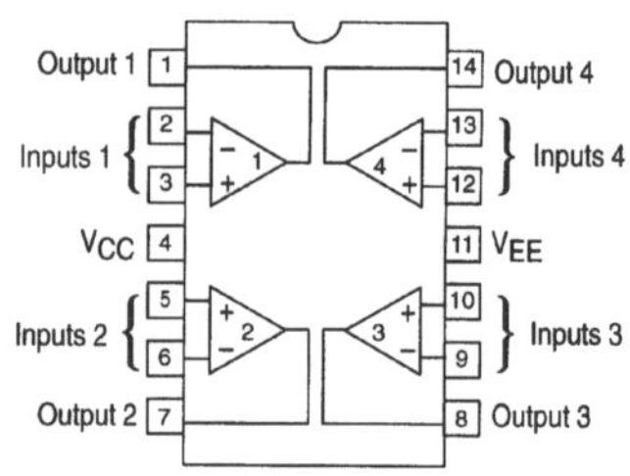

(Top View)

ORDERING INFORMATION

Op Amp Function	Device	Operating Temperature Range	Package
Single	MC33171D MC33171P	T_A = –40° to +85°C T_A = –40° to +85°C	SO–8 Plastic DIP
Dual	MC33172D MC33172P	T_A = –40° to +85°C T_A = –40° to +85°C	SO–8 Plastic DIP
Quad	MC33174D MC33174P	T_A = –40° to +85°C T_A = –40° to +85°C	SO–14 Plastic DIP

Appendix 4.1-2

MC33171 MC33172 MC33174

MAXIMUM RATINGS

Rating	Symbol	Value	Unit
Supply Voltage	V_{CC}/V_{EE}	±22	V
Input Differential Voltage Range	V_{IDR}	(Note 1)	V
Input Voltage Range	V_{IR}	(Note 1)	V
Output Short Circuit Duration (Note 2)	t_{SC}	Indefinite	sec
Operating Ambient Temperature Range	T_A	–40 to +85	°C
Operating Junction Temperature	T_J	+150	°C
Storage Temperature Range	T_{stg}	–65 to +150	°C

NOTES: 1. Either or both input voltages must not exceed the magnitude of V_{CC} or V_{EE}.
2. Power dissipation must be considered to ensure maximum junction temperature (T_J) is not exceeded.

Representative Schematic Diagram
(Each Amplifier)

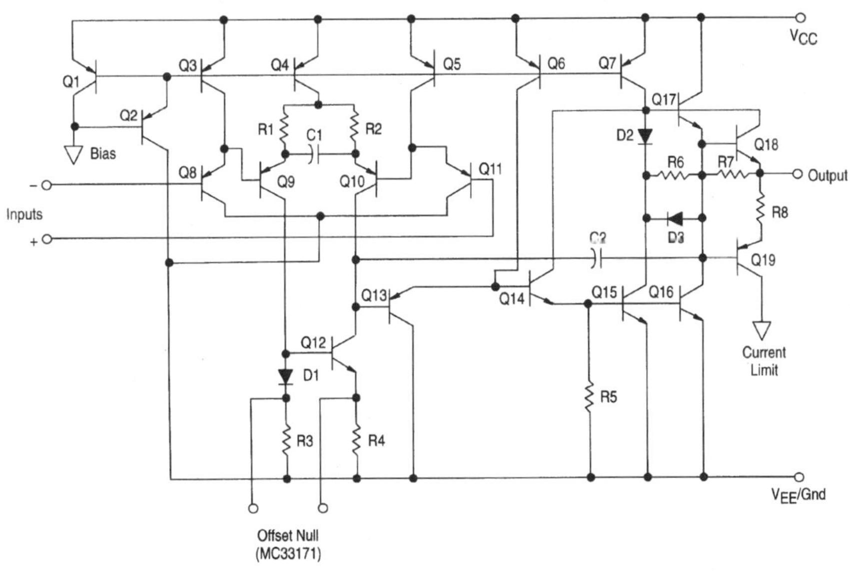

Appendix 4.1-3

MC33171 MC33172 MC33174

DC ELECTRICAL CHARACTERISTICS (V_{CC} = +15 V, V_{EE} = –15 V, R_L connected to ground, T_A = T_{low} to T_{high} [Note 3], unless otherwise noted.)

Characteristics	Symbol	Min	Typ	Max	Unit
Input Offset Voltage (V_{CM} = 0 V)	V_{IO}				mV
V_{CC} = +15 V, V_{EE} = –15 V, T_A = +25°C		—	2.0	4.5	
V_{CC} = +5.0 V, V_{EE} = 0 V, T_A = +25°C		—	2.5	5.0	
V_{CC} = +15 V, V_{EE} = –15 V, T_A = T_{low} to T_{high}		—	—	6.5	
Average Temperature Coefficient of Offset Voltage	$\Delta V_{IO}/\Delta T$	—	10	—	µV/°C
Input Bias Current (V_{CM} = 0 V)	I_{IB}				nA
T_A = +25°C		—	20	100	
T_A = T_{low} to T_{high}		—	—	200	
Input Offset Current (V_{CM} = 0 V)	I_{IO}				nA
T_A = +25°C		—	5.0	20	
T_A = T_{low} to T_{high}		—	—	40	
Large Signal Voltage Gain (V_O = ±10 V< R_L = 10 k)	A_{VOL}				V/mV
T_A = +25°C		50	500	—	
T_A = T_{low} to T_{high}		25	—	—	
Output Voltage Swing	V_{OH}				V
V_{CC} = +5.0 V, V_{EE} = 0 V, R_L = 10 k, T_A = +25°C		3.5	4.3	—	
V_{CC} = +15 V, V_{EE} = –15 V, R_L = 10 k, T_A = +25°C		13.6	14.2	—	
V_{CC} = +15 V, V_{EE} = –15 V, R_L = 10 k, T_A = T_{low} to T_{high}		13.3	—	—	
V_{CC} = +5.0 V, V_{EE} = 0 V, R_L = 10 k, T_A = +25°C	V_{OL}	—	0.05	0.15	
V_{CC} = +15 V, V_{EE} = –15 V, R_L = 10 k, T_A = +25°C		—	–14.2	–13.6	
V_{CC} = +15 V, V_{EE} = –15 V, R_L = 10 k, T_A = T_{low} to T_{high}		—	—	–13.3	
Output Short Circuit (T_A = +25°C)	I_{SC}				mA
Input Overdrive = 1.0 V, Output to Ground					
Source		3.0	5.0	—	
Sink		15	27	—	
Input Common Mode Voltage Range	V_{ICR}				V
T_A = +25°C		V_{EE} to (V_{CC} –1.8)			
T_A = T_{low} to T_{high}		V_{EE} to (V_{CC} –2.2)			
Common Mode Rejection Ratio (R_S ≤ 10 k) T_A = +25°C	CMRR	80	90	—	dB
Power Supply Rejection Ratio (R_S = 100 Ω) T_A = +25°C	PSRR	80	100	—	dB
Power Supply Current (Per Amplifier)	I_D				µA
V_{CC} = +5.0 V, V_{EE} = 0 V, T_A = +25°C		—	180	250	
V_{CC} = +15 V, V_{EE} = –15 V, T_A = +25°C		—	220	250	
V_{CC} = +15 V, V_{EE} = –15 V, T_A = T_{low} to T_{high}		—	—	300	

NOTE: 3. T_{low} = –40°C T_{high} = +85°C

Appendix 4.1-4

MC33171 MC33172 MC33174

AC ELECTRICAL CHARACTERISTICS (V_{CC} = +15 V, V_{EE} = −15 V, R_L connected to ground, T_A = +25°C, unless otherwise noted.)

Characteristics	Symbol	Min	Typ	Max	Unit
Slew Rate (V_{in} = −10 V to +10 V, R_L = 10 k, C_L = 100 pF) A_V +1 A_V −1	SR	 1.6 —	 2.1 2.1	 — —	V/μs
Gain Bandwidth Product (f = 100 kHz)	GBW	1.4	1.8	—	MHz
Power Bandwidth A_V = +1.0 R_L = 10 k, V_O = 20 V_{pp}, THD = 5%	BWp	—	35	—	kHz
Phase Margin R_L = 10 k R_L = 10 k, C_L = 100 pF	ϕm	 — —	 60 45	 — —	Degrees
Gain Margin R_L = 10 k R_L = 10 k, C_L = 100 pF	A_m	 — —	 15 5.0	 — —	dB
Equivalent Input Noise Voltage R_S = 100 Ω, f = 1.0 kHz	e_n	—	32	—	nV/$\sqrt{Hz}$
Equivalent Input Noise Current (f = 1.0 kHz)	I_n	—	0.2	—	pA/$\sqrt{Hz}$
Differential Input Resistance V_{cm} = 0 V	R_{in}	—	300	—	MΩ
Input Capacitance	C_i	—	0.8	—	pF
Total Harmonic Distortion A_V = +10, R_L = 10 k, 2.0 V_{pp} ≤ V_O ≤ 20 V_{pp}, f = 10 kHz	THD	—	0.03	—	%
Channel Separation (f = 10 kHz)	CS	—	120	—	dB
Open Loop Output Impedance (f = 1.0 MHz)	z_o	—	100	—	Ω

Figure 1. Input Common Mode Voltage Range versus Temperature

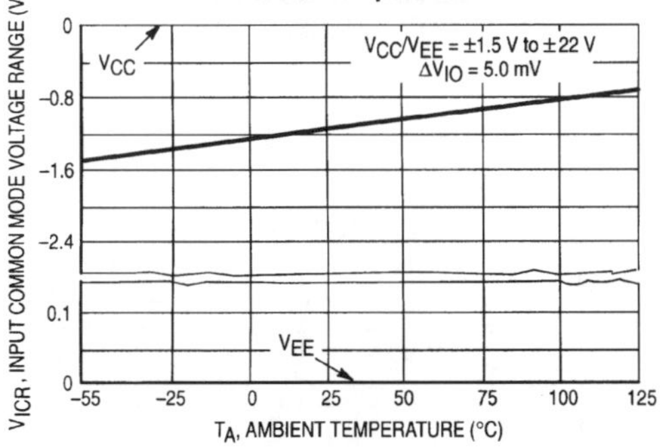

Figure 2. Split Supply Output Saturation versus Load Current

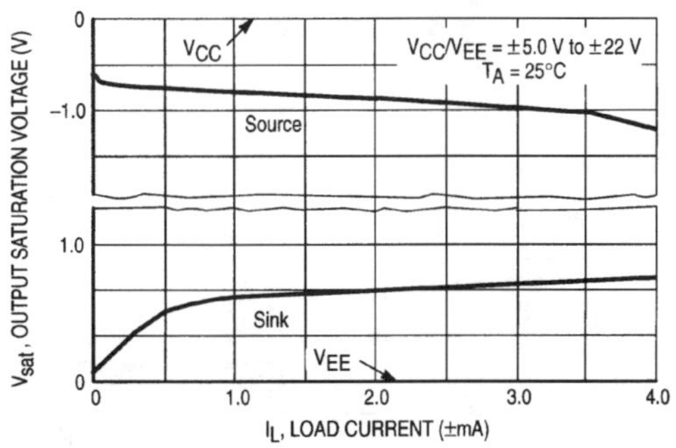

Appendix 4.1-5

MC33171 MC33172 MC33174

Figure 3. Open Loop Voltage Gain and Phase versus Frequency

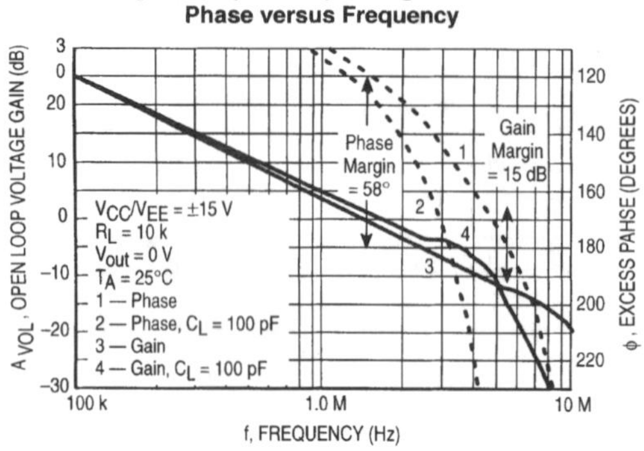

Figure 4. Phase Margin and Percent Overshoot versus Load Capacitance

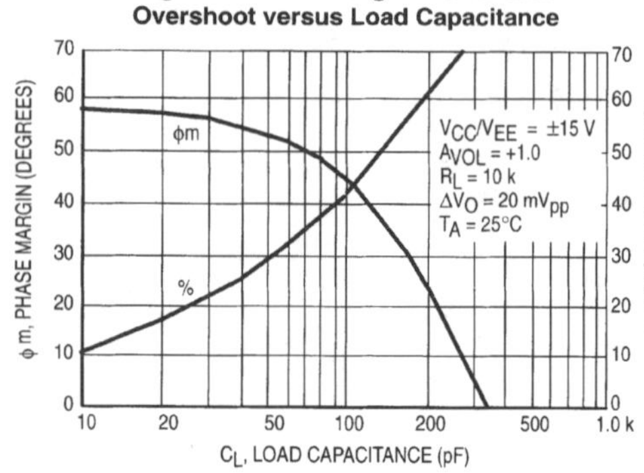

Figure 5. Normalized Gain Bandwidth Product and Slew Rate versus Temperature

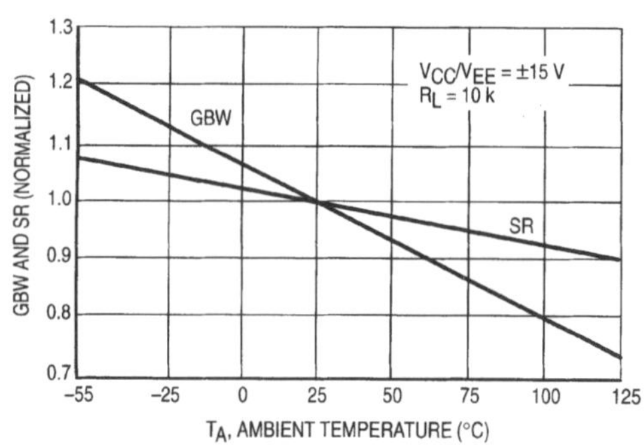

Figure 6. Small and Large Signal Transient Response

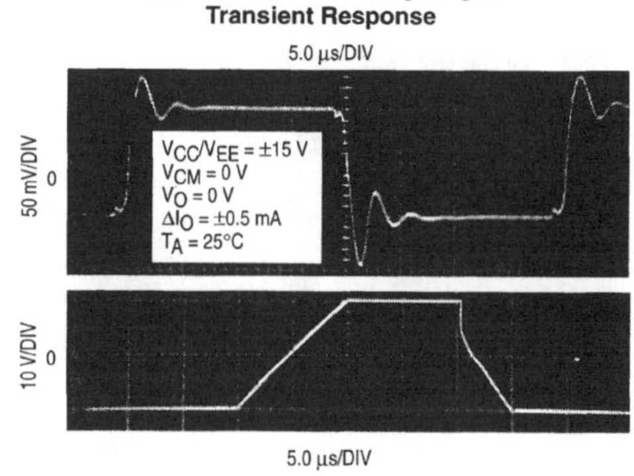

Figure 7. Output Impedance and Frequency

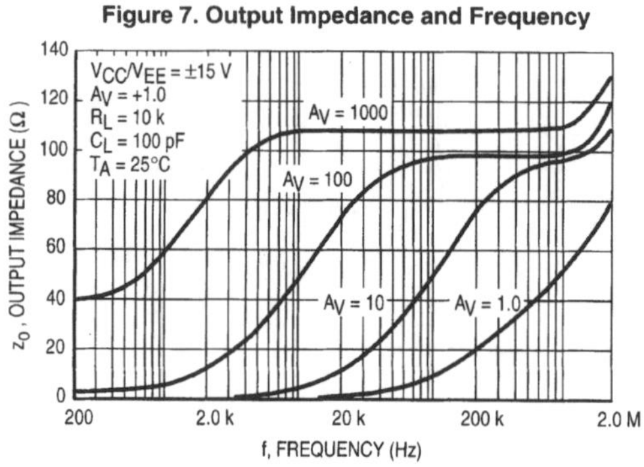

Figure 8. Supply Current versus Supply Voltage

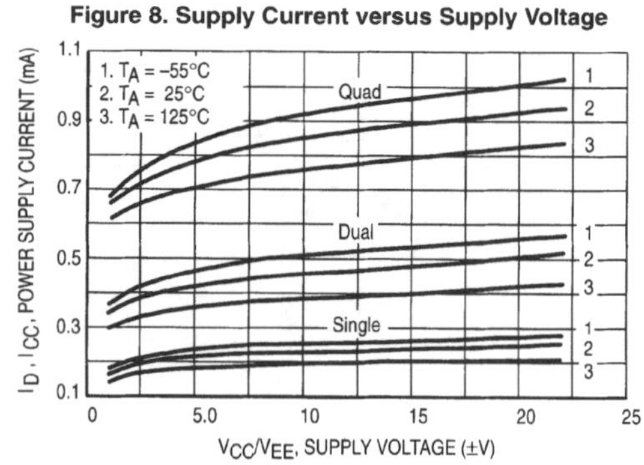

Appendix 4.1-6

MC33171 MC33172 MC33174

APPLICATIONS INFORMATION – CIRCUIT DESCRIPTION/PERFORMANCE FEATURES

Although the bandwidth, slew rate, and settling time of the MC33171/72/74 amplifier family is similar to low power op amp products utilizing JFET input devices, these amplifiers offer additional advantages as a result of the PNP transistor differential inputs and an all NPN transistor output stage.

Because the input common mode voltage range of this input stage includes the V_{EE} potential, single supply operation is feasible to as low as 3.0 V with the common mode input voltage at ground potential.

The input stage also allows differential input voltages up to ±44 V, provided the maximum input voltage range is not exceeded. Specifically, the input voltages must range between V_{CC} and V_{EE} supply voltages as shown by the maximum rating table. In practice, although *not recommended,* the input voltages can exceed the V_{CC} voltage by approximately 3.0 V and decrease below the V_{EE} voltage by 0.3 V without causing product damage, although output phase reversal may occur. It is also possible to source up to 5.0 mA of current from V_{EE} through either inputs' clamping diode without damage or latching, but phase reversal may again occur. If at least one input is within the common mode input voltage range and the other input is within the maximum input voltage range, no phase reversal will occur. If both inputs exceed the upper common mode input voltage limit, the output will be forced to its lowest voltage state.

Since the input capacitance associated with the small geometry input device is substantially lower (0.8 pF) than that of a typical JFET (3.0 pF), the frequency response for a given input source resistance is greatly enhanced. This becomes evident in D–to–A current to voltage conversion applications where the feedback resistance can form a pole with the input capacitance of the op amp. This input pole creates a 2nd Order system with the single pole op amp and is therefore detrimental to its settling time. In this context, lower input capacitance is desirable especially for higher values of feedback resistances (lower current DACs). This input pole can be compensated for by creating a feedback zero with a capacitance across the feedback resistance, if necessary, to reduce overshoot. For 10 kΩ of feedback resistance, the MC33171/72/74 family can typically settle to within 1/2 LSB of 8 bits in 4.2 µs, and within 1/2 LSB of 12 bits in 4.8 µs for a 10 V step. In a standard inverting unity gain fast settling configuration, the symmetrical slew rate is typically ±2.1 V/µs. In the classic noninverting unity gain configuration the typical output positive slew rate is also 2.1 V/µs, and the corresponding negative slew rate will usually exceed the positive slew rate as a function of the fall time of the input waveform.

The all NPN output stage, shown in its basic form on the equivalent circuit schematic, offers unique advantages over the more conventional NPN/PNP transistor Class AB output stage. A 10 kΩ load resistance can typically swing within 0.8 V of the positive rail (V_{CC}) and negative rail (V_{EE}), providing a 28.4 Vpp swing from ±15 V supplies. This large output swing becomes most noticeable at lower supply voltages.

The positive swing is limited by the saturation voltage of the current source transistor Q7, the V_{BE} of the NPN pull–up transistor Q17, and the voltage drop associated with the short circuit resistance, R5. For sink currents less than 0.4 mA, the negative swing is limited by the saturation voltage of the pull–down transistor Q15, and the voltage drop across R4 and R5. For small valued sink currents, the above voltage drops are negligible, allowing the negative swing voltage to approach within millivolts of V_{EE}. For sink currents (> 0.4 mA), diode D3 clamps the voltage across R4. Thus the negative swing is limited by the saturation voltage of Q15, plus the forward diode drop of D3 ($\approx V_{EE}$ +1.0 V). Therefore an unprecedented peak–to–peak output voltage swing is possible for a given supply voltage as indicated by the output swing specifications.

If the load resistance is referenced to V_{CC} instead of ground for single supply applications, the maximum possible output swing can be achieved for a given supply voltage. For light load currents, the load resistance will pull the output to V_{CC} during the positive swing and the output will pull the load resistance near ground during the negative swing. The load resistance value should be much less than that of the feedback resistance to maximize pull–up capability.

Because the PNP output emitter–follower transistor has been eliminated, the MC33171/72/74 family offers a 15 mA minimum current sink capability, typically to an output voltage of (V_{EE} +1.8 V). In single supply applications the output can directly source or sink base current from a common emitter NPN transistor for current switching applications.

In addition, the all NPN transistor output stage is inherently faster than PNP types, contributing to the bipolar amplifier's improved gain bandwidth product. The associated high frequency low output impedance (200 Ω typ @ 1.0 MHz) allows capacitive drive capability from 0 pF to 400 pF without oscillation in the noninverting unity gain configuration. The 60°C phase margin and 15 dB gain margin, as well as the general gain and phase characteristics, are virtually independent of the source/sink output swing conditions. This allows easier system phase compensation, since output swing will not be a phase consideration. The AC characteristics of the MC33171/72/74 family also allow excellent active filter capability, especially for low voltage single supply applications.

Although the single supply specification is defined at 5.0 V, these amplifiers are functional to at least 3.0 V @ 25°C. However slight changes in parametrics such as bandwidth, slew rate, and DC gain may occur.

If power to this integrated circuit is applied in reverse polarity, or if the IC is installed backwards in a socket, large unlimited current surges will occur through the device that may result in device destruction.

As usual with most high frequency amplifiers, proper lead dress, component placement and PC board layout should be exercised for optimum frequency performance. For example, long unshielded input or output leads may result in unwanted input/output coupling. In order to preserve the relatively low input capacitance associated with these amplifiers, resistors connected to the inputs should be immediately adjacent to the input pin to minimize additional stray input capacitance. This not only minimizes the input pole for optimum frequency response, but also minimizes extraneous "pick up" at this node. Supply decoupling with adequate capacitance immediately adjacent to the supply pin is also important, particularly over temperature, since many types of decoupling capacitors exhibit great impedance changes over temperature.

The output of any one amplifier is current limited and thus protected from a direct short to ground. However, under such conditions, it is important not to allow the device to exceed the maximum junction temperature rating. Typically for ±15 V supplies, any one output can be shorted continuously to ground without exceeding the maximum temperature rating.

Appendix 4.1-7

MC33171 MC33172 MC33174

Figure 9. AC Coupled Noninverting Amplifier with Single +5.0 V Supply

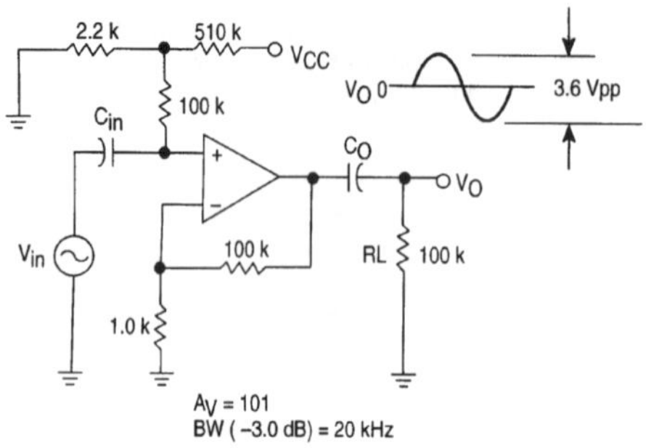

Figure 10. AC Coupled Inverting Amplifier with Single +5.0 V Supply

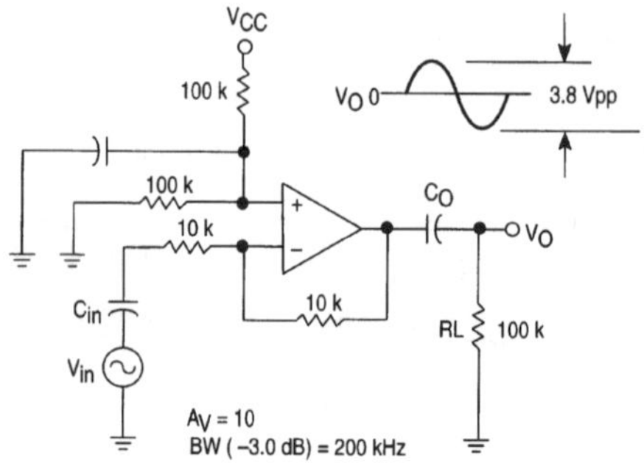

Figure 11. DC Coupled Inverting Amplifier Maximum Output Swing with Single +5.0 V Supply

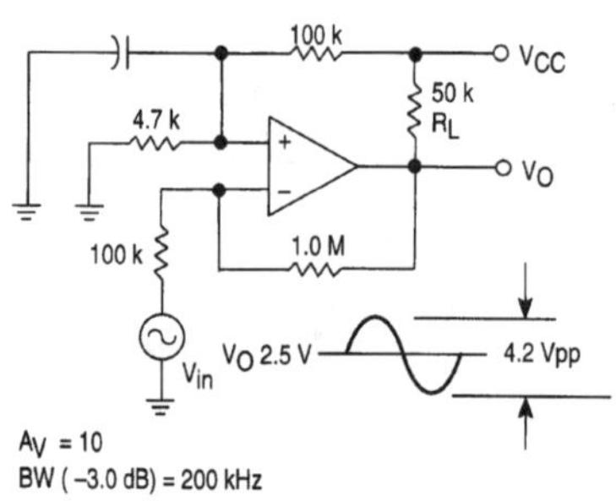

Figure 12. Offset Nulling Circuit

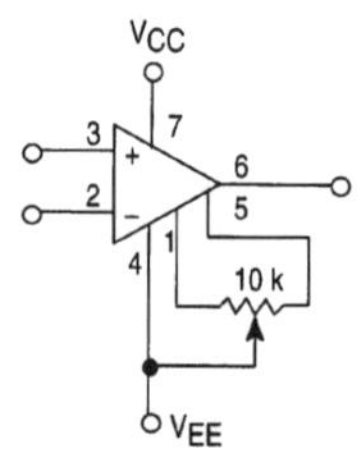

Offset Nulling range is approximately ±80 mV with a 10 k potentiometer, MC33171 only.

Figure 13. Active High–Q Notch Filter

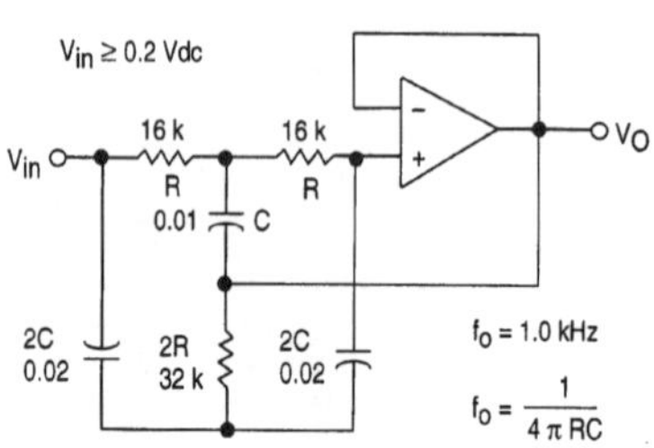

Figure 14. Active Bandpass Filter

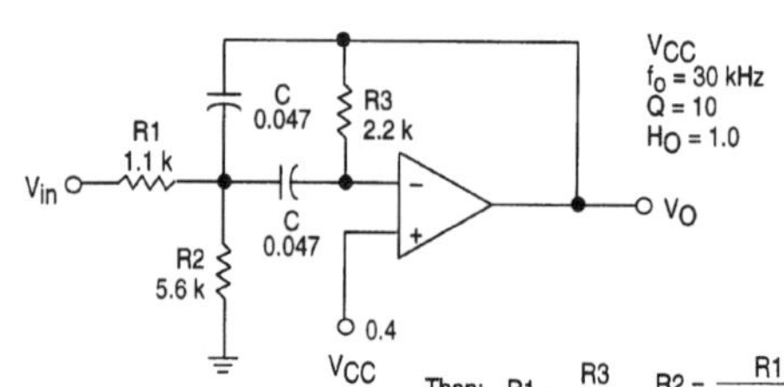

Then: $R1 = \frac{R3}{2\,H_O}$ $\quad R2 = \frac{R1\,R3}{4Q^2R1 - R3}$

$R3 = \frac{Q}{\pi f_o C}$ $\quad \frac{Q_o\, f_o}{GBW} < 0.1$

Given f_o = center frequency
A_o = Gain at center frequency
Choose Value f_o, Q, A_o, C
For less than 10% error for operational amplifier, where f_o and GBW are expressed in Hz.

Appendix 4.2-1

MOTOROLA

MC34080 thru MC34085

High Slew Rate, Wide Bandwidth, JFET Input Operational Amplifiers

HIGH PERFORMANCE JFET INPUT OPERATIONAL AMPLIFIERS

These devices are a new generation of high speed JFET input monolithic operational amplifiers. Innovative design concepts along with JFET technology provide wide gain bandwidth product and high slew rate. Well–matched JFET input devices and advanced trim techniques ensure low input offset errors and bias currents. The all NPN output stage features large output voltage swing, no deadband crossover distortion, high capacitive drive capability, excellent phase and gain margins, low open loop output impedance, and symmetrical source/sink AC frequency response.

This series of devices is available in fully compensated or decompensated ($A_{VCL} \leq 2$) and is specified over a commercial temperature range. They are pin compatible with existing Industry standard operational amplifiers, and allow the designer to easily upgrade the performance of existing designs.

- Wide Gain Bandwidth: 8.0 MHz for Fully Compensated Devices
  16 MHz for Decompensated Devices
- High Slew Rate: 25 V/μs for Fully Compensated Devices
  50 V/μs for Decompensated Devices
- High Input Impedance: $10^{12}\Omega$
- Input Offset Voltage: 0.5 mV Maximum (Single Amplifier)
- Large Output Voltage Swing: −14.7 V to +14 V for $V_{CC}/V_{EE} = \pm 15$ V
- Low Open Loop Output Impedance: 30 Ω @ 1.0 MHz
- Low THD Distortion: 0.01%
- Excellent Phase/Gain Margins: 55°/7.6 dB for Fully Compensated Devices

P SUFFIX
PLASTIC PACKAGE
CASE 626

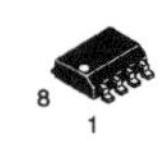

D SUFFIX
PLASTIC PACKAGE
CASE 751
(SO–8)

PIN CONNECTIONS

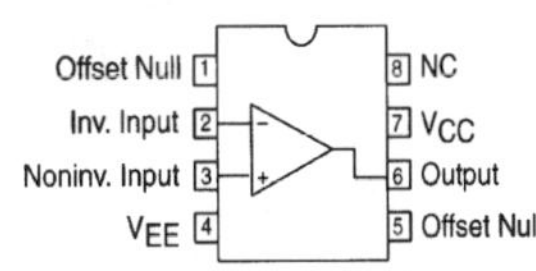

(Single, Top View)

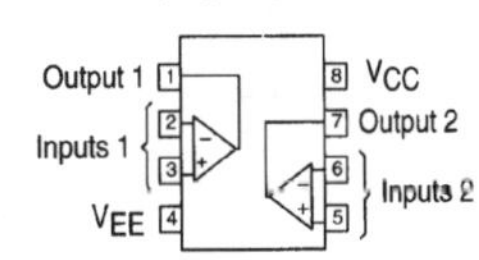

(Dual, Top View)

ORDERING INFORMATION

Op Amp Function	Fully Compen-sated	$A_{VCL} \geq 2$ Compensated	Operating Temperature Range	Package
Single	MC34081BD	MC34080BD	T_A = 0° to +70°C	SO–8
	MC34081BP	MC34080BP		Plastic DIP
Dual	MC34082P	MC34083BP		Plastic DIP
Quad	MC34084DW	MC34085BDW	T_A = 0° to +70°C	SO–16L
	MC34084P	MC34085BP		Plastic DIP

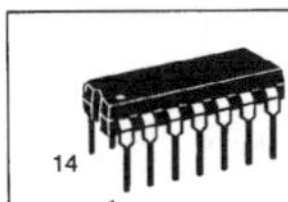

P SUFFIX
PLASTIC PACKAGE
CASE 646

DW SUFFIX
PLASTIC PACKAGE
CASE 751G
(SO–16L)

PIN CONNECTIONS

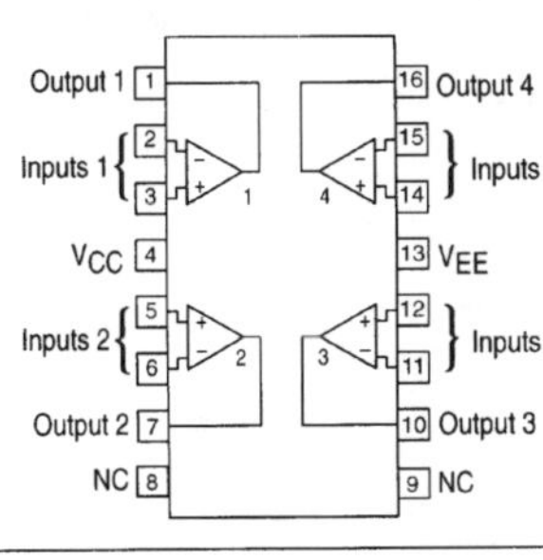

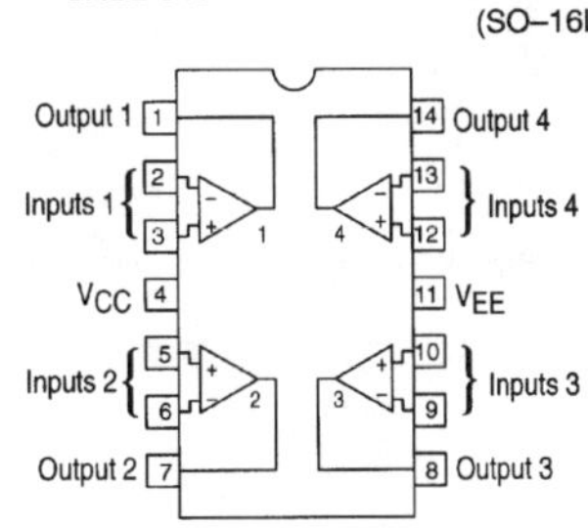

(Quad, Top View)

Appendix 4.2-2

MC34080 thru MC34085

MAXIMUM RATINGS

Rating	Symbol	Value	Unit
Supply Voltage (from V_{CC} to V_{EE})	V_S	+44	V
Input Differential Voltage Range	V_{IDR}	(Note 1)	V
Input Voltage Range	V_{IR}	(Note 1)	V
Output Short Circuit Duration (Note 2)	t_{SC}	Indefinite	sec
Operating Ambient Temperature Range	T_A	0 to +70	°C
Operating Junction Temperature	T_J	+125	°C
Storage Temperature Range	T_{stg}	− 65 to +165	°C

NOTES: 1. Either or both input voltages must not exceed the magnitude of V_{CC} or V_{EE}.
2. Power dissipation must be considered to ensure maximum junction temperature (T_J) is not exceeded.

Representative Schematic Diagram
(Each Amplifier)

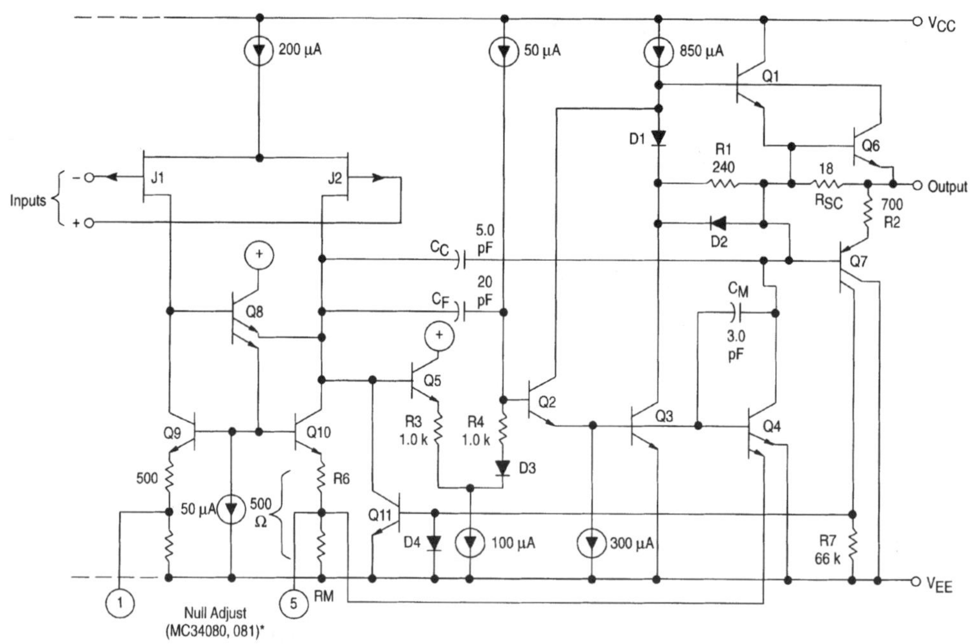

*Pins 1 & 5 (MC34080,081) should *not* be directly grounded or connected to V_{CC}.

Appendix 4.2-3

MC34080 thru MC34085

DC ELECTRICAL CHARACTERISTICS (V_{CC} = +15 V, V_{EE} = – 15 V, T_A = T_{low} to T_{high} [Note 3], unless otherwise noted.)

Characteristics	Symbol	Min	Typ	Max	Unit
Input Offset Voltage (Note 4)	V_{IO}				mV
Single					
T_A = +25°C		—	0.5	2.0	
T_A = 0° to +70°C (MC34080B, MC34081B)		—	—	4.0	
Dual					
T_A = +25°C		—	1.0	3.0	
T_A = 0° to +70°C (MC34082, MC34083)		—	—	5.0	
Quad					
T_A = +25°C		—	6.0	12	
T_A = 0° to +70°C (MC34084, MC34085)		—	—	14	
Average Temperature Coefficient of Offset Voltage	$\Delta V_{IO}/\Delta T$	—	10	—	μV/°C
Input Bias Current (V_{CM} = 0 Note 5)	I_{IB}				
T_A = +25°C		—	0.06	0.2	nA
T_A = 0° to +70°C		—	—	4.0	
Input Offset Current (V_{CM} = 0 Note 5)	I_{IO}				
T_A = +25°C		—	0.02	0.1	nA
T_A = 0° to +70°C		—	—	2.0	
Large Signal Voltage Gain (V_O = ±10 V, R_L = 2.0 k)	A_{VOL}				V/mV
T_A = +25°C		25	80	—	
T_A = T_{low} to T_{high}		15	—	—	
Output Voltage Swing	V_{OH}				V
R_L = 2.0 k, T_A = +25°C		13.2	13.7	—	
R_L = 10 k, T_A = +25°C		13.4	13.9	—	
R_L = 10 k, T_A = T_{low} to T_{high}		13.4	—	—	
R_L = 2.0 k, T_A = +25°C	V_{OL}	—	–14.1	–13.5	
R_L = 10 k, T_A = +25°C		—	–14.7	–14.1	
R_L = 10 k, T_A = T_{low} to T_{high}		—	—	–14.0	
Output Short Circuit Current (T_A = +25°C)	I_{SC}				mA
Input Overdrive = 1.0 V, Output to Ground					
Source		20	31	—	
Sink		20	28	—	
Input Common Mode Voltage Range	V_{ICR}	(V_{EE} +4.0) to			V
T_A = +25°C		(V_{CC} – 2.0)			
Common Mode Rejection Ratio (R_S ≤ 10 k, T_A = +25°C)	CMRR	70	90	—	dB
Power Supply Rejection Ratio (R_S = 100 Ω, T_A = 25°C)	PSRR	70	86	—	dB
Power Supply Current	I_D				mA
Single					
T_A = +25°C		—	2.5	3.4	
T_A = T_{low} to T_{high}		—	—	4.2	
Dual					
T_A = +25°C		—	4.9	6.0	
T_A = T_{low} to T_{high}		—	—	7.5	
Quad					
T_A = +25°C		—	9.7	11	
T_A = T_{low} to T_{high}		—	—	13	

NOTES: (continued)

3. T_{low} = 0°C for MC34080B, MC34081B, MC34084, MC34085 — T_{high} = +70°C for MC34080B, MC34081B, MC34084, MC34085
4. See application information for typical changes in input offset voltage due to solderability and temperature cycling.
5. Limits at T_A = +25°C are guaranteed by high temperature (T_{high}) testing.

Appendix 4.2-4

MC34080 thru MC34085

AC ELECTRICAL CHARACTERISTICS (V_{CC} = +15 V, V_{EE} = – 15 V, T_A = +25°C, unless otherwise noted.)

Characteristics	Symbol	Min	Typ	Max	Unit
Slew Rate (V_{in} = –10 V to +10 V, R_L = 2.0 kΩ, C_L = 100 pF)	SR				V/μs
Compensated A_V = +1.0		20	25	—	
A_V = –1.0		—	30	—	
Decompensated A_V = +2.0		35	50	—	
A_V = –1.0		—	50	—	
Settling Time (10 V Step, A_V = –1.0)	t_s				μs
To 0.10% (±1/2 LSB of 9–Bits)		—	0.72	—	
To 0.01% (±1/2 LSB of 12–Bits)		—	1.6	—	
Gain Bandwidth Product (f = 200 kHz)	GBW				MHz
Compensated		6.0	8.0	—	
Decompensated		12	16	—	
Power Bandwidth (R_L = 2.0 k, V_O = 20 V_{pp}, THD = 5.0%)	BWp				kHz
Compensated A_V = +1.0		—	400	—	
Decompensated A_V = – 1.0		—	800	—	
Phase Margin (Compensated)	ϕ_m				Degrees
R_L = 2.0 k		—	55	—	
R_L = 2.0 k, C_L = 100 pF		—	39	—	
Gain Margin (Compensated)	A_m				dB
R_L = 2.0 k		—	7.6	—	
R_L = 2.0 k, C_L = 100 pF		—	4.5	—	
Equivalent Input Noise Voltage R_S = 100 Ω, f = 1.0 kHz	e_n	—	30	—	nV/√Hz
Equivalent Input Noise Current (f = 1.0 kHz)	I_n	—	0.01	—	pA/√Hz
Input Capacitance	C_i	—	5.0	—	pF
Input Resistance	r_i	—	10^{12}	—	Ω
Total Harmonic Distortion A_V = +10, R_L = 2.0 k, 2.0 ≤ V_O ≤ 20 V_{pp}, f = 10 kHz	THD	—	0.05	—	%
Channel Separation (f = 10 kHz)	—	—	120	—	dB
Open Loop Output Impedance (f = 1.0 MHz)	Z_o	—	35	—	Ω

Figure 1. Input Common Mode Voltage Range versus Temperature

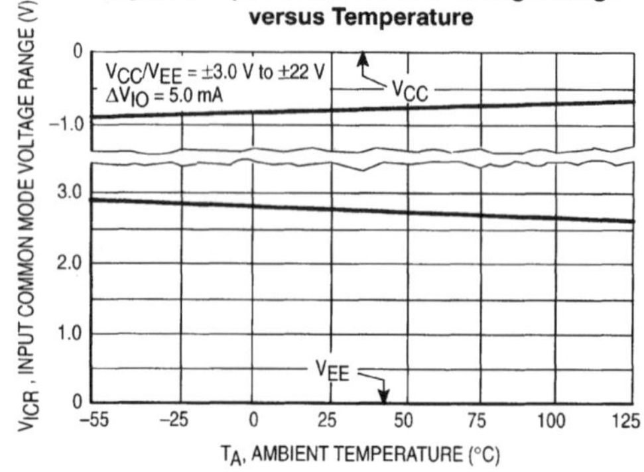

Figure 2. Input Bias Current versus Temperature

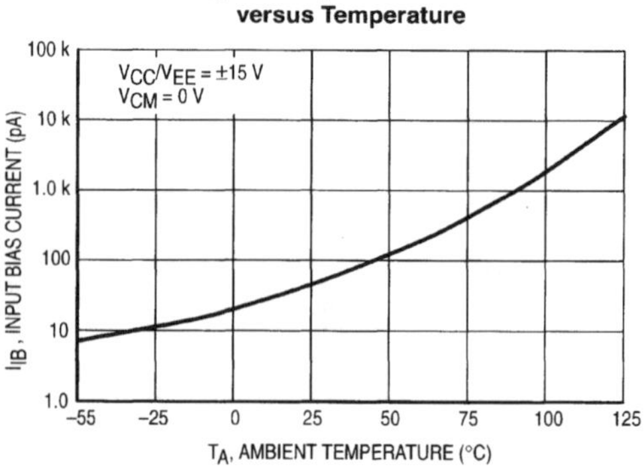

Appendix 4.2-5

MC34080 thru MC34085

Figure 3. Input Bias Current versus Input Common Mode Voltage

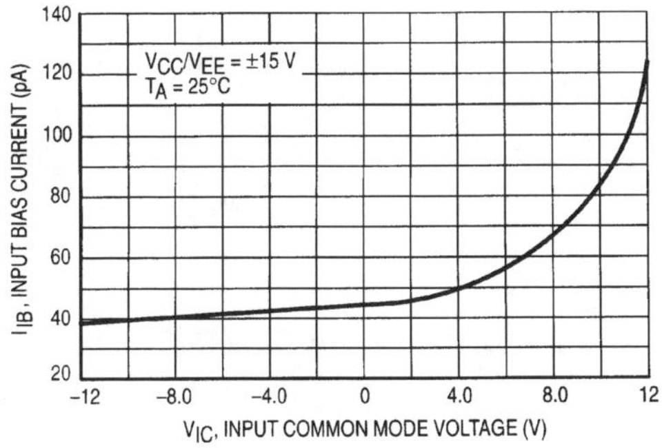

Figure 4. Output Voltage Swing versus Supply Voltage

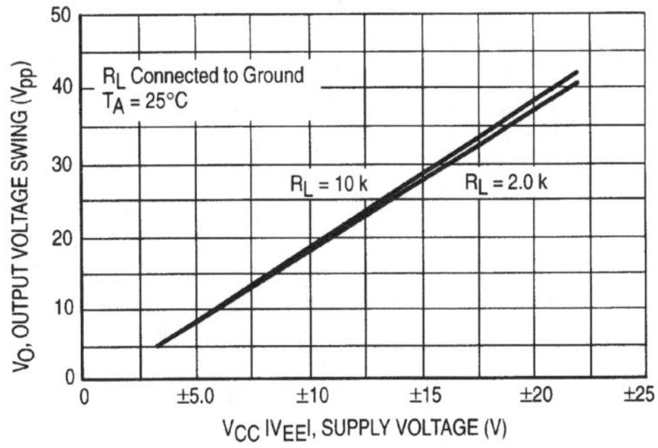

Figure 5. Output Saturation versus Load Current

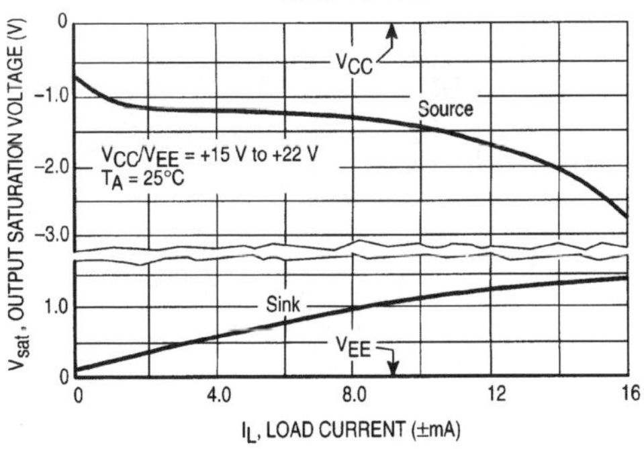

Figure 6. Output Saturation vesus Load Resistance to Ground

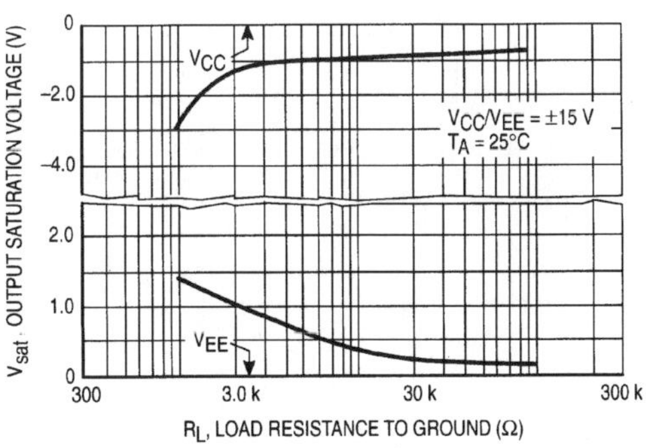

Figure 7. Output Saturation versus Load Resistance to V_{CC}

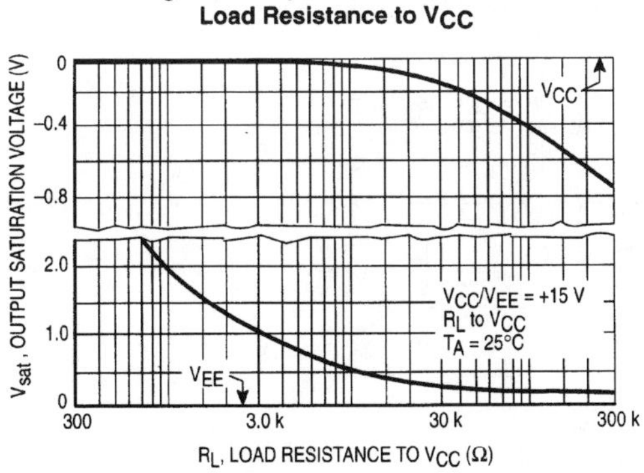

Figure 8. Output Short Circuit Current versus Temperature

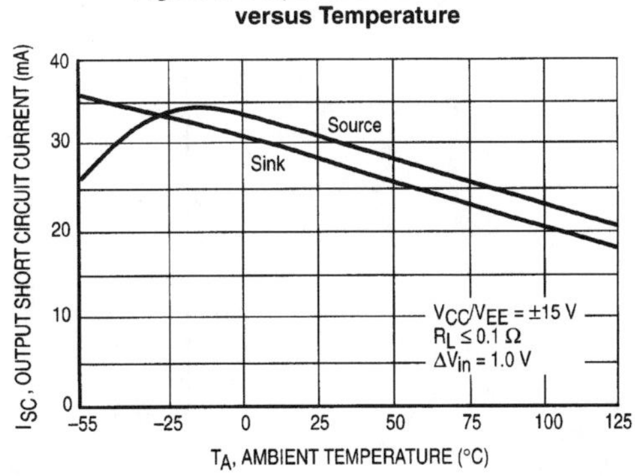

Appendix 4.2-6

MC34080 thru MC34085

Figure 9. Output Impedance versus Frequency

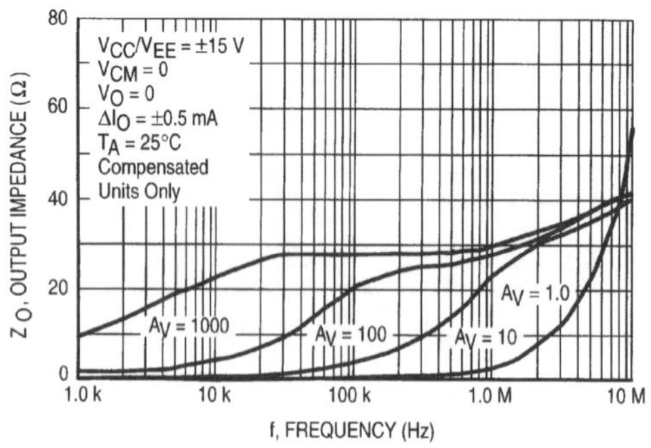

Figure 10. Output Impedance versus Frequency

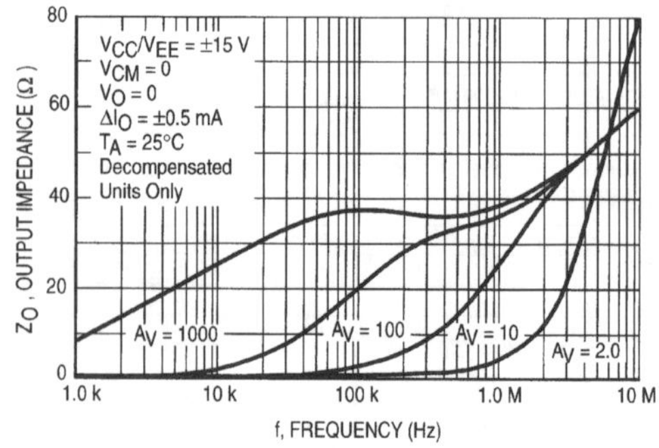

Figure 11. Output Voltage Swing versus Frequency

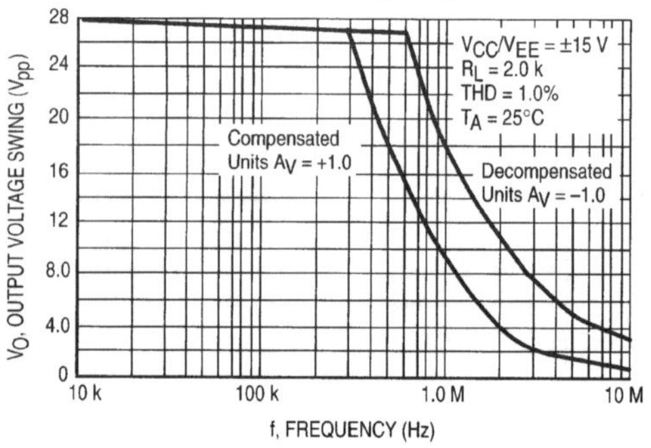

Figure 12. Output Distortion versus Frequency

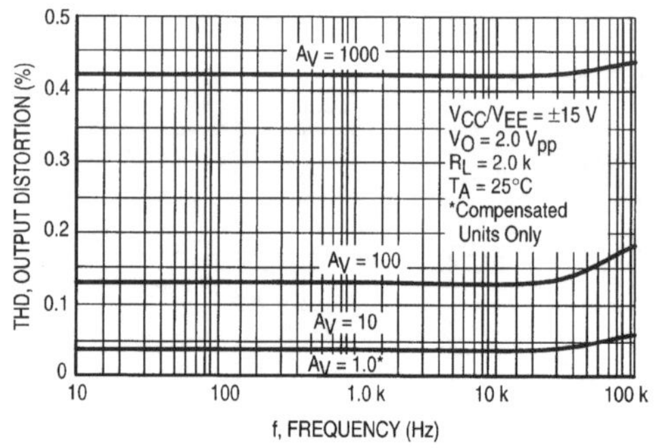

Figure 13. Open Loop Voltage Gain versus Temperature

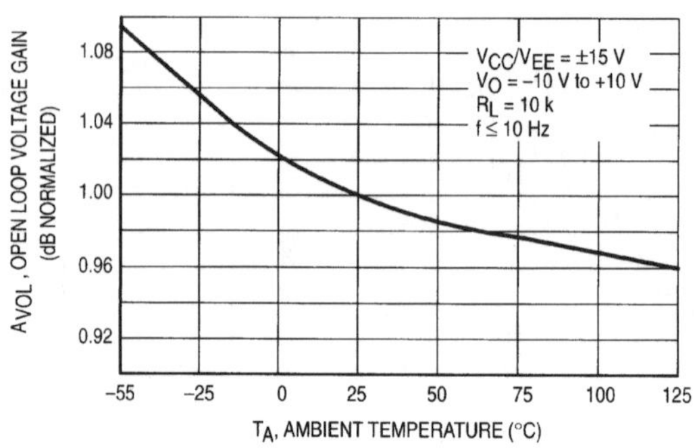

Appendix 4.2-7

MC34080 thru MC34085

Figure 14. Open Loop Voltage Gain and Phase versus Frequency

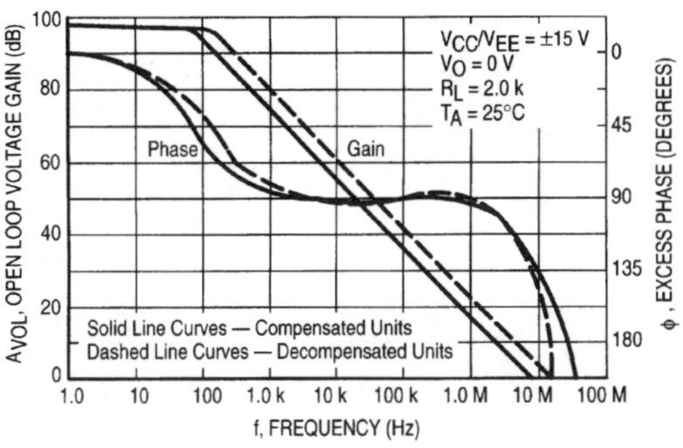

Figure 15. Open Loop Voltage Gain and Phase versus Frequency

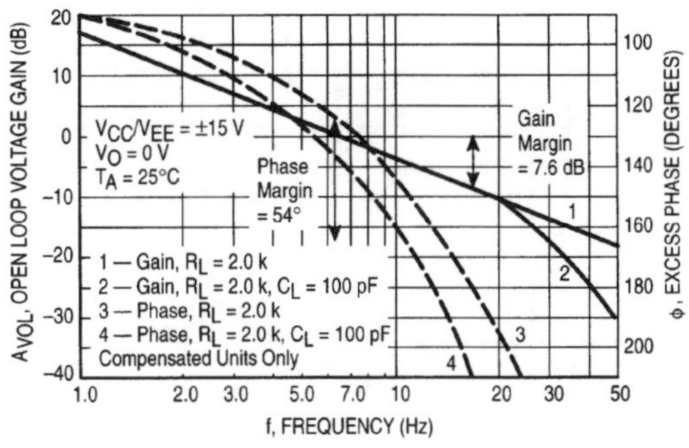

Figure 16. Open Loop Voltage Gain and Phase versus Frequency

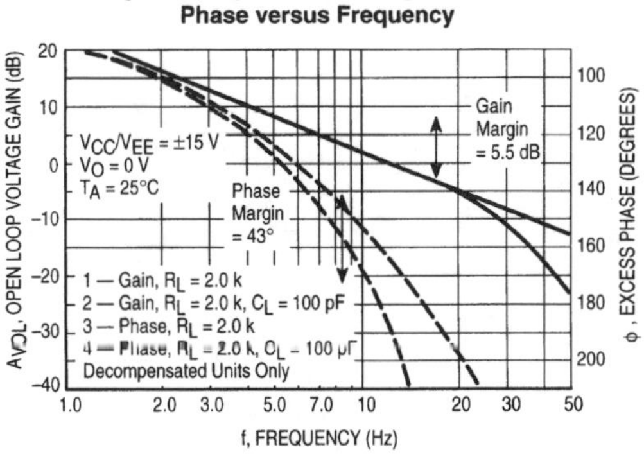

Figure 17. Normalized Gain Bandwidth Product versus Temperature

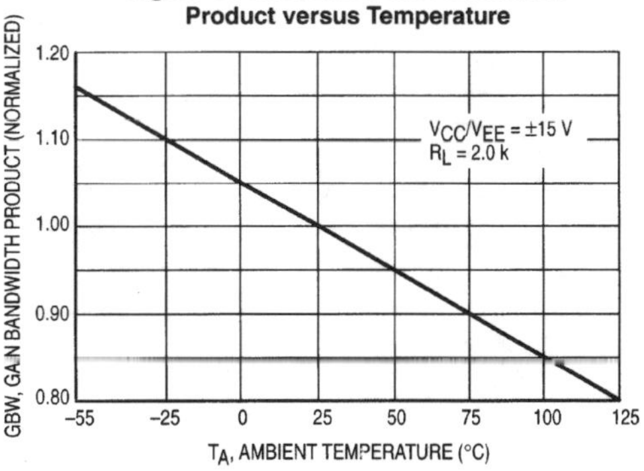

Figure 18. Percent Overshoot versus Load Capacitance

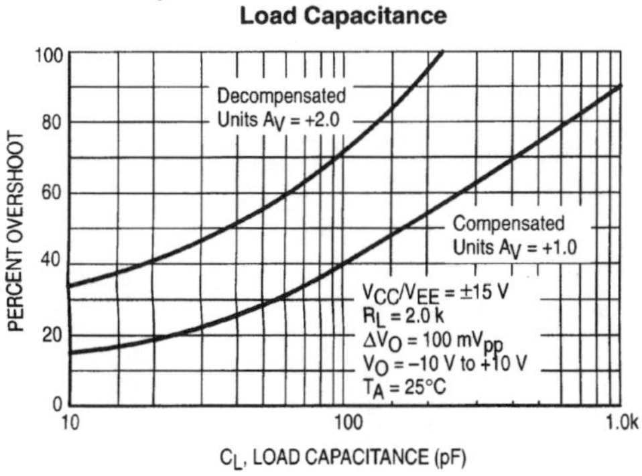

Figure 19. Phase Margin versus Load Capacitance

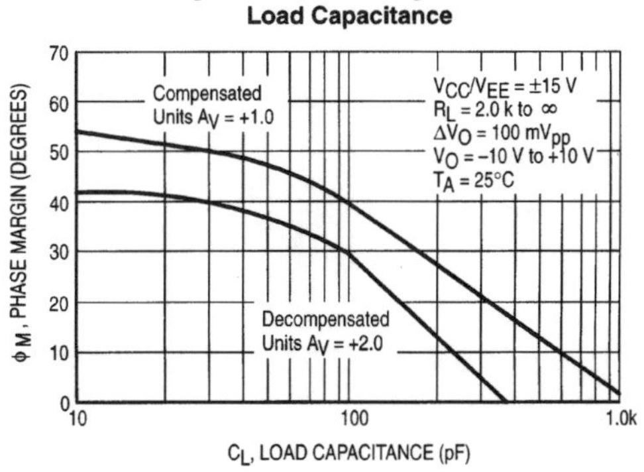

Appendix 4.2-8

MC34080 thru MC34085

Figure 20. Gain Margin versus Load Capacitance

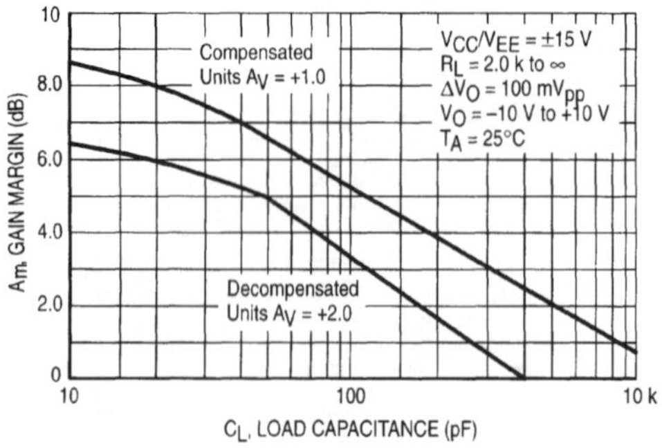

Figure 21. Phase Margin versus Temperature

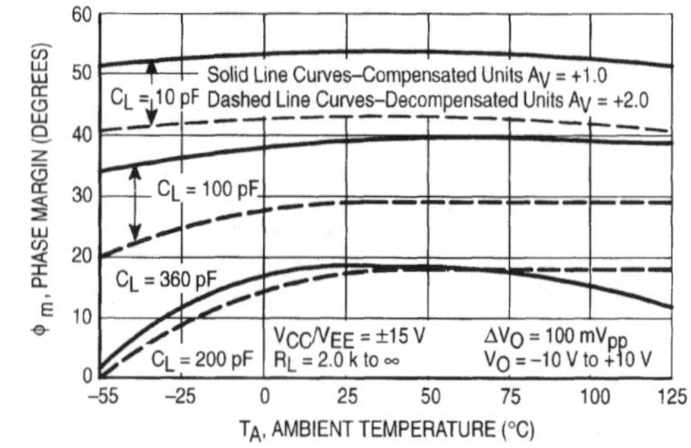

Figure 22. Gain Margin versus Temperature

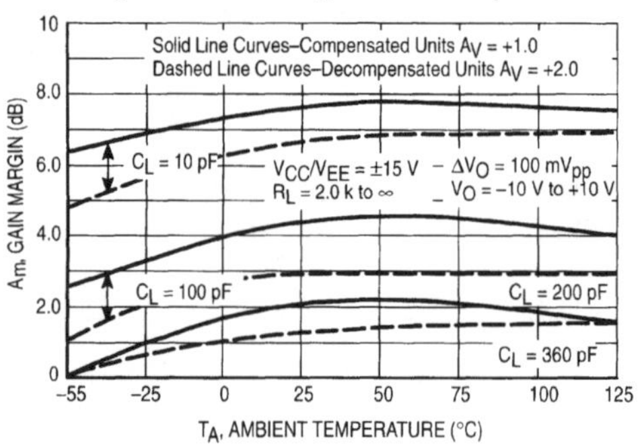

Figure 23. Normalized Slew Rate versus Temperature

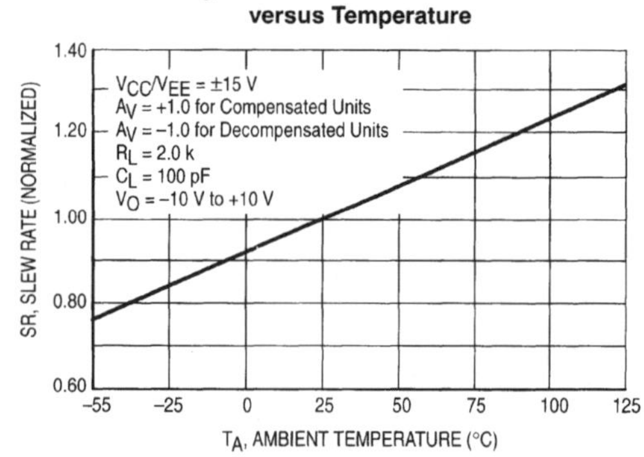

Appendix 4.2-9

MC34080 thru MC34085

MC34084 Transient Response

$A_V = +1.0$, $R_L = 2.0$ k, $V_{CC}/V_{EE} = \pm 15$ V, $T_A = 25°C$

Figure 24. Small Signal

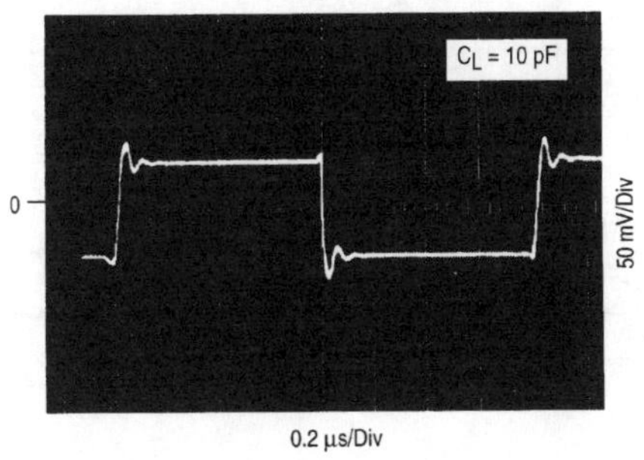

Figure 25. Large Signal

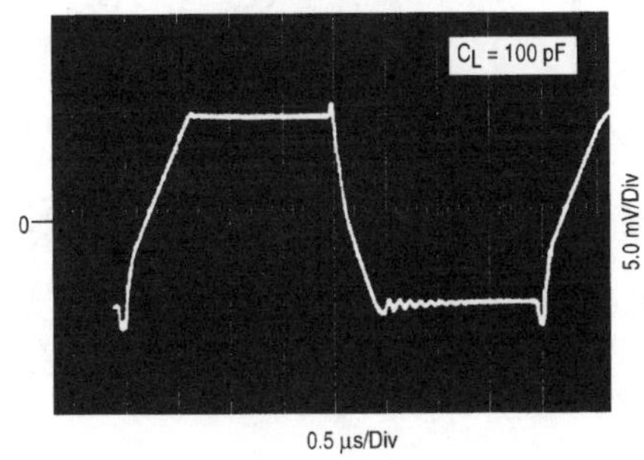

MC34085 Transient Response

$A_V = +2.0$, $R_L = 2.0$ k, $V_{CC}/V_{EE} = \pm 15$ V, $T_A = 25°C$

Figure 26. Small Signal

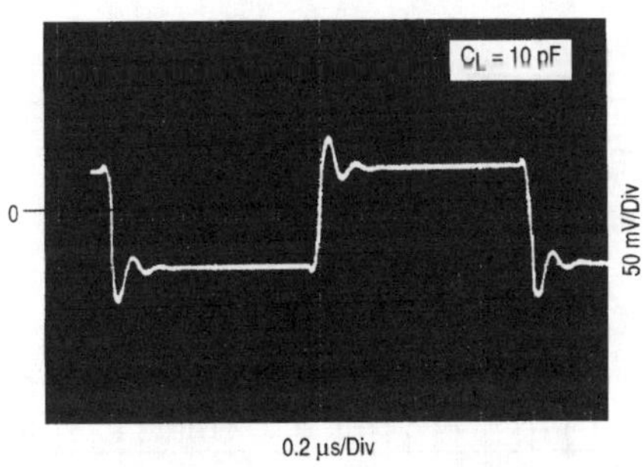

Figure 27. Large Signal

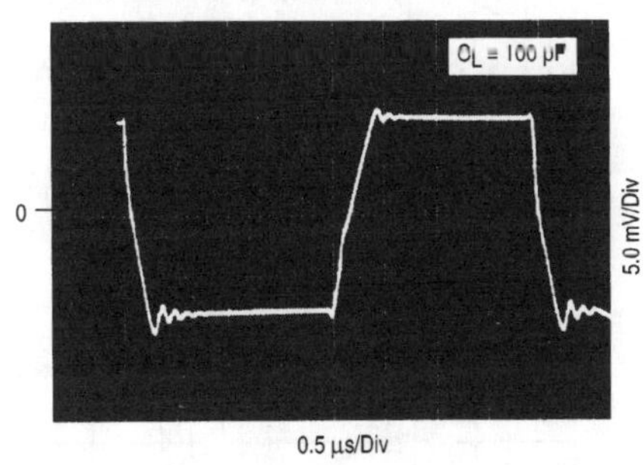

Appendix 4.2-10

MC34080 thru MC34085

Figure 28. Common Mode Rejection Ratio versus Frequency

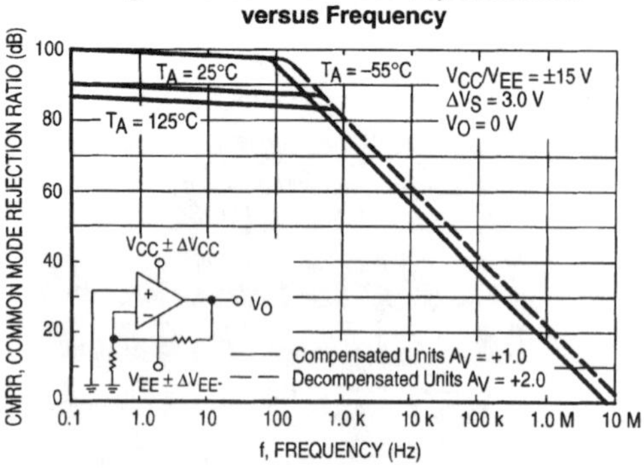

Figure 29. Power Supply Rejection Ratio versus Frequency

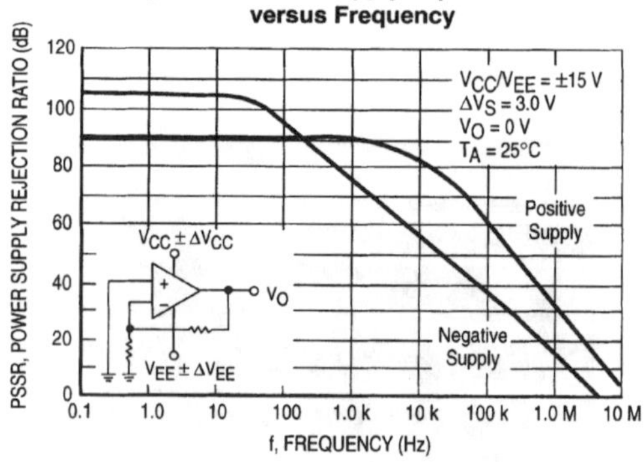

Figure 30. Power Supply Rejection Ratio versus Temperature

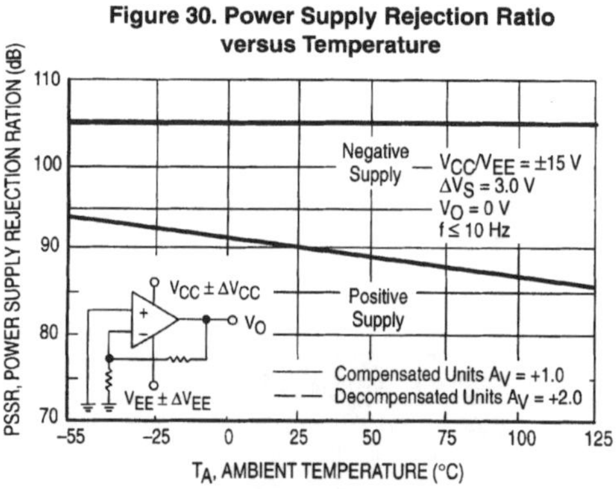

Figure 31. Normalized Supply Current versus Supply Voltage

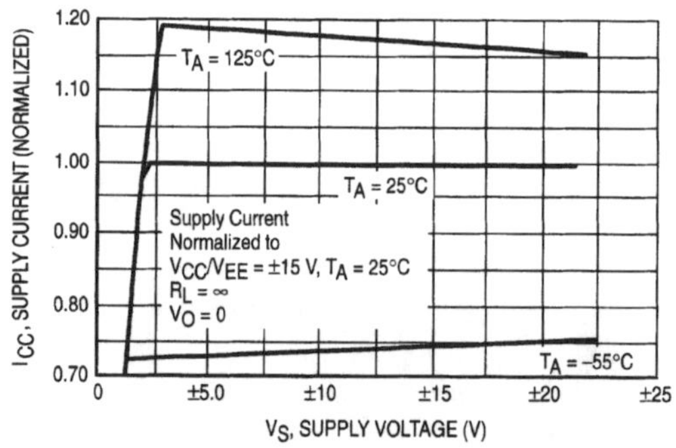

Figure 32. Channel Separation versus Frequency

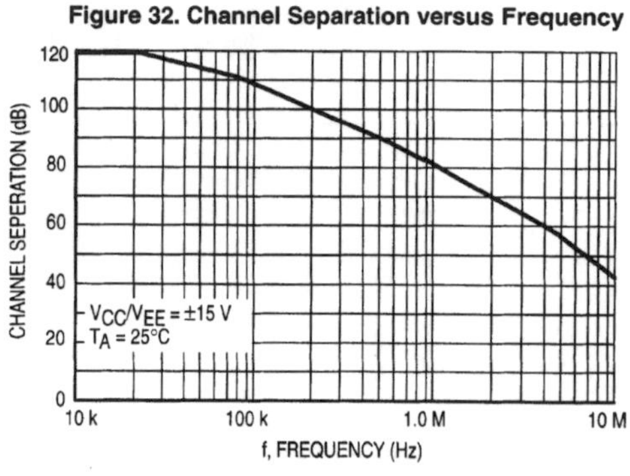

Figure 33. Spectral Noise Density

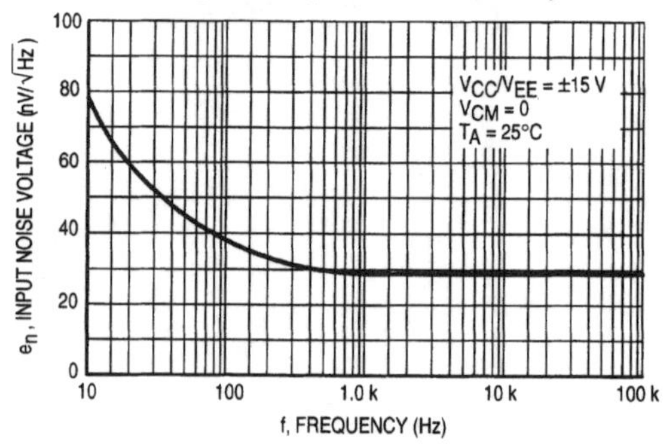

Appendix 4.2-11

MC34080 thru MC34085

APPLICATIONS INFORMATION

The bandwidth and slew rate of the MC34080 series is nearly double that of currently available general purpose JFET op–amps. This improvement in AC performance is due to the P–channel JFET differential input stage driving a compensated miller integration amplifier in conjunction with an all NPN output stage.

The all NPN output stage offers unique advantages over the more conventional NPN/PNP transistor Class AB output stage. With a 10 k load resistance, the op amp can typically swing within 1.0 V of the positive rail (V_{CC}), and within 0.3 V of the negative rail (V_{EE}), providing a 28.7 p–p swing from ±15 V supplies. This large output swing becomes most noticeable at lower supply voltages. If the load resistance is referenced to V_{CC} instead of ground, the maximum possible output swing can be achieved for a given supply voltage. For light load currents, the load resistance will pull the output to V_{CC} during the positive swing and the NPN output transistor will pull the output very near V_{EE} during the negative swing. The load resistance value should be much less than that of the feedback resistance to maximize pull–up capability.

The all NPN transistor output stage is also inherently fast, contributing to the operation amplifier's high gain–bandwidth product and fast settling time. The associated high frequency output impedance is 50 Ω (typical) at 8.0 MHz. This allows driving capacitive loads from 0 pF to 300 pF without oscillations over the military temperature range, and over the full range of output swing. The 55°C phase margin and 7.6 dB gain margin as well as the general gain and phase characteristics are virtually independent of the sink/source output swing conditions. The high frequency characteristics of the MC34080 series is especially useful for active filter applications.

The common mode input range is from 2.0 V below the positive rail (V_{CC}) to 4.0 V above the negative rail (V_{EE}). The amplifier remains active if the inputs are biased at the positive rail. This may be useful for some applications in that single supply operation is possible with a single negative supply. However, a degradation of offset voltage and voltage gain may result.

Phase reversal does not occur if either the inverting or noninverting input (or both) exceeds the positive common mode limit. If either input (or both) exceeds the negative common mode limit, the output will be in the high state. The input stage also allows a differential up to ±44 V, provided the maximum input voltage range is not exceeded. The supply voltage operating range is from ±5.0 V to ±22 V.

For optimum frequency performance and stability, careful component placement and printed circuit board layout should be exercised. For example, long unshielded input or output leads may result in unwanted input–output coupling. In order to reduce the input capacitance, resistors connected to the input pins should be physically close to these pins. This not only minimizes the input pole for optimum frequency response, but also minimizes extraneous "pickup" at this node.

Supply decoupling with adequate capacitance close to the supply pin is also important, particularly over temperature, since many types of decoupling capacitors exhibit large impedance changes over temperature.

Primarily due to the JFET inputs of the op amp, the input offset voltage may change due to temperature cycling and board soldering. After 20 temperature cycles (– 55° to 165°C), the typical standard deviation for input offset voltage is 559 μV in the plastic packages. With respect to board soldering (260°C, 10 seconds), the typical standard deviation for input offset voltage is 525 μV in the plastic package. Socketed devices should be used over a minimal temperature range for optimum input offset voltage performance.

Figure 34. Offset Nulling Circuit

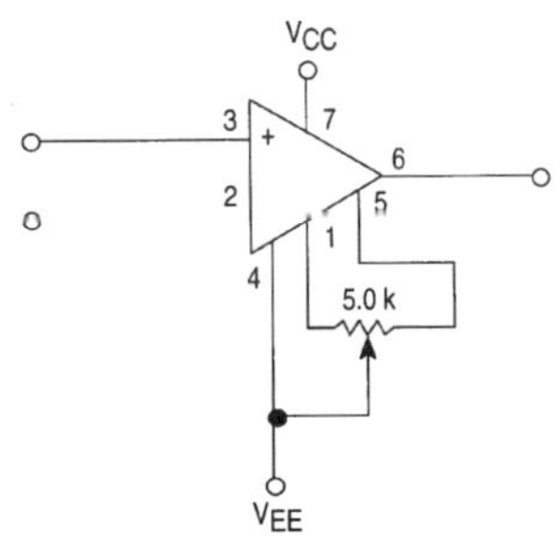

Appendix 4.3-1

MOTOROLA

MC3405

Dual Operational Amplifier and Dual Comparator

DUAL OPERATIONAL AMPLIFIER / DUAL VOLTAGE COMPARATOR

SEMICONDUCTOR TECHNICAL DATA

The MC3405 contains two differential–input operational amplifiers and two comparators, each set capable of single supply operation. This operational amplifier–comparator circuit fulfills its applications as a general purpose product for automotive and consumer circuits as well as an industrial building block.

The MC3405 is specified over the commercial operating temperature range of 0° to +70°C.

- Operational Amplifier Equivalent in Performance to MC3403
- Comparator Similar in Performance to LM339
- Single Supply Operation: 3.0 V to 36 V
- Split Supply Operation: ±1.5 V to ±18 V
- Low Supply Current Drain
- Operational Amplifier is Internally Frequency Compensated
- Comparator TTL and CMOS Compatible

P SUFFIX
PLASTIC PACKAGE
CASE 646

PIN CONNECTIONS

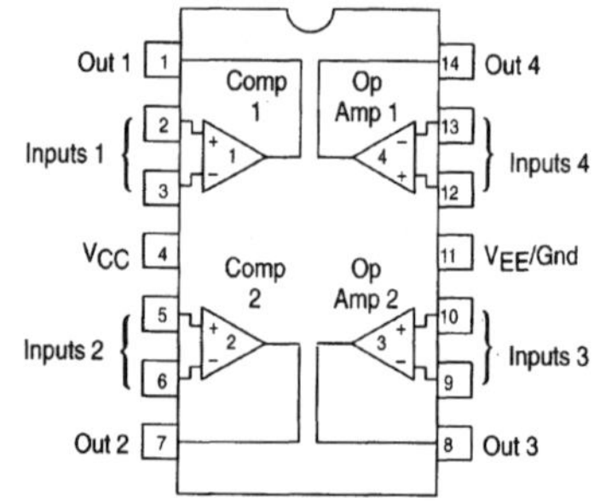

(Top View)

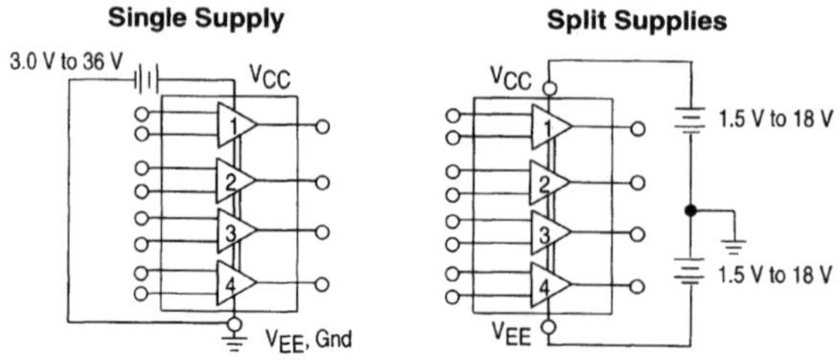

ORDERING INFORMATION

Device	Operating Temperature Range	Package
MC3405P	T_A = 0° to +70°C	Plastic DIP

Appendix 4.3-2

MC3405

OPERATIONAL AMPLIFIER SECTION

MAXIMUM RATINGS

Rating	Symbol	Value	Unit
Power Supply Voltage – Single Supply Split Supplies	V_{CC} V_{CC}, V_{EE}	36 ±18	Vdc
Input Differential Voltage Range	V_{IDR}	±36	Vdc
Input Common Mode Voltage Range	V_{ICR}	±18	Vdc
Operating Ambient Temperature Range	T_A	0 to +70	°C
Storage Temperature Range	Tstg	–55 to +125	°C
Operating Junction Temperature Range	T_J	150	°C

ELECTRICAL CHARACTERISTICS (V_{CC} = 5.0 V, V_{EE} = Gnd, T_A = 25°C, unless otherwise noted.)

Characteristic	Symbol	Min	Typ	Max	Unit
Input Offset Voltage	V_{IO}	–	2.0	10	mV
Input Offset Current	I_{IO}	–	30	50	nA
Input Bias Current	I_{IB}	–	–200	–500	nA
Large–Signal, Open Loop Voltage Gain (R_L = 2.0 kΩ)	A_{VOL}	20	200	–	V/mV
Power Supply Rejection	PSR	–	–	150	µV/V
Output Voltage Range (Note 1) (R_L = 10 kΩ, V_{CC} = 5.0 V) (R_L = 10 kΩ, 5.0 V ≤ V_{CC} ≤ 30 V)	V_{OR}	 3.3 V_{CC}–2.0	 3.5 V_{CC} –1.7	 – –	V_{pp}
Power Supply Current (Notes 2 and 3)	I_{CC}	–	2.5	7.0	mA
Channel Separation, f = 1.0 kHz to 20 kHz (Input Referenced)	–	–	–120	–	dB

ELECTRICAL CHARACTERISTICS (V_{CC} = +15 V, V_{EE} = –15 V, T_A = 25°C, unless otherwise noted.)

Characteristic	Symbol	Min	Typ	Max	Unit
Input Offset Voltage (T_A = T_{low} + T_{high}) (Note 4)	V_{IO}	– –	2.0 –	10 12	mV
Average Temperature Coefficient of Input Offset Voltage	$\Delta V_{IO}/\Delta T$	–	15	–	µV/°C
Input Offset Current (T_A = T_{low} to T_{high}) (Note 4)	I_{IO}	– –	– –	50 200	nA
Input Bias Current (T_A = T_{low} to T_{high}) (Note 4)	I_{IB}	– –	–200 –	–500 –800	nA
Input Common Mode Voltage Range	V_{ICR}	+13 –V_{EE}	–	–	Vdc
Large Signal, Open Loop Voltage Gain (V_O = ±10 V, R_L = 2.0 kΩ) (T_A = T_{low} to T_{high}) (Note 4)	A_{VOL}	 20 15	 200 100	 – –	V/mV
Common Mode Rejection	CMR	70	90	–	dB
Power Supply Rejection Ratio	PSRR	–	30	150	µV/V
Output Voltage (R_L = 10 kΩ) (R_L = 2.0 kΩ) (R_L = 2.0 kΩ, T_A = T_{low} to T_{high}) (Note 4)	V_O	 ±12 ±10 ±10	 ±13.5 ±13 –	 – – –	Vdc
Output Short Circuit Current	I_{SC}	±10	±20	±45	mA
Power Supply Current (Notes 2 and 3)	I_{CC}, I_{EE}	–	2.8	7.0	mA
Phase Margin	ϕm	–	60	–	Degrees
Small–Signal Bandwidth (A_V = 1, R_L = 10 kΩ, V_O = 50 mV)	BW	–	1.0	–	MHz

NOTES: 1. Output will swing to ground.
2. Not to exceed maximum package power dissipation.
3. For operational amplifier and comparator.
4. T_{low} = 0°C, T_{high} = +70°C

Appendix 4.3-3

MC3405

ELECTRICAL CHARACTERISTICS (V_{CC} = +15 V, V_{EE} = −15 V, T_A = 25°C, unless otherwise noted.)

Characteristic	Symbol	Min	Typ	Max	Unit
Power Bandwidth (A_V = 1, R_L = 2.0 kΩ, V_O = 20 V_{pp}, THD = 5%)	BWp	–	9.0	–	kHz
Rise Time/Fall Time	t_{TLH}, t_{THL}	–	0.35	–	μs
Overshoot (A_V = 1, R_L = 10 kΩ, V_O = 50 mV)	os	–	20	–	%
Slew Rate	SR	–	0.6	–	V/μs

COMPARATOR SECTION

MAXIMUM RATINGS

Rating	Symbol	Value	Unit
Power Supply Voltage – Single Supply Split Supplies	V_{CC} V_{CC}, V_{EE}	36 ±18	Vdc
Input Differential Voltage Range	V_{IDR}	±36	Vdc
Input Common Mode Voltage Range	V_{ICR}	−0.3 to +36	Vdc
Sink Current	I_{Sink}	20	mA
Operating Ambient Temperature Range	T_A	0 to +70	°C
Storage Temperature Range	Tstg	−55 to +125	°C
Operating Junction Temperature Range	T_J	150	°C

ELECTRICAL CHARACTERISTICS (V_{CC} = 5.0 V, V_{EE} = Gnd, T_A = 25°C, unless otherwise noted.)

Characteristic	Symbol	Min	Typ	Max	Unit
Input Offset Voltage (T_A = T_{low} to T_{high}) (Notes 1 and 2)	V_{IO}	– –	2.0 –	10 12	mV
Average Temperature Coefficient of Input Offset Voltage	$\Delta V_{IO}/\Delta T$	–	15	–	μV/°C
Input Offset Current (T_A = T_{low} to T_{high}) (Note 1)	I_{IO}	– –	50 –	100 200	nA
Input Bias Current (T_A = T_{low} to T_{high}) (Note 1)	I_{IB}	– –	−125 –	−500 −800	nA
Input Common Mode Voltage Range (T_A = T_{low} to T_{high}) (Note 1)	V_{ICR}	0 0	V_{CC} −1.5 V_{CC} −1.7	V_{CC} −1.7 V_{CC} −2.0	Vpp
Input Differential Voltage (All V_{in} ≥ 0 Vdc)	V_{ID}	–	–	36	V
Large–Signal, Open Loop Voltage Gain (R_L = 15 kΩ)	A_{VOL}	–	200	–	V/mV
Output Sink Current (−V_{in} ≥ 1.0 Vdc, +V_{in}= 0, V_O ≤ 1.5 V)	I_{Sink}	6.0	16	–	mA
Low Level Output Voltage (+V_{in}= 0 V, −V_{in}= 1.0 V, I_{Sink} = 4.0 mA) (T_A = T_{low} to T_{high}) (Note 1)	V_{OL}	 – –	 350 –	 500 700	μA
Output Leakage Current (+V_{in} ≥ 1.0 Vdc, −V_{in}= 0, V_O = 5.0 Vdc) (T_A = T_{low} to T_{high}) (Note 1)	I_{OL}	 – –	 0.1 0.1	 1.0 1.0	μA
Large–Signal Response	–	–	300	–	ns
Response Time (Note 3) (V_{RL} = 5.0 Vdc, R_L = 5.1 kΩ)	–	–	1.3	–	μs

NOTES: 1. T_{low} = 0°C, T_{high} = +70°C
2. V_O ≅ 1.4 V, R_S = 0 Ω with V_{CC} from 5.0 Vdc to 30 Vdc, and over the input common mode range 0 to V_{CC} −1.7 V.
3. The response time specified is for a 100 mV input step with 5.0 mV overdrive. For larger signals 300 ns is typical.

Appendix 4.3-4

MC3405

Representative Schematic Diagram
(1/2 of Circuit Shown)

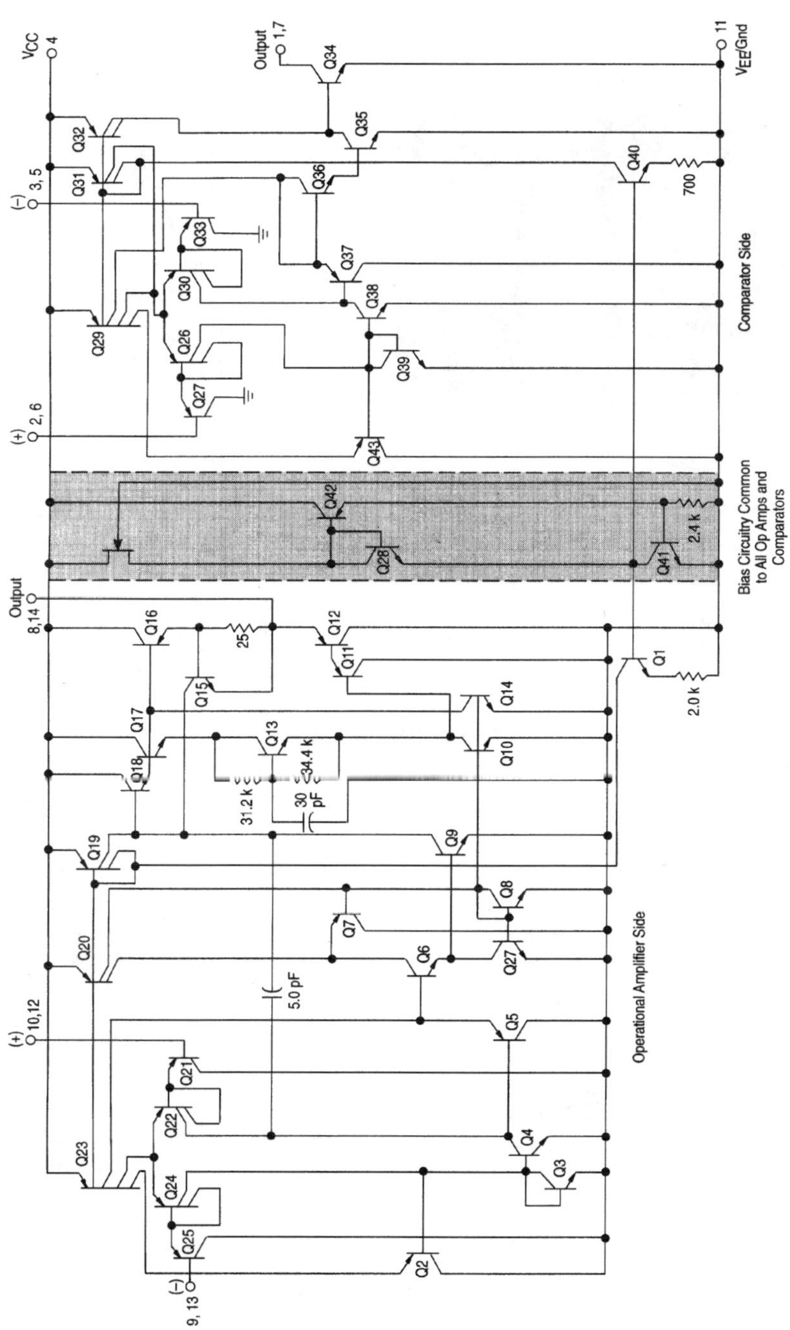

Appendix 4.3-5

MC3405

OPERATIONAL AMPLIFIER SECTION

Figure 1. Sine Wave Response

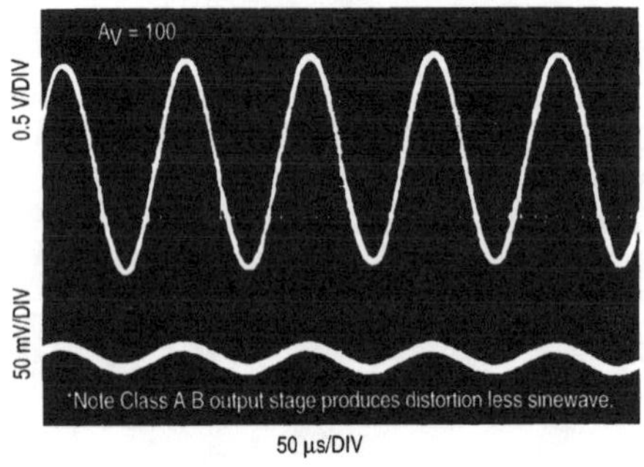

Figure 2. Open Loop Frequency Response

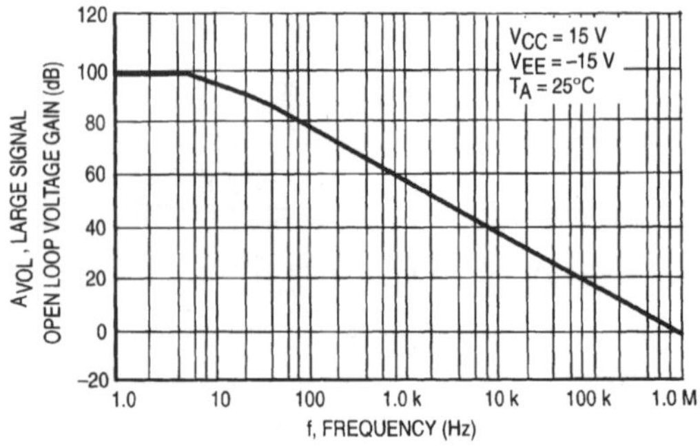

Figure 3. Power Bandwidth

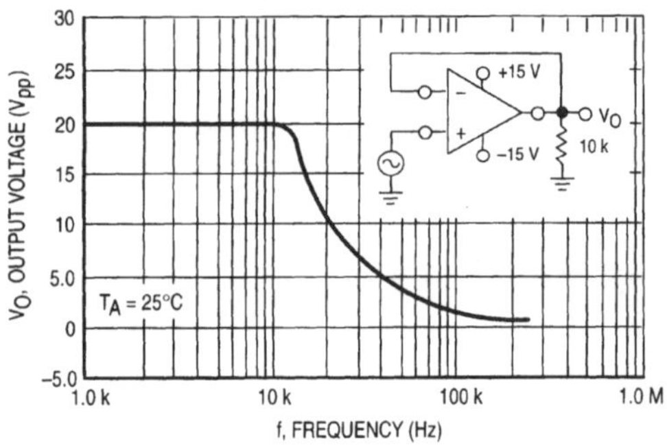

Figure 4. Output Swing versus Supply Voltage

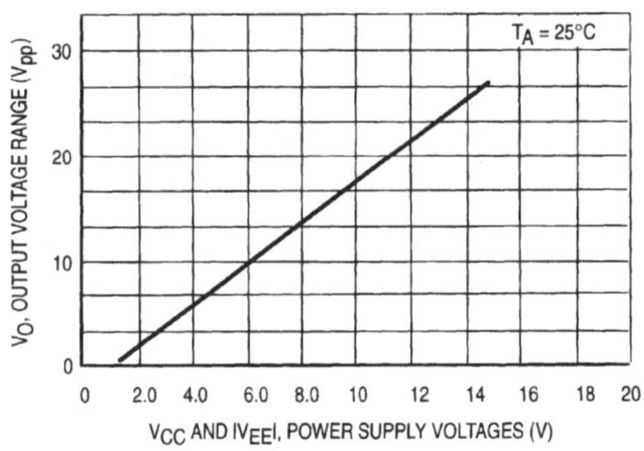

Figure 5. Input Bias Current versus Temperature

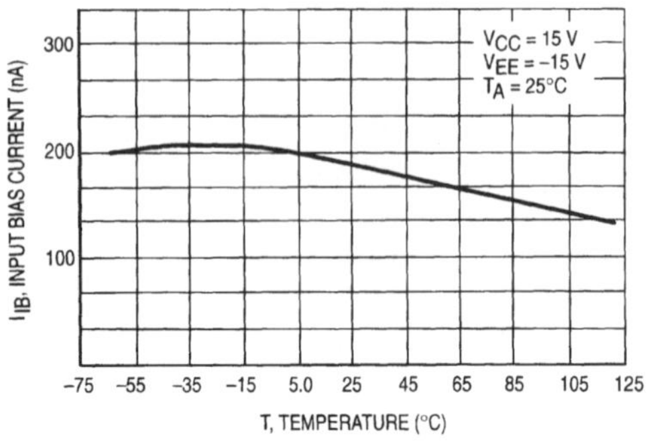

Figure 6. Input Bias Current versus Supply Voltage

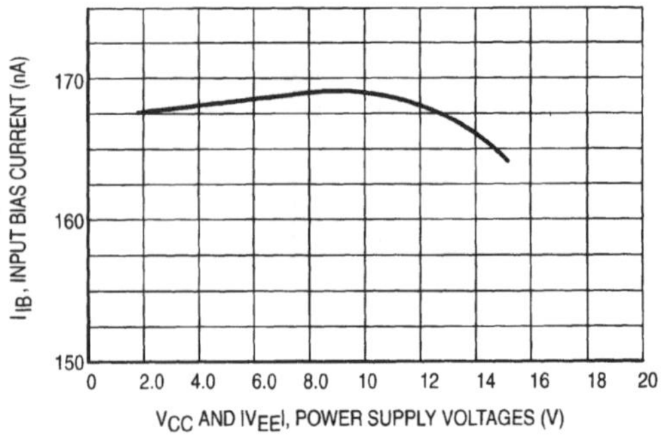

Appendix 4.3-6

MC3405

COMPARATOR SECTION

Figure 7. Normalized Input Offset Voltage

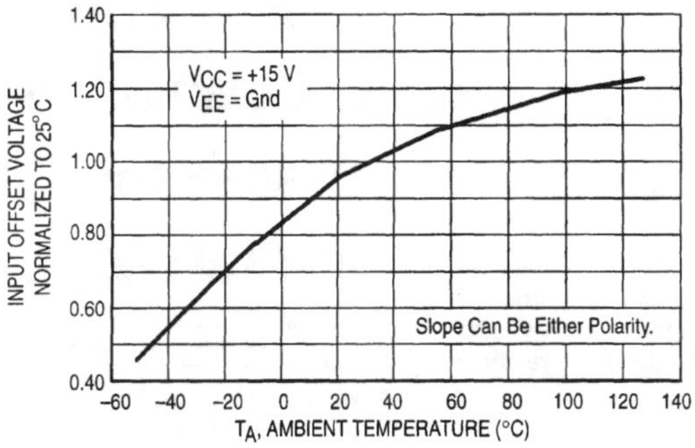

Figure 8. Input Bias Current

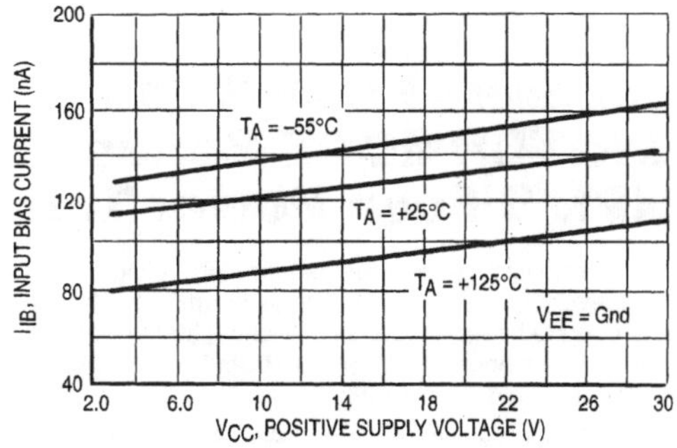

Figure 9. Normalized Input Offset Current

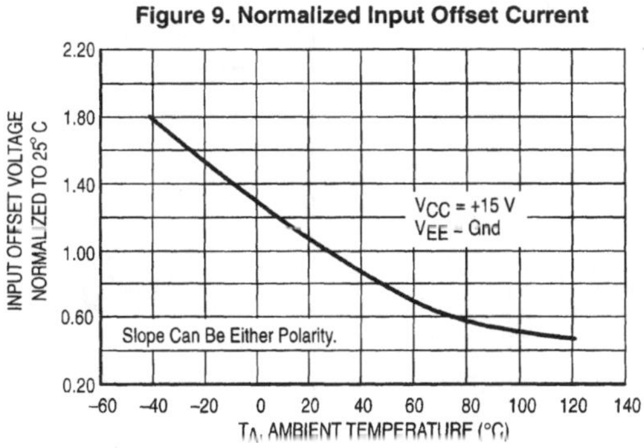

Figure 10. Output Sink Current versus Output Voltage

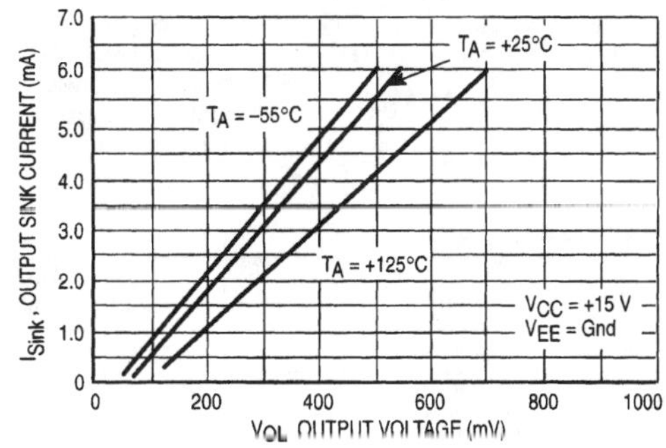

Figure 11. Pulse Width Modulator Schematic and Waveforms

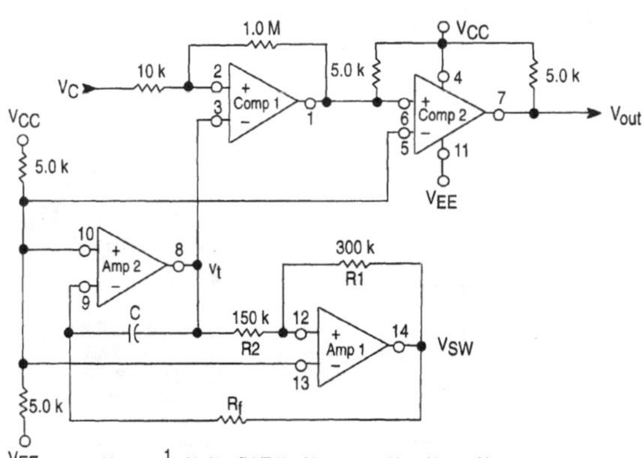

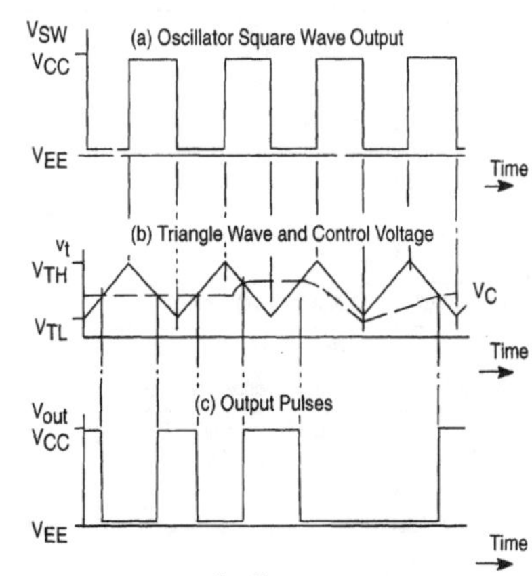

$V_{TH} = \frac{1}{2} V_S(1 + R2/R1) + V_{EE}$ $\quad V_S = V_{CC} - V_{EE}$

$V_{TL} = \frac{1}{2} V_S(1 - R2/R1) + V_{EE}$

Oscillator Frequency $f = \frac{R1}{4R_fCR2}$

Pulse Width $= \left(\frac{1}{f}\right)\left(\frac{V_C - V_{TL}}{V_{TH} - V_{TL}}\right)$ when: $V_{TL} < V_C < V_{TH}$

Duty Cycle in % $= \left(\frac{V_C - V_{TL}}{V_{TH} - V_{TL}}\right)(100)$

Appendix 4.4-1

MOTOROLA
SEMICONDUCTOR TECHNICAL DATA

MJ16010
MJW16010
MJ16012*
MJW16012*

*Motorola Preferred Device

Designer's™ Data Sheet
SWITCHMODE Series
NPN Silicon Power Transistors

These transistors are designed for high–voltage, high–speed, power switching in inductive circuits where fall time is critical. They are particularly suited for line–operated switchmode applications. The MJ16012 and MJW16012 are selected high gain versions of the MJ16010 and MJW16010 for applications where drive current is limited.

- Switching Regulators
- Inverters
- Solenoids
- Relay Drivers
- Motor Controls
- Deflection Circuits

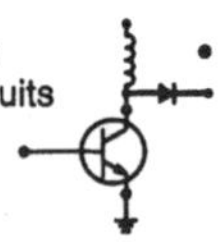

- Fast Turn–Off Times — $T_C = 100°C$
  50 ns Inductive Fall Time (Typ)
  90 ns Inductive Crossover Time (Typ)
  800 ns Inductive Storage Time (Typ)
- 100°C Performance Specified for:
  Reverse–Biased SOA with Inductive Loads
  Switching Times with Inductive Loads
  Saturation Voltages
  Leakage Currents

15 AMPERE
NPN SILICON
POWER TRANSISTORS
450 VOLTS
135 AND 175 WATTS

CASE 1–07
TO–204AA
(TO–3)
MJ16010
MJ16012

CASE 340F–03
TO–247AE
MJW16010
MJW16012

MAXIMUM RATINGS

Rating	Symbol	MJ16010 MJ16012	MJW16010 MJW16012	Unit
Collector–Emitter Voltage	V_{CEO}	450		Vdc
Collector–Emitter Voltage	V_{CEV}	850		Vdc
Emitter–Base Voltage	V_{EB}	6.0		Vdc
Collector Current — Continuous — Peak (1)	I_C I_{CM}	15 20		Adc
Base Current — Continuous — Peak (1)	I_B I_{BM}	10 15		Adc
Total Device Dissipation @ $T_C = 25°C$ @ $T_C = 100°C$ Derate above 25°C	P_D	 175 100 1.0	 135 53.8 1.11	Watts W/°C
Operating and Storage Junction Temperature Range	T_J, T_{stg}	–65 to 200	–55 to 150	°C

THERMAL CHARACTERISTICS

Characteristic	Symbol	Max		Unit
Thermal Resistance, Junction to Case	$R_{\theta JC}$	1.0	0.93	°C/W
Lead Temperature for Soldering Purposes, 1/8″ from Case for 5 Seconds	T_L	275		°C

(1) Pulse Test: Pulse Width ≤ 50 μs, Duty Cycle ≥ 10%

Designer's Data for "Worst Case" Conditions — The Designer's Data Sheet permits the design of most circuits entirely from the information presented. SOA Limit curves — representing boundaries on device characteristics — are given to facilitate "worst case" design.

Preferred devices are Motorola recommended choices for future use and best overall value.

Appendix 4.4-2

MJ16010 MJW16010 MJ16012 MJW16012

ELECTRICAL CHARACTERISTICS (T_C = 25°C unless otherwise noted)

Characteristic	Symbol	Min	Typ	Max	Unit
OFF CHARACTERISTICS					
Collector–Emitter Sustaining Voltage (Table 2) (I_C = 100 mA, I_B = 0)	$V_{CEO(sus)}$	450	—	—	Vdc
Collector Cutoff Current (V_{CEV} = 850 Vdc, $V_{BE(off)}$ = 1.5 Vdc) (V_{CEV} = 850 Vdc, $V_{BE(off)}$ = 1.5 Vdc, T_C = 100°C)	I_{CEV}	 — —	 — —	 0.25 1.5	mAdc
Collector Cutoff Current (V_{CE} = 850 Vdc, R_{BE} = 50 Ω, T_C = 100°C)	I_{CER}	—	—	2.5	mAdc
Emitter Cutoff Current (V_{EB} = 6.0 Vdc, I_C = 0)	I_{EBO}	—	—	10	mAdc
SECOND BREAKDOWN					
Second Breakdown Collector Current with Base Forward Biased	$I_{S/b}$	See Figure 15			
Clamped Inductive SOA with Base Reverse Biased	RBSOA	See Figure 16			
ON CHARACTERISTICS (1)					
Collector–Emitter Saturation Voltage (I_C = 5.0 Adc, I_B = 0.7 Adc) (I_C = 10 Adc, I_B = 1.3 Adc) (I_C = 10 Adc, I_B = 1.3 Adc, T_C = 100°C)	$V_{CE(sat)}$	 — — —	 — — —	 2.5 3.0 3.0	Vdc
Base–Emitter Saturation Voltage (I_C = 10 Adc, I_B = 1.3 Adc) (I_C = 10 Adc, I_B = 1.3 Adc, T_C = 100°C)	$V_{BE(sat)}$	 — —	 — —	 1.5 1.5	Vdc
DC Current Gain (I_C = 15 Adc, V_{CE} = 5.0 Vdc)	h_{FE}	5.0	—	—	—
DYNAMIC CHARACTERISTICS					
Output Capacitance (V_{CB} = 10 Vdc, I_E = 0, f_{test} = 1.0 kHz)	C_{ob}	—	—	400	pF

SWITCHING CHARACTERISTICS

Characteristic	Conditions		Symbol	Min	Typ	Max	Unit
Resistive Load (Table 1)							
Delay Time	(I_C = 10 Adc, V_{CC} = 250 Vdc, I_{B1} = 1.3 Adc, PW = 30 μs, Duty Cycle ≤ 2.0%)	(I_{B2} = 2.6 Adc, R_{B2} = 1.6 Ω)	t_d	—	20	—	ns
Rose Time			t_r	—	200	—	
Storage Time			t_s	—	1200	—	
Fall Time			t_f	—	200	—	
Storage Time		($V_{BE(off)}$ = 5.0 Vdc)	t_s	—	650	—	
Fall Time			t_f	—	80	—	
Inductive Load (Table 2)							
Storage Time	(I_C = 10 Adc, I_{B1} = 1.3 Adc, $V_{BE(off)}$ = 5.0 Vdc, $V_{CE(pk)}$ = 400 Vdc)	(T_C = 100°C)	t_{sv}	—	800	1800	ns
Fall Time			t_{fi}	—	50	200	
Crossover Time			t_c	—	90	250	
Storage Time		(T_C = 150°C)	t_{sv}	—	1050	—	
Fall Time			t_{fi}	—	70	—	
Crossover Time			t_c	—	120	—	

(1) Pulse Test: Pulse Width = 300 μs, Duty Cycle ≤ 2.0%

Appendix 4.4-3

MJ16010 MJW16010 MJ16012 MJW16012

ELECTRICAL CHARACTERISTICS (T_C = 25°C unless otherwise noted)

Characteristic	Symbol	Min	Typ	Max	Unit
OFF CHARACTERISTICS					
Collector–Emitter Sustaining Voltage (Table 2) (I_C = 100 mA, I_B = 0)	$V_{CEO(sus)}$	450	—	—	Vdc
Collector Cutoff Current (V_{CEV} = 850 Vdc, $V_{BE(off)}$ = 1.5 Vdc) (V_{CEV} = 850 Vdc, $V_{BE(off)}$ = 1.5 Vdc, T_C = 100°C)	I_{CEV}	— —	— —	0.25 1.5	mAdc
Collector Cutoff Current (V_{CE} = 850 Vdc, R_{BE} = 50 Ω, T_C = 100°C)	I_{CER}	—	—	2.5	mAdc
Emitter Cutoff Current (V_{EB} = 6.0 Vdc, I_C = 0)	I_{EBO}	—	—	10	mAdc
SECOND BREAKDOWN					
Second Breakdown Collector Current with Base Forward Biased	$I_{S/b}$	See Figure 15			
Clamped Inductive SOA with Base Reverse Biased	RBSOA	See Figure 16			
ON CHARACTERISTICS (1)					
Collector–Emitter Saturation Voltage (I_C = 5.0 Adc, I_B = 0.7 Adc) (I_C = 10 Adc, I_B = 1.0 Adc) (I_C = 10 Adc, I_B = 1.0 Adc, T_C = 100°C)	$V_{CE(sat)}$	— — —	— — —	2.5 3.0 3.0	Vdc
Base–Emitter Saturation Voltage (I_C = 10 Adc, I_B = 1.0 Adc) (I_C = 10 Adc, I_B = 1.0 Adc, T_C = 100°C)	$V_{BE(sat)}$	— —	— —	1.5 1.5	Vdc
DC Current Gain (I_C = 15 Adc, V_{CE} = 5.0 Vdc)	h_{FE}	7.0	—	—	—
DYNAMIC CHARACTERISTICS					
Output Capacitance (V_{CB} = 10 Vdc, I_E = 0, f_{test} = 1.0 kHz)	C_{ob}	—	—	400	pF

SWITCHING CHARACTERISTICS

Characteristic	Conditions		Symbol	Min	Typ	Max	Unit
Resistive Load (Table 1)							
Delay Time	(I_C = 10 Adc, V_{CC} = 250 Vdc, I_{B1} = 1.0 Adc, PW = 30 μs, Duty Cycle ≤ 2.0%)	(I_{B2} = 2.0 Adc, R_{B2} = 1.6 Ω)	t_d	—	20	—	ns
Rose Time			t_r	—	200	—	
Storage Time			t_s	—	900	—	
Fall Time			t_f	—	150	—	
Storage Time		($V_{BE(off)}$ = 5.0 Vdc)	t_s	—	500	—	
Fall Time			t_f	—	40	—	
Inductive Load (Table 2)							
Storage Time	(I_C = 10 Adc, I_{B1} = 1.0 Adc, $V_{BE(off)}$ = 5.0 Vdc, $V_{CE(pk)}$ = 400 Vdc)	(T_C = 100°C)	t_{sv}	—	650	1500	ns
Fall Time			t_{fi}	—	30	150	
Crossover Time			t_c	—	50	200	
Storage Time		(T_C = 150°C)	t_{sv}	—	850	—	
Fall Time			t_{fi}	—	30	—	
Crossover Time			t_c	—	70	—	

(1) Pulse Test: Pulse Width = 300 μs, Duty Cycle ≤ 2.0%

Appendix 4.4-4

MJ16010 MJW16010 MJ16012 MJW16012

TYPICAL STATIC CHARACTERISTICS

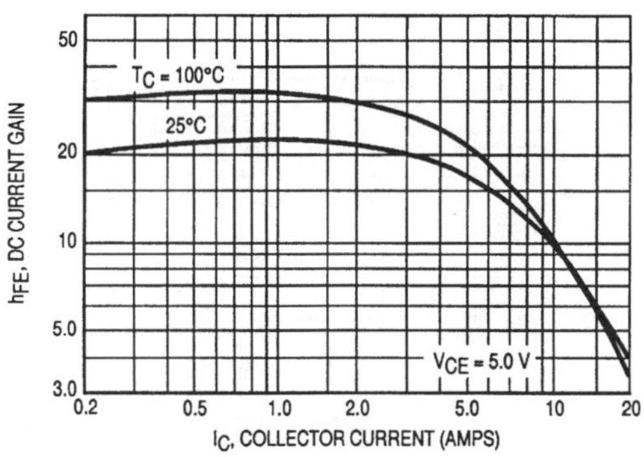

Figure 1. DC Current Gain

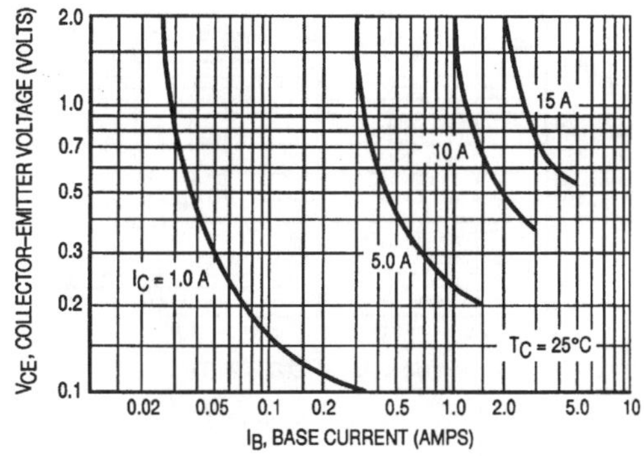

Figure 2. Collector Saturation Region

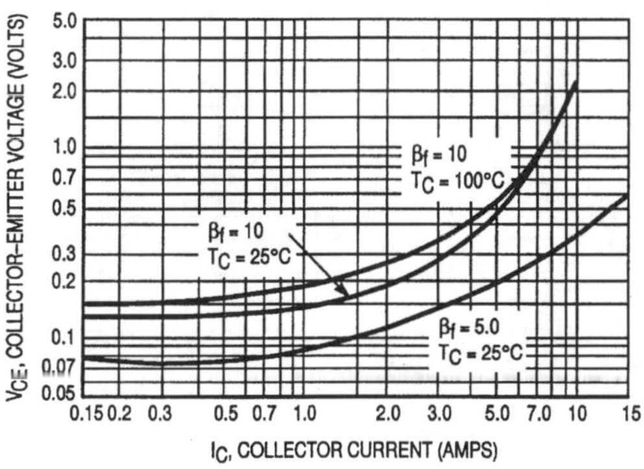

Figure 3. Collector–Emitter Saturation Voltage

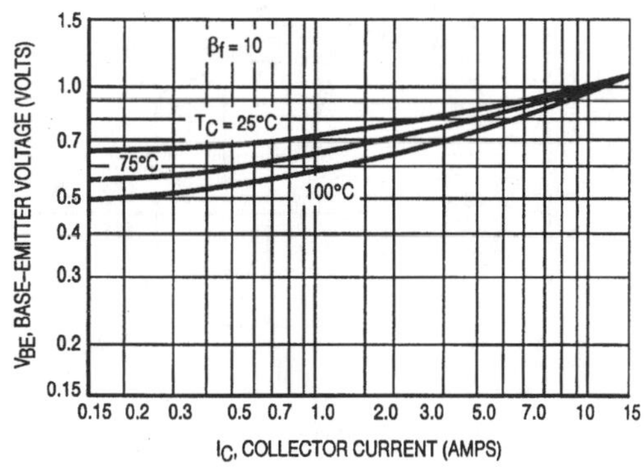

Figure 4. Base–Emitter Voltage

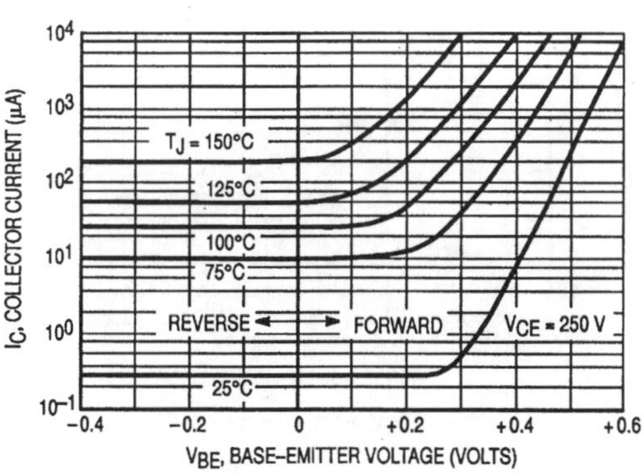

Figure 5. Collector Cutoff Region

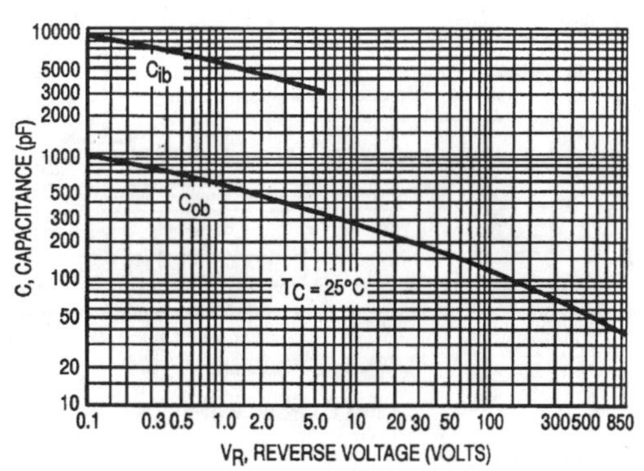

Figure 6. Capacitance

Appendix 4.4-5

MJ16010 MJW16010 MJ16012 MJW16012

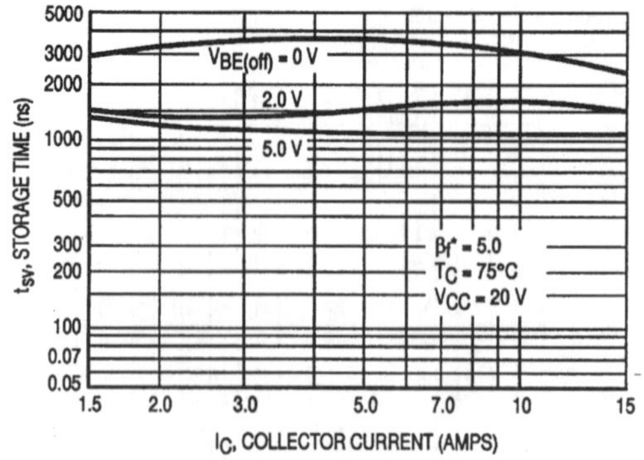

Figure 7. Storage Time

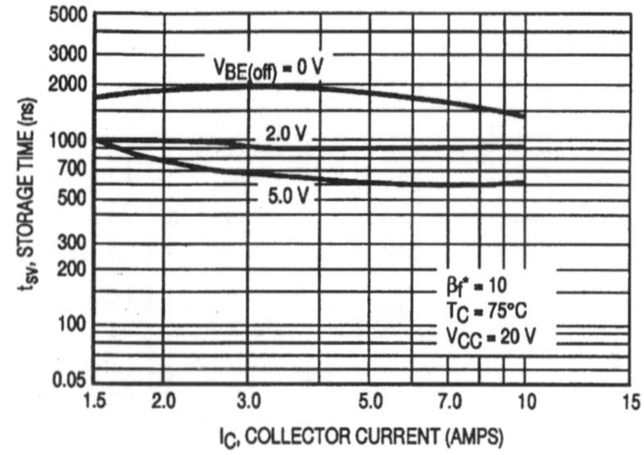

Figure 8. Storage Time

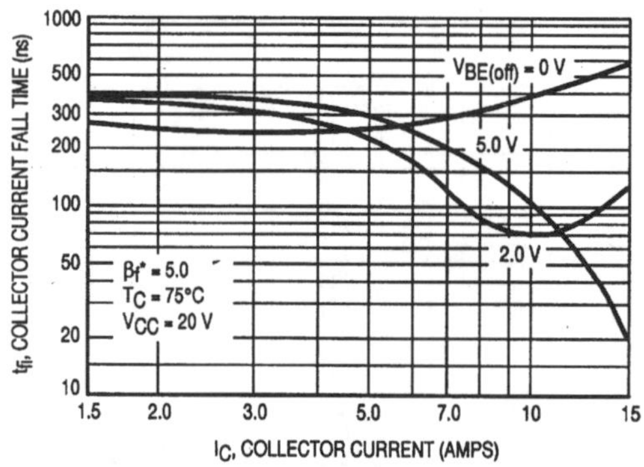

Figure 9. Collector Current Fall Time

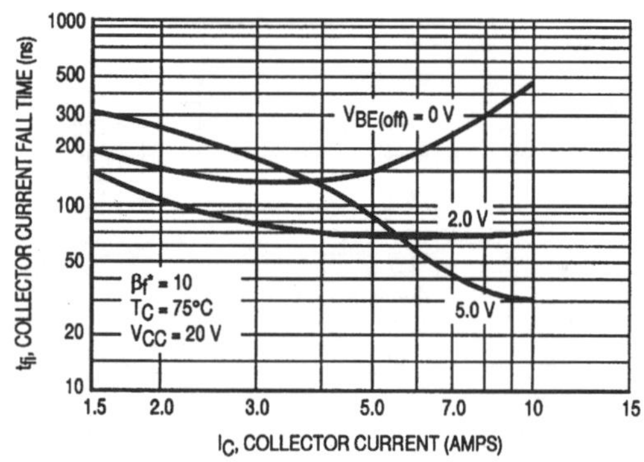

Figure 10. Collector Current Fall Time

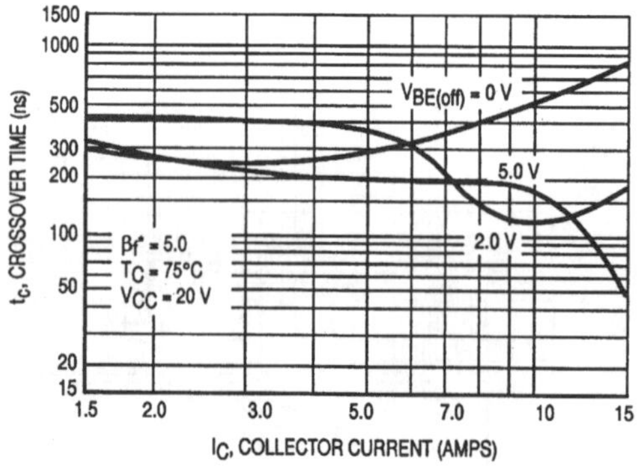

Figure 11. Crossover Time

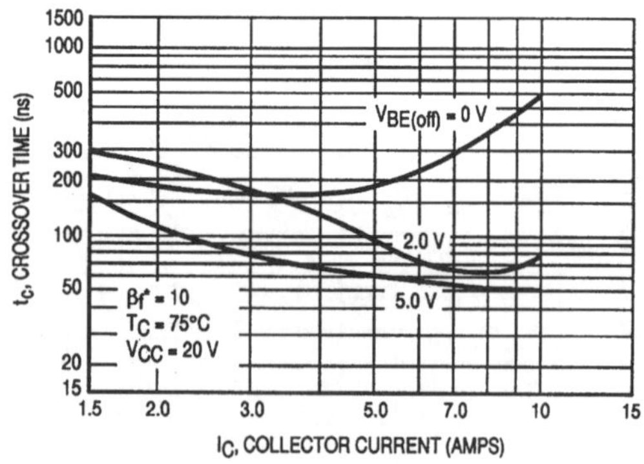

Figure 12. Crossover Time

$^*\beta_f = \frac{I_C}{I_{B1}}$

Appendix 4.4-6

MJ16010 MJW16010 MJ16012 MJW16012

GUARANTEED SAFE OPERATING AREA LIMITS

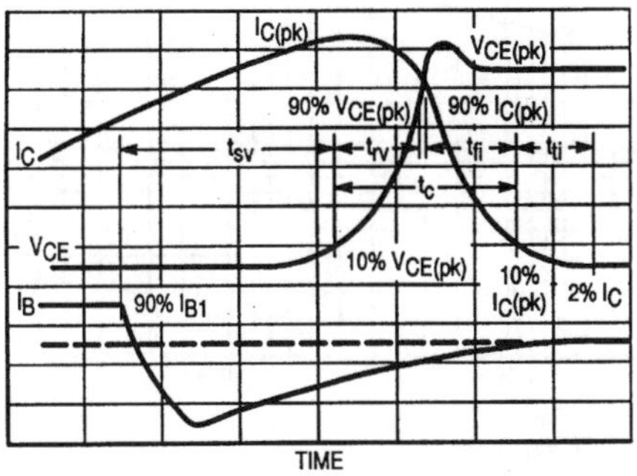

Figure 13. Inductive Switching Measurements

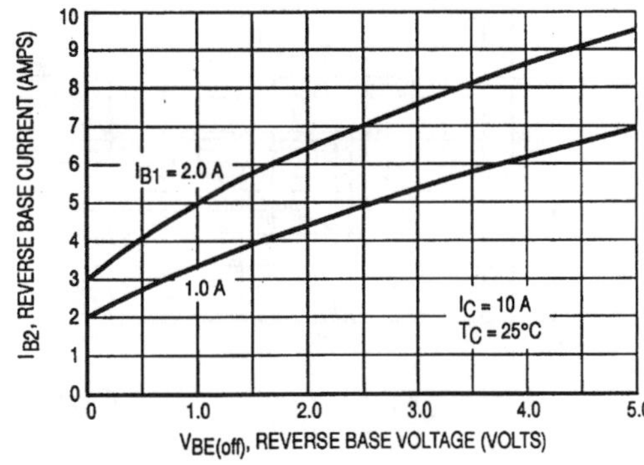

Figure 14. Peak Reverse Base Current

SAFE OPERATING AREA INFORMATION

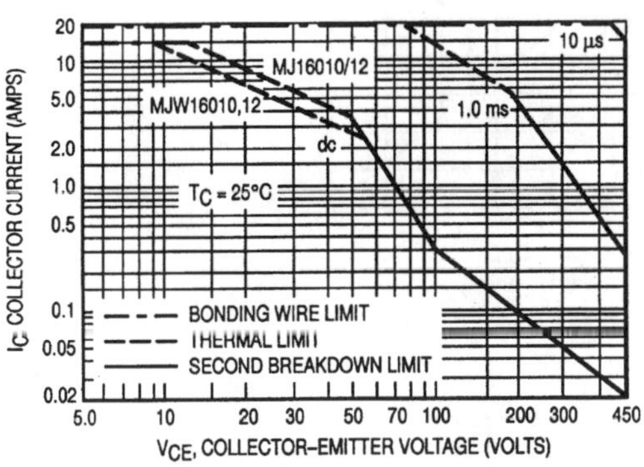

Figure 15. Maximum Forward Bias Safe Operating Area

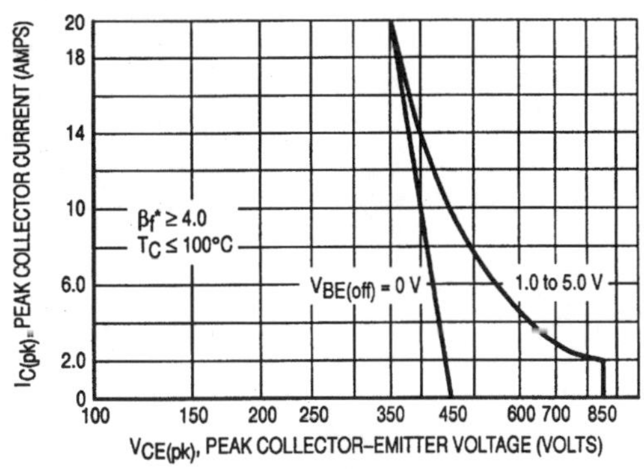

Figure 16. Maximum Reverse Bias Safe Operating Area

$^*\beta_f = \frac{I_C}{I_{B1}}$

FORWARD BIAS

There are two limitations on the power handling ability of a transistor: average junction temperature and second breakdown. Safe operating area curves indicate $I_C - V_{CE}$ limits of the transistor that must be observed for reliable operation; i.e., the transistor must not be subjected to greater dissipation than the curves indicate.

The data of Figure 15 is based on T_C = 25°C; $T_{J(pk)}$ is variable depending on power level. Second breakdown pulse limits are valid for duty cycles to 10% but must be derated when $T_C \geq 25°C$. Second breakdown limitations do not derate the same as thermal limitations. Allowable current at the voltages shown on Figure 15 may be found at any case temperature by using the appropriate curve on Figure 18.

$T_{J(pk)}$ may be calculated from the data in Figure 17. At high case temperatures, thermal limitations will reduce the power that can be handled to values less than the limitations imposed by second breakdown.

REVERSE BIAS

For inductive loads, high voltage and high current must be sustained simultaneously during turn–off, in most cases, with the base–to–emitter junction reverse biased. Under these conditions the collector voltage must be held to a safe level at or below a specific value of collector current. This can be accomplished by several means such as active clamping, RC snubbing, load line shaping, etc. The safe level for these devices is specified as Reverse Bias Safe Operating Area and represents the voltage current condition allowable during reverse biased turnoff. This rating is verified under clamped conditions so that the device is never subjected to an avalanche mode. Figure 16 gives the RBSOA characteristics.

Appendix 4.4-7

MJ16010 MJW16010 MJ16012 MJW16012

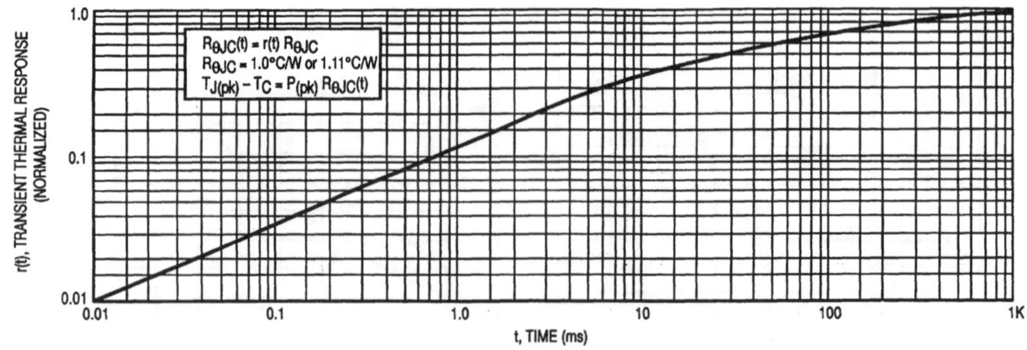

Figure 17. Thermal Response

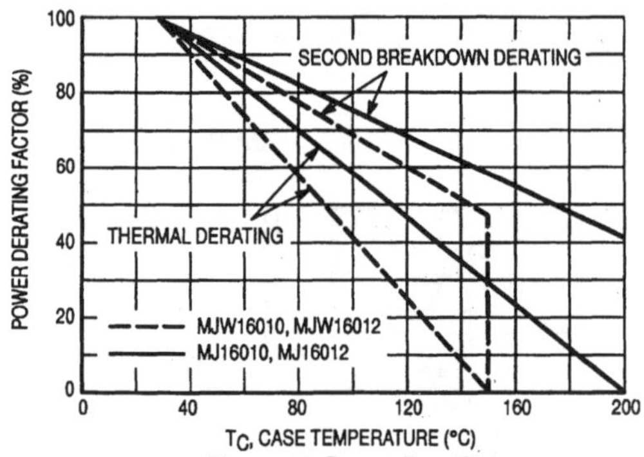

Figure 18. Power Derating

Table 1. Resistive Load Switching

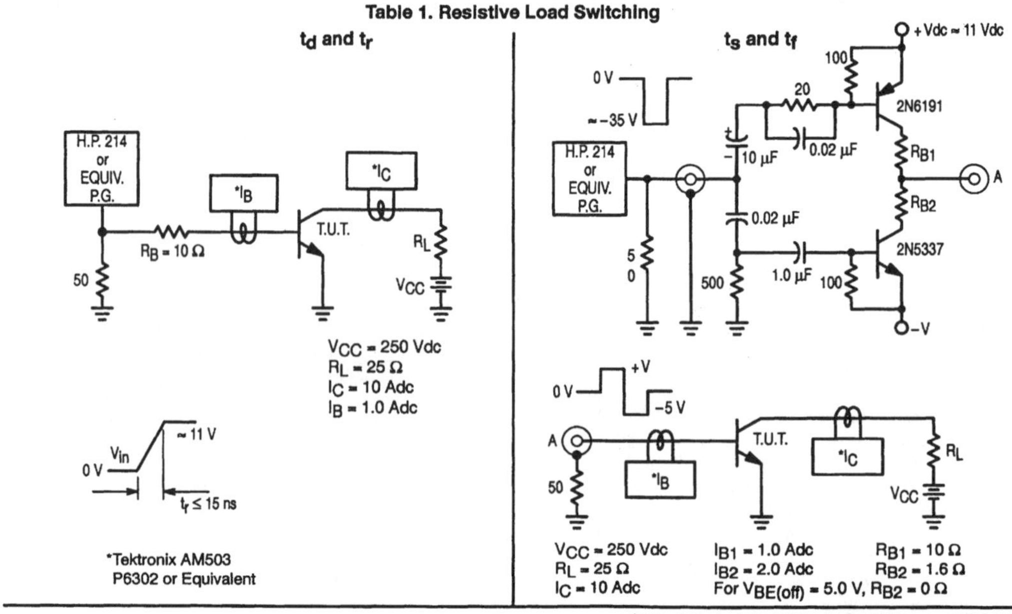

Note: Adjust –V to obtain desired $V_{BE(off)}$ at Point A.

Appendix 4.4-8

MJ16010 MJW16010 MJ16012 MJW16012

Table 2. Inductive Load Switching

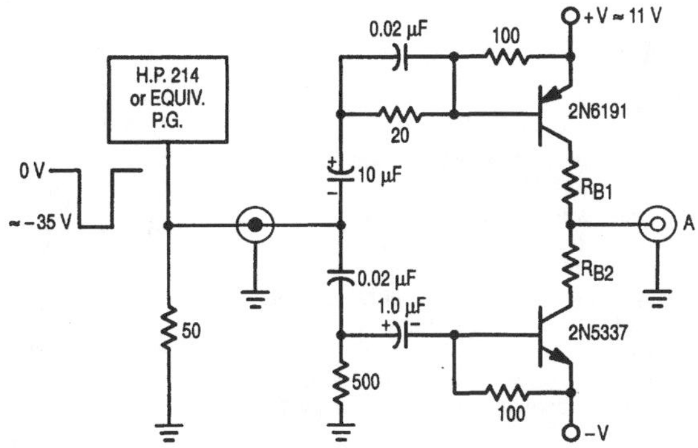

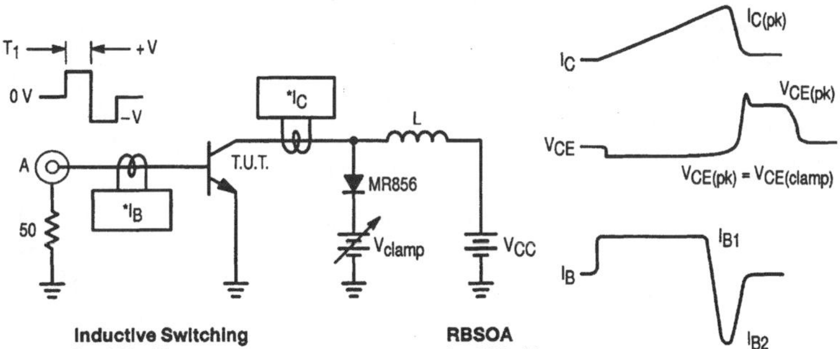

$$t_1 \approx \frac{L_{coil}\ (I_{Cpk})}{V_{CC}}$$

T_1 adjusted to obtain $I_{C(pk)}$

$V_{CEO(sus)}$
$L = 10$ mH
$R_{B2} = \infty$
$V_{CC} = 20$ V lts

*Tektronix AM503
P6302 or Equivalent

Inductive Switching
$L = 200$ μH
$R_{B2} = 0$
$V_{CC} = 20$ V
R_{B1} selected for desired I_{B1}

Scope — Tektronix
7403 or Equivalent

RBSOA
$L = 200$ μH
$R_{B2} = 0$
$V_{CC} = 20$ V
R_{B1} selected for desired I_{B1}

Note: Adjust –V to obtain desired $V_{BE(off)}$ at Point A.

TYPICAL INDUCTIVE SWITCHING WAVEFORMS

t_{sv}

$I_{C(pk)} = 10$ A
$I_{B1} = 1.0$ A
$V_{BE(off)} = 5.0$ V
$V_{CE(pk)} = 400$ V
$T_C = 25°C$
Time Base = 100 ns/cm

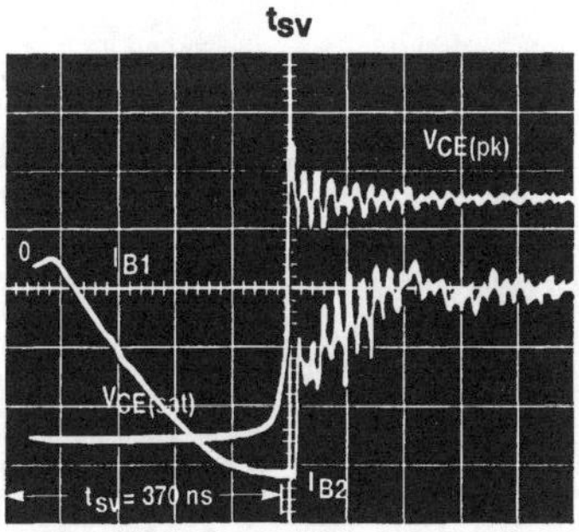

t_{fi}, t_c

$I_{C(pk)} = 10$ A
$I_{B1} = 1.0$ A
$V_{BE(off)} = 5.0$ V
$V_{CE(pk)} = 400$ V
$T_C = 25°C$
Time Base = 20 ns/cm

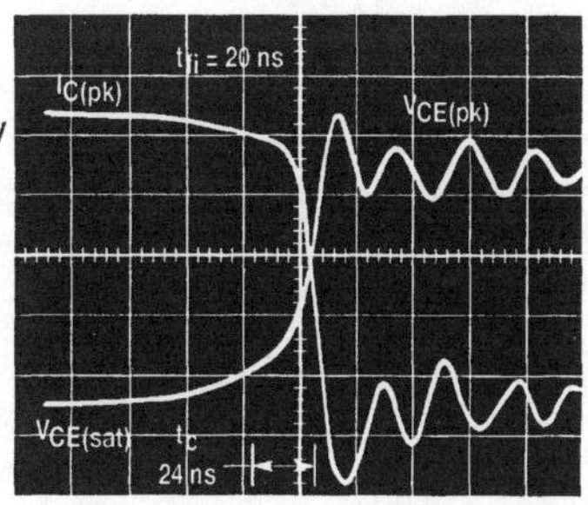

Appendix 4.5-1

MOTOROLA

MC34152
MC33152

High Speed Dual MOSFET Drivers

HIGH SPEED DUAL MOSFET DRIVERS

SEMICONDUCTOR TECHNICAL DATA

The MC34152/MC33152 are dual noninverting high speed drivers specifically designed for applications that require low current digital signals to drive large capacitive loads with high slew rates. These devices feature low input current making them CMOS/LSTTL logic compatible, input hysteresis for fast output switching that is independent of input transition time, and two high current totem pole outputs ideally suited for driving power MOSFETs. Also included is an undervoltage lockout with hysteresis to prevent system erratic operation at low supply voltages.

Typical applications include switching power supplies, dc–to–dc converters, capacitor charge pump voltage doublers/inverters, and motor controllers.

This device is available in dual–in–line and surface mount packages.

- Two Independent Channels with 1.5 A Totem Pole Outputs
- Output Rise and Fall Times of 15 ns with 1000 pF Load
- CMOS/LSTTL Compatible Inputs with Hysteresis
- Undervoltage Lockout with Hysteresis
- Low Standby Current
- Efficient High Frequency Operation
- Enhanced System Performance with Common Switching Regulator Control ICs

P SUFFIX
PLASTIC PACKAGE
CASE 626

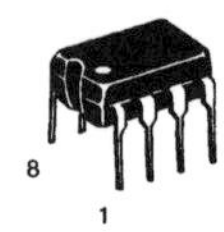

D SUFFIX
PLASTIC PACKAGE
CASE 751
(SO–8)

Representative Diagram

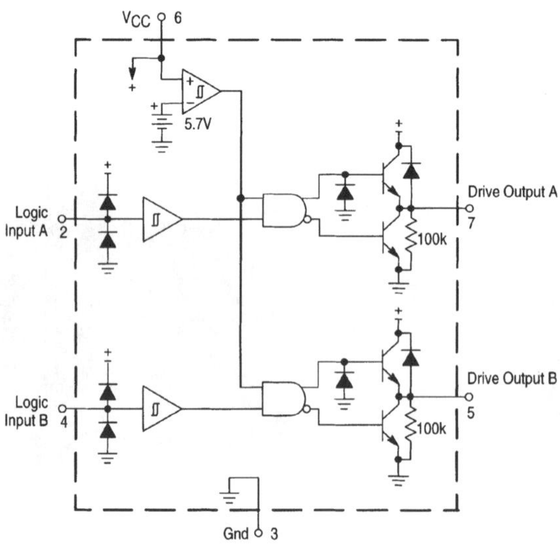

PIN CONNECTIONS

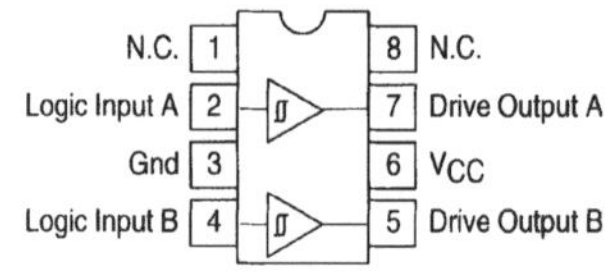

(Top View)

ORDERING INFORMATION

Device	Operating Temperature Range	Package
MC34152D	$T_A = 0°$ to $+70°C$	SO–8
MC34152P		Plastic DIP
MC33152D	$T_A = -40°$ to $+85°C$	SO–8
MC33152P		Plastic DIP

Appendix 4.5-2

MC34152 MC33152

MAXIMUM RATINGS

Rating	Symbol	Value	Unit
Power Supply Voltage	V_{CC}	20	V
Logic Inputs (Note 1)	V_{in}	−0.3 to $+V_{CC}$	V
Drive Outputs (Note 2) Totem Pole Sink or Source Current Diode Clamp Current (Drive Output to V_{CC})	 I_O $I_{O(clamp)}$	 1.5 1.0	A
Power Dissipation and Thermal Characteristics D Suffix, Plastic Package Case 751 Maximum Power Dissipation @ T_A = 50°C Thermal Resistance, Junction–to–Air P Suffix, Plastic Package, Case 626 Maximum Power Dissipation @ T_A = 50°C Thermal Resistance, Junction–to–Air	 P_D $R_{\theta JA}$ P_D $R_{\theta JA}$	 0.56 180 1.0 100	 W °C/W W °C/W
Operating Junction Temperature	T_J	+150	°C
Operating Ambient Temperature MC34152 MC33152	T_A	0 to +70 −40 to +85	°C
Storage Temperature Range	T_{stg}	−65 to +150	°C

ELECTRICAL CHARACTERISTICS (V_{CC} = 12 V, for typical values T_A = 25°C, for min/max values T_A is the operating ambient temperature range that applies [Note 3], unless otherwise noted.)

Characteristics	Symbol	Min	Typ	Max	Unit
LOGIC INPUTS					
Input Threshold Voltage High State Logic 1 Low State Logic 0	 V_{IH} V_{IL}	 2.6 –	 1.75 1.58	 – 0.9	V
Input Current High State (V_{IH} = 2.6 V) Low State (V_{IL} = 0.8 V)	 I_{IH} I_{IL}	 – –	 100 20	 300 100	μA
DRIVE OUTPUT					
Output Voltage Low State (I_{sink} = 10 mA) (I_{sink} = 50 mA) (I_{sink} = 400 mA) High State (I_{source} = 10 mA) (I_{source} = 50 mA) (I_{source} = 400 mA)	 V_{OL} V_{OH}	 – – – 10.5 10.4 10	 0.8 1.1 1.8 11.2 11.1 10.8	 1.2 1.5 2.5 – – –	V
Output Pull–Down Resistor	R_{PD}	–	100	–	kΩ
SWITCHING CHARACTERISTICS (T_A = 25°C)					
Propagation Delay (C_L = 1.0 nF) Logic Input to: Drive Output Rise (10% Input to 10% Output) Drive Output Fall (90% Input to 90% Output)	 t_{PLH} (IN/OUT) t_{PHL} (IN/OUT)	 – –	 55 40	 120 120	ns
Drive Output Rise Time (10% to 90%) C_L = 1.0 nF C_L = 2.5 nF	t_r	– –	14 36	30 –	ns
Drive Output Fall Time (90% to 10%) C_L = 1.0 nF C_L = 2.5 nF	t_f	– –	15 32	30 –	ns
TOTAL DEVICE					
Power Supply Current Standby (Logic Inputs Grounded) Operating (C_L = 1.0 nF Drive Outputs 1 and 2, f = 100 kHz)	I_{CC}	 – –	 6.0 10.5	 8.0 15	mA
Operating Voltage	V_{CC}	6.5	–	18	V

NOTES: 1. For optimum switching speed, the maximum input voltage should be limited to 10 V or V_{CC}, whichever is less.
2. Maximum package power dissipation limits must be observed.
3. Low duty cycle pulse techniques are used during test to maintain junction temperature as close to ambient as possible.
T_{low} = 0°C for MC34152 T_{high} = +70°C for MC34152
= −40°C for MC33152 = +85°C for MC33152

Appendix 4.5-3

MC34152 MC33152

Figure 1. Switching Characteristics Test Circuit

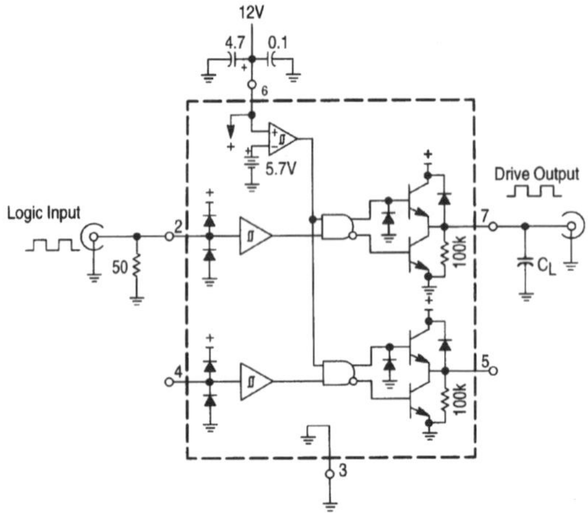

Figure 2. Switching Waveform Definitions

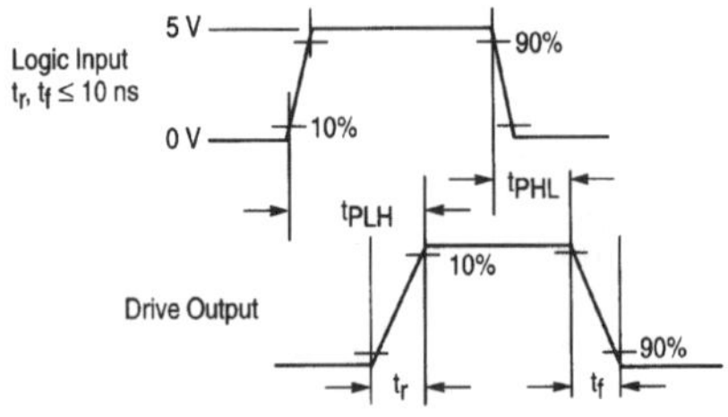

Figure 3. Logic Input Current versus Input Voltage

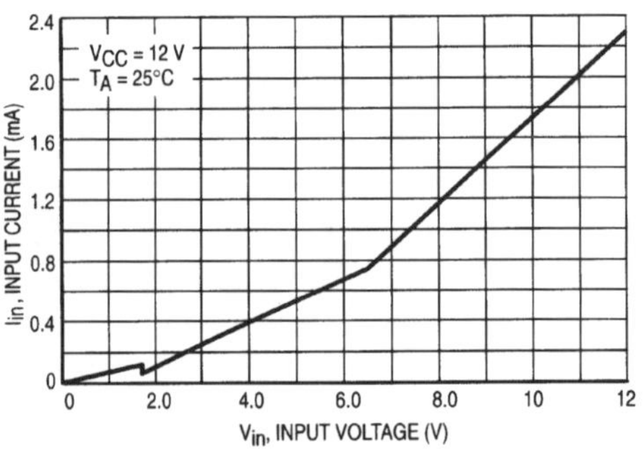

Figure 4. Logic Input Threshold Voltage versus Temperature

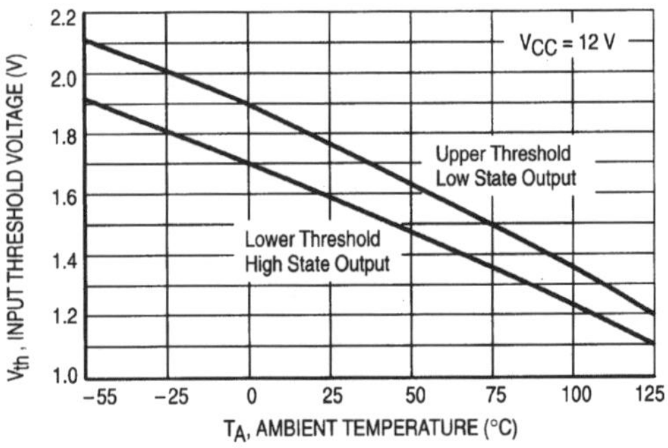

Figure 5. Drive Output High to Low Propagation Delay versus Logic Input Overdrive Voltage

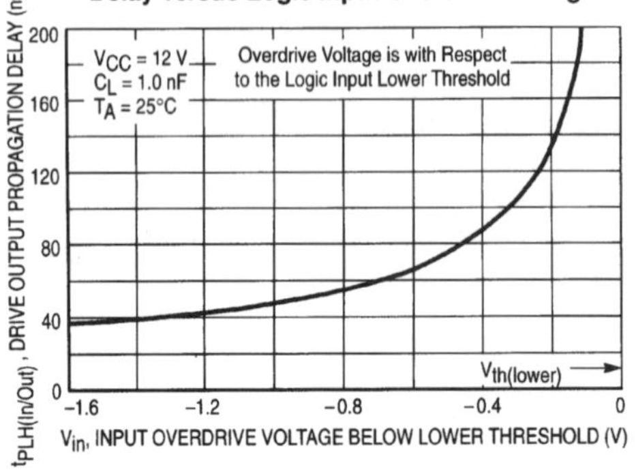

Figure 6. Drive Output Low to High Propagation Delay versus Logic Input Overdrive Voltage

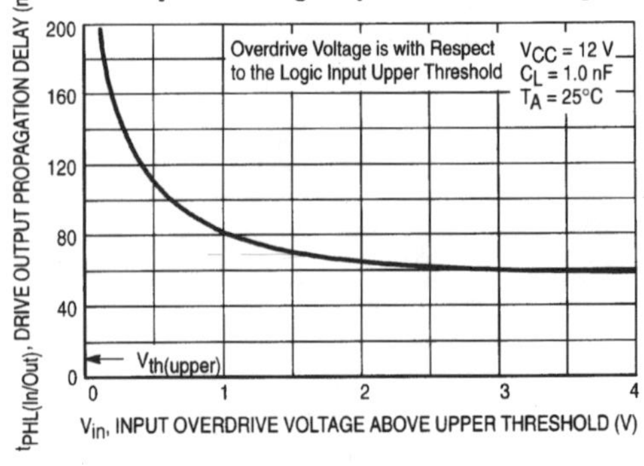

Appendix 4.5-4

MC34152 MC33152

Figure 7. Propagation Delay

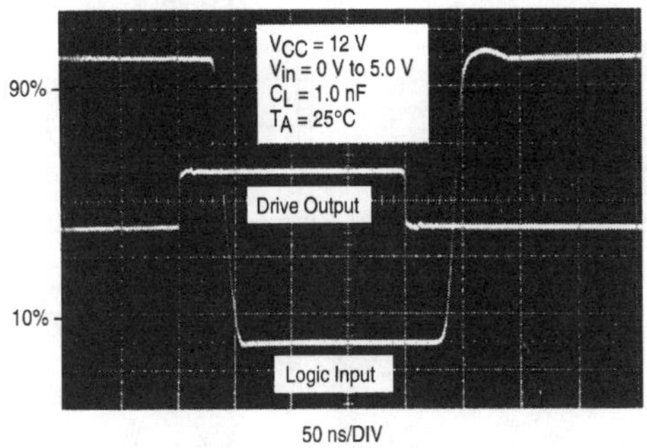

Figure 8. Drive Output Clamp Voltage versus Clamp Current

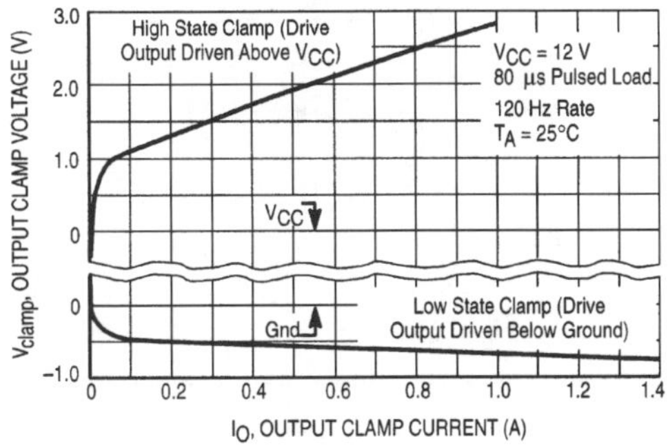

Figure 9. Drive Output Saturation Voltage versus Load Current

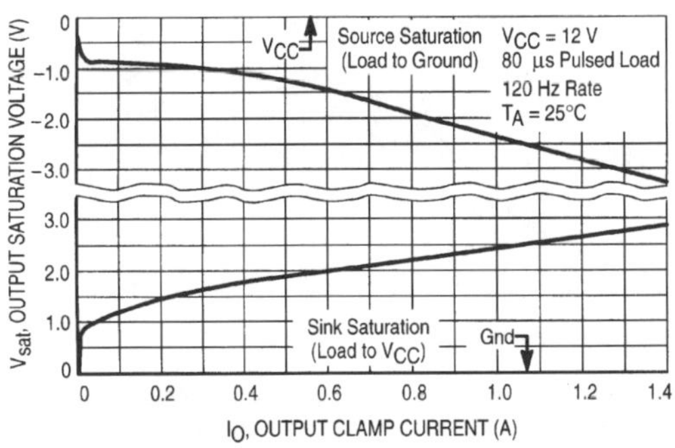

Figure 10. Drive Output Saturation Voltage versus Temperature

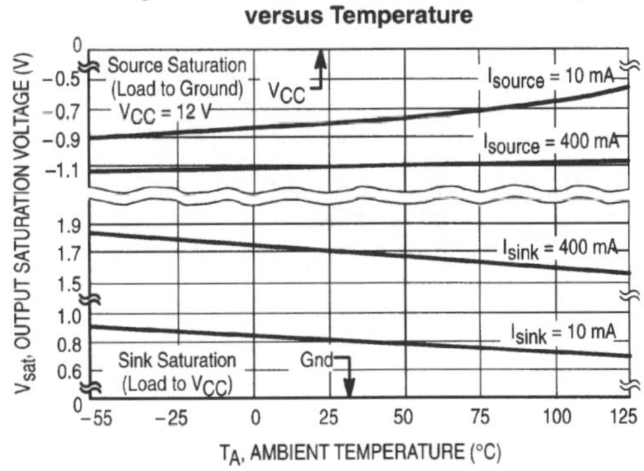

Figure 11. Drive Output Rise Time

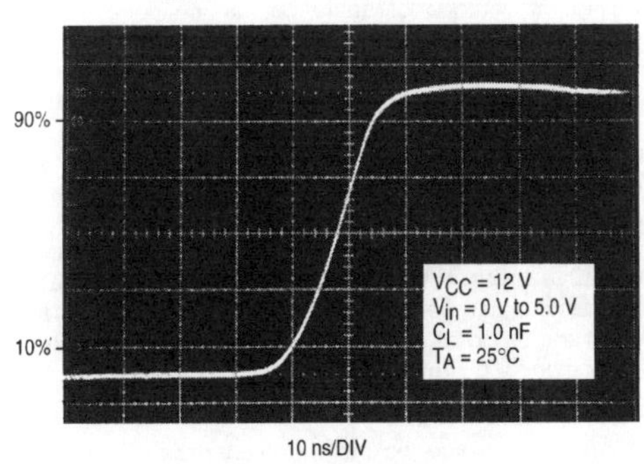

Figure 12. Drive Output Fall Time

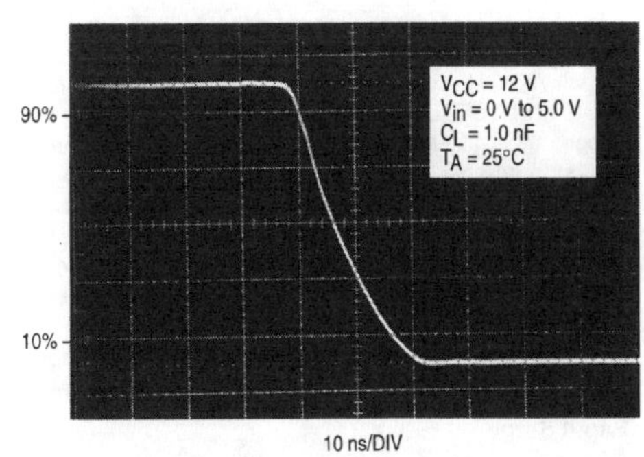

Appendix 4.5-5

MC34152 MC33152

Figure 13. Drive Output Rise and Fall Time versus Load Capacitance

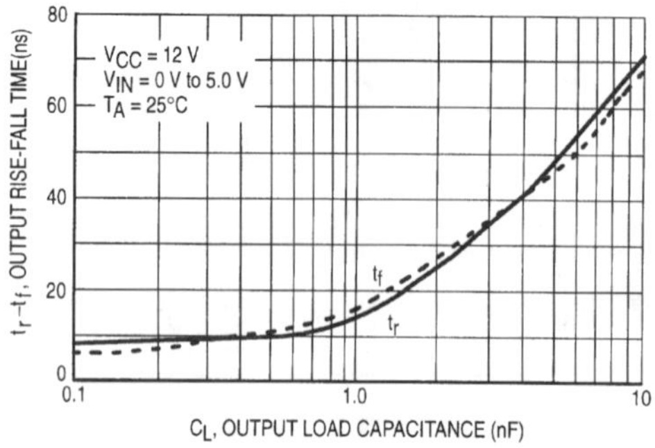

Figure 14. Supply Current versus Drive Output Load Capacitance

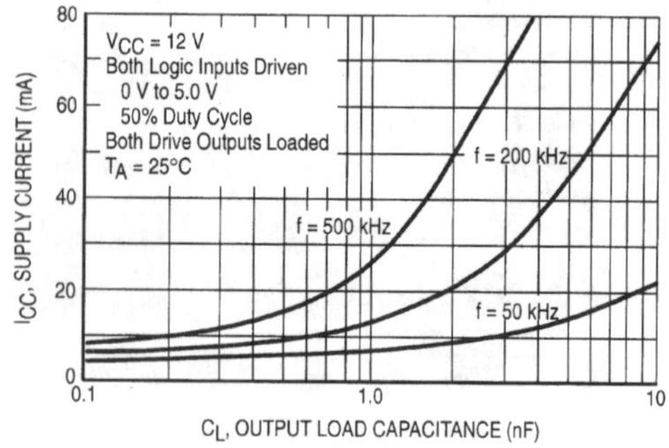

Figure 15. Supply Current versus Input Frequency

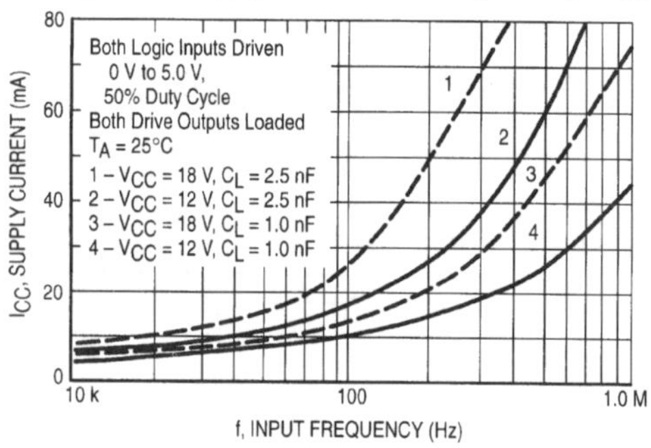

Figure 16. Supply Current versus Supply Voltage

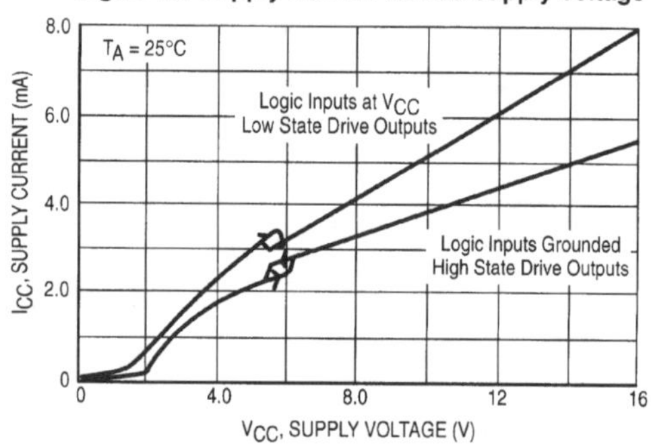

APPLICATIONS INFORMATION

Description

The MC34152 is a dual noninverting high speed driver specifically designed to interface low current digital circuitry with power MOSFETs. This device is constructed with Schottky clamped Bipolar Analog technology which offers a high degree of performance and ruggedness in hostile industrial environments.

Input Stage

The Logic Inputs have 170 mV of hysteresis with the input threshold centered at 1.67 V. The input thresholds are insensitive to V_{CC} making this device directly compatible with CMOS and LSTTL logic families over its entire operating voltage range. Input hysteresis provides fast output switching that is independent of the input signal transition time, preventing output oscillations as the input thresholds are crossed. The inputs are designed to accept a signal amplitude ranging from ground to V_{CC}. This allows the output of one channel to directly drive the input of a second channel for master–slave operation. Each input has a 30 kΩ pull–down resistor so that an unconnected open input will cause the associated Drive Output to be in a known low state.

Output Stage

Each totem pole Drive Output is capable of sourcing and sinking up to 1.5 A with a typical 'on' resistance of 2.4 Ω at 1.0 A. The low 'on' resistance allows high output currents to be attained at a lower V_{CC} than with comparative CMOS drivers. Each output has a 100 kΩ pull–down resistor to keep the MOSFET gate low when V_{CC} is less than 1.4 V. No over current or thermal protection has been designed into the device, so output shorting to V_{CC} or ground must be avoided.

Parasitic inductance in series with the load will cause the driver outputs to ring above V_{CC} during the turn–on transition, and below ground during the turn–off transition. With CMOS drivers, this mode of operation can cause a destructive output latch–up condition. The MC34152 is immune to output latch–up. The Drive Outputs contain an internal diode to V_{CC} for clamping positive voltage transients. When operating with V_{CC} at 18 V, proper power supply bypassing must be observed to prevent the output ringing from exceeding the maximum 20 V device rating. Negative output transients are clamped by the internal NPN pull–up transistor. Since full supply voltage is applied across the NPN pull–up during the negative output transient, power dissipation at high frequencies can become excessive. Figures 19, 20, and 21 show a method of using external Schottky diode clamps to reduce driver power dissipation.

Undervoltage Lockout

An undervoltage lockout with hysteresis prevents erratic system operation at low supply voltages. The UVLO forces the Drive Outputs into a low state as V_{CC} rises from 1.4 V to

Appendix 4.5-6

MC34152 MC33152

the 5.8 V upper threshold. The lower UVLO threshold is 5.3 V, yielding about 500 mV of hysteresis.

Power Dissipation

Circuit performance and long term reliability are enhanced with reduced die temperature. Die temperature increase is directly related to the power that the integrated circuit must dissipate and the total thermal resistance from the junction to ambient. The formula for calculating the junction temperature with the package in free air is:

$$T_J = T_A + P_D (R_{\theta JA})$$

where: T_J = Junction Temperature
T_A = Ambient Temperature
P_D = Power Dissipation
$R_{\theta JA}$ = Thermal Resistance Junction to Ambient

There are three basic components that make up total power to be dissipated when driving a capacitive load with respect to ground. They are:

$$P_D = P_Q + P_C + P_T$$

where: P_Q = Quiescent Power Dissipation
P_C = Capacitive Load Power Dissipation
P_T = Transition Power Dissipation

The quiescent power supply current depends on the supply voltage and duty cycle as shown in Figure 16. The device's quiescent power dissipation is:

$$P_Q = V_{CC} (I_{CCL} [1-D] + I_{CCH} [D])$$

where: I_{CCL} = Supply Current with Low State Drive Outputs
I_{CCH} = Supply Current with High State Drive Outputs
D = Output Duty Cycle

The capacitive load power dissipation is directly related to the load capacitance value, frequency, and Drive Output voltage swing. The capacitive load power dissipation per driver is:

$$P_C = V_{CC} (V_{OH} - V_{OL}) C_L f$$

where: V_{OH} = High State Drive Output Voltage
V_{OL} = Low State Drive Output Voltage
C_L = Load Capacitance
f = Frequency

When driving a MOSFET, the calculation of capacitive load power P_C is somewhat complicated by the changing gate to source capacitance C_{GS} as the device switches. To aid in this calculation, power MOSFET manufacturers provide gate charge information on their data sheets. Figure 17 shows a curve of gate voltage versus gate charge for the Motorola MTM15N50. Note that there are three distinct slopes to the curve representing different input capacitance values. To completely switch the MOSFET 'on,' the gate must be brought to 10 V with respect to the source. The graph shows that a gate charge Q_g of 110 nC is required when operating the MOSFET with a drain to source voltage V_{DS} of 400 V.

Figure 17. Gate–to–Source Voltage versus Gate charge

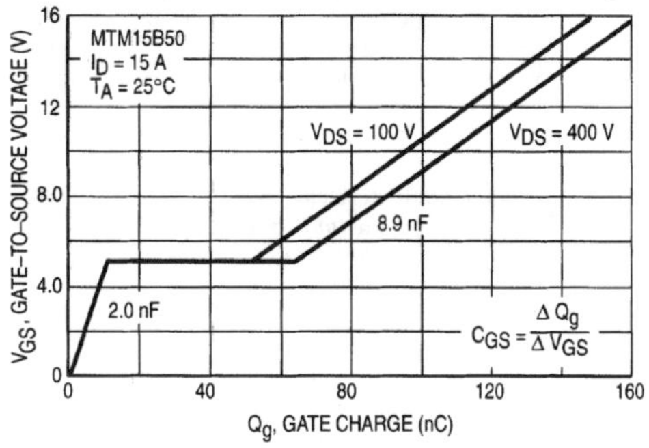

The capacitive load power dissipation is directly related to the required gate charge, and operating frequency. The capacitive load power dissipation per driver is:

$$P_{C(MOSFET)} = V_{CC} Q_g f$$

The flat region from 10 nC to 55 nC is caused by the drain–to–gate Miller capacitance, occurring while the MOSFET is in the linear region dissipating substantial amounts of power. The high output current capability of the MC34152 is able to quickly deliver the required gate charge for fast power efficient MOSFET switching. By operating the MC34152 at a higher V_{CC}, additional charge can be provided to bring the gate above 10 V. This will reduce the 'on' resistance of the MOSFET at the expense of higher driver dissipation at a given operating frequency.

The transition power dissipation is due to extremely short simultaneous conduction of internal circuit nodes when the Drive Outputs change state. The transition power dissipation per driver is approximately:

$$P_T \approx V_{CC} (1.08\ V_{CC}\ C_L\ f - 8 \times 10^{-4})$$

P_T must be greater than zero.

Switching time characterization of the MC34152 is performed with fixed capacitive loads. Figure 13 shows that for small capacitance loads, the switching speed is limited by transistor turn–on/off time and the slew rate of the internal nodes. For large capacitance loads, the switching speed is limited by the maximum output current capability of the integrated circuit.

Appendix 4.5-7

MC34152 MC33152

LAYOUT CONSIDERATIONS

High frequency printed circuit layout techniques are imperative to prevent excessive output ringing and overshoot. **Do not attempt to construct the driver circuit on wire–wrap or plug–in prototype boards.** When driving large capacitive loads, the printed circuit board must contain a low inductance ground plane to minimize the voltage spikes induced by the high ground ripple currents. All high current loops should be kept as short as possible using heavy copper runs to provide a low impedance high frequency path. For optimum drive performance, it is recommended that the initial circuit design contains dual power supply bypass capacitors connected with short leads as close to the V_{CC} pin and ground as the layout will permit. Suggested capacitors are a low inductance 0.1 μF ceramic in parallel with a 4.7 μF tantalum. Additional bypass capacitors may be required depending upon Drive Output loading and circuit layout.

Proper printed circuit board layout is extremely critical and cannot be over emphasized.

Figure 18. Enhanced System Performance with Common Switching Regulators

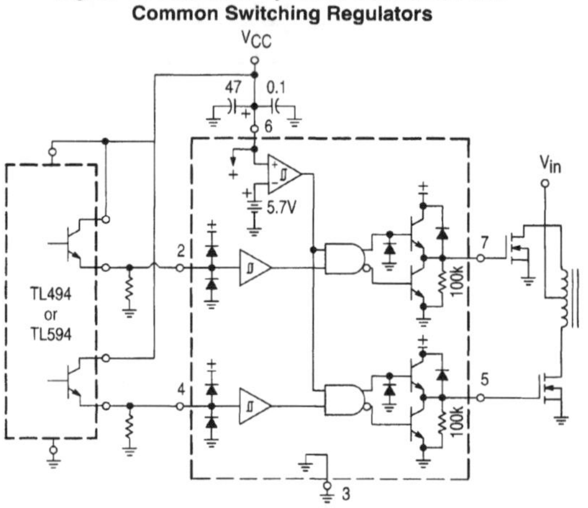

The MC34152 greatly enhances the drive capabilities of common switching regulators and CMOS/TTL logic devices.

Figure 19. MOSFET Parasitic Oscillations

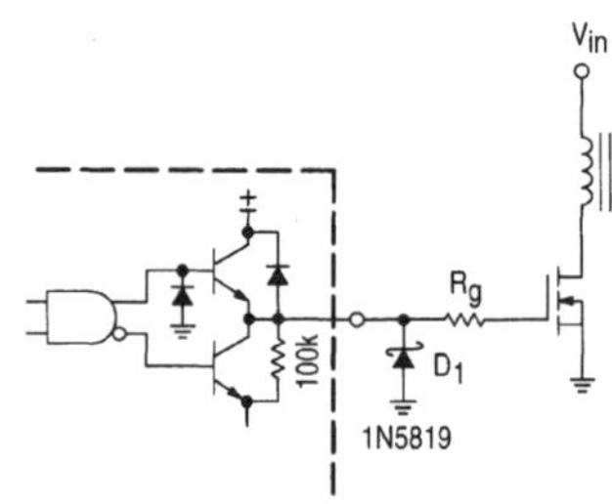

Series gate resistor R_g may be needed to damp high frequency parasitic oscillations caused by the MOSFET input capacitance and any series wiring inductance in the gate–source circuit. R_g will decrease the MOSFET switching speed. Schottky diode D_1 can reduce the driver's power dissipation due to excessive ringing, by preventing the output pin from being driven below ground.

Figure 20. Direct Transformer Drive

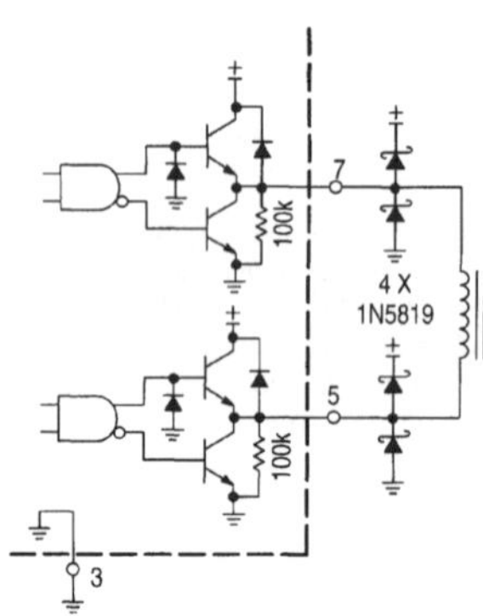

Output Schottky diodes are recommended when driving inductive loads at high frequencies. The diodes reduce the driver's power dissipation by preventing the output pins from being driven above V_{CC} and below ground.

Figure 21. Isolated MOSFET Drive

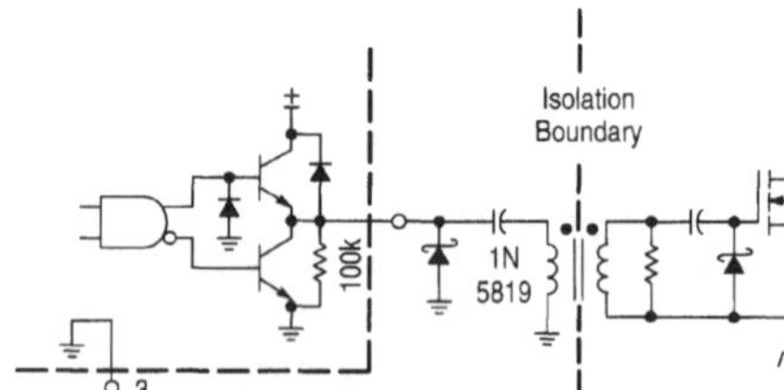

Appendix 4.5-8

MC34152 MC33152

Figure 22. Controlled MOSFET Drive

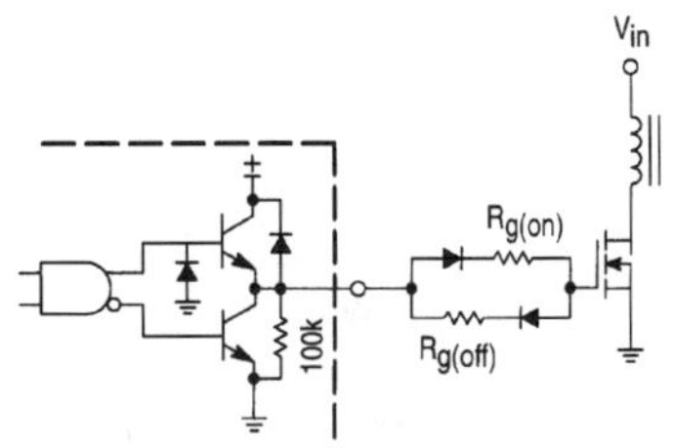

In noise sensitive applications, both conducted and radiated EMI can be reduced significantly by controlling the MOSFET's turn–on and turn–off times.

Figure 23. Bipolar Transistor Drive

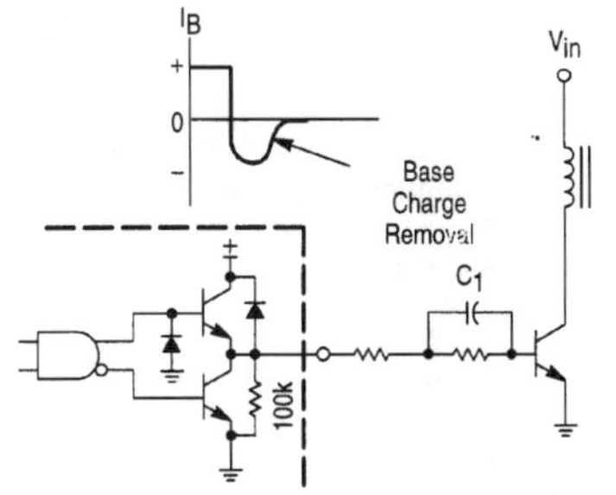

The totem–pole outputs can furnish negative base current for enhanced transistor turn–off, with the addition of capacitor C_1.

Figure 24. Dual Charge Pump Converter

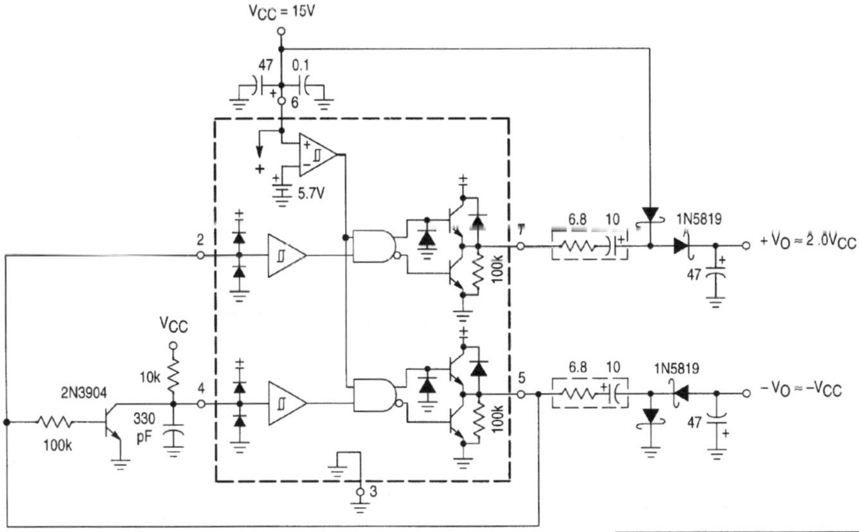

The capacitor's equivalent series resistance limits the Drive Output Current to 1.5 A. An additional series resistor may be required when using tantalum or other low ESR capacitors.

Output Load Regulation

I_O (mA)	$+V_O$ (V)	$-V_O$ (V)
0	27.7	–13.3
1.0	27.4	–12.9
10	26.4	–11.9
20	25.5	–11.2
30	24.6	–10.5
50	22.6	–9.4

Appendix 4.6-1

MOTOROLA

MC33153

Single IGBT Gate Driver

SINGLE IGBT GATE DRIVER

SEMICONDUCTOR TECHNICAL DATA

The MC33153 is specifcally designed as an IGBT driver for high power applications that include ac induction motor control, brushless dc motor control and uninterruptable power supplies. Although designed for driving discrete and module IGBTs, this device offers a cost effective solution for driving power MOSFETs and Bipolar Transistors. Device protection features include the choice of desaturation or overcurrent sensing and undervoltage detection. These devices are available in dual–in–line and surface mount packages and include the following features:

- High Current Output Stage: 1.0 A Source/2.0 A Sink
- Protection Circuits for Both Conventional and Sense IGBT's
- Programmable Fault Blanking Time
- Protection against Overcurrent and Short Circuit
- Undervoltage Lockout Optimzed for IGBT's
- Negative Gate Drive Capability
- Cost Effectively Drives Power MOSFETs and Bipolar Transistors

P SUFFIX
PLASTIC PACKAGE
CASE 626

D SUFFIX
PLASTIC PACKAGE
CASE 751
(SO–8)

PIN CONNECTIONS

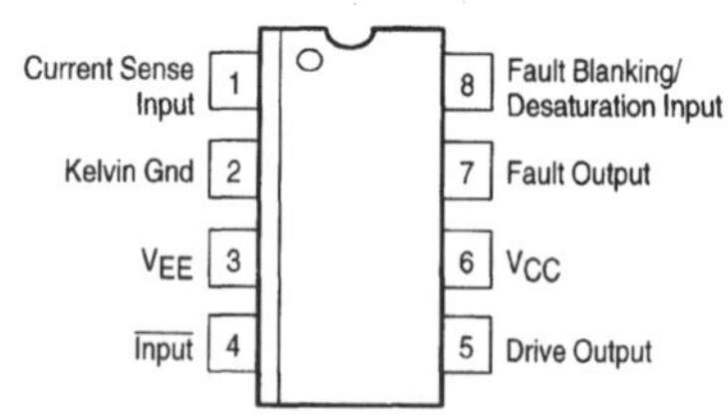

(Top View)

Representative Block Diagram

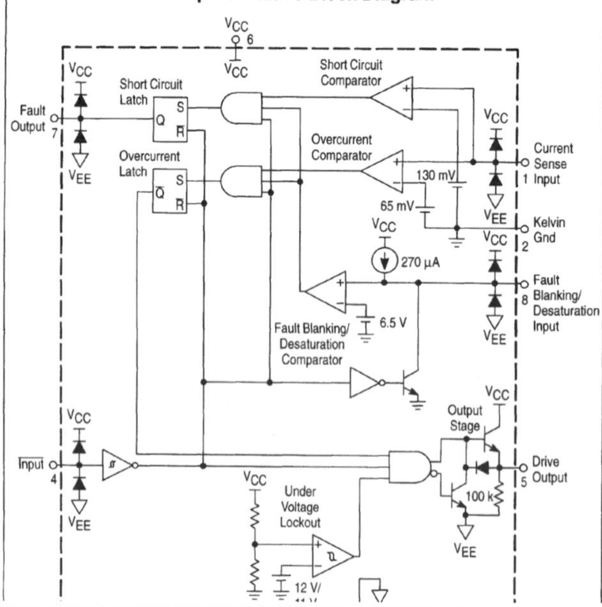

ORDERING INFORMATION

Device	Operating Temperature Range	Package

Appendix 4.6-2

MC33153

MAXIMUM RATINGS

Rating	Symbol	Value	Unit
Power Supply Voltage			V
V_{CC} to V_{EE}	$V_{CC} - V_{EE}$	23	
Kelvin Ground to V_{EE} (Note 1)	KGnd – V_{EE}	23	
Logic Input	V_{in}	V_{EE} –0.3 to V_{CC}	V
Current Sense Input	V_S	–0.3 to V_{CC}	V
Blanking/Desaturation Input	V_{BD}	–0.3 to V_{CC}	V
Gate Drive Output	I_O		A
Source Current		1.0	
Sink Current		2.0	
Diode Clamp Current		1.0	
Fault Output	I_{FO}		mA
Source Current		25	
Sink Curent		10	
Power Dissipation and Thermal Characteristics			
D Suffix SO–8 Package, Case 751			
Maximum Power Dissipation @ T_A = 50°C	P_D	0.56	W
Thermal Resistance, Junction–to–Air	$R_{\theta JA}$	180	°C/W
P Suffix DIP–8 Package, Case 626			
Maximum Power Dissipation @ T_A = 50°C	P_D	1.0	W
Thermal Resistance, Junction–to–Air	$R_{\theta JA}$	100	°C/W
Operating Junction Temperature	T_J	+150	°C
Operating Ambient Temperature	T_A	–40 to +105	°C
Storage Temperature Range	T_{stg}	–65 to +150	°C

ELECTRICAL CHARACTERISTICS (V_{CC} = 15 V, V_{EE} = 0 V, Kelvin Gnd connected to V_{EE}. For typical values T_A = 25°C, for min/max values T_A is the operating ambient temperature range that applies (Note 2), unless otherwise noted.)

Characteristic	Symbol	Min	Typ	Max	Unit
LOGIC INPUT					
Input Threshold Voltage					V
High State (Logic 1)	V_{IH}	–	2.70	3.2	
Low State (Logic 0)	V_{IL}	1.2	2.30	–	
Input Current					µA
High State (V_{IH} = 3.0 V)	I_{IH}	–	130	500	
Low State (V_{IL} = 1.2 V)	I_{IL}	–	50	100	
DRIVE OUTPUT					
Output Voltage					V
Low State (I_{Sink} = 1.0 A)	V_{OL}	–	2.0	2.5	
High State (I_{Source} = 500 mA)	V_{OH}	12	13.9	–	
Output Pull–Down Resistor	R_{PD}	–	100	200	kΩ
FAULT OUTPUT					
Output voltage					V
Low State (I_{Sink} = 5.0 mA)	V_{FL}	–	0.2	1.0	
High State (I_{Source} = 20 mA)	V_{FH}	12	13.3	–	
SWITCHING CHARACTERISTICS					
Propagation Delay (50% Input to 50% Output C_L = 1.0 nF)					ns
Logic Input to Drive Output Rise	$t_{PLH(in/out)}$	–	80	300	
Logic Input to Drive Output Fall	$t_{PHL\ (in/out)}$	–	120	300	
Drive Output Rise Time (10% to 90%) C_L = 1.0 nF	t_r	–	17	55	ns
Drive Output Fall Time (90% to 10%) C_L = 1.0 nF	t_f	–	17	55	ns
Propagation Delay					µs
Current Sense Input to Drive Output	$t_{P(OC)}$	–	0.3	1.0	
Fault Blanking/Desaturation Input to Drive Output	$t_{P(FLT)}$	–	0.3	1.0	

NOTE: 1. Kelvin Ground must always be between V_{EE} and V_{CC}.
2. Low duty cycle pulse techniques are used during test to maintain the junction temperature as close to ambient as possible.

Appendix 4.6-3

MC33153

ELECTRICAL CHARACTERISTICS (continued) (V_{CC} = 15 V, V_{EE} = 0 V, Kelvin Gnd connected to V_{EE}. For typical values T_A = 25°C, for min/max values T_A is the operating ambient temperature range that applies (Note 2), unless otherwise noted.)

Characteristic	Symbol	Min	Typ	Max	Unit
UVLO					
Startup Voltage	$V_{CC\ start}$	11.3	12	12.6	V
Disable Voltage	$V_{CC\ dis}$	10.4	11	11.7	V
COMPARATORS					
Overcurrent Threshold Voltage (V_{Pin8} > 7.0 V)	V_{SOC}	50	65	80	mV
Short Circuit Threshold Voltage (V_{Pin8} > 7.0 V)	V_{SSC}	100	130	160	mV
Fault Blanking/Desaturation Threshold (V_{Pin1} > 100 mV)	$V_{th(FLT)}$	6.0	6.5	7.0	V
Current Sense Input Current (V_{SI} = 0 V)	I_{SI}	–	–1.4	–10	μA
FAULT BLANKING/DESATURATION INPUT					
Current Source (V_{Pin8} = 0 V, V_{Pin4} = 0 V)	I_{chg}	–200	–270	–300	μA
Discharge Current (V_{Pin8} = 15 V, V_{Pin4} = 5.0 V)	I_{dschg}	1.0	2.5	–	mA
TOTAL DEVICE					
Power Supply Current Standby ($V_{Pin\ 4}$ = V_{CC}, Output Open) Operating (C_L = 1.0 nF, f = 20 kHz)	I_{CC}	 – –	 7.2 7.9	 14 20	mA

NOTE: 1. Kelvin Ground must always be between V_{EE} and V_{CC}.
2. Low duty cycle pulse techniques are used during test to maintain the junction temperature as close to ambient as possible.
T_{low} = –40°C for MC33153 T_{high} = +105°C for MC33153

Figure 1. Input Current versus Input Voltage

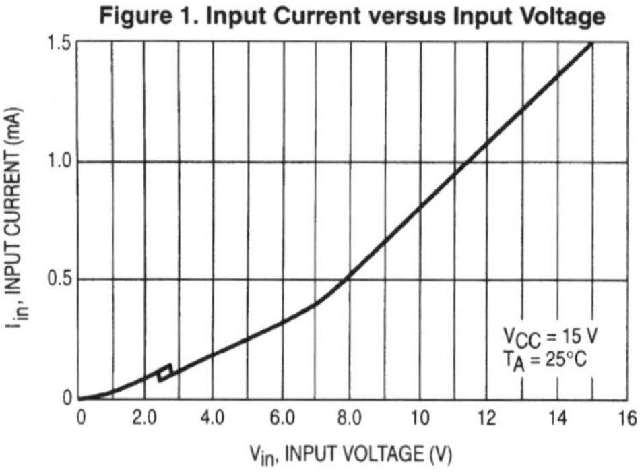

Figure 2. Output Voltage versus Input Voltage

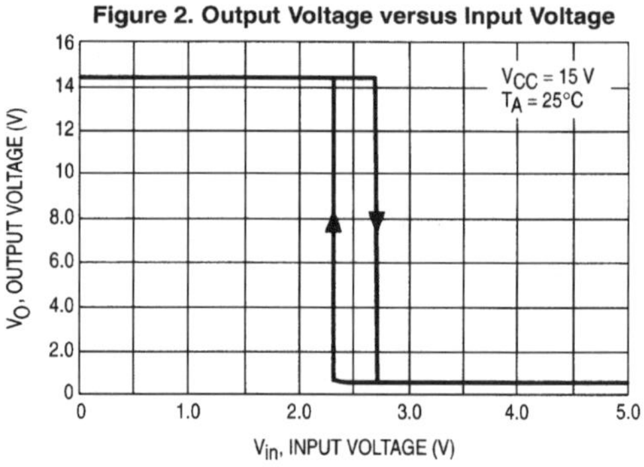

Appendix 4.6-4

MC33153

Figure 3. Input Threshold Voltage versus Temperature

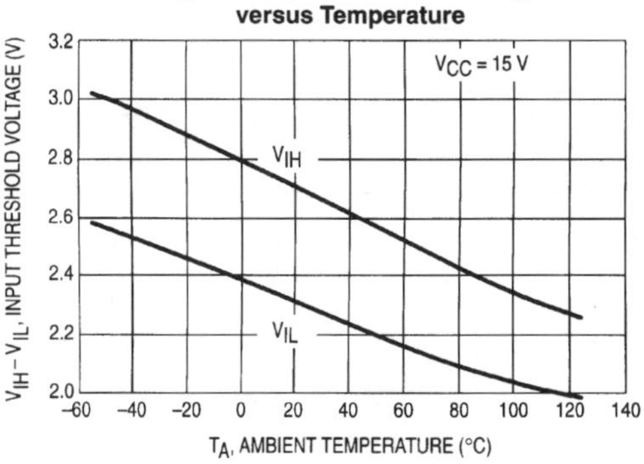

Figure 4. Input Threshold Voltage versus Supply Voltage

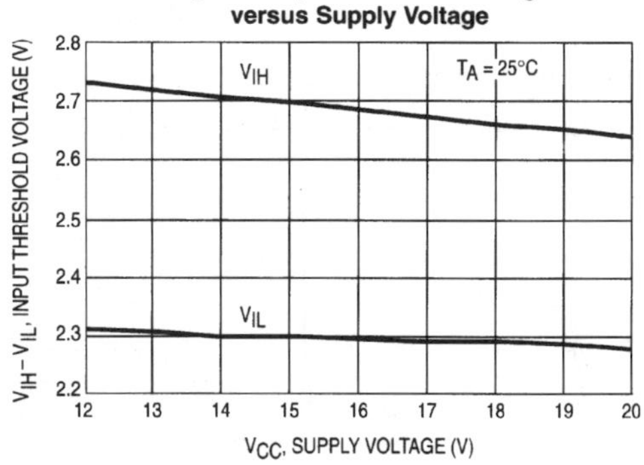

Figure 5. Drive Output Low State Voltage versus Temperature

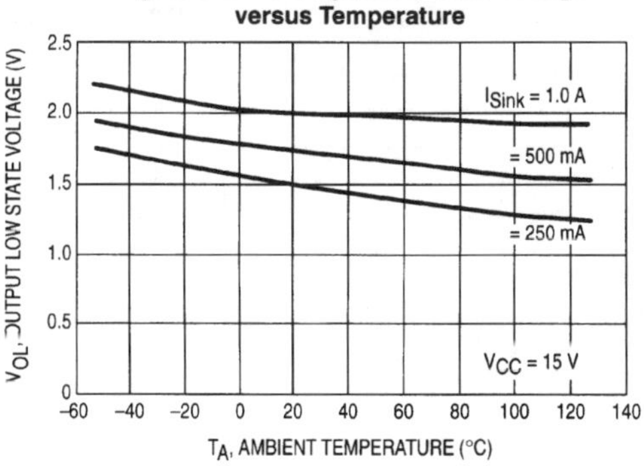

Figure 6. Drive Output Low State Voltage versus Sink Current

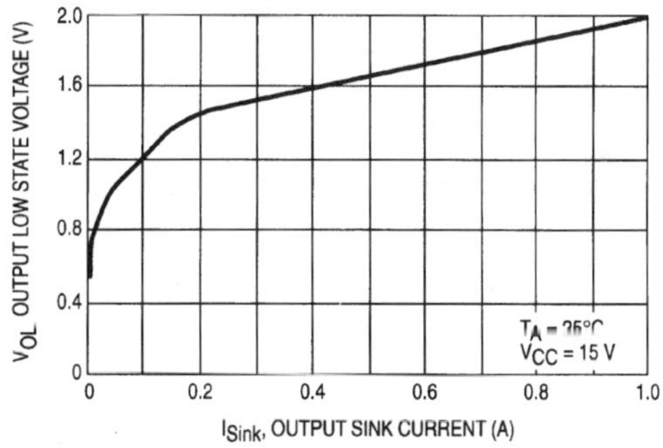

Figure 7. Drive Output High State Voltage versus Temperature

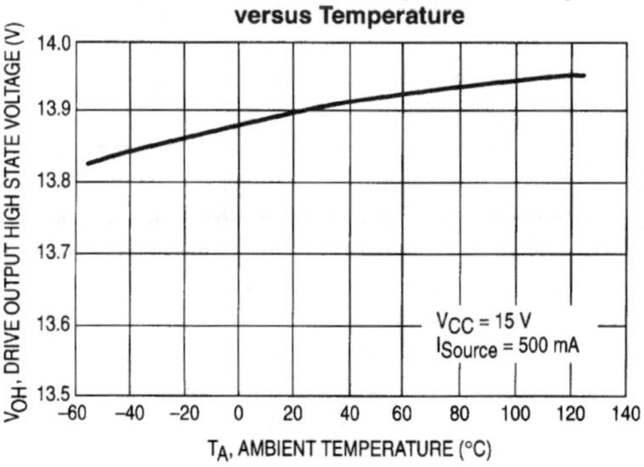

Figure 8. Drive Output High State Voltage versus Source Current

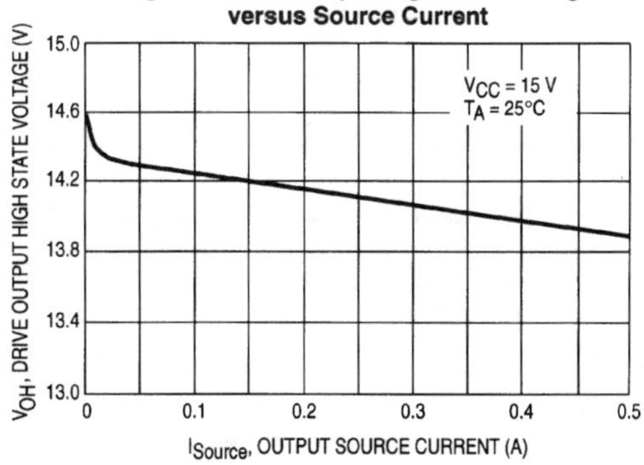

Appendix 4.6-5

MC33153

Figure 9. Drive Output Voltage versus Current Sense Input Voltage

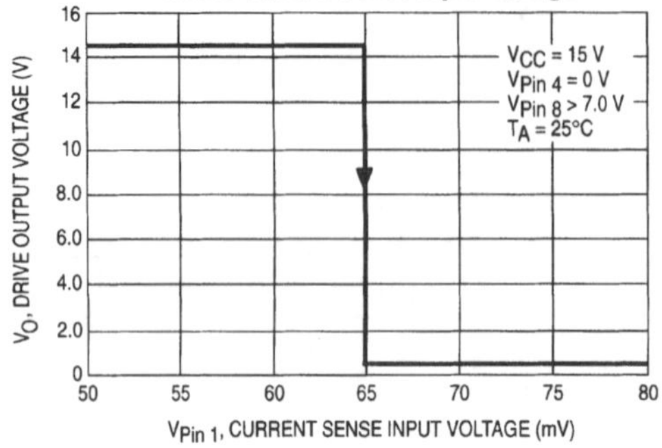

Figure 10. Fault Output Voltage versus Current Sense Input Voltage

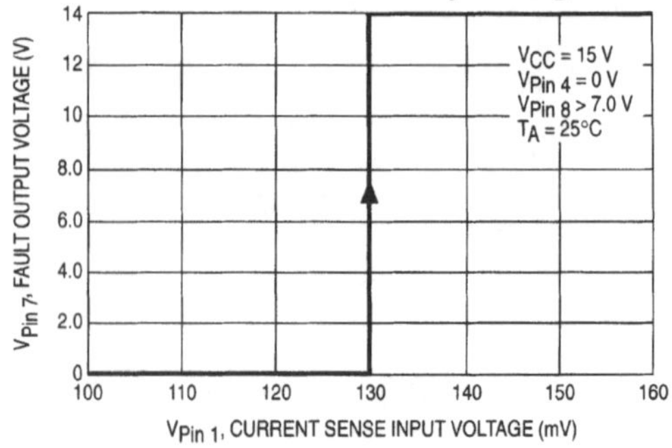

Figure 11. Overcurrent Protection Threshold Voltage versus Temperature

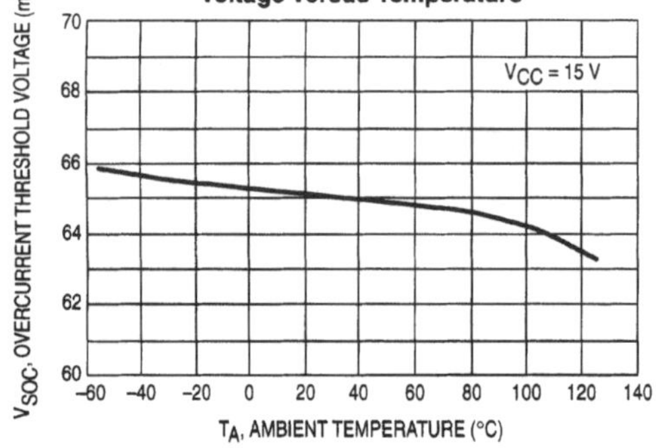

Figure 12. Overcurrent Protection Threshold Voltage versus Supply Voltage

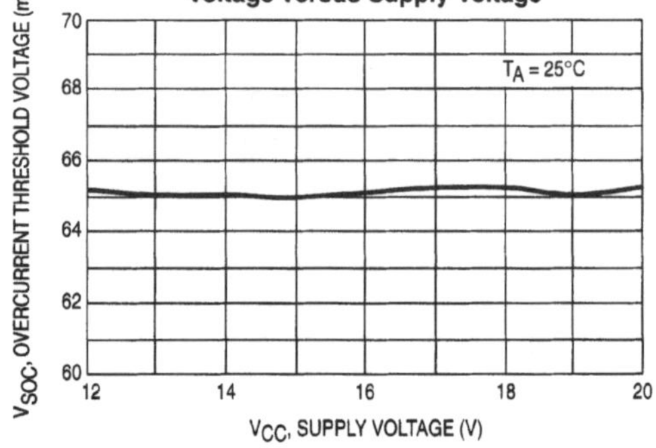

Figure 13. Short Circuit Comparator Threshold Voltage versus Temperature

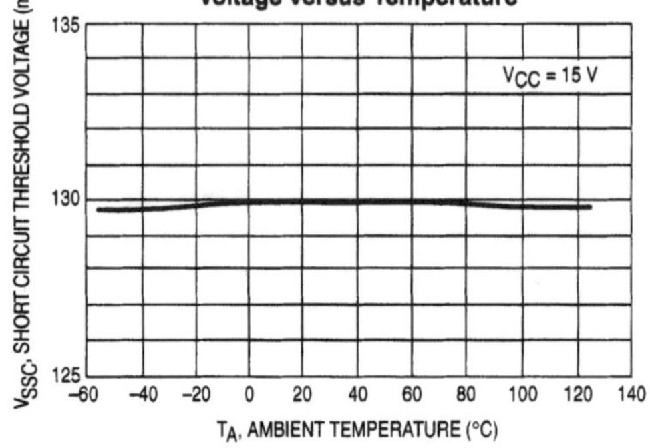

Figure 14. Short Circuit Comparator Threshold Voltage versus Supply Voltage

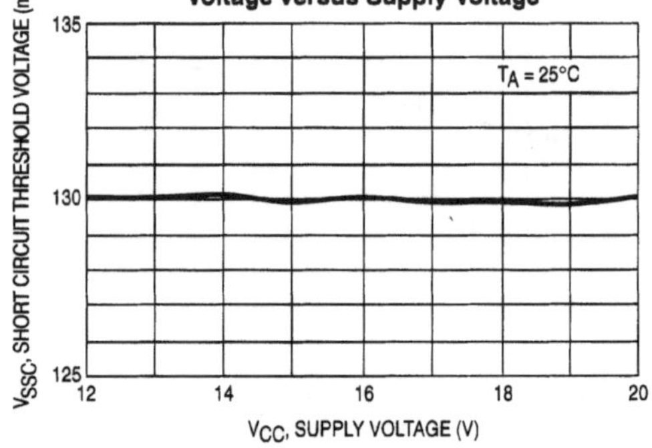

Appendix 4.6-6

MC33153

Figure 15. Current Sense Input Current versus Voltage

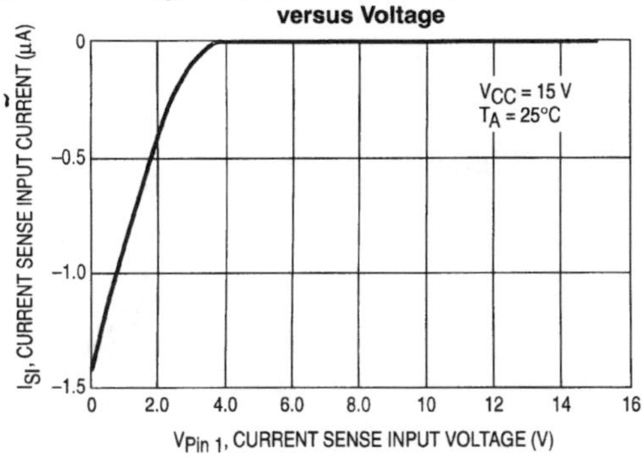

Figure 16. Drive Output Voltage versus Fault Blanking/Desaturation Input Voltage

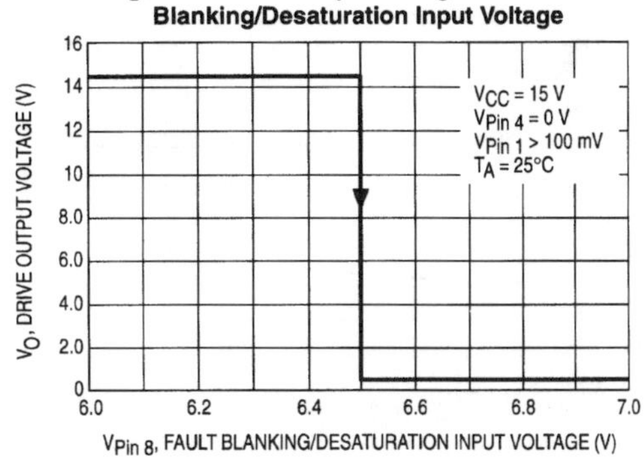

Figure 17. Fault Blanking/Desaturation Comparator Threshold Voltage versus Temperature

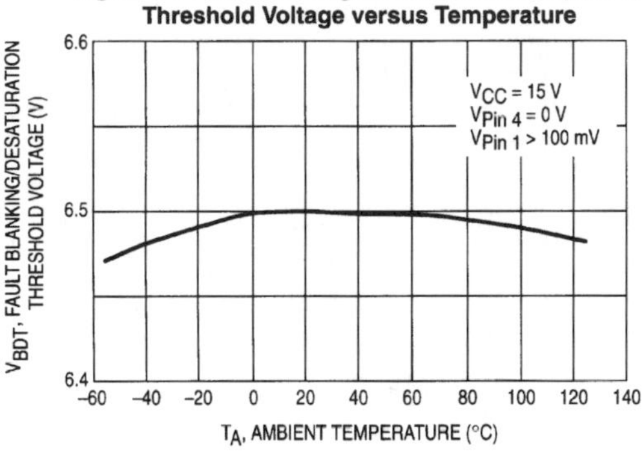

Figure 18. Fault Blanking/Desaturation Comparator Threshold Voltage versus Supply Voltage

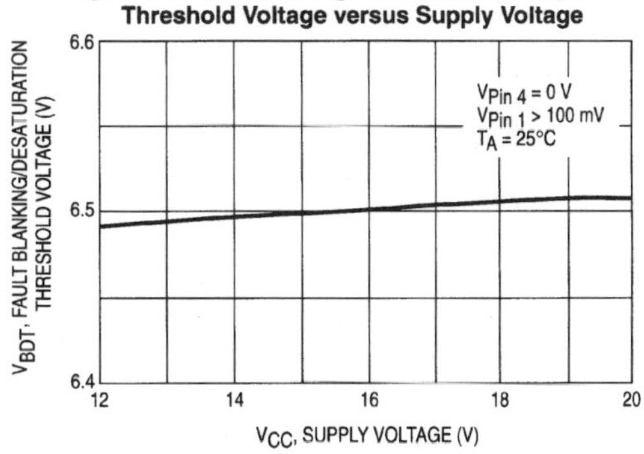

Figure 19. Fault Blanking/Desaturation Current Source versus Temperature

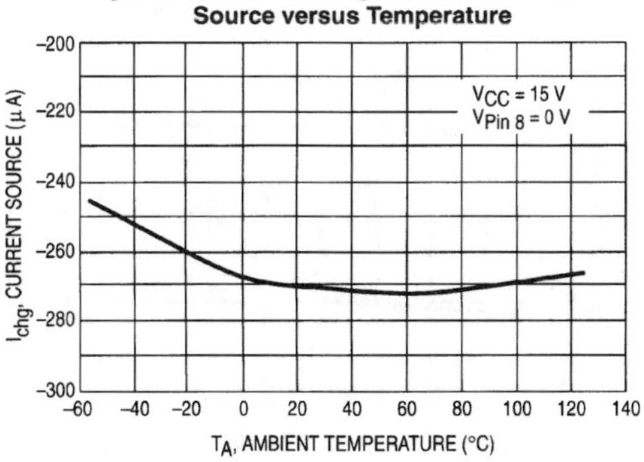

Figure 20. Fault Blanking/Desaturation Current Source versus Supply Voltage

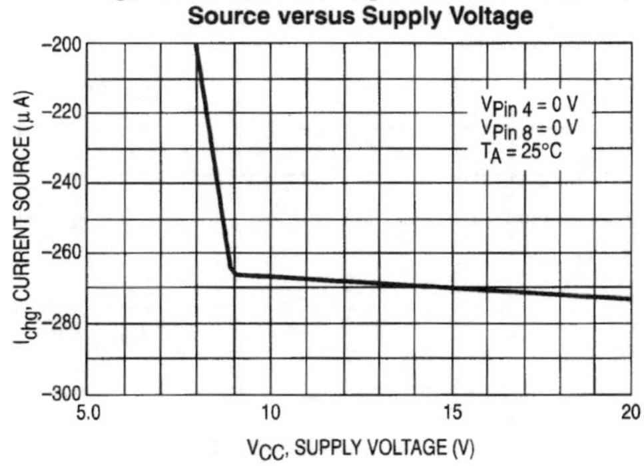

Appendix 4.6-7

MC33153

Figure 21. Fault Blanking/Desaturation Current Source versus Input Voltage

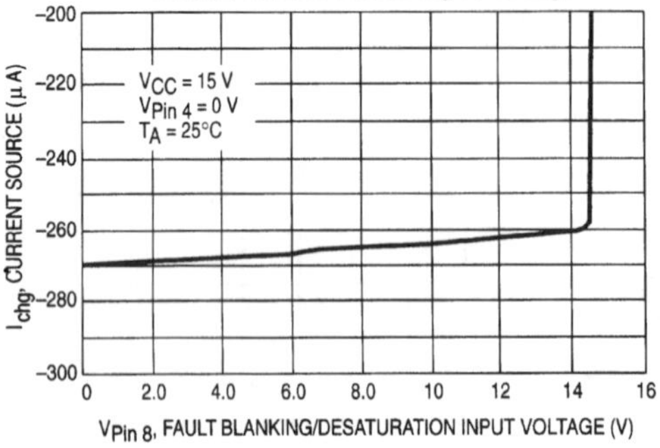

Figure 22. Fault Blanking/Desaturation Discharge Current versus Input Voltage

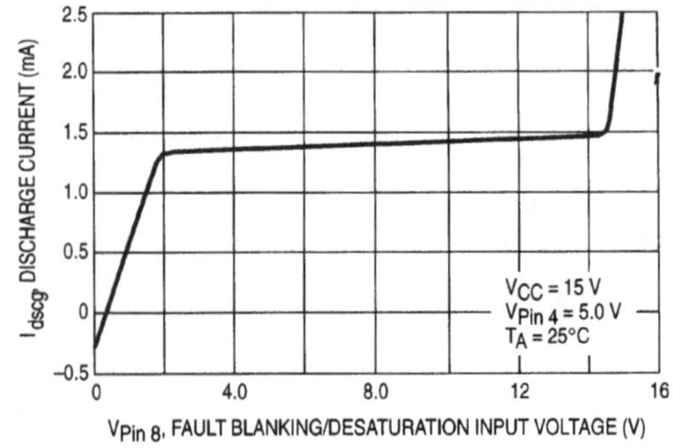

Figure 23. Fault Output Low State Voltage versus Sink Current

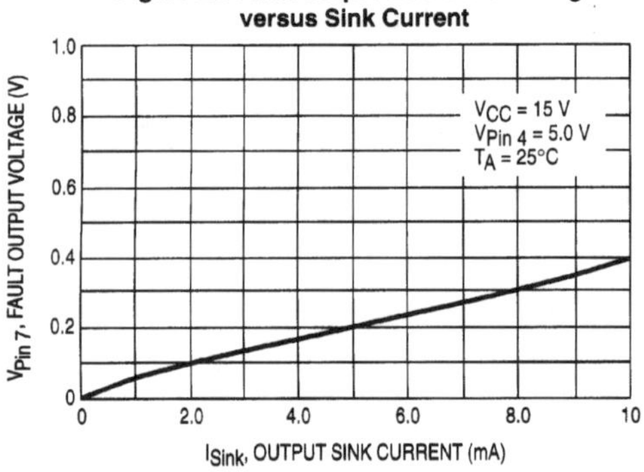

Figure 24. Fault Output High State Voltage versus Source Current

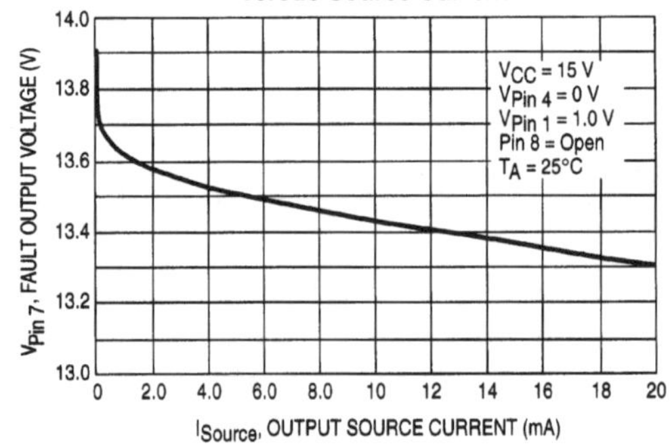

Figure 25. Drive Output Voltage versus Supply Voltage

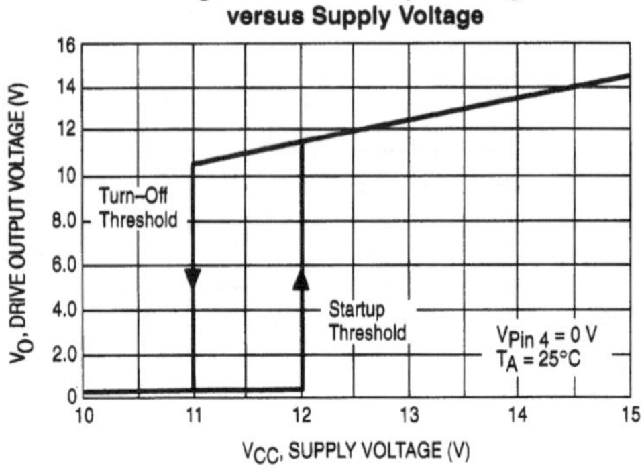

Figure 26. UVLO Thresholds versus Temperature

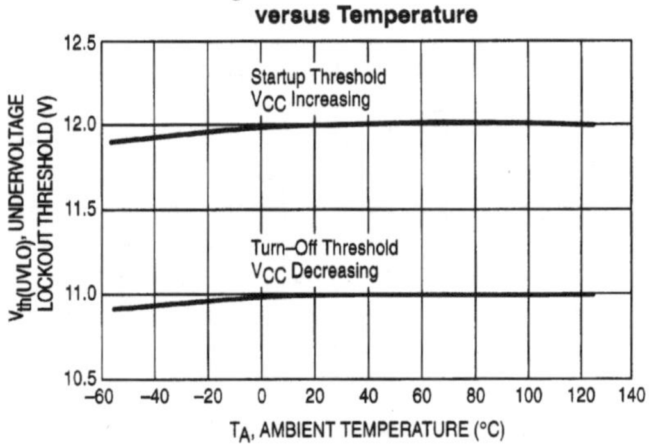

Appendix 4.6-8

MC33153

Figure 27. Supply Current versus Supply Voltage

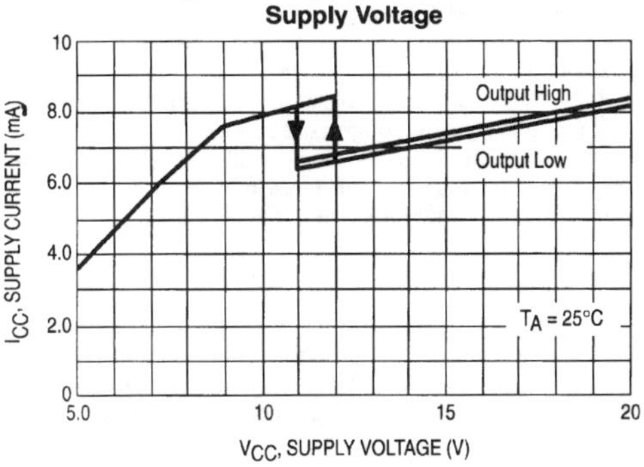

Figure 28. Supply Current versus Temperature

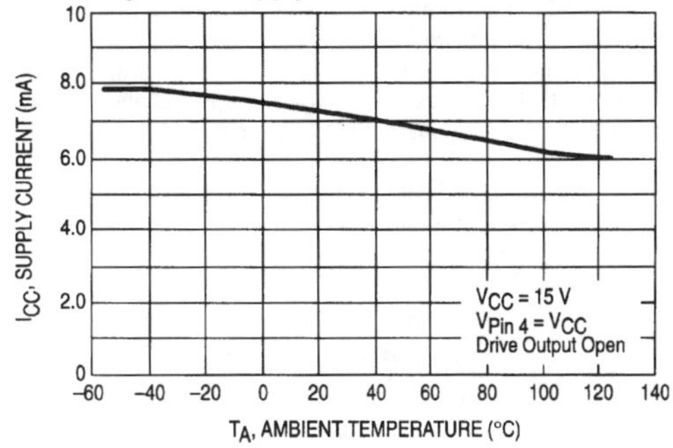

Figure 29. Supply Current versus Input Frequency

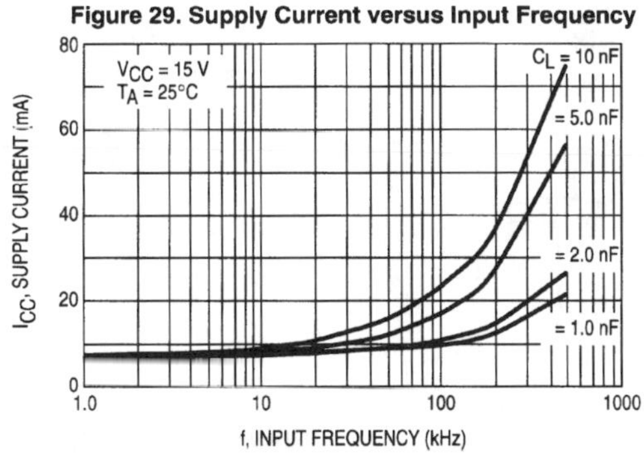

OPERATING DESCRIPTION

GATE DRIVE

Controlling Switching Times

The most important design aspect of an IGBT gate drive is optimization of the switching characteristics. The switching characteristics are especially important in motor control applications in which PWM transistors are used in a bridge configuration. In these applications, the gate drive circuit components should be selected to optimize turn–on, turn–off and off–state impedance. A single resistor may be used to control both turn–on and turn–off as shown in Figure 30. However, the resistor value selected must be a compromise in turn–on abruptness and turn–off losses. Using a single resistor is normally suitable only for very low frequency PWM. An optimized gate drive output stage is shown in Figure 31. This circuit allows turn–on and turn–off to be optimized separately. The turn–on resistor, R_{on}, provides control over the IGBT turn–on speed. In motor control circuits, the resistor sets the turn–on di/dt that controls how fast the free–wheel diode is cleared. The interaction of the IGBT and free–wheeling diode determines the turn–on dv/dt. Excessive turn–on dv/dt is a common problem in half–bridge circuits. The turn–off resistor, R_{off}, controls the turn–off speed and ensures that the IGBT remains off under commutation stresses. Turn–off is critical to obtain low switching losses. While IGBTs exhibit a fixed minimum loss due to minority carrier recombination, a slow gate drive will dominate the turn–off losses. This is particularly true for fast IGBTs. It is also possible to turn–off an IGBT too fast. Excessive turn–off speed will result in large overshoot voltages. Normally, the turn–off resistor is a small fraction of the turn–on resistor.

The MC33153 contains a bipolar totem pole output stage that is capable of sourcing 1.0 amp and sinking 2.0 amps peak. This output also contains a pull down resistor to ensure that the IGBT is off whenever there is insufficient V_{CC} to the MC33153.

In a PWM inverter, IGBTs are used in a half–bridge configuration. Thus, at least one device is always off. While the IGBT is in the off–state, it will be subjected to changes in voltage caused by the other devices. This is particularly a problem when the opposite transistor turns on.

Appendix 4.6-9

MC33153

When the lower device is turned on, clearing the upper diode, the turn–on dv/dt of the lower device appears across the collector emitter of the upper device. To eliminate shoot–through currents, it is necessary to provide a low sink impedance to the device that is in the off–state. In most applications the turn–off resistor can be made small enough to hold off the device that is under commutation without causing excessively fast turn–off speeds.

Figure 30. Using a Single Gate Resistor

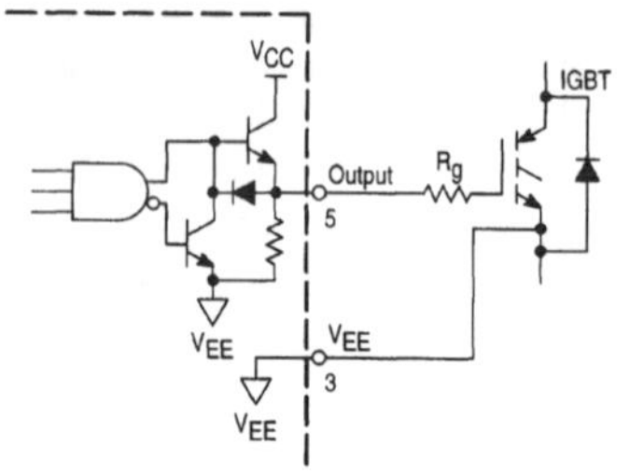

Figure 31. Using Separate Resistors for Turn–On and Turn–Off

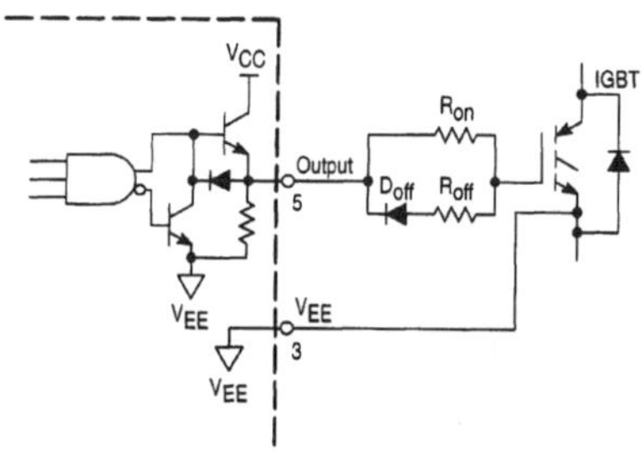

A negative bias voltage can be used to drive the IGBT into the off–state. This is a practice carried over from bipolar Darlington drives and is generally not required for IGBTs. However, a negative bias will reduce the possibility of shoot–through. The MC33153 has separate pins for V_{EE} and Kelvin Ground. This permits operation using a +15/–5.0 V supply.

INTERFACING WITH OPTOISOLATORS

Isolated Input

The MC33153 may be used with an optically isolated input. The optoisolator can be used to provide level shifting, and if desired, isolation from ac line voltages. An optoisolator with a very high dv/dt capability should be used, such as the Hewlett Packard HCPL4053. The IGBT gate turn–on resistor should be set large enough to ensure that the opto's dv/dt capability is not exceeded. Like most optoisolators, the HCPL4053 has an active low open–collector output. Thus, when the LED is on, the output will be low. The MC33153 has an inverting input pin to interface directly with an optoisolator using a pull up resistor. The input may also be interfaced directly to 5.0 V CMOS logic or a microcontroller.

Optoisolator Output Fault

The MC33153 has an active high fault output. The fault output may be easily interfaced to an optoisolator. While it is important that all faults are properly reported, it is equally important that no false signals are propagated. Again, a high dv/dt optoisolator should be used.

The LED drive provides a resistor programmable current of 10 to 20 mA when on, and provides a low impedance path when off. An active high output, resistor, and small signal diode provide an excellent LED driver. This circuit is shown in Figure 32.

Figure 32. Output Fault Optoisolator

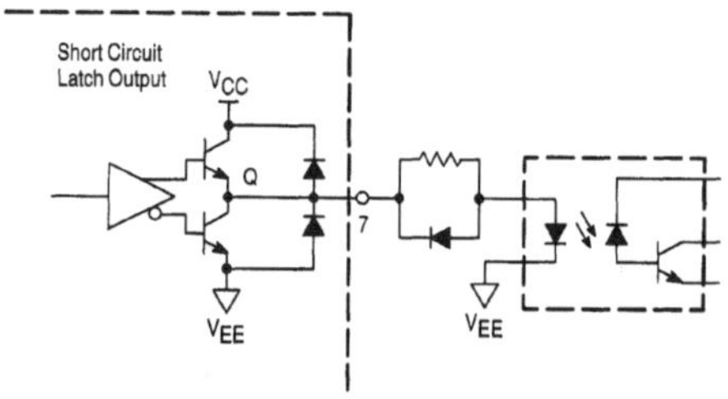

UNDERVOLTAGE LOCKOUT

It is desirable to protect an IGBT from insufficient gate voltage. IGBTs require 15 V on the gate to achieve the rated on–voltage. At gate voltages below 13 V, the on–voltage increases dramatically, especially at higher currents. At very low gate voltages, below 10 V, the IGBT may operate in the linear region and quickly overheat. Many PWM motor drives use a bootstrap supply for the upper gate drive. The UVLO provides protection for the IGBT in case the bootstrap capacitor discharges.

The MC33153 will typically start up at about 12 V. The UVLO circuit has about 1.0 V of hysteresis and will disable the output if the supply voltage falls below about 11V.

Appendix 4.6-10

MC33153

PROTECTION CIRCUITRY

Desaturation Protection

Bipolar Power circuits have commonly used what is known as "Desaturation Detection". This involves monitoring the collector voltage and turning off the device if this voltage rises above a certain limit. A bipolar transistor will only conduct a certain amount of current for a given base drive. When the base is overdriven, the device is in saturation. When the collector current rises above the knee, the device pulls out of saturation. The maximum current the device will conduct in the linear region is a function of the base current and the dc current gain (h_{FE}) of the transistor.

The output characteristics of an IGBT are similar to a Bipolar device. However, the output current is a function of gate voltage instead of current. The maximum current depends on the gate voltage and the device type. IGBTs tend to have a very high transconductance and a much higher current density under a short circuit than a bipolar device. Motor control IGBTs are designed for a lower current density under shorted conditions and a longer short circuit survival time.

The best method for detecting desaturation is the use of a high voltage clamp diode and a comparator. The MC33153 has a Fault Blanking/Desaturation Comparator which senses the collector voltage and provides an output indicating when the device is not fully saturated. Diode D1 is an external high voltage diode with a rated voltage comparable to the power device. When the IGBT is "on" and saturated, D1 will pull down the voltage on the Fault Blanking/Desaturation Input. When the IGBT pulls out of saturation or is "off", the current source will pull up the input and trip the comparator. The comparator threshold is 6.5 V, allowing a maximum on–voltage of about 5.8 V.

A fault exists when the gate input is high and V_{CE} is greater than the maximum allowable $V_{CE(sat)}$. The output of the Desaturation Comparator is ANDed with the gate Input signal and fed into the Short Circuit and Overcurrent Latches. The Overcurrent Latch will turn–off the IGBT for the remainder of the cycle when a fault is detected. When input goes high, both latches are reset. The reference voltage is tied to the Kelvin Ground instead of the V_{EE} to make the threshold independent of negative gate bias. Note that for proper operation of the Desaturation Comparator and the Fault Output, the Current Sense Input must be biased above the Overcurrent and Short Circuit Comparator thresholds. This can be accomplished by connecting Pin 1 to V_{CC}.

Figure 33. Desaturation Detection

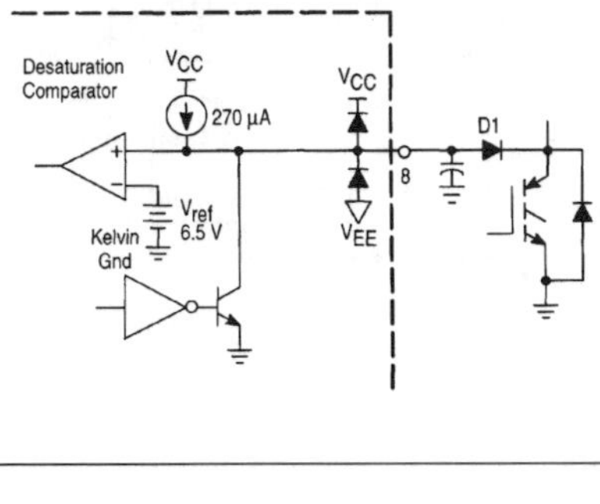

The MC33153 also features a programmable fault blanking time. During turn–on, the IGBT must clear the opposing free–wheeling diode. The collector voltage will remain high until the diode is cleared. Once the diode has been cleared, the voltage will come down quickly to the $V_{CE(sat)}$ of the device. Following turn–on, there is normally considerable ringing on the collector due to the C_{OSS} capacitance of the IGBTs and the parasitic wiring inductance. The fault signal from the Desaturation Comparator must be blanked sufficiently to allow the diode to be cleared and the ringing to settle out.

The blanking function uses an NPN transistor to clamp the comparator input when the gate input is low. When the input is switched high, the clamp transistor will turn "off", allowing the internal current source to charge the blanking capacitor. The time required for the blanking capacitor to charge up from the on–voltage of the internal NPN transistor to the trip voltage of the comparator is the blanking time.

If a short circuit occurs after the IGBT is turned on and saturated, the delay time will be the time required for the current source to charge up the blanking capacitor from the $V_{CE(sat)}$ level of the IGBT to the trip voltage of the comparator. Fault blanking can be disabled by leaving Pin 8 unconnected.

Sense IGBT Protection

Another approach to protecting the IGBTs is to sense the emitter current using a current shunt or Sense IGBTs. This method has the advantage of being able to use high gain IGBTs which do not have any inherent short circuit capability. Current sense IGBTs work as well as current sense MOSFETs in most circumstances. However, the basic problem of working with very low sense voltages still exists. Sense IGBTs sense current through the channel and are therefore linear with respect to the collector current. Because IGBTs have a very low incremental on–resistance, sense IGBTs behave much like low–on resistance current sense MOSFETs. The output voltage of a properly terminated sense IGBT is very low, normally less than 100 mV.

The sense IGBT approach requires fault blanking to prevent false tripping during turn–on. The sense IGBT also requires that the sense signal is ignored while the gate is low. This is because the mirror output normally produces large transient voltages during both turn–on and turn–off due to the collector to mirror capacitance. With non–sensing types of IGBTs, a low resistance current shunt (5.0 to 50 mΩ) can be used to sense the emitter current. When the output is an actual short circuit, the inductance will be very low. Since the blanking circuit provides a fixed minimum on–time, the peak current under a short circuit can be very high. A short circuit discern function is implemented by the second comparator which has a higher trip voltage. The short circuit signal is latched and appears at the Fault Output. When a short circuit is detected, the IGBT should be turned–off for several milliseconds allowing it to cool down before it is turned back on. The sense circuit is very similar to the desaturation circuit. It is possible to build a combination circuit that provides protection for both Short Circuit capable IGBTs and Sense IGBTs.

Appendix 4.6-11

MC33153

APPLICATION INFORMATION

Figure 34 shows a basic IGBT driver application. When driven from an optoisolator, an input pull up resistor is required. This resistor value should be set to bias the output transistor at the desired current. A decoupling capacitor should be placed close to the IC to minimize switching noise.

A bootstrap diode may be used for a floating supply. If the protection features are not required, then both the Fault Blanking/Desaturation and Current Sense Inputs should both be connected to the Kelvin Ground (Pin 2). When used with a single supply, the Kelvin Ground and V_{EE} pins should be connected together. Separate gate resistors are recommended to optimize the turn–on and turn–off drive.

Figure 34. Basic Application

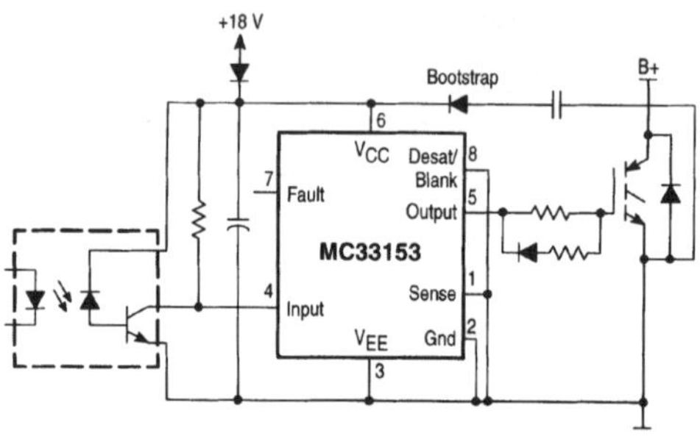

Figure 35. Dual Supply Application

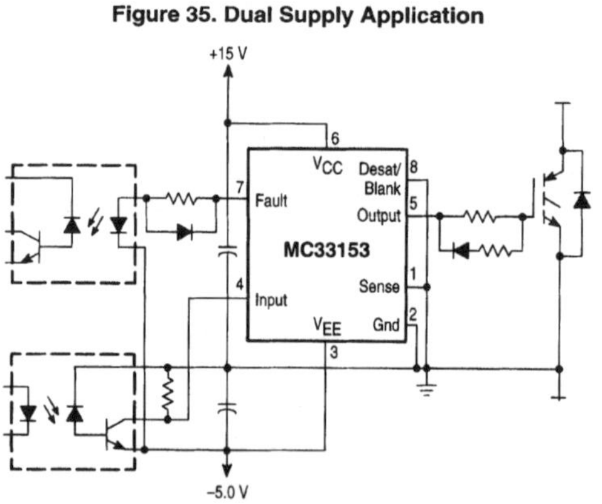

When used in a dual supply application as in Figure 35, the Kelvin Ground should be connected to the emitter of the IGBT. If the protection features are not used, then both the Fault Blanking/Desaturation and the Current Sense Inputs should be connected to Ground. The input optoisolator should always be referenced to V_{EE}.

If desaturation protection is desired, a high voltage diode is connected to the Fault Blanking/Desaturation pin. The blanking capacitor should be connected from the Desaturation pin to the V_{EE} pin. If a dual supply is used, the blanking capacitor should be connected to the Kelvin Ground. The Current Sense Input should be tied high because the two comparator outputs are ANDed together. Although the reverse voltage on collector of the IGBT is clamped to the emitter by the free–wheeling diode, there is normally considerable inductance within the package itself. A small resistor in series with the diode can be used to protect the IC from reverse voltage transients.

Figure 36. Desaturation Application

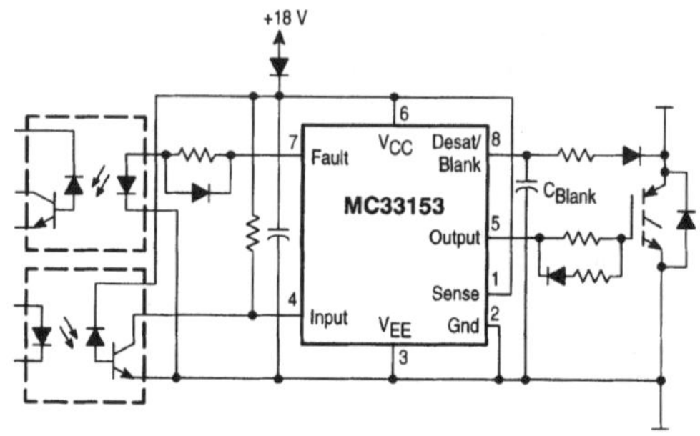

When using sense IGBTs or a sense resistor, the sense voltage is applied to the Current Sense Input. The sense trip voltages are referenced to the Kelvin Ground pin. The sense voltage is very small, typically about 65 mV, and sensitive to noise. Therefore, the sense and ground return conductors should be routed as a differential pair. An RC filter is useful in filtering any high frequency noise. A blanking capacitor is connected from the blanking pin to V_{EE}. The stray capacitance on the blanking pin provides a very small level of blanking if left open. The blanking pin should not be grounded when using current sensing, that would disable the sense. The blanking pin should never be tied high, that would short out the clamp transistor.

Figure 37. Sense IGBT Application

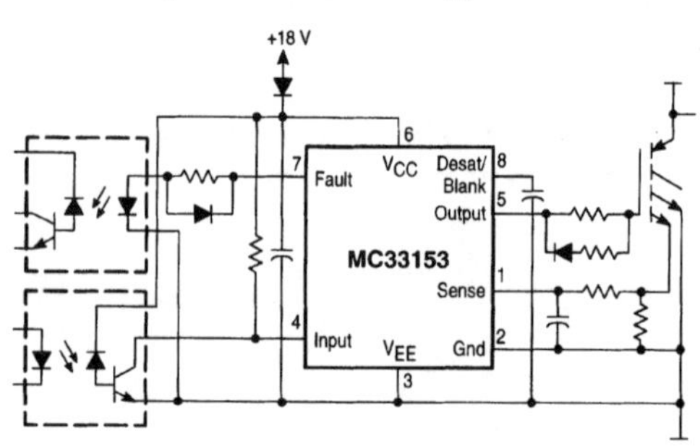

Appendix 4.7-1

MOTOROLA

LM2575

Advance Information

Easy Switcher™ 1.0 A Step-Down Voltage Regulator

The LM2575 series of regulators are monolithic integrated circuits ideally suited for easy and convenient design of a step–down switching regulator (buck converter). All circuits of this series are capable of driving a 1.0 A load with excellent line and load regulation. These devices are available in fixed output voltages of 3.3 V, 5.0 V, 12 V, 15 V, and an adjustable output version.

These regulators were designed to minimize the number of external components to simplify the power supply design. Standard series of inductors optimised for use with the LM2575 are offered by several different inductor manufacturers.

Since the LM2575 converter is a switch–mode power supply, its efficiency is significantly higher in comparison with popular three–terminal linear regulators, especially with higher input voltages. In many cases, the power dissipated by the LM2575 regulator is so low, that no heatsink is required or its size could be reduced dramatically.

The LM2575 features include a guaranteed ±4% tolerance on output voltage within specified input voltages and output load conditions, and ±10% on the oscillator frequency (±2% over 0°C to 125°C). External shutdown is included, featuring 80 μA typical standby current. The output switch includes cycle–by–cycle current limiting, as well as thermal shutdown for full protection under fault conditions.

Features

- 3.3 V, 5.0 V, 12 V, 15 V, and Adjustable Output Versions
- Adjustable Version Output Voltage Range of 1.23 V to 37 V ±4% Maximum Over Line and Load Conditions
- Guaranteed 1.0 A Output Current
- Wide Input Voltage Range: 4.75 V to 40 V
- Requires Only 4 External Components
- 52 kHz Fixed Frequency Internal Oscillator
- TTL Shutdown Capability, Low Power Standby Mode
- High Efficiency
- Uses Readily Available Standard Inductors
- Thermal Shutdown and Current Limit Protection

Applications

- Simple and High–Efficiency Step–Down (Buck) Regulators
- Efficient Pre–Regulator for Linear Regulators
- On–Card Switching Regulators
- Positive to Negative Converters (Buck–Boost)
- Negative Step–Up Converters
- Power Supply for Battery Chargers

EASY SWITCHER™
1.0 A STEP–DOWN
VOLTAGE REGULATOR

SEMICONDUCTOR
TECHNICAL DATA

T SUFFIX
PLASTIC PACKAGE
CASE 314D

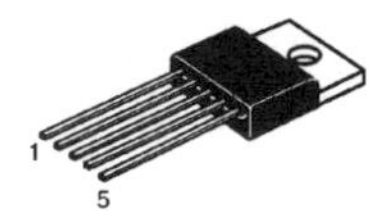

Pin 1. V_{in}
2. Output
3. Ground
4. Feedback
5. $\overline{\text{ON}}$/OFF

TV SUFFIX
PLASTIC PACKAGE
CASE 314B

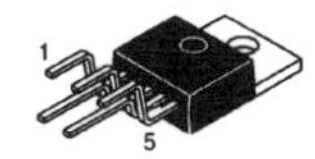

Heatsink surface connected to Pin 3.

D2T SUFFIX
PLASTIC PACKAGE
CASE 936A
(D^2PAK)

Heatsink surface (shown as terminal 6 in case outline drawing) is connected to Pin 3.

DEVICE TYPE/NOMINAL OUTPUT VOLTAGE

LM2575–3.3	3.3 V
LM2575–5	5.0 V
LM2575–12	12 V
LM2575–15	15 V
LM2575–Adj	1.23 V to 37 V

ORDERING INFORMATION

Device	Operating Temperature Range	Package
LM2575T–**		Straight Lead
LM2575TV–**	T_J = –40° to +125°C	Vertical Mount
LM2575D2T–**		Surface Mount

** = Voltage Option, ie. 3.3, 5.0, 12, 15 V and Adjustable Output.

Appendix 4.7-2

LM2575

Figure 1. Block Diagram and Typical Application

Typical Application (Fixed Output Voltage Versions)

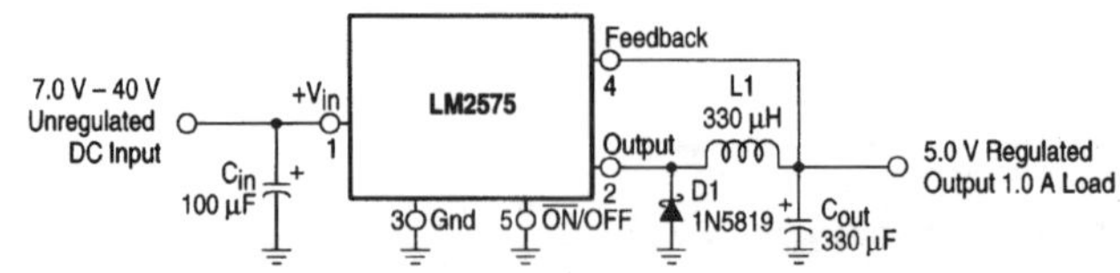

Representative Block Diagram and Typical Application

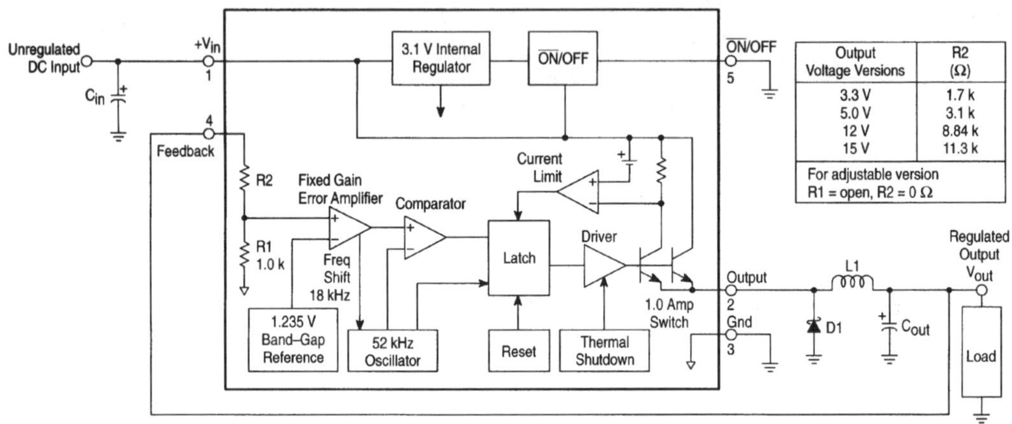

This device contains 162 active transistors.

ABSOLUTE MAXIMUM RATINGS (Absolute Maximum Ratings indicate limits beyond which damage to the device may occur.)

Rating	Symbol	Value	Unit
Maximum Supply Voltage	V_{in}	45	V
$\overline{\text{ON}}$/OFF Pin Input Voltage	–	$-0.3\ V \le V \le +V_{in}$	V
Output Voltage to Ground (Steady–State)	–	−1.0	V
Power Dissipation Case 314B and 314D (TO–220, 5–Lead) Thermal Resistance, Junction–to–Ambient Thermal Resistance, Junction–to–Case Case 936A (D^2PAK) Thermal Resistance, Junction–to–Ambient (Figure 34) Thermal Resistance, Junction–to–Case	 P_D $R_{\theta JA}$ $R_{\theta JC}$ P_D $R_{\theta JA}$ $R_{\theta JC}$	 Internally Limited 65 5.0 Internally Limited 70 5.0	 W °C/W °C/W W °C/W °C/W
Storage Temperature Range	T_{stg}	−65 to +150	°C
Minimum ESD Rating (Human Body Model: C = 100 pF, R = 1.5 kΩ)	–	3.0	kV
Lead Temperature (Soldering, 10 s)	–	260	°C
Maximum Junction Temperature	T_J	150	°C

NOTE: ESD data available upon request.

Appendix 4.7-3

LM2575

OPERATING RATINGS (Operating Ratings indicate conditions for which the device is intended to be functional, but do not guarantee specific performance limits. For guaranteed specifications and test conditions, see the Electrical Characteristics.)

Rating	Symbol	Value	Unit
Operating Junction Temperature Range	T_J	–40 to +125	°C
Supply Voltage	V_{in}	40	V

SYSTEM PARAMETERS ([Note 1] Test Circuit Figure 14)

ELECTRICAL CHARACTERISTICS (Unless otherwise specified, V_{in} = 12 V for the 3.3 V, 5.0 V, and Adjustable version, V_{in} = 25 V for the 12 V version, and V_{in} = 30 V for the 15 V version. I_{Load} = 200 mA. For typical values T_J = 25°C, for min/max values T_J is the operating junction temperature range that applies [Note 2], unless otherwise noted.)

Characteristics	Symbol	Min	Typ	Max	Unit
LM2575–3.3 ([Note 1] Test Circuit Figure 14)					
Output Voltage (V_{in} = 12 V, I_{Load} = 0.2 A, T_J = 25°C)	V_{out}	3.234	3.3	3.366	V
Output Voltage (4.75 V ≤ V_{in} ≤ 40 V, 0.2 A ≤ I_{Load} ≤ 1.0 A) T_J = 25°C T_J = –40 to +125°C	V_{out}	 3.168 3.135	 3.3 –	 3.432 3.465	V
Efficiency (V_{in} = 12 V, I_{Load} = 1.0 A)	η	–	75	–	%
LM2575–5 ([Note 1] Test Circuit Figure 14)					
Output Voltage (V_{in} = 12 V, I_{Load} = 0.2 A, T_J = 25°C)	V_{out}	4.9	5.0	5.1	V
Output Voltage (8.0 V ≤ V_{in} ≤ 40 V, 0.2 A ≤ I_{Load} ≤ 1.0 A) T_J = 25°C T_J = –40 to +125°C	V_{out}	 4.8 4.75	 5.0 –	 5.2 5.25	V
Efficiency (V_{in} = 12 V, I_{Load} = 1.0 A)	η	–	77	–	%
LM2575–12 ([Note 1] Test Circuit Figure 14)					
Output Voltage (V_{in} = 25 V, I_{Load} = 0.2 A, T_J = 25°C)	V_{out}	11.76	12	12.24	V
Output Voltage (15 V ≤ V_{in} ≤ 40 V, 0.2 A ≤ I_{Load} ≤ 1.0 A) T_J = 25°C T_J = –40 to +125°C	V_{out}	 11.52 11.4	 12 –	 12.48 12.6	V
Efficiency (V_{in} = 15V, I_{Load} = 1.0 A)	η	–	88	–	%
LM2575–15 ([Note 1] Test Circuit Figure 14)					
Output Voltage (V_{in} = 30 V, I_{Load} = 0.2 A, T_J = 25°C)	V_{out}	14.7	15	15.3	V
Output Voltage (18 V ≤ V_{in} ≤ 40 V, 0.2 A ≤ I_{Load} ≤ 1.0 A) T_J = 25°C T_J = –40 to +125°C	V_{out}	 14.4 14.25	 15 –	 15.6 15.75	V
Efficiency (V_{in} = 18 V, I_{Load} = 1.0 A)	η	–	88	–	%
LM2575 ADJUSTABLE VERSION ([Note 1] Test Circuit Figure 14)					
Feedback Voltage (V_{in} = 12 V, I_{Load} = 0.2 A, V_{out} = 5.0 V, T_J = 25°C)	V_{FB}	1.217	1.23	1.243	V
Feedback Voltage (8.0 V ≤ V_{in} ≤ 40 V, 0.2 A ≤ I_{Load} ≤ 1.0 A, V_{out} = 5.0 V) T_J = 25°C T_J = –40 to +125°C	V_{FB}	 1.193 1.18	 1.23 –	 1.267 1.28	V
Efficiency (V_{in} = 12 V, I_{Load} = 1.0 A, V_{out} = 5.0 V)	η	–	77	–	%

NOTES: 1. External components such as the catch diode, inductor, input and output capacitors can affect switching regulator system performance. When the LM2575 is used as shown in the Figure 14 test circuit, system performance will be as shown in system parameters section.

2. Tested junction temperature range for the LM2575: T_{low} = –40°C T_{high} = +125°C

Appendix 4.7-4

LM2575

DEVICE PARAMETERS

ELECTRICAL CHARACTERISTICS (Unless otherwise specified, V_{in} = 12 V for the 3.3 V, 5.0 V, and Adjustable version, V_{in} = 25 V for the 12 V version, and V_{in} = 30 V for the 15 V version. I_{Load} = 200 mA. For typical values T_J = 25°C, for min/max values T_J is the operating junction temperature range that applies [Note 2], unless otherwise noted.)

Characteristics	Symbol	Min	Typ	Max	Unit
ALL OUTPUT VOLTAGE VERSIONS					
Feedback Bias Current (V_{out} = 5.0 V [Adjustable Version Only])	I_b				nA
T_J = 25°C		–	25	100	
T_J = –40 to +125°C		–	–	200	
Oscillator Frequency [Note 3]	f_{osc}				kHz
T_J = 25°C		–	52	–	
T_J = 0 to +125°C		47	–	58	
T_J = –40 to +125°C		42	–	63	
Saturation Voltage (I_{out} = 1.0 A [Note 4])	V_{sat}				V
T_J = 25°C		–	1.0	1.2	
T_J = –40 to +125°C		–	–	1.3	
Max Duty Cycle ("on") [Note 5]	DC	94	98	–	%
Current Limit (Peak Current [Notes 4 and 3])	I_{CL}				A
T_J = 25°C		1.7	2.3	3.0	
T_J = –40 to +125°C		1.4	–	3.2	
Output Leakage Current [Notes 6 and 7], T_J = 25°C	I_L				mA
Output = 0 V		–	0.8	2.0	
Output = –1.0 V		–	6.0	20	
Quiescent Current [Note 6]	I_Q				mA
T_J = 25°C		–	5.0	9.0	
T_J = –40 to +125°C		–	–	11	
Standby Quiescent Current ($\overline{ON}$/OFF Pin = 5.0 V ("off"))	I_{stby}				µA
T_J = 25°C		–	80	200	
T_J = –40 to +125°C		–	–	400	
$\overline{ON}$/OFF Pin Logic Input Level (Test Circuit Figure 14)					V
V_{out} = 0 V	V_{IH}				
T_J = 25°C		2.2	1.4	–	
T_J = –40 to +125°C		2.4	–	–	
V_{out} = Nominal Output Voltage	V_{IL}				
T_J = 25°C		–	1.2	1.0	
T_J = –40 to +125°C		–	–	0.8	
$\overline{ON}$/OFF Pin Input Current (Test Circuit Figure 14)					µA
$\overline{ON}$/OFF Pin = 5.0 V ("off"), T_J = 25°C	I_{IH}	–	15	30	
$\overline{ON}$/OFF Pin = 0 V ("on"), T_J = 25°C	I_{IL}	–	0	5.0	

NOTES: 3. The oscillator frequency reduces to approximately 18 kHz in the event of an output short or an overload which causes the regulated output voltage to drop approximately 40% from the nominal output voltage. This self protection feature lowers the average dissipation of the IC by lowering the minimum duty cycle from 5% down to approximately 2%.

4. Output (Pin 2) sourcing current. No diode, inductor or capacitor connected to output pin.
5. Feedback (Pin 4) removed from output and connected to 0 V.
6. Feedback (Pin 4) removed from output and connected to +12 V for the Adjustable, 3.3 V, and 5.0 V versions, and +25 V for the 12 V and 15 V versions, to force the output transistor "off".
7. V_{in} = 40 V.

Appendix 4.7-5

LM2575

TYPICAL PERFORMANCE CHARACTERISTICS (Circuit of Figure 14)

Figure 2. Normalized Output Voltage

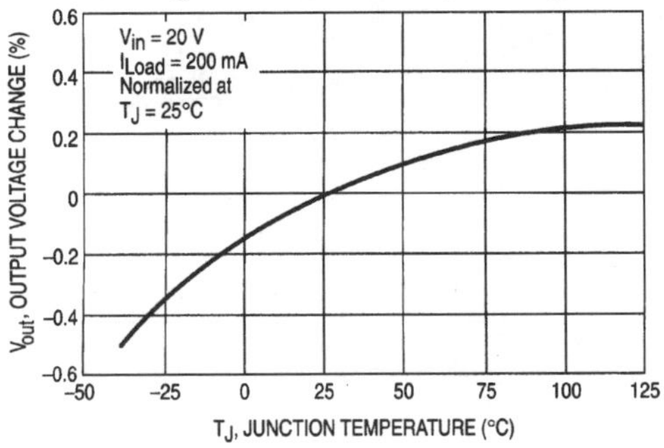

Figure 3. Line Regulation

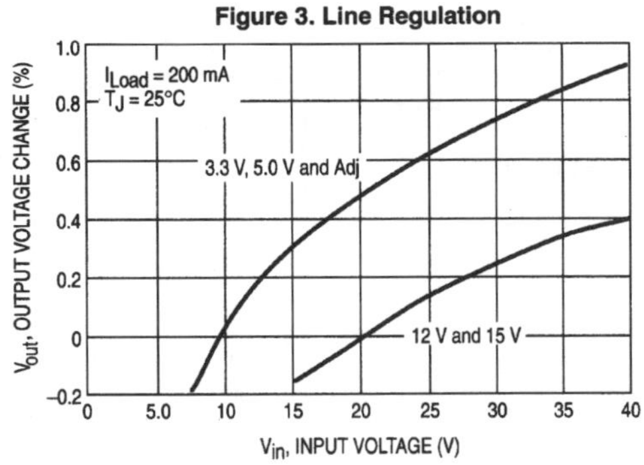

Figure 4. Switch Saturation Voltage

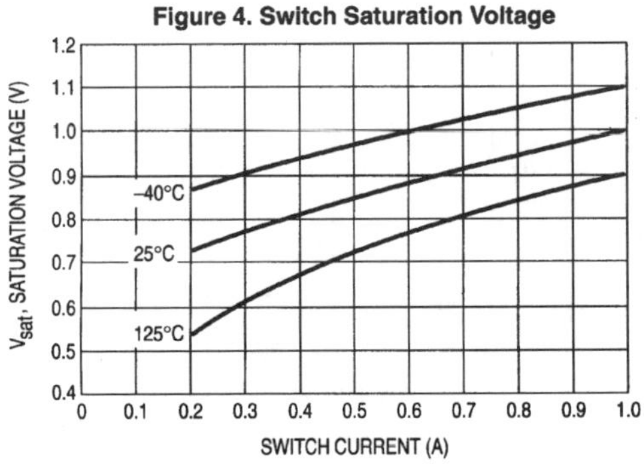

Figure 5. Current Limit

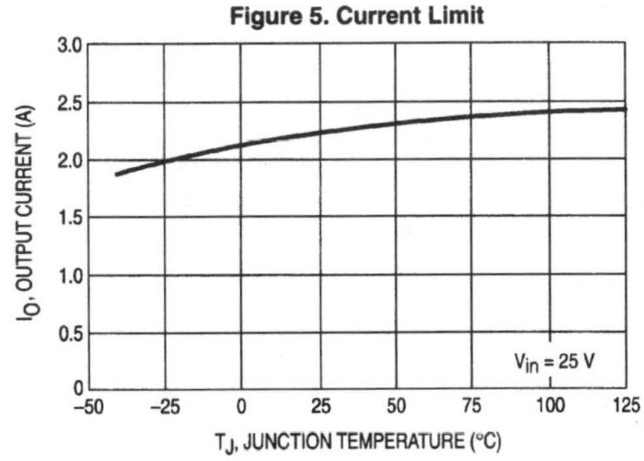

Figure 6. Dropout Voltage

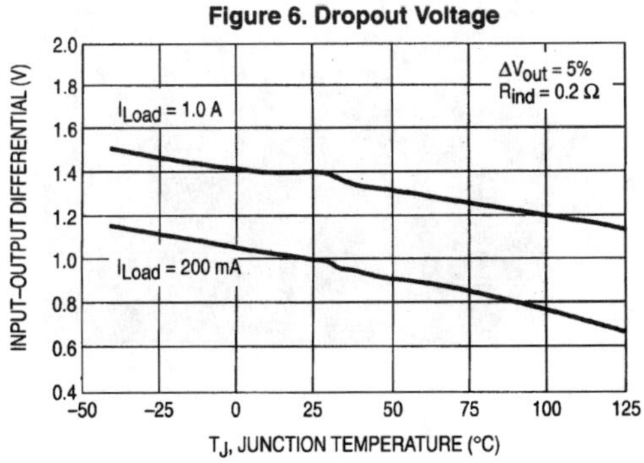

Figure 7. Quiescent Current

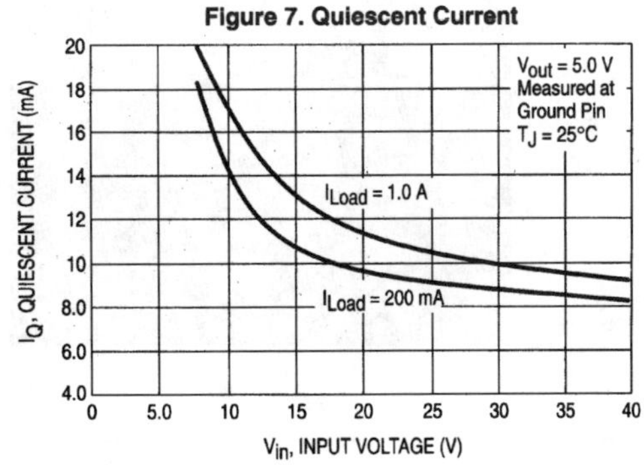

Appendix 4.7-6

LM2575

Figure 8. Standby Quiescent Current

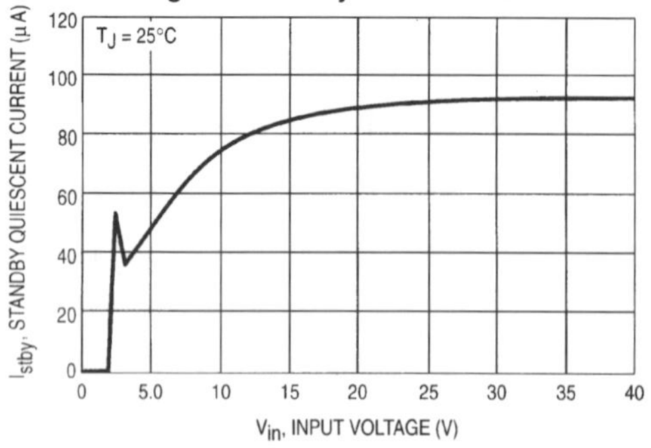

Figure 9. Standby Quiescent Current

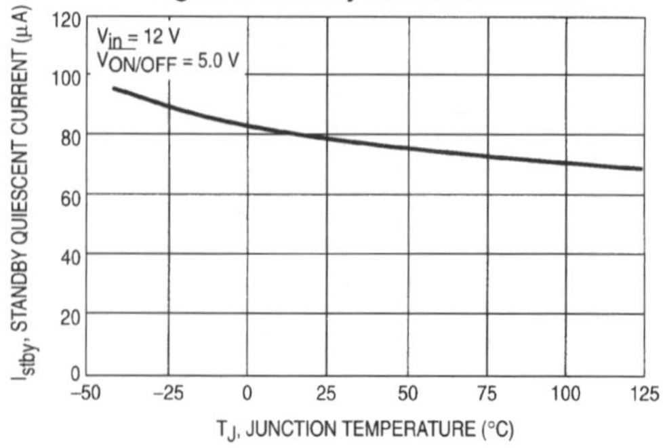

Figure 10. Oscillator Frequency

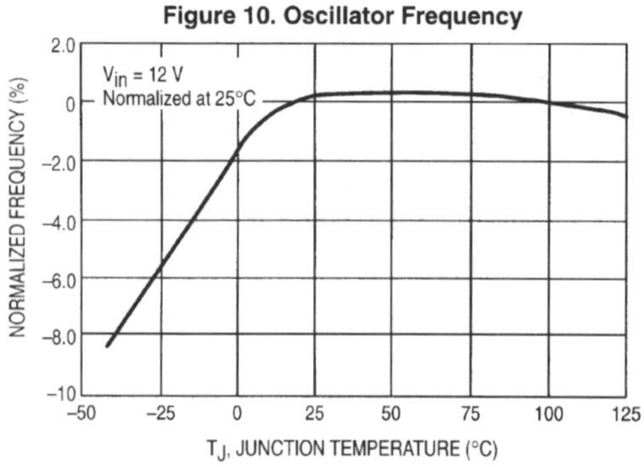

Figure 11. Feedback Pin Current

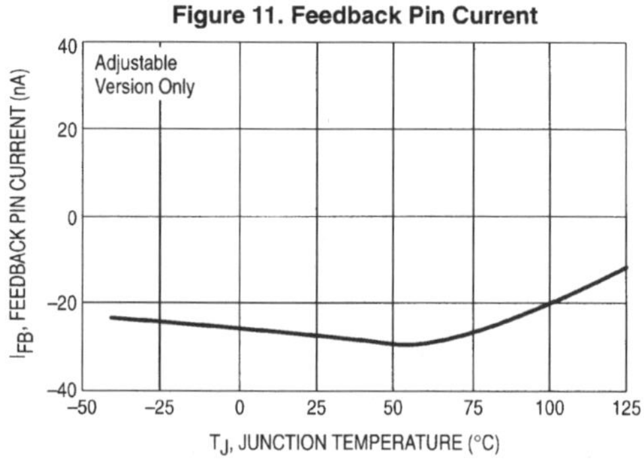

Figure 12. Switching Waveforms

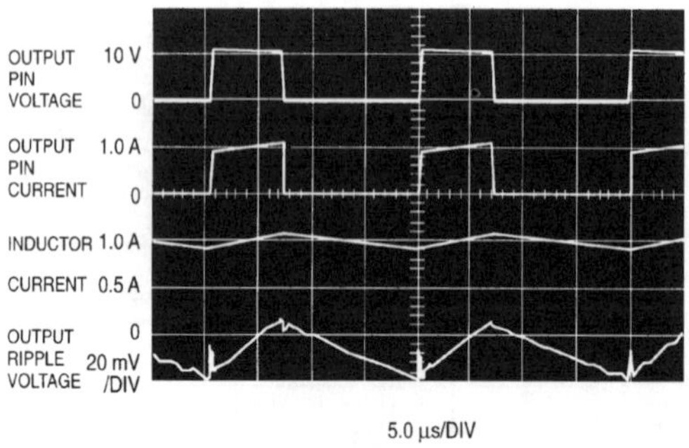

Figure 13. Load Transient Response

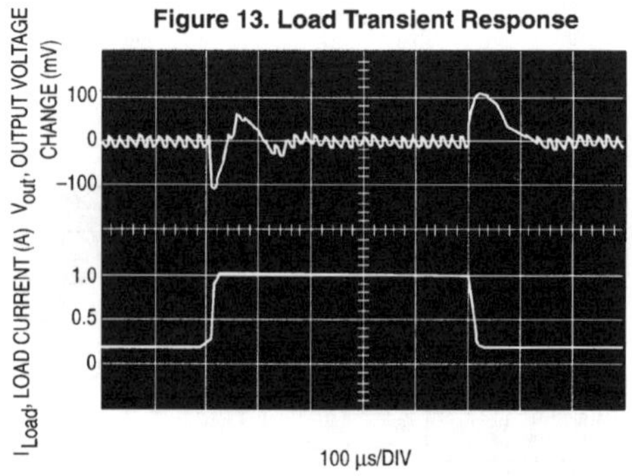

Appendix 4.7-7

LM2575

Figure 14. Typical Test Circuit

5.0 Output Voltage Versions

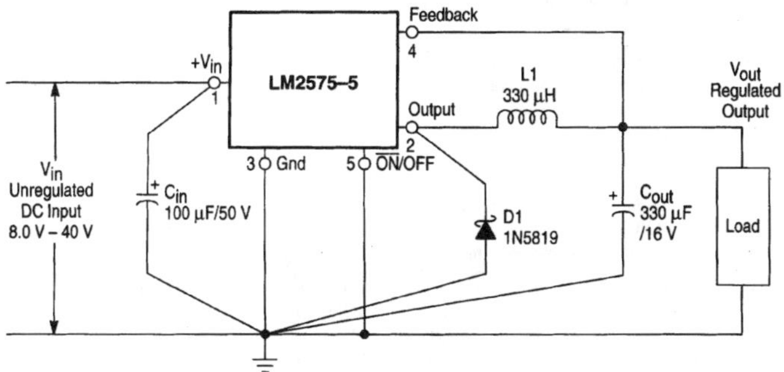

C_{in}	100 μF, 50 V, Aluminium Electrolytic
C_{out}	330 μF, 16 V, Aluminium Electrolytic
D1	Schottky, 1N5819
L1	330 μF, PE–52627 (for 5.0 V in, 3.3 V out, use 100 μH, PE–92108)

Adjustable Output Voltage Versions

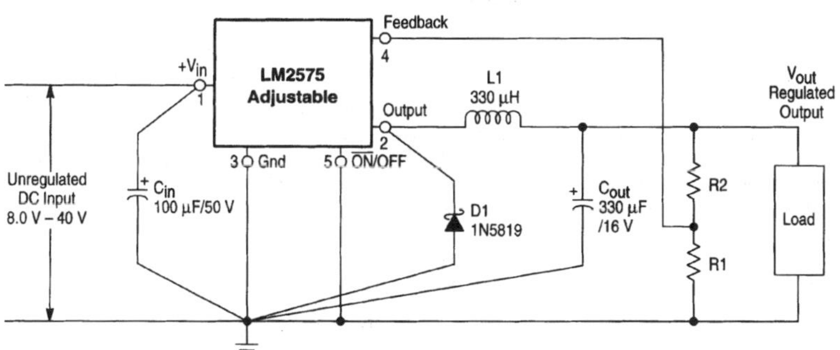

$$V_{out} = V_{ref}\left(1 + \frac{R2}{R1}\right)$$

$$R2 = R1\left(\frac{V_{out}}{V_{ref}} - 1\right)$$

Where V_{ref} = 1.23 V, R1 between 1.0 kΩ and 5.0 kΩ

PCB LAYOUT GUIDELINES

As in any switching regulator, the layout of the printed circuit board is very important. Rapidly switching currents associated with wiring inductance, stray capacitance and parasitic inductance of the printed circuit board traces can generate voltage transients which can generate electromagnetic interferences (EMI) and affect the desired operation. As indicated in the Figure 14, to minimize inductance and ground loops, the length of the leads indicated by heavy lines should be kept as short as possible. For best results, single–point grounding (as indicated) or ground plane construction should be used.

On the other hand, the PCB area connected to the Pin 2 (emitter of the internal switch) of the LM2575 should be kept to a minimum in order to minimize coupling to sensitive circuitry.

Another sensitive part of the circuit is the feedback. It is important to keep the sensitive feedback wiring short. To assure this, physically locate the programming resistors near to the regulator, when using the adjustable version of the LM2575 regulator.

Appendix 4.7-8

LM2575

PIN FUNCTION DESCRIPTION

Pin	Symbol	Description (Refer to Figure 1)
1	V_{in}	This pin is the positive input supply for the LM2575 step–down switching regulator. In order to minimize voltage transients and to supply the switching currents needed by the regulator, a suitable input bypass capacitor must be present (C_{in} in Figure 1).
2	Output	This is the emitter of the internal switch. The saturation voltage V_{sat} of this output switch is typically 1.0 V. It should be kept in mind that the PCB area connected to this pin should be kept to a minimum in order to minimize coupling to sensitive circuitry.
3	Gnd	Circuit ground pin. See the information about the printed circuit board layout.
4	Feedback	This pin senses regulated output voltage to complete the feedback loop. The signal is divided by the internal resistor divider network R2, R1 and applied to the non–inverting input of the internal error amplifier. In the Adjustable version of the LM2575 switching regulator this pin is the direct input of the error amplifier and the resistor network R2, R1 is connected externally to allow programming of the output voltage.
5	$\overline{\text{ON}}$/OFF	It allows the switching regulator circuit to be shut down using logic level signals, thus dropping the total input supply current to approximately 80 µA. The input threshold voltage is typically 1.4 V. Applying a voltage above this value (up to $+V_{in}$) shuts the regulator off. If the voltage applied to this pin is lower than 1.4 V or if this pin is connected to ground, the regulator will be in the "on" condition.

DESIGN PROCEDURE

Buck Converter Basics

The LM2575 is a "Buck" or Step–Down Converter which is the most elementary forward–mode converter. Its basic schematic can be seen in Figure 15.

The operation of this regulator topology has two distinct time periods. The first one occurs when the series switch is on, the input voltage is connected to the input of the inductor.

The output of the inductor is the output voltage, and the rectifier (or catch diode) is reverse biased. During this period, since there is a constant voltage source connected across the inductor, the inductor current begins to linearly ramp upwards, as described by the following equation:

$$I_{L(on)} = \frac{(V_{in} - V_{out})\, t_{on}}{L}$$

During this "on" period, energy is stored within the core material in the form of magnetic flux. If the inductor is properly designed, there is sufficient energy stored to carry the requirements of the load during the "off" period.

Figure 15. Basic Buck Converter

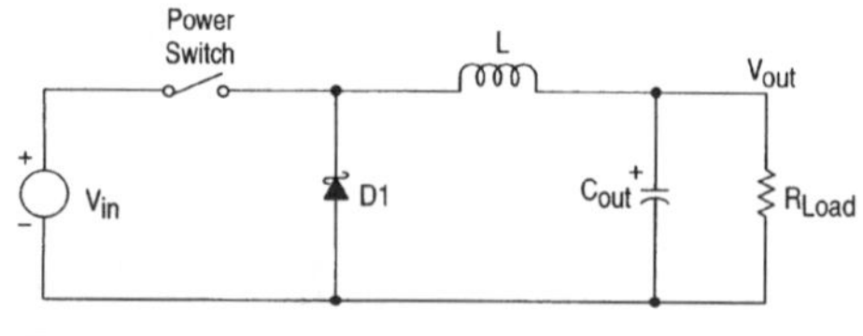

The next period is the "off" period of the power switch. When the power switch turns off, the voltage across the inductor reverses its polarity and is clamped at one diode voltage drop below ground by catch dioded. Current now flows through the catch diode thus maintaining the load current loop. This removes the stored energy from the inductor. The inductor current during this time is:

$$I_{L(off)} = \frac{(V_{out} - V_D)\, t_{off}}{L}$$

This period ends when the power switch is once again turned on. Regulation of the converter is accomplished by varying the duty cycle of the power switch. It is possible to describe the duty cycle as follows:

$d = \frac{t_{on}}{T}$, where T is the period of switching.

For the buck converter with ideal components, the duty cycle can also be described as:

$$d = \frac{V_{out}}{V_{in}}$$

Figure 16 shows the buck converter idealized waveforms of the catch diode voltage and the inductor current.

Figure 16. Buck Converter Idealized Waveforms

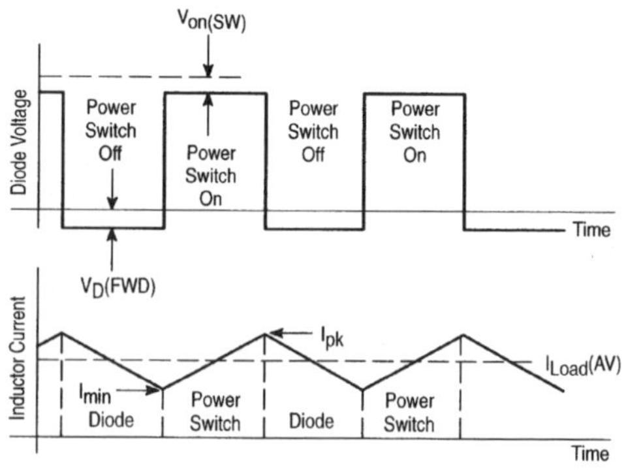

Appendix 4.7-9

LM2575

Procedure (Fixed Output Voltage Version) In order to simplify the switching regulator design, a step–by–step design procedure and example is provided.

Procedure	Example
Given Parameters: V_{out} = Regulated Output Voltage (3.3 V, 5.0 V, 12 V or 15 V) $V_{in(max)}$ = Maximum DC Input Voltage $I_{Load(max)}$ = Maximum Load Current	**Given Parameters:** V_{out} = 5.0 V $V_{in(max)}$ = 20 V $I_{Load(max)}$ = 0.8 A
1. Controller IC Selection According to the required input voltage, output voltage and current, select the appropriate type of the controller IC output voltage version.	**1. Controller IC Selection** According to the required input voltage, output voltage, current polarity and current value, use the LM2575–5 controller IC
2. Input Capacitor Selection (C_{in}) To prevent large voltage transients from appearing at the input and for stable operation of the converter, an aluminium or tantalum electrolytic bypass capacitor is needed between the input pin +V_{in} and ground pin Gnd. This capacitor should be located close to the IC using short leads. This capacitor should have a low ESR (Equivalent Series Resistance) value.	**2. Input Capacitor Selection (C_{in})** A 47 μF, 25 V aluminium electrolytic capacitor located near to the input and ground pins provides sufficient bypassing.
3. Catch Diode Selection (D1) **A.** Since the diode maximum peak current exceeds the regulator maximum load current the catch diode current rating must be at least 1.2 times greater than the maximum load current. For a robust design the diode should have a current rating equal to the maximum current limit of the LM2575 to be able to withstand a continuous output short **B.** The reverse voltage rating of the diode should be at least 1.25 times the maximum input voltage.	**3. Catch Diode Selection (D1)** **A.** For this example the current rating of the diode is 1.0 A. **B.** Use a 30 V 1N5818 Schottky diode, or any of the suggested fast recovery diodes shown in the Table 4.
4. Inductor Selection (L1) **A.** According to the required working conditions, select the correct inductor value using the selection guide from Figures 17 to 21. **B.** From the appropriate inductor selection guide, identify the inductance region intersected by the Maximum Input Voltage line and the Maximum Load Current line. Each region is identified by an inductance value and an inductor code. **C.** Select an appropriate inductor from the several different manufacturers part numbers listed in Table 1 or Table 2. When using Table 2 for selecting the right inductor the designer must realize that the inductor current rating must be higher than the maximum peak current flowing through the inductor. This maximum peak current can be calculated as follows: $I_{p(max)} = I_{Load(max)} + \frac{(V_{in} - V_{out})\, t_{on}}{2L}$ where t_{on} is the "on" time of the power switch and $t_{on} = \frac{V_{out}}{V_{in}} \times \frac{1}{f_{osc}}$ For additional information about the inductor, see the inductor section in the "External Components" section of this data sheet.	**4. Inductor Selection (L1)** **A.** Use the inductor selection guide shown in Figures 17 to 21. **B.** From the selection guide, the inductance area intersected by the 20 V line and 0.8 A line is L330. **C.** Inductor value required is 330 μH. From the Table 1 or Table 2, choose an inductor from any of the listed manufacturers.

Appendix 4.7-10

LM2575

Procedure (Fixed Output Voltage Version) (continued) In order to simplify the switching regulator design, a step–by–step design procedure and example is provided.

Procedure	Example
5. Output Capacitor Selection (C_{out}) **A.** Since the LM2575 is a forward–mode switching regulator with voltage mode control, its open loop 2–pole–2–zero frequency characteristic has the dominant pole–pair determined by the output capacitor and inductor values. For stable operation and an acceptable ripple voltage, (approximately 1% of the output voltage) a value between 100 μF and 470 μF is recommended. **B.** Due to the fact that the higher voltage electrolytic capacitors generally have lower ESR (Equivalent Series Resistance) numbers, the output capacitor's voltage rating should be at least 1.5 times greater than the output voltage. For a 5.0 V regulator, a rating at least 8V is appropriate, and a 10 V or 16 V rating is recommended.	**5. Output Capacitor Selection (C_{out})** **A.** C_{out} = 100 μF to 470 μF standard aluminium electrolytic. **B.** Capacitor voltage rating = 16 V.

Procedure (Adjustable Output Version: LM2575–Adj)

Procedure	Example
Given Parameters: V_{out} = Regulated Output Voltage $V_{in(max)}$ = Maximum DC Input Voltage $I_{Load(max)}$ = Maximum Load Current	**Given Parameters:** V_{out} = 8.0 V $V_{in(max)}$ = 12 V $I_{Load(max)}$ = 1.0 A
1. Programming Output Voltage To select the right programming resistor R1 and R2 value (see Figure 14) use the following formula: $V_{out} = V_{ref}\left(1 + \frac{R2}{R1}\right)$ where V_{ref} = 1.23 V Resistor R1 can be between 1.0 k and 5.0 kΩ. (For best temperature coefficient and stability with time, use 1% metal film resistors). $R2 = R1\left(\frac{V_{out}}{V_{ref}} - 1\right)$	**1. Programming Output Voltage** (selecting R1 and R2) Select R1 and R2: $V_{out} = 1.23\left(1 + \frac{R2}{R1}\right)$ Select R1 = 1.8 kΩ $R2 = R1\left(\frac{V_{out}}{V_{ref}} - 1\right) = 1.8\ k\left(\frac{8.0\ V}{1.23\ V} - 1\right)$ R2 = 9.91 kΩ, choose a 9.88 k metal film resistor.
2. Input Capacitor Selection (C_{in}) To prevent large voltage transients from appearing at the input and for stable operation of the converter, an aluminium or tantalum electrolytic bypass capacitor is needed between the input pin $+V_{in}$ and ground pin Gnd This capacitor should be located close to the IC using short leads. This capacitor should have a low ESR (Equivalent Series Resistance) value. For additional information see input capacitor section in the "External Components" section of this data sheet.	**2. Input Capacitor Selection (C_{in})** A 100 μF aluminium electrolytic capacitor located near the input and ground pin provides sufficient bypassing.
3. Catch Diode Selection (D1) **A.** Since the diode maximum peak current exceeds the regulator maximum load current the catch diode current rating must be at least 1.2 times greater than the maximum load current. For a robust design, the diode should have a current rating equal to the maximum current limit of the LM2575 to be able to withstand a continuous output short. **B.** The reverse voltage rating of the diode should be at least 1.25 times the maximum input voltage.	**3. Catch Diode Selection (D1)** **A.** For this example, a 3.0 A current rating is adequate. **B.** Use a 20 V 1N5820 or MBR320 Schottky diode or any suggested fast recovery diode in the Table 4.

Appendix 4.7-11

LM2575

Procedure (Adjustable Output Version: LM2575–Adj) (continued)

Procedure	Example
4. Inductor Selection (L1) **A.** Use the following formula to calculate the inductor Volt x microsecond [V x µs] constant: $$E \times T = \left(V_{in} - V_{out}\right) \frac{V_{out}}{V_{on}} \times \frac{10^6}{F[Hz]} \; [V \times \mu s]$$	**4. Inductor Selection (L1)** **A.** Calculate E x T [V x µs] constant: $$E \times T = (12 - 8.0) \times \frac{8.0}{12} \times \frac{1000}{52} = 51 \; [V \times \mu s]$$
B. Match the calculated E x T value with the corresponding number on the vertical axis of the Inductor Value Selection Guide shown in Figure 21. This E x T constant is a measure of the energy handling capability of an inductor and is dependent upon the type of core, the core area, the number of turns, and the duty cycle.	**B.** E x T = 51 [V x µs]
C. Next step is to identify the inductance region intersected by the E x T value and the maximum load current value on the horizontal axis shown in Figure 21.	**C.** $I_{Load(max)}$ = 1.0 A Inductance Region = L220
D. From the inductor code, identify the inductor value. Then select an appropriate inductor from the Table 1 or Table 2. The inductor chosen must be rated for a switching frequency of 52 kHz and for a current rating of 1.15 x I_{load}. The inductor current rating can also be determined by calculating the inductor peak current: $$I_{p(max)} = I_{Load(max)} + \frac{\left(V_{in} - V_{out}\right) t_{on}}{2L}$$ where t_{on} is the "on" time of the power switch and $$t_{on} = \frac{V_{out}}{V_{in}} \times \frac{1}{f_{osc}}$$ For additional information about the inductor, see the inductor section in the "External Components" section of this data sheet.	**D.** Proper inductor value = 220 µH Choose the inductor from the Table 1 or Table 2.
5. Output Capacitor Selection (C_{out}) **A.** Since the LM2575 is a forward–mode switching regulator with voltage mode control, its open loop 2–pole–2–zero frequency characteristic has the dominant pole pair determined by the output capacitor and inductor values. For stable operation, the capacitor must satisfy the following requirement: $$C_{out} \geq 7.785 \frac{V_{in(max)}}{V_{out} \times L \, [\mu H]} \; [\mu F]$$ **B.** Capacitor values between 10 µF and 2000 µF will satisfy the loop requirements for stable operation. To achieve an acceptable output ripple voltage and transient response, the output capacitor may need to be several times larger than the above formula yields. **C.** Due to the fact that the higher voltage electrolytic capacitors generally have lower ESR (Equivalent Series Resistance) numbers, the output capacitor's voltage rating should be at least 1.5 times greater than the output voltage. For a 5.0 V regulator, a rating of at least 8V is appropriate, and a 10 V or 16 V rating is recommended.	**5. Output Capacitor Selection (C_{out})** **A.** $$C_{out} \geq 7.785 \frac{12}{8.220} = 53 \; \mu F$$ To achieve an acceptable ripple voltage, select C_{out} = 100 µF electrolytic capacitor.

Appendix 4.7-12

LM2575

INDUCTOR VALUE SELECTION GUIDE

Figure 17. LM2575–3.3

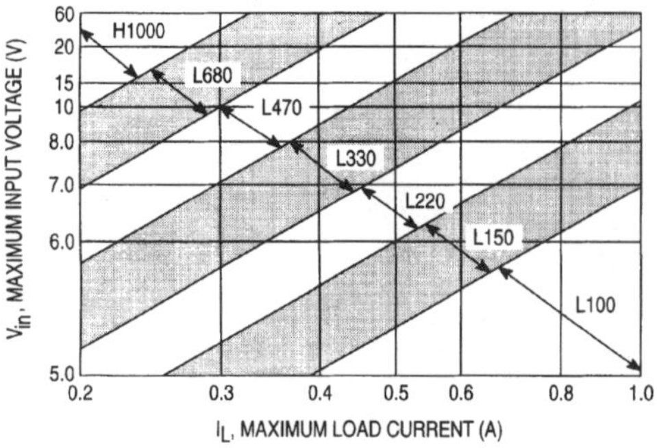

Figure 18. LM2575–5.0

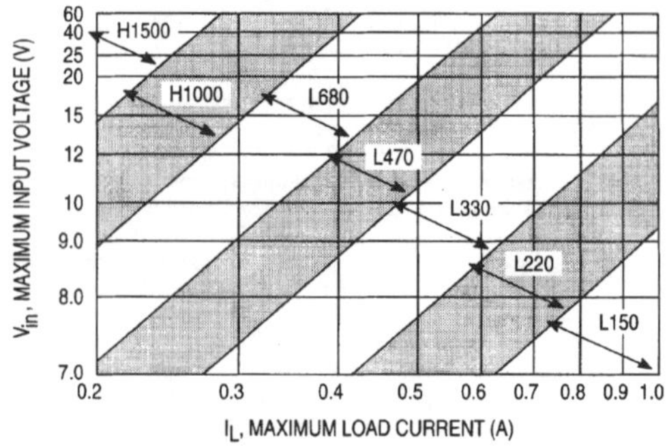

Figure 19. LM2575–12

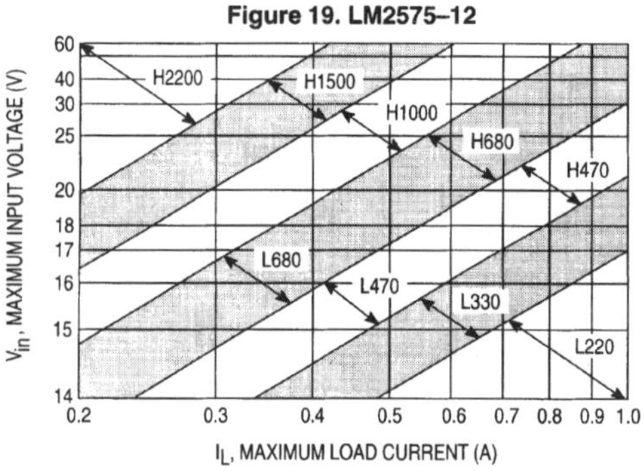

Figure 20. LM2575–15

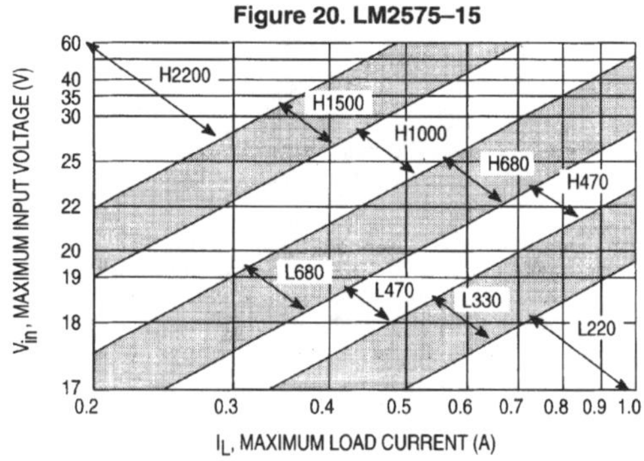

Figure 21. LM2575–Adj

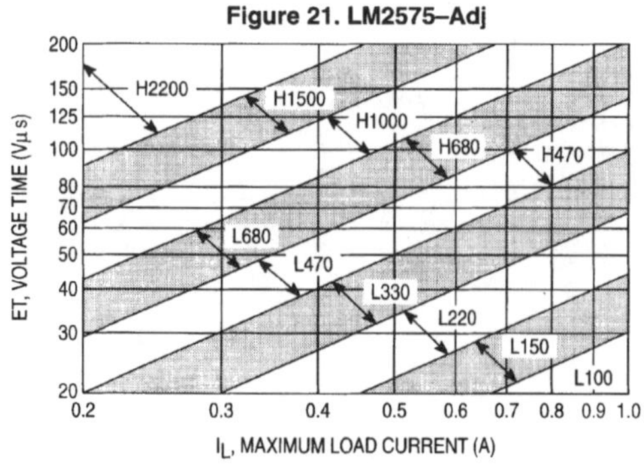

NOTE: This Inductor Value Selection Guide is applicable for continuous mode only.

Appendix 4.7-13

LM2575

Table 1. Inductor Selection Guide

Inductor Code	Inductor Value	Pulse Eng	Renco	AIE	Tech 39
L100	100 μH	PE–92108	RL2444	415–0930	77 308 BV
L150	150 μH	PE–53113	RL1954	415–0953	77 358 BV
L220	220 μH	PE–52626	RL1953	415–0922	77 408 BV
L330	330 μH	PE–52627	RL1952	415–0926	77 458 BV
L470	470 μH	PE–53114	RL1951	415–0927	–
L680	680 μH	PE–52629	RL1950	415–0928	77 508 BV
H150	150 μH	PE–53115	RL2445	415–0936	77 368 BV
H220	220 μH	PE–53116	RL2446	430–0636	77 410 BV
H330	330 μH	PE–53117	RL2447	430–0635	77 460 BV
H470	470 μH	PE–53118	RL1961	430–0634	–
H680	680 μH	PE–53119	RL1960	415–0935	77 510 BV
H1000	1000 μH	PE–53120	RL1959	415–0934	77 558 BV
H1500	1500 μH	PE–53121	RL1958	415–0933	–
H2200	2200 μH	PE–53122	RL2448	415–0945	77 610 BV

Table 2. Inductor Selection Guide

Inductance	Current	Schott		Renco		Pulse Engineering		Coilcraft
(μH)	(A)	THT	SMT	THT	SMT	THT	SMT	SMT
68	0.32	67143940	67144310	RL–1284–68–43	RL1500–68	PE–53804	PE–53804–S	DO1608–68
	0.58	67143990	67144360	RL–5470–6	RL1500–68	PE–53812	PE–53812–S	DO3308–683
	0.99	67144070	67144450	RL–5471–5	RL1500–68	PE–53821	PE–53821–S	DO3316–683
	1.78	67144140	67144520	RL–5471–5	–	PE–53830	PE–53830–S	DO5022P–683
100	0.48	67143980	67144350	RL–5470–5	RL1500–100	PE–53811	PE–53811–S	DO3308–104
	0.82	67144060	67144440	RL–5471–4	RL1500–100	PE–53820	PE–53820–S	DO3316–104
	1.47	67144130	67144510	RL–5471–4	–	PE–53829	PE–53829–S	DO5022P–104
150	0.39	–	67144340	RL–5470–4	RL1500–150	PE–53810	PE–53810–S	DO3308–154
	0.66	67144050	67144430	RL–5471–3	RL1500–150	PE–53819	PE–53819–S	DO3316–154
	1.20	67144120	67144500	RL–5471–3	–	PE–53828	PE–53828–S	DO5022P–154
220	0.32	67143960	67144330	RL–5470–3	RL1500–220	PE–53809	PE–53809–S	DO3308–224
	0.55	67144040	67144420	RL–5471–2	RL1500–220	PE–53818	PE–53818–S	DO3316–224
	1.00	67144110	67144490	RL–5471–2	–	PE–53827	PE–53827–S	DO5022P–224
330	0.42	67144030	67144410	RL–5471–1	RL1500–330	PE–53817	PE–53817–S	DO3316–334
	0.80	67144100	67144480	RL–5471–1	–	PE–53826	PE–53826–S	DO5022P–334

NOTE: Table 1 and Table 2 of this Indicator Selection Guide shows some examples of different manufacturer products suitable for design with the LM2575.

Appendix 4.7-14

LM2575

Table 3. Example of Several Inductor Manufacturers Phone/Fax Numbers

Pulse Engineering Inc.	Phone Fax	+ 1–619–674–8100 + 1–619–674–8262
Pulse Engineering Inc. Europe	Phone Fax	+ 353 93 24 107 + 353 93 24 459
Renco Electronics Inc.	Phone Fax	+ 1–516–645–5828 + 1–516–586–5562
AIE Magnetics	Phone Fax	+ 1–813–347–2181
Coilcraft Inc.	Phone Fax	+ 1–708–322–2645 + 1–708–639–1469
Coilcraft Inc., Europe	Phone Fax	+ 44 1236 730 595 + 44 1236 730 627
Tech 39	Phone Fax	+ 33 8425 2626 + 33 8425 2610
Schott Corp.	Phone Fax	+ 1–612–475–1173 + 1–612–475–1786

Table 4. Diode Selection Guide gives an overview about both surface–mount and through–hole diodes for an effective design. Device listed in bold are available from Motorola.

	Schottky				**Ultra–Fast Recovery**			
	1.0 A		**3.0 A**		**1.0 A**		**3.0 A**	
V_R	**SMT**	**THT**	**SMT**	**THT**	**SMT**	**THT**	**SMT**	**THT**
20 V	SK12	**1N5817** SR102	SK32 **MBRD320**	**1N5820** **MBR320** SR302				
30 V	**MBRS130LT3** SK13	**1N5818** SR103 11DQ03	SK33 **MBRD330**	**1N5821** **MBR330** SR303 31DQ03	**MURS120T3**	**MUR120** 11DF1 HER102	**MURS320T3**	
40 V	**MBRS140T3** SK14 10BQ040 10MQ040	**1N5819** SR104 11DQ04	**MBRS340T3** **MBRD340** 30WQ04 SK34	**1N5822** **MBR340** SR304 31DQ04	10BF10		**MURD320**	**MUR320** 30WF10 **MUR420**
50 V	**MBRS150** 10BQ050	**MBR150** SR105 11DQ05	**MBRD350** SK35 30WQ05	**MBR350** SR305 11DQ05			31DF1 HER302	

Appendix 4.7-15

LM2575

EXTERNAL COMPONENTS

Input Capacitor (C_{in})

The Input Capacitor Should Have a Low ESR

For stable operation of the switch mode converter a low ESR (Equivalent Series Resistance) aluminium or solid tantalum bypass capacitor is needed between the input pin and the ground pin to prevent large voltage transients from appearing at the input. It must be located near the regulator and use short leads. With most electrolytic capacitors, the capacitance value decreases and the ESR increases with lower temperatures. For reliable operation in temperatures below −25°C larger values of the input capacitor may be needed. Also paralleling a ceramic or solid tantalum capacitor will increase the regulator stability at cold temperatures.

RMS Current Rating of C_{in}

The important parameter of the input capacitor is the RMS current rating. Capacitors that are physically large and have large surface area will typically have higher RMS current ratings. For a given capacitor value, a higher voltage electrolytic capacitor will be physically larger than a lower voltage capacitor, and thus be able to dissipate more heat to the surrounding air, and therefore will have a higher RMS current rating. The consequence of operating an electrolytic capacitor above the RMS current rating is a shortened operating life. In order to assure maximum capacitor operating lifetime, the capacitor's RMS ripple current rating should be:

$$I_{rms} > 1.2 \times d \times I_{Load}$$

where d is the duty cycle, for a buck regulator

$$d = \frac{t_{on}}{T} = \frac{V_{out}}{V_{in}}$$

and $d = \dfrac{t_{on}}{T} = \dfrac{|V_{out}|}{|V_{out}| + V_{in}}$ for a buck–boost regulator.

Output Capacitor (C_{out})

For low output ripple voltage and good stability, low ESR output capacitors are recommended. An output capacitor has two main functions: it filters the output and provides regulator loop stability. The ESR of the output capacitor and the peak–to–peak value of the inductor ripple current are the main factors contributing to the output ripple voltage value.Standard aluminium electrolytics could be adequate for some applications but for quality design low ESR types are recommended.

An aluminium electrolytic capacitor's ESR value is related to many factors such as the capacitance value, the voltage rating, the physical size and the type of construction. In most cases, the higher voltage electrolytic capacitors have lower ESR value. Often capacitors with much higher voltage ratings may be needed to provide low ESR values that are required for low output ripple voltage.

The Output Capacitor Requires an ESR Value That Has an Upper and Lower Limit

As mentioned above, a low ESR value is needed for low output ripple voltage, typically 1% to 2% of the output voltage. But if the selected capacitor's ESR is extremely low (below 0.05 Ω), there is a possibility of an unstable feedback loop, resulting in oscillation at the output. This situation can occur when a tantalum capacitor, that can have a very low ESR, is used as the only output capacitor.

At Low Temperatures, Put in Parallel Aluminium Electrolytic Capacitors with Tantalum Capacitors

Electrolytic capacitors are not recommended for temperatures below −25°C. The ESR rises dramatically at cold temperatures and typically rises 3 times at −25°C and as much as 10 times at −40°C. Solid tantalum capacitors have much better ESR spec at cold temperatures and are recommended for temperatures below −25°C. They can be also used in parallel with aluminium electrolytics. The value of the tantalum capacitor should be about 10% or 20% of the total capacitance. The output capacitor should have at least 50% higher RMS ripple current rating at 52 kHz than the peak–to–peak inductor ripple current.

Catch Diode

Locate the Catch Diode Close to the LM2575

The LM2575 is a step–down buck converter; it requires a fast diode to provide a return path for the inductor current when the switch turns off. This diode must be located close to the LM2575 using short leads and short printed circuit traces to avoid EMI problems.

Use a Schottky or a Soft Switching Ultra–Fast Recovery Diode

Since the rectifier diodes are very significant source of losses within switching power supplies, choosing the rectifier that best fits into the converter design is an important process. Schottky diodes provide the best performance because of their fast switching speed and low forward voltage drop.

They provide the best efficiency especially in low output voltage applications (5.0 V and lower). Another choice could be Fast–Recovery, or Ultra–Fast Recovery diodes. It has to be noted, that some types of these diodes with an abrupt turnoff characteristic may cause instability or EMI troubles.

A fast–recovery diode with soft recovery characteristics can better fulfill a quality, low noise design requirements. Table 4 provides a list of suitable diodes for the LM2575 regulator. Standard 50/60 Hz rectifier diodes such as the 1N4001 series or 1N5400 series are **NOT** suitable.

Inductor

The magnetic components are the cornerstone of all switching power supply designs. The style of the core and the winding technique used in the magnetic component's design has a great influence on the reliability of the overall power supply.

Using an improper or poorly designed inductor can cause high voltage spikes generated by the rate of transitions in current within the switching power supply, and the possibility of core saturation can arise during an abnormal operational mode. Voltage spikes can cause the semiconductors to enter avalanche breakdown and the part can instantly fail if enough energy is applied. It can also cause significant RFI (Radio Frequency Interference) and EMI (Electro–Magnetic Interference) problems.

Continuous and Discontinuous Mode of Operation

The LM2575 step–down converter can operate in both the continuous and the discontinuous modes of operation. The regulator works in the continuous mode when loads are relatively heavy, the current flows through the inductor continuously and never falls to zero. Under light load

Appendix 4.7-16

LM2575

conditions, the circuit will be forced to the discontinuous mode when inductor current falls to zero for certain period of time (see Figure 22 and Figure 23). Each mode has distinctively different operating characteristics, which can affect the regulator performance and requirements. In many cases the preferred mode of operation is the continuous mode. It offers greater output power, lower peak currents in the switch, inductor and diode, and can have a lower output ripple voltage. On the other hand it does require larger inductor values to keep the inductor current flowing continuously, especially at low output load currents and/or high input voltages.

To simplify the inductor selection process, an inductor selection guide for the LM2575 regulator was added to this data sheet (Figures 17 through 21). This guide assumes that the regulator is operating in the continuous mode, and selects an inductor that will allow a peak–to–peak inductor ripple current to be a certain percentage of the maximum design load current. This percentage is allowed to change as different design load currents are selected. For light loads (less than approximately 200 mA) it may be desirable to operate the regulator in the discontinuous mode, because the inductor value and size can be kept relatively low. Consequently, the percentage of inductor peak–to–peak current increases. This discontinuous mode of operation is perfectly acceptable for this type of switching converter. Any buck regulator will be forced to enter discontinuous mode if the load current is light enough.

Figure 22. Continuous Mode Switching Current Waveforms

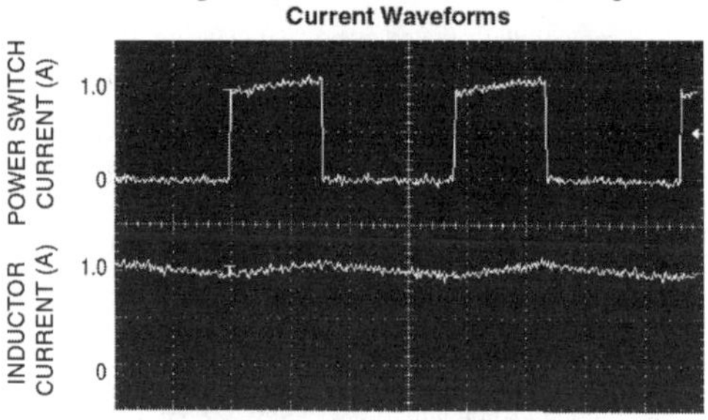

HORTIZONTAL TIME BASE: 5.0 μs /DIV

Selecting the Right Inductor Style

Some important considerations when selecting a core type are core material, cost, the output power of the power supply, the physical volume the inductor must fit within, and the amount of EMI (Electro–Magnetic Interference) shielding that the core must provide. The inductor selection guide covers different styles of inductors, such as pot core, E–core, toroid and bobbin core, as well as different core materials such as ferrites and powdered iron from different manufacturers.

For high quality design regulators the toroid core seems to be the best choice. Since the magnetic flux is completely contained within the core, it generates less EMI, reducing noise problems in sensitive circuits. The least expensive is the bobbin core type, which consists of wire wound on a ferrite rod core. This type of inductor generates more EMI due to the fact that its core is open, and the magnetic flux is not completely contained within the core.

When multiple switching regulators are located on the same printed circuit board, open core magnetics can cause interference between two or more of the regulator circuits, especially at high currents due to mutual coupling. A toroid, pot core or E–core (closed magnetic structure) should be used in such applications.

Do Not Operate an Inductor Beyond its Maximum Rated Current

Exceeding an inductor's maximum current rating may cause the inductor to overheat because of the copper wire losses, or the core may saturate. Core saturation occurs when the flux density is too high and consequently the cross sectional area of the core can no longer support additional lines of magnetic flux.

This causes the permeability of the core to drop, the inductance value decreases rapidly and the inductor begins to look mainly resistive. It has only the dc resistance of the winding. This can cause the switch current to rise very rapidly and force the LM2575 internal switch into cycle–by–cycle current limit, thus reducing the dc output load current. This can also result in overheating of the inductor and/or the LM2575. Different inductor types have different saturation characteristics, and this should be kept in mind when selecting an inductor.

Figure 23. Discontinuous Mode Switching Current Waveforms

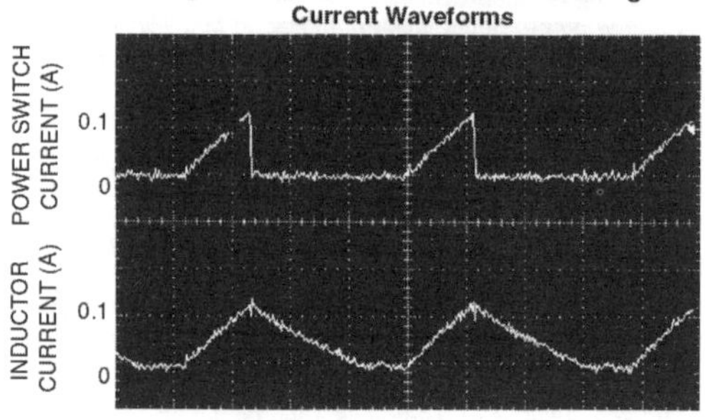

HORTIZONTAL TIME BASE: 5.0 μs /DIV

Appendix 4.7-17

LM2575

GENERAL RECOMMENDATIONS

Output Voltage Ripple and Transients

Source of the Output Ripple

Since the LM2575 is a switch mode power supply regulator, its output voltage, if left unfiltered, will contain a sawtooth ripple voltage at the switching frequency. The output ripple voltage value ranges from 0.5% to 3% of the output voltage. It is caused mainly by the inductor sawtooth ripple current multiplied by the ESR of the output capacitor.

Short Voltage Spikes and How to Reduce Them

The regulator output voltage may also contain short voltage spikes at the peaks of the sawtooth waveform (see Figure 24). These voltage spikes are present because of the fast switching action of the output switch, and the parasitic inductance of the output filter capacitor. There are some other important factors such as wiring inductance, stray capacitance, as well as the scope probe used to evaluate these transients, all these contribute to the amplitude of these spikes. To minimise these voltage spikes, low inductance capacitors should be used, and their lead lengths must be kept short. The importance of quality printed circuit board layout design should also be highlighted.

Figure 24. Output Ripple Voltage Waveforms

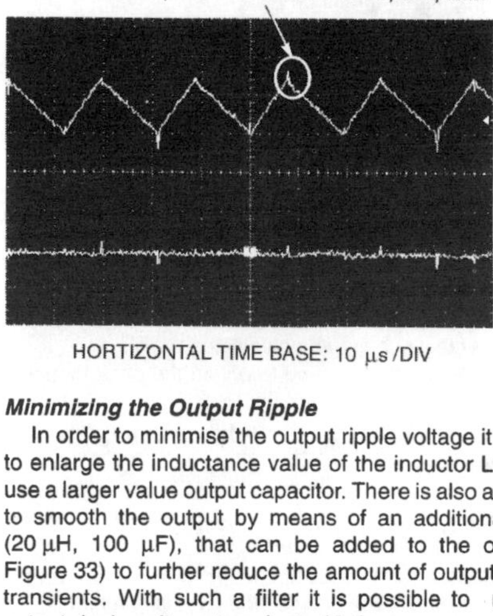

Minimizing the Output Ripple

In order to minimise the output ripple voltage it is possible to enlarge the inductance value of the inductor L1 and/or to use a larger value output capacitor. There is also another way to smooth the output by means of an additional LC filter (20 μH, 100 μF), that can be added to the output (see Figure 33) to further reduce the amount of output ripple and transients. With such a filter it is possible to reduce the output ripple voltage transients 10 times or more. Figure 24 shows the difference between filtered and unfiltered output waveforms of the regulator shown in Figure 33.

The upper waveform is from the normal unfiltered output of the converter, while the lower waveform shows the output ripple voltage filtered by an additional LC filter.

Heatsinking and Thermal Considerations

The Through–Hole Package TO–220

The LM2575 is available in two packages, a 5–pin TO–220(T, TV) and a 5–pin surface mount D2PAK(D2T). There are many applications that require no heatsink to keep the LM2575 junction temperature within the allowed operating range. The TO–220 package can be used without a heatsink for ambient temperatures up to approximately 50°C (depending on the output voltage and load current). Higher ambient temperatures require some heatsinking, either to the printed circuit (PC) board or an external heatsink.

The Surface Mount Package D2PAK and Its Heatsinking

The other type of package, the surface mount D2PAK, is designed to be soldered to the copper on the PC board. The copper and the board are the heatsink for this package and the other heat producing components, such as the catch diode and inductor. The PC board copper area that the package is soldered to should be at least 0.4 in^2 (or 100 mm^2) and ideally should have 2 or more square inches (1300 mm^2) of 0.0028 inch copper. Additional increasing of copper area beyond approximately 3.0 in^2 (2000 mm^2) will not improve heat dissipation significantly. If further thermal improvements are needed, double sided or multilayer PC boards with large copper areas should be considered.

Thermal Analysis and Design

The following procedure must be performed to determine whether or not a heatsink will be required. First determine:

1. $P_{D(max)}$ maximum regulator power dissipation in the application.
2. $T_{A(max)}$ maximum ambient temperature in the application.
3. $T_{J(max)}$ maximum allowed junction temperature (125°C for the LM2575). For a conservative design, the maximum junction temperature should not exceed 110°C to assure safe operation. For every additional 10°C temperature rise that the junction must withstand, the estimated operating lifetime of the component is halved.
4. $R_{\theta JC}$ package thermal resistance junction–case.
5. $R_{\theta JA}$ package thermal resistance junction–ambient.

(Refer to Absolute Maximum Ratings in this data sheet or $R_{\theta JC}$ and $R_{\theta JA}$ values).

The following formula is to calculate the total power dissipated by the LM2575:

$$P_D = (V_{in} \times I_Q) + d \times I_{Load} \times V_{sat}$$

where d is the duty cycle and for buck converter

$$d = \frac{t_{on}}{T} = \frac{V_O}{V_{in}},$$

I_Q (quiescent current) and V_{sat} can be found in the LM2575 data sheet,

V_{in} is minimum input voltage applied,

V_O is the regulator output voltage,

I_{Load} is the load current.

The dynamic switching losses during turn–on and turn–off can be neglected if proper type catch diode is used.

Packages Not on a Heatsink (Free–Standing)

For a free–standing application when no heatsink is used, the junction temperature can be determined by the following expression:

$$T_J = (R_{\theta JA})(P_D) + T_A$$

where $(R_{\theta JA})(P_D)$ represents the junction temperature rise caused by the dissipated power and T_A is the maximum ambient temperature.

Appendix 4.7-18

LM2575

Packages on a Heatsink

If the actual operating junction temperature is greater than the selected safe operating junction temperature determined in step 3, than a heatsink is required. The junction temperature will be calculated as follows:

$$T_J = P_D (R_{\theta JA} + R_{\theta CS} + R_{\theta SA}) + T_A$$

where $R_{\theta JC}$ is the thermal resistance junction–case,
$R_{\theta CS}$ is the thermal resistance case–heatsink,
$R_{\theta SA}$ is the thermal resistance heatsink–ambient.

If the actual operating temperature is greater than the selected safe operating junction temperature, then a larger heatsink is required.

Some Aspects That can Influence Thermal Design

It should be noted that the package thermal resistance and the junction temperature rise numbers are all approximate, and there are many factors that will affect these numbers, such as PC board size, shape, thickness, physical position, location, board temperature, as well as whether the surrounding air is moving or still.

Other factors are trace width, total printed circuit copper area, copper thickness, single– or double–sided, multilayer board, the amount of solder on the board or even colour of the traces.

The size, quantity and spacing of other components on the board can also influence its effectiveness to dissipate the heat.

Figure 25. Inverting Buck–Boost Regulator Using the LM2575–12 Develops –12 V @ 0.35 A

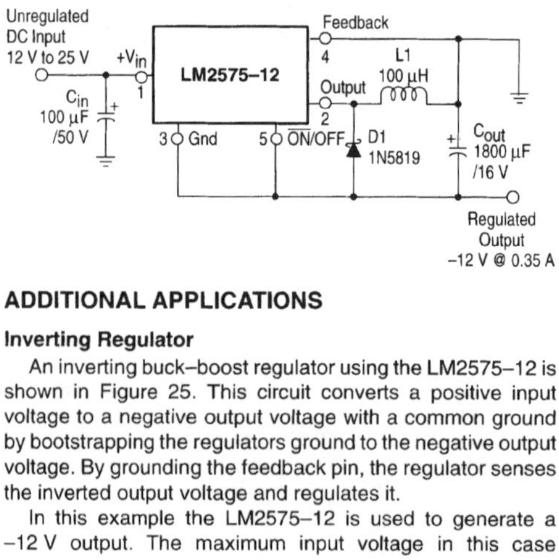

ADDITIONAL APPLICATIONS

Inverting Regulator

An inverting buck–boost regulator using the LM2575–12 is shown in Figure 25. This circuit converts a positive input voltage to a negative output voltage with a common ground by bootstrapping the regulators ground to the negative output voltage. By grounding the feedback pin, the regulator senses the inverted output voltage and regulates it.

In this example the LM2575–12 is used to generate a –12 V output. The maximum input voltage in this case cannot exceed +28 V because the maximum voltage appearing across the regulator is the absolute sum of the input and output voltages and this must be limited to a maximum of 40 V.

This circuit configuration is able to deliver approximately 0.35 A to the output when the input voltage is 12 V or higher. At lighter loads the minimum input voltage required drops to approximately 4.7 V, because the buck–boost regulator topology can produce an output voltage that, in its absolute value, is either greater or less than the input voltage.

Since the switch currents in this buck–boost configuration are higher than in the standard buck converter topology, the available output current is lower.

This type of buck–boost inverting regulator can also require a larger amount of startup input current, even for light loads. This may overload an input power source with a current limit less than 1.5 A.

Such an amount of input startup current is needed for at least 2.0 ms or more. The actual time depends on the output voltage and size of the output capacitor.

Because of the relatively high startup currents required by this inverting regulator topology, the use of a delayed startup or an undervoltage lockout circuit is recommended.

Using a delayed startup arrangement, the input capacitor can charge up to a higher voltage before the switch–mode regulator begins to operate.

The high input current needed for startup is now partially supplied by the input capacitor C_{in}.

Design Recommendations:

The inverting regulator operates in a different manner than the buck converter and so a different design procedure has to be used to select the inductor L1 or the output capacitor C_{out}.

The output capacitor values must be larger than is normally required for buck converter designs. Low input voltages or high output currents require a large value output capacitor (in the range of thousands of µF).

The recommended range of inductor values for the inverting converter design is between 68 µH and 220 µH. To select an inductor with an appropriate current rating, the inductor peak current has to be calculated.

The following formula is used to obtain the peak inductor current:

$$I_{peak} \approx \frac{I_{Load}(V_{in} + |V_O|)}{V_{in}} + \frac{V_{in} \times t_{on}}{2L_1}$$

where $t_{on} = \frac{|V_O|}{V_{in} + |V_O|} \times \frac{1}{f_{osc}}$, and $f_{osc} = 52$ kHz.

Under normal continuous inductor current operating conditions, the worst case occurs when V_{in} is minimal.

Note that the voltage appearing across the regulator is the absolute sum of the input and output voltage, and must not exceed 40 V.

Appendix 4.7-19

LM2575

Figure 26. Inverting Buck–Boost Regulator with Delayed Startup

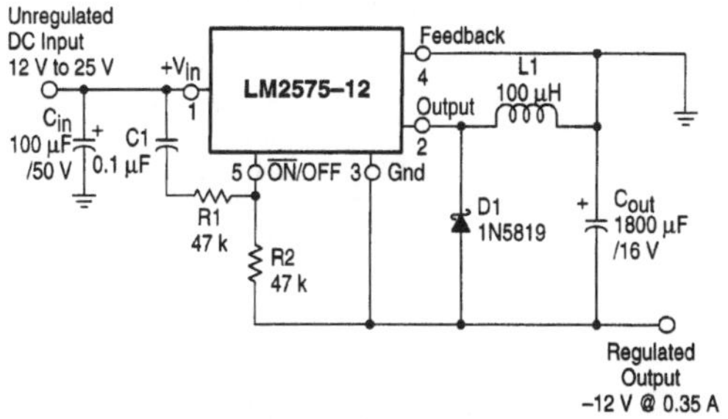

It has been already mentioned above, that in some situations, the delayed startup or the undervoltage lockout features could be very useful. A delayed startup circuit applied to a buck–boost converter is shown in Figure 26. Figure 31 in the "Undervoltage Lockout" section describes an undervoltage lockout feature for the same converter topology.

Figure 27. Inverting Buck–Boost Regulator Shut Down Circuit Using an Optocoupler

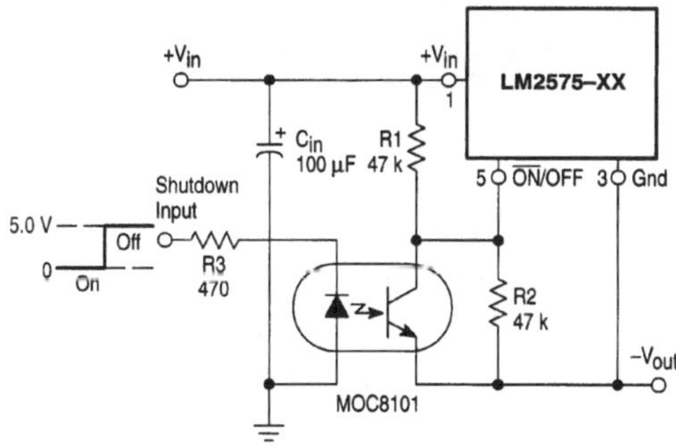

NOTE: This picture does not show the complete circuit.

With the inverting configuration, the use of the $\overline{\text{ON}}$/OFF pin requires some level shifting techniques. This is caused by the fact, that the ground pin of the converter IC is no longer at ground. Now, the $\overline{\text{ON}}$/OFF pin threshold voltage (1.4 V approximately) has to be related to the negative output voltage level. There are many different possible shut down methods, two of them are shown in Figures 27 and 28.

Figure 28. Inverting Buck–Boost Regulator Shut Down Circuit Using a PNP Transistor

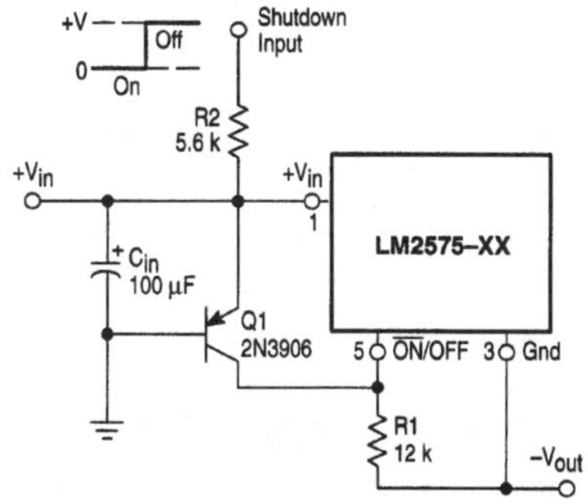

NOTE: This picture does not show the complete circuit.

Negative Boost Regulator

This example is a variation of the buck–boost topology and is called a negative boost regulator. This regulator experiences relatively high switch current, especially at low input voltages. The internal switch current limiting results in lower output load current capability.

The circuit in Figure 29 shows the negative boost configuration. The input voltage in this application ranges from −5.0 V to −12 V and provides a regulated −12 V output. If the input voltage is greater than −12 V, the output will rise above −12 V accordingly, but will not damage the regulator.

Figure 29. Negative Boost Regulator

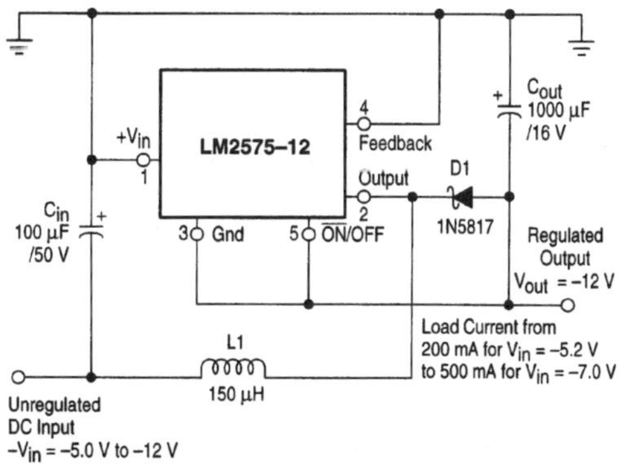

Appendix 4.7-20

LM2575

Design Recommendations:

The same design rules as for the previous inverting buck–boost converter can be applied. The output capacitor C_{out} must be chosen larger than would be required for a standard buck converter. Low input voltages or high output currents require a large value output capacitor (in the range of thousands of μF). The recommended range of inductor values for the negative boost regulator is the same as for inverting converter design.

Another important point is that these negative boost converters cannot provide current limiting load protection in the event of a short in the output so some other means, such as a fuse, may be necessary to provide the load protection.

Delayed Startup

There are some applications, like the inverting regulator already mentioned above, which require a higher amount of startup current. In such cases, if the input power source is limited, this delayed startup feature becomes very useful.

To provide a time delay between the time the input voltage is applied and the time when the output voltage comes up, the circuit in Figure 30 can be used. As the input voltage is applied, the capacitor C1 charges up, and the voltage across the resistor R2 falls down. When the voltage on the $\overline{\text{ON}}$/OFF pin falls below the threshold value 1.4 V, the regulator starts up. Resistor R1 is included to limit the maximum voltage applied to the $\overline{\text{ON}}$/OFF pin, reduces the power supply noise sensitivity, and also limits the capacitor C1 discharge current, but its use is not mandatory.

When a high 50 Hz or 60 Hz (100 Hz or 120 Hz respectively) ripple voltage exists, a long delay time can cause some problems by coupling the ripple into the $\overline{\text{ON}}$/OFF pin, the regulator could be switched periodically on and off with the line (or double) frequency.

Figure 30. Delayed Startup Circuitry

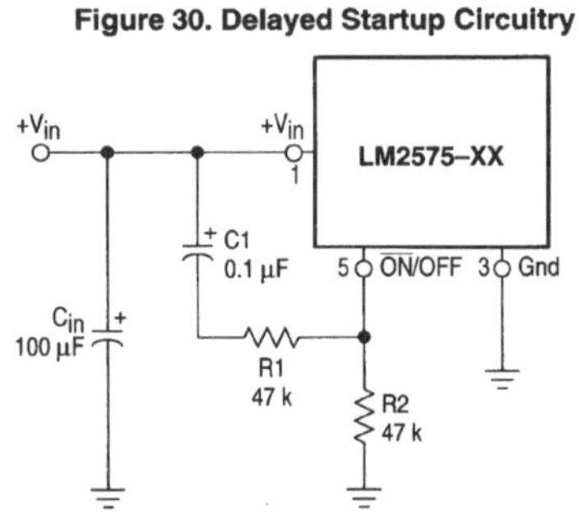

NOTE: This picture does not show the complete circuit.

Undervoltage Lockout

Some applications require the regulator to remain off until the input voltage reaches a certain threshold level. Figure 31 shows an undervoltage lockout circuit applied to a buck regulator. A version of this circuit for buck–boost converter is shown in Figure 32. Resistor R3 pulls the $\overline{\text{ON}}$/OFF pin high and keeps the regulator off until the input voltage reaches a predetermined threshold level, which is determined by the following expression:

$$V_{th} \approx V_{Z1} + \left(1 + \frac{R2}{R1}\right) V_{BE} \text{ (Q1)}$$

Figure 31. Undervoltage Lockout Circuit for Buck Converter

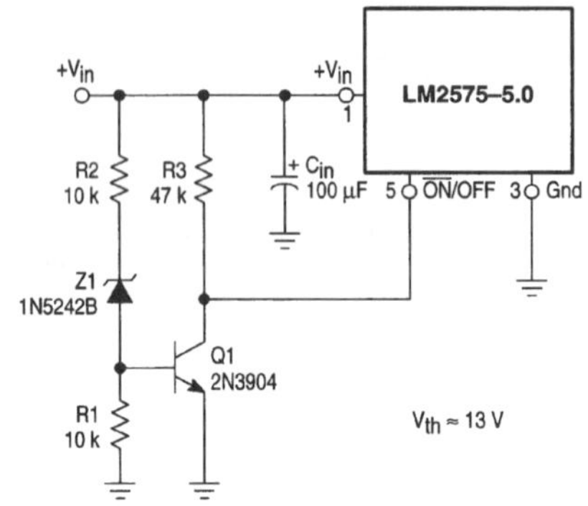

NOTE: This picture does not show the complete circuit.

Figure 32. Undervoltage Lockout Circuit for Buck–Boost Converter

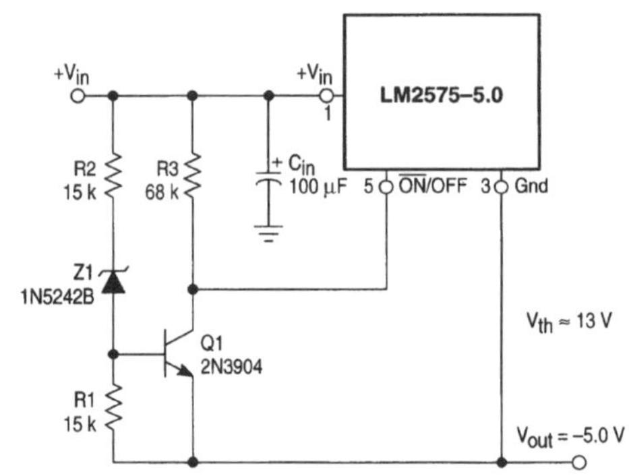

NOTE: This picture does not show the complete circuit.

Adjustable Output, Low–Ripple Power Supply

A 1.0 A output current capability power supply that features an adjustable output voltage is shown in Figure 33.

This regulator delivers 1.0 A into 1.2 V to 35 V output. The input voltage ranges from roughly 8.0 V to 40 V. In order to achieve a 10 or more times reduction of output ripple, an additional L–C filter is included in this circuit.

Appendix 4.7-21

LM2575

Figure 33. Adjustable Power Supply with Low Ripple Voltage

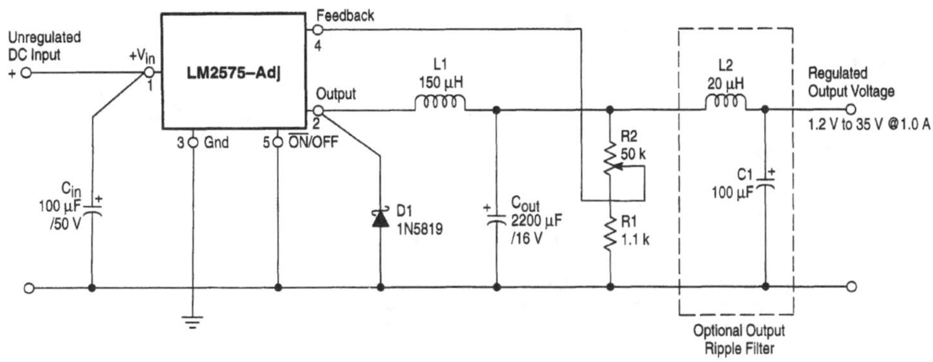

Figure 34. D²PAK Thermal Resistance and Maximum Power Dissipation versus P.C.B. Copper Length

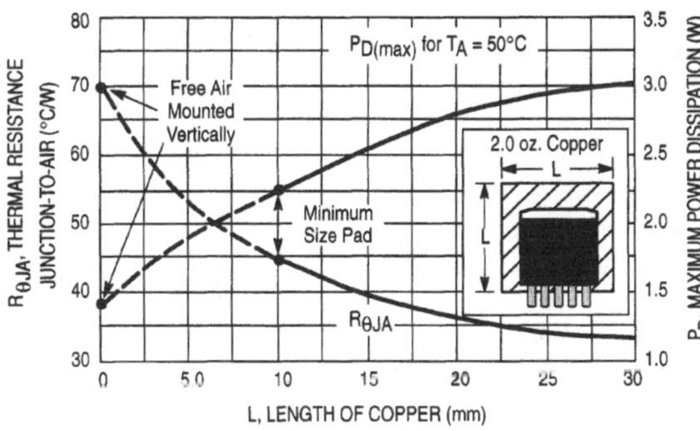

Appendix 4.8-1

MOTOROLA

MC34025
MC33025

High Speed Double-Ended PWM Controller

The MC34025 series are high speed, fixed frequency, double–ended pulse width modulator controllers optimized for high frequency operation. They are specifically designed for Off–Line and DC–to–DC converter applications offering the designer a cost effective solution with minimal external components. These integrated circuits feature an oscillator, a temperature compensated reference, a wide bandwidth error amplifier, a high speed current sensing comparator, steering flip–flop, and dual high current totem pole outputs ideally suited for driving power MOSFETs.

Also included are protective features consisting of input and reference undervoltage lockouts each with hysteresis, cycle–by–cycle current limiting, and a latch for single pulse metering.

The flexibility of this series allows it to be easily configured for either current mode or voltage mode control.

- 50 ns Propagation Delay to Outputs
- Dual High Current Totem Pole Outputs
- Wide Bandwidth Error Amplifier
- Fully–Latched Logic with Double Pulse Suppression
- Latching PWM for Cycle–By–Cycle Current Limiting
- Soft–Start Control with Latched Overcurrent Reset
- Input Undervoltage Lockout with Hysteresis
- Low Start–Up Current (500 μA Typ)
- Internally Trimmed Reference with Undervoltage Lockout
- 90% Maximum Duty Cycle (Externally Adjustable)
- Precision Trimmed Oscillator
- Voltage or Current Mode Operation to 1.0 MHz
- Functionally Similar to the UC3825

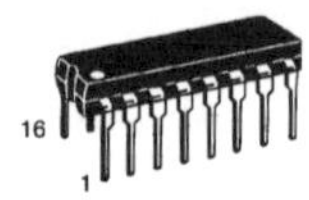

P SUFFIX
PLASTIC PACKAGE
CASE 648

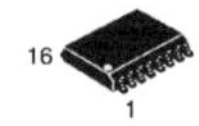

DW SUFFIX
PLASTIC PACKAGE
CASE 751G
(SO–16L)

PIN CONNECTIONS

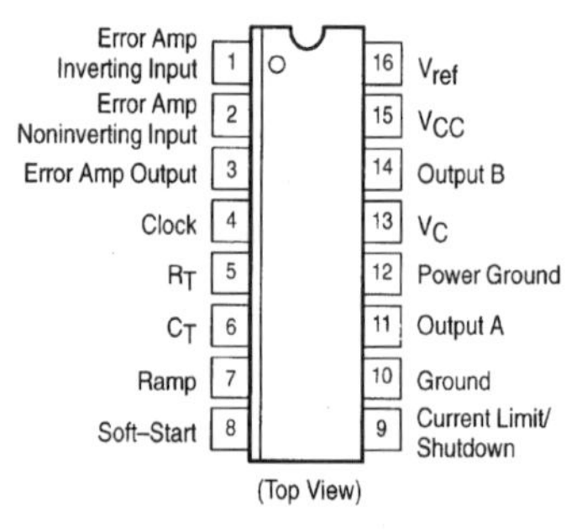

Simplified Application

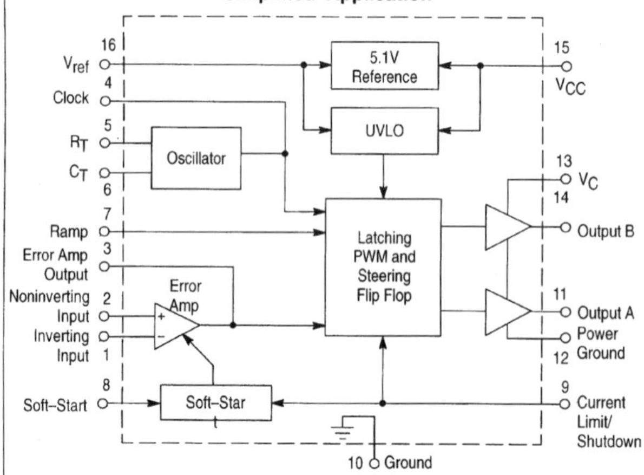

This device contains 227 active transistors.

ORDERING INFORMATION

Device	Operating Temperature Range	Package
MC33025DW	T_A = –40° to +105°C	SO–16L
MC33025P		Plastic DIP
MC34025DW	T_A = 0° to +70°C	SO–16L
MC34025P		Plastic DIP

Appendix 4.8-2

MC34025 MC33025

MAXIMUM RATINGS

Rating	Symbol	Value	Unit
Power Supply Voltage	V_{CC}	30	V
Output Driver Supply Voltage	V_C	20	V
Output Current, Source or Sink (Note 1) DC Pulsed (0.5 μs)	I_O	 0.5 2.0	A
Current Sense, Soft–Start, Ramp, and Error Amp Inputs	V_{in}	−0.3 to +7.0	V
Error Amp Output and Soft–Start Sink Current	I_O	10	mA
Clock and R_T Output Current	I_{CO}	5.0	mA
Power Dissipation and Thermal Characteristics SO–16 Package (Case 751G) Maximum Power Dissipation @ T_A = +25°C Thermal Resistance, Junction–to–Air DIP Package (Case 648) Maximum Power Dissipation @ T_A = +25°C Thermal Resistance, Junction–to–Air	 P_D $R_{\theta JA}$ P_D $R_{\theta JA}$	 862 145 1.25 100	 mW °C/W W °C/W
Operating Junction Temperature	T_J	+150	°C
Operating Ambient Temperature (Note 2) MC34025 MC33025	T_A	 0 to +70 −40 to +105	°C
Storage Temperature Range	T_{stg}	−55 to +150	°C

ELECTRICAL CHARACTERISTICS (V_{CC} = 15 V, R_T = 3.65 kΩ, C_T = 1.0 nF, for typical values T_A = +25°C, for min/max values T_A is the operating ambient temperature range that applies [Note 2], unless otherwise noted.)

Characteristic	Symbol	Min	Typ	Max	Unit
REFERENCE SECTION					
Reference Output Voltage (I_O = 1.0 mA, T_J = +25°C)	V_{ref}	5.05	5.1	5.15	V
Line Regulation (V_{CC} = 10 V to 30 V)	Reg_{line}	–	2.0	15	mV
Load Regulation (I_O = 1.0 mA to 10 mA)	Reg_{load}	–	2.0	15	mV
Temperature Stability	T_S	–	0.2	–	mV/°C
Total Output Variation over Line, Load, and Temperature	V_{ref}	4.95	–	5.25	V
Output Noise Voltage (f = 10 Hz to 10 kHz, T_J = +25°C)	V_n	–	50	–	μV
Long Term Stability (T_A = +125°C for 1000 Hours)	S	–	5.0	–	mV
Output Short Circuit Current	I_{SC}	–30	–65	–100	mA
OSCILLATOR SECTION					
Frequency T_J = +25°C Line (V_{CC} = 10 V to 30 V) and Temperature (T_A = T_{low} to T_{high})	f_{osc}	 380 370	 400 400	 420 430	kHz
Frequency Change with Voltage (V_{CC} = 10 V to 30 V)	$\Delta f_{osc}/\Delta V$	–	0.2	1.0	%
Frequency Change with Temperature (T_A = T_{low} to T_{high})	$\Delta f_{osc}/\Delta T$	–	2.0	–	%
Sawtooth Peak Voltage	V_P	2.6	2.8	3.0	V
Sawtooth Valley Voltage	V_V	0.7	1.0	1.25	V
Clock Output Voltage High State Low State	 V_{OH} V_{OL}	 3.9 –	 4.5 2.3	 – 2.9	V

NOTES: 1. Maximum package power dissipation limits must be observed.
2. Low duty cycle pulse techniques are used during test to maintain junction temperature as close to ambient as possible.
T_{low} = 0°C for MC34025 T_{high} = +70°C for MC34025
= – 40°C for MC33025 = +105°C for MC33025

Appendix 4.8-3

MC34025 MC33025

ELECTRICAL CHARACTERISTICS (V_{CC} = 15 V, R_T = 3.65 kΩ, C_T = 1.0 nF, for typical values T_A = +25°C, for min/max values T_A is the operating ambient temperature range that applies [Note 2], unless otherwise noted.)

Characteristic	Symbol	Min	Typ	Max	Unit
ERROR AMPLIFIER SECTION					
Input Offset Voltage	V_{IO}	–	–	15	mV
Input Bias Current	I_{IB}	–	0.6	3.0	μA
Input Offset Current	I_{IO}	–	0.1	1.0	μA
Open–Loop Voltage Gain (V_O = 1.0 V to 4.0 V)	A_{VOL}	60	95	–	dB
Gain Bandwidth Product (T_J = +25°C)	GBW	4.0	8.3	–	MHz
Common Mode Rejection Ratio (V_{CM} = 1.5 V to 5.5 V)	CMRR	75	95	–	dB
Power Supply Rejection Ratio (V_{CC} = 10 V to 30 V)	PSRR	85	110	–	dB
Output Current, Source (V_O = 4.0 V) Sink (V_O = 1.0 V)	I_{Source} I_{Sink}	0.5 1.0	3.0 3.6	– –	mA
Output Voltage Swing, High State (I_O = – 0.5 mA) Low State (I_O = 1.0 mA)	V_{OH} V_{OL}	4.5 0	4.75 0.4	5.0 1.0	V
Slew Rate	SR	6.0	12	–	V/μs
PWM COMPARATOR SECTION					
Ramp Input Bias Current	I_{IB}	–	–0.5	–5.0	μA
Duty Cycle, Maximum Minimum	$DC_{(max)}$ $DC_{(min)}$	80 –	90 –	– 0	%
Zero Duty Cycle Threshold Voltage Pin 3(4) (Pin 7(9) = 0 V)	V_{th}	1.1	1.25	1.4	V
Propagation Delay (Ramp Input to Output, T_J = +25°C)	$t_{PLH(in/out)}$	–	60	100	ns
SIFT–START SECTION					
Charge Current ($V_{Soft–Start}$ = 0.5 V)	I_{chg}	3.0	9.0	20	μA
Discharge Current ($V_{Soft–Start}$ = 1.5 V)	I_{dischg}	1.0	4.0	–	mA
CURRENT SENSE SECTION					
Input Bias Current (Pin 9(12) = 0 V to 4.0 V)	I_{IB}	–	–	15	μA
Current Limit Comparator Threshold Shutdown Comparator Threshold	V_{th} V_{th}	0.9 1.25	1.0 1.40	1.10 1.55	V
Propagation Delay (Current Limit/Shutdown to Output, T_J = +25°C)	$t_{PLH(in/out)}$	–	50	80	ns
OUTPUT SECTION					
Output Voltage Low State (I_{Sink} = 20 mA) (I_{Sink} = 200 mA) High State (I_{Source} = 20 mA) (I_{Source} = 200 mA)	 V_{OL} V_{OH} 	 – – 13 12	 0.25 1.2 13.5 13	 0.4 2.2 – –	V
Output Voltage with UVLO Activated (V_{CC} = 6.0 V, I_{Sink} = 0.5 mA)	$V_{OL(UVLO)}$	–	0.25	1.0	V
Output Leakage Current (V_C = 20 V)	I_L	–	100	500	μA
Output Voltage Rise Time (C_L = 1.0 nF, T_J = +25°C)	t_r	–	30	60	ns
Output Voltage Fall Time (C_L = 1.0 nF, T_J = +25°C)	t_f	–	30	60	ns
UNDERVOLTAGE LOCKOUT SECTION					
Start–Up Threshold (V_{CC} Increasing)	$V_{th(on)}$	8.8	9.2	9.6	V
UVLO Hysteresis Voltage (V_{CC} Decreasing After Turn–On)	V_H	0.4	0.8	1.2	V
TOTAL DEVICE					
Power Supply Current Start–Up (V_{CC} = 8.0 V) Operating	I_{CC}	 – –	 0.5 25	 1.2 35	mA

NOTES: 1. Maximum package power dissipation limits must be observed.
2. Low duty cycle pulse techniques are used during test to maintain junction temperature as close to ambient as possible.
T_{low} = 0°C for MC34025 T_{high} = +70°C for MC34025
= – 40°C for MC33025 = +105°C for MC33025

Appendix 4.8-4

MC34025 MC33025

Figure 1. Timing Resistor versus Oscillator Frequency

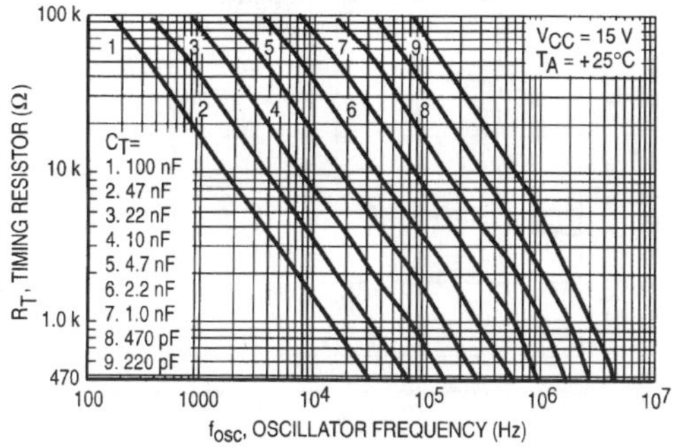

Figure 2. Oscillator Frequency versus Temperature

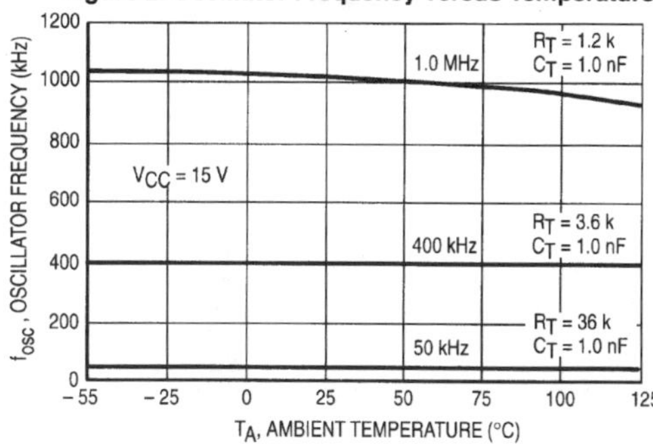

Figure 3. Error Amp Open Loop Gain and Phase versus Frequency

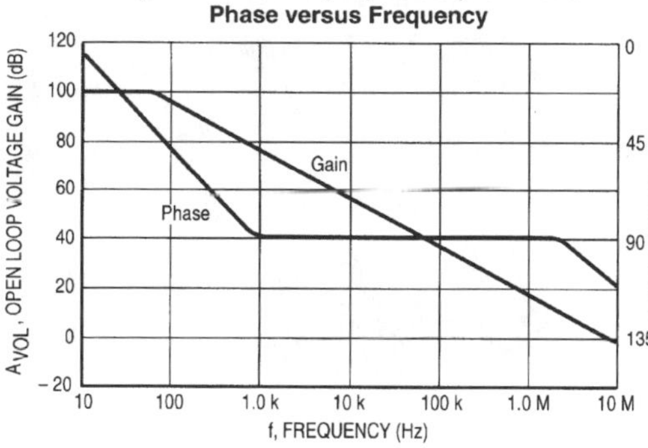

Figure 4. PWM Comparator Zero Duty Cycle Threshold Voltage versus Temperature

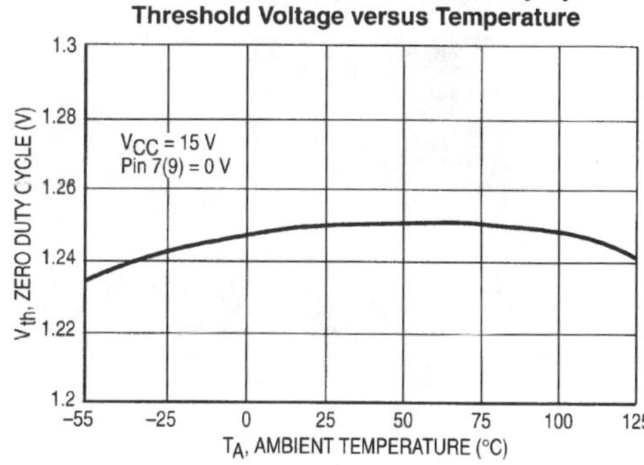

Figure 5. Error Amp Small Signal Transient Response

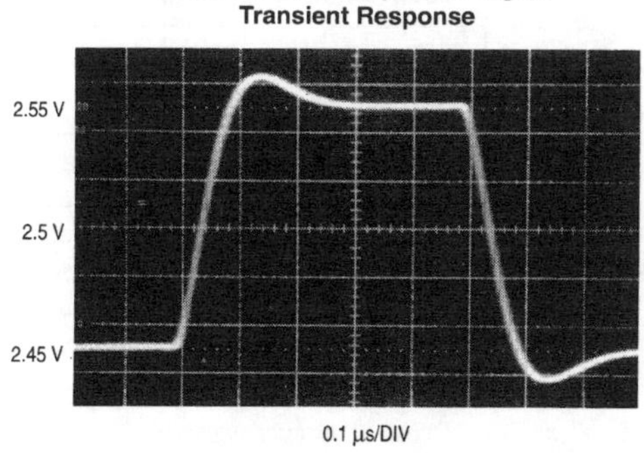

Figure 6. Error Amp Large Signal Transient Response

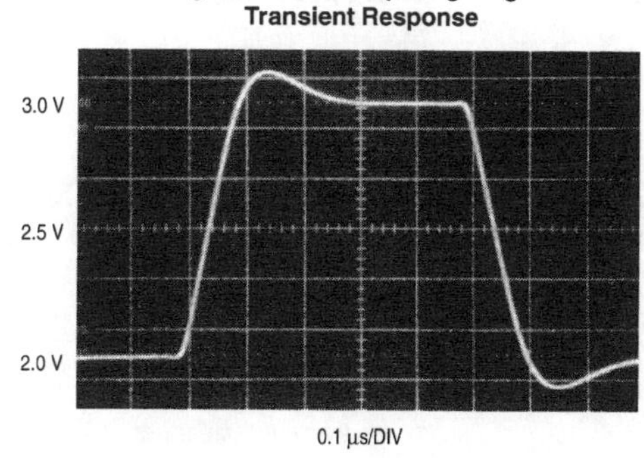

Appendix 4.8-5

MC34025 MC33025

Figure 7. Reference Voltage Change versus Source Current

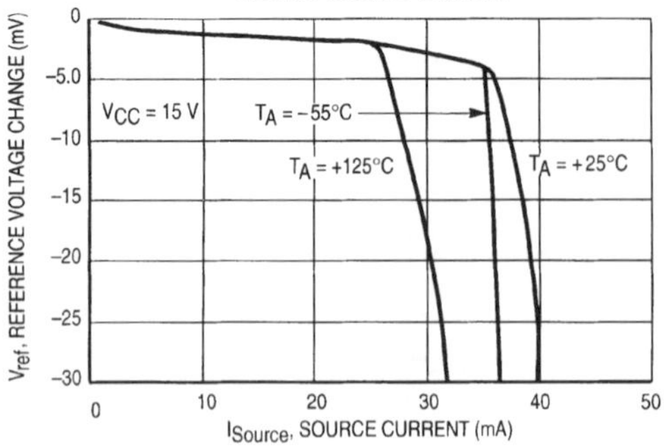

Figure 8. Reference Short Circuit Current versus Temperature

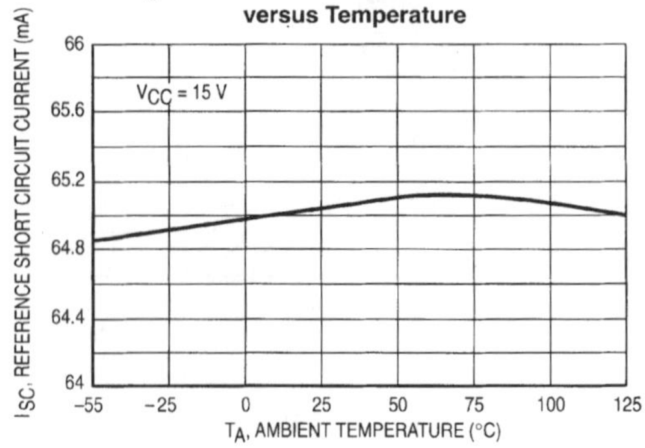

Figure 9. Reference Line Regulation

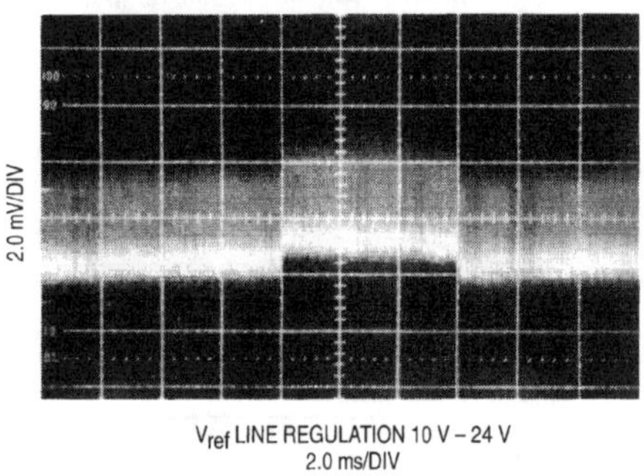

Figure 10. Reference Load Regulation

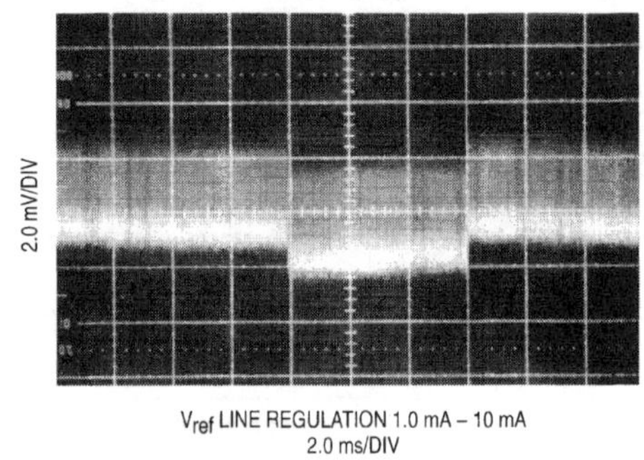

Figure 11. Current Limit Comparator Threshold Change versus Temperature

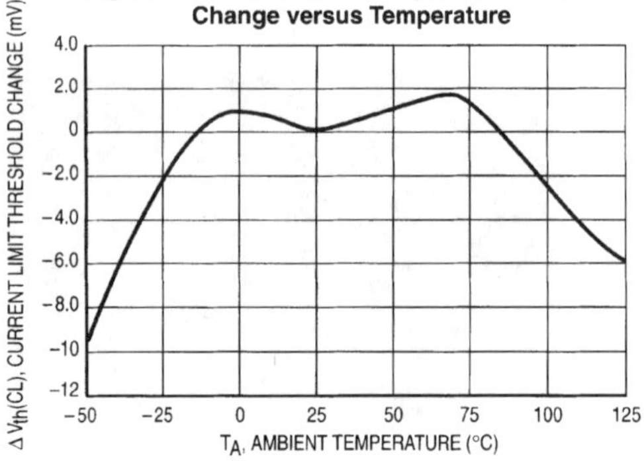

Figure 12. Shutdown Comparator Threshold Voltage versus Temperature

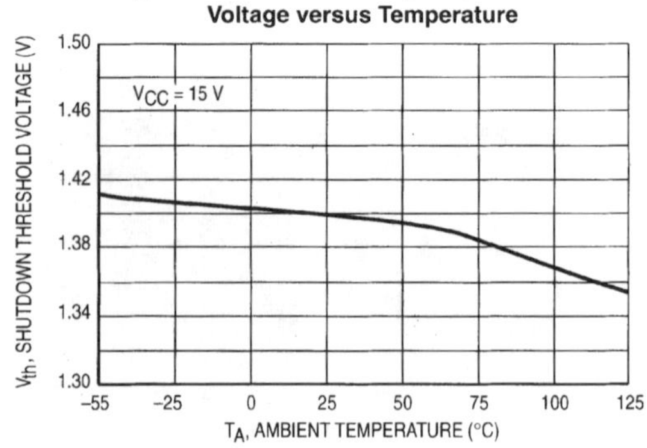

Appendix 4.8-6

MC34025 MC33025

Figure 13. Soft–Start Charge Current versus Temperature

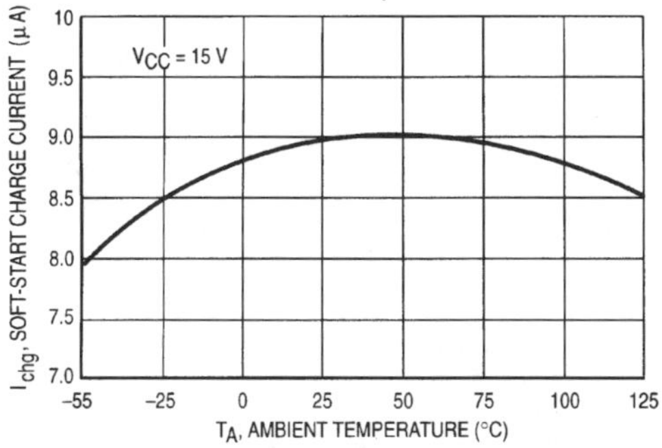

Figure 14. Output Saturation Voltage versus Load Current

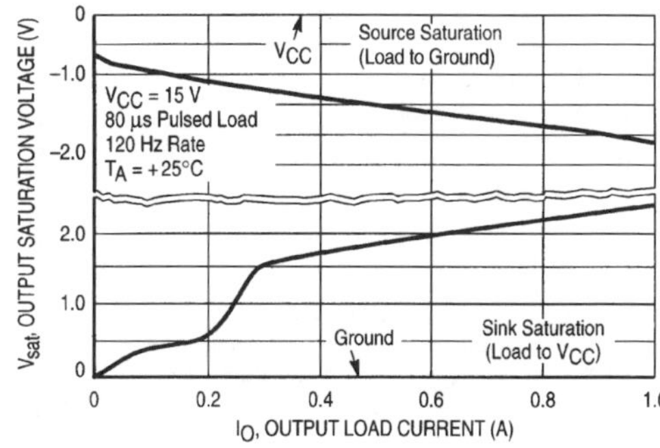

Figure 15. Drive Output Rise and Fall Time

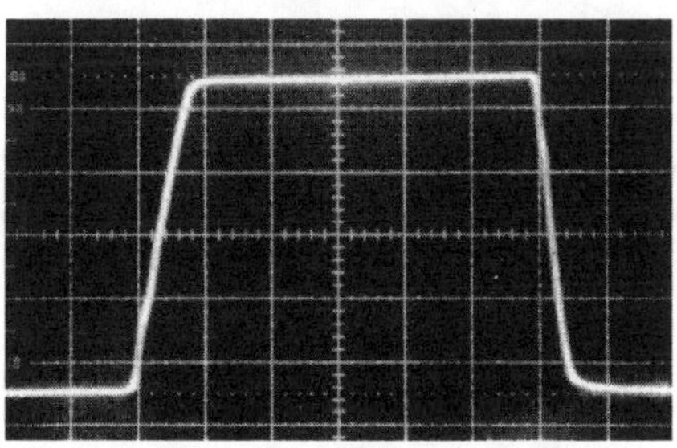

OUTPUT RISE & FALL TIME 1.0 nF LOAD
50 ns/DIV

Figure 16. Drive Output Rise and Fall Time

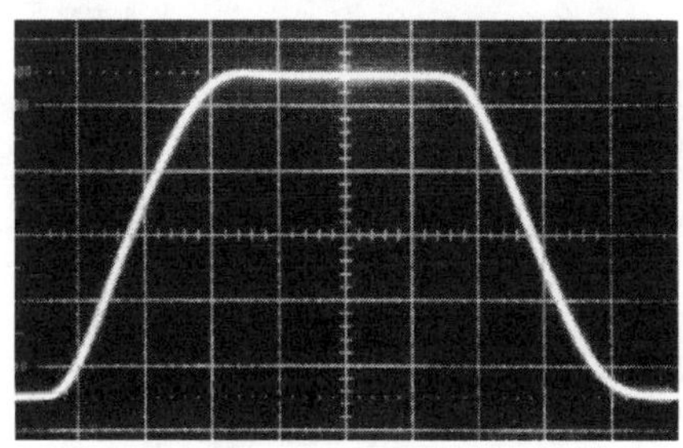

OUTPUT RISE & FALL TIME 10.0 nF LOAD
50 ns/DIV

Figure 17. Supply Voltage versus Supply Current

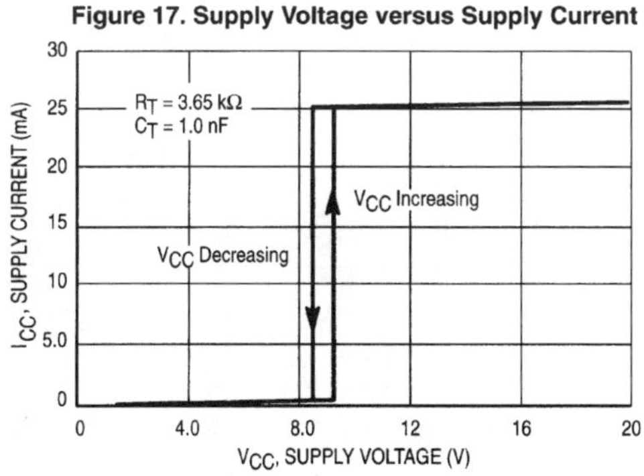

Appendix 4.8-7

MC34025 MC33025

Figure 18. Representative Block Diagram

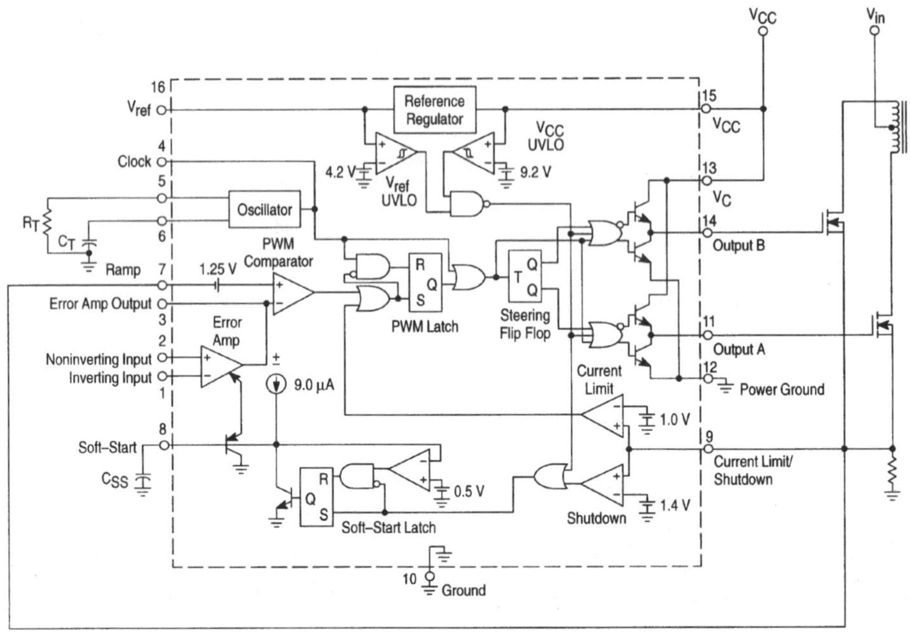

Figure 19. Current Limit Operating Waveforms

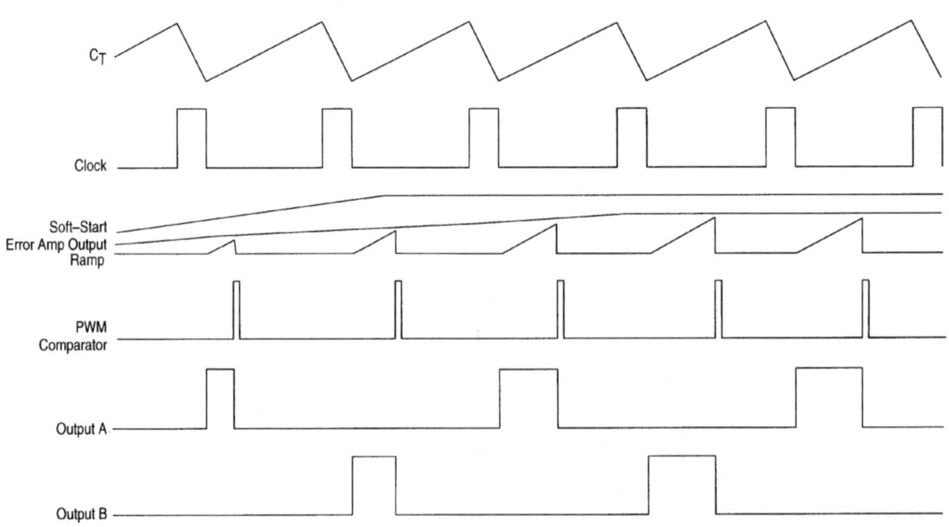

Appendix 4.8-8

MC34025 MC33025

OPERATING DESCRIPTION

The MC33025 and MC34025 series are high speed, fixed frequency, double–ended pulse width modulator controllers optimized for high frequency operation. They are specifically designed for Off–Line and DC–to–DC converter applications offering the designer a cost effective solution with minimal external components. A representative block diagram is shown in Figure 18.

Oscillator

The oscillator frequency is programmed by the values selected for the timing components R_T and C_T. The R_T pin is set to a temperature compensated 3.0 V. By selecting the value of R_T, the charge current is set through a current mirror for the timing capacitor C_T. This charge current runs continuously through C_T. The discharge current is ratioed to be 10 times the charge current, which yields the maximum duty cycle of 90%. C_T is charged to 2.8 V and discharged to 1.0 V. During the discharge of C_T, the oscillator generates an internal blanking pulse that resets the PWM Latch, inhibits the outputs, and toggles the steering flip–flop. The threshold voltages on the oscillator comparator is trimmed to guarantee an oscillator accuracy of 5.0% at 25°C.

Additional dead time can be added by externally increasing the charge current to C_T as shown in Figure 23. This changes the charge to discharge ratio of C_T which is set internally to $I_{charge}/10\ I_{charge}$. The new charge to discharge ratio will be:

$$\%\ \text{Deadtime} = \frac{I_{additional} + I_{charge}}{10\ (I_{charge})}$$

A bidirectional clock pin is provided for synchronization or for master/slave operation. As a master, the clock pin provides a positive output pulse during the discharge of C_T. As a slave, the clock pin is an input that resets the PWM latch and blanks the drive output, but does not discharge C_T. Therefore, the oscillator is not synchronized by driving the clock pin alone. Figures 29 and 30 provide suggested synchronization.

Error Amplifier

A fully compensated Error Amplifier is provided. It features a typical DC voltage gain of 95 dB and a gain bandwidth product of 8.3 MHz with 75 degrees of phase margin (Figure 3). Typical application circuits will have the noninverting input tied to the reference. The inverting input will typically be connected to a feedback voltage generated from the output of the switching power supply. Both inputs have a Common Mode Voltage (V_{CM}) input range of 1.5 V to 5.5 V. The Error Amplifier Output is provided for external loop compensation.

Soft–Start Latch

Soft–Start is accomplished in conjunction with an external capacitor. The soft start capacitor is charged by an internal 9.0 µA current source. This capacitor clamps the output of the error amplifier to less than its normal output voltage, thus limiting the duty cycle.

The time it takes for a capacitor to reach full charge is given by:

$$t \approx (4.5 \bullet 10^5)\ C_{Soft\text{-}Start}$$

A Soft–Start latch is incorporated to prevent erratic operation of this circuitry. Two conditions can cause the Soft–Start circuit to latch so that the Soft–Start capacitor stays discharged. The first condition is activation of an undervoltage lockout of either V_{CC} or V_{ref}. The second condition is when current sense input exceeds 1.4 V. Since this latch is "set dominant", it cannot be reset until either of these signals is removed, and the voltage at $C_{Soft–Start}$ is less than 0.5 V.

PWM Comparator and Latch

A PWM circuit typically compares an error voltage with a ramp signal. The outcome of this comparison determines the state of the output. In voltage mode operation the ramp signal is the voltage ramp of the timing capacitor. In current mode operation the ramp signal is the voltage ramp induced in a current sensing element. The ramp input of the PWM comparator is pinned out so that the user can decide which mode of operation best suits the application requirements. The ramp input has a 1.25 V offset such that whenever the voltage at this pin exceeds the Error Amplifier Output voltage minus 1.25 V, the PWM comparator will cause the PWM latch to set, disabling the outputs. Once the PWM latch is set, only a blanking pulse by the oscillator can reset it, thus initiating the next cycle.

A toggle flip flop connected to the output of the PWM latch controls which output is active. The flip flop is pulsed by an OR gate that gets its inputs from the oscillator clock and the output of the PWM latch. A pulse from either one will cause the flip flop to enable the other output.

Current Limiting and Shutdown

A pin is provided to perform current limiting and shutdown operations. Two comparators are connected to the input of this pin. When the voltage at this pin exceeds 1.0 V, one of the comparators is activated. The output of this comparator sets the PWM latch, which disables the output. In this way cycle–by–cycle current limiting is accomplished. If a current limit resistor is used in series with the power devices, the value of the resistor is found by:

$$R_{Sense} = \frac{1.0\ V}{I_{pk\ (switch)}}$$

If the voltage at this pin exceeds 1.4 V, the second comparator is activated. This comparator sets a latch which, in turn, causes the Soft–Start capacitor to be discharged. In this way a "hiccup" mode of recovery is possible in the case of output short circuits. If a current limit resistor is used in series with the output devices, the peak current at which the controller will enter a "hiccup" mode is given by:

$$I_{shutdown} = \frac{1.4\ V}{R_{Sense}}$$

Appendix 4.8-9

MC34025 MC33025

Undervoltage Lockout

There are two undervoltage lockout circuits within the IC. The first senses V_{CC} and the second V_{ref}. During power–up, V_{CC} must exceed 9.2 V and V_{ref} must exceed 4.2 V before the outputs can be enabled and the Soft–Start latch released. If V_{CC} falls below 8.4 V or V_{ref} falls below 3.6 V, the outputs are disabled and the Soft–Start latch is activated. When the UVLO is active, the part is in a low current standby mode allowing the IC to have an off–line bootstrap start–up circuit. Typical start–up current is 500 μA.

Output

The MC34025 has two high current totem pole outputs specifically designed for direct drive of power MOSFETs. They are capable of up to ±2.0 A peak drive current with a typical rise and fall time of 30 ns driving a 1.0 nF load.

Separate pins for V_C and Power Ground are provided. With proper implementation, a significant reduction of switching transient noise imposed on the control circuitry is possible. The separate V_C supply input also allows the designer added flexibility in tailoring the drive voltage independent of V_{CC}.

Reference

A 5.1 V bandgap reference is pinned out and is trimmed to an initial accuracy of ±1.0% at 25°C. This reference has short circuit protection and can source in excess of 10 mA for powering additional control system circuitry.

Design Considerations

Do not attempt to construct the converter on wire–wrap or plug–in prototype boards. With high frequency, high power, switching power supplies it is imperative to have separate current loops for the signal paths and for the power paths. The printed circuit layout should contain a ground plane with low current signal and high current switch and output grounds returning on separate paths back to the input filter capacitor. All bypass capacitors and snubbers should be connected as close as possible to the specific part in question. The PC board lead lengths must be less than 0.5 inches for effective bypassing or snubbing.

Instabilities

In current mode control, an instability can be encountered at any given duty cycle. The instability is caused by the current feedback loop. It has been shown that the instability is caused by a double pole at half the switching frequency. If an external ramp (S_e) is added to the on–time ramp (S_n) of the current–sense waveform, stability can be achieved (see Figure 20).

One must be careful not to add too much ramp compensation. If too much is added, the system will start to perform like a voltage mode regulator. All benefits of current mode control will be lost. Figures 28A and 28B show examples of two different ways in which external ramp compensation can be implemented.

Figure 20. Ramp Compensation

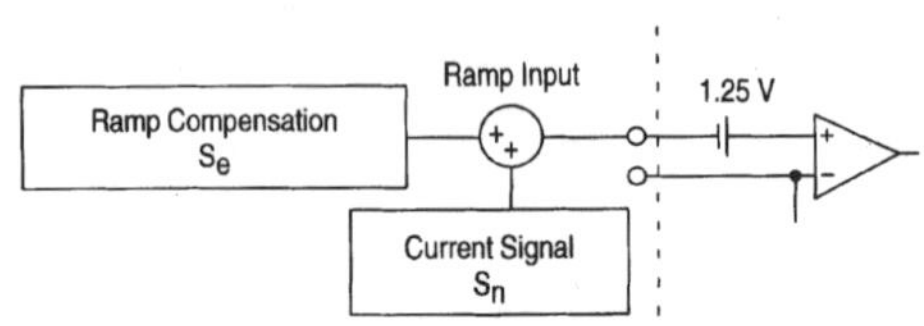

A simple equation can be used to calculate the amount of external ramp necessary to add that will achieve stability in the current loop. For the following equations, the calculated values for the application circuit in Figure 36 are also shown.

$$S_e = \frac{V_O}{L}\left(\frac{N_S}{N_P}\right)(R_S)A_i$$

where:
- V_O = DC output voltage
- N_P, N_S = number of power transformer primary or secondary turns
- A_i = gain of the current sense network (see Figures 25, 26 and 27)
- L = output inductor
- R_S = current sense resistance

For the application circuit: $S_e = \frac{5}{1.8\ \mu}\left(\frac{4}{16}\right)(0.3)(0.55)$

$= 0.115\ V/\mu s$

Appendix 4.8-10

MC34025 MC33025

PIN FUNCTION DESCRIPTION

Pin No. DIP/SOIC	Function	Description
1	Error Amp Inverting Input	This pin is usually used for feedback from the output of the power supply.
2	Error Amp Noninverting Input	This pin is used to provide a reference in which an error signal can be produced on the output of the error amp. Usually this is connected to V_{ref}, however an external reference can also be used.
3	Error Amp Output	This pin is provided for compensating the error amp for poles and zeros encountered in the power supply system, mostly the output LC filter.
4	Clock	This is a bidirectional pin used for synchronization.
5	R_T	The value of R_T sets the charge current through timing Capacitor, C_T.
6	C_T	In conjunction with R_T, the timing Capacitor sets the switching frequency. Because this part is a push–pull output, each output runs at one–half the frequency set at this pin.
7	Ramp Input	For voltage mode operation this pin is connected to C_T. For current mode operation this pin is connected through a filter to the current sensing element.
8	Soft–Start	A capacitor at this pin sets the Soft–Start time.
9	Current Limit/Shutdown	This pin has two functions. First, it provides cycle–by–cycle current limiting. Second, if the current is excessive, this pin will reinitiate a Soft–Start cycle.
10	Ground	This pin is the ground for the control circuitry.
11	Output A	This is a high current totem pole output.
12	Power Ground	This is a separate power ground return that is connected back to the power source. It is used to reduce the effects of switching transient noise on the control circuitry.
13	V_C	This is a separate power source connection for the outputs that is connected back to the power source input. With a separate power source connection, it can reduce the effects of switching transient noise on the control circuitry.
14	Output B	This is a high current totem pole output.
15	V_{CC}	This pin is the positive supply of the control IC.
16	V_{ref}	This is a 5.1 V reference. It is usually connected to the noninverting input of the error amplifier.

Figure 21. Voltage Mode Operation

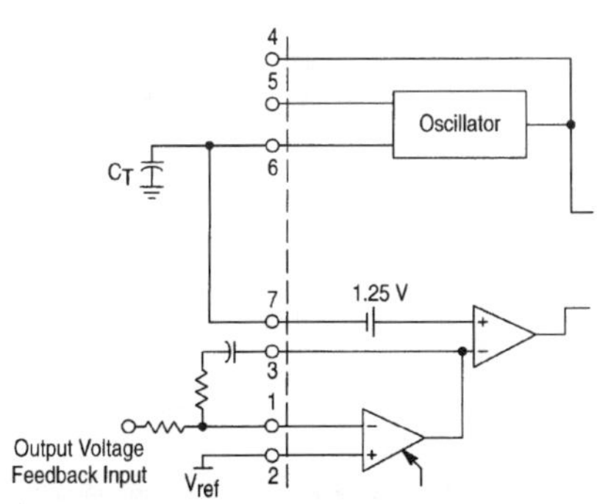

In voltage mode operation, the control range on the output of the Error Amplifier from 0% to 90% duty cycle is from 2.25 V to 4.05 V.

Figure 22. Current Mode Operation

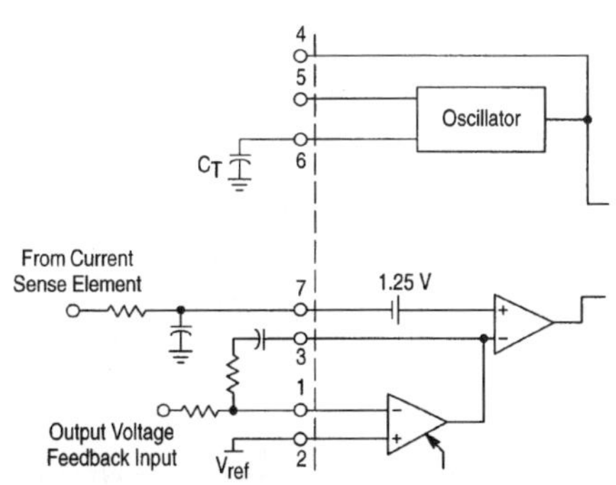

In current mode control, an RC filter should be placed at the ramp input to filter the leading edge spike caused by turn–on of a power MOSFET.

Appendix 4.8-11

MC34025 MC33025

Figure 23. Dead Time Addition

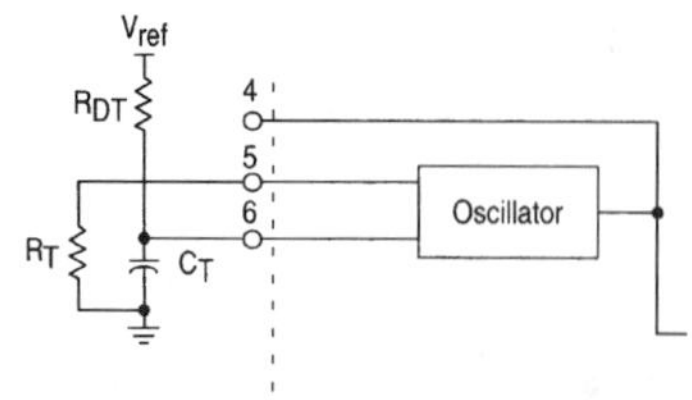

Additional dead time can be added by the addition of a dead time resistor from V_{ref} to C_T. See text on oscillator section for more information.

Figure 24. External Clock Synchronization

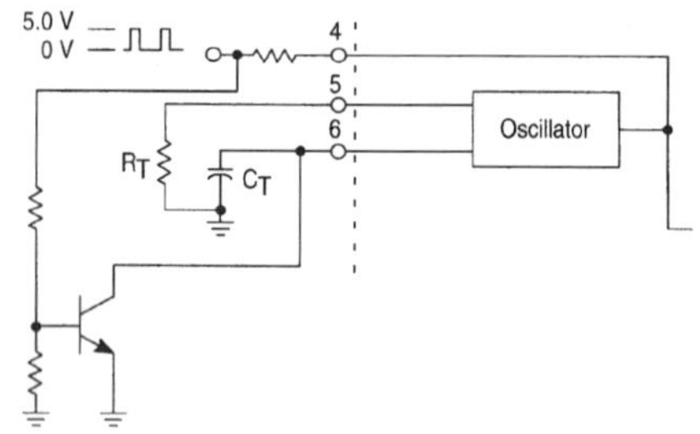

The sync pulse fed into the clock pin must be at least 3.9 V. R_T and C_T need to be set 10% slower than the sync frequency. This circuit is also used in voltage mode operation for master/slave operation. The clock signal would be coming from the master which is set at the desired operating frequency, while the slave is set 10% slower.

Figure 25. Resistive Current Sensing

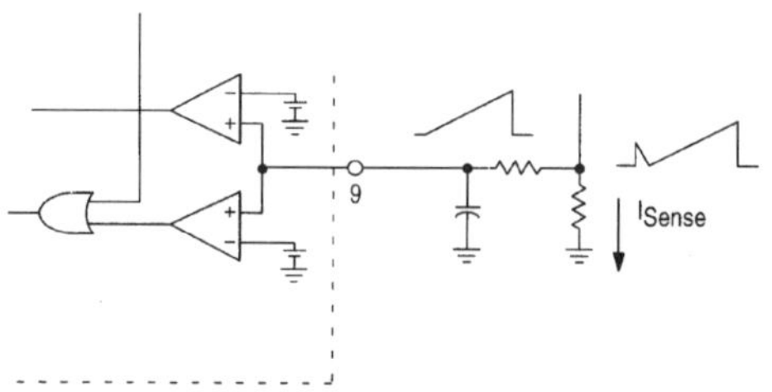

The addition of an RC filter will eliminate instability caused by the leading edge spike on the current waveform. This sense signal can also be used at the ramp input pin for current mode control. For ramp compensation it is necessary to know the gain of the current feedback loop. If a transformer is used, the gain can be calculated by:

$$A_i = \frac{R_{Sense}}{\text{turns ratio}}$$

Figure 26. Primary Side Current Sensing

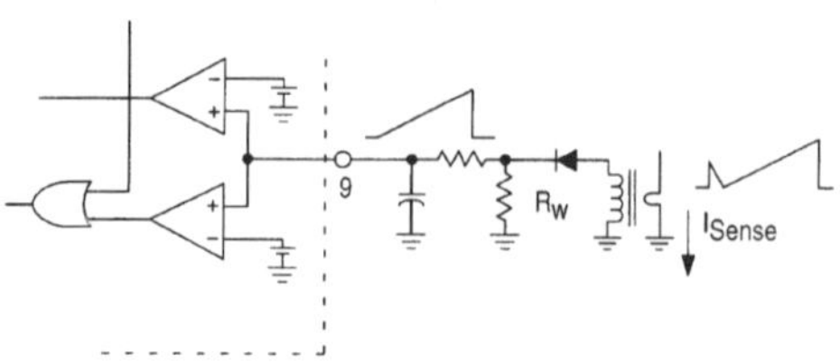

Figure 27. Primary or Secondary Side Current Sensing

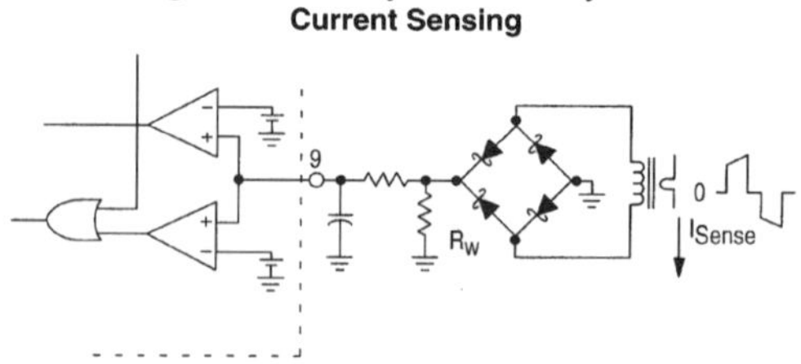

The addition of an RC filter will eliminate instability caused by the leading edge spike on the current waveform. This sense signal can also be used at the ramp input pin for current mode control. For ramp compensation it is necessary to know the gain of the current feedback loop. The gain can be calculated by:

$$A_i = \frac{R_W}{\text{turns ratio}}$$

Appendix 4.8-12

MC34025 MC33025

Figure 28A. Slope Compensation (Noise Sensitive)

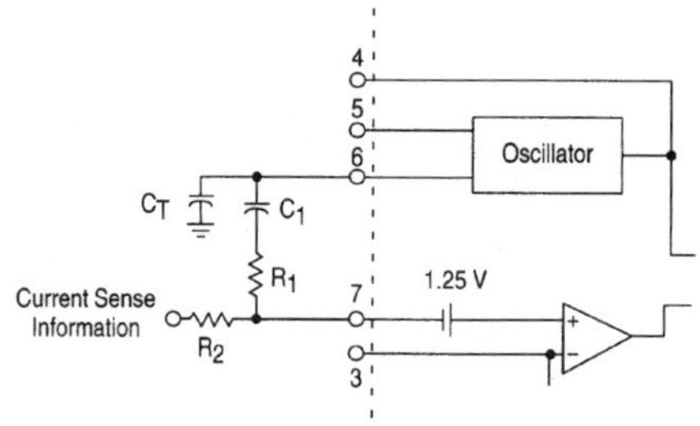

This method of slope compensation is easy to implement, however, it is noise sensitive. Capacitor C_1 provides AC coupling. The oscillator signal is added to the current signal by a voltage divider consisting of resistors R_1 and R_2.

Figure 28B. Slope Compensation (Noise Immune)

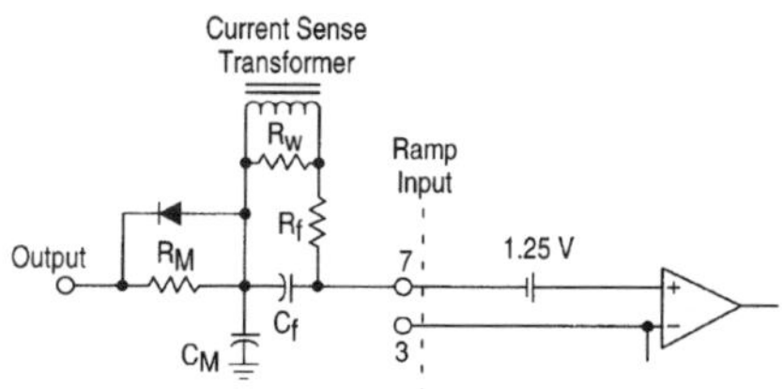

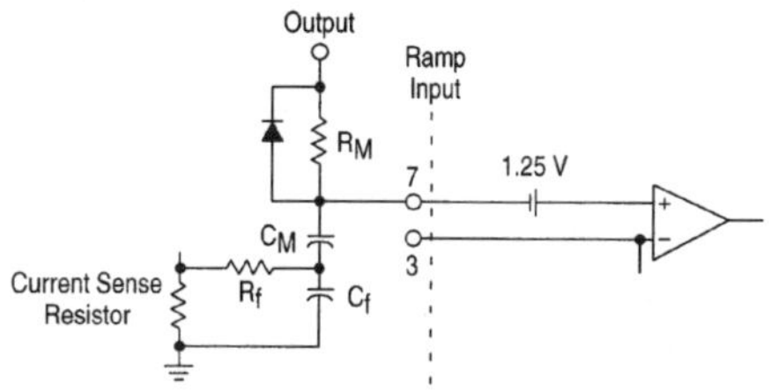

When only one output is used, this method of slope compensation can be used and it is relatively noise immune. Resistor R_M and capacitor C_M provide the added slope necessary. By choosing R_M and C_M with a larger time constant than the switching frequency, you can assume that its charge is linear. First choose C_M, then R_M can be adjusted to achieve the required slope. The diode provides a reset pulse at the ramp input at the end of every cycle. The charge current I_M can be calculated by $I_M = C_M S_e$. Then R_M can be calculated by $R_M = V_{CC}/I_M$.

Figure 29. Current Mode Master/Slave Operation Over Short Distances

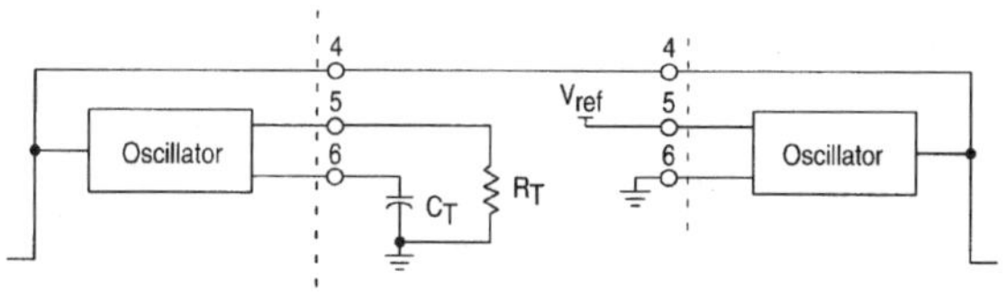

Appendix 4.8-13

MC34025 MC33025

Figure 30. Synchronization Over Long Distances

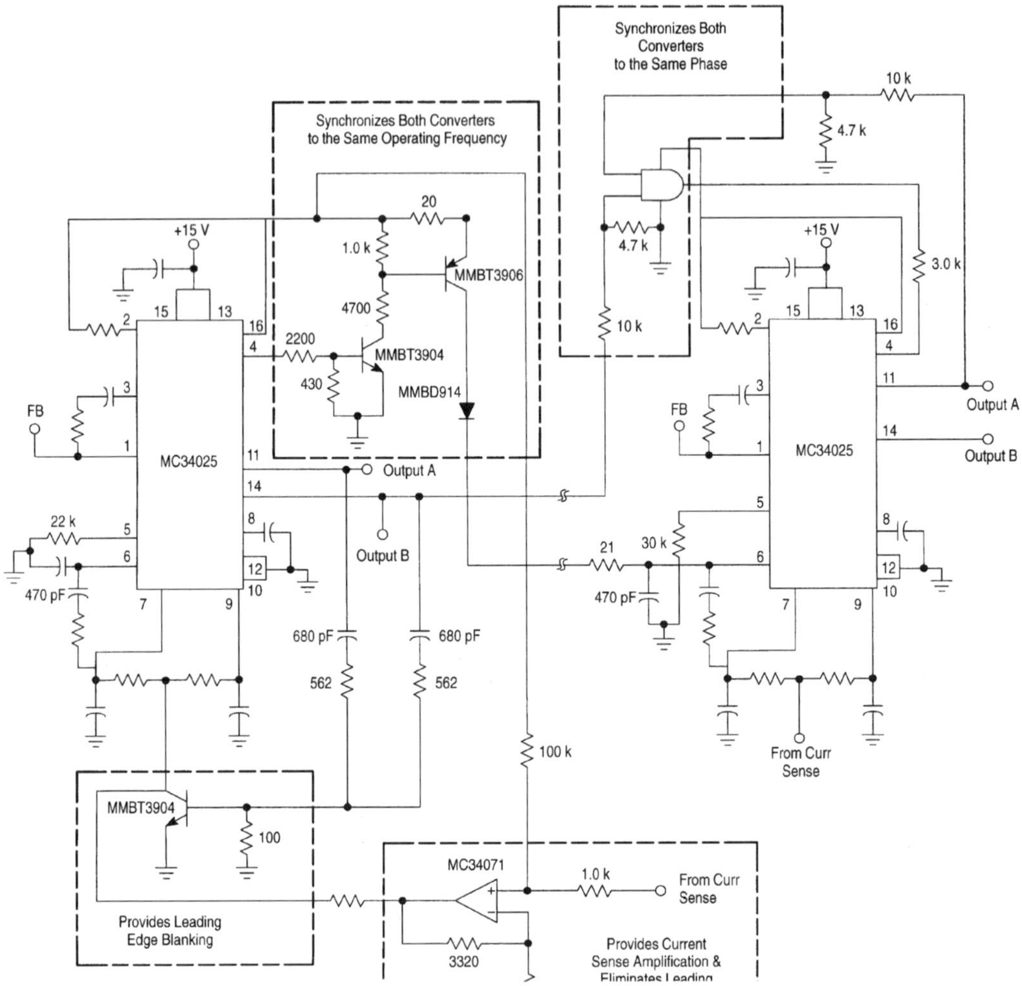

Appendix 4.8-14

MC34025 MC33025

Figure 31. Buffered Maximum Clamp Level

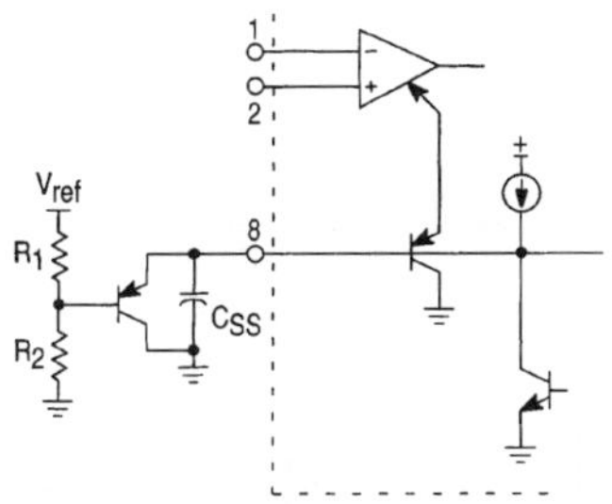

In voltage mode operation, the maximum duty cycle can be clamped. By the addition of a PNP transistor to buffer the clamp voltage, the Soft–Start current is not affected by R_1.

The new equation for Soft–Start is $t \approx \frac{V_{clamp} + 0.6}{9.0\ \mu A}\left(C_{SS}\right)$

In current mode operation, this circuit will limit the maximum voltage allowed at the ramp input to end a cycle.

Figure 32. Bipolar Transistor Drive

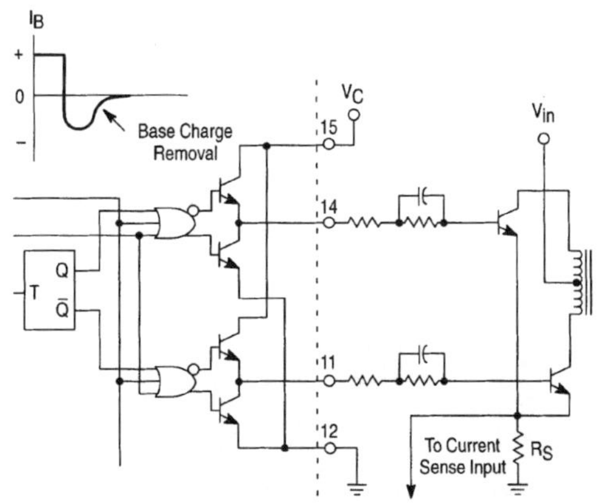

The totem pole output can furnish negative base current for enhanced transistor turn–off, with the addition of the capacitor in series with the base.

Figure 33. Isolated MOSFET Drive

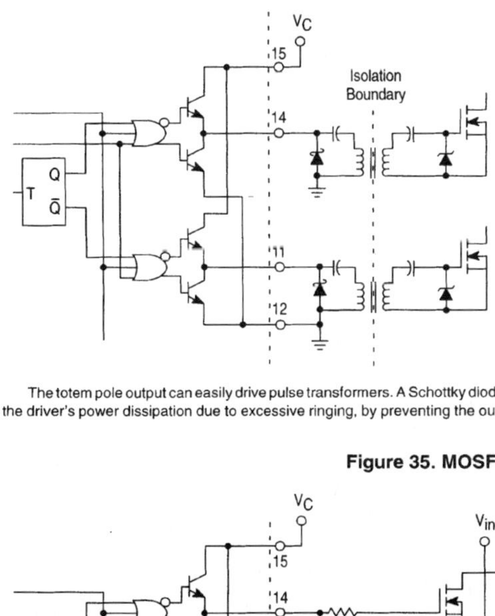

Figure 34. Direct Transformer Drive

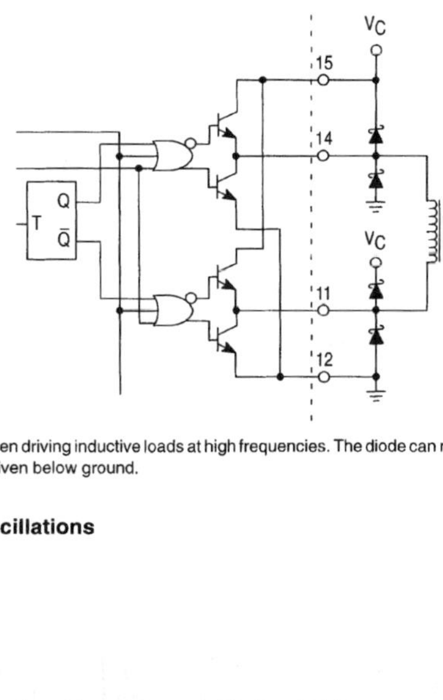

The totem pole output can easily drive pulse transformers. A Schottky diode is recommended when driving inductive loads at high frequencies. The diode can reduce the driver's power dissipation due to excessive ringing, by preventing the output pin from being driven below ground.

Figure 35. MOSFET Parasitic Oscillations

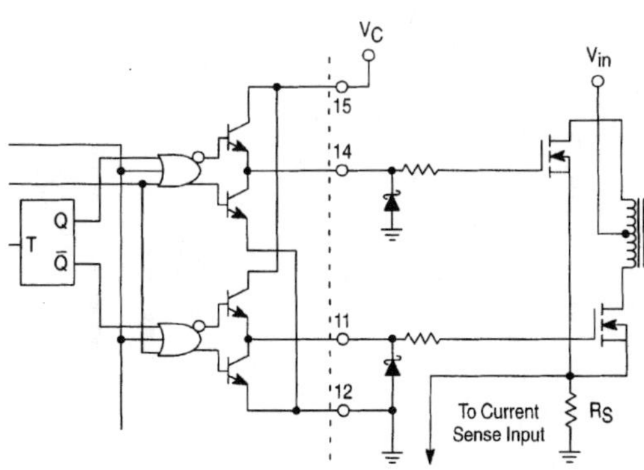

A series gate resistor may be needed to damp high frequency parasitic oscillation caused by a MOSFET's input capacitance and any series wiring inductance in the gate–source circuit. The series resistor will also decrease the MOSFET's switching speed. A Schottky diode can reduce the driver's power dissipation due to excessive ringing, by preventing the output pin from being driven below ground. The Schottky diode also prevents substrate injection when the output pin is driven below ground.

Appendix 4.8-15

MC34025 MC33025

Figure 36. Application Circuit

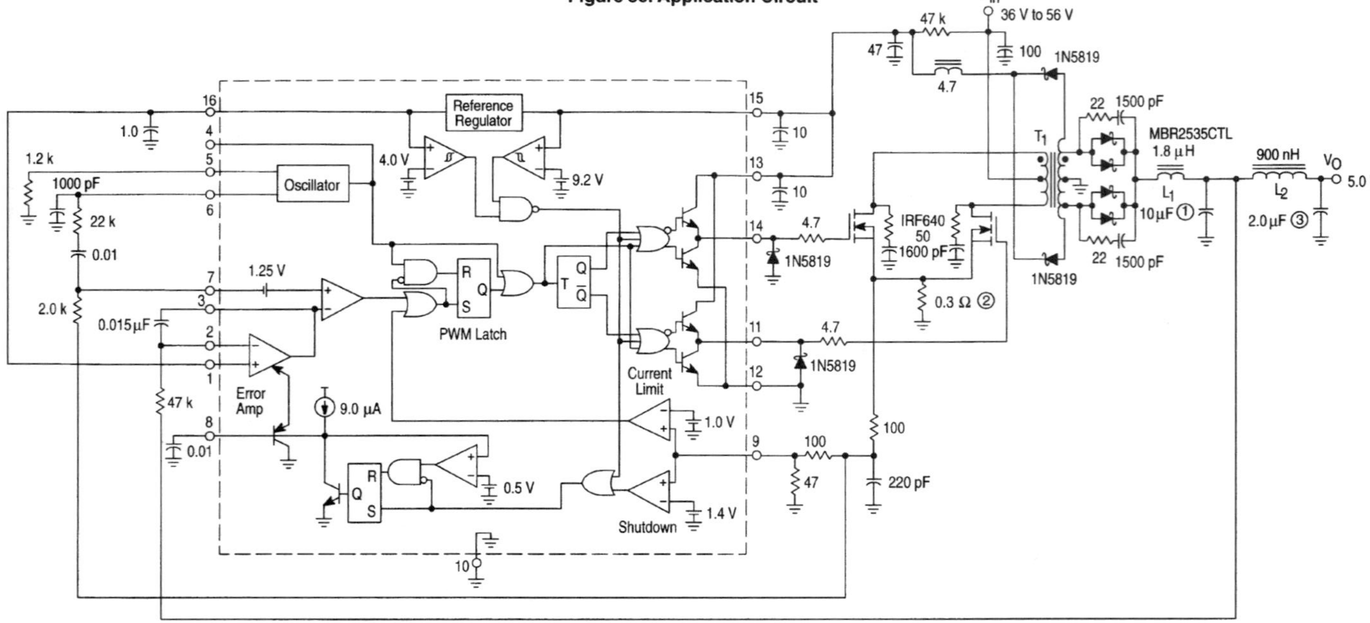

T1 – Primary: 16 turns center tapped #48 AWG (1300 strands litz wire)
Secondary: 4 turns center tapped 0.003" (2 layers) copper foil
Bootstrap: 1 turn added to each secondary output #36 AWG
Core: Philips 3F3, part #4312 020 4124
Bobbin: Philips part #4322 021 3525
Coilcraft P3269–A

L1 – 2 turns #48 AWG (1300 strands litz wire)
Core: Philips 3F3, part #EP10–3F3
Bobbin: Philips part #EP10PCB1–8
L = 1.8 µH
Coilcraft P3270–A

L2 – 7 turns #18 AWG, 1/2" diameter air core
Coilcraft P3271–A

Heatsinks – Power FET: AAVID Heatsink #533902B02554 with clip
Output Rectifiers: AAVID Heatsink #533402B02552 with clip

Insulators – All power devices are insulated with Berquist Sil–Pad 1500

① – 10 (1.0 µF) ceramic capacitors in parallel
② – 5 (1.5 Ω) resistors in parallel
③ – 2 (1.0 µF) cearmic capacitors in parallel

Test	Condition	Result
Line Regulation	V_{in} = 40 V to 56 V, I_O = 15 A	14 mV = ±0.275%
Load Regulation	V_{in} = 48 V, I_O = 8.0 V to 15 A	54 mV = ±1.0%
Output Ripple	V_{in} = 48 V, I_O = 15 A	50 mVp–p
Efficiency	V_{in} = 48 V, I_O = 15 A	71.2%

Appendix 4.9-1

MOTOROLA

MC34063A
MC33063A

DC-to-DC Converter Control Circuits

The MC34063A Series is a monolithic control circuit containing the primary functions required for DC–to–DC converters. These devices consist of an internal temperature compensated reference, comparator, controlled duty cycle oscillator with an active current limit circuit, driver and high current output switch. This series was specifically designed to be incorporated in Step–Down and Step–Up and Voltage–Inverting applications with a minimum number of external components. Refer to Application Notes AN920A/D and AN954/D for additional design information.

- Operation from 3.0 V to 40 V Input
- Low Standby Current
- Current Limiting
- Output Switch Current to 1.5 A
- Output Voltage Adjustable
- Frequency Operation to 100 kHz
- Precision 2% Reference

DC–to–DC CONVERTER CONTROL CIRCUITS

SEMICONDUCTOR TECHNICAL DATA

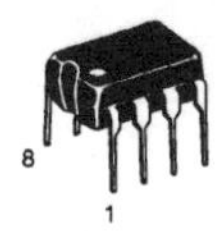

P, P1 SUFFIX
PLASTIC PACKAGE
CASE 626

D SUFFIX
PLASTIC PACKAGE
CASE 751
(SO–8)

Representative Schematic Diagram

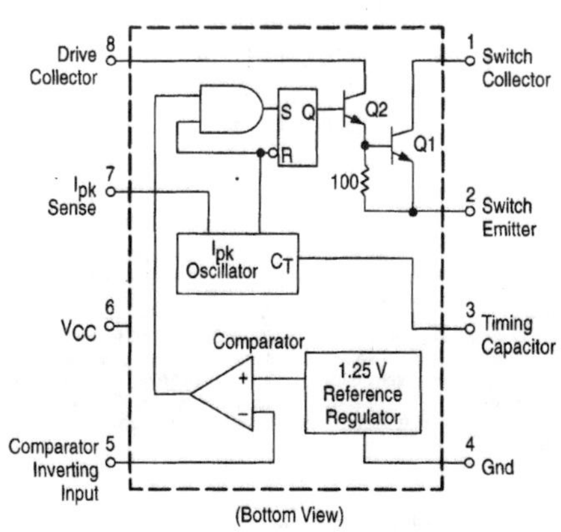

(Bottom View)

This device contains 51 active transistors.

PIN CONNECTIONS

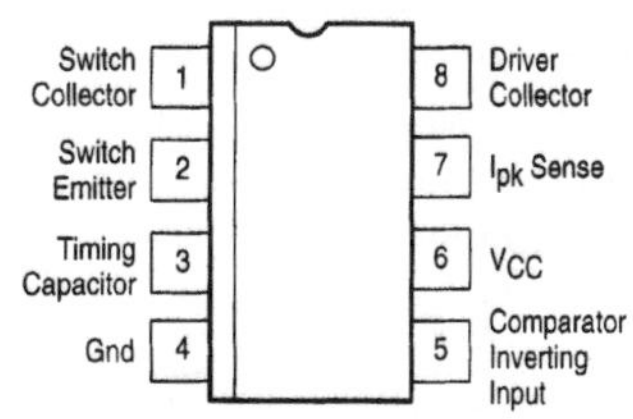

(Top View)

ORDERING INFORMATION

Device	Operating Temperature Range	Package
MC33063AD	T_A = – 40° to +85°C	SO–8
MC33063AP1		Plastic DIP
MC33063AVD	T_A = –40° to +125°C	SO–8
MC33063AVP		Plastic DIP
MC34063AD	T_A = 0° to +70°C	SO–8
MC34063AP1		Plastic DIP

Appendix 4.9-2

MC34063A MC33063A

MAXIMUM RATINGS

Rating	Symbol	Value	Unit
Power Supply Voltage	V_{CC}	40	Vdc
Comparator Input Voltage Range	V_{IR}	−0.3 to +40	Vdc
Switch Collector Voltage	$V_{C(switch)}$	40	Vdc
Switch Emitter Voltage ($V_{Pin\ 1}$ = 40 V)	$V_{E(switch)}$	40	Vdc
Switch Collector to Emitter Voltage	$V_{CE(switch)}$	40	Vdc
Driver Collector Voltage	$V_{C(driver)}$	40	Vdc
Driver Collector Current (Note 1)	$I_{C(driver)}$	100	mA
Switch Current	I_{SW}	1.5	A
Power Dissipation and Thermal Characteristics Plastic Package, P, P1 Suffix T_A = 25°C Thermal Resistance SOIC Package, D Suffix T_A = 25°C Thermal Resistance	 P_D $R_{\theta JA}$ P_D $R_{\theta JA}$	 1.25 100 625 160	 W °C/W W °C/W
Operating Junction Temperature	T_J	+150	°C
Operating Ambient Temperature Range MC34063A MC33063AV MC33063A	T_A	 0 to +70 −40 to +125 −40 to +85	°C
Storage Temperature Range	T_{stg}	−65 to +150	°C

NOTES: 1. Maximum package power dissipation limits must be observed.
2. ESD data available upon request.

ELECTRICAL CHARACTERISTICS (V_{CC} = 5.0 V, T_A = T_{low} to T_{high} [Note 3], unless otherwise specified.)

Characteristics	Symbol	Min	Typ	Max	Unit
OSCILLATOR					
Frequency ($V_{Pin\ 5}$ = 0 V, C_T = 1.0 nF, T_A = 25°C)	f_{osc}	24	33	42	kHz
Charge Current (V_{CC} = 5.0 V to 40 V, T_A = 25°C)	I_{chg}	24	35	42	μA
Discharge Current (V_{CC} = 5.0 V to 40 V, T_A = 25°C)	I_{dischg}	140	220	260	μA
Discharge to Charge Current Ratio (Pin 7 to V_{CC}, T_A = 25°C)	I_{dischg}/I_{chg}	5.2	6.5	7.5	–
Current Limit Sense Voltage (I_{chg} = I_{dischg}, T_A = 25°C)	$V_{ipk(sense)}$	250	300	350	mV
OUTPUT SWITCH (Note 4)					
Saturation Voltage, Darlington Connection (Note 5) (I_{SW} = 1.0 A, Pins 1, 8 connected)	$V_{CE(sat)}$	–	1.0	1.3	V
Saturation Voltage, Darlington Connection (I_{SW} = 1.0 A, $R_{Pin\ 8}$ = 82 Ω to V_{CC}, Forced β ≃ 20)	$V_{CE(sat)}$	–	0.45	0.7	V
DC Current Gain (I_{SW} = 1.0 A, V_{CE} = 5.0 V, T_A = 25°C)	h_{FE}	50	75	–	–
Collector Off–State Current (V_{CE} = 40 V)	$I_{C(off)}$	–	0.01	100	μA

NOTES: 3. T_{low} = 0°C for MC34063A, −40°C for MC33063A, AV T_{high} = +70°C for MC34063A, +85°C for MC33063A, +125°C for MC33063AV
4. Low duty cycle pulse techniques are used during test to maintain junction temperature as close to ambient temperature as possible.
5. If the output switch is driven into hard saturation (non–Darlington configuration) at low switch currents (≤ 300 mA) and high driver currents (≥ 30 mA), it may take up to 2.0 μs for it to come out of saturation. This condition will shorten the off time at frequencies ≥ 30 kHz, and is magnified at high temperatures. This condition does not occur with a Darlington configuration, since the output switch cannot saturate. If a non–Darlington configuration is used, the following output drive condition is recommended:

$$\text{Forced } \beta \text{ of output switch}: \frac{I_C \text{ output}}{I_C \text{ driver} - 7.0 \text{ mA}^*} \geq 10$$

*The 100 Ω resistor in the emitter of the driver device requires about 7.0 mA before the output switch conducts.

Appendix 4.9-3

MC34063A MC33063A

ELECTRICAL CHARACTERISTICS (continued) (V_{CC} = 5.0 V, $T_A = T_{low}$ to T_{high} [Note 3], unless otherwise specified.)

Characteristics	Symbol	Min	Typ	Max	Unit
COMPARATOR					
Threshold Voltage	V_{th}				V
T_A = 25°C		1.225	1.25	1.275	
$T_A = T_{low}$ to T_{high}		1.21	–	1.29	
Threshold Voltage Line Regulation (V_{CC} = 3.0 V to 40 V)	Reg_{line}				mV
MC33063A, MC34063A		–	1.4	5.0	
MC33363AV		–	1.4	6.0	
Input Bias Current (V_{in} = 0 V)	I_{IB}	–	–20	–400	nA
TOTAL DEVICE					
Supply Current (V_{CC} = 5.0 V to 40 V, C_T = 1.0 nF, Pin 7 = V_{CC}, $V_{Pin\ 5} > V_{th}$, Pin 2 = Gnd, remaining pins open)	I_{CC}	–	–	4.0	mA

NOTES: 3. T_{low} = 0°C for MC34063A, –40°C for MC33063A, AV T_{high} = +70°C for MC34063A, +85°C for MC33063A, +125°C for MC33063AV
4. Low duty cycle pulse techniques are used during test to maintain junction temperature as close to ambient temperature as possible.
5. If the output switch is driven into hard saturation (non–Darlington configuration) at low switch currents (≤ 300 mA) and high driver currents (≥ 30 mA), it may take up to 2.0 µs for it to come out of saturation. This condition will shorten the off time at frequencies ≥ 30 kHz, and is magnified at high temperatures. This condition does not occur with a Darlington configuration, since the output switch cannot saturate. If a non–Darlington configuration is used, the following output drive condition is recommended:

$$\text{Forced } \beta \text{ of output switch}: \frac{I_C \text{ output}}{I_C \text{ driver} - 7.0 \text{ mA}^*} \geq 10$$

*The 100 Ω resistor in the emitter of the driver device requires about 7.0 mA before the output switch conducts.

Figure 1. Output Switch On–Off Time versus Oscillator Timing Capacitor

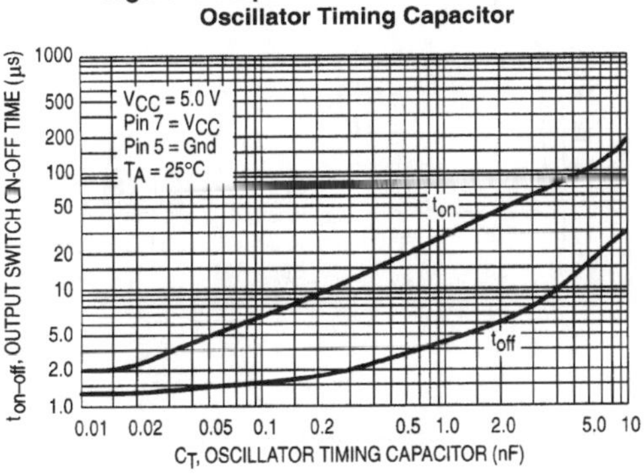

Figure 2. Timing Capacitor Waveform

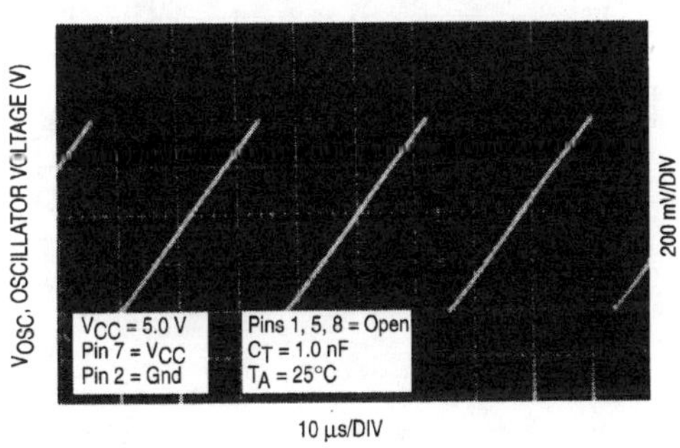

Appendix 4.9-4

MC34063A MC33063A

Figure 3. Emitter Follower Configuration Output Saturation Voltage versus Emitter Current

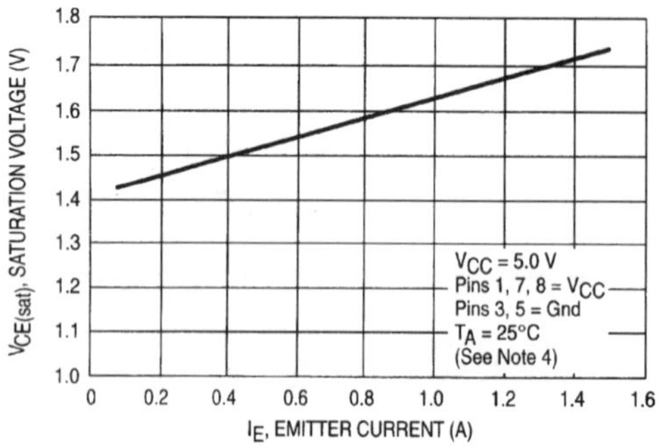

Figure 4. Common Emitter Configuration Output Switch Saturation Voltage versus Collector Current

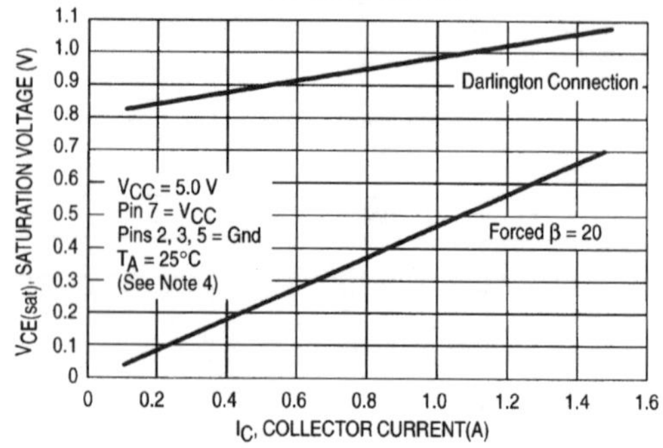

Figure 5. Current Limit Sense Voltage versus Temperature

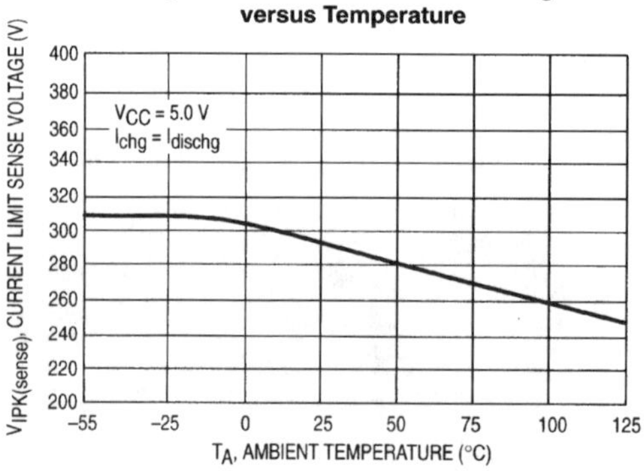

Figure 6. Standby Supply Current versus Supply Voltage

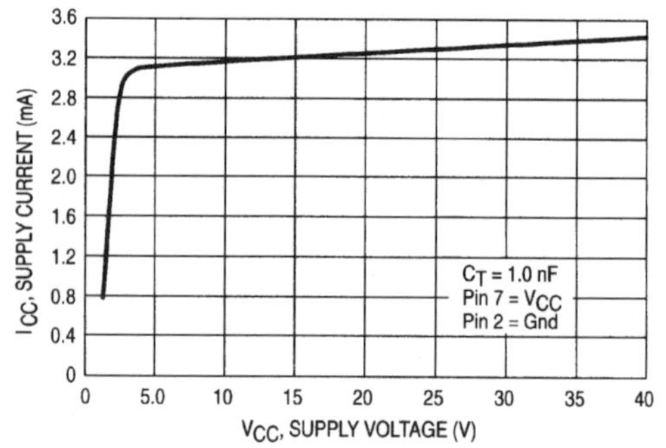

NOTE: 4. Low duty cycle pulse techniques are used during test to maintain junction temperature as close to ambient temperature as possible.

Appendix 4.9-5

MC34063A MC33063A

Figure 7. Step–Up Converter

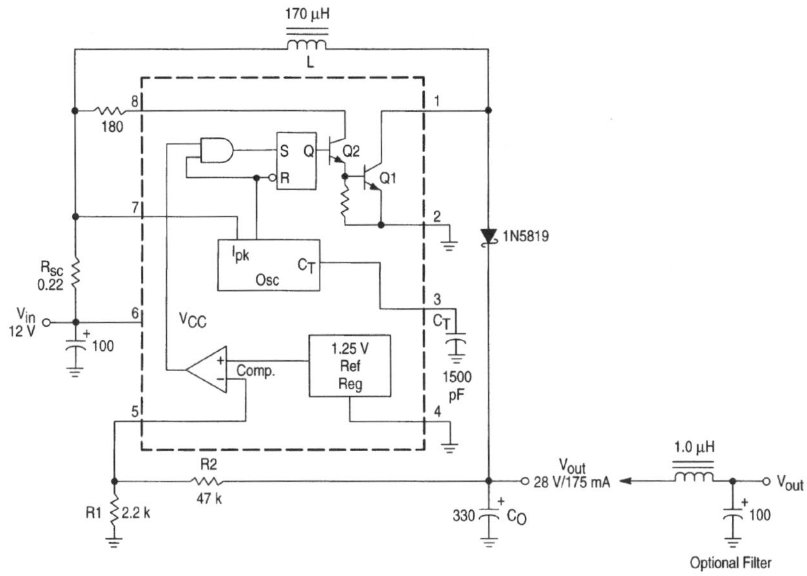

Test	Conditions	Results
Line Regulation	V_{in} = 8.0 V to 16 V, I_O = 175 mA	30 mV = ±0.05%
Load Regulation	V_{in} = 12 V, I_O = 75 mA to 175 mA	10 mV = ±0.017%
Output Ripple	V_{in} = 12 V, I_O = 175 mA	400 mVpp
Efficiency	V_{in} = 12 V, I_O = 175 mA	87.7%
Output Ripple With Optional Filter	V_{in} = 12 V, I_O = 175 mA	40 mVpp

Figure 8. External Current Boost Connections for I_C Peak Greater than 1.5 A

8a. External NPN Switch

8b. External NPN Saturated Switch

(See Note 5)

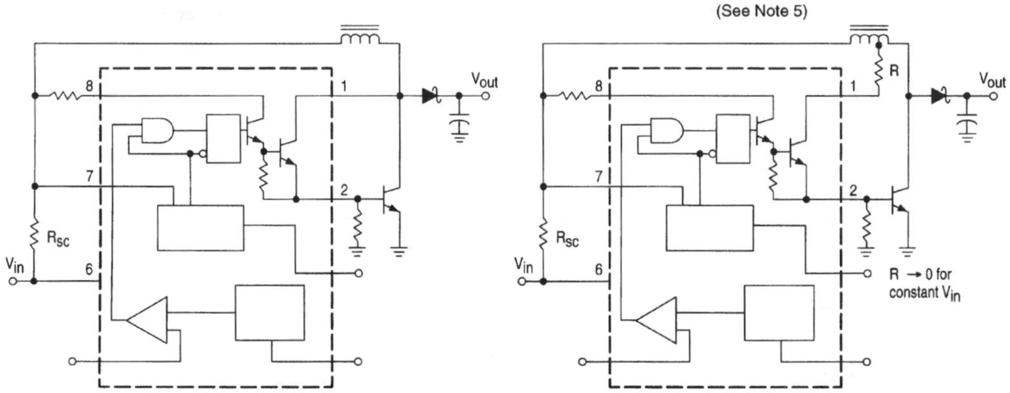

NOTE: 5. If the output switch is driven into hard saturation (non–Darlington configuration) at low switch currents (≤ 300 mA) and high driver currents (≥ 30 mA), it may take up to 2.0 μs to come out of saturation. This condition will shorten the off time at frequencies ≥ 30 kHz, and is magnified at high temperatures. This condition does not occur with a Darlington configuration, since the output switch cannot saturate. If a non–Darlington configuration is used, the following output drive condition is recommended.

Appendix 4.9-6

MC34063A MC33063A

Figure 9. Step–Down Converter

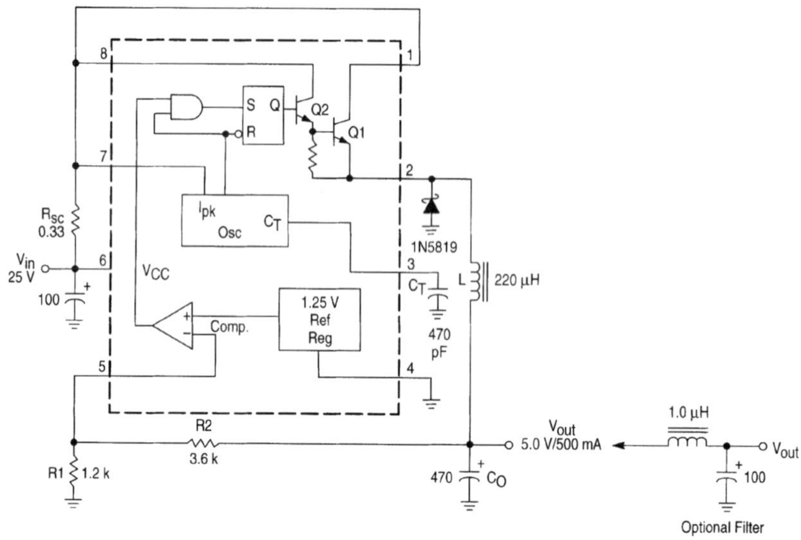

Test	Conditions	Results
Line Regulation	V_{in} = 15 V to 25 V, I_O = 500 mA	12 mV = ±0.12%
Load Regulation	V_{in} = 25 V, I_O = 50 mA to 500 mA	3.0 mV = ±0.03%
Output Ripple	V_{in} = 25 V, I_O = 500 mA	120 mVpp
Short Circuit Current	V_{in} = 25 V, R_L = 0.1 Ω	1.1 A
Efficiency	V_{in} = 25 V, I_O = 500 mA	83.7%
Output Ripple With Optional Filter	V_{in} = 25 V, I_O = 500 mA	40 mVpp

Figure 10. External Current Boost Connections for I_C Peak Greater than 1.5 A

10a. External NPN Switch **10b. External PNP Saturated Switch**

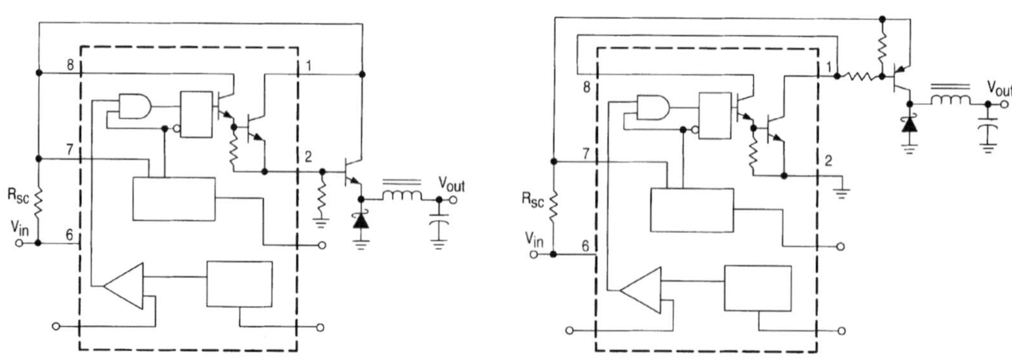

Appendix 4.9-7

MC34063A MC33063A

Figure 11. Voltage Inverting Converter

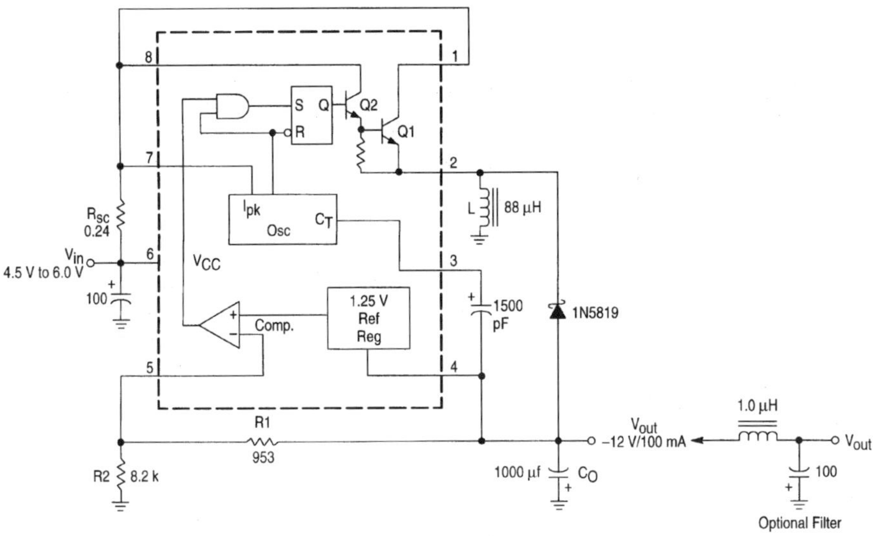

Test	Conditions	Results
Line Regulation	V_{in} = 4.5 V to 6.0 V, I_O = 100 mA	3.0 mV = ±0.012%
Load Regulation	V_{in} = 5.0 V, I_O = 10 mA to 100 mA	0.022 V = ±0.09%
Output Ripple	V_{in} = 5.0 V, I_O = 100 mA	500 mVpp
Short Circuit Current	V_{in} = 5.0 V, R_L = 0.1 Ω	910 mA
Efficiency	V_{in} = 5.0 V, I_O = 100 mA	62.2%
Output Ripple With Optional Filter	V_{in} = 5.0 V, I_O = 100 mA	70 mVpp

Figure 12. External Current Boost Connections for I_C Peak Greater than 1.5 A

12a. External NPN Switch

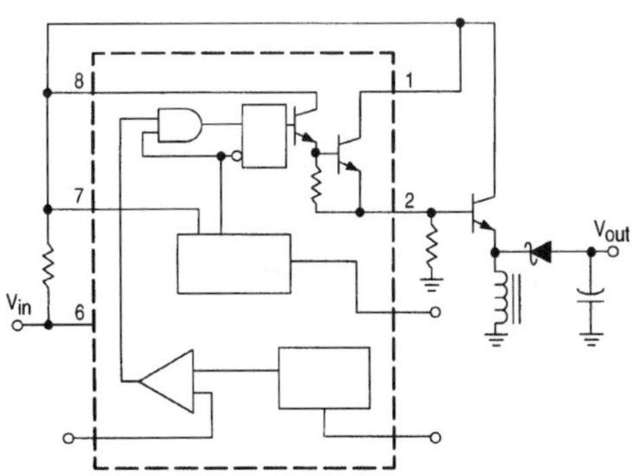

12b. External PNP Saturated Switch

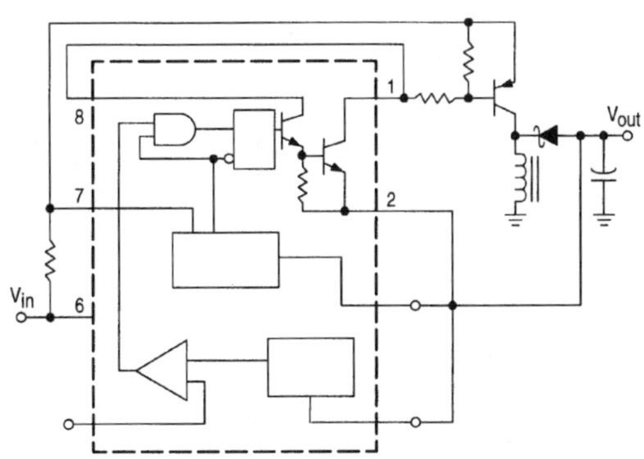

chapter five

Direct-current machines

List of basic variables and symbols used in this chapter

i_a and i_f are the currents in the armature and field windings
i_s and i_r are the currents in the stator and rotor windings
u_a and u_f are the applied voltages to the armature and field windings
u_s and u_r are the applied voltages to the stator and rotor windings
ψ_a and ψ_f are the armature and field flux linkages
ω_r and θ_r are the angular velocity and angular displacement of the rotor
E_a is the *electromotive force*
T_e and T_L are the electromagnetic and load torques
B_m is the viscous friction coefficient
J is the equivalent moment of inertial of the rotor and attached load
r_a and r_f are the resistances of the armature and field windings
r_s and r_r are the resistances of the stator and rotor windings
L_a and L_f are the self-inductances of the armature and field windings
L_{af} is the mutual inductance between the armature and field windings
L_s and L_r are the self-inductances of the stator and rotor windings
L_{sr} is the mutual inductance between the stator and rotor windings
$\Re_m$ is the magnetizing reluctance
N_s and N_r are the number of turns in the stator and rotor windings
P is the number of poles

5.1 Introduction

Electric motors and generators are rotating energy-transfer electromechanical motion devices, called electric machines, that convert energy by means of rotational motions. Electric machines, as the major part of electromechanical systems, should be studied along with the driving power electronics and converters. In addition to power converters, mechanical systems and mechanical coupling must be integrated into analysis and design.

Electric motors convert electrical energy to mechanical energy, while generators convert mechanical energy to electrical energy. It is worth mentioning that many electric machines can be used as motors and generators. Hence, the energy conversion is reversible, and conventional generators can be operated as motors and vice versa. There are two types of electric machines used; namely, AC and DC machines. Electric machines have

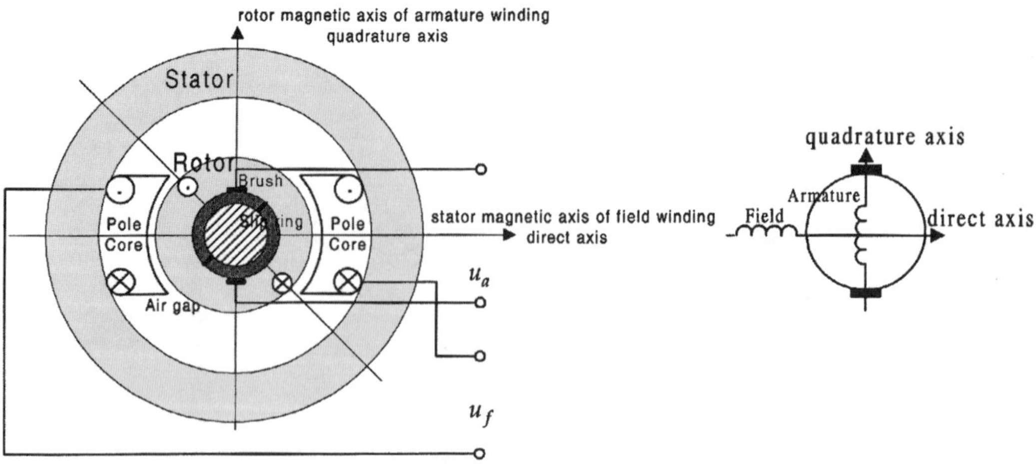

Figure 5.1 Elementary two-pole commutator machine.

certain basic features; however, one should distinguish direct-current, induction, and synchronous machines. In this chapter, we will study DC machines; in particular, separately excited, shunt-connected, series-connected, compound-connected, and permanent-magnet DC motors and generators will be thoroughly covered.

All electric machines have stationary and rotating members, separated by an air gap. The armature winding is placed in the rotor slots and connected to a rotating commutator, which rectifies the induced voltage (see Figure 5.1). One supplies the armature and excitation voltages, or feeds the armature and excitation currents to the armature (rotor) and field (stator) windings. These stator and rotor windings are coupled magnetically.

The brushes, which are connected to the armature windings, ride on the commutator. The armature winding consists of identical coils carried in slots that are uniformly distributed around the periphery of the rotor. The stator has salient poles, and conventional DC machines are excited by direct current; in particular, if a voltage-fed converter is used, a dc voltage u_f is supplied to the stationary field windings. Hence, the excitation magnetic field is produced by the field coils. It should be emphasized that in permanent-magnet machines, the magnetic field is established by permanent magnets. Due to the commutator (circular conducting segments), armature and field windings produce stationary *magnetomotive forces* that are displaced by 90 electrical degrees. Separately excited, permanent-magnet, shunt-, series-, and compound-connected DC machines are widely used. These machines have different steady-state and dynamic characteristics due to different operating, torque production, and excitation features. However, all of these machines have a commutator. The standard general-purpose DC machine is shown in Figure 5.2. In electromechanical systems, the so-called *servo*-motors are used. In direct-drive applications, motors are commonly assembled with gears, and the optical encoder is used to measure the angular velocity of the motor. Figure 5.3 illustrates the permanent-magnet servo-motor.

The armature magnetic force is along the *quadrature* (rotor) magnetic axis, while the *direct* axis stands for a field magnetic axis. The electromagnetic torque is produced as a result of the interaction of these stationary *magnetomotive forces.*

The *electromotive force,* which results due to rotation, should be thoroughly studied. The induced (generated) voltage E_a is a function of the flux per pole and the angular velocity of the rotor, and E_a opposes the current produced by the voltage applied (see Lenz's law). In particular, for one revolution $\theta_r \in [0 \quad 2\pi]$, one obtains

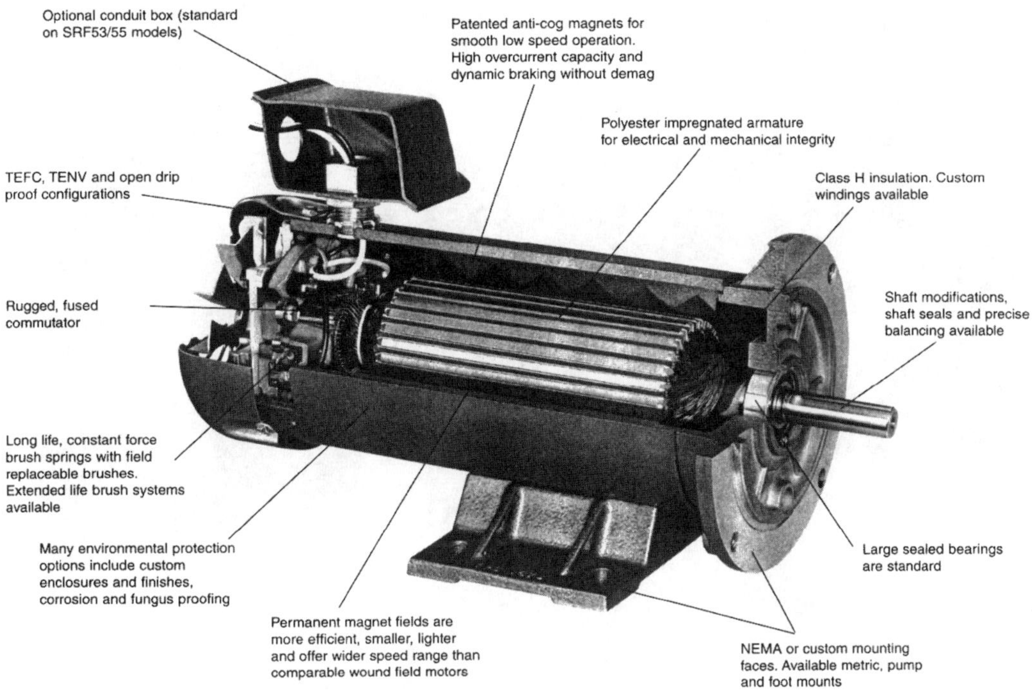

Figure 5.2 Cutaway view of a general-purpose permanent-magnet DC machine. (Courtesy of Pacific Scientific.)

$$E_{a\,av} = \frac{1}{2\pi}\int_0^{2\pi} \frac{d\psi_p(\theta_r)}{dt}\, d\theta_r$$

Then, we have

$$E_a = \frac{2PN_a\psi_p\omega_r}{a} = L_{af}i_f\omega_r \tag{5.1}$$

where ψ_p is the magnetic flux per pole, $\psi_p = \frac{N_f i_f}{\Re_m}$; N_a is the number of turns; and a is the number of parallel current paths in the armature winding.

From Kirchhoff's law, one obtains the following steady-state equation for the armature voltage for electric motors (the armature current opposes the induced *electromotive force*)

$$u_a - E_a = i_a r_a$$

For generators, the armature current is in the same direction as the generated *electromotive force*, and we have

$$u_a - E_a = -i_a r_a$$

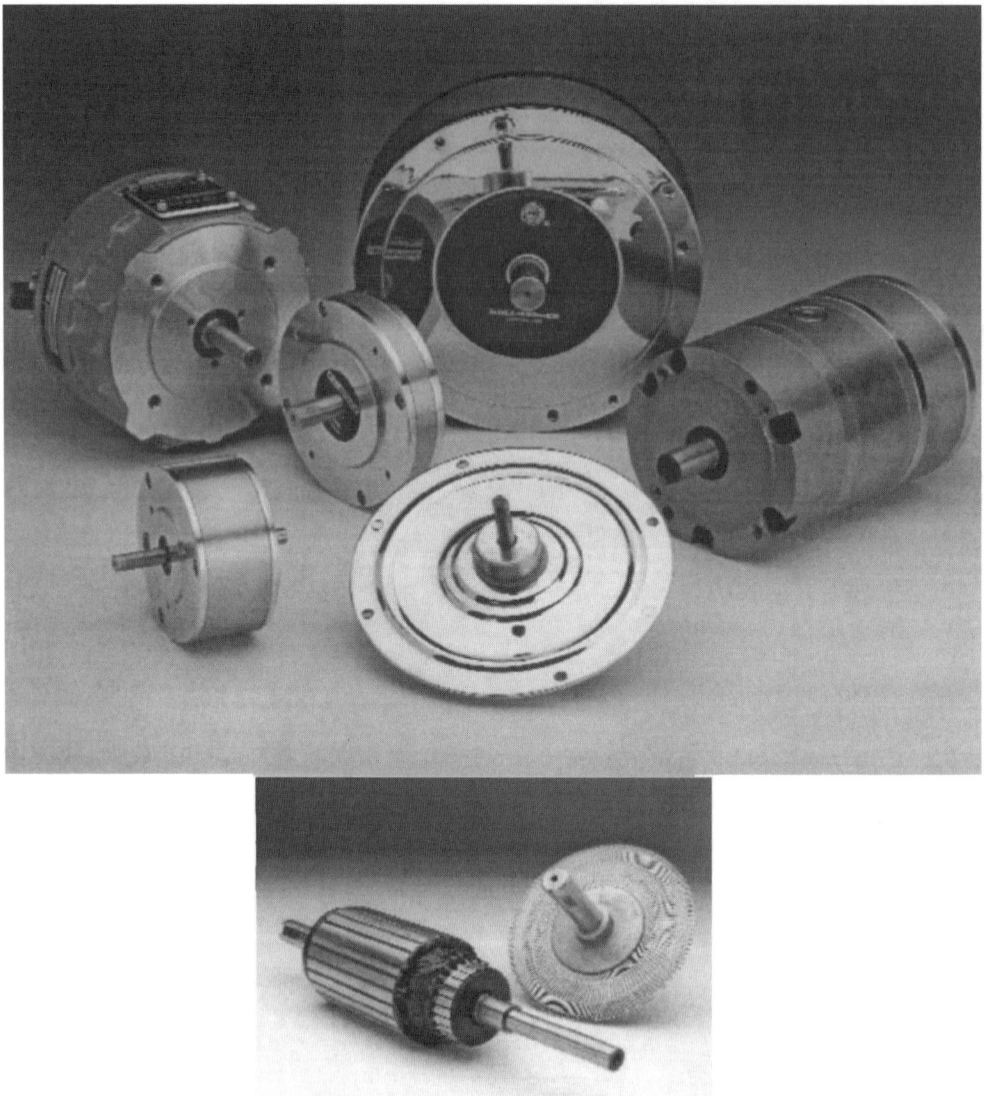

Figure 5.3 Cutaway view of a permanent-magnet servo-motor with gear and attached optical encoder. (Courtesy of Maxon Precision Motors, Inc.)

The difference between the applied voltage and the induced *electromotive force* is the voltage drop across the internal armature resistance r_a. One concludes that electric machines rotate at an angular velocity at which the *electromotive force* generated in the armature winding balances the armature voltage. If an electric machine operates as a motor, the induced *electromotive force* is less than the voltage applied to the windings. If an electric machine operates as a generator, the generated (induced) *electromotive force* is greater than the terminal voltage.

Kirchhoff's second law is extensively used in the development of mathematical models of electric machines. The equivalence of results derived by applying the Lagrange concept and Kirchhoff's law has already been shown. To demonstrate an indispensable link between these different avenues, recall that using the Lagrange equations of motion, the mathematical model was built for a simple, but quite general, electromechanical motion device. In particular, the doubly excited transducer was thoroughly studied in Chapter 2.

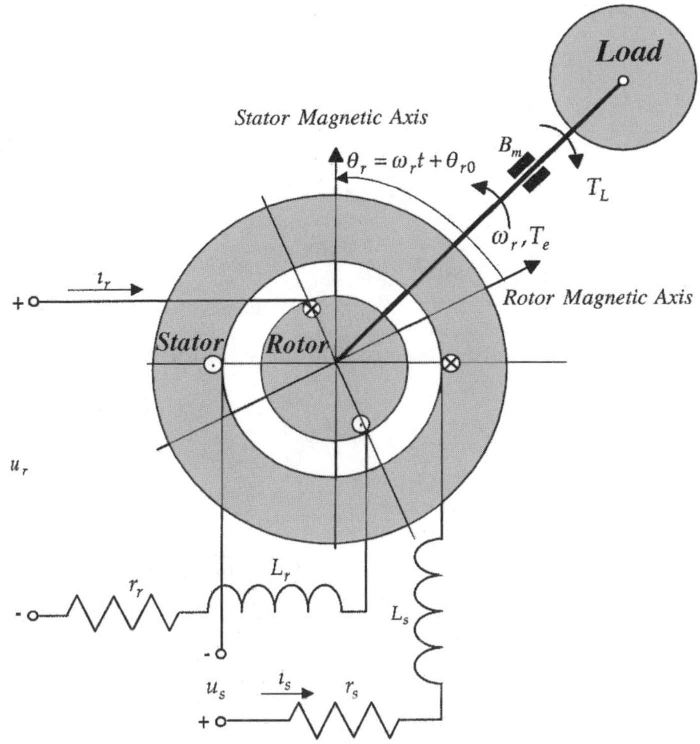

Figure 5.4 Transducer with stator and rotor windings.

5.2 Doubly excited transducer and introduction to fundamentals of alternating-current and direct-current electric machines

Transducers are devices that convert energy. As an important example in the analysis of rotating electric machines, consider a transducer with two independently excited stator and rotor windings (see Example 2.9). Figure 5.4 shows the configuration for the doubly excited transducer studied.

It was demonstrated in Example 2.9 that by using the electric charges in the stator and rotor windings and the angular rotor displacement as the generalized coordinates q_1, q_2, and q_3 $\left(q_1 = \frac{i_s}{s},\ q_2 = \frac{i_r}{s},\ q_3 = \theta_r,\ \dot{q}_1 = i_s,\ \dot{q}_2 = i_r,\ \dot{q}_3 = \omega_r\right)$, and assigning the applied voltages to the stator and rotor windings and the load torque to be the generalized forces Q_1, Q_2, and Q_3 ($Q_1 = u_s$, $Q_2 = u_r$, and $Q_3 = -T_L$), the Lagrange equations of motion

$$\frac{d}{dt}\left(\frac{\partial \Gamma}{\partial \dot{q}_1}\right) - \frac{\partial \Gamma}{\partial q_1} + \frac{\partial D}{\partial \dot{q}_1} + \frac{\partial \Pi}{\partial q_1} = Q_1$$

$$\frac{d}{dt}\left(\frac{\partial \Gamma}{\partial \dot{q}_2}\right) - \frac{\partial \Gamma}{\partial q_2} + \frac{\partial D}{\partial \dot{q}_2} + \frac{\partial \Pi}{\partial q_2} = Q_2$$

$$\frac{d}{dt}\left(\frac{\partial \Gamma}{\partial \dot{q}_3}\right) - \frac{\partial \Gamma}{\partial q_3} + \frac{\partial D}{\partial \dot{q}_3} + \frac{\partial \Pi}{\partial q_3} = Q_3$$

are straightforwardly applied to find the resulting differential equations as given by

$$L_s \frac{di_s}{dt} + L_M \cos\theta_r \frac{di_r}{dt} - L_M i_r \sin\theta_r \frac{d\theta_r}{dt} + r_s i_s = u_s$$

$$L_r \frac{di_r}{dt} + L_M \cos\theta_r \frac{di_s}{dt} - L_M i_s \sin\theta_r \frac{d\theta_r}{dt} + r_r i_r = u_r$$

$$J \frac{d^2\theta_r}{dt^2} + L_M i_s i_r \sin\theta_r + B_m \frac{d\theta_r}{dt} = -T_L$$

Using the rotor angular velocity ω_r, the mathematical model in Cauchy's form is found to be

$$\frac{di_s}{dt} = \frac{1}{L_s L_r - L_M^2 \cos^2\theta_r}$$
$$\left(-r_s L_r i_s - \frac{1}{2} L_M^2 i_s \omega_r \sin 2\theta_r + r_r L_M i_r \cos\theta_r + L_r L_M i_r \omega_r \sin\theta_r + L_r u_s - L_M \cos\theta_r u_r\right)$$

$$\frac{di_r}{dt} = \frac{1}{L_s L_r - L_M^2 \cos^2\theta_r}$$
$$\left(r_s L_M i_s \cos\theta_r + L_s L_M i_s \omega_r \sin\theta_r - r_r L_s i_r - \frac{1}{2} L_M^2 i_r \omega_r \sin 2\theta_r - L_M \cos\theta_r u_s + L_s u_r\right) \tag{5.2}$$

$$\frac{d\omega_r}{dt} = \frac{1}{J}(-L_M i_s i_r \sin\theta_r - B_m \omega_r - T_L)$$

$$\frac{d\theta_r}{dt} = \omega_r$$

We have omitted a lot of details in deviations of the transducer dynamics that were given in Example 2.9.

Taking note of

$$\frac{d\omega_r}{dt} = \frac{1}{J}(-L_M i_s i_r \sin\theta_r - B_m \omega_r - T_L)$$

one immediately obtains the expression for the electromagnetic torque T_e developed. In particular, we have

$$T_e = -L_M i_s i_r \sin\theta_r \tag{5.3}$$

This equation allows one to calculate the developed electromagnetic torque, which is a function of the rotor angular displacement as well as the stator and rotor currents.

It is of great importance to note that the model developed is valid for single-phase AC and DC electric machines, and the corresponding mathematical models can be developed. There are several interesting features in the analysis of a doubly excited transducer. One can now study the torque production features. The most important is that the armature and field windings can be supplied by AC and DC voltages to guarantee electromagnetic torque development. Clearly, if direct current flows in both stator and rotor windings, the rotor turns from the initial position until the magnetic axis of the rotor comes in line with the magnetic axis of the stator; see equation (5.3) to justify this result (for $i_s = const$ and $i_r = const$, we have $T_e = 0$ as $\theta_r = 0, \pi, 2\pi, \ldots$). However, if an alternating current flows in the rotor or stator winding, the rotor will continue to rotate because an electromagnetic torque is produced by the magnetic fields (assume that $i_s = -\sin\theta_r$ and $i_r = \frac{1}{L_M} = const$, then $T_e = \sin^2\theta_r$). These important features in torque production are implemented in practice using the commutator in DC electric machines and supplying ac voltages to the phase windings of AC machines. The significance of the doubly excited transducer studied is that this electromechanical motion device gives an indispensable example for analysis and understanding of the basic features in operation and torque production of the commutator-type DC electric machines, as well as induction and synchronous machines. Furthermore, the model in (5.2) provides the fundamental equations of motion for rotating single-phase AC machines, as well as allows one to study steady-state and dynamic characteristics.

5.2.1 *Analysis of commutator machines*

Conventional commutator-type DC electric machines differ in several respects from the transducer due to 90 electrical degrees displacement of the stator and rotor magnetic axes. In fact, for commutator DC machines, by using the fact that the armature and field windings produce stationary *magnetomotive forces* that are displaced by 90 electrical degrees, the developed nonlinear model (5.2) and the torque production equation (5.3) should be rewritten. The most important feature is that due to the commutator action, one obtains the following expression for the electromagnetic torque

$$T_e = -L_{af} i_a i_f \sin(-90°) = L_{af} i_a i_f$$

Hence, the electromagnetic torque is developed if the armature and field currents have dc quantities.

Separately excited, shunt-connected, series-connected, compound-connected, and permanent-magnet DC machines will be covered in this chapter. One can demonstrate that the results derived using the Lagrange equations of motion can be applied to find the differential equations of the separately excited DC motor. By applying the differential equations obtained to map the transient dynamics and steady-state operation of the doubly excited transducer (5.2), one easily finds the resulting mathematical model. In particular, for a separately excited machine, which is shown in Figure 5.5, denoting $i_a = i_r$, $i_f = i_s$, $u_a = u_r$, $u_f = u_s$, $r_a = r_r$, $r_f = r_s$, $L_a = L_r$, $L_f = L_s$, and $L_{af} = L_M$, we have

$$\frac{di_a}{dt} = \frac{1}{L_a}(-r_a i_a - L_{af} i_f \omega_r + u_a)$$

$$\frac{di_f}{dt} = \frac{1}{L_f}(-r_f i_f + u_f) \qquad (5.4)$$

$$\frac{d\omega_r}{dt} = \frac{1}{J}(L_{af} i_a i_f - B_m \omega_r - T_L)$$

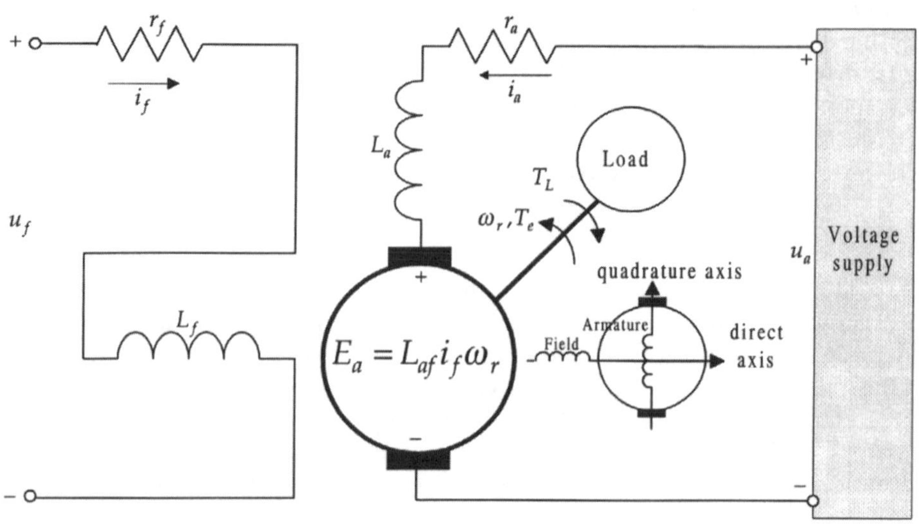

Figure 5.5 Separately excited DC motors.

The *electromotive force* is expressed as $E_a = -L_{af} i_f \omega_r$.

5.2.2 *Analysis of induction machines*

In induction machines, the stator and rotor windings are excited by AC voltages. The current in the stator winding has a *line* frequency, while the current in the rotor winding has a *slip* frequency. In squirrel-cage induction machines, alternating current in the rotor short-circuited winding results due to electromagnetic induction. When the rotor rotates, the induced voltage in the rotor winding depends on the angular velocity of the rotor and the frequency of the voltage applied to the stator winding. The angular velocity of the rotor is found as

$$\omega_r = \omega_e (1 - slip)$$

One can find this relationship using the following formula for the *slip*,

$$slip = \frac{\omega_e - \omega_r}{\omega_e}$$

where the synchronous angular velocity ω_e is found using the frequency of the applied voltage f and the number of poles P as

$$\omega_e = \frac{4\pi f}{P}$$

In motor operation, the angular velocity of the rotor is less than the synchronous angular velocity; that is, $\omega_e > \omega_r$ and $0 < slip < 1$. For generator action, the *slip* is negative because $\omega_e < \omega_r$. The current flowing in the stator winding produces the air gap *magnetomotive force*, which rotates at the synchronous angular velocity ω_e relative to the stator. The induced voltage generates alternating current in the rotor winding, and the resulting rotor magnetomotive force rotates at $(\omega_e - \omega_r)$, as observed from the rotor. The electromagnetic torque is produced by these two magnetomotive forces. Furthermore, it appears that the electromagnetic torque is zero if the rotor rotates at the synchronous angular velocity because the current in the rotor winding is not induced.

The equations of motion for a single-phase induction motor are given by (5.2). In viewing formula (5.3), one concludes that single-phase squirrel-cage induction motors (the rotor windings are short-circuited) do not develop a starting electromagnetic torque because no rotor current is induced when $\omega_r = 0$; that is, $i_r = 0$ if $\omega_r = 0$. To start single-phase induction motors, an auxiliary winding is used. Observe that for squirrel-cage induction machines, $u_r = 0$. The comprehensive analysis and model developments for two- and three-phase induction machines is given in Chapter 6.

5.2.3 *Analysis of synchronous machines*

Using a doubly excited transducer, we will briefly discuss the basic principles of operation of synchronous machines. The goal is to study a case when one feeds a direct current to one winding and alternating current to the other. In particular, it is assumed that the field winding on the rotor is excited by the direct current supplied by means of carbon brushes bearing on slip rings. As the rotor rotates, the induced *electromotive force* appears. This generated voltage is a function of the rotor angular velocity and number of poles. The angular frequency of the induced voltage is $\omega_e = \frac{1}{2}P\omega_{rm}$, where ω_{rm} denotes the mechanical angular velocity of the rotor. The direct current in the rotor winding creates a magnetic flux in the air gap that rotates at the synchronous angular velocity. The electromagnetic torque is developed due to the interaction of the rotating air gap magnetomotive force, produced by alternating current flowing in the armature winding on the stator, and the rotor air gap magnetomotive force. In particular, the electromagnetic torque results from the tendency of two magnetomotive forces to align. Hence, in the steady-state, the air gap magnetomotive force produced by the armature (stator) winding, rotates with synchronous angular velocity. The magnetic field, created by the current flowing in the armature (stator) winding, rotates at the same angular velocity as the magnetic field created by the current flowing in the field (rotor) winding. Here, one assumes that the stator windings are connected to a balanced voltage set, displaced by 180 and 120 electrical degrees for two- and three-phase synchronous machines, respectively.

A doubly excited transducer is a viable example to study synchronous machines. A single-phase synchronous machine has one winding on the stator (armature winding) and another winding on the rotor (field winding). The field winding is excited by the direct current, while the armature winding is excited by the alternating current. Using the results derived for a transducer, one can show that alternating and direct currents must be supplied to the stator and rotor windings to develop the electromagnetic torque. The goal is to find the expression for T_e if the armature current is a sinusoidal function of the rotor angular displacement θ_r and the dc field current is fed to the rotor winding; that is,

$$i_s = -I_s \sin\theta_r, \text{ and } i_r = I_r = const$$

From equation (5.3), one obtains

$$T_{eav} = \frac{L_M I_s I_r}{2\pi} \int_0^{2\pi} \sin^2\theta_r d\theta_r = \frac{L_M I_s I_r}{2\pi} \left[\frac{1}{2}\theta_r - \frac{1}{4}\sin 2\theta_r \right]_0^{-2\pi} = \frac{L_M I_s I_r}{2}$$

This result leads one to an important conclusion: the electromagnetic torque is developed only if the applied voltage to the stator winding is a function of the rotor angular displacement.

Single-phase synchronous machines are mapped by nonlinear differential equations (5.2). It is worth mentioning that the differential equations for a variety of three-phase synchronous machines are derived in Chapter 7, based on the Lagrange concept and Kirchhoff's law.

5.3 Separately excited direct-current machines

Conventional separately excited DC electric machines have stator and rotor windings excited by direct currents. The rotor has a commutator, and the dc voltage to the armature winding is supplied through the brushes, which establish electric contact with the commutator. The brushes are fixed with respect to the stator, and they are placed in the specified angular displacement to produce the maximum electromagnetic torque. In particular, to maximize the electromagnetic torque, the stator and rotor magnetic axes are displaced by 90 electrical degrees using a commutator. Expression (5.3) for the electromagnetic torque $T_e = -L_{sr}i_si_r \sin\theta_r$ indicates that T_e can be developed only if there exists a displacement between the armature (rotor) and field (stator) magnetic axes.

An electric machine can be either a motor or a generator, depending on whether it drives a load or it is driven by a prime mover. The direction of the armature current is reversed when an electric machine changes from motor to generator operation. However, there is no change of line voltage polarity, and the direction of rotation and field current are the same. These facts are essential, and Kirchhoff's voltage law can be applied to study the steady-state operation. In particular, by using the Kirchhoff law, one obtains

$$i_a = \frac{u_a - E_a}{r_a}$$

If u_a is greater than E_a, the armature current is positive, and the machine operates as a motor. An electric machine operates as a generator if E_a is greater than u_a, and i_a is negative. Schematic diagrams of separately excited motors and generators are shown in Figures 5.6 and 5.7.

Using Kirchhoff's second law, one can find the armature and field voltage equations to study the transient dynamics of electric motors. Assuming that $r_{ar} = r_{fr} = 0$, one obtains the following differential equations for a motor (shown in Figure 5.6).

$$\begin{aligned} u_a - L_{af}i_f\omega_r &= r_ai_a + \frac{d\psi_a}{dt} = r_ai_a + L_a\frac{di_a}{dt} \\ u_f &= r_fi_f + \frac{d\psi_f}{dt} = r_fi_f + L_f\frac{di_f}{dt} \end{aligned} \tag{5.5}$$

A generator is driven by a prime mover that rotates with angular velocity ω_{pm}. When an electric machine is used as a generator, the output quantity of interest is the voltage induced, while in the motor application, the output is the angular velocity. When the electric machine studied is operated as a generator, the direction of the armature current is reversed compared with the motor operation because $E_a > u_a$. Then, we have the following differential equations of motion for stator and rotor circuitry

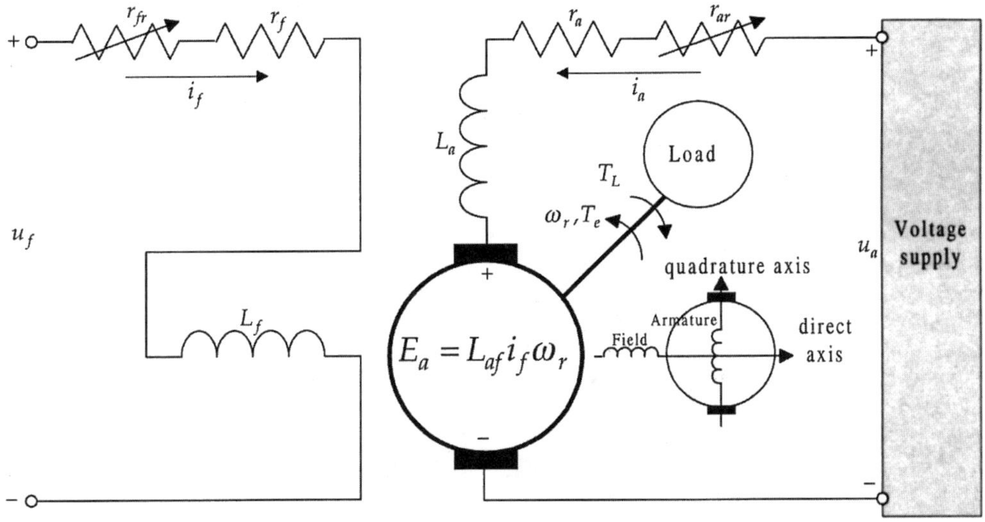

Figure 5.6 Equivalent circuit for separately excited DC motors.

$$\begin{aligned} u_a - L_{af} i_f \omega_r &= -r_a i_a - L_a \frac{di_a}{dt} \\ u_f &= r_f i_f + L_f \frac{di_f}{dt} \end{aligned} \tag{5.6}$$

The steady-state operating conditions for a generator are

$$u_a - L_{af} i_f \omega_r = -r_a i_a$$
$$u_f = r_f i_f$$

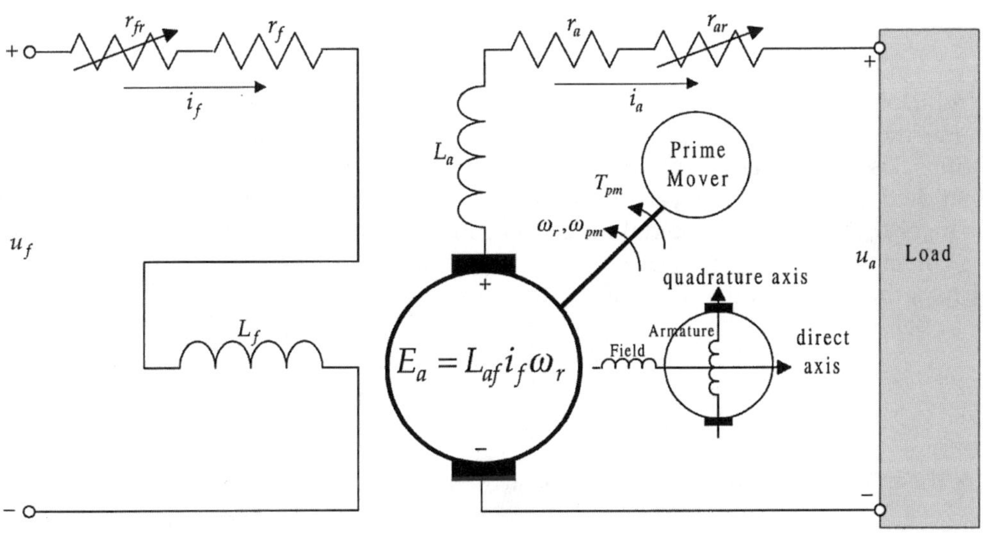

Figure 5.7 Equivalent circuit for separately excited DC generators.

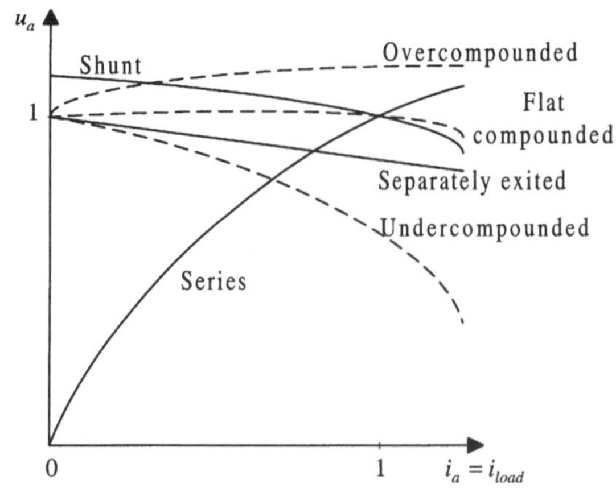

Figure 5.8 Volt-ampere characteristics of *self-* and *separately* excited generators.

Equation $u_a - L_{af}i_f\omega_r = -r_ai_a$ gives the relationship between the generated (induced) *electromotive force* E_a and the voltage at the load terminal u_a. One concludes that the terminal voltage is equal to the induced *electromotive force* minus the voltage drop in the armature resistance, and $u_a = L_{af}i_f\omega_r - r_ai_a$. Hence, for separately excited DC generators, the generated voltage is a function of the angular velocity and the currents flowing in the field winding. It is important to emphasize that, in general, generators can be classified as *self-* (shunt- and compound-connected) and *separately* excited. Assuming that the angular velocity is constant, the steady-state per-unit volt-ampere characteristics of *self-* and *separately* excited generators are shown in Figure 5.8.

Now one can focus on separately excited electric motors to study the steady-state quantities. To find the steady-state torque-speed characteristics, one should derive the expression for the electromagnetic torque developed. This was accomplished in Chapter 2 using the coenergy concept. Assuming that the magnetic system is linear, the coenergy is found as

$$W_C(i_a, i_f, \theta_r) = \frac{1}{2}L_ai_a^2 + L_{sr}(\theta_r)i_ai_f + \frac{1}{2}L_fi_f^2$$

Inductances L_a and L_f are independent on the angular displacement θ_r. We should examine the mutual inductance $L_{sr}(\theta_r)$ between the stator (field) and rotor (armature) windings. Using the results obtained from Example 2.9 where a doubly excited transducer was studied, one has $L_{sr} = L_M \cos\theta_r = L_{af}\cos\theta_r$. Recall that the mutual inductance of the transducer $L_{sr}(\theta_r)$ is a cosinusoidal function of the rotor displacement because the magnetizing reluctance is a function of θ_r.

Then, the electromagnetic torque is found to be

$$T_e = \frac{\partial W_C(i_a, i_f, \theta_r)}{\partial \theta_r} = -L_{af}i_ai_f \sin\theta_r$$

where the mutual inductance between the armature and field windings L_{af} is expressed by

$$L_{af} = L_M = L_{sr\ \max} = \frac{N_aN_f}{\Re_m(90°)}$$

It should be emphasized that the same formula is derived for the electromagnetic torque developed using the Lagrange equations of motion (see equation (5.3)).

In conventional commutator DC machines, the *quadrature* (armature) and *direct* (field) magnetic axes are displaced by 90 electrical degrees and, thus, the magnetizing reluctance $\Re_m(90°)$ is constant. Therefore, the mutual inductance is maximized, and L_{af} is constant. Hence, using the fact that the *quadrature* and *direct* magnetic axes are displaced by 90 electrical degrees, we have

$$T_e = L_{af} i_a i_f \tag{5.7}$$

This equation for the electromagnetic torque can be verified by studying the balance of energy. The electromagnetic power developed is

$$P_{em} = E_a i_a$$

and the mechanical energy is expressed as

$$P_{mec} = T_e \omega_r$$

Therefore, the consideration of the power balance indicates that

$$T_e = \frac{E_a i_a}{\omega_r}$$

Using the formula for the *electromotive force*, which is given as $E_a = L_{af} i_f \omega_r$, equation (5.7) for T_e results. That is, it is found that $T_e = L_{af} i_a i_f$.

From (5.7) and setting the current derivatives to zero in (5.5), one obtains an explicit expression for the angular velocity as a function of the electromagnetic torque. In particular, using $u_a - L_{af} i_f \omega_r = r_a i_a$ and $T_e = L_{af} i_a i_f$, the torque-speed characteristic curves are given by

$$\omega_r = \frac{u_a - r_a i_a}{L_{af} i_f} = \frac{u_a}{L_{af} i_f} - \frac{r_a}{(L_{af} i_f)^2} T_e \tag{5.8}$$

This equation describes the steady-state relationship between the angular velocity and the electromagnetic torque.

Analysis of (5.8) indicates that the angular velocity of the separately excited motor can be regulated by changing the applied voltages to the armature u_a and field u_f windings. One should distinguish the armature voltage control and field weakening regulation. It is clear that the angular velocity of the motor can be controlled by varying the applied armature and field voltages. The flux is a function of the field current in the stator winding, and higher angular velocity can be achieved by field weakening by reducing the stator current i_f (see equation (5.8)). However, there exists a mechanical limit imposed on the maximum angular velocity. The maximum allowed (rated) armature current is specified as well, and using the equation $T_e = L_{af} i_a i_f$, one concludes that the electromagnetic torque is bounded. Electric machines operate in four quadrants of the torque-speed plane. Figure 5.9 illustrates motoring and generating regions. To reverse the direction of rotation, the armature or field current must be reversed by changing the polarity of the voltage applied to the rotor or stator winding, but not both. Usually, the field current is less than the

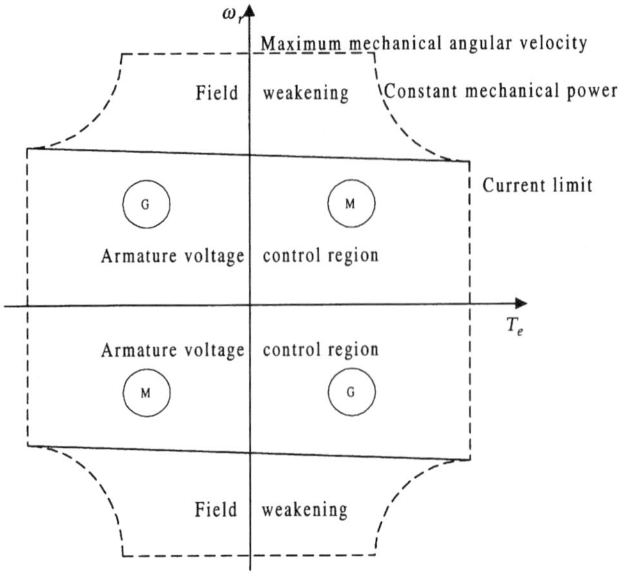

Figure 5.9 Typical operating regions of separately excited DC machines.

armature current. Therefore, it is easier to change the direction of the magnetic flux by changing the polarity of the applied voltage to the field winding.

If a separately excited machine is used as a motor, a family of torque-speed characteristic curves can be plotted, as shown in Figure 5.10.

The load torque, which is usually a function of the angular velocity $T_L = f(\omega_r)$, can be estimated. In most cases, it is possible to define the maximum load conditions to select a

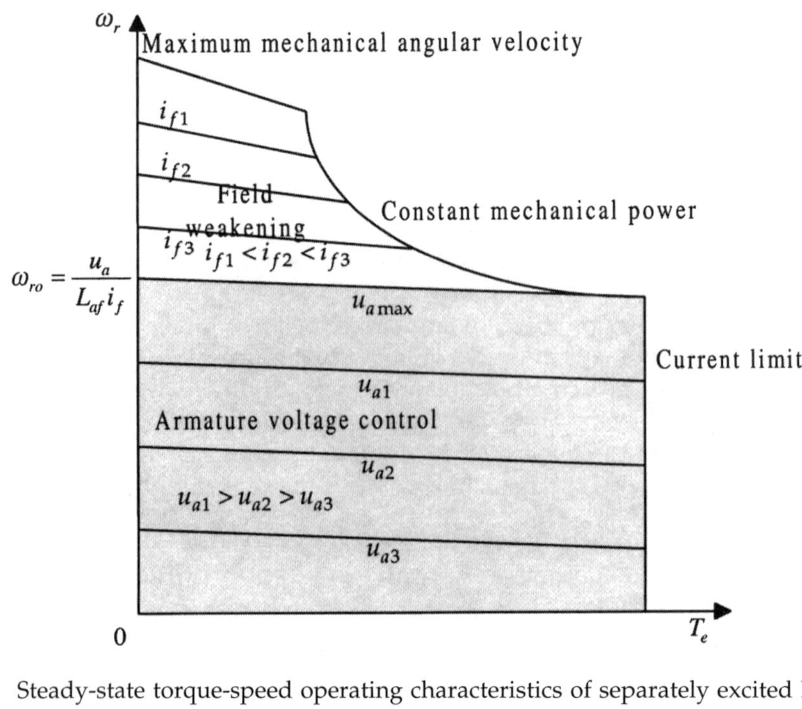

Figure 5.10 Steady-state torque-speed operating characteristics of separately excited DC motors.

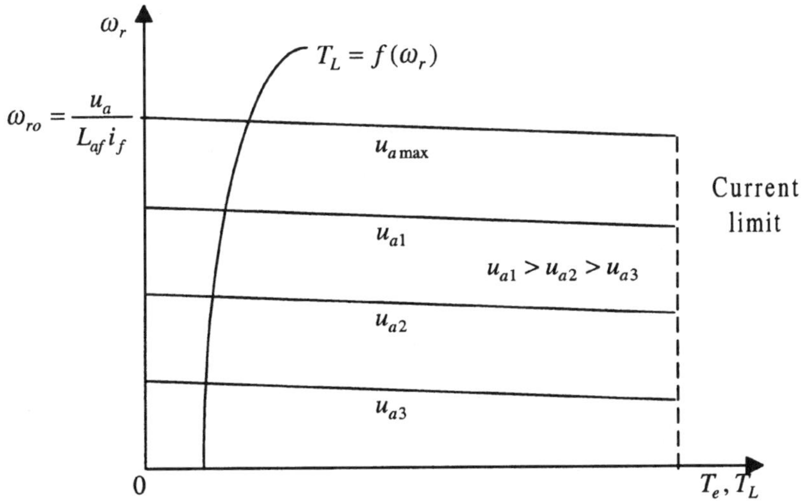

Figure 5.11 Torque-speed characteristic curves for $i_f = const$ and the load characteristic.

motor by analyzing the rated (allowed) continuous and peak electromagnetic torques. In the steady-state, we have $T_e = T_L$. Hence, the intersection of a torque-speed characteristic curve with a load characteristic gives an operating point, and the value of ω_r results. The field weakening regulation leads one to a limited developed electromagnetic torque. This drawback restricts the application of the field weakening concept in speed control. By changing the supplied armature voltage u_a, and keeping the stator voltage at the rated value ($i_f = const$), a family of torque-speed characteristic curves for a separately excited motor controlled by varying the armature voltage can be drawn (see Figure 5.11).

In many applications, small electric motors (up to a few kilowatts) can be started by applying the rated armature and field voltages. If one attempts to reduce the starting current, which depends on the voltages applied to the motor terminals and the load torque, *rheostat* resistances r_{ar} and r_{fr} can be inserted if the armature and field voltages cannot be regulated (it is assumed that the motor is connected directly to unregulated voltage-fed converters). These *rheostat* resistances are inserted into the armature and field circuits, and using Kirchhoff's law, one obtains the following differential equations for armature and field circuits

$$u_a - L_{af} i_f \omega_r = (r_a + r_{ar}) i_a + L_a \frac{di_a}{dt}$$

$$u_f = (r_f + r_{fr}) i_f + L_f \frac{di_f}{dt}$$

The steady-state relationship between the angular velocity and electromagnetic torque is given by

$$\omega_r = \frac{u_a}{L_{af} i_f} - \frac{r_a + r_{ar}}{(L_{af} i_f)^2} T_e$$

It is concluded that the angular velocity can be regulated by inserting the rheostats in the armature and field circuits; a spectrum of the torque-speed characteristic curves is

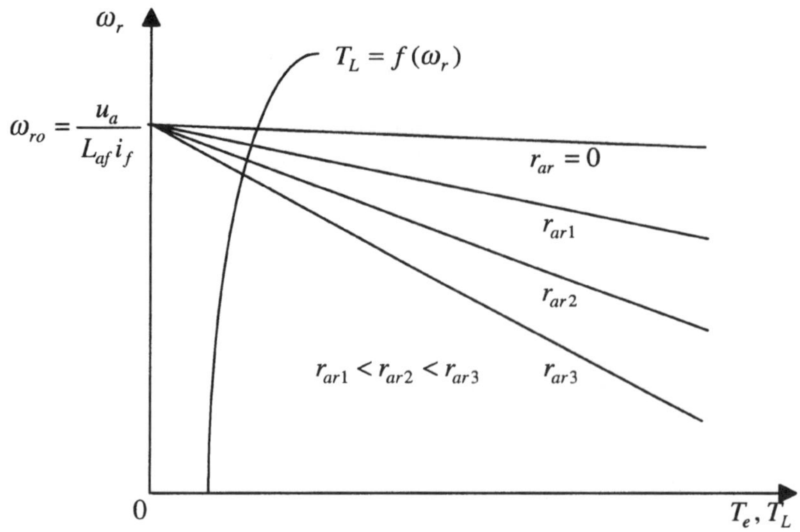

Figure 5.12 Steady-state torque-speed characteristics when the armature rheostat r_{ar} is inserted.

plotted in Figure 5.12 assuming that the field current is constant. However, this simple method is extremely inefficient, and has limited practical value. After the motor is started, these additional *rheostat* resistances must be removed (disconnected) to avoid heat losses. This leads to additional circuitry. Furthermore, by inserting r_{ar} and r_{fr}, one reduces the starting electromagnetic torque. This is a serious drawback in achieving good performance. Therefore, this concept, in many cases, is unacceptable.

Let us analyze the separately excited electric machine dynamics. Using the circuitry dynamics, as given by (5.5), and taking note of the *torsional-mechanical* equation, which is obtained using Newton's second law,

$$\frac{d\omega_r}{dt} = \frac{1}{J}(T_e - B_m\omega_r - T_L)$$

we derive three highly coupled nonlinear differential equations. In particular, separately excited DC motors are modeled using

$$\begin{aligned}
\frac{di_a}{dt} &= -\frac{r_a}{L_a}i_a - \frac{L_{af}}{L_a}i_f\omega_r + \frac{1}{L_a}u_a \\
\frac{di_f}{dt} &= -\frac{r_f}{L_f}i_f + \frac{1}{L_f}u_f \\
\frac{d\omega_r}{dt} &= \frac{L_{af}}{J}i_a i_f - \frac{B_m}{J}\omega_r - \frac{1}{J}T_L
\end{aligned} \tag{5.9}$$

Differential equations (5.9), which were found using Kirchhoff's voltage law and Newton's second law of motion, as well as the Lagrange equations of motion, provide an augmented electromechanical model for separately excited DC motors. It is evident that the developed model in (5.9) matches the derived equations of motion (5.4). These non-

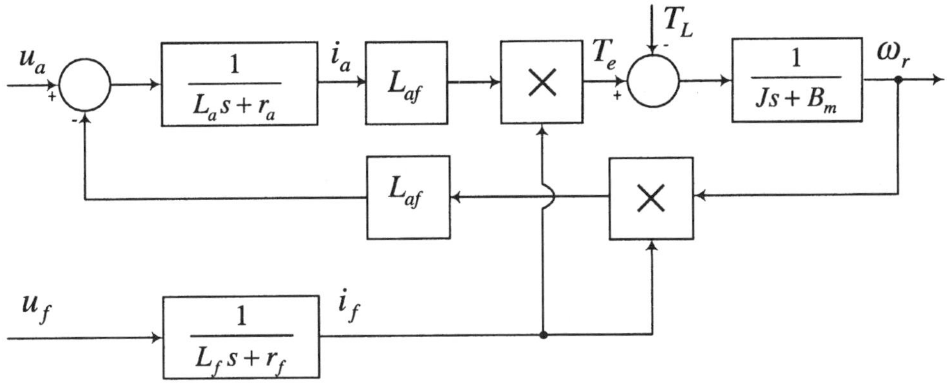

Figure 5.13 Block diagram of separately excited DC motors.

linear differential equations cannot be linearized, and the designer must employ the complete nonlinear dynamics, as given by (5.9). An *s*-domain block diagram of separately excited DC motors is shown in Figure 5.13.

To find the equations of motion for generator, the circuitry equations (5.6) are used. Taking note of Newton's second law, which is given by

$$-T_e - B_m\omega_r + T_{pm} = J\frac{d\omega_r}{dt}$$

one obtains a set of nonlinear differential equations to map the generator dynamics

$$\frac{di_a}{dt} = -\frac{r_a}{L_a}i_a + \frac{L_{af}}{L_a}i_f\omega_r - \frac{1}{L_a}u_a$$

$$\frac{di_f}{dt} = -\frac{r_f}{L_f}i_f + \frac{1}{L_f}u_f \tag{5.10}$$

$$\frac{d\omega_r}{dt} = -\frac{1}{J}L_{af}i_a i_f - \frac{B_m}{J}\omega_r + \frac{1}{J}T_{pm}$$

It should be emphasized that the expression for the voltage at the load terminal u_a must be used. For example, for the resistive load with R_L, we have $u_a = R_L i_a$.

Example 5.1

A separately excited, 2-kW DC motor with rated armature current 20 A and angular velocity 200 rad-s^{-1} operates at the constant voltages u_a = 100 V and u_f = 20 V. The motor parameters are: r_a = 0.18 ohm, r_f = 3.5 ohm, L_{af} = 0.1 H, and B_m = 0.007 N-m-s-rad^{-1}. Calculate the steady-state angular velocity and armature current at the minimum and maximum load conditions, $T_{L\,\min}$ = 0 N-m and $T_{L\,\max}$ = 10 N-m. Analyzing the rated motor data and load conditions, make conclusions regarding the application of this motor based upon the steady-state and dynamic analysis.

Solution

The angular velocity should be calculated using (5.8). We have

$$\omega_r = \frac{u_a}{L_{af}i_f} - \frac{r_a}{(L_{af}i_f)^2}T_e$$

For steady-state operation, $T_e = T_L$.

Hence, having obtained the field current as $i_f = \frac{u_f}{r_f} = 5.7$ A, one can find the angular velocity. Using the viscous friction coefficient, for $T_{L\,\min} = 0$ N-m, we have

$$\omega_r = \frac{100}{0.1\times 5.7} - \frac{0.18}{(0.1\times 5.7)^2}0.007\omega_r$$

and $\omega_r = 174.7$ rad-s^{-1}. For $T_{L\,\max} = 10$ N-m, one obtains

$$\omega_r = \frac{100}{0.1\times 5.7} - \frac{0.18}{(0.1\times 5.7)^2}(10 + 0.007\omega_r)$$

Hence, $\omega_r = 169.2$ rad-s^{-1}.

Using equation (5.7) $T_e = L_{af}i_a i_f$, the armature current at the steady-state can be calculated as

$$i_a = \frac{T_e}{L_{af}i_f} = \frac{T_L + B_m\omega_r}{L_{af}i_f}$$

Then, if $T_{L\,\min} = 0$ N-m, we have $i_a = 2.15$ A.

For $T_{L\,\max} = 10$ N-m, one obtains $i_a = 19.6$ A. Hence, the calculated angular velocities and armature currents conform with the rated motor data, and one might tentatively conclude that the motor can be used for the assigned application.

The basic steady-state analysis of achieving preliminary results with the least expenditure might be used. But in the last few years, the evaluation of results has changed. To view transient behavior, incorporate the analysis of dynamics, and capture all essential features, comprehensive transient analysis must be performed. To ensure the reliability of results, one should study the dynamics. For the motor considered, the self-inductances are $L_a = 0.0062$ H and $L_f = 0.0095$ H, while the equivalent moment of inertia is $J = 0.04$ kg-m^2.

The MATLAB environment provides a great opportunity to implement the advanced modeling software. Using differential equations (5.9), the motor dynamics are studied, and two MATLAB m-files are:

MATLAB script 1

```
% The transient dynamics of a separately excited direct-current motor
t0=0; tfinal=0.4; tol=1e-7; trace=1e-7; y0=[0 0 0]';
[t,y]=ode45('script2',t0,tfinal,y0,tol,trace);
subplot(2,3,1); plot(t,y(:,1),'-');
xlabel('Time (second)'); title('Armature current, ia [A]'); pause;
subplot(2,3,2); plot(t,y(:,2),'-');
xlabel('Time (second)'); title('Field current, if [A]'); pause;
```

```
subplot(2,3,3); plot(t,y(:,3),'-');
xlabel('Time (second)'); title('Angular velocity, wr [rad/sec]'); pause;
clear; disp('End');
```

MATLAB script 2

```
% Transient dynamics of a separately excited direct-current motor
function yprime=difer(t,y);
% Motor parameters
ra=0.18; rf=3.5; La=0.0062; Lf=0.0095; Laf=0.1; J=0.04; Bm=0.007;
% Load torques are 0 and 10
Tl=0;
% Rated armature and field applied voltages
ua=100; uf=20;
% Third-order differential equations for a separately excited motor
yprime=[(-ra*y(1,:)-Laf*y(2,:)*y(3,:)+ua)/La;...
(-rf*y(2,:)+uf)/Lf;...
(Laf*y(1,:)*y(2,:)-Bm*y(3,:)-Tl)/J];
```

The transient dynamics for the motor states are calculated if the motor, which is loaded by $T_{L\,\min} = 0$ N-m and $T_{L\,\max} = 10$ N-m, starts from stall (see Figure 5.14).

SIMULINK can be efficiently used to analyze the dynamic behavior. A SIMULINK diagram that allows one to study the nonlinear dynamics of separately excited DC motors is given in Figure 5.15. It should be emphasized that the developed SIMULINK diagram is built using nonlinear differential equations (5.9) and the *s*-domain block diagram documented in Figure 5.13.

The transient dynamics, as well as the waveforms for the electromagnetic and load torques, are illustrated in Figure 5.16 if $T_L = -10$ N-m for $t \in [0 \quad 0.5)$ s and $t \in [1 \quad 1.5]$ s, and $T_L = 10$ N-m for $t \in [0.5 \quad 1)$ s.

One can appreciate the simplicity of nonlinear modeling, which allows us to perform a complete analysis of transient dynamics and study steady-state operation by applying advanced software tools such as MATLAB. In fact, it is understood that nonlinear analysis becomes a necessary ingredient whenever one needs to successfully perform a reasonable spectrum of analysis objectives as benchmarks for reference (e.g., the studied motor cannot be operated with the armature current 270 A; that is, the motor studied cannot be started by applying the rated armature and field voltages). It appears that the nonlinear concept developed ensures the evolution from traditional steady-state viewpoints or linear analysis to an extremely urgent nonlinear transient and steady-state analysis in order to explicitly account for all essential effects and features, as well as to initiate and fulfill the relevant performance adjustments needed.

□

Example 5.2

Plot the torque-speed characteristic curves for a separately excited, 2-kW DC motor if the rated (maximum) armature voltage is $u_{a\max} = 100$ V and the field voltage is $u_f = 20$ V. The motor parameters are: $r_a = 0.18$ ohm, $r_f = 3.5$ ohm, $L_{af} = 0.1$ H, and $B_m = 0.007$ N-m-s-rad^{-1}. Plot the load characteristic if $T_L = T_{L0} + B_m\omega_r$, $T_{L0} = 5$ N-m.

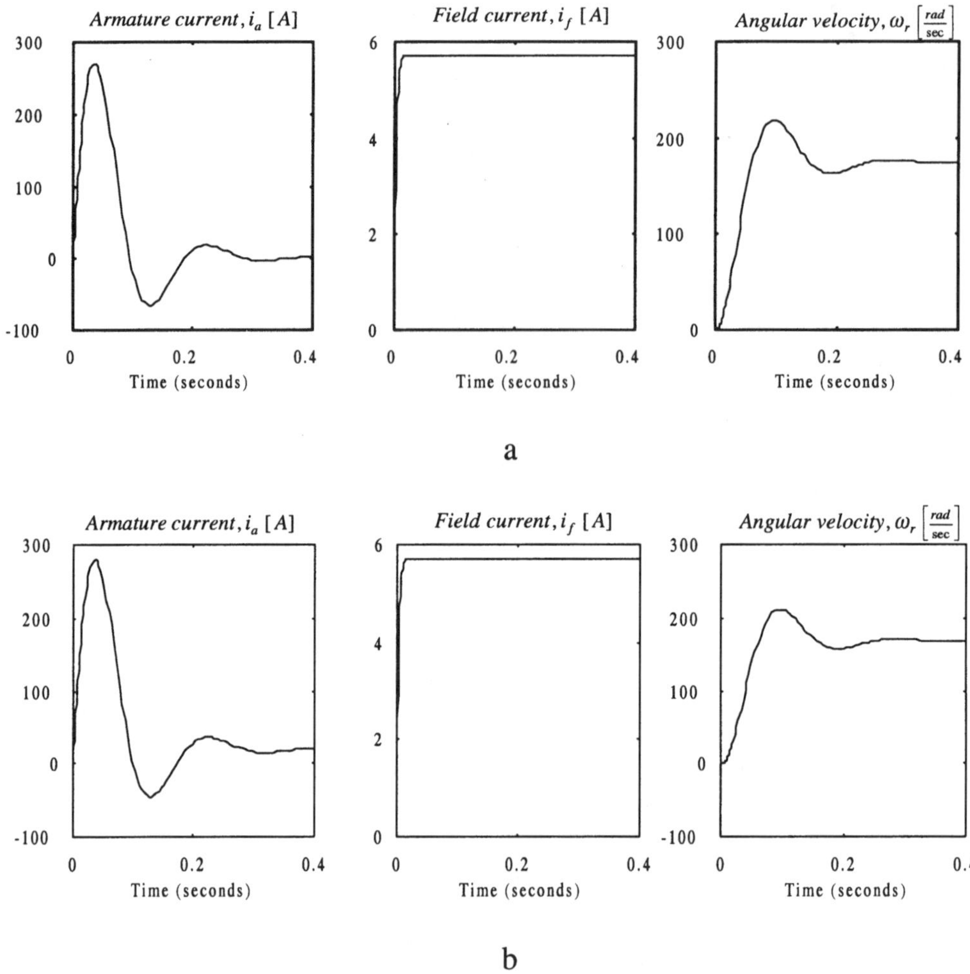

Figure 5.14 Motor dynamics: (a) $T_{L\,min} = 0$ N-m; (b) $T_{L\,max} = 10$ N-m.

Solution

A family of the torque-speed characteristic curves are found by using formula (5.8). Applying

$$\omega_r = \frac{u_a}{L_{af} i_f} - \frac{r_a}{(L_{af} i_f)^2} T_e$$

and having found the field current to be $i_f = \dfrac{u_f}{r_f} = 5.7$ A, the steady-state torque-speed characteristic curves $\omega_r = \Omega_T(T_e)$ are calculated and plotted in Figure 5.17.

The load characteristic $T_L = T_{L0} + B_m \omega_r$ is of particular interest, and the plot for the load torque is given in Figure 5.17. The intersection of the torque-speed and load characteristics gives the values for the motor angular velocity for the applied armature voltage specified. The MATLAB m-file to calculate and plot the steady-state torque-speed and characteristics is given.

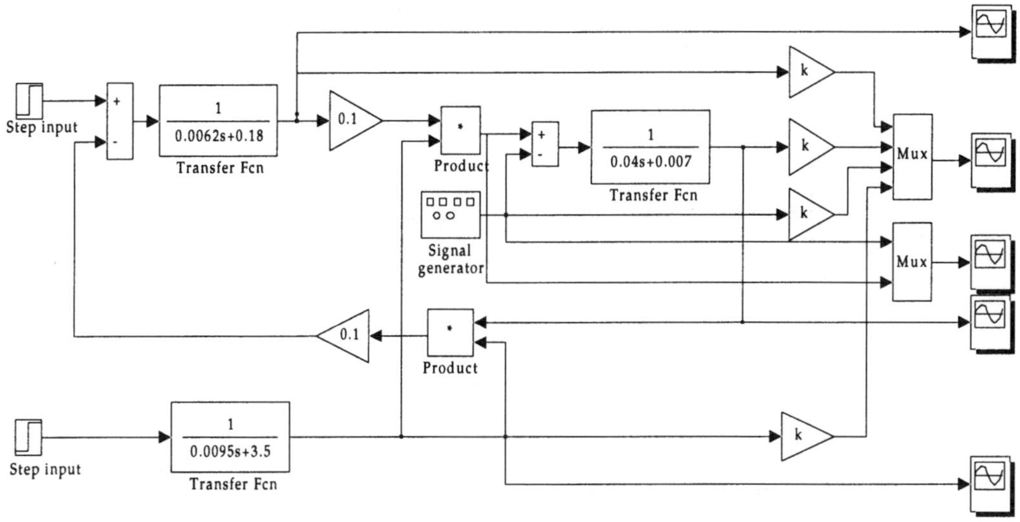

Figure 5.15 SIMULINK diagram of a separately excited DC motor.

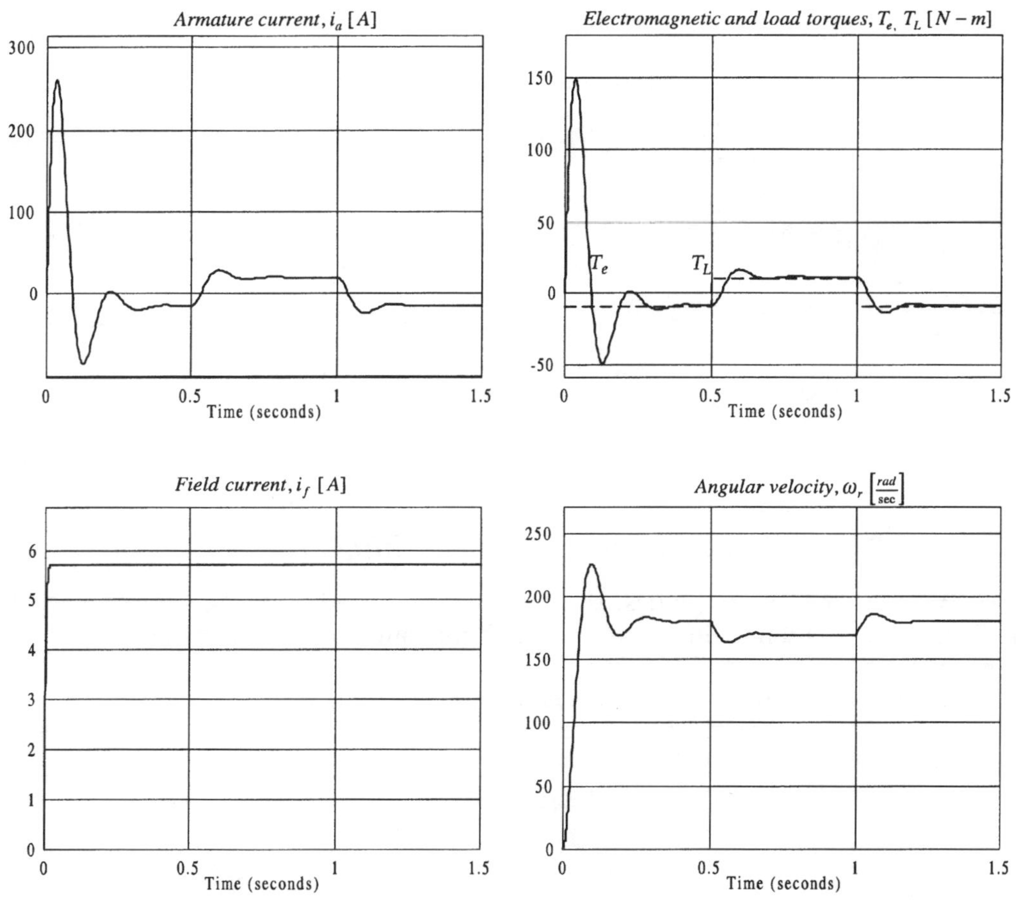

Figure 5.16 Dynamic behavior of a separately excited motor.

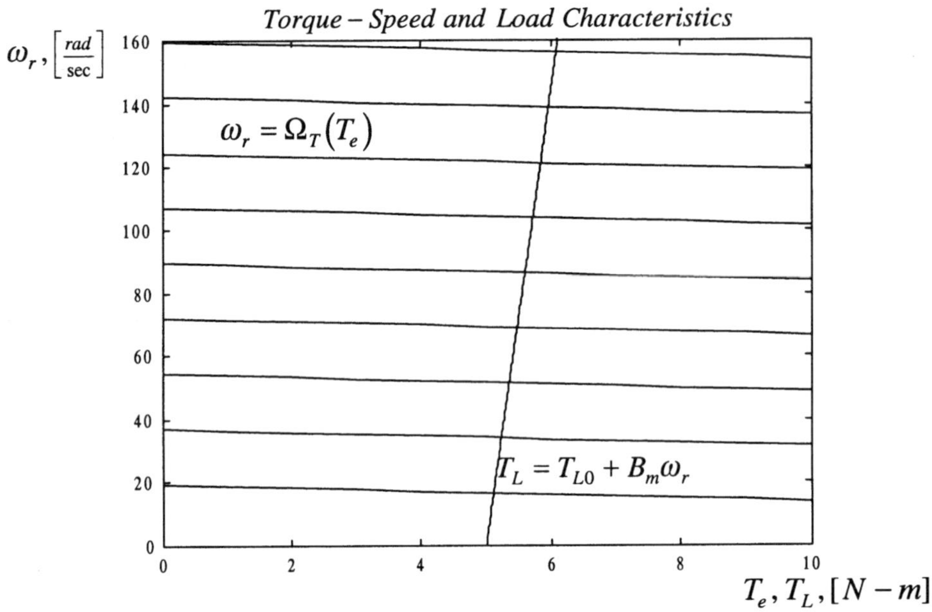

Figure 5.17 Torque-speed and load characteristics for a separately excited motor.

MATLAB script (ch_5_01.m)

```
% parameters of a separately excited motor
ra=0.18; Laf=0.1; Bm=0.007; If=5.7; Tl0=5;
Te=0:1:10;
for ua=11:10:100;
wr=ua/(Laf*If)-(ra/((Laf*If)^2))*Te;
wrl=0:1:200; Tl=Tl0+Bm*wrl;
plot(Te,wr,'-',Tl,wrl,'-');hold on;
axis([0, 10, 0, 160]);
end; disp('End')
```

□

5.4 Shunt-connected direct-current machines

In shunt-connected DC machines, the armature and field windings are connected in parallel. Self-excitation is in place, and a shunt-connected DC machine is shown in Figure 5.18.

Applying Kirchhoff's voltage law and assigning $r_{ar} = r_{fr} = 0$, one obtains

$$u_a - L_{af} i_f \omega_r = r_a i_a + L_a \frac{di_a}{dt}$$

$$u_f = r_f i_f + L_f \frac{di_f}{dt}$$

where $u_a = u_f$.

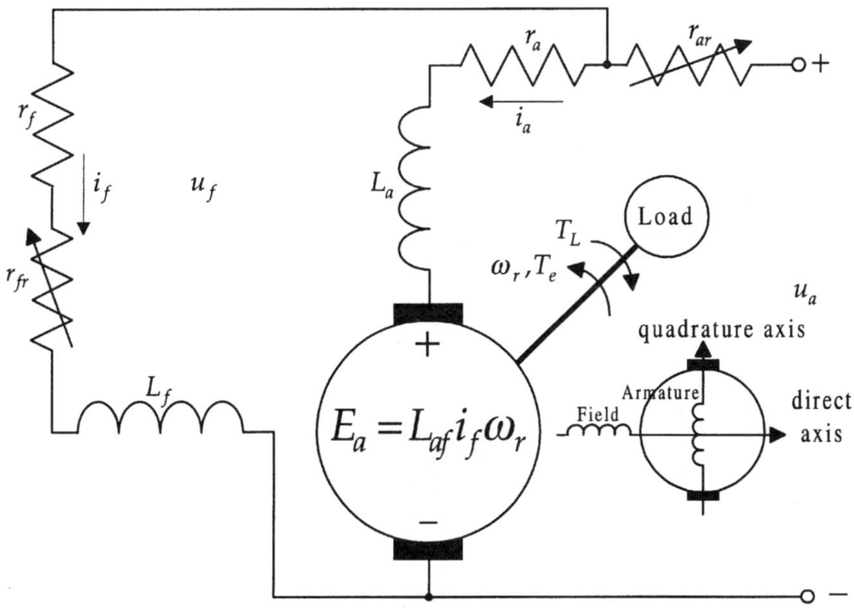

Figure 5.18 Schematic diagram of shunt-connected DC machines (the armature current direction corresponds to motor operation).

Using these Kirchhoff's equations for circuitry and applying Newton's second law, the following differential equations can be found and used to model the nonlinear transient dynamics for shunt-connected DC motors

$$\frac{di_a}{dt} = -\frac{r_a}{L_a} i_a - \frac{L_{af}}{L_a} i_f \omega_r + \frac{1}{L_a} u_a$$

$$\frac{di_f}{dt} = -\frac{r_f}{L_f} i_f + \frac{1}{L_f} u_a, \quad u_a = u_f \tag{5.11}$$

$$\frac{d\omega_r}{dt} = \frac{L_{af}}{J} i_a i_f - \frac{B_m}{J} \omega_r - \frac{1}{J} T_L$$

Under steady-state conditions, since $u_a = u_f$, we have

$$i_a = \frac{u_a - L_{af} i_f \omega_r}{r_a} \quad \text{and} \quad i_f = \frac{u_a}{r_f}$$

The substitution of these currents into the torque equation gives

$$T_e = L_{af} i_a i_f = \frac{L_{af}}{r_a r_f} \left(1 - \frac{L_{af} \omega_r}{r_f} \right) u_a^2 \tag{5.12}$$

The analysis of (5.12) shows that the electromagnetic torque is a linear function of the angular velocity, and T_e varies as the square of the armature voltage applied. In shunt-

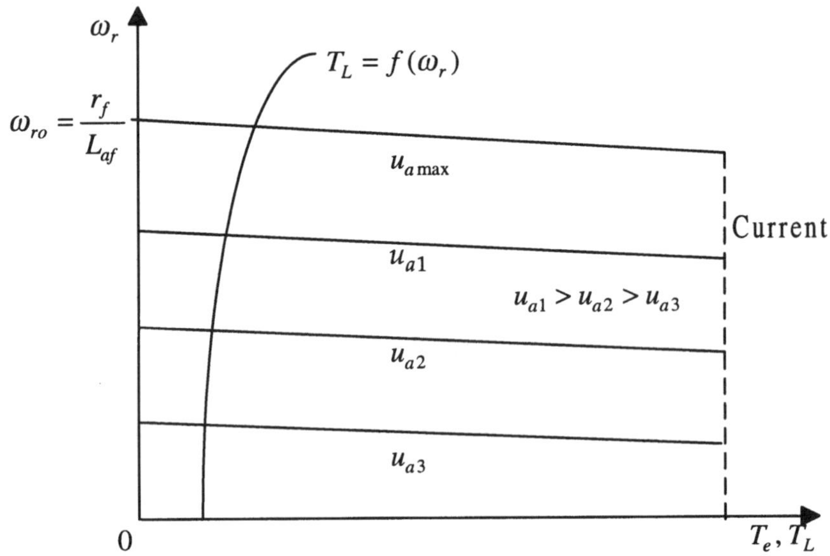

Figure 5.19 Steady-state torque-speed for shunt-connected DC motors.

connected DC machines, the armature and field windings are connected across a voltage source. An external resistance r_{ar} can be used to control the angular velocity and, theoretically, r_{fr} can be inserted as well. However, a regulated source of voltage is usually used. To change the angular velocity, the applied voltage is varied, and a family of steady-state torque-speed characteristic curves can be derived from (5.12). The torque-speed characteristics with slope $-\frac{L_{af}^2}{r_a r_f^2} u_a^2$ are shown in Figure 5.19.

Example 5.3

A shunt-connected motor ($r_a = 0.12$ ohm, $r_f = 23$ ohm, and $L_{af} = 0.15$ H, $r_{ar} = r_{fr} = 0$) drives a fan. When one applies $u_a = 100$ V, the angular velocity is 150 rad-s^{-1}. For steady-state operating conditions and assuming that the viscous friction is negligibly small, find the developed electromagnetic torque and currents in the armature and field windings.

Solution

Using formula (5.12), one obtains the developed electromagnetic torque. In particular,

$$T_e = \frac{L_{af}}{r_a r_f}\left(1 - \frac{L_{af}\omega_r}{r_f}\right)u_a^2 = \frac{0.15}{0.12 \times 23}\left(1 - \frac{0.15 \times 150}{23}\right)100^2 = 11.8 \text{ N-m}$$

The current in the field winding is found as

$$i_f = \frac{u_f}{r_f} = \frac{100}{23} = 4.35 \text{ A}$$

Then, the armature current is obtained by

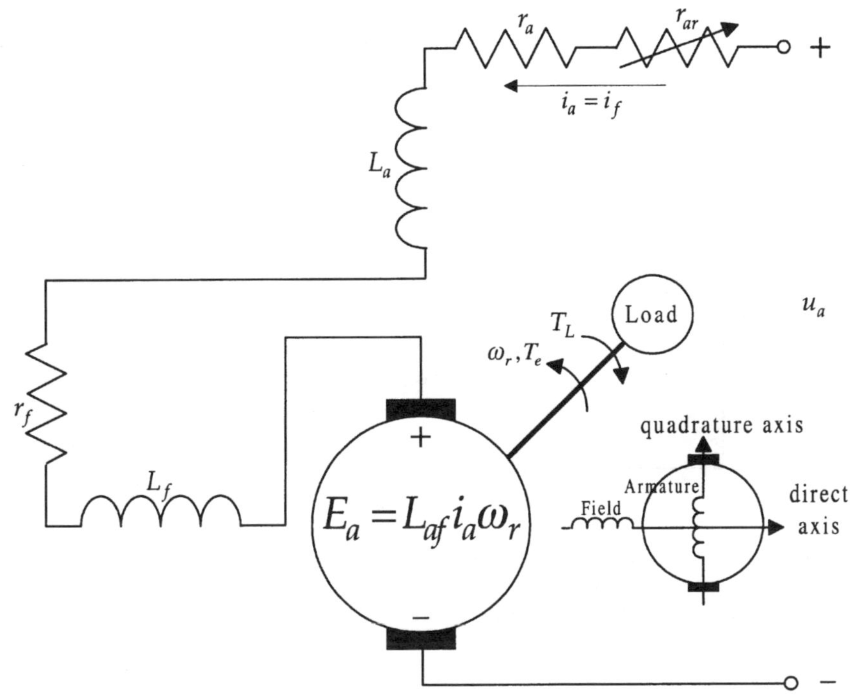

Figure 5.20 Schematic diagram of series-connected DC electric machines (current direction corresponds to motor operation).

$$i_a = \frac{T_e}{L_{af} i_f} = \frac{11.8}{0.15 \times 4.35} = 18.1 \text{ A}$$ □

5.5 Series-connected direct-current machines

Series-connected DC machines have their field winding in series with the armature winding. Therefore, the armature current ensures the field excitation (see the schematic diagram in Figure 5.20). As the armature current increases, the flux also increases.

Consider the motor magnetic system. The flux linkages in the armature and field windings are produced by the armature current. According to Kirchhoff's voltage law, the applied voltage is expressed as ($r_{ar} = 0$)

$$u_a - L_{af} i_a \omega_r = (r_a + r_f) i_a + (L_a + L_f) \frac{di_a}{dt} \tag{5.13}$$

Using differential equation (5.13) and equating $\frac{di_a}{dt}$ to zero (for steady-state analysis), we have the following expression for the armature current

$$i_a = \frac{u_a}{L_{af}\omega_r + r_a + r_f}$$

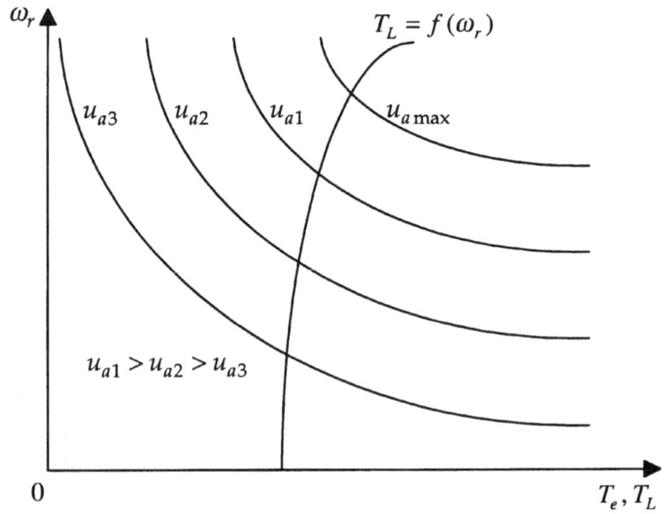

Figure 5.21 Steady-state torque-speed operating characteristics for series-connected DC motors.

Substituting this equation into the expression for electromagnetic torque $T_e = L_{af} i_a^2$, one obtains

$$T_e = L_{af} i_a^2 = \frac{L_{af}}{(L_{af}\omega_r + r_a + r_f)^2} u_a^2 \tag{5.14}$$

The torque-speed characteristic curves of separately excited and shunt-connected motors are entirely different when compared with series-connected electric machines. Assuming that the magnetic system is linear, one concludes that because the flux is proportional to the armature current, the developed electromagnetic torque is proportional to the square of the current. However, the saturation effect should be taken into account. The torque-speed characteristics can be measured experimentally, and typical curves are shown in Figure 5.21.

Using the differential equation for the armature circuitry (5.13), as well as the *torsional-mechanical* dynamics, one obtains the following nonlinear differential equations to map and analyze the transient behavior of series-connected motors

$$\begin{aligned} \frac{di_a}{dt} &= -\frac{r_a + r_f}{L_a + L_f} i_a - \frac{L_{af}}{L_a + L_f} i_a \omega_r + \frac{1}{L_a + L_f} u_a \\ \frac{d\omega_r}{dt} &= \frac{L_{af}}{J} i_a^2 - \frac{B_m}{J}\omega_r - \frac{1}{J} T_L \end{aligned} \tag{5.15}$$

Series-connected motors have been widely used in electric vehicles and other industrial electric drives when high starting torques are required (see the torque-speed characteristic curves illustrated in Figure 5.21). At stall, the starting electromagnetic torque is limited by the augmented armature resistance; see equation (5.14). Series-connected motors can be used if the corresponding load varies in the particular region. In particular, the load torque cannot be less than 10 to 15% compared with the rated electromagnetic torque. Furthermore, the load must be permanently attached to the motor (the angular velocity of series-connected motors is extremely high for unloaded motors because the values of the internal armature r_a and field r_f resistances are small). Therefore, series-connected motors are used in cars, locomotives, mills, ships, and trains. The angular velocity is regulated by varying the applied armature voltage. Clearly, to limit the armature current and protect the motor, the armature *rheostat* can be used. However, r_{ar} must be removed from the circuit because rheostat regulation is clearly inefficient for speed control.

It should be emphasized that series-connected motors develop electromagnetic torque and rotate if the alternating voltage is supplied.

Example 5.4

Find an equation to calculate the average electromagnetic torque if a sinusoidal voltage of frequency $\frac{\omega_f}{2\pi}$ is applied to the series-connected DC motor. Here, ω_f is the angular frequency of the supplied ac voltage. Assume that 10 A current (*rms* value), shifted by 30° behind the voltage with frequency 60 Hz $\left(\frac{\omega_f}{2\pi} = 60 \text{ Hz}\right)$, is fed to the series-connected DC motor. Calculate the average value of the developed electromagnetic torque if $L_{af} = 0.01$ H.

Solution

The armature current is expressed as

$$i_a = I_a \sin(\omega_f t - \varphi)$$

Observe that the *rms* (root-mean-square) value of the current is $I_{rms} = \frac{I_a}{\sqrt{2}}$.

The period is related to the frequency, and is given as $\frac{2\pi}{\omega_f}$. Then, the average electromagnetic torque can be calculated. In particular, we have

$$T_{e\,av} = L_{af}\frac{\omega_f}{2\pi} I_a^2 \int_0^{\frac{2\pi}{\omega_f}} \sin^2(\omega_f t - \varphi)dt = L_{af}\frac{\omega_f}{\pi} I_{rms}^2 \int_0^{\frac{2\pi}{\omega_f}} \sin^2(\omega_f t - \varphi)dt$$

For a given frequency and *rms* value of the armature current, the average electromagnetic torque developed can be found. The integral $\int_0^{\frac{2\pi}{\omega_f}} \sin^2(\omega_f t - \varphi)dt$ can be easily found. Using the table of integrals, one has

$$\int_a^b \sin^2 x dx = \left[\frac{1}{2}x - \frac{1}{4}\sin 2x\right]_a^b$$

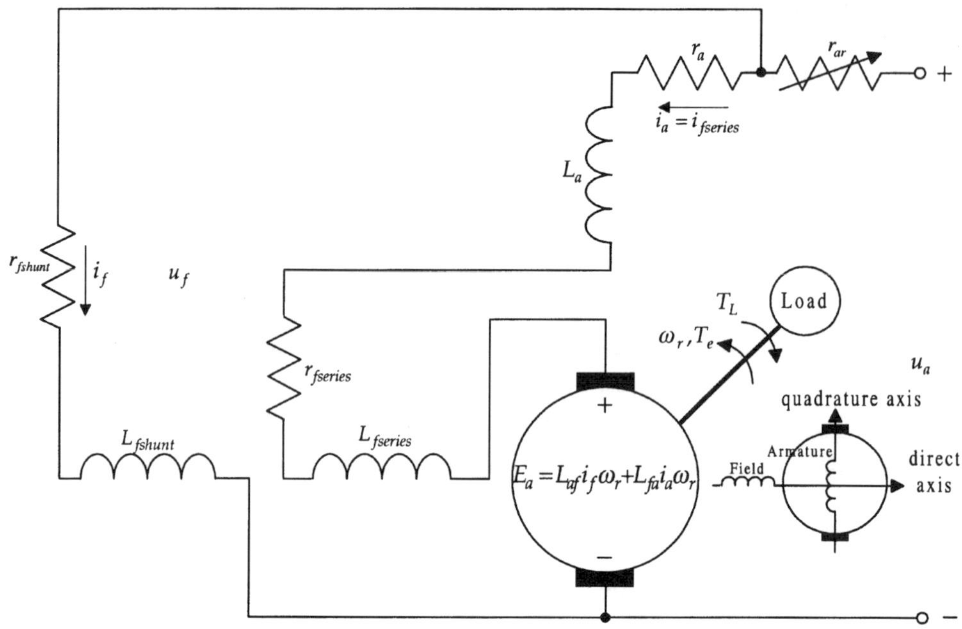

Figure 5.22 Schematic diagram of cumulatively compounded DC machines (armature current direction corresponds to motor operation).

Applying the formula for $T_{e\,av}$ derived, we have

$$T_{e\,av} = L_{af} 200 \times 60 \int_0^{\frac{1}{60}} \sin^2\left(2\pi 60t - \frac{1}{12}\pi\right) dt = 1 \qquad \square$$

5.6 Compound-connected direct-current electric machines

Consider compound-connected electric machines with series-field and shunt-field windings that are placed on the main field poles. When the resulting flux linkages, produced by the currents in series and shunt windings, are in the same direction, this machine is said to be *cumulatively compounded*. If the flux produced by the current in the series winding opposes the flux produced by the current in the shunt winding, the machine is said to be *differentially compounded*. Figures 5.22 and 5.23 illustrate the cumulatively compounded and differentially compounded DC motors, respectively.

The characteristics of compound-connected DC electric machines combine the basic features of shunt- and series-connected machines. Kirchhoff's voltage law leads one to the differential equations to map the circuitry dynamics, and for *cumulatively compounded* motors assigning $r_{ar} = 0$, one obtains

$$u_a - L_{fa} i_a \omega_r - L_{af} i_f \omega_r = (r_a + r_{fseries}) i_a + (L_a + L_{fseries}) \frac{di_a}{dt} + L_{ff} \frac{di_f}{dt}$$

$$u_f = r_{fshunt} i_f + L_{fshunt} \frac{di_f}{dt} + L_{ff} \frac{di_a}{dt} \qquad (5.16)$$

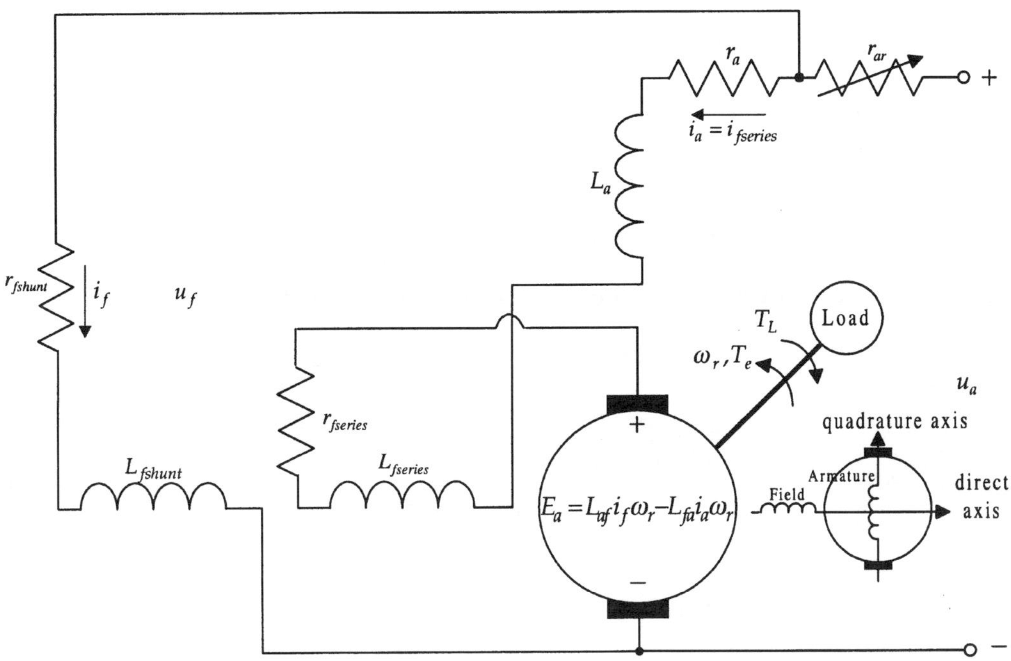

Figure 5.23 Schematic diagram of differentially compounded DC machines (armature current direction corresponds to motor operation).

where L_{ff} is the mutual inductance between the shunt and series windings; L_{af} is the mutual inductance between the armature and shunt windings; L_{fa} is the mutual inductance between the armature and series windings; and $L_{fseries}$ and L_{fshunt} are the self-inductances of the series and shunt windings, respectively.

Hence, one distinguishes the mutual inductances between the armature and shunt windings L_{af}, and the armature and series windings L_{fa}.

It is of a great interest to find the steady-state torque-speed characteristic curves using the expression for the developed electromagnetic torque. For *cumulatively compounded* machines, we have

$$T_e = L_{af} i_a i_f + L_{fa} i_a^2$$

The steady-state solution of equations (5.16) is found, and the resulting equation is

$$T_e = \frac{L_{af}\left(1 - \dfrac{L_{af}\omega_r}{r_{fshunt}}\right)}{r_{fshunt}(r_a + r_{fseries} + L_{fa}\omega_r)} u_a^2 + \frac{L_{af}\left(1 - \dfrac{L_{af}\omega_r}{r_{fshunt}}\right)^2}{(r_a + r_{fseries} + L_{fa}\omega_r)^2} u_a^2$$

Using Newton's second law and (5.16), the differential equations to model cumulatively compounded motors are found to be

$$
\begin{aligned}
&(L_a + L_{fseries})\frac{di_a}{dt} + L_{ff}\frac{di_f}{dt} = -(r_a + r_{fseries})i_a - L_{fa}i_a\omega_r - L_{af}i_f\omega_r + u_a \\
&L_{fshunt}\frac{di_f}{dt} + L_{ff}\frac{di_a}{dt} = -r_{fshunt}i_f + u_a \\
&\frac{d\omega_r}{dt} = \frac{L_{af}i_a i_f + L_{fa}i_a^2}{J} - \frac{B_m}{J}\omega_r - \frac{1}{J}T_L
\end{aligned} \tag{5.17}
$$

From (5.17), Cauchy's form of the differential equations is found, and

$$
\begin{aligned}
&\frac{di_a}{dt} = \frac{-(r_a + r_{fseries})L_{fshunt}i_a + r_{fshunt}L_{ff}i_f - L_{fa}L_{fshunt}i_a\omega_r - L_{af}L_{fshunt}i_f\omega_r + (L_{fshunt} - L_{ff})u_a}{(L_a + L_{fseries})L_{fshunt} - L_{ff}^2} \\
&\frac{di_f}{dt} = \frac{(r_a + r_{fseries})L_{ff}i_a - r_{fshunt}(L_a + L_{fseries})i_f + L_{fa}L_{ff}i_a\omega_r + L_{af}L_{ff}i_f\omega_r + (L_a + L_{fseries} - L_{ff})u_a}{(L_a + L_{fseries})L_{fshunt} - L_{ff}^2} \\
&\frac{d\omega_r}{dt} = \frac{L_{af}i_a i_f + L_{fa}i_a^2}{J} - \frac{B_m}{J}\omega_r - \frac{1}{J}T_L
\end{aligned} \tag{5.18}
$$

Cumulatively compounded DC machines can be designed as overcompounded, flat compounded, and undercompounded.

Let us study differentially compounded electric machines. The straightforward application of Kirchhoff's second law leads one to the nonlinear differential equations to model the circuitry transient dynamics. In particular, making use of the circuitry configuration as given in Figure 5.23, we have

$$
\begin{aligned}
&u_a + L_{fa}i_a\omega_r - L_{af}i_f\omega_r = (r_a + r_{fseries})i_a + (L_a + L_{fseries})\frac{di_a}{dt} - L_{ff}\frac{di_f}{dt} \\
&u_f = r_{fshunt}i_f + L_{fshunt}\frac{di_f}{dt} - L_{ff}\frac{di_a}{dt}
\end{aligned} \tag{5.19}
$$

where $u_f = u_a$ if $r_{ar} = 0$.

The electromagnetic torque is expressed as

$$T_e = L_{af}i_a i_f - L_{fa}i_a^2$$

and, deriving the steady-state solution of (5.19), we have

$$T_e = \frac{L_{af}\left(1 - \dfrac{L_{af}\omega_r}{r_{fshunt}}\right)}{r_{fshunt}(r_a + r_{fseries} - L_{fa}\omega_r)}u_a^2 - \frac{L_{af}\left(1 - \dfrac{L_{af}\omega_r}{r_{fshunt}}\right)^2}{(r_a + r_{fseries} - L_{fa}\omega_r)^2}u_a^2$$

Using differential equations for the circuitry dynamics (5.19) as well as Newton's second law of motion, one obtains a set of three differential equations to map the dynamics of differentially compounded electric motors

$$
\begin{aligned}
&(L_a + L_{fseries})\frac{di_a}{dt} - L_{ff}\frac{di_f}{dt} = -(r_a + r_{fseries})i_a + L_{fa}i_a\omega_r - L_{af}i_f\omega_r + u_a \\
&L_{fshunt}\frac{di_f}{dt} - L_{ff}\frac{di_a}{dt} = -r_{fshunt}i_f + u_a \\
&\frac{d\omega_r}{dt} = \frac{L_{af}i_a i_f - L_{fa}i_a^2}{J} - \frac{B_m}{J}\omega_r - \frac{1}{J}T_L
\end{aligned}
\tag{5.20}
$$

From (5.20), we have the final form of a mathematical model of differentially compounded DC motors as given by

$$
\begin{aligned}
&\frac{di_a}{dt} = \frac{-(r_a + r_{fseries})L_{fshunt}i_a - r_{fshunt}L_{ff}i_f + L_{fa}L_{fshunt}i_a\omega_r - L_{af}L_{fshunt}i_f\omega_r + (L_{fshunt} + L_{ff})u_a}{(L_a + L_{fseries})L_{fshunt} - L_{ff}^2} \\
&\frac{di_f}{dt} = \frac{-(r_a + r_{fseries})L_{ff}i_a - r_{fshunt}(L_a + L_{fseries})i_f + L_{fa}L_{ff}i_a\omega_r - L_{af}L_{ff}i_f\omega_r + (L_a + L_{fseries} + L_{ff})u_a}{(L_a + L_{fseries})L_{fshunt} - L_{ff}^2} \\
&\frac{d\omega_r}{dt} = \frac{L_{af}i_a i_f - L_{fa}i_a^2}{J} - \frac{B_m}{J}\omega_r - \frac{1}{J}T_L
\end{aligned}
\tag{5.21}
$$

A set of highly nonlinear coupled differential equations (5.21) thus derived must be used to analyze and simulate *differentially compounded* electric motors.

5.7 Permanent-magnet direct-current machines

The constant magnetic flux in AC and DC machines can be produced by permanent-magnet poles, which allow one to establish a residual flux density up to 1.3 T. Electric machines with permanent-magnet poles are called permanent-magnet machines. Permanent-magnet motors and generators are widely used, and Figure 5.24 shows the basic structural features.

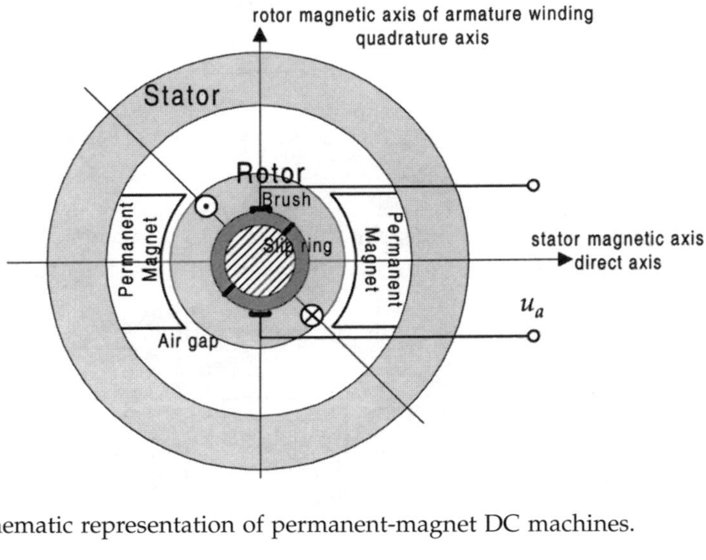

Figure 5.24 Schematic representation of permanent-magnet DC machines.

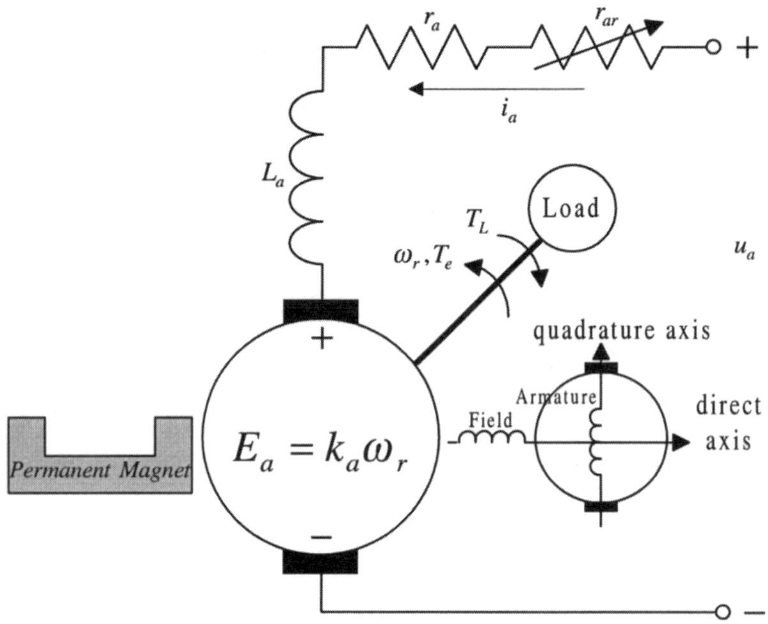

Figure 5.25 Schematic diagram of permanent-magnet electric machines (current direction corresponds to motor operation).

Permanent-magnet motors have been widely used in high-performance electric drives and servo-systems. For the purpose of deriving the equations of motion, a schematic diagram of permanent-magnet DC machines is illustrated in Figure 5.25.

Using Kirchhoff's voltage law and Newton's second law of motion, the differential equations for permanent-magnet DC machines can be easily derived. Assuming that the *susceptibility* is constant (in reality, Curie's constant varies as a function of temperature), one supposes that the flux, established by the permanent magnet poles, is constant. Then, denoting the *back emf* and *torque* constants as k_a, one obtains the following differential equations describing the transient behavior of the armature winding and *torsional-mechanical* dynamics

$$\begin{aligned}\frac{di_a}{dt} &= -\frac{r_a}{L_a}i_a - \frac{k_a}{L_a}\omega_r + \frac{1}{L_a}u_a\\ \frac{d\omega_r}{dt} &= \frac{k_a}{J}i_a - \frac{B_m}{J}\omega_r - \frac{1}{J}T_L\end{aligned} \tag{5.22}$$

Permanent-magnet electric machines do not have field windings. Therefore, by using the differential equations for separately excited electric machines (5.9) and denoting $k_a = L_{af}i_f$, one obtains equations (5.22).

Using (5.22), it is easy to find the model in matrix form. We have

$$\begin{bmatrix}\dfrac{di_a}{dt}\\ \dfrac{d\omega_r}{dt}\end{bmatrix} = \begin{bmatrix}-\dfrac{r_a}{L_a} & -\dfrac{k_a}{L_a}\\ \dfrac{k_a}{J} & -\dfrac{B_m}{J}\end{bmatrix}\begin{bmatrix}i_a\\ \omega_r\end{bmatrix} + \begin{bmatrix}\dfrac{1}{L_a}\\ 0\end{bmatrix}u_a - \begin{bmatrix}0\\ \dfrac{1}{J}\end{bmatrix}T_L$$

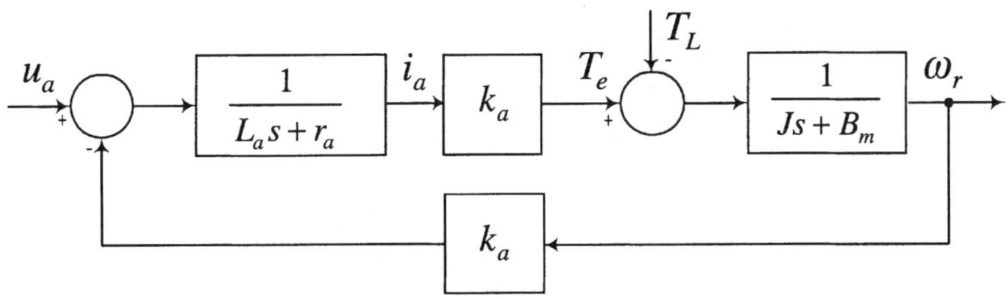

Figure 5.26 Block diagram of permanent-magnet DC motors.

An s-domain block diagram of permanent-magnet DC motors is illustrated in Figure 5.26.

The angular velocity can be reversed if the polarity of the applied voltage is changed (the direction of the field flux cannot be changed). The steady-state torque-speed characteristic curves obey the following equation

$$\omega_r = \frac{u_a - r_a i_a}{k_a} = \frac{u_a}{k_a} - \frac{r_a}{k_a^2} T_e \tag{5.23}$$

and a spectrum of the torque-speed characteristic curves is illustrated in Figure 5.27.

The steady-state torque-speed characteristic curves are available from catalogs. For example, the catalog data and torque-speed characteristics for the TT-2931 series Kollmorgen permanent-magnet DC motors are given in Figure 5.28.

Example 5.5

Perform the modeling of a permanent-magnet Clifton Precision motor JDH-2250 that has the following parameters: $r_a = 2.7$ ohm, $L_a = 0.004$ H, $k_a = 0.105$ V-s-rad^{-1}, $k_a = 0.105$ N-m-A^{-1}, $J = 0.0001$ kg-m^2, and $B_m = 0.0000093$ N-m-s-rad^{-1}.

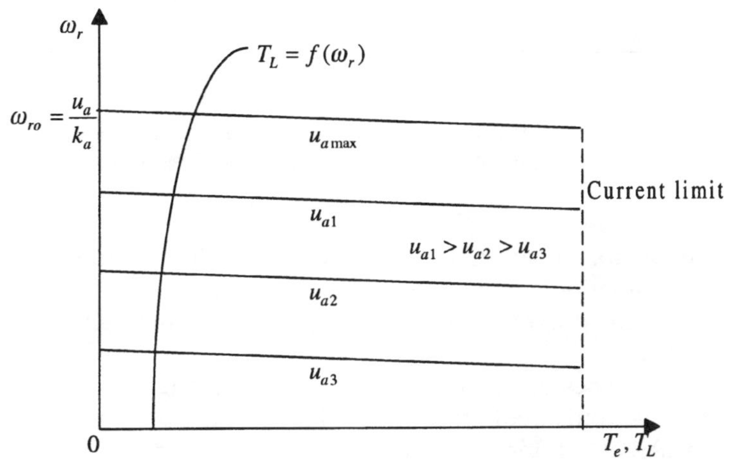

Figure 5.27 Torque-speed characteristic curves for permanent-magnet motors.

Motor Parameters	Symbols	Units	2931		2933				2934				2936				2937		
			AA	BA	AA	BA	CA	DA	AA	BA	CA	DA	AA	BA	CA	DA	AA	BA	CA
Horsepower	HP Rated	HP	1.2	0.8	1.4	1.4	0.8	1.5	1.7	1.7	1.2	2.0	2.5	2.0	1.6	2.5	2.0	2.2	3.0
Kilowatts	KW Rated	KW	0.90	0.60	1.04	1.04	0.60	1.12	1.3	1.3	0.9	1.5	1.9	1.5	1.2	1.9	1.5	1.6	2.2
Max Operating Speed	N Max	RPM	4500	2900	3500	2450	1400	4500	3000	2400	1500	4000	2700	2000	1500	3200	1400	1800	2500
Cont. Torque (stall)	TC	lb-ft	1.7	1.7	3.4	3.4	3.4	3.4	5.0	5.0	5.0	5.0	6.5	6.5	6.5	6.5	8.7	8.7	8.7
@ 40°C amb.		N-m	2.3	2.3	4.6	4.6	4.6	4.6	6.8	6.8	6.8	6.8	8.8	8.8	8.8	8.8	11.8	11.8	11.8
Current @ Cont. Torque	IC	amps	9.88	6.29	15.3	10.6	6.3	19.7	18.9	14.7	9.4	26.3	21.5	16.1	12.9	26.0	15.4	19.8	27.6
Peak Torque	TP	lb-ft	8.0	12.0	13.8	20.8	30	13	19.2	23.2	35.5	14.8	23.5	26.3	28.5	18.7	29	31	23.5
		N-m	10.8	16.2	18.6	28.2	40.5	17.6	26.0	31.3	47.9	19.9	31.8	35.1	38.5	25.2	39.2	41.6	31.7
Current @ Peak Torque	IP	amps	90	70	90	90	90	90	90	90	90	90	90	90	90	90	90	90	90
Max Terminal Voltage	VMax	Volts	115	115	115	115	115	115	115	115	115	115	115	115	115	115	115	115	115
Torque Sensitivity ± 10%	KT	lb-ft/amp	0172	0.270	0.222	0.321	0.542	0.173	0.265	0.340	0.530	0.190	0.302	0.403	0.504	0.250	0.566	0.440	0.315
		N-m/amp	0.232	0.365	0.299	0.433	0.732	0.233	0.359	0.461	0.716	0.257	0.409	0.546	0.680	0.338	0.767	0.596	0.425
Back EMF Constant	KB	V/K RPM	24.4	38.3	31.5	45.6	76.9	24.6	37.6	48.3	75.3	27.0	42.8	57.2	71.6	35.5	80.4	62.5	44.7
DC Resistance ± 12%	RT	ohms	0.685	1.652	0.366	0.715	2.295	0.229	0.273	0.456	1.255	0.159	0.257	0.432	0.707	0.195	0.646	0.396	0.239
Inductance ± 30%	LM	MH	1.8	4.4	1.6	2.5	9.5	0.78	2.0	1.21	8.0	1.0	1.2	2.1	3.2	0.83	2.5	1.5	1.4
Rotor Inertia	JM	lb-ft-sec²	0.000695	0.000695	0.0010	0.0010	0.0010	0.0010	0.0015	0.0015	0.0015	0.00015	0.00202	0.00202	0.00202	0.00202	0.00253	0.00253	0.00253
		Kg-m²	0.000942	0.000942	0.0014	0.0014	0.0014	0.0014	00.0020	0.0020	0.0020	0.00274	0.00274	0.00274	0.00274	0.00274	0.00343	0.00343	0.00343
Weight	WT	lb	14.8	14.8	18.3	18.3	18.3	18.3	23.3	23.3	23.3	23.3	28.4	28.4	28.4	28.4	32.4	32.4	32.4
		Kg	6.7	6.7	8.3	8.3	8.3	8.3	10.6	10.6	10.6	10.6	12.9	12.9	12.9	12.9	14.7	14.7	14.7
Tachometer Parameters	TGF-2030-147-A		1% Avg-pk ripple at 66 cy / Rev., 2.6K Min load resistance																
Voltage Sensitivity ± 10%	KG	V/K RPM	19	19	19	19	19	19	19	19	19	19	19	19	19	19	19	19	19

TT-2931-A

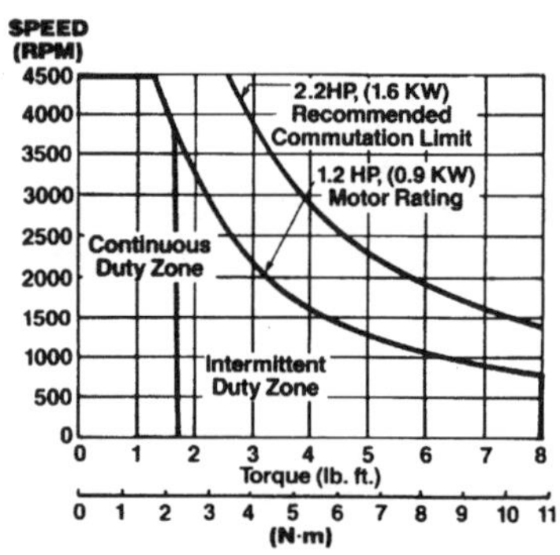

TT-2931-B

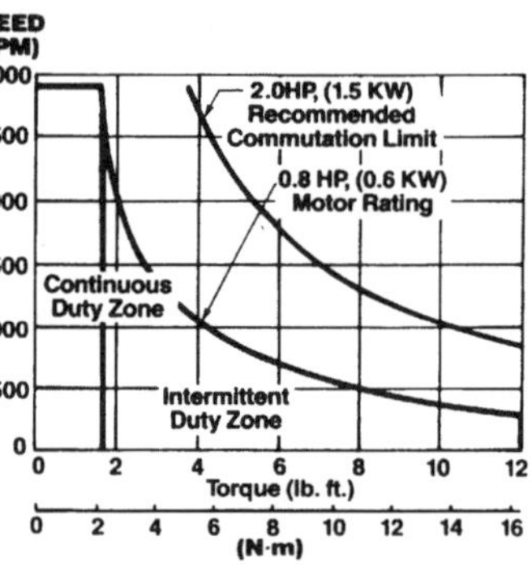

Figure 5.28 Performance data and torque-speed characteristic curves for TT-2931-A and TT-2931-B permanent-magnet DC motors. (Courtesy of the Kollmorgen Corporation.)

Solution

The analysis of the motor dynamics is straightforward using the model developed in the form of differential equations (5.22). Using the *s*-domain block diagram, as illustrated in Figure 5.26, a SIMULINK diagram to model a permanent-magnet motor is built (see Figure 5.29).

The transient responses for the armature current and angular velocity, as well as the waveforms of the electromagnetic torque developed and load torque applied, are illustrated in Figure 5.30. Performing the modeling, the applied armature voltage is 30 V, and the load torque is assigned to be

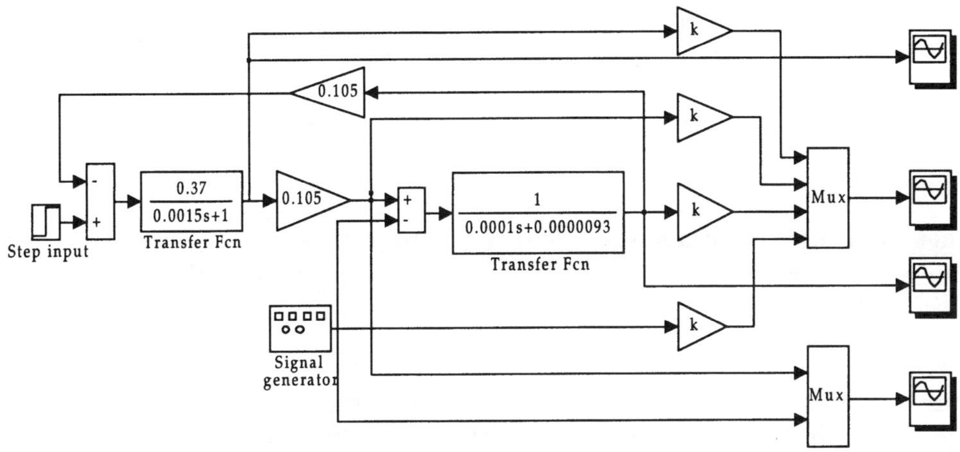

Figure 5.29 SIMULINK diagram.

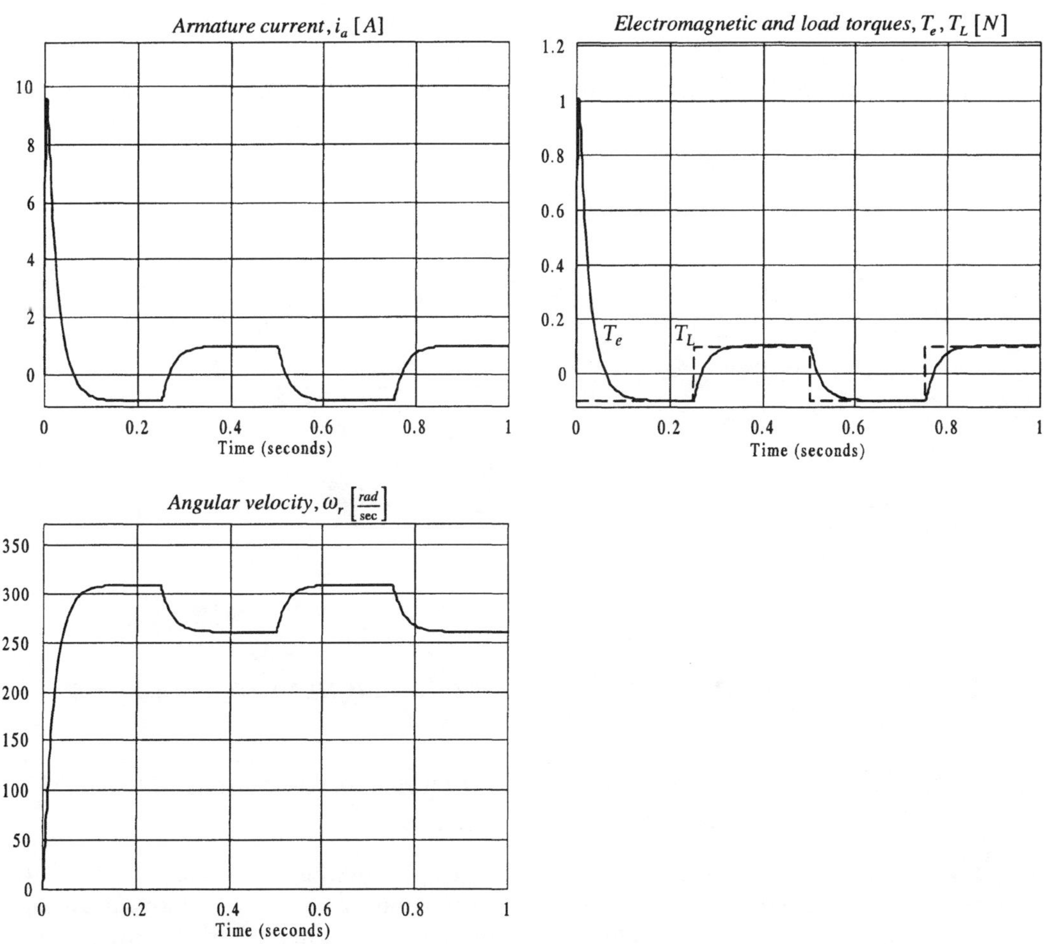

Figure 5.30 Permanent-magnet motor dynamics.

$$T_L = \begin{cases} 0.1 \text{ N}-m, & t \in [0.25 \quad 0.5) \text{ s} \\ -0.1 \text{ N}-m, & t \in [0 \quad 0.25), \; t \in [0.5 \quad 0.75] \text{ s} \end{cases}$$

□

Example 5.6

Calculate and plot the torque-speed characteristic curves for the permanent-magnet DC motor studied in Example 5.5. Plot the load characteristic if the load is given as $T_L = f(\omega_r) = 0.05 + 0.0001\,\omega_r^2$.

Solution

A family of torque-speed characteristic curves are found by applying (5.23). In particular, plugging the different values of the armature voltage u_a in

$$\omega_r = \frac{u_a - r_a i_a}{k_a} = \frac{u_a}{k_a} - \frac{r_a}{k_a^2} T_e$$

the steady-state characteristics are found and plotted in Figure 5.31.

The load characteristic, as given by $T_L = f(\omega_r) = 0.05 + 0.0001\,\omega_r^2$, is documented in Figure 5.31. The MATLAB m-file developed to calculate and plot the steady-state operating characteristics is given.

MATLAB script (`ch_5_02.m`)

```
% parameters of a permanent-magnet DC motor
ra=2.7; ka=0.105;
Te=0:0.01:1;
for ua=5:5:30;
wr=ua/ka-(ra/ka^2)*Te;
wrl=0:1:100; Tl=0.05+0.0001*wrl.^2;
plot(Te,wr,'-',Tl,wrl,'-');hold on;
axis([0, 1, 0, 300]);
end; disp('End')
```

□

5.8 Modeling and analysis of permanent-magnet direct-current generators

5.8.1 Modeling and analysis of permanent-magnet direct-current generators driven by permanent-magnet direct-current motors

In studying the permanent-magnet DC generator driven by the prime mover, the goal is to thoroughly study the transient dynamics and perform the steady-state analysis. To solve these problems, a mathematical model should be derived, and simulations must be performed, augmenting the generator and prime mover dynamics. Assume that the permanent-magnet DC motor is used as a prime mover, and the power generation system considered is illustrated in Figure 5.32.

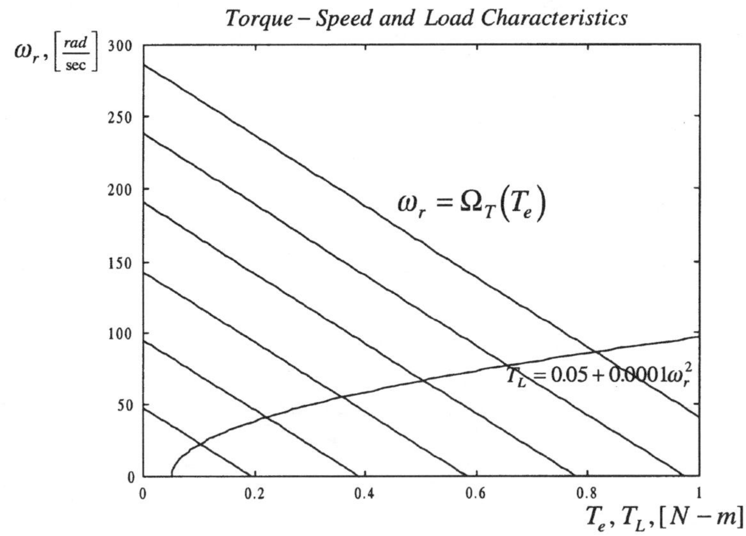

Figure 5.31 Torque-speed and load characteristics.

The mathematical model for permanent-magnet DC generator circuitry dynamics, if the resistive load is studied, is found using Kirchhoff's law

$$\frac{di_{ag}}{dt} = -\frac{r_{ag} + R_L}{L_{ag}} i_{ag} + \frac{k_{ag}}{L_{ag}} \omega_{rpm} \tag{5.24}$$

and Newton's second law of motion maps the *torsional-mechanical* dynamics of the generator–prime mover system

$$T_{epm} - (B_{mpm} + B_{mg})\omega_{rpm} - T_{eg} = (J_{pm} + J_g)\frac{d\omega_{rpm}}{dt} \tag{5.25}$$

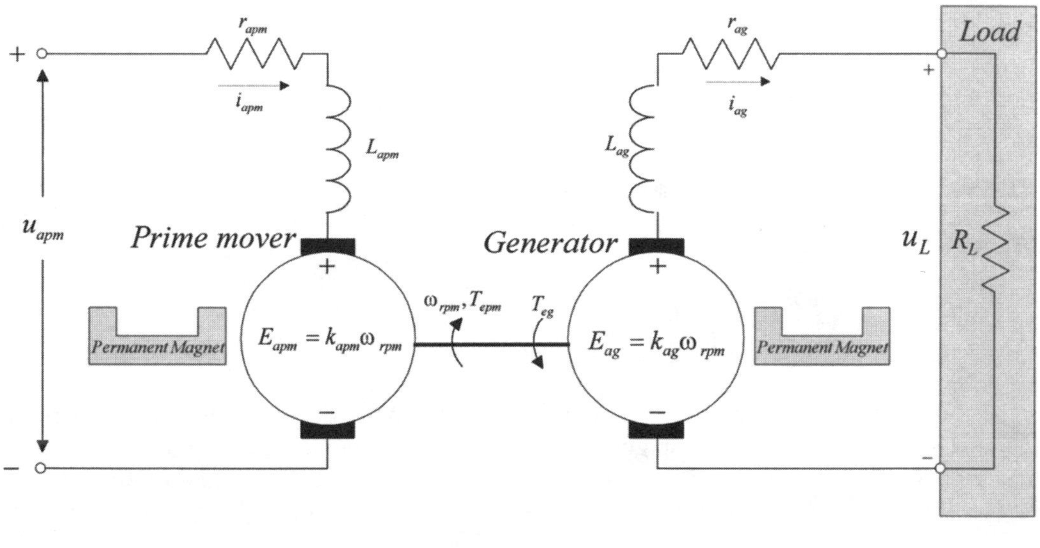

Figure 5.32 Permanent-magnet DC generator driven by DC motor.

where i_{ag} is the generator armature current; ω_{rpm} is the angular velocity of the prime mover and generator; r_{ag} and L_{ag} are the armature resistance and inductance of the generator rotor winding; R_L is the load resistance; k_{ag} is the *back emf* (*torque*) constant of the generator; B_{mpm} and B_{mg} are the viscous friction coefficients; and J_{pm} and J_g are the moments of inertia of the prime mover and generator.

The electromagnetic torque developed by the permanent-magnet DC motor (prime mover), denoted as T_{epm}, is

$$T_{epm} = k_{apm} i_{apm} \tag{5.26}$$

where i_{apm} is the armature current in the prime mover rotor winding; and k_{apm} is the *back emf* (*torque*) constant of the prime mover.

The load torque for the prime mover is the generator electromagnetic torque, and we have

$$T_{eg} = k_{ag} i_{ag} \tag{5.27}$$

From (5.25), (5.26), and (5.27), one obtains

$$\frac{d\omega_{rpm}}{dt} = \frac{k_{apm}}{J_{pm}+J_g} i_{apm} - \frac{B_{mpm}+B_{mg}}{J_{pm}+J_g} \omega_{rpm} - \frac{k_{ag}}{J_{pm}+J_g} i_{ag} \tag{5.28}$$

The mathematical model for the prime mover circuitry dynamics must be found. In particular, we have

$$\frac{di_{apm}}{dt} = -\frac{r_{apm}}{L_{apm}} i_{apm} - \frac{k_{apm}}{L_{apm}} \omega_{rpm} + \frac{1}{L_{apm}} u_{apm} \tag{5.29}$$

where u_{apm} is the voltage applied to the armature winding of the prime mover; and r_{apm} and L_{apm} are the armature resistance and inductance of the prime mover rotor winding.

Taking note of (5.24), (5.28), and (5.29), we have

$$\begin{aligned}
\frac{di_{apm}}{dt} &= -\frac{r_{apm}}{L_{apm}} i_{apm} - \frac{k_{apm}}{L_{apm}} \omega_{rpm} + \frac{1}{L_{apm}} u_{apm} \\
\frac{di_{ag}}{dt} &= -\frac{r_{ag}+R_L}{L_{ag}} i_{ag} + \frac{k_{ag}}{L_{ag}} \omega_{rpm} \\
\frac{d\omega_{rpm}}{dt} &= \frac{k_{apm}}{J_{pm}+J_g} i_{apm} - \frac{B_{mpm}+B_{mg}}{J_{pm}+J_g} \omega_{rpm} - \frac{k_{ag}}{J_{pm}+J_g} i_{ag}
\end{aligned} \tag{5.30}$$

In the MATLAB environment, the mathematical model derived for the permanent-magnet DC generator–DC motor generation system is used to perform simulations. In particular, a SIMULINK diagram is developed based on a set of differential equations (5.30). The resulting diagram is illustrated in Figure 5.33 (file name is `pm_pmg_1.mdl`).

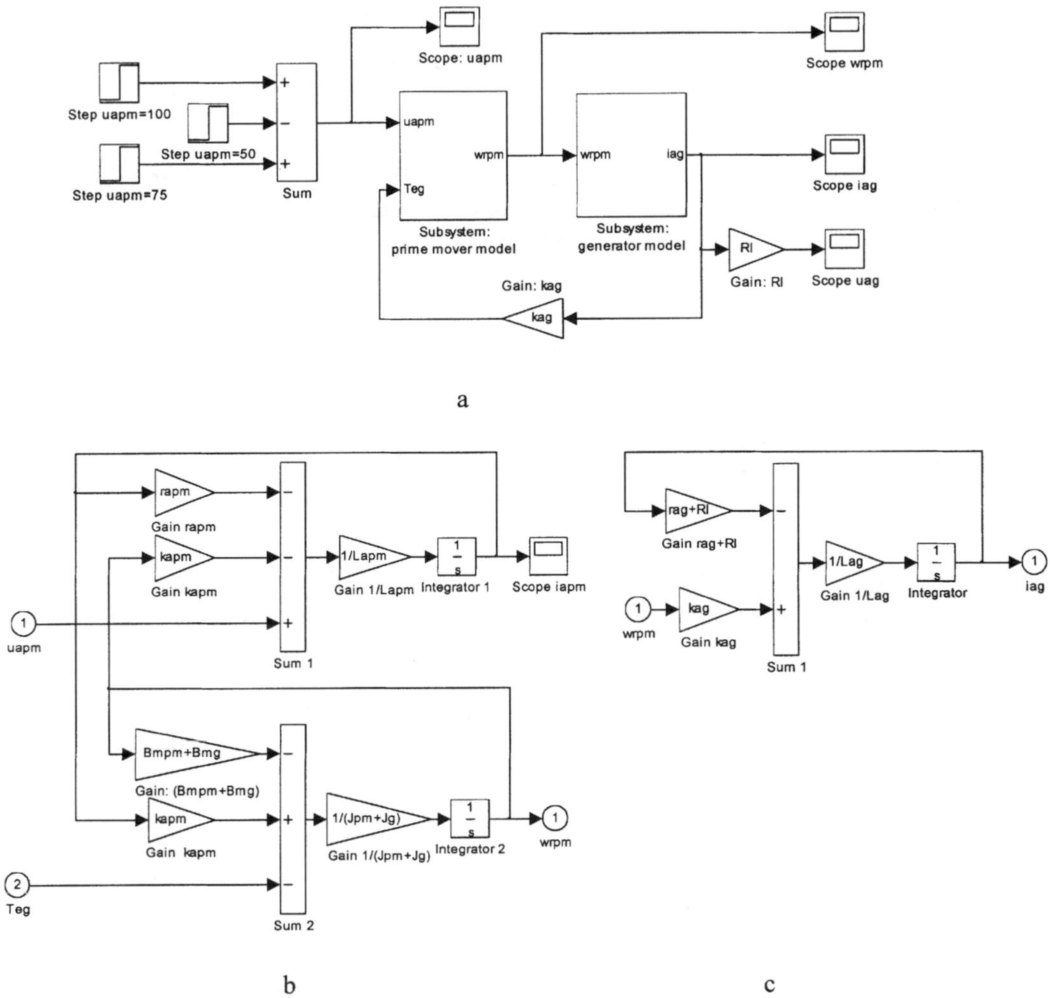

Figure 5.33 SIMULINK diagram to simulate the generation system: permanent-magnet DC generator–DC motor (a) augmented system model; (b) SIMULINK subsystem diagram to model permanent-magnet DC motors; (c) SIMULINK subsystem diagram to model permanent-magnet generators.

Example 5.7

Model the generation system with permanent-magnet DC generator and permanent-magnet DC motor. Study the transient dynamics and voltage induced by the generator for different resistive load and angular velocity of the prime mover. In particular, assign

$$u_{apm} = \begin{cases} 100\text{ V}, \ \forall t \in [0 \quad 2)\text{ s} \\ 50\text{ V}, \ \forall t \in [2 \quad 4)\text{ s} \\ 75\text{ V}, \ \forall t \in [4 \quad 6]\text{ s} \end{cases}$$

to change the angular velocity, and

$$R_L = \begin{cases} 100 \text{ ohm}, \ \forall t \in [0 \quad 2) \text{ s} \\ 5 \text{ ohm}, \ \forall t \in [2 \quad 4) \text{ s} \\ 25 \text{ ohm}, \ \forall t \in [4 \quad 6] \text{ s} \end{cases}$$

The parameters of the electric machines used are: $r_{apm} = 0.4$ ohm, $r_{ag} = 0.3$ ohm, $L_{apm} = 0.05$ H, $L_{ag} = 0.06$ H, $k_{apm} = 0.3$ V-s-rad^{-1} or $k_{apm} = 0.3$ N-m-A^{-1}, $k_{ag} = 0.25$ V-s-rad^{-1} or $k_{ag} = 0.25$ N-m-A^{-1}, $J_{pm} = 0.04$ kg-m^2, $J_g = 0.05$ kg-m^2, $B_{mpm} = 0.0007$ N-m-s-rad^{-1}, and $B_{mg} = 0.0008$ N-m-s-rad^{-1}.

Solution

The following m-file (pm_pmg_d.m) downloads the parameters of the generation system.

```
% Parameters of the prime mover
rapm=0.4; Lapm=0.05; kapm=0.3; Jpm=0.04; Bmpm=0.0007;
% Parameters of the permanent-magnet DC generator
rag=0.3; Lag=0.06; kag=0.25; Jg=0.05; Bmg=0.0008;
% Load resistance
Rl=5;
```

Inserting the distinct values of the load resistance ($R_L = 5$ ohm, $R_L = 25$ ohm, and $R_L = 100$ ohm) in the armature circuit of the permanent-magnet DC generator, and applying the following armature voltages to the motor (prime mover) armature winding

$$u_{apm} = \begin{cases} 100 \text{ V}, \ \forall t \in [0 \quad 2) \text{ s} \\ 50 \text{ V}, \ \forall t \in [2 \quad 4) \text{ s} \\ 75 \text{ V}, \ \forall t \in [4 \quad 6] \text{ s} \end{cases}$$

the dynamics of the permanent-magnet DC generator–prime mover generation system are studied. The transient behavior of the state variables, as well as the induced voltage waveforms, are documented in Figures 5.34 to 5.36.

One concludes that if $R_L = 5$ ohm, the settling time is 1.3 s, and for different angular velocities 320, 160, and 235 rad-s^{-1} of the prime mover, which depend on the armature voltage applied

$$u_{apm} = \begin{cases} 100 \text{ V}, \ \forall t \in [0 \quad 2) \text{ s} \\ 50 \text{ V}, \ \forall t \in [2 \quad 4) \text{ s} \\ 75 \text{ V}, \ \forall t \in [4 \quad 6] \text{ s} \end{cases}$$

the steady-state values of the voltage induced are 75, 37, and 55.5 V. The armature currents in the generator and prime mover are related to the load resistance inserted, and Figures 5.34 to 5.36 illustrate the transient dynamics for different R_L.

Supplying $u_{apm} = 50$ V, $u_{apm} = 75$ V, and $u_{apm} = 100$ V to the rotor winding of the permanent-magnet DC motor, the values of the load resistance inserted are assigned to be

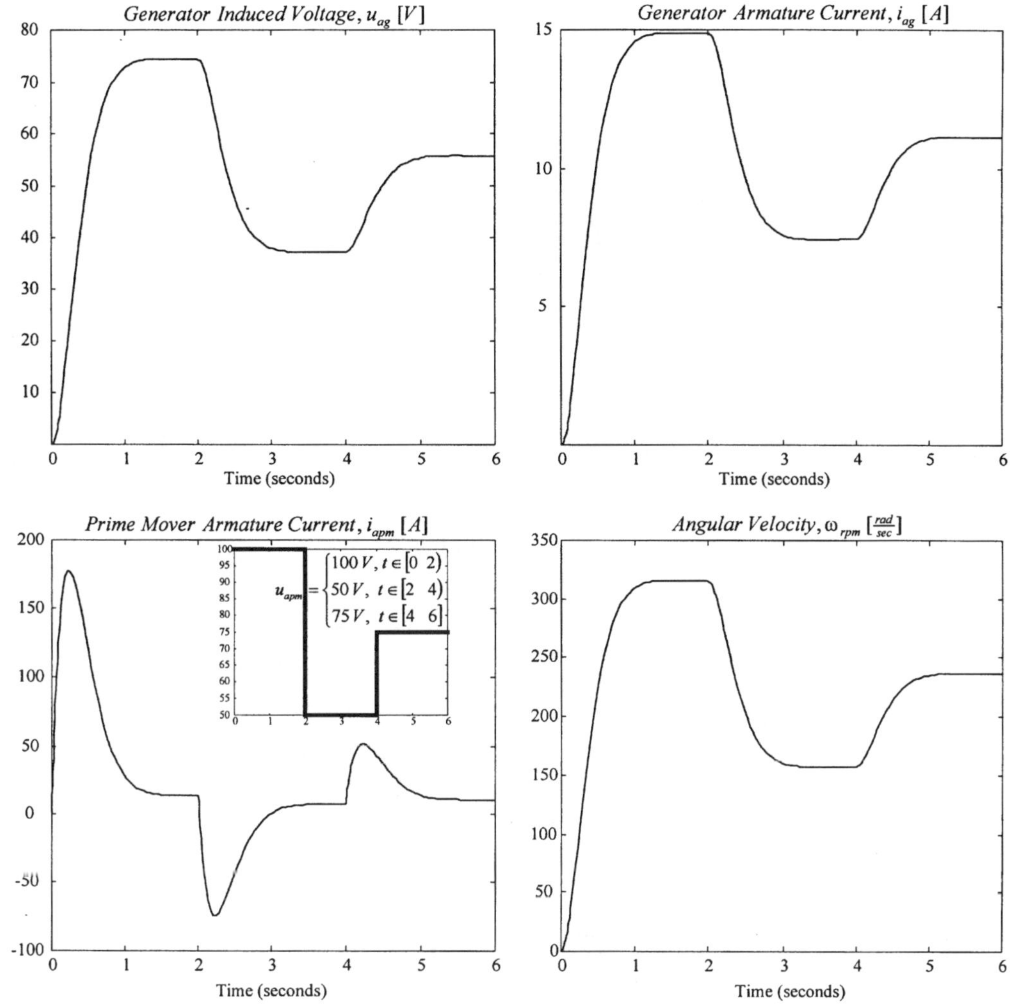

Figure 5.34 Generator and motor dynamics; R_L = 5 ohm and $u_{apm} = \begin{cases} 100\text{ V}, \ \forall t \in [0 \quad 2)\text{ s} \\ 50\text{ V}, \ \forall t \in [2 \quad 4)\text{ s} \\ 75\text{ V}, \ \forall t \in [4 \quad 6]\text{ s} \end{cases}$.

$$R_L = \begin{cases} 100\text{ ohm}, \ \forall t \in [0 \quad 2)\text{ s} \\ 5\text{ ohm}, \ \forall t \in [2 \quad 4)\text{ s} \\ 25\text{ ohm}, \ \forall t \in [4 \quad 6]\text{ s} \end{cases}$$

SIMULINK is used to perform simulations. A SIMULINK diagram is developed and documented in Figure 5.37 (`pm_pmg_2.mdl`).

The dynamics of the permanent-magnet DC generator and motor are studied. In particular, the state variables $i_{apm}(t)$, $\omega_{rpm}(t)$, $i_{ag}(t)$, as well as the voltage induced $u_{ag}(t)$, are plotted in Figures 5.38 to 5.40.

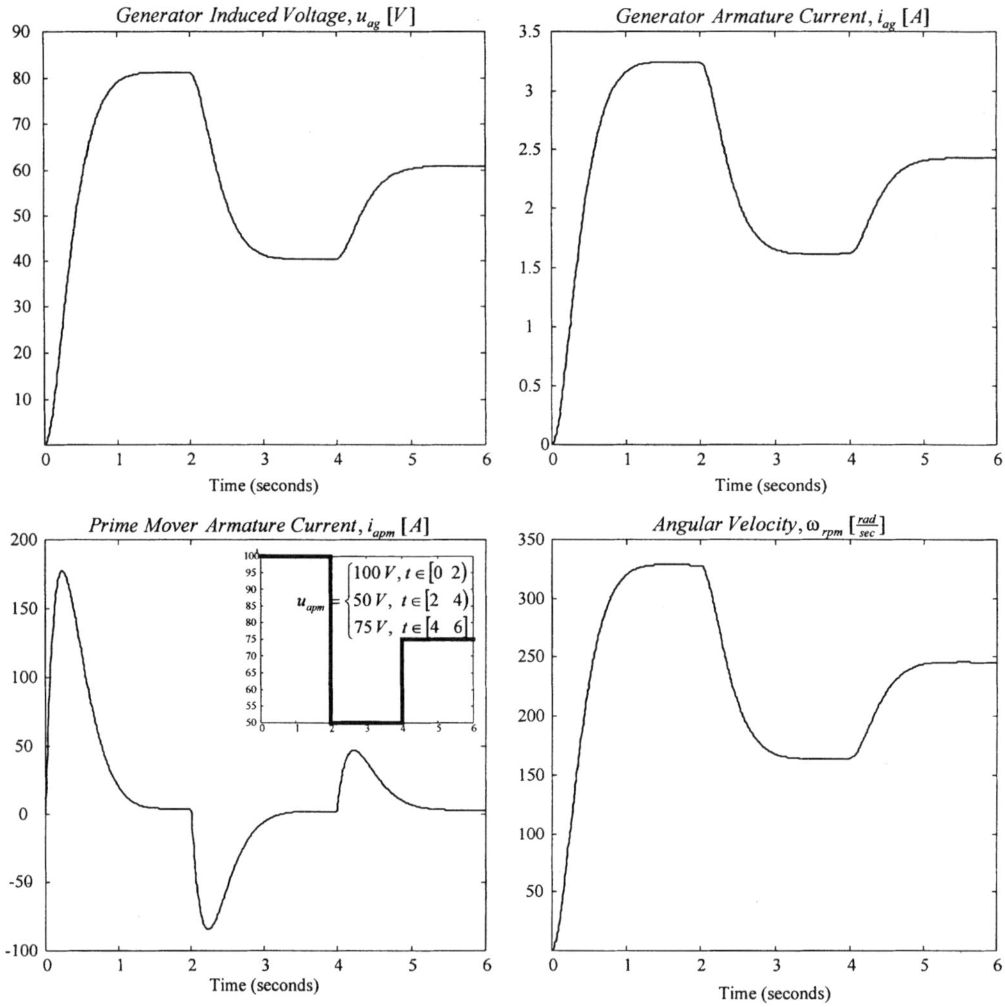

Figure 5.35 Generator and motor dynamics; R_L = 25 ohm and $u_{apm} = \begin{cases} 100\text{ V}, \forall t \in [0 \quad 2)\text{ s} \\ 50\text{ V}, \forall t \in [2 \quad 4)\text{ s} \\ 75\text{ V}, \forall t \in [4 \quad 6]\text{ s} \end{cases}$.

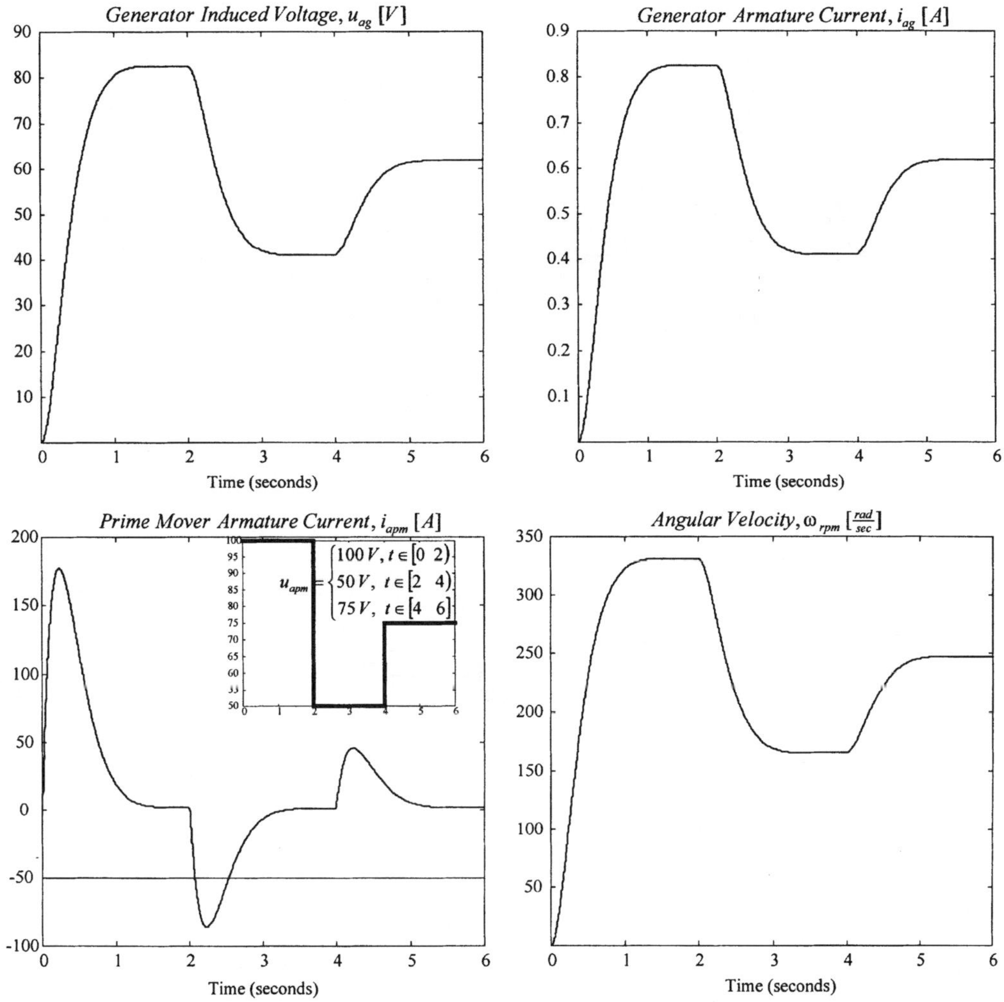

Figure 5.36 Generator and motor dynamics; $R_L = 100$ ohm and $u_{apm} = \begin{cases} 100\text{ V}, \forall t \in [0 \quad 2)\text{ s} \\ 50\text{ V}, \forall t \in [2 \quad 4)\text{ s} \\ 75\text{ V}, \forall t \in [4 \quad 6]\text{ s} \end{cases}$.

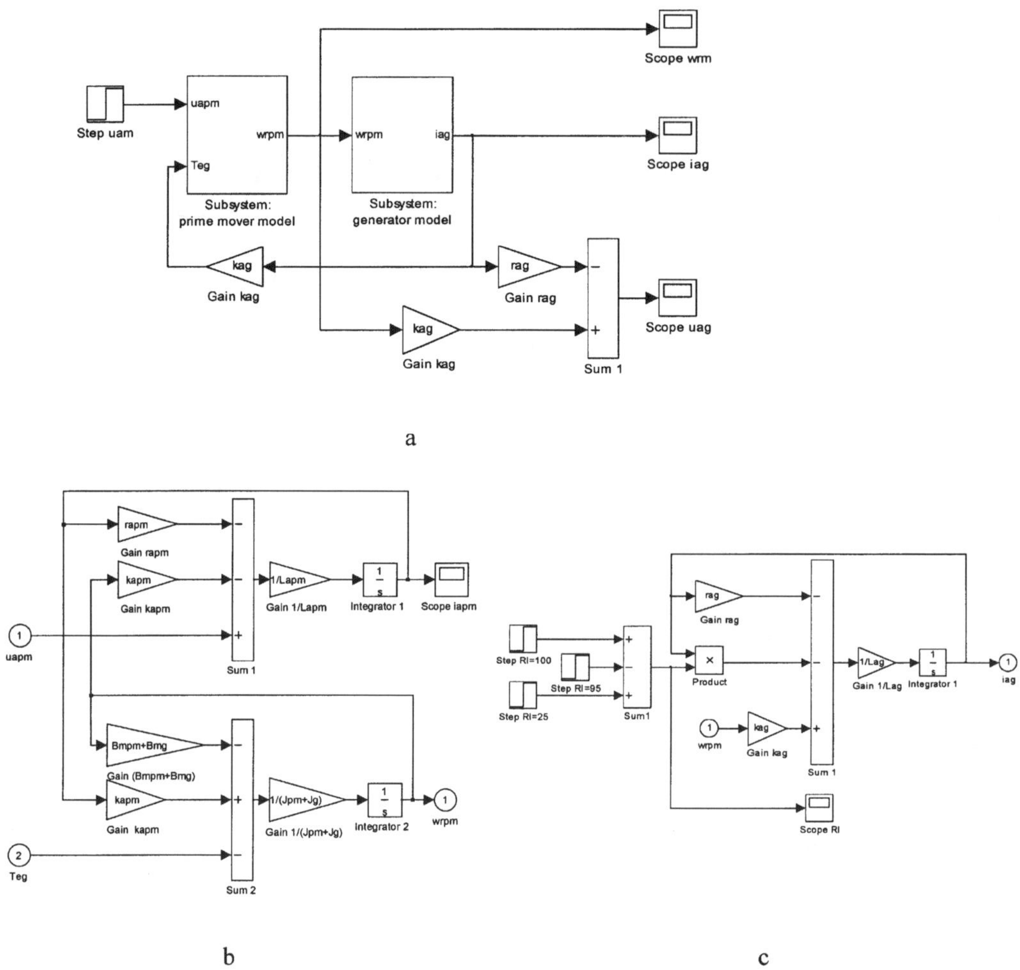

Figure 5.37 SIMULINK diagram to simulate the generation system: permanent-magnet DC generator–DC motor: (a) augmented system model; (b) SIMULINK subsystem diagram to model permanent-magnet DC motors; (c) SIMULINK subsystem diagram to model permanent-magnet generators.

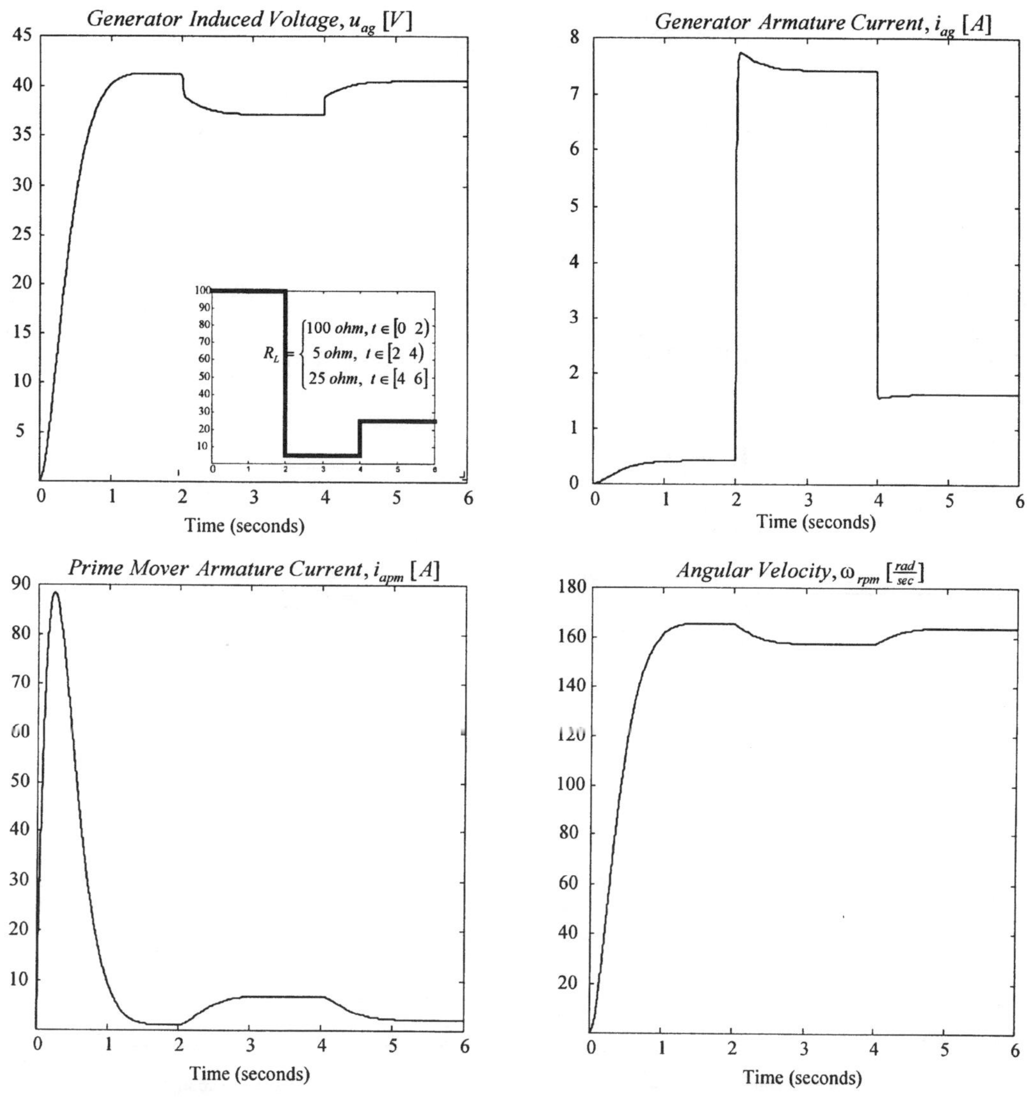

Figure 5.38 Generator and motor dynamics; $u_{apm} = 50$ V, $R_L = \begin{cases} 100 \text{ ohm}, \ \forall t \in [0 \quad 2) \text{ s} \\ 5 \text{ ohm}, \ \forall t \in [2 \quad 4) \text{ s} \\ 25 \text{ ohm}, \ \forall t \in [4 \quad 6] \text{ s} \end{cases}$.

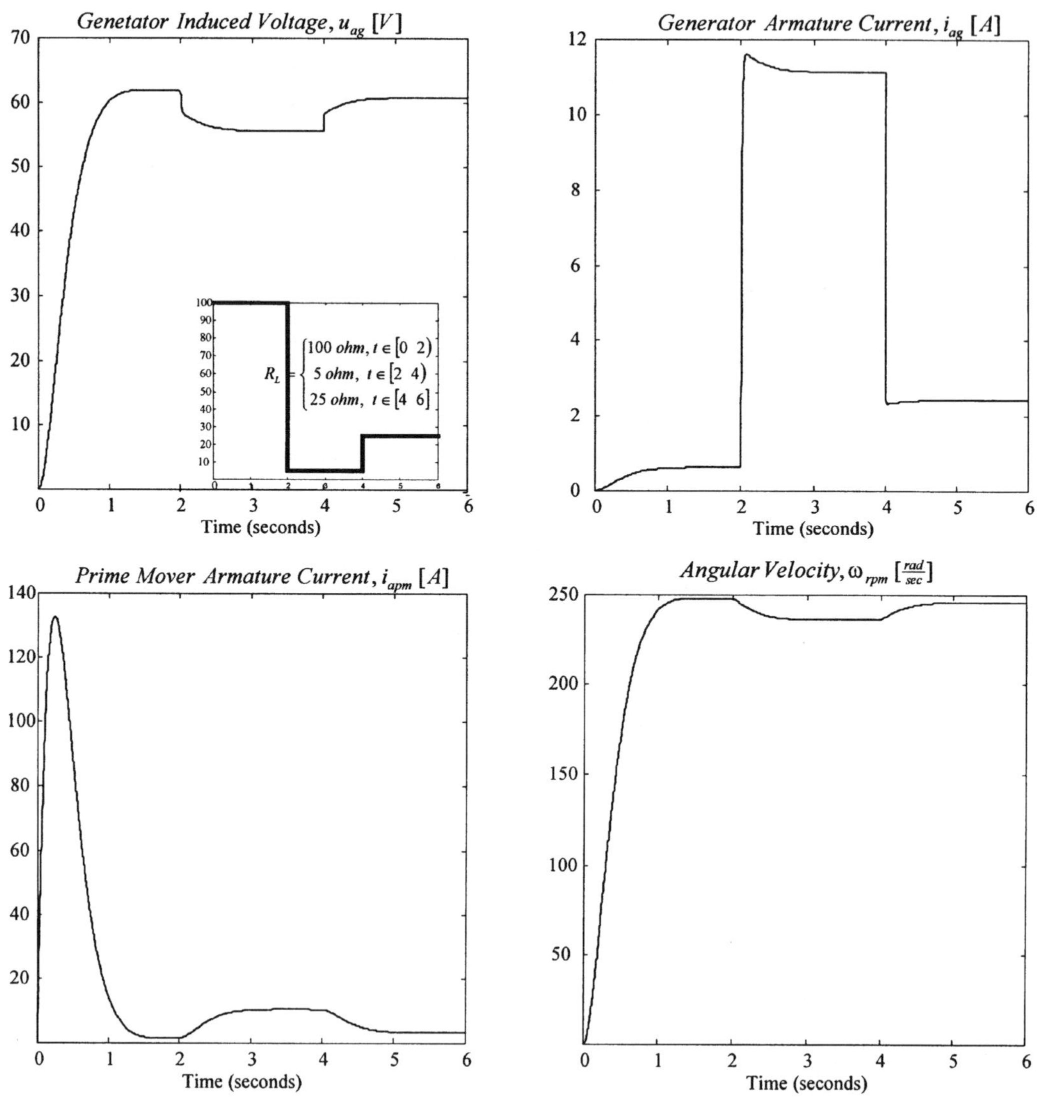

Figure 5.39 Generator and motor dynamics; $u_{apm} = 75$ V, $R_L = \begin{cases} 100 \text{ ohm}, \ \forall t \in [0 \quad 2) \text{ s} \\ 5 \text{ ohm}, \ \forall t \in [2 \quad 4) \text{ s} \\ 25 \text{ ohm}, \ \forall t \in [4 \quad 6] \text{ s} \end{cases}$.

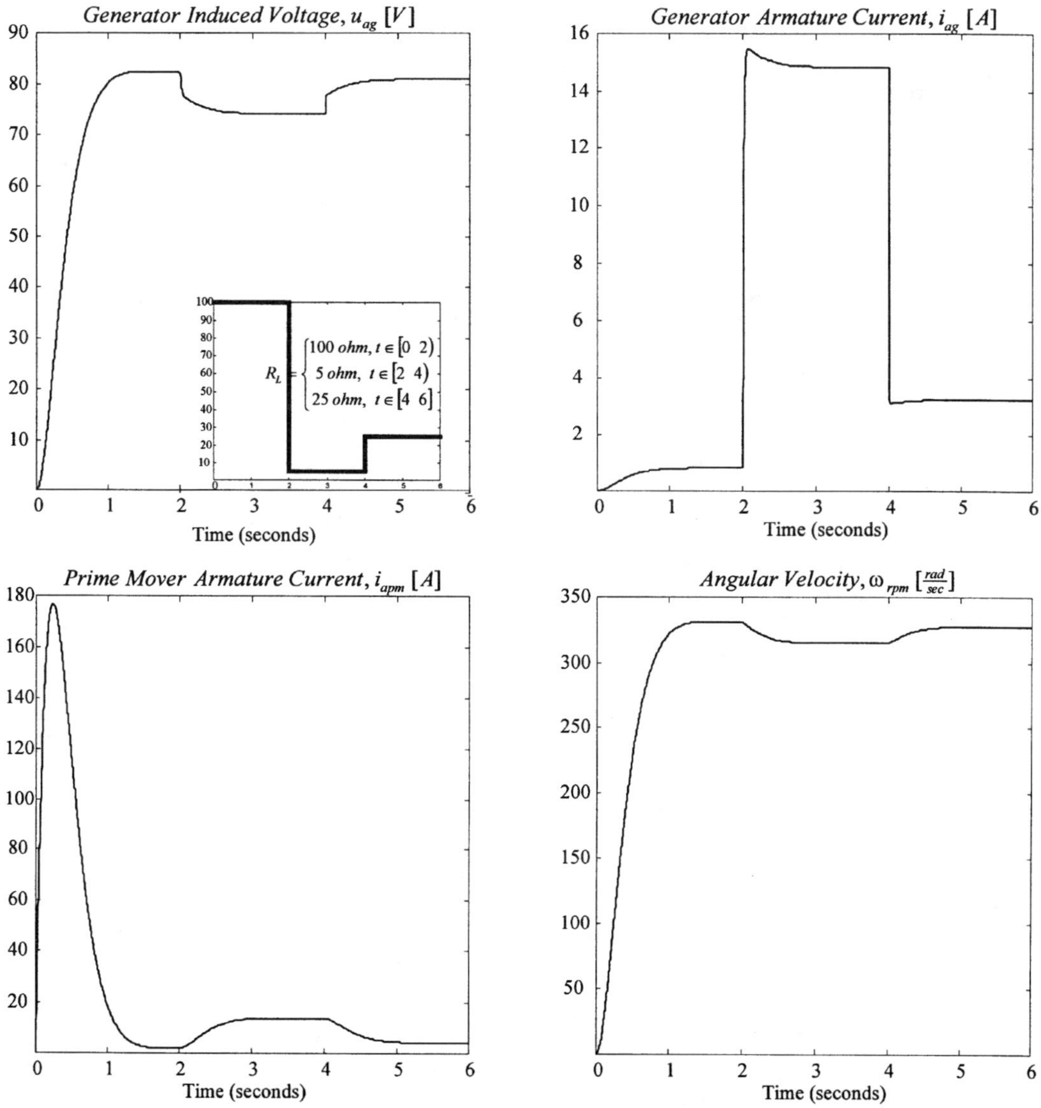

Figure 5.40 Generator and motor dynamics; $u_{apm} = 100$ V, $R_L = \begin{cases} 100 \text{ ohm}, \ \forall t \in [0 \quad 2) \text{ s} \\ 5 \text{ ohm}, \ \forall t \in [2 \quad 4) \text{ s} \\ 25 \text{ ohm}, \ \forall t \in [4 \quad 6] \text{ s} \end{cases}$.

5.8.2 Modeling and analysis of direct-current generators with permanent-magnet direct-current motors

Having derived the mathematical model for the prime mover–permanent-magnet DC generator, it should be emphasized that other DC electric machines can be used as generators. The differential equations for the generation systems can be found by augmenting the prime mover circuitry dynamics, with the *tosional-mechanical* dynamics.

Denoting

$$J_\Sigma = J_{pm} + J_g \text{ and } B_{m\Sigma} = B_{mpm} + B_{mg}$$

from the *tosional-mechanical* equation of motion

$$\frac{d\omega_{rpm}}{dt} = \frac{1}{J_{pm}+J_g}T_{epm} - \frac{B_{mpm}+B_{mg}}{J_{pm}+J_g}\omega_{rpm} - \frac{1}{J_{pm}+J_g}T_{eg}$$

we have

$$\frac{d\omega_{rpm}}{dt} = \frac{1}{J_\Sigma}T_{epm} - \frac{B_{m\Sigma}}{J_\Sigma}\omega_{rpm} - \frac{1}{J_\Sigma}T_{eg}$$

It is assumed that the permanent-magnet DC motor is used as a prime mover, and hence, the circuitry dynamics is mapped as

$$\frac{di_{apm}}{dt} = -\frac{r_{apm}}{L_{apm}}i_{apm} - \frac{k_{apm}}{L_{apm}}\omega_{rpm} + \frac{1}{L_{apm}}u_{apm}$$

and the electromagnetic torque is expressed as

$$T_{epm} = k_{apm}i_{apm}$$

Then, for different DC generators used in the generation systems illustrated in Figures 5.41 to 5.45, the augmented mathematical model can be easily derived using a set of

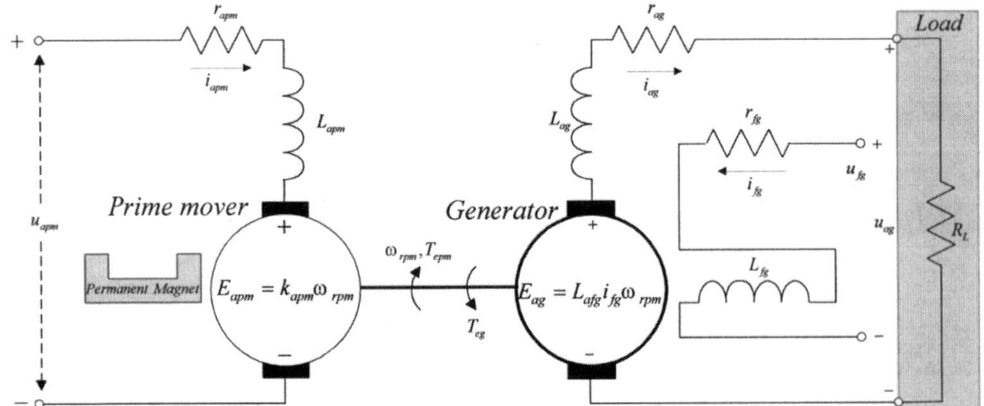

Figure 5.41 Separately excited DC generator driven by permanent-magnet DC motor.

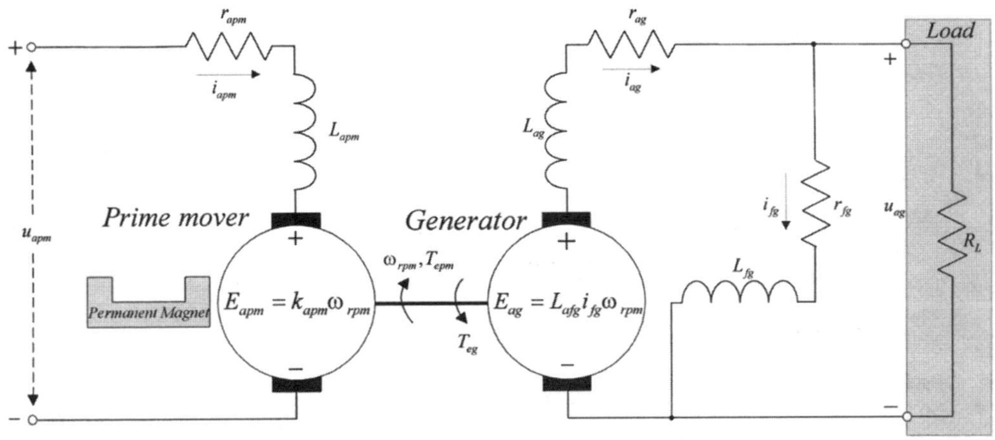

Figure 5.42 Shunt-connected DC generator driven by permanent-magnet DC motor.

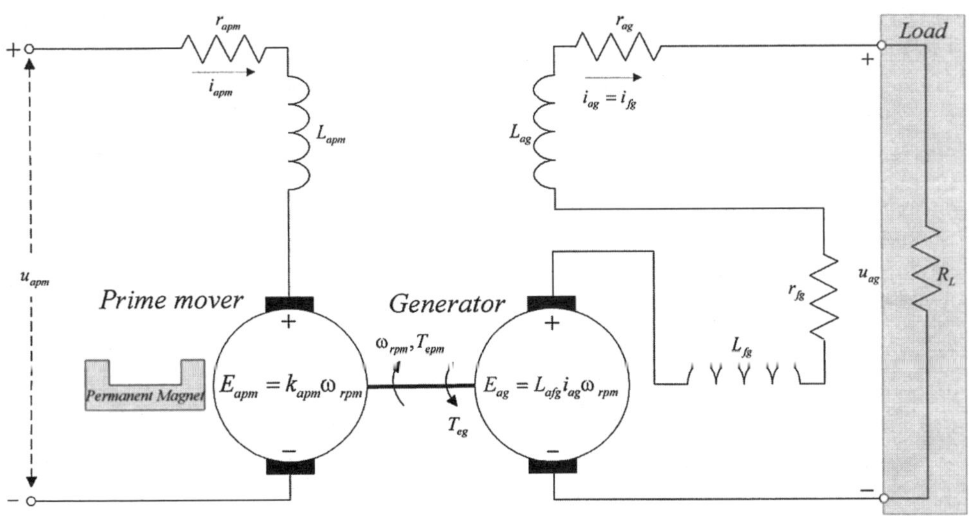

Figure 5.43 Series-connected DC generator driven by permanent-magnet DC motor.

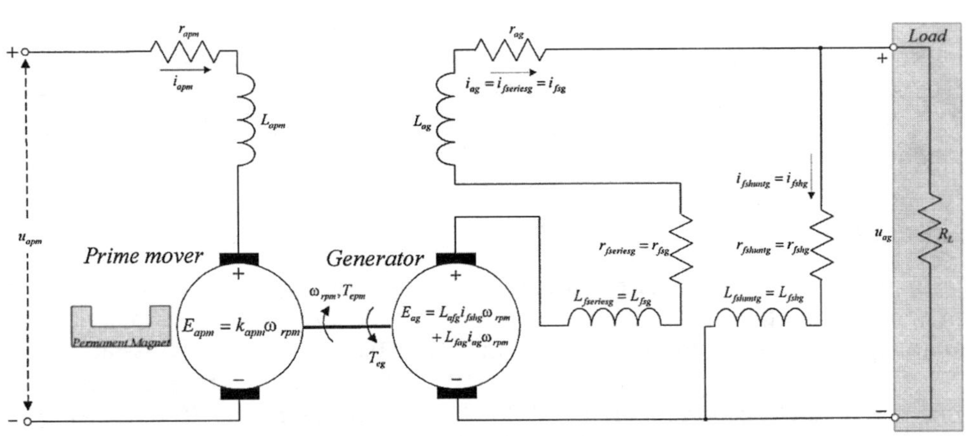

Figure 5.44 *Cumulatively compounded DC* generator driven by permanent-magnet DC motor.

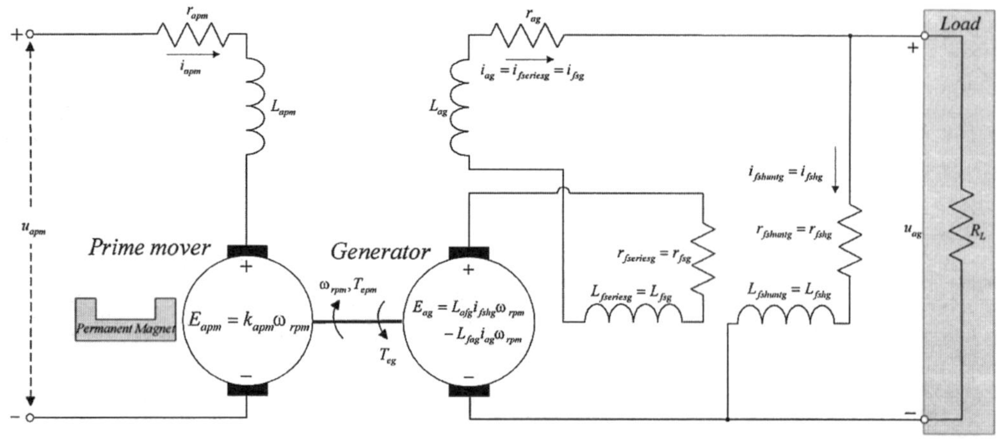

Figure 5.45 *Differentially compounded* DC generator driven by permanent-magnet DC motor.

differential equations given in Table 5.1. The corresponding equations for the load torque T_{eg} are documented.

Augmenting the differential equations derived, the mathematical models of the generation systems with permanent-magnet, separately excited, shunt- and series-connected, as well as cumulatively and differentially compounded DC generators, can be found.

For example, for differentially compounded DC generators driven by permanent-magnet DC motors, we have the following set of differential equations

$$\frac{di_{apm}}{dt} = -\frac{r_{apm}}{L_{apm}} i_{apm} - \frac{k_{apm}}{L_{apm}} \omega_{rpm} + \frac{1}{L_{apm}} u_{apm}$$

$$\frac{di_{ag}}{dt} = \frac{1}{(L_{ag} + L_{fsg})L_{fshg} - L_{ffg}^2} (-(r_{ag} + r_{fsg})L_{fshg} i_a - r_{fshg} L_{ffg} i_{fg}$$
$$- L_{fag} L_{fshg} i_{ag} \omega_{rpm} + L_{afg} L_{fsgh} i_{fg} \omega_{rpm} - (L_{fshg} + L_{ffg}) R_L i_{ag} + (L_{fshg} + L_{ffg}) R_L i_{fg})$$

$$\frac{di_{fg}}{dt} = \frac{1}{(L_{ag} + L_{fsg})L_{fshg} - L_{ffg}^2} ((r_{ag} + r_{fsg})L_{ffg} i_{ag} - r_{fshg}(L_{ag} + L_{fsg}) i_{fg}$$
$$+ L_{fag} L_{ffg} i_{ag} \omega_{rpm} - L_{afg} L_{ffg} i_{fg} \omega_{rpm} + (L_{ag} + L_{fsg} + L_{ffg}) R_L i_{ag} - (L_{ag} + L_{fsg} + L_{ffg}) R_L i_{fg})$$

$$\frac{d\omega_{rpm}}{dt} = \frac{k_{apm}}{J_\Sigma} i_{apm} - \frac{B_{m\Sigma}}{J_\Sigma} \omega_{rpm} - \frac{L_{afg}}{J_\Sigma} i_{ag} i_{fg} + \frac{L_{fag}}{J_\Sigma} i_{ag}^2$$

Example 5.8

Study the transient dynamics of the separately excited DC generator driven by a permanent-magnet DC motor. The parameters of the 2 kW, 100 V, 400 rad-s^{-1} permanent-magnet DC motor (prime mover) are: r_{apm} = 0.4 ohm, L_{apm} = 0.05 H, k_{apm} = 0.3 V-s-rad^{-1} or k_{apm} = 0.3 N-m-A^{-1}, J_{pm} = 0.01 kg-m^2, and B_{mpm} = 0.0007 N-m-s-rad^{-1}. The parameters of the separately excited DC generator are r_{ag} = 0.18 ohm, r_{fg} = 3.5 ohm, L_{ag} = 0.06 H, L_{fg} = 0.095 H, L_{afg} = 0.1 H, J_g = 0.01 kg-m^2, and B_{mg} = 0.001 N-m-s-rad^{-1}. Study the dynamics for different

Table 5.1 Differential Equations for the DC Generators

Generator Circuitry Dynamics	*Tosional-Mechanical* Dynamics
Permanent-magnet DC generator $\frac{di_{ag}}{dt}=-\frac{r_{ag}+R_L}{L_{ag}}i_{ag}+\frac{k_{ag}}{L_{ag}}\omega_{rpm}$	$\frac{d\omega_{rpm}}{dt}=\frac{1}{J_\Sigma}T_{epm}-\frac{B_{m\Sigma}}{J_\Sigma}\omega_{rpm}-\frac{1}{J_\Sigma}T_{eg}$ $T_{eg}=k_{ag}i_{ag}$
Separately excited DC generator $\frac{di_{ag}}{dt}=-\frac{r_{ag}+R_L}{L_{ag}}i_{ag}+\frac{L_{afg}}{L_{ag}}i_{fg}\omega_{rpm}$ $\frac{di_{fg}}{dt}=-\frac{r_{fg}}{L_{fg}}i_{fg}+\frac{1}{L_{fg}}u_{fg}$	$\frac{d\omega_{rpm}}{dt}=\frac{1}{J_\Sigma}T_{epm}-\frac{B_{m\Sigma}}{J_\Sigma}\omega_{rpm}-\frac{1}{J_\Sigma}T_{eg}$ $T_{eg}=L_{afg}i_{ag}i_{fg}$
Shunt-connected DC generator $\frac{di_{ag}}{dt}=-\frac{r_{ag}+R_L}{L_{ag}}i_{ag}+\frac{L_{afg}}{L_{ag}}i_{fg}\omega_{rpm}+\frac{R_L}{L_{ag}}i_{fg}$ $\frac{di_{fg}}{dt}=\frac{R_L}{L_{fg}}i_{ag}-\frac{r_{fg}+R_L}{L_{fg}}i_{fg}$	$\frac{d\omega_{rpm}}{dt}=\frac{1}{J_\Sigma}T_{epm}-\frac{B_{m\Sigma}}{J_\Sigma}\omega_{rpm}-\frac{1}{J_\Sigma}T_{eg}$ $T_{eg}=L_{afg}i_{ag}i_{fg}$
Series-connected DC generator $\frac{di_{ag}}{dt}=-\frac{r_{ag}+r_{fg}+R_L}{L_{ag}+L_{fg}}i_{ag}+\frac{L_{afg}}{L_{ag}+L_{fg}}i_{ag}\omega_{rpm}$	$\frac{d\omega_{rpm}}{dt}=\frac{1}{J_\Sigma}T_{epm}-\frac{B_{m\Sigma}}{J_\Sigma}\omega_{rpm}-\frac{1}{J_\Sigma}T_{eg}$ $T_{eg}=L_{afg}i_{ag}^2$

Table 5.1 Differential Equations for the DC Generators *(continued)*

Generator Circuitry Dynamics	*Tosional-Mechanical* Dynamics
Comulatively compounded DC generator	
$\frac{di_{ag}}{dt}=\frac{1}{(L_{ag}+L_{fsg})L_{fshg}-L_{ffg}^2}(-(r_{ag}+r_{fsg})L_{fshg}i_{ag}+r_{fshg}L_{ffg}i_{fg}$ $+L_{fag}L_{fshg}i_{ag}\omega_{rpm}+L_{afg}L_{fshg}i_{fg}\omega_{rpm}-(L_{fshg}-L_{ffg})R_Li_{ag}+(L_{fshg}-L_{ffg})R_Li_{fg})$	$\frac{d\omega_{rpm}}{dt}=\frac{1}{J_\Sigma}T_{epm}-\frac{B_{m\Sigma}}{J_\Sigma}\omega_{rpm}-\frac{1}{J_\Sigma}T_{eg}$ $T_{eg}=L_{afg}i_{ag}i_{fg}+L_{fag}i_{ag}^2$
$\frac{di_{fg}}{dt}=\frac{1}{(L_{ag}+L_{fsg})L_{fshg}-L_{ffg}^2}(-(r_{ag}+r_{fsh})L_{ffg}i_{ag}-r_{fshgg}(L_{ag}+L_{fsg})i_{fg}$ $-L_{fag}L_{ffg}i_{ag}\omega_{rpm}+L_{afg}L_{ffg}i_{fg}\omega_{rpm}+(L_{ag}+L_{fsg}-L_{ffg})R_Li_{ag}-(L_{ag}+L_{fsg}-L_{ffg})R_Li_{fg})$	
Differentially compounded DC generator	
$\frac{di_{ag}}{dt}=\frac{1}{(L_{ag}+L_{fsg})L_{fshg}-L_{ffg}^2}(-(r_{ag}+r_{fsg})L_{fshg}i_a-r_{fshg}L_{ffg}i_{fg}$ $-L_{fag}L_{fshg}i_{ag}\omega_{rpm}+L_{afg}L_{fshg}i_{fg}\omega_{rpm}-(L_{fshg}+L_{ffg})R_Li_{ag}+(L_{fshg}+L_{ffg})R_Li_{fg})$	$\frac{d\omega_{rpm}}{dt}=\frac{1}{J_\Sigma}T_{epm}-\frac{B_{m\Sigma}}{J_\Sigma}\omega_{rpm}-\frac{1}{J_\Sigma}T_{eg}$ $T_e=L_{afg}i_{ag}i_{fg}-L_{fag}i_{ag}^2$
$\frac{di_{fg}}{dt}=\frac{1}{(L_{ag}+L_{fsg})L_{fshg}-L_{ffg}^2}((r_{ag}+r_{fsg})L_{ffg}i_{ag}-r_{fshg}(L_{ag}+L_{fsg})i_{fg}$ $+L_{fag}L_{ffg}i_{ag}\omega_{rpm}-L_{afg}L_{ffg}i_{fg}\omega_{rpm}+(L_{ag}+L_{fsg}+L_{ffg})R_Li_{ag}-(L_{ag}+L_{fsg}+L_{ffg})R_Li_{fg})$	

angular velocities and resistive loads. Perform the analysis of the induced voltage for the different angular velocities, load conditions, and excitation voltages applied.

Solution

Using Table 5.1, one finds the following set of differential equations to be used in modeling:

$$\frac{di_{apm}}{dt} = -\frac{r_{apm}}{L_{apm}} i_{apm} - \frac{k_{apm}}{L_{apm}} \omega_{rpm} + \frac{1}{L_{apm}} u_{apm}$$

$$\frac{di_{ag}}{dt} = -\frac{r_{ag} + R_L}{L_{ag}} i_{ag} + \frac{L_{afg}}{L_{ag}} i_{fg} \omega_{rpm}$$

$$\frac{di_{fg}}{dt} = -\frac{r_{fg}}{L_{fg}} i_{fg} + \frac{1}{L_{fg}} u_{fg}$$

$$\frac{d\omega_{rpm}}{dt} = \frac{k_{apm}}{J_\Sigma} i_{apm} - \frac{B_{m\Sigma}}{J_\Sigma} \omega_{rpm} - \frac{L_{afg}}{J_\Sigma} i_{ag} i_{fg}$$

That is, the state variables are $i_{apm}(t)$, $i_{ag}(t)$, $i_{fg}(t)$, and $\omega_{rpm}(t)$; and for the generation system studied, the output is the voltage generated $u_{ag}(t)$.

To vary the angular velocity of the prime mover, one changes the armature voltage applied to the rotor winding of the prime mover. In particular, for $u_a = 50$ V, $u_a = 75$ V and $u_a = 100$ V, one obtains different steady-state values for the angular velocity ω_{rpm}. Different resistive loads R_L and the field voltages u_{fg} are used. In particular, let us study the generation system dynamics and steady-state performance, generation systems stability, and induced voltage waveforms if

$$R_L = \begin{cases} 100 \text{ ohm} \\ 5 \text{ ohm} \\ 25 \text{ ohm} \end{cases} \quad \text{and} \quad u_{fg} = \begin{cases} 30 \text{ V} \\ 10 \text{ V} \\ 20 \text{ V} \end{cases}$$

The dynamics of the state variables $i_{amp}(t)$, $i_{ag}(t)$, $i_{fg}(t)$, and $\omega_{rpm}(t)$, as well as the waveforms for the voltage induced $u_{ag}(t)$, are plotted in Figures 5.46 to 5.51. In particular, the steady-state values for the currents, angular velocity, and the generated voltage are found and illustrated by numerically solving the differential equations in the MATLAB environment. To perform the modeling, two MATLAB scripts are used (files `pm_seg_1.m` and `pm_seg_2.m`).

MATLAB script (pm_seg_1.m)

```
tspan=[0 4]; y0=[0 0 0 0]';
global rag Lafg;
options = odeset('RelTol',1e-4,'AbsTol',[1e-4 1e-4 1e-4 1e-4]);
[t,y] = ODE45('pm_seg_2',tspan,y0,options);
subplot(3,2,1);
% Induced voltage by separately excited DC generators
uag=Lafg*y(:,3).*y(:,4)-rag*y(:,2);
plot(t,uag);title('Generator Induced Voltage, uag [V]');
```

```
xlabel('Time (seconds)'); grid;
subplot(3,2,2);
plot(t,y(:,1));title('Prime Mover Armature Current, iapm [A]');
xlabel('Time (seconds)'); grid;
subplot(3,2,3);
plot(t,y(:,2));title('Generator Armature Current, iag [A]');
xlabel('Time (seconds)'); grid;
subplot(3,2,4);
plot(t,y(:,3));title('Generator Field Current, ifg [A]');
xlabel('Time (seconds)'); grid;
subplot(3,2,5);
plot(t,y(:,4));title('Angular Velocity, wrpm [rad/sec]');
xlabel('Time (seconds)'); grid;
disp('End')
```

MATLAB script (pm_seg_2.m)

```
function yprime=difer(t,y);
global rag Lafg;
% Parameters of the prime mover
rapm=0.4;Lapm=0.05;kapm=0.3;Jpm=0.01;Bmpm=0.0007;
% Applied voltage to the prime mover
uapm=100;
% Parameters of the separately excited DC generator
rag=0.18;rfg=3.5;Lag=0.06;Lfg=0.095;Lafg=0.1;Jg=0.01;Bmg=0.001;
% Applied voltage to the generator field winding
if t<=2, ufg=30;
else ufg=10;
end
if t>=3, ufg=20;
end
% Load resistance
Rl=25;
% Differential equations of the generation system
yprime=[(-rapm*y(1,:)-kapm*y(4,:)+uapm)/Lapm;...
(-(rag+Rl)*y(2,:)+Lafg*y(3,:).*y(4,:))/Lag;...
(-rfg*y(3,:)+ufg)/Lfg;...
(kapm*y(1,:)-Lafg*y(2,:).*y(3,:)-(Bmpm+Bmg)*y(4,:))/(Jpm+Jg)];
```

Example 5.9

Study the transient dynamics and perform the analysis of the steady-state performance, including the voltage generation, for a shunt-connected DC generator driven by a permanent-magnet DC motor. The parameters of the 2 kW, 100 V, 400 rad-s^{-1} permanent-magnet DC motor are: r_{apm} = 0.4 ohm, L_{apm} = 0.05 H, k_{apm} = 0.3 V-s-rad^{-1} or k_{apm} = 0.3 N-m-A^{-1}, J_{pm} = 0.01 kg-m^2, and B_{mpm} = 0.0007 N-m-s-rad^{-1}. The parameters of the shunt-connected DC generator are r_{ag} = 0.1 ohm, r_{fg} = 75 ohm, L_{ag} = 0.06 H, L_{fg} = 150 H, L_{afg} = 1 H, J_g = 0.01 kg-m^2, and B_{mg} = 0.001 N-m-s-rad^{-1}.

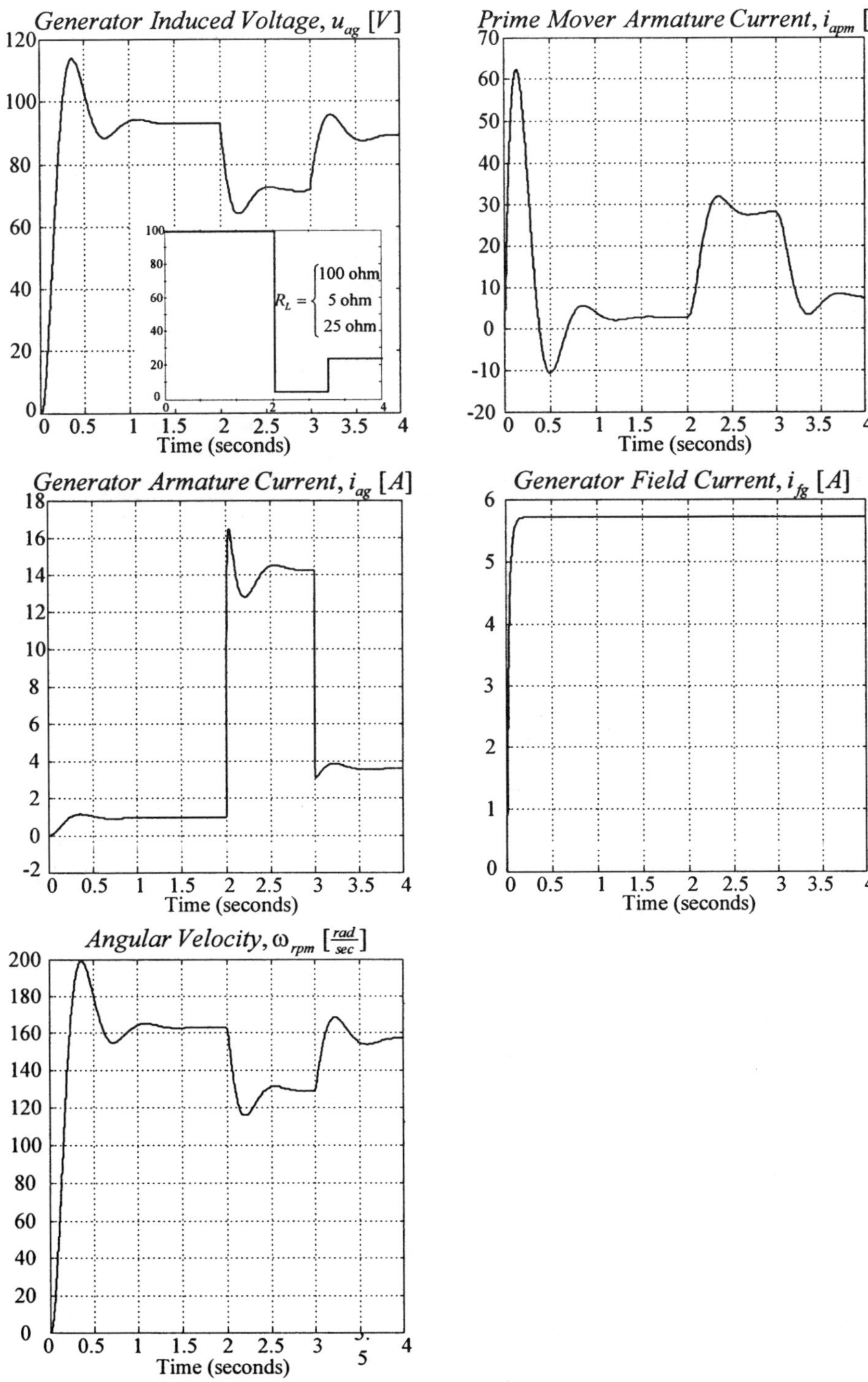

Figure 5.46 Generator and motor dynamics; $u_{apm} = 50$ V, $u_{fg} = 20$ V, and $R_L = \begin{cases} 100\text{ ohm}, \ \forall t \in [0 \quad 2)\text{ s} \\ 5\text{ ohm}, \ \forall t \in [2 \quad 3)\text{ s} \\ 25\text{ ohm}, \ \forall t \in [3 \quad 4)\text{ s} \end{cases}$.

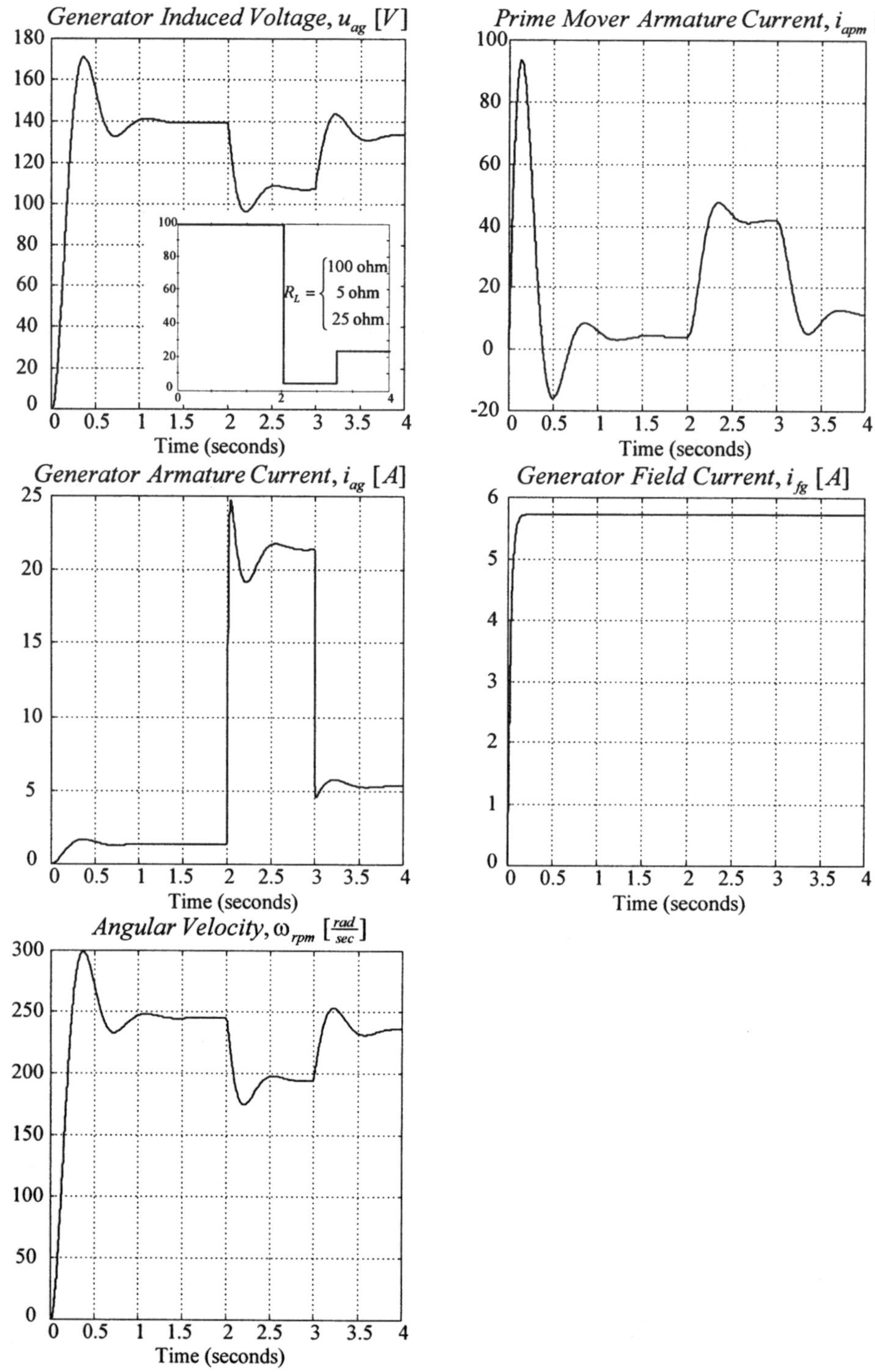

Figure 5.47 Generator and motor dynamics; $u_{apm} = 75$ V, $u_{fg} = 20$ V, and $R_L = \begin{cases} 100 \text{ ohm}, \ \forall t \in [0 \quad 2) \text{ s} \\ 5 \text{ ohm}, \ \forall t \in [2 \quad 3) \text{ s} \\ 25 \text{ ohm}, \ \forall t \in [3 \quad 4) \text{ s} \end{cases}$.

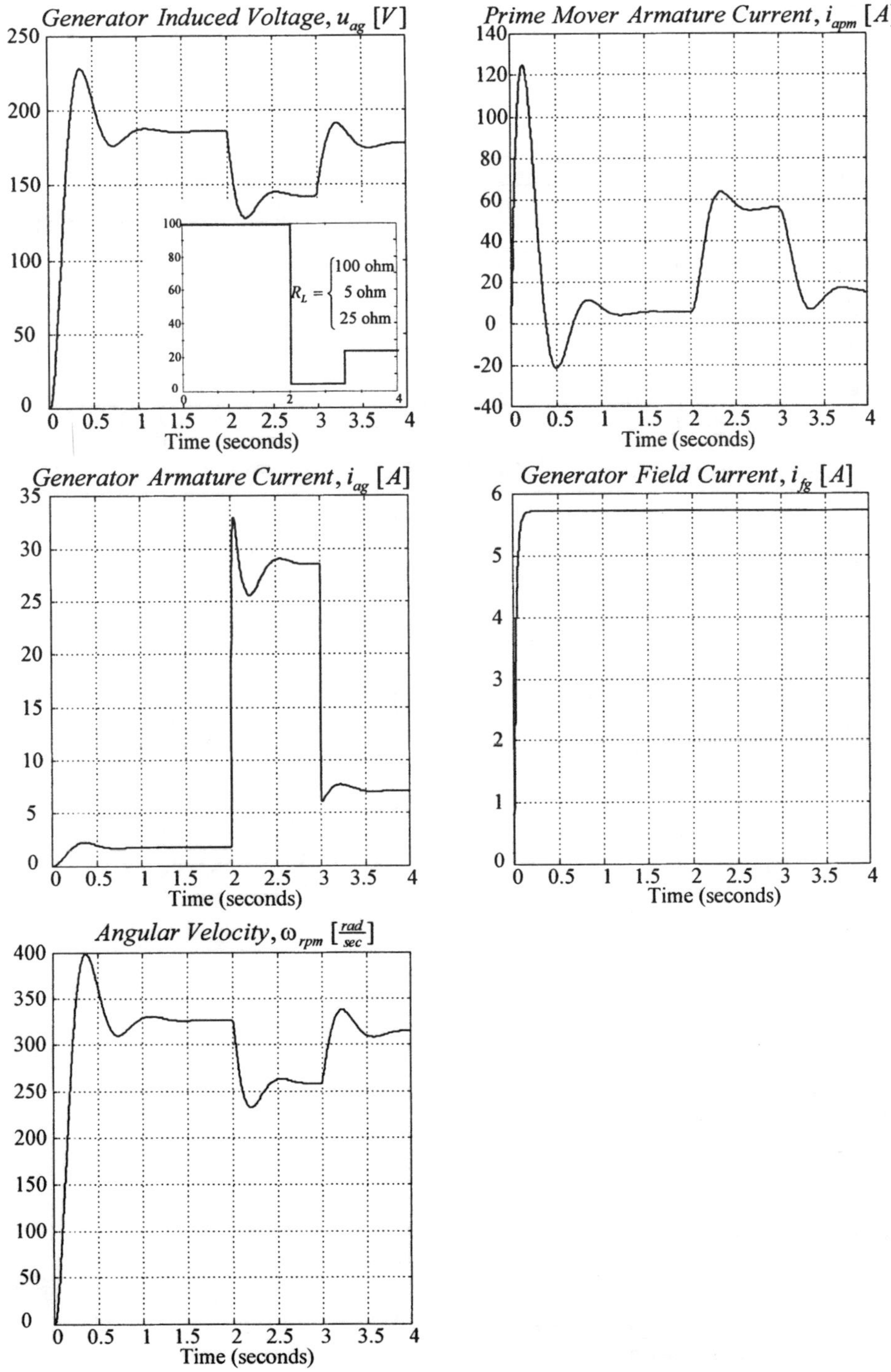

Figure 5.48 Generator and motor dynamics; $u_{apm} = 100$ V, $u_{fg} = 20$ V, and $R_L = \begin{cases} 100\ \text{ohm},\ \forall t \in [0 \quad 2)\ \text{s} \\ 5\ \text{ohm},\ \forall t \in [2 \quad 3)\ \text{s} \\ 25\ \text{ohm},\ \forall t \in [3 \quad 4)\ \text{s} \end{cases}$.

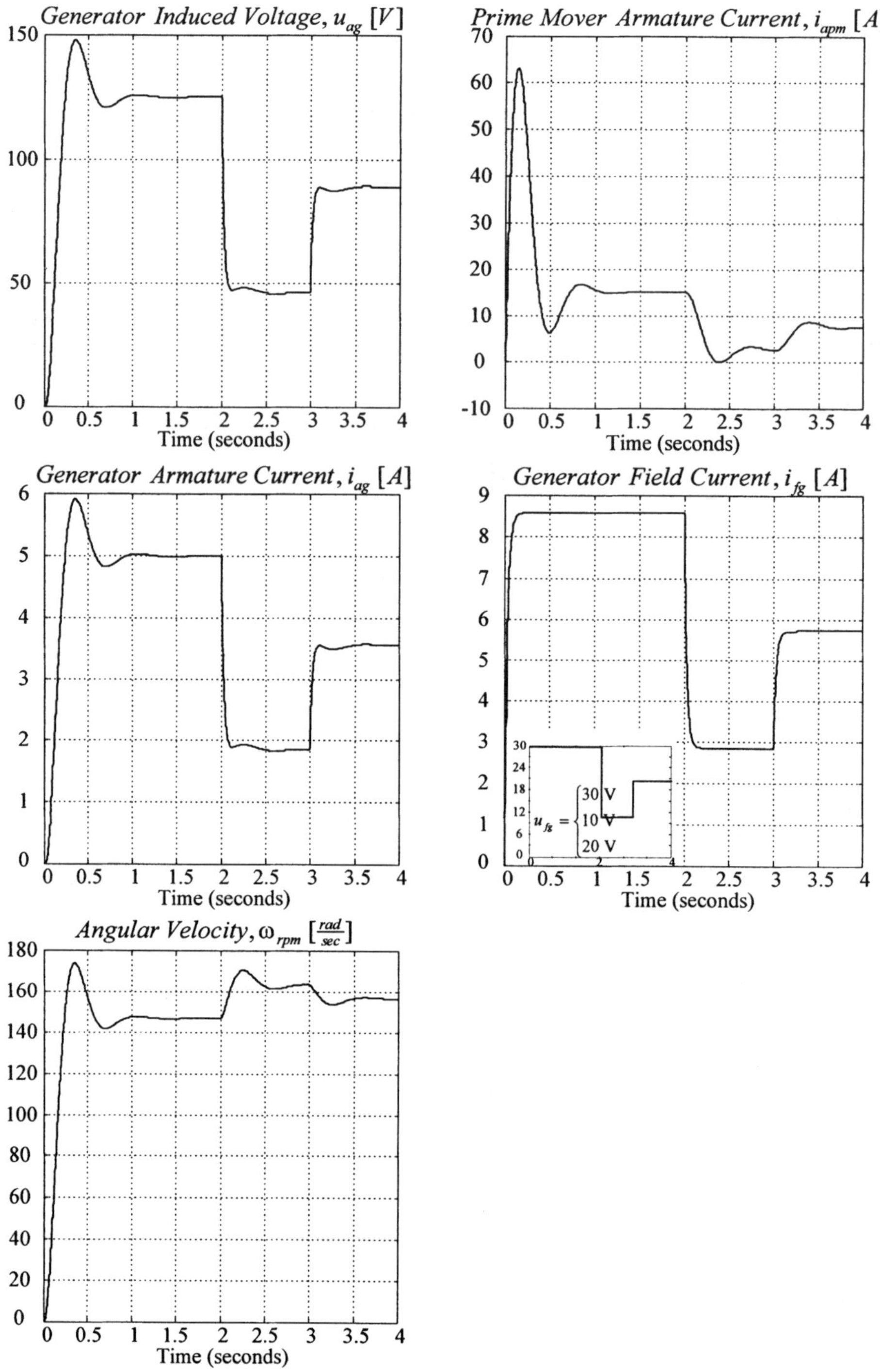

Figure 5.49 Generator and motor dynamics; $u_{apm} = 50$ V, $R_L = 25$ ohm, and $u_{fg} = \begin{cases} 30\text{ V}, \ \forall t \in [0 \quad 2)\text{ s} \\ 10\text{ V}, \ \forall t \in [2 \quad 3)\text{ s}. \\ 20\text{ V}, \ \forall t \in [3 \quad 4)\text{ s} \end{cases}$

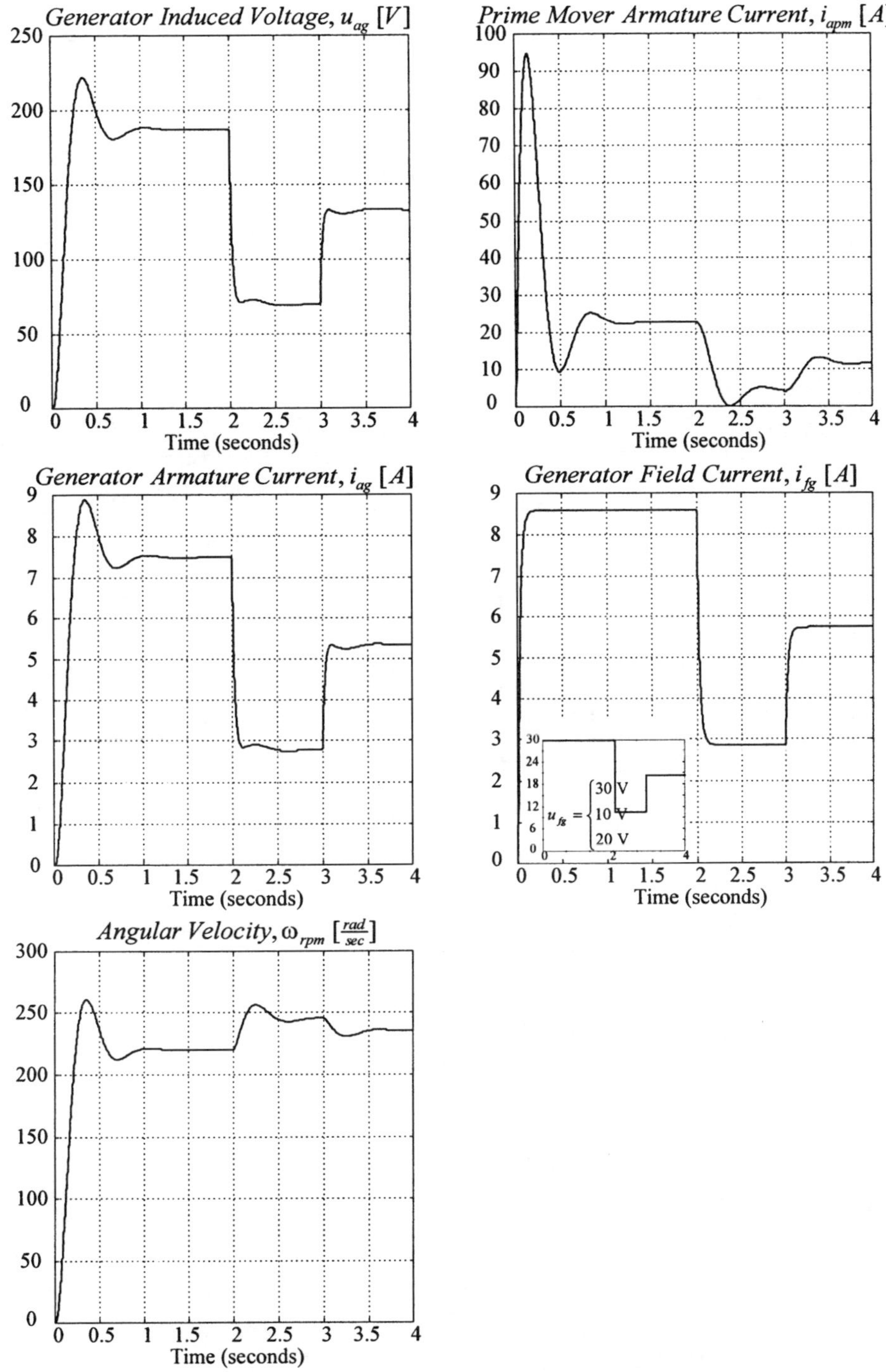

Figure 5.50 Generator and motor dynamics; u_{apm} = 75 V, R_L = 25 ohm, and $u_{fg} = \begin{cases} 30\text{ V}, \ \forall t \in [0 \quad 2)\text{ s} \\ 10\text{ V}, \ \forall t \in [2 \quad 3)\text{ s}. \\ 20\text{ V}, \ \forall t \in [3 \quad 4)\text{ s} \end{cases}$

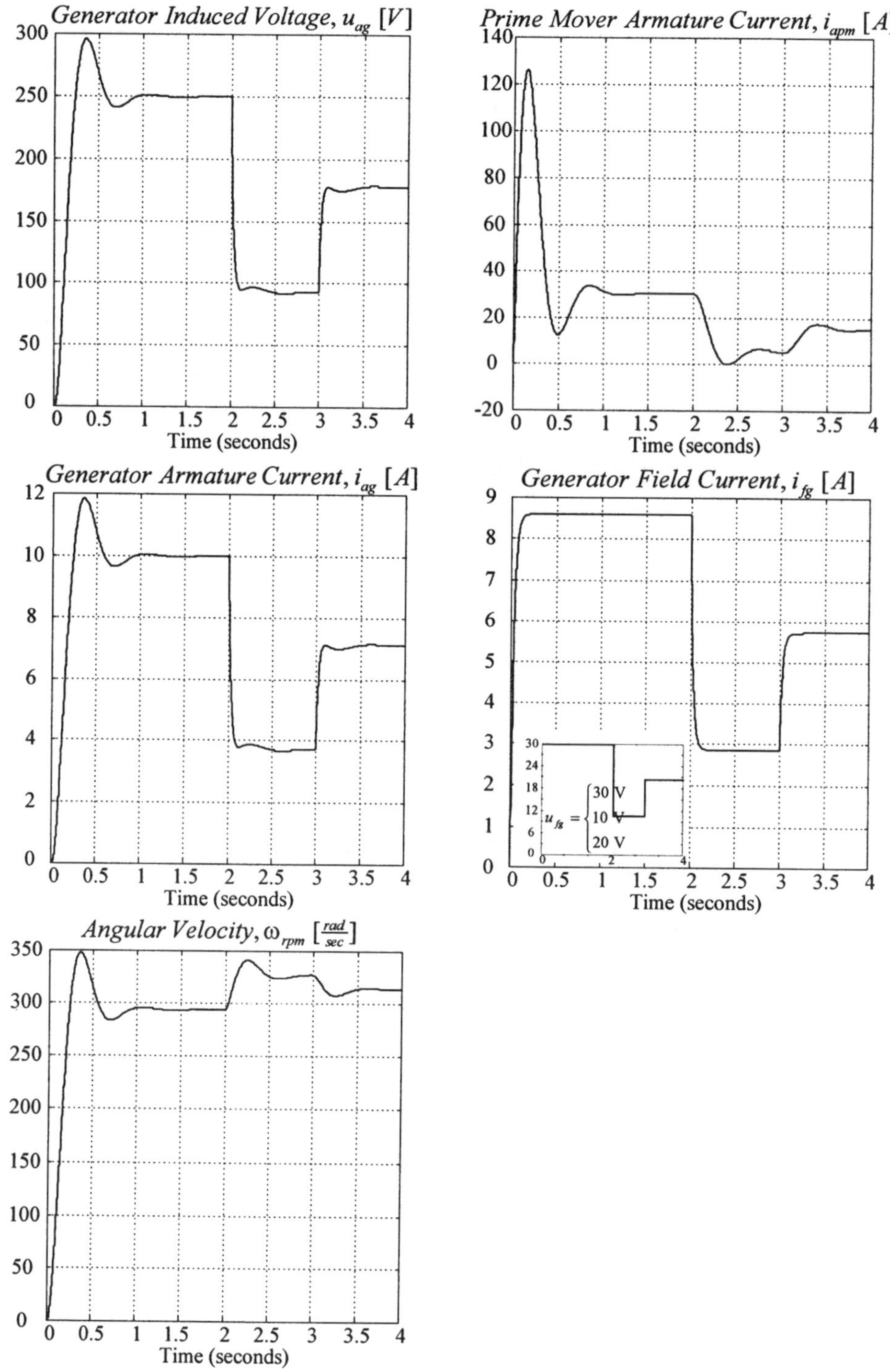

Figure 5.51 Generator and motor dynamics; $u_{apm} = 100$ V, $R_L = 25$ ohm, and $u_{fg} = \begin{cases} 30\text{ V}, & \forall t \in [0 \quad 2)\text{ s} \\ 10\text{ V}, & \forall t \in [2 \quad 3)\text{ s}. \\ 20\text{ V}, & \forall t \in [3 \quad 4)\text{ s} \end{cases}$

Solution

The dynamics and steady-state performance of the shunt-connected DC generator driven by the permanent-magnet DC motor are studied. Using Table 5.1, the differential equations result to map the transient behavior. The state variables are $i_{apm}(t)$, $i_{ag}(t)$, $i_{fg}(t)$, and $\omega_{rpm}(t)$, and the voltage induced by the generator $u_{ag}(t)$ is the output. Two m-files are developed to solve a set of highly nonlinear differential equations, as given by

$$\frac{di_{apm}}{dt} = -\frac{r_{apm}}{L_{apm}} i_{apm} - \frac{k_{apm}}{L_{apm}} \omega_{rpm} + \frac{1}{L_{apm}} u_{apm}$$

$$\frac{di_{ag}}{dt} = -\frac{r_{ag} + R_L}{L_{ag}} i_{ag} + \frac{L_{afg}}{L_{ag}} i_{fg} \omega_{rpm} + \frac{R_L}{L_{ag}} i_{fg}$$

$$\frac{di_{fg}}{dt} = \frac{R_L}{L_{fg}} i_{ag} - \frac{r_{fg} + R_L}{L_{fg}} i_{fg}$$

$$\frac{d\omega_{rpm}}{dt} = \frac{k_{apm}}{J_\Sigma} i_{apm} - \frac{B_{m\Sigma}}{J_\Sigma} \omega_{rpm} - \frac{L_{afg}}{J_\Sigma} i_{ag} i_{fg}$$

The resulting transient dynamics are plotted in Figure 5.52 if

$$u_{apm} = 50 \text{ V}$$

and

$$R_L = \begin{cases} 50 \text{ ohm}, \ \forall t \in [0 \quad 2) \text{ s} \\ 10 \text{ ohm}, \ \forall t \in [2 \quad 3) \text{ s} \\ 25 \text{ ohm}, \ \forall t \in [3 \quad 4) \text{ s} \end{cases}$$

It should be emphasized that shunt-connected DC generators belong to the class of self-excited generators. These generators are sensitive to variations of the parameters, as well as to load changes and variations of the angular velocity. Furthermore, to build the voltage, the self-excited generators should be initially excited. In our simulations, the following initial conditions are assigned $i_{fg}(0) = 3$ A and $\omega_{rpm}(0) = 50$ rad-s^{-1}.

MATLAB script (`pm_shg_1.m`)

```
tspan=[0 4]; y0=[0 0 3 50]';
global rag Lafg;
options = odeset('RelTol',1e-4,'AbsTol',[1e-4 1e-4 1e-4 1e-4]);
[t,y] = ODE45('pm_shg_2',tspan,y0,options);
subplot(3,2,1);
% Induced voltage by shunt-connected DC generators
uag=Lafg*y(:,3).*y(:,4)-rag*y(:,2);
plot(t,uag);title('Generator Induced Voltage, uag [V]');
xlabel('Time (seconds)'); grid;
```

```
subplot(3,2,2);
plot(t,y(:,1));title('Prime Mover Armature Current, iapm [A]');
xlabel('Time (seconds)'); grid;
subplot(3,2,3);
plot(t,y(:,2));title('Generator Armature Current, iag [A]');
xlabel('Time (seconds)'); grid;
subplot(3,2,4);
plot(t,y(:,3));title('Generator Field Current, ifg [A]');
xlabel('Time (seconds)'); grid;
subplot(3,2,5);
plot(t,y(:,4));title('Angular Velocity, wrpm [rad/sec]');
xlabel('Time (seconds)');
disp('End')
```

MATLAB script (pm_shg_2.m)

```
function yprime=difer(t,y);
global rag Lafg;
% Parameters of the prime mover
rapm=0.4;Lapm=0.05;kapm=0.3;Jpm=0.01;Bmpm=0.0007;
% Applied voltage to the prime mover
uapm=50;
% Parameters of the shunt-connected DC generator
rag=0.1;rfg=75;Lag=0.06;Lfg=150;Lafg=1;Jg=0.01;Bmg=0.001;
% Load resistance
if t<=2, Rl=50;
else Rl=10;
end
if t>=3, Rl=25;
end
% Differential equations of the generation system
yprime=[(-rapm*y(1,:)-kapm*y(4,:)+uapm)/Lapm;...
(-(rag+Rl)*y(2,:)+Lafg*y(3,:).*y(4,:)+Rl*y(3,:))/Lag;...
(Rl*y(2,:)-(rfg+Rl)*y(3,:))/Lfg;...
(kapm*y(1,:)-Lafg*y(2,:).*y(3,:)-(Bmpm+Bmg)*y(4,:))/(Jpm+Jg)];
```

Example 5.10

Model the series-connected DC generator driven by the 2 kW, 100 V, 400 rad-s^{-1} permanent-magnet DC motor with the following parameters: r_{apm} = 0.4 ohm, L_{apm} = 0.05 H, k_{apm} = 0.3 V-s-rad^{-1} or k_{apm} 0.3 N-m-A^{-1}, J_{pm} = 0.01 kg-m^2, and B_{mpm} = 0.0007 N-m-s-rad^{-1}. The parameters of a series-connected DC generator are: r_{ag} = 0.1 ohm, r_{fg} = 2 ohm, L_{ag} = 0.06 H, L_{fg} = 6 H, L_{afg} = 0.71 H, J_g = 0.01 kg-m^2, and B_{mg} = 0.001 N-m-s-rad^{-1}. Study the dynamics and analyze the voltage induced by the generator for different angular velocity and load resistance.

Solution

The differential equations, needed to be used and to be numerically solved, are given in Table 5.1. The state variables of interest are $i_{apm}(t)$, $i_{ag}(t)$, and $\omega_{rpm}(t)$. The voltage induced by the generator is denoted as $u_{ag}(t)$. Two m-files are developed, and the resulting transient

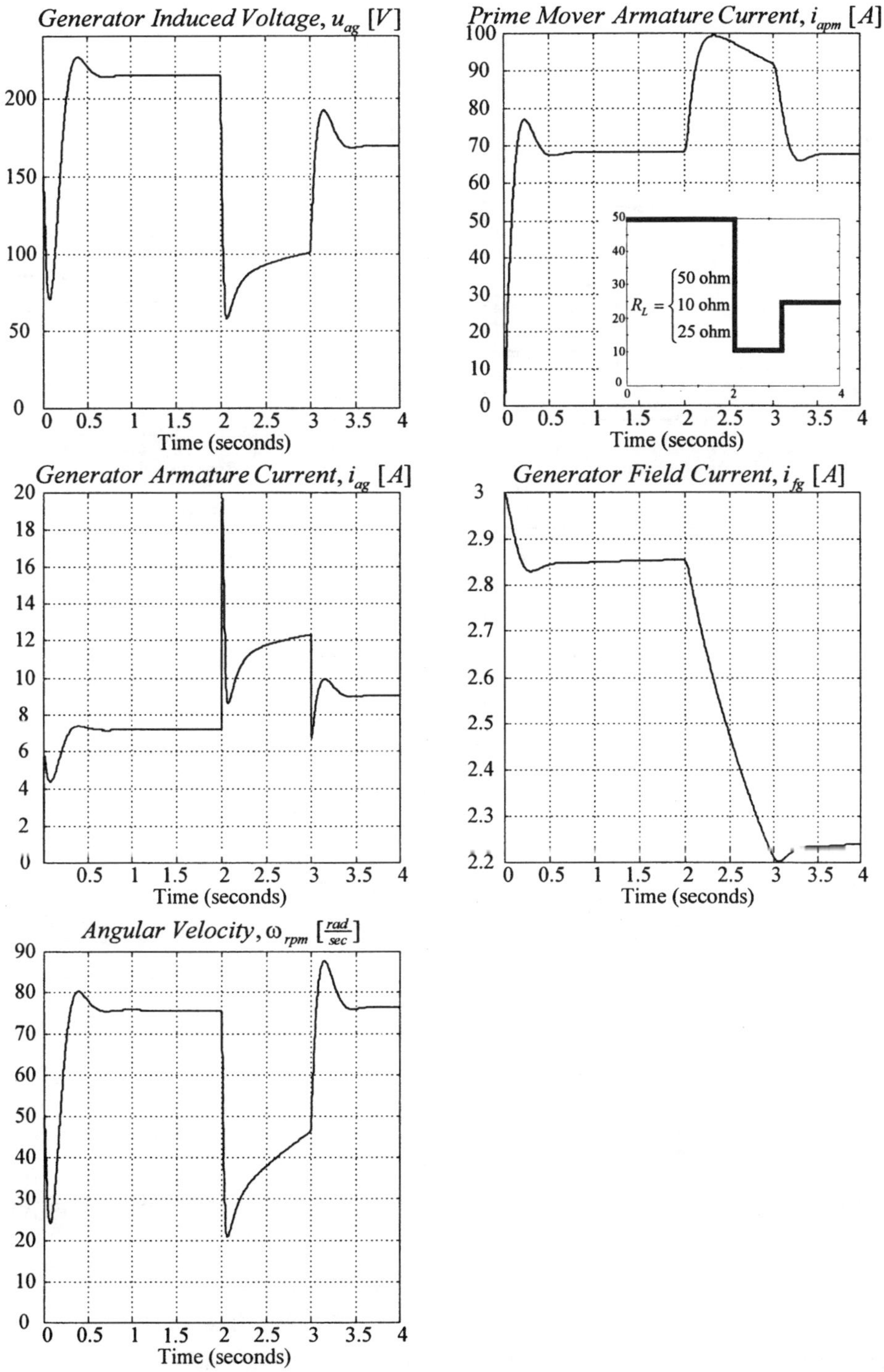

Figure 5.52 Generator and motor dynamics; $u_{apm} = 50$ V, $R_L = \begin{cases} 50 \text{ ohm}, \ \forall t \in [0 \quad 2) \text{ s} \\ 10 \text{ ohm}, \ \forall t \in [2 \quad 3) \text{ s}. \\ 25 \text{ ohm}, \ \forall t \in [3 \quad 4] \text{ s} \end{cases}$

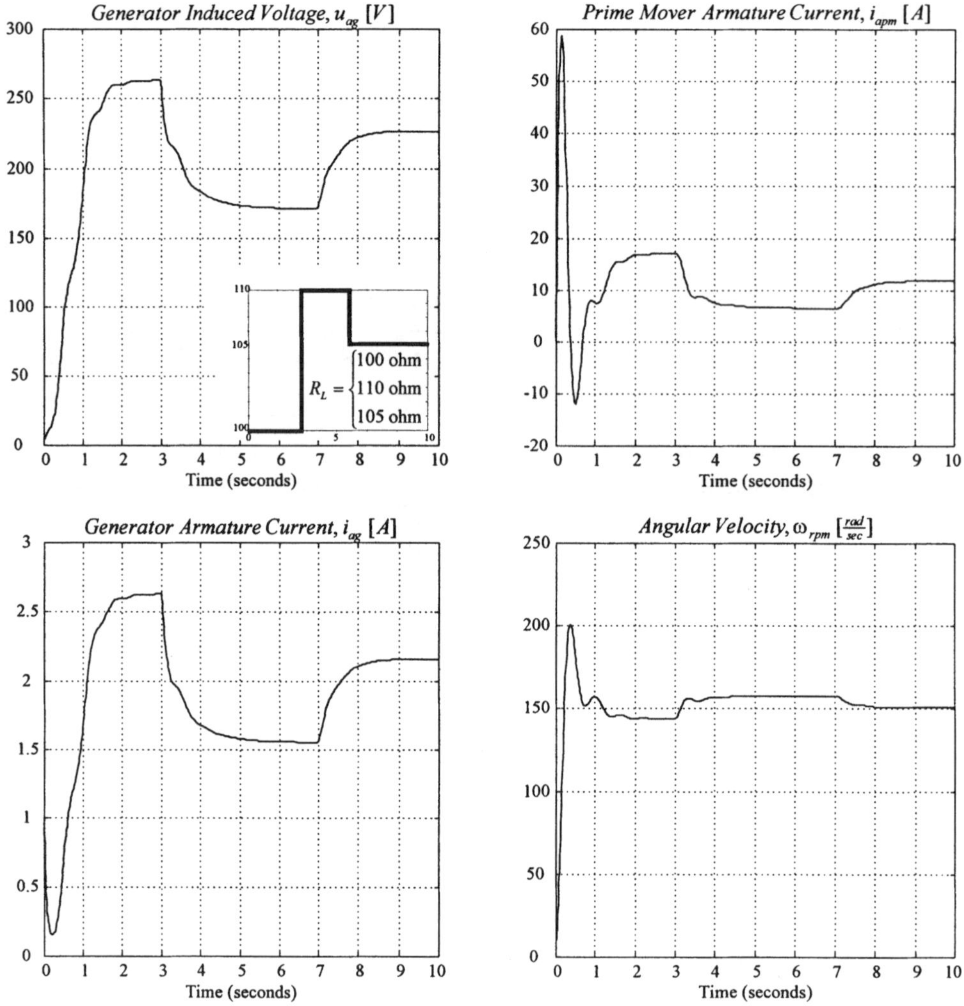

Figure 5.53 Generator and motor dynamics; $u_{apm} = 50$ V, $R_L = \begin{cases} 100 \text{ ohm}, \ \forall t \in [0 \quad 3) \text{ s} \\ 110 \text{ ohm}, \ \forall t \in [3 \quad 7) \text{ s} \\ 105 \text{ ohm}, \ \forall t \in [7 \quad 10] \text{ s} \end{cases}$.

dynamics for the states and the waveforms for voltage induced are plotted in Figures 5.53 and 5.54 if $u_{apm} = 50$ V, $u_{apm} = 60$ V, and $R_L = \begin{cases} 100 \text{ ohm}, \ \forall t \in [0 \quad 3) \text{ s} \\ 110 \text{ ohm}, \ \forall t \in [3 \quad 7) \text{ s} \\ 105 \text{ ohm}, \ \forall t \in [7 \quad 10] \text{ s} \end{cases}$.

For the self-excited series-connected generator under consideration, the following initial conditions are assigned: $i_{ag}(0) = 1$ A and $\omega_{rpm}(0) = 10$ rad-s^{-1}. The analysis of the steady-state and dynamic characteristics of the generation system shows that the series-connected generators are sensitive to the parametric variations, as well as to angular velocity and load changes.

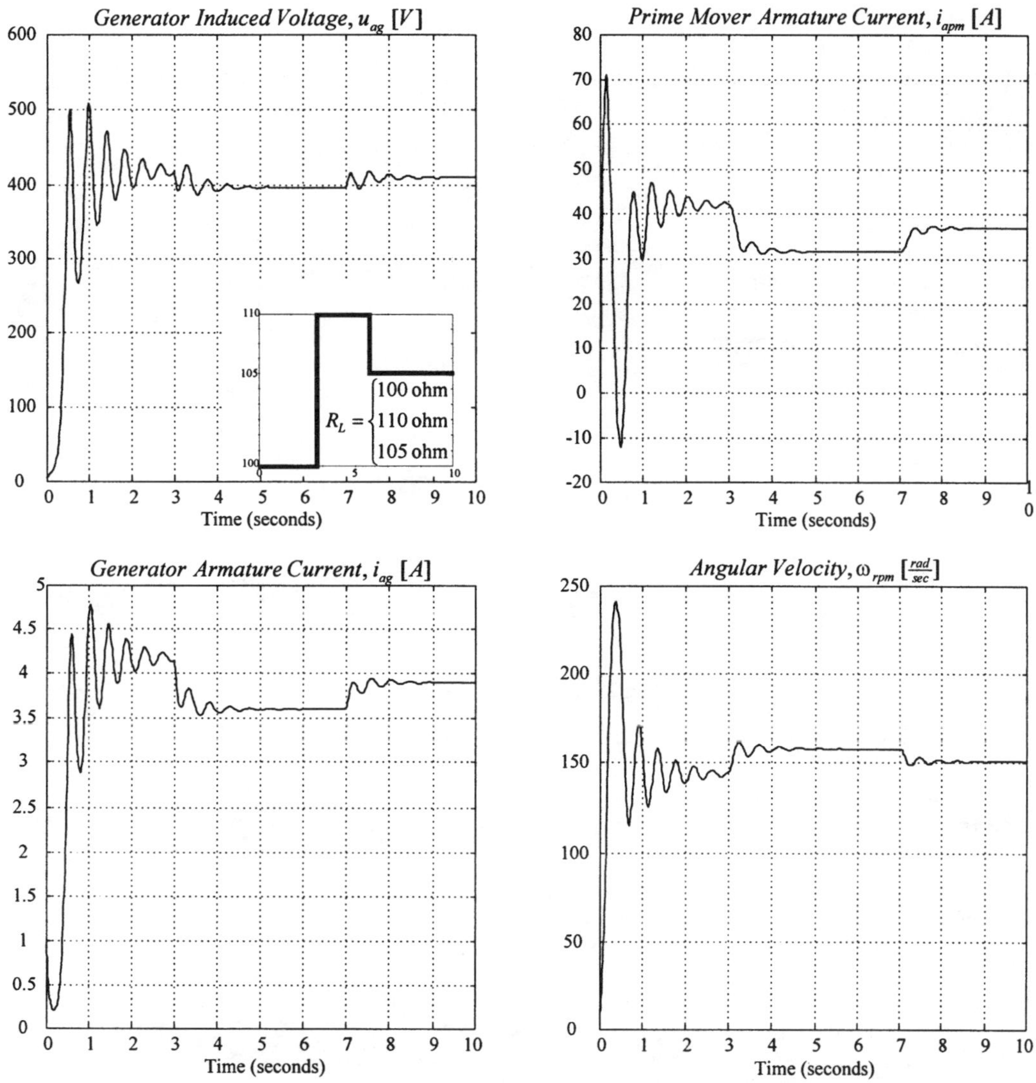

Figure 5.54 Generator and motor dynamics; $u_{apm} = 60$ V, $R_L = \begin{cases} 100\text{ ohm}, \forall t \in [0 \quad 3)\text{ s} \\ 110\text{ ohm}, \forall t \in [3 \quad 7)\text{ s} \\ 105\text{ ohm}, \forall t \in [7 \quad 10]\text{ s} \end{cases}$.

MATLAB script (pm_sg_1.m)

```
tspan=[0 10]; y0=[0 1 10]';
global rag rfg Lafg ;
options = odeset('RelTol',1e-4,'AbsTol',[1e-4 1e-4 1e-4]);
[t,y] = ODE45('pm_sg_2',tspan,y0,options);
subplot(2,2,1);
% Induced voltage by series-connected DC generators
uag=Lafg*y(:,2).*y(:,3)-(rag+rfg)*y(:,2);
plot(t,uag);title('Generator Induced Voltage, uag [V]');
xlabel('Time (seconds)'); grid;
subplot(2,2,2);
plot(t,y(:,1));title('Prime Mover Armature Current, iapm [A]');
xlabel('Time (seconds)'); grid;
subplot(2,2,3);
plot(t,y(:,2));title('Generator Armature Current, iag [A]');
xlabel('Time (seconds)'); grid;
subplot(2,2,4);
plot(t,y(:,3));title('Angular Velocity, wrpm [rad/sec]');
xlabel('Time (seconds)'); grid;
disp('End')
```

MATLAB script (pm_sg_2.m)

```
function yprime=difer(t,y);
global rag rfg Lafg;
% Parameters of the prime mover
rapm=0.4;Lapm=0.05;kapm=0.3;Jpm=0.01;Bmpm=0.0007;
% Applied voltage to the prime mover
uapm=60;
% Parameters of the shunt-connected DC generator
rag=0.1;rfg=2;Lag=0.06;Lfg=6;Lafg=0.71;Jg=0.01;Bmg=0.001;
% Load resistance
if t<=3, Rl=100;
else Rl=110;
end
if t>=7, Rl=105;
end
% Differential equations of the generation system
yprime=[(-rapm*y(1,:)-kapm*y(3,:)+uapm)/Lapm;...
(-(rag+rfg+Rl)*y(2,:)+Lafg*y(2,:).*y(3,:))/(Lag+Lfg);...
(kapm*y(1,:)-Lafg*y(2,:).*y(2,:)-(Bmpm+Bmg)*y(3,:))/(Jpm+Jg)];
```

5.9 *Model development and analysis of the Ward-Leonard systems with direct-current electric machines*

5.9.1 *Ward-Leonard system with permanent-magnet direct-current motors and generators*

The permanent-magnet DC generator is driven by the permanent-magnet DC motor, and the voltage, induced by the generator, is supplied to the armature winding of the permanent-magnet traction DC motor (see Figure 5.55).

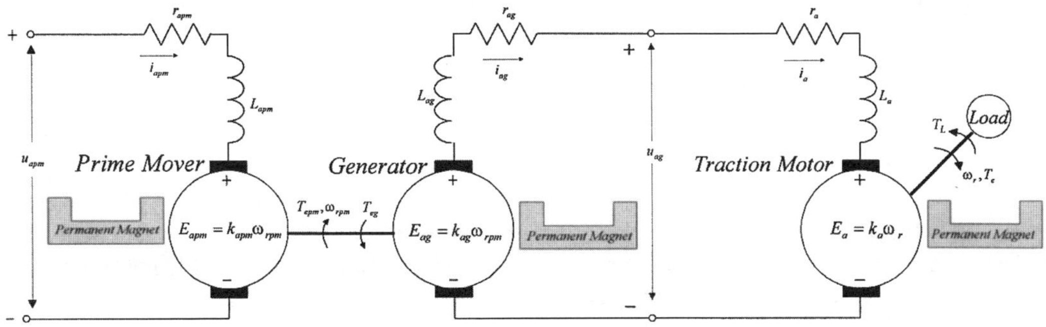

Figure 5.55 Permanent-magnet DC generator, driven by the prime mover, feeds the traction permanent-magnet DC motor.

Newton's law for the considered generator–prime mover *torsional-mechanical* coupling gives

$$T_{epm} - (B_{mpm} + B_{mg})\omega_{rpm} - T_{eg} = (J_{pm} + J_g)\frac{d\omega_{rpm}}{dt} \tag{5.31}$$

where ω_{rpm} is the angular velocity of the prime mover; B_{mpm} and B_{mg} are the viscous friction coefficients of the prime mover and generator; and J_{pm} and J_g are the moments of inertia of the prime mover and generator.

The electromagnetic torque T_{epm}, developed by the permanent-magnet DC motor (prime mover), is found as

$$T_{epm} = k_{apm} i_{apm} \tag{5.32}$$

The load torque for the prime mover is the generator electromagnetic torque, and

$$T_{eg} = k_{ag} i_{ag} \tag{5.33}$$

Here, i_{apm} and i_{ag} are the armature currents in the prime mover and generator rotor windings; and k_{apm} and k_{ag} are the *back emf* (*torque*) constants of the prime mover and generator.

Using Kirchhoff's voltage law and augmenting (5.31), (5.32), and (5.33), the following differential equations for the prime mover (permanent-magnet DC motor) are derived:

$$\begin{aligned}
\frac{di_{apm}}{dt} &= -\frac{r_{apm}}{L_{apm}} i_{apm} - \frac{k_{apm}}{L_{apm}} \omega_{rpm} + \frac{1}{L_{apm}} u_{apm} \\
\frac{d\omega_{rpm}}{dt} &= \frac{k_{apm}}{J_{pm} + J_g} i_{apm} - \frac{B_{mpm} + B_{mg}}{J_{pm} + J_g} \omega_{rpm} - \frac{k_{ag}}{J_{pm} + J_g} i_{ag}
\end{aligned} \tag{5.34}$$

where u_{apm} is the voltage applied to the armature winding of the prime mover; and r_{apm} and L_{apm} are the resistance and self-inductance of the prime mover rotor winding, respectively.

The armature current of the generator i_{ag} is equal to the armature current of the traction motor i_a; that is, $i_{ag} = i_a$. Hence, the differential equation for the augmented circuitry, which integrates the rotor windings of the permanent-magnet DC generator and traction motor, is found as

$$\frac{di_{ag}}{dt} = -\frac{r_{ag} + r_a}{L_{ag} + L_a} i_{ag} + \frac{k_{ag}}{L_{ag} + L_a} \omega_{rpm} - \frac{k_a}{L_{ag} + L_a} \omega_r \tag{5.35}$$

Here, ω_r is the angular velocity of the traction motor; r_{ag}, r_a, L_{ag}, and L_a are the armature resistances and self-inductances of the generator and traction motor; and k_a is the *back emf* (*torque*) constant of the traction motor.

The *torsional-mechanical* dynamics of the traction motor is mapped by

$$\frac{d\omega_r}{dt} = \frac{k_a}{J} i_{ag} - \frac{B_m}{J} \omega_r - \frac{1}{J} T_L \tag{5.36}$$

where B_m is the viscous friction coefficient; and J is the equivalent moment of inertia of the traction motor and the load attached.

From (5.35) and (5.36), we have

$$\begin{aligned} \frac{di_{ag}}{dt} &= -\frac{r_{ag} + r_a}{L_{ag} + L_a} i_{ag} + \frac{k_{ag}}{L_{ag} + L_a} \omega_{rpm} - \frac{k_a}{L_{ag} + L_a} \omega_r \\ \frac{d\omega_r}{dt} &= \frac{k_a}{J} i_{ag} - \frac{B_m}{J} \omega_r - \frac{1}{J} T_L \end{aligned} \tag{5.37}$$

Using (5.34) and (5.37), a set of differential equations for the Ward-Leonard system considered is found to be

$$\begin{aligned} \frac{di_{apm}}{dt} &= -\frac{r_{apm}}{L_{apm}} i_{apm} - \frac{k_{apm}}{L_{apm}} \omega_{rpm} + \frac{1}{L_{apm}} u_{apm} \\ \frac{di_{ag}}{dt} &= -\frac{r_{ag} + r_a}{L_{ag} + L_a} i_{ag} + \frac{k_{ag}}{L_{ag} + L_a} \omega_{rpm} - \frac{k_a}{L_{ag} + L_a} \omega_r \\ \frac{d\omega_{rpm}}{dt} &= \frac{k_{apm}}{J_{pm} + J_g} i_{apm} - \frac{B_{mpm} + B_{mg}}{J_{pm} + J_g} \omega_{rpm} - \frac{k_{ag}}{J_{pm} + J_g} i_{ag} \\ \frac{d\omega_r}{dt} &= \frac{k_a}{J} i_{ag} - \frac{B_m}{J} \omega_r - \frac{1}{J} T_L \end{aligned} \tag{5.38}$$

The voltage, induced by the generator, is expressed as

$$u_{ag} = E_{ag} - r_{ag} i_{ag} = k_{ag} \omega_{rpm} - r_{ag} i_{ag} \tag{5.39}$$

For the Ward-Leonard system studied, using differential equations (5.38), as well as the expression for the voltage built by the permanent-magnet generator u_{ag} (5.39), a SIMULINK diagram develops and is shown in Figure 5.56 (`wlpm_1.mdl`).

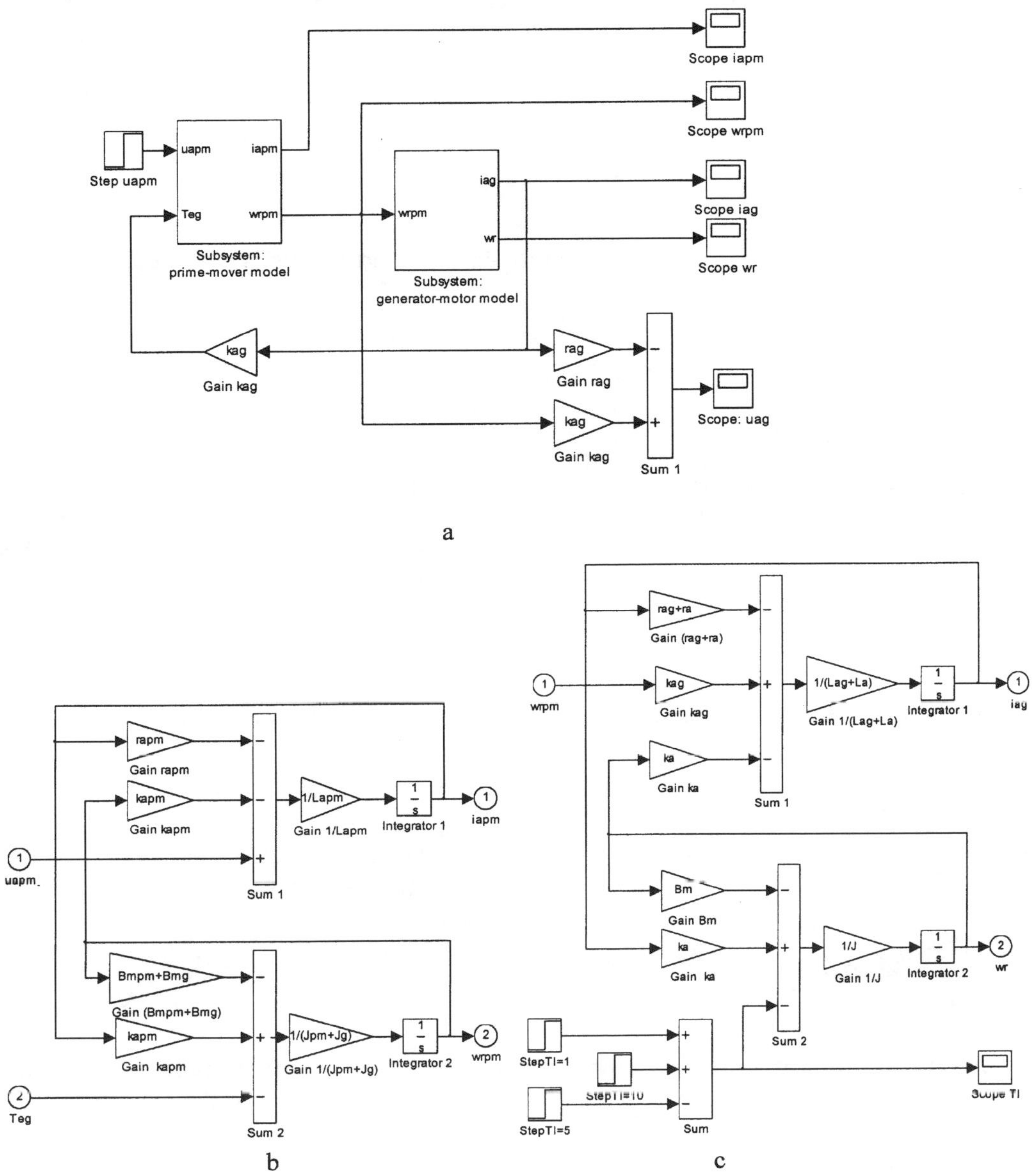

Figure 5.56 SIMULINK diagrams to model the Ward-Leonard system: (a) augmented Ward-Leonard system model; (b) SIMULINK subsystem diagram to model permanent-magnet DC motors; (c) SIMULINK subsystem diagram to model DC generators and traction motors.

In performing the modeling, it is convenient to use the symbolic parameters of electric machines. First, one downloads the parameters of the Ward-Leonard system, and then, differential equations (5.38) can be numerically solved using the SIMULINK diagram developed and illustrated in Figure 5.56.

Example 5.11

Model the Ward-Leonard system with a permanent-magnet DC generator and permanent-magnet DC motors (prime mover and traction motor). Study the transient dynamics for different angular velocities of the prime mover and load torque to the traction motor. In particular, use

$$u_{apm} = \begin{cases} 100\text{ V}, \ \forall t \in [0 \quad 10)\text{ s} \\ 50\text{ V}, \ \forall t \in [10 \quad 20)\text{ s} \\ 75\text{ V}, \ \forall t \in [20 \quad 30]\text{ s} \end{cases}$$

to vary the angular velocity of the prime mover. The load torque is assigned as

$$T_L = \begin{cases} 1\text{ N-m}, \ \forall t \in [0 \quad 10)\text{ s} \\ 10\text{ N-m}, \ \forall t \in [10 \quad 20)\text{ s} \\ 5\text{ N-m}, \ \forall t \in [20 \quad 30]\text{ s} \end{cases}$$

The parameters of the electric machines used are: r_{apm} = 0.4 ohm, r_{ag} = 0.3 ohm, r_a = 0.2 ohm, L_{apm} = 0.05 H, L_{ag} = 0.06 H, L_a = 0.07 H, k_{apm} = 0.3 V-s-rad^{-1} or k_{apm} = 0.3 N-m-A^{-1}, k_{ag} = 0.25 V-s-rad^{-1} or k_{ag} = 0.25 N-m-A^{-1}; k_a = 0.2 V-s-rad^{-1} or k_a = 0.2 N-m-A^{-1}, J_{pm} = 0.04 kg-m^2, J_g = 0.05 kg-m^2, J = 0.06 kg-m^2, B_{mpm} = 0.0007 N-m-s-rad^{-1}, B_{mg} = 0.0008 N-m-s-rad^{-1}, and B_m 0.0009 N-m-s-rad^{-1}.

Solution

The following m-file is used (`wlpm_d.m`) to download the parameters of the Ward-Leonard system:

```
% Parameters of the prime mover
rapm=0.4; Lapm=0.05; kapm=0.3; Jpm=0.04; Bmpm=0.0007;
% Parameters of the permanent-magnet DC generator
rag=0.3; Lag=0.06; kag=0.25; Jg=0.05; Bmg=0.0008;
% Parameters of the permanent-magnet traction motor
ra=0.2; La=0.07; ka=0.2; J=0.06; Bm=0.0009;
```

We supply three distinct armature voltages to the permanent-magnet DC motor: u_{apm} = 50 V, u_{apm} = 75 V, and u_{apm} = 100 V. The load torque, applied to the traction motor, is assigned as

$$T_L = \begin{cases} 1\text{ N-m}, \ \forall t \in [0 \quad 10)\text{ s} \\ 10\text{ N-m}, \ \forall t \in [10 \quad 20)\text{ s} \\ 5\text{ N-m}, \ \forall t \in [20 \quad 30]\text{ s} \end{cases}$$

The dynamics of the prime mover, permanent-magnet generator, and traction motor are found by numerically solving a set of differential equations (5.38) and using the SIMULINK diagram documented in Figure 5.56. The state variables are plotted in Figure 5.57 to 5.59. The settling time is 6 s; and for different loads, the steady-state values of the angular velocities and currents in the rotor windings of the prime mover and traction motor are different. The voltage, induced by the permanent-magnet DC generator, varies as well. Recall that $u_{ag} = E_{ag} - r_{ag}i_{ag} = k_{ag}\omega_{rpm} - r_{ag}i_{ag}$ (see (5.39)). For example, for

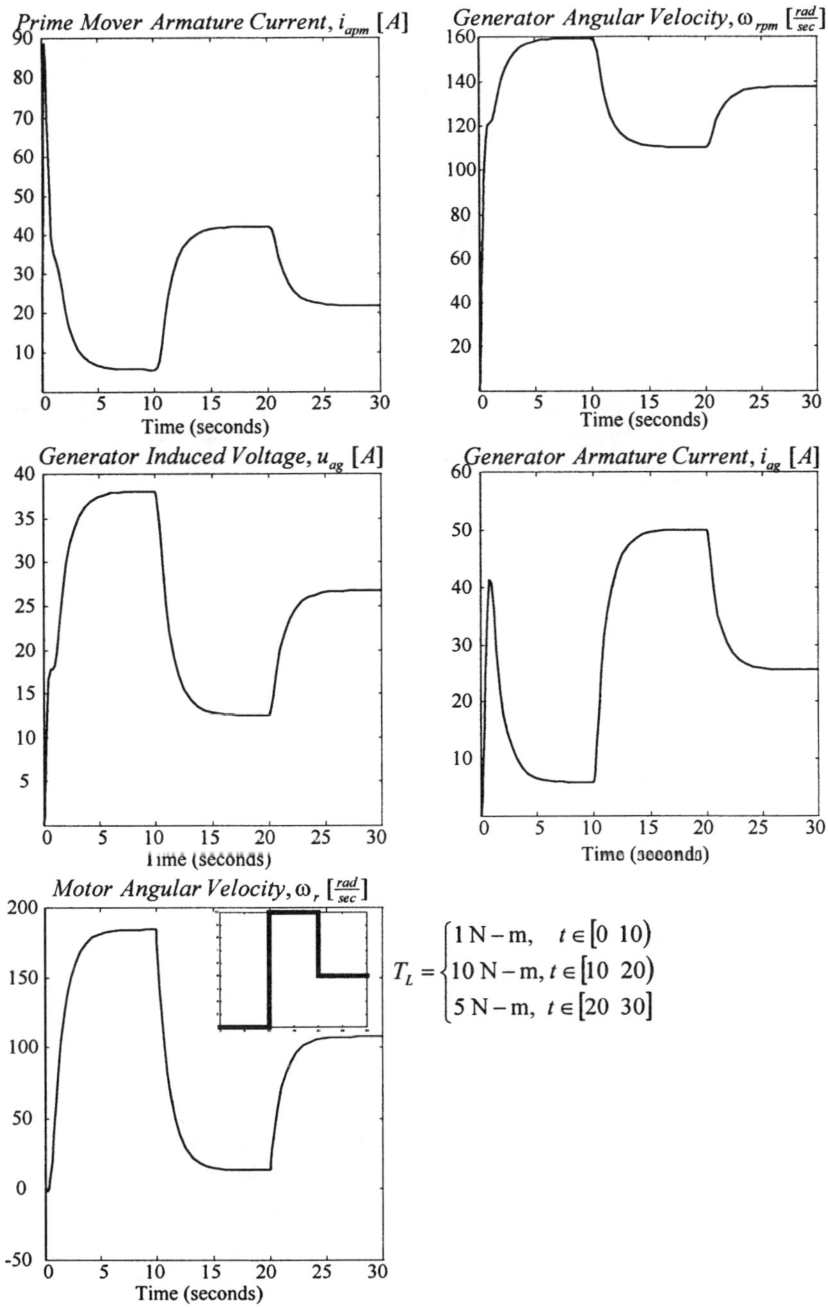

Figure 5.57 Dynamics of the Ward-Leonard system; $T_L = \begin{cases} 1\text{ N-m}, \ \forall t \in [0 \quad 10)\text{ s} \\ 10\text{ N-m}, \ \forall t \in [10 \quad 20)\text{ s} \\ 5\text{ N-m}, \ \forall t \in [20 \quad 30]\text{ s} \end{cases}$.

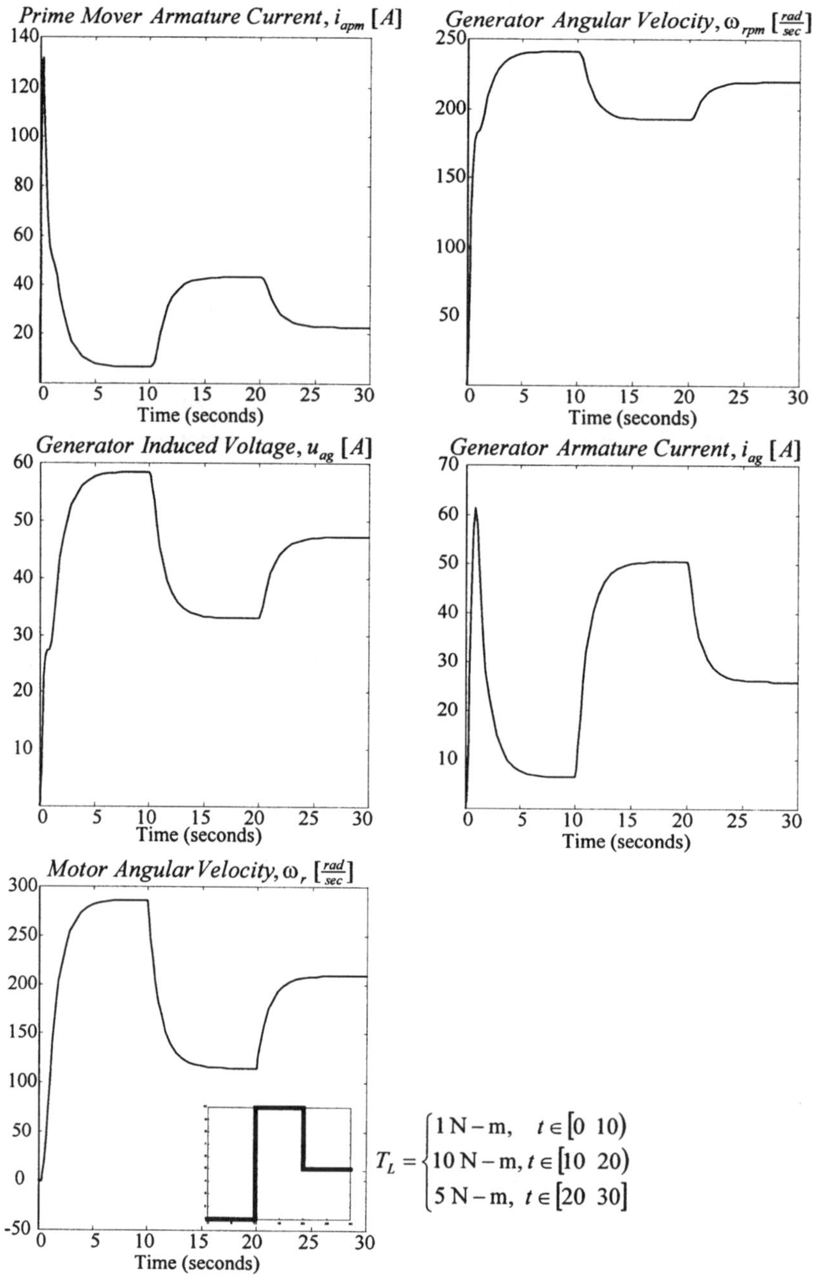

Figure 5.58 Dynamics of the Ward-Leonard system; $u_{apm} = 75$ V and $T_L = \begin{cases} 1\ \text{N-m}, & \forall t \in [0 \quad 10)\ \text{s} \\ 10\ \text{N-m}, & \forall t \in [10 \quad 20)\ \text{s} \\ 5\ \text{N-m}, & \forall t \in [20 \quad 30]\ \text{s} \end{cases}$.

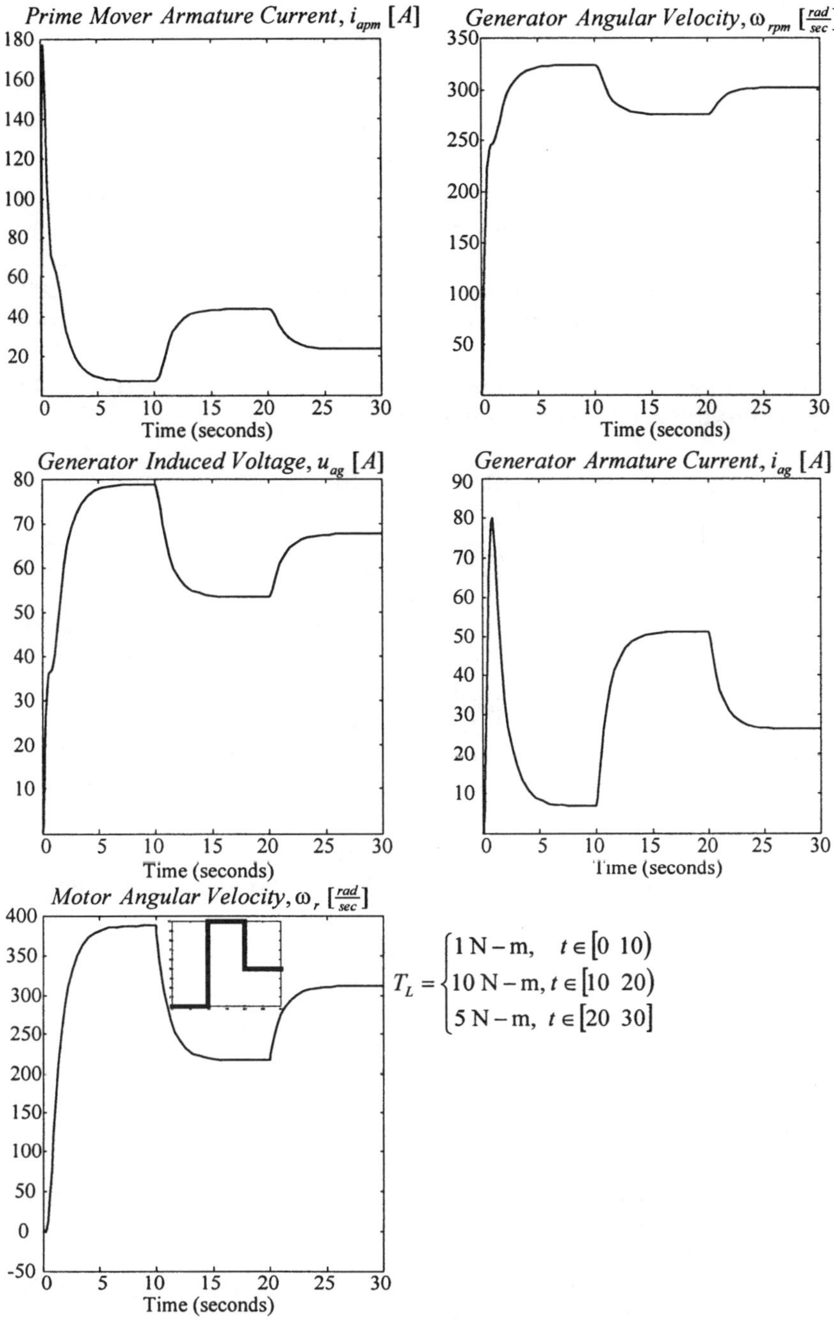

Figure 5.59 Dynamics of the Ward-Leonard system; $u_{apm} = 100$ V and $T_L = \begin{cases} 1\text{ N-m}, \ \forall t \in [0 \quad 10)\text{ s} \\ 10\text{ N-m}, \ \forall t \in [10 \quad 20)\text{ s} \\ 5\text{ N-m}, \ \forall t \in [20 \quad 30]\text{ s} \end{cases}$.

$$T_L = \begin{cases} 1\ \text{N-m},\ \forall t \in [0 \quad 10)\ \text{s} \\ 10\ \text{N-m},\ \forall t \in [10 \quad 20)\ \text{s} \\ 5\ \text{N-m},\ \forall t \in [20 \quad 30]\ \text{s} \end{cases}$$

supplying u_{apm} = 50 V, the steady-state values of the angular velocity of the prime mover are 160, 110, and 138 rad-s^{-1}. From Figure 5.57, it is easy to see that the steady-state values of the voltage induced by the permanent-magnet DC generator are 38, 12.5, and 27 V. The angular velocity of the traction motor, as well as the voltage induced by the generator, can be viewed as the output. Using the modeling results documented in Figures 5.58 and 5.59, if the applied voltages to the prime mover are u_{apm} = 75 V and u_{apm} = 100 V, one can analyze the steady-state and dynamic characteristics, and the studies performed have a great degree of merit.

For the load torque T_L = 1 N-m applied to the traction motor, the values of the armature voltage supplied to the prime mover are assigned as

$$u_{apm} = \begin{cases} 100\ \text{V},\ \forall t \in [0 \quad 10)\ \text{s} \\ 50\ \text{V},\ \forall t \in [10 \quad 20)\ \text{s} \\ 75\ \text{V},\ \forall t \in [20 \quad 30]\ \text{s} \end{cases}$$

A Simulink diagram (`wlpm_2.mdl`) is developed and shown in Figure 5.60.

The dynamics of the prime mover, permanent-magnet DC generator, and traction motor are of particular interest. The resulting transient behavior for the state variables $i_{apm}(t)$, $i_{ag}(t)$, $\omega_{rpm}(t)$, and $\omega_r(t)$, as well as the waveform of the voltage induced by the generator $u_{ag}(t)$, are plotted in Figure 5.61. The analysis of the steady-state and transient dynamics shows that as one increases the applied voltage supplied to the prime mover armature winding, one increases the angular velocity of the prime mover. Hence, the voltage induced by the generator increases. For the load torque T_L = 1 N-m, which is applied at t = 0 s, the analysis of the steady-state and dynamic performances is accomplished for

$$u_{apm} = \begin{cases} 100\ \text{V},\ \forall t \in [0 \quad 10)\ \text{s} \\ 50\ \text{V},\ \forall t \in [10 \quad 20)\ \text{s} \\ 75\ \text{V},\ \forall t \in [20 \quad 30]\ \text{s} \end{cases}$$

See the results documented in Figure 5.61.

5.9.2 Ward-Leonard system with separately exited direct-current machines

The separately excited DC generator is driven by the permanent-magnet DC motor, and the voltage, induced by the generator, is supplied to the armature winding of the separately excited traction DC motor (see Figure 5.62).

The *torsional-mechanical* dynamics of the prime mover and generator are mapped as

$$T_{epm} - (B_{mpm} + B_{mg})\omega_{rpm} - T_{eg} = (J_{pm} + J_g)\frac{d\omega_{rpm}}{dt} \tag{5.40}$$

The electromagnetic torque, developed by the permanent-magnet DC motor, is

$$T_{epm} = k_{apm} i_{apm} \tag{5.41}$$

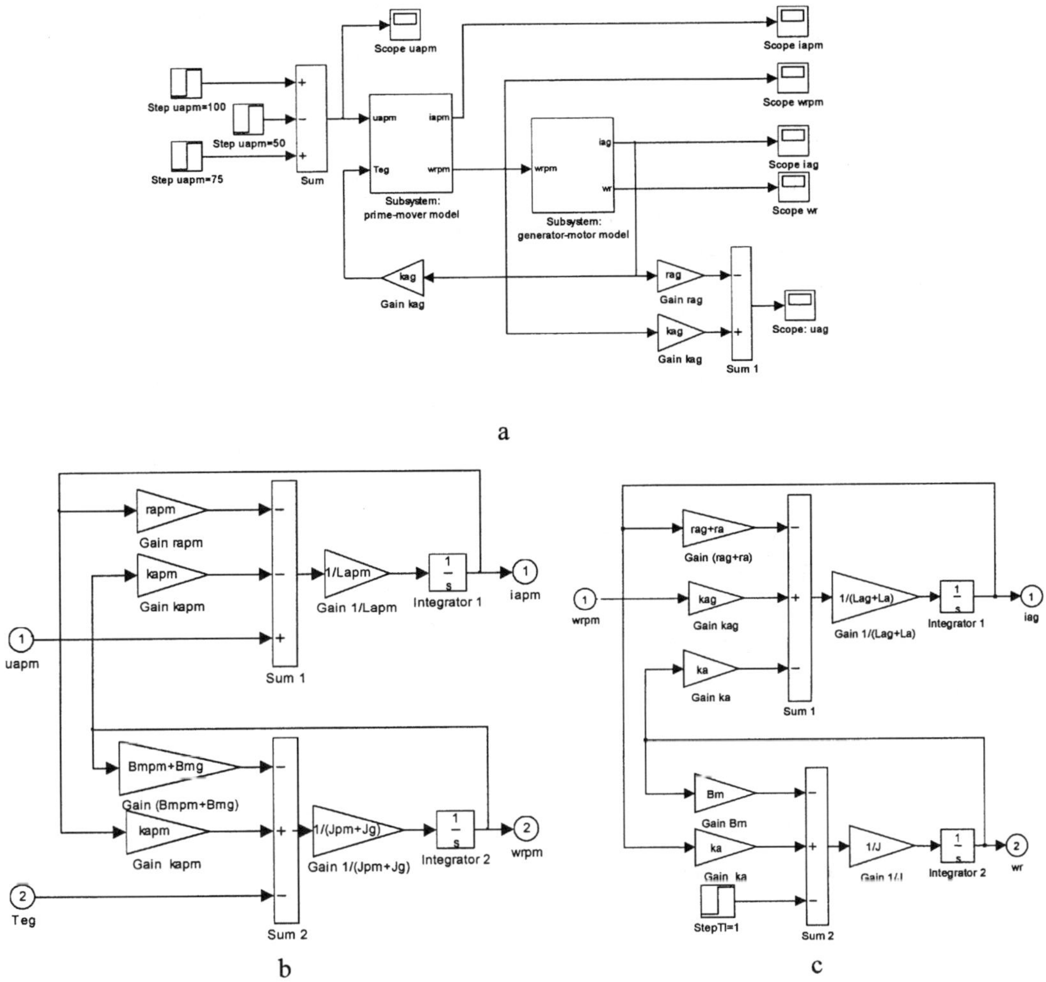

Figure 5.60 SIMULINK diagrams to model the Ward-Leonard system: (a) augmented Ward-Leonard system model; (b) SIMULINK subsystem diagram to model permanent-magnet DC motors; (c) SIMULINK subsystem diagram to model DC generators and traction motors.

and the load torque for the prime mover is expressed as

$$T_{eg} = L_{afg} i_{ag} i_{fg} \tag{5.42}$$

Applying Kirchhoff's voltage law and using (5.40) to (5.42), the following differential equations for the prime mover result:

$$\begin{aligned} \frac{di_{apm}}{dt} &= -\frac{r_{apm}}{L_{apm}} i_{apm} - \frac{k_{apm}}{L_{apm}} \omega_{rpm} + \frac{1}{L_{apm}} u_{apm} \\ \frac{d\omega_{rpm}}{dt} &= \frac{k_{apm}}{J_{pm}+J_g} i_{apm} - \frac{B_{mpm}+B_{mg}}{J_{pm}+J_g} \omega_{rpm} - \frac{L_{afg}}{J_{pm}+J_g} i_{ag} i_{fg} \end{aligned} \tag{5.43}$$

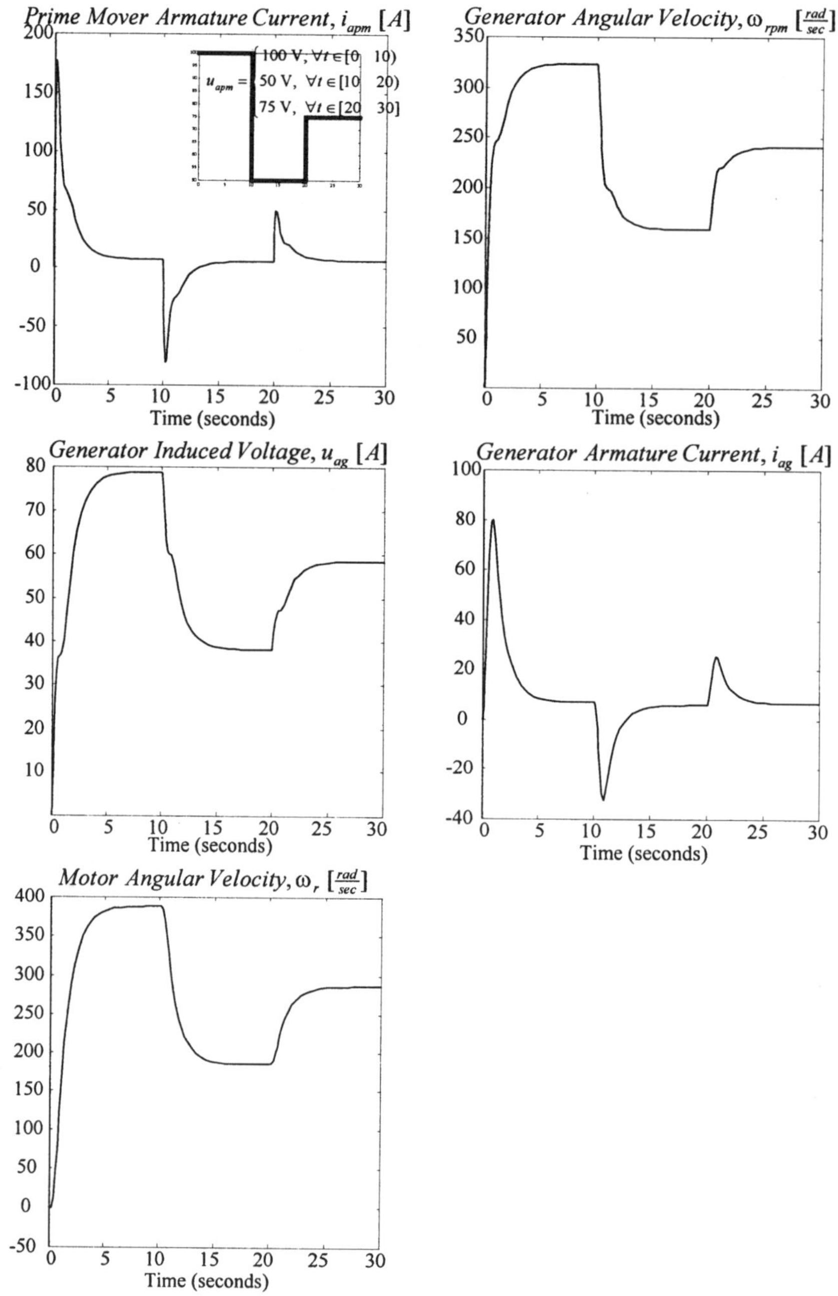

Figure 5.61 Dynamics of the Ward-Leonard system; $T_L = 1$ N-m and $u_{apm} = \begin{cases} 100\text{ V}, \ \forall t \in [0 \quad 10)\text{ s} \\ 50\text{ V}, \ \forall t \in [10 \quad 20)\text{ s}. \\ 75\text{ V}, \ \forall t \in [20 \quad 30]\text{ s} \end{cases}$

The armature current of the generator i_{ag} is equal to the armature current of the traction motor i_a. The differential equations for the circuitry, which map the separately exited DC generator and traction motor windings dynamic behavior, are

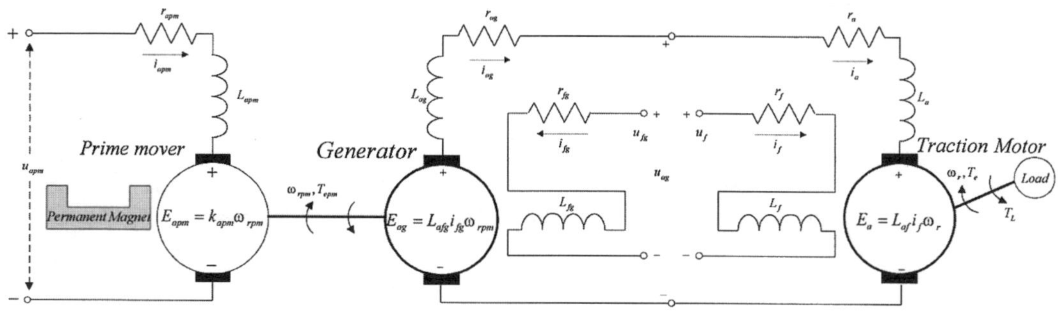

***Figure* 5.62** Separately excited DC generator, driven by the prime mover, feeds the traction separately excited DC motor.

$$\frac{di_{ag}}{dt} = -\frac{r_{ag}+r_a}{L_{ag}+L_a}i_{ag} + \frac{L_{afg}}{L_{ag}+L_a}\omega_{rpm} - \frac{L_{af}}{L_{ag}+L_a}\omega_r$$

$$\frac{di_{fg}}{dt} = -\frac{r_{fg}}{L_{fg}}i_{fg} + \frac{1}{L_{fg}}u_{fg} \tag{5.44}$$

$$\frac{di_f}{dt} = -\frac{r_f}{L_f}i_f + \frac{1}{L_f}u_f$$

and the *torsional-mechanical* dynamics of the traction motor is

$$\frac{d\omega_r}{dt} = \frac{L_{af}}{J}i_{ag}i_{fg} - \frac{B_m}{J}\omega_r - \frac{1}{J}T_L \tag{5.45}$$

From (5.44) and (5.45), we have

$$\begin{aligned}
\frac{di_{ag}}{dt} &= -\frac{r_{ag}+r_a}{L_{ag}+L_a}i_{ag} + \frac{L_{afg}}{L_{ag}+L_a}\omega_{rpm} - \frac{L_{af}}{L_{ag}+L_a}\omega_r \\
\frac{di_{fg}}{dt} &= -\frac{r_{fg}}{L_{fg}}i_{fg} + \frac{1}{L_{fg}}u_{fg} \\
\frac{di_f}{dt} &= -\frac{r_f}{L_f}i_f + \frac{1}{L_f}u_f \\
\frac{d\omega_r}{dt} &= \frac{L_{af}}{J}i_{ag}i_{fg} - \frac{B_m}{J}\omega_r - \frac{1}{J}T_L
\end{aligned} \tag{5.46}$$

Using (5.43) and (5.46), a set of differential equations for the Ward-Leonard system considered is found to be

$$\frac{di_{apm}}{dt} = -\frac{r_{apm}}{L_{apm}} i_{apm} - \frac{k_{apm}}{L_{apm}} \omega_{rpm} + \frac{1}{L_{apm}} u_{apm}$$

$$\frac{di_{ag}}{dt} = -\frac{r_{ag} + r_a}{L_{ag} + L_a} i_{ag} + \frac{L_{afg}}{L_{ag} + L_a} \omega_{rpm} - \frac{L_{af}}{L_{ag} + L_a} \omega_r$$

$$\frac{di_{fg}}{dt} = -\frac{r_{fg}}{L_{fg}} i_{fg} + \frac{1}{L_{fg}} u_{fg}$$

$$\frac{di_f}{dt} = -\frac{r_f}{L_f} i_f + \frac{1}{L_f} u_f \tag{5.47}$$

$$\frac{d\omega_r}{dt} = \frac{L_{af}}{J} i_{ag} i_{fg} - \frac{B_m}{J} \omega_r - \frac{1}{J} T_L$$

$$\frac{d\omega_{rpm}}{dt} = \frac{k_{apm}}{J_{pm} + J_g} i_{apm} - \frac{B_{mpm} + B_{mg}}{J_{pm} + J_g} \omega_{rpm} - \frac{L_{afg}}{J_{pm} + J_g} i_{ag} i_{fg}$$

The voltage, induced by the generator, is expressed as

$$u_{ag} = E_{ag} - r_{ag} i_{ag} = L_{afg} i_{fg} \omega_{rpm} - r_{ag} i_{ag} \tag{5.48}$$

Example 5.12

Model the Ward-Leonard system with separately excited generator and traction motor. The generator is driven by the permanent-magnet DC motor. Study the transient dynamics for different angular velocities of the prime mover, assigning the armature voltage to be 50, 75, and 100 V; that is, $u_{apm} = 50$ V, $u_{apm} = 75$ V, and $u_{apm} = 100$ V. The load torque is assumed to be constant and equal to 5 N-m. Perform the analysis supplying

$$u_{fg} = \begin{cases} 30 \text{ V}, \ \forall t \in [0 \quad 4) \text{ s} \\ 10 \text{ V}, \ \forall t \in [4 \quad 7) \text{ s} \\ 20 \text{ V}, \ \forall t \in [7 \quad 10] \text{ s} \end{cases} \text{ and } u_f = \begin{cases} 30 \text{ V}, \ \forall t \in [0 \quad 4) \text{ s} \\ 10 \text{ V}, \ \forall t \in [4 \quad 7) \text{ s} \\ 20 \text{ V}, \ \forall t \in [7 \quad 10] \text{ s} \end{cases}$$

to the field windings of the generator and traction motor. The parameters of the electric machines used are: permanent-magnet DC motor: $r_{apm} = 0.4$ ohm, $L_{apm} = 0.05$ H, $k_{apm} = 0.3$ V-s-rad^{-1} or $k_{apm} = 0.3$ N-m-A^{-1}, $J_{pm} = 0.01$ kg-m^2, and $B_{mpm} = 0.0007$ N-m-s-rad^{-1}; separately excited DC generator: $r_{ag} = 0.18$ ohm, $r_{fg} = 3.5$ ohm, $L_{ag} = 0.06$ H, $L_{fg} = 0.095$ H, $L_{afg} = 0.1$ H, $J_g = 0.01$ kg-m^2, and $B_{mg} = 0.001$ N-m-s-rad^{-1}; separately excited DC traction motor: $r_a = 0.15$ ohm, $r_f = 3$ ohm, $L_a = 0.05$ H, $L_f = 0.08$ H, $L_{af} = 0.09$ H, $J = 0.0095$ kg-m^2, and $B_m = 0.00095$ N-m-s-rad^{-1}.

Solution

Using differential equations (5.47), as well as the expression for the voltage built u_{ag} (5.48), two m-files develop.

MATLAB script 1 (wl_seg_1.m)

```
tspan=[0 6]; y0=[0 0 0 0 0 0]';
global rag Lafg;
options = odeset('RelTol',1e-4,'AbsTol',[1e-4 1e-4 1e-4 1e-4 1e-4 1e-4]);
[t,y] = ODE45('wl_seg_2',tspan,y0,options);
subplot(4,2,1);
% Induced voltage by separately excited DC generators
uag=Lafg*y(:,3).*y(:,6)-rag*y(:,2);
plot(t,uag);title('Generator Induced Voltage, uag [V]');
xlabel('Time (seconds)'); grid;
subplot(4,2,2);
plot(t,y(:,1));title('Prime Mover Armature Current, iapm [A]');
xlabel('Time (seconds)'); grid;
subplot(4,2,3);
plot(t,y(:,2));title('Generator Armature Current, iag [A]');
xlabel('Time (seconds)'); grid;
subplot(4,2,4);
plot(t,y(:,3));title('Generator Field Current, ifg [A]');
xlabel('Time (seconds)'); grid;
subplot(4,2,5);
plot(t,y(:,4));title('Traction Motor Field Current, if [A]');
xlabel('Time (seconds)'); grid;
subplot(4,2,6);
plot(t,y(:,5));title('Traction Motor Angular velocity, wr [rad/sec]');
xlabel('Time (seconds)'); grid;
subplot(4,2,7);
plot(t,y(:,6));title('Angular Velocity, wrpm [rad/sec]');
xlabel('Time (seconds)'); grid;
disp('End')
```

MATLAB script 2 (wl_seg_2.m)

```
function yprime=difer(t,y);
global rag Lafg;
% Parameters of the prime mover
rapm=0.4;Lapm=0.05;kapm=0.3;Jpm=0.01;Bmpm=0.0007;
% Applied voltage to the prime mover
uapm=50;
% Parameters of the separately excited DC generator
rag=0.18;rfg=3.5;Lag=0.06;Lfg=0.095;Lafg=0.1;Jg=0.01;Bmg=0.001;
% Parameters of the traction motor
ra=0.15;rf=3;La=0.05;Lf=0.08;Laf=0.09;J=0.0095;Bm=0.00095;
% Applied voltage to traction motor field windings
uf=20;
% Applied voltage to the generator field winding
if t<=2, ufg=30;
else ufg=10;
end
if t>=4, ufg=20;
end
% Load torque
Tl=5;
```

```
function yprime=difer(t,y);
% Differential equations of the generation system
yprime=[(-rapm*y(1,:)-kapm*y(6,:)+uapm)/Lapm;...
(-(rag+ra)*y(2,:)+Lafg*y(3,:).*y(6,:)-Laf*y(4,:)*y(5,:))/(Lag+La);...
(-rfg*y(3,:)+ufg)/Lfg;...
(-rf*y(4,:)+uf)/Lf;...
(Laf*y(2,:).*y(4,:)-Bm*y(5,:)-Tl)/J;...
(kapm*y(1,:)-Lafg*y(2,:).*y(3,:)-(Bmpm+Bmg)*y(6,:))/(Jpm+Jg)];
```

We supply three distinct values of the armature voltage to the permanent-magnet DC motor: $u_{apm} = 50$ V, $u_{apm} = 75$ V, and $u_{apm} = 100$ V. The applied voltage to the generator field winding is specified as

$$u_{fg} = \begin{cases} 30\text{ V}, \ \forall t \in [0 \quad 4)\text{ s} \\ 10\text{ V}, \ \forall t \in [4 \quad 7)\text{ s} \\ 20\text{ V}, \ \forall t \in [7 \quad 10]\text{ s} \end{cases}$$

and the applied voltage to the traction motor field winding is $u_f = 20$ V. Modeling is performed if the load torque to the traction motor is $T_L = 5$ N-m. The dynamics of the prime mover, separately excited generator, and traction motor are plotted in Figures 5.63 to 5.65. The settling time is 6 s; and for different applied armature voltages to the prime mover, the angular velocity varies. The changes in the field voltage supplied significantly influence the induced voltage of the generator (the voltage, induced by the separately excited DC generator is found using Eq. (5.48) as $u_{ag} = E_{ag} - r_{ag}i_{ag} = L_{afg}i_{fg}\omega_{rpm} - r_{ag}i_{ag}$). Hence, the armature voltage applied to the traction motor varies, and the angular velocity of the separately excited traction motor changes. For

$$u_{fg} = \begin{cases} 30\text{ V}, \ \forall t \in [0 \quad 4)\text{ s} \\ 10\text{ V}, \ \forall t \in [4 \quad 7)\text{ s} \\ 20\text{ V}, \ \forall t \in [7 \quad 10]\text{ s} \end{cases}$$

supplying $u_{apm} = 50$ V and $u_f = 20$ V, the steady-state values of the angular velocity of the prime mover are 133, 154, and 144 rad-s^{-1}. From Figure 5.63, it is easy to see that the steady-state values of the voltage induced by the separately excited DC generator are 112, 43, and 81 V. The angular velocity of the traction motor, as well as the voltage induced by the separately excited generator, can be considered as the outputs. However, a thorough analysis of armature currents, which are the state variables, must be performed because the rated armature currents are specified. The corresponding transient dynamics for the state variables and the waveforms for the voltage induced u_{apm} are documented in Figures 5.63 to 5.65. The steady-state and dynamic characteristics, as well as the system performance, are analyzed using the modeling performed.

For the load torque 5 N-m applied to the separately excited traction motor, the armature voltages supplied to the prime mover are assigned as $u_{apm} = 50$ V, $u_{apm} = 75$, and $u_{apm} = 100$ V. The field voltage applied to the traction motor field winding is

$$u_f = \begin{cases} 30\ \text{V}, \ \forall t \in [0 \quad 4)\ \text{s} \\ 10\ \text{V}, \ \forall t \in [4 \quad 7)\ \text{s} \\ 20\ \text{V}, \ \forall t \in [7 \quad 10]\ \text{s} \end{cases}$$

We supply 20 V to the generator field winding; that is, u_{fg} = 20 V. The dynamics of the prime mover, separately excited DC generator, and traction motor are of particular interest. The resulting transient behavior of the state variables $i_{apm}(t)$, $i_{ag}(t)$, $i_{fg}(t)$, $i_f(t)$, $\omega_{rpm}(t)$, $\omega_r(t)$, as well as the waveforms for the voltage induced by the generator $u_{ag}(t)$, are plotted in Figures 5.66 to 5.68. The analysis of the steady-state and transient dynamics shows that by increasing the applied voltage supplied to the prime mover armature winding, one increases the angular velocity of the prime mover. Hence, the voltage induced by the DC generator increases. For the load T_L = 5 N-m, which is applied at t = 0 s, the analysis of the steady-state and dynamic performances is accomplished for

$$u_f = \begin{cases} 30\ \text{V}, \ \forall t \in [0 \quad 4)\ \text{s} \\ 10\ \text{V}, \ \forall t \in [4 \quad 7)\ \text{s} \\ 20\ \text{V}, \ \forall t \in [7 \quad 10]\ \text{s} \end{cases} \quad \text{and} \quad u_{fg} = 20\ \text{V}$$

See Figures 5.66 to 5.68.

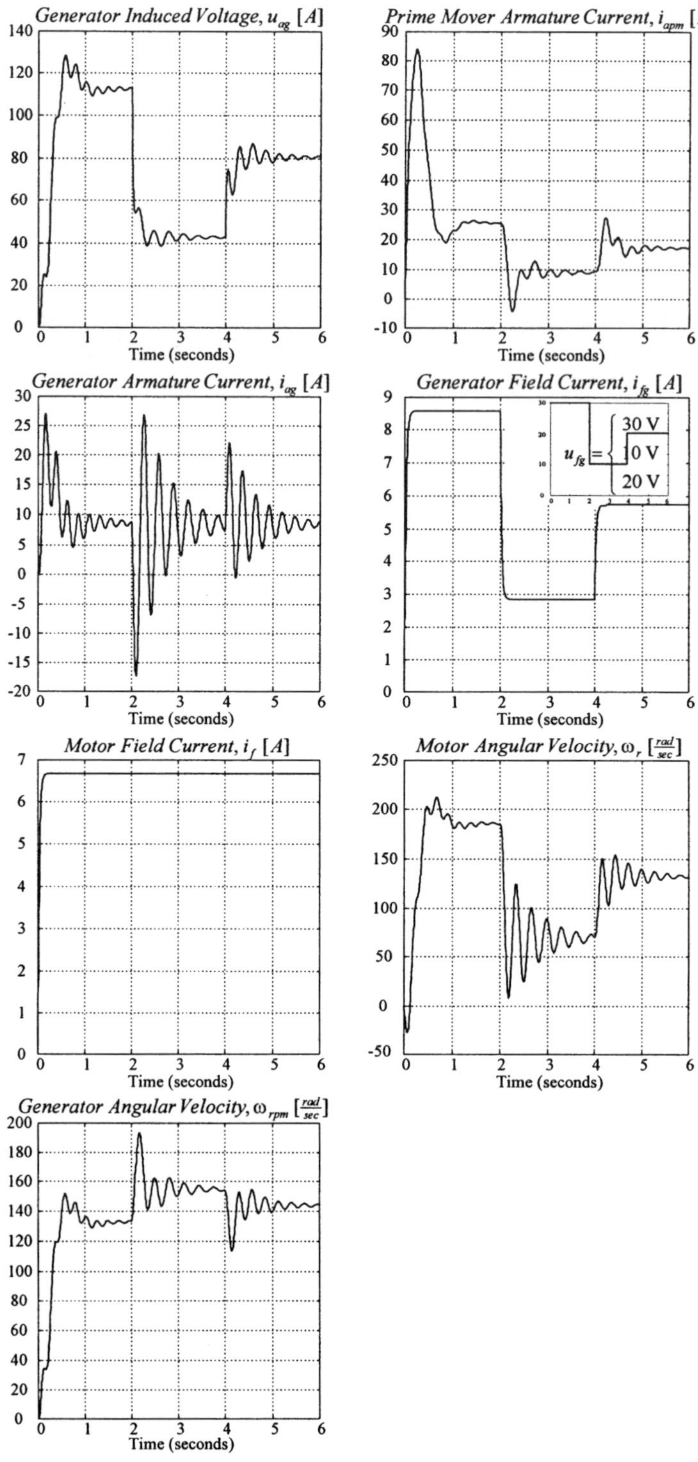

Figure 5.63 Dynamics of the Ward-Leonard system; $u_{apm} = 50$ V, $u_f = 20$ V, and $u_{fg} = \begin{cases} 30\text{ V}, & \forall t \in [0 \quad 4)\text{ s} \\ 10\text{ V}, & \forall t \in [4 \quad 7)\text{ s} \\ 20\text{ V}, & \forall t \in [7 \quad 10]\text{ s} \end{cases}$.

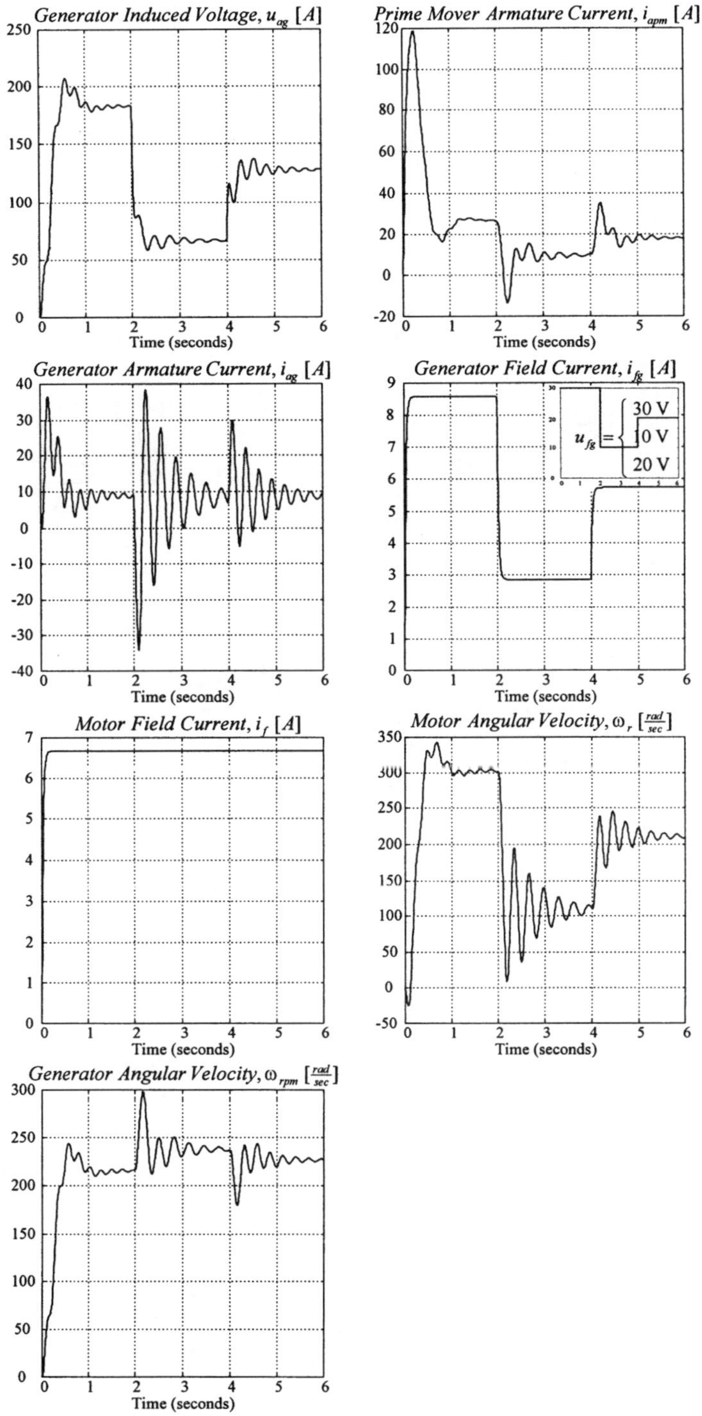

Figure 5.64 Dynamics of the Ward-Leonard system; $u_{apm} = 70$ V, $u_f = 20$ V, and $u_{fg} = \begin{cases} 30\text{ V}, & \forall t \in [0 \quad 4)\text{ s} \\ 10\text{ V}, & \forall t \in [4 \quad 7)\text{ s} \\ 20\text{ V}, & \forall t \in [7 \quad 10]\text{ s} \end{cases}$.

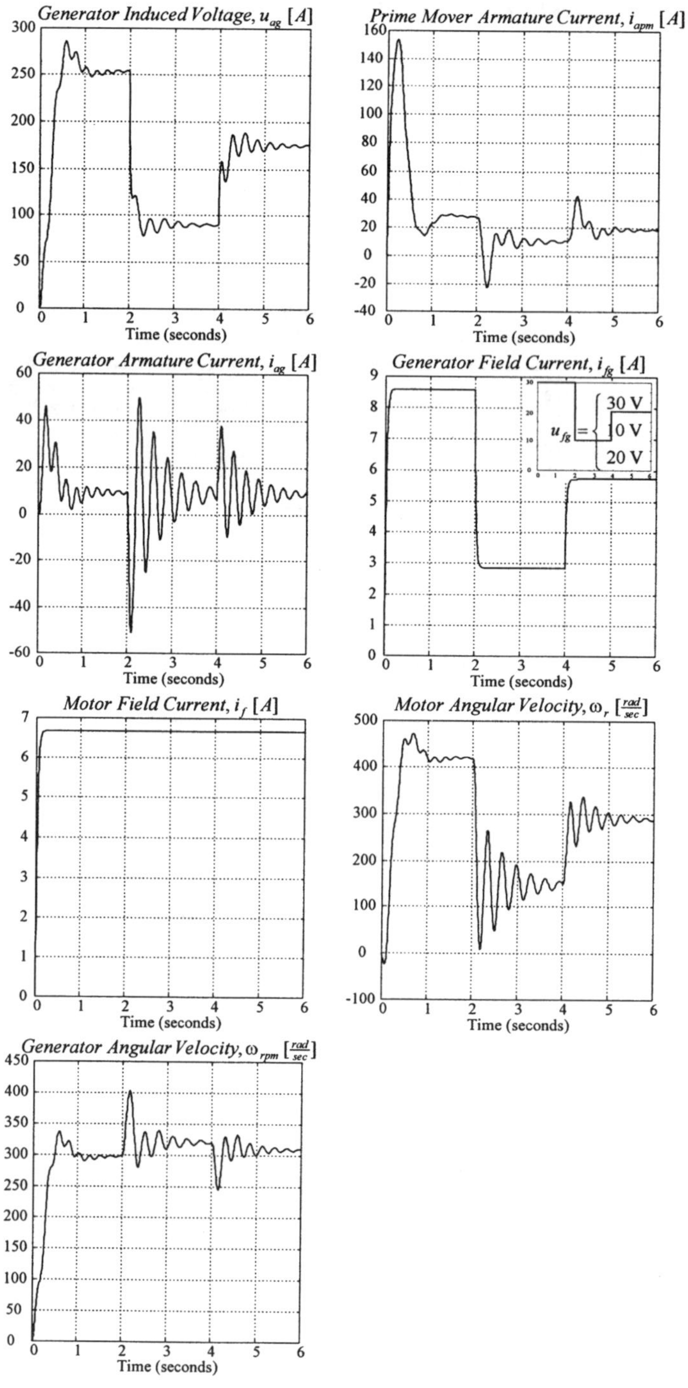

Figure 5.65 Dynamics of the Ward-Leonard system; $u_{apm} = 100$ V, $u_f = 20$ V, and $u_{fg} = \begin{cases} 30\ \text{V},\ \forall t \in [0 \quad 4)\ \text{s} \\ 10\ \text{V},\ \forall t \in [4 \quad 7)\ \text{s} \\ 20\ \text{V},\ \forall t \in [7 \quad 10]\ \text{s} \end{cases}$.

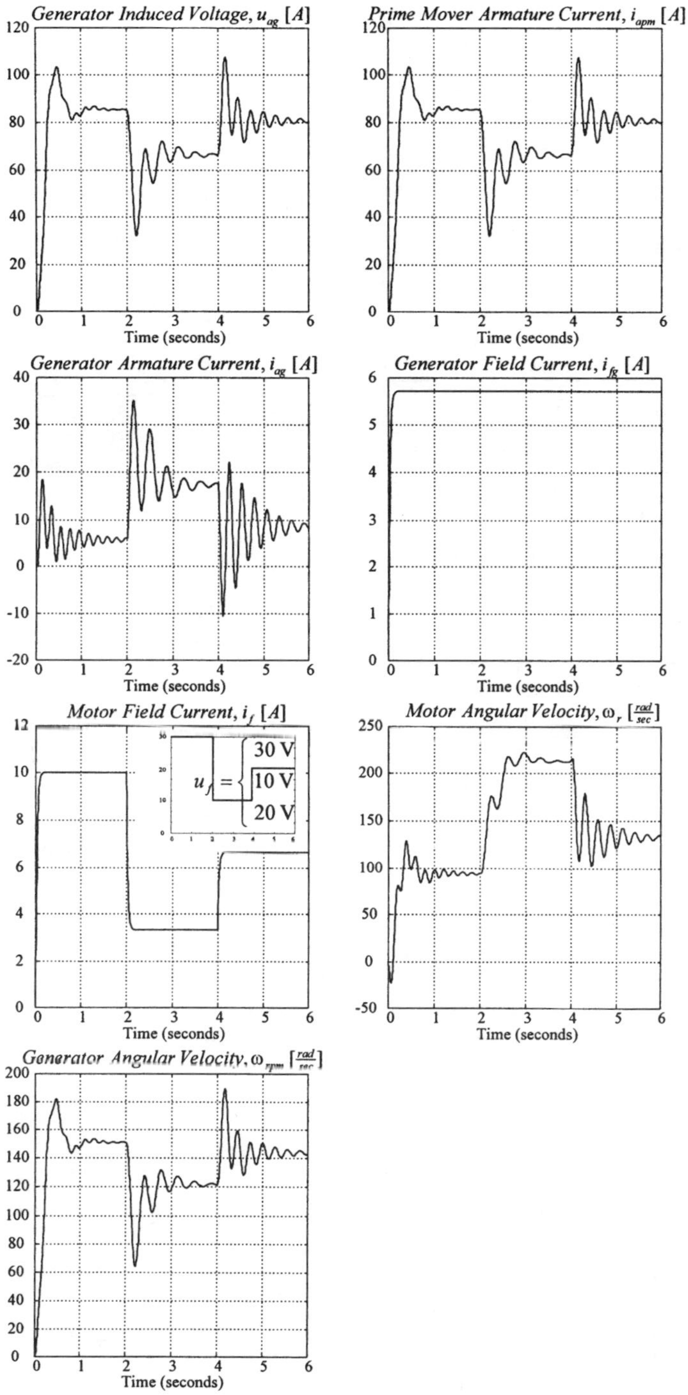

Figure 5.66 Dynamics of the Ward-Leonard system; $u_{apm} = 50$ V, $u_{fg} = 20$ V, and $u_f = \begin{cases} 30\text{ V}, & \forall t \in [0 \quad 4)\text{ s} \\ 10\text{ V}, & \forall t \in [4 \quad 7)\text{ s} \\ 20\text{ V}, & \forall t \in [7 \quad 10]\text{ s} \end{cases}$.

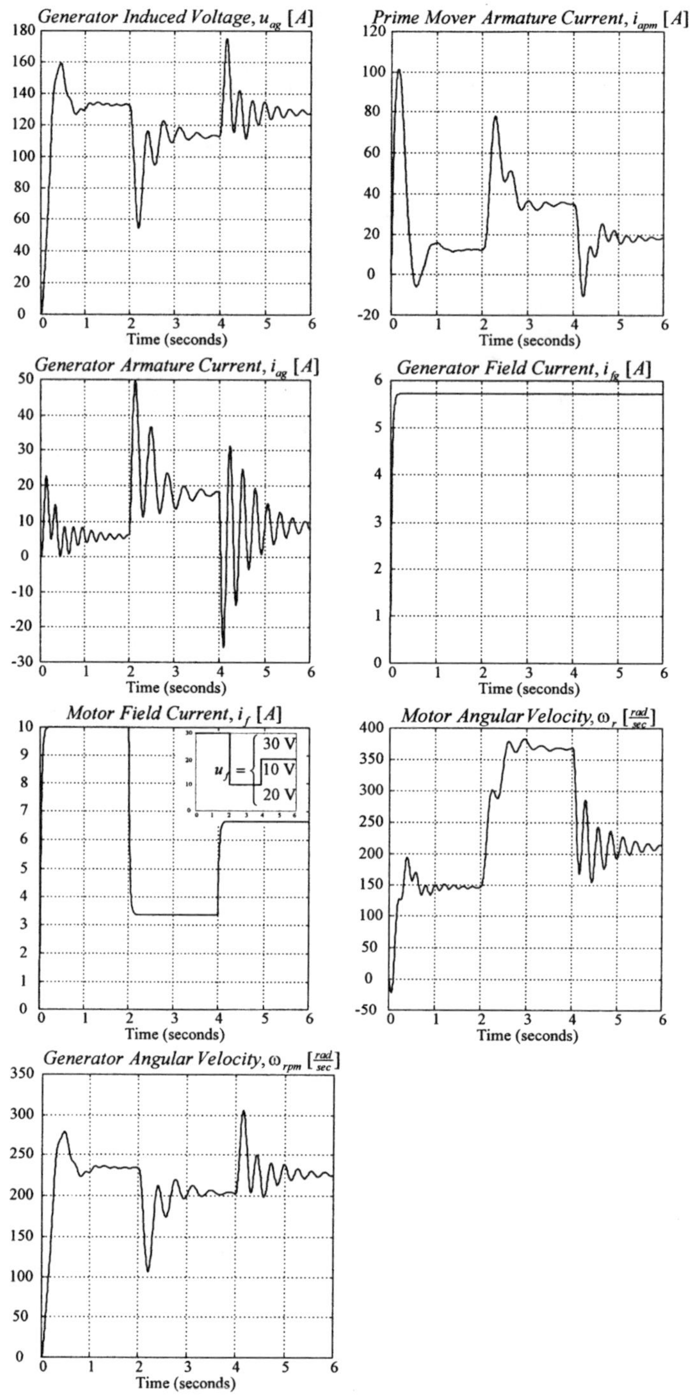

Figure 5.67 Dynamics of the Ward-Leonard system; $u_{apm} = 75$ V, $u_{fg} = 20$ V, and $u_f = \begin{cases} 30\text{ V}, \ \forall t \in [0 \quad 4)\text{ s} \\ 10\text{ V}, \ \forall t \in [4 \quad 7)\text{ s} \\ 20\text{ V}, \ \forall t \in [7 \quad 10]\text{ s} \end{cases}$.

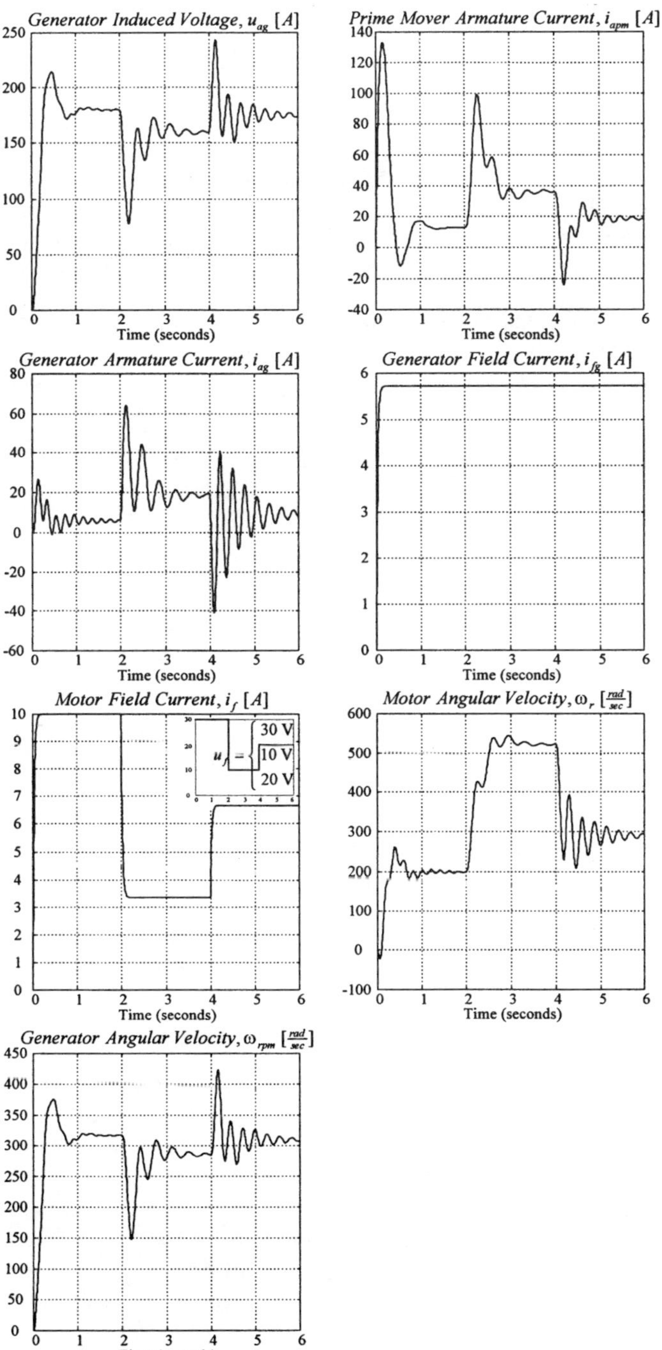

Figure 5.68 Dynamics of the Ward-Leonard system; $u_{apm} = 100$ V, $u_{fg} = 20$ V, and $u_f = \begin{cases} 30\text{ V}, & \forall t \in [0 \quad 4)\text{ s} \\ 10\text{ V}, & \forall t \in [4 \quad 7)\text{ s} \\ 20\text{ V}, & \forall t \in [7 \quad 10]\text{ s} \end{cases}$.

References

Chapman, S. J., *Electric Machinery Fundamentals,* McGraw-Hill, New York, 1999.
Fitzgerald, A. E., Kingsley, C., and Umans, S. D., *Electric Machinery,* McGraw-Hill, New York, 1990.
Krause, P. C., Wasynczuk, O., and Sudhoff, S. D., *Analysis of Electric Machinery,* IEEE Press, New York, 1995.
Krause, P. C. and Wasynczuk, O., *Electromechanical Motion Devices,* McGraw-Hill, New York, 1989.
Leonhard, W., *Control of Electrical Drives,* Springer, Berlin, 1996.
Ong, C.-M., *Dynamic Simulation of Electric Machines,* Prentice-Hall, Upper Saddle River, NJ, 1998.
Sarma, M. S., *Electric Machines: Steady-State Theory and Dynamic Performance,* PWS Press, New York, 1996.
Sen, P. C., *Principles of Electric Machines and Power Electronics,* John Wiley & Sons, New York, 1989.
Slemon, G. R., *Electric Machines and Drives,* Addison-Wesley, Reading, MA, 1992.
White, D. C. and Woodson, H. H., *Electromechanical Energy Conversion,* John Wiley & Sons, New York, 1959.

chapter six

Induction machines

List of basic variables and symbols used in this chapter

u_{as}, u_{bs}, and u_{cs} are the phase voltages in the stator windings *as, bs,* and *cs*
u_{qs}, u_{ds}, and u_{os} are the *quadrature-, direct-,* and *zero*-axis components of stator voltages
i_{as}, i_{bs}, and i_{cs} are the phase currents in the stator windings *as, bs,* and *cs*
i_{qs}, i_{ds}, and i_{os} are the *quadrature-, direct-,* and *zero*-axis components of stator currents
ψ_{as}, ψ_{bs}, and ψ_{cs} are the stator flux linkages
ψ_{qs}, ψ_{ds}, and ψ_{os} are the *quadrature-, direct-,* and *zero*-axis components of stator flux linkages
u_{ar}, u_{br}, and u_{cr} are the voltages in the rotor windings *ar, br,* and *cr*
u_{qr}, u_{dr}, and u_{or} are the *quadrature-, direct-,* and *zero*-axis components of rotor voltages
i_{ar}, i_{br}, and i_{cr} are the currents in the rotor windings *ar, br,* and *cr*
i_{qr}, i_{dr}, and i_{or} are the *quadrature-, direct-,* and *zero*-axis components of rotor currents
ψ_{ar}, ψ_{br}, and ψ_{cr} are the rotor flux linkages
ψ_{qr}, ψ_{dr}, and ψ_{or} are the *quadrature-, direct-,* and *zero*-axis components of rotor flux linkages
ω_r is the electrical angular velocity
ω_{rm} is the angular velocity of the rotor (the mechanical angular velocity)
θ_r is the electrical angular displacement
θ_{rm} is the angular displacement of the rotor (the mechanical angular displacement)
T_e is the electromagnetic torque developed by the motor
T_L is the load torque applied
r_s is the resistance of the stator windings
r_r is the resistance of the rotor windings
L_{ss} is the self-inductance of the stator windings
L_{rr} is the self-inductance of the rotor windings
L_{ms} is the stator magnetizing inductance
L_{ls} is the stator leakage inductance
L_{lr} is the rotor leakage inductance
N_s and N_r are the number of turns of the stator and rotor windings
P is the number of poles
B_m is the viscous friction coefficient
J is the equivalent moment of inertia
ω is the angular velocity of the reference frame
θ is the angular displacement of the reference frame

6.1 Introduction

Induction and synchronous machines were invented by Nikola Tesla, and high-efficiency machines have been manufactured in a wide range of rated power, torque, and angular velocity. Nikola Tesla was born in Croatia on July 9, 1856. In 1884, Tesla immigrated to the United States and worked with Thomas Edison. The principle of the rotating magnetic field, which is the basis of alternating-current electric machines, was invented and patented by Tesla in 1888. He discovered that two coils can be rotated due to the magnetic field if the currents supplied have 90° difference in phase. George Westinghouse purchased the patent, and the first induction motors were manufactured by the Westinghouse Motor Company in the 1890s. In addition to electric machinery, Tesla performed pioneering research in high-voltage power systems, transmission lines, and wireless communications.

Induction machines are used in the majority of industrial electric drives. Integrated with advanced variable-frequency power converters, sensors, and control systems, induction motors are increasingly used in high-performance variable-speed drives. Induction machines are rigid and reliable, have low weight-to-torque and power-to-torque ratios, operate at high angular velocity, have low cost and maintenance, can be used in harsh environments, etc.

Squirrel-cage induction machines have a cylindrical rotor with short-circuited rotor windings, and the voltage is supplied to stator windings. The rotor of a squirrel-cage induction machine, which does not have a brush and commutator, consists of conducting bars that are placed into the rotor slots, and these bars are shorted by the shorting rings. Aluminum and copper conductors are placed in slots around the outer periphery of the rotor core, and sheet-steel lamination technology is used to fabricate rotors. In contrast, the wound-rotor induction machines have wye-connected rotor windings and, in addition to the voltage applied to the stator windings, the phase voltages can be supplied to the rotor windings using brushes and slip rings. That is, one can shape the torque-speed characteristics of wound-rotor induction motors by supplying the voltage to the rotor windings or by inserting the external resistances into the rotor. The efficiency and maximal angular velocity of wound-rotor induction machines are lower compared with squirrel-cage design; and due to high weight-to-torque and power-to-torque ratios, and lower raggedness and reliability, wound-rotor induction machines have not been widely used. In contrast, for years, squirrel-cage induction machines have been used in high-performance electric drives and electromechanical systems.

Studying the doubly excited transducer, it has been shown that the applied voltage to the stator winding produces a magnetic field that induces the currents in the rotor winding. The electromagnetic torque is developed due to the interaction of the magnetic fields. These basic features were demonstrated by studying a doubly excited transducer, which was thoroughly examined to perform the analysis of single-phase induction machines. The main objectives of this chapter are to analyze the steady-state characteristics and dynamic performance of two- and three-phase induction machines, to develop mathematical models, to perform nonlinear simulations, as well as to introduce the power converters needed to control induction motors. The analysis of induction machines will be performed using the so-called *machine* (*abc*) variables and an *arbitrary* reference frame. The arbitrary reference frame allows one to examine the steady-state and dynamic behaviors of induction machines using the *quadrature-*, *direct-*, and *zero*-axis (*qd0*) quantities. In particular, the *quadrature-*, *direct-*, and *zero*-axis components of stator and rotor voltages, currents, and flux linkages are used. The arbitrary reference frame permits one to study induction machines in the stationary, rotor, and synchronous reference frames by assigning the angular velocity of the reference frame ω to be equal to zero, $\omega = 0$; equal to the rotor angular velocity $\omega = \omega_r$; or equal to the synchronous angular velocity, $\omega = \omega_e$.

Figure 6.1 Photograph of an induction motor. (Courtesy of Baldor Electric Company.)

A typical induction motor and assembly of induction machines are shown in Figures 6.1 to 6.4.

6.2 *Voltage, flux linkages, and torque equations for two-phase induction machines: dynamics in the **machine** variables*

To develop a mathematical model of a two-phase induction motor, which is shown in Figure 6.5, one can map the stator and rotor circuitry dynamics using the voltages applied

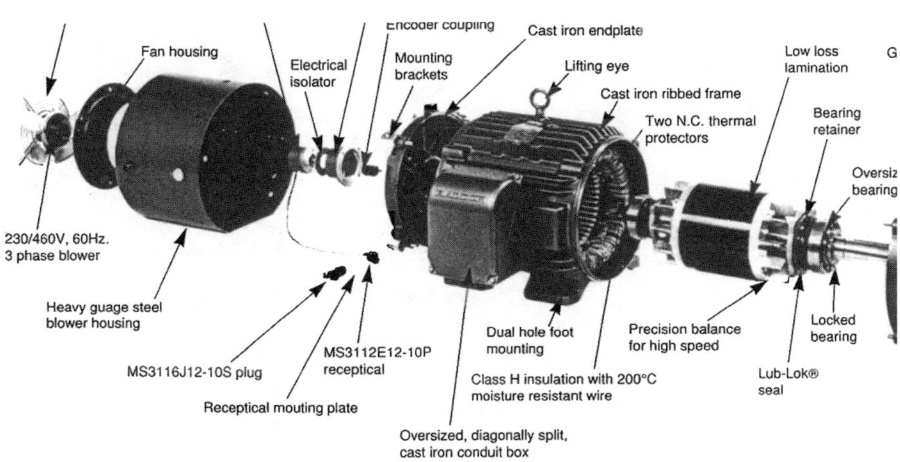

Figure 6.2 Assembly of induction machines. (Courtesy of Baldor Electric Company.)

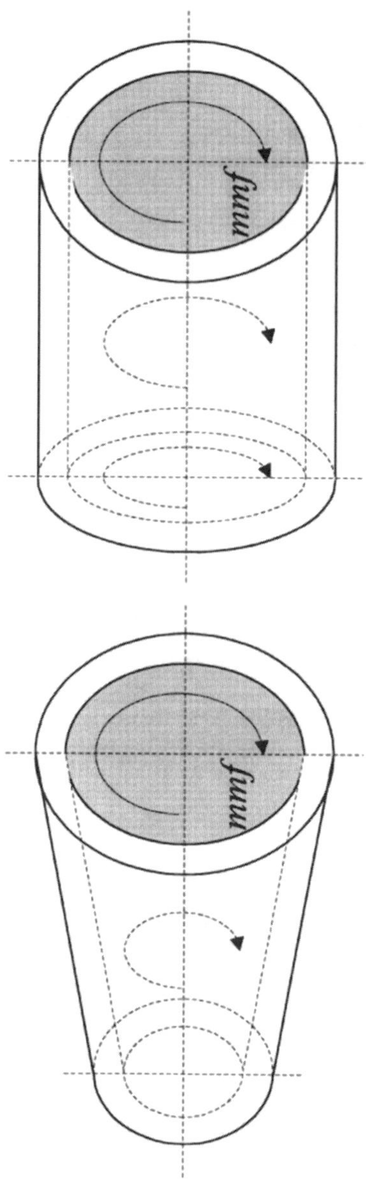

Figure 6.3 Conical and cylindrical geometry of induction machines: endless magnetic system.

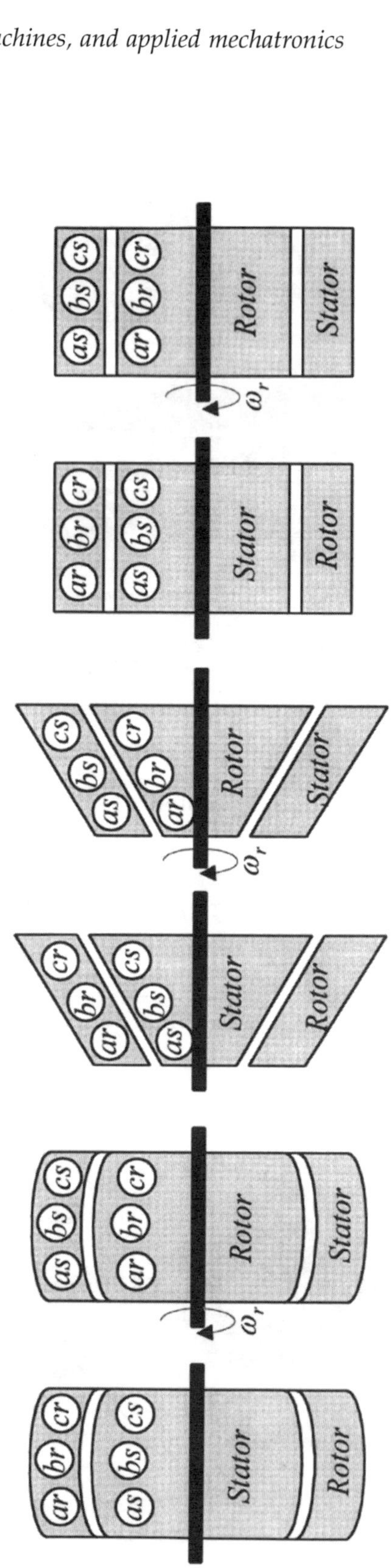

Figure 6.4 Three-phase induction machines: spherical, conical, and cylindrical geometry.

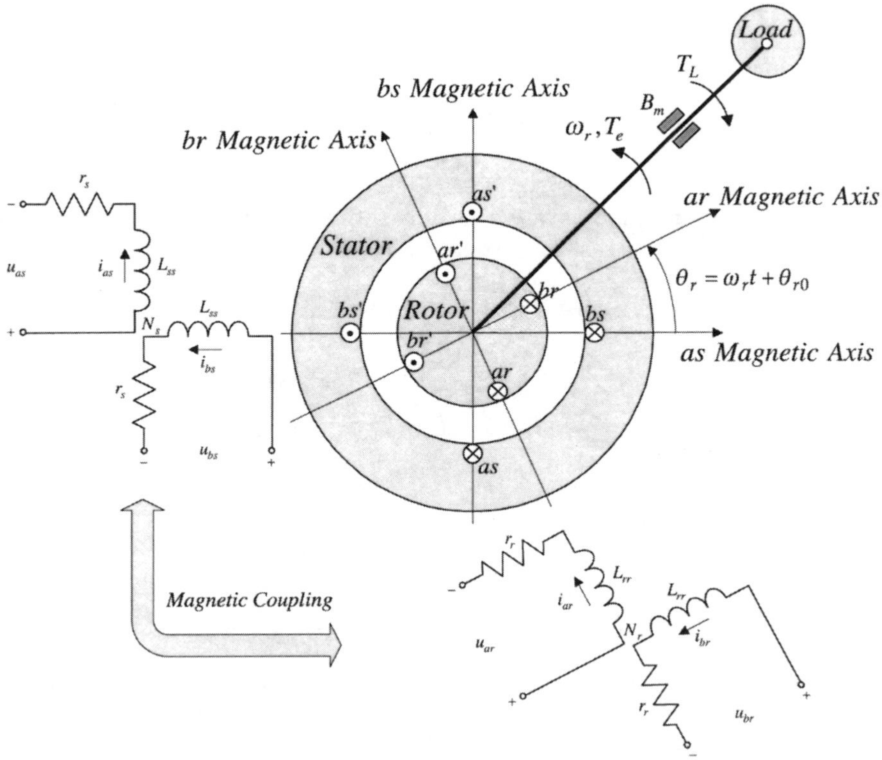

Figure 6.5 Two-phase symmetrical induction motor.

to the stator (*as* and *bs*) and rotor (*ar* and *br*) windings, as well as the stator and rotor currents and flux linkages.

Using Kirchhoff's voltage law, four differential equations result for two-phase induction motors circuitry:

$$u_{as} = r_s i_{as} + \frac{d\psi_{as}}{dt}$$

$$u_{bs} = r_s i_{bs} + \frac{d\psi_{bs}}{dt}$$

$$u_{ar} = r_r i_{ar} + \frac{d\psi_{ar}}{dt}$$

$$u_{br} = r_r i_{br} + \frac{d\psi_{br}}{dt}$$

and in matrix form, we have

$$\mathbf{u}_{abs} = \mathbf{r}_s \mathbf{i}_{abs} + \frac{d\boldsymbol{\psi}_{abs}}{dt}$$
$$\mathbf{u}_{abr} = \mathbf{r}_r \mathbf{i}_{abr} + \frac{d\boldsymbol{\psi}_{abr}}{dt} \quad (6.1)$$

where $\mathbf{u}_{abs} = \begin{bmatrix} u_{as} \\ u_{bs} \end{bmatrix}$, $\mathbf{u}_{abr} = \begin{bmatrix} u_{ar} \\ u_{br} \end{bmatrix}$, $\mathbf{i}_{abs} = \begin{bmatrix} i_{as} \\ i_{bs} \end{bmatrix}$, $\mathbf{i}_{abr} = \begin{bmatrix} i_{ar} \\ i_{br} \end{bmatrix}$, $\boldsymbol{\psi}_{abs} = \begin{bmatrix} \psi_{as} \\ \psi_{bs} \end{bmatrix}$, and $\boldsymbol{\psi}_{abr} = \begin{bmatrix} \psi_{ar} \\ \psi_{br} \end{bmatrix}$ are the phase voltages, currents, and flux linkages; $\mathbf{r}_s = \begin{bmatrix} r_s & 0 \\ 0 & r_s \end{bmatrix}$ and $\mathbf{r}_r = \begin{bmatrix} r_r & 0 \\ 0 & r_r \end{bmatrix}$ are the matrices of the stator and rotor resistances.

Assuming that the magnetic system is linear, the flux linkages are expressed using the self- and mutual inductances. In particular,

$$\psi_{as} = L_{asas} i_{as} + L_{asbs} i_{bs} + L_{asar} i_{ar} + L_{asbr} i_{br}$$

$$\psi_{bs} = L_{bsas} i_{as} + L_{bsbs} i_{bs} + L_{bsar} i_{ar} + L_{bsbr} i_{br}$$

$$\psi_{ar} = L_{aras} i_{as} + L_{arbs} i_{bs} + L_{arar} i_{ar} + L_{arbr} i_{br}$$

$$\psi_{br} = L_{bras} i_{as} + L_{brbs} i_{bs} + L_{brar} i_{ar} + L_{brbr} i_{br}$$

The stator and rotor self-inductances are denoted as L_{ss} and L_{rr}. One obtains

$$L_{ss} = L_{asas} = L_{bsbs} \text{ and } L_{rr} = L_{arar} = L_{brbr}$$

The stator and rotor windings are orthogonal. Therefore, the coupling between the *as* and *bs* stator windings, as well as between the *ar* and *br* rotor windings, does not exist. Thus, the mutual inductances between the stator windings are

$$L_{asbs} = L_{bsas} = 0$$

and the mutual inductances between the rotor windings are given as

$$L_{arbr} = L_{brar} = 0$$

The magnetic coupling between stator and rotor windings should be studied. Varying mutual inductances, which are functions of θ_r, have minimum and maximum values. In fact, the stator-rotor mutual inductances vary sinusoidally as functions of the rotor displacement (recall that the stator is the stationary member, and the rotor rotates with the electrical angular velocity ω_r; see the coupling shown in Figure 6.5). Assuming that these variations obey the cosine law, one obtains the following expressions for the mutual inductances between stator and rotor windings:

$$L_{asar} = L_{aras} = L_{sr} \cos\theta_r$$

$$L_{asbr} = L_{bras} = -L_{sr} \sin\theta_r$$

$$L_{bsar} = L_{arbs} = L_{sr} \sin\theta_r$$

$$L_{bsbr} = L_{brbs} = L_{sr} \cos\theta_r$$

The following matrix equation for the flux linkages results

$$\begin{bmatrix} \boldsymbol{\psi}_{abs} \\ \boldsymbol{\psi}_{abr} \end{bmatrix} = \begin{bmatrix} \mathbf{L}_s & \mathbf{L}_{sr} \\ \mathbf{L}_{sr}^T & \mathbf{L}_r \end{bmatrix} \begin{bmatrix} \mathbf{i}_{abs} \\ \mathbf{i}_{abr} \end{bmatrix}$$

where $\mathbf{L}_s$ is the matrix of the stator inductances, $\mathbf{L}_s = \begin{bmatrix} L_{ss} & 0 \\ 0 & L_{ss} \end{bmatrix}$, $L_{ss} = L_{ls} + L_{ms}$, $L_{ms} = \dfrac{N_s^2}{\Re_m}$;

$\mathbf{L}_r$ is the matrix of the rotor inductances, $\mathbf{L}_r = \begin{bmatrix} L_{rr} & 0 \\ 0 & L_{rr} \end{bmatrix}$, $L_{rr} = L_{lr} + L_{mr}$, $L_{mr} = \dfrac{N_r^2}{\Re_m}$;

$\mathbf{L}_{sr}$ is the matrix of the stator-rotor mutual inductances, $\mathbf{L}_{sr} = \begin{bmatrix} L_{sr}\cos\theta_r & -L_{sr}\sin\theta_r \\ L_{sr}\sin\theta_r & L_{sr}\cos\theta_r \end{bmatrix}$,

$L_{sr} = \dfrac{N_s N_r}{\Re_m}$.

Using the number of turns in the stator and rotor windings, we have

$$\mathbf{i}'_{abr} = \frac{N_r}{N_s}\mathbf{i}_{abr}$$

$$\mathbf{u}'_{abr} = \frac{N_s}{N_r}\mathbf{u}_{abr}$$

and

$$\boldsymbol{\psi}'_{abr} = \frac{N_s}{N_r}\boldsymbol{\psi}_{abr}$$

Then, the flux linkages are written in matrix form as

$$\begin{bmatrix} \psi_{abs} \\ \psi'_{abr} \end{bmatrix} = \begin{bmatrix} \mathbf{L}_s & \mathbf{L}'_{sr} \\ \mathbf{L}'^{T}_{sr} & \mathbf{L}'_r \end{bmatrix} \begin{bmatrix} \mathbf{i}_{abs} \\ \mathbf{i}'_{abr} \end{bmatrix} \tag{6.2}$$

where $\mathbf{L}'_r = \left(\dfrac{N_s}{N_r}\right)^2 \mathbf{L}_r = \begin{bmatrix} L'_{rr} & 0 \\ 0 & L'_{rr} \end{bmatrix}$, $L'_{rr} = L'_{lr} + L'_{mr}$;

$$\mathbf{L}'_{sr} = \left(\frac{N_s}{N_r}\right)\mathbf{L}_{sr} = L_{ms}\begin{bmatrix} \cos\theta_r & -\sin\theta_r \\ \sin\theta_r & \cos\theta_r \end{bmatrix},\quad L_{ms} = \frac{N_s}{N_r}L_{sr}.$$

Observe that $L'_{mr} = \left(\dfrac{N_s}{N_r}\right)^2 L_{mr}$ and $L'_{mr} = L_{ms} = \dfrac{N_s}{N_r}L_{sr}$. Thus, $L'_{rr} = L'_{lr} + L_{ms}$.

Substituting the matrices for self- and mutual inductance, $\mathbf{L}_s$, $\mathbf{L}'_r$ and $\mathbf{L}'_{sr}$ in (6.2), one obtains

$$\begin{bmatrix} \psi_{as} \\ \psi_{bs} \\ \psi'_{ar} \\ \psi'_{br} \end{bmatrix} = \begin{bmatrix} L_{ss} & 0 & L_{ms}\cos\theta_r & -L_{ms}\sin\theta_r \\ 0 & L_{ss} & L_{ms}\sin\theta_r & L_{ms}\cos\theta_r \\ L_{ms}\cos\theta_r & L_{ms}\sin\theta_r & L'_{rr} & 0 \\ -L_{ms}\sin\theta_r & L_{ms}\cos\theta_r & 0 & L'_{rr} \end{bmatrix} \begin{bmatrix} i_{as} \\ i_{bs} \\ i'_{ar} \\ i'_{br} \end{bmatrix}$$

In expanded form, the flux linkages are found to be

$$\begin{aligned} \psi_{as} &= L_{ss}i_{as} + L_{ms}i'_{ar}\cos\theta_r - L_{ms}i'_{br}\sin\theta_r \\ \psi_{bs} &= L_{ss}i_{bs} + L_{ms}i'_{ar}\sin\theta_r + L_{ms}i'_{br}\cos\theta_r \\ \psi'_{ar} &= L_{ms}i_{as}\cos\theta_r + L_{ms}i_{bs}\sin\theta_r + L'_{rr}i_{ar} \\ \psi'_{br} &= -L_{ms}i_{as}\sin\theta_r + L_{ms}i_{bs}\cos\theta_r + L'_{rr}i'_{br} \end{aligned} \tag{6.3}$$

The circuitry differential equations (6.1) are rewritten as

$$\begin{aligned} \mathbf{u}_{abs} &= \mathbf{r}_s\mathbf{i}_{abs} + \frac{d\psi_{abs}}{dt} \\ \mathbf{u}'_{abr} &= \mathbf{r}'_r\mathbf{i}'_{abr} + \frac{d\psi'_{abr}}{dt} \end{aligned} \tag{6.4}$$

where $\mathbf{r}'_r = \frac{N_s^2}{N_r^2}\mathbf{r}_r = \frac{N_s^2}{N_r^2}\begin{bmatrix} r'_r & 0 \\ 0 & r'_r \end{bmatrix}$.

Making use of Eq. (6.4) we have

$$\begin{aligned} \frac{d\psi_{as}}{dt} &= -r_si_{as} + u_{as} \\ \frac{d\psi_{bs}}{dt} &= -r_si_{bs} + u_{bs} \\ \frac{d\psi'_{ar}}{dt} &= -r'_ri'_{ar} + u'_{ar} \\ \frac{d\psi'_{br}}{dt} &= -r'_ri'_{br} + u'_{br} \end{aligned}$$

Assuming that the self- and mutual inductances L_{ss}, L'_{rr} , and L_{ms} are time-invariant, and using the expressions for the flux linkages (6.3), one obtains a set of highly coupled nonlinear differential equations to map the circuitry dynamics:

$$\begin{aligned} L_{ss}\frac{di_{as}}{dt} + L_{ms}\frac{d(i'_{ar}\cos\theta_r)}{dt} - L_{ms}\frac{d(i'_{br}\sin\theta_r)}{dt} &= -r_si_{as} + u_{as} \\ L_{ss}\frac{di_{bs}}{dt} + L_{ms}\frac{d(i'_{ar}\sin\theta_r)}{dt} + L_{ms}\frac{d(i'_{br}\cos\theta_r)}{dt} &= -r_si_{bs} + u_{bs} \end{aligned}$$

$$L_{ms}\frac{d(i_{as}\cos\theta_r)}{dt}+L_{ms}\frac{d(i_{bs}\sin\theta_r)}{dt}+L'_{rr}\frac{di'_{ar}}{dt}=-r'_r i'_{ar}+u'_{ar}$$

$$-L_{ms}\frac{d(i_{as}\sin\theta_r)}{dt}+L_{ms}\frac{d(i_{bs}\cos\theta_r)}{dt}+L'_{rr}\frac{di'_{br}}{dt}=-r'_r i'_{br}+u'_{br}$$

Cauchy's form of these differential equations is found. In particular, we have

$$\frac{di_{as}}{dt}=-\frac{L'_{rr}r_s}{L_{ss}L'_{rr}-L^2_{ms}}i_{as}+\frac{L^2_{ms}}{L_{ss}L'_{rr}-L^2_{ms}}i_{bs}\omega_r+\frac{L_{ms}L'_{rr}}{L_{ss}L'_{rr}-L^2_{ms}}i'_{ar}\left(\omega_r\sin\theta_r+\frac{r'_r}{L'_{rr}}\cos\theta_r\right)$$
$$+\frac{L_{ms}L'_{rr}}{L_{ss}L'_{rr}-L^2_{ms}}i'_{br}\left(\omega_r\cos\theta_r-\frac{r'_r}{L'_{rr}}\sin\theta_r\right)+\frac{L'_{rr}}{L_{ss}L'_{rr}-L^2_{ms}}u_{as}-\frac{L_{ms}}{L_{ss}L'_{rr}-L^2_{ms}}\cos\theta_r u'_{ar}$$
$$+\frac{L_{ms}}{L_{ss}L'_{rr}-L^2_{ms}}\sin\theta_r u'_{br}$$

$$\frac{di_{bs}}{dt}=-\frac{L'_{rr}r_s}{L_{ss}L'_{rr}-L^2_{ms}}i_{bs}-\frac{L^2_{ms}}{L_{ss}L'_{rr}-L^2_{ms}}i_{as}\omega_r-\frac{L_{ms}L'_{rr}}{L_{ss}L'_{rr}-L^2_{ms}}i'_{ar}\left(\omega_r\cos\theta_r-\frac{r'_r}{L'_{rr}}\sin\theta_r\right)$$
$$+\frac{L_{ms}L'_{rr}}{L_{ss}L'_{rr}-L^2_{ms}}i'_{br}\left(\omega_r\sin\theta_r+\frac{r'_r}{L'_{rr}}\cos\theta_r\right)+\frac{L'_{rr}}{L_{ss}L'_{rr}-L^2_{ms}}u_{bs}-\frac{L_{ms}}{L_{ss}L'_{rr}-L^2_{ms}}\sin\theta_r u'_{ar}$$
$$-\frac{L_{ms}}{L_{ss}L'_{rr}-L^2_{ms}}\cos\theta_r u'_{br}$$

$$\frac{di'_{ar}}{dt}=-\frac{L_{ss}r'_r}{L_{ss}L'_{rr}-L^2_{ms}}i'_{ar}+\frac{L_{ms}L_{ss}}{L_{ss}L'_{rr}-L^2_{ms}}i_{as}\left(\omega_r\sin\theta_r+\frac{r_s}{L_{ss}}\cos\theta_r\right)-\frac{L_{ms}L_{ss}}{L_{ss}L'_{rr}-L^2_{ms}}i_{bs}\left(\omega_r\cos\theta_r-\frac{r_s}{L_{ss}}\sin\theta_r\right)$$
$$-\frac{L^2_{ms}}{L_{ss}L'_{rr}-L^2_{ms}}i'_{br}\omega_r-\frac{L_{ms}}{L_{ss}L'_{rr}-L^2_{ms}}\cos\theta_r u_{as}-\frac{L_{ms}}{L_{ss}L'_{rr}-L^2_{ms}}\sin\theta_r u_{bs}+\frac{L_{ss}}{L_{ss}L'_{rr}-L^2_{ms}}u'_{ar}$$

$$\frac{di'_{br}}{dt}=-\frac{L_{ss}r'_r}{L_{ss}L'_{rr}-L^2_{ms}}i'_{br}+\frac{L_{ms}L_{ss}}{L_{ss}L'_{rr}-L^2_{ms}}i_{as}\left(\omega_r\cos\theta_r-\frac{r_s}{L_{ss}}\sin\theta_r\right)+\frac{L_{ms}L_{ss}}{L_{ss}L'_{rr}-L^2_{ms}}i_{bs}\left(\omega_r\sin\theta_r+\frac{r_s}{L_{ss}}\cos\theta_r\right)$$
$$+\frac{L^2_{ms}}{L_{ss}L'_{rr}-L^2_{ms}}i'_{ar}\omega_r+\frac{L_{ms}}{L_{ss}L'_{rr}-L^2_{ms}}\sin\theta_r u_{as}-\frac{L_{ms}}{L_{ss}L'_{rr}-L^2_{ms}}\cos\theta_r u_{bs}+\frac{L_{ss}}{L_{ss}L'_{rr}-L^2_{ms}}u'_{br}\tag{6.5}$$

The derived differential equations map the augmented stator-rotor circuitry dynamics for two-phase induction motors, and the electrical angular velocity ω_r and displacement θ_r are used in (6.5) as the state variables. Hence, the *torsional-mechanical* equation of motion must be incorporated to map the evolution of ω_r and θ_r. Using Newton's second law, we have

$$\sum T=T_e-B_m\omega_{rm}-T_L=J\frac{d\omega_{rm}}{dt}$$

$$\frac{d\theta_{rm}}{dt}=\omega_{rm}$$

The mechanical angular velocity ω_{rm} must be expressed using the electrical angular velocity ω_r and the number of poles P. In particular,

$$\omega_{rm} = \frac{2}{P}\omega_r$$

The mechanical and electrical angular displacements θ_{rm} and θ_r are related as

$$\theta_{rm} = \frac{2}{P}\theta_r$$

Therefore, taking note of Newton's second law, one obtains

$$\frac{d\omega_r}{dt} = \frac{P}{2J}T_e - \frac{B_m}{J}\omega_r - \frac{P}{2J}T_L$$
$$\frac{d\theta_r}{dt} = \omega_r$$

To find the expression for the electromagnetic torque developed by two-phase induction motors, the equation for coenergy must be used. For linear magnetic system, we have

$$W_c = W_f = \frac{1}{2}\mathbf{i}_{abs}^T(\mathbf{L}_s - L_{ls}\mathbf{I})\mathbf{i}_{abs} + \mathbf{i}_{abs}^T\mathbf{L}'_{sr}\mathbf{i}'_{abr} + \frac{1}{2}\mathbf{i}'^T_{abr}(\mathbf{L}'_r - L'_{lr}\mathbf{I})\mathbf{i}'_{abr}$$

The self-inductances L_{ss} and L'_{rr}, as well as the leakage inductances L_{ls} and L'_{lr}, are not functions of the angular displacement θ_r, while the following expression for the matrix of stator-rotor mutual inductances $\mathbf{L}'_{sr}$ was derived $\mathbf{L}'_{sr} = L_{ms}\begin{bmatrix}\cos\theta_r & -\sin\theta_r \\ \sin\theta_r & \cos\theta_r\end{bmatrix}$. Then, the electromagnetic torque developed by P-pole two-phase induction motors is found to be

$$T_e = \frac{P}{2}\frac{\partial W_c(\mathbf{i}_{abs}, \mathbf{i}'_{abr}, \theta_r)}{\partial \theta_r} = \frac{P}{2}\mathbf{i}_{abs}^T\frac{\partial \mathbf{L}'_{sr}(\theta_r)}{\partial \theta_r}\mathbf{i}'_{abr} = \frac{P}{2}L_{ms}[i_{as} \quad i_{bs}]\begin{bmatrix}-\sin\theta_r & -\cos\theta_r \\ \cos\theta_r & -\sin\theta_r\end{bmatrix}\begin{bmatrix}i'_{ar} \\ i'_{br}\end{bmatrix}$$
$$= -\frac{P}{2}L_{ms}[(i_{as}i'_{ar} + i_{bs}i'_{br})\sin\theta_r + (i_{as}i'_{br} - i_{bs}i'_{ar})\cos\theta_r] \quad (6.6)$$

Using expression (6.6) for the electromagnetic torque T_e in the *torsional-mechanical* equations of motion, one obtains

$$\frac{d\omega_r}{dt} = -\frac{P^2}{4J}L_{ms}[(i_{as}i'_{ar} + i_{bs}i'_{br})\sin\theta_r + (i_{as}i'_{br} - i_{bs}i'_{ar})\cos\theta_r] - \frac{B_m}{J}\omega_r - \frac{P}{2J}T_L$$
$$\frac{d\theta_r}{dt} = \omega_r \quad (6.7)$$

It is evident that

$$\theta_r = \int_{t_0}^{t} \omega_r(\tau)d\tau + \theta_{r0} \tag{6.8}$$

It is concluded that the mathematical model of two-phase induction motors is found using the stator and rotor currents, as well as the electrical angular velocity and displacement as the state variables. Augmenting differential equations (6.5) and (6.7), the following set of highly nonlinear differential equations results

$$\frac{di_{as}}{dt} = -\frac{L'_{rr}r_s}{L_\Sigma}i_{as} + \frac{L_{ms}^2}{L_\Sigma}i_{bs}\omega_r + \frac{L_{ms}L'_{rr}}{L_\Sigma}i'_{ar}\left(\omega_r\sin\theta_r + \frac{r'_r}{L'_{rr}}\cos\theta_r\right) + \frac{L_{ms}L'_{rr}}{L_\Sigma}i'_{br}\left(\omega_r\cos\theta_r - \frac{r'_r}{L'_{rr}}\sin\theta_r\right)$$
$$+\frac{L'_{rr}}{L_\Sigma}u_{as} - \frac{L_{ms}}{L_\Sigma}\cos\theta_r u'_{ar} + \frac{L_{ms}}{L_\Sigma}\sin\theta_r u'_{br}$$

$$\frac{di_{bs}}{dt} = -\frac{L'_{rr}r_s}{L_\Sigma}i_{bs} - \frac{L_{ms}^2}{L_\Sigma}i_{as}\omega_r - \frac{L_{ms}L'_{rr}}{L_\Sigma}i'_{ar}\left(\omega_r\cos\theta_r - \frac{r'_r}{L'_{rr}}\sin\theta_r\right) + \frac{L_{ms}L'_{rr}}{L_\Sigma}i'_{br}\left(\omega_r\sin\theta_r + \frac{r'_r}{L'_{rr}}\cos\theta_r\right)$$
$$+\frac{L'_{rr}}{L_\Sigma}u_{bs} - \frac{L_{ms}}{L_\Sigma}\sin\theta_r u'_{ar} - \frac{L_{ms}}{L_\Sigma}\cos\theta_r u'_{br}$$

$$\frac{di'_{ar}}{dt} = -\frac{L_{ss}r'_r}{L_\Sigma}i'_{ar} + \frac{L_{ms}L_{ss}}{L_\Sigma}i_{as}\left(\omega_r\sin\theta_r + \frac{r_s}{L_{ss}}\cos\theta_r\right) - \frac{L_{ms}L_{ss}}{L_\Sigma}i_{bs}\left(\omega_r\cos\theta_r - \frac{r_s}{L_{ss}}\sin\theta_r\right) - \frac{L_{ms}^2}{L_\Sigma}i'_{br}\omega_r$$
$$-\frac{L_{ms}}{L_\Sigma}\cos\theta_r u_{as} - \frac{L_{ms}}{L_\Sigma}\sin\theta_r u_{bs} + \frac{L_{ss}}{L_\Sigma}u'_{ar}$$

$$\frac{di'_{br}}{dt} = -\frac{L_{ss}r'_r}{L_\Sigma}i'_{br} + \frac{L_{ms}L_{ss}}{L_\Sigma}i_{as}\left(\omega_r\cos\theta_r - \frac{r_s}{L_{ss}}\sin\theta_r\right) + \frac{L_{ms}L_{ss}}{L_\Sigma}\left(\omega_r\sin\theta_r - \frac{r_s}{L_{ss}}\cos\theta_r\right) + \frac{L_{ms}^2}{L_\Sigma}i'_{ar}\omega_r$$
$$+\frac{L_{ms}}{L_\Sigma}\sin\theta_r u_{as} - \frac{L_{ms}}{L_\Sigma}\cos\theta_r u_{bs} + \frac{L_{ss}}{L_\Sigma}u'_{br}$$

$$\frac{d\omega_r}{dt} = -\frac{P^2}{4J}L_{ms}[(i_{as}i'_{ar} + i_{bs}i'_{br})\sin\theta_r + (i_{as}i'_{br} - i_{bs}i'_{ar})\cos\theta_r] - \frac{B_m}{J}\omega_r - \frac{P}{2J}T_L$$

$$\frac{d\theta_r}{dt} = \omega_r \tag{6.9}$$

where $L_\Sigma = L_{ss}L'_{rr} - L_{ms}^2$.

In matrix form, a set of six highly coupled nonlinear differential equations (6.9) is written as

$$\begin{bmatrix} \frac{di_{as}}{dt} \\ \frac{di_{bs}}{dt} \\ \frac{di'_{ar}}{dt} \\ \frac{di'_{br}}{dt} \\ \frac{d\omega_r}{dt} \\ \frac{d\theta_r}{dt} \end{bmatrix} = \begin{bmatrix} -\frac{L'_{rr}r_s}{L_\Sigma} & 0 & 0 & 0 & 0 & 0 \\ 0 & -\frac{L'_{rr}r_s}{L_\Sigma} & 0 & 0 & 0 & 0 \\ 0 & 0 & -\frac{L'_{ss}r'_r}{L_\Sigma} & 0 & 0 & 0 \\ 0 & 0 & 0 & -\frac{L'_{ss}r'_{rr}}{L_\Sigma} & 0 & 0 \\ 0 & 0 & 0 & 0 & -\frac{B_m}{J} & 0 \\ 0 & 0 & 0 & 0 & 1 & 0 \end{bmatrix} \begin{bmatrix} i_{as} \\ i_{bs} \\ i'_{ar} \\ i'_{br} \\ \omega_r \\ \theta_r \end{bmatrix}$$

$$+ \begin{bmatrix} \frac{L_{ms}^2}{L_\Sigma} i_{bs}\omega_r + \frac{L_{ms}L'_{rr}}{L_\Sigma} i'_{ar}\left(\omega_r \sin\theta_r + \frac{r'_r}{L'_{rr}}\cos\theta_r\right) + \frac{L_{ms}L'_{rr}}{L_\Sigma} i'_{br}\left(\omega_r \cos\theta_r - \frac{r'_r}{L'_{rr}}\sin\theta_r\right) \\ -\frac{L_{ms}^2}{L_\Sigma} i_{as}\omega_r - \frac{L_{ms}L'_{rr}}{L_\Sigma} i'_{ar}\left(\omega_r \cos\theta_r - \frac{r'_r}{L'_{rr}}\sin\theta_r\right) + \frac{L_{ms}L'_{rr}}{L_\Sigma} i'_{br}\left(\omega_r \sin\theta_r + \frac{r'_r}{L'_{rr}}\cos\theta_r\right) \\ \frac{L_{ms}L_{ss}}{L_\Sigma} i_{as}\left(\omega_r \sin\theta_r + \frac{r_s}{L_{ss}}\cos\theta_r\right) - \frac{L_{ms}L_{ss}}{L_\Sigma} i_{bs}\left(\omega_r \cos\theta_r - \frac{r_s}{L_{ss}}\sin\theta_r\right) - \frac{L_{ms}^2}{L_\Sigma} i'_{br}\omega_r \\ \frac{L_{ms}L_{ss}}{L_\Sigma} i_{as}\left(\omega_r \cos\theta_r - \frac{r_s}{L_{ss}}\sin\theta_r\right) + \frac{L_{ms}L_{ss}}{L_\Sigma} i_{bs}\left(\omega_r \sin\theta_r + \frac{r_s}{L_{ss}}\cos\theta_r\right) + \frac{L_{ms}^2}{L_\Sigma} i'_{ar}\omega_r \\ -\frac{P^2}{4J} L_{ms}[(i_{as}i'_{ar} + i_{bs}i'_{br})\sin\theta_r + (i_{as}i'_{br} - i_{bs}i'_{ar})\cos\theta_r] \\ 0 \end{bmatrix}$$

$$+ \begin{bmatrix} \frac{L'_{rr}}{L_\Sigma} & 0 & 0 & 0 \\ 0 & \frac{L'_{rr}}{L_\Sigma} & 0 & 0 \\ 0 & 0 & \frac{L_{ss}}{L_\Sigma} & 0 \\ 0 & 0 & 0 & \frac{L_{ss}}{L_\Sigma} \\ 0 & 0 & 0 & 0 \\ 0 & 0 & 0 & 0 \end{bmatrix} \begin{bmatrix} u_{as} \\ u_{bs} \\ u'_{ar} \\ u'_{br} \end{bmatrix} + \begin{bmatrix} -\frac{L_{ms}}{L_\Sigma}\cos\theta_r u'_{ar} + \frac{L_{ms}}{L_\Sigma}\sin\theta_r u'_{br} \\ -\frac{L_{ms}}{L_\Sigma}\sin\theta_r u'_{ar} - \frac{L_{ms}}{L_\Sigma}\cos\theta_r u'_{br} \\ -\frac{L_{ms}}{L_\Sigma}\cos\theta_r u_{as} - \frac{L_{ms}}{L_\Sigma}\sin\theta_r u_{bs} \\ \frac{L_{ms}}{L_\Sigma}\sin\theta_r u_{as} - \frac{L_{ms}}{L_\Sigma}\cos\theta_r u_{bs} \\ 0 \\ 0 \end{bmatrix} - \begin{bmatrix} 0 \\ 0 \\ 0 \\ 0 \\ \frac{P}{2J} \\ 0 \end{bmatrix} T_L$$

The application of the Lagrange equations of motion in the modeling of two-phase induction motors

The mathematical model of two-phase induction machines can be derived by using the Lagrange equations of motion. Assigning the generalized coordinates and the generalized forces to be

$$q_1 = \frac{i_{as}}{s},\ q_2 = \frac{i_{bs}}{s},\ q_3 = \frac{i'_{ar}}{s},\ q_4 = \frac{i'_{br}}{s},\ q_5 = \theta_r$$

and

$$Q_1 = u_{as},\ Q_2 = u_{bs},\ Q_3 = u'_{ar},\ Q_4 = u'_{br},\ Q_5 = -T_L$$

the following five equations result

$$\frac{d}{dt}\left(\frac{\partial \Gamma}{\partial \dot{q}_1}\right) - \frac{\partial \Gamma}{\partial q_1} + \frac{\partial D}{\partial \dot{q}_1} + \frac{\partial \Pi}{\partial q_1} = Q_1$$

$$\frac{d}{dt}\left(\frac{\partial \Gamma}{\partial \dot{q}_2}\right) - \frac{\partial \Gamma}{\partial q_2} + \frac{\partial D}{\partial \dot{q}_2} + \frac{\partial \Pi}{\partial q_2} = Q_2$$

$$\frac{d}{dt}\left(\frac{\partial \Gamma}{\partial \dot{q}_3}\right) - \frac{\partial \Gamma}{\partial q_3} + \frac{\partial D}{\partial \dot{q}_3} + \frac{\partial \Pi}{\partial q_3} = Q_3$$

$$\frac{d}{dt}\left(\frac{\partial \Gamma}{\partial \dot{q}_4}\right) - \frac{\partial \Gamma}{\partial q_4} + \frac{\partial D}{\partial \dot{q}_4} + \frac{\partial \Pi}{\partial q_4} = Q_4$$

$$\frac{d}{dt}\left(\frac{\partial \Gamma}{\partial \dot{q}_5}\right) - \frac{\partial \Gamma}{\partial q_5} + \frac{\partial D}{\partial \dot{q}_5} + \frac{\partial \Pi}{\partial q_5} = Q_5$$

The expressions for the total kinetic, potential, and dissipated energies are given as

$$\Gamma = \frac{1}{2}L_{ss}\dot{q}_1^2 + L_{ms}\dot{q}_1\dot{q}_3\cos q_5 - L_{ms}\dot{q}_1\dot{q}_4\sin q_5 + \frac{1}{2}L_{ss}\dot{q}_2^2 + L_{ms}\dot{q}_2\dot{q}_3\sin q_5 + L_{ms}\dot{q}_2\dot{q}_4\cos q_5$$
$$+ \frac{1}{2}L'_{rr}\dot{q}_3^2 + \frac{1}{2}L'_{rr}\dot{q}_4^2 + \frac{1}{2}J\dot{q}_5^2$$

$$\Pi = 0$$

$$D = \frac{1}{2}(r_s\dot{q}_1^2 + r_s\dot{q}_2^2 + r'_r\dot{q}_3^2 + r'_r\dot{q}_4^2 + B_m\dot{q}_5^2)$$

Hence, we have

$$\frac{\partial \Gamma}{\partial q_1} = 0,\quad \frac{\partial \Gamma}{\partial \dot{q}_1} = L_{ss}\dot{q}_1 + L_{ms}\dot{q}_3\cos q_5 - L_{ms}\dot{q}_4\sin q_5$$

$$\frac{\partial \Gamma}{\partial q_2} = 0,\quad \frac{\partial \Gamma}{\partial \dot{q}_2} = L_{ss}\dot{q}_2 + L_{ms}\dot{q}_3\sin q_5 + L_{ms}\dot{q}_4\cos q_5$$

$$\frac{\partial \Gamma}{\partial q_3} = 0,\quad \frac{\partial \Gamma}{\partial \dot{q}_3} = L'_{rr}\dot{q}_3 + L_{ms}\dot{q}_1\cos q_5 + L_{ms}\dot{q}_2\sin q_5$$

$$\frac{\partial \Gamma}{\partial q_4} = 0,\quad \frac{\partial \Gamma}{\partial \dot{q}_4} = L'_{rr}\dot{q}_4 - L_{ms}\dot{q}_1\sin q_5 + L_{ms}\dot{q}_2\cos q_5$$

$$\frac{\partial \Gamma}{\partial q_5} = -L_{ms}\dot{q}_1\dot{q}_3\sin q_5 - L_{ms}\dot{q}_1\dot{q}_4\cos q_5 + L_{ms}\dot{q}_2\dot{q}_3\cos q_5 - L_{ms}\dot{q}_2\dot{q}_4\sin q_5$$
$$= -L_{ms}[(\dot{q}_1\dot{q}_3 + \dot{q}_2\dot{q}_4)\sin q_5 + (\dot{q}_1\dot{q}_4 - \dot{q}_2\dot{q}_3)\cos q_5]$$

$$\frac{\partial \Gamma}{\partial \dot{q}_5} = J\dot{q}_5$$

$$\frac{\partial \Pi}{\partial q_1} = 0, \quad \frac{\partial \Pi}{\partial q_2} = 0, \quad \frac{\partial \Pi}{\partial q_3} = 0, \quad \frac{\partial \Pi}{\partial q_4} = 0, \quad \frac{\partial \Pi}{\partial q_5} = 0$$

$$\frac{\partial D}{\partial \dot{q}_1} = r_s\dot{q}_1, \quad \frac{\partial D}{\partial \dot{q}_2} = r_s\dot{q}_2, \quad \frac{\partial D}{\partial \dot{q}_3} = r_r'\dot{q}_3, \quad \frac{\partial D}{\partial \dot{q}_4} = r_r'\dot{q}_4, \quad \frac{\partial D}{\partial \dot{q}_5} = B_m\dot{q}_5$$

Taking note of $\dot{q}_1 = i_{as}$, $\dot{q}_2 = i_{bs}$, $\dot{q}_3 = i'_{ar}$, $\dot{q}_4 = i'_{br}$, and $\dot{q}_5 = \omega_r$, we have

$$L_{ss}\frac{di_{as}}{dt} + L_{ms}\frac{d(i'_{ar}\cos\theta_r)}{dt} - L_{ms}\frac{d(i'_{br}\sin\theta_r)}{dt} + r_s i_{as} = u_{as}$$

$$L_{ss}\frac{di_{bs}}{dt} + L_{ms}\frac{d(i'_{ar}\sin\theta_r)}{dt} + L_{ms}\frac{d(i'_{br}\cos\theta_r)}{dt} + r_s i_{bs} = u_{bs}$$

$$L_{ms}\frac{d(i_{as}\cos\theta_r)}{dt} + L_{ms}\frac{d(i_{bs}\sin\theta_r)}{dt} + L'_{rr}\frac{di'_{ar}}{dt} + r'_r i'_{ar} = u'_{ar}$$

$$-L_{ms}\frac{d(i_{as}\sin\theta_r)}{dt} + L_{ms}\frac{d(i_{bs}\cos\theta_r)}{dt} + L'_{rr}\frac{di'_{br}}{dt} + r'_r i'_{br} = u'_{br}$$

$$J\frac{d^2\theta_r}{dt^2} + L_{ms}[(i_{as}i'_{ar} + i_{bs}i'_{br})\sin\theta_r + (i_{as}i'_{br} - i_{bs}i'_{ar})\cos\theta_r] + B_m\frac{d\theta_r}{dt} = -T_l$$

For P-pole induction motors, by making use of $\frac{d\theta_r}{dt} = \omega_r$ six differential equations, as found in Cauchy's form (6.9), result.

The power, simplicity, and versatility of the Lagrangian concept have been demonstrated, and the augmented mathematical model of induction motors has been derived. It is shown that there is no need to use Kirchhoff's voltage law, Newton's second law of motion, and coenergy concept to find the circuitry and *torsional-mechanical* dynamics, as well as the electromagnetic torque developed. Unified, straightforward results have been obtained using the Lagrange equations of motion.

Torque-speed characteristics and control of induction motors

The angular velocity of induction motors must be controlled, and the torque-speed characteristic curves should be examined. The electromagnetic torque developed by two-phase induction motors, as a function of the stator and rotor currents and rotor displacement, was found. In particular, equation (6.6) gives the explicit expression for the electromagnetic torque, and

$$T_e = -\frac{P}{2}L_{ms}[(i_{as}i'_{ar} + i_{bs}i'_{br})\sin\theta_r + (i_{as}i'_{br} - i_{bs}i'_{ar})\cos\theta_r]$$

To guarantee the balanced operating condition for two-phase induction motors, one supplies the following phase voltages to the stator windings

$$u_{as}(t) = \sqrt{2}u_M \cos(\omega_f t)$$

$$u_{bs}(t) = \sqrt{2}u_M \sin(\omega_f t)$$

and the sinusoidal steady-state phase currents are

$$i_{as}(t) = \sqrt{2}i_M \cos(\omega_f t - \varphi_i)$$

$$i_{bs}(t) = \sqrt{2}i_M \sin(\omega_f t - \varphi_i)$$

Here, u_M is the magnitude of the voltages applied to the *as* and *bs* stator windings, and u_M cannot exceed the admissible voltage $u_{M\,\max}$ specified by the manufacturer in the motor data sheet, $u_{M\,\min} \le u_M \le u_{M\,\max}$; i_M is the magnitude of the *as* and *bs* stator currents; ω_f is the angular frequency of the applied phase voltages, and $\omega_f = 2\pi f$; f is the frequency of the supplied voltage; and φ_i is the phase difference in radians.

It is important to emphasize that the synchronous angular velocity ω_e is found using the number of poles as

$$\omega_e = \frac{4\pi f}{P}$$

and ω_e can be changed by changing the frequency f.

To regulate the angular velocity, one varies the magnitude of the applied voltages as well as the frequency of the voltages supplied. The torque-speed characteristic curves of induction motors must be thoroughly studied. Performing the transient analysis by solving the derived differential equations (6.9), one can find the steady-state curves $\omega_r = \Omega_T(T_e)$ by plotting the angular velocity versus the electromagnetic torque. The alternative way to find the torque-speed characteristics is to study the equivalent motor circuitry. The National Electric Manufacturers Association (NEMA) in the U.S. and the International Electromechanical Commission (IEC) in Europe have defined four basic classes of induction machines. In particular, A-, B-, C-, and D-class induction machines were designated. For different classes, typical torque-speed characteristic curves in the motor region are plotted in Figure 6.6, where the *slip* is found as

$$slip = \frac{\omega_e - \omega_r}{\omega_e}$$

and $\omega_r = (1 - slip)\omega_e$.

The standard industrial induction motors are designed to be the A- or B-class machines. These motors have normal starting torque and low *slip*, which is usually 0.05. In contrast, the C-class induction motors have higher starting torque due to double-rotor design, and the *slip* can be greater than 0.05. The D-class motors have high rotor resistance, and high starting torque results. The high losses and low efficiency prevent the application of these induction motors. Typically, the *slip* of D-class induction motors is in the range from 0.5 to 0.9. The E- and F-class induction motors have very low starting torque, and the rotor bars are deeply buried and high leakage inductance results.

Figure 6.7 illustrates the torque-speed characteristic of A-class induction machines in the motor, generator, and braking regions.

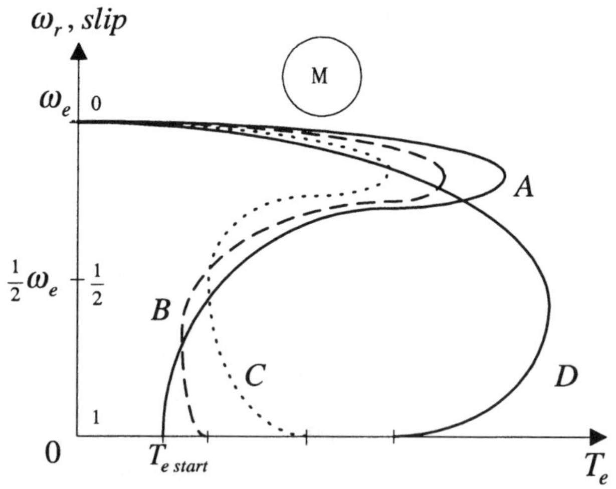

Figure 6.6 Typical torque-speed characteristic curves of induction motors.

Having derived the expression for the electromagnetic torque developed

$$T_e = -\frac{P}{2} L_{ms}[(i_{as}i'_{ar} + i_{bs}i'_{br})\sin\theta_r + (i_{as}i'_{br} - i_{bs}i'_{ar})\cos\theta_r]$$

and using Newton's second law of motion

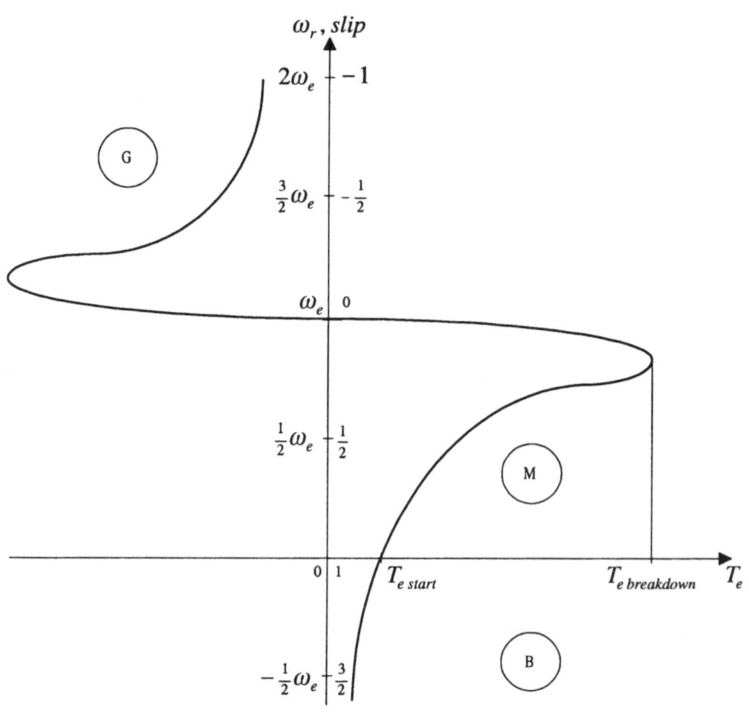

Figure 6.7 Torque-speed characteristic curves in the motor (M), generator (G), and breaking (B) regions.

$$\frac{d\omega_r}{dt} = \frac{P}{2J}T_e - \frac{B_m}{J}\omega_r - \frac{P}{2J}T_L$$

it evident that the angular velocity of induction motors is regulated by changing the magnitude u_M and the frequency f of the voltages applied to the *as* and *bs* windings. A balanced two-phase set (a set of equal magnitude sinusoidal voltages displaced by 90°) is

$$u_{as}(t) = \sqrt{2}u_M \cos(\omega_f t)$$

$$u_{bs}(t) = \sqrt{2}u_M \sin(\omega_f t)$$

The following principles are used to control the angular velocity.

Voltage control. By changing the magnitude u_M of the applied phase voltages to the stator windings, the angular velocity can be regulated in the stable operating region (see Figure 6.8a). It is emphasized that $u_{M\,\min} < u_M < u_{M\,\max}$, where $u_{M\,\max}$ is the maximum allowed (rated) voltage.

Frequency control. The magnitude of the applied phase voltages is assigned to be constant $u_M^{constant}$, and the angular velocity is regulated above and below the synchronous angular velocity ω_e, $\omega_e = \frac{4\pi f}{P}$ by changing the frequency of the supplied voltages. The speed of general-purpose induction motors can be regulated from $0.05\omega_e$ to $2\omega_e$ (the mechanical limit is imposed on the maximum speed). The torque-speed characteristic curves for different values of the frequency are shown in Figure 6.8b.

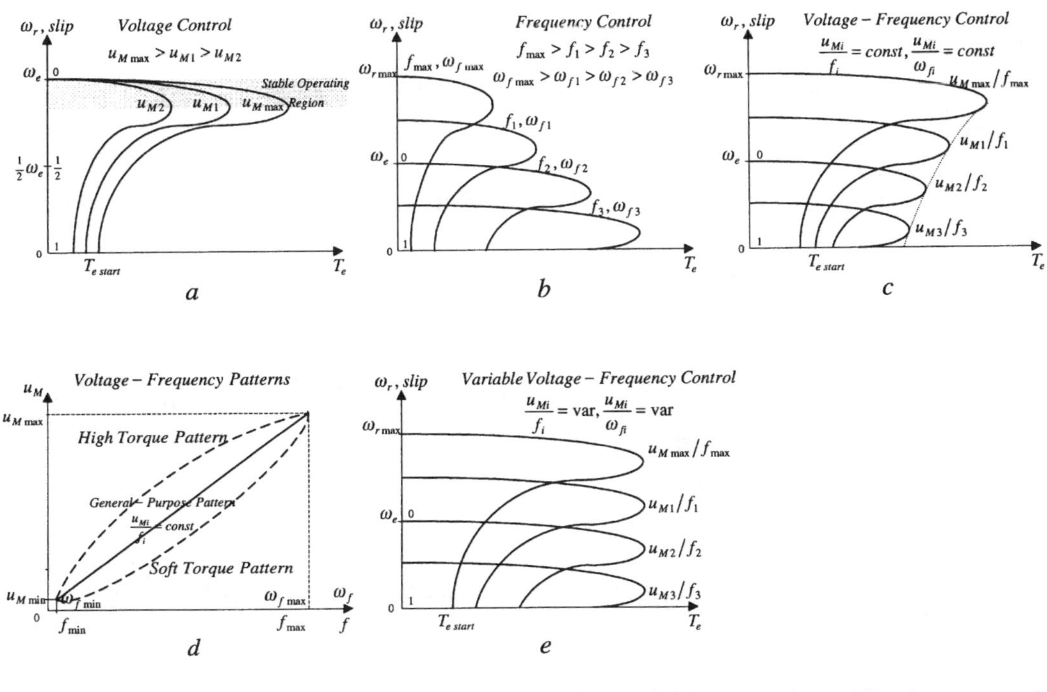

Figure 6.8 Torque-speed characteristic curves $\omega_r = \Omega_T(T_e)$ of induction motors: (a) voltage control; (b) frequency control; (c) voltage-frequency control: *constant volts per hertz* control; (d) voltage-frequency patterns; (e) variable voltage-frequency control.

Voltage-frequency control. The angular frequency ω_f is proportional to the frequency of the supplied voltages, and $\omega_f = 2\pi f$. The torque-speed characteristic curves are shown in Figure 6.8b if one changes the frequency. To minimize losses, the magnitude u_M of the applied voltages should be regulated if the frequency is changed. In particular, the magnitude of phase voltages applied to the stator windings can be decreased linearly by decreasing the frequency. That is, one maintains the following relationship $\frac{u_{Mi}}{f_i} = const$ or $\frac{u_{Mi}}{\omega_{fi}} = const$ to attain the so-called *constant volts per hertz* control. The corresponding torque-speed characteristics are plotted in Figure 6.8c. It must be emphasized that by regulating the voltage-frequency patterns, one shapes the torque-speed curves. For example, the following relation $\sqrt{\frac{u_{Mi}}{f_i}} = const$ is commonly used to adjust the magnitude u_M and frequency f. The standard (general-purpose) voltage-frequency pattern is illustrated in Figure 6.8d (the solid line). For electric drives and electromechanical systems, to guarantee the acceleration and settling time specified, overshoot and rise time needed, stability and robustness in the operating envelope, the general-purpose-, soft- and high-starting torque patterns are implemented based on the requirements and criteria imposed (see the soft- and high-torque patterns as shown in Figure 6.8d by the dashed lines). In particular, assigning $\omega_f = \varphi(u_M)$ with domain $u_{M\,\min} < u_M < u_{M\,\max}$ and range $\omega_{f\min} < \omega_f < \omega_{f\max}$, one maintains $\frac{u_{Mi}}{f_i} = \text{var}$ or $\frac{u_{Mi}}{\omega_{fi}} = \text{var}$. For example, the specified torque-speed characteristic curves for general-purpose electric drives, as plotted in Figure 6.8e, can be easily achieved.

S-domain block diagram of two-phase induction motors

To perform the analysis of dynamics, to control induction machines, as well as to visualize the results, it is imperative to develop s-domain block diagrams. For squirrel-cage induction motors, the rotor windings are short-circuited, and therefore $u'_{ar} = u'_{br} = 0$. The block diagram for two-phase squirrel-cage induction motors using the stator and rotor currents, electrical angular velocity, and electrical angular displacement as the state variables can be built employing differential equations (6.9). In particular, substituting $s = \frac{d}{dt}$, we have

$$i_{as} = \frac{L_{ms}}{L_\Sigma s + L'_{rr} r_s}\left[L_{ms} i_{bs}\omega_r + L'_{rr} i'_{ar}\left(\omega_r \sin\theta_r + \frac{r'_r}{L'_{rr}}\cos\theta_r\right) + L'_{rr} i'_{br}\left(\omega_r \cos\theta_r - \frac{r'_r}{L'_{rr}}\sin\theta_r\right)\right.$$
$$\left. + \frac{L'_{rr}}{L_{ms}} u_{as} - (\cos\theta_r u'_{ar} - \sin\theta_r u'_{br})\right]$$

$$i_{bs} = \frac{L_{ms}}{L_\Sigma s + L'_{rr} r_s}\left[-L_{ms} i_{as}\omega_r - L'_{rr} i'_{ar}\left(\omega_r \cos\theta_r - \frac{r'_r}{L'_{rr}}\sin\theta_r\right) + L'_{rr} i'_{br}\left(\omega_r \sin\theta_r + \frac{r'_r}{L'_{rr}}\cos\theta_r\right)\right.$$
$$\left. + \frac{L'_{rr}}{L_{ms}} u_{bs} - (\sin\theta_r u'_{ar} + \cos\theta_r u'_{br})\right]$$

$$i'_{ar} = \frac{L_{ms}}{L_\Sigma s + L_{ss} r'_r}\left[L_{ss} i_{as}\left(\omega_r \sin\theta_r + \frac{r_s}{L_{ss}}\cos\theta_r\right) - L_{ss} i_{bs}\left(\omega_r \cos\theta_r - \frac{r_s}{L_{ss}}\sin\theta_r\right) - L_{ms} i'_{br}\omega_r \right.$$

$$\left. -(\cos\theta_r u_{as} + \sin\theta_r u_{bs}) + \frac{L_{ss}}{L_{ms}} u'_{ar}\right]$$

$$i'_{br} = \frac{L_{ms}}{L_\Sigma s + L_{ss} r'_{rr}}\left[L_{ss} i_{as}\left(\omega_r \cos\theta_r - \frac{r_s}{L_{ss}}\sin\theta_r\right) + L_{ss} i_{bs}\left(\omega_r \sin\theta_r + \frac{r_s}{L_{ss}}\cos\theta_r\right) + L_{ms} i'_{ar}\omega_r \right.$$

$$\left. +(\sin\theta_r u_{as} - \cos\theta_r u_{bs}) + \frac{L_{ss}}{L_{ms}} u'_{br}\right]$$

$$\omega_r = \frac{P}{2}\frac{1}{(Js + B_m)}\left\{-\frac{P}{2} L_{ms}[(i_{as} i'_{ar} + i_{bs} i'_{br})\sin\theta_r + (i_{as} i'_{br} - i_{bs} i'_{ar})\cos\theta_r] - T_L\right\}$$

$$\theta_r = \frac{1}{s}\omega_r$$

The resulting *s*-domain block diagram is shown in Figure 6.9.

Example 6.1

Using the block diagram for squirrel-cage induction motors documented in Figure 6.9 develop the SIMULINK diagram and perform simulations for two-phase, four-pole, 115 V (*rms*), 60 Hz induction motors. Plot the torque-speed characteristic curves. Perform the analysis of the steady-state and dynamic characteristics of induction motors to show that the balanced set of phase voltages must be supplied to the stator windings. The parameters of A- and D-class motors to be simulated are given in Table 6.1.

Solution

To guarantee the balanced operating condition, the voltages to by applied are

$$u_{as}(t) = \sqrt{2} u_M \cos(\omega_f t), \quad u_{bs}(t) = \sqrt{2} u_M \sin(\omega_f t)$$

Supplying the rated voltage, we have

$$u_{as}(t) = \sqrt{2}115\cos(377t) \text{ and } u_{bs}(t) = \sqrt{2}115\sin(377t)$$

No load and load conditions are examined, and the load torques are assigned to be 0 and 0.5 N-m. The SIMULINK diagram, which is illustrated in Figure 6.10, is developed using the *s*-domain diagram of two-phase induction motors shown in Figure 6.9. Nonlinear simulations were performed, and transient dynamics of the stator and rotor currents in the *as*, *bs*, *ar*, and *br* windings $i_{as}(t)$, $i_{bs}(t)$, $i'_{ar}(t)$, and $i'_{br}(t)$, as well as the rotor angular velocity $\omega_{rm}(t)$, are plotted in Figures 6.11 to 6.14.

Figures 6.11 and 6.12 illustrate the transient dynamics for the A-class motor. The motor accelerates from stall, and the load torques 0 and 0.5 N-m are applied at $t = 0$ s. The acceleration capabilities of the D-class motor are documented in Figures 6.13 and 6.14; see the waveforms for the rotor angular velocity. The A-class induction motor reaches the

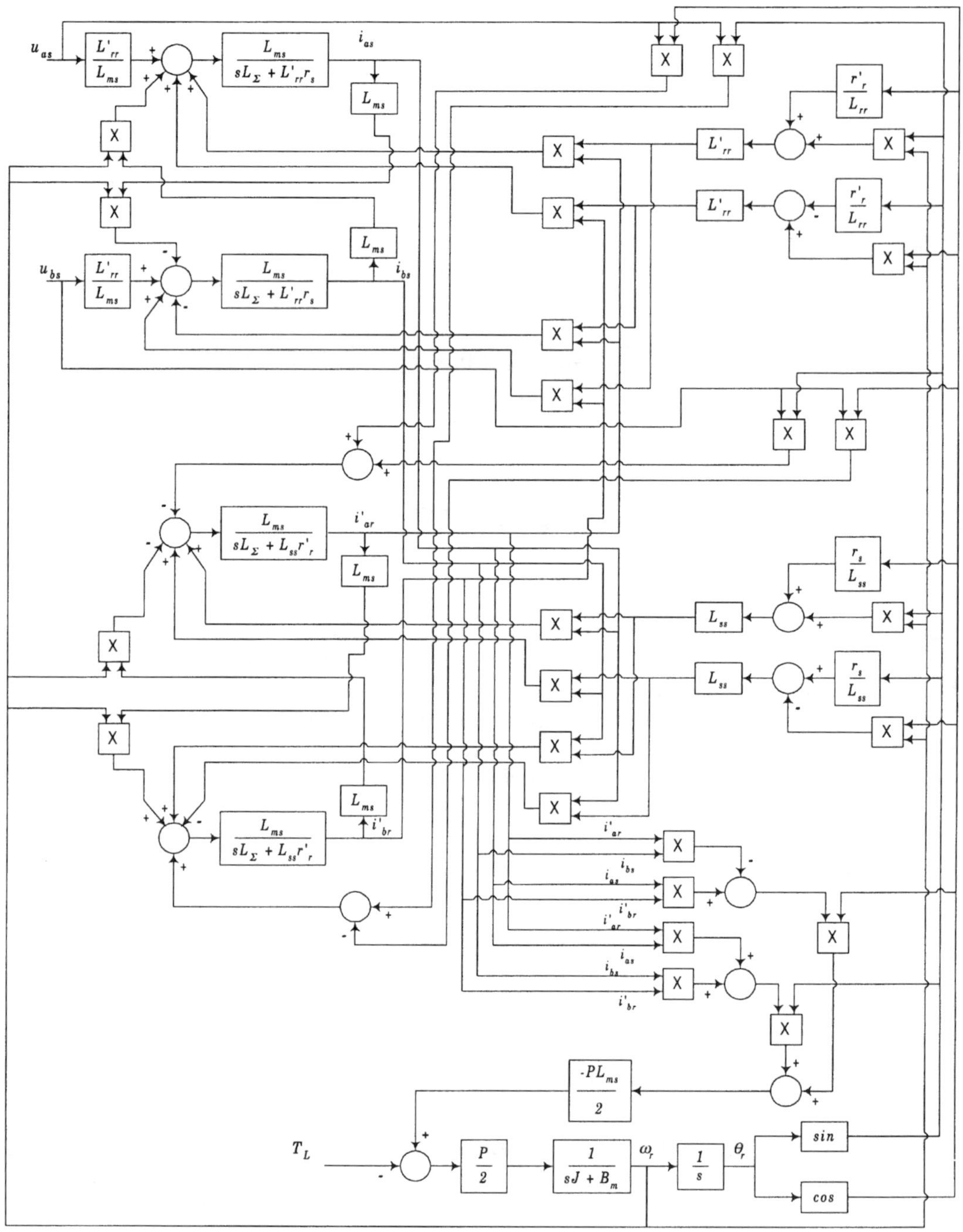

Figure 6.9 Block diagram of two-phase squirrel-cage induction motors.

Table 6.1 Parameters of Two-Phase Squirrel-Cage Induction Motors

Motor Parameters	A-Class	D-Class [1]
Stator resistance, r_s [ohm]	1.2	24.5
Rotor resistance, r_r [ohm]	1.5	23
Stator magnetizing inductance, L_{ms} [H]	0.16	0.27
Stator leakage inductance, L_{ls} [H]	0.02	0.027
Rotor leakage inductance, L_{lr} [H]	0.02	0.027
Moment of inertia, J [kg-m^2]	0.005	0.001
Number of poles, P	4	4

steady-state angular velocity within 0.8 s (with no load), and the settling time is 1.4 s if the load torque 0.5 N-m is applied at $t = 0$ s. Better acceleration capabilities are observed for the D-class motor (0.5 and 0.8 s for no load and load conditions). These results can be easily justified. Analyzing the torque-speed characteristics, it is emphasized that the D-class induction motors develop higher starting electromagnetic torque compared with the A- and B-class motors (see Figure 6.6).

It is important to thoroughly examine the torque-speed characteristic curves for the induction motors studied. In fact, in addition to analysis of transient dynamics, the torque-speed curves $\omega_r = \Omega_T(T_e)$ are of particular interest. The waveforms for the electromagnetic torque developed $T_e(t)$ are documented in Figures 6.11 to 6.14. The resulting characteristics $\omega_{rm} = \Omega_T(T_e)$ are obtained by plotting the angular velocity versus the electromagnetic torque. In fact, the torque-speed characteristics can be thoroughly examined using the results of nonlinear modeling of motor dynamics. It must be emphasized that steady-state analysis can be accomplished by performing the analysis of transient dynamics, but not vice versa. Figures 6.11 and 6.12 document the torque-speed curves $\omega_{rm} = \Omega_T(T_e)$ of the A-class motor; and Figures 6.13 and 6.14 illustrate the torque-speed characteristics for the D-class motor. It is evident that the torque-speed characteristic curves are obtained by solving the differential equations (6.9), which map the motor dynamics, allowing one to accomplish more comprehensive analysis compared with the steady-state studies. It is concluded that the torque-speed characteristic curves, as plotted in Figures 6.6 and 6.8, do not allow one to thoroughly examine the electromagnetic torque waveforms as induction motors accelerate. Hence, the steady-sate analysis of the torque-speed characteristics can be used in preliminary studies. That is, in general, to guarantee the detailed analysis needed in high-performance electric drives and electromechanical systems, the induction motor transient dynamics must be studied.

To analyze the unbalanced operation of the A- and D-class induction motors considered, we supply an unbalanced two-phase voltage set. In particular, the voltage supplied to the *bs* winding $u_{bs}(t)$ is 10% higher than the rated voltage applied to the *as* winding $u_{as}(t)$. That is, one supplies

$$u_{as}(t) = \sqrt{2}115\cos(377t) \text{ and } u_{bs}(t) = \sqrt{2}126\sin(377t)$$

The transient responses and the resulting torque-speed characteristics obtained are documented in Figures 6.15 and 6.16 for the A-class motor for different load conditions. Figures 6.17 and 6.18 illustrate the state transients if the D-class induction motor starts from stall with 0 and 0.5 N-m load torques. The chattering of the stator and rotor currents $i_{as}(t)$, $i_{bs}(t)$, $i'_{ar}(t)$, and $i'_{br}(t)$, as well as the electromagnetic torque $T_e(t)$, is evident (see Figures 6.15 to 6.18). The torque-speed characteristics if an unbalanced two-phase

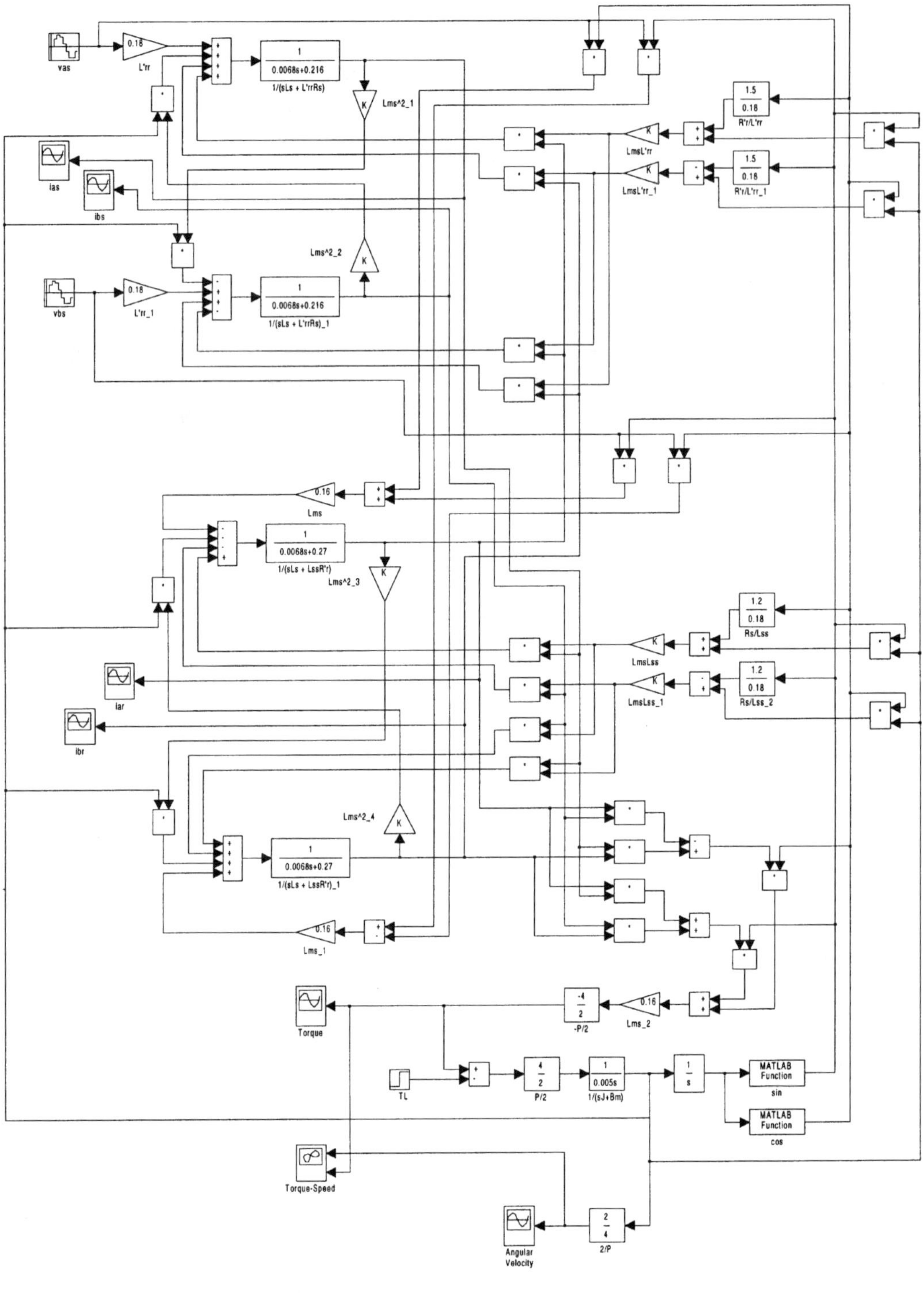

Figure 6.10 SIMULINK diagram to model the A-class squirrel-cage induction motor.

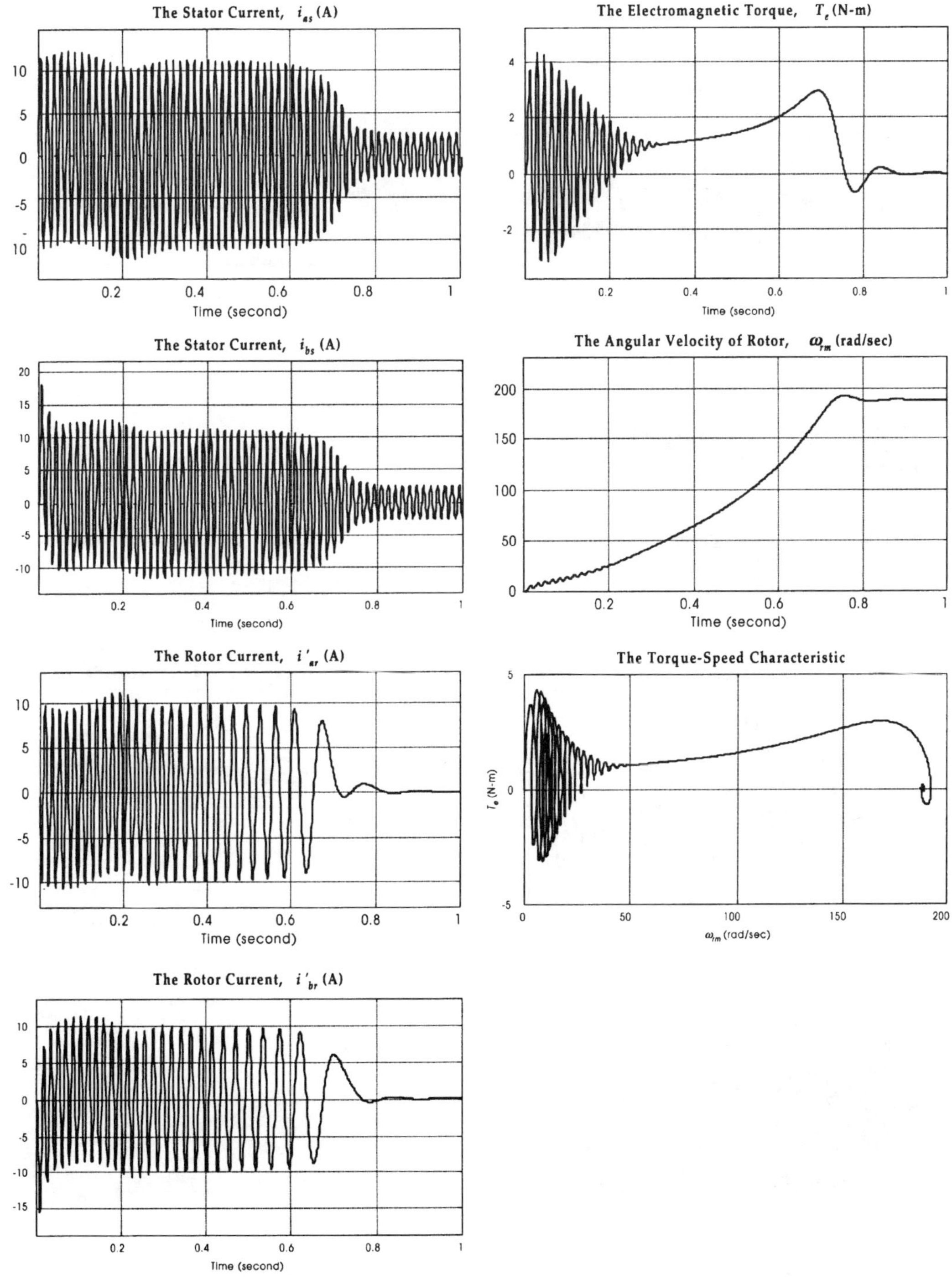

Figure 6.11 Dynamics and torque-speed characteristic of the A-class induction motor: $T_L = 0$ N-m, $u_{as}(t) = \sqrt{2}115\cos(377t)$, and $u_{bs}(t) = \sqrt{2}115\sin(377t)$.

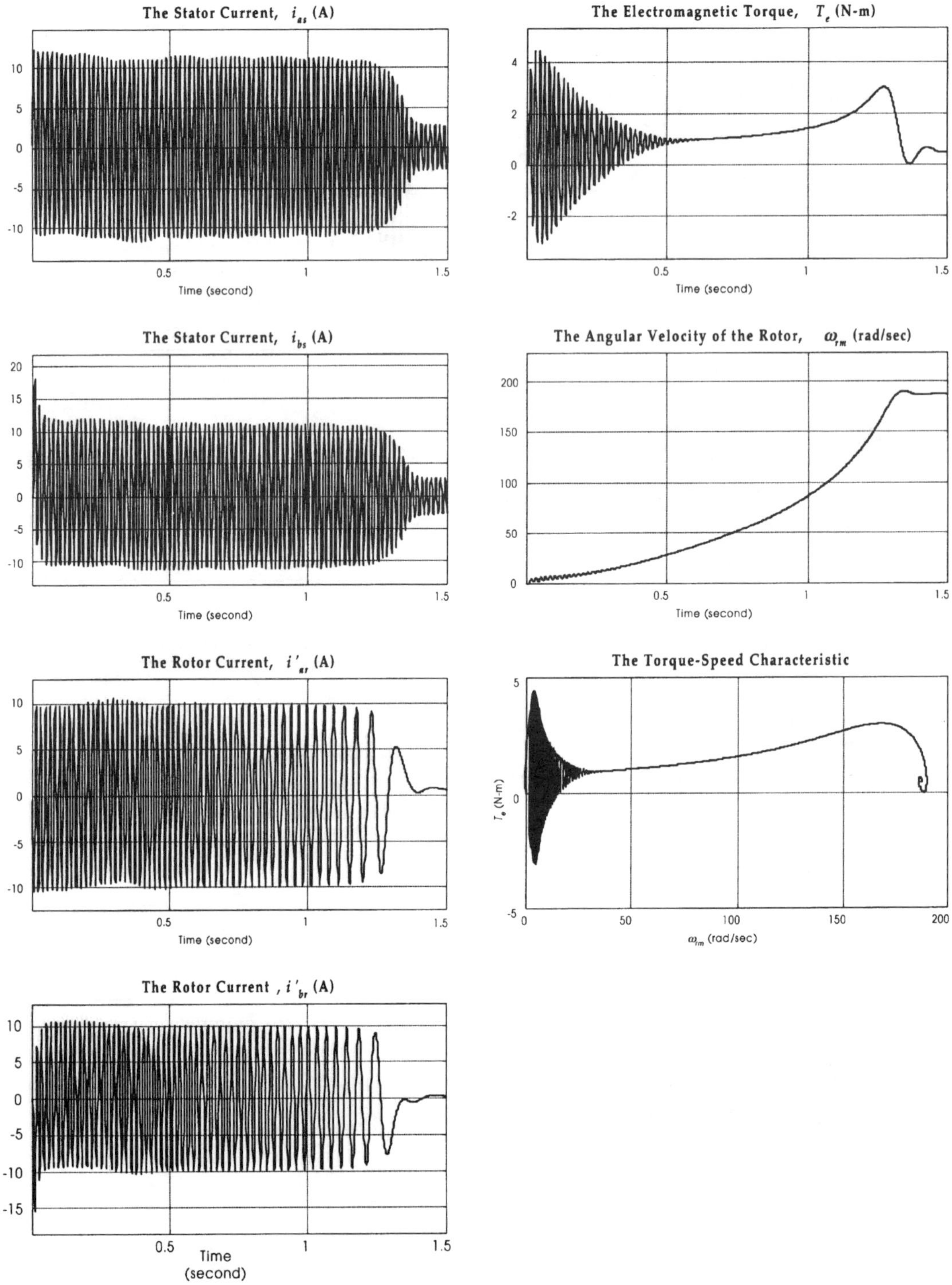

Figure 6.12 Dynamics and torque-speed characteristic of the A-class induction motor: $T_L = 0.5$ N-m, $u_{as}(t) = \sqrt{2}115\cos(377t)$, and $u_{bs}(t) = \sqrt{2}115\sin(377t)$.

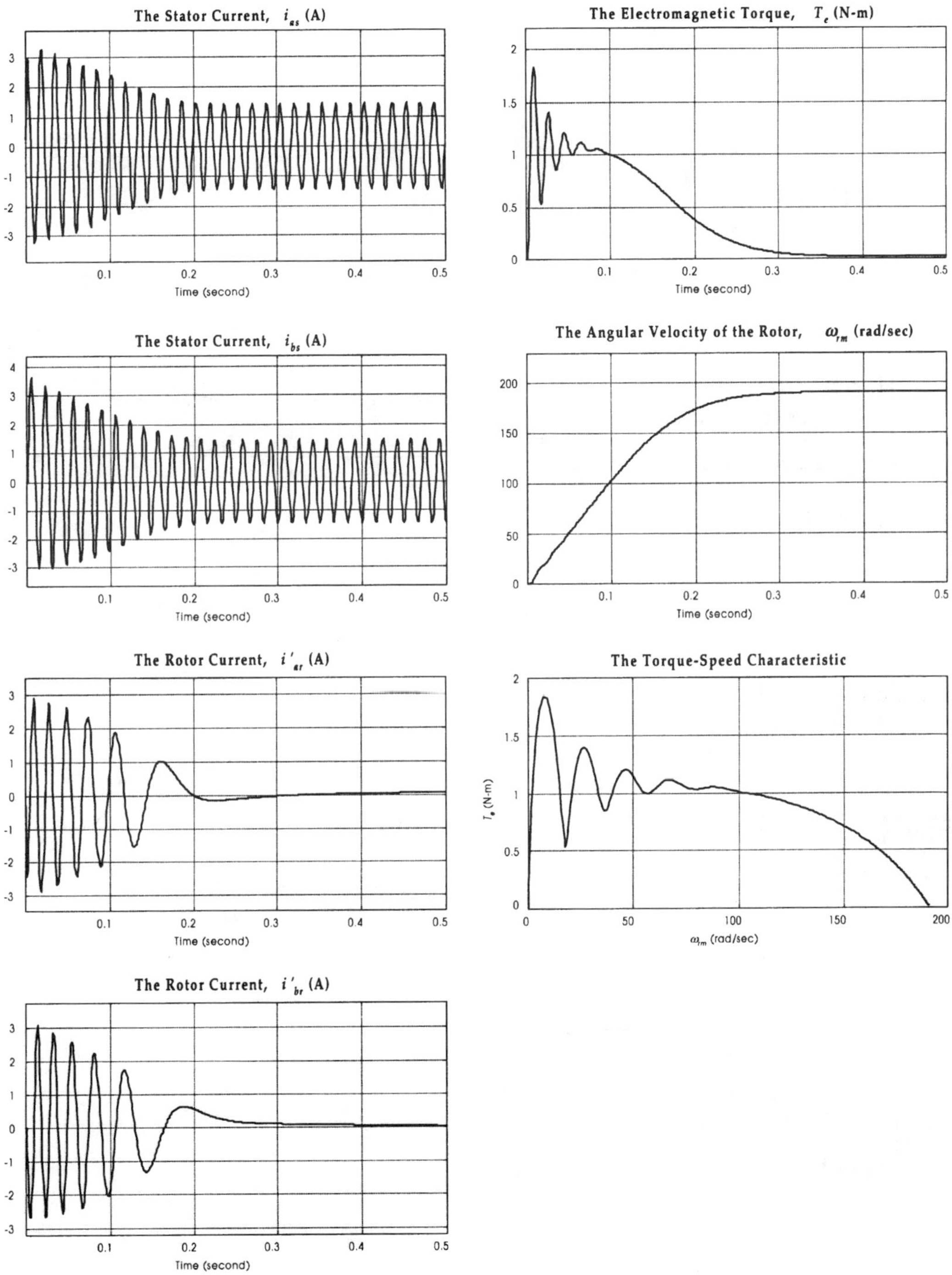

Figure 6.13 Dynamics and torque-speed characteristic of the D-class induction motor: $T_L = 0$ N-m, $u_{as}(t) = \sqrt{2}115\cos(377t)$, and $u_{bs}(t) = \sqrt{2}115\sin(377t)$.

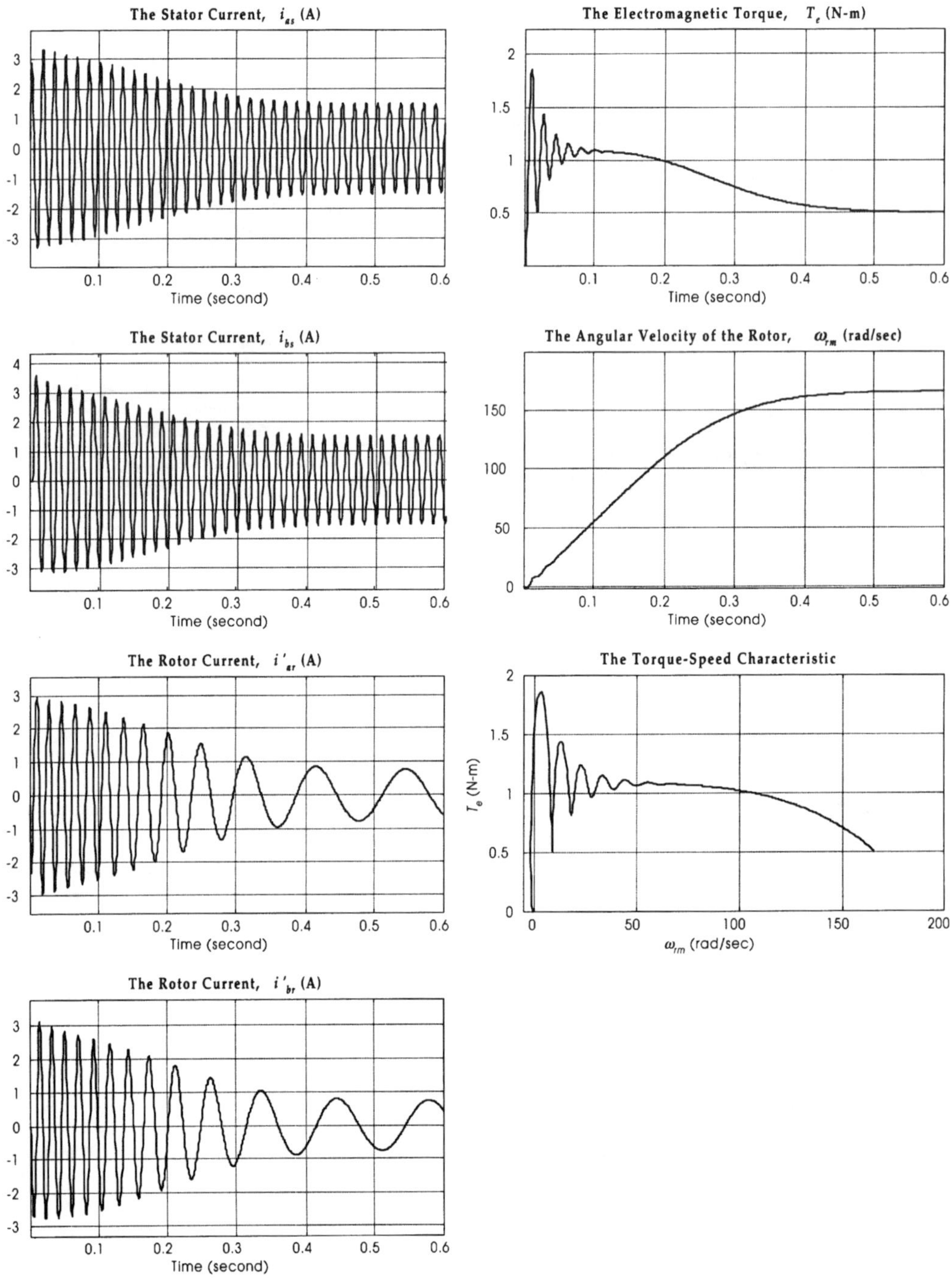

Figure 6.14 Dynamics and torque-speed characteristic of the D-class induction motor: $T_L = 0.5$ N-m, $u_{as}(t) = \sqrt{2}115\cos(377t)$, and $u_{bs}(t) = \sqrt{2}115\sin(377t)$.

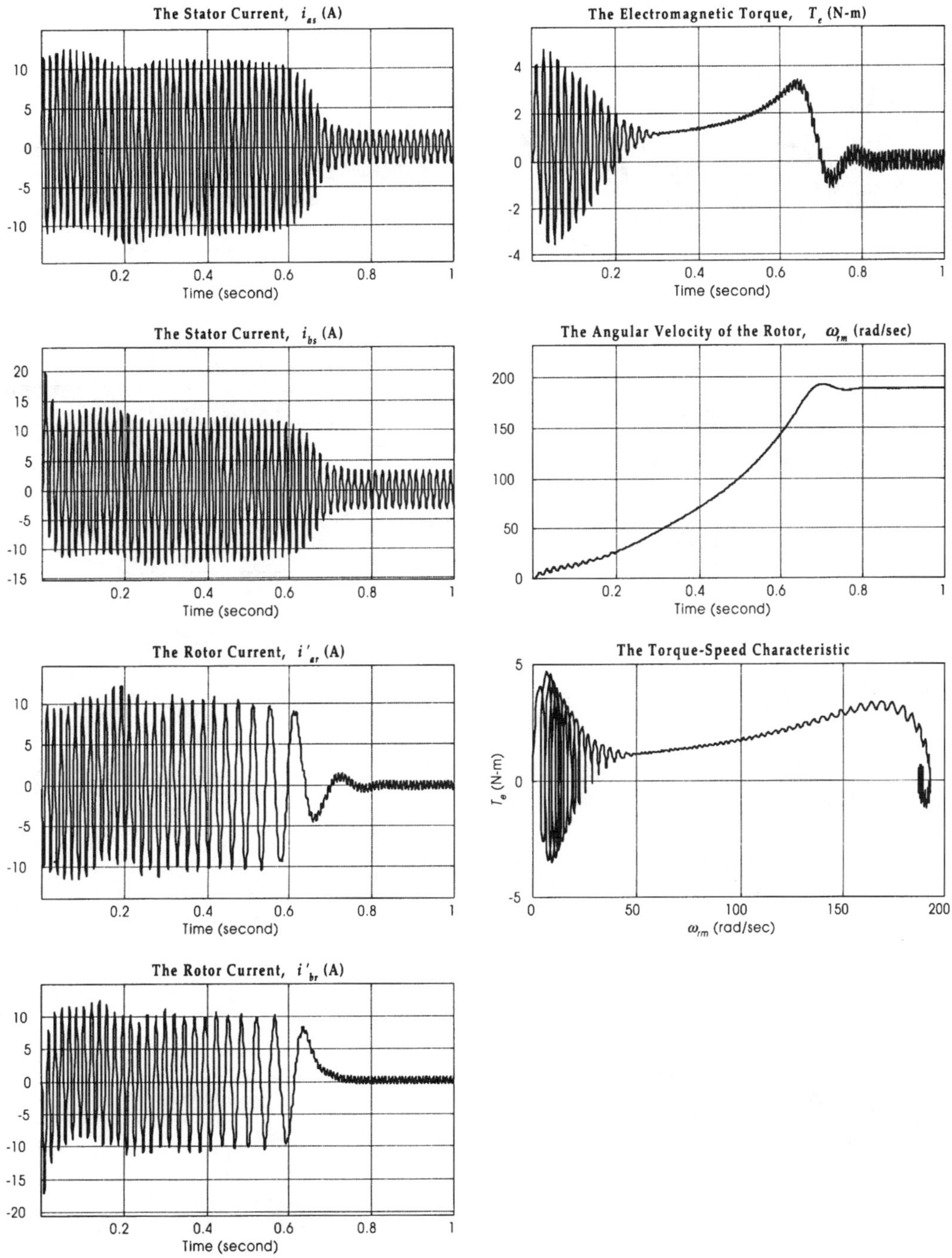

Figure 6.15 Dynamics and torque-speed characteristic of the A-class induction motor: $T_L = 0$ N-m, $u_{as}(t) = \sqrt{2}115\cos(377t)$, and $u_{bs}(t) = \sqrt{2}126\sin(377t)$.

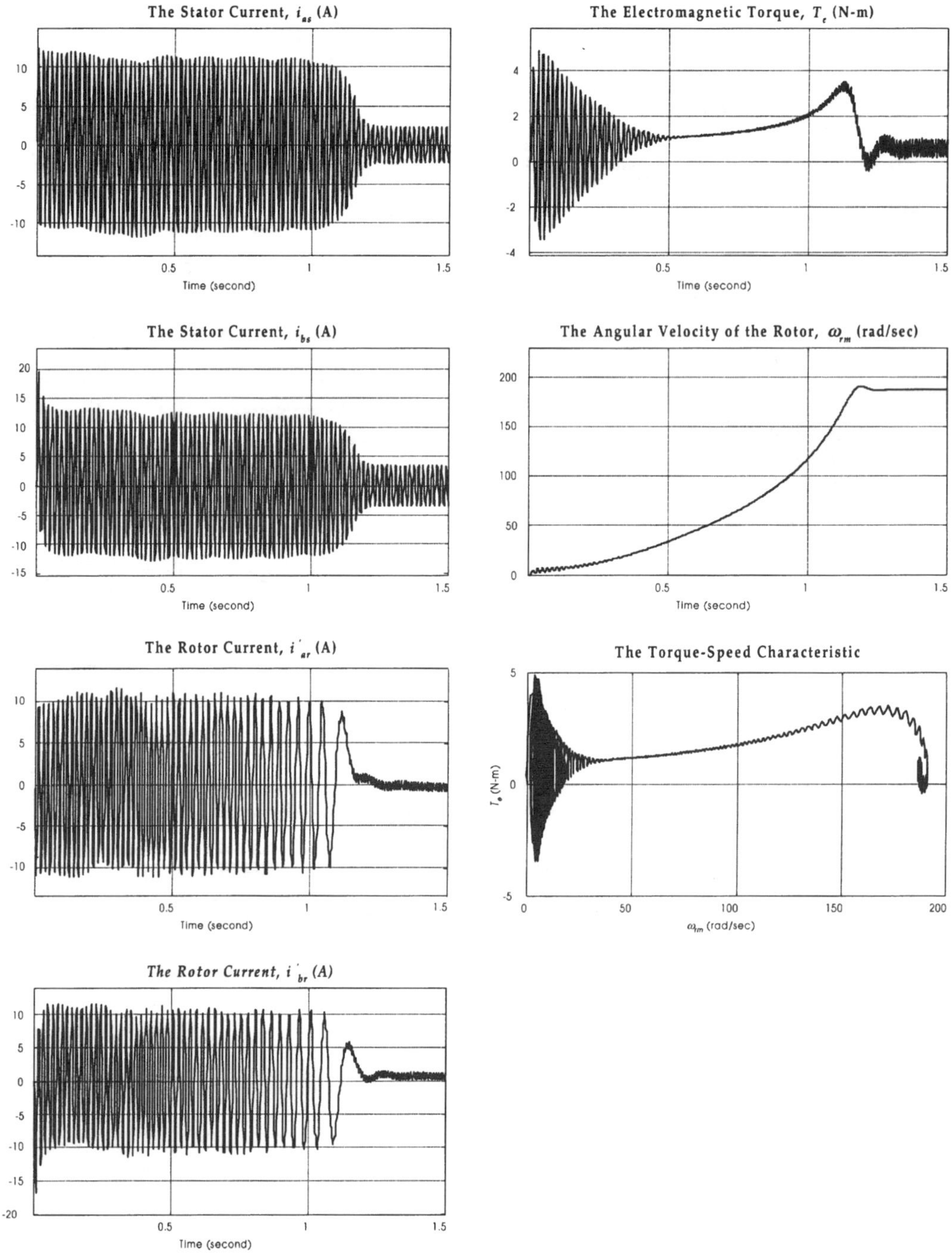

Figure 6.16 Dynamics and torque-speed characteristic of the A-class induction motor: $T_L = 0.5$ N-m, $u_{as}(t) = \sqrt{2}115\cos(377t)$, and $u_{bs}(t) = \sqrt{2}126\sin(377t)$.

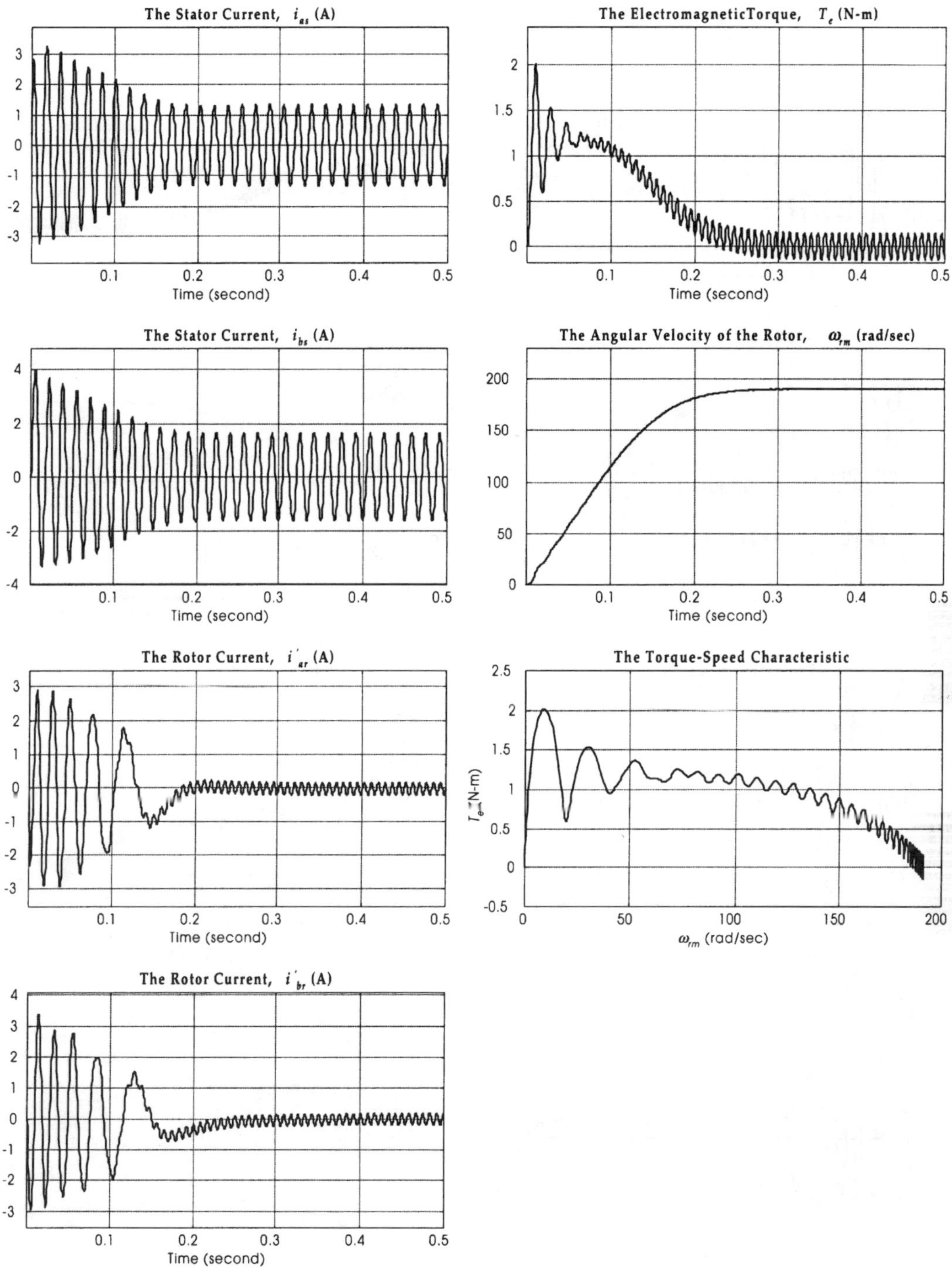

Figure 6.17 Dynamics and torque-speed characteristic of the D-class induction motor: $T_L = 0$ N-m, $u_{as}(t) = \sqrt{2}115\cos(377t)$, and $u_{bs}(t) = \sqrt{2}126\sin(377t)$.

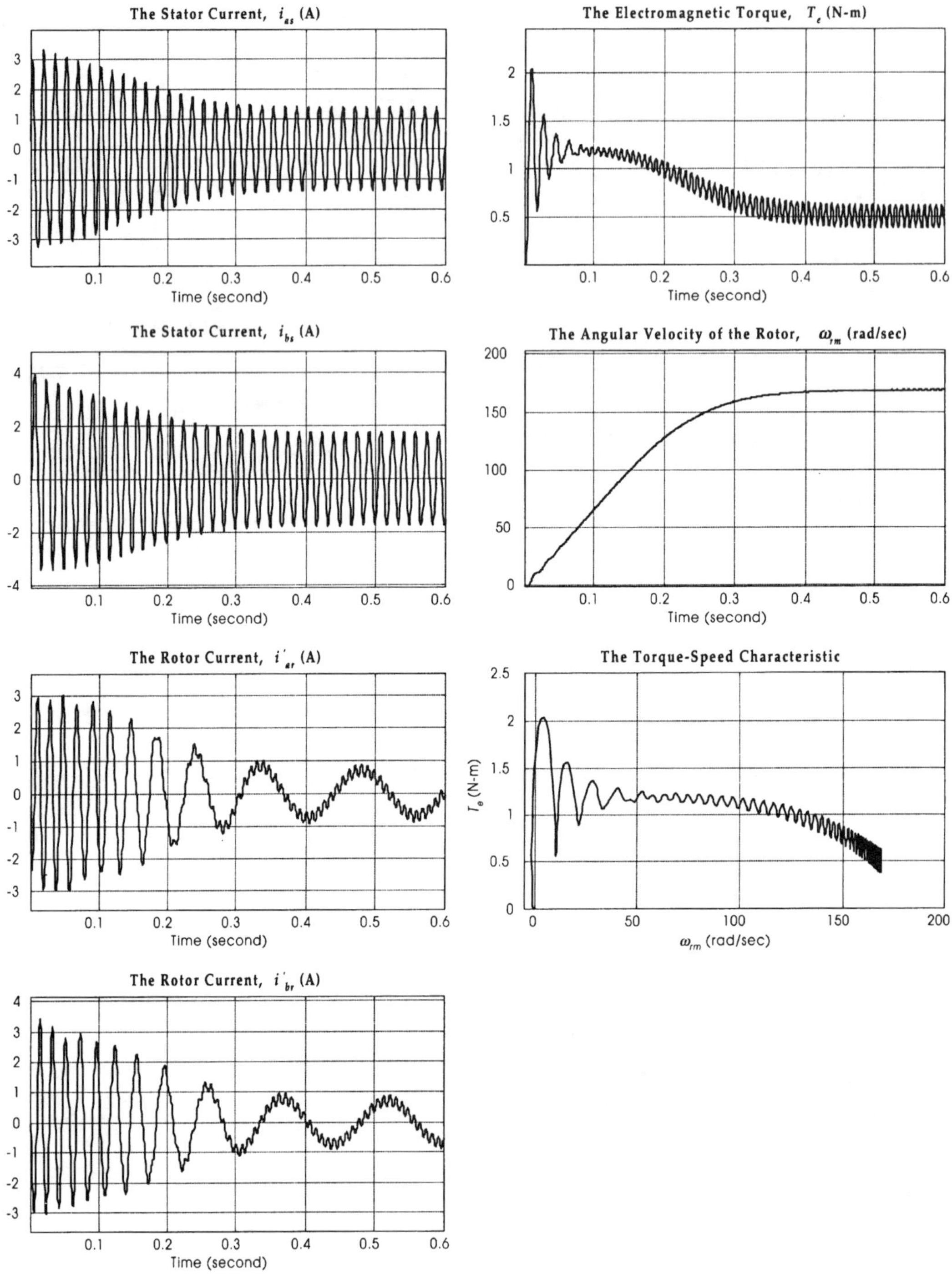

Figure 6.18 Dynamics and torque-speed characteristic of the D-class induction motor: $T_L = 0.5$ N-m, $u_{as}(t) = \sqrt{2}115\cos(377t)$, and $u_{bs}(t) = \sqrt{2}126\sin(377t)$.

voltage set is applied are plotted in Figures 6.15 to 6.18. One concludes that the applied voltages must have the same magnitude. It is not difficult to verify that the frequency must be the same. The difference in the stator and rotor resistances or inductances lead to the chattering effect even if the magnitude and frequency of the applied voltages are the same. In fact, supplying $u_{as}(t)=\sqrt{2}u_M\cos(\omega_f t)$ and $u_{bs}(t)=\sqrt{2}u_M\sin(\omega_f t)$, the magnitudes of the sinusoidal steady-state phase currents in the a and b windings are different if the phase resistances are not equal. If the inductances of the a and b windings are different, the phase is different. Therefore, the motor windings must be identical.

Dynamics of two-phase induction generators

Having derived and analyzed the mathematical model for two-phase induction motors, which is given by equations (6.9), it must be emphasized that the corresponding differential equations for two-phase induction generators can be easily obtained. In particular, to analyze the generator dynamics, one should use the following Kirchhoff voltage equations for stator and rotor windings

$$\mathbf{u}_{abs}=-\mathbf{r}_s\mathbf{i}_{abs}+\frac{d\boldsymbol{\psi}_{abs}}{dt}$$

$$\mathbf{u}'_{abr}=\mathbf{r}'_r\mathbf{i}'_{abr}+\frac{d\boldsymbol{\psi}'_{abr}}{dt}$$

where the flux linkages are given as

$$\begin{bmatrix}\boldsymbol{\psi}_{abs}\\ \boldsymbol{\psi}'_{abr}\end{bmatrix}=\begin{bmatrix}\mathbf{L}_s & \mathbf{L}'_{sr}\\ \mathbf{L}'^{T}_{sr} & \mathbf{L}_r\end{bmatrix}\begin{bmatrix}-\mathbf{i}_{abs}\\ \mathbf{i}'_{abr}\end{bmatrix}$$

Augmenting four differential equations, which map the stator and rotor circuitry dynamics, with the *torsional-mechanical* dynamics

$$\frac{d\omega_r}{dt}=\frac{P^2}{4J}L_{ms}[(i_{as}i'_{ar}+i_{bs}i'_{br})\sin\theta_r+(i_{as}i'_{br}-i_{bs}i'_{ar})\cos\theta_r]-\frac{B_m}{J}\omega_r+\frac{P}{2J}T_{pm}$$

$$\frac{d\theta_r}{dt}=\omega_r$$

a set of nonlinear differential equations to map the dynamics of two-phase induction generators can be found.

The application of the Lagrange equations of motion in the modeling of two-phase induction generators

To develop the mathematical model of two-phase induction generators, the Lagrangian concept can be straightforwardly applied. The model development problem was solved for two-phase induction motors. Since induction generators are under consideration, one assigns the following generalized coordinates $q_1=-\frac{i_{as}}{s}$, $q_2=-\frac{i_{bs}}{s}$, $q_3=\frac{i'_{ar}}{s}$, $q_4=\frac{i'_{br}}{s}$, $q_5=\theta_r$, and the generalized forces $Q_1=u_{as}$, $Q_2=u_{bs}$, $Q_3=u'_{ar}$, $Q_4=u'_{br}$, $Q_5=T_{pm}$ to be used in

$$\frac{d}{dt}\left(\frac{\partial \Gamma}{\partial \dot{q}_1}\right)-\frac{\partial \Gamma}{\partial q_1}+\frac{\partial D}{\partial \dot{q}_1}+\frac{\partial \Pi}{\partial q_1}=Q_1$$

$$\frac{d}{dt}\left(\frac{\partial \Gamma}{\partial \dot{q}_2}\right)-\frac{\partial \Gamma}{\partial q_2}+\frac{\partial D}{\partial \dot{q}_2}+\frac{\partial \Pi}{\partial q_2}=Q_2$$

$$\frac{d}{dt}\left(\frac{\partial \Gamma}{\partial \dot{q}_3}\right)-\frac{\partial \Gamma}{\partial q_3}+\frac{\partial D}{\partial \dot{q}_3}+\frac{\partial \Pi}{\partial q_3}=Q_3$$

$$\frac{d}{dt}\left(\frac{\partial \Gamma}{\partial \dot{q}_4}\right)-\frac{\partial \Gamma}{\partial q_4}+\frac{\partial D}{\partial \dot{q}_4}+\frac{\partial \Pi}{\partial q_4}=Q_4$$

$$\frac{d}{dt}\left(\frac{\partial \Gamma}{\partial \dot{q}_5}\right)-\frac{\partial \Gamma}{\partial q_5}+\frac{\partial D}{\partial \dot{q}_5}+\frac{\partial \Pi}{\partial q_5}=Q_5$$

Taking note of the generalized coordinates used, the total kinetic, potential, and dissipated energies are found to be

$$\Gamma=\frac{1}{2}L_{ss}\dot{q}_1^2+L_{ms}\dot{q}_1\dot{q}_3\cos q_5-L_{ms}\dot{q}_1\dot{q}_4\sin q_5+\frac{1}{2}L_{ss}\dot{q}_2^2+L_{ms}\dot{q}_2\dot{q}_3\sin q_5+L_{ms}\dot{q}_2\dot{q}_4\cos q_5$$
$$+\frac{1}{2}L'_{rr}\dot{q}_3^2+\frac{1}{2}L'_{rr}\dot{q}_4^2+\frac{1}{2}J\dot{q}_5^2$$

$$\Pi=0$$

$$D=\frac{1}{2}(r_s\dot{q}_1^2+r_s\dot{q}_2^2+r'_r\dot{q}_3^2+r'_r\dot{q}_4^2+B_m\dot{q}_5^2)$$

Hence, we have the following expressions for the partial derivatives

$$\frac{\partial \Gamma}{\partial q_1}=0,\quad \frac{\partial \Gamma}{\partial \dot{q}_1}=L_{ss}\dot{q}_1+L_{ms}\dot{q}_3\cos q_5-L_{ms}\dot{q}_4\sin q_5$$

$$\frac{\partial \Gamma}{\partial q_2}=0,\quad \frac{\partial \Gamma}{\partial \dot{q}_2}=L_{ss}\dot{q}_2+L_{ms}\dot{q}_3\sin q_5+L_{ms}\dot{q}_4\cos q_5$$

$$\frac{\partial \Gamma}{\partial q_3}=0,\quad \frac{\partial \Gamma}{\partial \dot{q}_3}=L'_{rr}\dot{q}_3+L_{ms}\dot{q}_1\cos q_5+L_{ms}\dot{q}_2\sin q_5$$

$$\frac{\partial \Gamma}{\partial q_4}=0,\quad \frac{\partial \Gamma}{\partial \dot{q}_4}=L'_{rr}\dot{q}_4-L_{ms}\dot{q}_1\sin q_5+L_{ms}\dot{q}_2\cos q_5$$

$$\frac{\partial \Gamma}{\partial q_5}=-L_{ms}\dot{q}_1\dot{q}_3\sin q_5-L_{ms}\dot{q}_1\dot{q}_4\cos q_5+L_{ms}\dot{q}_2\dot{q}_3\cos q_5-L_{ms}\dot{q}_2\dot{q}_4\sin q_5$$
$$=-L_{ms}[(\dot{q}_1\dot{q}_3+\dot{q}_2\dot{q}_4)\sin q_5+(\dot{q}_1\dot{q}_4-\dot{q}_2\dot{q}_3)\cos q_5]$$

$$\frac{\partial \Gamma}{\partial \dot{q}_5}=J\dot{q}_5$$

$$\frac{\partial \Pi}{\partial q_1}=0,\quad \frac{\partial \Pi}{\partial q_2}=0,\quad \frac{\partial \Pi}{\partial q_3}=0,\quad \frac{\partial \Pi}{\partial q_4}=0,\quad \frac{\partial \Pi}{\partial q_5}=0$$

$$\frac{\partial D}{\partial \dot q_1}=r_s\dot q_1,\quad \frac{\partial D}{\partial \dot q_2}=r_s\dot q_2,\quad \frac{\partial D}{\partial \dot q_3}=r_r'\dot q_3,\quad \frac{\partial D}{\partial \dot q_4}=r_r'\dot q_4,\quad \frac{\partial D}{\partial \dot q_5}=B_m\dot q_5$$

Using the currents in the stator and rotor windings, which are related to the generalized coordinates as $\dot q_1=-i_{as}$, $\dot q_2=-i_{bs}$, $\dot q_3=i_{ar}'$, $\dot q_4=i_{br}'$, as well as the angular velocity $\dot q_5=\omega_r$ and the angular displacement θ_r, one obtains

$$-L_{ss}\frac{di_{as}}{dt}+L_{ms}\frac{d(i_{ar}'\cos\theta_r)}{dt}-L_{ms}\frac{d(i_{br}'\sin\theta_r)}{dt}-r_si_{as}=u_{as}$$

$$-L_{ss}\frac{di_{bs}}{dt}+L_{ms}\frac{d(i_{ar}'\sin\theta_r)}{dt}+L_{ms}\frac{d(i_{br}'\cos\theta_r)}{dt}-r_si_{bs}=u_{bs}$$

$$-L_{ms}\frac{d(i_{as}\cos\theta_r)}{dt}-L_{ms}\frac{d(i_{bs}\sin\theta_r)}{dt}+L_{rr}'\frac{di_{br}'}{dt}+r_r'i_{ar}'=u_{ar}'$$

$$L_{ms}\frac{d(i_{as}\sin\theta_r)}{dt}-L_{ms}\frac{d(i_{bs}\cos\theta_r)}{dt}+L_{rr}'\frac{di_{br}'}{dt}+r_r'i_{br}'=u_{br}'$$

$$J\frac{d^2\theta_r}{dt^2}-L_{ms}[(i_{as}i_{ar}'+i_{bs}i_{br}')\sin\theta_r+(i_{as}i_{br}'-i_{bs}i_{ar}')\cos\theta_r]+B_m\frac{d\theta_r}{dt}=T_{pm}$$

It is concluded that the application of the Lagrange equations of motion allows one to straightforwardly derive the resulting mathematical model for induction generators.

Analysis of induction machines using the flux linkages

The mathematical model of induction motors using stator and rotor currents, as given by the set of six highly nonlinear differential equations (6.9), was developed. The circuitry dynamics of induction machines using the stator currents and rotor flux linkages as the state variables are of interest. To develop the mathematical model, one can use the relationships between the rotor flux linkages $\frac{}{dt}=\omega_r$ and the currents i_{ar}', i_{br}'. In particular, applying (6.3), we have

$$\psi_{ar}'=L_{ms}i_{as}\cos\theta_r+L_{ms}i_{bs}\sin\theta_r+L_{rr}'i_{ar}'$$

$$\psi_{br}'=-L_{ms}i_{as}\sin\theta_r+L_{ms}i_{bs}\cos\theta_r+L_{rr}'i_{br}'$$

Hence,

$$i_{ar}'=\frac{1}{L_{rr}'}(\psi_{ar}'-L_{ms}i_{as}\cos\theta_r-L_{ms}i_{bs}\sin\theta_r)$$
$$i_{br}'=\frac{1}{L_{rr}'}(\psi_{br}'+L_{ms}i_{as}\sin\theta_r-L_{ms}i_{bs}\cos\theta_r) \qquad (6.10)$$

Taking note of (6.5) and (6.10), one obtains

$$\frac{di_{as}}{dt}=-\frac{L_{rr}'^2 r_s+L_{ms}^2 r_r'}{L_{rr}'(L_{ss}L_{rr}'-L_{ms}^2)}i_{as}+\frac{L_{ms}}{L_{ss}L_{rr}'-L_{ms}^2}\psi_{ar}'\left(\omega_r\sin\theta_r+\frac{r_r'}{L_{rr}'}\cos\theta_r\right)$$
$$+\frac{L_{ms}}{L_{ss}L_{rr}'-L_{ms}^2}\psi_{br}'\left(\omega_r\cos\theta_r-\frac{r_r'}{L_{rr}'}\sin\theta_r\right)+\frac{L_{rr}'}{L_{ss}L_{rr}'-L_{ms}^2}u_{as}$$
$$-\frac{L_{ms}}{L_{ss}L_{rr}'-L_{ms}^2}\cos\theta_r u_{ar}'+\frac{L_{ms}}{L_{ss}L_{rr}'-L_{ms}^2}\sin\theta_r u_{br}'$$
$$\frac{di_{bs}}{dt}=-\frac{L_{rr}'^2 r_s+L_{ms}^2 r_r'}{L_{rr}'(L_{ss}L_{rr}'-L_{ms}^2)}i_{bs}-\frac{L_{ms}}{L_{ss}L_{rr}'-L_{ms}^2}\psi_{ar}'\left(\omega_r\cos\theta_r-\frac{r_r'}{L_{rr}'}\sin\theta_r\right)$$
$$+\frac{L_{ms}}{L_{ss}L_{rr}'-L_{ms}^2}\psi_{br}'\left(\omega_r\sin\theta_r+\frac{r_r'}{L_{rr}'}\cos\theta_r\right)+\frac{L_{rr}'}{L_{ss}L_{rr}'-L_{ms}^2}u_{bs} \qquad (6.11)$$
$$-\frac{L_{ms}}{L_{ss}L_{rr}'-L_{ms}^2}\sin\theta_r u_{ar}'-\frac{L_{ms}}{L_{ss}L_{rr}'-L_{ms}^2}\cos\theta_r u_{br}'$$
$$\frac{d\psi_{ar}'}{dt}=-\frac{r_r'}{L_{rr}'}\psi_{ar}'+\frac{L_{ms}r_r'}{L_{rr}'}(i_{as}\cos\theta_r+i_{bs}\sin\theta_r)+u_{ar}'$$
$$\frac{d\psi_{br}'}{dt}=-\frac{r_r'}{L_{rr}'}\psi_{br}'-\frac{L_{ms}r_s'}{L_{rr}'}(i_{as}\sin\theta_r-i_{bs}\cos\theta_r)+u_{br}'$$

Equation (6.6) for the electromagnetic torque, as given by

$$T_e=-\frac{P}{2}L_{ms}[(i_{as}i_{ar}'+i_{bs}i_{br}')\sin\theta_r+(i_{as}i_{br}'-i_{bs}i_{ar}')\cos\theta_r]$$

should be rewritten in terms of the rotor flux linkages, which are used as the state variables. Applying (6.10), one obtains

$$T_e=-\frac{P}{2}\frac{L_{ms}}{L_{rr}'}[(i_{as}\psi_{ar}'+i_{bs}\psi_{br}')\sin\theta_r+(i_{as}\psi_{br}'-i_{bs}\psi_{ar}')\cos\theta_r] \qquad (6.12)$$

Substituting the derived expression for the electromagnetic torque (6.12) into the *torsional-mechanical* equations, we have

$$\frac{d\omega_r}{dt}=-\frac{P^2}{4J}\frac{L_{ms}}{L_{rr}'}[(i_{as}\psi_{ar}'+i_{bs}\psi_{br}')\sin\theta_r+(i_{as}\psi_{br}'-i_{bs}\psi_{ar}')\cos\theta_r]-\frac{B_m}{J}\omega_r-\frac{P}{2J}T_L \qquad (6.13)$$
$$\frac{d\theta_r}{dt}=\omega_r$$

Making use of equations (6.11), (6.12), and (6.13), the resulting mathematical model for two-phase induction motors, using the stator currents and rotor flux linkages, the electrical angular velocity, and the electrical angular displacement as the state variables, is found to be

$$
\begin{aligned}
\frac{di_{as}}{dt} &= -\frac{L_{rr}'^2 r_s + L_{ms}^2 r_r'}{L_{rr}' L_\Sigma} i_{as} + \frac{L_{ms}}{L_\Sigma}\psi_{ar}'\left(\omega_r \sin\theta_r + \frac{r_r'}{L_{rr}'}\cos\theta_r\right) + \frac{L_{ms}}{L_\Sigma}\psi_{br}'\left(\omega_r \cos\theta_r - \frac{r_r'}{L_{rr}'}\sin\theta_r\right) \\
&\quad + \frac{L_{rr}'}{L_\Sigma}u_{as} - \frac{L_{ms}}{L_\Sigma}\cos\theta_r u_{ar}' + \frac{L_{ms}}{L_\Sigma}\sin\theta_r u_{br}' \\
\frac{di_{bs}}{dt} &= -\frac{L_{rr}'^2 r_s + L_{ms}^2 r_r'}{L_{rr}' L_\Sigma} i_{bs} - \frac{L_{ms}}{L_\Sigma}\psi_{ar}'\left(\omega_r \cos\theta_r - \frac{r_r'}{L_{rr}'}\sin\theta_r\right) + \frac{L_{ms}}{L_\Sigma}\psi_{br}'\left(\omega_r \sin\theta_r + \frac{r_r'}{L_{rr}'}\cos\theta_r\right) \\
&\quad + \frac{L_{rr}'}{L_\Sigma}u_{bs} - \frac{L_{ms}}{L_\Sigma}\sin\theta_r u_{ar}' - \frac{L_{ms}}{L_\Sigma}\cos\theta_r u_{br}' \\
\frac{d\psi_{ar}'}{dt} &= -\frac{r_r'}{L_{rr}'}\psi_{ar}' + \frac{L_{ms} r_r'}{L_{rr}'}(i_{as}\cos\theta_r + i_{bs}\sin\theta_r) + u_{ar}' \\
\frac{d\psi_{br}'}{dt} &= -\frac{r_r'}{L_{rr}'}\psi_{br}' - \frac{L_{ms} r_r'}{L_{rr}'}(i_{as}\sin\theta_r - i_{bs}\cos\theta_r) + u_{br}' \\
\frac{d\omega_r}{dt} &= -\frac{P^2}{4J}\frac{L_{ms}}{L_{rr}'}[(i_{as}\psi_{ar}' + i_{bs}\psi_{br}')\sin\theta_r + (i_{as}\psi_{br}' - i_{bs}\psi_{ar}')\cos\theta_r] - \frac{B_m}{J}\omega_r - \frac{P}{2J}T_L \\
\frac{d\theta_r}{dt} &= \omega_r
\end{aligned}
\tag{6.14}
$$

Here, $L_\Sigma = L_{ss}L_{rr}' - L_{ms}^2$.

These nonlinear differential equations can be written in matrix form as

$$
\begin{bmatrix} \frac{di_{as}}{dt} \\ \frac{di_{bs}}{dt} \\ \frac{d\psi_{ar}'}{dt} \\ \frac{d\psi_{br}'}{dt} \\ \frac{d\omega_r}{dt} \\ \frac{d\theta_r}{dt} \end{bmatrix} =
\begin{bmatrix}
-\frac{L_{rr}'^2 r_s + L_{ms}^2 r_r'}{L_{rr}' L_\Sigma} & 0 & 0 & 0 & 0 & 0 \\
0 & -\frac{L_{rr}'^2 r_s + L_{ms}^2 r_r'}{L_{rr}' L_\Sigma} & 0 & 0 & 0 & 0 \\
0 & 0 & -\frac{r_r'}{L_{rr}'} & 0 & 0 & 0 \\
0 & 0 & 0 & -\frac{r_r'}{L_{rr}'} & 0 & 0 \\
0 & 0 & 0 & 0 & -\frac{B_m}{J} & 0 \\
0 & 0 & 0 & 0 & 1 & 0
\end{bmatrix}
\begin{bmatrix} i_{as} \\ i_{bs} \\ \psi_{ar}' \\ \psi_{br}' \\ \omega_r \\ \theta_r \end{bmatrix}
$$

$$
+\begin{bmatrix}
\frac{L_{ms}}{L_\Sigma}\psi'_{ar}\left(\omega_r\sin\theta_r+\frac{r'_r}{L'_{rr}}\cos\theta_r\right)+\frac{L_{ms}}{L_\Sigma}\psi'_{br}\left(\omega_r\cos\theta_r-\frac{r'_r}{L'_{rr}}\sin\theta_r\right)\\
-\frac{L_{ms}}{L_\Sigma}\psi'_{ar}\left(\omega_r\cos\theta_r-\frac{r'_r}{L'_{rr}}\sin\theta_r\right)+\frac{L_{ms}}{L_\Sigma}\psi'_{br}\left(\omega_r\sin\theta_r+\frac{r'_r}{L'_{rr}}\cos\theta_r\right)\\
\frac{L_{ms}r'_r}{L'_{rr}}(i_{as}\cos\theta_r+i_{bs}\sin\theta_r)\\
-\frac{L_{ms}r'_r}{L'_{rr}}(i_{as}\sin\theta_r-i_{bs}\cos\theta_r)\\
-\frac{P^2}{4J}\frac{L_{ms}}{L'_{rr}}[(i_{as}\psi'_{ar}+i_{bs}\psi'_{br})\sin\theta_r+(i_{as}\psi'_{br}-i_{bs}\psi'_{ar})\cos\theta_r]\\
0
\end{bmatrix}
$$

$$
+\begin{bmatrix}
\frac{L'_{rr}}{L_\Sigma} & 0 & 0 & 0\\
0 & \frac{L'_{rr}}{L_\Sigma} & 0 & 0\\
0 & 0 & 1 & 0\\
0 & 0 & 0 & 1\\
0 & 0 & 0 & 0\\
0 & 0 & 0 & 0
\end{bmatrix}
\begin{bmatrix} u_{as}\\ u_{bs}\\ u'_{ar}\\ u'_{br}\end{bmatrix}
+\begin{bmatrix}
-\frac{L_{ms}}{L_\Sigma}(\cos\theta_r u'_{ar}-\sin\theta_r u'_{br})\\
-\frac{L_{ms}}{L_\Sigma}(\sin\theta_r u'_{ar}+\cos\theta_r u'_{br})\\
0\\ 0\\ 0\\ 0
\end{bmatrix}
-\begin{bmatrix}0\\0\\0\\0\\ \frac{P}{2J}\\0\end{bmatrix}T_L
\tag{6.15}
$$

Mathematical models of electromechanical systems (electric drives and servomechanisms augmented with power electronics and sensors) can be expressed in expanded and matrix forms. These forms are equivalent; see, for example, induction machine dynamics as given by (6.14) and (6.15). Matrix notations are commonly used to attain better visualization.

The s-domain block diagrams are commonly used in analysis and modeling of induction machines. The differential equations, which map the induction machines dynamics, can be visualized using block diagrams. To develop the block diagram, one can use the stator currents and rotor flux linkages, as well as the electrical angular velocity and displacement, as the state variables. The resulting block diagram is developed from differential equations (6.14) by substituting the Laplace operator $s=\frac{d}{dt}$. In particular, we have

$$
i_{as}=\frac{L'_{rr}L_{ms}}{L'_{rr}L_\Sigma s+L'^2_{rr}r_s+L^2_{ms}r'_r}\left[\psi'_{ar}\left(\omega_r\sin\theta_r+\frac{r'_r}{L'_{rr}}\cos\theta_r\right)+\psi'_{br}\left(\omega_r\cos\theta_r-\frac{r'_r}{L'_{rr}}\sin\theta_r\right)\right.
$$

$$
\left.+\frac{L'_{rr}}{L_{ms}}u_{as}-\cos\theta_r u'_{ar}+\sin\theta_r u'_{br}\right]
$$

$$
i_{bs}=\frac{L'_{rr}L_{ms}}{L'_{rr}L_\Sigma s+L'^2_{rr}r_s+L^2_{ms}r'_r}\left[-\psi'_{ar}\left(\omega_r\cos\theta_r-\frac{r'_r}{L'_{rr}}\sin\theta_r\right)+\psi'_{br}\left(\omega_r\sin\theta_r+\frac{r'_r}{L'_{rr}}\cos\theta_r\right)\right.
$$

$$
\left.+\frac{L'_{rr}}{L_{ms}}u_{bs}-\sin\theta_r u'_{ar}-\cos\theta_r u'_{br}\right]
$$

$$
\begin{aligned}
\psi'_{ar} &= \frac{1}{L'_{rr}s + r'_r}[L_{ms}r'_r(i_{as}\cos\theta_r + i_{bs}\sin\theta_r) + L'_{rr}u'_{ar}] \\
\psi'_{br} &= \frac{1}{L'_{rr}s + r'_r}[-L_{ms}r'_r(i_{as}\sin\theta_r - i_{bs}\cos\theta_r) + L'_{rr}u'_{br}] \\
\omega_r &= \frac{P}{2}\frac{1}{(Js + B_m)}\left\{-\frac{P}{2}\frac{L_{ms}}{L'_{rr}}[(i_{as}\psi'_{ar} + i_{bs}\psi'_{br})\sin\theta_r + (i_{as}\psi'_{br} - i_{bs}\psi'_{ar})\cos\theta_r] - T_L\right\} \\
\theta_r &= \frac{1}{s}\omega_r
\end{aligned}
\tag{6.16}
$$

Using (6.16), the block diagram of two-phase squirrel-cage induction motors is built. This block diagram is illustrated in Figure 6.19.

Example 6.2

Using an *s*-domain block diagram, as given by Figure 6.19, develop the SIMULINK diagram and simulate the A- and D-class two-phase squirrel-cage induction motors considered in Example 6.1. Analyze the torque-speed characteristics and transient dynamics.

Solution

Modeling of the squirrel-cage motors under consideration is performed using the developed SIMULINK diagram shown in Figure 6.20. The A- and D class induction motors are studied under the different load torques T_L, and balanced and unbalanced two-phase voltage sets are supplied. The transient dynamics and torque-speed characteristic curves are plotted in Figures 6.21 to 6.24 if the motors start from stall. The phase voltages applied to the stator windings are

$$u_{as}(t) = \sqrt{2}115\cos(377t) \text{ and } u_{bs}(t) = \sqrt{2}115\sin(377t)$$

The evolution of the stator currents and angular velocity are identical, as was examined in Example 6.1 (see Figures 6.11 to 6.14). The behavior of the rotor flux linkages is of interest, and the waveforms for $\psi'_{ar}(t)$ and $\psi'_{br}(t)$ are documented in Figures 6.21 to 6.24.

Figures 6.25 to 6.28 illustrate the transient dynamics as well as the torque-speed characteristic curves $\omega_{rm} = \Omega_T(T_e)$ if the stator voltages supplied have different magnitudes. In particular, the motors were fed by the following *as* and *bs* voltages

$$u_{as}(t) = \sqrt{2}115\cos(377t) \text{ and } u_{bs}(t) = \sqrt{2}126\sin(377t).$$

6.3 Mathematical models of two-phase induction machines in the arbitrary, stationary, rotor, and synchronous reference frames

Different concepts have been applied to analyze electric machines, and the *arbitrary* reference frame is used as a benchmark to study induction and synchronous machines. The steady-state and transient analysis of induction machines can be performed in the assigned frame of reference. The *arbitrary* reference frame, which rotates with unspecified angular velocity ω, is of great importance. Assigning the angular velocity of the frame, induction

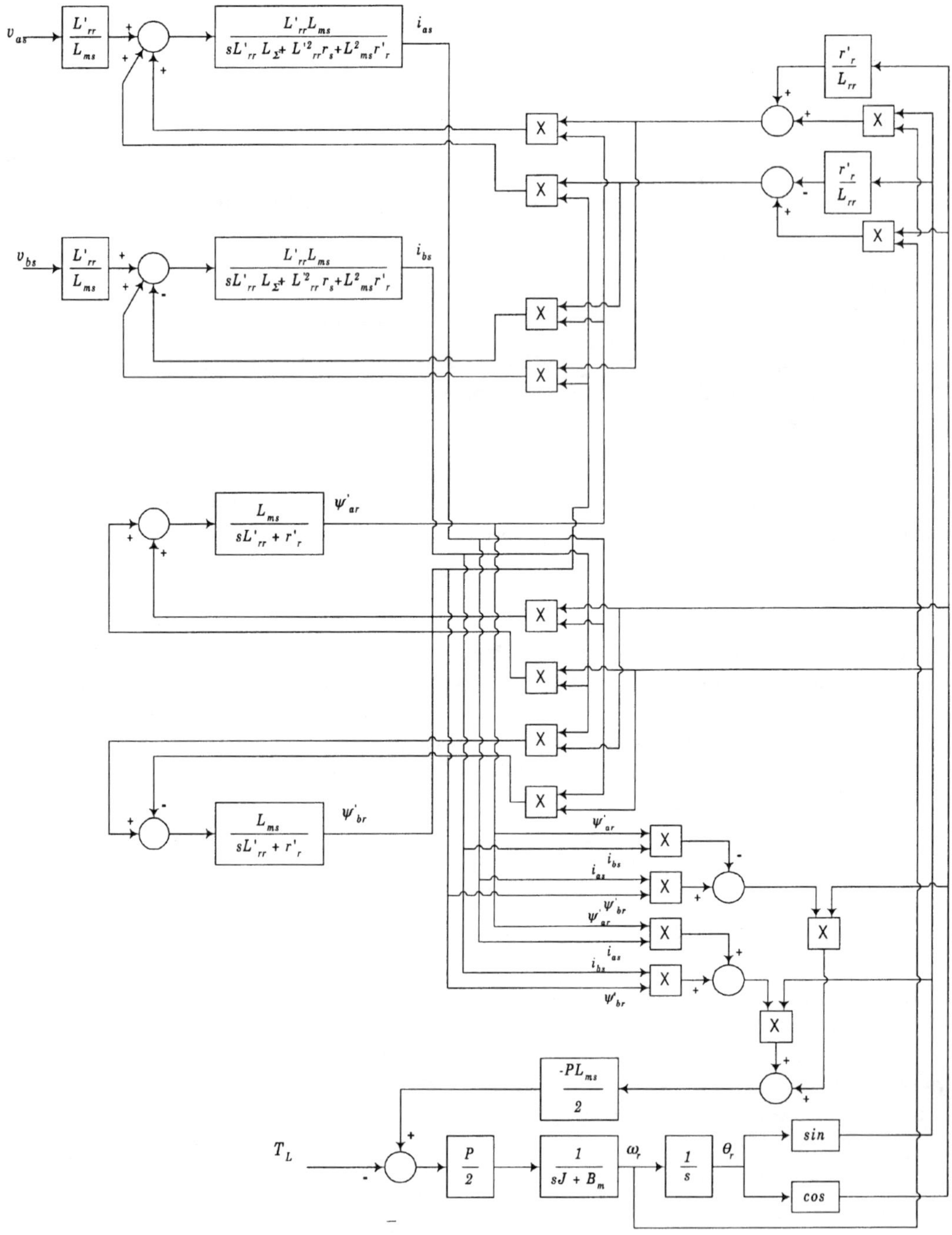

Figure 6.19 Block diagram of two-phase squirrel-cage induction motors.

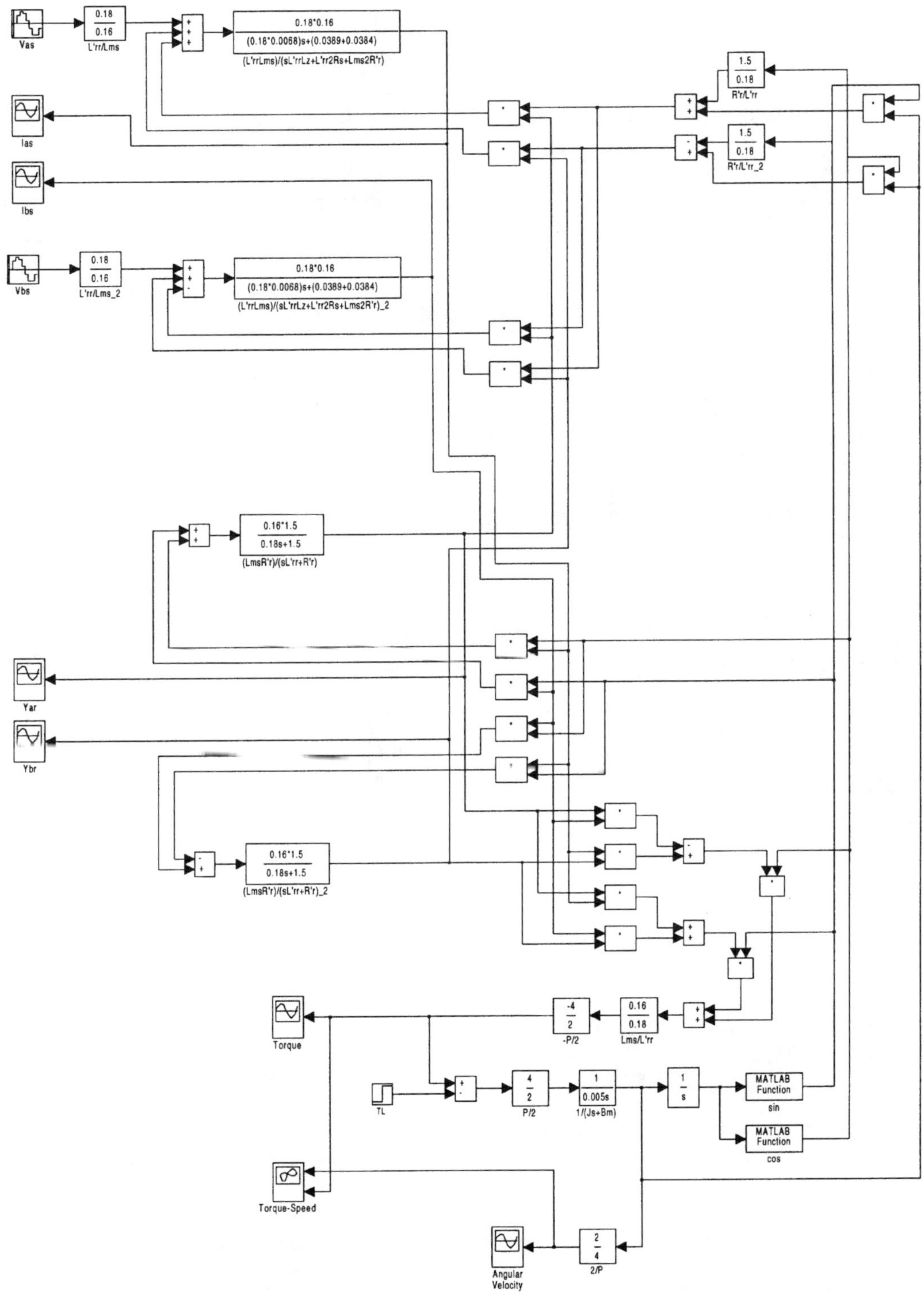

Figure 6.20 SIMULINK diagram to model the transient dynamics and plot the torque-speed characteristics of the A-class squirrel-cage motor.

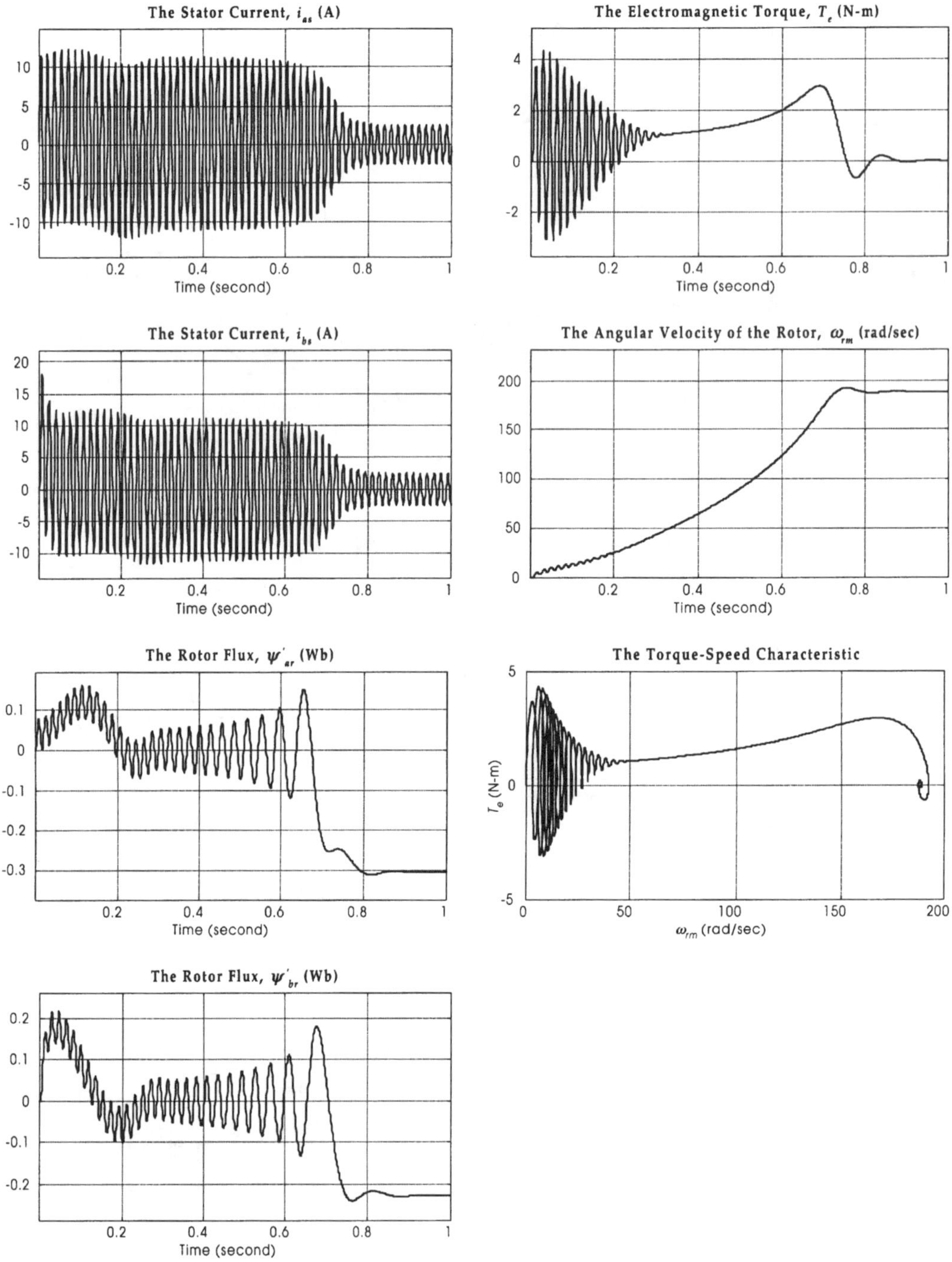

Figure 6.21 Dynamics and torque-speed characteristic of the A-class induction motor: $T_L = 0$ N-m, $u_{as}(t) = \sqrt{2}115\cos(377t)$, and $u_{bs}(t) = \sqrt{2}115\sin(377t)$.

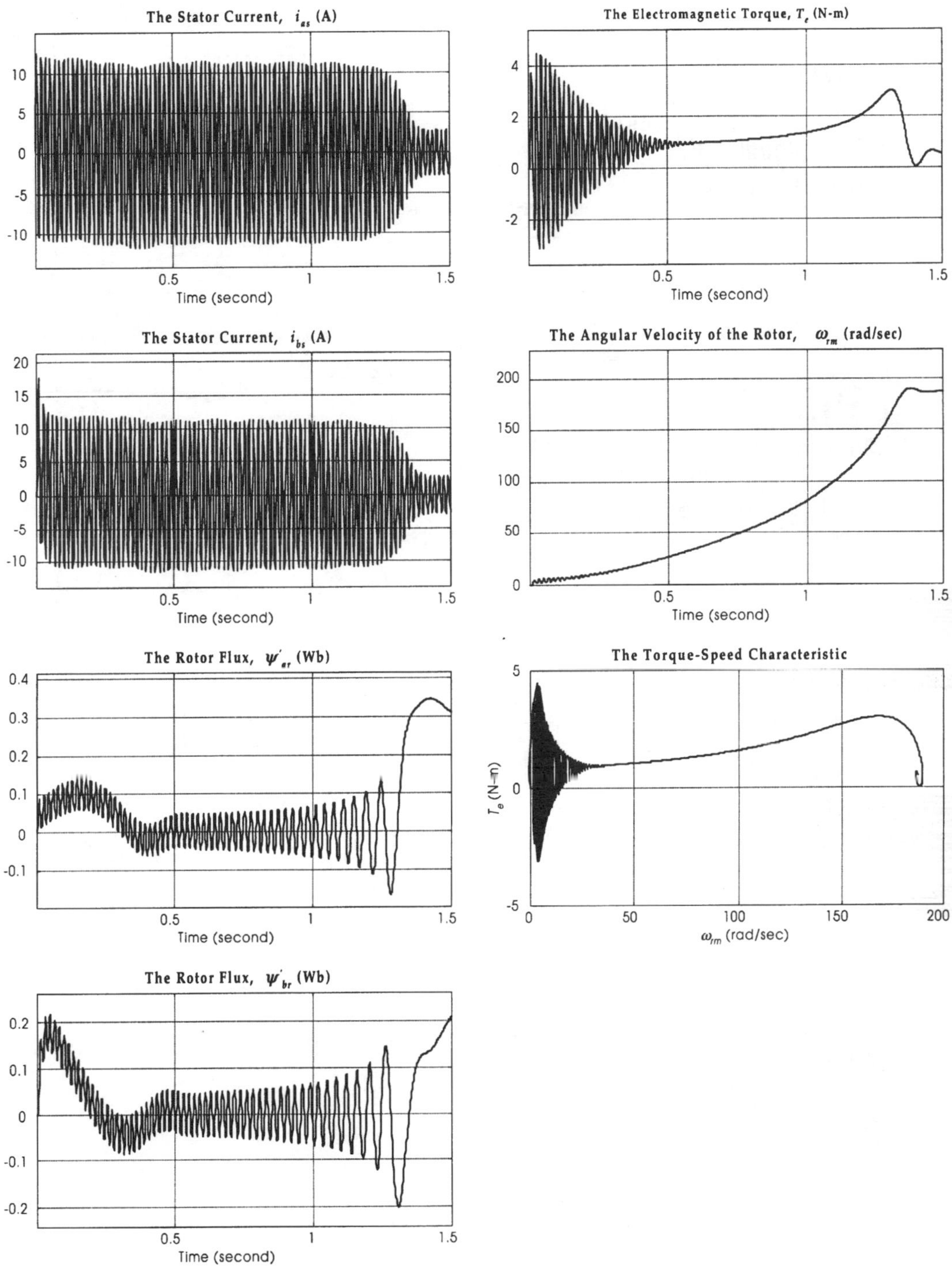

Figure 6.22 Dynamics and torque-speed characteristic of the A-class induction motor: $T_L = 0.5$ N-m, $u_{as}(t) = \sqrt{2}115\cos(377t)$, and $u_{bs}(t) = \sqrt{2}115\sin(377t)$.

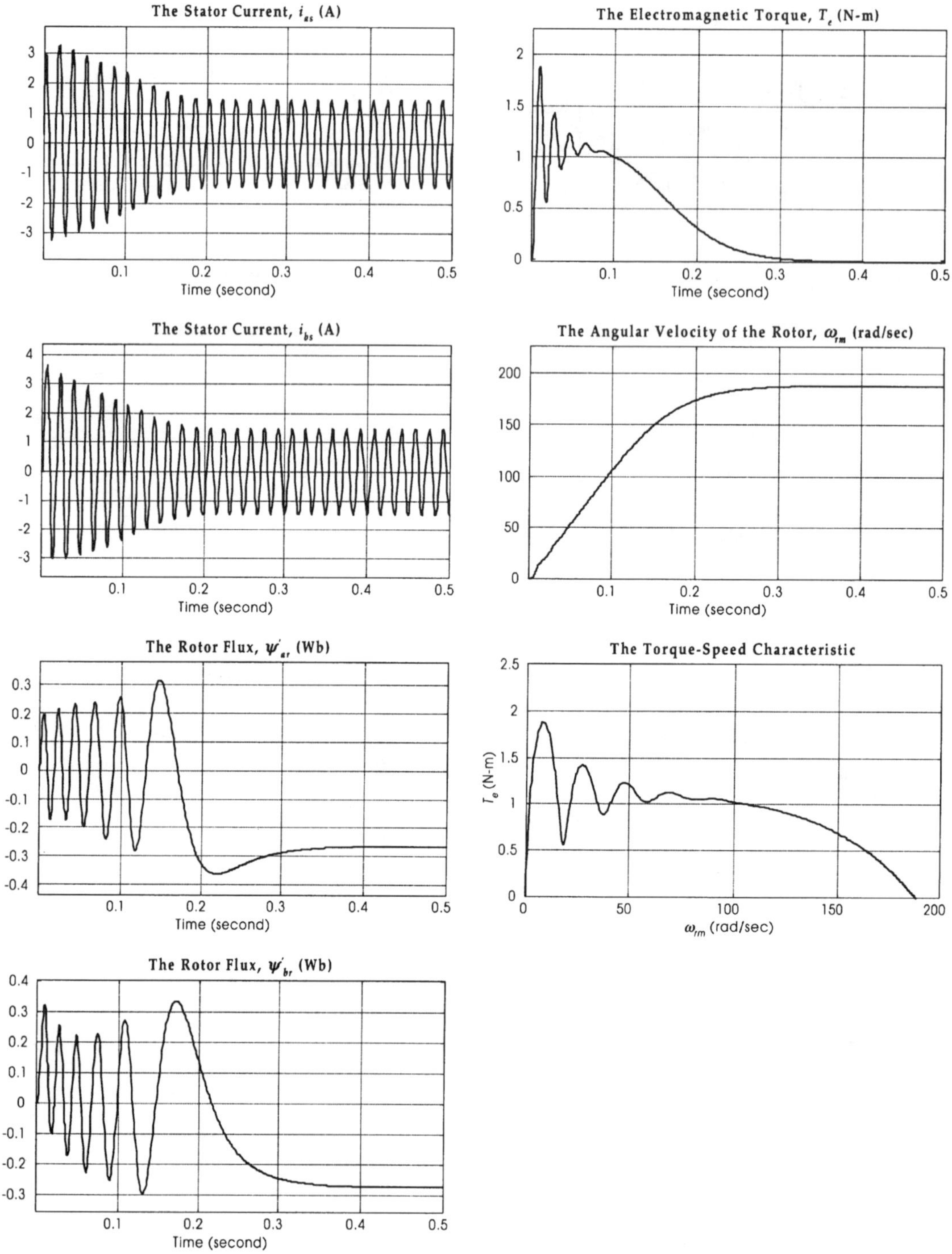

Figure 6.23 Dynamics and torque-speed characteristic of the D-class induction motor: $T_L = 0$ N-m, $u_{as}(t) = \sqrt{2}115\cos(377t)$, and $u_{bs}(t) = \sqrt{2}115\sin(377t)$.

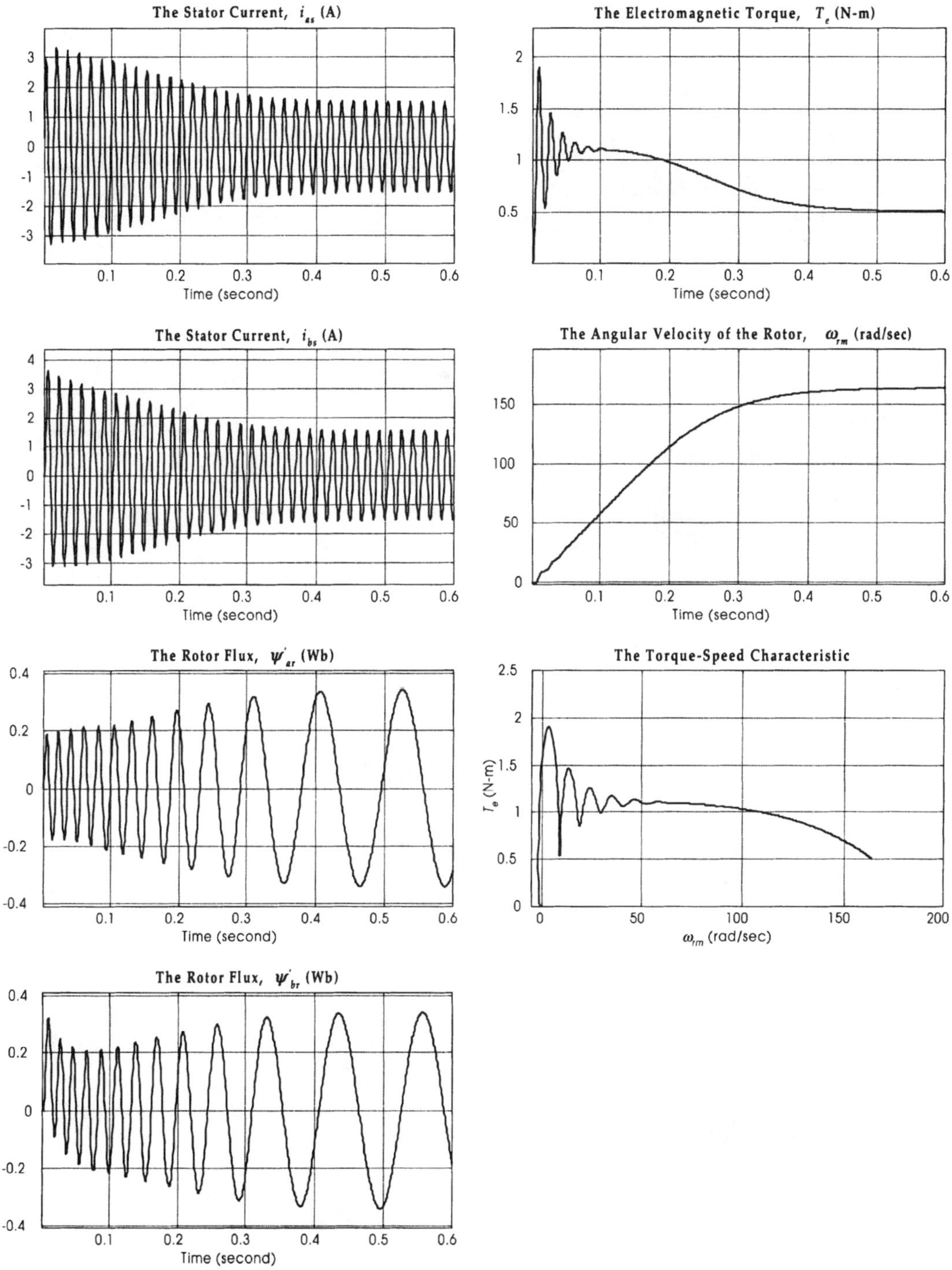

Figure 6.24 Dynamics and torque-speed characteristic of the D-class induction motor: $T_L = 0.5$ N-m, $u_{as}(t) = \sqrt{2}115\cos(377t)$, and $u_{bs}(t) = \sqrt{2}115\sin(377t)$.

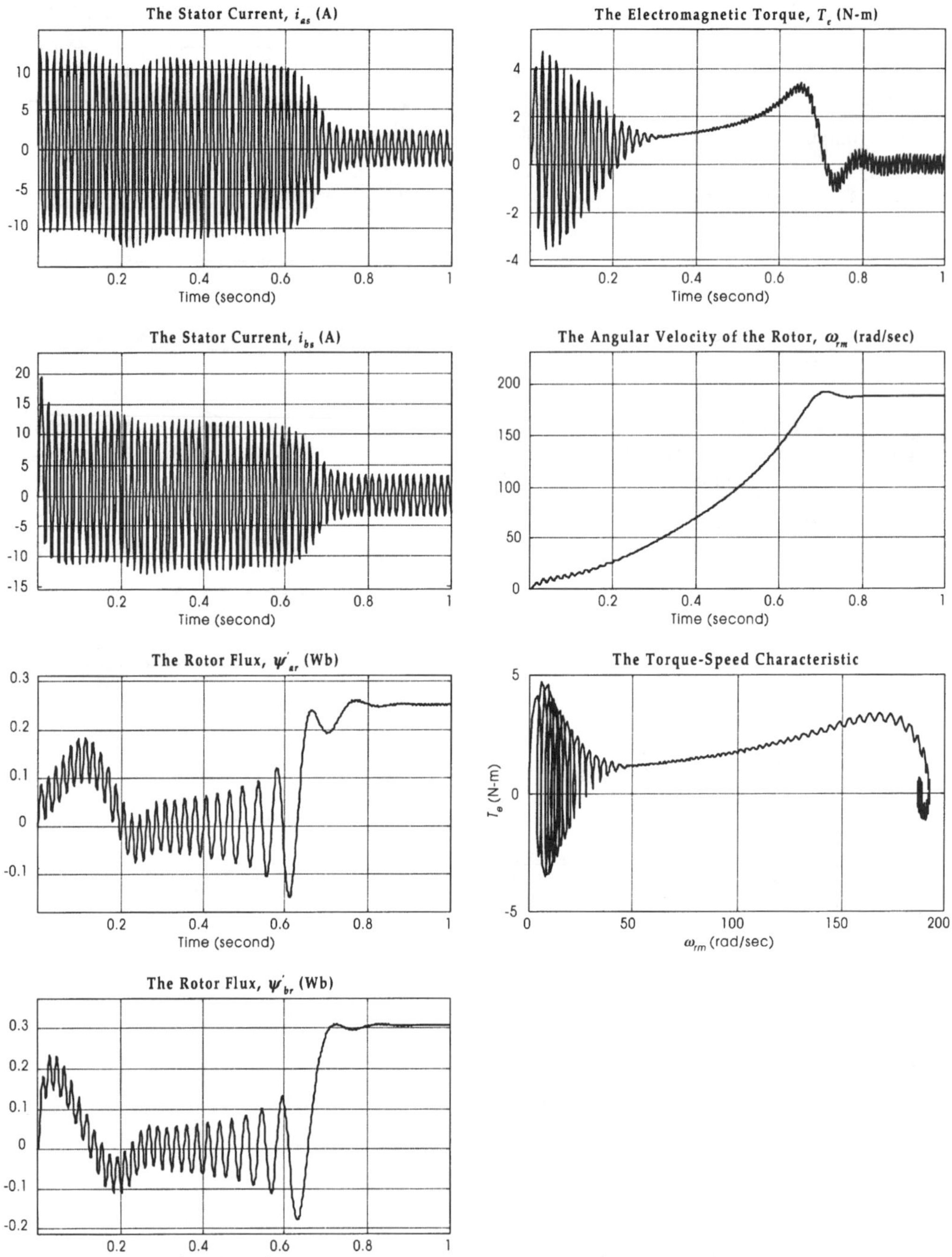

Figure 6.25 Dynamics and torque-speed characteristic of the A-class induction motor: $T_L = 0$ N-m, $u_{as}(t) = \sqrt{2}115\cos(377t)$, and $u_{bs}(t) = \sqrt{2}126\sin(377t)$.

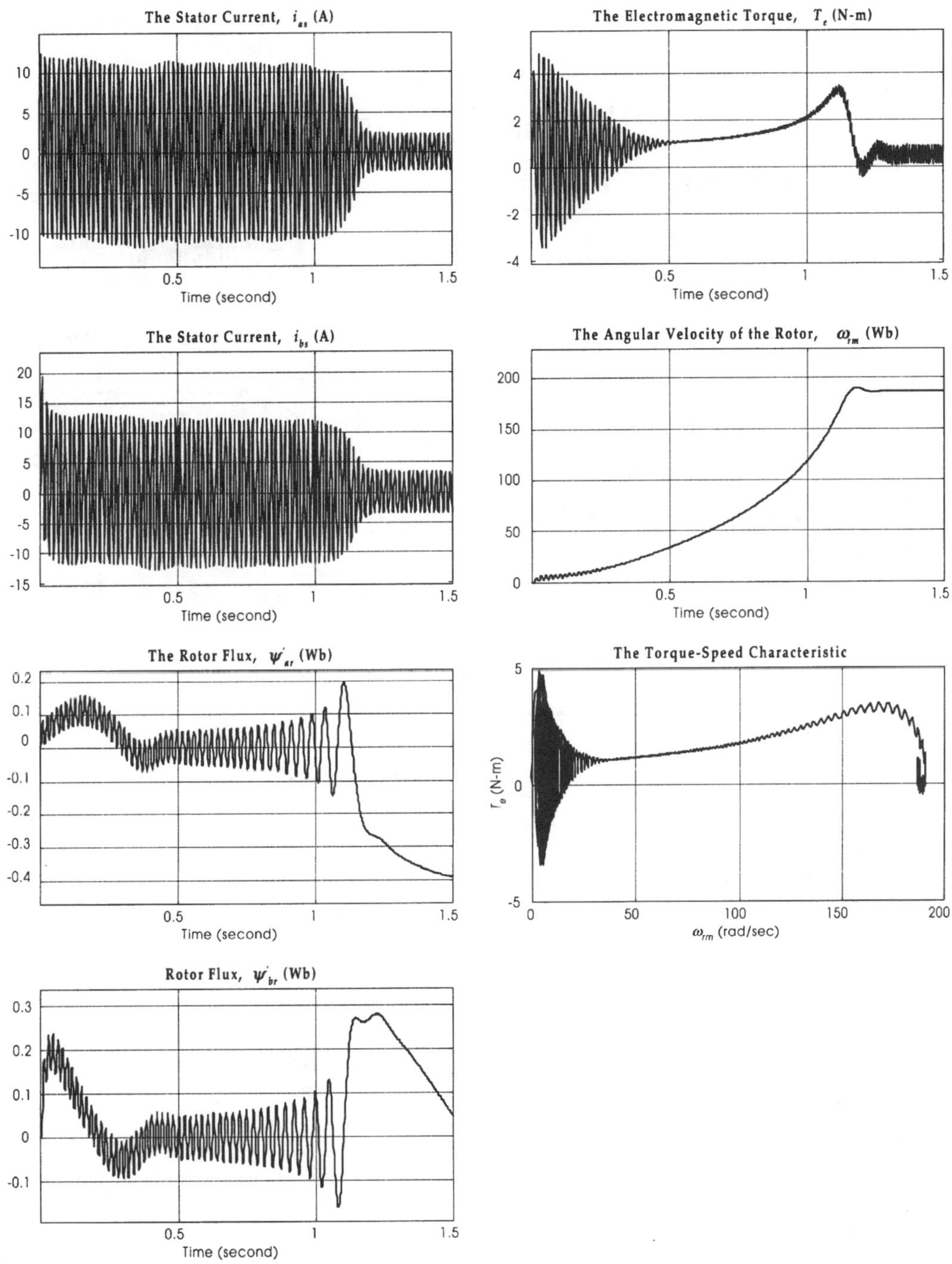

Figure 6.26 Dynamics and torque-speed characteristic of the A-class induction motor: $T_L = 0.5$ N-m, $u_{as}(t) = \sqrt{2}115\cos(377t)$, and $u_{bs}(t) = \sqrt{2}126\sin(377t)$.

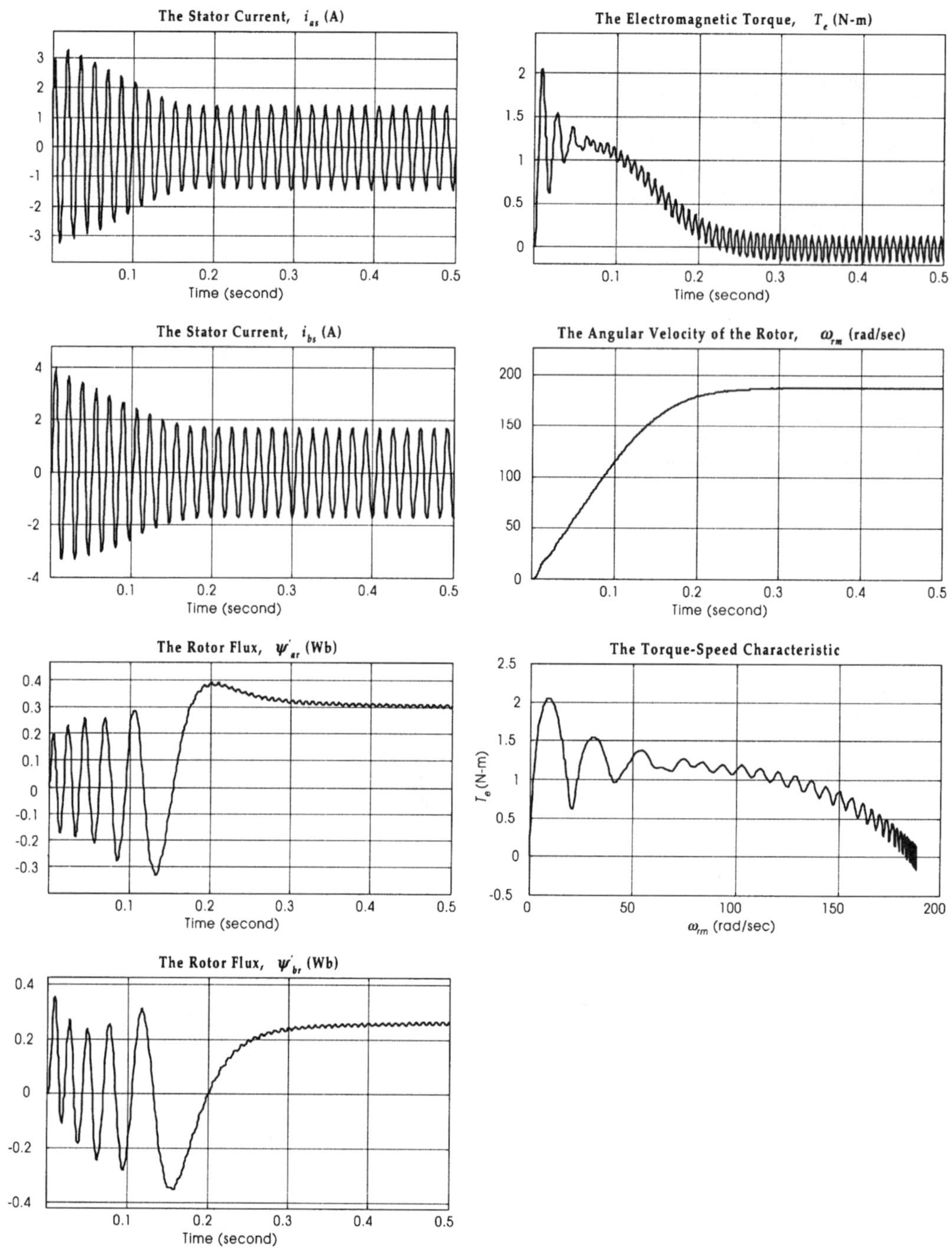

Figure 6.27 Dynamics and torque-speed characteristic of the D-class induction motor: $T_L = 0$ N-m, $u_{as}(t) = \sqrt{2}115\cos(377t)$, and $u_{bs}(t) = \sqrt{2}126\sin(377t)$.

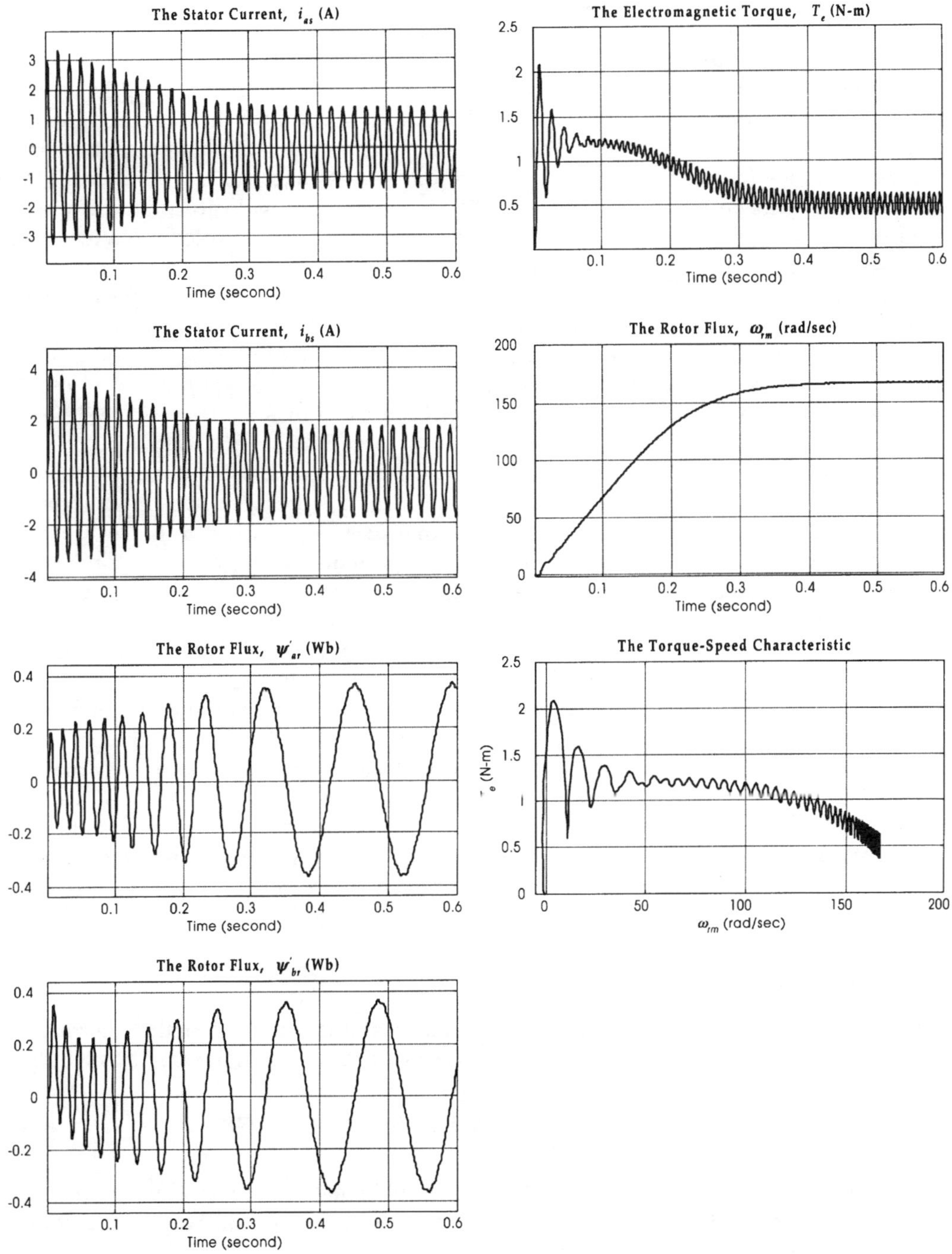

Figure 6.28 Dynamics and torque-speed characteristic of the D-class induction motor: $T_L = 0.5$ N-m, $u_{as}(t) = \sqrt{2}115\cos(377t)$, and $u_{bs}(t) = \sqrt{2}126\sin(377t)$.

and synchronous machines can be studied in three commonly used reference frames. In particular, in the stationary reference frame, one assigns $\omega = 0$, the rotor reference frame results if $\omega = \omega_r$, and the synchronous reference frame is in place if the angular velocity of the reference frame is equal to the synchronous angular velocity $\omega = \omega_e$. In the most general case, the induction machine dynamics should be found if the angular velocity of the reference frame is not specified. That is, as one derives the mathematical model in the *arbitrary* reference frame, the corresponding equations to map the machine steady-state and transients can be easily derived specifying the angular velocity of the frame; for example, $\omega = 0$ for the stationary reference frame, $\omega = \omega_r$ for the rotor reference frame, and $\omega = \omega_e$ for the synchronous reference frame.

Using the Park transformations, we change the *machine* (*abc*) variables to the *quadrature*, *direct*, and *zero* (*qd0*) quantities. It should be emphasized that the developed differential equations in the *machine* variables (voltages, currents, and flux linkages), as given by (6.5) or (6.11), as well as the equations for the electromagnetic torque (6.6) or (6.12), must be employed as the starting point to map the machine dynamics in the *arbitrary* reference frame using the *quadrature*- and *direct*-axis components of stator and rotor voltages, currents, and flux linkages. That is, at the beginning, we use the models of induction machines given in terms of the phase voltages, currents, and flux linkages. For two-phase induction machines, using the *direct* Park transformation matrices, we transform the *as*, *bs*, *ar*, and *br* variables to the *quadrature*- and *direct*-axis components of stator and rotor voltages, currents, and flux linkages. Using the *inverse* Park transformation matrices, the *quadrature*- and *direct*-axis quantities can be transformed back to the *ab* variables. For three-phase induction machines, the *abc* variables are transformed to the *quadrature-*, *direct-*, and *zero*-axis components of stator and rotor voltages, currents, and flux linkages. These transformations can be performed to reduce the complexity of the differential equations needed to be used in analysis and design. The figure, given in Table 6.2 illustrates the orientation of the *machine* stator and rotor magnetic axes, as well as the *quadrature* and *direct* magnetic axes. In addition, the basic transformation equations to transform the *machine* variables to the *quadrature* and *direct* quantities and vice versa are given in Table 6.2 by making use of the *direct* and *inverse* Park transformation matrices. Letting the angular velocity of the reference frame be ω, $\omega = 0$, $\omega = \omega_r$, and $\omega = \omega_e$, one has the corresponding transformations in the *arbitrary*, stationary, rotor, and synchronous reference frames.

Illustrative Example 6.1

Find the *quadrature*- and *direct*-axis components of stator voltages in the *arbitrary* and stationary reference frames if the following voltages are supplied to the *as* and *bs* windings

$$u_{as}(t) = 100\cos(377t) \ \text{ and } \ u_{bs}(t) = 100\sin(377t)$$

Find $u^e_{qs}(t)$, $u^e_{ds}(t)$ and $u^s_{qs}(t)$, $u^s_{ds}(t)$ in the synchronous and stationary reference frames, assigning $\omega = 377$ rad-s^{-1} and $\omega = 0$. Prove that $u_{qs}(t)$ and $u_{ds}(t)$ are the dc voltages if the angular velocity of the frame is $\omega = 377$ rad-s^{-1}.

Solution

To solve the problem, one applies the Park transformation matrix

$$\mathbf{K}_s = \begin{bmatrix} \cos\theta & \sin\theta \\ \sin\theta & -\cos\theta \end{bmatrix}$$

Table 6.2 Transformations of the *Machine* and *qd* Variables

<table>
<tr><th>Stator, rotor, and reference frame magnetic axes</th><th>Transformation matrices</th><th>Transformation of variables</th></tr>
<tr><td rowspan="2">

***Arbitrary* reference frame**

ω unspecified

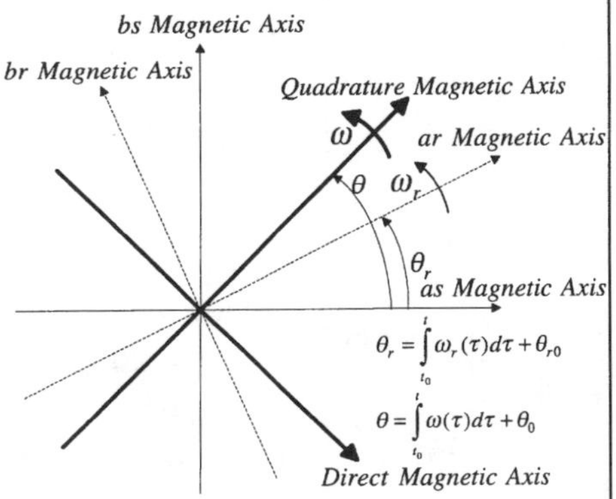

</td><td>

Stator transformation matrices

$\mathbf{K}_s = \begin{bmatrix} \cos\theta & \sin\theta \\ \sin\theta & -\cos\theta \end{bmatrix}$

$\mathbf{K}_s^{-1} = \begin{bmatrix} \cos\theta & \sin\theta \\ \sin\theta & -\cos\theta \end{bmatrix}$

</td><td>

Direct transformation

$\mathbf{u}_{qds} = \mathbf{K}_s \mathbf{u}_{abs}$

$\mathbf{i}_{qds} = \mathbf{K}_s \mathbf{i}_{abs}$

$\boldsymbol{\psi}_{qds} = \mathbf{K}_s \boldsymbol{\psi}_{abs}$

Inverse transformation

$\mathbf{u}_{abs} = \mathbf{K}_s^{-1} \mathbf{u}_{qds}$

$\mathbf{i}_{abs} = \mathbf{K}_s^{-1} \mathbf{i}_{qds}$

$\boldsymbol{\psi}_{abs} = \mathbf{K}_s^{-1} \boldsymbol{\psi}_{qds}$

</td></tr>
<tr><td>

Rotor transformation matrices

$\mathbf{K}_r = \begin{bmatrix} \cos(\theta-\theta_r) & \sin(\theta-\theta_r) \\ \sin(\theta-\theta_r) & -\cos(\theta-\theta_r) \end{bmatrix}$

$\mathbf{K}_r^{-1} = \begin{bmatrix} \cos(\theta-\theta_r) & \sin(\theta-\theta_r) \\ \sin(\theta-\theta_r) & -\cos(\theta-\theta_r) \end{bmatrix}$

</td><td>

Direct transformation

$\mathbf{u}_{qdr} = \mathbf{K}_r \mathbf{u}_{abr}$

$\mathbf{i}_{qdr} = \mathbf{K}_r \mathbf{i}_{abr}$

$\boldsymbol{\psi}_{qdr} = \mathbf{K}_r \boldsymbol{\psi}_{abr}$

Inverse transformation

$\mathbf{u}_{abr} = \mathbf{K}_r^{-1} \mathbf{u}_{qdr}$

$\mathbf{i}_{abs} = \mathbf{K}_r^{-1} \mathbf{i}_{qds}$

$\boldsymbol{\psi}_{abs} = \mathbf{K}_r^{-1} \boldsymbol{\psi}_{qds}$

</td></tr>
<tr><td rowspan="2">

Stationary reference frame, $\omega = 0$

</td><td>

Stator transformation matrices

$\mathbf{K}_s^s = \begin{bmatrix} 1 & 0 \\ 0 & -1 \end{bmatrix},$

$\mathbf{K}_s^{s^{-1}} = \begin{bmatrix} 1 & 0 \\ 0 & -1 \end{bmatrix}$

</td><td>

Direct transformation

$\mathbf{u}_{qds}^s = \mathbf{K}_s^s \mathbf{u}_{abs}$

$\mathbf{i}_{qds}^s = \mathbf{K}_s^s \mathbf{i}_{abs}$

$\boldsymbol{\psi}_{qds}^s = \mathbf{K}_s^s \boldsymbol{\psi}_{abs}$

Inverse transformation

$\mathbf{u}_{abs} = \mathbf{K}_s^{s^{-1}} \mathbf{u}_{qds}^s$

$\mathbf{i}_{abs} = \mathbf{K}_s^{s^{-1}} \mathbf{i}_{qds}^s$

$\boldsymbol{\psi}_{abs} = \mathbf{K}_s^{s^{-1}} \boldsymbol{\psi}_{qds}^s$

</td></tr>
<tr><td>

Rotor transformation matrices

$\mathbf{K}_r^s = \begin{bmatrix} \cos\theta_r & \sin\theta_r \\ \sin\theta_r & -\cos\theta_r \end{bmatrix},$

$\mathbf{K}_r^{r^{-1}} = \begin{bmatrix} \cos\theta_r & \sin\theta_r \\ \sin\theta_r & -\cos\theta_r \end{bmatrix}$

</td><td>

Direct transformation

$\mathbf{u}_{qdr}^s = \mathbf{K}_r^s \mathbf{u}_{abr}$

$\mathbf{i}_{qdr}^s = \mathbf{K}_r^s \mathbf{i}_{abr}$

$\boldsymbol{\psi}_{qdr}^s = \mathbf{K}_r^s \boldsymbol{\psi}_{abr}$

Inverse transformation

$\mathbf{u}_{abr} = \mathbf{K}_r^{s^{-1}} \mathbf{u}_{qdr}^s$

$\mathbf{i}_{abr} = \mathbf{K}_r^{s^{-1}} \mathbf{i}_{qdr}^s$

$\boldsymbol{\psi}_{abr} = \mathbf{K}_r^{s^{-1}} \boldsymbol{\psi}_{qdr}^s$

</td></tr>
</table>

Rotor reference frame, $\omega = \omega_r$	**Stator** transformation matrices $\mathbf{K}_s^r = \begin{bmatrix} \cos\theta_r & \sin\theta_r \\ \sin\theta_r & -\cos\theta_r \end{bmatrix}$ $\mathbf{K}_s^{r^{-1}} = \begin{bmatrix} \cos\theta_r & \sin\theta_r \\ \sin\theta_r & -\cos\theta_r \end{bmatrix}$	*Direct* transformation $\mathbf{u}_{qds}^r = \mathbf{K}_s^r \mathbf{u}_{abs}$ $\mathbf{i}_{qds}^r = \mathbf{K}_s^r \mathbf{i}_{abs}$ $\boldsymbol{\psi}_{qds}^r = \mathbf{K}_s^r \boldsymbol{\psi}_{abs}$ *Inverse* transformation $\mathbf{u}_{abs} = \mathbf{K}_s^{r^{-1}} \mathbf{u}_{qds}^r$ $\mathbf{i}_{abs} = \mathbf{K}_s^{r^{-1}} \mathbf{i}_{qds}^r$ $\boldsymbol{\psi}_{abs} = \mathbf{K}_s^{r^{-1}} \boldsymbol{\psi}_{qds}^r$
	Rotor transformation matrices $\mathbf{K}_r^r = \begin{bmatrix} 1 & 0 \\ 0 & -1 \end{bmatrix}$ $\mathbf{K}_r^{r^{-1}} = \begin{bmatrix} 1 & 0 \\ 0 & -1 \end{bmatrix}$	*Direct* transformation $\mathbf{u}_{qdr}^r = \mathbf{K}_r^r \mathbf{u}_{abr}$ $\mathbf{i}_{qdr}^r = \mathbf{K}_r^r \mathbf{i}_{abr}$ $\boldsymbol{\psi}_{qdr}^r = \mathbf{K}_r^r \boldsymbol{\psi}_{abr}$ *Inverse* transformation $\mathbf{u}_{abr} = \mathbf{K}_r^{r^{-1}} \mathbf{u}_{qdr}^r$ $\mathbf{i}_{abr} = \mathbf{K}_r^{r^{-1}} \mathbf{i}_{qdr}^r$ $\boldsymbol{\psi}_{abr} = \mathbf{K}_r^{r^{-1}} \boldsymbol{\psi}_{qdr}^r$
Synchronous reference frame, $\omega = \omega_e$	**Stator** transformation matrices $\mathbf{K}_s^e = \begin{bmatrix} \cos\theta_e & \sin\theta_e \\ \sin\theta_e & -\cos\theta_e \end{bmatrix}$ $\mathbf{K}_s^{e^{-1}} = \begin{bmatrix} \cos\theta_e & \sin\theta_e \\ \sin\theta_e & -\cos\theta_e \end{bmatrix}$	*Direct* transformation $\mathbf{u}_{qds}^e = \mathbf{K}_s^e \mathbf{u}_{abs}$ $\mathbf{i}_{qds}^e = \mathbf{K}_s^e \mathbf{i}_{abs}$ $\boldsymbol{\psi}_{qds}^e = \mathbf{K}_s^e \boldsymbol{\psi}_{abs}$ *Inverse* transformation $\mathbf{u}_{abs} = \mathbf{K}_s^{e^{-1}} \mathbf{u}_{qds}^e$ $\mathbf{i}_{abs} = \mathbf{K}_s^{e^{-1}} \mathbf{i}_{qds}^e$ $\boldsymbol{\psi}_{abs} = \mathbf{K}_s^{e^{-1}} \boldsymbol{\psi}_{qds}^e$
	Rotor transformation matrices $\mathbf{K}_r^e = \begin{bmatrix} \cos(\theta_e - \theta_r) & \sin(\theta_e - \theta_r) \\ \sin(\theta_e - \theta_r) & -\cos(\theta_e - \theta_r) \end{bmatrix}$ $\mathbf{K}_r^{e^{-1}} = \begin{bmatrix} \cos(\theta_e - \theta_r) & \sin(\theta_e - \theta_r) \\ \sin(\theta_e - \theta_r) & -\cos(\theta_e - \theta_r) \end{bmatrix}$	*Direct* transformation $\mathbf{u}_{qdr}^e = \mathbf{K}_r^e \mathbf{u}_{abr}$ $\mathbf{i}_{qdr}^e = \mathbf{K}_r^e \mathbf{i}_{abr}$ $\boldsymbol{\psi}_{qdr}^e = \mathbf{K}_r^e \boldsymbol{\psi}_{abr}$ *Inverse* transformation $\mathbf{u}_{abr} = \mathbf{K}_r^{e^{-1}} \mathbf{u}_{qdr}^e$ $\mathbf{i}_{abr} = \mathbf{K}_r^{e^{-1}} \mathbf{i}_{qdr}^e$ $\boldsymbol{\psi}_{abr} = \mathbf{K}_r^{e^{-1}} \boldsymbol{\psi}_{qdr}^e$

Taking note of

$$\mathbf{u}_{qds} = \mathbf{K}_s \mathbf{u}_{abs}, \quad \begin{bmatrix} u_{qs} \\ u_{ds} \end{bmatrix} = \begin{bmatrix} \cos\theta & \sin\theta \\ \sin\theta & -\cos\theta \end{bmatrix} \begin{bmatrix} u_{as} \\ u_{bs} \end{bmatrix}$$

we have

$$u_{qs}(t) = \cos\theta u_{as}(t) + \sin\theta u_{bs}(t)$$

$$u_{ds}(t) = \sin\theta u_{as}(t) - \cos\theta u_{bs}(t)$$

Thus, in the *arbitrary* reference frame, the *quadrature*- and *direct*-axis components of voltages are found to be

$$u_{qs}(t) = 100(\cos\theta\cos(377t) + \sin\theta\sin(377t))$$

$$u_{ds}(t) = 100(\sin\theta\cos(377t) - \cos\theta\sin(377t))$$

Specifying the angular velocity of the reference frame, one obtains the *quadrature*- and *direct*-axis components of voltages in the stationary, rotor, and synchronous reference frames. The synchronous reference frame rotates with the constant angular velocity $\omega = 377$ rad-s^{-1} and $\theta_0 = 0$. Thus, we have $\theta = \omega t = 377t$.

Hence,

$$u_{qs}^e(t) = 100(\cos^2(377t) + \sin^2(377t)) = 100$$

$$u_{ds}^e(t) = 100(\sin(377t)\cos(377t) - \cos(377)\sin(377t)) = 0$$

One concludes that in the synchronous reference frame, the *quadrature*- and *direct*-axis components $u_{qs}^e(t)$ and $u_{ds}^e(t)$ are the dc voltages (recall that it is assigned that the angular frequency of the supplied voltages and the reference frame angular velocity are equal). This result cannot be generalized to all reference frames used. For example, one finds that in the stationary reference frame, $u_{qs}^s(t)$ and $u_{ds}^s(t)$ are the ac voltages.

In the stationary reference frame, the frame angular velocity is assigned to be zero. Making use of $\omega = 0$, we plug $\theta = 0$ in $\mathbf{K}_s = \begin{bmatrix} \cos\theta & \sin\theta \\ \sin\theta & -\cos\theta \end{bmatrix}$. Hence, the following transformation matrix needs to be used in the stationary reference frame

$$\mathbf{K}_s^s = \begin{bmatrix} 1 & 0 \\ 0 & -1 \end{bmatrix}$$

Therefore, the following expressions result

$$\mathbf{u}_{qds}^s = \mathbf{K}_s^s \mathbf{u}_{abs}, \quad \begin{bmatrix} u_{qs}^s \\ u_{ds}^s \end{bmatrix} = \begin{bmatrix} 1 & 0 \\ 0 & -1 \end{bmatrix} \begin{bmatrix} u_{as} \\ u_{bs} \end{bmatrix}$$

Thus, in expanded form, one obtains

$$u_{qs}^s(t) = u_{as}(t), \quad u_{ds}^s(t) = -u_{bs}(t)$$

Hence, making use of the expressions for the phase voltages, we have

$$u_{qs}^s(t) = 100\cos(377t)$$

$$u_{ds}^s(t) = -100\sin(377t)$$

□

Consider the two-phase induction motor shown in Figure 6.29. The stator is the stationary member. Hence, the *as* and *bs* magnetic axes are stationary. The rotor rotates with electrical angular velocity ω_r, and the angular displacement of the *ar* magnetic axis with respect to the *as* magnetic axis is found as $\theta_r = \int_{t_0}^{t} \omega_r(\tau)d\tau + \theta_{r0}$. The *qd* reference frame (see the *quadrature* and *direct* magnetic axes shown in Figure 6.29), rotates with unspecified

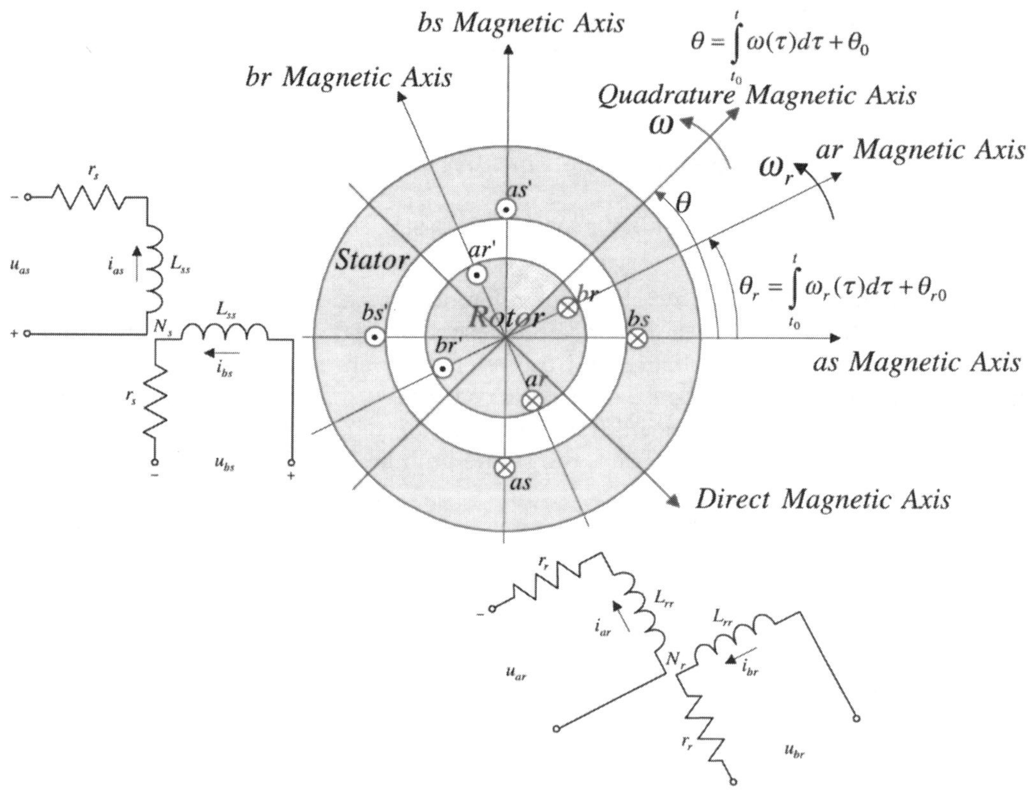

Figure 6.29 Two-phase symmetrical induction motor with *quadrature* and *direct* axes.

angular velocity ω. Hence, the angular displacement between the *quadrature* and *as* magnetic axes is

$$\theta = \int_{t_0}^{t} \omega(\tau)d\tau + \theta_0$$

For two-phase induction machines, the transformation of the stator voltages, currents, and flux linkages, as given in the *machine* (*ab*) variables, to the *quadrature* and *direct* quantities is performed using the stator transformation matrix $\mathbf{K}_s$, which is expressed by

$$\mathbf{K}_s = \begin{bmatrix} \cos\theta & \sin\theta \\ \sin\theta & -\cos\theta \end{bmatrix}$$

In particular, the *quadrature-* and *direct*-axis components of stator voltages, currents, and flux linkages are found using the matrix $\mathbf{K}_s$ as

$$\mathbf{u}_{qds} = \mathbf{K}_s \mathbf{u}_{abs}, \quad \begin{bmatrix} u_{qs} \\ u_{ds} \end{bmatrix} = \begin{bmatrix} \cos\theta & \sin\theta \\ \sin\theta & -\cos\theta \end{bmatrix} \begin{bmatrix} u_{as} \\ u_{bs} \end{bmatrix}$$

$$\mathbf{i}_{qds} = \mathbf{K}_s \mathbf{i}_{abs}, \quad \begin{bmatrix} i_{qs} \\ i_{ds} \end{bmatrix} = \begin{bmatrix} \cos\theta & \sin\theta \\ \sin\theta & -\cos\theta \end{bmatrix} \begin{bmatrix} i_{as} \\ i_{bs} \end{bmatrix}$$

and

$$\boldsymbol{\psi}_{qds} = \mathbf{K}_s \boldsymbol{\psi}_{abs}, \quad \begin{bmatrix} \psi_{qs} \\ \psi_{ds} \end{bmatrix} = \begin{bmatrix} \cos\theta & \sin\theta \\ \sin\theta & -\cos\theta \end{bmatrix} \begin{bmatrix} \psi_{as} \\ \psi_{bs} \end{bmatrix}$$

It should be emphasized that the *machine* variables can be found using the *quadrature-* and *direct*-axis quantities. In particular, using the inverse matrix $\mathbf{K}_s^{-1}$, as given by

$$\mathbf{K}_s^{-1} = \begin{bmatrix} \cos\theta & \sin\theta \\ \sin\theta & -\cos\theta \end{bmatrix}$$

we have

$$\mathbf{u}_{abs} = \mathbf{K}_s^{-1} \mathbf{u}_{qds}, \quad \begin{bmatrix} u_{as} \\ u_{bs} \end{bmatrix} = \begin{bmatrix} \cos\theta & \sin\theta \\ \sin\theta & -\cos\theta \end{bmatrix} \begin{bmatrix} u_{qs} \\ u_{ds} \end{bmatrix}$$

$$\mathbf{i}_{abs} = \mathbf{K}_s^{-1} \mathbf{i}_{qds}, \quad \begin{bmatrix} i_{as} \\ i_{bs} \end{bmatrix} = \begin{bmatrix} \cos\theta & \sin\theta \\ \sin\theta & -\cos\theta \end{bmatrix} \begin{bmatrix} i_{qs} \\ i_{ds} \end{bmatrix}$$

and

$$\boldsymbol{\psi}_{abs} = \mathbf{K}_s^{-1} \boldsymbol{\psi}_{qds}, \quad \begin{bmatrix} \psi_{as} \\ \psi_{bs} \end{bmatrix} = \begin{bmatrix} \cos\theta & \sin\theta \\ \sin\theta & -\cos\theta \end{bmatrix} \begin{bmatrix} \psi_{qs} \\ \psi_{ds} \end{bmatrix}$$

We have demonstrated how to perform the *direct* and *inverse* Park transformations for the stator voltages, currents, and flux linkages. The angular velocity of the reference frame is ω, and the angular displacement of the *quadrature* magnetic axis with respect to the stationary *machine as* magnetic axis is found as

$$\theta = \int_{t_0}^{t} \omega(\tau)\,d\tau + \theta_0$$

The rotor rotates with the electrical angular velocity ω_r, and the angular displacement is

$$\theta_r = \int_{t_0}^{t} \omega_r(\tau)d\tau + \theta_{r0}$$

The difference between the angular displacements θ and θ_r is employed to find the rotor transformation matrix needed to be applied to transform the *ab* rotor voltages, currents, and flux linkages to the corresponding *qd* variables. In particular, for the rotor variables, the transformation matrix $\mathbf{K}_r$ is given by

$$\mathbf{K}_r = \begin{bmatrix} \cos(\theta-\theta_r) & \sin(\theta-\theta_r) \\ \sin(\theta-\theta_r) & -\cos(\theta-\theta_r) \end{bmatrix}$$

For two-phase induction machines, the transformation of the rotor voltages, currents, and flux linkages to the corresponding *quadrature*- and *direct*-axis quantities is accomplished as

$$\mathbf{u}_{qdr} = \mathbf{K}_r \mathbf{u}_{abr}, \quad \begin{bmatrix} u_{qr} \\ u_{dr} \end{bmatrix} = \begin{bmatrix} \cos(\theta-\theta_r) & \sin(\theta-\theta_r) \\ \sin(\theta-\theta_r) & -\cos(\theta-\theta_r) \end{bmatrix} \begin{bmatrix} u_{ar} \\ u_{br} \end{bmatrix}$$

$$\mathbf{i}_{qdr} = \mathbf{K}_r \mathbf{i}_{abr}, \quad \begin{bmatrix} i_{qr} \\ i_{dr} \end{bmatrix} = \begin{bmatrix} \cos(\theta-\theta_r) & \sin(\theta-\theta_r) \\ \sin(\theta-\theta_r) & -\cos(\theta-\theta_r) \end{bmatrix} \begin{bmatrix} i_{ar} \\ i_{br} \end{bmatrix}$$

and

$$\boldsymbol{\psi}_{qdr} = \mathbf{K}_r \boldsymbol{\psi}_{abr}, \quad \begin{bmatrix} \psi_{qr} \\ \psi_{dr} \end{bmatrix} = \begin{bmatrix} \cos(\theta-\theta_r) & \sin(\theta-\theta_r) \\ \sin(\theta-\theta_r) & -\cos(\theta-\theta_r) \end{bmatrix} \begin{bmatrix} \psi_{ar} \\ \psi_{br} \end{bmatrix}$$

The *machine* variables can be found using the *quadrature*- and *direct*-axis components of rotor voltages, currents, and flux linkages. Applying the inverse matrix $\mathbf{K}_r^{-1}$, which is given as

$$\mathbf{K}_r^{-1} = \begin{bmatrix} \cos(\theta-\theta_r) & \sin(\theta-\theta_r) \\ \sin(\theta-\theta_r) & -\cos(\theta-\theta_r) \end{bmatrix}$$

we have

$$\mathbf{u}_{abr} = \mathbf{K}_r^{-1} \mathbf{u}_{qdr}, \quad \begin{bmatrix} u_{ar} \\ u_{br} \end{bmatrix} = \begin{bmatrix} \cos(\theta-\theta_r) & \sin(\theta-\theta_r) \\ \sin(\theta-\theta_r) & -\cos(\theta-\theta_r) \end{bmatrix} \begin{bmatrix} u_{qr} \\ u_{dr} \end{bmatrix}$$

$$\mathbf{i}_{abr} = \mathbf{K}_r^{-1} \mathbf{i}_{qdr}, \quad \begin{bmatrix} i_{ar} \\ i_{br} \end{bmatrix} = \begin{bmatrix} \cos(\theta-\theta_r) & \sin(\theta-\theta_r) \\ \sin(\theta-\theta_r) & -\cos(\theta-\theta_r) \end{bmatrix} \begin{bmatrix} i_{qr} \\ i_{dr} \end{bmatrix}$$

and

$$\boldsymbol{\psi}_{abr} = \mathbf{K}_r^{-1}\boldsymbol{\psi}_{qdr}, \quad \begin{bmatrix} \psi_{ar} \\ \psi_{br} \end{bmatrix} = \begin{bmatrix} \cos(\theta-\theta_r) & \sin(\theta-\theta_r) \\ \sin(\theta-\theta_r) & -\cos(\theta-\theta_r) \end{bmatrix} \begin{bmatrix} \psi_{qr} \\ \psi_{dr} \end{bmatrix}$$

Let us apply the Park transformation to find the mathematical model of induction motors in the *arbitrary* reference frame. To map the circuitry dynamics of two-phase induction motors, matrix equation (6.4) was used. In particular,

$$\mathbf{u}_{abs} = \mathbf{r}_s\mathbf{i}_{abs} + \frac{d\boldsymbol{\psi}_{abs}}{dt}$$

$$\mathbf{u}'_{abr} = \mathbf{r}'_r\mathbf{i}'_{abr} + \frac{d\boldsymbol{\psi}'_{abr}}{dt}$$

Using the stator voltage equation

$$\mathbf{u}_{abs} = \mathbf{r}_s\mathbf{i}_{abs} + \frac{d\boldsymbol{\psi}_{abs}}{dt}$$

we transform the *ab* variables to the *qd* variables. By multiplying the stator voltages, currents, and flux linkages by the inverse transformation matrix $\mathbf{K}_s^{-1} = \begin{bmatrix} \cos\theta & \sin\theta \\ \sin\theta & -\cos\theta \end{bmatrix}$, one obtains

$$\mathbf{K}_s^{-1}\mathbf{u}_{qds} = \mathbf{r}_s\mathbf{K}_s^{-1}\mathbf{i}_{qds} + \frac{d(\mathbf{K}_s^{-1}\boldsymbol{\psi}_{qds})}{dt}$$

Multiplying both sides of the above equation by $\mathbf{K}_s$ and differentiating $\frac{d(\mathbf{K}_s^{-1}\boldsymbol{\psi}_{qds})}{dt}$, we have

$$\mathbf{u}_{qds} = \mathbf{r}_s\mathbf{i}_{qds} + \mathbf{K}_s\frac{d(\mathbf{K}_s^{-1}\boldsymbol{\psi}_{qds})}{dt} = \mathbf{r}_s\mathbf{i}_{qds} + \mathbf{K}_s\frac{d\mathbf{K}_s^{-1}}{dt}\boldsymbol{\psi}_{qds} + \mathbf{K}_s\mathbf{K}_s^{-1}\frac{d\boldsymbol{\psi}_{qds}}{dt}$$

Observe that

$$\mathbf{K}_s\mathbf{K}_s^{-1} = \begin{bmatrix} \cos\theta & \sin\theta \\ \sin\theta & -\cos\theta \end{bmatrix}\begin{bmatrix} \cos\theta & \sin\theta \\ \sin\theta & -\cos\theta \end{bmatrix} = \begin{bmatrix} 1 & 0 \\ 0 & 1 \end{bmatrix}$$

Taking note of

$$\mathbf{K}_s\frac{d\mathbf{K}_s^{-1}}{dt} = \omega\begin{bmatrix} \cos\theta & \sin\theta \\ \sin\theta & -\cos\theta \end{bmatrix}\begin{bmatrix} -\sin\theta & \cos\theta \\ \cos\theta & \sin\theta \end{bmatrix} = \omega\begin{bmatrix} 0 & 1 \\ -1 & 0 \end{bmatrix}$$

one obtains the following matrix equation to map the stator circuitry dynamics using the *quadrature*- and *direct*-axis components of stator voltages, currents, and flux linkages

$$\mathbf{u}_{qds} = \mathbf{r}_s \mathbf{i}_{qds} + \omega \begin{bmatrix} 0 & 1 \\ -1 & 0 \end{bmatrix} \boldsymbol{\psi}_{qds} + \frac{d\boldsymbol{\psi}_{qds}}{dt} \tag{6.17}$$

From (6.4), the rotor circuitry dynamics in the *machine* variables is given by

$$\mathbf{u}'_{abr} = \mathbf{r}'_r \mathbf{i}'_{abr} + \frac{d\boldsymbol{\psi}'_{abr}}{dt}$$

By multiplying the voltages, currents, and flux linkages by $\mathbf{K}_r^{-1}$, one obtains

$$\mathbf{K}_r^{-1} \mathbf{u}'_{qdr} = \mathbf{r}'_r \mathbf{K}_r^{-1} \mathbf{i}'_{qdr} + \frac{d(\mathbf{K}_r^{-1} \boldsymbol{\psi}'_{qdr})}{dt}$$

Multiplying both sides of this equation by $\mathbf{K}_r$, we have

$$\mathbf{u}'_{qdr} = \mathbf{r}'_r \mathbf{i}'_{qdr} + \mathbf{K}_r \frac{d(\mathbf{K}_r^{-1} \boldsymbol{\psi}'_{qdr})}{dt} = \mathbf{r}'_r \mathbf{i}'_{qdr} + \mathbf{K}_r \frac{d\mathbf{K}_r^{-1}}{dt} \boldsymbol{\psi}'_{qdr} + \mathbf{K}_r \mathbf{K}_r^{-1} \frac{d\boldsymbol{\psi}'_{qdr}}{dt}$$

Taking note of

$$\mathbf{K}_r \mathbf{K}_r^{-1} = \begin{bmatrix} 1 & 0 \\ 0 & 1 \end{bmatrix}$$

and

$$\mathbf{K}_r \frac{d\mathbf{K}_r^{-1}}{dt} = (\omega - \omega_r) \begin{bmatrix} 0 & 1 \\ -1 & 0 \end{bmatrix}$$

the following matrix equation in the *qd* variables results

$$\mathbf{u}'_{qdr} = \mathbf{r}'_r \mathbf{i}'_{qdr} + (\omega - \omega_r) \begin{bmatrix} 0 & 1 \\ -1 & 0 \end{bmatrix} \boldsymbol{\psi}'_{qdr} + \frac{d\boldsymbol{\psi}'_{qdr}}{dt} \tag{6.18}$$

Using (6.17) and (6.18), for two-phase induction motors, we have

$$\begin{aligned} \mathbf{u}_{qds} &= \mathbf{r}_s \mathbf{i}_{qds} + \begin{bmatrix} 0 & \omega \\ -\omega & 0 \end{bmatrix} \boldsymbol{\psi}_{qds} + \frac{d\boldsymbol{\psi}_{qds}}{dt} \\ \mathbf{u}'_{qdr} &= \mathbf{r}'_r \mathbf{i}'_{qdr} + \begin{bmatrix} 0 & \omega - \omega_r \\ -\omega + \omega_r & 0 \end{bmatrix} \boldsymbol{\psi}'_{qdr} + \frac{d\boldsymbol{\psi}'_{qdr}}{dt} \end{aligned} \tag{6.19}$$

Matrix differential equations (6.19) in expanded form are given by

$$u_{qs} = r_s i_{qs} + \omega \psi_{ds} + \frac{d\psi_{qs}}{dt}$$

$$u_{ds} = r_s i_{ds} - \omega \psi_{qs} + \frac{d\psi_{ds}}{dt}$$

$$u'_{qr} = r'_r i'_{qr} + (\omega - \omega_r)\psi'_{dr} + \frac{d\psi'_{qr}}{dt} \tag{6.20}$$

$$u'_{dr} = r'_r i'_{dr} - (\omega - \omega_r)\psi'_{qr} + \frac{d\psi'_{dr}}{dt}$$

The *quadrature*- and *direct*-axis components of stator and rotor flux linkages can be found by applying the Park transformation matrices $\mathbf{K}_s$ and $\mathbf{K}_r$. The flux linkages equations were obtained for two-phase induction machines as

$$\begin{bmatrix} \psi_{as} \\ \psi_{bs} \\ \psi'_{ar} \\ \psi'_{br} \end{bmatrix} = \begin{bmatrix} L_{ss} & 0 & L_{ms}\cos\theta_r & -L_{ms}\sin\theta_r \\ 0 & L_{ss} & L_{ms}\sin\theta_r & L_{ms}\cos\theta_r \\ L_{ms}\cos\theta_r & L_{ms}\sin\theta_r & L'_{rr} & 0 \\ -L_{ms}\sin\theta_r & L_{ms}\cos\theta_r & 0 & L'_{rr} \end{bmatrix} \begin{bmatrix} i_{as} \\ i_{bs} \\ i'_{ar} \\ i'_{br} \end{bmatrix}$$

Using Eq. (6.2)

$$\begin{bmatrix} \boldsymbol{\psi}_{abs} \\ \boldsymbol{\psi}'_{abr} \end{bmatrix} = \begin{bmatrix} \mathbf{L}_s & \mathbf{L}'_{sr} \\ \mathbf{L}'^{T}_{sr} & \mathbf{L}'_r \end{bmatrix} \begin{bmatrix} \mathbf{i}_{abs} \\ \mathbf{i}'_{abr} \end{bmatrix}$$

the stator and rotor flux linkages are expressed as

$$\boldsymbol{\psi}_{abs} = \mathbf{L}_s \mathbf{i}_{abs} + \mathbf{L}'_{sr}\mathbf{i}'_{abr}$$

$$\boldsymbol{\psi}'_{abr} = \mathbf{L}'^{T}_{sr}\mathbf{i}_{abs} + \mathbf{L}'_r \mathbf{i}'_{abr}$$

Multiplying the stator and rotor currents and flux linkages by $\mathbf{K}_s^{-1}$ and $\mathbf{K}_r^{-1}$, respectively, we obtain

$$\mathbf{K}_s^{-1}\boldsymbol{\psi}_{qds} = \mathbf{L}_s \mathbf{K}_s^{-1}\mathbf{i}_{qds} + \mathbf{L}'_{sr}\mathbf{K}_r^{-1}\mathbf{i}'_{qdr}$$

$$\mathbf{K}_r^{-1}\boldsymbol{\psi}'_{qdr} = \mathbf{L}'^{T}_{sr}\mathbf{K}_s^{-1}\mathbf{i}_{qds} + \mathbf{L}'_r \mathbf{K}_r^{-1}\mathbf{i}'_{qdr}$$

Both sides of these matrix equations should be multiplied by $\mathbf{K}_s$ and $\mathbf{K}_r$. One has

$$\boldsymbol{\psi}_{qds} = \mathbf{K}_s \mathbf{L}_s \mathbf{K}_s^{-1}\mathbf{i}_{qds} + \mathbf{K}_s \mathbf{L}'_{sr}\mathbf{K}_r^{-1}\mathbf{i}'_{qdr}$$

$$\boldsymbol{\psi}'_{qdr} = \mathbf{K}_r \mathbf{L}'^{T}_{sr}\mathbf{K}_s^{-1} i_{qds} + \mathbf{K}_r \mathbf{L}'_r \mathbf{K}_r^{-1}\mathbf{i}'_{qdr}$$

The matrices $\mathbf{L}_s$ and $\mathbf{L}'_r$ are diagonal. Hence,

$$\mathbf{K}_s\mathbf{L}_s\mathbf{K}_s^{-1} = \mathbf{L}_s = \begin{bmatrix} L_{ss} & 0 \\ 0 & L_{ss} \end{bmatrix}$$

and

$$\mathbf{K}_r\mathbf{L}'_r\mathbf{K}_r^{-1} = \mathbf{L}'_r = \begin{bmatrix} L'_{rr} & 0 \\ 0 & L'_{rr} \end{bmatrix}$$

Terms $\mathbf{K}_s\mathbf{L}'_{sr}\mathbf{K}_r^{-1}$ and $\mathbf{K}_r\mathbf{L}'^{T}_{sr}\mathbf{K}_s^{-1}$ are found to be

$$\mathbf{K}_s\mathbf{L}'_{sr}\mathbf{K}_r^{-1} = L_{ms}\begin{bmatrix} \cos\theta & \sin\theta \\ \sin\theta & -\cos\theta \end{bmatrix}\begin{bmatrix} \cos\theta_r & -\sin\theta_r \\ \sin\theta_r & \cos\theta_r \end{bmatrix}\begin{bmatrix} \cos(\theta-\theta_r) & \sin(\theta-\theta_r) \\ \sin(\theta-\theta_r) & -\cos(\theta-\theta_r) \end{bmatrix}$$

$$= L_{ms}\begin{bmatrix} \cos\theta\cos\theta_r + \sin\theta\sin\theta_r & \sin\theta\cos\theta_r - \cos\theta\sin\theta_r \\ \sin\theta\cos\theta_r - \cos\theta\sin\theta_r & -\sin\theta\sin\theta_r - \cos\theta\cos\theta_r \end{bmatrix}\begin{bmatrix} \cos(\theta-\theta_r) & \sin(\theta-\theta_r) \\ \sin(\theta-\theta_r) & -\cos(\theta-\theta_r) \end{bmatrix}$$

$$= L_{ms}\begin{bmatrix} 1 & 0 \\ 0 & 1 \end{bmatrix} = \begin{bmatrix} L_{ms} & 0 \\ 0 & L_{ms} \end{bmatrix}$$

and

$$\mathbf{K}_r\mathbf{L}'^{T}_{sr}\mathbf{K}_s^{-1} = L_{ms}\begin{bmatrix} \cos(\theta-\theta_r) & \sin(\theta-\theta_r) \\ \sin(\theta-\theta_r) & -\cos(\theta-\theta_r) \end{bmatrix}\begin{bmatrix} \cos\theta_r & \sin\theta_r \\ -\sin\theta_r & \cos\theta_r \end{bmatrix}\begin{bmatrix} \cos\theta & \sin\theta \\ \sin\theta & -\cos\theta \end{bmatrix}$$

$$= L_{ms}\begin{bmatrix} \cos(\theta-\theta_r) & \sin(\theta-\theta_r) \\ \sin(\theta-\theta_r) & -\cos(\theta-\theta_r) \end{bmatrix}\begin{bmatrix} \cos\theta\cos\theta_r + \sin\theta\sin\theta_r & \sin\theta\cos\theta_r - \cos\theta\sin\theta_r \\ \sin\theta\cos\theta_r - \cos\theta\sin\theta_r & -\sin\theta\sin\theta_r - \cos\theta\cos\theta_r \end{bmatrix}$$

$$= L_{ms}\begin{bmatrix} 1 & 0 \\ 0 & 1 \end{bmatrix} = \begin{bmatrix} L_{ms} & 0 \\ 0 & L_{ms} \end{bmatrix}$$

Hence, the *quadrature*- and *direct*-axis components of stator and rotor flux linkages are

$$\begin{aligned} \boldsymbol{\psi}_{qds} &= \mathbf{L}_s\mathbf{i}_{qds} + \mathbf{L}_{ms}\mathbf{i}'_{qdr} \\ \boldsymbol{\psi}'_{qdr} &= \mathbf{L}_{ms}\mathbf{i}_{qds} + \mathbf{L}'_r\mathbf{i}'_{qdr} \end{aligned} \tag{6.21}$$

where $\mathbf{L}_{ms} = \begin{bmatrix} L_{ms} & 0 \\ 0 & L_{ms} \end{bmatrix}$.

Matrix equations (6.21) are written in expanded form as

$$\begin{aligned}\psi_{qs} &= L_{ss}i_{qs} + L_{ms}i'_{qr}\\ \psi_{ds} &= L_{ss}i_{ds} + L_{ms}i'_{dr}\\ \psi'_{qr} &= L_{ms}i_{qs} + L'_{rr}i'_{qr}\\ \psi'_{dr} &= L_{ms}i_{ds} + L'_{rr}i'_{dr}\end{aligned} \tag{6.22}$$

The goal is to develop the mathematical model for two-phase induction motors in the *arbitrary* reference frame. Using the differential equations (6.20) and incorporating the expressions for the flux linkages (6.22), we find the mathematical model for two-phase induction motors using the *quadrature*- and *direct*-axis components of stator and rotor voltages and currents. From (6.20) and (6.22), one has

$$u_{qs} = r_s i_{qs} + (L_{ss}i_{ds} + L_{ms}i'_{dr})\omega + \frac{d(L_{ss}i_{qs} + L_{ms}i'_{qr})}{dt}$$

$$u_{ds} = r_s i_{ds} - (L_{ss}i_{qs} + L_{ms}i'_{qr})\omega + \frac{d(L_{ss}i_{ds} + L_{ms}i'_{dr})}{dt}$$

$$u'_{qr} = r'_r i'_{qr} + (L_{ms}i_{ds} + L'_{rr}i'_{dr})(\omega - \omega_r) + \frac{d(L_{ms}i_{qs} + L'_{rr}i'_{qr})}{dt}$$

$$u'_{dr} = r'_r i'_{dr} - (L_{ms}i_{qs} + L'_{rr}i'_{qr})(\omega - \omega_r) + \frac{d(L_{ms}i_{ds} + L'_{rr}i'_{dr})}{dt}$$

Using these equations, Cauchy's form of the differential equations is found. In particular, the stator and rotor circuitry dynamics are given by

$$\frac{di_{qs}}{dt} = \frac{1}{L_\Sigma}[-L'_{rr}r_s i_{qs} + L_{ms}r'_r i'_{qr} - L'_{rr}(L_{ss}i_{ds} + L_{ms}i'_{dr})\omega + L_{ms}(L_{ms}i_{ds} + L'_{rr}i'_{dr})(\omega - \omega_r) + L'_{rr}u_{qs} - L_{ms}u'_{qr}]$$

$$\frac{di_{ds}}{dt} = \frac{1}{L_\Sigma}[-L'_{rr}r_s i_{ds} + L_{ms}r'_r i'_{dr} + L'_{rr}(L_{ss}i_{qs} + L_{ms}i'_{qr})\omega - L_{ms}(L_{ms}i_{qs} + L'_{rr}i'_{qr})(\omega - \omega_r) + L'_{rr}u_{ds} - L_{ms}u'_{dr}]$$

$$\frac{di'_{qr}}{dt} = \frac{1}{L_\Sigma}[L_{ms}r_s i_{qs} - L_{ss}r'_r i'_{qr} + L_{ms}(L_{ss}i_{ds} + L_{ms}i'_{dr})\omega - L_{ss}(L_{ms}i_{ds} + L'_{rr}i'_{dr})(\omega - \omega_r) - L_{ms}u_{qs} + L_{ss}u'_{qr}]$$

$$\frac{di'_{dr}}{dt} = \frac{1}{L_\Sigma}[L_{ms}r_s i_{ds} - L_{ss}r'_r i'_{dr} - L_{ms}(L_{ss}i_{qs} + L_{ms}i'_{qr})\omega + L_{ss}(L_{ms}i_{qs} + L'_{rr}i'_{qr})(\omega - \omega_r) - L_{ms}u_{ds} + L_{ss}u'_{dr}]$$

(6.23)

Newton's second law of motion must be applied to derive the dynamics for mechanical system, and the *torsional-mechanical* equations are

$$T_e - B_m\omega_m - T_L = J\frac{d\omega_{rm}}{dt}$$

$$\frac{d\theta_{rm}}{dt} = \omega_{rm}$$

Using the electrical angular velocity and displacement, we have

$$\frac{d\omega_r}{dt} = \frac{P}{2J}T_e - \frac{B_m}{J}\omega_r - \frac{P}{2J}T_L$$
$$\frac{d\theta_r}{dt} = \omega_r \tag{6.24}$$

The expression for electromagnetic torque needs to be derived using the *quadrature-* and *direct*-axis components of stator and rotor currents. Using the matrix equation for coenergy

$$W_c = \frac{1}{2}\mathbf{i}_{abs}^T(\mathbf{L}_s - L_{ls}\mathbf{I})\mathbf{i}_{abs} + \mathbf{i}_{abs}^T\mathbf{L}'_{sr}\mathbf{i}'_{abr} + \frac{1}{2}\mathbf{i}'^T_{abr}(\mathbf{L}'_r - L'_{lr}\mathbf{I})\mathbf{i}'_{abr}$$

the electromagnetic torque is found as

$$T_e = \frac{P}{2}\frac{\partial W_c(\mathbf{i}_{abs}, \mathbf{i}'_{abr}, \theta_r)}{\partial\theta_r} = \frac{P}{2}\mathbf{i}_{abs}^T\frac{\partial \mathbf{L}'_{sr}(\theta_r)}{\partial\theta_r}\mathbf{i}'_{abr}$$

Applying the Park transformation matrices $\mathbf{K}_s^{-1}$ and $\mathbf{K}_r^{-1}$, we obtain

$$T_e = \frac{P}{2}(\mathbf{K}_s^{-1}\mathbf{i}_{qds})^T\frac{\partial \mathbf{L}'_{sr}(\theta_r)}{\partial\theta_r}\mathbf{K}_r^{-1}\mathbf{i}'_{qdr}$$

$$= \frac{P}{2}L_{ms}[i_{qs} \quad i_{ds}]\begin{bmatrix}\cos\theta & \sin\theta \\ \sin\theta & -\cos\theta\end{bmatrix}\begin{bmatrix}-\sin\theta_r & -\cos\theta_r \\ \cos\theta_r & -\sin\theta_r\end{bmatrix}\begin{bmatrix}\cos(\theta-\theta_r) & \sin(\theta-\theta_r) \\ \sin(\theta-\theta_r) & -\cos(\theta-\theta_r)\end{bmatrix}\begin{bmatrix}i'_{qr} \\ i'_{dr}\end{bmatrix}$$

$$= \frac{P}{2}L_{ms}\begin{bmatrix}i_{qs}\cos\theta + i_{ds}\sin\theta \\ i_{qs}\sin\theta - i_{ds}\cos\theta\end{bmatrix}^T\begin{bmatrix}-\sin\theta_r & -\cos\theta_r \\ \cos\theta_r & -\sin\theta_r\end{bmatrix}\begin{bmatrix}i'_{qr}\cos(\theta-\theta_r) + i'_{dr}\sin(\theta-\theta_r) \\ i'_{qr}\sin(\theta-\theta_r) - i'_{dr}\cos(\theta-\theta_r)\end{bmatrix}$$

$$= \frac{P}{2}L_{ms}\begin{bmatrix}-\sin\theta_r(i_{qs}\cos\theta + i_{ds}\sin\theta) + \cos\theta_r(i_{qs}\sin\theta - i_{ds}\cos\theta) \\ -\cos\theta_r(i_{qs}\cos\theta + i_{ds}\sin\theta) - \sin\theta_r(i_{qs}\sin\theta - i_{ds}\cos\theta)\end{bmatrix}^T\begin{bmatrix}i'_{qr}\cos(\theta-\theta_r) + i'_{dr}\sin(\theta-\theta_r) \\ i'_{qr}\sin(\theta-\theta_r) - i'_{dr}\cos(\theta-\theta_r)\end{bmatrix}$$

$$= \frac{P}{2}L_{ms}\begin{bmatrix}i_{qs}(\sin\theta\cos\theta_r - \cos\theta\sin\theta_r) - i_{ds}(\cos\theta\cos\theta_r + \sin\theta\sin\theta_r) \\ -i_{qs}(\cos\theta\cos\theta_r + \sin\theta\sin\theta_r) - i_{ds}(\sin\theta\cos\theta_r - \cos\theta\sin\theta_r)\end{bmatrix}^T\begin{bmatrix}i'_{qr}\cos(\theta-\theta_r) + i'_{dr}\sin(\theta-\theta_r) \\ i'_{qr}\sin(\theta-\theta_r) - i'_{dr}\cos(\theta-\theta_r)\end{bmatrix}$$

$$= \frac{P}{2}L_{ms}[i_{qs}\sin(\theta-\theta_r) - i_{ds}\cos(\theta-\theta_r) \quad -i_{qs}\cos(\theta-\theta_r) - i_{ds}\sin(\theta-\theta_r)]\begin{bmatrix}i'_{qr}\cos(\theta-\theta_r) + i'_{dr}\sin(\theta-\theta_r) \\ i'_{qr}\sin(\theta-\theta_r) - i'_{dr}\cos(\theta-\theta_r)\end{bmatrix}$$

$$= \frac{P}{2}L_{ms}[i_{qs}i'_{qr}\sin(\theta-\theta_r)\cos(\theta-\theta_r) + i_{qs}i'_{dr}\sin^2(\theta-\theta_r) - i_{ds}i'_{qr}\cos^2(\theta-\theta_r) - i_{ds}i'_{dr}\cos(\theta-\theta_r)\sin(\theta-\theta_r)$$
$$- i_{qs}i'_{qr}\cos(\theta-\theta_r)\sin(\theta-\theta_r) + i_{qs}i'_{dr}\cos^2(\theta-\theta_r) - i_{ds}i'_{qr}\sin^2(\theta-\theta_r) + i_{ds}i'_{dr}\sin(\theta-\theta_r)\cos(\theta-\theta_r)]$$

$$= \frac{P}{2}L_{ms}\{i_{qs}i'_{dr}[\sin^2(\theta-\theta_r) + \cos^2(\theta-\theta_r)] - i_{ds}i'_{qr}[\sin^2(\theta-\theta_r) + \cos^2(\theta-\theta_r)]\}$$

Hence, the electromagnetic torque is expressed as

$$T_e = \frac{P}{2} L_{ms}(i_{qs}i'_{dr} - i_{ds}i'_{qr}) \tag{6.25}$$

Using equations (6.24) and (6.25), one obtains

$$\begin{aligned}
\frac{d\omega_r}{dt} &= \frac{P^2}{4J} L_{ms}(i_{qs}i'_{dr} - i_{ds}i'_{qr}) - \frac{B_m}{J}\omega_r - \frac{P}{2J}T_L \\
\frac{d\theta_r}{dt} &= \omega_r
\end{aligned} \tag{6.26}$$

Augmenting differential equations (6.23) and (6.26), the mathematical model is found for two-phase induction motors using the *quadrature*- and *direct*-axis components of stator and rotor voltages and currents. In particular, we have

$$\begin{aligned}
\frac{di_{qs}}{dt} &= \frac{1}{L_\Sigma}[-L'_{rr}r_s i_{qs} + L_{ms}r'_r i'_{qr} - L'_{rr}(L_{ss}i_{ds} + L_{ms}i'_{dr})\omega + L_{ms}(L_{ms}i_{ds} + L'_{rr}i'_{dr})(\omega - \omega_r) + L'_{rr}u_{qs} - L_{ms}u'_{qr}] \\
\frac{di_{ds}}{dt} &= \frac{1}{L_\Sigma}[L'_{rr}r_s i_{ds} + L_{ms}r'_r i'_{dr} + L'_{rr}(L_{ss}i_{qs} + L_{ms}i'_{qr})\omega - L_{ms}(L_{ms}i_{qs} + L'_{rr}i'_{qr})(\omega - \omega_r) + L'_{rr}u_{ds} - L_{ms}u'_{dr}] \\
\frac{di'_{qr}}{dt} &= \frac{1}{L_\Sigma}[L_{ms}r_s i_{qs} - L_{ss}r'_r i'_{qr} + L_{ms}(L_{ss}i_{ds} + L_{ms}i'_{dr})\omega - L_{ss}(L_{ms}i_{ds} + L'_{rr}i'_{dr})(\omega - \omega_r) - L_{ms}u_{qs} + L_{ss}u'_{qr}] \\
\frac{di'_{dr}}{dt} &= \frac{1}{L_\Sigma}[L_{ms}r_s i_{ds} - L_{ss}r'_r i'_{dr} - L_{ms}(L_{ss}i_{qs} + L_{ms}i'_{qr})\omega + L_{ss}(L_{ms}i_{qs} + L'_{rr}i'_{qr})(\omega - \omega_r) - L_{ms}u_{ds} + L_{ss}u'_{dr}] \\
\frac{d\omega_r}{dt} &= \frac{P^2}{4J} L_{ms}(i_{qs}i'_{dr} - i_{ds}i'_{qr}) - \frac{B_m}{J}\omega_r - \frac{P}{2J}T_L \\
\frac{d\theta_r}{dt} &= \omega_r
\end{aligned} \tag{6.27}$$

A set of nonlinear differential equations (6.27) is given in matrix form as

$$\begin{bmatrix} \frac{di_{qs}}{dt} \\ \frac{di_{ds}}{dt} \\ \frac{di'_{qr}}{dt} \\ \frac{di'_{dr}}{dt} \\ \frac{d\omega_r}{dt} \\ \frac{d\theta_r}{dt} \end{bmatrix} = \begin{bmatrix} -\frac{L'_{rr}r_s}{L_\Sigma} & -\omega & \frac{L_{ms}r'_r}{L_\Sigma} & 0 & 0 & 0 \\ \omega & -\frac{L'_{rr}r_s}{L_\Sigma} & 0 & \frac{L_{ms}r'_r}{L_\Sigma} & 0 & 0 \\ \frac{L_{ms}r_s}{L_\Sigma} & 0 & -\frac{L_{ss}r'_r}{L_\Sigma} & -\omega & 0 & 0 \\ 0 & \frac{L_{ms}r_s}{L_\Sigma} & \omega & -\frac{L_{ss}r'_r}{L_\Sigma} & 0 & 0 \\ 0 & 0 & 0 & 0 & -\frac{B_m}{J} & 0 \\ 0 & 0 & 0 & 0 & 1 & 0 \end{bmatrix} \begin{bmatrix} i_{qs} \\ i_{ds} \\ i'_{qr} \\ i'_{dr} \\ \omega_r \\ \theta_r \end{bmatrix} + \begin{bmatrix} -\frac{L_{ms}}{L_\Sigma}(L_{ms}i_{ds} + L'_{rr}i'_{dr})\omega_r \\ \frac{L_{ms}}{L_\Sigma}(L_{ms}i_{qs} + L'_{rr}i'_{qr})\omega_r \\ \frac{L_{ss}}{L_\Sigma}(L_{ms}i_{ds} + L'_{rr}i'_{dr})\omega_r \\ -\frac{L_{ss}}{L_\Sigma}(L_{ms}i_{qs} + L'_{rr}i'_{qr})\omega_r \\ \frac{P^2 L_{ms}}{4J}(i_{qs}i'_{dr} - i_{ds}i'_{qr}) \\ 0 \end{bmatrix}$$

$$
+\begin{bmatrix} \frac{L'_{rr}}{L_\Sigma} & 0 & -\frac{L_{ms}}{L_\Sigma} & 0 \\ 0 & \frac{L'_{rr}}{L_\Sigma} & 0 & -\frac{L_{ms}}{L_\Sigma} \\ -\frac{L_{ms}}{L_\Sigma} & 0 & \frac{L_{ss}}{L_\Sigma} & 0 \\ 0 & -\frac{L_{ms}}{L_\Sigma} & 0 & \frac{L_{ss}}{L_\Sigma} \\ 0 & 0 & 0 & 0 \\ 0 & 0 & 0 & 0 \end{bmatrix} \begin{bmatrix} u_{qs} \\ u_{ds} \\ u'_{qr} \\ u'_{dr} \end{bmatrix} - \begin{bmatrix} 0 \\ 0 \\ 0 \\ 0 \\ \frac{P}{2J} \\ 0 \end{bmatrix} T_L
$$

To develop an *s*-domain block diagram for two-phase squirrel-cage induction motors, one can use the *quadrature*- and *direct*-axis components of stator and rotor currents, as well as the electrical angular velocity, as the state variables. That is, the system states are i_{as}, i_{bs}, i'_{qr}, i'_{dr}, and ω_r. The block diagram is built using differential equations (6.27). In particular, we have

$$
i_{qs} = \frac{1}{L_\Sigma s + L'_{rr} r_s} [L_{ms} r'_r i'_{qr} + (L^2_{ms} - L_{ss} L'_{rr}) \omega i_{ds} - L_{ms}(L_{ms} i_{ds} + L'_{rr} i'_{dr}) \omega_r + L'_{rr} u_{qs} - L_{ms} u'_{qr}]
$$

$$
i_{ds} = \frac{1}{L_\Sigma s + L'_{rr} r_s} [-(L^2_{ms} - L_{ss} L'_{rr}) \omega i_{qs} + L_{ms} r'_r i'_{dr} + L_{ms}(L_{ms} i_{qs} + L'_{rr} i'_{qr}) \omega_r + L'_{rr} u_{ds} - L_{ms} u'_{dr}]
$$

$$
i'_{qr} = \frac{1}{L_\Sigma s + L_{ss} r'_r} [L_{ms} r_s i_{qs} + (L^2_{ms} - L_{ss} L'_{rr}) \omega i'_{dr} + L_{ss}(L_{ms} i_{ds} + L'_{rr} i'_{dr}) \omega_r - L_{ms} u_{qs} + L_{ss} u'_{qr}]
$$

$$
i_{dr} = \frac{1}{L_\Sigma s + L_{ss} r'_r} [L_{ms} r_s i_{ds} - (L^2_{ms} - L_{ss} L'_{rr}) \omega i'_{qr} - L_{ss}(L_{ms} i_{qs} + L'_{rr} i'_{qr}) \omega_r - L_{ms} u_{ds} + L_{ss} u'_{dr}]
$$

$$
\omega_r = \frac{P}{2}\left(\frac{1}{Js + B_m}\right)\left[\frac{P}{2J} L_{ms}(i_{qs} i'_{dr} - i_{ds} i'_{qr}) - T_L\right]
$$

For squirrel-cage induction motors, the voltages are supplied only to the stator windings, and the resulting *s*-domain diagram is shown in Figure 6.30.

Analysis of induction machines using flux linkages

The mathematical model of two-phase induction machines in the *arbitrary* reference frame can be developed using the *quadrature*- and *direct*-axis components of stator currents and rotor flux linkages, as well as the electrical angular velocity and displacement. That is, the system states can be chosen as i_{as}, i_{bs}, ψ'_{qr}, ψ'_{dr}, ω_r, and θ_r. We use the voltage equations (6.20)

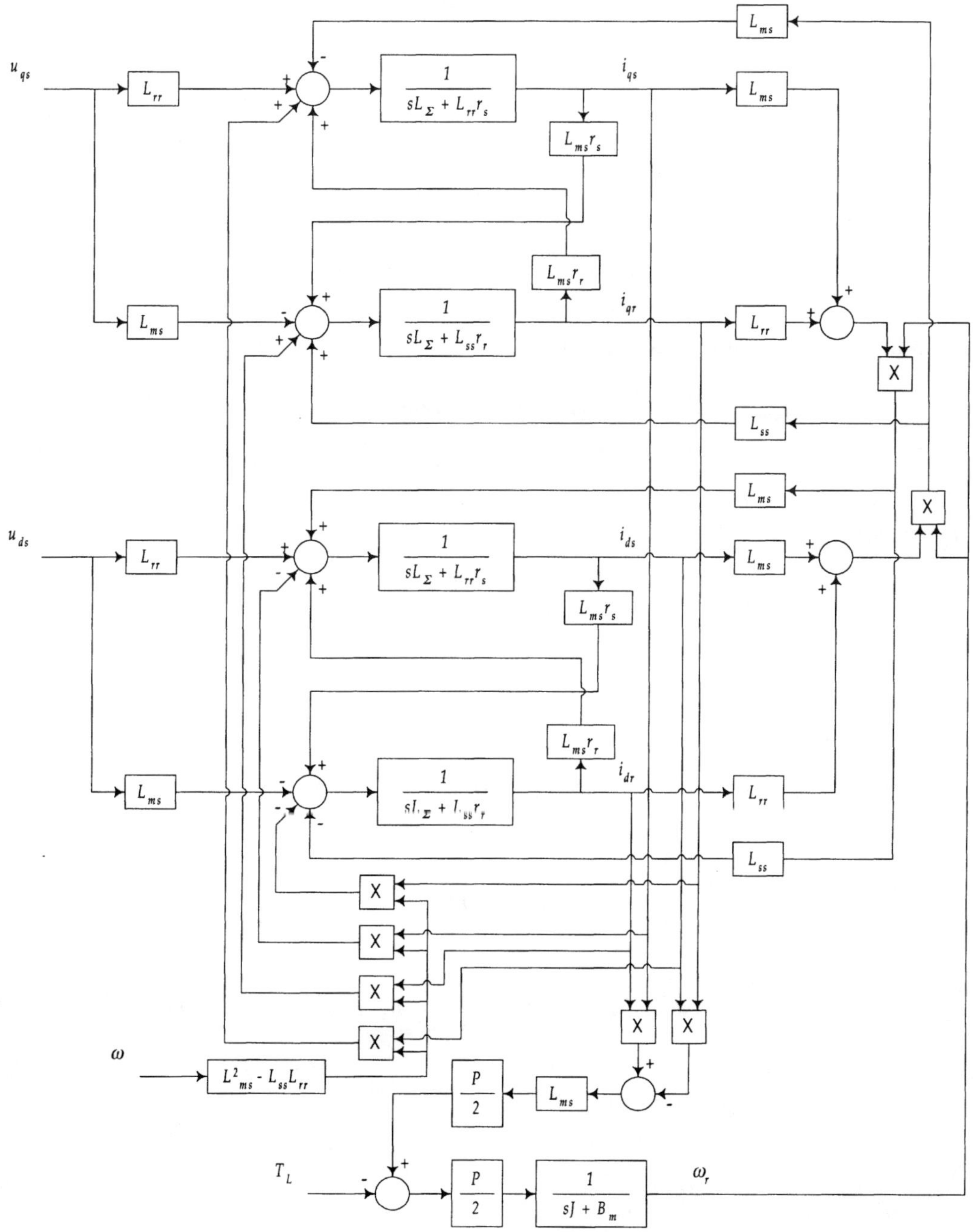

Figure 6.30 Block diagram of two-phase squirrel-cage induction motors in the *arbitrary* reference frame.

$$u_{qs} = r_s i_{qs} + \omega\psi_{ds} + \frac{d\psi_{qs}}{dt}$$

$$u_{ds} = r_s i_{ds} - \omega\psi_{qs} + \frac{d\psi_{ds}}{dt}$$

$$u'_{qr} = r'_r i'_{qr} + (\omega - \omega_r)\psi'_{dr} + \frac{d\psi'_{qr}}{dt}$$

$$u'_{dr} = r'_r i'_{dr} - (\omega - \omega_r)\psi'_{qr} + \frac{d\psi'_{dr}}{dt}$$

Taking note of relationships (6.22)

$$\psi'_{qr} = L_{ms} i_{qs} + L'_{rr} i'_{qr}$$

$$\psi'_{dr} = L_{ms} i_{ds} + L'_{rr} i'_{dr}$$

one expresses i'_{qr} and i'_{dr} as

$$i'_{qr} = \frac{1}{L'_{rr}}(\psi'_{qr} - L_{ms} i_{qs})$$

$$i'_{dr} = \frac{1}{L'_{rr}}(\psi'_{dr} - L_{ms} i_{ds})$$

Using the expressions derived for i'_{qr} and i'_{dr}, as well as equations (6.20) and (6.23), we have

$$\frac{di_{qs}}{dt} = \frac{1}{L_\Sigma}\left[-L'_{rr} r_s i_{qs} + \frac{L_{ms} r'_r}{L'_{rr}}(\psi'_{qr} - L_{ms} i_{qs}) - (L_\Sigma \omega + L_{ms}^2 \omega_r) i_{ds} - L_{ms}(\psi'_{dr} - L_{ms} i_{ds})\omega_r + L'_{rr} u_{qs} - L_{ms} u'_{qr}\right]$$

$$\frac{di_{ds}}{dt} = \frac{1}{L_\Sigma}\left[-L'_{rr} r_s i_{ds} + \frac{L_{ms} r'_r}{L'_{rr}}(\psi'_{dr} - L_{ms} i_{ds}) + (L_\Sigma \omega + L_{ms}^2 \omega_r) i_{qs} + L_{ms}(\psi'_{qr} - L_{ms} i_{qs})\omega_r + L'_{rr} u_{ds} - L_{ms} u'_{dr}\right]$$

$$\frac{d\psi'_{qr}}{dt} = \frac{L_{ms} r'_r}{L'_{rr}} i_{qs} - \frac{r'_r}{L'_{rr}}\psi'_{qr} - (\omega - \omega_r)\psi'_{dr} + u'_{qr}$$

$$\frac{d\psi'_{dr}}{dt} = \frac{L_{ms} r'_r}{L'_{rr}} i_{ds} + (\omega - \omega_r)\psi'_{qr} - \frac{r'_r}{L'_{rr}}\psi'_{dr} + u'_{dr}$$

Simplifying these differential equations, one obtains

$$\frac{di_{qs}}{dt}=\frac{1}{L_\Sigma}\left(-\frac{L_{rr}'^2 r_s+L_{ms}^2 r_r'}{L_{rr}'}i_{qs}-L_\Sigma\omega i_{ds}+\frac{L_{ms}r_r'}{L_{rr}'}\psi_{qr}'-L_{ms}\psi_{dr}'\omega_r+L_{rr}'u_{qs}-L_{ms}u_{qr}'\right)$$

$$\frac{di_{ds}}{dt}=\frac{1}{L_\Sigma}\left(-\frac{L_{rr}'^2 r_s+L_{ms}^2 r_r'}{L_{rr}'}i_{ds}+\frac{L_{ms}r_r'}{L_{rr}'}\psi_{dr}'+L_\Sigma\omega i_{qs}+L_{ms}\psi_{qr}'\omega_r+L_{rr}'u_{ds}-L_{ms}u_{dr}'\right)$$

$$\frac{d\psi_{qr}'}{dt}=\frac{L_{ms}r_r'}{L_{rr}'}i_{qs}-\frac{r_r'}{L_{rr}'}\psi_{qr}'-(\omega-\omega_r)\psi_{dr}'+u_{qr}' \tag{6.28}$$

$$\frac{d\psi_{dr}'}{dt}=\frac{L_{ms}r_r'}{L_{rr}'}i_{ds}+(\omega-\omega_r)\psi_{qr}'-\frac{r_r'}{L_{rr}'}\psi_{dr}'+u_{dr}'$$

From

$$i_{qr}'=\frac{1}{L_{rr}'}(\psi_{qr}'-L_{ms}i_{qs})$$

and

$$i_{dr}'=\frac{1}{L_{rr}'}(\psi_{dr}'-L_{ms}i_{ds})$$

the equation for electromagnetic torque (6.25) $T_e=\frac{P}{2}L_{ms}(i_{qs}i_{dr}'-i_{ds}i_{qr}')$ is found to be

$$T_e=\frac{P}{2}\frac{L_{ms}}{L_{rr}'}[i_{qs}(\psi_{dr}'-L_{ms}i_{ds})-i_{ds}(\psi_{qr}'-L_{ms}i_{qs})]=\frac{P}{2}\frac{L_{ms}}{L_{rr}'}(i_{qs}\psi_{dr}'-i_{ds}\psi_{qr}') \tag{6.29}$$

Using Newton's equations of motion (6.24) and (6.29), one finds the *torsional-mechanical* dynamics, and the following two differential equations result

$$\frac{d\omega_r}{dt}=\frac{P^2}{4J}\frac{L_{ms}}{L_{rr}'}(i_{qs}\psi_{dr}'-i_{ds}\psi_{qr}')-\frac{B_m}{J}\omega_r-\frac{P}{2J}T_L$$

$$\frac{d\theta_r}{dt}=\omega_r \tag{6.30}$$

Augmenting (6.28) and (6.30), the resulting model in the *arbitrary* reference frame for induction motors is found. In particular,

$$
\begin{aligned}
\frac{di_{qs}}{dt} &= \frac{1}{L_\Sigma}\left(-\frac{L'^2_{rr}r_s + L^2_{ms}r'_r}{L'_{rr}}i_{qs} - L_\Sigma \omega i_{ds} + \frac{L_{ms}r'_r}{L'_{rr}}\psi'_{qr} - L_{ms}\psi'_{dr}\omega_r + L'_{rr}u_{qs} - L_{ms}u'_{qr}\right)\\
\frac{di_{ds}}{dt} &= \frac{1}{L_\Sigma}\left(-\frac{L'^2_{rr}r_s + L^2_{ms}r'_r}{L'_{rr}}i_{ds} + \frac{L_{ms}r'_r}{L'_{rr}}\psi'_{dr} + L_\Sigma \omega i_{qs} + L_{ms}\psi'_{qr}\omega_r + L'_{rr}u_{ds} - L_{ms}u'_{dr}\right)\\
\frac{d\psi'_{qr}}{dt} &= \frac{L_{ms}r'_r}{L'_{rr}}i_{qs} - \frac{r'_r}{L'_{rr}}\psi'_{qr} - (\omega - \omega_r)\psi'_{dr} + u'_{qr}\\
\frac{d\psi'_{dr}}{dt} &= \frac{L_{ms}r'_r}{L'_{rr}}i_{ds} + (\omega - \omega_r)\psi'_{qr} - \frac{r'_r}{L'_{rr}}\psi'_{dr} + u'_{dr}\\
\frac{d\omega_r}{dt} &= \frac{P^2}{4J}\frac{L_{ms}}{L'_{rr}}(i_{qs}\psi'_{dr} - i_{ds}\psi'_{qr}) - \frac{B_m}{J}\omega_r - \frac{P}{2J}T_L\\
\frac{d\theta_r}{dt} &= \omega_r
\end{aligned}
\tag{6.31}
$$

In matrix form, a set of differential equations (6.31) is given as

$$
\begin{bmatrix}\dfrac{di_{qs}}{dt}\\ \dfrac{di_{ds}}{dt}\\ \dfrac{d\psi'_{qr}}{dt}\\ \dfrac{d\psi'_{dr}}{dt}\\ \dfrac{d\omega_r}{dt}\\ \dfrac{d\theta_r}{dt}\end{bmatrix}
=
\begin{bmatrix}
-\dfrac{L'^2_{rr}r_s + L^2_{ms}r'_r}{L_\Sigma L'_{rr}} & -\omega & \dfrac{L_{ms}r'_r}{L_\Sigma L'_{rr}} & 0 & 0 & 0\\
\omega & -\dfrac{L'_{rr}r_s + L^2_{ms}r'_r}{L_\Sigma L'_{rr}} & 0 & \dfrac{L_{ms}r'_r}{L_\Sigma L'_{rr}} & 0 & 0\\
\dfrac{L_{ms}r'_r}{L'_{rr}} & 0 & -\dfrac{r'_r}{L'_{rr}} & -\omega & 0 & 0\\
0 & \dfrac{L_{ms}r'_r}{L'_{rr}} & \omega & -\dfrac{r'_r}{L'_{rr}} & 0 & 0\\
0 & 0 & 0 & 0 & -\dfrac{B_m}{J} & 0\\
0 & 0 & 0 & 0 & 1 & 0
\end{bmatrix}
\begin{bmatrix} i_{qs}\\ i_{ds}\\ \psi'_{qr}\\ \psi'_{dr}\\ \omega_r\\ \theta_r\end{bmatrix}
$$

$$
+\begin{bmatrix}
-\dfrac{L_{ms}}{L_\Sigma}\psi'_{dr}\omega_r\\
\dfrac{L_{ms}}{L_\Sigma}\psi'_{qr}\omega_r\\
\psi'_{dr}\omega_r\\
-\psi'_{qr}\omega_r\\
\dfrac{P^2}{4J}\dfrac{L_{ms}}{L'_{rr}}(i_{qs}\psi'_{dr} - i_{ds}\psi'_{qr})\\
0
\end{bmatrix}
+\begin{bmatrix}
\dfrac{L'_{rr}}{L_\Sigma} & 0 & -\dfrac{L_{ms}}{L_\Sigma} & 0\\
0 & \dfrac{L'_{rr}}{L_\Sigma} & 0 & -\dfrac{L_{ms}}{L_\Sigma}\\
0 & 0 & 1 & 0\\
0 & 0 & 0 & 1\\
0 & 0 & 0 & 0\\
0 & 0 & 0 & 0
\end{bmatrix}
\begin{bmatrix} u_{qs}\\ u_{ds}\\ u'_{qr}\\ u'_{dr}\end{bmatrix}
-\begin{bmatrix}0\\0\\0\\0\\ \dfrac{P}{2J}\\ 0\end{bmatrix}T_L
$$

One will find that it is of great importance to build the block diagram for two-phase induction machines. As the state variables, we use the *quadrature-* and *direct*-axis components of stator currents and rotor flux linkages i_{as}, i_{bs}, ψ'_{qr}, ψ'_{dr}, as well as the electrical angular velocity ω_r. Using the Laplace operator s, from (6.31), we have

$$i_{qs}=\frac{L'_{rr}}{L_\Sigma L'_{rr}s+L'^2_{rr}r_s+L^2_{ms}r'_r}\left(-L_\Sigma\omega i_{ds}+\frac{L_{ms}r'_r}{L'_{rr}}\psi'_{qr}-L_{ms}\psi'_{dr}\omega_r+L'_{rr}u_{qs}-L_{ms}u'_{qr}\right)$$

$$i_{ds}=\frac{L'_{rr}}{L_\Sigma L'_{rr}s+L'^2_{rr}r_s+L^2_{ms}r'_r}\left(\frac{L_{ms}r'_r}{L'_{rr}}\psi'_{dr}+L_\Sigma\omega i_{qs}+L_{ms}\psi'_{qr}\omega_r+L'_{rr}u_{ds}-L_{ms}u'_{dr}\right)$$

$$\psi'_{qr}=\frac{L'_{rr}}{L'_{rr}s+r'_r}\left(\frac{L_{ms}r'_r}{L'_{rr}}i_{qs}-\omega\psi'_{dr}+\psi'_{dr}\omega_r+u'_{qr}\right)$$

$$\psi'_{dr}=\frac{L'_{rr}}{L'_{rr}s+r'_r}\left(\frac{L_{ms}r'_r}{L'_{rr}}i_{ds}+\omega\psi'_{qr}-\psi'_{qr}\omega_r+u'_{dr}\right)$$

$$\omega_r=\frac{P}{2}\left(\frac{1}{Js+B_m}\right)\left[\frac{P}{2}\frac{L_{ms}}{L'_{rr}}(i_{qs}\psi'_{dr}-i_{ds}\psi'_{qr})-T_L\right]$$

For two-phase squirrel-cage motors, the resulting block diagram is documented in Figure 6.31.

6.3.1 *Development of the mathematical model of two-phase induction motors in the stationary reference frame*

The mathematical model of induction motors in the stationary reference frame is found by substituting $\omega = 0$ into the differential equations that map the motor transient dynamics in the *arbitrary* reference frame. That is, differential equations (6.27) are applied. The superscript s denotes the stationary frame of reference, and taking note of (6.27) and letting $\omega = 0$, we have

$$\begin{aligned}
\frac{di^s_{qs}}{dt}&=\frac{1}{L_\Sigma}[-L'_{rr}r_si^s_{qs}+L_{ms}r'_ri'^s_{qr}-L_{ms}(L_{ms}i^s_{ds}+L'_{rr}i'^s_{dr})\omega_r+L'_{rr}u^s_{qs}-L_{ms}u'^s_{qr}]\\
\frac{di^s_{ds}}{dt}&=\frac{1}{L_\Sigma}[-L'_{rr}r_si^s_{ds}+L_{ms}r'_ri'^s_{dr}+L_{ms}(L_{ms}i^s_{qs}+L'_{rr}i'^s_{qr})\omega_r+L'_{rr}u^s_{ds}-L_{ms}u'^s_{dr}]\\
\frac{di'^s_{qr}}{dt}&=\frac{1}{L_\Sigma}[L_{ms}r_si^s_{qs}-L_{ss}r'_ri'^s_{qr}+L_{ss}(L_{ms}i^s_{ds}+L'_{rr}i'^s_{dr})\omega_r-L_{ms}u^s_{qs}+L_{ss}u'^s_{qr}]\\
\frac{di'^s_{dr}}{dt}&=\frac{1}{L_\Sigma}[L_{ms}r_si^s_{ds}-L_{ss}r'_ri'^s_{dr}-L_{ss}(L_{ms}i^s_{qs}+L'_{rr}i'^s_{qr})\omega_r-L_{ms}u^s_{ds}+L_{ss}u'^s_{dr}]\\
\frac{d\omega_r}{dt}&=\frac{P^2L_{ms}}{4J}(i^s_{qs}i'^s_{dr}-i^s_{ds}i'^s_{qr})-\frac{B_m}{J}\omega_r-\frac{P}{2J}T_L\\
\frac{d\theta_r}{dt}&=\omega_r
\end{aligned}\tag{6.32}$$

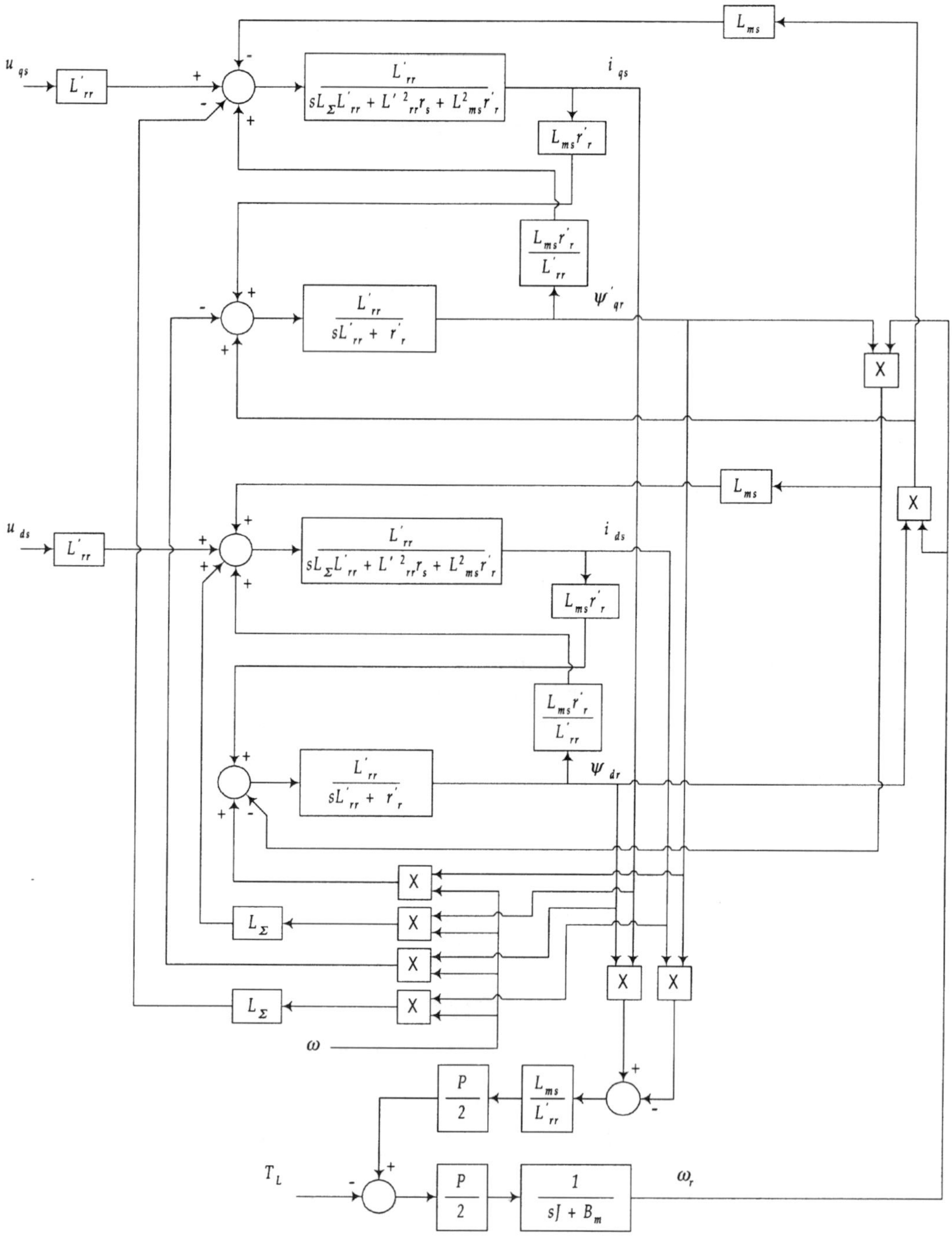

Figure 6.31 Block diagram of two-phase squirrel-cage induction motors in the *arbitrary* reference frame.

From (6.32), one obtains

$$\begin{bmatrix} \frac{di_{qs}^s}{dt} \\ \frac{di_{ds}^s}{dt} \\ \frac{di_{qr}^{\prime s}}{dt} \\ \frac{di_{dr}^{\prime s}}{dt} \\ \frac{d\omega_r}{dt} \\ \frac{d\theta_r}{dt} \end{bmatrix} = \begin{bmatrix} -\frac{L'_{rr}r_s}{L_\Sigma} & 0 & \frac{L_{ms}r'_r}{L_\Sigma} & 0 & 0 & 0 \\ 0 & -\frac{L'_{rr}r_s}{L_\Sigma} & 0 & \frac{L_{ms}r'_r}{L_\Sigma} & 0 & 0 \\ \frac{L_{ms}r_s}{L_\Sigma} & 0 & -\frac{L_{ss}r'_r}{L_\Sigma} & 0 & 0 & 0 \\ 0 & \frac{L_{ms}r_s}{L_\Sigma} & 0 & -\frac{L_{ss}r'_r}{L_\Sigma} & 0 & 0 \\ 0 & 0 & 0 & 0 & -\frac{B_m}{J} & 0 \\ 0 & 0 & 0 & 0 & 1 & 0 \end{bmatrix} \begin{bmatrix} i_{qs}^s \\ i_{ds}^s \\ i_{qr}^{\prime s} \\ i_{dr}^{\prime s} \\ \omega_r \\ \theta_r \end{bmatrix} + \begin{bmatrix} -\frac{L_{ms}}{L_\Sigma}(L_{ms}i_{ds}^s + L'_{rr}i_{dr}^{\prime s})\omega_r \\ \frac{L_{ms}}{L_\Sigma}(L_{ms}i_{qs}^s + L'_{rr}i_{qr}^{\prime s})\omega_r \\ \frac{L_{ss}}{L_\Sigma}(L_{ms}i_{ds}^s + L'_{rr}i_{dr}^{\prime s})\omega_r \\ -\frac{L_{ss}}{L_\Sigma}(L_{ms}i_{qs}^s + L'_{rr}i_{qr}^{\prime s})\omega_r \\ \frac{P^2L_{ms}}{4J}(i_{qs}^s i_{dr}^{\prime s} - i_{ds}^s i_{qr}^{\prime s}) \\ 0 \end{bmatrix}$$

$$+ \begin{bmatrix} \frac{L'_{rr}}{L_\Sigma} & 0 & -\frac{L_{ms}}{L_\Sigma} & 0 \\ 0 & \frac{L'_{rr}}{L_\Sigma} & 0 & -\frac{L_{ms}}{L_\Sigma} \\ -\frac{L_{ms}}{L_\Sigma} & 0 & \frac{L_{ss}}{L_\Sigma} & 0 \\ 0 & -\frac{L_{ms}}{L_\Sigma} & 0 & \frac{L_{ss}}{L_\Sigma} \\ 0 & 0 & 0 & 0 \\ 0 & 0 & 0 & 0 \end{bmatrix} \begin{bmatrix} u_{qs}^s \\ u_{ds}^s \\ u_{qr}^{\prime s} \\ u_{dr}^{\prime s} \end{bmatrix} - \begin{bmatrix} 0 \\ 0 \\ 0 \\ 0 \\ \frac{P}{2J} \\ 0 \end{bmatrix} T_L$$

To guarantee balanced operating conditions, the voltages to be supplied to the stator windings must be found. The *quadrature*- and *direct*-axis components of voltages are obtained using the Park transformation matrix. In particular, in the *arbitrary* reference frame to transform the *machine* stator variables, we have to apply the following transformation matrix

$$\mathbf{K}_s = \begin{bmatrix} \cos\theta & \sin\theta \\ \sin\theta & -\cos\theta \end{bmatrix}$$

and using this $\mathbf{K}_s$, one finds

$$\mathbf{u}_{qds} = \mathbf{K}_s \mathbf{u}_{abs}$$

In the stationary reference frame, $\omega = 0$. Recalling that the angular displacement is found as $\theta = \int_{t_0}^{t} \omega(\tau)d\tau + \theta_0$ and assuming $\theta_0 = 0$, we have $\theta = 0$. Substituting $\theta = 0$ in $\mathbf{K}_s$, one obtains the Park transformation matrix to be used if induction and synchronous machines are studied in the stationary reference frame. In particular, we have

$$\mathbf{K}_s^s = \begin{bmatrix} 1 & 0 \\ 0 & -1 \end{bmatrix}$$

Therefore,

$$\mathbf{u}_{qds}^s = \mathbf{K}_s^s \mathbf{u}_{abs}$$

and

$$\begin{bmatrix} u_{qs}^s \\ u_{ds}^s \end{bmatrix} = \begin{bmatrix} 1 & 0 \\ 0 & -1 \end{bmatrix} \begin{bmatrix} u_{as} \\ u_{bs} \end{bmatrix}$$

Hence, the following explicit expressions for the *quadrature*- and *direct*-axis components of stator voltages result

$$u_{qs}^s(t) = u_{as}(t)$$

$$u_{ds}^s(t) = -u_{bs}(t)$$

Taking note of a balanced two-phase voltage set, as given by

$$u_{as}(t) = \sqrt{2}u_M \cos(\omega_f t)$$

$$u_{bs}(t) = \sqrt{2}u_M \sin(\omega_f t)$$

one obtains the expressions for the *quadrature*- and *direct*-axis components of voltages $u_{qs}^s(t)$ and $u_{ds}^s(t)$. In particular, we have

$$u_{qs}^s(t) = \sqrt{2}u_M \cos(\omega_f t)$$

$$u_{ds}^s(t) = -\sqrt{2}u_M \sin(\omega_f t)$$

The block diagram for two-phase induction motors modeled in the stationary reference frame is developed using (6.32). Taking note of

$$i_{qs}^{s} = \frac{L_{ms}}{L_{\Sigma}s + L'_{rr}r_s}\left[r'_r i'^{s}_{qr} - (L_{ms}i_{ds}^{s} + L'_{rr}i'^{s}_{dr})\omega_r + \frac{L'_{rr}}{L_{ms}}u_{qs}^{s} - u'^{s}_{qr}\right]$$

$$i_{ds}^{s} = \frac{L_{ms}}{L_{\Sigma}s + L'_{rr}r_s}\left[r'_r i'^{s}_{dr} + (L_{ms}i_{qs}^{s} + L'_{rr}i'^{s}_{qr})\omega_r + \frac{L'_{rr}}{L_{ms}}u_{ds}^{s} - u'^{s}_{dr}\right]$$

$$i'^{s}_{qr} = \frac{1}{L_{\Sigma}s + L_{ss}r'_r}[L_{ms}r_s i_{qs}^{s} + L_{ss}(L_{ms}i_{ds}^{s} + L'_{rr}i'^{s}_{dr})\omega_r - L_{ms}u_{qs}^{s} + L_{ss}u'^{s}_{qr}]$$

$$i'^{s}_{dr} = \frac{1}{L_{\Sigma}s + L_{ss}r'_r}[L_{ms}r_s i_{ds}^{s} - L_{ss}(L_{ms}i_{qs}^{s} + L'_{rr}i'^{s}_{qr})\omega_r - L_{ms}u_{ds}^{s} + L_{ss}u'^{s}_{dr}]$$

$$\omega_r = \frac{P}{2}\left(\frac{1}{Js + B_m}\right)\left[\frac{P}{2}L_{ms}(i_{qs}^{s}i'^{s}_{dr} - i_{ds}^{s}i'^{s}_{qr}) - T_L\right]$$

the block diagram for squirrel-cage induction motors is shown in Figure 6.32.

Example 6.3

Perform the modeling of the A- and D-class squirrel-cage induction motors examined in Examples 6.1 and 6.2, in the stationary reference frame. Develop the SIMULINK diagram.

Solution

The SIMULINK diagram is built using the developed block diagram illustrated in Figure 6.32. To guarantee balanced operating conditions, the phase voltages to be supplied to the stator windings are found. In particular, the *quadrature-* and *direct*-axis components of stator voltages $u_{qs}^{s}(t)$ and $u_{ds}^{s}(t)$ are derived using the expressions for a balanced two-phase voltage set. Applying the rated voltages, we have

$$u_{as}(t) = \sqrt{2}115\cos(377t) \text{ and } u_{bs}(t) = \sqrt{2}115\sin(377t)$$

Making use of

$$u_{qs}^{s} = u_{as} \text{ and } u_{ds}^{s} = -u_{bs}$$

one finds

$$u_{qs}^{s}(t) = \sqrt{2}115\cos(377t) \text{ and } u_{ds}^{s}(t) = -\sqrt{2}115\sin(377t)$$

The results of simulations when induction motors are modeled using the stationary reference frame are shown in Figures 6.33 to 6.40. Transient dynamics and torque-speed characteristic curves for the A-class induction motor are illustrated in Figures 6.33 and 6.34. The load torques applied are $T_L = 0$ and $T_L = 0.5$ N-m. Figures 6.35 and 6.36 illustrate the acceleration of the D-class motor.

Under the unbalanced operating condition, supplying

$$u_{qs}^{s}(t) = \sqrt{2}115\cos(377t) \text{ and } u_{ds}^{s}(t) = -\sqrt{2}126\sin(377t)$$

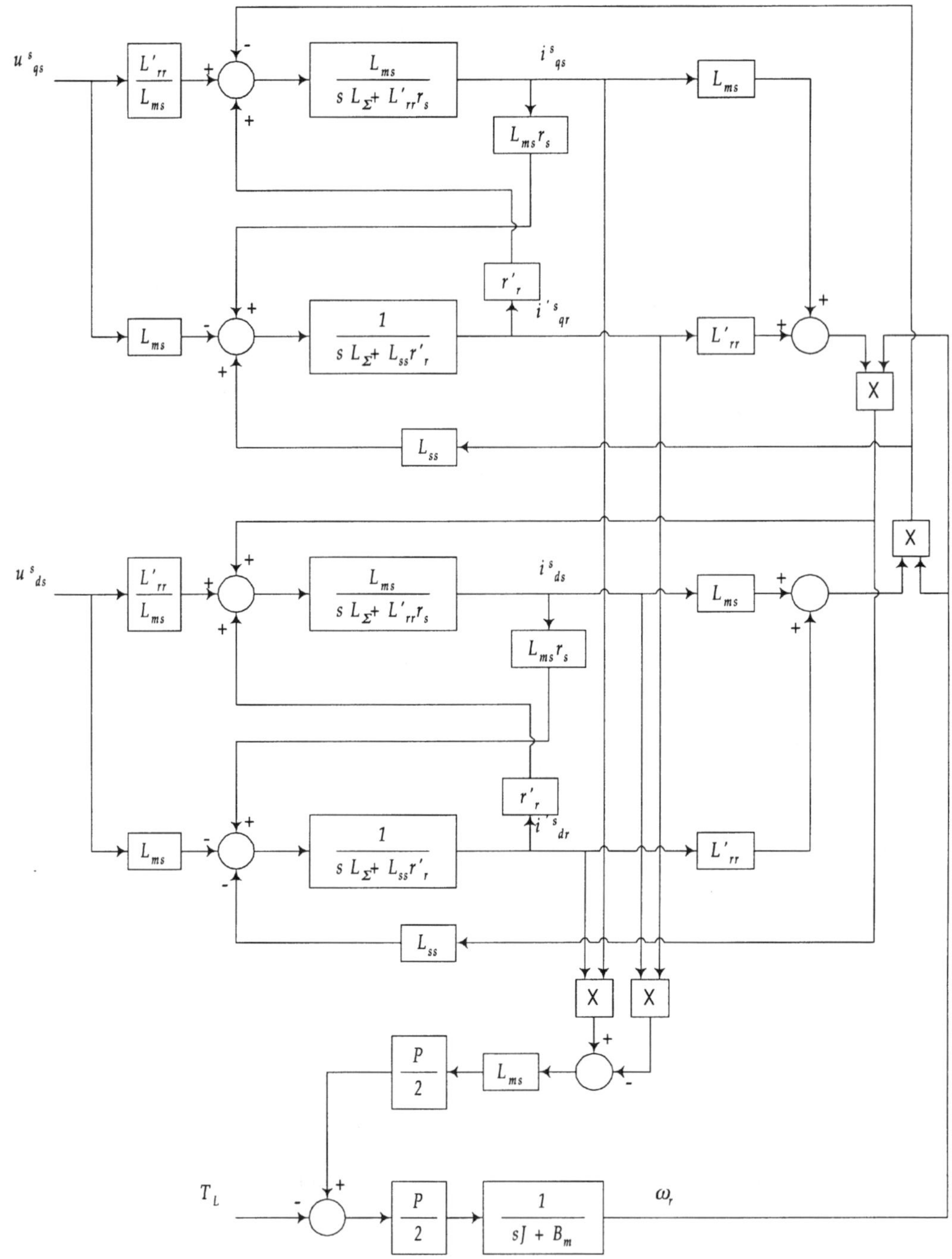

Figure 6.32 Block diagram of squirrel-cage induction motors modeled in the stationary reference frame.

the chattering is observed, and Figures 6.37 to 6.40 document the A- and D-class motor transient dynamics. The torque-speed characteristics are plotted.

The SIMULINK diagram used to simulate the induction motors is documented in Figure 6.41. The parameters used correspond to the A-class motor studied.

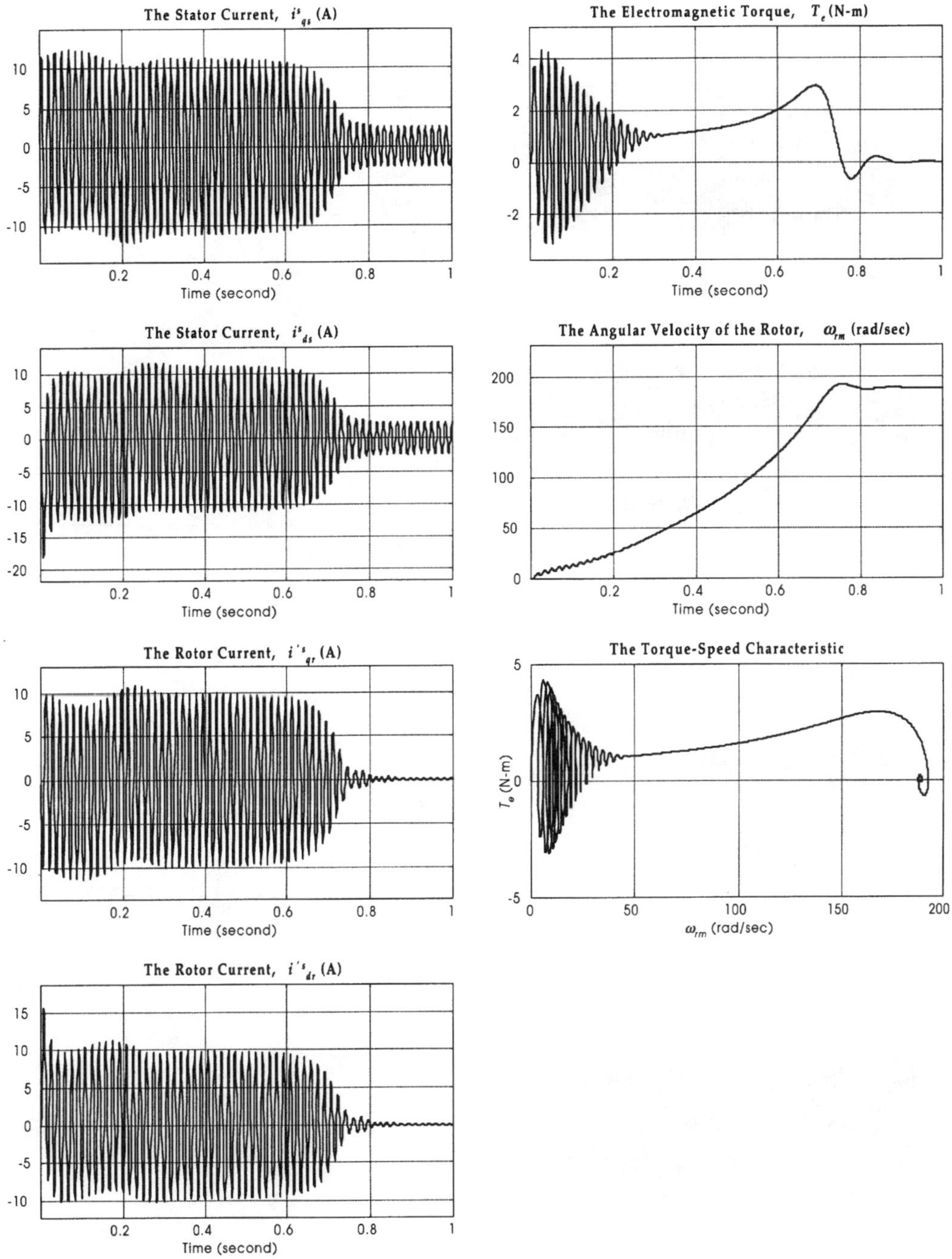

Figure 6.33 Dynamics and torque-speed characteristic of the A-class induction motor modeled in the stationary reference frame; $T_L = 0$ N-m, $u^s_{qs}(t) = \sqrt{2}115\cos(377t)$, and $u^s_{ds}(t) = -\sqrt{2}115\sin(377t)$.

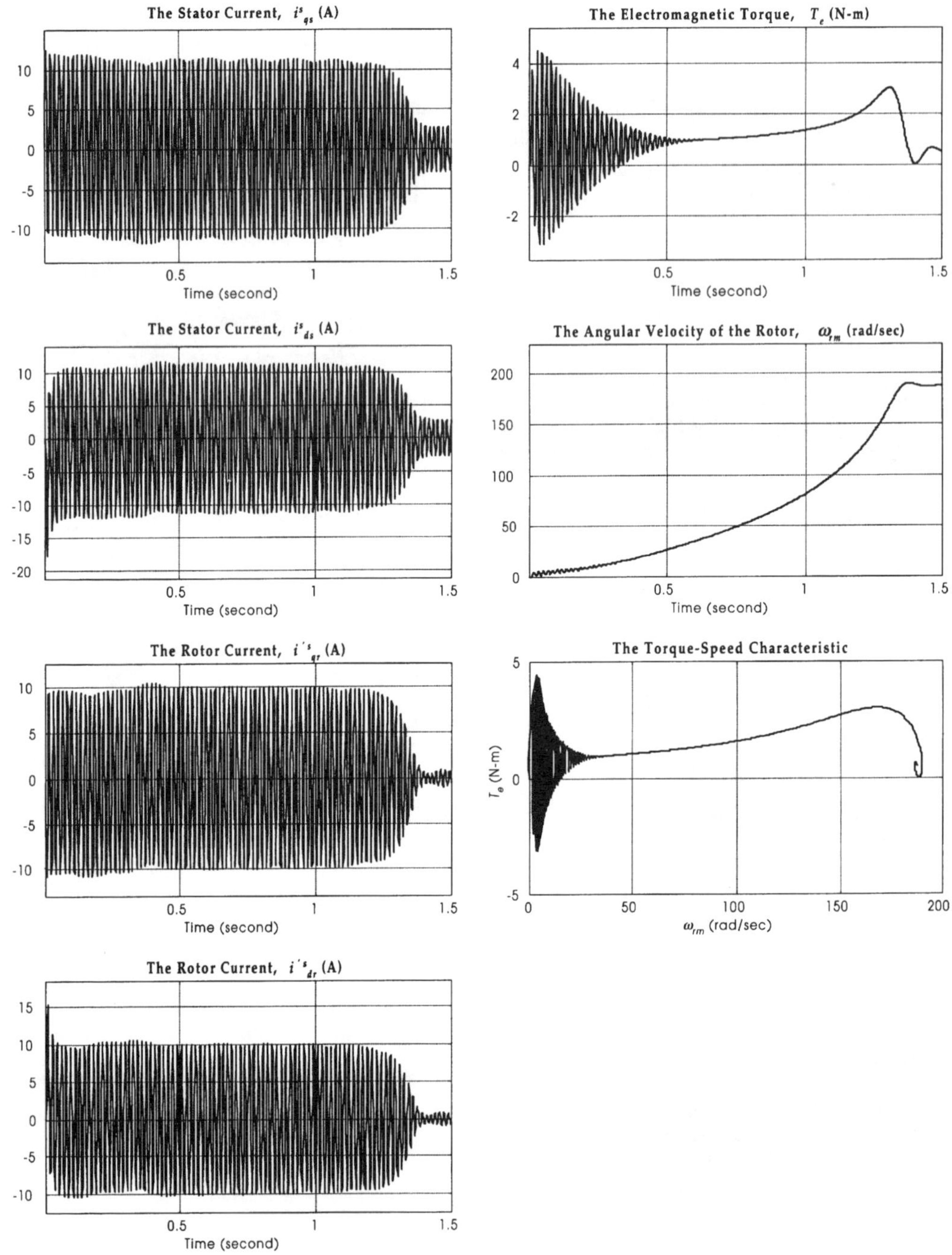

Figure 6.34 Dynamics and torque-speed characteristic of the A-class induction motor modeled in the stationary reference frame; $T_L = 0.5$ N-m, $u_{qs}^s(t) = \sqrt{2}115\cos(377t)$, and $u_{ds}^s(t) = -\sqrt{2}115\sin(377t)$.

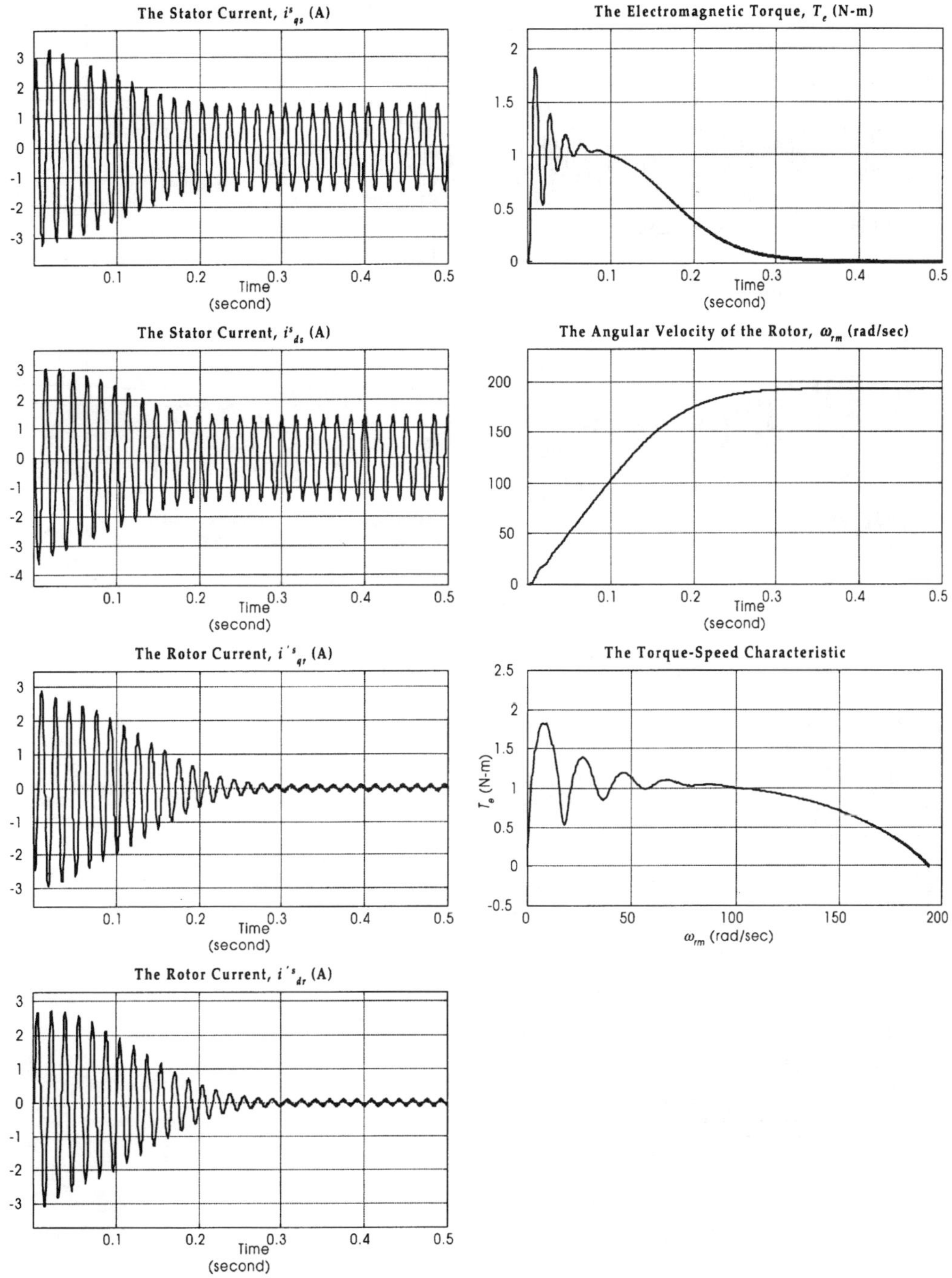

Figure 6.35 Dynamics and torque-speed characteristic of the D-class induction motor modeled in the stationary reference frame; $T_L = 0$ N-m, $u^s_{qs}(t) = \sqrt{2}115\cos(377t)$, and $u^s_{ds}(t) = -\sqrt{2}115\sin(377t)$.

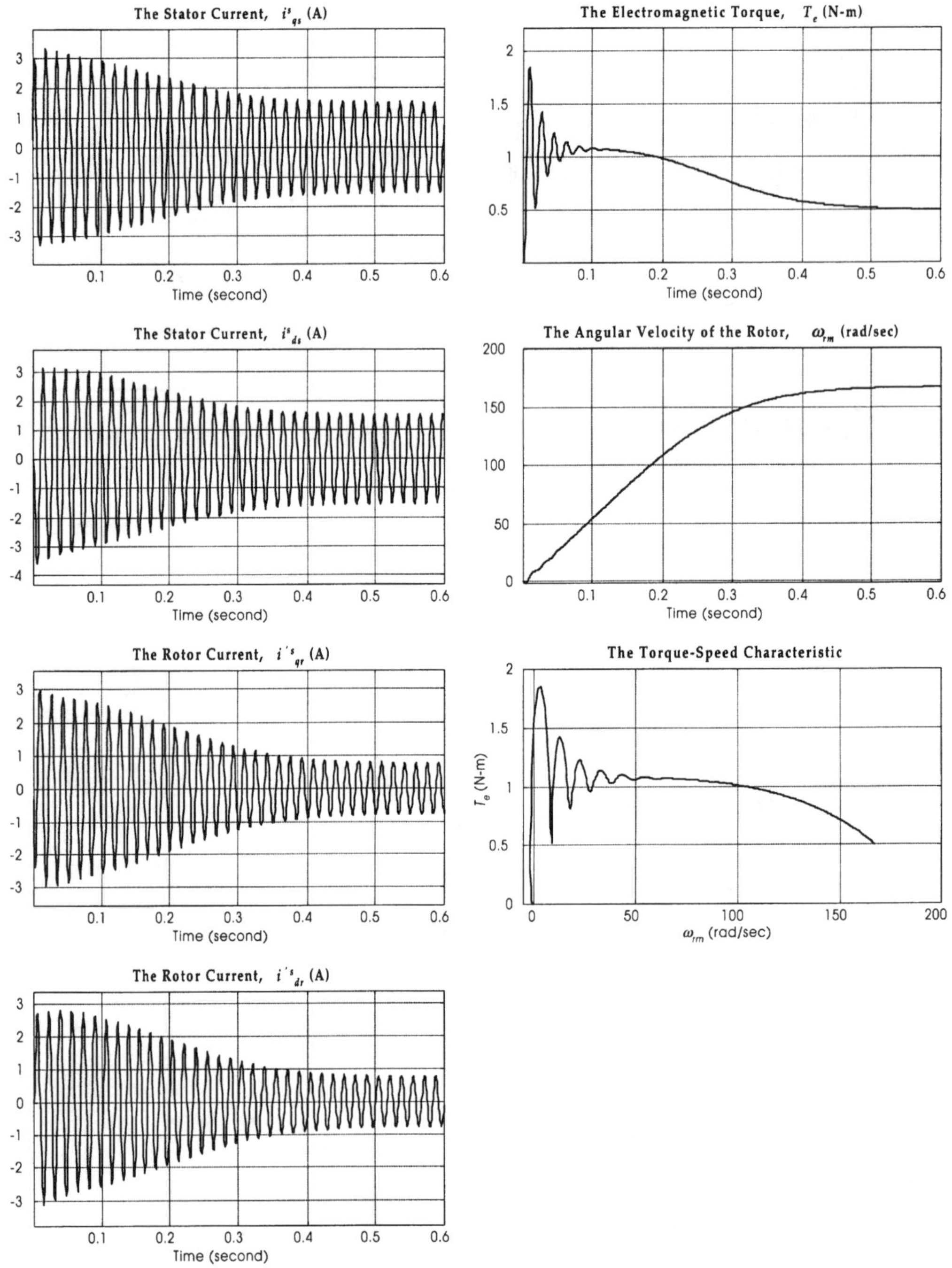

Figure 6.36 Dynamics and torque-speed characteristic of the D-class induction motor modeled in the stationary reference frame; $T_L = 0.5$ N-m, $u^s_{qs}(t) = \sqrt{2}115\cos(377t)$, and $u^s_{ds}(t) = -\sqrt{2}115\sin(377t)$.

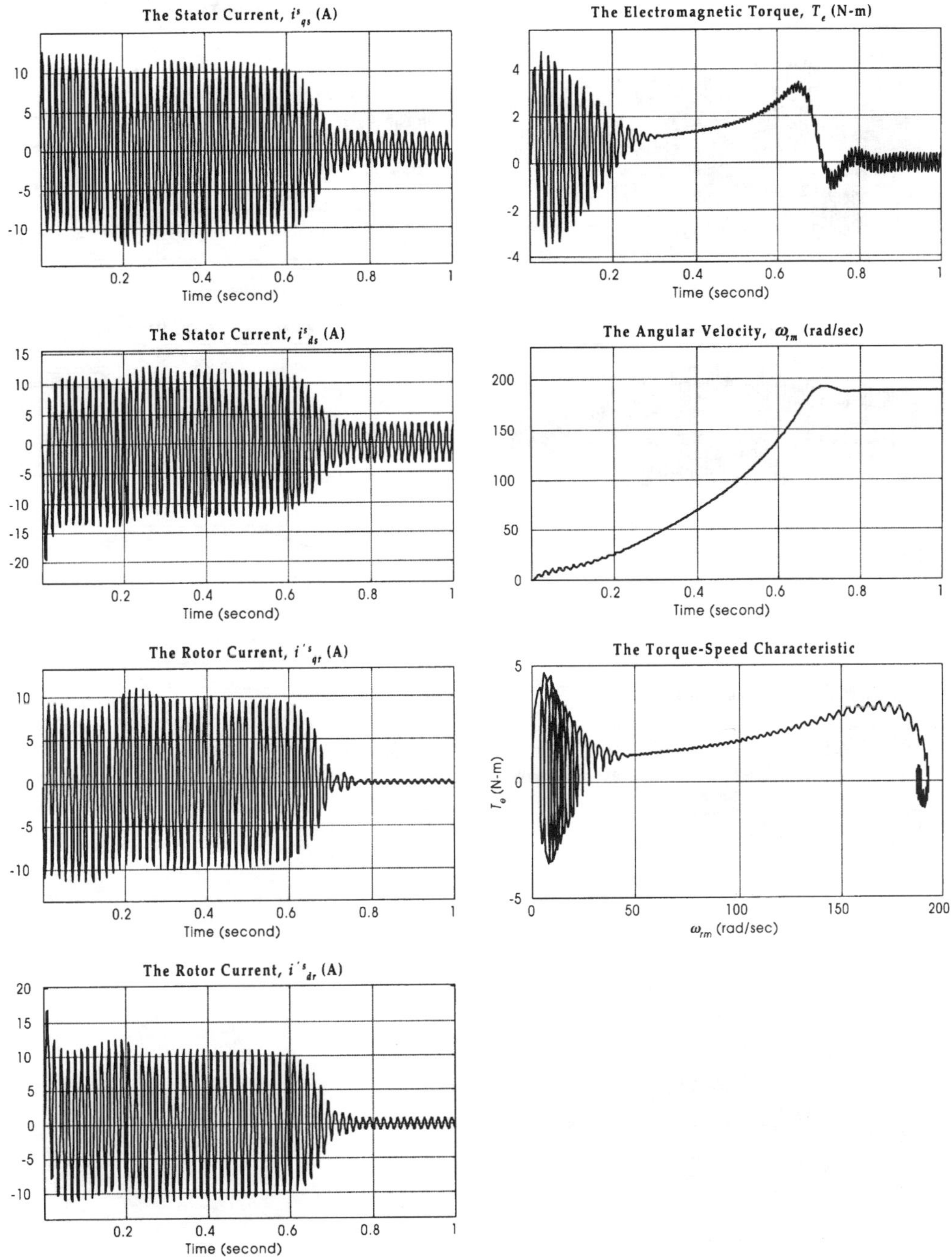

Figure 6.37 Dynamics and torque-speed characteristic of the A-class induction motor modeled in the stationary reference frame; $T_L = 0$ N-m, $u^s_{qs}(t) = \sqrt{2}115\cos(377t)$, and $u^s_{ds}(t) = -\sqrt{2}126\sin(377t)$.

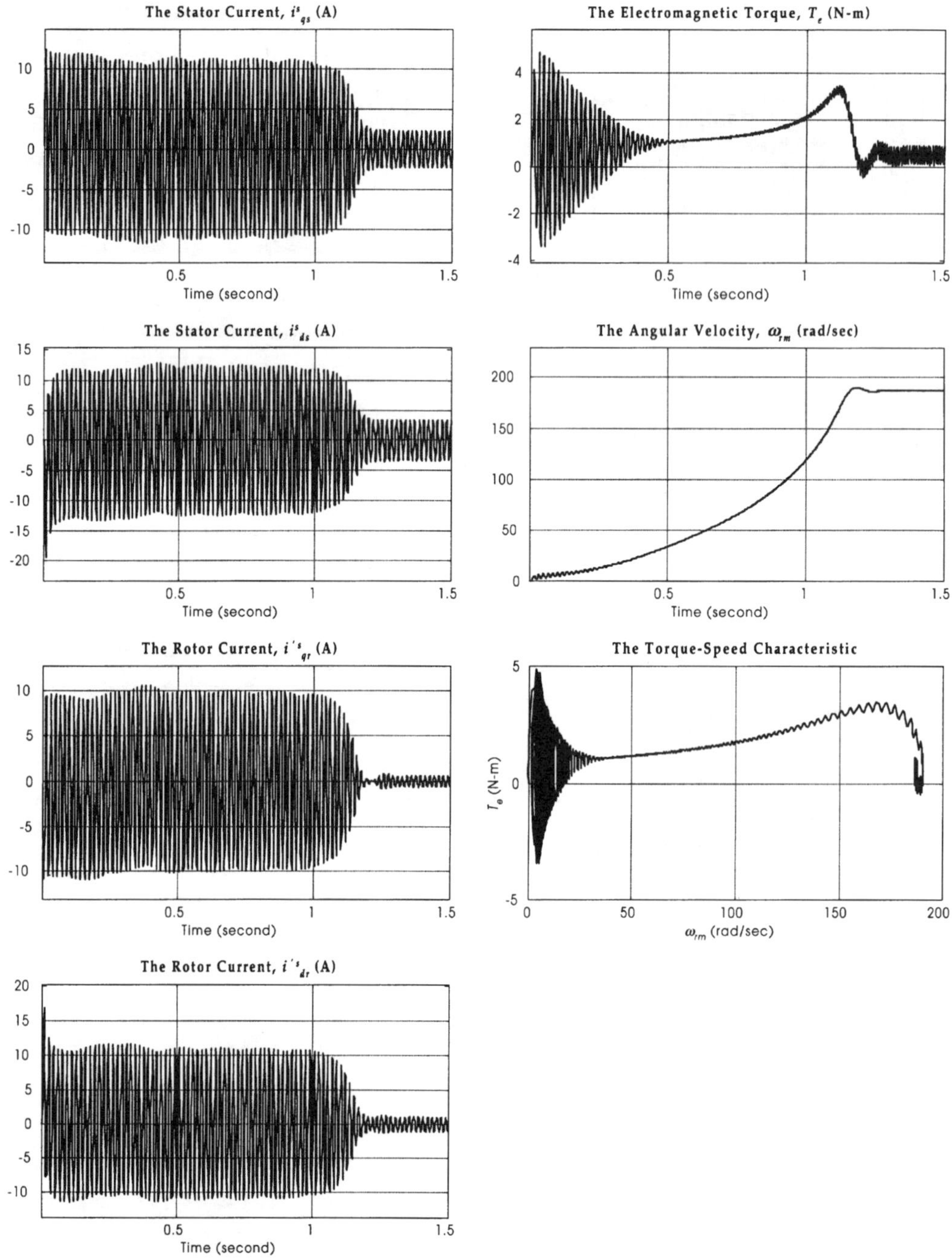

Figure 6.38 Dynamics and torque-speed characteristic of the A-class induction motor modeled in the stationary reference frame; $T_L = 0.5$ N-m, $u^s_{qs}(t) = \sqrt{2}115\cos(377t)$, and $u^s_{ds}(t) = -\sqrt{2}126\sin(377t)$.

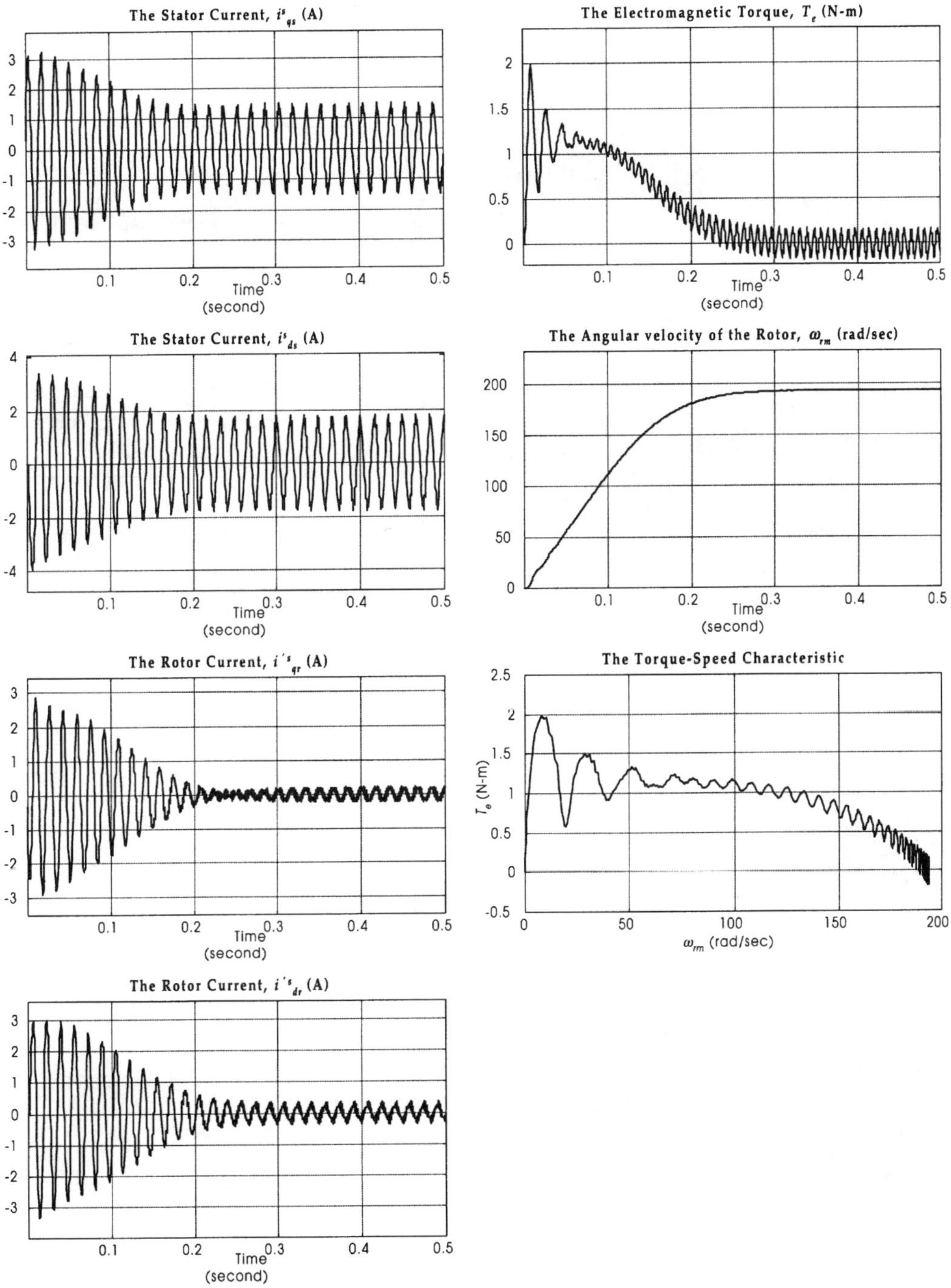

Figure 6.39 Dynamics and torque-speed characteristic of the D-class induction motor modeled in the stationary reference frame; $T_L = 0$ N-m, $u^s_{qs}(t) = \sqrt{2}115\cos(377t)$, and $u^s_{ds}(t) = -\sqrt{2}126\sin(377t)$.

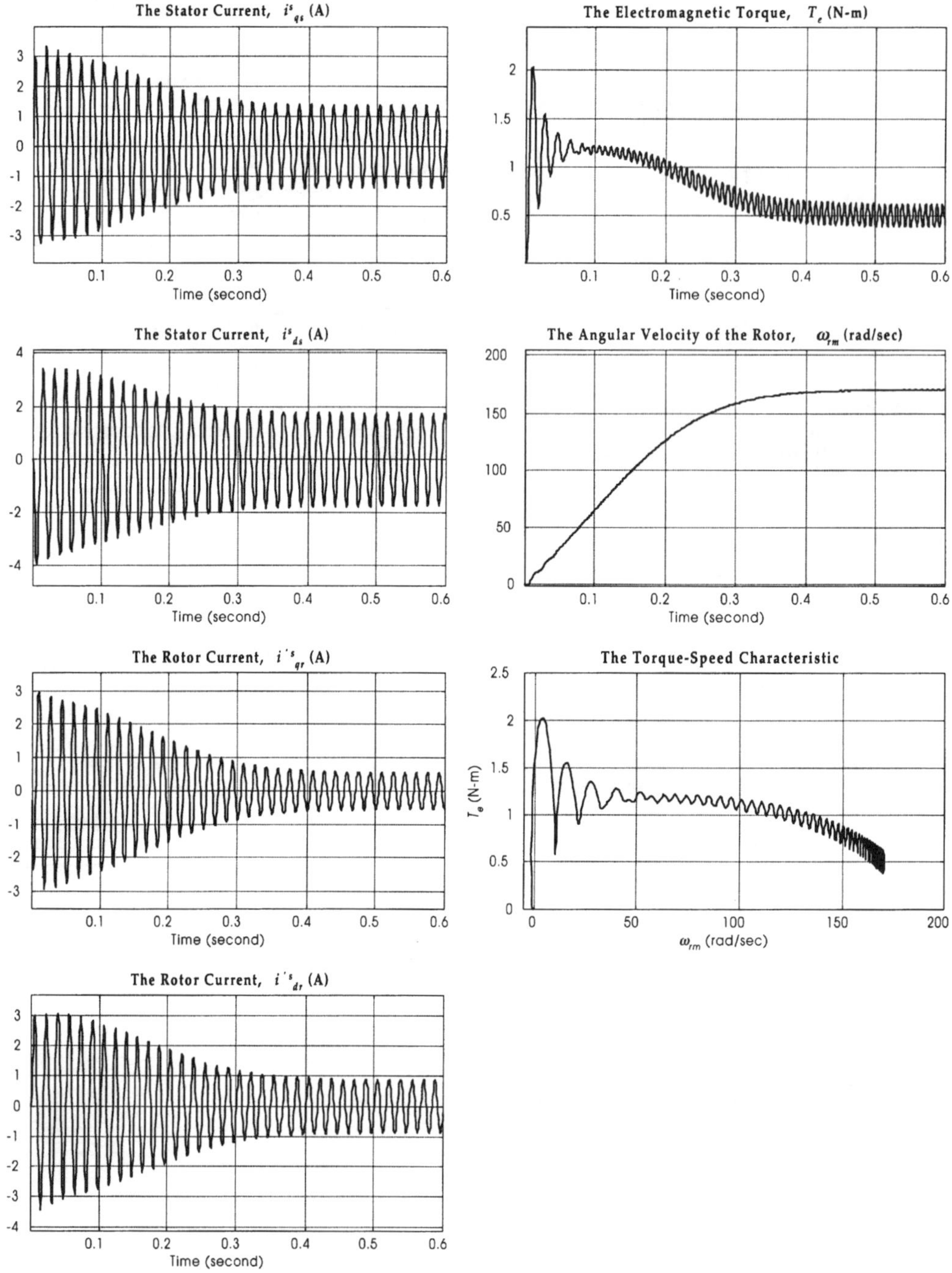

Figure 6.40 Dynamics and torque-speed characteristic of the D-class induction motor modeled in the stationary reference frame; T_L = 0.5 N-m, $u^s_{qs}(t) = \sqrt{2}115\cos(377t)$, and $u^s_{ds}(t) = -\sqrt{2}126\sin(377t)$.

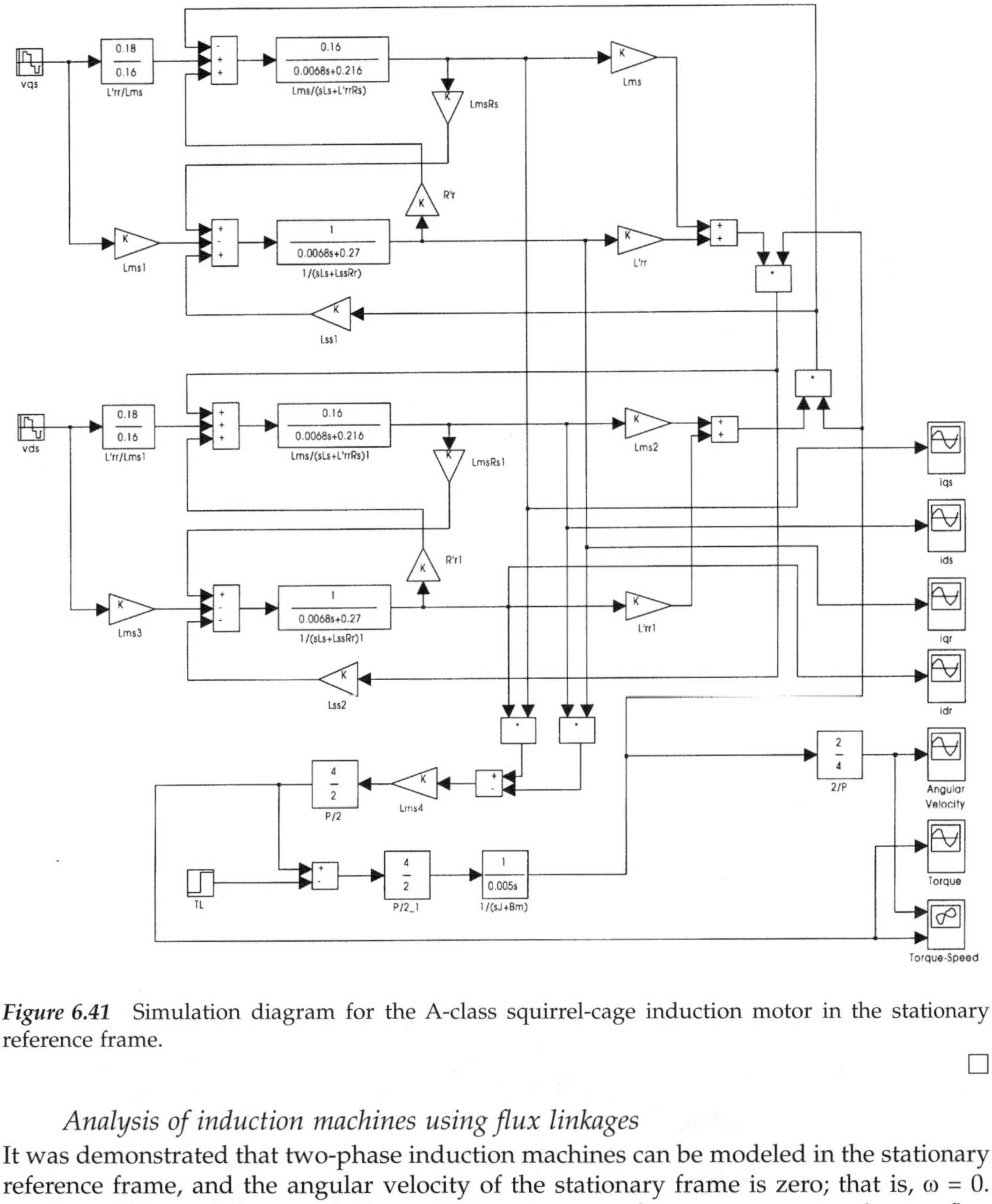

Figure 6.41 Simulation diagram for the A-class squirrel-cage induction motor in the stationary reference frame.

□

Analysis of induction machines using flux linkages

It was demonstrated that two-phase induction machines can be modeled in the stationary reference frame, and the angular velocity of the stationary frame is zero; that is, $\omega = 0$. Applying the *quadrature*- and *direct*-axis components of stator currents and rotor flux linkages, one can develop the corresponding differential equations to map the transient dynamics. The induction motor states are chosen to be i_{qs}^s, i_{ds}^s, $\psi_{qr}'^s$, $\psi_{dr}'^s$, ω_r, and θ_r. Plugging $\omega = 0$ into (6.31), one obtains the model if the stationary reference frame is in place. In particular, we have

$$\frac{di_{qs}^s}{dt} = \frac{1}{L_\Sigma}\left(-\frac{L_{rr}'^2 r_s + L_{ms}^2 r_r'}{L_{rr}'} i_{qs}^s + \frac{L_{ms} r_r'}{L_{rr}'} \psi_{qr}'^s - L_{ms}\psi_{dr}'^s\omega_r + L_{rr}' u_{qs}^s - L_{ms} u_{qr}'^s\right)$$

$$\frac{di_{ds}^s}{dt}=\frac{1}{L_\Sigma}\left(-\frac{L_{rr}'^2 r_s+L_{ms}^2 r_r'}{L_{rr}'}i_{ds}^s+\frac{L_{ms}r_r'}{L_{rr}'}\psi_{dr}'^s+L_{ms}\psi_{qr}'^s\omega_r+L_{rr}'u_{ds}^s-L_{ms}u_{dr}'^s\right)$$

$$\frac{d\psi_{qr}'^s}{dt}=\frac{L_{ms}r_r'}{L_{rr}'}i_{qs}^s-\frac{r_r'}{L_{rr}'}\psi_{qr}'^s+\omega_r\psi_{dr}'^s+u_{qr}'^s$$

$$\frac{d\psi_{dr}'^s}{dt}=\frac{L_{ms}r_r'}{L_{rr}'}i_{ds}^s-\omega_r\psi_{qr}'^s-\frac{r_r'}{L_{rr}'}\psi_{dr}'^s+u_{dr}'^s \tag{6.33}$$

$$\frac{d\omega_r}{dt}=\frac{P^2}{4J}\frac{L_{ms}}{L_{rr}'}(i_{qs}^s\psi_{dr}'^s-i_{ds}^s\psi_{qr}'^s)-\frac{B_m}{J}\omega_r-\frac{P}{2J}T_L$$

$$\frac{d\theta_r}{dt}=\omega_r$$

Differential equations (6.33) in matrix form are given as

$$\begin{bmatrix}\frac{di_{qs}^s}{dt}\\ \frac{di_{ds}^s}{dt}\\ \frac{d\psi_{qr}'^s}{dt}\\ \frac{d\psi_{dr}'^s}{dt}\\ \frac{d\omega_r}{dt}\\ \frac{d\theta_r}{dt}\end{bmatrix}=\begin{bmatrix}-\frac{L_{rr}'^2 r_s+L_{ms}^2 r_r'}{L_\Sigma L_{rr}'} & 0 & \frac{L_{ms}r_r'}{L_\Sigma L_{rr}'} & 0 & 0 & 0\\ 0 & -\frac{L_{rr}'^2 r_s+L_{ms}^2 r_r'}{L_\Sigma L_{rr}'} & 0 & \frac{L_{ms}r_r'}{L_\Sigma L_{rr}'} & 0 & 0\\ \frac{L_{ms}r_r'}{L_{rr}'} & 0 & -\frac{r_r'}{L_{rr}'} & 0 & 0 & 0\\ 0 & \frac{L_{ms}r_r'}{L_{rr}'} & 0 & -\frac{r_r'}{L_{rr}'} & 0 & 0\\ 0 & 0 & 0 & 0 & -\frac{B_m}{J} & 0\\ 0 & 0 & 0 & 0 & 1 & 0\end{bmatrix}\begin{bmatrix}i_{qs}^s\\ i_{ds}^s\\ \psi_{qr}'^s\\ \psi_{dr}'^s\\ \omega_r\\ \theta_r\end{bmatrix}$$

$$+\begin{bmatrix}-\frac{L_{ms}}{L_\Sigma}\psi_{dr}'^s\omega_r\\ \frac{L_{ms}}{L_\Sigma}\psi_{qr}'^s\omega_r\\ \psi_{dr}'^s\omega_r\\ -\psi_{qr}'^s\omega_r\\ \frac{P^2}{4J}\frac{L_{ms}}{L_{rr}'}(i_{qs}^s\psi_{dr}'^s-i_{ds}^s\psi_{qr}'^s)\\ 0\end{bmatrix}+\begin{bmatrix}\frac{L_{rr}'}{L_\Sigma} & 0 & -\frac{L_{ms}}{L_\Sigma} & 0\\ 0 & \frac{L_{rr}'}{L_\Sigma} & 0 & -\frac{L_{ms}}{L_\Sigma}\\ 0 & 0 & 1 & 0\\ 0 & 0 & 0 & 1\\ 0 & 0 & 0 & 0\\ 0 & 0 & 0 & 0\end{bmatrix}\begin{bmatrix}u_{qs}^s\\ u_{ds}^s\\ u_{qr}'^s\\ u_{dr}'^s\end{bmatrix}-\begin{bmatrix}0\\ 0\\ 0\\ 0\\ \frac{P}{2J}\\ 0\end{bmatrix}T_L$$

In the stationary reference frame, the block diagram for two-phase squirrel-cage induction motors using the *quadrature*- and *direct*-axis components of stator currents and rotor flux linkages is developed by rewriting a set of differential equations (6.33) as

$$i_{qs}^s=\frac{L_{rr}'}{L_\Sigma L_{rr}'s+L_{rr}'^2 r_s+L_{ms}^2 r_r'}\left(\frac{L_{ms}r_r'}{L_{rr}'}\psi_{qr}'^s-L_{ms}\psi_{dr}'^s\omega_r+L_{rr}'u_{qs}^s-L_{ms}u_{qr}'^s\right)$$

$$i_{ds}^{s}=\frac{L'_{rr}}{L_{\Sigma}L'_{rr}s+L'^{2}_{rr}r_{s}+L^{2}_{ms}r'_{r}}\left(\frac{L_{ms}r'_{r}}{L'_{rr}}\psi'^{s}_{dr}+L_{ms}\psi'^{s}_{qr}\omega_{r}+L'_{rr}u^{s}_{ds}-L_{ms}u'^{s}_{dr}\right)$$

$$\psi'^{s}_{qr}=\frac{1}{L'_{rr}s+r'_{r}}(L_{ms}r'_{r}i^{s}_{qs}+L'_{rr}\psi'^{s}_{dr}\omega_{r}+L'_{rr}u'^{s}_{qr})$$

$$\psi'^{s}_{dr}=\frac{1}{L'_{rr}s+r'_{r}}(L_{ms}r'_{r}i^{s}_{ds}-L'_{rr}\psi'^{s}_{qr}\omega_{r}+L'_{rr}u'^{s}_{dr})$$

$$\omega_{r}=\frac{P}{2}\left(\frac{1}{Js+B_{m}}\right)\left[\frac{P}{2}\frac{L_{ms}}{L'_{rr}}(i^{s}_{qs}\psi'^{s}_{dr}-i^{s}_{ds}\psi'^{s}_{qr})-T_{L}\right]$$

From this set of equations in the *s*-domain, the block diagram of two-phase squirrel-cage induction motors modeled in the stationary reference frame is developed and shown in Figure 6.42.

Example 6.4

Using i^{s}_{qs}, i^{s}_{ds}, ψ'^{s}_{qr}, ψ'^{s}_{dr}, and ω_{r} as the state variables, perform the nonlinear simulation of the A- and D-class squirrel-cage induction motors, parameters of which are given in Table 6.1, in the stationary reference frame. Develop the SIMULINK diagram to be used in modeling.

Solution

The SIMULINK diagram, shown in Figure 6.43 using A-class motor parameters, is developed by applying the diagram as documented in Figure 6.42. It was shown in Example 6.3 that the *qd* voltages

$$u^{s}_{qs}(t)=\sqrt{2}115\cos(377t)\ \text{ and }\ u^{s}_{ds}(t)=-\sqrt{2}115\sin(377t)$$

guarantee the balanced operating conditions.

The simulation results are documented in Figures 6.44 and 6.45 for the A-class motor, while Figures 6.46 and 6.47 illustrate the transient behavior and the torque-speed characteristic curves of the D-class motor.

Under the unbalanced operating condition, supplying

$$u^{s}_{qs}(t)=\sqrt{2}115\cos(377t)\ \text{ and }\ u^{s}_{ds}(t)=-\sqrt{2}126\sin(377t)$$

the dynamics and the torque-speed characteristics of two-phase induction motors studied are demonstrated in Figures 6.48 to 6.51.

6.3.2 *Development of the mathematical model of two-phase induction motors in the rotor reference frame*

Using the stator and rotor currents as the state variables, the mathematical model of induction motors can be developed in the rotor reference frame. Substituting the angular velocity of the reference frame in the differential equations (6.27), which were derived in the *arbitrary* reference frame when the angular velocity of the frame of reference was not specified, the resulting set of nonlinear differential equations is obtained. In particular, plugging in $\omega = \omega_r$, one finds the following six highly nonlinear differential equations

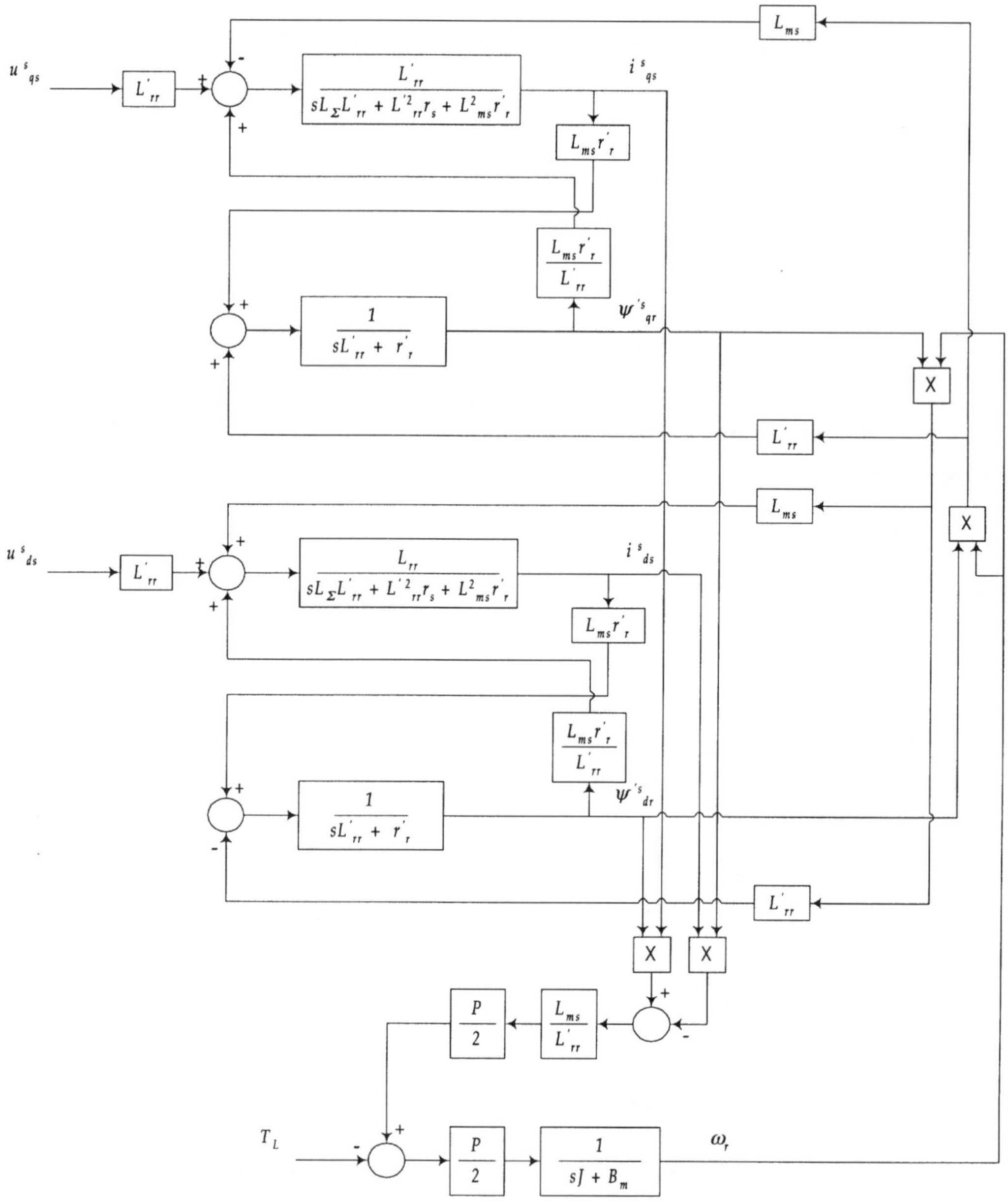

Figure 6.42 Block diagram of two-phase squirrel-cage induction motors modeled in the stationary reference frame.

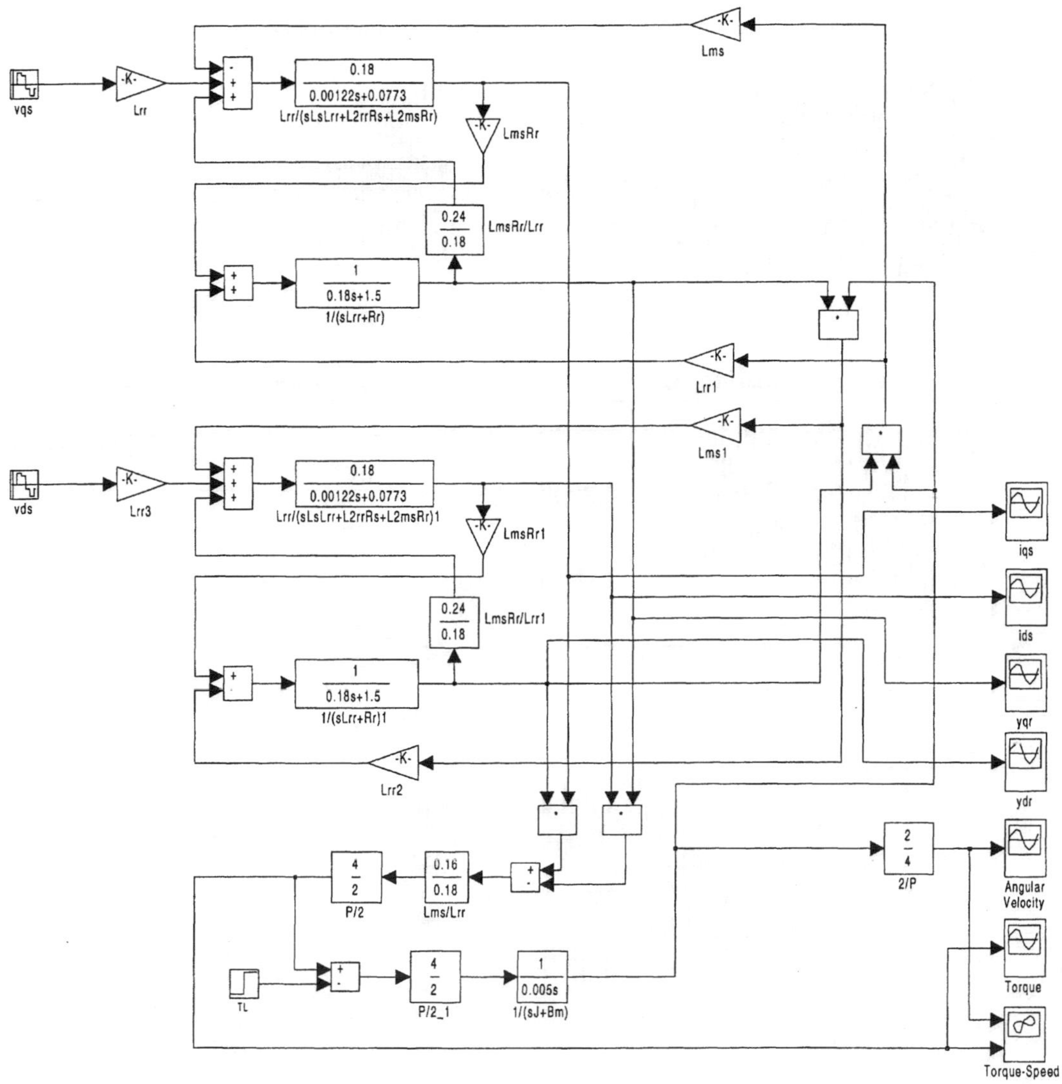

Figure 6.43 SIMULINK diagram to model the A-class squirrel-cage induction motor in the stationary reference frame.

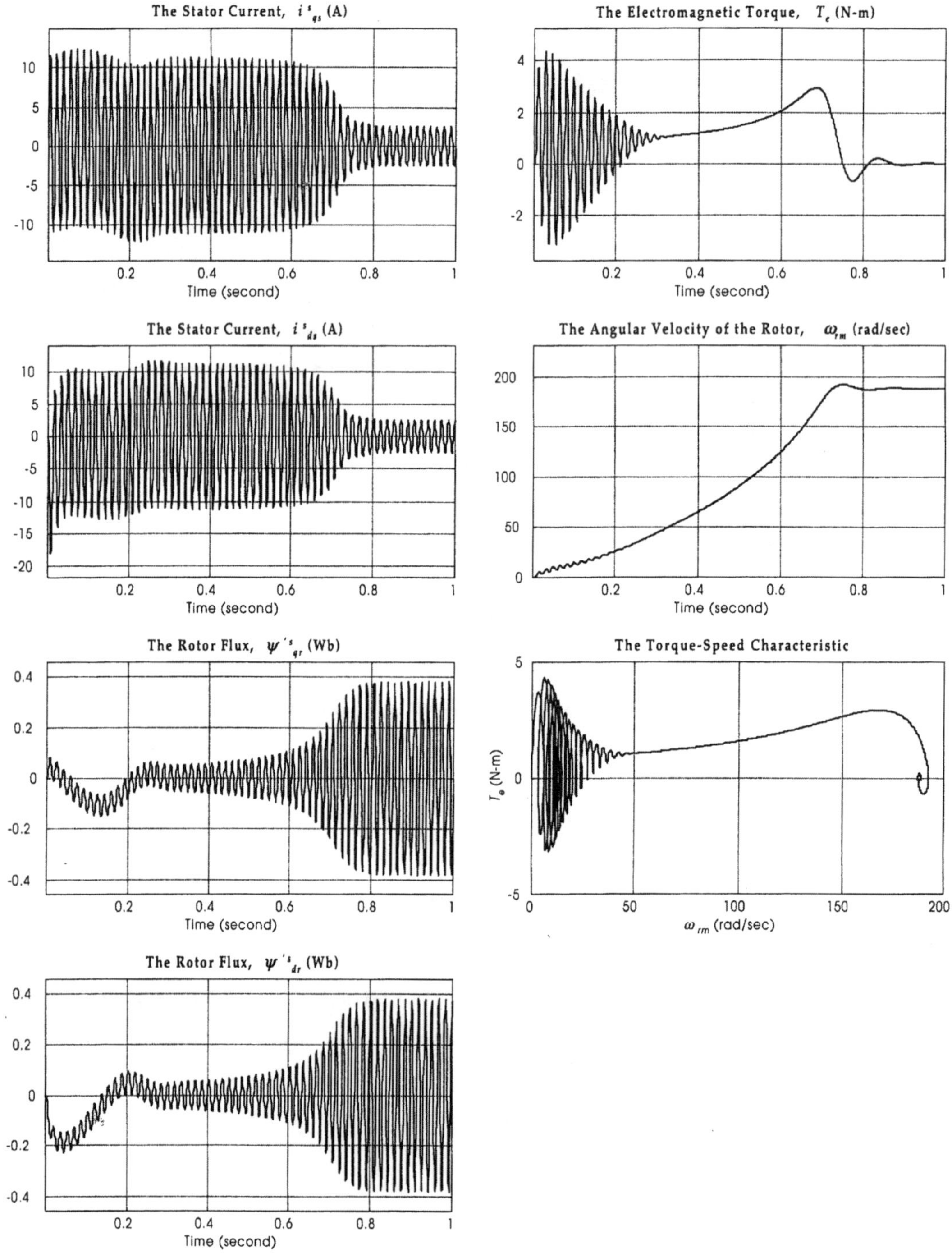

Figure 6.44 Dynamics and torque-speed characteristic of the A-clsss induction motor modeled in the stationary reference frame; $T_L = 0$ N-m, $u^s_{qs}(t) = \sqrt{2}115\cos(377t)$, and $u^s_{ds}(t) = -\sqrt{2}115\sin(377t)$.

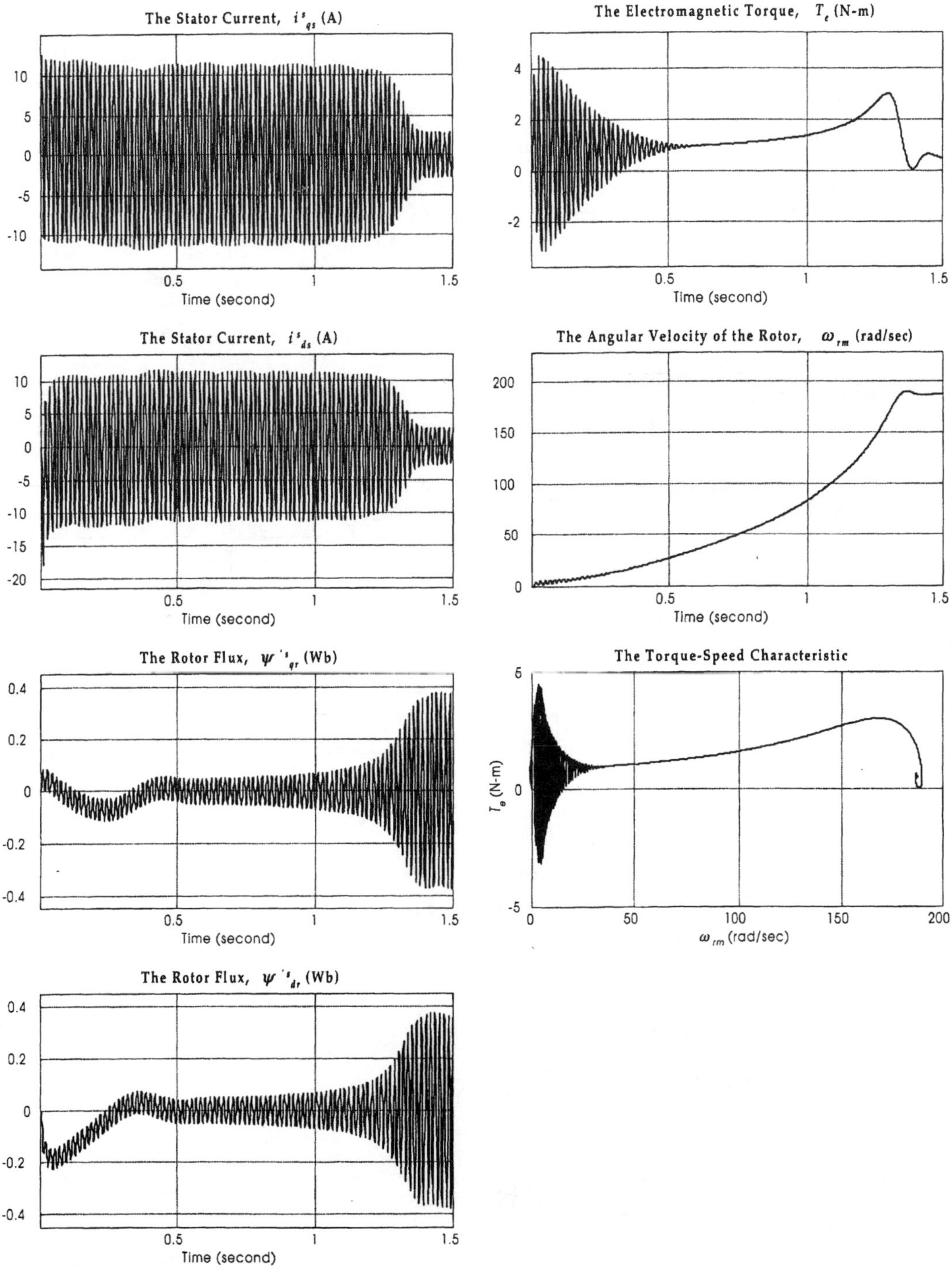

Figure 6.45 Dynamics and torque-speed characteristic of the A-clsss induction motor modeled in the stationary reference frame; $T_L = 0.5$ N-m, $u^s_{qs}(t) = \sqrt{2}115\cos(377t)$, and $u^s_{ds}(t) = -\sqrt{2}115\sin(377t)$.

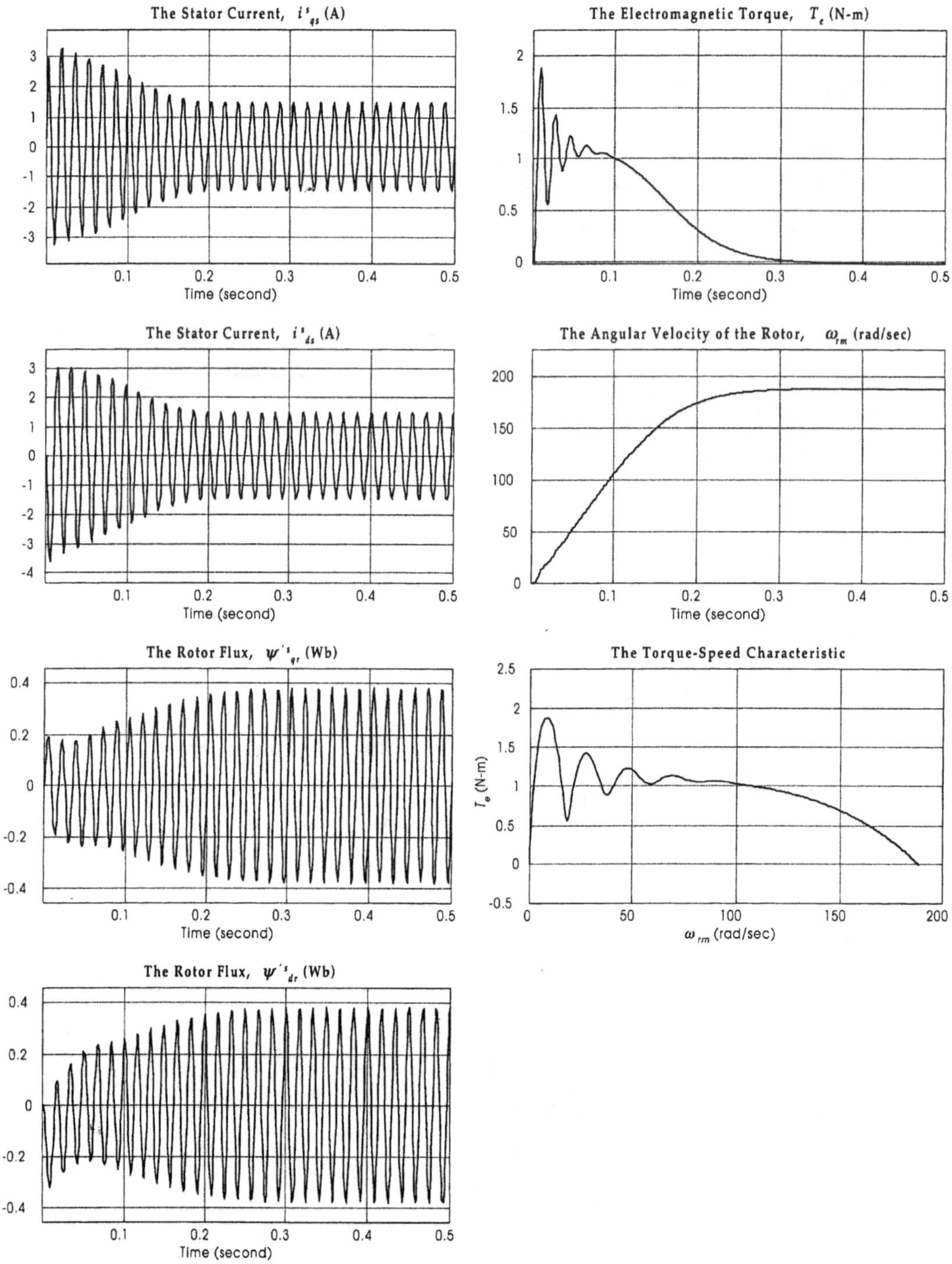

Figure 6.46 Dynamics and torque-speed characteristic of the D-clsss induction motor modeled in the stationary reference frame; $T_L = 0$ N-m, $u_{qs}^s(t) = \sqrt{2}115\cos(377t)$, and $u_{ds}^s(t) = -\sqrt{2}115\sin(377t)$.

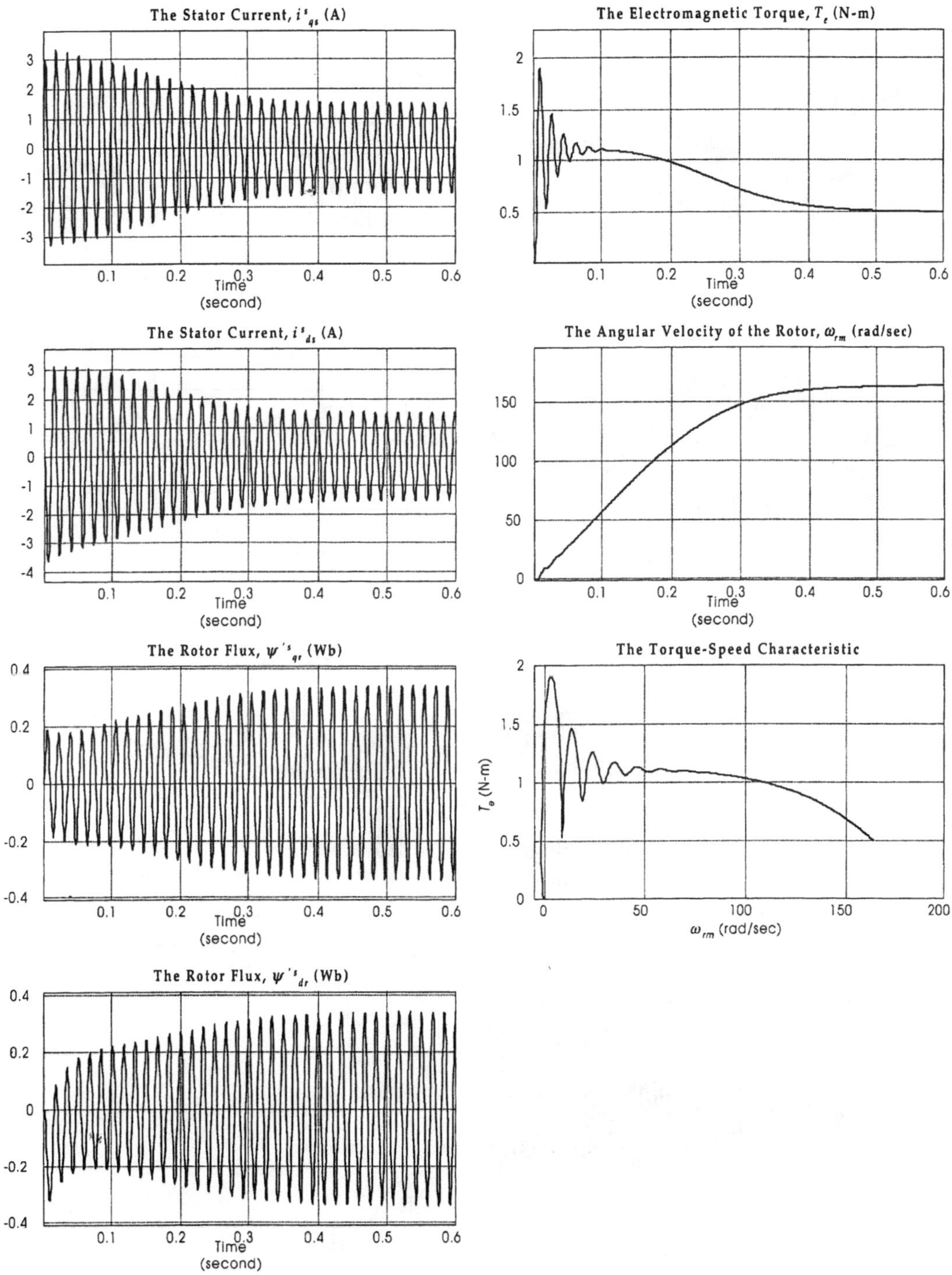

Figure 6.47 Dynamics and torque-speed characteristic of the D-clsss induction motor modeled in the stationary reference frame; $T_L = 0.5$ N-m, $u_{qs}^s(t) = \sqrt{2}115\cos(377t)$, and $u_{ds}^s(t) = -\sqrt{2}115\sin(377t)$.

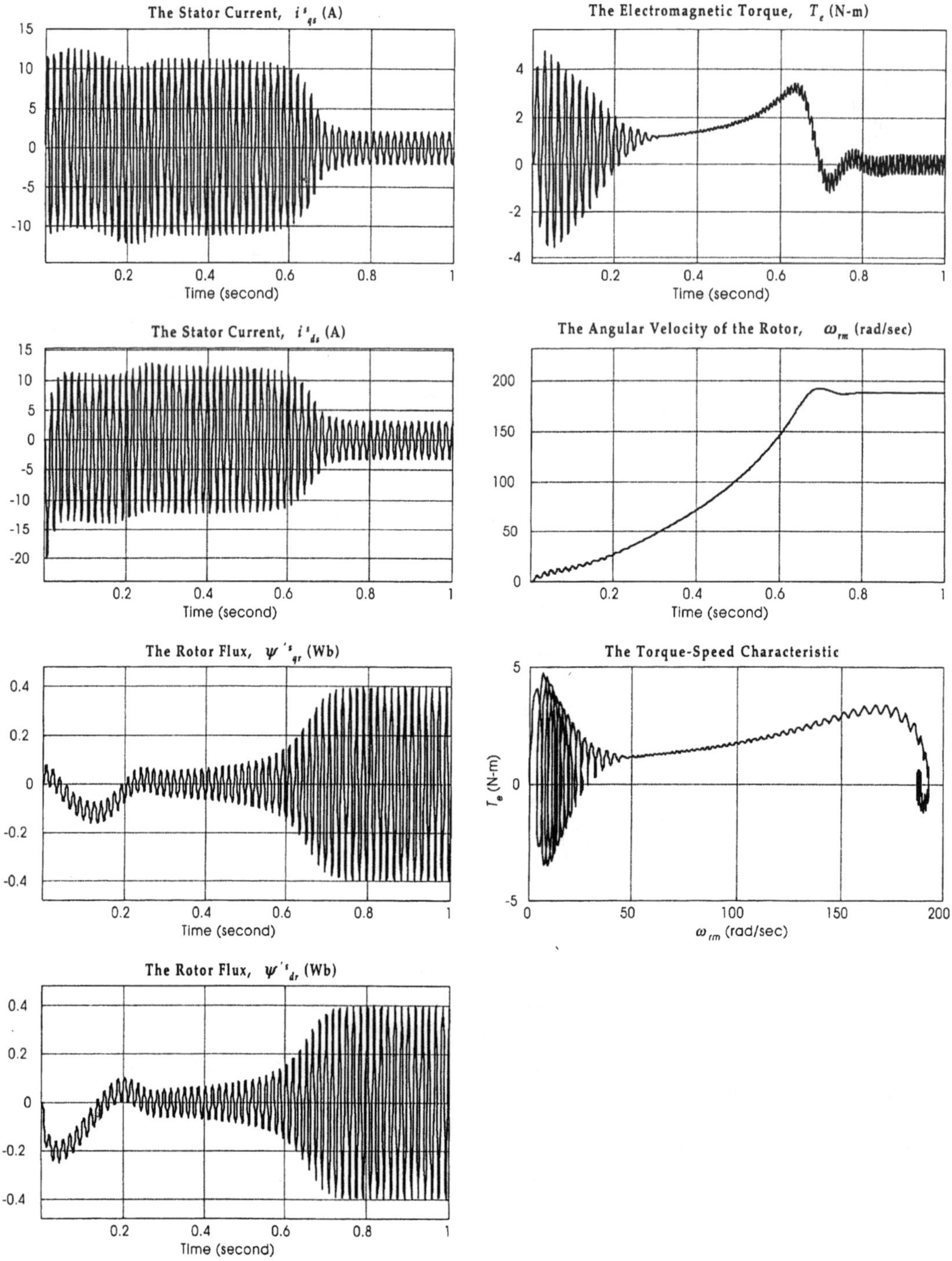

Figure 6.48 Dynamics and torque-speed characteristic of the A-clsss induction motor modeled in the stationary reference frame; $T_L = 0$ N-m, $u^s_{qs}(t) = \sqrt{2}115\cos(377t)$, and $u^s_{ds}(t) = -\sqrt{2}126\sin(377t)$.

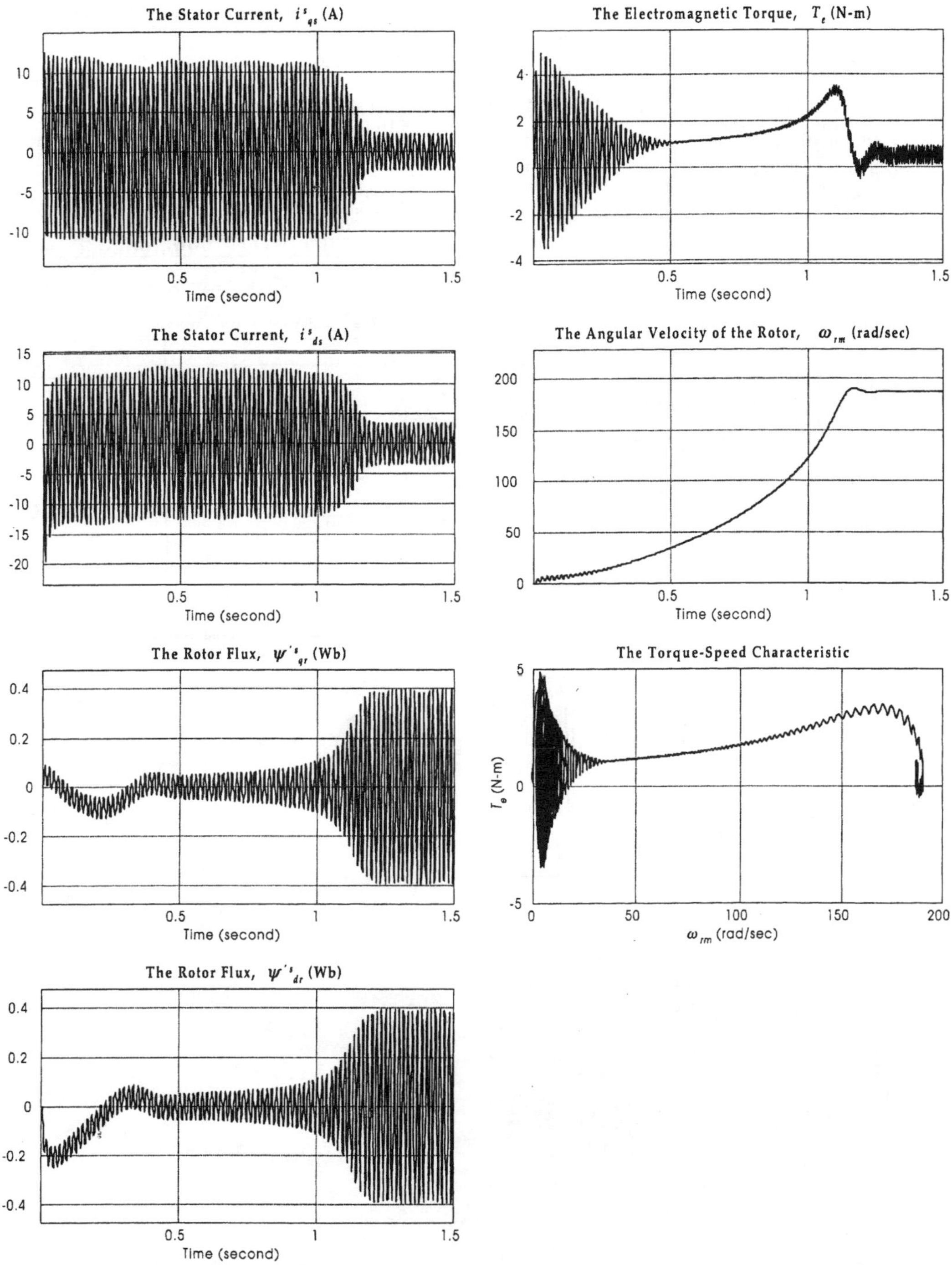

Figure 6.49 Dynamics and torque-speed characteristic of the A-clsss induction motor modeled in the stationary reference frame; $T_L = 0.5$ N-m, $u^s_{qs}(t) = \sqrt{2}115\cos(377t)$, and $u^s_{ds}(t) = -\sqrt{2}126\sin(377t)$.

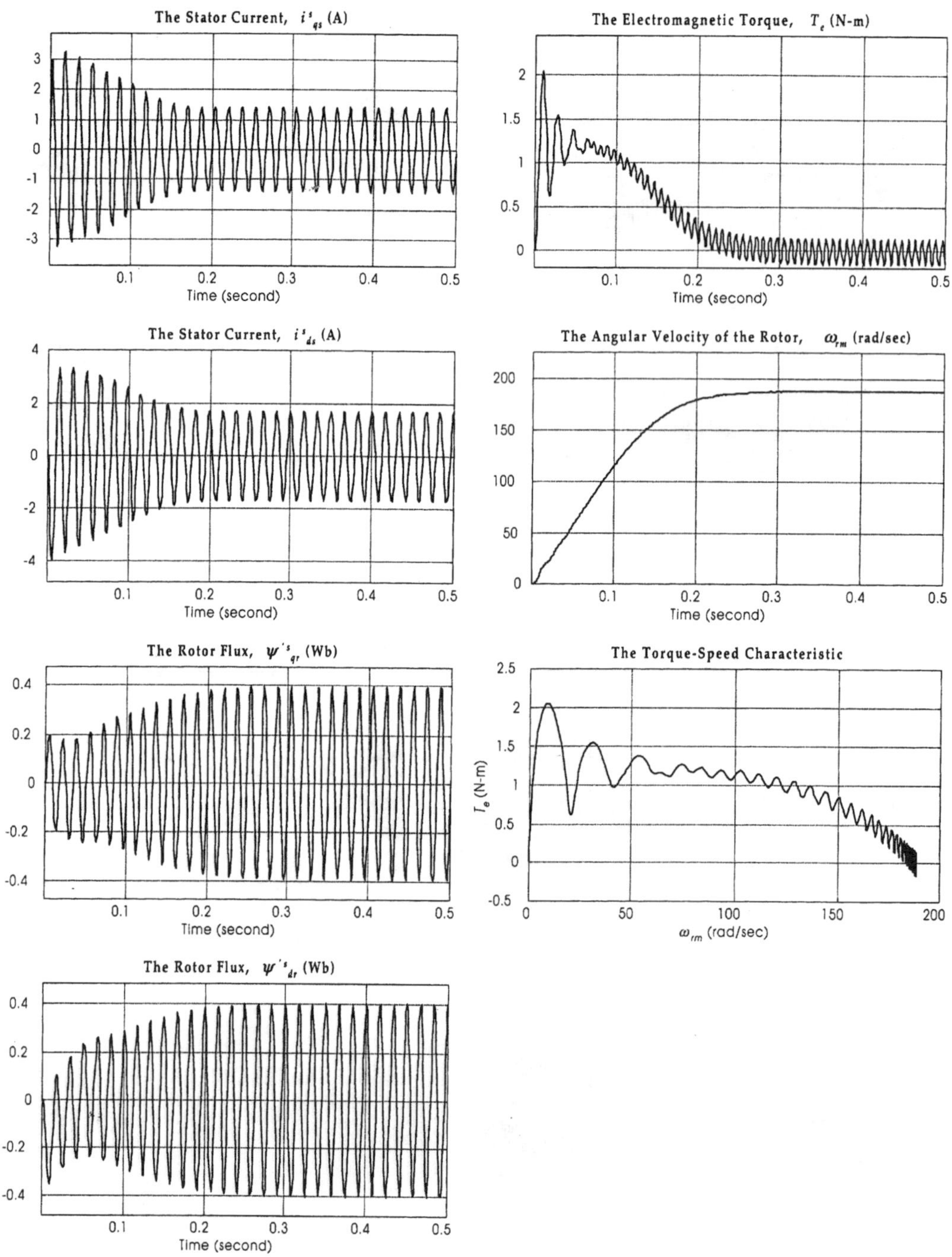

Figure 6.50 Dynamics and torque-speed characteristic of the D-clsss induction motor modeled in the stationary reference frame; $T_L = 0$ N-m, $u^s_{qs}(t) = \sqrt{2}115\cos(377t)$, and $u^s_{ds}(t) = -\sqrt{2}126\sin(377t)$.

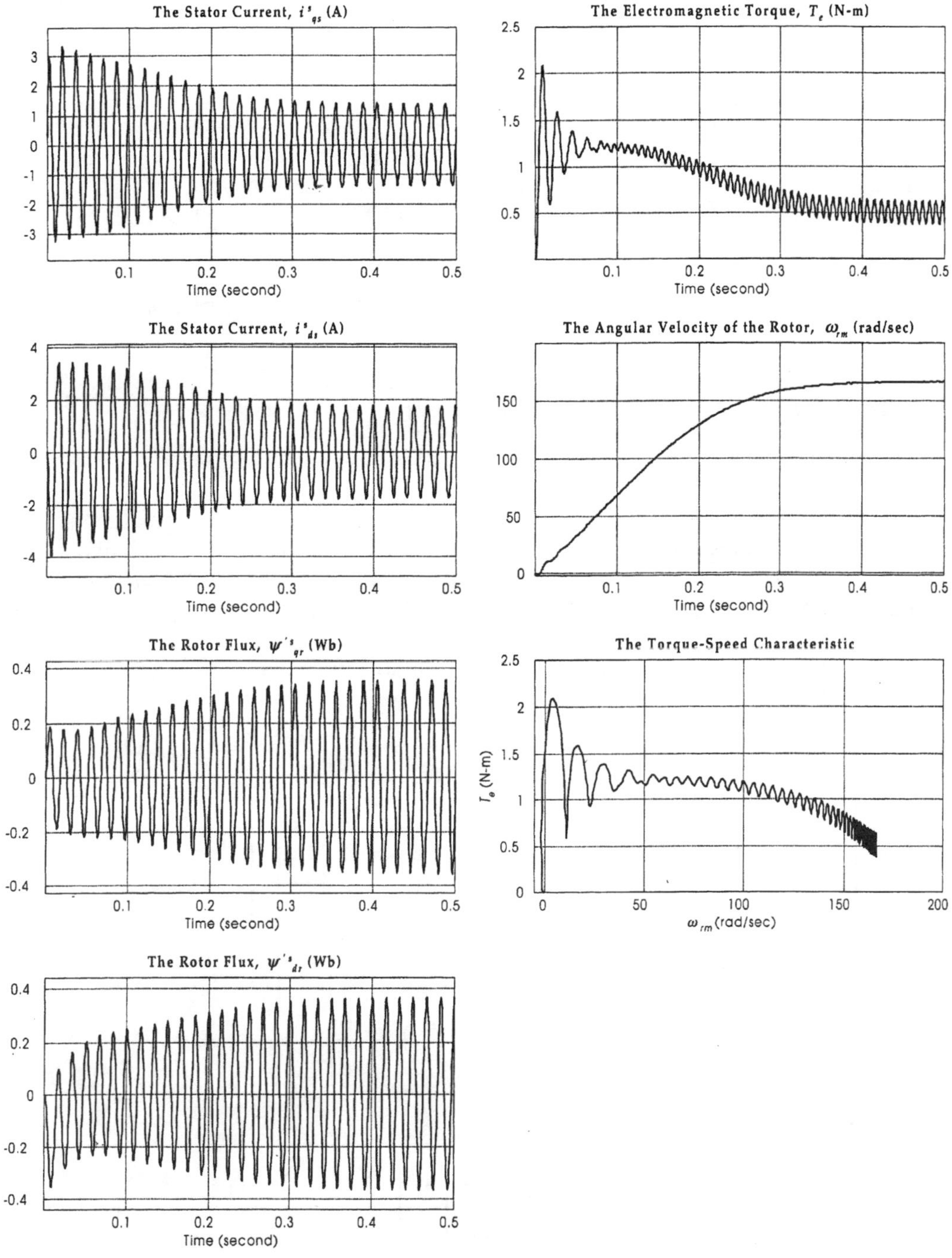

Figure 6.51 Dynamics and torque-speed characteristic of the D-clsss induction motor modeled in the stationary reference frame; $T_L = 0.5$ N-m, $u^s_{qs}(t) = \sqrt{2}115\cos(377t)$, and $u^s_{ds}(t) = -\sqrt{2}126\sin(377t)$.

$$\frac{di_{qs}^r}{dt} = \frac{1}{L_\Sigma}[-L'_{rr} r_s i_{qs}^r + L_{ms} r'_r i'^r_{qr} - L'_{rr}(L_{ss} i_{ds}^r + L_{ms} i'^r_{dr})\omega_r + L'_{rr} u_{qs}^r - L_{ms} u'^r_{qr}]$$

$$\frac{di_{ds}^r}{dt} = \frac{1}{L_\Sigma}[-L'_{rr} r_s i_{ds}^r + L_{ms} r'_r i'^r_{dr} + L'_{rr}(L_{ss} i_{qs}^r + L_{ms} i'^r_{qr})\omega_r + L'_{rr} u_{ds}^r - L_{ms} u'^r_{dr}]$$

$$\frac{di'^r_{qr}}{dt} = \frac{1}{L_\Sigma}[L_{ms} r_s i_{qs}^r - L_{ss} r'_r i'^r_{qr} + L_{ms}(L_{ss} i_{ds}^r + L_{ms} i'^r_{dr})\omega_r - L_{ms} u_{qs}^r + L_{ss} u'^r_{qr}] \qquad (6.34)$$

$$\frac{di'^r_{dr}}{dt} = \frac{1}{L_\Sigma}[L_{ms} r_s i_{ds}^r - L_{ss} r'_r i'^r_{dr} - L_{ms}(L_{ss} i_{qs}^r + L_{ms} i'^r_{qr})\omega_r - L_{ms} u_{ds}^r + L_{ss} u'^r_{dr}]$$

$$\frac{d\omega_r}{dt} = \frac{P^2 L_{ms}}{4J}(i_{qs}^r i'^r_{dr} - i_{ds}^r i'^r_{qr}) - \frac{B_m}{J}\omega_r - \frac{P}{2J}T_L$$

$$\frac{d\theta_r}{dt} = \omega_r$$

In the differential equations (6.34), which map the induction motor transient dynamics in the rotor reference frame, the superscript r designates the fact that the *quadrature*- and *direct*-axis components of stator and rotor currents and voltages are given in the rotor reference frame.

Differential equations (6.34) are rewritten in matrix form as

$$\begin{bmatrix} \frac{di_{qs}^r}{dt} \\ \frac{di_{ds}^r}{dt} \\ \frac{di'^r_{qr}}{dt} \\ \frac{di'^r_{dr}}{dt} \\ \frac{d\omega_r}{dt} \\ \frac{d\theta_r}{dt} \end{bmatrix} = \begin{bmatrix} -\frac{L'_{rr} r_s}{L_\Sigma} & 0 & \frac{L_{ms} r'_r}{L_\Sigma} & 0 & 0 & 0 \\ 0 & -\frac{L'_{rr} r_s}{L_\Sigma} & 0 & \frac{L_{ms} r'_r}{L_\Sigma} & 0 & 0 \\ \frac{L_{ms} r_s}{L_\Sigma} & 0 & -\frac{L_{ss} r'_r}{L_\Sigma} & 0 & 0 & 0 \\ 0 & \frac{L_{ms} r_s}{L_\Sigma} & 0 & -\frac{L_{ss} r'_r}{L_\Sigma} & 0 & 0 \\ 0 & 0 & 0 & 0 & -\frac{B_m}{J} & 0 \\ 0 & 0 & 0 & 0 & 1 & 0 \end{bmatrix} \begin{bmatrix} i_{qs}^r \\ i_{ds}^r \\ i'^r_{qr} \\ i'^r_{dr} \\ \omega_r \\ \theta_r \end{bmatrix} + \begin{bmatrix} -\frac{L'_{rr}}{L_\Sigma}(L_{ss} i_{ds}^r + L_{ms} i'^r_{dr})\omega_r \\ \frac{L'_{rr}}{L_\Sigma}(L_{ss} i_{qs}^r + L_{ms} i'^r_{qr})\omega_r \\ \frac{L_{ms}}{L_\Sigma}(L_{ss} i_{ds}^r + L_{ms} i'^r_{dr})\omega_r \\ -\frac{L_{ms}}{L_\Sigma}(L_{ss} i_{qs}^r + L_{ms} i'^r_{qr})\omega_r \\ \frac{P^2 L_{ms}}{4J}(i_{qs}^r i'^r_{dr} - i_{ds}^r i'^r_{qr}) \\ 0 \end{bmatrix}$$

$$+ \begin{bmatrix} \frac{L'_{rr}}{L_\Sigma} & 0 & -\frac{L_{ms}}{L_\Sigma} & 0 \\ 0 & \frac{L'_{rr}}{L_\Sigma} & 0 & -\frac{L_{ms}}{L_\Sigma} \\ -\frac{L_{ms}}{L_\Sigma} & 0 & \frac{L_{ss}}{L_\Sigma} & 0 \\ 0 & -\frac{L_{ms}}{L_\Sigma} & 0 & \frac{L_{ss}}{L_\Sigma} \\ 0 & 0 & 0 & 0 \\ 0 & 0 & 0 & 0 \end{bmatrix} \begin{bmatrix} u_{qs}^r \\ u_{ds}^r \\ u'^r_{qr} \\ u'^r_{dr} \end{bmatrix} - \begin{bmatrix} 0 \\ 0 \\ 0 \\ 0 \\ \frac{P}{2J} \\ 0 \end{bmatrix} T_L$$

It is of great importance to find the *quadrature-* and *direct*-axis components of voltages to control induction motors. To obtain the explicit expressions for u_{qs}^r and u_{ds}^r, we use the Park transformation. Applying the transformation matrix for the stator variables, as given by $\mathbf{K}_s = \begin{bmatrix} \cos\theta & \sin\theta \\ \sin\theta & -\cos\theta \end{bmatrix}$, and substituting $\theta = \theta_r$, one obtains the following transformation matrix for the rotor reference frame

$$\mathbf{K}_s^r = \begin{bmatrix} \cos\theta_r & \sin\theta_r \\ \sin\theta_r & -\cos\theta_r \end{bmatrix}$$

Hence,

$$\mathbf{u}_{qds}^r = \mathbf{K}_s^r \mathbf{u}_{abs}$$

and therefore,

$$\begin{bmatrix} u_{qs}^r \\ u_{ds}^r \end{bmatrix} = \begin{bmatrix} \cos\theta_r & \sin\theta_r \\ \sin\theta_r & -\cos\theta_r \end{bmatrix} \begin{bmatrix} u_{as} \\ u_{bs} \end{bmatrix}$$

In expanded form, we have

$$u_{qs}^r = u_{as}\cos\theta_r + u_{bs}\sin\theta_r$$

$$u_{ds}^r = u_{as}\sin\theta_r - u_{bs}\cos\theta_r$$

Taking note of a balanced two-phase set, as given by

$$u_{as}(t) = \sqrt{2}u_M \cos(\omega_f t)$$

$$u_{bs}(t) = \sqrt{2}u_M \sin(\omega_f t)$$

the following equations for u_{qs}^r and u_{ds}^r result

$$u_{qs}^r(t) = \sqrt{2}u_M \cos(\omega_f t - \theta_r)$$
$$u_{ds}^r(t) = -\sqrt{2}u_M \sin(\omega_f t - \theta_r) \qquad (6.35)$$

Equations (6.35) guide one to the expressions for the *quadrature-* and *direct*-axis components of stator voltages needed to be supplied to guarantee balanced operating conditions for squirrel-cage induction motors. The electrical displacement θ_r is used to find u_{qs}^r and u_{ds}^r. Hence, in the rotor reference frame, one must measure or estimate the angular displacement θ_r to apply the voltages needed to guarantee the balanced conditions.

As one derives the differential equations, the block diagram for two-phase induction motors can be developed to visualize the model and to build the SIMULINK diagram to analyze the steady-state and transient characteristics. Using (6.34), the corresponding block diagram is built and documented in Figure 6.52.

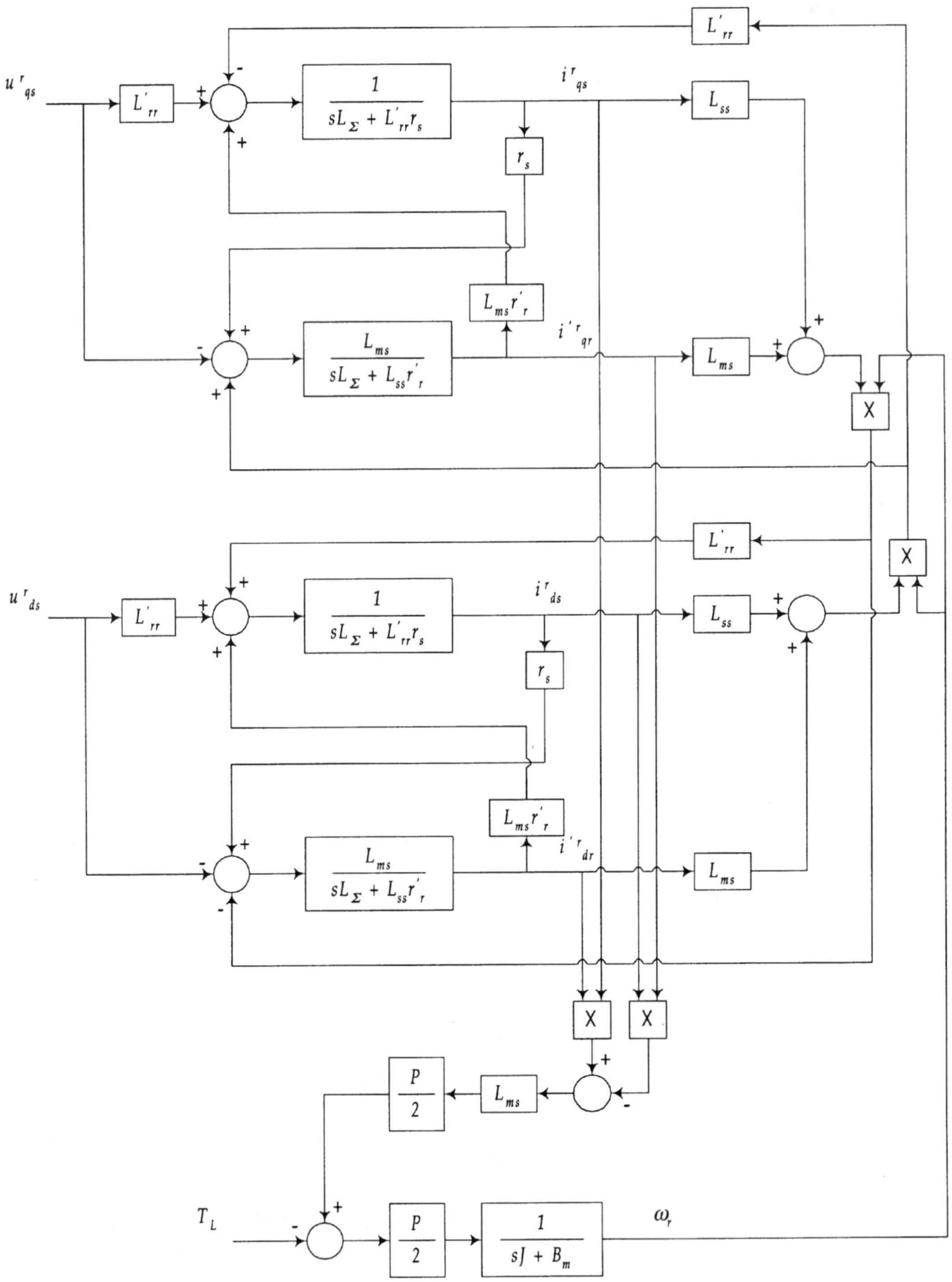

Figure 6.52 Block diagram of two-phase squirrel-cage induction motors modeled using the rotor reference frame.

Example 6.5

Using the rotor reference frame, model the A- and D-class squirrel-cage induction motors examined in Examples 6.1 to 6.4. Develop the SIMULINK diagram.

Solution

The SIMULINK diagram, as illustrated in Figure 6.53 using the parameters of the A-class motor, is built using differential equations (6.34) as well as the block diagram documented in Figure 6.52. To guarantee balanced operating conditions, the *quadrature*- and *direct*-axis components of stator voltages are obtained using the Park transformation. In particular, from (6.35), we have the following expressions for u^r_{qs} and u^r_{ds}

$$u^r_{qs}(t) = \sqrt{2}u_M \cos(\omega_f t - \theta_r)$$

$$u^r_{ds}(t) = -\sqrt{2}u_M \sin(\omega_f t - \theta_r)$$

Applying the rated voltages and using a two-phase balanced set

$$u_{as}(t) = \sqrt{2}115\cos(377t) \text{ and } u_{bs}(t) = \sqrt{2}115\sin(377t)$$

one finds

$$u^r_{qs}(t) = \sqrt{2}115\cos(377t - \theta_r) \text{ and } u^r_{ds}(t) = -\sqrt{2}115\sin(377t - \theta_r)$$

Motors start from stall, and as the rated voltage is applied, the goal is to examine the starting capabilities, dynamic performance, and torque-speed characteristics. Figures 6.54 and 6.55 illustrate the transient dynamics and torque-speed characteristic curves for the A-class squirrel-cage induction motor if $T_L = 0$ N-m and $T_L = 0.5$ N-m. The D-class motor dynamics and torque-speed characteristics for the unloaded and loaded motor are plotted in Figures 6.56 and 6.57.

Analysis of induction motors using flux linkages

The mathematical model in the rotor reference frame, if i^r_{qs}, i^r_{ds}, ψ'^r_{qr}, ψ'^r_{dr}, ω_r, and θ_r are assigned as the state variables, can be easily developed using the differential equations that map the induction motors dynamics in the *arbitrary* reference frame (6.31). In particular, substituting $\omega = \omega_r$, the following differential equations are found to be

$$\frac{di^r_{qs}}{dt} = \frac{1}{L_\Sigma}\left[-\frac{L'^2_{rr}r_s + L^2_{ms}r'_r}{L'_{rr}}i^r_{qs} + \frac{L_{ms}r'_r}{L'_{rr}}\psi'^r_{qr} - (L_\Sigma i^r_{ds} + L_{ms}\psi'^r_{dr})\omega_r + L'_{rr}u^r_{qs} - L_{ms}u'^r_{qr}\right]$$

$$\frac{di^r_{ds}}{dt} = \frac{1}{L_\Sigma}\left[-\frac{L'^2_{rr}r_s + L^2_{ms}r'_r}{L'_{rr}}i^r_{ds} + \frac{L_{ms}r'_r}{L'_{rr}}\psi'^r_{dr} + (L_\Sigma i^r_{qs} + L_{ms}\psi'^r_{qr})\omega_r + L'_{rr}u^r_{ds} - L_{ms}u'^r_{dr}\right]$$

$$\frac{d\psi'^r_{qr}}{dt} = \frac{L_{ms}r'_r}{L'_{rr}}i^r_{qs} - \frac{r'_r}{L'_{rr}}\psi'^r_{qr} + u'^r_{qr}$$

$$\frac{d\psi'^r_{dr}}{dt} = \frac{L_{ms}r'_r}{L'_{rr}}i^r_{ds} - \frac{r'_r}{L'_{rr}}\psi'^r_{dr} + u'^r_{dr}$$

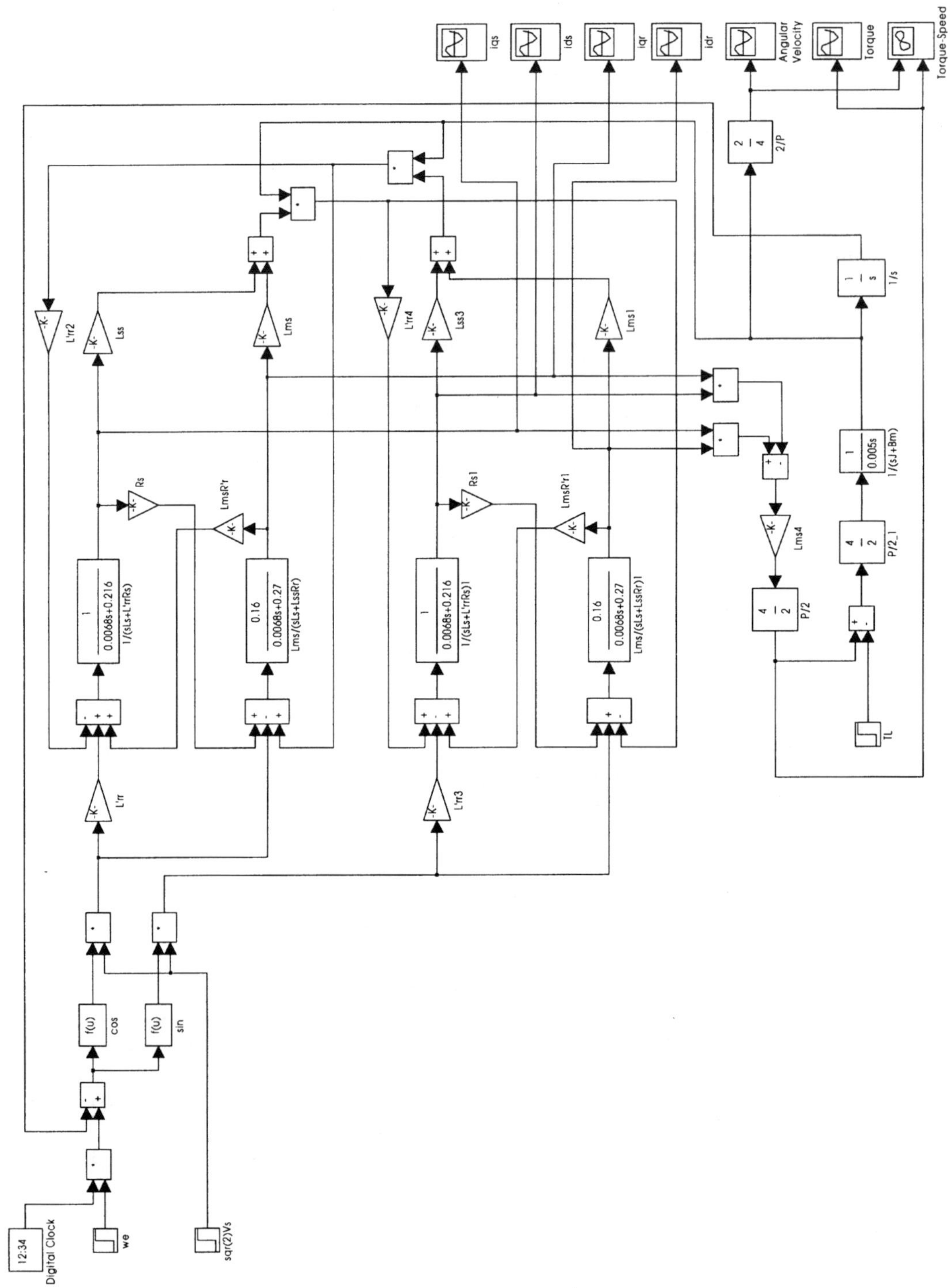

Figure 6.53 SIMULINK diagram of the A-class squirrel-cage induction motor in the rotor reference frame.

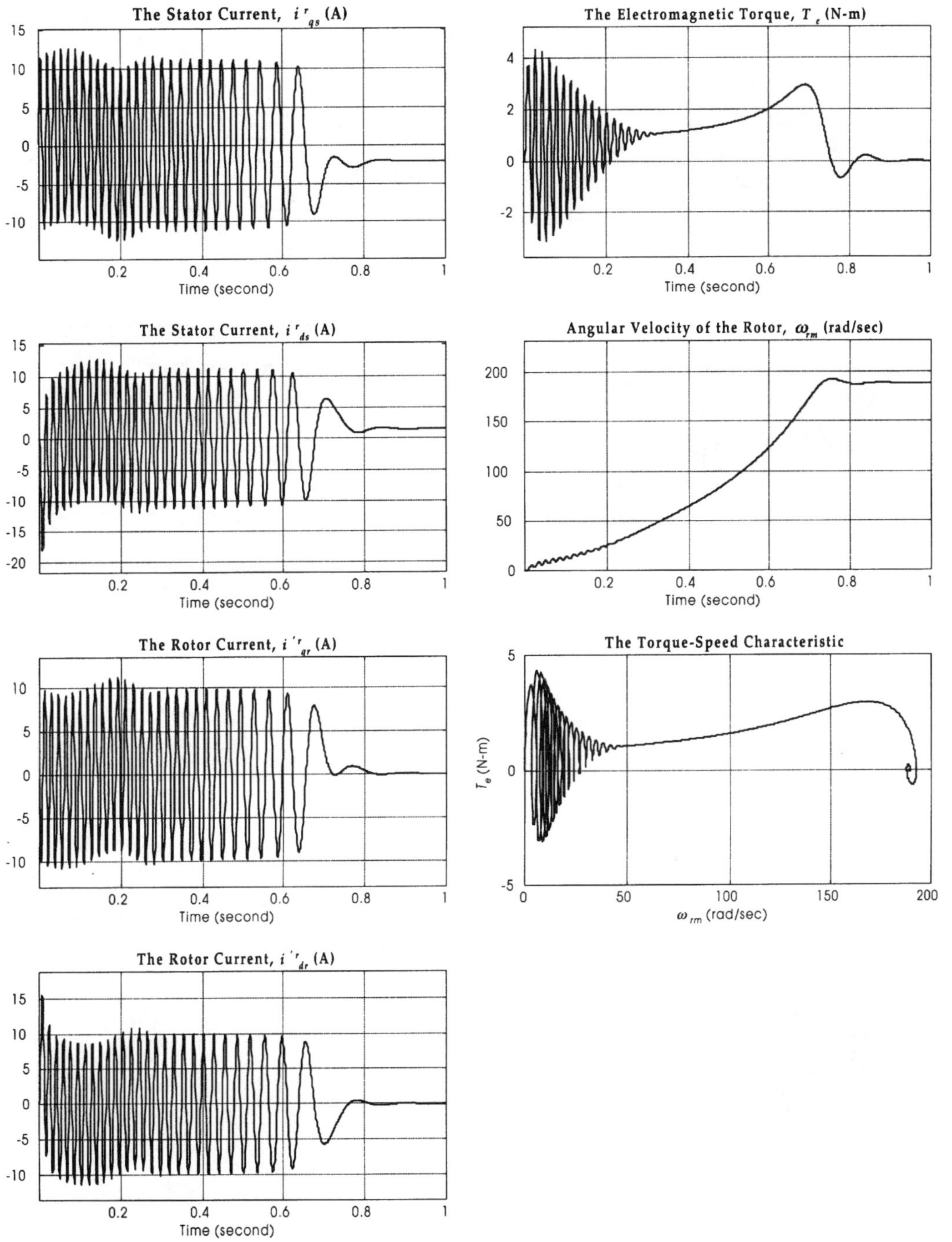

Figure 6.54 Dynamics and torque-speed characteristic of the A-class induction motor modeled in the rotor reference frame; $T_L = 0$ N-m, $u^{r}_{qs}(t) = \sqrt{2}115\cos(377t - \theta_r)$, and $u^{r}_{ds}(t) = -\sqrt{2}115\sin(377t - \theta_r)$.

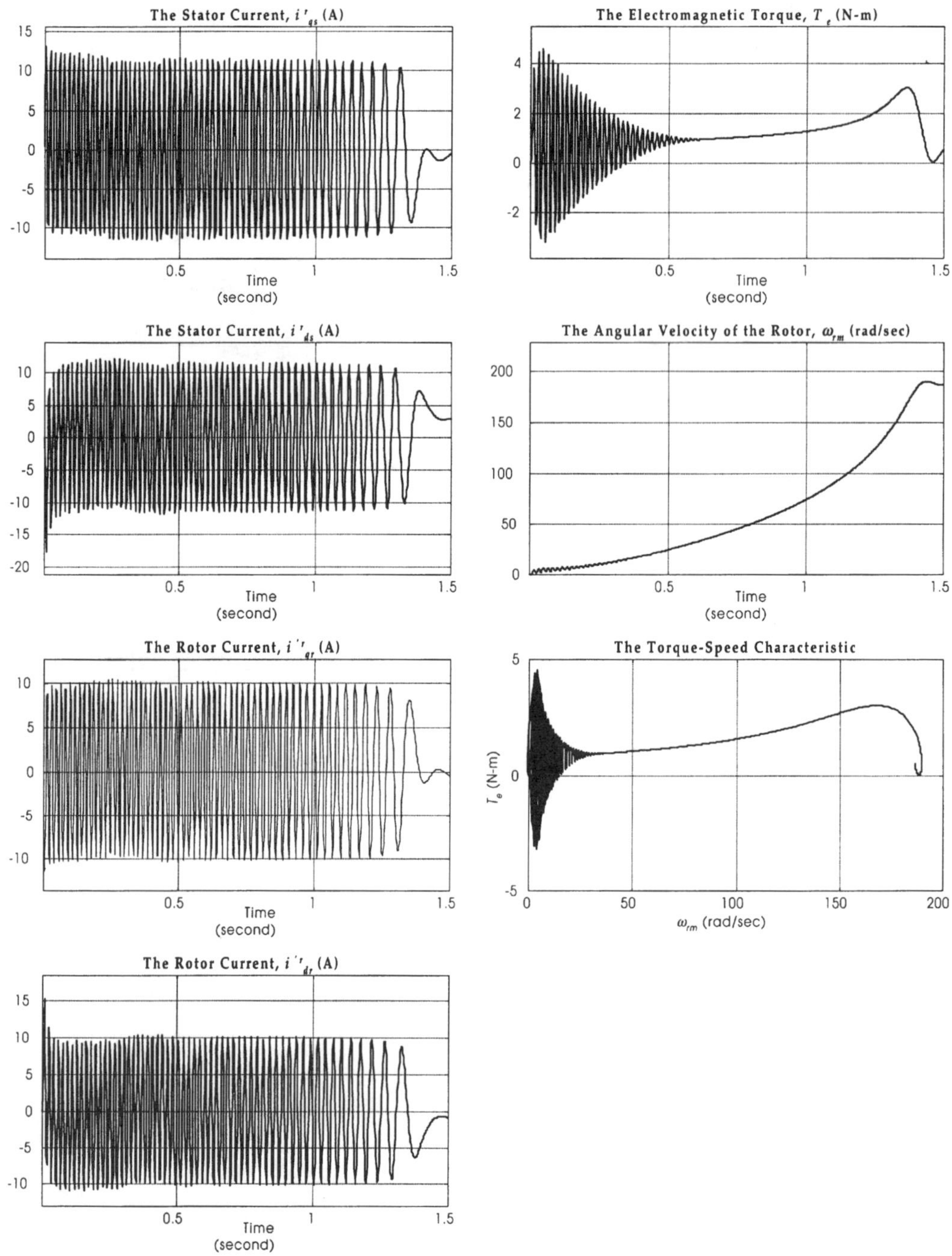

Figure 6.55 Dynamics and torque-speed characteristic of the A-class induction motor modeled in the rotor reference frame; $T_L = 0.5$ N-m, $u_{qs}^{r}(t) = \sqrt{2}115\cos(377t - \theta_r)$, and $u_{ds}^{r}(t) = -\sqrt{2}115\sin(377t - \theta_r)$.

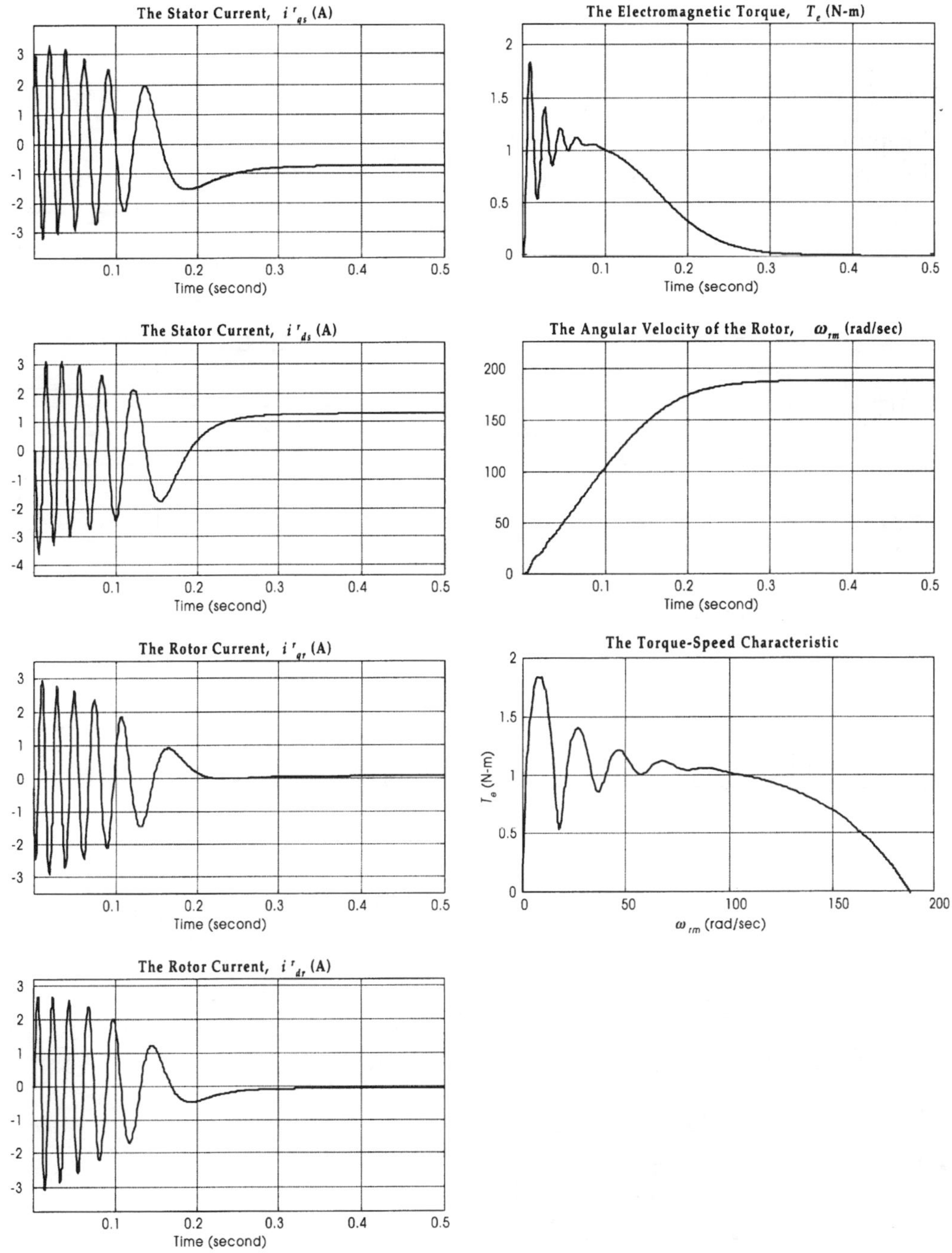

Figure 6.56 Dynamics and torque-speed characteristic of the D-class induction motor modeled in the rotor reference frame; $T_L = 0$ N-m, $u^r_{qs}(t) = \sqrt{2}115\cos(377t - \theta_r)$, and $u^r_{ds}(t) = -\sqrt{2}115\sin(377t - \theta_r)$.

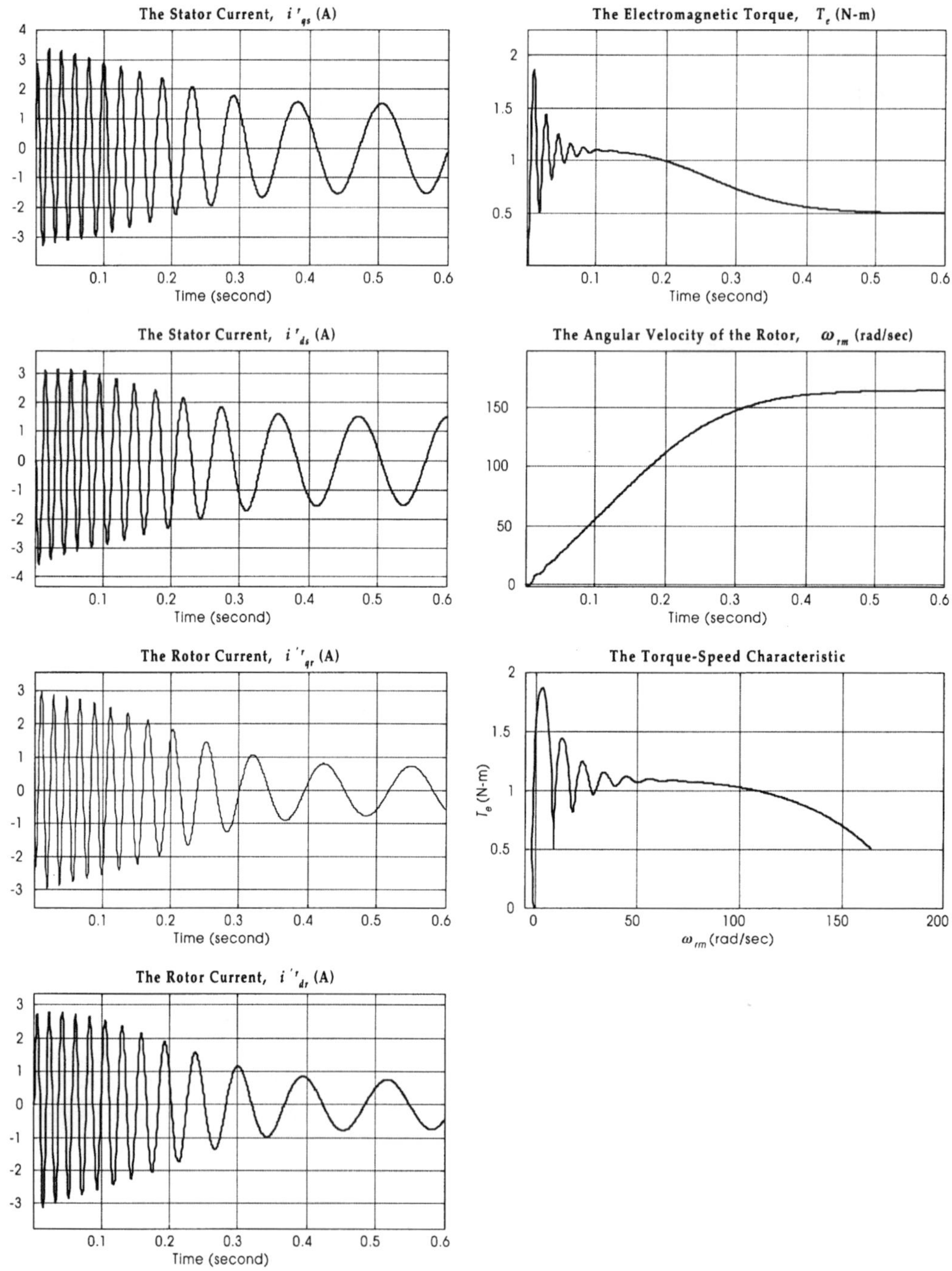

Figure 6.57 Dynamics and torque-speed characteristic of the D-class induction motor modeled in the rotor reference frame; $T_L = 0.5$ N-m, $u^r_{qs}(t) = \sqrt{2}115\cos(377t - \theta_r)$, and $u^r_{ds}(t) = -\sqrt{2}115\sin(377t - \theta_r)$.

$$\frac{d\omega_r}{dt} = \frac{P^2}{4J}\frac{L_{ms}}{L'_{rr}}(i^r_{qs}\psi'^r_{dr} - i^r_{ds}\psi'^r_{qr}) - \frac{B_m}{J}\omega_r - \frac{P}{2J}T_L$$

$$\frac{d\theta_r}{dt} = \omega_r$$

In matrix form, this set of differential equations is rewritten as

$$\begin{bmatrix} \frac{di^r_{qs}}{dt} \\ \frac{di^r_{ds}}{dt} \\ \frac{d\psi'^r_{qr}}{dt} \\ \frac{d\psi'^r_{dr}}{dt} \\ \frac{d\omega_r}{dt} \\ \frac{d\theta_r}{dt} \end{bmatrix} = \begin{bmatrix} -\frac{L'^2_{rr}r_s + L^2_{ms}r'_r}{L_\Sigma L'_{rr}} & 0 & \frac{L_{ms}r'_r}{L_\Sigma L'_{rr}} & 0 & 0 & 0 \\ 0 & -\frac{L'^2_{rr}r_s + L^2_{ms}r'_r}{L_\Sigma L'_{rr}} & 0 & \frac{L_{ms}r'_r}{L_\Sigma L'_{rr}} & 0 & 0 \\ \frac{L_{ms}r'_r}{L'_{rr}} & 0 & -\frac{r'_r}{L'_{rr}} & 0 & 0 & 0 \\ 0 & \frac{L_{ms}r'_r}{L'_{rr}} & 0 & -\frac{r'_r}{L'_{rr}} & 0 & 0 \\ 0 & 0 & 0 & 0 & -\frac{B_m}{J} & 0 \\ 0 & 0 & 0 & 0 & 1 & 0 \end{bmatrix} \begin{bmatrix} i^r_{qs} \\ i^r_{ds} \\ \psi'^r_{qr} \\ \psi'^r_{dr} \\ \omega_r \\ \theta_r \end{bmatrix}$$

$$+ \begin{bmatrix} -\frac{(L_\Sigma i^r_{ds} + L_{ms}\psi'^r_{dr})\omega_r}{L_\Sigma} \\ \frac{(L_\Sigma i^r_{qs} + L_{ms}\psi'^r_{qr})\omega_r}{L_\Sigma} \\ 0 \\ 0 \\ \frac{P^2}{4J}\frac{L_{ms}}{L'_{rr}}(i^r_{qs}\psi'^r_{dr} - i^r_{ds}\psi'^r_{qr}) \\ 0 \end{bmatrix} + \begin{bmatrix} \frac{L'_{rr}}{L_\Sigma} & 0 & \frac{-L_{ms}}{L_\Sigma} & 0 \\ 0 & \frac{L'_{rr}}{L_\Sigma} & 0 & \frac{-L_{ms}}{L_\Sigma} \\ 0 & 0 & 1 & 0 \\ 0 & 0 & 0 & 1 \\ 0 & 0 & 0 & 0 \\ 0 & 0 & 0 & 0 \end{bmatrix} \begin{bmatrix} u^r_{qs} \\ u^r_{ds} \\ u'^r_{qr} \\ u'^r_{dr} \end{bmatrix} - \begin{bmatrix} 0 \\ 0 \\ 0 \\ 0 \\ \frac{P}{2J} \\ 0 \end{bmatrix} T_L$$

For completeness of the analysis, the block diagram of two-phase induction motors in the rotor reference frame is developed and illustrated in Figure 6.58.

Example 6.6

Using the following state variables i^r_{qs}, i^r_{ds}, ψ'^r_{qr}, ψ'^r_{dr}, ω_r, and θ_r, develop the SIMULINK diagram and perform the modeling of two-phase squirrel-cage motors studied in Examples 6.1 to 6.5 in the rotor reference frame.

Solution

An s-domain block diagram for two-phase induction motors was built and documented in Figure 6.58. Using the parameters for an A-class motor, the corresponding SIMULINK diagram is developed (see Figure 6.59). The nonlinear modeling is performed for balanced operating conditions by supplying the following *quadrature* and *direct* stator voltages

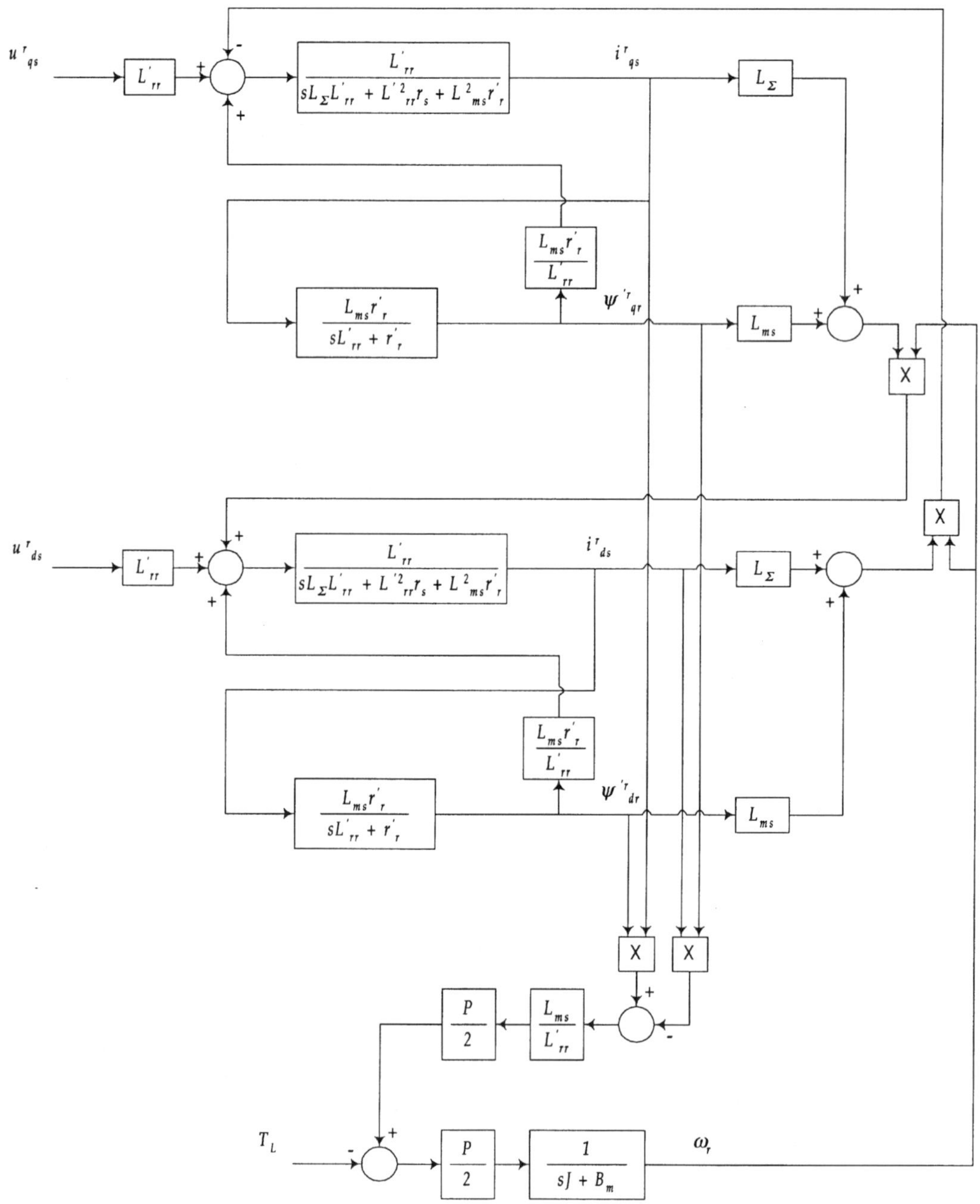

Figure 6.58 Block diagram of two-phase squirrel-cage induction motors modeled in the rotor reference frame.

$$u_{qs}^{r}(t) = \sqrt{2}115\cos(377t - \theta_r)$$

$$u_{ds}^{r}(t) = -\sqrt{2}115\sin(377t - \theta_r)$$

Figures 6.60 to 6.63 document the transient dynamics of the A- and D-class induction motors that are started from stall. The torque-speed characteristic curves are plotted.

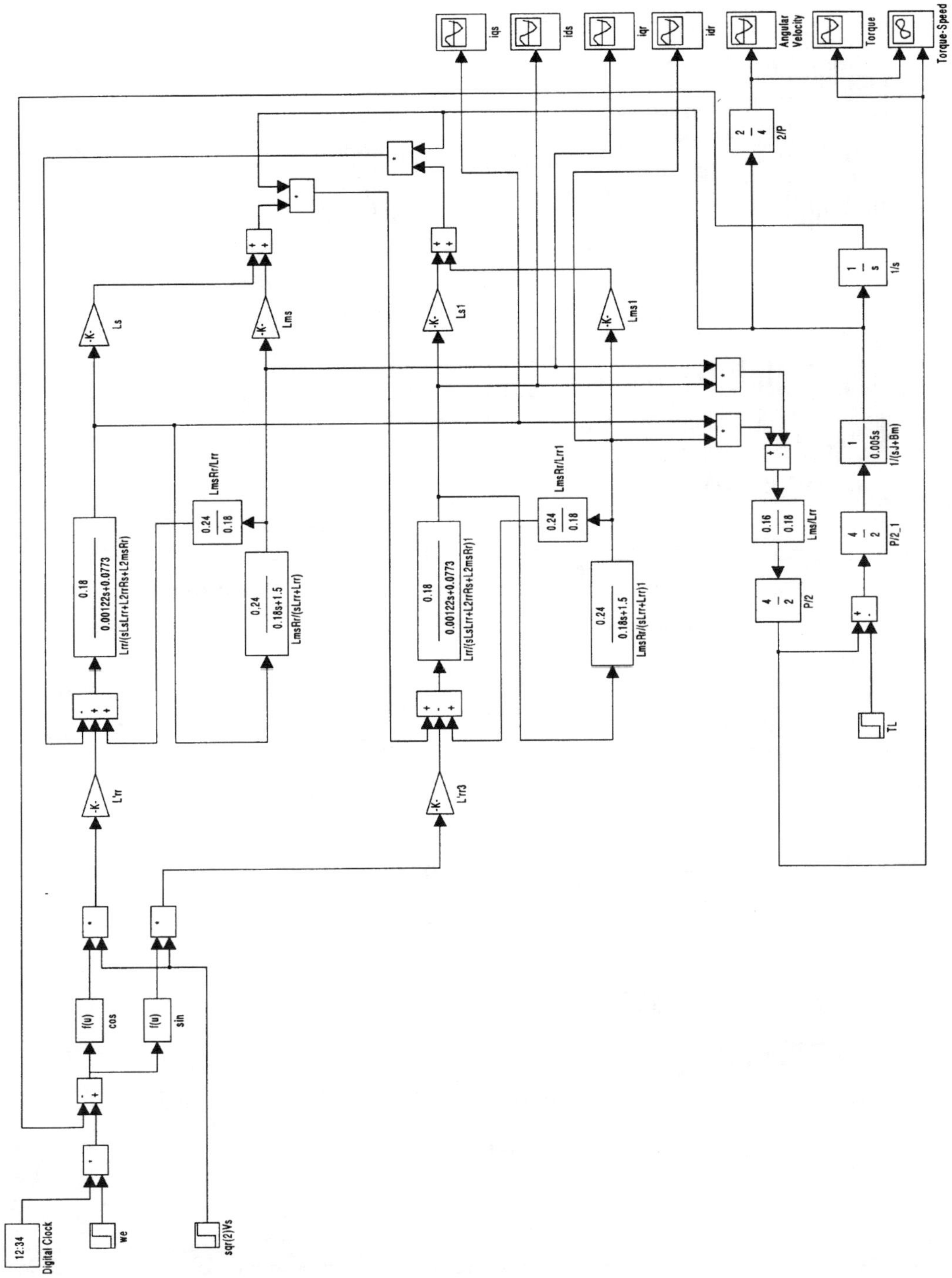

Figure 6.59 Simulink diagram for the A-class induction motor modeled in the rotor reference frame.

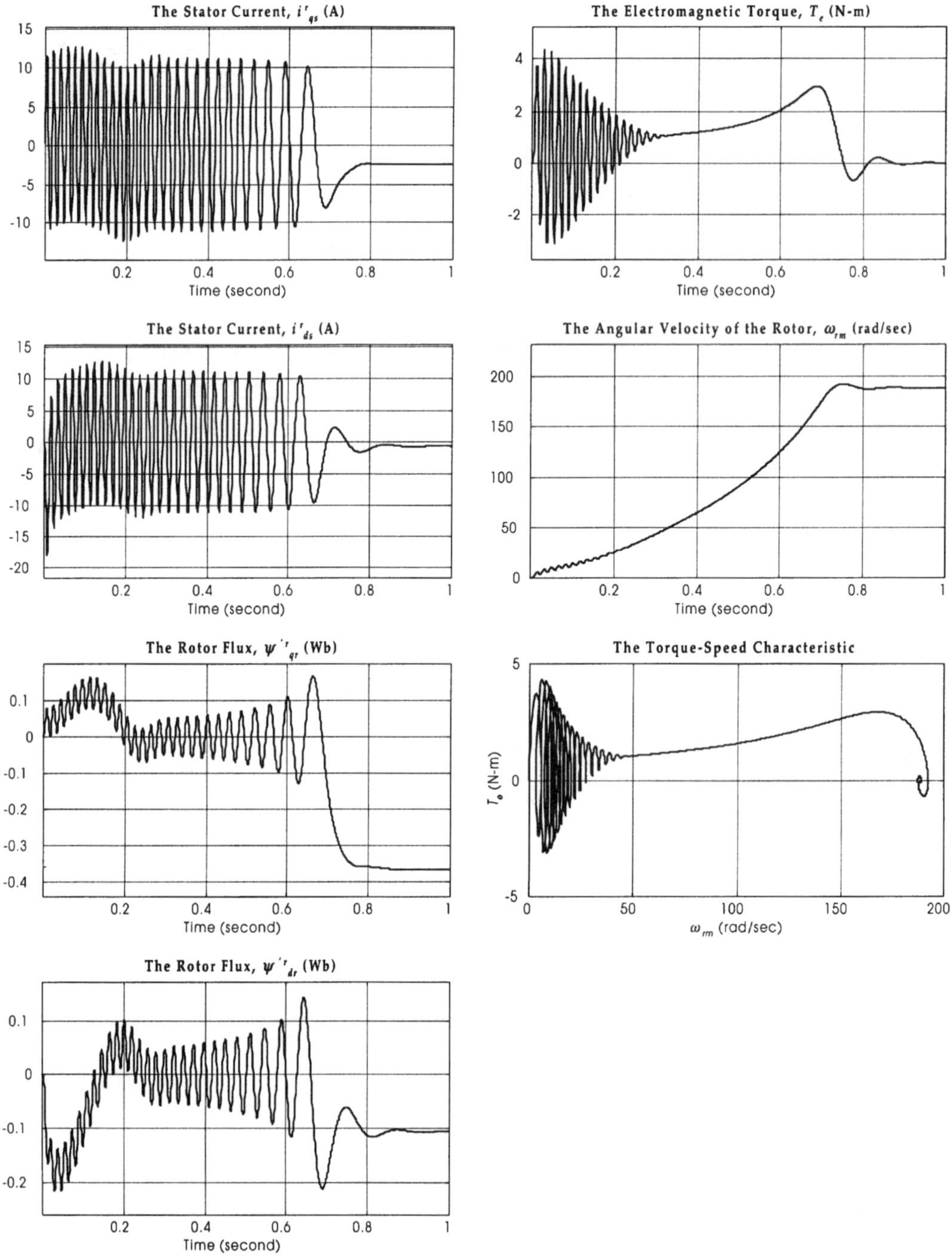

Figure 6.60 Dynamics and torque-speed characteristic of the A-class induction motor modeled in the rotor reference frame; $T_L = 0$ N-m, $u^r_{qs}(t) = \sqrt{2}115\cos(377t - \theta_r)$, and $u^r_{ds}(t) = -\sqrt{2}115\sin(377t - \theta_r)$.

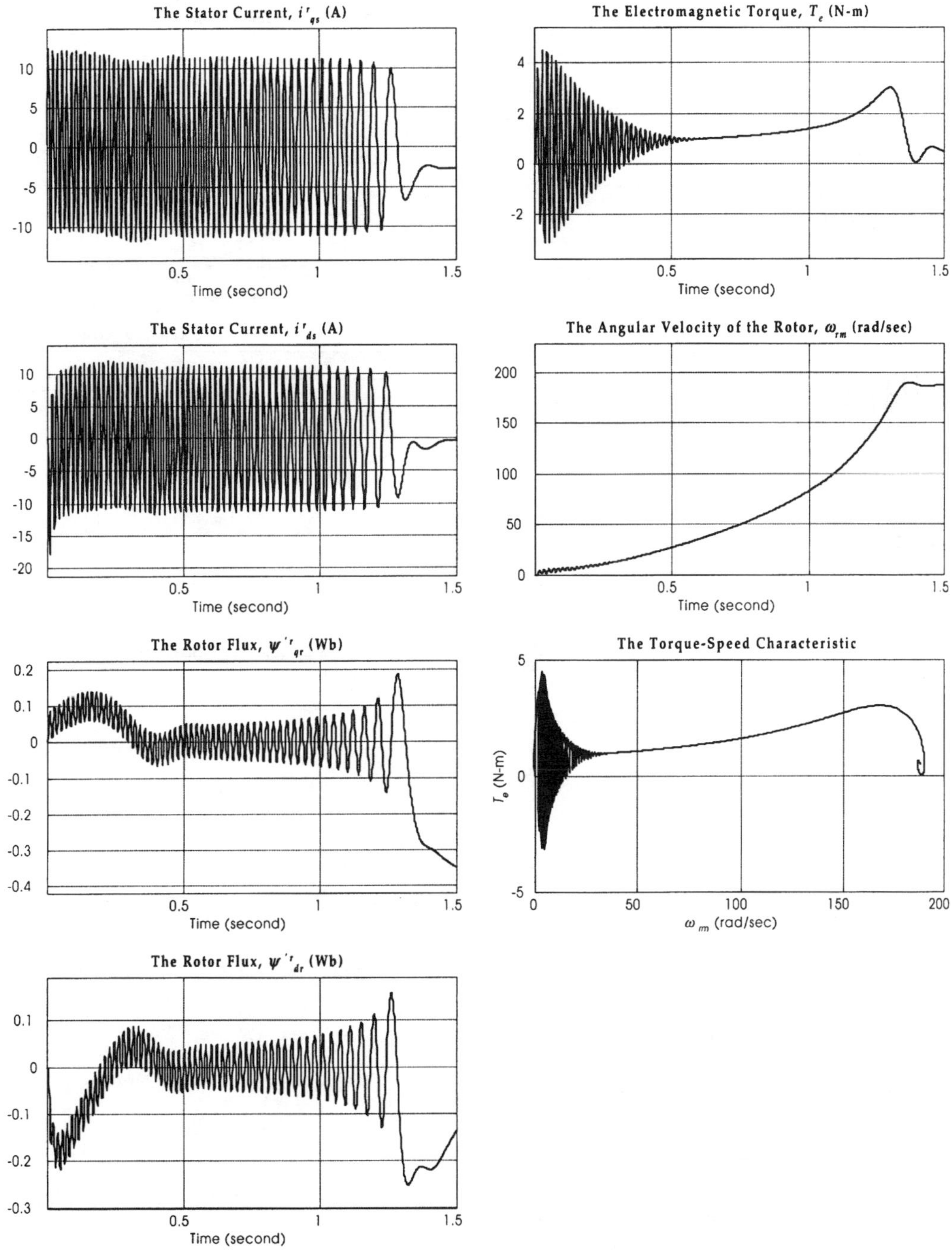

Figure 6.61 Dynamics and torque-speed characteristic of the A-class induction motor modeled in the rotor reference frame; $T_L = 0.5$ N-m, $u^r_{qs}(t) = \sqrt{2}115\cos(377t - \theta_r)$, and $u^r_{ds}(t) = -\sqrt{2}115\sin(377t - \theta_r)$.

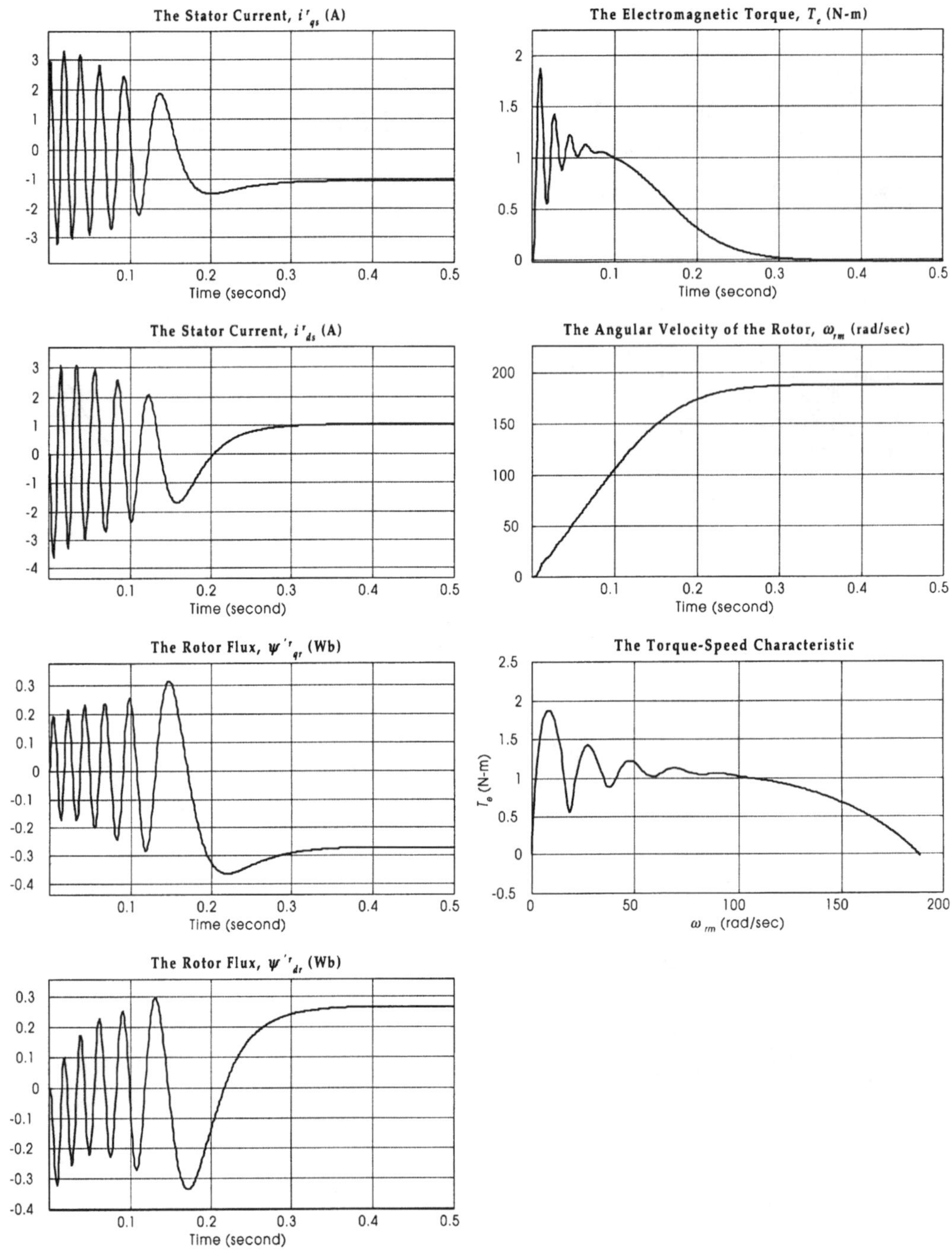

Figure 6.62 Dynamics and torque-speed characteristic of the D-class induction motor modeled in the rotor reference frame; $T_L = 0$ N-m, $u^r_{qs}(t) = \sqrt{2}115\cos(377t - \theta_r)$, and $u^r_{ds}(t) = -\sqrt{2}115\sin(377t - \theta_r)$.

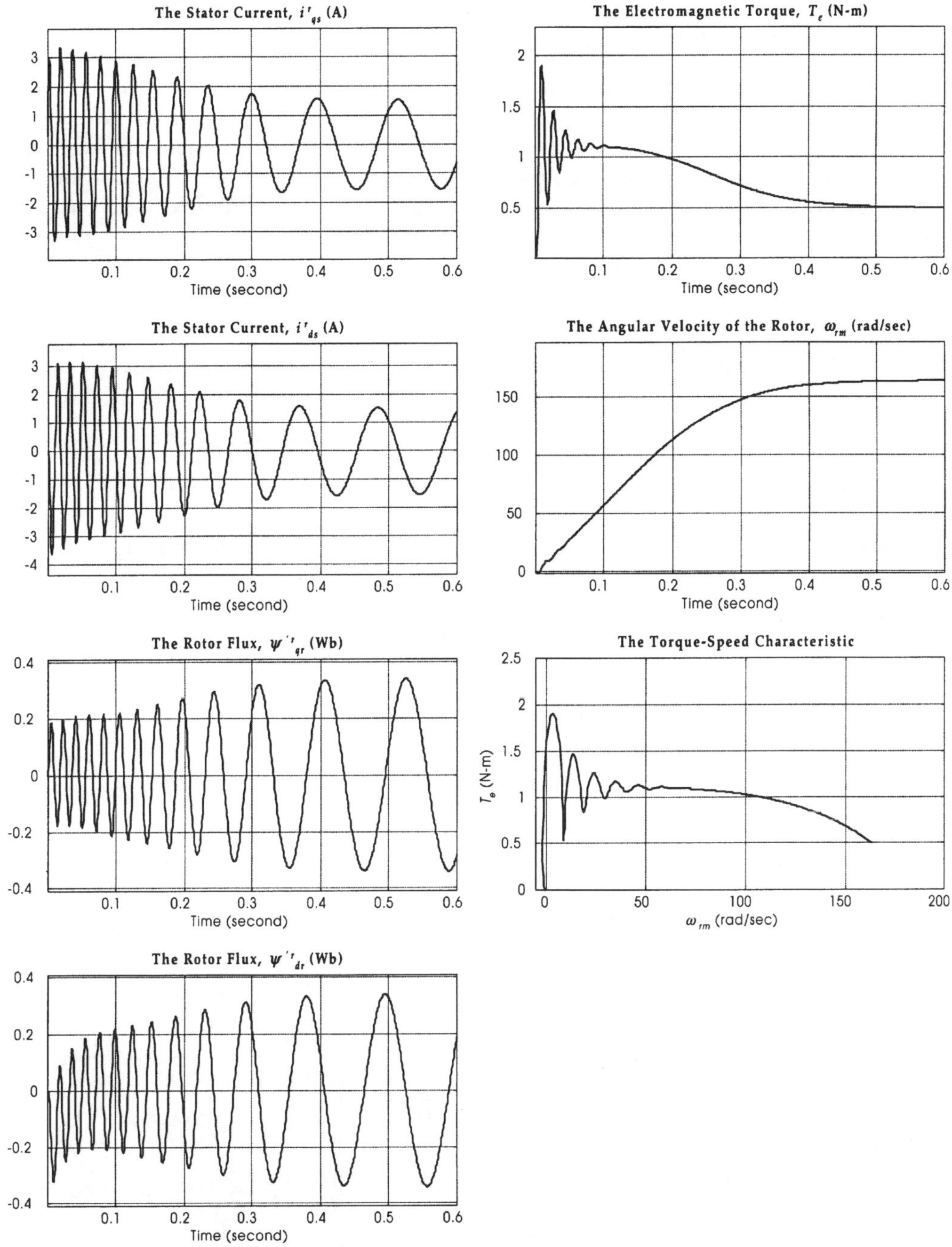

Figure 6.63 Dynamics and torque-speed characteristic of the D-class induction motor modeled in the rotor reference frame; $T_L = 0.5$ N-m, $u^r_{qs}(t) = \sqrt{2}115\cos(377t - \theta_r)$, and $u^r_{ds}(t) = -\sqrt{2}115\sin(377t - \theta_r)$.

6.3.3 Development of the mathematical model of two-phase induction motors in the synchronous reference frame

The mathematical model of induction motors in the synchronous reference frame is developed by substituting the corresponding reference frame angular velocity in the differential equations (6.27). That is, by letting $\omega = \omega_e$, the model in the synchronous reference frame results, and the superscript e indicates that the *quadrature*- and *direct*-axis components of voltages, currents, and flux linkages are given in the synchronous frame of reference. One obtains

$$\begin{aligned}
\frac{di_{qs}^e}{dt} &= \frac{1}{L_\Sigma}[-L'_{rr}r_s i_{qs}^e + L_{ms}r'_r i'^e_{qr} - L_\Sigma\omega_e i_{ds}^e - L_{ms}(L_{ms}i_{ds}^e + L'_{rr}i'^e_{dr})\omega_r + L'_{rr}u_{qs}^e - L_{ms}u'^e_{qr}]\\
\frac{di_{ds}^e}{dt} &= \frac{1}{L_\Sigma}[L_\Sigma\omega_e i_{qs}^e - L'_{rr}r_s i_{ds}^e + L_{ms}r'_r i'^e_{dr} + L_{ms}(L_{ms}i_{qs}^e + L'_{rr}i'^e_{qr})\omega_r + L'_{rr}u_{ds}^e - L_{ms}u'^e_{dr}]\\
\frac{di'^e_{qr}}{dt} &= \frac{1}{L_\Sigma}[L_{ms}r_s i_{qs}^e - L_{ss}r'_r i'^e_{qr} - L_\Sigma\omega_e i'^e_{dr} + L_{ss}(L_{ms}i_{ds}^e + L'_{rr}i'^e_{dr})\omega_r - L_{ms}u_{qs}^e + L_{ss}u'^e_{qr}]\\
\frac{di'^e_{dr}}{dt} &= \frac{1}{L_\Sigma}[L_{ms}r_s i_{ds}^e + L_\Sigma\omega_e i'^e_{qr} - L_{ss}r'_r i'^e_{dr} - L_{ss}(L_{ms}i_{qs}^e + L'_{rr}i'^e_{qr})\omega_r - L_{ms}u_{ds}^e + L_{ss}u'^e_{dr}]\\
\frac{d\omega_r}{dt} &= \frac{P^2}{4J}L_{ms}(i_{qs}^e i'^e_{dr} - i_{ds}^e i'^e_{qr}) - \frac{B_m}{J}\omega_r - \frac{P}{2J}T_L\\
\frac{d\theta_r}{dt} &= \omega_r
\end{aligned} \tag{6.36}$$

Hence, in matrix form, we have

$$\begin{bmatrix}\frac{di_{qs}^e}{dt}\\ \frac{di_{ds}^e}{dt}\\ \frac{di'^e_{qr}}{dt}\\ \frac{di'^e_{dr}}{dt}\\ \frac{d\omega_r}{dt}\\ \frac{d\theta_r}{dt}\end{bmatrix} = \begin{bmatrix} -\frac{L'_{rr}r_s}{L_\Sigma} & -\omega_e & \frac{L_{ms}r'_r}{L_\Sigma} & 0 & 0 & 0\\ \omega_e & -\frac{L'_{rr}r_s}{L_\Sigma} & 0 & \frac{L_{ms}r'_r}{L_\Sigma} & 0 & 0\\ \frac{L_{ms}r_s}{L_\Sigma} & 0 & -\frac{L_{ss}r'_r}{L_\Sigma} & -\omega_e & 0 & 0\\ 0 & \frac{L_{ms}r_s}{L_\Sigma} & \omega_e & -\frac{L_{ss}r'_r}{L_\Sigma} & 0 & 0\\ 0 & 0 & 0 & 0 & -\frac{B_m}{J} & 0\\ 0 & 0 & 0 & 0 & 1 & 0\end{bmatrix}\begin{bmatrix} i_{qs}^e\\ i_{ds}^e\\ i'^e_{qr}\\ i'^e_{dr}\\ \omega_r\\ \theta_r\end{bmatrix} + \begin{bmatrix} -\frac{L_{ms}}{L_\Sigma}(L_{ms}i_{ds}^e + L'_{rr}i'^e_{dr})\omega_r\\ \frac{L_{ms}}{L_\Sigma}(L_{ms}i_{qs}^e + L'_{rr}i'^e_{qr})\omega_r\\ \frac{L_{ss}}{L_\Sigma}(L_{ms}i_{ds}^e + L'_{rr}i'^e_{dr})\omega_r\\ -\frac{L_{ss}}{L_\Sigma}(L_{ms}i_{qs}^e + L'_{rr}i'^e_{qr})\omega_r\\ \frac{P^2L_{ms}}{4J}(i_{qs}^e i'^e_{dr} - i_{ds}^e i'^e_{qr})\\ 0\end{bmatrix}$$

$$+\begin{bmatrix} \frac{L'_{rr}}{L_\Sigma} & 0 & -\frac{L_{ms}}{L_\Sigma} & 0 \\ 0 & \frac{L'_{rr}}{L_\Sigma} & 0 & -\frac{L_{ms}}{L_\Sigma} \\ -\frac{L_{ms}}{L_\Sigma} & 0 & \frac{L_{ss}}{L_\Sigma} & 0 \\ 0 & -\frac{L_{ms}}{L_\Sigma} & 0 & \frac{L_{ss}}{L_\Sigma} \\ 0 & 0 & 0 & 0 \\ 0 & 0 & 0 & 0 \end{bmatrix}\begin{bmatrix} u^e_{qs} \\ u^e_{ds} \\ u'^e_{qr} \\ u'^e_{dr} \end{bmatrix} - \begin{bmatrix} 0 \\ 0 \\ 0 \\ 0 \\ \frac{P}{2J} \\ 0 \end{bmatrix} T_L$$

To guarantee balanced operating conditions, the phase voltages to be supplied to the stator windings must be found. The *quadrature*- and *direct*-axis components of voltages are obtained using the Park transformation matrices. In particular, in the *arbitrary* reference frame, to transform the *machine* stator variables, we have to apply the following transformation matrix

$$\mathbf{K}_s = \begin{bmatrix} \cos\theta & \sin\theta \\ \sin\theta & -\cos\theta \end{bmatrix}$$

and using $\mathbf{K}_s$ one finds

$$\mathbf{u}_{qds} = \mathbf{K}_s \mathbf{u}_{abs}$$

To find the *quadrature*- and *direct*-axis components of stator voltages u^e_{qs} and u^e_{ds} that need to be supplied to guarantee the balanced operating conditions, we use the corresponding Park transformation matrix as obtained using $\theta = \theta_e$. That is, in the synchronous reference frame,

$$\mathbf{K}^e_s = \begin{bmatrix} \cos\theta_e & \sin\theta_e \\ \sin\theta_e & -\cos\theta_e \end{bmatrix}$$

For the stator voltages, the following formula is applied

$$\mathbf{u}^e_{qds} = \mathbf{K}^e_s \mathbf{u}_{abs}$$

and

$$\begin{bmatrix} u^e_{qs} \\ u^e_{ds} \end{bmatrix} = \begin{bmatrix} \cos\theta_e & \sin\theta_e \\ \sin\theta_e & -\cos\theta_e \end{bmatrix}\begin{bmatrix} u_{as} \\ u_{bs} \end{bmatrix}$$

Thus, one has

$$u^e_{qs} = u_{as}\cos\theta_e + u_{bs}\sin\theta_e$$

$$u^e_{ds} = u_{as}\sin\theta_e - u_{bs}\cos\theta_e$$

To guarantee the balanced operation of two-phase induction motors, the phase voltages that need to be supplied are expressed as

$$u_{as}(t)=\sqrt{2}u_M\cos(\omega_f t)$$

$$u_{bs}(t)=\sqrt{2}u_M\sin(\omega_f t)$$

Hence, one obtains the following expressions for the *quadrature*- and *direct*-axis components of voltages $u_{qs}^e(t)$ and $u_{ds}^e(t)$ in the synchronous reference frame

$$u_{qs}^e(t)=\sqrt{2}u_M\cos(\omega_f t-\theta_e)$$

$$u_{ds}^s(t)=-\sqrt{2}u_M\sin(\omega_f t-\theta_e)$$

Making use of $\theta_e=\int_{t_0}^{t}\omega_e(\tau)d\tau+\theta_e(t_0)$ and assigning $\theta_e(t_0)=0$, we have $\theta_e=\omega_e t$. Taking note that $\omega_f t=\omega_e t$, one concludes that the *quadrature*- and *direct*-axis components of stator voltages are given as

$$u_{qs}^e=\sqrt{2}u_M$$
$$u_{ds}^e=0 \tag{6.37}$$

The most important feature observed in the synchronous reference frame is the fact that the *quadrature*- and *direct*-axis components of voltages, currents, and flux linkages have dc quantities. This can be easily justified using the Park transformation. It has been shown that one applies the dc *quadrature* stator voltage u_{qs}^e (see equation (6.37)). Please also recall the results given in Illustrative Example 6.1. For a complete exposition of motor dynamics, one is advised to examine the transient responses for the *quadrature*- and *direct*-axis components of stator and rotor currents and flux linkages as given in Examples 6.7 and 6.8.

Taking note of (6.36) and using the Laplace operator s, we have

$$i_{qs}^e=\frac{1}{L_\Sigma s+L'_{rr}r_s}[L_{ms}r'_r i'^e_{qr}-L_\Sigma\omega_e i_{ds}^e-L_{ms}(L_{ms}i_{ds}^e+L'_{rr}i'^e_{dr})\omega_r+L'_{rr}u_{qs}^e-L_{ms}u'^e_{qr}]$$

$$i_{ds}^e=\frac{1}{L_\Sigma s+L'_{rr}r_s}[L_\Sigma\omega_e i_{qs}^e+L_{ms}r'_r i'^e_{dr}+L_{ms}(L_{ms}i_{qs}^e+L'_{rr}i'^e_{qr})\omega_r+L'_{rr}u_{ds}^e-L_{ms}u'^e_{dr}]$$

$$i'^e_{qr}=\frac{1}{L_\Sigma s+L_{ss}r'_r}[L_{ms}r_s i_{qs}^e-L_\Sigma\omega_e i'^e_{dr}+L_{ss}(L_{ms}i_{ds}^e+L'_{rr}i'^e_{dr})\omega_r-L_{ms}u_{qs}^e+L_{ss}u'^e_{qr}]$$

$$i_{dr}^e=\frac{1}{L_\Sigma s+L_{ss}r'_r}[L_{ms}r_s i_{ds}^e+L_\Sigma\omega_e i'^e_{qr}-L_{ss}(L_{ms}i_{qs}^e+L'_{rr}i'^e_{qr})\omega_r-L_{ms}u_{ds}^e+L_{ss}u'^e_{dr}]$$

$$\omega_r=\frac{P}{2}\left(\frac{1}{Js+B_m}\right)\left[\frac{P}{2J}L_{ms}(i_{qs}^e i'^e_{dr}-i_{ds}^e i'^e_{dr})-T_L\right]$$

and the block diagram of squirrel-cage induction motors is shown in Figure 6.64.

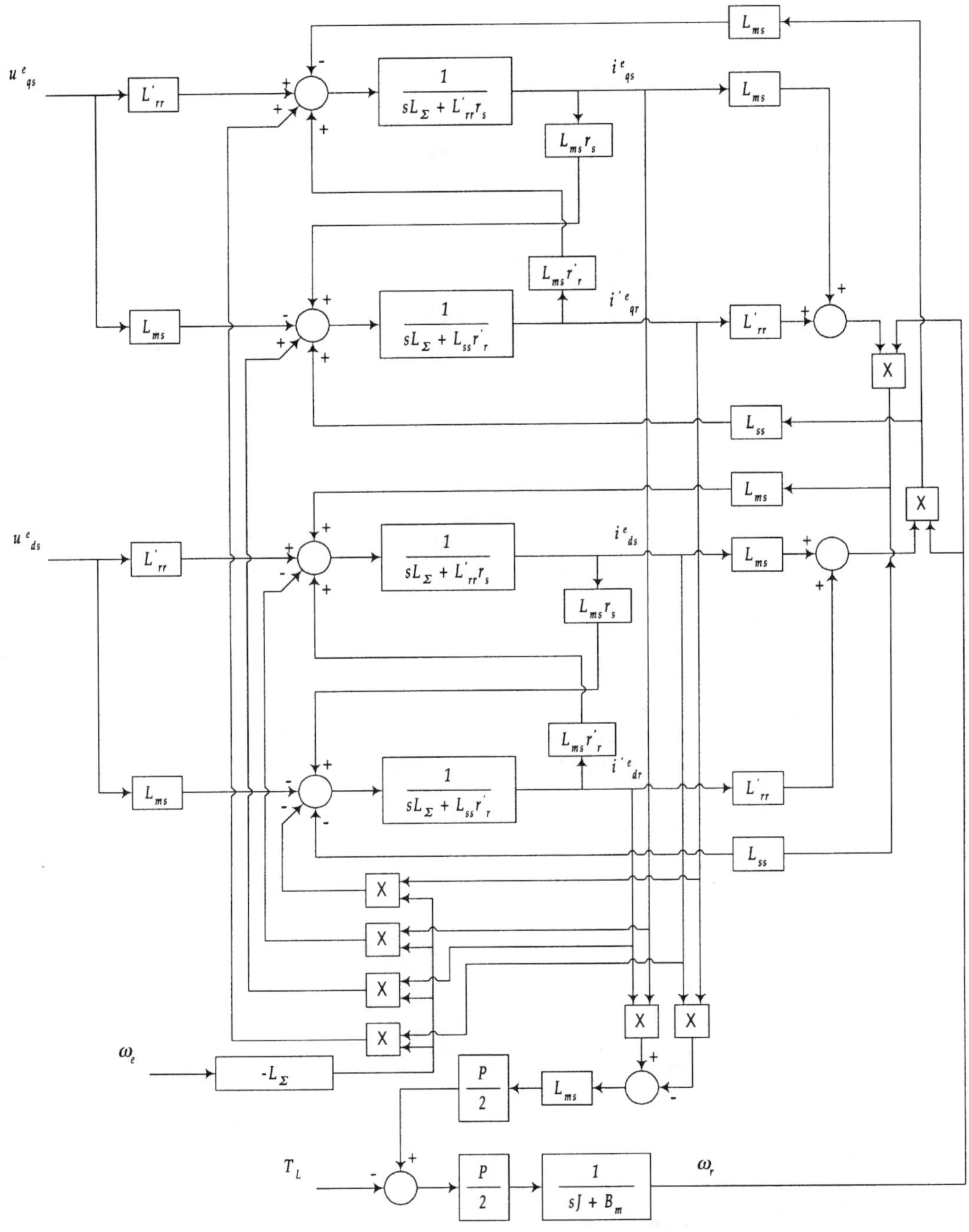

Figure 6.64 Block diagram of two-phase squirrel-cage induction motors in the synchronous reference frame.

Example 6.7

Using the parameters of two-phase squirrel-cage motors, as given in Table 6.1, perform the modeling of induction motors in the synchronous reference frame.

Solution

The SIMULINK diagram is developed to simulate nonlinear transient dynamics and plot the torque-speed characteristic curves (see Figure 6.65). Assume that the rated voltage is applied. Hence, for the balanced operating conditions, we have

$$u_{qs}^{e}=\sqrt{2}115,\ u_{ds}^{e}=0$$

The A- and D-class induction motors are modeled, and the dynamics are illustrated in Figures 6.66 to 6.69 for different load conditions. It must be emphasized that the *quadrature*- and *direct*-axis components of stator and rotor voltages and currents have dc quantities.

Analysis of induction motors using flux linkages

It was demonstrated that the rotor flux linkages can be used to map the induction machine dynamics. Using the differential equations in the *arbitrary* reference frame (6.31), and substituting $\omega = \omega_e$, we have a set of highly coupled differential equations to map the induction motor dynamics in the synchronous reference frame. The resulting differential equations are

$$\begin{aligned}
\frac{di_{qs}^{e}}{dt}&=\frac{1}{L_{\Sigma}}\left(-\frac{L_{rr}'^{2}r_{s}+L_{ms}^{2}r_{r}'}{L_{rr}'}i_{qs}^{e}-L_{\Sigma}\omega_{e}i_{ds}^{e}+\frac{L_{ms}r_{r}'}{L_{rr}'}\psi_{qr}'^{e}-L_{ms}\psi_{dr}'^{e}\omega_{r}+L_{rr}'u_{qs}^{e}-L_{ms}u_{qr}'^{e}\right)\\
\frac{di_{ds}^{e}}{dt}&=\frac{1}{L_{\Sigma}}\left(-\frac{L_{rr}'^{2}r_{s}+L_{ms}^{2}r_{r}'}{L_{rr}'}i_{ds}^{e}+\frac{L_{ms}r_{r}'}{L_{rr}'}\psi_{dr}'^{e}+L_{\Sigma}\omega_{e}i_{qs}^{e}+L_{ms}\psi_{qr}'^{e}\omega_{r}+L_{rr}'u_{ds}^{e}-L_{ms}u_{dr}'^{e}\right)\\
\frac{d\psi_{qr}'^{e}}{dt}&=\frac{L_{ms}r_{r}'}{L_{rr}'}i_{qs}^{e}-\frac{r_{r}'}{L_{rr}'}\psi_{qr}'^{e}-\omega_{e}\psi_{dr}'^{e}+\psi_{dr}'^{e}\omega_{r}+u_{qr}'^{e}\\
\frac{d\psi_{dr}'^{e}}{dt}&=\frac{L_{ms}r_{r}'}{L_{rr}'}i_{ds}^{e}+\omega_{e}\psi_{qr}'^{e}-\psi_{qr}'^{e}\omega_{r}-\frac{r_{r}'}{L_{rr}'}\psi_{dr}'^{e}+u_{dr}'^{e}\\
\frac{d\omega_{r}}{dt}&=\frac{P^{2}}{4J}\frac{L_{ms}}{L_{rr}'}(i_{qs}^{e}\psi_{dr}'^{e}-i_{ds}^{e}\psi_{qr}'^{e})-\frac{B_{m}}{J}\omega_{r}-\frac{P}{2J}T_{L}\\
\frac{d\theta_{r}}{dt}&=\omega_{r}
\end{aligned}\tag{6.38}$$

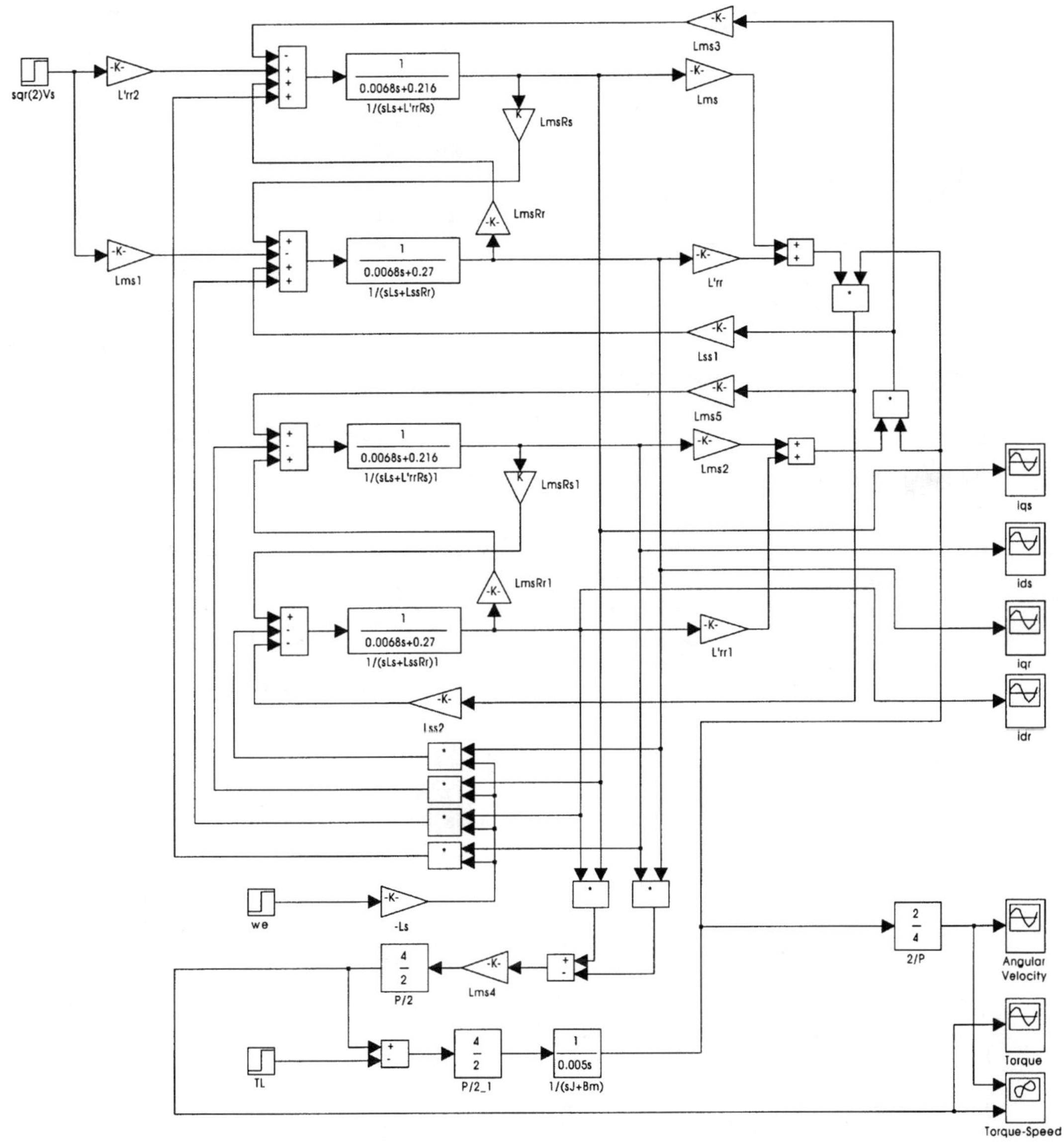

Figure 6.65 SIMULINK diagram for the A-class squirrel-cage induction motor modeled in the synchronous reference frame.

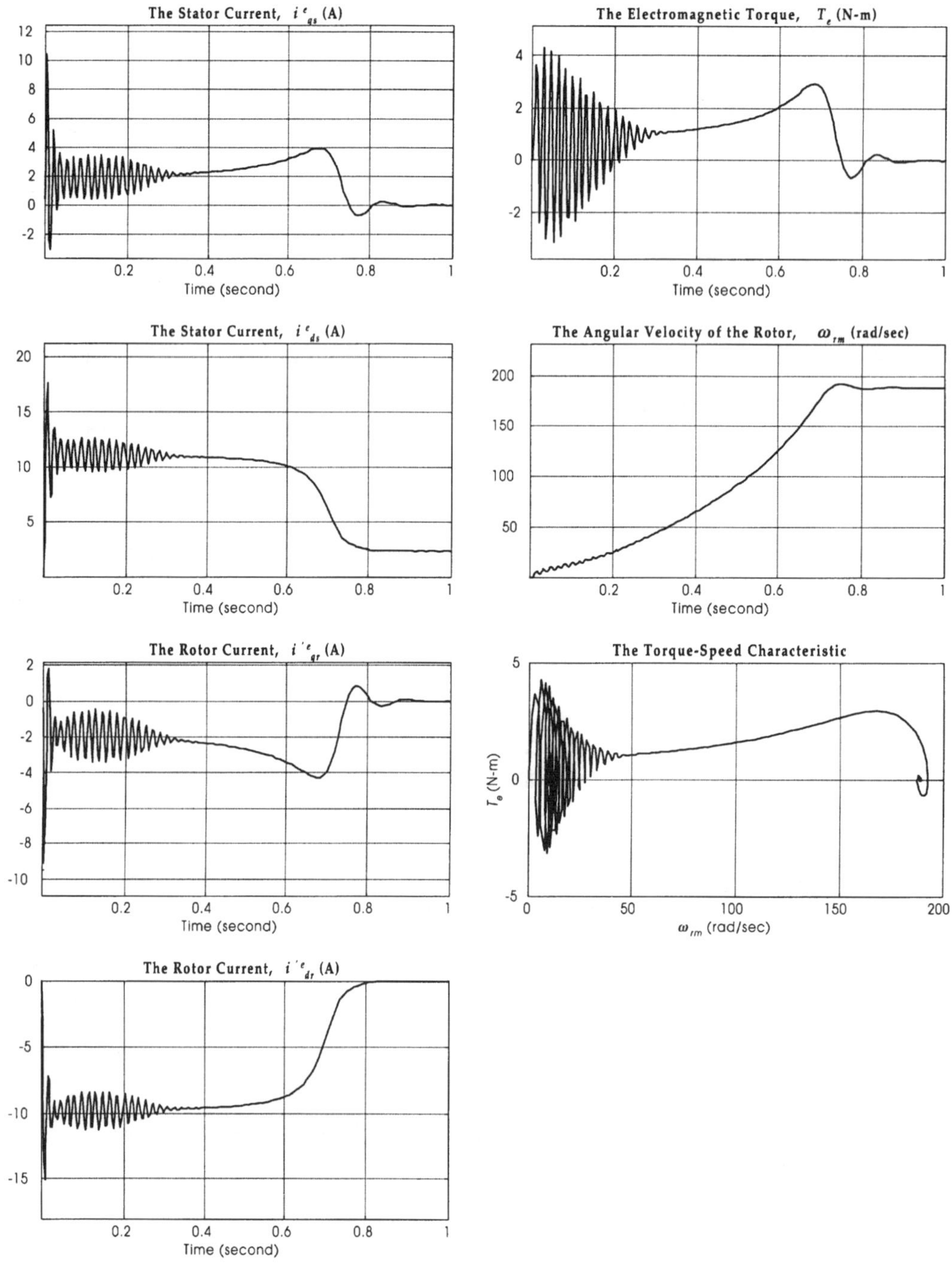

Figure 6.66 Dynamics and torque-speed characteristic of the A-class induction motor modeled in the synchronous reference frame; $T_L = 0$ N-m, $u^e_{qs}(t) = \sqrt{2}115$, and $u^e_{ds}(t) = 0$.

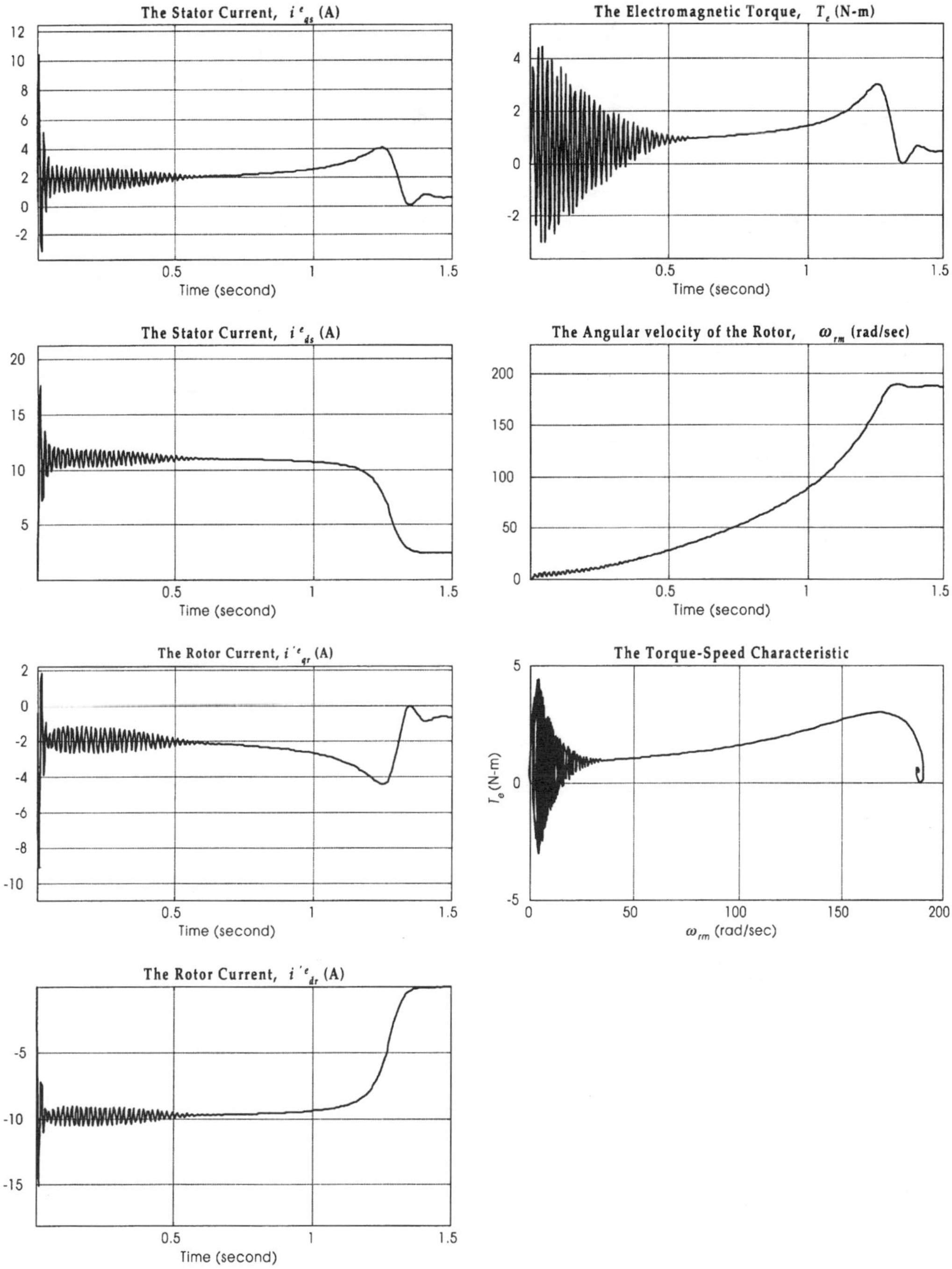

Figure 6.67 Dynamics and torque-speed characteristic of the A-class induction motor modeled in the synchronous reference frame; $T_L = 0.5$ N-m, $u^e_{qs}(t) = \sqrt{2}115$, and $u^e_{ds}(t) = 0$.

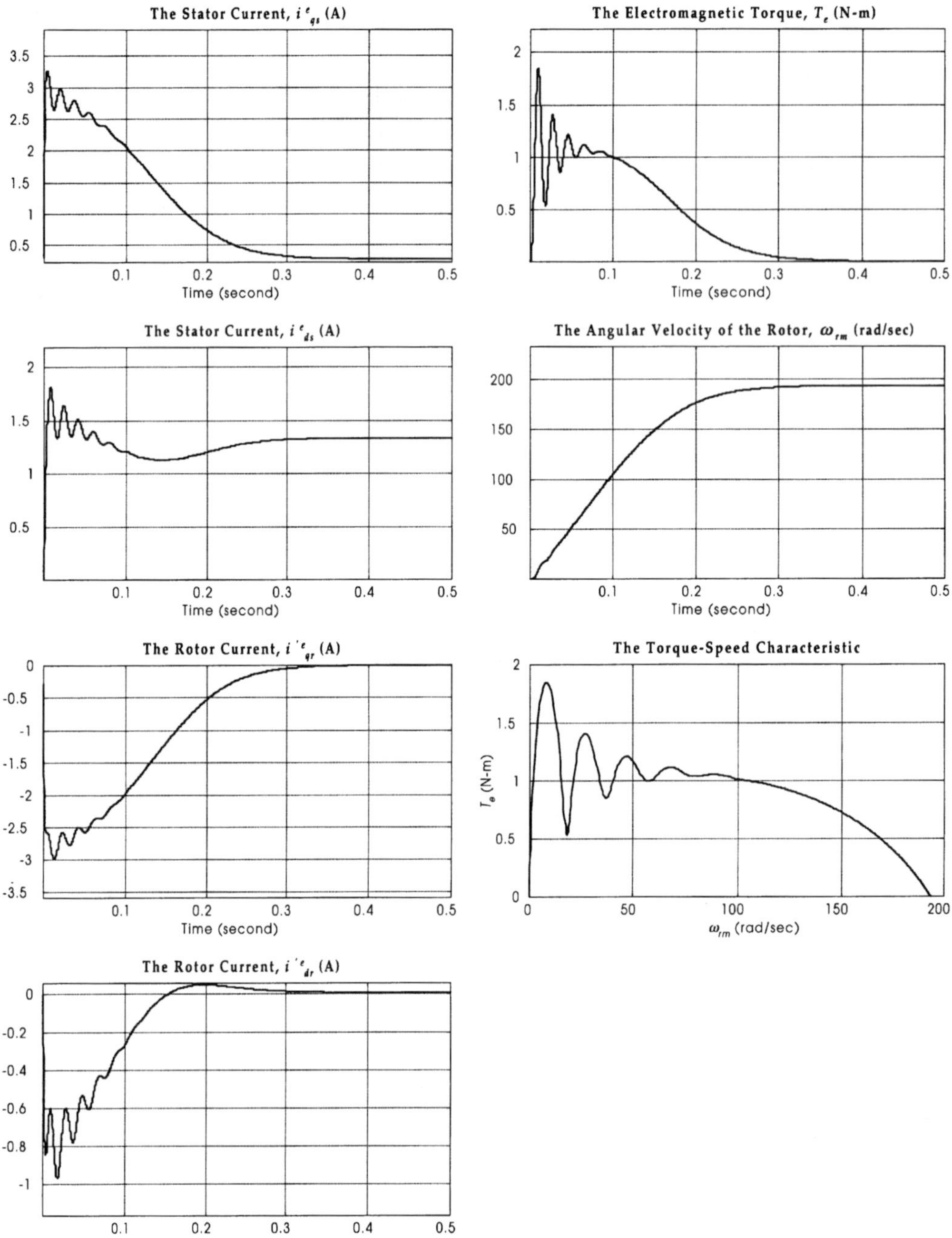

Figure 6.68 Dynamics and torque-speed characteristic of the D-class induction motor modeled in the synchronous reference frame; $T_L = 0$ N-m, $u^e_{qs}(t) = \sqrt{2}115$, and $u^e_{ds}(t) = 0$.

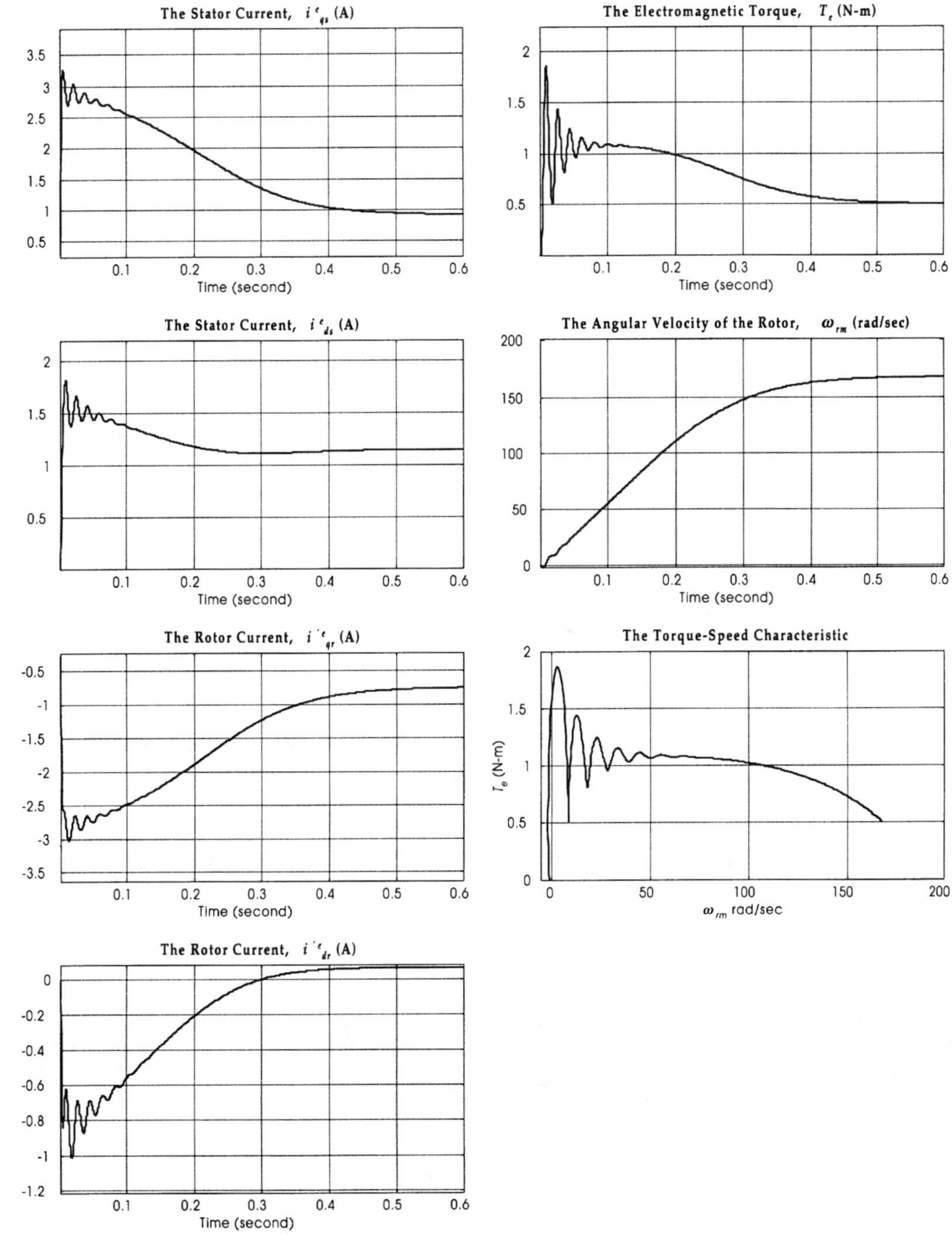

Figure 6.69 Dynamics and torque-speed characteristic of the D-class induction motor modeled in the synchronous reference frame; $T_L = 0.5$ N-m, $u^e_{qs}(t) = \sqrt{2}115$, and $u^e_{ds}(t) = 0$.

From (6.38), one obtains the resulting model in matrix form. In particular,

$$\begin{bmatrix} \frac{di_{qs}^e}{dt} \\ \frac{di_{ds}^e}{dt} \\ \frac{d\psi_{qr}'^e}{dt} \\ \frac{d\psi_{dr}'^e}{dt} \\ \frac{d\omega_r}{dt} \\ \frac{d\theta_r}{dt} \end{bmatrix} = \begin{bmatrix} -\frac{L_{rr}'^2 r_s + L_{ms}^2 r_r'}{L_\Sigma L_{rr}'} & -\omega_e & \frac{L_{ms} r_r'}{L_\Sigma L_{rr}'} & 0 & 0 & 0 \\ \omega_e & -\frac{L_{rr}'^2 r_s + L_{ms}^2 r_r'}{L_\Sigma L_{rr}'} & 0 & \frac{L_{ms} r_r'}{L_\Sigma L_{rr}'} & 0 & 0 \\ \frac{L_{ms} r_r'}{L_{rr}'} & 0 & -\frac{r_r'}{L_{rr}'} & -\omega_e & 0 & 0 \\ 0 & \frac{L_{ms} r_r'}{L_{rr}'} & \omega_e & -\frac{r_r'}{L_{rr}'} & 0 & 0 \\ 0 & 0 & 0 & 0 & -\frac{B_m}{J} & 0 \\ 0 & 0 & 0 & 0 & 1 & 0 \end{bmatrix} \begin{bmatrix} i_{qs}^e \\ i_{ds}^e \\ \psi_{qr}'^e \\ \psi_{dr}'^e \\ \omega_r \\ \theta_r \end{bmatrix}$$

$$+ \begin{bmatrix} -\frac{L_{ms}}{L_\Sigma}\psi_{dr}'^e\omega_r \\ \frac{L_{ms}}{L_\Sigma}\psi_{qr}'^e\omega_r \\ \psi_{dr}'^e\omega_r \\ -\psi_{qr}'^e\omega_r \\ \frac{P^2}{4J}\frac{L_{ms}}{L_{rr}'}(i_{qs}^e\psi_{dr}'^e - i_{ds}^e\psi_{qr}'^e) \\ 0 \end{bmatrix} + \begin{bmatrix} \frac{L_{rr}'}{L_\Sigma} & 0 & -\frac{L_{ms}}{L_\Sigma} & 0 \\ 0 & \frac{L_{rr}'}{L_\Sigma} & 0 & -\frac{L_{ms}}{L_\Sigma} \\ 0 & 0 & 1 & 0 \\ 0 & 0 & 0 & 1 \\ 0 & 0 & 0 & 0 \\ 0 & 0 & 0 & 0 \end{bmatrix} \begin{bmatrix} u_{qs}^e \\ u_{ds}^e \\ u_{qr}'^e \\ u_{dr}'^e \end{bmatrix} - \begin{bmatrix} 0 \\ 0 \\ 0 \\ 0 \\ \frac{P}{2J} \\ 0 \end{bmatrix} T_L$$

The block diagram, which is shown in Figure 6.70, was developed using (6.38).

Example 6.8

In SIMULINK, perform nonlinear modeling of the A- and D-class induction motors using i_{qs}^e, i_{ds}^e, $\psi_{qr}'^e$, $\psi_{dr}'^e$, and ω_r as the state variables. The parameters of the motors studied are given in Table 6.1.

Solution

The SIMULINK diagram is built and documented in Figure 6.71. Assuming that the rated phase voltages are applied, we have the following *quadrature* and *direct* voltages to guarantee the balanced operating conditions

$$u_{qs}^e = \sqrt{2}115, \quad u_{ds}^e = 0$$

The dynamics of the A- and D-class motors are illustrated in Figures 6.72 to 6.75, and the torque-speed characteristic curves are documented as well.

It should be emphasized that in electric drives, the mechanical angular velocity is usually considered as the output variable and, therefore, ω_{rm} is plotted in Figures 6.72 to 6.75. Recall that the mechanical and electrical angular velocities are related as $\omega_{rm} = \frac{2}{P}\omega_r$.

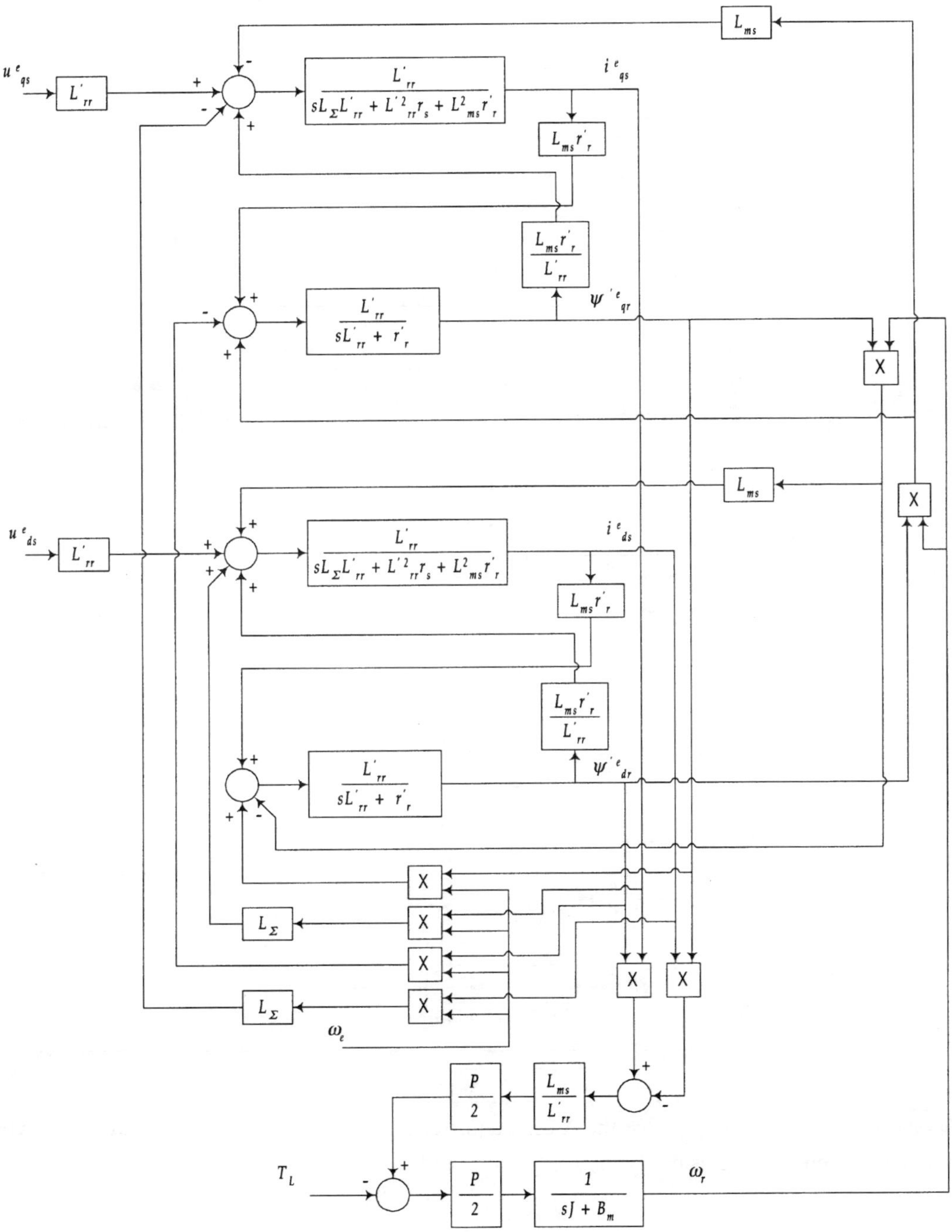

Figure 6.70 Block diagram of two-phase squirrel-cage induction motors modeled in the synchronous reference frame.

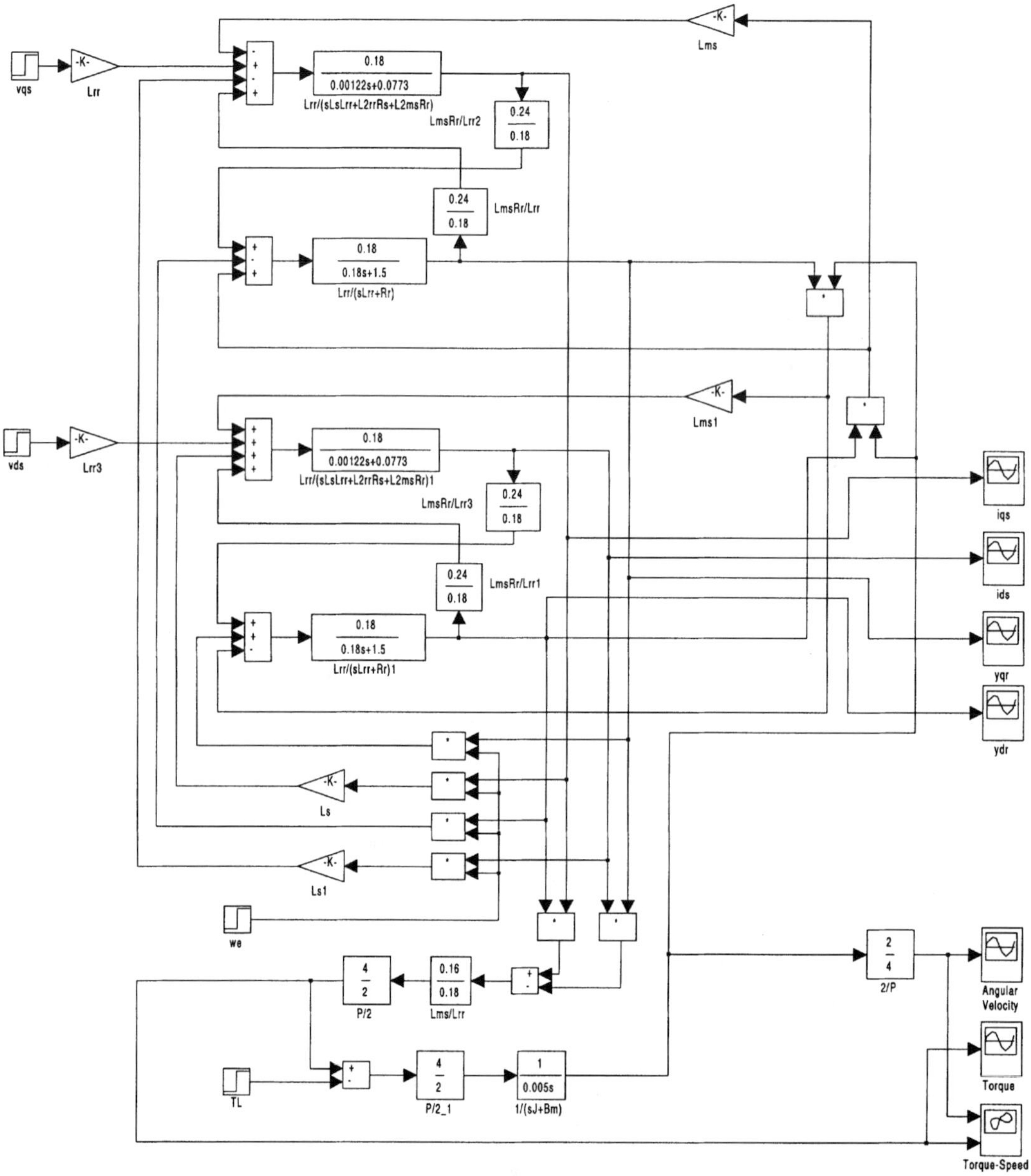

Figure 6.71 SIMULINK diagram for the A-class squirrel-cage induction motor modeled in the synchronous reference frame using i_{qs}^{e}, i_{ds}^{e}, $\psi_{qr}'^{e}$, $\psi_{dr}'^{e}$, and ω_r.

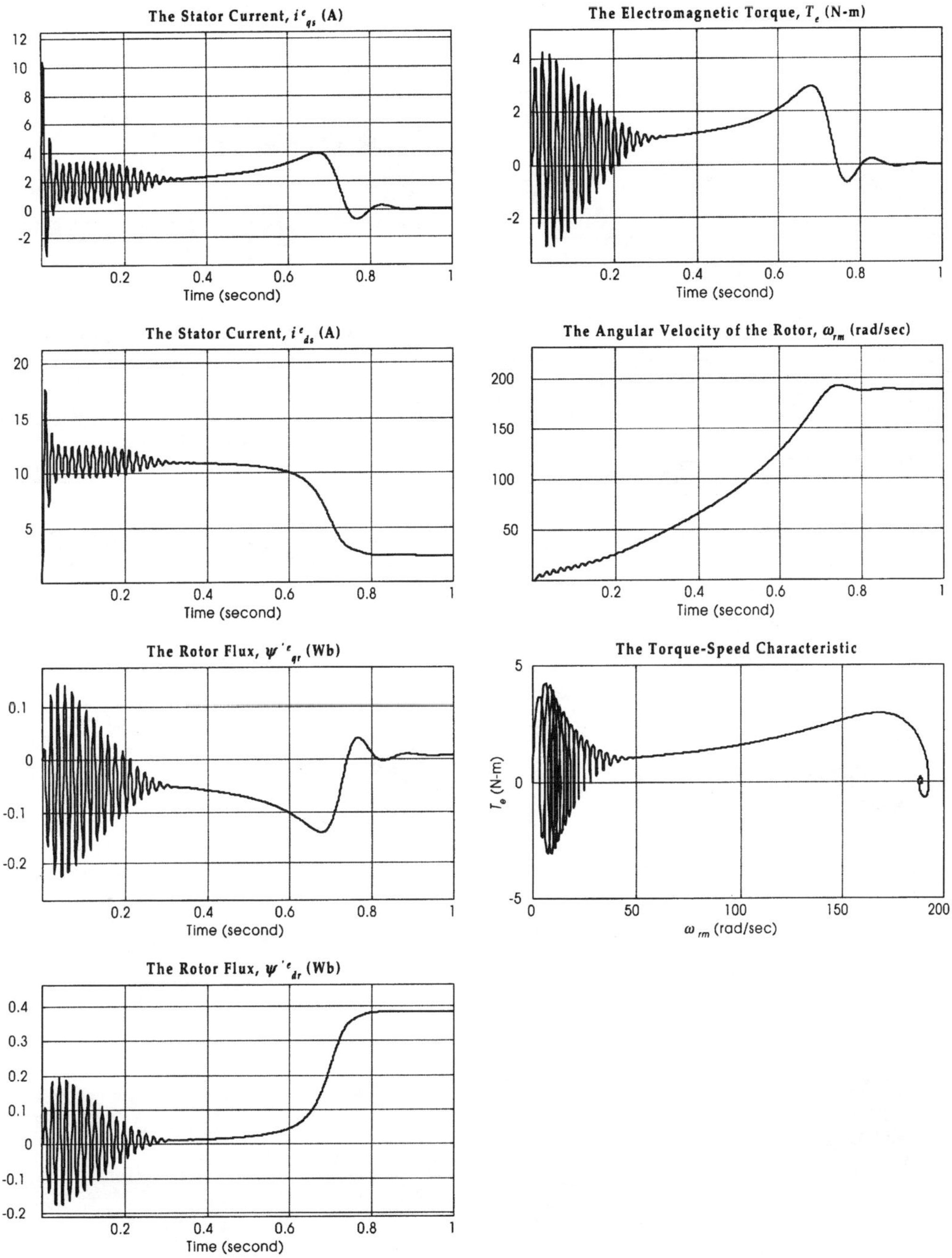

Figure 6.72 Dynamics and torque-speed characteristic of the A-class induction motor modeled in the synchronous reference frame; $T_L = 0$ N-m, $u^e_{qs}(t) = \sqrt{2}115$, and $u^e_{ds}(t) = 0$.

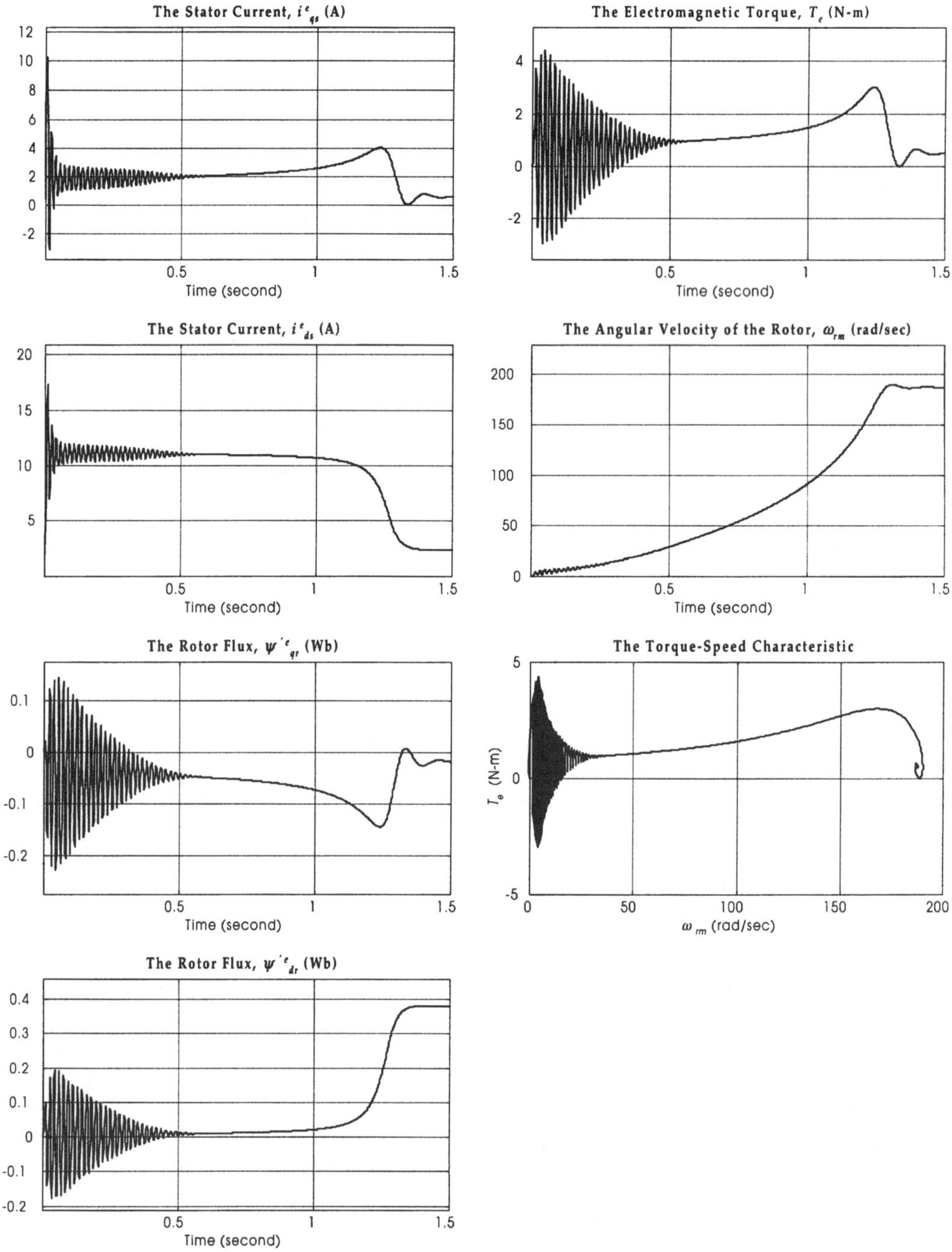

Figure 6.73 Dynamics and torque-speed characteristic of the A-class induction motor modeled in the synchronous reference frame; $T_L = 0.5$ N-m, $u^e_{qs}(t) = \sqrt{2}115$, and $u^e_{ds}(t) = 0$.

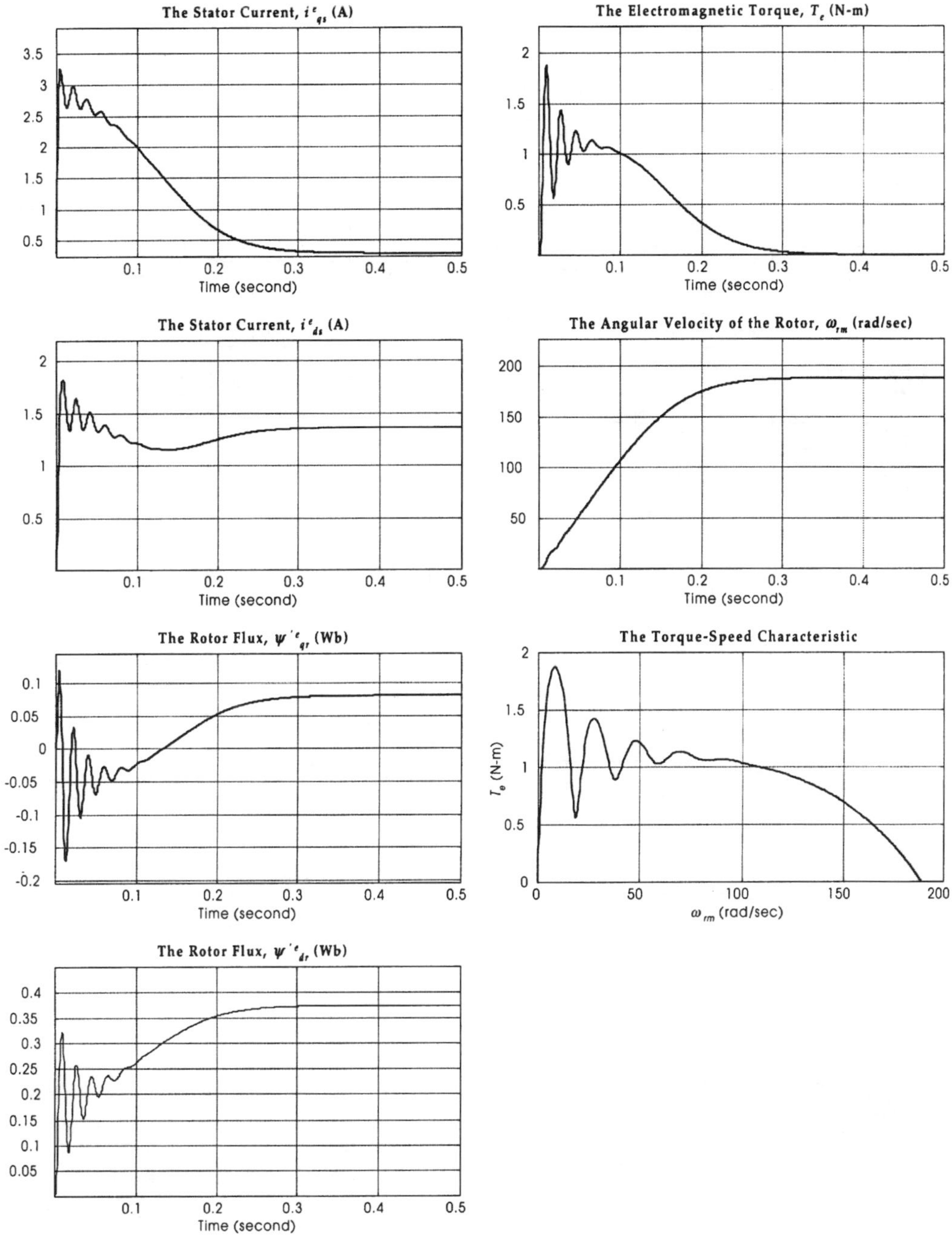

Figure 6.74 Dynamics and torque-speed characteristic of the D-class induction motor modeled in the synchronous reference frame; $T_L = 0$ N-m, $u^e_{qs}(t) = \sqrt{2}115$, and $u^e_{ds}(t) = 0$.

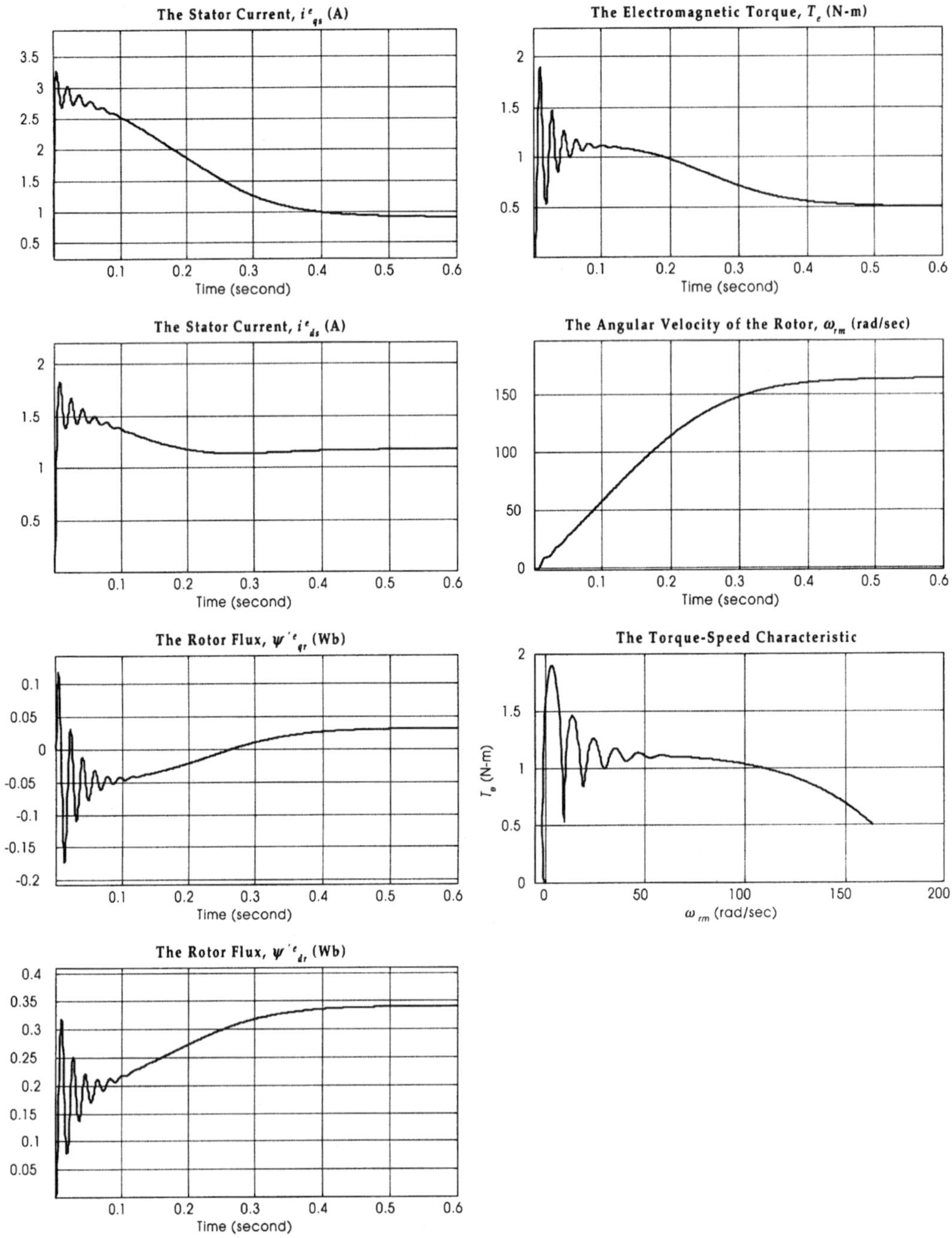

Figure 6.75 Dynamics and torque-speed characteristic of the D-class induction motor modeled in the synchronous reference frame; $T_L = 0.5$ N-m, $u^e_{qs}(t) = \sqrt{2}115$, and $u^e_{ds}(t) = 0$.

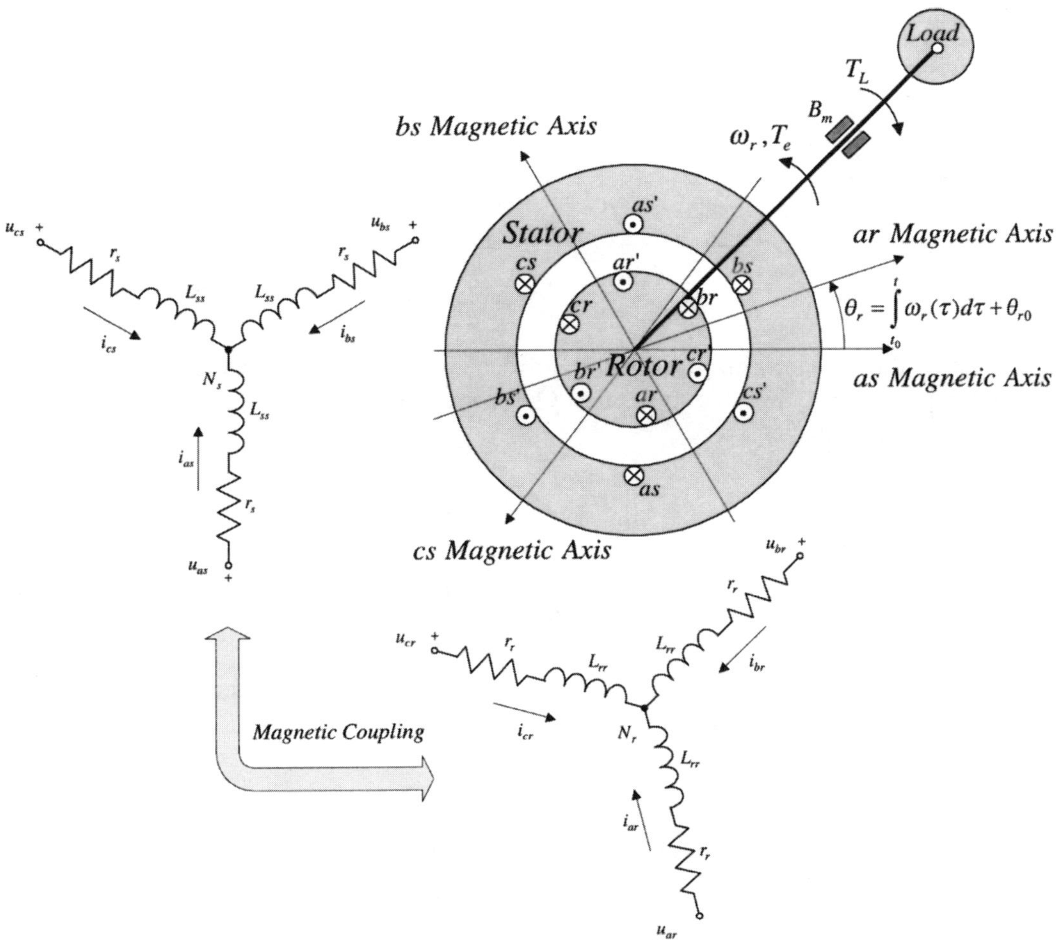

Figure 6.76 Three-phase symmetrical induction motor.

6.4 Voltage, flux linkages, and torque equations for three-phase induction machines: dynamics in the **machine** variables

This section develops the mathematical model of three-phase induction machines using *machine* variables. The goal is to find a set of differential equations to map the dynamics of induction machines in order to perform a thorough analysis of the transient behavior and steady-state performance. A three-phase induction motor is illustrated in Figure 6.76.

Kirchhoff's voltage law gives the differential equations for the voltages, applied to the *abc* stator and rotor windings, the *abc* stator and rotor currents, and flux linkages. In particular, we have

$$u_{as} = r_s i_{as} + \frac{d\psi_{as}}{dt}$$

$$u_{bs} = r_s i_{bs} + \frac{d\psi_{bs}}{dt}$$

$$u_{cs} = r_s i_{cs} + \frac{d\psi_{cs}}{dt}$$

$$u_{ar} = r_r i_{ar} + \frac{d\psi_{ar}}{dt}$$

$$u_{br} = r_r i_{br} + \frac{d\psi_{br}}{dt}$$

$$u_{cr} = r_r i_{cr} + \frac{d\psi_{cr}}{dt}$$

That is, the *abc* stator and rotor voltages, currents, and flux linkages are used as the variables to map the circuitry dynamics. One concludes that there is a set of two matrix differential equations

$$\mathbf{u}_{abcs} = \mathbf{r}_s \mathbf{i}_{abcs} + \frac{d\boldsymbol{\psi}_{abcs}}{dt}$$
$$\mathbf{u}_{abcr} = \mathbf{r}_r \mathbf{i}_{abcr} + \frac{d\boldsymbol{\psi}_{abcr}}{dt} \tag{6.39}$$

where the *abc* stator and rotor voltages, currents, and flux linkages are given in vector form, and

$$\mathbf{u}_{abcs} = \begin{bmatrix} u_{as} \\ u_{bs} \\ u_{cs} \end{bmatrix}, \ \mathbf{u}_{abcr} = \begin{bmatrix} u_{ar} \\ u_{br} \\ u_{cr} \end{bmatrix}, \ \mathbf{i}_{abcs} = \begin{bmatrix} i_{as} \\ i_{bs} \\ i_{cs} \end{bmatrix}, \ \mathbf{i}_{abcr} = \begin{bmatrix} i_{ar} \\ i_{br} \\ i_{cr} \end{bmatrix}, \ \boldsymbol{\psi}_{abcs} = \begin{bmatrix} \psi_{as} \\ \psi_{bs} \\ \psi_{cs} \end{bmatrix}, \text{ and } \boldsymbol{\psi}_{abcr} = \begin{bmatrix} \psi_{ar} \\ \psi_{br} \\ \psi_{cr} \end{bmatrix}$$

In (6.39), the diagonal matrices of the stator and rotor resistances are expressed as

$$\mathbf{r}_s = \begin{bmatrix} r_s & 0 & 0 \\ 0 & r_s & 0 \\ 0 & 0 & r_s \end{bmatrix} \text{ and } \mathbf{r}_r = \begin{bmatrix} r_r & 0 & 0 \\ 0 & r_r & 0 \\ 0 & 0 & r_r \end{bmatrix}$$

The flux linkages are found using the corresponding currents in the stator and rotor windings and self- and mutual inductances. In particular, the analysis of the stator and rotor magnetically coupled system, which can be accomplished using Figure 6.76, leads one to the following equations

$$\psi_{as} = L_{asas} i_{as} + L_{asbs} i_{bs} + L_{ascs} i_{cs} + L_{asar} i_{ar} + L_{asbr} i_{br} + L_{ascr} i_{cr}$$

$$\psi_{bs} = L_{bsas} i_{as} + L_{bsbs} i_{bs} + L_{bscs} i_{cs} + L_{bsar} i_{ar} + L_{bsbr} i_{br} + L_{bscr} i_{cr}$$

$$\psi_{cs} = L_{csas} i_{as} + L_{csbs} i_{bs} + L_{cscs} i_{cs} + L_{csar} i_{ar} + L_{csbr} i_{br} + L_{cscr} i_{cr}$$

$$\psi_{ar} = L_{aras} i_{as} + L_{arbs} i_{bs} + L_{arcs} i_{cs} + L_{arar} i_{ar} + L_{arbr} i_{br} + L_{arcr} i_{cr}$$

$$\psi_{br} = L_{bras}i_{as} + L_{brbs}i_{bs} + L_{brcs}i_{cs} + L_{brar}i_{ar} + L_{brbr}i_{br} + L_{brcr}i_{cr}$$

$$\psi_{cr} = L_{cras}i_{as} + L_{crbs}i_{bs} + L_{crcs}i_{cs} + L_{crar}i_{ar} + L_{crbr}i_{br} + L_{crcr}i_{cr}$$

The stator and rotor *abc* windings, which are identical, are displaced magnetically by $\frac{2}{3}\pi$. Hence, there exists a coupling between the *abc* stator and rotor windings. The mutual inductances between the stator windings are equal, and we have

$$L_{asbs} = L_{ascs} = L_{bscs} = L_{ms}\cos\left(\frac{2}{3}\pi\right) = -\frac{1}{2}L_{ms}$$

where $L_{ms} = \frac{N_s^2}{\Re_m}$.

Taking note that the rotor windings are displaced by 120°, one has the following formula to express the mutual inductances between the rotor windings

$$L_{arbr} = L_{arcr} = L_{brcr} = L_{mr}\cos\left(\frac{2}{3}\pi\right) = -\frac{1}{2}L_{mr}$$

where $L_{mr} = \frac{N_r^2}{N_s^2}L_{ms}$.

The stator and rotor self-inductances are

$$L_{ss} = L_{ls} + L_{ms} \text{ and } L_{rr} = L_{lr} + L_{mr}$$

and the matrices of self- and mutual inductances $\mathbf{L}_s$ and $\mathbf{L}_r$ are found to be

$$\mathbf{L}_s = \begin{bmatrix} L_{ls}+L_{ms} & -\frac{1}{2}L_{ms} & -\frac{1}{2}L_{ms} \\ -\frac{1}{2}L_{ms} & L_{ls}+L_{ms} & -\frac{1}{2}L_{ms} \\ -\frac{1}{2}L_{ms} & -\frac{1}{2}L_{ms} & L_{ls}+L_{ms} \end{bmatrix} \text{ and } \mathbf{L}_r = \begin{bmatrix} L_{lr}+L_{mr} & -\frac{1}{2}L_{mr} & -\frac{1}{2}L_{mr} \\ -\frac{1}{2}L_{mr} & L_{lr}+L_{mr} & -\frac{1}{2}L_{mr} \\ -\frac{1}{2}L_{mr} & -\frac{1}{2}L_{mr} & L_{lr}+L_{mr} \end{bmatrix}$$

Assume that the mutual inductances between the stator and rotor windings are sinusoidal functions of the rotor angular displacement θ_r. Hence,

$$L_{asar} = L_{aras} = L_{sr}\cos\theta_r,\ L_{asbr} = L_{bras} = L_{sr}\cos\left(\theta_r + \frac{2}{3}\pi\right),\ L_{ascr} = L_{cras} = L_{sr}\cos\left(\theta_r - \frac{2}{3}\pi\right)$$

$$L_{bsar} = L_{arbs} = L_{sr}\cos\left(\theta_r - \frac{2}{3}\pi\right),\ L_{bsbr} = L_{brbs} = L_{sr}\cos\theta_r,\ L_{bscr} = L_{crbs} = L_{sr}\cos\left(\theta_r + \frac{2}{3}\pi\right)$$

$$L_{csar} = L_{arcs} = L_{sr}\cos\left(\theta_r + \frac{2}{3}\pi\right),\ L_{csbr} = L_{brcs} = L_{sr}\cos\left(\theta_r - \frac{2}{3}\pi\right),\ L_{cscr} = L_{crcs} = L_{sr}\cos\theta_r$$

where $L_{sr} = \dfrac{N_s N_r}{\Re_m}$.

Therefore, the following matrix for the stator-rotor mutual inductances results

$$\mathbf{L}_{sr} = L_{sr}\begin{bmatrix} \cos\theta_r & \cos\left(\theta_r + \frac{2}{3}\pi\right) & \cos\left(\theta_r - \frac{2}{3}\pi\right) \\ \cos\left(\theta_r - \frac{2}{3}\pi\right) & \cos\theta_r & \cos\left(\theta_r + \frac{2}{3}\pi\right) \\ \cos\left(\theta_r + \frac{2}{3}\pi\right) & \cos\left(\theta_r - \frac{2}{3}\pi\right) & \cos\theta_r \end{bmatrix}$$

One obtains the following matrix equation

$$\begin{bmatrix} \boldsymbol{\psi}_{abcs} \\ \boldsymbol{\psi}_{abcr} \end{bmatrix} = \begin{bmatrix} \mathbf{L}_s & \mathbf{L}_{sr} \\ \mathbf{L}_{sr}^T & \mathbf{L}_r \end{bmatrix} \begin{bmatrix} \mathbf{i}_{abcs} \\ \mathbf{i}_{abcr} \end{bmatrix}. \tag{6.40}$$

Using the number of turns N_s and N_r, one finds

$$\mathbf{u}'_{abcr} = \frac{N_s}{N_r}\mathbf{u}_{abcr},\quad \mathbf{i}'_{abcr} = \frac{N_r}{N_s}\mathbf{i}_{abcr},\quad \text{and } \boldsymbol{\psi}'_{abcr} = \frac{N_s}{N_r}\boldsymbol{\psi}'_{abcr}$$

The inductances are expressed as

$$L_{ms} = \frac{N_s}{N_r}L_{sr},\quad L_{sr} = \frac{N_s N_r}{\Re_m},\quad \text{and } L_{ms} = \frac{N_s^2}{\Re_m}$$

Then, we have

$$\mathbf{L}'_{sr} = \frac{N_s}{N_r}\mathbf{L}_{sr} = L_{ms}\begin{bmatrix} \cos\theta_r & \cos\left(\theta_r + \frac{2}{3}\pi\right) & \cos\left(\theta_r - \frac{2}{3}\pi\right) \\ \cos\left(\theta_r - \frac{2}{3}\pi\right) & \cos\theta_r & \cos\left(\theta_r + \frac{2}{3}\pi\right) \\ \cos\left(\theta_r + \frac{2}{3}\pi\right) & \cos\left(\theta_r - \frac{2}{3}\pi\right) & \cos\theta_r \end{bmatrix}$$

and

$$\mathbf{L}'_r = \frac{N_s^2}{N_r^2}\mathbf{L}_r = \begin{bmatrix} L'_{lr} + L_{ms} & -\frac{1}{2}L_{ms} & -\frac{1}{2}L_{ms} \\ -\frac{1}{2}L_{ms} & L'_{lr} + L_{ms} & -\frac{1}{2}L_{ms} \\ -\frac{1}{2}L_{ms} & -\frac{1}{2}L_{ms} & L'_{lr} + L_{ms} \end{bmatrix}$$

where $L'_{lr} = \dfrac{N_s^2}{N_r^2}L_{lr}$.

Therefore, the equation for flux linkages (6.40) is rewritten as

$$\begin{bmatrix} \boldsymbol{\Psi}_{abcs} \\ \boldsymbol{\Psi}'_{abcr} \end{bmatrix} = \begin{bmatrix} \mathbf{L}_s & \mathbf{L}'_{sr} \\ \mathbf{L}'^{T}_{sr} & \mathbf{L}'_r \end{bmatrix} \begin{bmatrix} \mathbf{i}_{abcs} \\ \mathbf{i}'_{abcr} \end{bmatrix} \tag{6.41}$$

and substituting the matrices $\mathbf{L}_s$, $\mathbf{L}'_{sr}$, and $\mathbf{L}'_r$, we have

$$\begin{bmatrix} \psi_{as} \\ \psi_{bs} \\ \psi_{cs} \\ \psi'_{ar} \\ \psi'_{br} \\ \psi'_{cr} \end{bmatrix} = \begin{bmatrix} L_{ls}+L_{ms} & -\frac{1}{2}L_{ms} & -\frac{1}{2}L_{ms} & L_{ms}\cos\theta_r & L_{ms}\cos\left(\theta_r+\frac{2}{3}\pi\right) & L_{ms}\cos\left(\theta_r-\frac{2}{3}\pi\right) \\ -\frac{1}{2}L_{ms} & L_{ls}+L_{ms} & -\frac{1}{2}L_{ms} & L_{ms}\cos\left(\theta_r-\frac{2}{3}\pi\right) & L_{ms}\cos\theta_r & L_{ms}\cos\left(\theta_r+\frac{2}{3}\pi\right) \\ -\frac{1}{2}L_{ms} & -\frac{1}{2}L_{ms} & L_{ls}+L_{ms} & L_{ms}\cos\left(\theta_r+\frac{2}{3}\pi\right) & L_{ms}\cos\left(\theta_r-\frac{2}{3}\pi\right) & L_{ms}\cos\theta_r \\ L_{ms}\cos\theta_r & L_{ms}\cos\left(\theta_r-\frac{2}{3}\pi\right) & L_{ms}\cos\left(\theta_r+\frac{2}{3}\pi\right) & L'_{lr}+L_{ms} & -\frac{1}{2}L_{ms} & -\frac{1}{2}L_{ms} \\ L_{ms}\cos\left(\theta_r+\frac{2}{3}\pi\right) & L_{ms}\cos\theta_r & L_{ms}\cos\left(\theta_r-\frac{2}{3}\pi\right) & -\frac{1}{2}L_{ms} & L'_{lr}+L_{ms} & -\frac{1}{2}L_{ms} \\ L_{ms}\cos\left(\theta_r-\frac{2}{3}\pi\right) & L_{ms}\cos\left(\theta_r+\frac{2}{3}\pi\right) & L_{ms}\cos\theta_r & -\frac{1}{2}L_{ms} & -\frac{1}{2}L_{ms} & L'_{lr}+L_{ms} \end{bmatrix}$$

$$\times \begin{bmatrix} i_{as} \\ i_{bs} \\ i_{cs} \\ i'_{ar} \\ i'_{br} \\ i'_{cr} \end{bmatrix}$$

In expanded form, one has

$$\begin{aligned}
\psi_{as} &= (L_{ls}+L_{ms})i_{as} - \frac{1}{2}L_{ms}i_{bs} - \frac{1}{2}L_{ms}i_{cs} + L_{ms}\cos\theta_r i'_{ar} + L_{ms}\cos\left(\theta_r+\frac{2}{3}\pi\right)i'_{br} + L_{ms}\cos\left(\theta_r-\frac{2}{3}\pi\right)i'_{cr} \\
\psi_{bs} &= -\frac{1}{2}L_{ms}i_{as} + (L_{ls}+L_{ms})i_{bs} - \frac{1}{2}L_{ms}i_{cs} + L_{ms}\cos\left(\theta_r-\frac{2}{3}\pi\right)i'_{ar} + L_{ms}\cos\theta_r i'_{br} + L_{ms}\cos\left(\theta_r+\frac{2}{3}\pi\right)i'_{cr} \\
\psi_{cs} &= -\frac{1}{2}L_{ms}i_{as} - \frac{1}{2}L_{ms}i_{bs} + (L_{ls}+L_{ms})i_{cs} + L_{ms}\cos\left(\theta_r+\frac{2}{3}\pi\right)i'_{ar} + L_{ms}\cos\left(\theta_r-\frac{2}{3}\pi\right)i'_{br} + L_{ms}\cos\theta_r i'_{cr} \\
\psi'_{ar} &= L_{ms}\cos\theta_r i_{as} + L_{ms}\cos\left(\theta_r-\frac{2}{3}\pi\right)i_{bs} + L_{ms}\cos\left(\theta_r+\frac{2}{3}\pi\right)i_{cs} + (L'_{lr}+L_{ms})i'_{ar} - \frac{1}{2}L_{ms}i'_{br} - \frac{1}{2}L_{ms}i'_{cr} \\
\psi'_{br} &= L_{ms}\cos\left(\theta_r+\frac{2}{3}\pi\right)i_{as} + L_{ms}\cos\theta_r i_{bs} + L_{ms}\cos\left(\theta_r-\frac{2}{3}\pi\right)i_{cs} - \frac{1}{2}L_{ms}i'_{ar} + (L'_{lr}+L_{ms})i'_{br} - \frac{1}{2}L_{ms}i'_{cr} \\
\psi'_{cr} &= L_{ms}\cos\left(\theta_r-\frac{2}{3}\pi\right)i_{as} + L_{ms}\cos\left(\theta_r+\frac{2}{3}\pi\right)i_{bs} + L_{ms}\cos\theta_r i_{cs} - \frac{1}{2}L_{ms}i'_{ar} - \frac{1}{2}L_{ms}i'_{br} + (L'_{lr}+L_{ms})i'_{cr}
\end{aligned} \tag{6.42}$$

From (6.42), it follows that one can find explicit expressions for derivatives

$$\frac{d\boldsymbol{\Psi}_{abcs}}{dt} \text{ and } \frac{d\boldsymbol{\Psi}'_{abcr}}{dt}$$

Using (6.39) and (6.41), one obtains

$$\mathbf{u}_{abcs} = \mathbf{r}_s \mathbf{i}_{abcs} + \frac{d\boldsymbol{\psi}_{abcs}}{dt} = \mathbf{r}_s \mathbf{i}_{abcs} + \mathbf{L}_s \frac{d\mathbf{i}_{abcs}}{dt} + \frac{d(\mathbf{L}'_{sr}\mathbf{i}'_{abcr})}{dt}$$
$$\mathbf{u}'_{abcr} = \mathbf{r}'_r \mathbf{i}'_{abcr} + \frac{d\boldsymbol{\psi}'_{abcr}}{dt} = \mathbf{r}'_r \mathbf{i}'_{abcr} + \mathbf{L}'_r \frac{d\mathbf{i}'_{abcr}}{dt} + \frac{d(\mathbf{L}'^{T}_{sr}\mathbf{i}'_{abcs})}{dt} \quad (6.43)$$

where $\mathbf{r}'_r = \frac{N_s^2}{N_r^2}\mathbf{r}_r$.

Matrix equations (6.43) in expanded form using (6.42) are written as

$$u_{as} = r_s i_{as} + (L_{ls} + L_{ms})\frac{di_{as}}{dt} - \frac{1}{2}L_{ms}\frac{di_{bs}}{dt} - \frac{1}{2}L_{ms}\frac{di_{cs}}{dt} + L_{ms}\frac{d(i'_{ar}\cos\theta_r)}{dt} + L_{ms}\frac{d\left(i'_{br}\cos\left(\theta_r + \frac{2\pi}{3}\right)\right)}{dt} + L_{ms}\frac{d\left(i'_{cr}\cos\left(\theta_r - \frac{2\pi}{3}\right)\right)}{dt}$$

$$u_{bs} = r_s i_{bs} - \frac{1}{2}L_{ms}\frac{di_{as}}{dt} + (L_{ls} + L_{ms})\frac{di_{bs}}{dt} - \frac{1}{2}L_{ms}\frac{di_{cs}}{dt} + L_{ms}\frac{d\left(i'_{ar}\cos\left(\theta_r - \frac{2\pi}{3}\right)\right)}{dt} + L_{ms}\frac{d(i'_{br}\cos\theta_r)}{dt} + L_{ms}\frac{d\left(i'_{cr}\cos\left(\theta_r + \frac{2\pi}{3}\right)\right)}{dt}$$

$$u_{cs} = r_s i_{cs} - \frac{1}{2}L_{ms}\frac{di_{as}}{dt} - \frac{1}{2}L_{ms}\frac{di_{bs}}{dt} + (L_{ls} + L_{ms})\frac{di_{cs}}{dt} + L_{ms}\frac{d\left(i'_{ar}\cos\left(\theta_r + \frac{2\pi}{3}\right)\right)}{dt} + L_{ms}\frac{d\left(i'_{br}\cos\left(\theta_r - \frac{2\pi}{3}\right)\right)}{dt} + L_{ms}\frac{d(i'_{cr}\cos\theta_r)}{dt}$$

$$u'_{ar} = r'_r i'_{ar} + L_{ms}\frac{d(i_{as}\cos\theta_r)}{dt} + L_{ms}\frac{d\left(i_{bs}\cos\left(\theta_r - \frac{2\pi}{3}\right)\right)}{dt} + L_{ms}\frac{d\left(i_{cs}\cos\left(\theta_r + \frac{2\pi}{3}\right)\right)}{dt} + (L'_{lr} + L_{ms})\frac{di'_{ar}}{dt} - \frac{1}{2}L_{ms}\frac{di'_{br}}{dt} - \frac{1}{2}L_{ms}\frac{di'_{cr}}{dt}$$

$$u'_{br} = r'_r i'_{br} + L_{ms}\frac{d\left(i_{as}\cos\left(\theta_r + \frac{2\pi}{3}\right)\right)}{dt} + L_{ms}\frac{d(i_{bs}\cos\theta_r)}{dt} + L_{ms}\frac{d\left(i_{cs}\cos\left(\theta_r - \frac{2\pi}{3}\right)\right)}{dt} - \frac{1}{2}L_{ms}\frac{di'_{ar}}{dt} + (L'_{lr} + L_{ms})\frac{di'_{br}}{dt} - \frac{1}{2}L_{ms}\frac{di'_{cr}}{dt}$$

$$u'_{cr} = r'_r i'_{cr} + L_{ms}\frac{d\left(i_{as}\cos\left(\theta_r - \frac{2\pi}{3}\right)\right)}{dt} + L_{ms}\frac{d\left(i_{bs}\cos\left(\theta_r + \frac{2\pi}{3}\right)\right)}{dt} + L_{ms}\frac{d(i_{cs}\cos\theta_r)}{dt} - \frac{1}{2}L_{ms}\frac{di'_{ar}}{dt} - \frac{1}{2}L_{ms}\frac{di'_{br}}{dt} + (L'_{lr} + L_{ms})\frac{di'_{cr}}{dt}$$

Hence, one obtains the following set of voltage equations:

$$
\begin{aligned}
u_{as} &= r_s i_{as} + (L_{ls} + L_{ms})\frac{di_{as}}{dt} - \frac{1}{2}L_{ms}\frac{di_{bs}}{dt} - \frac{1}{2}L_{ms}\frac{di_{cs}}{dt} + L_{ms}\cos\theta_r\frac{di'_{ar}}{dt} + L_{ms}\cos\left(\theta_r + \frac{2\pi}{3}\right)\frac{di'_{br}}{dt} \\
&\quad + L_{ms}\cos\left(\theta_r - \frac{2\pi}{3}\right)\frac{di'_{cr}}{dt} - L_{ms}\left[i'_{ar}\sin\theta_r + i'_{br}\sin\left(\theta_r + \frac{2\pi}{3}\right) + i'_{cr}\sin\left(\theta_r - \frac{2\pi}{3}\right)\right]\omega_r \\
u_{bs} &= r_s i_{bs} - \frac{1}{2}L_{ms}\frac{di_{as}}{dt} + (L_{ls} + L_{ms})\frac{di_{bs}}{dt} - \frac{1}{2}L_{ms}\frac{di_{cs}}{dt} + L_{ms}\cos\left(\theta_r - \frac{2\pi}{3}\right)\frac{di'_{ar}}{dt} + L_{ms}\cos\theta_r\frac{di'_{br}}{dt} \\
&\quad + L_{ms}\cos\left(\theta_r + \frac{2\pi}{3}\right)\frac{di'_{cr}}{dt} - L_{ms}\left[i'_{ar}\sin\left(\theta_r - \frac{2\pi}{3}\right) + i'_{br}\sin\theta_r + i'_{cr}\sin\left(\theta_r + \frac{2\pi}{3}\right)\right]\omega_r \\
u_{cs} &= r_s i_{cs} - \frac{1}{2}L_{ms}\frac{di_{as}}{dt} - \frac{1}{2}L_{ms}\frac{di_{bs}}{dt} + (L_{ls} + L_{ms})\frac{di_{cs}}{dt} + L_{ms}\cos\left(\theta_r + \frac{2\pi}{3}\right)\frac{di'_{ar}}{dt} + L_{ms}\cos\left(\theta_r - \frac{2\pi}{3}\right)\frac{di'_{br}}{dt} \\
&\quad + L_{ms}\cos\theta_r\frac{di'_{cr}}{dt} - L_{ms}\left[i'_{ar}\sin\left(\theta_r + \frac{2\pi}{3}\right) + i'_{br}\sin\left(\theta_r - \frac{2\pi}{3}\right) + i'_{cr}\sin\theta_r\right]\omega_r \\
u'_{ar} &= r'_r i'_{ar} + L_{ms}\cos\theta_r\frac{di_{as}}{dt} + L_{ms}\cos\left(\theta_r - \frac{2\pi}{3}\right)\frac{di_{bs}}{dt} + L_{ms}\cos\left(\theta_r + \frac{2\pi}{3}\right)\frac{di_{cs}}{dt} + (L'_{lr} + L_{ms})\frac{di'_{ar}}{dt} \\
&\quad - \frac{1}{2}L_{ms}\frac{di'_{br}}{dt} - \frac{1}{2}L_{ms}\frac{di'_{cr}}{dt} - L_{ms}\left[i_{as}\sin\theta_r + i_{bs}\sin\left(\theta_r - \frac{2\pi}{3}\right) + i_{cs}\sin\left(\theta_r + \frac{2\pi}{3}\right)\right]\omega_r \\
u'_{br} &= r'_r i'_{br} + L_{ms}\cos\left(\theta_r + \frac{2\pi}{3}\right)\frac{di_{as}}{dt} + L_{ms}\cos\theta_r\frac{di_{bs}}{dt} + L_{ms}\cos\left(\theta_r - \frac{2\pi}{3}\right)\frac{di_{cs}}{dt} - \frac{1}{2}L_{ms}\frac{di'_{ar}}{dt} \\
&\quad + (L'_{lr} + L_{ms})\frac{di'_{br}}{dt} - \frac{1}{2}L_{ms}\frac{di'_{cr}}{dt} - L_{ms}\left[i_{as}\sin\left(\theta_r + \frac{2\pi}{3}\right) + i_{bs}\sin\theta_r + i_{cs}\sin\left(\theta_r - \frac{2\pi}{3}\right)\right]\omega_r \\
u'_{cr} &= r'_r i'_{cr} + L_{ms}\left(\theta_r - \frac{2\pi}{3}\right)\frac{di_{as}}{dt} + L_{ms}\cos\left(\theta_r + \frac{2\pi}{3}\right)\frac{di_{bs}}{dt} + L_{ms}\cos\theta_r\frac{di_{cs}}{dt} - \frac{1}{2}L_{ms}\frac{di'_{ar}}{dt} \\
&\quad - \frac{1}{2}L_{ms}\frac{di'_{br}}{dt} + (L'_{lr} + L_{ms})\frac{di'_{cr}}{dt} - L_{ms}\left[i_{as}\sin\left(\theta_r - \frac{2\pi}{3}\right) + i_{bs}\sin\left(\theta_r + \frac{2\pi}{3}\right) + i_{cs}\sin\theta_r\right]\omega_r
\end{aligned}
\tag{6.44}
$$

Taking note of (6.44), Cauchy's form of the differential equations, given in matrix form, are found to be:

$$
\begin{bmatrix} \frac{di_{as}}{dt} \\ \frac{di_{bs}}{dt} \\ \frac{di_{cs}}{dt} \\ \frac{di'_{ar}}{dt} \\ \frac{di'_{br}}{dt} \\ \frac{di'_{cr}}{dt} \end{bmatrix}
= \frac{1}{L_{\Sigma L}}
\begin{bmatrix}
-r_s L_{\Sigma m} & -\frac{1}{2}r_s L_{ms} & -\frac{1}{2}r_s L_{ms} & 0 & 0 & 0 \\
-\frac{1}{2}r_s L_{ms} & -r_s L_{\Sigma m} & -\frac{1}{2}r_s L_{ms} & 0 & 0 & 0 \\
-\frac{1}{2}r_s L_{ms} & -\frac{1}{2}r_s L_{ms} & -r_s L_{\Sigma m} & 0 & 0 & 0 \\
0 & 0 & 0 & -r_r L_{\Sigma m} & -\frac{1}{2}r_r L_{ms} & -\frac{1}{2}r_r L_{ms} \\
0 & 0 & 0 & -\frac{1}{2}r_r L_{ms} & -r_r L_{\Sigma m} & -\frac{1}{2}r_r L_{ms} \\
0 & 0 & 0 & -\frac{1}{2}r_r L_{ms} & -\frac{1}{2}r_r L_{ms} & -r_r L_{\Sigma m}
\end{bmatrix}
\begin{bmatrix} i_{as} \\ i_{bs} \\ i_{cs} \\ i'_{ar} \\ i'_{br} \\ i'_{cr} \end{bmatrix}
$$

$$
+\frac{1}{L_{\Sigma L}}\begin{bmatrix}
0 & 0 & 0 & r_r L_{ms}\cos\theta_r & r_r L_{ms}\cos\left(\theta_r+\frac{2}{3}\pi\right) & r_r L_{ms}\cos\left(\theta_r-\frac{2}{3}\pi\right)\\
0 & 0 & 0 & r_r L_{ms}\cos\left(\theta_r-\frac{2}{3}\pi\right) & r_r L_{ms}\cos\theta_r & r_r L_{ms}\cos\left(\theta_r+\frac{2}{3}\pi\right)\\
0 & 0 & 0 & r_r L_{ms}\cos\left(\theta_r+\frac{2}{3}\pi\right) & r_r L_{ms}\cos\left(\theta_r-\frac{2}{3}\pi\right) & r_r L_{ms}\cos\theta_r\\
r_s L_{ms}\cos\theta_r & r_s L_{ms}\cos\left(\theta_r-\frac{2}{3}\pi\right) & r_s L_{ms}\cos\left(\theta_r+\frac{2}{3}\pi\right) & 0 & 0 & 0\\
r_s L_{ms}\cos\left(\theta_r+\frac{2}{3}\pi\right) & r_s L_{ms}\cos\theta_r & r_s L_{ms}\cos\left(\theta_r-\frac{2}{3}\pi\right) & 0 & 0 & 0\\
r_s L_{ms}\cos\left(\theta_r-\frac{2}{3}\pi\right) & r_s L_{ms}\cos\left(\theta_r+\frac{2}{3}\pi\right) & r_s L_{ms}\cos\theta_r & 0 & 0 & 0
\end{bmatrix}\begin{bmatrix} i_{as}\\ i_{bs}\\ i_{cs}\\ i'_{ar}\\ i'_{br}\\ i'_{cr}\end{bmatrix}
$$

$$
+\frac{1}{L_{\Sigma L}}\begin{bmatrix}
0 & 1.3L_{ms}^2\omega_r & -1.3L_{ms}^2\omega_r & L_{\Sigma ms}\omega_r\sin\theta_r & L_{\Sigma ms}\omega_r\sin\left(\theta_r+\frac{2}{3}\pi\right) & L_{\Sigma ms}\omega_r\sin\left(\theta_r-\frac{2}{3}\pi\right)\\
-1.3L_{ms}^2\omega_r & 0 & 1.3L_{ms}^2\omega_r & L_{\Sigma ms}\omega_r\sin\left(\theta_r-\frac{2}{3}\pi\right) & L_{\Sigma ms}\omega_r\sin\theta_r & L_{\Sigma ms}\omega_r\sin\left(\theta_r+\frac{2}{3}\pi\right)\\
1.3L_{ms}^2\omega_r & -1.3L_{ms}^2\omega_r & 0 & L_{\Sigma ms}\omega_r\sin\left(\theta_r+\frac{2}{3}\pi\right) & L_{\Sigma ms}\omega_r\sin\left(\theta_r-\frac{2}{3}\pi\right) & L_{\Sigma ms}\omega_r\sin\theta_r\\
L_{\Sigma ms}\omega_r\sin\theta_r & L_{\Sigma ms}\omega_r\sin\left(\theta_r-\frac{2}{3}\pi\right) & L_{\Sigma ms}\omega_r\sin\left(\theta_r+\frac{2}{3}\pi\right) & 0 & -1.3L_{ms}^2\omega_r & 1.3L_{ms}^2\omega_r\\
L_{\Sigma ms}\omega_r\sin\left(\theta_r+\frac{2}{3}\pi\right) & L_{\Sigma ms}\omega_r\sin\theta_r & L_{\Sigma ms}\omega_r\sin\left(\theta_r-\frac{2}{3}\pi\right) & 1.3L_{ms}^2\omega_r & 0 & -1.3L_{ms}^2\omega_r\\
L_{\Sigma ms}\omega_r\sin\left(\theta_r-\frac{2}{3}\pi\right) & L_{\Sigma ms}\omega_r\sin\left(\theta_r+\frac{2}{3}\pi\right) & L_{\Sigma ms}\omega_r\sin\theta_r & -1.3L_{ms}^2\omega_r & 1.3L_{ms}^2\omega_r & 0
\end{bmatrix}\begin{bmatrix} i_{as}\\ i_{bs}\\ i_{cs}\\ i'_{ar}\\ i'_{br}\\ i'_{cr}\end{bmatrix}
$$

$$
+\frac{1}{L_{\Sigma L}}\begin{bmatrix}
2L_{ms}+L'_{lr} & \frac{1}{2}L_{ms} & \frac{1}{2}L_{ms} & -L_{ms}\cos\theta_r & -L_{ms}\cos\left(\theta_r+\frac{2}{3}\pi\right) & -L_{ms}\cos\left(\theta_r-\frac{2}{3}\pi\right)\\
\frac{1}{2}L_{ms} & 2L_{ms}+L'_{lr} & \frac{1}{2}L_{ms} & -L_{ms}\cos\left(\theta_r-\frac{2}{3}\pi\right) & -L_{ms}\cos\theta_r & -L_{ms}\cos\left(\theta_r+\frac{2}{3}\pi\right)\\
\frac{1}{2}L_{ms} & \frac{1}{2}L_{ms} & 2L_{ms}+L'_{lr} & -L_{ms}\cos\left(\theta_r+\frac{2}{3}\pi\right) & -L_{ms}\cos\left(\theta_r-\frac{2}{3}\pi\right) & -L_{ms}\cos\theta_r\\
-L_{ms}\cos\theta_r & -L_{ms}\cos\left(\theta_r-\frac{2}{3}\pi\right) & -L_{ms}\cos\left(\theta_r+\frac{2}{3}\pi\right) & 2L_{ms}+L'_{lr} & \frac{1}{2}L_{ms} & \frac{1}{2}L_{ms}\\
-L_{ms}\cos\left(\theta_r+\frac{2}{3}\pi\right) & -L_{ms}\cos\theta_r & -L_{ms}\cos\left(\theta_r-\frac{2}{3}\pi\right) & \frac{1}{2}L_{ms} & 2L_{ms}+L'_{lr} & \frac{1}{2}L_{ms}\\
-L_{ms}\cos\left(\theta_r-\frac{2}{3}\pi\right) & -L_{ms}\cos\left(\theta_r+\frac{2}{3}\pi\right) & -L_{ms}\cos\theta_r & \frac{1}{2}L_{ms} & \frac{1}{2}L_{ms} & 2L_{ms}+L'_{lr}
\end{bmatrix}\begin{bmatrix} u_{as}\\ u_{bs}\\ u_{cs}\\ u'_{ar}\\ u'_{br}\\ u'_{cr}\end{bmatrix}
\tag{6.45}
$$

Here, $L_{\Sigma L}=(3L_{ms}+L'_{lr})L'_{lr}$, $L_{\Sigma m}=2L_{ms}+L'_{lr}$, and $L_{\Sigma ms}=\frac{3}{2}L_{ms}^2+L_{ms}L'_{lr}$.

Newton's second law of motion is applied to derive the *torsional-mechanical* equations. To find the *net* torque, the expression for the electromagnetic torque developed by induction motors must be obtained. The electromagnetic torque for P-pole three-phase induction machines is found using the expression for coenergy $W_c(\mathbf{i}_{abcs}, \mathbf{i}'_{abcr}, \theta_r)$. In particular,

$$T_e=\frac{P}{2}\frac{\partial W_c(\mathbf{i}_{abcs},\mathbf{i}'_{abcr},\theta_r)}{\partial\theta_r}$$

where

$$W_c=W_f=\frac{1}{2}\mathbf{i}_{abcs}^T(\mathbf{L}_s-L_{ls}\mathbf{I})\mathbf{i}_{abcs}+\mathbf{i}_{abcs}^T\mathbf{L}'_{sr}(\theta_r)\mathbf{i}'_{abcr}+\frac{1}{2}\mathbf{i}'^T_{abcr}(\mathbf{L}'_r-L'_{lr}\mathbf{I})\mathbf{i}'_{abcr}$$

Here, $\mathbf{L}_s$, L_{ls}, $\mathbf{L}'_r$, and L'_{lr} are not functions of the electrical displacement θ_r. Then, the expression for the electromagnetic torque becomes

$$T_e = \frac{P}{2}\mathbf{i}_{abcs}^T \frac{\partial \mathbf{L}'_{sr}(\theta_r)}{\partial \theta_r}\mathbf{i}'_{abcr}$$

$$= -\frac{P}{2}L_{ms}[i_{as} \quad i_{bs} \quad i_{cs}]\begin{bmatrix} \sin\theta_r & \sin\left(\theta_r + \frac{2}{3}\pi\right) & \sin\left(\theta_r - \frac{2}{3}\pi\right) \\ \sin\left(\theta_r - \frac{2}{3}\pi\right) & \sin\theta_r & \sin\left(\theta_r + \frac{2}{3}\pi\right) \\ \sin\left(\theta_r + \frac{2}{3}\pi\right) & \sin\left(\theta_r - \frac{2}{3}\pi\right) & \sin\theta_r \end{bmatrix}\begin{bmatrix} i'_{ar} \\ i'_{br} \\ i'_{cr} \end{bmatrix}$$

$$= -\frac{P}{2}L_{ms}\left\{\left[i_{as}\left(i'_{ar} - \frac{1}{2}i'_{br} - \frac{1}{2}i'_{cr}\right) + i_{bs}\left(i'_{br} - \frac{1}{2}i'_{ar} - \frac{1}{2}i'_{cr}\right) + i_{cs}\left(i'_{cr} - \frac{1}{2}i'_{br} - \frac{1}{2}i'_{ar}\right)\right]\sin\theta_r \right.$$

$$\left. + \frac{\sqrt{3}}{2}[i_{as}(i'_{br} - i'_{cr}) + i_{bs}(i'_{cr} - i'_{ar}) + i_{cs}(i'_{ar} - i'_{br})]\cos\theta_r \right\} \tag{6.46}$$

Using Newton's second law and (6.46), the *tortional-mechanical* equations are written as

$$\frac{d\omega_r}{dt} = \frac{P}{2J}T_e - \frac{B_m}{J}\omega_r - \frac{P}{2J}T_L$$

$$= -\frac{P^2}{4J}L_{ms}\left\{\left[i_{as}\left(i'_{ar} - \frac{1}{2}i'_{br} - \frac{1}{2}i'_{cr}\right) + i_{bs}\left(i'_{br} - \frac{1}{2}i'_{ar} - \frac{1}{2}i'_{cr}\right) + i_{cs}\left(i'_{cr} - \frac{1}{2}i'_{br} - \frac{1}{2}i'_{ar}\right)\right]\sin\theta_r \right.$$

$$\left. + \frac{\sqrt{3}}{2}[i_{as}(i'_{br} - i'_{cr}) + i_{bs}(i'_{cr} - i'_{ar}) + i_{cs}(i'_{ar} - i'_{br})]\cos\theta_r \right\} - \frac{B_m}{J}\omega_r - \frac{P}{2J}T_L$$

$$\frac{d\theta_r}{dt} = \omega_r \tag{6.47}$$

Augmenting differential equations (6.45) and (6.47), the model for three-phase induction motors modeled in *machine* variables results.

The derived nonlinear model allows one to perform analysis of steady-state and dynamic characteristics, as well as to design control algorithms. Induction motors must be controlled; and to regulate the angular velocity, one changes the magnitude of the applied phase voltages to the *abc* windings. The torque-speed characteristic curves of three-phase induction motors are shaped by changing the magnitude and frequency of the phase voltages. For squirrel-cage induction motors, the following three-phase voltage set must be supplied to the *as*, *bs*, and *cs* windings to guarantee balanced operation

$$u_{as}(t) = \sqrt{2}u_M \cos(\omega_f t)$$

$$u_{bs}(t) = \sqrt{2}u_M \cos\left(\omega_f t - \frac{2}{3}\pi\right)$$

$$u_{cs}(t) = \sqrt{2}u_M \cos\left(\omega_f t + \frac{2}{3}\pi\right)$$

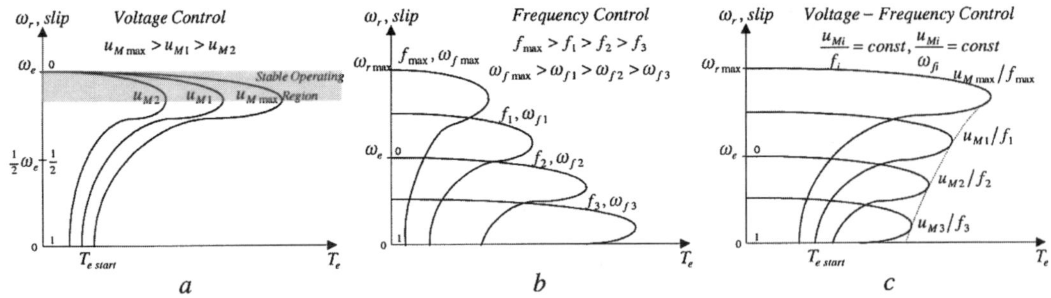

Figure 6.77 Torque-speed characteristic curves of induction motors, $\omega_r = \Omega_T(T_e)$: (a) voltage control; (b) frequency control; (c) voltage-frequency (constant volts per hertz) control.

where the angular frequency of the applied voltage is $\omega_f = 2\pi f$.

The sinusoidal steady-state currents

$$i_{as}(t) = \sqrt{2}i_M \cos(\omega_f t - \varphi_i),\ i_{bs}(t) = \sqrt{2}i_M \cos\left(\omega_f t - \frac{2}{3}\pi - \varphi_i\right),\ i_{cs}(t) = \sqrt{2}i_M \cos\left(\omega_f t + \frac{2}{3}\pi - \varphi_i\right)$$

ensure balanced operation.

Figure 6.77a illustrates the torque-speed characteristic curves if one changes the magnitude u_M of the applied phase voltages. The frequency of the applied voltage f can be changed to control the angular velocity above or below the synchronous angular velocity ω_e (see Figure 6.77b). It should be emphasized that the angular velocity of general-purpose induction motors can be controlled in the region $[0.05\omega_e \quad 2\omega_e]$. Voltage-frequency control is widely used, and the corresponding torque-speed characteristic curves are plotted in Figure 6.77c. In conventional electric drives, the voltage-frequency pattern $\frac{u_{Mi}}{f_i}$ $= const$ is usually assigned. That is, the magnitude and the frequency are adjusted to guarantee $\frac{u_{Mi}}{f_i} = const$. It was demonstrated in Section 6.2 (see Figures 6.8d and 6.8e) that using $\frac{u_{Mi}}{f_i} = \text{var}$, one shapes the torque-speed characteristic curves, and the soft- and high-starting torque patterns can be implemented.

The torque-speed characteristic curves can be obtained by performing nonlinear modeling of induction motors. To approach the preliminary analysis, using the stator and rotor resistances and reactances X_s and X'_r, as well as the magnetizing reactance X_M, one can apply the following formula to plot the torque-speed characteristics

$$T_e = \frac{3\left(u_M \dfrac{X_M}{X_s + X_M}\right)^2 \dfrac{r'_r}{slip}}{\omega_e\left[\left(r_s\left(\dfrac{X_M}{X_s + X_M}\right)^2 + \dfrac{r'_r}{slip}\right)^2 + (X_s + X'_r)^2\right]}$$

Taking note of the following expression for the *slip*

$$slip = \frac{\omega_e - \omega_r}{\omega_e}, \quad \omega_e = \frac{4\pi f}{P}$$

the torque-speed characteristic curves $\omega_r = \Omega_T(T_e)$ can be found for different values of the magnitude and frequency of the phase voltage supplied.

Example 6.9

Find and plot the torque-speed characteristic curve for a three-phase two-pole induction motor that has the following parameters: $r_s = 0.27$ ohm, $r_r' = 0.15$ ohm, $X_s = 0.9$ ohm, $X_r' = 0.41$ ohm, and $X_M = 17$ ohm, if $u_M = 220$ V and $f = 60$ Hz.

Solution

For the assigned voltage and frequency, one calculates the synchronous angular velocity by applying the formula $\omega_e = \frac{4\pi f}{P}$. Then, using the following formulas

$$T_e = \frac{3\left(u_M \dfrac{X_M}{X_s + X_M}\right)^2 \dfrac{r_r'}{slip}}{\omega_e\left[\left(r_s\left(\dfrac{X_M}{X_s + X_M}\right)^2 + \dfrac{r_r'}{slip}\right)^2 + (X_s + X_r')^2\right]} \quad \text{and} \quad slip = \frac{\omega_e - \omega_r}{\omega_e}$$

the torque-speed characteristic curve can be found by assigning different values for the angular velocity. The following MATLAB file is developed to calculate and plot $\omega_r = \Omega_T(T_e)$ for the induction motor studied.

MATLAB script (ch_6_01.m)

```
% parameters of induction motor
rs=0.25; rr=0.14; Xs=0.9; Xr=0.41; Xm=17;
ull=220; f=60; P=2;
we=4*pi*f/P;
% torque-speed characteristic
for wr=[1:10:340 340:1:4*pi*f/P];
slip=(we-wr)/we;
Te=3*(ull*Xm/(Xs+Xm))^2*(rr/slip)...
/(we*((rs*(Xm/(Xs+Xm))^2+rr/slip)^2+(Xs+Xr)^2));
plot(Te,wr,'o');hold on;
end; disp('End')
```

The resulting torque-speed characteristic curve is documented in Figure 6.78. □

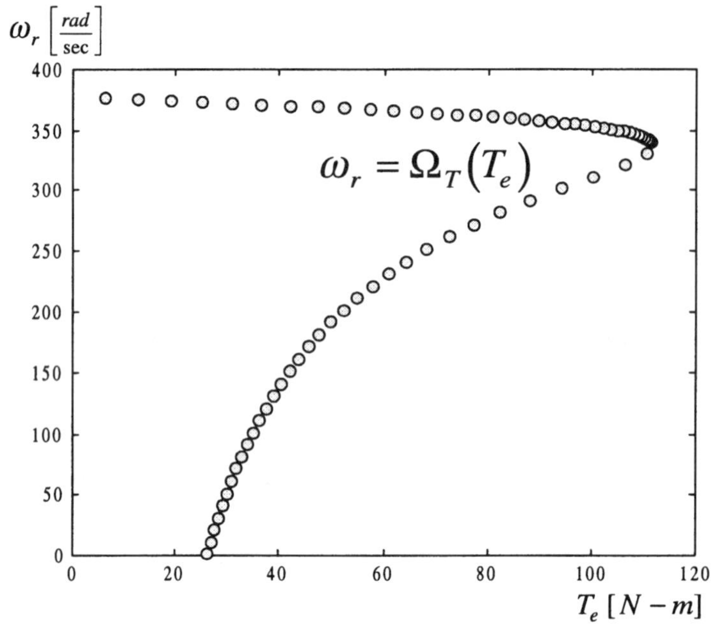

Figure 6.78 Torque-speed characteristic curve for a three-phase induction motor.

6.5 Mathematical models of three-phase induction machines in the arbitrary, stationary, rotor, and synchronous reference frames

The differential equations to comprehensively study the steady-state and dynamic performance of induction machines were developed in the *machine* variables using Kirchhoff's second law and Newton's law of motion. That is, the motor steady-state and transient behavior are described using the *abc* stator and rotor quantities. It has been shown that the *quadrature* and *direct* (*qd*) quantities can be applied to reduce the complexity of the resulting differential equations that map the dynamics of two-phase induction machines. For three-phase induction and synchronous machines, the transformations of the *machine* stator and rotor voltages, currents, and flux linkages to the *quadrature-*, *direct-*, and *zero*-axis components of stator and rotor voltages, currents, and flux linkages can be performed using Park transformations. In the most general case, the model is developed in the *arbitrary* reference frame, and the frame angular velocity ω is not specified. Assigning the frame angular velocity to be $\omega = 0$, $\omega = \omega_r$, and $\omega = \omega_e$, the mathematical models in three reference frames, commonly used in practice, can be easily found. In particular, the stationary, rotor, and synchronous reference frames are widely used. Table 6.3. documents the transformations of the *machine* variables to the *quadrature, direct,* and *zero* (*qd0*) quantities, and vice versa.

Illustrative Example 6.2

Find the *quadrature-*, *direct-*, and *zero*-axis components of voltages in the *arbitrary* reference frame if the following voltages are supplied to the *as*, *bs*, and *cs* windings

$$u_{as}(t) = 100\cos(377t),\quad u_{bs}(t) = 100\cos\left(377t - \frac{2}{3}\pi\right),\quad \text{and}\quad u_{cs}(t) = 100\left(377t + \frac{2}{3}\pi\right)$$

Table 6.3 Transformations of the *machine* and *qd0* Variables

Stator, rotor, *quadrature* and *direct* magnetic axes	Transformation of variables using transformation matrices
Arbitrary reference frame ω unspecified 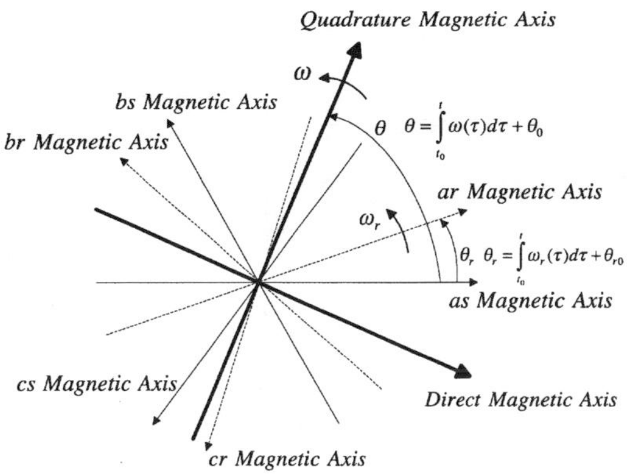	*Direct* transformation $\mathbf{u}_{qdos}=\mathbf{K}_s\mathbf{u}_{abcs}$, $\mathbf{i}_{qdos}=\mathbf{K}_s\mathbf{i}_{abcs}$, $\boldsymbol{\psi}_{qdos}=\mathbf{K}_s\boldsymbol{\psi}_{abcs}$ *Inverse* transformation $\mathbf{u}_{abcs}=\mathbf{K}_s^{-1}\mathbf{u}_{qdos}$, $\mathbf{i}_{abcs}=\mathbf{K}_s^{-1}\mathbf{i}_{qdos}$, $\boldsymbol{\psi}_{abcs}=\mathbf{K}_s^{-1}\boldsymbol{\psi}_{qdos}$ **Stator** transformation matrices $\mathbf{K}_s=\frac{2}{3}\begin{bmatrix}\cos\theta & \cos\left(\theta-\frac{2}{3}\pi\right) & \cos\left(\theta+\frac{2}{3}\pi\right)\\ \sin\theta & \sin\left(\theta-\frac{2}{3}\pi\right) & \sin\left(\theta+\frac{2}{3}\pi\right)\\ \frac{1}{2} & \frac{1}{2} & \frac{1}{2}\end{bmatrix}$, $\mathbf{K}_s^{-1}=\begin{bmatrix}\cos\theta & \sin\theta & 1\\ \cos\left(\theta-\frac{2}{3}\pi\right) & \sin\left(\theta-\frac{2}{3}\pi\right) & 1\\ \cos\left(\theta+\frac{2}{3}\pi\right) & \sin\left(\theta+\frac{2}{3}\pi\right) & 1\end{bmatrix}$ *Direct* transformation $\mathbf{u}_{qdor}=\mathbf{K}_r\mathbf{u}_{abcr}$, $\mathbf{i}_{qdor}=\mathbf{K}_r\mathbf{i}_{abcr}$, $\boldsymbol{\psi}_{qdor}=\mathbf{K}_r\boldsymbol{\psi}_{abcr}$ *Inverse* transformation $\mathbf{u}_{abcr}=\mathbf{K}_r^{-1}\mathbf{u}_{qdor}$, $\mathbf{i}_{abcs}=\mathbf{K}_r^{-1}\mathbf{i}_{qdos}$, $\boldsymbol{\psi}_{abcs}=\mathbf{K}_r^{-1}\boldsymbol{\psi}_{qdos}$ **Rotor** transformation matrices $\mathbf{K}_r=\frac{2}{3}\begin{bmatrix}\cos(\theta-\theta_r) & \cos\left(\theta-\theta_r-\frac{2}{3}\pi\right) & \cos\left(\theta-\theta_r+\frac{2}{3}\pi\right)\\ \sin(\theta-\theta_r) & \sin\left(\theta-\theta_r-\frac{2}{3}\pi\right) & \sin\left(\theta-\theta_r+\frac{2}{3}\pi\right)\\ \frac{1}{2} & \frac{1}{2} & \frac{1}{2}\end{bmatrix}$, $\mathbf{K}_r^{-1}=\begin{bmatrix}\cos(\theta-\theta_r) & \sin(\theta-\theta_r) & 1\\ \cos\left(\theta-\theta_r-\frac{2}{3}\pi\right) & \sin\left(\theta-\theta_r-\frac{2}{3}\pi\right) & 1\\ \cos\left(\theta-\theta_r+\frac{2}{3}\pi\right) & \sin\left(\theta-\theta_r+\frac{2}{3}\pi\right) & 1\end{bmatrix}$
Stationary reference frame, $\omega=0$	*Direct* transformation $\mathbf{u}_{qdos}^s=\mathbf{K}_s^s\mathbf{u}_{abcs}$, $\mathbf{i}_{qdos}^s=\mathbf{K}_s^s\mathbf{i}_{abcs}$, $\boldsymbol{\psi}_{qdos}^s=\mathbf{K}_s^s\boldsymbol{\psi}_{abcs}$ *Inverse* transformation $\mathbf{u}_{abcs}=\mathbf{K}_s^{s^{-1}}\mathbf{u}_{qdos}^s$, $\mathbf{i}_{abcs}=\mathbf{K}_s^{s^{-1}}\mathbf{i}_{qdos}^s$, $\boldsymbol{\psi}_{abcs}=\mathbf{K}_s^{s^{-1}}\boldsymbol{\psi}_{qdos}^s$ *Direct* transformation $\mathbf{u}_{qdor}^s=\mathbf{K}_r^s\mathbf{u}_{abcr}$, $\mathbf{i}_{qdor}^s=\mathbf{K}_r^s\mathbf{i}_{abcr}$, $\boldsymbol{\psi}_{qdor}^s=\mathbf{K}_r^s\boldsymbol{\psi}_{abcr}$ *Inverse* transformation $\mathbf{u}_{abcr}=\mathbf{K}_r^{s^{-1}}\mathbf{u}_{qdor}^s$, $\mathbf{i}_{abcr}=\mathbf{K}_r^{s^{-1}}\mathbf{i}_{qdor}^s$, $\boldsymbol{\psi}_{abcr}=\mathbf{K}_r^{s^{-1}}\boldsymbol{\psi}_{qdor}^s$
Rotor reference frame, $\omega=\omega_r$	*Direct* transformation $\mathbf{u}_{qdos}^r=\mathbf{K}_s^r\mathbf{u}_{abcs}$, $\mathbf{i}_{qdos}^r=\mathbf{K}_s^r\mathbf{i}_{abcs}$, $\boldsymbol{\psi}_{qdos}^r=\mathbf{K}_s^r\boldsymbol{\psi}_{abcs}$ *Inverse* transformation $\mathbf{u}_{abcs}=\mathbf{K}_s^{r^{-1}}\mathbf{u}_{qdos}^r$, $\mathbf{i}_{abcs}=\mathbf{K}_s^{r^{-1}}\mathbf{i}_{qdos}^r$, $\boldsymbol{\psi}_{abcs}=\mathbf{K}_s^{r^{-1}}\boldsymbol{\psi}_{qdos}^r$ *Direct* transformation $\mathbf{u}_{qdor}^r=\mathbf{K}_r^r\mathbf{u}_{abcr}$, $\mathbf{i}_{qdor}^r=\mathbf{K}_r^r\mathbf{i}_{abcr}$, $\boldsymbol{\psi}_{qdor}^r=\mathbf{K}_r^r\boldsymbol{\psi}_{abcr}$ *Inverse* transformation $\mathbf{u}_{abcr}=\mathbf{K}_r^{r^{-1}}\mathbf{u}_{qdor}^r$, $\mathbf{i}_{abcr}=\mathbf{K}_r^{r^{-1}}\mathbf{i}_{qdor}^r$, $\boldsymbol{\psi}_{abcr}=\mathbf{K}_r^{r^{-1}}\boldsymbol{\psi}_{qdor}^r$
Synchronous reference frame, $\omega=\omega_e$	*Direct* transformation $\mathbf{u}_{qdos}^e=\mathbf{K}_s^e\mathbf{u}_{abcs}$, $\mathbf{i}_{qdos}^e=\mathbf{K}_s^e\mathbf{i}_{abcs}$, $\boldsymbol{\psi}_{qdos}^e=\mathbf{K}_s^e\boldsymbol{\psi}_{abcs}$ *Inverse* transformation $\mathbf{u}_{abcs}=\mathbf{K}_s^{e^{-1}}\mathbf{u}_{qdos}^e$, $\mathbf{i}_{abcs}=\mathbf{K}_s^{e^{-1}}\mathbf{i}_{qdos}^e$, $\boldsymbol{\psi}_{abcs}=\mathbf{K}_s^{e^{-1}}\boldsymbol{\psi}_{qdos}^e$ *Direct* transformation $\mathbf{u}_{qdor}^e=\mathbf{K}_r^e\mathbf{u}_{abcr}$, $\mathbf{i}_{qdor}^e=\mathbf{K}_r^e\mathbf{i}_{abcr}$, $\boldsymbol{\psi}_{qdor}^e=\mathbf{K}_r^e\boldsymbol{\psi}_{abcr}$ *Inverse* transformation $\mathbf{u}_{abcr}=\mathbf{K}_r^{e^{-1}}\mathbf{u}_{qdor}^e$, $\mathbf{i}_{abcr}=\mathbf{K}_r^{e^{-1}}\mathbf{i}_{qdor}^e$, $\boldsymbol{\psi}_{abcr}=\mathbf{K}_r^{e^{-1}}\boldsymbol{\psi}_{qdor}^e$

Assigning the frame angular velocities to be $\omega = 377$ rad-s^{-1} and $\omega = 0$ (the synchronous and stationary reference frames, respectively), obtain the corresponding *qd0* voltages.

Solution

Using the *direct* Park transformation, as given by

$$\mathbf{u}_{qdos} = \mathbf{K}_s \mathbf{u}_{abcs}$$

where the transformation matrix is

$$\mathbf{K}_s = \frac{2}{3}\begin{bmatrix} \cos\theta & \cos\left(\theta - \frac{2}{3}\pi\right) & \cos\left(\theta + \frac{2}{3}\pi\right) \\ \sin\theta & \sin\left(\theta - \frac{2}{3}\pi\right) & \sin\left(\theta + \frac{2}{3}\pi\right) \\ \frac{1}{2} & \frac{1}{2} & \frac{1}{2} \end{bmatrix}$$

we have

$$\begin{bmatrix} u_{qs} \\ u_{ds} \\ u_{os} \end{bmatrix} = \frac{2}{3}\begin{bmatrix} \cos\theta & \cos\left(\theta - \frac{2}{3}\pi\right) & \cos\left(\theta + \frac{2}{3}\pi\right) \\ \sin\theta & \sin\left(\theta - \frac{2}{3}\pi\right) & \sin\left(\theta + \frac{2}{3}\pi\right) \\ \frac{1}{2} & \frac{1}{2} & \frac{1}{2} \end{bmatrix}\begin{bmatrix} u_{as} \\ u_{bs} \\ u_{cs} \end{bmatrix}$$

and in expanded form, the following expressions for the *quadrature-*, *direct-*, and *zero*-axis components of voltages result

$$u_{qs}(t) = \frac{2}{3}\left(\cos\theta u_{as}(t) + \cos\left(\theta - \frac{2}{3}\pi\right)u_{bs}(t) + \cos\left(\theta + \frac{2}{3}\pi\right)u_{cs}(t)\right)$$

$$u_{ds}(t) = \frac{2}{3}\left(\sin\theta u_{as}(t) + \sin\left(\theta - \frac{2}{3}\pi\right)u_{bs}(t) + \sin\left(\theta + \frac{2}{3}\pi\right)u_{cs}(t)\right)$$

$$u_{os}(t) = \frac{1}{3}(u_{as}(t) + u_{bs}(t) + u_{cs}(t))$$

Taking note of

$$u_{as}(t) = 100\cos(377t),\ \ u_{bs}(t) = 100\cos\left(377t - \frac{2}{3}\pi\right),\ \text{and}\ \ u_{cs}(t) = 100\cos\left(377t + \frac{2}{3}\pi\right)$$

one obtains the equations for the *quadrature-*, *direct-*, and *zero*-axis components of voltages in the *arbitrary* reference frame as

$$u_{qs}(t)=\frac{200}{3}\left(\cos\theta\cos(377t)+\cos\left(\theta-\frac{2}{3}\pi\right)\cos\left(377t-\frac{2}{3}\pi\right)+\cos\left(\theta+\frac{2}{3}\pi\right)\cos\left(377t+\frac{2}{3}\pi\right)\right)$$

$$u_{ds}(t)=\frac{200}{3}\left(\sin\theta\cos(377t)+\sin\left(\theta-\frac{2}{3}\pi\right)\cos\left(377t-\frac{2}{3}\pi\right)+\sin\left(\theta+\frac{2}{3}\pi\right)\cos\left(377t+\frac{2}{3}\pi\right)\right)$$

$$u_{os}(t)=\frac{200}{3}\left(\cos(377t)+\cos\left(377t-\frac{2}{3}\pi\right)+\cos\left(377t+\frac{2}{3}\pi\right)\right)$$

Thus, in the *arbitrary* reference frame, the *quadrature-*, *direct-*, and *zero*-axis components of voltages are found.

Letting the angular velocity of the frame be ω = 377 rad-s^{-1}, one concludes that the synchronous reference frame is in place because the angular frequency of the supplied voltages is 377 rad-s^{-1}. Assuming that $\theta_0 = 0$, we have

$$\theta=\omega t=377t$$

Hence, using the trigonometric identities, one finds

$$u^e_{qs}(t)=\frac{200}{3}\left(\cos^2(377t)+\cos^2\left(377t-\frac{2}{3}\pi\right)+\cos^2\left(377t+\frac{2}{3}\pi\right)\right)=\frac{200}{3}\frac{3}{2}=100$$

$$u^e_{do}(t)=\frac{200}{3}\left(\sin(377t)\cos(377t)+\sin\left(377t-\frac{2}{3}\pi\right)\cos\left(377t-\frac{2}{3}\pi\right)\right.$$

$$\left.+\sin\left(377t+\frac{2}{3}\pi\right)\cos\left(377t+\frac{2}{3}\right)\right)=0$$

$$u^e_{os}(t)=\frac{200}{3}\left(\cos(377t)+\cos\left(377t-\frac{2}{3}\pi\right)+\cos\left(377t+\frac{2}{3}\pi\right)\right)=0$$

For a balanced three-phase voltage set (a set of equal-amplitude sinusoidal voltages displaced by 120°), as given by

$$u_{as}(t)=100\cos(377t),\;\; u_{bs}(t)=100\sin\left(377t-\frac{2}{3}\pi\right),\;\text{and}\;\; u_{cs}(t)=100\sin\left(377t+\frac{2}{3}\pi\right)$$

the resulting *qd0* components $u^e_{qs}(t)$, $u^e_{ds}(t)$, and $u^e_{os}(t)$ are found to be dc voltages, and furthermore $u^e_{ds}(t)=0$ and $u^e_{os}(t)=0$.

In the stationary reference frame, one assigns $\omega = 0$, and thus, $\theta = 0$. Therefore, using

$$\mathbf{K}_s=\frac{2}{3}\begin{bmatrix}\cos\theta & \cos\left(\theta-\frac{2}{3}\pi\right) & \cos\left(\theta+\frac{2}{3}\pi\right)\\ \sin\theta & \sin\left(\theta-\frac{2}{3}\pi\right) & \sin\left(\theta+\frac{2}{3}\pi\right)\\ \frac{1}{2} & \frac{1}{2} & \frac{1}{2}\end{bmatrix}$$

we can find the matrix that needs to be applied in the stationary reference frame. In particular,

$$\mathbf{K}_s^s = \frac{2}{3}\begin{bmatrix} \cos\theta & \cos\left(\theta - \frac{2}{3}\pi\right) & \cos\left(\theta + \frac{2}{3}\pi\right) \\ \sin\theta & \sin\left(\theta - \frac{2}{3}\pi\right) & \sin\left(\theta + \frac{2}{3}\pi\right) \\ \frac{1}{2} & \frac{1}{2} & \frac{1}{2} \end{bmatrix}_{\theta=0} = \frac{2}{3}\begin{bmatrix} 1 & -\frac{1}{2} & -\frac{1}{2} \\ 0 & -\frac{\sqrt{3}}{2} & \frac{\sqrt{3}}{2} \\ \frac{1}{2} & \frac{1}{2} & \frac{1}{2} \end{bmatrix} = \begin{bmatrix} \frac{2}{3} & -\frac{1}{3} & -\frac{1}{3} \\ 0 & -\frac{1}{\sqrt{3}} & \frac{1}{\sqrt{3}} \\ \frac{1}{3} & \frac{1}{3} & \frac{1}{3} \end{bmatrix}$$

Hence, from

$$\mathbf{u}_{qdos}^s = \mathbf{K}_s^s \mathbf{u}_{abcs}, \quad \begin{bmatrix} u_{qs}^s \\ u_{ds}^s \\ u_{os}^s \end{bmatrix} = \begin{bmatrix} \frac{2}{3} & -\frac{1}{3} & -\frac{1}{3} \\ 0 & -\frac{1}{\sqrt{3}} & \frac{1}{\sqrt{3}} \\ \frac{1}{3} & \frac{1}{3} & \frac{1}{3} \end{bmatrix}\begin{bmatrix} u_{as} \\ u_{bs} \\ u_{cs} \end{bmatrix}$$

one finds

$$u_{qs}^s(t) = \frac{2}{3}u_{as}(t) - \frac{1}{3}u_{bs}(t) - \frac{1}{3}u_{cs}(t)$$

$$u_{ds}^s(t) = -\frac{1}{\sqrt{3}}u_{bs}(t) + \frac{1}{\sqrt{3}}u_{cs}(t)$$

$$u_{os}^s(t) = \frac{1}{3}u_{as}(t) + \frac{1}{3}u_{bs}(t) + \frac{1}{3}u_{cs}(t)$$

Therefore,

$$u_{qs}^s(t) = \frac{200}{3}\cos(377t) - \frac{100}{3}\cos\left(377t - \frac{2}{3}\pi\right) - \frac{100}{3}\cos\left(377t + \frac{2}{3}\pi\right) = 100\cos(377t)$$

$$u_{ds}^s(t) = -\frac{100}{\sqrt{3}}\cos\left(377t - \frac{2}{3}\pi\right) + \frac{100}{\sqrt{3}}\cos\left(377t + \frac{2}{3}\pi\right) = -100\sin(377t)$$

$$u_{os}^s(t) = \frac{100}{3}\cos(377t) + \frac{100}{3}\cos\left(377t - \frac{2}{3}\pi\right) + \frac{100}{3}\cos\left(377t + \frac{2}{3}\pi\right) = 0$$

□

In order to develop the mathematical model of three-phase induction machines in the *arbitrary* reference frame, we consider three-phase induction motors with *quadrature* and *direct* magnetic axes as illustrated in Figure 6.79.

To transform the *machine* (*abc*) stator voltages, currents, and flux linkages to the *quadrature-*, *direct-*, and *zero-*axis components of stator voltages, currents, and flux linkages, the following transformations are used

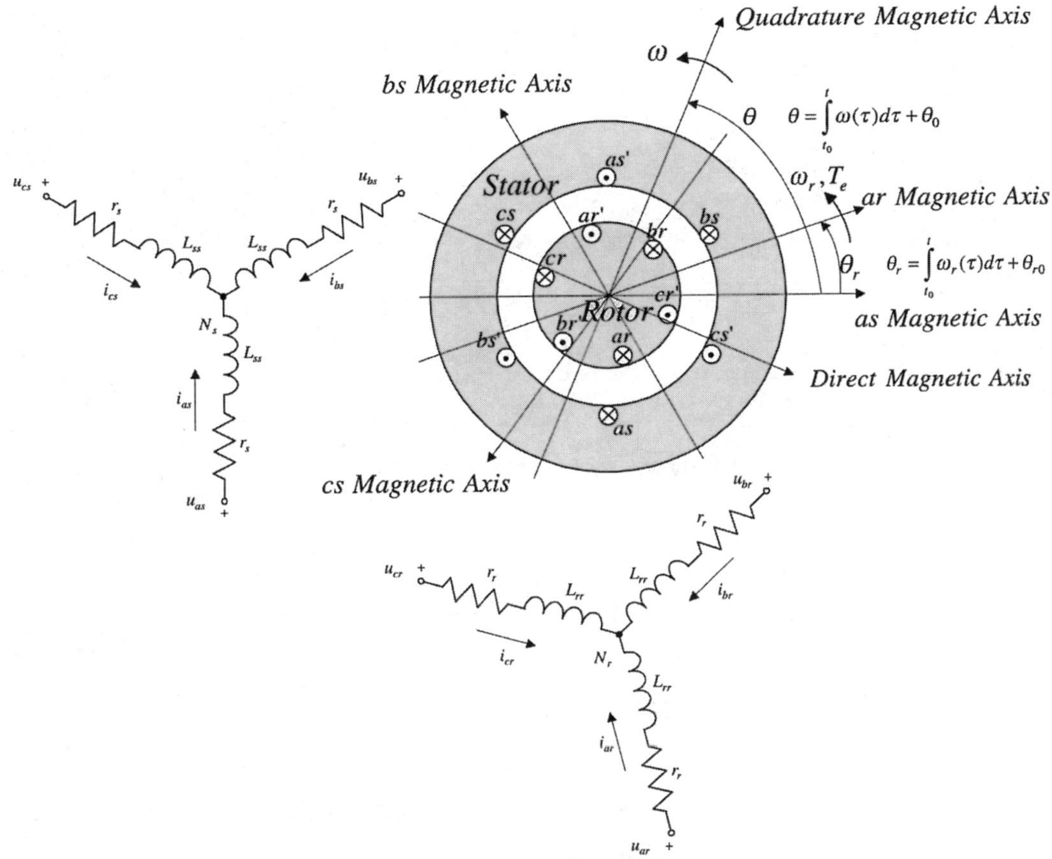

Figure 6.79 Three-phase symmetrical induction motor with *quadrature* and *direct* axes.

$$\mathbf{u}_{qdos} = \mathbf{K}_s \mathbf{u}_{abcs}$$
$$\mathbf{i}_{qdos} = \mathbf{K}_s \mathbf{i}_{abcs} \tag{6.48}$$
$$\boldsymbol{\Psi}_{qdos} = \mathbf{K}_s \boldsymbol{\Psi}_{abcs}$$

where the stator transformation matrix $\mathbf{K}_s$ is give as

$$\mathbf{K}_s = \frac{2}{3}\begin{bmatrix} \cos\theta & \cos\left(\theta - \frac{2}{3}\pi\right) & \cos\left(\theta + \frac{2}{3}\pi\right) \\ \sin\theta & \sin\left(\theta - \frac{2}{3}\pi\right) & \sin\left(\theta + \frac{2}{3}\pi\right) \\ \frac{1}{2} & \frac{1}{2} & \frac{1}{2} \end{bmatrix} \tag{6.49}$$

Here, the angular displacement of the reference frame is found as $\theta = \int_{t_0}^{t} \omega(\tau)d\tau + \theta_0$.

The *abc* rotor variables must be transformed to the *quadrature*, *direct*, and *zero* quantities as well. In particular, the *quadrature-*, *direct-*, and *zero*-axis components of rotor voltages, currents, and flux linkages are found using the *abc* rotor voltages, currents, and flux linkages. Taking note of the rotor transformations matrix $\mathbf{K}_r$, we have

$$\begin{aligned}\mathbf{u}'_{qdor} &= \mathbf{K}_r\mathbf{u}'_{abcr}\\ \mathbf{i}'_{qdor} &= \mathbf{K}_r\mathbf{i}'_{abcr}\\ \boldsymbol{\psi}'_{qdor} &= \mathbf{K}_r\boldsymbol{\psi}'_{abcr}\end{aligned} \tag{6.50}$$

Here, the rotor transformation matrix is expressed as

$$\mathbf{K}_r = \frac{2}{3}\begin{bmatrix}\cos(\theta-\theta_r) & \cos\left(\theta-\theta_r-\frac{2}{3}\pi\right) & \cos\left(\theta-\theta_r+\frac{2}{3}\pi\right)\\ \sin(\theta-\theta_r) & \sin\left(\theta-\theta_r-\frac{2}{3}\pi\right) & \sin\left(\theta-\theta_r+\frac{2}{3}\pi\right)\\ \frac{1}{2} & \frac{1}{2} & \frac{1}{2}\end{bmatrix} \tag{6.51}$$

From differential equations (6.43) that map the stator and rotor circuitry dynamics of three-phase induction motors and are given by

$$\begin{aligned}\mathbf{u}_{abcs} &= \mathbf{r}_s\mathbf{i}_{abcs} + \frac{d\boldsymbol{\psi}_{abcs}}{dt}\\ \mathbf{u}'_{abcr} &= \mathbf{r}'_r\mathbf{i}'_{abcr} + \frac{d\boldsymbol{\psi}'_{abcr}}{dt}\end{aligned}$$

and taking note of the inverse Park transformation matrices and $\mathbf{K}_s^{-1}$ and $\mathbf{K}_r^{-1}$, we have

$$\begin{aligned}\mathbf{K}_s^{-1}\mathbf{u}_{qdos} &= \mathbf{r}_s\mathbf{K}_s^{-1}\mathbf{i}_{qdos} + \frac{d(\mathbf{K}_s^{-1}\boldsymbol{\psi}_{qdos})}{dt}\\ \mathbf{K}_r^{-1}\mathbf{u}'_{qdor} &= \mathbf{r}'_r\mathbf{K}_r^{-1}\mathbf{i}'_{qdor} + \frac{d(\mathbf{K}_r^{-1}\boldsymbol{\psi}'_{qdor})}{dt}\end{aligned} \tag{6.52}$$

Making use of (6.49) and (6.51), one finds inverse matrices $\mathbf{K}_s^{-1}$ and $\mathbf{K}_r^{-1}$ to be

$$\mathbf{K}_s^{-1} = \begin{bmatrix}\cos\theta & \sin\theta & 1\\ \cos\left(\theta-\frac{2}{3}\pi\right) & \sin\left(\theta-\frac{2}{3}\pi\right) & 1\\ \cos\left(\theta+\frac{2}{3}\pi\right) & \sin\left(\theta+\frac{2}{3}\pi\right) & 1\end{bmatrix}$$

and

$$\mathbf{K}_r^{-1} = \begin{bmatrix} \cos(\theta-\theta_r) & \sin(\theta-\theta_r) & 1 \\ \cos\left(\theta-\theta_r-\frac{2}{3}\pi\right) & \sin\left(\theta-\theta_r-\frac{2}{3}\pi\right) & 1 \\ \cos\left(\theta-\theta_r+\frac{2}{3}\pi\right) & \sin\left(\theta-\theta_r+\frac{2}{3}\pi\right) & 1 \end{bmatrix}$$

Multiplying left and right sides of equations (6.52) by $\mathbf{K}_s$ and $\mathbf{K}_r$, we have

$$\begin{aligned} \mathbf{u}_{qdos} &= \mathbf{K}_s\mathbf{r}_s\mathbf{K}_s^{-1}\mathbf{i}_{qdos} + \mathbf{K}_s\frac{d\mathbf{K}_s^{-1}}{dt}\boldsymbol{\psi}_{qdos} + \mathbf{K}_s\mathbf{K}_s^{-1}\frac{d\boldsymbol{\psi}_{qdos}}{dt} \\ \mathbf{u}'_{qdor} &= \mathbf{K}_r\mathbf{r}'_r\mathbf{K}_r^{-1}\mathbf{i}'_{qdor} + \mathbf{K}_r\frac{d\mathbf{K}_r^{-1}}{dt}\boldsymbol{\psi}_{qdor} + \mathbf{K}_r\mathbf{K}_r^{-1}\frac{d\boldsymbol{\psi}'_{qdor}}{dt} \end{aligned} \tag{6.53}$$

The matrices of the stator and rotor resistances $\mathbf{r}_s$ and $\mathbf{r}'_r$ are diagonal and, hence,

$$\mathbf{K}_s\mathbf{r}_s\mathbf{K}_s^{-1} = \mathbf{r}_s \quad \text{and} \quad \mathbf{K}_r\mathbf{r}'_r\mathbf{K}_r^{-1} = \mathbf{r}'_r$$

Performing differentiation, one finds

$$\frac{d\mathbf{K}_s^{-1}}{dt} = \omega\begin{bmatrix} -\sin\theta & \cos\theta & 0 \\ -\sin\left(\theta-\frac{2}{3}\pi\right) & \cos\left(\theta-\frac{2}{3}\pi\right) & 0 \\ -\sin\left(\theta+\frac{2}{3}\pi\right) & \cos\left(\theta+\frac{2}{3}\pi\right) & 0 \end{bmatrix}$$

and

$$\frac{d\mathbf{K}_r^{-1}}{dt} = (\omega-\omega_r)\begin{bmatrix} -\sin(\theta-\theta_r) & \cos(\theta-\theta_r) & 0 \\ -\sin\left(\theta-\theta_r-\frac{2}{3}\pi\right) & \cos\left(\theta-\theta_r-\frac{2}{3}\pi\right) & 0 \\ -\sin\left(\theta-\theta_r+\frac{2}{3}\pi\right) & \cos\left(\theta-\theta_r+\frac{2}{3}\pi\right) & 0 \end{bmatrix}$$

Therefore, using $\mathbf{K}_s$ and $\mathbf{K}_r$, the expressions for $\mathbf{K}_s\frac{d\mathbf{K}_s^{-1}}{dt}$ and $\mathbf{K}_r\frac{d\mathbf{K}_r^{-1}}{dt}$ in (6.53) are found to be

$$\mathbf{K}_s\frac{d\mathbf{K}_s^{-1}}{dt} = \omega\begin{bmatrix} 0 & 1 & 0 \\ -1 & 0 & 0 \\ 0 & 0 & 0 \end{bmatrix}$$

and

$$\mathbf{K}_r \frac{d\mathbf{K}_r^{-1}}{dt} = (\omega - \omega_r)\begin{bmatrix} 0 & 1 & 0 \\ -1 & 0 & 0 \\ 0 & 0 & 0 \end{bmatrix}$$

Finally, we obtain the voltage equations for stator and rotor circuits in the *arbitrary* reference frame. In particular, from (6.53), the following matrix differential equations result

$$\mathbf{u}_{qdos} = \mathbf{r}_s \mathbf{i}_{qdos} + \begin{bmatrix} 0 & \omega & 0 \\ -\omega & 0 & 0 \\ 0 & 0 & 0 \end{bmatrix} \boldsymbol{\psi}_{qdos} + \frac{d\boldsymbol{\psi}_{qdos}}{dt}$$

$$\mathbf{u}'_{qdor} = \mathbf{r}'_r \mathbf{i}'_{qdor} + \begin{bmatrix} 0 & \omega - \omega_r & 0 \\ -\omega + \omega_r & 0 & 0 \\ 0 & 0 & 0 \end{bmatrix} \boldsymbol{\psi}'_{qdor} + \frac{d\boldsymbol{\psi}'_{qdor}}{dt} \tag{6.54}$$

From (6.54), six differential equations in expanded form are found to map the circuitry dynamics. In particular,

$$u_{qs} = r_s i_{qs} + \omega \psi_{ds} + \frac{d\psi_{qs}}{dt}$$

$$u_{ds} = r_s i_{ds} - \omega \psi_{qs} + \frac{d\psi_{ds}}{dt}$$

$$u_{os} = r_s i_{os} + \frac{d\psi_{os}}{dt}$$

$$u'_{qr} = r'_r i'_{qr} + (\omega - \omega_r)\psi'_{dr} + \frac{d\psi'_{qr}}{dt} \tag{6.55}$$

$$u'_{dr} = r'_r i'_{dr} - (\omega - \omega_r)\psi'_{qr} + \frac{d\psi'_{dr}}{dt}$$

$$u'_{or} = r'_r i'_{or} + \frac{d\psi'_{or}}{dt}$$

One must derive the expressions for the *quadrature-*, *direct-*, and *zero*-axis components of stator and rotor flux linkages. Using the matrix equation for flux linkages in the *machine* variables (6.41)

$$\begin{bmatrix} \boldsymbol{\psi}_{abcs} \\ \boldsymbol{\psi}'_{abcr} \end{bmatrix} = \begin{bmatrix} \mathbf{L}_s & \mathbf{L}'_{sr} \\ \mathbf{L}'^{T}_{sr} & \mathbf{L}'_r \end{bmatrix} \begin{bmatrix} \mathbf{i}_{abcs} \\ \mathbf{i}'_{abcr} \end{bmatrix}$$

we have

$$\boldsymbol{\psi}_{abcs} = \mathbf{L}_s \mathbf{i}_{abcs} + \mathbf{L}'_{sr} \mathbf{i}'_{abcr}$$
$$\boldsymbol{\psi}'_{abcr} = \mathbf{L}'^{T}_{sr} \mathbf{i}_{abcs} + \mathbf{L}'_r \mathbf{i}'_{abcr}$$

These equations can be represented using the *quadrature*, *direct*, and *zero* quantities of flux linkages by employing the Park transformation matrices. In particular, from

$$\mathbf{K}_s^{-1} \boldsymbol{\psi}_{qdos} = \mathbf{L}_s \mathbf{K}_s^{-1} \mathbf{i}_{qdos} + \mathbf{L}'_{sr} \mathbf{K}_r^{-1} \mathbf{i}'_{qdor}$$
$$\mathbf{K}_r^{-1} \boldsymbol{\psi}'_{qdor} = \mathbf{L}'^{T}_{sr} \mathbf{K}_s^{-1} \mathbf{i}_{qdos} + \mathbf{L}'_r \mathbf{K}_r^{-1} \mathbf{i}'_{qdor}$$

we have

$$\begin{aligned} \boldsymbol{\psi}_{qdos} &= \mathbf{K}_s \mathbf{L}_s \mathbf{K}_s^{-1} \mathbf{i}_{qdos} + \mathbf{K}_s \mathbf{L}'_{sr} \mathbf{K}_r^{-1} \mathbf{i}'_{qdor} \\ \boldsymbol{\psi}'_{qdor} &= \mathbf{K}_r \mathbf{L}'^{T}_{sr} \mathbf{K}_s^{-1} i_{qdos} + \mathbf{K}_r \mathbf{L}'_r \mathbf{K}_r^{-1} \mathbf{i}'_{qdor} \end{aligned} \tag{6.56}$$

Taking note of the Park transformation matrices $\mathbf{K}_s$, $\mathbf{K}_s^{-1}$, $\mathbf{K}_r$, $\mathbf{K}_r^{-1}$, and applying the derived expressions for matrices of inductances

$$\mathbf{L}_s = \begin{bmatrix} L_{ls} + L_{ms} & -\frac{1}{2}L_{ms} & -\frac{1}{2}L_{ms} \\ -\frac{1}{2}L_{ms} & L_{ls} + L_{ms} & -\frac{1}{2}L_{ms} \\ -\frac{1}{2}L_{ms} & -\frac{1}{2}L_{ms} & L_{ls} + L_{ms} \end{bmatrix}$$

$$\mathbf{L}'_{sr} = L_{ms} \begin{bmatrix} \cos\theta_r & \cos\left(\theta_r + \frac{2}{3}\pi\right) & \cos\left(\theta_r - \frac{2}{3}\pi\right) \\ \cos\left(\theta_r - \frac{2}{3}\pi\right) & \cos\theta_r & \cos\left(\theta_r + \frac{2}{3}\pi\right) \\ \cos\left(\theta_r + \frac{2}{3}\pi\right) & \cos\left(\theta_r - \frac{2}{3}\pi\right) & \cos\theta_r \end{bmatrix}$$

and

$$\mathbf{L}'_r = \begin{bmatrix} L'_{lr} + L_{ms} & -\frac{1}{2}L_{ms} & -\frac{1}{2}L_{ms} \\ -\frac{1}{2}L_{ms} & L'_{lr} + L_{ms} & -\frac{1}{2}L_{ms} \\ -\frac{1}{2}L_{ms} & -\frac{1}{2}L_{ms} & L'_{lr} + L_{ms} \end{bmatrix}$$

multiplying the matrices results in

$$\mathbf{K}_s\mathbf{L}_s\mathbf{K}_s^{-1} = \begin{bmatrix} L_{ls}+M & 0 & 0 \\ 0 & L_{ls}+M & 0 \\ 0 & 0 & L_{ls} \end{bmatrix}$$

$$\mathbf{K}_s\mathbf{L}'_{sr}\mathbf{K}_r^{-1} = \mathbf{K}_r\mathbf{L}'^{T}_{sr}\mathbf{K}_s^{-1} = \begin{bmatrix} M & 0 & 0 \\ 0 & M & 0 \\ 0 & 0 & 0 \end{bmatrix}$$

and

$$\mathbf{K}_r\mathbf{L}'_r\mathbf{K}_r^{-1} = \begin{bmatrix} L'_{lr}+M & 0 & 0 \\ 0 & L'_{lr}+M & 0 \\ 0 & 0 & L'_{lr} \end{bmatrix}$$

where $M = \frac{3}{2}L_{ms}$.

In expanded form, the flux linkage equations (6.56) are written as

$$\begin{aligned}
\psi_{qs} &= L_{ls}i_{qs} + Mi_{qs} + Mi'_{qr} \\
\psi_{ds} &= L_{ls}i_{ds} + Mi_{ds} + Mi'_{dr} \\
\psi_{os} &= L_{ls}i_{os} \\
\psi'_{qr} &= L'_{lr}i'_{qr} + Mi_{qs} + Mi'_{qr} \\
\psi'_{dr} &= L'_{lr}i'_{dr} + Mi_{ds} + Mi'_{dr} \\
\psi'_{or} &= L'_{lr}i'_{or}
\end{aligned} \tag{6.57}$$

Using the expressions obtained for the *qd0* flux linkages (6.57) in (6.55), the following differential equations result to map three-phase induction motor circuitry dynamics in the *arbitrary* reference frame

$$u_{qs} = r_s i_{qs} + \omega(L_{ls}i_{ds} + Mi_{ds} + Mi'_{dr}) + \frac{d(L_{ls}i_{qs} + Mi_{qs} + Mi'_{qr})}{dt}$$

$$u_{ds} = r_s i_{ds} - \omega(L_{ls}i_{qs} + Mi_{qs} + Mi'_{qr}) + \frac{d(L_{ls}i_{ds} + Mi_{ds} + Mi'_{dr})}{dt}$$

$$u_{os} = r_s i_{os} + \frac{d(L_{ls}i_{os})}{dt}$$

$$u'_{qr} = r'_r i'_{qr} + (\omega - \omega_r)(L'_{lr}i'_{dr} + Mi_{ds} + Mi'_{dr}) + \frac{d(L'_{lr}i'_{qr} + Mi_{qs} + Mi'_{qr})}{dt}$$

$$u'_{dr} = r'_r i'_{dr} - (\omega - \omega_r)(L'_{lr}i'_{qr} + Mi_{qs} + Mi'_{qr}) + \frac{d(L'_{lr}i'_{dr} + Mi_{ds} + Mi'_{dr})}{dt}$$

$$u'_{or} = r'_r i'_{or} + \frac{d(L'_{lr}i'_{or})}{dt}$$

Cauchy's form of these differential equations needs to be found. We have

$$\frac{di_{qs}}{dt} = \frac{1}{L_{SM}L_{RM} - M^2}[-L_{RM}r_s i_{qs} - (L_{SM}L_{RM} - M^2)\omega i_{ds} + Mr_r' i_{qr}' - M(Mi_{ds} + L_{RM} i_{dr}')\omega_r + L_{RM}u_{qs} - Mu_{qr}']$$

$$\frac{di_{ds}}{dt} = \frac{1}{L_{SM}L_{RM} - M^2}[(L_{SM}L_{RM} - M^2)\omega i_{qs} - L_{RM}r_s i_{ds} + Mr_r' i_{dr}' + M(Mi_{qs} + L_{RM} i_{qr}')\omega_r + L_{RM}u_{ds} - Mu_{dr}']$$

$$\frac{di_{os}}{dt} = \frac{1}{L_{ls}}(-r_s i_{os} + u_{os})$$

$$\frac{di_{qr}'}{dt} = \frac{1}{L_{SM}L_{RM} - M^2}[Mr_s i_{qs} - L_{SM}r_r' i_{qr}' - (L_{SM}L_{RM} - M^2)\omega i_{dr}' + L_{SM}(Mi_{ds} + L_{RM} i_{dr}')\omega_r - Mu_{qs} + L_{SM}u_{qr}'] \tag{6.58}$$

$$\frac{di_{dr}'}{dt} = \frac{1}{L_{SM}L_{RM} - M^2}[Mr_s i_{ds} + (L_{SM}L_{RM} - M^2)\omega i_{qr}' - L_{SM}r_r' i_{dr}' - L_{SM}(Mi_{qs} + L_{RM} i_{qr}')\omega_r - Mu_{ds} + L_{SM}u_{dr}']$$

$$\frac{di_{or}'}{dt} = \frac{1}{L_{lr}'}(-r_r' i_{or}' + u_{or}')$$

where $L_{SM} = L_{ls} + M = L_{ls} + \frac{3}{2}L_{ms}$ and $L_{RM} = L_{lr}' + M = L_{lr}' + \frac{3}{2}L_{ms}$.

As the nonlinear differential equations are found to describe the circuitry transient behavior, to complete the model developments, the *torsional-mechanical* equations,

$$T_e - B_m\omega_{rm} - T_L = J\frac{d\omega_{rm}}{dt}$$
$$\frac{d\theta_{rm}}{dt} = \omega_{rm} \tag{6.59}$$

must be used. The equation for the electromagnetic torque developed by three-phase induction motors must be obtained using the *quadrature-* and *direct*-axis components of stator and rotor currents. Using the formula for coenergy

$$W_c = \frac{1}{2}\mathbf{i}_{abcs}^T(\mathbf{L}_s - L_{ls}\mathbf{I})\mathbf{i}_{abcs} + \mathbf{i}_{abcs}^T\mathbf{L}_{sr}'(\theta_r)\mathbf{i}_{abcr}' + \frac{1}{2}\mathbf{i}_{abcr}'^T(\mathbf{L}_r' - L_{lr}'\mathbf{I})\mathbf{i}_{abcr}'$$

one finds that

$$T_e = \frac{P}{2}\frac{\partial W_c(\mathbf{i}_{abcs}, \mathbf{i}_{abcr}', \theta_r)}{\partial\theta_r} = \frac{P}{2}\mathbf{i}_{abcs}^T\frac{\partial\mathbf{L}_{sr}'(\theta_r)}{\partial\theta_r}\mathbf{i}_{abcr}'$$

Hence,

$$T_e = \frac{P}{2}(\mathbf{K}_s^{-1}\mathbf{i}_{qdos})^T \frac{\partial \mathbf{L}'_{sr}(\theta_r)}{\partial \theta_r}\mathbf{K}_r^{-1}\mathbf{i}'_{qdor} = \frac{P}{2}\mathbf{i}_{qdos}^T \mathbf{K}_s^{-1T} \frac{\partial \mathbf{L}'_{sr}(\theta_r)}{\partial \theta_r}\mathbf{K}_r^{-1}\mathbf{i}'_{qdor}$$

Using the following expressions

$$\mathbf{i}_{qdos}^T \mathbf{K}_s^{-1^T} = \begin{bmatrix} i_{qs} & i_{ds} & i_{os} \end{bmatrix} \begin{bmatrix} \cos\theta & \cos\left(\theta - \frac{2}{3}\pi\right) & \cos\left(\theta + \frac{2}{3}\pi\right) \\ \sin\theta & \sin\left(\theta - \frac{2}{3}\pi\right) & \sin\left(\theta + \frac{2}{3}\pi\right) \\ 1 & 1 & 1 \end{bmatrix}$$

$$= \begin{bmatrix} i_{qs}\cos\theta + i_{ds}\sin\theta + i_{os} \\ i_{qs}\cos\left(\theta - \frac{2}{3}\pi\right) + i_{ds}\sin\left(\theta - \frac{2}{3}\pi\right) + i_{os} \\ i_{qs}\cos\left(\theta + \frac{2}{3}\pi\right) + i_{ds}\sin\left(\theta + \frac{2}{3}\pi\right) + i_{os} \end{bmatrix}^T$$

$$\mathbf{K}_r^{-1}\mathbf{i}'_{qdor} = \begin{bmatrix} \cos(\theta - \theta_r) & \sin(\theta - \theta_r) & 1 \\ \cos\left(\theta - \theta_r - \frac{2}{3}\pi\right) & \sin\left(\theta - \theta_r - \frac{2}{3}\pi\right) & 1 \\ \cos\left(\theta - \theta_r + \frac{2}{3}\pi\right) & \sin\left(\theta - \theta_r + \frac{2}{3}\right) & 1 \end{bmatrix} \begin{bmatrix} i'_{qr} \\ i'_{dr} \\ i'_{or} \end{bmatrix}$$

$$= \begin{bmatrix} i'_{qr}\cos(\theta - \theta_r) + i'_{dr}\sin(\theta - \theta_r) + i'_{or} \\ i'_{qr}\cos\left(\theta - \theta_r - \frac{2}{3}\pi\right) + i'_{dr}\sin\left(\theta - \theta_r - \frac{2}{3}\pi\right) + i'_{or} \\ i'_{qr}\cos\left(\theta - \theta_r + \frac{2}{3}\pi\right) + i'_{dr}\sin\left(\theta - \theta_r + \frac{2}{3}\pi\right) + i'_{or} \end{bmatrix}$$

and

$$\frac{\partial \mathbf{L}'_{sr}(\theta_r)}{\partial \theta_r} = -L_{ms} \begin{bmatrix} \sin\theta_r & \sin\left(\theta_r + \frac{2}{3}\pi\right) & \sin\left(\theta_r - \frac{2}{3}\pi\right) \\ \sin\left(\theta_r - \frac{2}{3}\pi\right) & \sin\theta_r & \sin\left(\theta_r + \frac{2}{3}\pi\right) \\ \sin\left(\theta_r + \frac{2}{3}\pi\right) & \sin\left(\theta_r - \frac{2}{3}\pi\right) & \sin\theta_r \end{bmatrix}$$

one obtains

$$T_e = -\frac{PL_{ms}}{2}\begin{bmatrix} i_{qs}\cos\theta + i_{ds}\sin\theta + i_{os} \\ i_{qs}\cos\left(\theta - \frac{2}{3}\pi\right) + i_{ds}\sin\left(\theta - \frac{2}{3}\pi\right) + i_{os} \\ i_{qs}\cos\left(\theta + \frac{2}{3}\pi\right) + i_{ds}\sin\left(\theta + \frac{2}{3}\pi\right) + i_{os} \end{bmatrix}^T$$

$$\times\begin{bmatrix} \sin\theta_r & \sin\left(\theta_r + \frac{2}{3}\pi\right) & \sin\left(\theta_r - \frac{2}{3}\pi\right) \\ \sin\left(\theta_r - \frac{2}{3}\pi\right) & \sin\theta_r & \sin\left(\theta_r + \frac{2}{3}\pi\right) \\ \sin\left(\theta_r + \frac{2}{3}\pi\right) & \sin\left(\theta_r - \frac{2}{3}\pi\right) & \sin\theta_r \end{bmatrix}$$

$$\times\begin{bmatrix} i'_{qr}\cos(\theta - \theta_r) + i'_{dr}\sin(\theta - \theta_r) + i'_{or} \\ i'_{qr}\cos\left(\theta - \theta_r - \frac{2}{3}\pi\right) + i'_{dr}\sin\left(\theta - \theta_r - \frac{2}{3}\pi\right) + i'_{or} \\ i'_{qr}\cos\left(\theta - \theta_r + \frac{2}{3}\pi\right) + i'_{dr}\sin\left(\theta - \theta_r + \frac{2}{3}\pi\right) + i'_{or} \end{bmatrix}$$

By performing multiplication of matrices, the following formula for the electromagnetic torque results

$$T_e = \frac{3P}{4}M(i_{qs}i'_{dr} - i_{ds}i'_{qr}) \tag{6.60}$$

From (6.59) and (6.60), one concludes that the *torsional-mechanical* equations are given as

$$\begin{aligned} \frac{d\omega_r}{dt} &= \frac{3P^2}{8J}M(i_{qs}i'_{dr} - i_{ds}i'_{qr}) - \frac{B_m}{J}\omega_r - \frac{P}{2J}T_L \\ \frac{d\theta_r}{dt} &= \omega_r \end{aligned} \tag{6.61}$$

Augmenting the circuitry and *torsional-mechanical* dynamics, as given by (6.58) and (6.61), the model for three-phase induction motors in the *arbitrary* reference frame results. In particular, we have a set of eight highly coupled nonlinear differential equations

$$\frac{di_{qs}}{dt} = \frac{1}{L_{SM}L_{RM} - M^2}[-L_{RM}r_s i_{qs} - (L_{SM}L_{RM} - M^2)\omega i_{ds} + Mr'_r i'_{qr} - M(Mi_{ds} + L_{RM}i'_{dr})\omega_r$$
$$+ L_{RM}u_{qs} - Mu'_{qr}],$$

$$\frac{di_{ds}}{dt} = \frac{1}{L_{SM}L_{RM} - M^2}[(L_{SM}L_{RM} - M^2)\omega i_{qs} - L_{RM}r_s i_{ds} + Mr'_r i'_{dr} + M(Mi_{qs} + L_{RM}i'_{qr})\omega_r$$
$$+ L_{RM}u_{ds} - Mu'_{dr}]$$

$$\frac{di_{os}}{dt} = \frac{1}{L_{ls}}(-r_s i_{os} + u_{os})$$

$$\frac{di'_{qr}}{dt} = \frac{1}{L_{SM}L_{RM} - M^2}[Mr_s i_{qs} - L_{SM}r'_r i'_{qr} - (L_{SM}L_{RM} - M^2)\omega i'_{dr} + L_{SM}(Mi_{ds} + L_{RM}i'_{dr})\omega_r - Mu_{qs} + L_{SM}u'_{qr}]$$

$$\frac{di'_{dr}}{dt} = \frac{1}{L_{SM}L_{RM} - M^2}[Mr_s i_{ds} + (L_{SM}L_{RM} - M^2)\omega i'_{qr} - L_{SM}r'_r i'_{dr} - L_{SM}(Mi_{qs} + L_{RM}i'_{qr})\omega_r - Mu_{ds} + L_{SM}u'_{dr}], \tag{6.62}$$

$$\frac{di'_{or}}{dt} = \frac{1}{L'_{lr}}(-r'_r i'_{or} + u'_{or})$$

$$\frac{d\omega_r}{dt} = \frac{3P^2}{8J}M(i_{qs}i'_{dr} - i_{ds}i'_{qr}) - \frac{B_m}{J}\omega_r - \frac{P}{2J}T_L$$

$$\frac{d\theta_r}{dt} = \omega_r$$

One concludes that the mathematical model was developed. It is important to emphasize that in the *arbitrary* reference frame, the angular displacement is not used in the differential equations that map the stator and rotor circuitry dynamics. Furthermore, the electromagnetic torque is not a function of θ_r. One concludes that the last differential equation $\frac{d\theta_r}{dt} = \omega_r$ in (6.62) can be omitted if induction motors are used in the electric drive application. That is, in electric drives, the following set of equations should be used

$$\frac{di_{qs}}{dt} = \frac{1}{L_{SM}L_{RM} - M^2}[-L_{RM}r_s i_{qs} - (L_{SM}L_{RM} - M^2)\omega i_{ds} + Mr'_r i'_{qr} - M(Mi_{ds} + L_{RM}i'_{dr})\omega_r + L_{RM}u_{qs} - Mu'_{qr}]$$

$$\frac{di_{ds}}{dt} = \frac{1}{L_{SM}L_{RM} - M^2}[(L_{SM}L_{RM} - M^2)\omega i_{qs} - L_{RM}r_s i_{ds} + Mr'_r i'_{dr} + M(Mi_{qs} + L_{RM}i'_{qr})\omega_r + L_{RM}u_{ds} - Mu'_{dr}]$$

$$\frac{di_{os}}{dt} = \frac{1}{L_{ls}}(-r_s i_{os} + u_{os})$$

$$\frac{di'_{qr}}{dt} = \frac{1}{L_{SM}L_{RM} - M^2}[Mr_s i_{qs} - L_{SM}r'_r i'_{qr} - (L_{SM}L_{RM} - M^2)\omega i'_{dr} + L_{SM}(Mi_{ds} + L_{RM}i'_{dr})\omega_r - Mu_{qs} + L_{SM}u'_{qr}] \tag{6.63}$$

$$\frac{di'_{dr}}{dt} = \frac{1}{L_{SM}L_{RM} - M^2}[Mr_s i_{ds} + (L_{SM}L_{RM} - M^2)\omega i'_{qr} - L_{SM}r'_r i'_{dr} - L_{SM}(Mi_{qs} + L_{RM}i'_{qr})\omega_r - Mu_{ds} + L_{SM}u'_{dr}]$$

$$\frac{di'_{or}}{dt} = \frac{1}{L'_{lr}}(-r'_r i'_{or} + u'_{or})$$

$$\frac{d\omega_r}{dt} = \frac{3P^2}{8J}M(i_{qs}i'_{dr} - i_{ds}i'_{qr}) - \frac{B_m}{J}\omega_r - \frac{P}{2J}T_L$$

In matrix form, (6.63) is written as

$$\begin{bmatrix} \frac{di_{qs}}{dt} \\ \frac{di_{ds}}{dt} \\ \frac{di_{os}}{dt} \\ \frac{di'_{qr}}{dt} \\ \frac{di'_{dr}}{dt} \\ \frac{di'_{or}}{dt} \\ \frac{d\omega_r}{dt} \end{bmatrix} = \begin{bmatrix} -\frac{L_{RM}r_s}{L_{SM}L_{RM}-M^2} & -\omega & 0 & \frac{Mr'_r}{L_{SM}L_{RM}-M^2} & 0 & 0 & 0 \\ \omega & -\frac{L_{RM}r_s}{L_{SM}L_{RM}-M^2} & 0 & 0 & \frac{Mr'_r}{L_{SM}L_{RM}-M^2} & 0 & 0 \\ 0 & 0 & -\frac{r_s}{L_{ls}} & 0 & 0 & 0 & 0 \\ \frac{Mr_s}{L_{SM}L_{RM}-M^2} & 0 & 0 & -\frac{L_{SM}r'_r}{L_{SM}L_{RM}-M^2} & -\omega & 0 & 0 \\ 0 & \frac{Mr_s}{L_{SM}L_{RM}-M^2} & 0 & \omega & -\frac{L_{SM}r'_r}{L_{SM}L_{RM}-M^2} & 0 & 0 \\ 0 & 0 & 0 & 0 & 0 & -\frac{r'_r}{L'_{lr}} & 0 \\ 0 & 0 & 0 & 0 & 0 & 0 & -\frac{B_m}{J} \end{bmatrix}$$

$$\times \begin{bmatrix} i_{qs} \\ i_{ds} \\ i_{os} \\ i'_{qr} \\ i'_{dr} \\ i'_{or} \\ \omega_r \end{bmatrix} + \begin{bmatrix} -\frac{M(Mi_{ds}+L_{RM}i'_{dr})\omega_r}{L_{SM}L_{RM}-M^2} \\ \frac{M(Mi_{qs}+L_{RM}i'_{qr})\omega_r}{L_{SM}L_{RM}-M^2} \\ 0 \\ \frac{L_{SM}(Mi_{ds}+L_{RM}i'_{dr})\omega_r}{L_{SM}L_{RM}-M^2} \\ -\frac{L_{SM}(Mi_{qs}+L_{RM}i'_{qr})\omega_r}{L_{SM}L_{RM}-M^2} \\ 0 \\ \frac{3P^2}{8J}M(i_{qs}i'_{dr}-i_{ds}i'_{qr}) \end{bmatrix}$$

$$+ \begin{bmatrix} \frac{L_{RM}}{L_{SM}L_{RM}-M^2} & 0 & 0 & -\frac{M}{L_{SM}L_{RM}-M^2} & 0 & 0 \\ 0 & \frac{L_{RM}}{L_{SM}L_{RM}-M^2} & 0 & 0 & -\frac{M}{L_{SM}L_{RM}-M^2} & 0 \\ 0 & 0 & \frac{1}{L_{ls}} & 0 & 0 & 0 \\ -\frac{M}{L_{SM}L_{RM}-M^2} & 0 & 0 & \frac{L_{SM}}{L_{SM}L_{RM}-M^2} & 0 & 0 \\ 0 & -\frac{M}{L_{SM}L_{RM}-M^2} & 0 & 0 & \frac{L_{SM}}{L_{SM}L_{RM}-M^2} & 0 \\ 0 & 0 & 0 & 0 & 0 & \frac{1}{L'_{lr}} \\ 0 & 0 & 0 & 0 & 0 & 0 \end{bmatrix} \begin{bmatrix} u_{qs} \\ u_{ds} \\ u_{os} \\ u'_{qr} \\ u'_{dr} \\ u'_{or} \end{bmatrix} - \begin{bmatrix} 0 \\ 0 \\ 0 \\ 0 \\ 0 \\ 0 \\ \frac{P}{2J} \end{bmatrix} T_L$$

The block diagram for three-phase induction motors modeled in the *arbitrary* reference frame can be developed using differential equations (6.63). In particular, using the Laplace operator, we have

$$i_{qs} = \frac{1}{(L_{SM}L_{RM}-M^2)s+L_{RM}r_s}[-(L_{SM}L_{RM}-M^2)\omega i_{ds} + Mr'_r i'_{qr} - M(Mi_{ds}+L_{RM}i'_{dr})\omega_r$$
$$+L_{RM}u_{qs} - Mu'_{qr}]$$

$$i_{ds} = \frac{1}{(L_{SM}L_{RM} - M^2)s + L_{RM}r_s}[(L_{SM}L_{RM} - M^2)\omega i_{qs} + Mr_r' i_{dr}' + M(Mi_{qs} + L_{RM}i_{qr}')\omega_r$$
$$+ L_{RM}u_{ds} - Mu_{dr}']$$

$$i_{os} = \frac{1}{L_{ls}s + r_s}u_{os}$$

$$i_{qr}' = \frac{1}{(L_{SM}L_{RM} - M^2)s + L_{SM}r_r'}[Mr_s i_{qs} - (L_{SM}L_{RM} - M^2)\omega i_{dr}' + L_{SM}(Mi_{ds} + L_{RM}i_{dr}')\omega_r$$
$$- Mu_{qs} + L_{SM}u_{qr}']$$

$$i_{dr}' = \frac{1}{(L_{SM}L_{RM} - M^2)s + L_{SM}r_r'}[Mr_s i_{ds} + (L_{SM}L_{RM} - M^2)\omega i_{qr}' - L_{SM}(Mi_{qs} + L_{RM}i_{qr}')\omega_r$$
$$- Mu_{ds} + L_{SM}u_{dr}']$$

$$i_{or}' = \frac{1}{L_{lr}'s + r_r'}u_{or}'$$

$$\omega_r = \frac{P}{2(Js + B_m)}\left[\frac{3P}{4}M(i_{qs}i_{dr}' - i_{ds}i_{qr}') - T_L\right]$$

Figure 6.80 illustrates the block diagram built for three-phase squirrel-cage induction motors modeled in the *arbitrary* reference frame.

The *qd0* voltages to be supplied to three-phase induction motors need to be found. It has been emphasized that the majority of induction motors are squirrel-cage motors. In squirrel-cage induction machines, the rotor windings are short-circuited. To guarantee balanced operating conditions, one supplies the following balanced three-phase voltage set to the *as, bs,* and *cs* windings

$$u_{as}(t) = \sqrt{2}u_M \cos(\omega_f t)$$

$$u_{bs}(t) = \sqrt{2}u_M \cos\left(\omega_f t - \frac{2}{3}\pi\right)$$

$$u_{cs}(t) = \sqrt{2}u_M \cos\left(\omega_f t + \frac{2}{3}\pi\right)$$

where the frequency of the applied voltage is $\omega_f = 2\pi f$.

The *quadrature-*, *direct-*, and *zero*-axis components of stator voltages can be obtained using the stator Park transformation matrix. In particular,

$$\mathbf{u}_{qdos} = \mathbf{K}_s \mathbf{u}_{abcs}, \quad \mathbf{K}_s = \frac{2}{3}\begin{bmatrix} \cos\theta & \cos\left(\theta - \frac{2}{3}\pi\right) & \cos\left(\theta + \frac{2}{3}\pi\right) \\ \sin\theta & \sin\left(\theta - \frac{2}{3}\pi\right) & \sin\left(\theta - \frac{2}{3}\pi\right) \\ \frac{1}{2} & \frac{1}{2} & \frac{1}{2} \end{bmatrix}$$

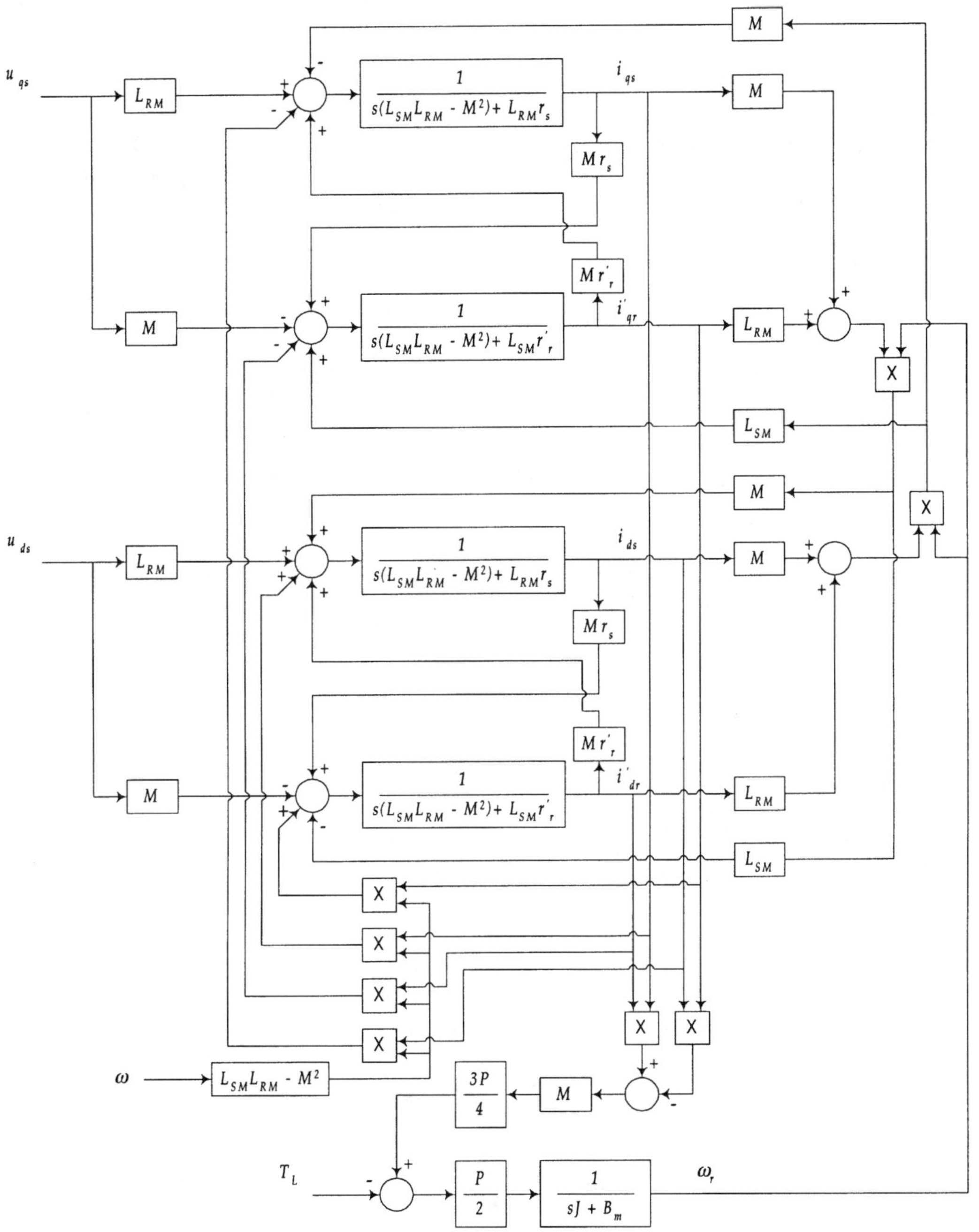

Figure 6.80 Block diagram of three-phase squirrel-cage induction motors in the *arbitrary* reference frame.

For stationary, rotor, and synchronous reference frames, assigning the frame angular velocity to be $\omega = 0$, $\omega = \omega_r$, and $\omega = \omega_e$, the corresponding angular displacements result; and for zero initial conditions, $\theta = 0$, $\theta = \theta_r$, and $\theta = \theta_e$. Hence, the *quadrature-*, *direct-*, and *zero*-axis components of voltages can be obtained to guarantee the balanced operation of induction motors.

6.5.1 Development of the mathematical model of three-phase induction motors in the stationary reference frame

The mathematical model for three-phase induction motors using the *qd0* stator and rotor currents, as well as the angular velocity and displacement, as the state variables can be found in the stationary reference frame by assigning $\omega = 0$. In particular, using the differential equations for the *arbitrary* reference frame (6.62) and plugging in $\omega = 0$, we have the following set of nonlinear differential equations to map the motor dynamics

$$
\begin{aligned}
\frac{di_{qs}^s}{dt} &= \frac{1}{L_{SM}L_{RM}-M^2}[-L_{RM}r_s i_{qs}^s + Mr_r' i_{qr}'^s - M(Mi_{ds}^s + L_{RM}i_{dr}'^s)\omega_r + L_{RM}u_{qs}^s - Mu_{qr}'^s] \\
\frac{di_{ds}^s}{dt} &= \frac{1}{L_{SM}L_{RM}-M^2}[-L_{RM}r_s i_{ds}^s + Mr_r' i_{dr}'^s + M(Mi_{qs}^s + L_{RM}i_{qr}'^s)\omega_r + L_{RM}u_{ds}^s - Mu_{dr}'^s] \\
\frac{di_{os}^s}{dt} &= \frac{1}{L_{ls}}(-r_s i_{os}^s + u_{os}^s) \\
\frac{di_{qr}'^s}{dt} &= \frac{1}{L_{SM}L_{RM}-M^2}[Mr_s i_{qs}^s - L_{SM}r_r' i_{qr}'^s + L_{SM}(Mi_{ds}^s + L_{RM}i_{dr}'^s)\omega_r - Mu_{qs}^s + L_{SM}u_{qr}'^s] \\
\frac{di_{dr}'^s}{dt} &= \frac{1}{L_{SM}L_{RM}-M^2}[Mr_s i_{ds}^s - L_{SM}r_r' i_{dr}'^s - L_{SM}(Mi_{qs}^s + L_{RM}i_{qr}'^s)\omega_r - Mu_{ds}^s + L_{SM}u_{dr}'^s] \\
\frac{di_{or}'^s}{dt} &= \frac{1}{L_{lr}'}(-r_r' i_{or}'^s + u_{or}'^s) \\
\frac{d\omega_r}{dt} &= \frac{3P^2}{8J}M(i_{qs}^s i_{dr}'^s - i_{ds}^s i_{qr}'^s) - \frac{B_m}{J}\omega_r - \frac{P}{2J}T_L \\
\frac{d\theta_r}{dt} &= \omega_r
\end{aligned}
\tag{6.64}
$$

For electric drives driven by three-phase induction motors, we have

$$
\begin{bmatrix} \frac{di_{qs}^s}{dt} \\ \frac{di_{ds}^s}{dt} \\ \frac{di_{os}^s}{dt} \\ \frac{di_{qr}'^s}{dt} \\ \frac{di_{dr}'^s}{dt} \\ \frac{di_{or}'^s}{dt} \\ \frac{d\omega_r}{dt} \end{bmatrix} =
\begin{bmatrix}
-\frac{L_{RM}r_s}{L_{SM}L_{RM}-M^2} & 0 & 0 & \frac{Mr_r'}{L_{SM}L_{RM}-M^2} & 0 & 0 & 0 \\
0 & -\frac{L_{RM}r_s}{L_{SM}L_{RM}-M^2} & 0 & 0 & \frac{Mr_r'}{L_{SM}L_{RM}-M^2} & 0 & 0 \\
0 & 0 & -\frac{r_s}{L_{ls}} & 0 & 0 & 0 & 0 \\
\frac{Mr_s}{L_{SM}L_{RM}-M^2} & 0 & 0 & -\frac{L_{SM}r_r'}{L_{SM}L_{RM}-M^2} & 0 & 0 & 0 \\
0 & \frac{Mr_s}{L_{SM}L_{RM}-M^2} & 0 & 0 & -\frac{L_{SM}r_r'}{L_{SM}L_{RM}-M^2} & 0 & 0 \\
0 & 0 & 0 & 0 & 0 & -\frac{r_r'}{L_{lr}'} & 0 \\
0 & 0 & 0 & 0 & 0 & 0 & -\frac{B_m}{J}
\end{bmatrix}
$$

$$
\times\begin{bmatrix} i_{qs}^s \\ i_{ds}^s \\ i_{os}^s \\ i_{qr}'^s \\ i_{dr}'^s \\ i_{or}'^s \\ \omega_r \end{bmatrix} + \begin{bmatrix} -\dfrac{M(Mi_{ds}^s + L_{RM}i_{dr}'^s)\omega_r}{L_{SM}L_{RM}-M^2} \\ \dfrac{M(Mi_{qs}^s + L_{RM}i_{qr}'^s)\omega_r}{L_{SM}L_{RM}-M^2} \\ 0 \\ \dfrac{L_{SM}(Mi_{ds}^s + L_{RM}i_{dr}'^s)\omega_r}{L_{SM}L_{RM}-M^2} \\ -\dfrac{L_{SM}(Mi_{qs}^s + L_{RM}i_{qr}'^s)\omega_r}{L_{SM}L_{RM}-M^2} \\ 0 \\ \dfrac{3P^2}{8J}M(i_{qs}^s i_{dr}'^s - i_{ds}^s i_{qr}'^s) \end{bmatrix}
$$

$$
+\begin{bmatrix} \dfrac{L_{RM}}{L_{SM}L_{RM}-M^2} & 0 & 0 & -\dfrac{M}{L_{SM}L_{RM}-M^2} & 0 & 0 \\ 0 & \dfrac{L_{RM}}{L_{SM}L_{RM}-M^2} & 0 & 0 & -\dfrac{M}{L_{SM}L_{RM}-M^2} & 0 \\ 0 & 0 & \dfrac{1}{L_{ls}} & 0 & 0 & 0 \\ -\dfrac{M}{L_{SM}L_{RM}-M^2} & 0 & 0 & \dfrac{L_{SM}}{L_{SM}L_{RM}-M^2} & 0 & 0 \\ 0 & -\dfrac{M}{L_{SM}L_{RM}-M^2} & 0 & 0 & \dfrac{L_{SM}}{L_{SM}L_{RM}-M^2} & 0 \\ 0 & 0 & 0 & 0 & 0 & \dfrac{1}{L_{lr}'} \\ 0 & 0 & 0 & 0 & 0 & 0 \end{bmatrix} \begin{bmatrix} u_{qs}^s \\ u_{ds}^s \\ u_{os}^s \\ u_{qr}'^s \\ u_{dr}'^s \\ u_{or}'^s \end{bmatrix} - \begin{bmatrix} 0 \\ 0 \\ 0 \\ 0 \\ 0 \\ \dfrac{P}{2J} \end{bmatrix} T_L
$$

The *quadrature-*, *direct-*, and *zero*-axis components of voltages u_{qs}^s, u_{ds}^s, and u_{os}^s are obtained using the stator Park transformation matrices. In particular, plugging $\theta = 0$ into the transformation matrix

$$
\mathbf{K}_s = \frac{2}{3}\begin{bmatrix} \cos\theta & \cos\left(\theta - \frac{2}{3}\pi\right) & \cos\left(\theta + \frac{2}{3}\pi\right) \\ \sin\theta & \sin\left(\theta - \frac{2}{3}\pi\right) & \sin\left(\theta + \frac{2}{3}\pi\right) \\ \frac{1}{2} & \frac{1}{2} & \frac{1}{2} \end{bmatrix}
$$

results in

$$
\mathbf{u}_{qdos}^s = \mathbf{K}_s^s \mathbf{u}_{abcs}, \quad \mathbf{K}_s^s = \begin{bmatrix} \frac{2}{3} & -\frac{1}{3} & -\frac{1}{3} \\ 0 & -\frac{1}{\sqrt{3}} & \frac{1}{\sqrt{3}} \\ \frac{1}{3} & \frac{1}{3} & \frac{1}{3} \end{bmatrix}
$$

Hence,

$$\begin{bmatrix} u_{qs}^s \\ u_{ds}^s \\ u_{os}^s \end{bmatrix} = \begin{bmatrix} \frac{2}{3} & -\frac{1}{3} & -\frac{1}{3} \\ 0 & -\frac{1}{\sqrt{3}} & \frac{1}{\sqrt{3}} \\ \frac{1}{3} & \frac{1}{3} & \frac{1}{3} \end{bmatrix} \begin{bmatrix} u_{as} \\ u_{bs} \\ u_{cs} \end{bmatrix}$$

Therefore, in expanded form, one obtains

$$u_{qs}^s(t) = \frac{2}{3}u_{as}(t) - \frac{1}{3}u_{bs}(t) - \frac{1}{3}u_{cs}(t)$$

$$u_{ds}^s(t) = -\frac{1}{\sqrt{3}}u_{bs}(t) + \frac{1}{\sqrt{3}}u_{cs}(t)$$

$$u_{os}^s(t) = \frac{1}{3}u_{as}(t) + \frac{1}{3}u_{bs}(t) + \frac{1}{3}u_{cs}(t)$$

Using a balanced three-phase voltage set

$$u_{as}(t) = \sqrt{2}u_M \cos(\omega_f t)$$

$$u_{bs}(t) = \sqrt{2}u_M \cos\left(\omega_f t - \frac{2}{3}\pi\right)$$

$$u_{cs}(t) = \sqrt{2}u_M \cos\left(\omega_f t + \frac{2}{3}\pi\right)$$

one obtains the following expressions for the *qd0* voltages:

$$\begin{aligned} u_{qs}^s(t) &= \sqrt{2}u_M \cos(\omega_f t) \\ u_{ds}^s(t) &= -\sqrt{2}u_M \sin(\omega_f t) \\ u_{os}^s(t) &= 0 \end{aligned} \tag{6.65}$$

It was shown that the dynamics of three-phase induction motors in the stationary reference frame are mapped by a set of differential equations (6.64). Making use of the Laplace operator, the corresponding mathematical model in the *s*-domain can be found. The block diagram for three-phase squirrel-cage induction motors modeled in the stationary reference frame is developed and documented in Figure 6.81.

Example 6.10

In the stationary reference frame, perform nonlinear simulations of a three-phase, two-pole, 220 V (*rms*), 60 Hz squirrel-cage induction motor and plot the torque-speed characteristic curves for different load conditions. The parameters of the motor are given in Table 6.4.

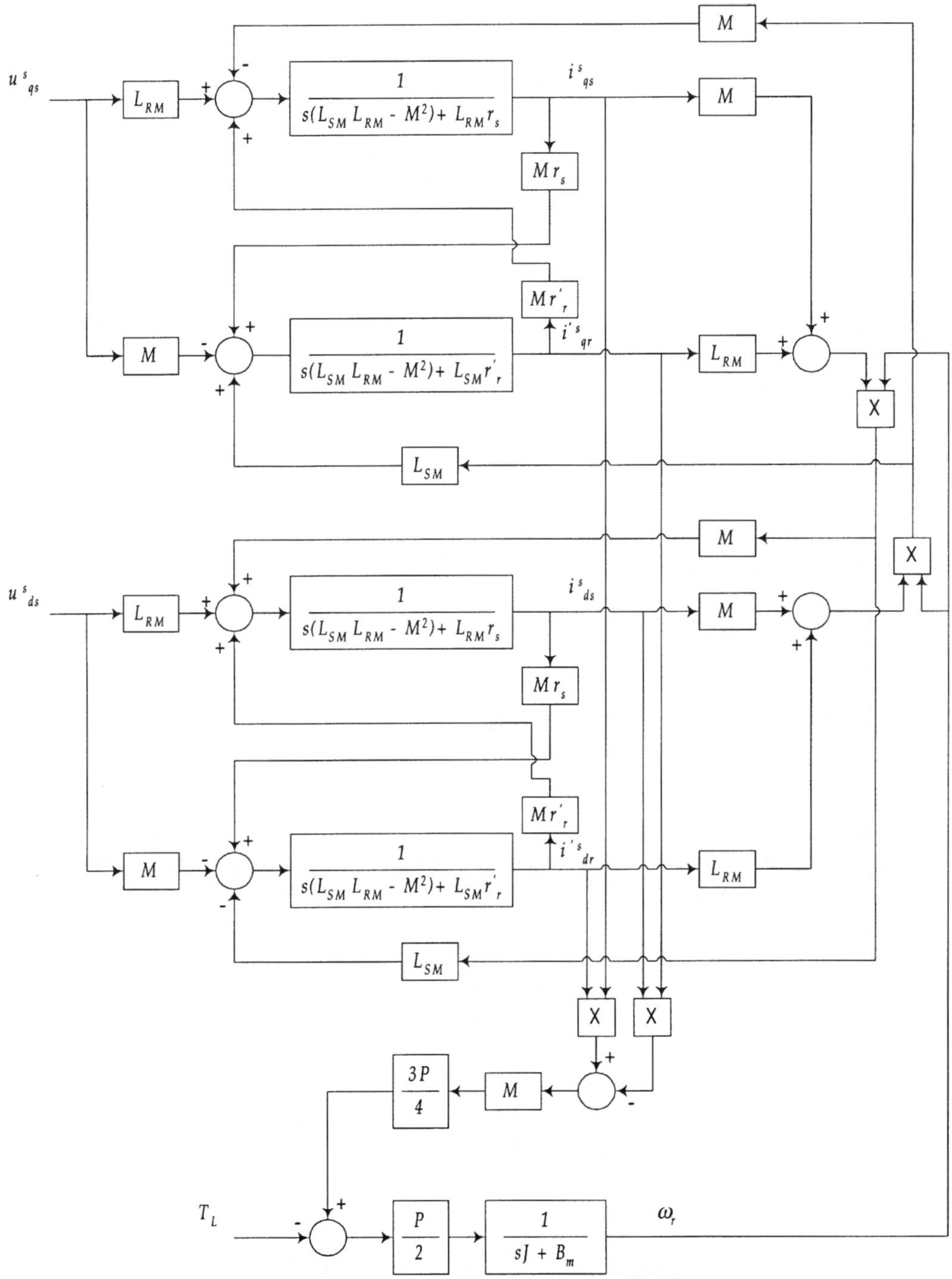

Figure 6.81 Block diagram of three-phase induction motors in the stationary reference frame.

Solution

Using the mathematical model developed to map the three-phase induction motor dynamics in the stationary reference frame, as given by the set of differential equations (6.64), a block diagram is developed (see Figure 6.81). The SIMULINK diagram is built and illustrated

Table 6.4 Parameters of a Three-Phase Squirrel-Cage Induction Motor

Parameters	Value
Rotor resistance, r_r [ohm]	7.1
Stator resistant, r_s [ohm]	9.4
Stator magnetizing inductance, L_{ms} [H]	0.37
Stator leakage inductance, L_{ls} [H]	0.032
Rotor leakage inductance, L_{lr} [H]	0.032
Moment of inertia, J [kg-m^2]	0.001
Number of poles, P	2

in Figure 6.82. Using the rated value for the phase voltages, the following *qd0* voltages are found

$$u_{qs}^s(t)=\sqrt{2}220\cos(377t),\ \ u_{ds}^s(t)=-\sqrt{2}220\sin(377t),\ \text{and}\ \ u_{os}^s(t)=0$$

The motor transients, plotted in Figures 6.83 and 6.84, illustrate the states' evolution (waveforms for the stator and rotor currents and angular velocity) if the motor accelerates from stall. The load torques applied at $t = 0$ s are 0 N-m and 2.5 N-m. For $T_L = 0$ N-m and $T_L = 2.5$ N-m, the motor reaches the steady-state angular velocity within 0.12 and 0.23 s, respectively.

6.5.2 *Development of the mathematical model of three-phase induction motors in the rotor reference frame*

The mathematical model of three-phase induction motors in the rotor reference frame can be easily found by substituting the angular velocity of the frame, which is equal to the electrical angular velocity of motor, into differential equations (6.62). That is, taking note that $\omega = \omega_r$, from (6.62), we have

$$\frac{di_{qs}^r}{dt}=\frac{1}{L_{SM}L_{RM}-M^2}[-L_{RM}r_s i_{qs}^s+Mr_r' i_{qr}'^r-L_{RM}(L_{SM}i_{ds}^r+Mi_{dr}'^r)\omega_r+L_{RM}u_{qs}^r-Mu_{qr}'^r]$$

$$\frac{di_{ds}^r}{dt}=\frac{1}{L_{SM}L_{RM}-M^2}[-L_{RM}r_s i_{ds}^r+Mr_r' i_{dr}'^r+L_{RM}(L_{SM}i_{qs}^r+Mi_{qr}'^r)\omega_r+L_{RM}u_{ds}^r-Mu_{dr}'^r]$$

$$\frac{di_{os}^r}{dt}=\frac{1}{L_{ls}}(-r_s i_{os}^r+u_{os}^r)$$

$$\frac{di_{qr}'^r}{dt}=\frac{1}{L_{SM}L_{RM}-M^2}[Mr_s i_{qs}^r-L_{SM}r_r' i_{qr}'^r+M(L_{SM}i_{ds}^r+Mi_{dr}'^r)\omega_r-Mu_{qs}^r+L_{SM}u_{qr}'^r]$$

$$\frac{di_{dr}'^r}{dt}=\frac{1}{L_{SM}L_{RM}-M^2}[Mr_s i_{ds}^r-L_{SM}r_r' i_{dr}'^r-M(L_{SM}i_{qs}^r+Mi_{qr}'^r)\omega_r-Mu_{ds}^r+L_{SM}u_{dr}'^r]$$

$$\frac{di_{or}'^r}{dt}=\frac{1}{L_{lr}'}(-r_r' i_{or}'^r+u_{or}'^r)$$

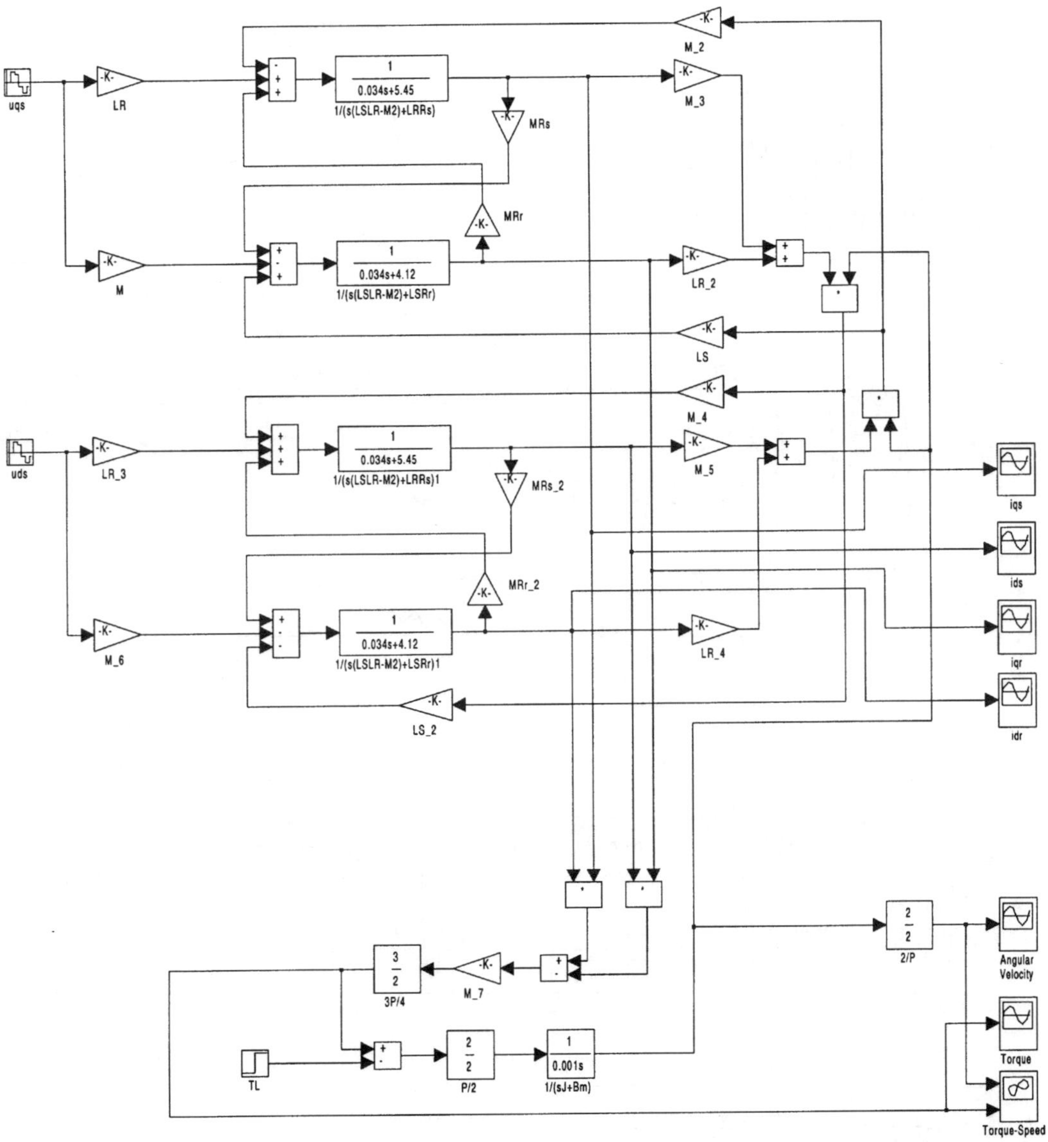

Figure 6.82 SIMULINK diagram to model a three-phase squirrel-cage motor in the stationary reference frame.

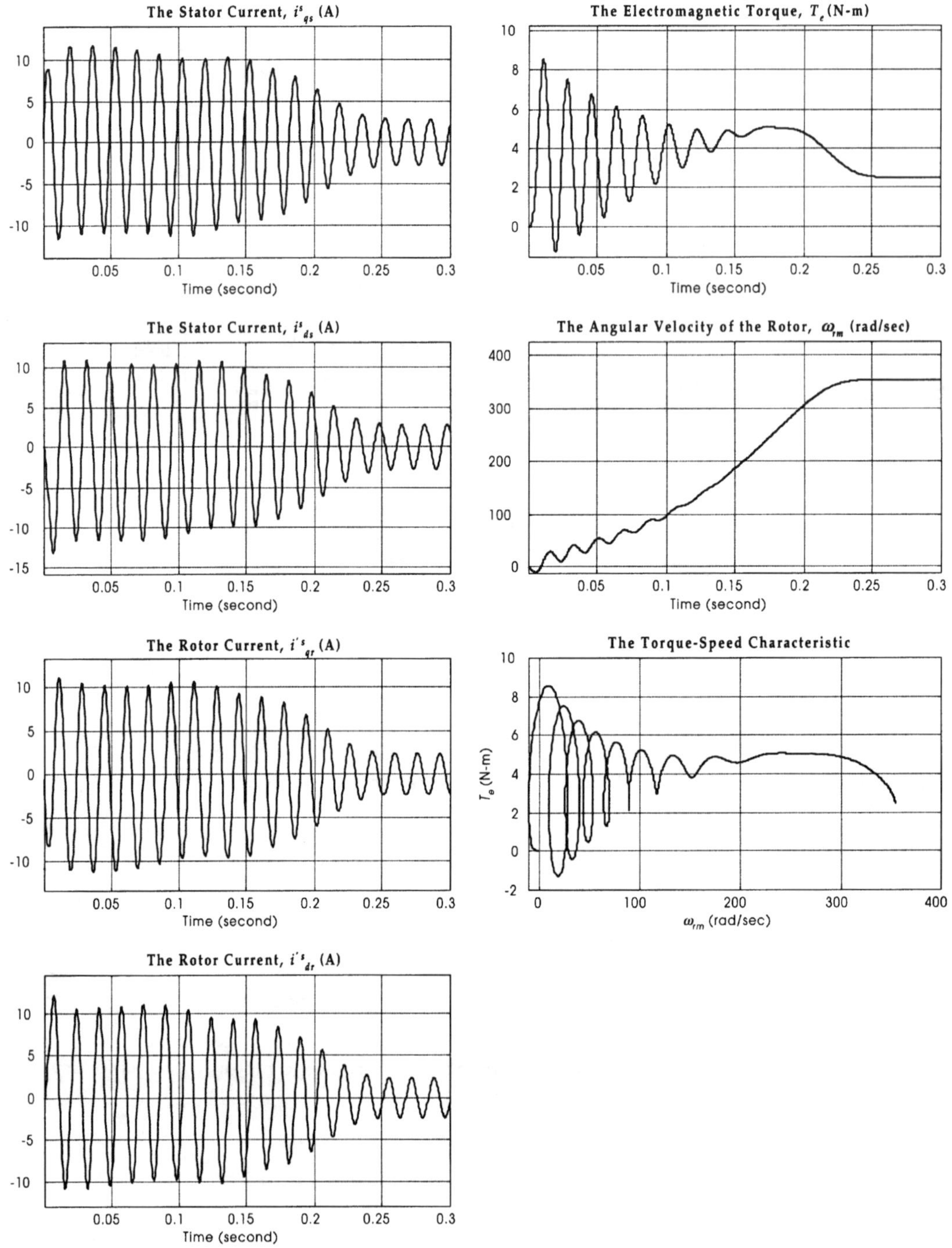

Figure 6.83 Dynamics and torque-speed characteristic of three-phase induction motor modeled in the stionary reference frame; T_L = 0 N-m, $u^s_{qs}(t) = \sqrt{2}220\cos(377t)$, $u^s_{ds}(t) = -\sqrt{2}220\sin(377t)$, and $u^s_{os}(t) = 0$.

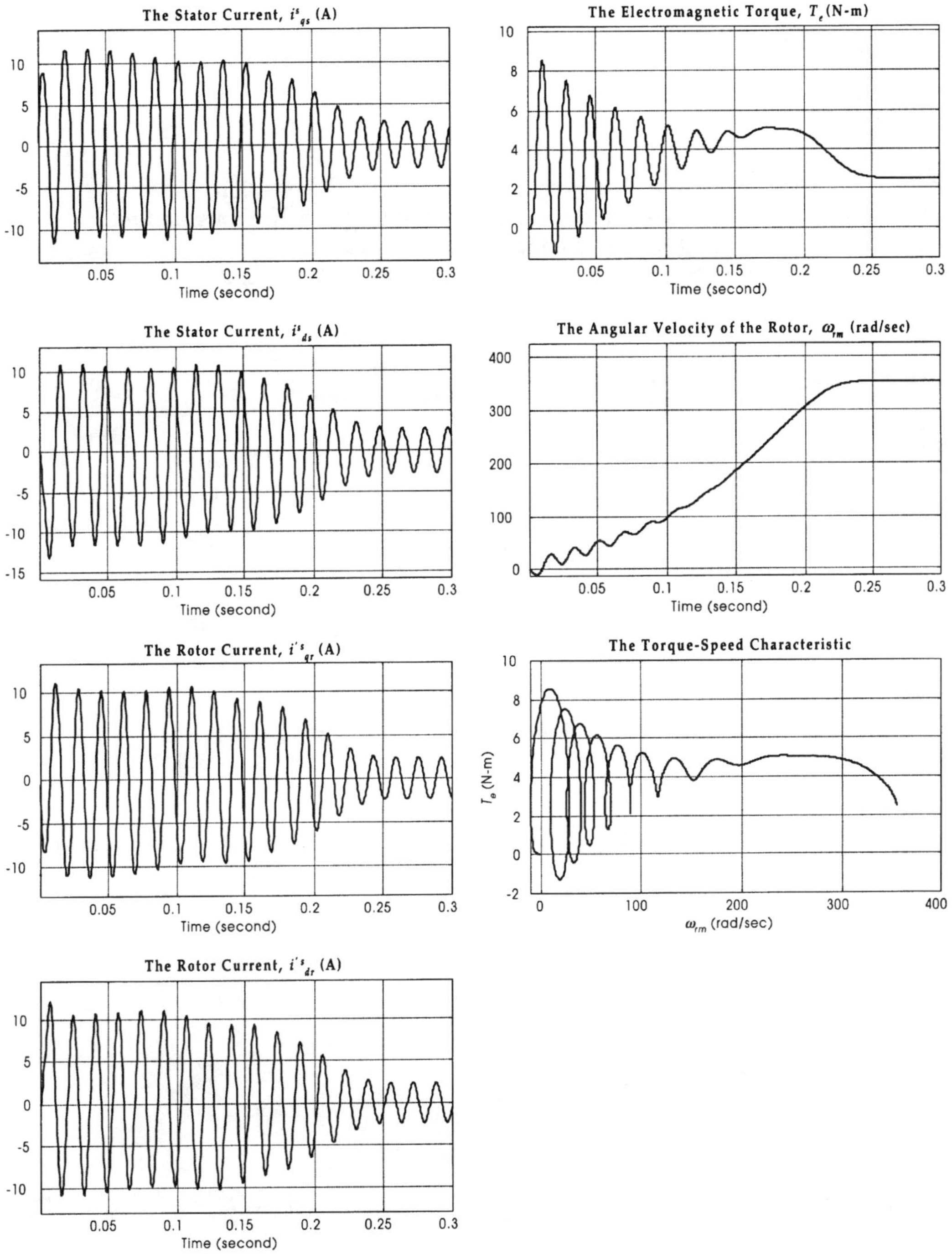

Figure 6.84 Dynamics and torque-speed characteristic of three-phase induction motor modeled in the stionary reference frame; T_L = 2.5 N-m, $u^s_{qs}(t) = \sqrt{2}220\cos(377t)$, $u^s_{ds}(t) = -\sqrt{2}220\sin(377t)$, and $u^s_{os}(t) = 0$.

$$\frac{d\omega_r}{dt} = \frac{3P^2}{8J} M(i_{qs}^r i_{dr}'^r - i_{ds}^r i_{qr}'^r) - \frac{B_m}{J}\omega_r - \frac{P}{2J}T_L$$
$$\frac{d\theta_r}{dt} = \omega_r \tag{6.66}$$

For electric drives, the displacement is usually not considered a state variable if induction motors are modeled in the *arbitrary*, stationary, synchronous, and rotor reference frames. Hence, the last differential equation in (6.66) can be omitted. In matrix form, for electric drives with induction motors, one obtains the following differential equation to map the transient dynamics.

$$\begin{bmatrix} \frac{di_{qs}^r}{dt} \\ \frac{di_{ds}^r}{dt} \\ \frac{di_{os}^r}{dt} \\ \frac{di_{qr}'^r}{dt} \\ \frac{di_{dr}'^r}{dt} \\ \frac{di_{or}'^r}{dt} \\ \frac{d\omega_r}{dt} \end{bmatrix} = \begin{bmatrix} -\frac{L_{RM}r_s}{L_{SM}L_{RM}-M^2} & 0 & 0 & \frac{Mr_r'}{L_{SM}L_{RM}-M^2} & 0 & 0 & 0 \\ 0 & -\frac{L_{RM}r_s}{L_{SM}L_{RM}-M^2} & 0 & 0 & \frac{Mr_r'}{L_{SM}L_{RM}-M^2} & 0 & 0 \\ 0 & 0 & -\frac{r_s}{L_{ls}} & 0 & 0 & 0 & 0 \\ \frac{Mr_s}{L_{SM}L_{RM}-M^2} & 0 & 0 & -\frac{L_{SM}r_r'}{L_{SM}L_{RM}-M^2} & 0 & 0 & 0 \\ 0 & \frac{Mr_s}{L_{SM}L_{RM}-M^2} & 0 & 0 & -\frac{L_{SM}r_r'}{L_{SM}L_{RM}-M^2} & 0 & 0 \\ 0 & 0 & 0 & 0 & 0 & -\frac{r_r'}{L_{lr}'} & 0 \\ 0 & 0 & 0 & 0 & 0 & 0 & -\frac{B_m}{J} \end{bmatrix}$$

$$\times \begin{bmatrix} i_{qs}^r \\ i_{ds}^r \\ i_{os}^r \\ i_{qr}'^r \\ i_{dr}'^r \\ i_{or}'^r \\ \omega_r \end{bmatrix} + \begin{bmatrix} -\frac{L_{RM}(L_{SM}i_{ds}^r + Mi_{dr}'^r)}{L_{SM}L_{RM}-M^2}\omega_r \\ \frac{L_{RM}(L_{SM}i_{qs}^r + Mi_{qr}'^r)}{L_{SM}L_{RM}-M^2}\omega_r \\ 0 \\ \frac{M(L_{SM}i_{ds}^r + Mi_{dr}'^r)}{L_{SM}L_{RM}-M^2}\omega_r \\ -\frac{M(L_{SM}i_{qs}^r + Mi_{qr}'^r)}{L_{SM}L_{RM}-M^2} \\ 0 \\ \frac{3P^2}{8J}M(i_{qs}^r i_{dr}'^r - i_{ds}^r i_{qr}'^r) \end{bmatrix}$$

$$+ \begin{bmatrix} \frac{L_{RM}}{L_{SM}L_{RM}-M^2} & 0 & 0 & -\frac{M}{L_{SM}L_{RM}-M^2} & 0 & 0 \\ 0 & \frac{L_{RM}}{L_{SM}L_{RM}-M^2} & 0 & 0 & -\frac{M}{L_{SM}L_{RM}-M^2} & 0 \\ 0 & 0 & \frac{1}{L_{ls}} & 0 & 0 & 0 \\ -\frac{M}{L_{SM}L_{RM}-M^2} & 0 & 0 & \frac{L_{SM}}{L_{SM}L_{RM}-M^2} & 0 & 0 \\ 0 & -\frac{M}{L_{SM}L_{RM}-M^2} & 0 & 0 & \frac{L_{SM}}{L_{SM}L_{RM}-M^2} & 0 \\ 0 & 0 & 0 & 0 & 0 & \frac{1}{L_{lr}'} \\ 0 & 0 & 0 & 0 & 0 & 0 \end{bmatrix} \begin{bmatrix} u_{qs}^r \\ u_{ds}^r \\ u_{os}^r \\ u_{qr}'^r \\ u_{dr}'^r \\ u_{or}'^r \end{bmatrix} - \begin{bmatrix} 0 \\ 0 \\ 0 \\ 0 \\ 0 \\ 0 \\ \frac{P}{2J} \end{bmatrix} T_L$$

To control induction motors, one must find explicit expressions for u_{qs}^r, u_{ds}^r, and u_{os}^r in order to guarantee balanced operation. The stator Park transformation matrix $\mathbf{K}_s^r$ that needs to be used to obtain the *quadrature-*, *direct-*, and *zero*-axis components of voltages

$$\mathbf{u}_{qdos}^r = \mathbf{K}_s^r \mathbf{u}_{abcs}$$

is found by plugging in $\theta = \theta_r$. That is, employing

$$\mathbf{K}_s = \frac{2}{3}\begin{bmatrix} \cos\theta & \cos\left(\theta - \frac{2}{3}\pi\right) & \cos\left(\theta + \frac{2}{3}\pi\right) \\ \sin\theta & \sin\left(\theta - \frac{2}{3}\pi\right) & \sin\left(\theta + \frac{2}{3}\pi\right) \\ \frac{1}{2} & \frac{1}{2} & \frac{1}{2} \end{bmatrix}$$

results in

$$\mathbf{K}_s^r = \frac{2}{3}\begin{bmatrix} \cos\theta_r & \cos\left(\theta_r - \frac{2}{3}\pi\right) & \cos\left(\theta_r + \frac{2}{3}\pi\right) \\ \sin\theta_r & \sin\left(\theta_r - \frac{2}{3}\pi\right) & \sin\left(\theta_r + \frac{2}{3}\pi\right) \\ \frac{1}{2} & \frac{1}{2} & \frac{1}{2} \end{bmatrix}$$

Hence,

$$\begin{bmatrix} u_{qs}^r \\ u_{ds}^r \\ u_{os}^r \end{bmatrix} = \frac{2}{3}\begin{bmatrix} \cos\theta_r & \cos\left(\theta_r - \frac{2}{3}\pi\right) & \cos\left(\theta_r + \frac{2}{3}\pi\right) \\ \sin\theta_r & \sin\left(\theta_r - \frac{2}{3}\pi\right) & \sin\left(\theta_r + \frac{2}{3}\pi\right) \\ \frac{1}{2} & \frac{1}{2} & \frac{1}{2} \end{bmatrix}\begin{bmatrix} u_{as} \\ u_{bs} \\ u_{cs} \end{bmatrix}$$

and in expanded form, the following *quadrature*, *direct*, and *zero* voltages result

$$u_{qs}^r(t) = \frac{2}{3}\left[u_{as}\cos\theta_r + u_{bs}\cos\left(\theta_r - \frac{2}{3}\pi\right) + u_{cs}\cos\left(\theta_r + \frac{2}{3}\pi\right)\right]$$

$$u_{ds}^r(t) = \frac{2}{3}\left[u_{as}\sin\theta_r + u_{bs}\sin\left(\theta_r - \frac{2}{3}\pi\right) + u_{cs}\sin\left(\theta_r + \frac{2}{3}\pi\right)\right]$$

$$u_{os}^r(t) = \frac{1}{3}(u_{as} + u_{bs} + u_{cs})$$

Taking note of a balanced three-phase voltage set

$$u_{as}(t) = \sqrt{2}u_M \cos(\omega_f t)$$

$$u_{bs}(t) = \sqrt{2}u_M \cos\left(\omega_f t - \frac{2}{3}\pi\right)$$

$$u_{cs}(t) = \sqrt{2}u_M \cos\left(\omega_f t + \frac{2}{3}\pi\right)$$

we have

$$u^r_{qs}(t) = \sqrt{2}u_M \cos(\omega_f t - \theta_r)$$

$$u^r_{ds}(t) = -\sqrt{2}u_M \sin(\omega_f t - \theta_r) \tag{6.67}$$

$$u^r_{os}(t) = 0$$

From (6.67), we conclude that to guarantee the balanced operation one must use the electrical angular displacement θ_r to find the *quadrature* and *direct* voltages.

Using a set of differential equations (6.66), one can develop a block diagram in the s-domain. This diagram is needed to visualize the results. Figure 6.85 illustrates the resulting block diagram for three-phase squirrel-cage induction motors.

Example 6.11

In SIMULINK, model a three-phase squirrel-cage induction motor, parameters of which are documented in Table 6.4. The rotor reference frame should be used.

Solution

Using differential equations (6.66) as well as the block diagram demonstrated in Figure 6.85, the SIMULINK diagram is developed and shown in Figure 6.86. Applying the rated voltage, and taking note of the necessity to implement a balanced voltage set, from (6.67) we obtain

$$u^r_{qs}(t) = \sqrt{2}220\cos(377t - \theta_r)$$

$$u^r_{ds}(t) = -\sqrt{2}220\sin(377t - \theta_r)$$

$$u^r_{os}(t) = 0$$

The transient behavior of state variables and the torque-speed characteristic curves are plotted in Figures 6.87 and 6.88 for unloaded and loaded induction motors, respectively.

6.5.3 *Development of the mathematical model of three-phase induction motors in the synchronous reference frame*

The mathematical model of three-phase induction motors in the synchronous reference frame is found by substituting the frame angular velocity in the differential equations obtained for the *arbitrary* reference frame (6.62). That is, using $\omega = \omega_e$ in (6.62), we have

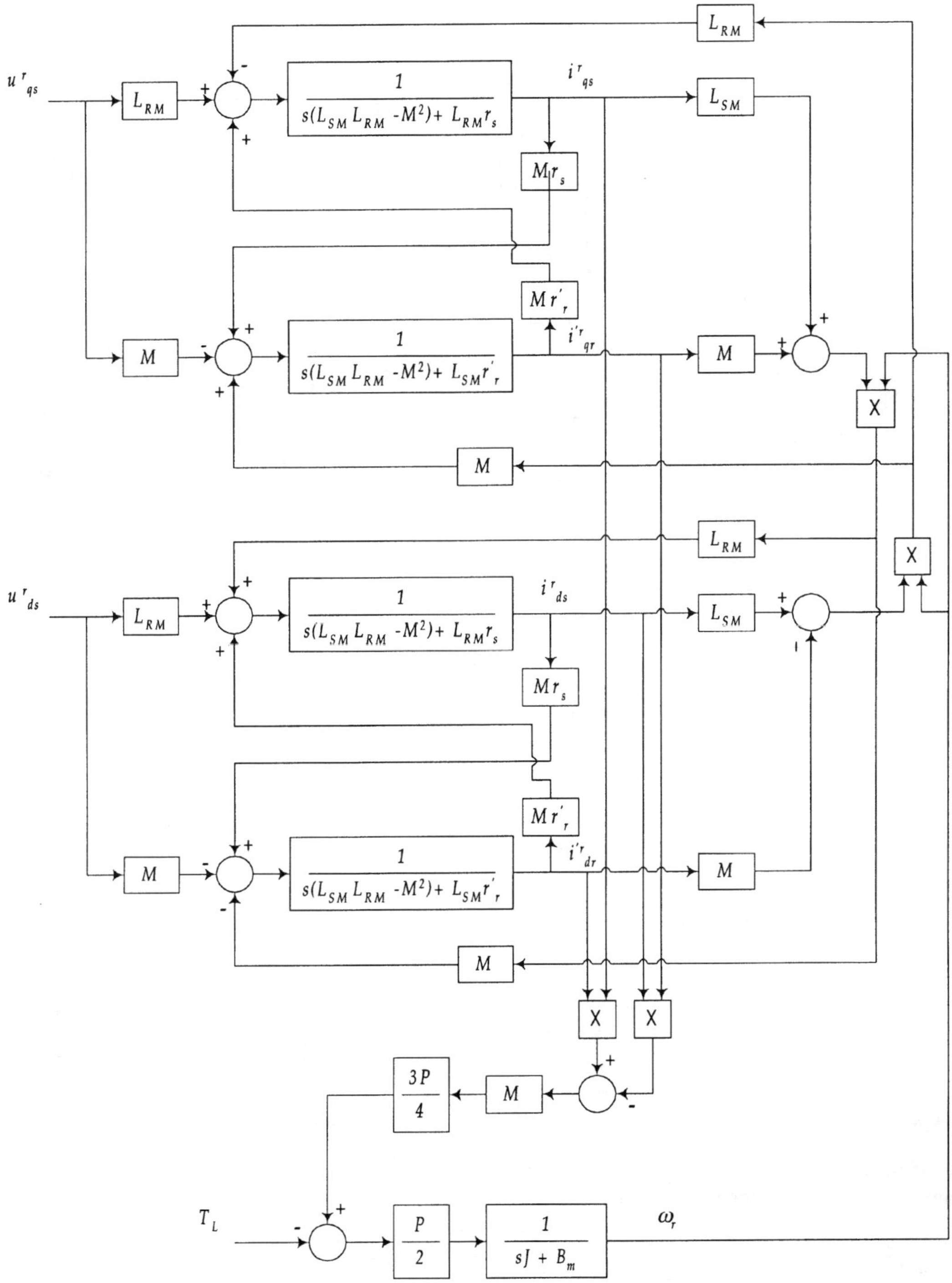

Figure 6.85 Block diagram of three-phase induction motors in the rotor reference frame.

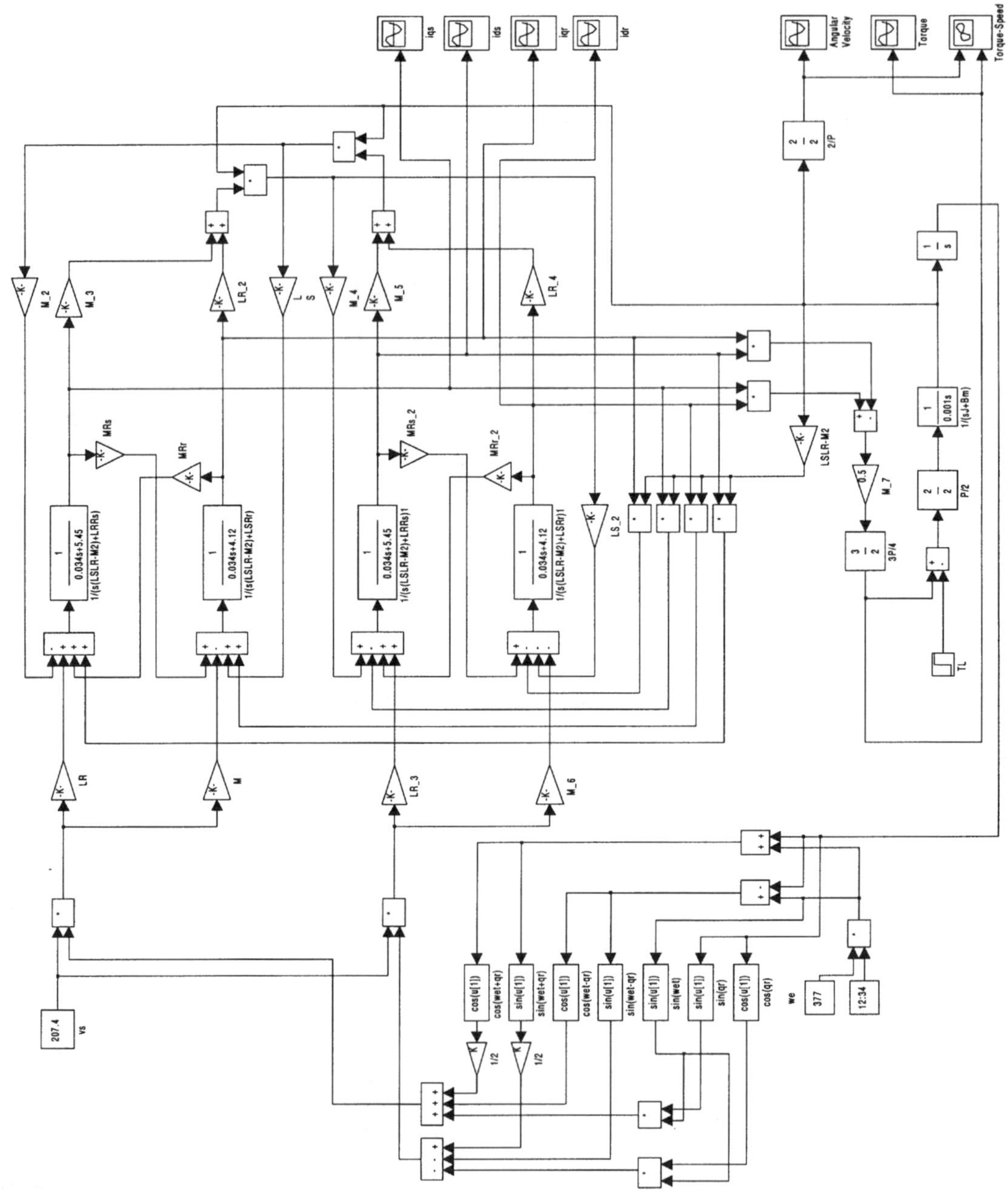

Figure 6.86 SIMULINK diagram to model the dynamics of a three-phase squirrel-cage induction motor in the rotor reference frame.

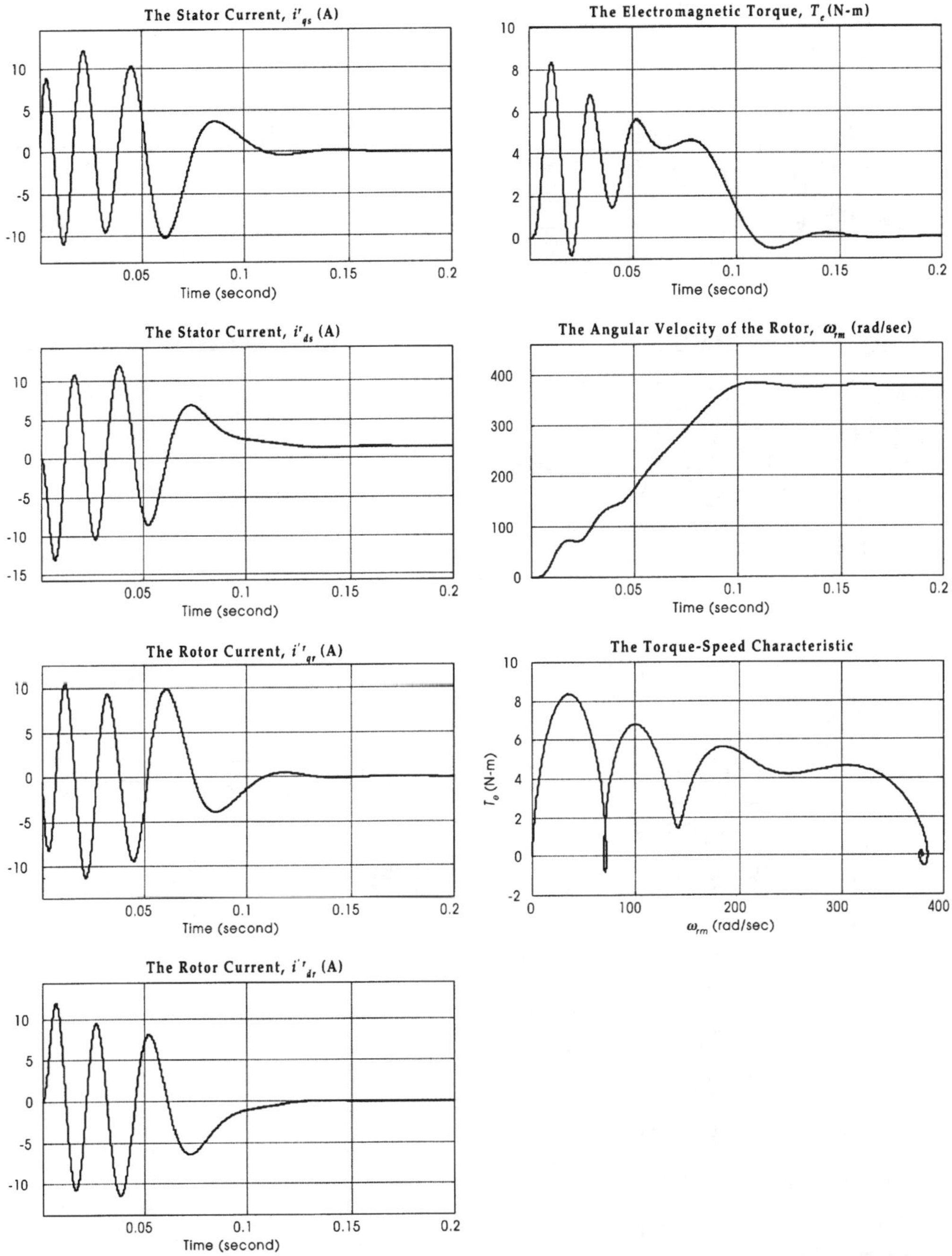

Figure 6.87 Dynamics and torque-speed characteristic of three-phase induction motor modeled in the rotor reference frame if $T_L = 0$ N-m.

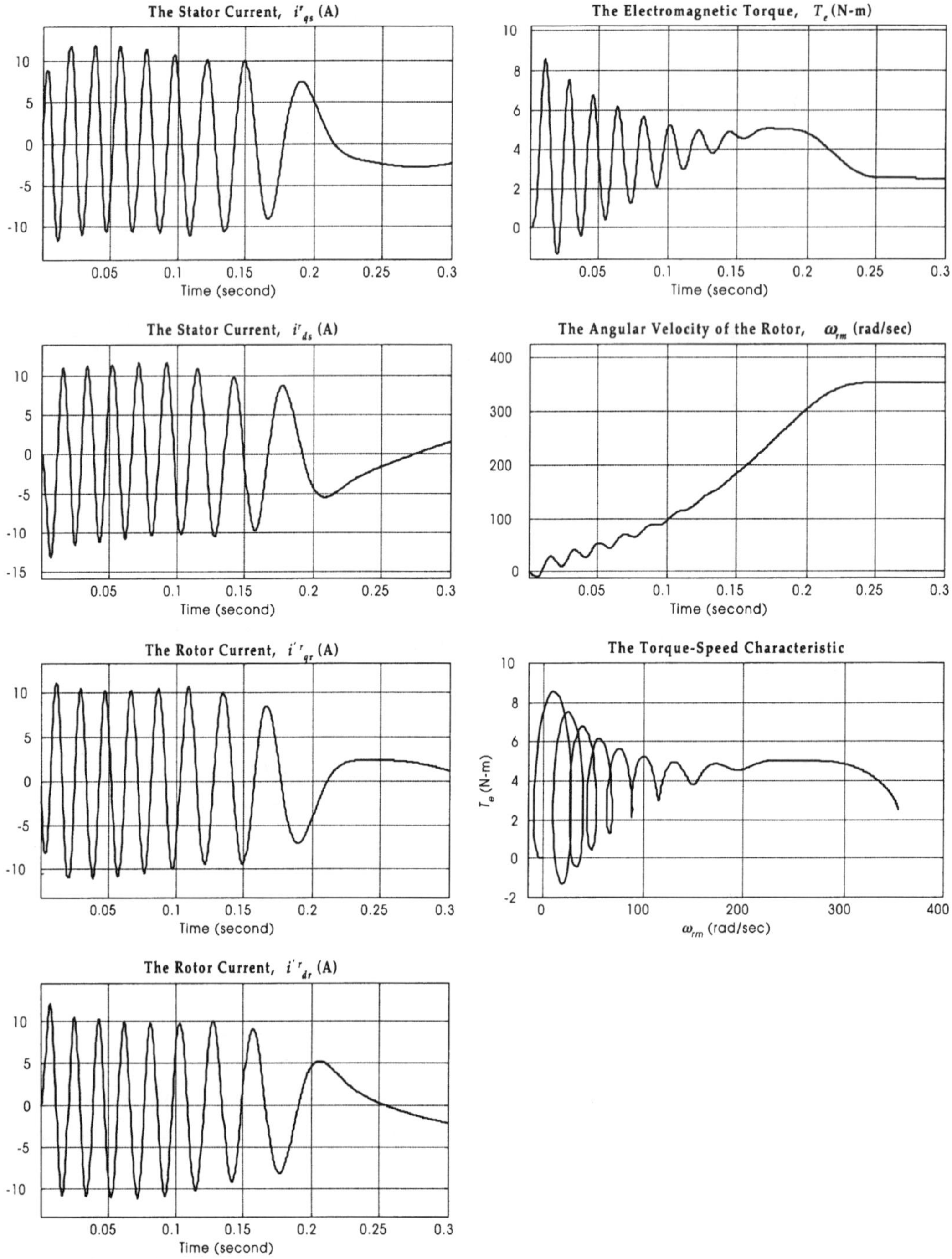

Figure 6.88 Dynamics and torque-speed characteristic of three-phase induction motor modeled in the rotor reference frame if T_L = 2.5 N-m.

$$
\begin{aligned}
\frac{di_{qs}^{e}}{dt} &= \frac{1}{L_{SM}L_{RM}-M^{2}}\left[-L_{RM}r_{s}i_{qs}^{e}-(L_{SM}L_{RM}-M^{2})\omega_{e}i_{ds}^{e}+Mr_{r}'i_{qr}'^{e}-M(Mi_{ds}^{e}+L_{RM}i_{dr}'^{e})\omega_{r}\right.\\
&\quad\left.+L_{RM}u_{qs}^{e}-Mu_{qr}'^{e}\right]\\
\frac{di_{ds}^{e}}{dt} &= \frac{1}{L_{SM}L_{RM}-M^{2}}\left[(L_{SM}L_{RM}-M^{2})\omega_{e}i_{qs}^{e}-L_{RM}r_{s}i_{ds}^{e}+Mr_{r}'i_{dr}'^{e}+M(Mi_{qs}^{e}+L_{RM}i_{qr}'^{e})\omega_{r}\right.\\
&\quad\left.+L_{RM}u_{ds}^{e}-Mu_{dr}'^{e}\right]\\
\frac{di_{os}^{e}}{dt} &= \frac{1}{L_{ls}}(-r_{s}i_{os}^{e}+u_{os}^{e})\\
\frac{di_{qr}'^{e}}{dt} &= \frac{1}{L_{SM}L_{RM}-M^{2}}\left[Mr_{s}i_{qs}^{e}-L_{SM}r_{r}'i_{qr}'^{e}-(L_{SM}L_{RM}-M^{2})\omega_{e}i_{dr}'^{e}+L_{SM}(Mi_{ds}^{e}+L_{RM}i_{dr}'^{e})\omega_{r}\right.\\
&\quad\left.-Mu_{qs}^{e}+L_{SM}u_{qr}'^{e}\right]\\
\frac{di_{dr}'^{e}}{dt} &= \frac{1}{L_{SM}L_{RM}-M^{2}}\left[Mr_{s}i_{ds}^{e}+(L_{SM}L_{RM}-M^{2})\omega_{e}i_{qr}'^{e}-L_{SM}r_{r}'i_{dr}'^{e}-L_{SM}(Mi_{qs}^{e}+L_{RM}i_{qr}'^{e})\omega_{r}\right.\\
&\quad\left.-Mu_{ds}^{e}+L_{SM}u_{dr}'^{e}\right]\\
\frac{di_{or}'^{e}}{dt} &= \frac{1}{L_{lr}'}(-r_{r}'i_{or}'^{e}+u_{or}'^{e})\\
\frac{d\omega_{r}}{dt} &= \frac{3P^{2}}{8J}M(i_{qs}^{e}i_{dr}'^{e}-i_{ds}^{e}i_{qr}'^{e})-\frac{B_{m}}{J}\omega_{r}-\frac{P}{2J}T_{L}\\
\frac{d\theta_{r}}{dt} &= \omega_{r}
\end{aligned}
\tag{6.68}
$$

The superscript e denotes the synchronous frame of reference.

In matrix form, using (6.68), we have the following differential equation for the electric drive application

$$
\begin{bmatrix}
\frac{di_{qs}^{e}}{dt}\\
\frac{di_{ds}^{e}}{dt}\\
\frac{di_{os}^{e}}{dt}\\
\frac{di_{qr}'^{e}}{dt}\\
\frac{di_{dr}'^{e}}{dt}\\
\frac{di_{or}'^{e}}{dt}\\
\frac{d\omega_{r}}{dt}
\end{bmatrix}
=
\begin{bmatrix}
-\frac{L_{RM}r_{s}}{L_{SM}L_{RM}-M^{2}} & -\omega_{e} & 0 & \frac{Mr_{r}'}{L_{SM}L_{RM}-M^{2}} & 0 & 0 & 0\\
\omega_{e} & -\frac{L_{RM}r_{s}}{L_{SM}L_{RM}-M^{2}} & 0 & 0 & \frac{Mr_{r}'}{L_{SM}L_{RM}-M^{2}} & 0 & 0\\
0 & 0 & -\frac{r_{s}}{L_{ls}} & 0 & 0 & 0 & 0\\
\frac{Mr_{s}}{L_{SM}L_{RM}-M^{2}} & 0 & 0 & -\frac{L_{SM}r_{r}'}{L_{SM}L_{RM}-M^{2}} & -\omega_{e} & 0 & 0\\
0 & \frac{Mr_{s}}{L_{SM}L_{RM}-M^{2}} & 0 & \omega_{e} & -\frac{L_{SM}r_{r}'}{L_{SM}L_{RM}-M^{2}} & 0 & 0\\
0 & 0 & 0 & 0 & 0 & -\frac{r_{r}'}{L_{lr}'} & 0\\
0 & 0 & 0 & 0 & 0 & 0 & -\frac{B_{m}}{J}
\end{bmatrix}
$$

$$
\times\begin{bmatrix} i_{qs}^e \\ i_{ds}^e \\ i_{os}^e \\ i_{qr}'^e \\ i_{dr}'^e \\ i_{or}'^e \\ \omega_r \end{bmatrix} + \begin{bmatrix} -\dfrac{M(Mi_{ds}^e + L_{RM} i_{dr}'^e)\omega_r}{L_{SM}L_{RM} - M^2} \\ \dfrac{M(Mi_{qs}^e + L_{RM} i_{qr}'^e)\omega_r}{L_{SM}L_{RM} - M^2} \\ 0 \\ \dfrac{L_{SM}(Mi_{ds}^e + L_{RM} i_{dr}'^e)\omega_r}{L_{SM}L_{RM} - M^2} \\ -\dfrac{L_{SM}(Mi_{qs}^e + L_{RM} i_{qr}'^e)\omega_r}{L_{SM}L_{RM} - M^2} \\ 0 \\ \dfrac{3P^2}{8J} M(i_{qs}^e i_{dr}'^e - i_{ds}^e i_{qr}'^e) \end{bmatrix}
$$

$$
+\begin{bmatrix} \dfrac{L_{RM}}{L_{SM}L_{RM} - M^2} & 0 & 0 & -\dfrac{M}{L_{SM}L_{RM} - M^2} & 0 & 0 \\ 0 & \dfrac{L_{RM}}{L_{SM}L_{RM} - M^2} & 0 & 0 & -\dfrac{M}{L_{SM}L_{RM} - M^2} & 0 \\ 0 & 0 & \dfrac{1}{L_{ls}} & 0 & 0 & 0 \\ -\dfrac{M}{L_{SM}L_{RM} - M^2} & 0 & 0 & \dfrac{L_{SM}}{L_{SM}L_{RM} - M^2} & 0 & 0 \\ 0 & -\dfrac{M}{L_{SM}L_{RM} - M^2} & 0 & 0 & \dfrac{L_{SM}}{L_{SM}L_{RM} - M^2} & 0 \\ 0 & 0 & 0 & 0 & 0 & \dfrac{1}{L_{lr}'} \\ 0 & 0 & 0 & 0 & 0 & 0 \end{bmatrix} \begin{bmatrix} u_{qs}^e \\ u_{ds}^e \\ u_{os}^e \\ u_{qr}'^e \\ u_{dr}'^e \\ u_{or}'^e \end{bmatrix} - \begin{bmatrix} 0 \\ 0 \\ 0 \\ 0 \\ 0 \\ 0 \\ \dfrac{P}{2J} \end{bmatrix} T_L
$$

The *quadrature, direct,* and *zero* voltages, u_{qs}^e, u_{ds}^e, and u_{os}^e, that need to be supplied to attain the balanced operation of induction motors are obtained using the Park transformation

$$
\mathbf{u}_{qdos}^e = \mathbf{K}_s^e \mathbf{u}_{abcs}
$$

Taking note that $\theta = \theta_e$ must be used in

$$
\mathbf{K}_s = \frac{2}{3}\begin{bmatrix} \cos\theta & \cos\left(\theta - \frac{2}{3}\pi\right) & \cos\left(\theta + \frac{2}{3}\pi\right) \\ \sin\theta & \sin\left(\theta - \frac{2}{3}\pi\right) & \sin\left(\theta + \frac{2}{3}\pi\right) \\ \frac{1}{2} & \frac{1}{2} & \frac{1}{2} \end{bmatrix}
$$

we obtain

$$
\mathbf{K}_s^e = \frac{2}{3}\begin{bmatrix} \cos\theta_e & \cos\left(\theta_e - \frac{2}{3}\pi\right) & \cos\left(\theta_e + \frac{2}{3}\pi\right) \\ \sin\theta_e & \sin\left(\theta_e - \frac{2}{3}\pi\right) & \sin\left(\theta_e + \frac{2}{3}\pi\right) \\ \frac{1}{2} & \frac{1}{2} & \frac{1}{2} \end{bmatrix}
$$

Therefore, one finds

$$\begin{bmatrix} u_{qs}^e \\ u_{ds}^e \\ u_{os}^e \end{bmatrix} = \frac{2}{3}\begin{bmatrix} \cos\theta_e & \cos\left(\theta_e - \frac{2}{3}\pi\right) & \cos\left(\theta_e + \frac{2}{3}\pi\right) \\ \sin\theta_e & \sin\left(\theta_e - \frac{2}{3}\pi\right) & \sin\left(\theta_e + \frac{2}{3}\pi\right) \\ \frac{1}{2} & \frac{1}{2} & \frac{1}{2} \end{bmatrix}\begin{bmatrix} u_{as} \\ u_{bs} \\ u_{cs} \end{bmatrix}$$

That is,

$$u_{qs}^e(t) = \frac{2}{3}\left[u_{as}\cos\theta_e + u_{bs}\cos\left(\theta_e - \frac{2}{3}\pi\right) + u_{cs}\cos\left(\theta_e + \frac{2}{3}\pi\right)\right]$$

$$u_{ds}^e(t) = \frac{2}{3}\left[u_{as}\sin\theta_e + u_{bs}\sin\left(\theta_e - \frac{2}{3}\pi\right) + u_{cs}\sin\left(\theta_e + \frac{2}{3}\pi\right)\right]$$

$$u_{os}^e(t) = \frac{1}{3}(u_{as} + u_{bs} + u_{cs})$$

where the following expressions for the *abc* applied phase voltages must be used

$$u_{as}(t) = \sqrt{2}u_M\cos(\omega_f t)$$

$$u_{bs}(t) = \sqrt{2}u_M\cos\left(\omega_f t - \frac{2}{3}\pi\right)$$

$$u_{cs}(t) = \sqrt{2}u_M\cos\left(\omega_f t + \frac{2}{3}\pi\right)$$

Assuming that the initial displacement of the *quadrature* magnetic axis is zero, and making use of the relationship $\theta_e = \omega_f t$, we have

$$u_{qs}^e(t) = \sqrt{2}u_M$$

$$u_{ds}^e(t) = 0 \tag{6.69}$$

$$u_{os}^e(t) = 0$$

It should be emphasized that the *quadrature-*, *direct-*, and *zero*-axis components of stator and rotor voltages, currents, and flux linkages have dc form. Furthermore, to control induction motors, only the dc *quadrature* voltage $u_{qs}^e(t)$ is regulated because $u_{ds}^e(t) = 0$ and $u_{os}^e(t) = 0$.

From (6.68), the block diagram can be developed. For three-phase induction motors modeled in the synchronous reference frame, the block diagram is built and plotted in Figure 6.89.

Example 6.12

Perform nonlinear modeling of a three-phase squirrel-cage motor studied in Examples 6.10 and 6.11. Develop the SIMULINK diagram to simulate squirrel-cage induction motors in the synchronous reference frame.

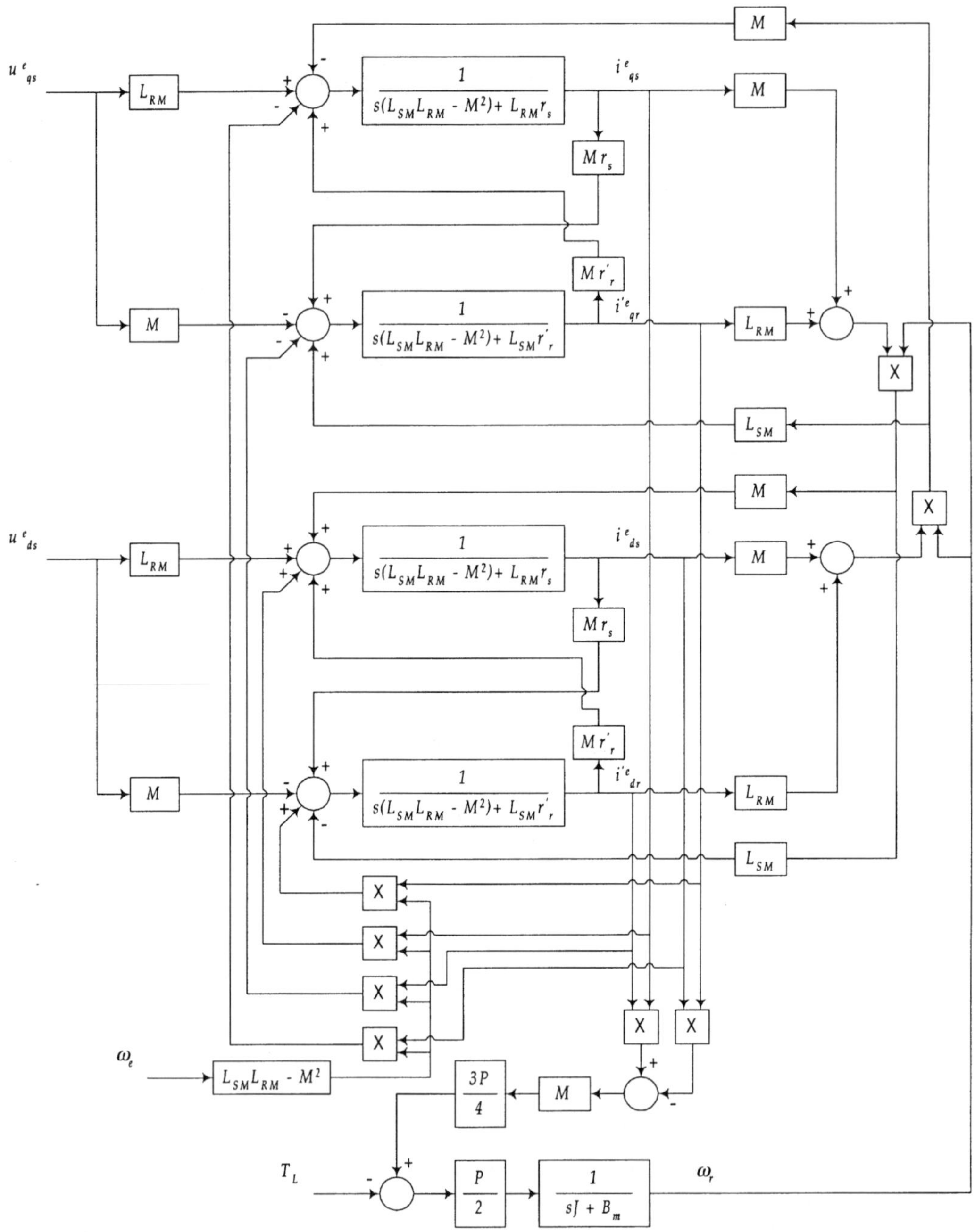

Figure 6.89 Block diagram for three-phase squirrel-cage induction motors modeled in the synchronous reference frame.

Solution

Using the block diagram developed for three-phase squirrel-cage induction motors (see Figure 6.89), the SIMULINK diagram is built and documented in Figure 6.90. The analysis of equation (6.69) shows that only *quadrature* voltage needs to be supplied, and this u_{qs}^e is the dc voltage. In particular, in modeling, we supply the following voltages

$$u_{qs}^e(t) = \sqrt{2}220$$

$$u_{ds}^e(t) = 0$$

$$u_{os}^e(t) = 0$$

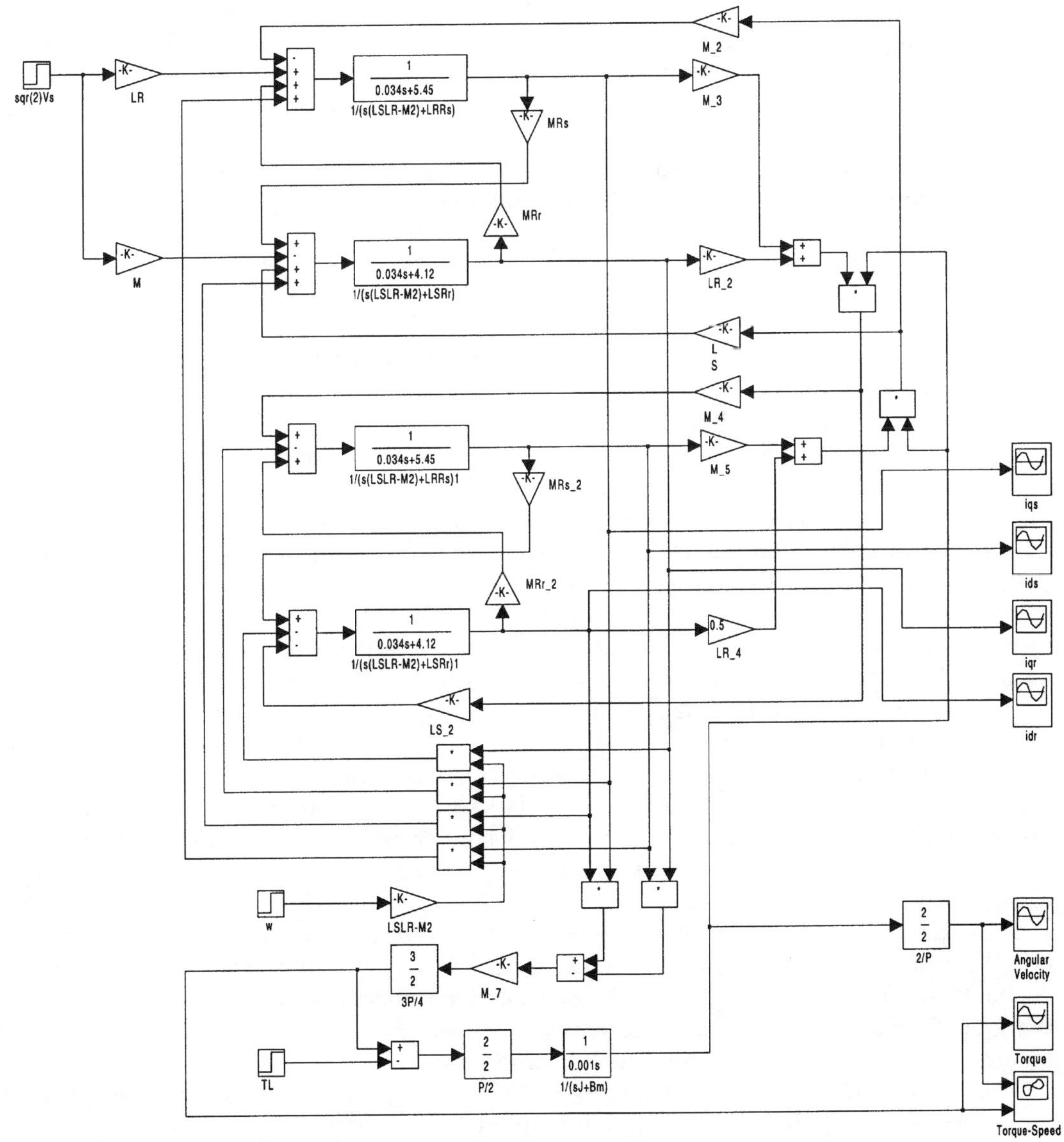

Figure 6.90 SIMULINK diagram for a three-phase squirrel-cage induction motor modeled in the synchronous reference frame.

The nonlinear simulations are performed for the motor studied for different load conditions. In particular, Figures 6.91 and 6.92 illustrate the motor dynamics and the torque-speed characteristic curves if $T_L = 0$ and $T_L = 2.5$ N-m. The analysis of the transient dynamics performed in the synchronous reference frame shows that the *quadrature*- and *direct*-axis components of stator and rotor currents have dc quantities.

6.6 Power converters and control of induction machines

Induction motors used in electric drives and servo-systems must be controlled. The angular velocity of squirrel-cage induction motors is regulated by changing the magnitude and frequency of the phase voltages applied to the stator windings, and power converters are used. The basic components of variable-frequency converters are rectifier, filter, and inverter. The simplest rectifiers are the single-phase half- and full-wave rectifiers. To control medium- and high-power induction motors, *polyphase* rectifiers are used. Polyphase rectifiers contain several ac sources, and the rectified voltage is summated at the output. The rectified voltage is filtered to reduce the harmonic content of the rectifier output voltage. Passive and active harmonic reduction, harmonic elimination, and harmonic cancellation can be attained using passive and active filters. To control the frequency, inverters are used. Inverters are electric circuits that convert the dc voltage or current to the ac voltage or current (voltage- and current-fed converters). Pulse-width modulation (PWM) reduces the *total harmonic distortion*. The PWM concept requires control circuitry to drive the switches with high frequency; however, the filtering requirements are significantly relaxed. Power converters provide an interface between the input energy source and the induction motor needed to be controlled (see Figure 6.93).

It has been emphasized that power converters produce sinusoidal voltages, applied to the induction motor windings, or sinusoidal currents fed to the motor windings, whose magnitude and frequency are regulated. The dc voltage is obtained by rectifying and filtering the line voltage, and the magnitude of the voltage can be controlled. The sinusoidal ac voltage or current with the regulated frequency, fed to the phase windings of induction motors, is obtained using DC-to-AC inverters. Design and deployment of high-performance electric drives and servo-systems have been directly related to the availability of power semiconductor devices. High-frequency switching power transistors, to be used in power converters for light-, medium-, and heavy-duty applications, are available. In Chapter 4, we provided the data for power transistors used in electric drives and servos. It is important to mention that specialized design is available; for example, IGBTs are integrated with diodes in the same package (Powerex and Toshiba have produced 3000 V and 1000 A IGBTs). Soft Switching Technology Corporation produces soft-switching resonant-link inverters up to 200 kVA with IGBTs switching at 70 kHz frequency. Electronic Power Conduction Inc. produces soft-switching converters up to 400 kW. The development of the gate turn-off (GTO) thyristor was the key to extending the power rating of many electric drive systems to the megawatt range. Power converters with GTO have found widespread application in traction drives (electric drivetrains in ships and locomotives). Gate turn-off thyristors are current-controlled devices that require a large gate current to enable turn-off of the anode current. Hence, large snubbers are needed to ensure turn-off without destructive failures. Consequently, the cost of power converters that use GTO increases because of the need for snubbers and control circuits.

There exist two basic types of inverters. Using voltage source inverters, one feeds the induction motor windings with variable-frequency phase voltages. In contrast, if current source inverters are used, the variable-frequency phase currents are fed to the induction motor windings. Figure 6.94 illustrates high-level schematic configurations of power converters, which include a pulse-width-modulated voltage source inverter with an unregu-

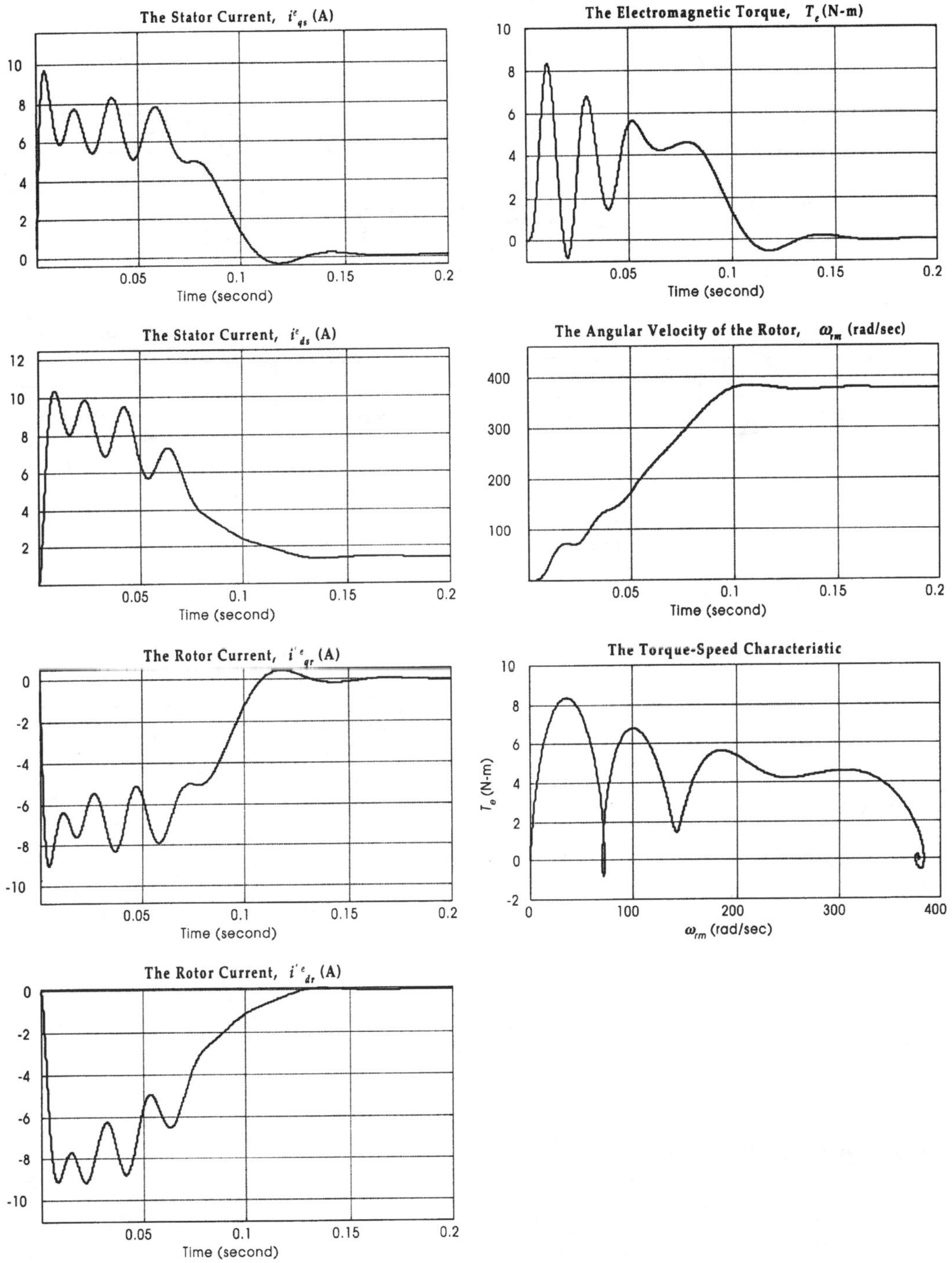

Figure 6.91 Dynamics and torque-speed characteristic of three-phase induction motor modeled in the synchronous reference frame; $T_L = 0$ N-m, $u^e_{qs}(t) = \sqrt{2}220$, $u^e_{ds}(t) = 0$, and $u^e_{os}(t) = 0$.

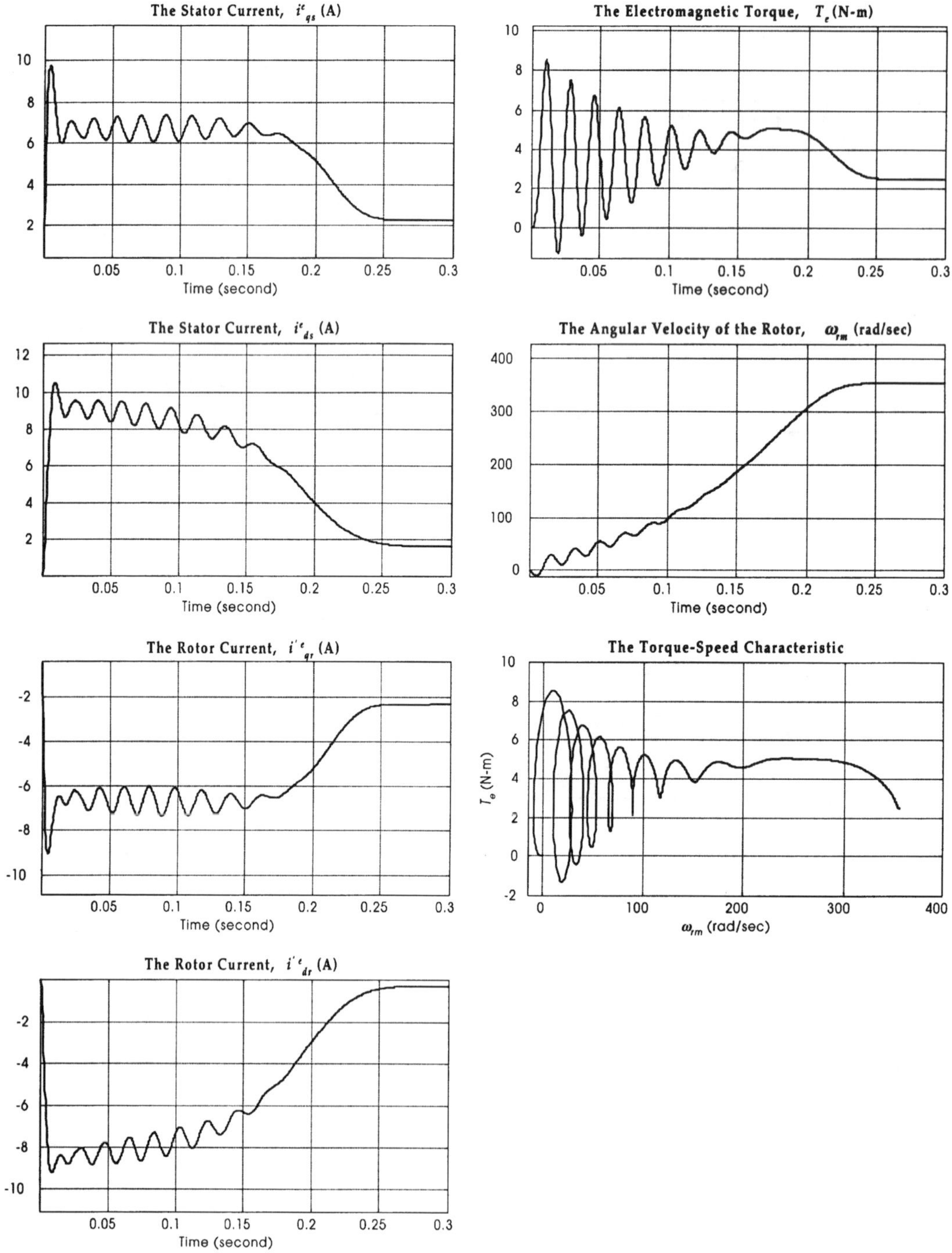

Figure 6.92 Dynamics and torque-speed characteristic of three-phase induction motor modeled in the synchronous reference frame; T_L = 2.5 N-m, $u^e_{qs}(t)=\sqrt{2}220,\ u^e_{ds}(t)=0,$ and $u^e_{os}(t)=0.$

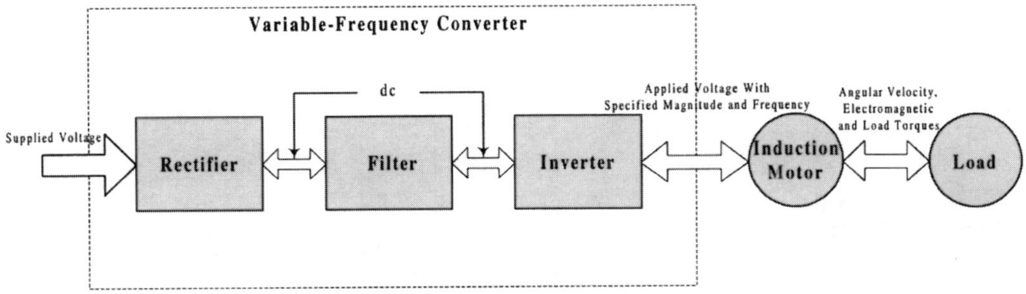

Figure 6.93 Variable-frequency power converter and induction motor.

lated rectifier, a square-wave voltage source inverter with a regulated rectifier, and a current source inverter with a regulated rectifier. These power converters allow one to control the magnitude and frequency of the phase voltages and currents.

Typical PWM power converter configurations consist of three legs, one for each phase, to control the frequency and the magnitude of the phase voltages applied to the motor

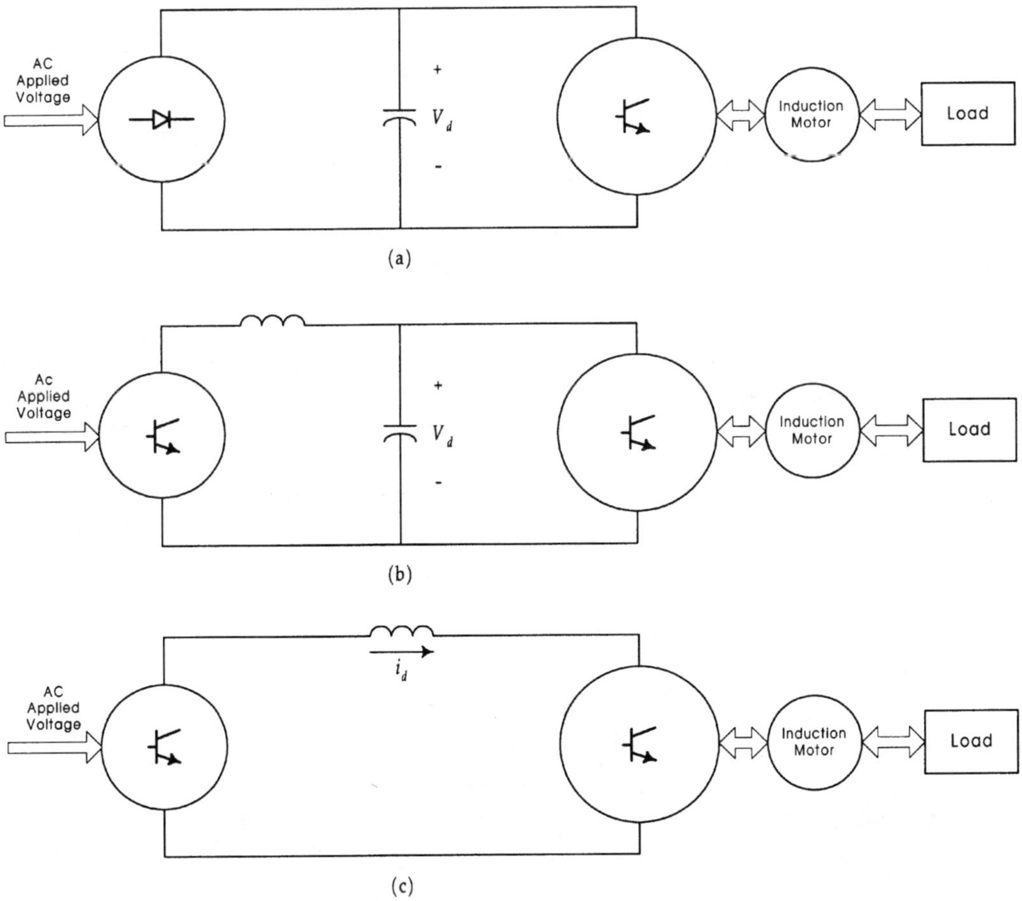

Figure 6.94 Variable-frequency power converters: (a) pulse-width-modulated voltage source inverter with an unregulated rectifier; (b) squire-wave voltage source inverter with a regulated rectifier; (c) current source inverter with a regulated rectifier.

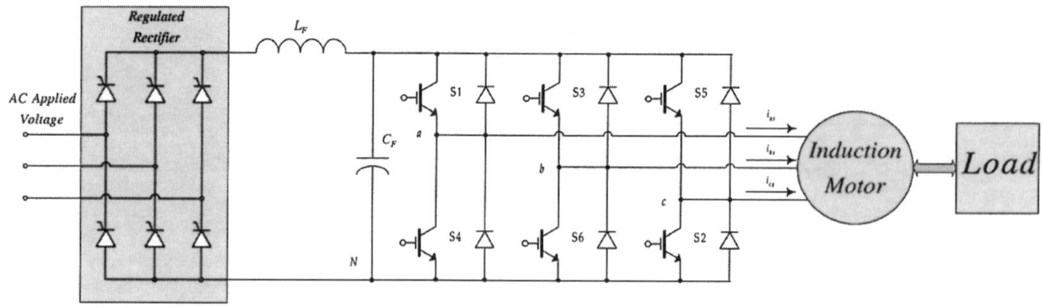

Figure 6.95 Power converter with three-phase hard-switching inverter.

windings (see Figures 6.95 and 6.96). The inverter converts the dc bus voltage into a *polyphase* ac voltage at the desired frequency to attain the desired torque-speed characteristics, transient behavior, efficiency, and starting capabilities, etc. Switching stresses and losses, high electromagnetic interference, and extended operating areas, among other drawbacks of the hard-switching inverters, led to the implementation of soft-switching technology. This enabling concept was developed in the last decade. Soft-switching using resonant-linked converters allows one to attain zero voltage across (current through) the switching device; that is, the device is switched when the voltage across it, or the current through it, is zero. Hence, low losses and electromagnetic interference, and high efficiency and switching capabilities, result compared with the hard-switching inverters.

For the hard- and soft-switching inverter configurations shown in Figures 6.95 and 6.96, the waveforms of the phase voltages supplied to the motor windings are illustrated in Figures 6.97 and 6.98.

Transistors and thyristors are current-controlled devices. Transistors require continuous drive signals, while thyristors need a momentary gate current to turn on and turn off. For example, a base current must be regulated to maintain the BJT in the conducting state, and the turn-on and turn-off times depend on how rapidly the charge needed to be supplied (to turn on) or removed (to turn off) can be delivered to the base region (turn-off switching speed is decreased by initially applying a spike of base current, and then reducing the current to the magnitude needed, while turn-off switching speed is decreased by initially applying a spike of negative base current). Control algorithms must be designed to drive high-frequency transistors, and the closed-loop system configuration for a three-phase induction motor with power converter is documented in Figure 6.99.

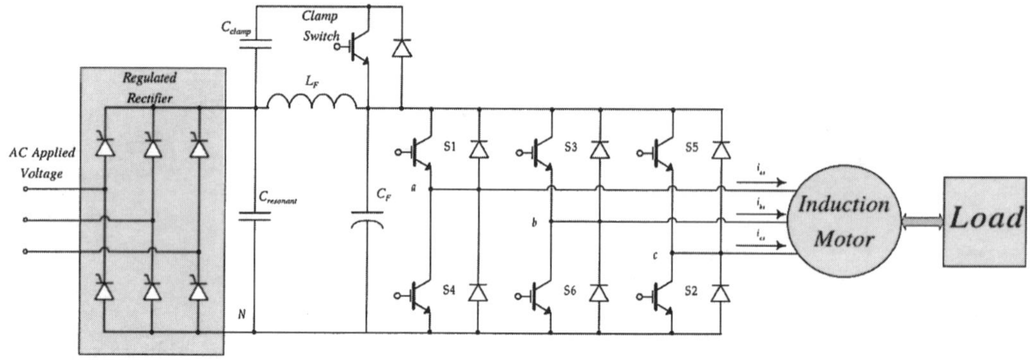

Figure 6.96 Power converter with three-phase soft-switching inverter.

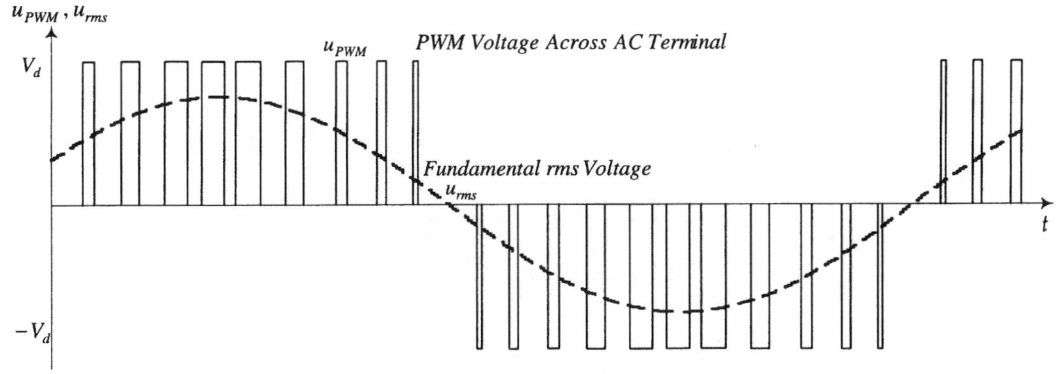

Figure 6.97 Phase voltage waveforms in the hard-switching PWM inverter.

Consider the hard-switching inverter configuration shown in Figures 6.95 and 6.99 with three switch pairs. To obtain three-phase balanced output voltages using the PWM concept, a triangular signal-level voltage is compared with three sinusoidal specified frequency control signals, and these control signals are shifted by 120° (see Figure 6.100a). High-frequency switches S1 and S4, S3 and S6, S5 and S2, close and open opposite to each other. That is, switches in each pair are turned on and off simultaneously. If, for example, S1 and S4 are closed at the same instant, the circuits is short-circuited across the source. The instantaneous voltages u_{aN}, u_{bN}, and u_{cN} are either equal to V_d or 0. In particular, the signal level voltages u_{ac}, u_{bc}, and u_{cc} are compared with the triangular signal u_t. If, for example, u_{ac} is greater than u_t, then S1 is closed, whereas S4 is opened. If the signal-level voltage u_{ac} is less than u_t, then S4 is closed, whereas S1 is opened. The resulting waveform for the phase voltage u_{aN} is shown in Figure 6.100b. In a similar manner, the phase voltages u_{bN} and u_{cN} are determined by comparing the signal-level voltages u_{bc} and u_{cc} with u_t to open or close switches S3/S6 and S5/S2. The resulting voltages u_{bN} and u_{cN} possess the same pattern as the *aN* voltage, except that u_{bN} and u_{cN} are shifted by 120° and 240°, respectively (see Figures 6.100c and 6.100d). It should be emphasized that the voltages u_{aN}, u_{bN}, and u_{cN} are measured with respect to the negative dc bus, and these dc components are canceled as one considers the line-to-line voltage, which is plotted in Figure 6.100e. In particular, the line-to-line voltage u_{ab} is obtained by subtracting voltage u_{bN} from u_{aN}. One can analyze the waveforms of the instantaneous and *rms* voltages shown in Figure 6.97 and 6.100e.

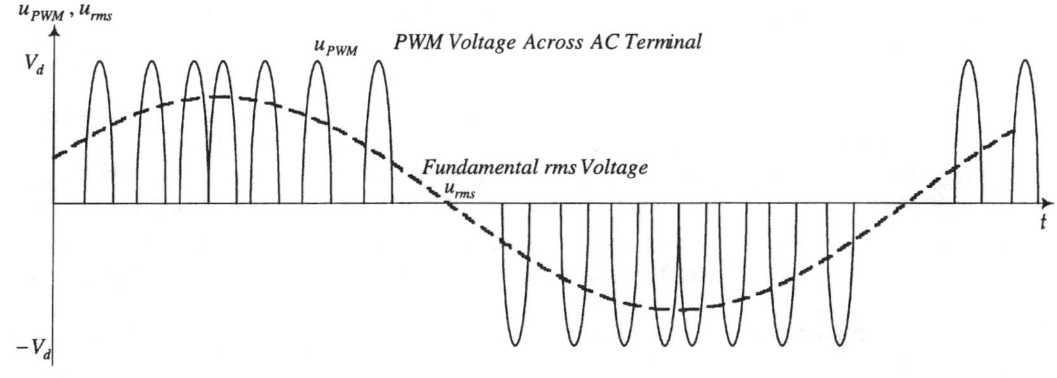

Figure 6.98 Phase voltage waveforms in the soft-switching PWM inverter.

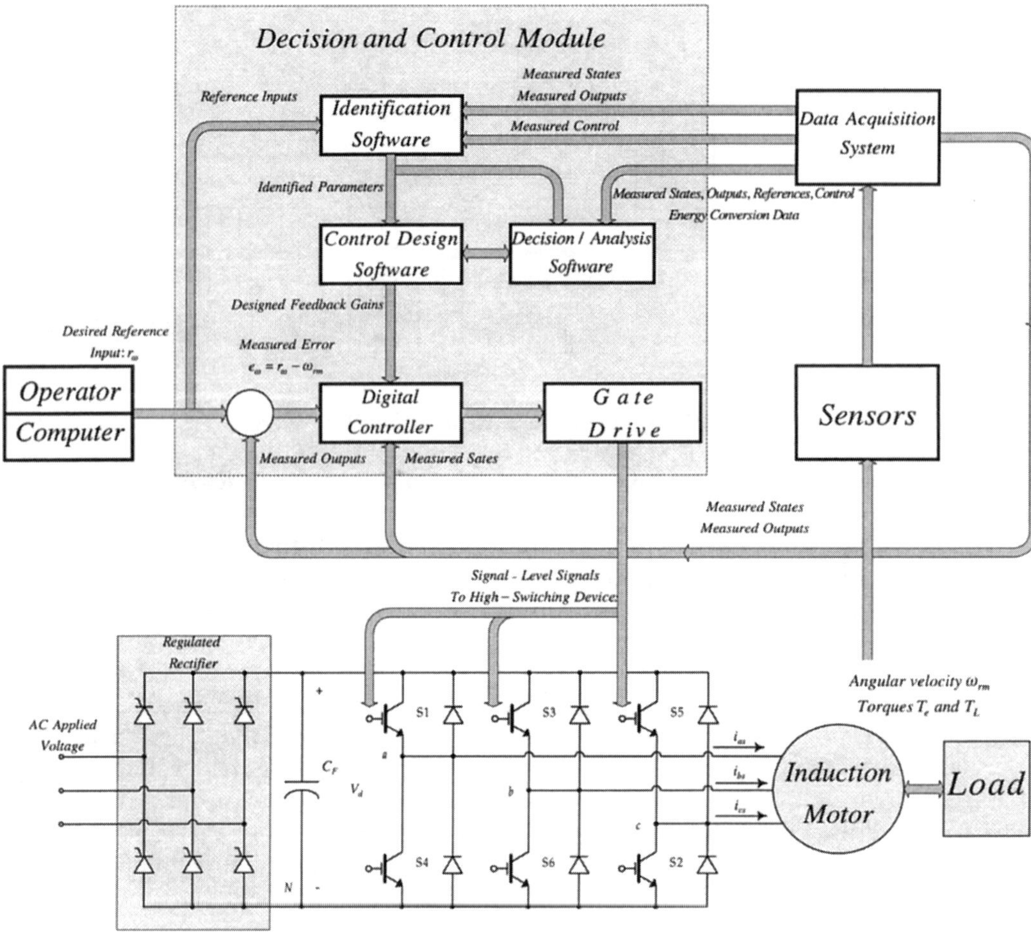

Figure 6.99 Three-phase squirrel-cage induction motor controlled by power converter: closed-loop system configuration.

Square-wave voltage source inverters, known as six-step inverters, are commonly used. A schematic diagram of a basic three-phase square-wave voltage source inverter bridge is shown in Figure 6.101. The rectifier rectifies the three-phase ac applied voltage, and a relatively large electrolytic capacitor is inserted to maintain constant dc voltage as well as to provide a path for the rapidly changing currents drawn by the inverter. The inductance is inserted to attenuate current spikes.

Assuming that the inverter consists of six ideal switches, consider the basic operation of the square-wave voltage inverter. Each switch is closed for 180° and is opened for the remaining 180° in a cyclic pattern where S3 is closed 120° after S1, S5 is closed 120° after S3, S4 is closed 180° after S1, S6 is closed 180° after S3, and S2 is closed 180° after S5 (see Figure 6.102). The result of this switching operation is that a combination of three switches is closed simultaneously for every 60° duration, as shown in Figure 6.102; that is, in three-phase sex-step inverters, the switching appears every 60° angle interval $\left(\frac{1}{6}T \text{ time interval}\right)$.

To determine the voltage waveforms applied to the *abc* windings, consider the six-step inverter and motor circuitry configurations as illustrated in Figure 6.101. During the

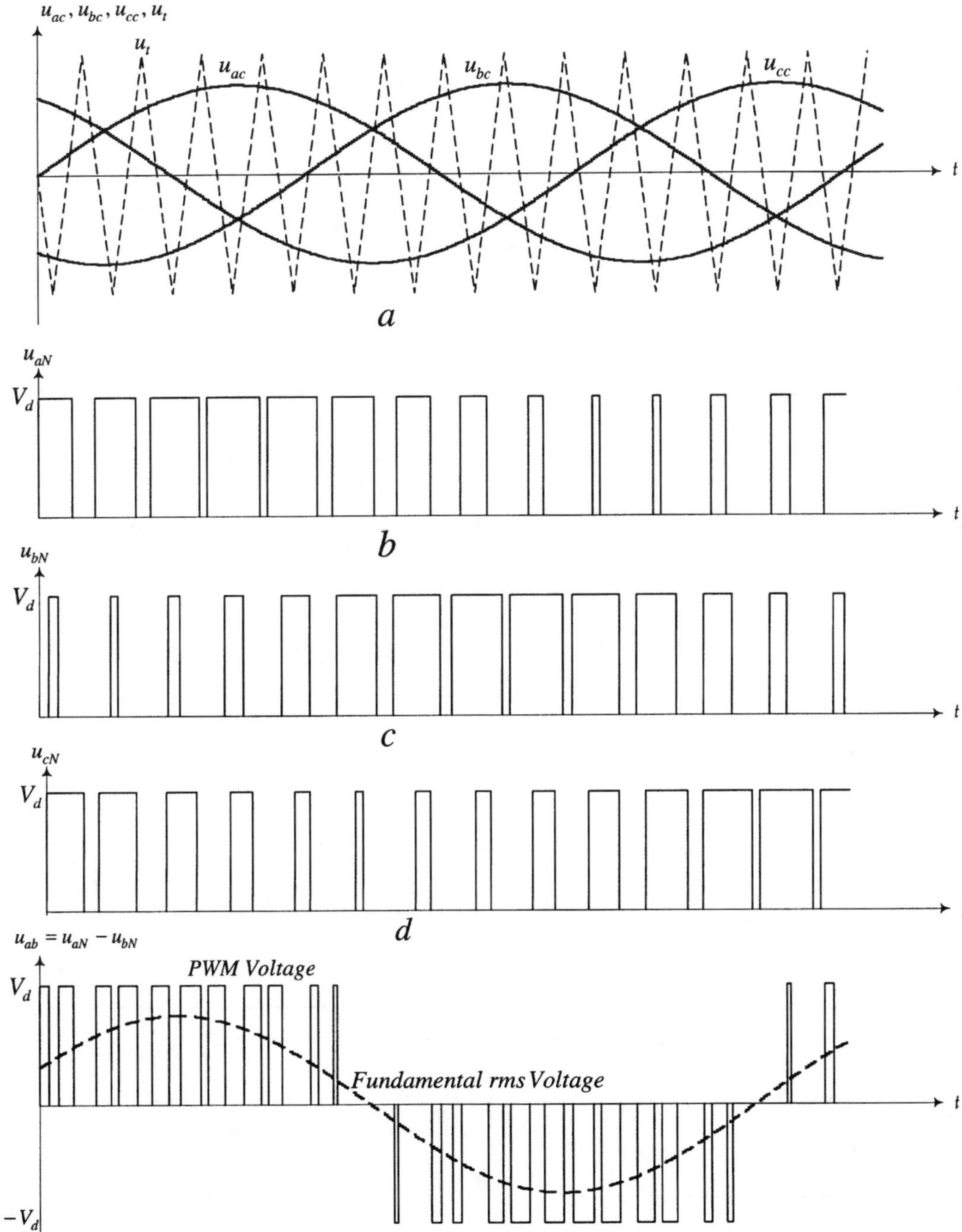

Figure 6.100 Voltage waveforms in three-phase hard-switching inverters.

interval from 0° to 60°, where switches S5, S6, and S1 are closed, phase a is in parallel with c, and they are connected to phase b in series, which is connected to the source via S6. As a result, the voltage waveforms, as shown in Figure 6.103, result. In particular, $u_{aN} = u_{cN} = V_d$ and $u_{bN} = 0$. Hence, $u_{ab} = V_d$, $u_{bc} = -V_d$, and $u_{ca} = 0$. Because phases a and c are connected in parallel, the apparent impedance, seen from the neutral of the motor (point

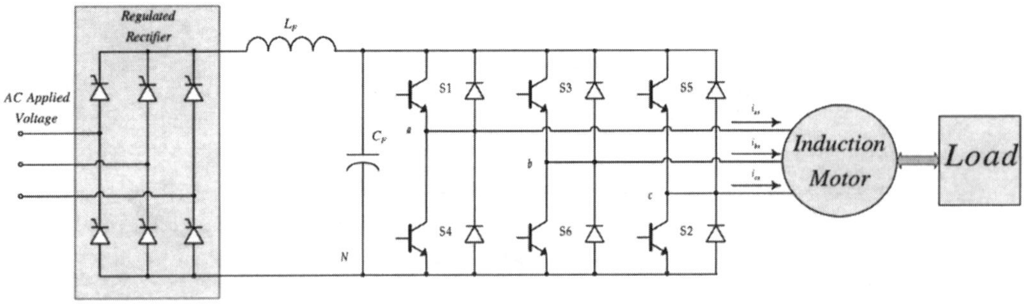

Figure 6.101 Three-phase square-wave (six-step) voltage source inverter.

N'), is halved. Hence, the voltage drop across phases *as* and *cs* is $\frac{2}{3}V_d$,, whereas the voltage drop across phase *bs* is $\frac{1}{3}V_d$. That is, $u_{as}=\frac{1}{3}V_d$, $u_{bs}=-\frac{2}{3}V_d$, and $u_{cs}=\frac{1}{3}V_d$. Hence, the voltage drop across the phase is always $\frac{1}{3}V_d$ or $\frac{2}{3}V_d$, depending on the

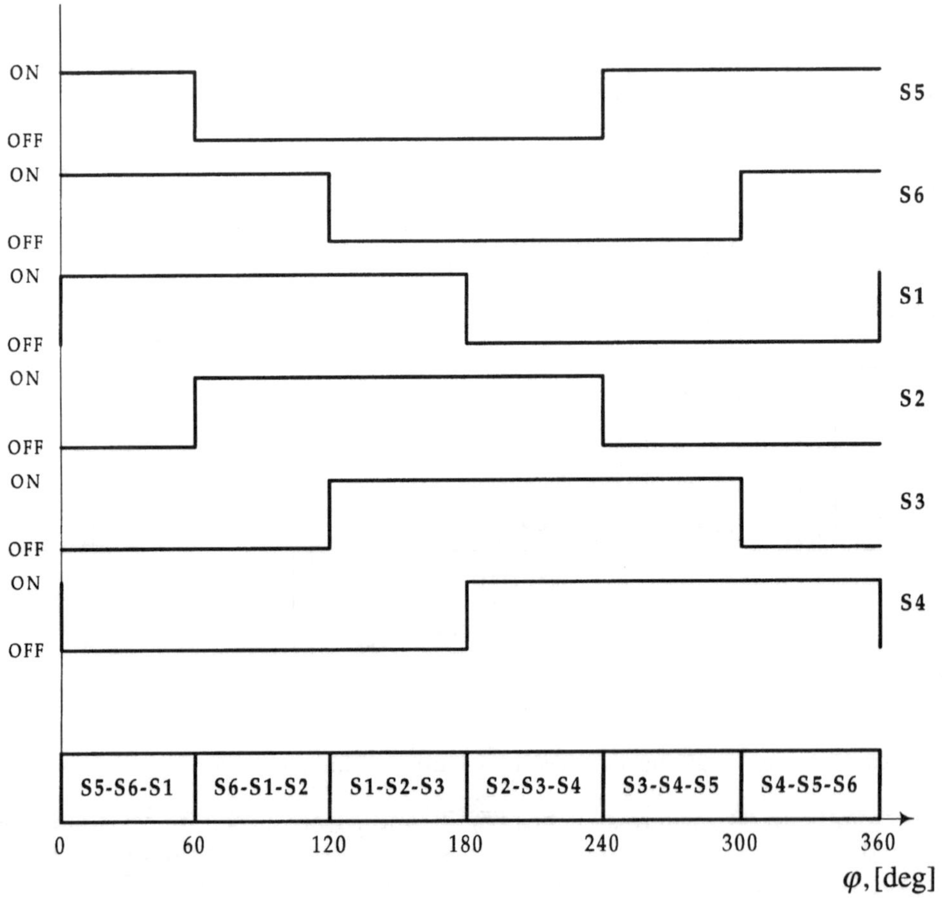

Figure 6.102 Switching pattern for three-phase six-step inverters.

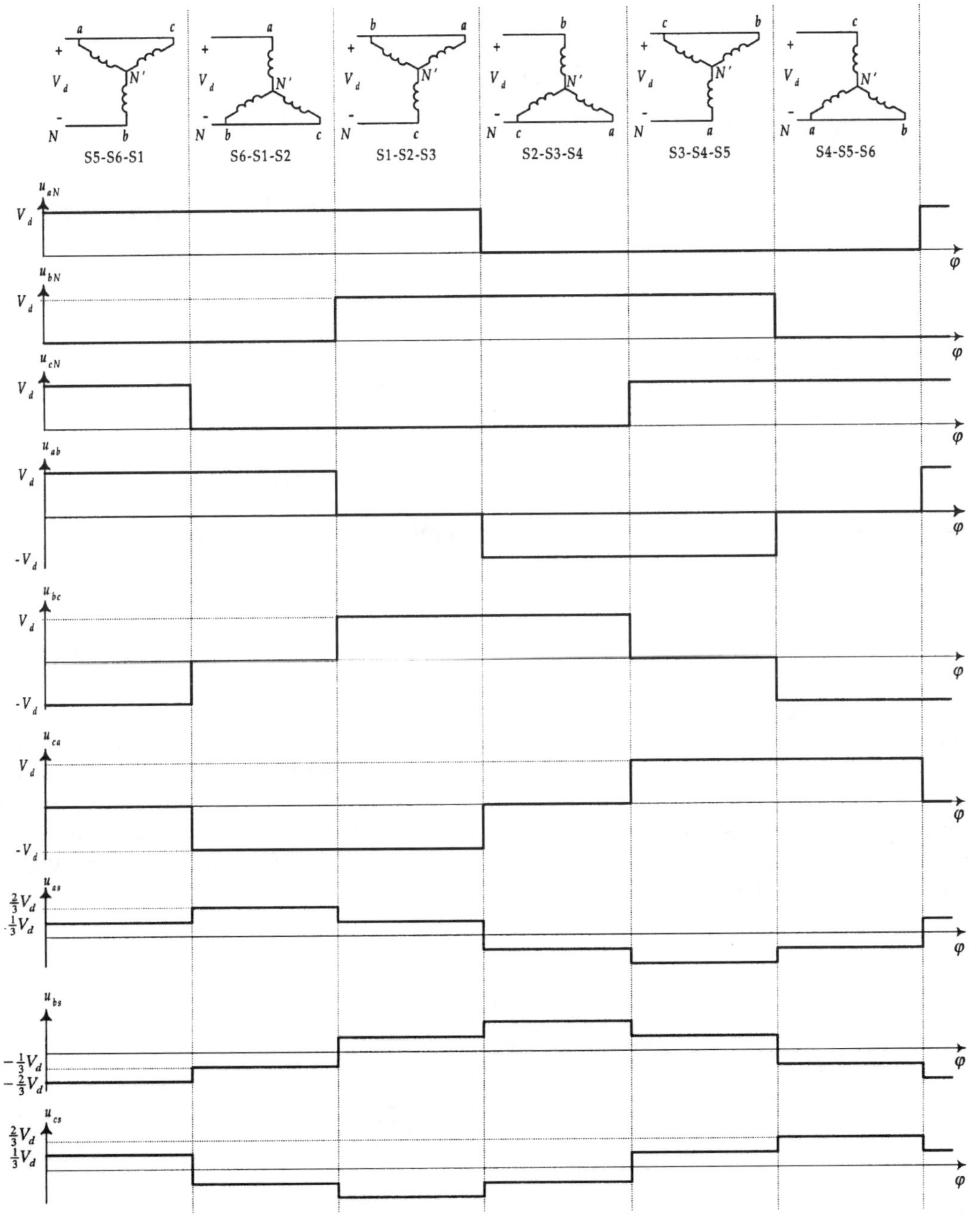

Figure 6.103 Voltage waveforms at the terminals *aN*, *bN*, and *cN*, line-to-line and line-to-neutral voltages applied to the induction motor.

connection of the phases (series or parallel). The waveforms for u_{as}, u_{bs}, and u_{cs} are shown in Figure 6.103.

Compared with voltage source inverters, power converter configurations — if the current source inverters are implemented — are different. In particular, the current source inverter is fed from a constant current source, which is generated by a controlled rectifier

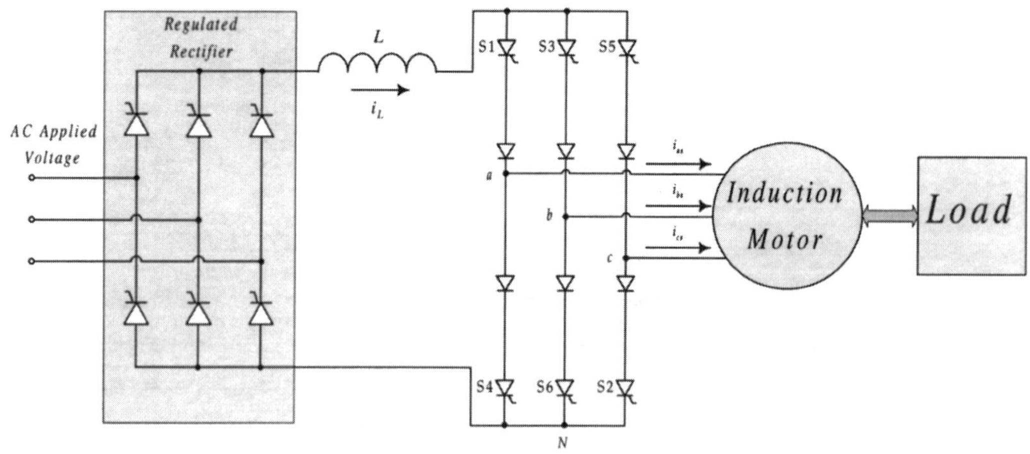

Figure 6.104 Power converter with a thyristor current source inverter.

with a large dc link inductor to smooth the generated current. The schematic of a current source inverter is shown in Figure 6.104.

At any time, only two thyristors conduct: one of the thyristors is connected to the positive dc link and the other is connected to the negative dc link. The current is switched sequentially into one of the phases of a three-phase induction motor by the top half of the inverter and returns from another phase to the dc link by the bottom half of the inverter. Because the current is constant, there will be a constant voltage drop across the stator winding of the motor and a zero voltage drop across the self-inductance of the winding. Hence, the motor terminal voltage is not set by the inverter, but by the resistance of the stator winding of the motor. Because the motor is wound with sinusoidally distributed windings, the voltages that appear on the terminals of the motor are sinusoidal. The current and voltage waveforms of the motor connected to a current source inverter are shown in Figure 6.105.

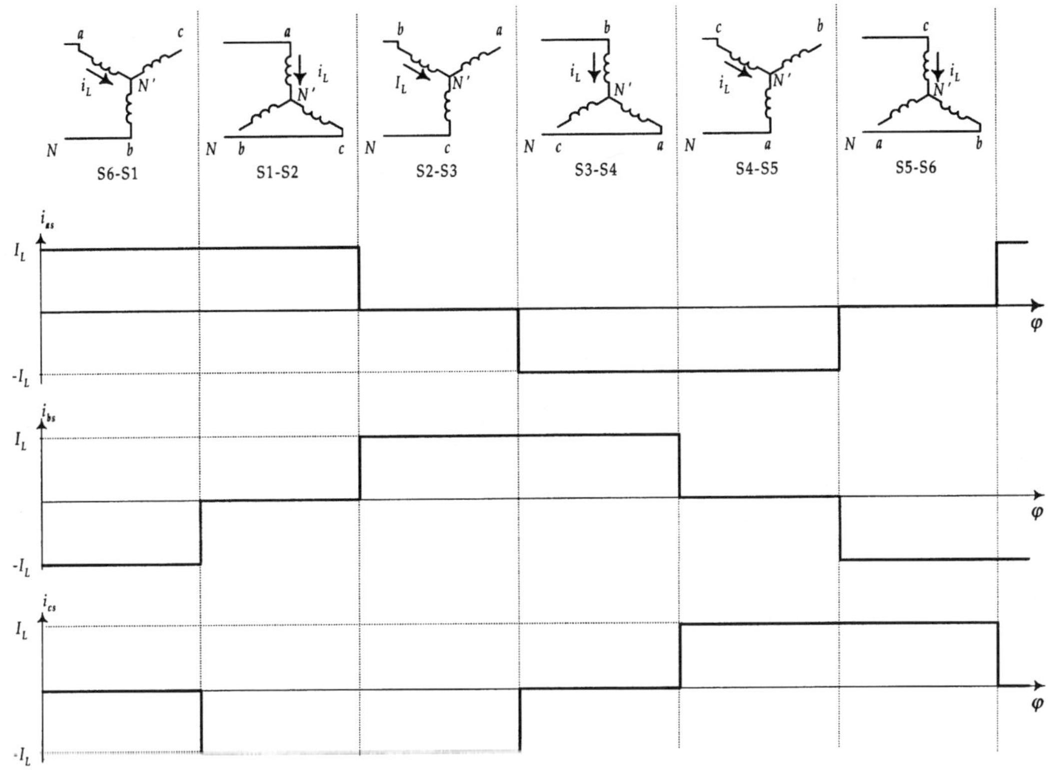

Figure 6.105 Typical phase currents in induction motors fed by a current source inverter.

References

Hart, D. W., *Introduction to Power Electronics,* Prentice-Hall, Upper Saddle River, NJ, 1997.

Kassakian, J. G., Schlecht, M. F., and Verghese, G. C., *Principles of Power Electronics,* Addison-Wesley Publishing, Reading, MA, 1991.

Krause, P. C., Wasynczuk, O., and Sudhoff, S. D., *Analysis of Electric Machinery,* IEEE Press, New York, 1995.

Krause, P. C. and Wasynczuk, O., *Electromechanical Motion Devices,* McGraw-Hill, New York, 1989.

Lorenz, R. D., Lipo, T. A., and Novotny, D. W., Motion Control with Induction Motors, in *Power Electronics and Variable Frequency Drives,* pp. 209–276, IEEE Press, New York, 1997.

Mohan, N., Undeland, T. M., and Robbins, W. P., *Power Electronics: Converters, Applications, and Design,* John Wiley & Sons, New York, 1995.

Novotny, D. W. and Lipo, T. A., *Vector Control and Dynamics of AC Drives,* Clarendon Press, Oxford, 1996.

chapter seven

Synchronous machines

List of basic variables and symbols used in this chapter

u_{as}, u_{bs}, and u_{cs} are the phase voltages in the stator windings *as, bs,* and *cs*, respectively
u_{qs}, u_{ds}, and u_{os} are the *quadrature-*, *direct-*, and *zero*-axis stator voltage components
i_{as}, i_{bs}, and i_{cs} are the phase currents in the stator windings *as, bs,* and *cs*, respectively
i_{qs}, i_{ds}, and i_{os} are the *quadrature-*, *direct-*, and *zero*-axis stator current components
ψ_{as}, ψ_{bs}, and ψ_{cs} are the stator flux linkages
ψ_{qs}, ψ_{ds}, and ψ_{0s} are the *quadrature-*, *direct-*, and *zero*-axis stator flux linkages components
u_{fr} is the excitation voltage applied to the rotor winding *fr*
i_{fr} is the current in the rotor winding
ψ_{fr} is the rotor flux linkage
ψ_m is the magnitude of the flux linkages established by the permanent-magnet
ω_r and ω_{rm} are the electrical and rotor angular velocities, respectively
θ_r and θ_{rm} are the electrical and rotor angular displacements, respectively
T_e is the electromagnetic torque developed
T_L is the load torque applied
B_m is the viscous friction coefficient
J is the equivalent moment of inertia
r_s and r_r are the resistances of the stator and rotor windings, respectively
L_{ss} and L_{rr} are the self-inductances of the stator and rotor windings, respectively
L_{ms} and L_{ls} are the stator magnetizing and leakage inductances, respectively
L_{lr} is the rotor leakage inductance
L_{mq} and L_{md} are the magnetizing inductances in the *quadrature* and *direct* axes
N_s and N_r are the number of turns of the stator and rotor windings, respectively
P is the number of poles
ω is the angular velocity of the reference frame
θ is the angular displacement of the reference frame

7.1 Introduction

Synchronous machines are widely used as motors and generators. When synchronous machines are used to convert mechanical energy to electrical energy, they are called synchronous generators. In contrast, synchronous motors convert electrical energy to mechanical energy. In conventional synchronous machines, a dc voltage that is applied to the rotor field winding produces a rotor magnetic field. If the synchronous generator is driven by the prime mover, the produced rotating magnetic field induces a three-phase set of voltages in the stator windings. The frequency of the phase voltages induced is locked with the electrical angular velocity. For two- and four-pole synchronous generators, 60 Hz frequency results if the angular velocity of the prime mover is 377 rad-s^{-1} and 188.5 rad-s^{-1}. To avoid brushes, needed to supply the dc voltage to the field winding, advanced high-speed synchronous generators have brushless exciters. A brushless exciter is a small AC generator that has field windings mounted on the stator and armature windings on the rotor. Three-phase induced voltage is rectified by the rectifier, which is mounted on the rotor, and the rectified dc voltage is fed to the field winding. A broad spectrum of synchronous electric machines is widely used in industrial electric drives, servos, and generation systems. In addition to the required steady-state and dynamic characteristics, particular features, such as power and torque density, efficiency and controllability, versatility and flexibility, absence of commutators and brushes, simplicity and ruggedness, reliability and cost, weight-to-torque and weight-to-power ratios, starting capabilities, and other criteria, are essential and commonly considered. The final decision involves a compromise between the variety of available options, and conventional synchronous machines are widely used in power generation systems.

Permanent-magnet synchronous machines are used in a wide range of industrial electric drives, servomechanisms, and power generation systems such as auxiliary power units. Permanent-magnet synchronous machines are simple electromechanical devices, and are the simplest AC-type motors from a control standpoint. In fact, the torque-speed characteristics fulfill the *controllability* criteria over an entire range of operation. In contrast, induction, switched-, and variable-reluctance motors require sophisticated control algorithms to achieve suitable torque-speed characteristics. Permanent-magnet synchronous motors and generators are used when severe starting conditions and variable loads occur, and these machines have superior advantages with regard to control capabilities in variable speed and load applications. As a result, these machines satisfy the industrial requirements and are used in various applications previously, exclusively associated with permanent-magnet DC machines (sparking reinforces the competitive position and restricts the specific areas of applicability of drives and servos based on direct-current technology). Commercially available permanent-magnet synchronous machines exceed the functional characteristics of brush-type DC motors. This strong competitive position of permanent-magnet synchronous machines demands further comprehensive studies from nonlinear analysis, modeling, and control perspectives. Innovative analytical, numerical, and experimental results are under development, and new comprehensive models need to be developed to improve existing models and to design advanced electric drives, servos, and power generation systems. This chapter develops nonlinear mathematical models, and performs nonlinear modeling and analysis of synchronous electric machines.

In direct-drive applications, permanent-magnet stepper motors, which are synchronous machines, are widely used. Cutaway views of a permanent-magnet synchronous machine and stepper motor are shown in Figure 7.1.

7.2 Synchronous reluctance motors

Synchronous reluctance motors are used in electric drives. The laminated rotor design results in different magnetizing reluctances $\Re_{md}$ and $\Re_{mq}$ in the *direct* and *quadrature* axes. The difference $(\Re_{md} - \Re_{mq})$, as well as the ratio between these reluctances, are important in order to optimize motor performance and to attain the desired torque-speed and dynamic characteristics. In particular, the optimum design can be achieved by optimizing

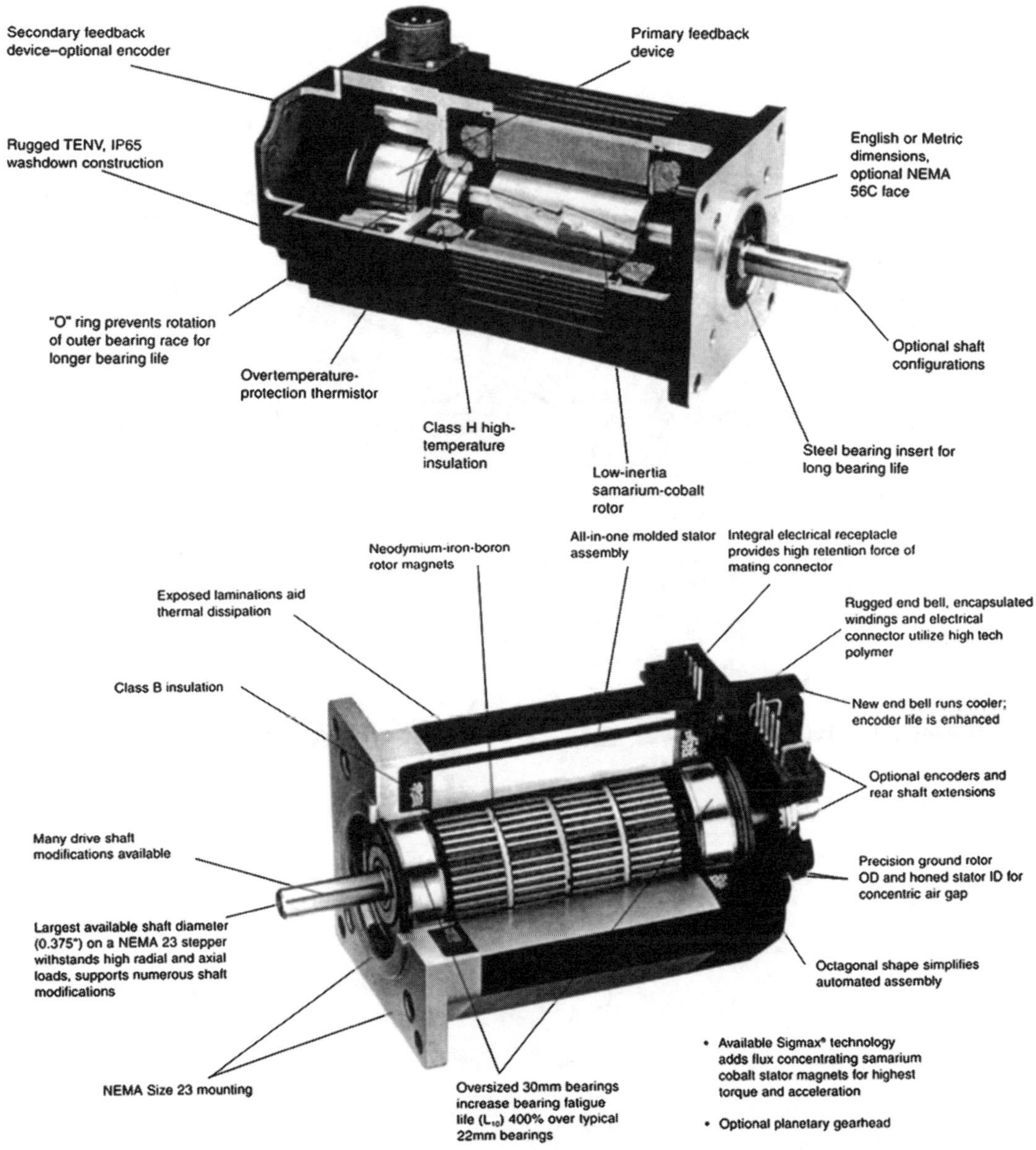

Figure 7.1 Synchronous machines: cutaway views of a permanent-magnet synchronous machine and stepper motor with encoders. (Courtesy of Pacific Scientific.)

the difference $(\Re_{md} - \Re_{mq})$ and the ratio $\frac{\Re_{md}}{\Re_{mq}}$. Various synchronous reluctance motor design and optimization concepts have been proposed to maximize $(\Re_{md} - \Re_{mq})$ and $\frac{\Re_{md}}{\Re_{mq}}$. Figures 7.2a and 7.2b show three-phase, two- and four-pole synchronous reluctance motors with axial-laminated rotors, respectively. The medium between the lamination is any nonmagnetic material such as aluminum, copper, or injected molded plastic. In this

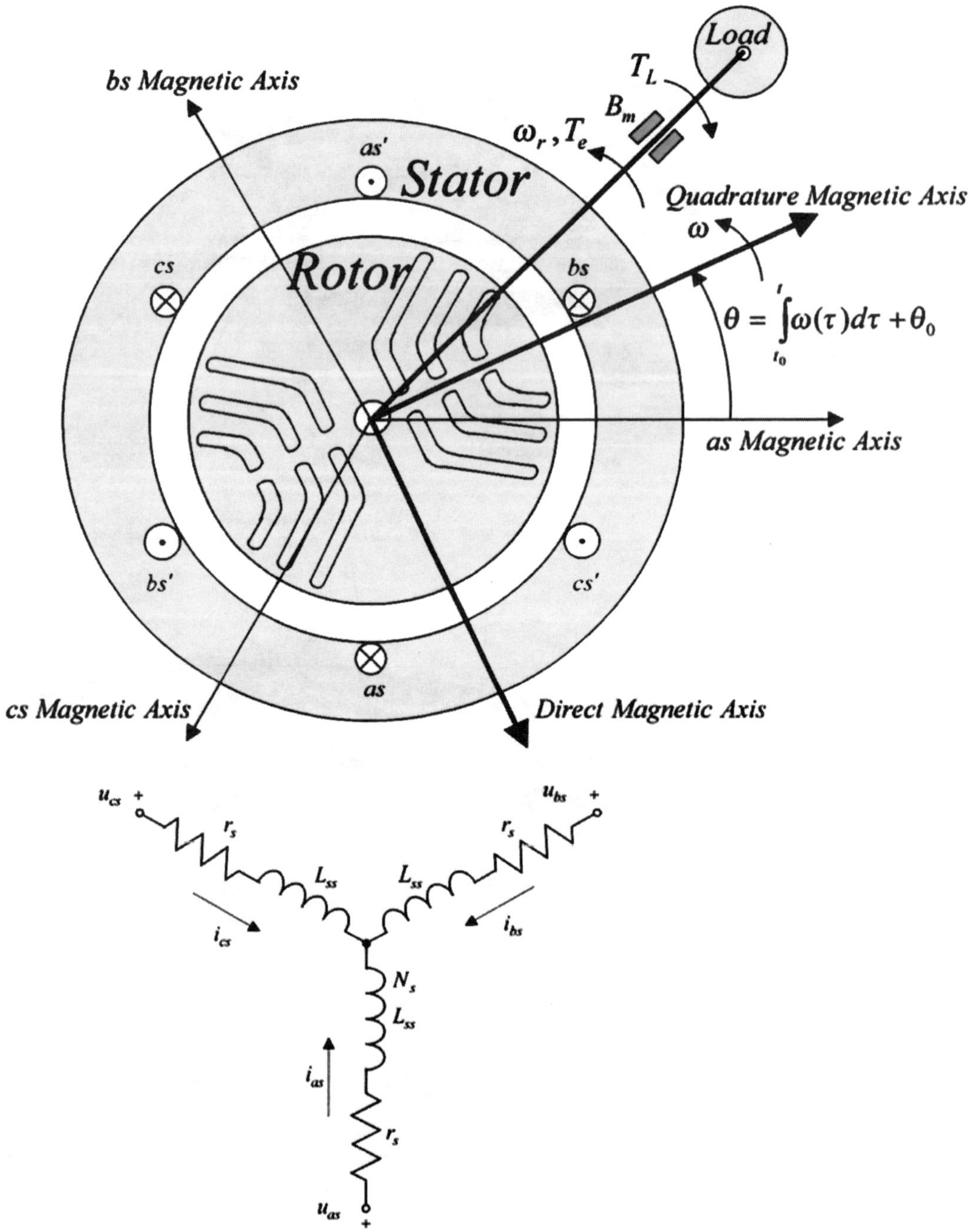

Figure 7.2a Three-phase, two-pole synchronous reluctance motor.

section, synchronous reluctance motor mathematical models in the *machine* (*abc*) and *quadrature-direct-zero* (*qd0*) variables are derived and thoroughly examined, and the analysis of the steady-state and dynamic performances is accomplished.

Three-phase, two- and four-pole synchronous reluctance motors are shown in Figures 7.2a and 7.2b. The stator windings, which are sinusoidally distributed, are identical and

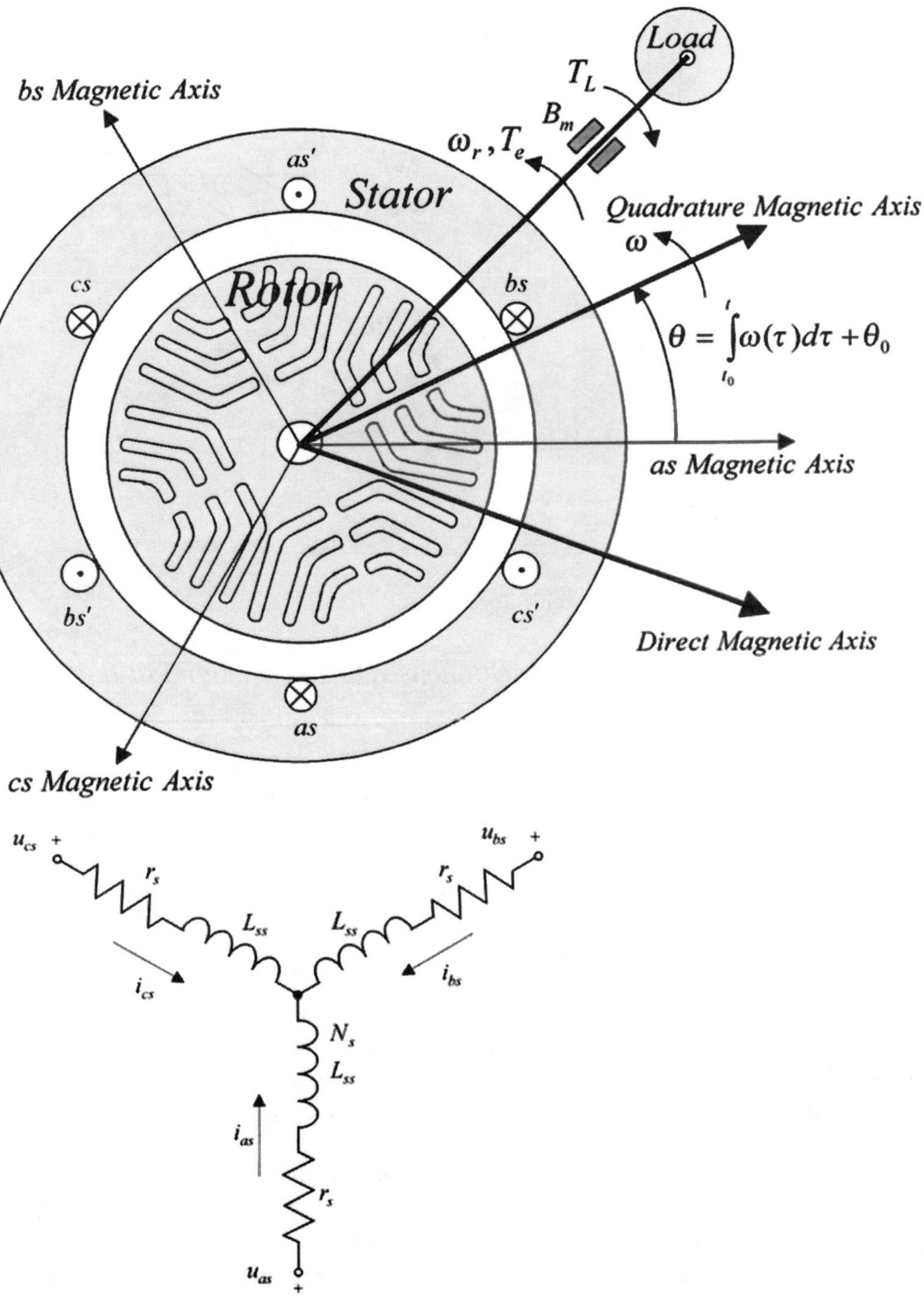

Figure 7.2b Three-phase, four-pole synchronous reluctance motor.

displaced by 120 electrical degrees. The rotating air gap *mmf* is established by symmetrical three-phase stator windings carrying balanced three-phase currents, and the electromagnetic torque is produced due to the interaction of *mmf*.

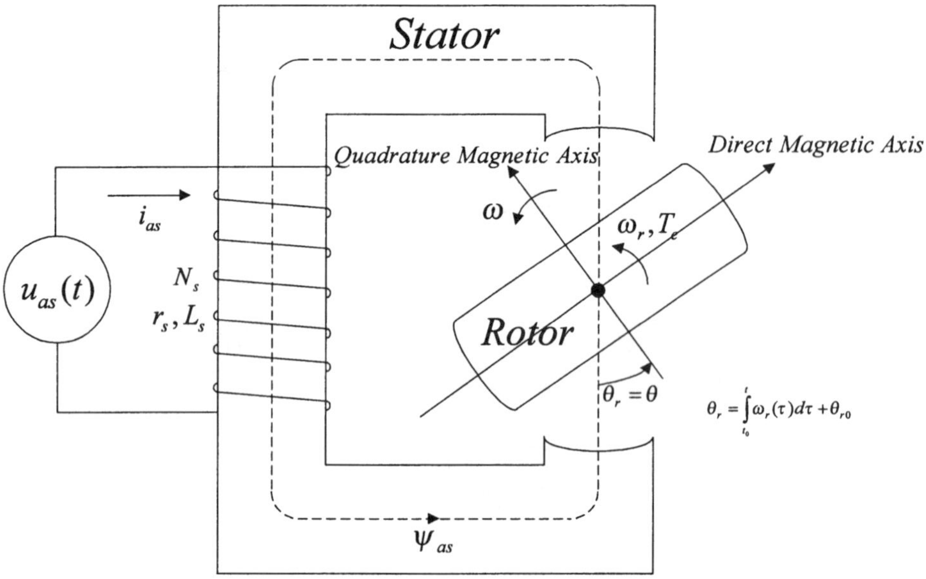

Figure 7.3 Elementary single-phase reluctance motor.

7.2.1 *Mathematical models of synchronous reluctance motors in the machine variables*

Single-phase reluctance motors

The major goal of this section is to develop a complete nonlinear mathematical model for synchronous reluctance motors using the *abc machine* variables. To approach the model development, analysis, and control of three-phase synchronous reluctance machines, an elementary single-phase reluctance machine is studied first. In particular, Figure 7.3 illustrates an elementary single-phase reluctance motor that rotates with angular velocity ω_r. The *quadrature* and *direct* magnetic axes are fixed with the rotor, as shown in Figure 7.3. These magnetic axes rotate with angular velocity ω, which is assigned to be equal to the rotor angular velocity ω_r; that is, $\omega = \omega_r$. It should be emphasized that under normal operation, the angular velocity of synchronous machines is equal to the synchronous angular velocity ω_e. That is, $\omega_r = \omega_e$, and hence $\omega = \omega_r = \omega_e$. Therefore, assuming that the initial conditions are zero, the angular displacement of the rotor θ_r and the angular displacement of the *quadrature* magnetic axis θ are equal, and $\theta_r = \theta = \int_{t_0}^{t} \omega_r(\tau)d\tau = \int_{t_0}^{t} \omega(\tau)d\tau$.

It is easy to see that the magnetizing reluctance $\Re_m$ is a function of the rotor angular displacement θ_r. Using the number of turns N_s, the magnetizing inductance is found as

$$L_m(\theta_r) = \frac{N_s^2}{\Re_m(\theta_r)}$$

This magnetizing inductance has minimum and maximum values. In particular, depending on θ_r, we have

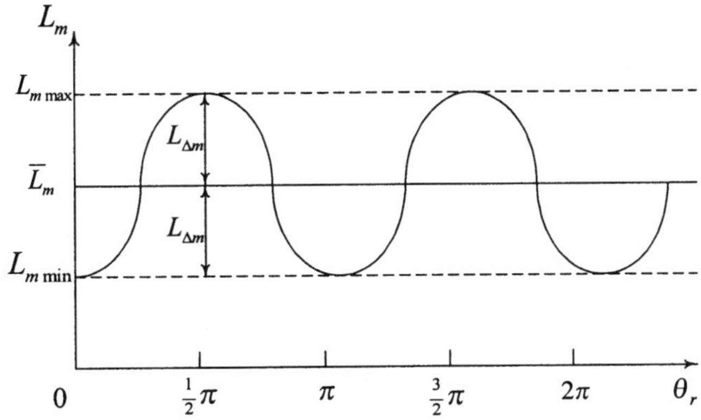

Figure 7.4 Plot of the magnetizing inductance $L_m(\theta_r)$ assuming that the initial rotor displacement is $\theta_{r0} = 0, \pi, 2\pi, \ldots$

$$L_{m\min} = \left.\frac{N_s^2}{\Re_{m\max}(\theta_r)}\right|_{\theta_r=0,\pi,2\pi,\ldots} \quad and \quad L_{m\max} = \left.\frac{N_s^2}{\Re_{m\min}(\theta_r)}\right|_{\theta_r=\frac{1}{2}\pi,\frac{3}{2}\pi,\frac{5}{2}\pi,\ldots}$$

The magnetizing inductance varies twice per one revolution of the rotor. Assuming that this variation is a sinusoidal function of the rotor angular displacement, which is proportional to the electrical angular displacement θ_r (for two-pole machines, $\theta_r = \theta_{rm}$), we have

$$L_m(\theta_r) = \bar{L}_m - L_{\Delta m}\cos 2\theta_r$$

and the plot for $L_m(\theta_r)$ is documented in Figure 7.4. Here, $\bar{L}_m$ is the average value of the magnetizing inductance, and $L_{\Delta m}$ is the half-amplitude of the sinusoidal variation of the magnetizing inductance.

The electromagnetic torque developed by an elementary reluctance motor is found using the coenergy $W_c(i_{as}, \theta_r)$. Taking note that

$$W_c(i_{as}, \theta_r) = \frac{1}{2}(L_{ls} + \bar{L}_m - L_{\Delta m}\cos 2\theta_r)i_{as}^2$$

we have

$$T_e = \frac{\partial W_c(i_{as}, \theta_r)}{\partial \theta_r} = \frac{\partial\left(\frac{1}{2}i_{as}^2(L_{ls} + \bar{L}_m - L_{\Delta m}\cos 2\theta_r)\right)}{\partial \theta_r} = L_{\Delta m} i_{as}^2 \sin 2\theta_r$$

This formula indicates a very important feature. One concludes that the electromagnetic torque is not developed by synchronous reluctance motors if one feeds the dc current or voltage to the motor winding. In fact, the average value of T_e is not equal to zero if and only if the current is a function of θ_r. For example, if one feeds the following current to the motor winding

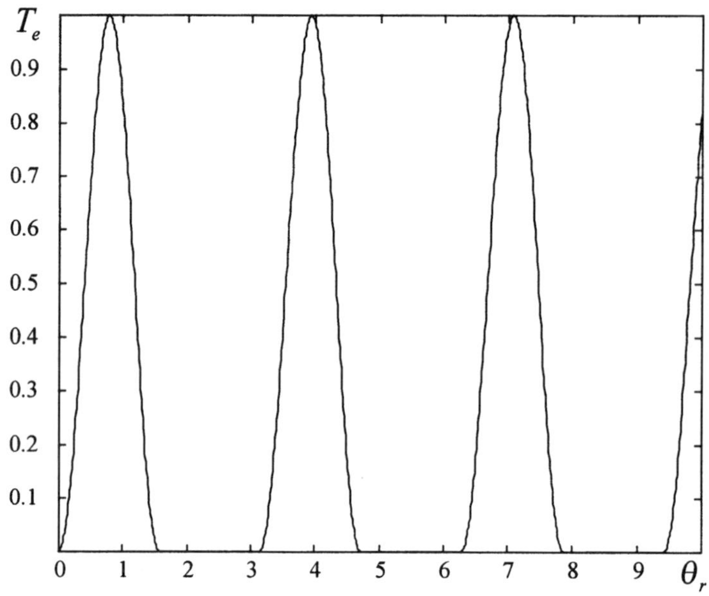

Figure 7.5 Waveforms for the electromagnetic torque developed by an elementary reluctance motor if $L_{\Delta m} = 1$ and $i_{as} = \mathrm{Re}\left(\sqrt{\sin 2\theta_r}\right)$.

$$i_{as} = i_M \,\mathrm{Re}\left(\sqrt{\sin 2\theta_r}\right)$$

one has

$$T_e = L_{\Delta m} i_{as}^2 \sin 2\theta_r = L_{\Delta m} i_M^2 \left(\mathrm{Re}\sqrt{\sin 2\theta_r}\right)^2 \sin 2\theta_r \neq 0$$

and

$$T_{eav} = \frac{1}{\pi}\int_0^{\pi} L_{\Delta m} i_{as}^2 \sin 2\theta_r d\theta_r = \frac{1}{4} L_{\Delta m} i_M^2$$

The electromagnetic torque developed can be plotted as a function of the rotor displacement. Assigning $L_{\Delta m} = 1$, $i_M = 1$, and taking note that $i_{as} = i_M \,\mathrm{Re}\left(\sqrt{\sin 2\theta_r}\right)$, we type in the MATLAB Command Window:

```
» th=0:0.01:10;  ias=real(sqrt(sin(2*th)));Te=ias.*ias.*sin(2*th);
plot(th,Te)
```

Pressing the Enter key, the plot for $T_e(\theta_r)$ results, and the average electromagnetic torque is 0.25 (see Figure 7.5).

The mathematical model of the reluctance motor studied is found using Kirchhoff's voltage law

$$u_{as} = r_s i_{as} + \frac{d\psi_{as}}{dt}, \quad \psi_{as} = (L_{ls} + \bar{L}_m - L_{\Delta m}\cos 2\theta_r)i_{as}$$

and Newton's second law

$$T_e - B_m\omega_r - L_L = J\frac{d^2\theta_r}{dt^2}$$

In particular, the following highly nonlinear differential equations result:

$$\frac{di_{as}}{dt} = -\frac{r_s}{L_{ls} + \bar{L}_m - L_{\Delta m}\cos 2\theta_r} i_{as} - \frac{2L_{\Delta m}}{L_{ls} + \bar{L}_m - L_{\Delta m}\cos 2\theta_r} i_{as}\omega_r \sin 2\theta_r + \frac{1}{L_{ls} + \bar{L}_m - L_{\Delta m}\cos 2\theta_r} u_{as}$$

$$\frac{d\omega_r}{dt} = \frac{1}{J}(L_{\Delta m} i_{as}^2 \sin 2\theta_r - B_m\omega_r - T_L)$$

$$\frac{d\theta_r}{dt} = \omega_r$$

Three-phase synchronous reluctance motors

Having performed the analysis of a single-phase reluctance motor, one can now concentrate on three-phase synchronous reluctance machines, as shown in Figures 7.2a and 7.2b. Using Kirchhoff's second law for the stator windings (see Figure 7.2a), three differential equations result

$$u_{as} = r_s i_{as} + \frac{d\psi_{as}}{dt}$$
$$u_{bs} = r_s i_{bs} + \frac{d\psi_{bs}}{dt} \tag{7.1}$$
$$u_{cs} = r_s i_{cs} + \frac{d\psi_{cs}}{dt}$$

The flux linkage equations are found to be

$$\psi_{as} = L_{asas} i_{as} + L_{asbs} i_{bs} + L_{ascs} i_{cs}$$
$$\psi_{bs} = L_{bsas} i_{as} + L_{bsbs} i_{bs} + L_{bscs} i_{cs}$$
$$\psi_{cs} = L_{csas} i_{as} + L_{csbs} i_{bs} + L_{cscs} i_{cs}$$

The self-inductances of the stator windings L_{asas}, L_{bsbs}, L_{cscs} and the mutual inductances between the stator windings L_{asbs}, L_{ascs}, L_{bsas}, L_{bscs}, L_{csas}, L_{csbs} are functions of the electrical angular displacement θ_r. The rotor has different magnetizing reluctances $\Re_{mq}$ and $\Re_{md}$ in the *quadrature* and *direct* axes, respectively. Therefore, the motor inductances are functions of the rotor angular displacement θ_{rm}. It should be emphasized that the electrical and rotor angular displacements are related. Using the number of poles P, we have

$$\theta_r = \frac{P}{2}\theta_{rm}$$

The stator windings, which are displaced by 120 electrical degrees, are identical and sinusoidally distributed. One can now examine the self-inductance L_{asas}. The expression for the self-inductance L_{asas} should be given as a function of the angular displacement θ_r. The *quadrature* and *direct* magnetic axes, fixed with the rotor, rotate with angular velocity ω, and $\omega = \omega_r$. That is, since the rotor of the two-pole machine studied rotates with synchronous angular velocity ω_e, one obtains $\omega = \omega_r = \omega_e$. Hence, assuming that $\theta_0 = \theta_{r0}$, we have $\theta = \theta_r$.

If $\theta_r = 0$, the *quadrature* axis (high-reluctance path with $\Re_{mq}$) is aligned with the *as* magnetic axis of the *as* winding, and one obtains

$$L_{asas}\Big|_{\theta_r=0} = L_{ls} + L_{mq}$$

For $\theta_r = \frac{1}{2}\pi$, the *direct* axis (low-reluctance path with $\Re_{md}$) is aligned with the *as* magnetic axis, and the following expression for the self-inductance L_{asas} results

$$L_{asas}\Big|_{\theta_r=\frac{1}{2}\pi} = L_{ls} + L_{md}$$

Equations for the magnetizing *quadrature* and *direct* inductances are

$$L_{mq} = \frac{N_s^2}{\Re_{mq}} \text{ and } L_{md} = \frac{N_s^2}{\Re_{md}}$$

The *quadrature* and *direct* magnetizing reluctances are different due to the laminated rotor design, and

$$\Re_{mq} > \Re_{md}$$

Hence, we have $L_{mq} < L_{md}$.

The minimum value of L_{asas} occurs periodically at $\theta_r = 0, \pi, 2\pi, \ldots$ The maximum value of L_{asas} occurs at $\theta_r = \frac{1}{2}\pi, \frac{3}{2}\pi, \frac{5}{2}\pi, \ldots$ One concludes that the self-inductance $L_{asas}(\theta_r)$, which is bounded as

$$L_{ls} + L_{mq} \le L_{asas} \le L_{ls} + L_{md}$$

is a periodic function of θ_r. Assuming that $L_{asas}(\theta_r)$ varies as a sine function with a dc component, we have

$$L_{asas} = L_{ls} + \bar{L}_m - L_{\Delta m}\cos 2\theta_r$$

where $\bar{L}_m$ is the average value of the magnetizing inductance; and $L_{\Delta m}$ is the half-amplitude of the sinusoidal variation of the magnetizing inductance.

The relationships between L_{mq}, L_{md}, and $\bar{L}_m$, $L_{\Delta m}$ must be found. For a single-phase reluctance machine, shown in Figure 7.3, the following results

$$L_{m\max} = L_{md} = \bar{L}_m + L_{\Delta m} \text{ and } L_{m\min} = L_{mq} = \bar{L}_m - L_{\Delta m}$$

For three-phase synchronous reluctance motors, one obtains

$$L_{mq} = \frac{3}{2}(\bar{L}_m - L_{\Delta m}) \text{ and } L_{md} = \frac{3}{2}(\bar{L}_m + L_{\Delta m})$$

Therefore,

$$\bar{L}_m = \frac{1}{3}(L_{mq} + L_{md}) \text{ and } L_{\Delta m} = \frac{1}{3}(L_{md} - L_{mq})$$

Using the expressions for L_{mq} and L_{md}, and taking note that $L_{mq} = \dfrac{N_s^2}{\Re_{mq}}$ and $L_{md} = \dfrac{N_s^2}{\Re_{md}}$, we have

$$\bar{L}_m = \frac{1}{3}\left(\frac{N_s^2}{\Re_{mq}} + \frac{N_s^2}{\Re_{md}}\right) \text{ and } L_{\Delta m} = \frac{1}{3}\left(\frac{N_s^2}{\Re_{md}} - \frac{N_s^2}{\Re_{mq}}\right)$$

Having derived the equation for L_{asas}, as given by

$$L_{asas} = L_{ls} + \bar{L}_m - L_{\Delta m}\cos(2\theta_r),\ \bar{L}_m = \frac{1}{3}(L_{mq} + L_{md}),\ L_{\Delta m} = \frac{1}{3}(L_{md} - L_{mq})$$

the self-inductances L_{bsbs} and L_{cscs} can be found. The *bs* and *cs* magnetic axes have 120 and 240 electrical degrees lag from the *as* magnetic axis. Therefore, we have

$$L_{bsbs} = L_{ls} + \bar{L}_m - L_{\Delta m}\cos 2\left(\theta_r - \frac{2}{3}\pi\right)$$

$$L_{cscs} = L_{ls} + \bar{L}_m - L_{\Delta m}\cos 2\left(\theta_r + \frac{2}{3}\pi\right)$$

The mutual inductances between sinusoidally distributed stator windings L_{asbs}, L_{ascs}, L_{bsas}, L_{bscs}, L_{csas}, and L_{csbs} are periodic functions of θ_r and have the average values (dc component). Assuming that the magnetic field is uniform, and making use of the fact that the magnetic axes are displaced by $\frac{2}{3}\pi$, one concludes that the dc component of L_{asbs}, L_{ascs}, L_{bsas}, L_{bscs}, L_{csas}, and L_{csbs} is

$$\bar{L}_{asbs} = \bar{L}_{ascs} = \bar{L}_{bsas} = \bar{L}_{bscs} = \bar{L}_{csas} = \bar{L}_{csbs} = \bar{L}_m \cos\left(\frac{2}{3}\pi\right) = -\frac{1}{2}\bar{L}_m$$

Considering the *as* and *bs* stator windings, maximum coupling between the windings occurs at $\theta_r = \frac{1}{3}\pi, \frac{4}{3}\pi, \frac{7}{3}\pi, \ldots,$ and minimum coupling results at $\theta_r = \frac{5}{6}\pi, \frac{11}{6}\pi, \frac{17}{6}\pi, \ldots$ That is, the variation can be expressed as a periodic function, and we have

$$L_{asbs} = L_{bsas} = -\frac{1}{2}\bar{L}_m - L_{\Delta m}\cos 2\left(\theta_r - \frac{1}{3}\pi\right)$$

For *as* and *cs* stator windings, maximum and minimum couplings occur at $\theta_r = \pi, 2\pi, 3\pi, \ldots$ and $\theta_r = \frac{1}{2}\pi, \frac{3}{2}\pi, \frac{3}{2}\pi, \ldots,$ respectively. Therefore,

$$L_{ascs} = L_{csas} = -\frac{1}{2}\bar{L}_m - L_{\Delta m}\cos 2\left(\theta_r + \frac{1}{3}\pi\right)$$

For *cs* and *bs* stator windings, maximum coupling occurs when $\theta_r = -\frac{1}{3}\pi, \frac{2}{3}\pi, \frac{5}{3}\pi, \ldots,$ and minimum coupling occurs at $\theta_r = \frac{1}{6}\pi, \frac{7}{6}\pi, \frac{13}{6}\pi, \ldots$ Hence,

$$L_{bscs} = L_{csbs} = -\frac{1}{2}\bar{L}_m - L_{\Delta m}\cos 2(\theta_r + \pi)$$

Therefore, having derived the explicit expressions for self- and mutual inductances, expanded equations for the flux linkages are found as

$$\psi_{as} = (L_{ls} + \bar{L}_m - L_{\Delta m}\cos 2\theta_r)i_{as} + \left(-\frac{1}{2}\bar{L}_m - L_{\Delta m}\cos 2\left(\theta_r - \frac{1}{3}\pi\right)\right)i_{bs} + \left(-\frac{1}{2}\bar{L}_m - L_{\Delta m}\cos 2\left(\theta_r + \frac{1}{3}\pi\right)\right)i_{cs}$$

$$\psi_{bs} = \left(-\frac{1}{2}\bar{L}_m - L_{\Delta m}\cos 2\left(\theta_r - \frac{1}{3}\pi\right)\right)i_{as} + \left(L_{ls} + \bar{L}_m - L_{\Delta m}\cos 2\left(\theta_r - \frac{2}{3}\pi\right)\right)i_{bs} + \left(-\frac{1}{2}\bar{L}_m - L_{\Delta m}\cos 2\theta_r\right)i_{cs}$$

$$\psi_{cs} = \left(-\frac{1}{2}\bar{L}_m - L_{\Delta m}\cos 2\left(\theta_r + \frac{1}{3}\pi\right)\right)i_{as} + \left(-\frac{1}{2}\bar{L}_m - L_{\Delta m}\cos 2\theta_r\right)i_{bs} + \left(L_{ls} + \bar{L}_m - L_{\Delta m}\cos 2\left(\theta_r + \frac{2}{3}\pi\right)\right)i_{cs}$$

In matrix form, we have

$$\boldsymbol{\psi}_{abcs} = \mathbf{L}_s \mathbf{i}_{absc} =$$

$$\begin{bmatrix} L_{ls} + \bar{L}_m - L_{\Delta m}\cos 2\theta_r & -\frac{1}{2}\bar{L}_m - L_{\Delta m}\cos 2\left(\theta_r - \frac{1}{3}\pi\right) & -\frac{1}{2}\bar{L}_m - L_{\Delta m}\cos 2\left(\theta_r + \frac{1}{3}\pi\right) \\ -\frac{1}{2}\bar{L}_m - L_{\Delta m}\cos 2\left(\theta_r - \frac{1}{3}\pi\right) & L_{ls} + \bar{L}_m - L_{\Delta m}\cos\left(\theta_r - \frac{2}{3}\pi\right) & -\frac{1}{2}\bar{L}_m - L_{\Delta m}\cos 2\theta_r \\ -\frac{1}{2}\bar{L}_m - L_{\Delta m}\cos\left(\theta_r + \frac{1}{3}\pi\right) & -\frac{1}{2}\bar{L}_m - L_{\Delta m}\cos 2\theta_r & L_{ls} + \bar{L}_m - L_{\Delta m}\cos\left(\theta_r + \frac{2}{3}\pi\right) \end{bmatrix}$$

$$\times \begin{bmatrix} i_{as} \\ i_{bs} \\ i_{cs} \end{bmatrix} \tag{7.2}$$

where the matrix $\mathbf{L}_s$ is

$$\mathbf{L}_s =$$

$$\begin{bmatrix} L_{ls} + \bar{L}_m - L_{\Delta m}\cos 2\theta_r & -\frac{1}{2}\bar{L}_m - L_{\Delta m}\cos 2\left(\theta_r - \frac{1}{3}\pi\right) & -\frac{1}{2}\bar{L}_m - L_{\Delta m}\cos 2\left(\theta_r + \frac{1}{3}\pi\right) \\ -\frac{1}{2}\bar{L}_m - L_{\Delta m}\cos 2\left(\theta_r - \frac{1}{3}\pi\right) & L_{ls} + \bar{L}_m - L_{\Delta m}\cos 2\left(\theta_r - \frac{2}{3}\pi\right) & -\frac{1}{2}\bar{L}_m - L_{\Delta m}\cos 2\theta_r \\ -\frac{1}{2}\bar{L}_m - L_{\Delta m}\cos 2\left(\theta_r + \frac{1}{3}\pi\right) & -\frac{1}{2}\bar{L}_m - L_{\Delta m}\cos 2\theta_r & L_{ls} + \bar{L}_m - L_{\Delta m}\cos 2\left(\theta_r + \frac{2}{3}\pi\right) \end{bmatrix}$$

Using (7.1) and (7.2), one finds

$$\mathbf{u}_{abcs} = \mathbf{r}_s \mathbf{i}_{abcs} + \frac{d\boldsymbol{\psi}_{abcs}}{dt} = \mathbf{r}_s \mathbf{i}_{abcs} + \frac{d(\mathbf{L}_s \mathbf{i}_{abcs})}{dt} \tag{7.3}$$

where $\boldsymbol{\psi}_{abcs}$ is the vector of the flux linkages, $\boldsymbol{\psi}_{abcs} = \mathbf{L}_s \mathbf{i}_{abcs}$; $\mathbf{r}_s$ is the matrix of the stator resistances, and $\mathbf{r}_s = \begin{bmatrix} r_s & 0 & 0 \\ 0 & r_s & 0 \\ 0 & 0 & r_s \end{bmatrix}$.

The *torsional-mechanical* dynamics must be studied. Using Newton's second law, we have the following differential equations for P-pole electric motors

$$T_e - \frac{2B_m}{P}\omega_r - T_L = \frac{2J}{P}\frac{d\omega_r}{dt}$$
$$\frac{d\theta_r}{dt} = \omega_r \tag{7.4}$$

To find the developed electromagnetic torque T_e, the coenergy concept is used. As the expression for the coenergy W_c is derived

$$W_c = \frac{1}{2}[i_{as} \quad i_{bs} \quad i_{cs}]\mathbf{L}_s \begin{bmatrix} i_{as} \\ i_{bs} \\ i_{cs} \end{bmatrix}$$

$$= \frac{1}{2}[i_{as} \quad i_{bs} \quad i_{cs}]$$

$$\times \begin{bmatrix} L_{ls} + \bar{L}_m - L_{\Delta m}\cos 2\theta_r & -\frac{1}{2}\bar{L}_m - L_{\Delta m}\cos 2\left(\theta_r - \frac{1}{3}\pi\right) & -\frac{1}{2}\bar{L}_m - L_{\Delta m}\cos 2\left(\theta_r + \frac{1}{3}\pi\right) \\ -\frac{1}{2}\bar{L}_m - L_{\Delta m}\cos 2\left(\theta_r - \frac{1}{3}\pi\right) & L_{ls} + \bar{L}_m - L_{\Delta m}\cos 2\left(\theta_r - \frac{2}{3}\pi\right) & -\frac{1}{2}\bar{L}_m - L_{\Delta m}\cos 2\theta_r \\ -\frac{1}{2}\bar{L}_m - L_{\Delta m}\cos 2\left(\theta_r + \frac{1}{3}\pi\right) & -\frac{1}{2}\bar{L}_m - L_{\Delta m}\cos 2\theta_r & L_{ls} + \bar{L}_m - L_{\Delta m}\cos 2\left(\theta_r + \frac{2}{3}\pi\right) \end{bmatrix}$$

$$\times \begin{bmatrix} i_{as} \\ i_{bs} \\ i_{cs} \end{bmatrix}$$

the formula to calculate the electromagnetic torque results, and

$$T_e = \frac{P}{2}\frac{\partial W_c}{\partial \theta_r} = \frac{P}{2}\frac{1}{2}[i_{as} \quad i_{bs} \quad i_{cs}]$$

$$\times \begin{bmatrix} 2L_{\Delta m}\sin 2\theta_r & 2L_{\Delta m}\sin 2\left(\theta_r - \frac{1}{3}\pi\right) & 2L_{\Delta m}\sin 2\left(\theta_r + \frac{1}{3}\pi\right) \\ 2L_{\Delta m}\sin 2\left(\theta_r - \frac{1}{3}\pi\right) & 2L_{\Delta m}\sin 2\left(\theta_r - \frac{2}{3}\pi\right) & 2L_{\Delta m}\sin 2\theta_r \\ 2L_{\Delta m}\sin 2\left(\theta_r + \frac{1}{3}\pi\right) & 2L_{\Delta m}\sin 2\theta_r & 2L_{\Delta m}\sin 2\left(\theta_r + \frac{2}{3}\pi\right) \end{bmatrix} \begin{bmatrix} i_{as} \\ i_{bs} \\ i_{cs} \end{bmatrix}$$

The electromagnetic torque developed is found to be

$$T_e = \frac{P}{2}L_{\Delta m}\left[i_{as}^2 \sin 2\theta_r + 2i_{as}i_{bs}\sin 2\left(\theta_r - \frac{1}{3}\pi\right) + 2i_{as}i_{cs}\sin 2\left(\theta_r + \frac{1}{3}\pi\right) + i_{bs}^2\sin 2\left(\theta_r - \frac{2}{3}\pi\right)\right.$$
$$\left. + 2i_{bs}i_{cs}\sin 2\theta_r + i_{cs}^2 \sin 2\left(\theta_r + \frac{2}{3}\pi\right)\right] \tag{7.5}$$

Using the trigonometric identities and recalling that

$$L_{\Delta m} = \frac{1}{3}(L_{md} - L_{mq}),\ L_{mq} = \frac{3}{2}(\bar{L}_m - L_{\Delta m}),\ \text{and } L_{md} = \frac{3}{2}(\bar{L}_m + L_{\Delta m})$$

we derive the following alternative formula for the developed electromagnetic torque

$$T_e = \frac{P(L_{md} - L_{mq})}{6}\left[\left(i_{as}^2 - \frac{1}{2}i_{bs}^2 - \frac{1}{2}i_{cs}^2 - i_{as}i_{bs} - i_{as}i_{cs} + 2i_{bs}i_{cs}\right)\sin 2\theta_r\right.$$
$$\left. + \frac{\sqrt{3}}{2}(i_{bs}^2 - i_{cs}^2 - 2i_{as}i_{bs} + 2i_{as}i_{cs})\cos 2\theta_r\right]$$

To perform the analysis, optimization, and design, the mathematical model of synchronous reluctance motors must be found in Cauchy's form. Otherwise, it is difficult to approach the design; modeling can be performed using SIMULINK. The resulting differential equations for synchronous reluctance motors in the *abc* variables are found using Kirchhoff's second law, which in matrix form is given by

$$\mathbf{u}_{abcs} = \mathbf{r}_s \mathbf{i}_{abcs} + \frac{d\boldsymbol{\psi}_{abcs}}{dt}, \quad \boldsymbol{\psi}_{abcs} = \mathbf{L}_s \mathbf{i}_{abcs}$$

and Newton's second law of motion

$$T_e - \frac{2B_m}{P}\omega_r - T_L = \frac{2J}{P}\frac{d\omega_r}{dt}$$

$$\frac{d\theta_r}{dt} = \omega_r$$

Hence, from (7.2), (7.3), and (7.4), we have

$$\frac{di_{as}}{dt} = \frac{1}{L_{ls} + \bar{L}_m - L_{\Delta m}\cos 2\theta_r}\left[-r_s i_{as} + u_{as} + \left(\frac{1}{2}\bar{L}_m + L_{\Delta m}\cos 2\left(\theta_r - \frac{1}{3}\pi\right)\right)\frac{di_{bs}}{dt} + \left(\frac{1}{2}\bar{L}_m + L_{\Delta m}\cos 2\left(\theta_r + \frac{1}{3}\pi\right)\right)\frac{di_{cs}}{dt}\right.$$
$$\left. 2L_{\Delta m}\omega_r\left(i_{as}\sin 2\theta_r + i_{bs}\sin 2\left(\theta_r - \frac{1}{3}\pi\right) + i_{cs}\sin 2\left(\theta_r + \frac{1}{3}\pi\right)\right)\right]$$

$$\frac{di_{bs}}{dt} = \frac{1}{L_{ls} + \bar{L}_m - L_{\Delta m}\cos 2\left(\theta_r - \frac{2}{3}\pi\right)}\left[-r_s i_{bs} + u_{bs} + \left(\frac{1}{2}\bar{L}_m + L_{\Delta m}\cos 2\left(\theta_r - \frac{1}{3}\pi\right)\right)\frac{di_{as}}{dt} + \left(\frac{1}{2}\bar{L}_m + L_{\Delta m}\cos 2\theta_r\right)\frac{di_{cs}}{dt}\right.$$
$$\left. -2L_{\Delta m}\omega_r\left(i_{as}\sin 2\left(\theta_r - \frac{1}{3}\pi\right) + i_{bs}\sin 2\left(\theta_r - \frac{2}{3}\pi\right) + i_{cs}\sin 2\theta_r\right)\right]$$

$$\frac{di_{cs}}{dt} = \frac{1}{L_{ls} + \bar{L}_m - L_{\Delta m}\cos 2\left(\theta_r + \frac{2}{3}\pi\right)}\left[-r_s i_{cs} + u_{cs} + \left(\frac{1}{2}\bar{L}_m + L_{\Delta m}\cos 2\left(\theta_r + \frac{1}{3}\pi\right)\right)\frac{di_{as}}{dt} + \left(\frac{1}{2}\bar{L}_m + L_{\Delta m}\cos 2\theta_r\right)\frac{di_{bs}}{dt}\right.$$
$$\left. -2L_{\Delta m}\omega_r\left(i_{as}\sin 2\left(\theta_r + \frac{1}{3}\pi\right) + i_{bs}\sin 2\theta_r + i_{cs}\sin 2\left(\theta_r + \frac{2}{3}\pi\right)\right)\right]$$

$$\frac{d\omega_r}{dt} = \frac{P^2}{4J}L_{\Delta m}\left[i_{as}^2\sin 2\theta_r + 2i_{as}i_{bs}\sin 2\left(\theta_r - \frac{1}{3}\pi\right) + 2i_{as}i_{cs}\sin 2\left(\theta_r + \frac{1}{3}\pi\right) + i_{bs}^2\sin 2\left(\theta_r - \frac{2}{3}\pi\right)\right.$$
$$\left. +2i_{bs}i_{cs}\sin 2\theta_r + i_{cs}^2\sin 2\left(\theta_r + \frac{2}{3}\pi\right)\right] - \frac{B_m}{J}\omega_r - \frac{P}{2J}T_L$$

$$\frac{d\theta_r}{dt} = \omega_r$$

In the MATLAB environment, by using SYMBOLIC MATH TOOLBOX, an m-file was developed to find a mathematical model for nonlinear circuitry dynamics in Cauchy's form. We use the following notations `Lbm` = $\bar{L}_m$, `Ldm` = $L_{\Delta m}$, `Lls` = L_{ls}, `rs` = r_s, `Bm` = B_m, `S1` = $\sin 2\theta_r$, `S2` = $\sin 2\left(\theta_r - \frac{1}{3}\pi\right)$, `S3` = $\sin 2\left(\theta_r + \frac{1}{3}\pi\right)$, `S4` = $\sin 2\left(\theta_r - \frac{2}{3}\pi\right)$, `S5` = $\sin 2(\theta_r + \pi)$, `S6` =

$\sin 2\left(\theta_r + \frac{2}{3}\pi\right)$, and C1 = $\cos 2\theta_r$, C2 = $\cos 2\left(\theta_r - \frac{1}{3}\pi\right)$, C3 = $\cos 2\left(\theta_r + \frac{1}{3}\pi\right)$, C4 = $\cos 2\left(\theta_r - \frac{2}{3}\pi\right)$, C5 = $\cos 2(\theta_r + \pi)$, C6 = $\cos 2\left(\theta_r + \frac{2}{3}\pi\right)$.

MATLAB script

```
L=sym('[Lls+Lbm-Ldm*C1,-Lbm/2-Ldm*C2,-Lbm/2-Ldm*C3,0;
     Lbm/2-Ldm*C2,Lls+Lbm-Ldm*C4,-Lbm/2-Ldm*C5,0;
     Lbm/2-Ldm*C3,-Lbm/2-Ldm*C5,Lls+Lbm-Ldm*C6,0;
     0,0,0,2*J/P]');
R=sym('[-rs 0 0 0;0 -rs 0 0;0 0 -rs 0;0 0 0 -Bm*(2/P)]');
I=sym('[Ias; Ibs; Ics; Wr]');
V=sym('[vas;vbs;vcs;-TL]');
K=sym('[Ldm*2*Wr*(S1*Ias+S2*Ibs+S3*Ics);Ldm*2*Wr*(S2*Ias+S4Ibs+
S5*Ics);
     Ldm*2*Wr*(S3*Ias+S5*Ibs+S6*Ics);Te]');
L1=inv(L); L2=simplify(L1)
FS1=L2*R*I; FS2=simplify(FS1)
FS3=L2*V; FS4=simplify(FS3)
FS5=L2*K; FS6=simplify(FS5)
FS7=FS2+FS4-FS6; FS=simplify(FS7)
disp('END')
```

Using the analytic results derived by running the MATLAB file and trigonometric identities, the following nonlinear differential equations are obtained to model the dynamics of synchronous reluctance motors in the *abc* variables:

$$\frac{di_{as}}{dt} = \frac{1}{L_{ls}(9L_{\Delta m}^2 - 4L_{ls}^2 - 12\bar{L}_m L_{ls} - 9\bar{L}_m^2} \Big[(r_s i_{as} - u_{as})(4L_{ls}^2 + 3\bar{L}_m^2 - 3L_{\Delta m}^2 + 8\bar{L}_m L_{ls} - 4L_{ls}L_{\Delta m}\cos 2\theta_r)$$

$$+ (r_s i_{bs} - u_{bs})\left(3\bar{L}_m^2 - 3L_{\Delta m}^2 + 2\bar{L}_m L_{ls} + 4L_{ls}L_{\Delta m}\cos 2\left(\theta_r - \frac{1}{3}\pi\right)\right) + (r_s i_{cs} - u_{cs})\left(3\bar{L}_m^2 - 3L_{\Delta m}^2 + 2\bar{L}_m L_{ls} + 4L_{ls}L_{\Delta m}\cos 2\left(\theta_r + \frac{1}{3}\pi\right)\right)$$

$$+6\sqrt{3}L_{\Delta m}^2 L_{ls}\omega_r(i_{cs} - i_{bs}) + (8L_{\Delta m}L_{ls}^2\omega_r + 12L_{\Delta m}\bar{L}_m L_{ls}\omega_r)\left(\sin 2\theta_r i_{as} + \sin 2\left(\theta_r - \frac{1}{3}\pi\right)i_{bs} + \sin 2\left(\theta_r + \frac{1}{3}\pi\right)i_{cs}\right)\Big]$$

$$\frac{di_{bs}}{dt} = \frac{1}{L_{ls}(9L_{\Delta m}^2 - 4L_{ls}^2 - 12\bar{L}_m L_{ls} - 9\bar{L}_m^2} \Big[(r_s i_{as} - u_{as})\left(3\bar{L}_m^2 - 3L_{\Delta m}^2 + 2\bar{L}_m L_{ls} + 4L_{ls}L_{\Delta m}\cos 2\left(\theta_r - \frac{1}{3}\pi\right)\right)$$

$$+ (r_s i_{bs} - u_{bs})\left(4L_{ls}^2 + 3\bar{L}_m^2 - 3L_{\Delta m}^2 + 8\bar{L}_m L_{ls} - 4L_{ls}L_{\Delta m}\cos 2\left(\theta_r + \frac{1}{3}\pi\right)\right) + (r_s i_{cs} - u_{cs})(3\bar{L}_m^2 - 3L_{\Delta m}^2 + 2\bar{L}_m L_{ls} + 4L_{ls}L_{\Delta m}\cos 2\theta_r)$$

$$+ 6\sqrt{3}L_{\Delta m}^2 L_{ls}\omega_r(i_{as} - i_{cs}) + (8L_{\Delta m}L_{ls}^2\omega_r + 12L_{\Delta m}\bar{L}_m L_{ls}\omega_r)\left(\sin 2\left(\theta_r - \frac{1}{3}\pi\right)i_{as} + \sin 2\left(\theta_r + \frac{1}{3}\pi\right)i_{bs} + \sin 2\theta_r i_{cs}\right)\Big]$$

$$\frac{di_{cs}}{dt} = \frac{1}{L_{ls}(9L_{\Delta m}^2 - 4L_{ls}^2 - 12\bar{L}_m L_{ls} - 9\bar{L}_m^2} \Big[(r_s i_{as} - u_{as})\left(3\bar{L}_m^2 - 3L_{\Delta m}^2 + 2\bar{L}_m L_{ls} + 4L_{ls}L_{\Delta m}\cos 2\left(\theta_r + \frac{1}{3}\pi\right)\right)$$

$$+ (r_s i_{bs} - u_{bs})(3\bar{L}_m^2 - 3L_{\Delta m}^2 + 2\bar{L}_m L_{ls} + 4L_{ls}L_{\Delta m}\cos 2\theta_r) + (r_s i_{cs} - u_{cs})\left(4L_{ls}^2 + 3\bar{L}_m^2 - 3L_{\Delta m}^2 + 8\bar{L}_m L_{ls} - 4L_{ls}L_{\Delta m}\cos 2\left(\theta_r - \frac{1}{3}\pi\right)\right)$$

$$+6\sqrt{3}L_{\Delta m}^2 L_{ls}\omega_r(i_{bs} - i_{as}) + (8L_{\Delta m}L_{ls}^2\omega_r + 12L_{\Delta m}\bar{L}_m L_{ls}\omega_r)\left(\sin 2\left(\theta_r + \frac{1}{3}\pi\right)i_{as} + \sin 2\theta_r i_{bs} + \sin 2\left(\theta_r - \frac{1}{3}\pi\right)i_{cs}\right)\Big]$$

$$\frac{d\omega_r}{dt} = \frac{P^2}{4J} L_{\Delta m} \left[i_{as}^2 \sin 2\theta_r + 2i_{as}i_{bs} \sin 2\left(\theta_r - \frac{1}{3}\pi\right) + 2i_{as}i_{cs} \sin 2\left(\theta_r + \frac{1}{3}\pi\right) + i_{bs}^2 \sin 2\left(\theta_r - \frac{2}{3}\pi\right) \right.$$

$$\left. + 2i_{bs}i_{cs} \sin 2\theta_r + i_{cs}^2 \sin 2\left(\theta_r + \frac{2}{3}\pi\right)\right] - \frac{B_m}{J}\omega_r - \frac{P}{2J}T_L$$

$$\frac{d\theta_r}{dt} = \omega_r \tag{7.6}$$

Hence, a set of five highly nonlinear differential equations in Cauchy's form is found, and these equations should be used in the analysis of dynamics, optimization, and control. In particular, the derived equations are employed to analyze the transient dynamics of synchronous reluctance motors.

Torque maximization problem

The expression for the developed electromagnetic torque was found; in particular,

$$T_e = \frac{P}{2} L_{\Delta m} \left[i_{as}^2 \sin 2\theta_r + 2i_{as}i_{bs} \sin 2\left(\theta_r - \frac{1}{3}\pi\right) + 2i_{as}i_{cs} \sin 2\left(\theta_r + \frac{1}{3}\pi\right) + i_{bs}^2 \sin 2\left(\theta_r - \frac{2}{3}\pi\right) \right.$$

$$\left. + 2i_{bs}i_{cs} \sin 2\theta_r + i_{cs}^2 \sin 2\left(\theta_r + \frac{2}{3}\pi\right)\right]$$

The goal is to maximize the electromagnetic torque produced in order to attain the optimum performance of the synchronous reluctance motors and ensure the specified steady-state and dynamic characteristics imposed in the full torque-speed envelope. Current- and voltage-fed converters are deployed to control synchronous reluctance motors. For an elementary single-phase reluctance motor, it was demonstrated that the current fed to the stator winding must be a function of the electrical angular displacement in order to develop the average electromagnetic torque. In particular, it was shown that if $i_{as} = i_M \operatorname{Re}\left(\sqrt{\sin 2\theta_r}\right)$, the T_e results (see Figure 7.5). For three-phase synchronous reluctance motors, the analysis of torque production is quite complex. From equation (7.5), one concludes that the electromagnetic torque T_e is a function of the *abc* phase currents i_{as}, i_{bs}, and i_{cs}, as well as the electrical angular displacement θ_r. Now our goal is to maximize the torque developed using the equation for T_e. Applying the trigonometric identities, one obtains the formulas for the currents to be fed to the *abc* windings in order to maximize the electromagnetic torque developed. A balanced three-phase sinusoidal current set that guarantees the maximum torque production is derived as

$$i_{as} = \sqrt{2}i_M \sin\left(\omega_r t + \frac{1}{3}\varphi_i\pi\right) = \sqrt{2}i_M \sin\left(\theta_r + \frac{1}{3}\varphi_i\pi\right)$$

$$i_{bs} = \sqrt{2}i_M \sin\left(\omega_r t - \frac{1}{3}(2-\varphi_i)\pi\right) = \sqrt{2}i_M \sin\left(\theta_r - \frac{1}{3}(2-\varphi_i)\pi\right) \tag{7.7}$$

$$i_{cs} = \sqrt{2}i_M \sin\left(\omega_r t + \frac{1}{3}(2+\varphi_i)\pi\right) = \sqrt{2}i_M \sin\left(\theta_r + \frac{1}{3}(2+\varphi_i)\pi\right)$$

where $\varphi_i = 0.3245$.

From (7.7), plugging a balanced three-phase current set, in which i_{as}, i_{bs}, and i_{cs} are the functions of θ_r, into the formula for electromagnetic torque (7.5), we have

$$T_e = PL_{\Delta m} i_M^2 \left[\sin^2\left(\theta_r + \frac{1}{3}\varphi_i\pi\right)\sin 2\theta_r + 2\sin\left(\theta_r + \frac{1}{3}\varphi_i\pi\right)\sin\left(\theta_r - \frac{1}{3}(2-\varphi_i)\pi\right)\sin 2\left(\theta_r - \frac{1}{3}\pi\right) \right.$$
$$+ 2\sin\left(\theta_r + \frac{1}{3}\varphi_i\pi\right)\sin\left(\theta_r + \frac{1}{3}(2+\varphi_i)\pi\right)\sin 2\left(\theta_r + \frac{1}{3}\pi\right) + \sin^2\left(\theta_r - \frac{1}{3}(2-\varphi_i)\pi\right)\sin 2\left(\theta_r - \frac{2}{3}\pi\right)$$
$$\left. + 2\sin\left(\theta_r - \frac{1}{3}(2-\varphi_i)\pi\right)\sin\left(\theta_r + \frac{1}{3}(2+\varphi_i)\pi\right)\sin 2\theta_r + \sin^2\left(\theta_r + \frac{1}{3}(2+\varphi_i)\pi\right)\sin 2\left(\theta_r + \frac{2}{3}\pi\right)\right]$$

That is,

$$T_e = \sqrt{2} PL_{\Delta m} i_M^2$$

For example, assigning $i_M = 1$ and $PL_{\Delta m} = 1$, we have $T_e = \sqrt{2}$.

A MATLAB m-file is developed to verify this result.

MATLAB script: Torque Maximization Problem

```
% Calculation of the Developed Electromagnetic Torque in the abc and
% qd0 variables: The Electromagnetic Torque Maximization Problem
th=0:0.01:10;                   % angular rotor displacement
phi=0.3245;                     % phase current angle
% balanced three-phase current set
Ia=sin(th+phi*pi/3);            % current fed to the as winding
Ib=sin(th-(2-phi)*pi/3);        % current fed to the bs winding
Ic=sin(th+(2+phi)*pi/3);        % current fed to the cs winding
% calculation of the electromagnetic torque developed
T=Ia.*(sin(2*th).*Ia+2*sin(2*th-2*pi/3).*Ib+2*sin(2*th+2*pi/3).*Ic)...
+Ib.*(sin(2*th-4*pi/3).*Ib+2*sin(2*th).*Ic)+Ic.*sin(2*th+4*pi/3).*Ic;
% plot the currents applied to abc windings
plot(th,Ia,'-',th,Ib,'--',th,Ic,'-.'); grid; pause;
% plot of the torque developed versus the angular displacement
plot(th,T); axis([0,10,0,1.5]); grid; pause;
disp 'End'
```

Using the balanced three-phase current set (7.7) and the MATLAB m-file developed, the steady-state *per-unit* electromagnetic torque T_e is calculated as a function of the electrical angular displacement θ_r (see Figure 7.6). The analysis shows that the maximum electromagnetic torque developed by three-phase synchronous reluctance motors fed by the following balanced three-phase current set

$$i_{as} = \sin\left(\theta_r + \frac{1}{3}\varphi_i\pi\right),\quad i_{bs} = \sin\left(\theta_r - \frac{1}{3}(2-\varphi_i)\pi\right),\quad i_{cs} = \sin\left(\theta_r + \frac{1}{3}(2+\varphi_i)\pi\right),\quad \varphi_i = 0.3245$$

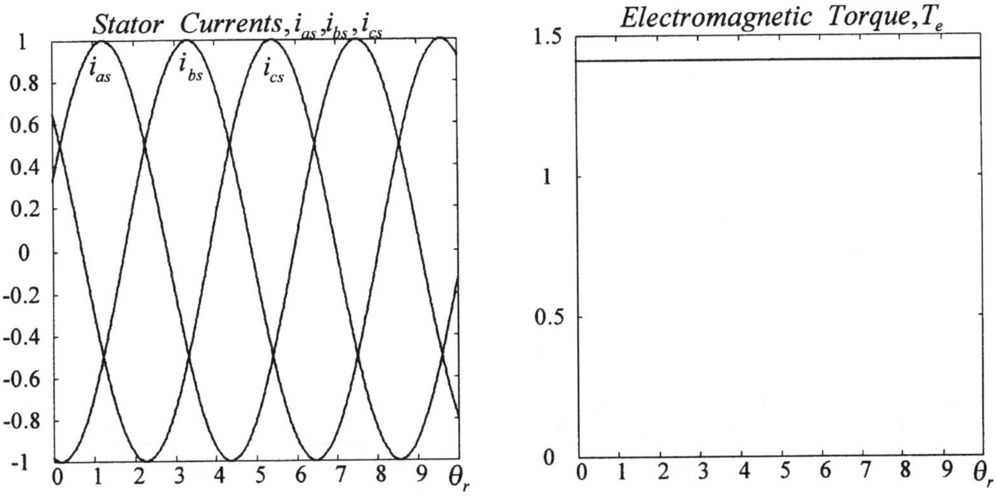

Figure 7.6 Phase currents and electromagnetic torque: *counterclockwise* rotation.

To reverse synchronous reluctance motors (which initially rotates in the *counterclockwise* direction because T_e is positive), the electromagnetic torque developed must be negative in order to attain the *clockwise* rotation. The maximization problem should be solved. To guarantee the *clockwise* rotation, the balanced three-phase current set to be implemented is

$$i_{as} = \sqrt{2}i_M \sin\left(\omega_r t + \frac{1}{3}\varphi_i\pi - \frac{1}{2}\pi\right) = \sqrt{2}i_M \sin\left(\theta_r + \frac{1}{3}\varphi_i\pi - \frac{1}{2}\pi\right)$$

$$i_{bs} = \sqrt{2}i_M \sin\left(\omega_r t - \frac{1}{3}(2-\varphi_i)\pi - \frac{1}{2}\pi\right) = \sqrt{2}i_M \sin\left(\theta_r - \frac{1}{3}(2-\varphi_i)\pi - \frac{1}{2}\pi\right)$$

$$i_{cs} = \sqrt{2}i_M \sin\left(\omega_r t + \frac{1}{3}(2+\varphi_i)\pi - \frac{1}{2}\pi\right) = \sqrt{2}i_M \sin\left(\theta_r + \frac{1}{3}(2+\varphi_i)\pi - \frac{1}{2}\pi\right)$$

$$\varphi_i = 0.3245$$

The resulting phase currents and torque waveforms are documented in Figure 7.7. It should be emphasized that negative electromagnetic torque is developed.

A balanced three-phase voltage set should be found to maximize the electromagnetic torque developed. The phase voltages to be applied to the stator windings are obtained using advanced phase shifting $\varphi_u\pi$. We have

$$u_{as} = \sqrt{2}u_M \sin\left(\omega_r t + \frac{1}{3}\varphi_i\pi + \varphi_u\pi\right) = \sqrt{2}u_M \sin\left(\theta_r + \frac{1}{3}\varphi_i\pi + \varphi_u\pi\right)$$

$$u_{bs} = \sqrt{2}u_M \sin\left(\omega_r t - \frac{1}{3}(2-\varphi_i)\pi + \varphi_u\pi\right) = \sqrt{2}u_M \sin\left(\theta_r - \frac{1}{3}(2-\varphi_i)\pi + \varphi_u\pi\right) \quad (7.8)$$

$$u_{cs} = \sqrt{2}u_M \sin\left(\omega_r t + \frac{1}{3}(2+\varphi_i)\pi + \varphi_u\pi\right) = \sqrt{2}u_M \sin\left(\theta_r + \frac{1}{3}(2+\varphi_i)\pi + \varphi_u\pi\right)$$

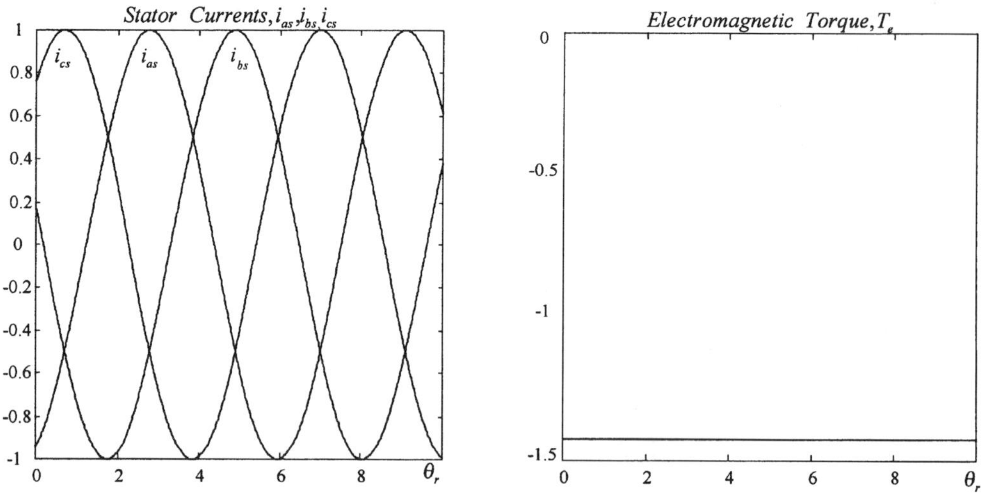

Figure 7.7 Phase currents and electromagnetic torque: *clockwise* rotation.

Example 7.1

Perform the nonlinear modeling using the differential equations (7.6) derived. The motor parameters are: $r_s = 0.01$ ohm, $L_{md} = 0.0012$ H, $L_{mq} = 0.0002$ H, $J = 0.6$ kg-m^2, and $B_m = 0.003$ N-m-s-rad^{-1}. The three-phase voltage set (7.8) is supplied. The magnitude of the phase voltages is $u_M = 110$ V, and $\varphi_u = 0.3882$. The electrical angular velocity is the state variable of interest.

Solution

The transient dynamics of the electrical angular velocity is of particular interest. For the three-phase synchronous reluctance motor studied, the transient dynamics are modeled in the MATLAB environment; and if the motor starts from stall, Figure 7.8 illustrates the transient dynamics for the angular velocity if a load torque of 10 N-m is applied at 1 s.

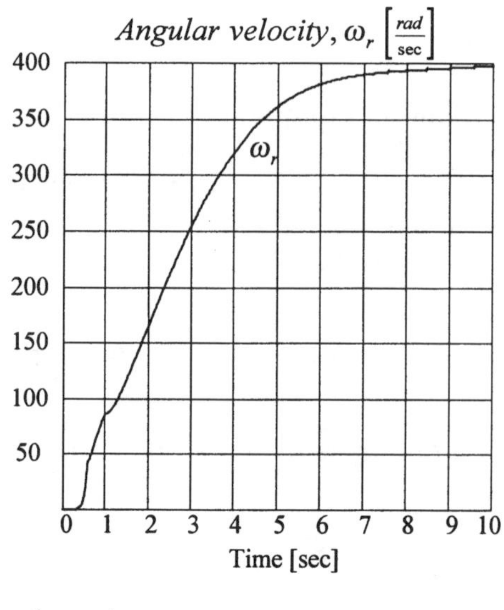

Figure 7.8 Open-loop motor dynamics.

□

7.2.2 *Mathematical models of synchronous reluctance motors in the rotor and synchronous reference frames*

To simplify the analysis, synchronous machines can be modeled using the *qd0* variables. We recall that the *arbitrary* reference frame already has been employed to study induction machines. The goal is to simplify steady-state and dynamic analysis. To avoid the complexity of mathematical deviations, one specifies the angular velocity of the reference frame. The common choice is the rotor reference frame, and $\omega = \omega_r$.

Remark. Under normal operation, the electrical angular velocity is equal to the synchronous angular velocity; that is, $\omega = \omega_e$. One concludes that the model developed in the rotor reference frame is identical to the model in the synchronous reference frame.

□

To transform the phase voltages, currents, and flux linkages (given in the *abc* variables) to the corresponding *quadrature-*, *direct-*, and *zero*-axis components of voltages, currents, and flux linkages, the Park transformation is applied. In particular,

$$\mathbf{u}^r_{qd0s} = \mathbf{K}^r_s \mathbf{u}_{absc}, \quad \begin{bmatrix} u^r_{qs} \\ u^r_{ds} \\ u^r_{0s} \end{bmatrix} = \mathbf{K}^r_s \begin{bmatrix} u_{as} \\ u_{bs} \\ u_{cs} \end{bmatrix}$$

$$\mathbf{i}^r_{qd0s} = \mathbf{K}^r_s \mathbf{i}_{abcs}, \quad \begin{bmatrix} i^r_{qs} \\ i^r_{ds} \\ i^r_{0s} \end{bmatrix} = \mathbf{K}^r_s \begin{bmatrix} i_{as} \\ i_{bs} \\ i_{cs} \end{bmatrix}$$

$$\boldsymbol{\psi}^r_{qd0s} = \mathbf{K}^r_s \boldsymbol{\psi}_{abcs}, \quad \begin{bmatrix} \psi^r_{qs} \\ \psi^r_{ds} \\ \psi^r_{0s} \end{bmatrix} = \mathbf{K}^r_s \begin{bmatrix} \psi_{as} \\ \psi_{bs} \\ \psi_{cs} \end{bmatrix}$$

where the stator Park transformation matrix $\mathbf{K}^r_s$ is

$$\mathbf{K}^r_s = \frac{2}{3} \begin{bmatrix} \cos\theta_r & \cos\left(\theta_r - \frac{2}{3}\pi\right) & \cos\left(\theta_r + \frac{2}{3}\pi\right) \\ \sin\theta_r & \sin\left(\theta_r - \frac{2}{3}\pi\right) & \sin\left(\theta_r + \frac{2}{3}\pi\right) \\ \frac{1}{2} & \frac{1}{2} & \frac{1}{2} \end{bmatrix}$$

The *inverse* Park transformation allows one to find the *abc* voltages, currents, and flux linkages using the *quadrature-*, *direct-*, and *zero*-axis components of the corresponding variables. That is, using the Park transformation matrix $\mathbf{K}^{r^{-1}}_s$, we have the following *inverse* transformations

$$\mathbf{u}_{abcs} = \mathbf{K}^{r^{-1}}_s \mathbf{u}^r_{qd0s}, \quad \begin{bmatrix} u_{as} \\ u_{bs} \\ u_{cs} \end{bmatrix} = \mathbf{K}^{r^{-1}}_s \begin{bmatrix} u^r_{qs} \\ u^r_{ds} \\ u^r_{0s} \end{bmatrix}$$

$$\mathbf{i}_{abcs} = \mathbf{K}_s^{r^{-1}} \mathbf{i}_{qd0s}^r, \begin{bmatrix} i_{as} \\ i_{bs} \\ i_{cs} \end{bmatrix} = \mathbf{K}_s^{r^{-1}} \begin{bmatrix} i_{qs}^r \\ i_{ds}^r \\ i_{0s}^r \end{bmatrix}$$

$$\boldsymbol{\psi}_{abcs} = \mathbf{K}_s^{r^{-1}} \boldsymbol{\psi}_{qd0s}^r, \begin{bmatrix} \psi_{as} \\ \psi_{bs} \\ \psi_{cs} \end{bmatrix} = \mathbf{K}_s^{r^{-1}} \begin{bmatrix} \psi_{qs}^r \\ \psi_{ds}^r \\ \psi_{0s}^r \end{bmatrix}$$

The inverse matrix $\mathbf{K}_s^{r^{-1}}$ used is

$$\mathbf{K}_s^{r^{-1}} = \begin{bmatrix} \cos\theta_r & \sin\theta_r & 1 \\ \cos\left(\theta_r - \frac{2}{3}\pi\right) & \sin\left(\theta_r - \frac{2}{3}\pi\right) & 1 \\ \cos\left(\theta_r + \frac{2}{3}\pi\right) & \sin\left(\theta_r + \frac{2}{3}\pi\right) & 1 \end{bmatrix}$$

One can now proceed with the development of synchronous reluctance motor dynamics in the rotor reference frame. Kirchhoff's voltage law is used, and equations (7.1) are applied. To avoid the complexity of notations, we use matrix notations, and equations (7.1) are

$$\mathbf{u}_{abcs} = \mathbf{r}_s \mathbf{i}_{abcs} + \frac{d\boldsymbol{\psi}_{abcs}}{dt}$$

Making use of the Park transformation, one performs the transformation of the *abc* variables to the *qd0* quantities by multiplying the left and right sides by $\mathbf{K}_s^{r^{-1}}$. We have

$$\mathbf{K}_s^{r^{-1}} \mathbf{u}_{qd0s}^r = \mathbf{r}_s \mathbf{K}_s^{r^{-1}} \mathbf{i}_{qd0s}^r + \frac{d(\mathbf{K}_s^{r^{-1}} \psi_{qd0s}^r}{dt}$$

By multiplying this equation by matrix $\mathbf{K}_s^r$, one obtains

$$\mathbf{K}_s^r \mathbf{K}_s^{r^{-1}} \mathbf{u}_{qd0s}^r = \mathbf{K}_s^r \mathbf{r}_s \mathbf{K}_s^{r^{-1}} \mathbf{i}_{qd0s}^r + \mathbf{K}_s^r \frac{d(\mathbf{K}_s^{r^{-1}} \boldsymbol{\psi}_{qd0s}^r)}{dt}$$

The following matrix identity is used: $\mathbf{K}_s^r \mathbf{K}_s^{r^{-1}} = \mathbf{I}$, where **I** is the identity matrix. Then,

$$\mathbf{K}_s^r \mathbf{K}_s^{r^{-1}} \mathbf{u}_{qd0s}^r = \mathbf{u}_{qd0s}^r$$

and

$$\mathbf{K}_s^r \mathbf{r}_s \mathbf{K}_s^{r^{-1}} \mathbf{i}_{qdos}^r = \mathbf{r}_s \mathbf{i}_{qdos}^r$$

Furthermore,

$$\mathbf{K}_s^r \frac{d(\mathbf{K}_s^{r^{-1}} \boldsymbol{\psi}_{qd0s}^r)}{dt} = \mathbf{K}_s^r \frac{d\mathbf{K}_s^{r^{-1}}}{dt} \boldsymbol{\psi}_{qd0s}^r + \mathbf{K}_s^r \mathbf{K}_s^{r^{-1}} \frac{d\boldsymbol{\psi}_{qd0s}^r}{dt}$$

$$= \frac{2}{3}\begin{bmatrix} \cos\theta_r & \cos\left(\theta_r - \frac{2}{3}\pi\right) & \cos\left(\theta_r + \frac{2}{3}\pi\right) \\ \sin\theta_r & \sin\left(\theta_r - \frac{2}{3}\pi\right) & \sin\left(\theta_r + \frac{2}{3}\pi\right) \\ \frac{1}{2} & \frac{1}{2} & \frac{1}{2} \end{bmatrix} \frac{d\theta_r}{dt} \begin{bmatrix} -\sin\theta_r & \cos\theta_r & 0 \\ -\sin\left(\theta_r - \frac{2}{3}\pi\right) & \cos\left(\theta_r - \frac{2}{3}\pi\right) & 0 \\ -\sin\left(\theta_r + \frac{2}{3}\pi\right) & \cos\left(\theta_r + \frac{2}{3}\pi\right) & 0 \end{bmatrix}$$

$$\times \begin{bmatrix} \psi_{qs}^r \\ \psi_{ds}^r \\ \psi_{0s}^r \end{bmatrix} + \begin{bmatrix} \frac{d\psi_{qs}^r}{dt} \\ \frac{d\psi_{ds}^r}{dt} \\ \frac{d\psi_{0s}^r}{dt} \end{bmatrix}$$

$$= \omega_r \begin{bmatrix} 0 & 1 & 0 \\ -1 & 0 & 0 \\ 0 & 0 & 0 \end{bmatrix} \begin{bmatrix} \psi_{qs}^r \\ \psi_{ds}^r \\ \psi_{0s}^r \end{bmatrix} + \begin{bmatrix} \frac{d\psi_{qs}^r}{dt} \\ \frac{d\psi_{ds}^r}{dt} \\ \frac{d\psi_{0s}^r}{dt} \end{bmatrix} = \begin{bmatrix} \omega_r \psi_{ds}^r \\ -\omega_r \psi_{qs}^r \\ 0 \end{bmatrix} + \begin{bmatrix} \frac{d\psi_{qs}^r}{dt} \\ \frac{d\psi_{ds}^r}{dt} \\ \frac{d\psi_{0s}^r}{dt} \end{bmatrix}$$

Therefore, one obtains

$$\begin{bmatrix} u_{qs}^r \\ u_{ds}^r \\ u_{0s}^r \end{bmatrix} = \begin{bmatrix} r_s & 0 & 0 \\ 0 & r_s & 0 \\ 0 & 0 & r_s \end{bmatrix} \begin{bmatrix} i_{qs}^r \\ i_{ds}^r \\ i_{0s}^r \end{bmatrix} + \begin{bmatrix} \omega_r \psi_{ds}^r \\ -\omega_r \psi_{qs}^r \\ 0 \end{bmatrix} + \begin{bmatrix} \frac{d\psi_{qs}^r}{dt} \\ \frac{d\psi_{ds}^r}{dt} \\ \frac{d\psi_{0s}^r}{dt} \end{bmatrix}$$

Making use of

$$\boldsymbol{\psi}_{abcs} = \mathbf{L}_s \mathbf{i}_{abcs}$$

we have $\mathbf{K}_s^r \mathbf{K}_s^{r^{-1}} \boldsymbol{\psi}_{qd0s}^r = \mathbf{K}_s^r \mathbf{L}_s \mathbf{K}_s^{r^{-1}} \mathbf{i}_{qd0s}^r$

Due to $\mathbf{K}_s^r \mathbf{K}_s^{r^{-1}} = \mathbf{I}$, the following results

$$\boldsymbol{\psi}_{qd0s}^r = \mathbf{K}_s^r \mathbf{L}_s \mathbf{K}_s^{r^{-1}} \mathbf{i}_{qd0s}^r$$

Here, the product of matrices $\mathbf{K}_s^r \mathbf{L}_s \mathbf{K}_s^{r^{-1}}$ must be found. We have

$$\mathbf{K}_s^r \mathbf{L}_s \mathbf{K}_s^{r^{-1}} = \frac{2}{3}\begin{bmatrix} \cos\theta_r & \cos\left(\theta_r - \frac{2}{3}\pi\right) & \cos\left(\theta_r + \frac{2}{3}\pi\right) \\ \sin\theta_r & \sin\left(\theta_r - \frac{2}{3}\pi\right) & \sin\left(\theta_r + \frac{2}{3}\pi\right) \\ \frac{1}{2} & \frac{1}{2} & \frac{1}{2} \end{bmatrix}$$

$$\times \begin{bmatrix} L_{ls} + \bar{L}_m - L_{\Delta m}\cos 2\theta_r & -\frac{1}{2}\bar{L}_m - L_{\Delta m}\cos 2\left(\theta_r - \frac{1}{3}\pi\right) & -\frac{1}{2}\bar{L}_m - L_{\Delta m}\cos 2\left(\theta_r + \frac{1}{3}\pi\right) \\ -\frac{1}{2}\bar{L}_m - L_{\Delta m}\cos 2\left(\theta_r - \frac{1}{3}\pi\right) & L_{ls} + \bar{L}_m - L_{\Delta m}\cos 2\left(\theta_r - \frac{2}{3}\pi\right) & -\frac{1}{2}\bar{L}_m - L_{\Delta m}\cos 2\theta_r \\ -\frac{1}{2}\bar{L}_m - L_{\Delta m}\cos 2\left(\theta_r + \frac{1}{3}\pi\right) & -\frac{1}{2}\bar{L}_m - L_{\Delta m}\cos 2\theta_r & L_{ls} + \bar{L}_m - L_{\Delta m}\cos 2\left(\theta_r + \frac{2}{3}\pi\right) \end{bmatrix}$$

$$\times \begin{bmatrix} \cos\theta_r & \sin\theta_r & 1 \\ \cos\left(\theta_r - \frac{2}{3}\pi\right) & \sin\left(\theta_r - \frac{2}{3}\pi\right) & 1 \\ \cos\left(\theta_r + \frac{2}{3}\pi\right) & \sin\left(\theta_r + \frac{2}{3}\pi\right) & 1 \end{bmatrix} = \begin{bmatrix} L_{ls} + L_{mq} & 0 & 0 \\ 0 & L_{ls} + L_{md} & 0 \\ 0 & 0 & L_{ls} \end{bmatrix}$$

Hence, one obtains

$$\begin{bmatrix} \psi_{qs}^r \\ \psi_{ds}^r \\ \psi_{0s}^r \end{bmatrix} = \begin{bmatrix} L_{ls} + L_{mq} & 0 & 0 \\ 0 & L_{ls} + L_{md} & 0 \\ 0 & 0 & L_{ls} \end{bmatrix} \begin{bmatrix} i_{qs}^r \\ i_{ds}^r \\ i_{0s}^r \end{bmatrix}$$

The following set of differential equations is found

$$u_{qs}^r = r_s i_{qs}^r + (L_{ls} + L_{mq})\frac{di_{qs}^r}{dt} + (L_{ls} + L_{md}) i_{ds}^r \omega_r$$

$$u_{ds}^r = r_s i_{ds}^r + (L_{ls} + L_{md})\frac{di_{ds}^r}{dt} - (L_{ls} + L_{mq}) i_{qs}^r \omega_r$$

$$u_{0s}^r = r_s i_{0s}^r + L_{ls}\frac{di_{0s}^r}{dt}$$

Hence, the circuitry dynamics of synchronous reluctance motors in the rotor reference frame is mapped by

$$\begin{aligned} \frac{di_{qs}^r}{dt} &= -\frac{r_s}{L_{ls} + L_{mq}} i_{qs}^r - \frac{L_{ls} + L_{md}}{L_{ls} + L_{mq}} i_{ds}^r \omega_r + \frac{1}{L_{ls} + L_{mq}} u_{qs}^r \\ \frac{di_{ds}^r}{dt} &= -\frac{r_s}{L_{ls} + L_{md}} i_{ds}^r + \frac{L_{ls} + L_{mq}}{L_{ls} + L_{md}} i_{qs}^r \omega_r + \frac{1}{L_{ls} + L_{md}} u_{ds}^r \\ \frac{di_{0s}^r}{dt} &= -\frac{r_s}{L_{ls}} i_{0s}^r + \frac{1}{L_{ls}} u_{0s}^r \end{aligned} \tag{7.9}$$

Using Newton's second law $T_e - \frac{2B_m}{P}\omega_r - T_L = \frac{2J}{P}\frac{d^2\theta_r}{dt^2}$, the *torsional-mechanical* dynamics are found to be

$$\frac{d\omega_r}{dt} = \frac{P}{2J}\left(T_e - \frac{2B_m}{P}\omega_r - T_L\right)$$
$$\frac{d\theta_r}{dt} = \omega_r \tag{7.10}$$

The electromagnetic torque, which was found to be

$$T_e = \frac{P}{2}\frac{1}{2}[i_{as} \quad i_{bs} \quad i_{cs}]\frac{\partial \mathbf{L}_s(\theta_r)}{\partial \theta_r}\begin{bmatrix} i_{as} \\ i_{bs} \\ i_{cs} \end{bmatrix}$$

must be derived using the *qd0* currents as variables. Applying the *inverse* Park transformation $\mathbf{i}_{abcs} = \mathbf{K}_s^{r^{-1}}\mathbf{i}_{qd0s}^r$, one obtains

$$T_e = \frac{P}{2}\frac{\partial W_c(\mathbf{i}_{qd0s}^r, \theta_r)}{\partial \theta_r} = \frac{P}{2}\frac{1}{2}[i_{qs}^r \quad i_{ds}^r \quad i_{0s}^r](\mathbf{K}_s^{r^{-1}})^T\frac{\partial \mathbf{L}_s(\theta_r)}{\partial \theta_r}\mathbf{K}_s^{r^{-1}}\begin{bmatrix} i_{qs}^r \\ i_{ds}^r \\ i_{0s}^r \end{bmatrix}$$

By multiplying the matrices in the expression $(\mathbf{K}_s^{r^{-1}})^T\frac{\partial \mathbf{L}_s(\theta_r)}{\partial \theta_r}\mathbf{K}_s^{r^{-1}}$, we have

$$(\mathbf{K}_s^{r^{-1}})^T\frac{\partial \mathbf{L}_s(\theta_r)}{\partial \theta_r}\mathbf{K}_s^{r^{-1}} = \begin{bmatrix} \cos\theta_r & \cos\left(\theta_r - \frac{2}{3}\pi\right) & \cos\left(\theta_r + \frac{2}{3}\pi\right) \\ \sin\theta_r & \sin\left(\theta_r - \frac{2}{3}\pi\right) & \sin\left(\theta_r + \frac{2}{3}\pi\right) \\ 1 & 1 & 1 \end{bmatrix}$$

$$\times\begin{bmatrix} 2L_{\Delta m}\sin 2\theta_r & 2L_{\Delta m}\sin 2\left(\theta_r - \frac{1}{3}\pi\right) & 2L_{\Delta m}\sin 2\left(\theta_r + \frac{1}{3}\pi\right) \\ 2L_{\Delta m}\sin 2\left(\theta_r - \frac{1}{3}\pi\right) & 2L_{\Delta m}\sin 2\left(\theta_r - \frac{2}{3}\pi\right) & 2L_{\Delta m}\sin 2\theta_r \\ 2L_{\Delta m}\sin 2\left(\theta_r + \frac{1}{3}\pi\right) & 2L_{\Delta m}\sin 2\theta_r & 2L_{\Delta m}\sin 2\left(\theta_r + \frac{2}{3}\pi\right) \end{bmatrix}$$

$$\times\begin{bmatrix} \cos\theta_r & \sin\theta_r & 1 \\ \cos\left(\theta_r - \frac{2}{3}\pi\right) & \sin\left(\theta_r - \frac{2}{3}\pi\right) & 1 \\ \cos\left(\theta_r + \frac{2}{3}\pi\right) & \sin\left(\theta_r + \frac{2}{3}\pi\right) & 1 \end{bmatrix} = 3L_{\Delta m}\begin{bmatrix} \sin\theta_r & \sin\left(\theta_r - \frac{2}{3}\pi\right) & \sin\left(\theta_r + \frac{2}{3}\pi\right) \\ \cos\theta_r & \cos\left(\theta_r - \frac{2}{3}\pi\right) & \cos\left(\theta_r + \frac{2}{3}\pi\right) \\ 0 & 0 & 0 \end{bmatrix}$$

$$\times\begin{bmatrix}\cos\theta_r & \sin\theta_r & 1\\ \cos\left(\theta_r-\frac{2}{3}\pi\right) & \sin\left(\theta_r-\frac{2}{3}\pi\right) & 1\\ \cos\left(\theta_r+\frac{2}{3}\pi\right) & \sin\left(\theta_r+\frac{2}{3}\pi\right) & 1\end{bmatrix}=3L_{\Delta m}\begin{bmatrix}0 & \frac{3}{2} & 0\\ \frac{3}{2} & 0 & 0\\ 0 & 0 & 0\end{bmatrix}$$

Therefore,

$$T_e=\frac{3P}{4}L_{\Delta m}[i_{qs}^r \quad i_{ds}^r \quad i_{0s}^r]\begin{bmatrix}0 & \frac{3}{2} & 0\\ \frac{3}{2} & 0 & 0\\ 0 & 0 & 0\end{bmatrix}\begin{bmatrix}i_{qs}^r\\ i_{ds}^r\\ i_{os}^r\end{bmatrix}=\frac{3P}{4}L_{\Delta m}[i_{qs}^r \quad i_{ds}^r \quad i_{0s}^r]\begin{bmatrix}\frac{3}{2}i_{ds}^r\\ \frac{3}{2}i_{qs}^r\\ 0\end{bmatrix}=\frac{9P}{4}L_{\Delta m}i_{qs}^r i_{ds}^r$$

where $L_{\Delta m}=\frac{1}{3}(L_{md}-L_{mq})$ can be found using the experimentally measured magnetizing *quadrature* and *direct* inductances $L_{mq}=\frac{3}{2}(\bar{L}_m-L_{\Delta m})$ and $L_{md}=\frac{3}{2}(\bar{L}_m+L_{\Delta m})$.

Then, we have the following expression for the electromagnetic torque developed

$$T_e=\frac{3P}{4}(L_{md}-L_{mq})i_{qs}^r i_{ds}^r$$

Plugging $T_e=\frac{3P}{4}(L_{md}-L_{mq})i_{qs}i_{ds}$ into (7.10), we have

$$\frac{d\omega_r}{dt}=\frac{3P^2}{8J}(L_{md}-L_{mq})i_{qs}^r i_{ds}^r-\frac{B_m}{J}\omega_r-\frac{P}{2J}T_L$$

$$\frac{d\theta_r}{dt}=\omega_r$$

Augmenting the *torsional-mechanical* equations of motion with circuitry dynamics (7.9), one obtains a set of nonlinear differential equations to be used to model synchronous reluctance motors in the rotor reference frame. In particular,

$$\frac{di_{qs}^r}{dt}=-\frac{r_s}{L_{ls}+L_{mq}}i_{qs}^r-\frac{L_{ls}+L_{md}}{L_{ls}+L_{mq}}i_{ds}^r\omega_r+\frac{1}{L_{ls}+L_{mq}}u_{qs}^r$$

$$\frac{di_{ds}^r}{dt}=-\frac{r_s}{L_{ls}+L_{md}}i_{ds}^r+\frac{L_{ls}+L_{mq}}{L_{ls}+L_{md}}i_{qs}^r\omega_r+\frac{1}{L_{ls}+L_{md}}u_{ds}^r$$

$$\frac{di_{0s}^r}{dt}=-\frac{r_s}{L_{ls}}i_{0s}^r+\frac{1}{L_{ls}}u_{0s}^r$$

$$\frac{d\omega_r}{dt} = \frac{3P^2}{8J}(L_{md} - L_{mq})i^r_{qs}i^r_{ds} - \frac{B_m}{J}\omega_r - \frac{P}{2J}T_L$$

$$\frac{d\theta_r}{dt} = \omega_r$$

This set of five nonlinear differential equations maps the synchronous reluctance motor dynamics in the rotor reference frame, and the angular velocity of the reference frame was assigned as $\omega = \omega_r$.

Model in the synchronous reference frame

In contrast, for the synchronous reference frame, one assigns the angular velocity of the reference frame equal to the synchronous angular velocity, $\omega = \omega_e$. Recall that the electrical angular velocity of synchronous machines is equal to the synchronous angular velocity; that, is $\omega_r = \omega_e$ and $\theta_r = \theta_e$. Hence, one finds that the mathematical models of synchronous machines in the rotor and synchronous frames are the same; the corresponding differential equations in the synchronous reference frame are

$$\frac{di^e_{qs}}{dt} = -\frac{r_s}{L_{ls} + L_{mq}}i^e_{qs} - \frac{L_{ls} + L_{md}}{L_{ls} + L_{mq}}i^e_{ds}\omega_r + \frac{1}{L_{ls} + L_{mq}}u^e_{qs}$$

$$\frac{di^e_{ds}}{dt} = -\frac{r_s}{L_{ls} + L_{md}}i^e_{ds} + \frac{L_{ls} + L_{mq}}{L_{ls} + L_{md}}i^e_{qs}\omega_r + \frac{1}{L_{ls} + L_{md}}u^e_{ds}$$

$$\frac{di^e_{0s}}{dt} = -\frac{r_s}{L_{ls}}i^e_{0s} + \frac{1}{L_{ls}}u^e_{0s}$$

$$\frac{d\omega_r}{dt} = \frac{3P^2}{8J}(L_{md} - L_{mq})i^e_{qs}i^e_{ds} - \frac{B_m}{J}\omega_r - \frac{P}{2J}T_L$$

$$\frac{d\theta_r}{dt} = \omega_r$$

That is, the mathematical models of synchronous reluctance motors in the rotor and synchronous reference frames were derived, and highly coupled nonlinear differential equations cannot be linearized or simplified.

Balanced operation of the synchronous reluctance motors

Balanced motor operation must be guaranteed in the full operating torque-speed envelope to attain optimum performance. To ensure balanced operation, the *quadrature-*, *direct-*, and *zero-*axis components of voltages and currents must be found. Using the *direct* Park transformation in the rotor reference frame, as given by

$$\mathbf{u}^r_{qd0s} = \mathbf{K}^r_s\mathbf{u}_{abcs} \text{ and } \mathbf{i}^r_{qd0s} = \mathbf{K}^r_s\mathbf{i}_{abcs}$$

we obtain

$$\begin{bmatrix} u_{qs}^r \\ u_{ds}^r \\ u_{0s}^r \end{bmatrix} = \frac{2}{3}\begin{bmatrix} \cos\theta_r & \cos\left(\theta_r - \frac{2}{3}\pi\right) & \cos\left(\theta_r + \frac{2}{3}\pi\right) \\ \sin\theta_r & \sin\left(\theta_r - \frac{2}{3}\pi\right) & \sin\left(\theta_r + \frac{2}{3}\pi\right) \\ \frac{1}{2} & \frac{1}{2} & \frac{1}{2} \end{bmatrix}\begin{bmatrix} u_{as} \\ u_{bs} \\ u_{cs} \end{bmatrix}$$

$$\begin{bmatrix} i_{qs}^r \\ i_{ds}^r \\ i_{0s}^r \end{bmatrix} = \frac{2}{3}\begin{bmatrix} \cos\theta_r & \cos\left(\theta_r - \frac{2}{3}\pi\right) & \cos\left(\theta_r + \frac{2}{3}\pi\right) \\ \sin\theta_r & \sin\left(\theta_r - \frac{2}{3}\pi\right) & \sin\left(\theta_r + \frac{2}{3}\pi\right) \\ \frac{1}{2} & \frac{1}{2} & \frac{1}{2} \end{bmatrix}\begin{bmatrix} i_{as} \\ i_{bs} \\ i_{cs} \end{bmatrix}$$

The balanced three-phase current set to be fed is found and given by equations (7.7). In particular,

$$i_{as} = \sqrt{2}i_M \sin\left(\theta_r + \frac{1}{3}\varphi_i\pi\right),\ i_{bs} = \sqrt{2}i_M \sin\left(\theta_r - \frac{1}{3}(2-\varphi_i)\pi\right),\ i_{cs} = \sqrt{2}i_M \sin\left(\theta_r + \frac{1}{3}(2+\varphi_i)\pi\right)$$

Using the Park transformation, we have

$$\begin{bmatrix} i_{qs}^r \\ i_{ds}^r \\ i_{0s}^r \end{bmatrix} = \frac{2}{3}\begin{bmatrix} \cos\theta_r & \cos\left(\theta_r - \frac{2}{3}\pi\right) & \cos\left(\theta_r + \frac{2}{3}\pi\right) \\ \sin\theta_r & \sin\left(\theta_r - \frac{2}{3}\pi\right) & \sin\left(\theta_r + \frac{2}{3}\pi\right) \\ \frac{1}{2} & \frac{1}{2} & \frac{1}{2} \end{bmatrix}\begin{bmatrix} \sqrt{2}i_M \sin\left(\theta_r + \frac{1}{3}\varphi_i\pi\right) \\ \sqrt{2}i_M \sin\left(\theta_r - \frac{1}{3}(2-\varphi_i)\pi\right) \\ \sqrt{2}i_M \sin\left(\theta_r + \frac{1}{3}(2+\varphi_i)\pi\right) \end{bmatrix} = \begin{bmatrix} 0.4714 i_M \\ 1.333 i_M \\ 0 \end{bmatrix}$$

Hence, in the rotor reference frame, the *quadrature, direct,* and *zero* currents are

$$i_{qs}^r = 0.4714 i_M,\ i_{ds}^r = 1.333 i_M,\ \text{and}\ i_{0s}^r = 0$$

Correspondingly, in the synchronous reference frame, we have

$$i_{qs}^e = 0.4714 i_M,\ i_{ds}^e = 1.333 i_M,\ \text{and}\ i_{0s}^e = 0$$

In the majority of electric drives, voltage-fed power converters are used, and one supplies the phase voltages to the motor windings. It was shown that by using advanced shifting, the corresponding voltage set (7.8) is $u_{as} = \sqrt{2}u_M \sin\left(\theta_r + \frac{1}{3}\varphi_i\pi + \varphi_u\pi\right)$, $u_{bs} = \sqrt{2}u_M \sin\left(\theta_r - \frac{1}{3}(2-\varphi_i)\pi + \varphi_u\pi\right)$, and $u_{cs} = \sqrt{2}u_M \sin\left(\theta_r + \frac{1}{3}(2+\varphi_i)\pi + \varphi_u\pi\right)$. Hence, in the rotor reference frame, the resulting *quadrature-*, *direct-*, and *zero*-axis components of voltages to be supplied are found as

$$\begin{bmatrix} u_{qs}^r \\ u_{ds}^r \\ u_{0s}^r \end{bmatrix} = \frac{2\sqrt{2}}{3} u_M \begin{bmatrix} \cos\theta_r & \cos\left(\theta_r - \frac{2}{3}\pi\right) & \cos\left(\theta_r + \frac{2}{3}\pi\right) \\ \sin\theta_r & \sin\left(\theta_r - \frac{2}{3}\pi\right) & \sin\left(\theta_r + \frac{2}{3}\pi\right) \\ \frac{1}{2} & \frac{1}{2} & \frac{1}{2} \end{bmatrix} \begin{bmatrix} \sin\left(\theta_r + \frac{1}{3}\varphi_i\pi + \varphi_u\pi\right) \\ \sin\left(\theta_r - \frac{1}{3}(2-\varphi_i)\pi + \varphi_u\pi\right) \\ \sin\left(\theta_r + \frac{1}{3}(2+\varphi_i)\pi + \varphi_u\pi\right) \end{bmatrix}$$

The $\varphi_u\pi$ can be found using the motor parameters, and the multiplication of matrices gives the *quadrature* and *direct* voltages u_{qs}^r and u_{ds}^r (the *zero* voltage is zero, $u_{0s}^r = 0$).

Example 7.2

Model the dynamics of a synchronous reluctance motor in the rotor reference frame. The motor parameters are the same as those used in Example 7.1.

Solution

To analyze the motor behavior, the differential equations derived to map the transient dynamics in the rotor reference frame,

$$\frac{di_{qs}^r}{dt} = -\frac{r_s}{L_{ls}+L_{mq}} i_{qs}^r - \frac{L_{ls}+L_{md}}{L_{ls}+L_{mq}} i_{ds}^r \omega_r + \frac{1}{L_{ls}+L_{mq}} u_{qs}^r$$

$$\frac{di_{ds}^r}{dt} = -\frac{r_s}{L_{ls}+L_{md}} i_{ds}^r + \frac{L_{ls}+L_{mq}}{L_{ls}+L_{md}} i_{qs}^r \omega_r + \frac{1}{L_{ls}+L_{md}} u_{ds}^r$$

$$\frac{di_{0s}^r}{dt} = -\frac{r_s}{L_{ls}} i_{0s}^r + \frac{1}{L_{ls}} u_{0s}^r$$

$$\frac{d\omega_r}{dt} = \frac{3P^2}{8J}(L_{md} - L_{mq}) i_{qs}^r i_{ds}^r - \frac{B_m}{J}\omega_r - \frac{P}{2J} T_L$$

must be numerically solved. The SIMULINK diagram is built to perform nonlinear simulations (see Figure 7.9), and we use the following expressions: $L_q = L_{ls} + L_{mq}$ and $L_d = L_{ls} + L_{md}$.

The *quadrature* and *direct* voltages are found to be $u_{qs}^r = 1.4u_M$ and $u_{ds}^r = 0.0024u_M$.

Small electric motors are started by applying the rated voltage. The synchronous reluctance motor studied, as well as other medium- and high-power electric motors are started by using the "ramp" or other starting voltage patterns. The transfer function $H(s) = \frac{1}{Ts^2 + 2\xi Ts + 1}$, $T = 3$, $\xi = 0.9$ is used in the SIMULINK diagram to assign the specified pattern for the voltages applied.

The motor dynamics, if T_L = 40 N-m is applied at 1 s, are illustrated in Figure 7.10, and transient responses of i_{qs}^r, i_{ds}^r, and ω_r are documented.

The *quadrature-*, *direct-*, and *zero*-axis components of voltages and currents can be transformed to the *abc* variables using the *inverse* Park transformation.

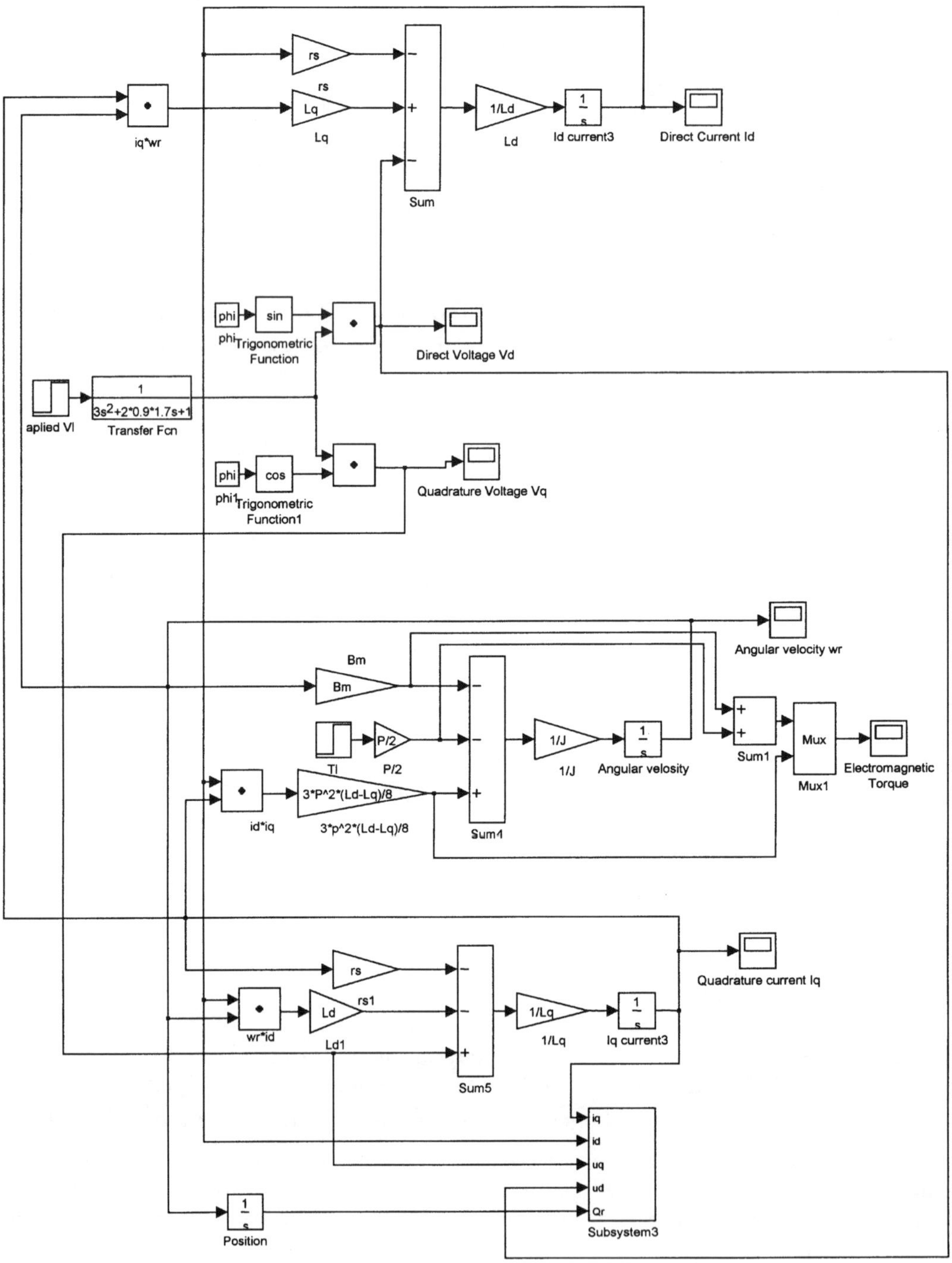

Figure 7.9 Simulink diagram to model synchronous reluctance motors dynamics in the rotor reference frame.

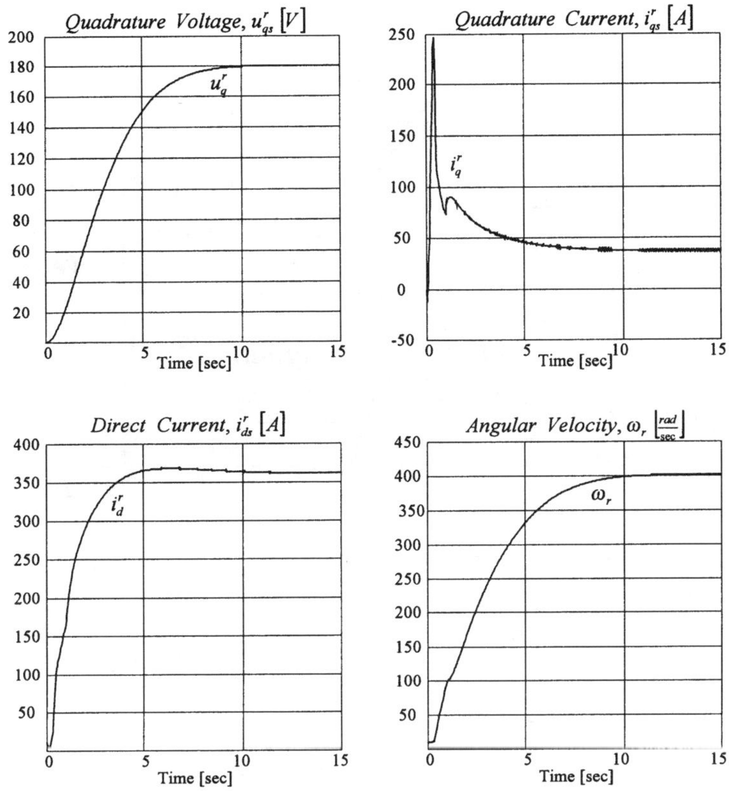

Figure 7.10 Synchronous reluctance motor dynamics in the rotor reference frame.

□

7.3 Permanent-magnet synchronous machines

It has been emphasized that permanent-magnet synchronous machines are widely used to drive electric drives, actuate servo-systems, and generate (induce) three-phase ac voltage in power generation systems. Permanent-magnet synchronous machines are brushless machines because the excitation flux is produced by permanent magnets. Commutatorless permanent-magnet synchronous motors are called brushless DC motors due to the similarity of torque-speed characteristics. Compared with DC, induction, and synchronous reluctance machines, permanent-magnet synchronous machines have lower weigh-to-torque and weight-to-power ratios (the angular velocity of general-purpose permanent-magnet synchronous motors and generators is from 1000 to 2000 rad-s^{-1}), higher efficiency, excellent controllability in the full torque-speed operating envelope, ruggedness and reliability, lower cost and maintenance, etc. For years, permanent-magnet synchronous machines have been the preferred choice for light- and medium-duty electric drives, servos, and power generation systems. There is much current interest in using brushless electronically-commutated servo-motors in high-performance electromechanical systems, and the application of neodymium-iron-boron ($Nd_2Fe_{14}B$) and samarium-cobalt (Sm_1Co_5 and Sm_2Co_{17}) rare-earth magnets results in high torque and power density, efficiency and controllability, reliability and ruggedness. Typical continuous and intermittent torque-speed envelopes for permanent-magnet synchronous motors are illustrated in Figure 7.11.

It must be emphasized that the ratio $\frac{T_{e\max}}{T_{rated}}$ can be 5 or greater.

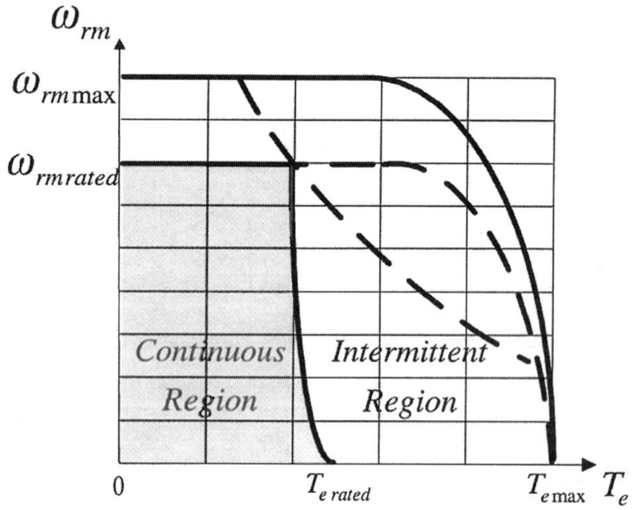

Figure 7.11 Typical torque-speed envelopes for permanent-magnet synchronous motors.

The torque-speed operating envelopes for 23 and 34 NEMA size permanent-magnet synchronous motors, produced by Kollmorgen, as well as the motor parameters are documented in Figure 7.12 and in Table 7.1.

7.3.1 *Dynamics of permanent-magnet synchronous machines in the* machine *variables*

This section approaches and thoroughly studies nonlinear analysis of permanent-magnet synchronous machines and develops nonlinear models. Three-phase, two- and four-pole permanent-magnet synchronous motors are illustrated in Figures 7.13 and 7.14, and a two-pole permanent-magnet synchronous generator is documented in Figure 7.15.

Using Kirchhoff's second law, one obtains three differential equations. In particular, for the *as*, *bs*, and *cs* stator windings, we have

$$u_{as} = r_s i_{as} + \frac{d\psi_{as}}{dt}$$
$$u_{bs} = r_s i_{bs} + \frac{d\psi_{bs}}{dt} \qquad (7.11)$$
$$u_{cs} = r_s i_{cs} + \frac{d\psi_{cs}}{dt}$$

where the flux linkages ψ_{as}, ψ_{bs}, and ψ_{cs} are given as

$$\psi_{as} = L_{asas} i_{as} + L_{asbs} i_{bs} + L_{ascs} i_{cs} + \psi_{asm}$$
$$\psi_{bs} = L_{bsas} i_{as} + L_{bsbs} i_{bs} + L_{bscs} i_{cs} + \psi_{bsm}$$
$$\psi_{cs} = L_{csas} i_{as} + L_{csbs} i_{bs} + L_{cscs} i_{cs} + \psi_{csm}$$

■ Motor: H-232-G ■ Amplifier: SO-4008 ■ Bus Voltage: 40

Performance Specification	Symbol	Units	
Cont. Torque at stall	Tc	oz-in	58.8
		N-m	0.415
Peak Torque at stall	Tp	oz-in	121.0
		N-m	0.85
Cont. Power	HP rated	HP	0.180
	W rated	Watts	134
Max. Speed	N	RPM	4150
Motor Inertia x 10^4	Jm	oz-in-s^2	24.4
		kg-cm^2	0.17
Motor Weight	Wt	lb	2.0
		kg	0.91

■ Motor: H-234-F ■ Amplifier: ROL-20004 ■ Bus Voltage: 150

Performance Specification	Symbol	Units	
Cont. Torque at stall	Tc	oz-in	111.0
		N-m	0.784
Peak Torque at stall	Tp	oz-in	272.0
		N-m	1.93
Cont. Power	HP rated	HP	0.40
	W rated	Watts	298
Max. Speed	N	RPM	4600
Motor Inertia x 10^4	Jm	oz-in-s^2	32.6
		kg-cm^2	0.23
Motor Weight	Wt	lb	2.8
		kg	1.27

■ Motor: H-234-G ■ Amplifier: SO-4008 ■ Bus Voltage: 40

Performance Specification	Symbol	Units	
Cont. Torque at stall	Tc	oz-in	106.0
		N-m	0.749
Peak Torque at stall	Tp	oz-in	210.0
		N-m	1.49
Cont. Power	HP rated	HP	0.22
	W rated	Watts	164
Max. Speed	N	RPM	2200
Motor Inertia x 10^4	Jm	oz-in-s^2	32.6
		kg-cm^2	0.23
Motor Weight	Wt	lb	2.8
		kg	1.27

■ Motor: H-234-H ■ Amplifier: ROL-20012 ■ Bus Voltage: 150

Performance Specification	Symbol	Units	
Cont. Torque at stall	Tc	oz-in	118.0
		N-m	0.833
Peak Torque at stall	Tp	oz-in	208
		N-m	1.47
Cont. Power	HP rated	HP	0.47
	W rated	Watts	351
Max. Speed	N	RPM	6000
Motor Inertia x 10^4	Jm	oz-in-s^2	32.6
		kg-cm^2	0.23
Motor Weight	Wt	lb	2.8
		kg	1.27

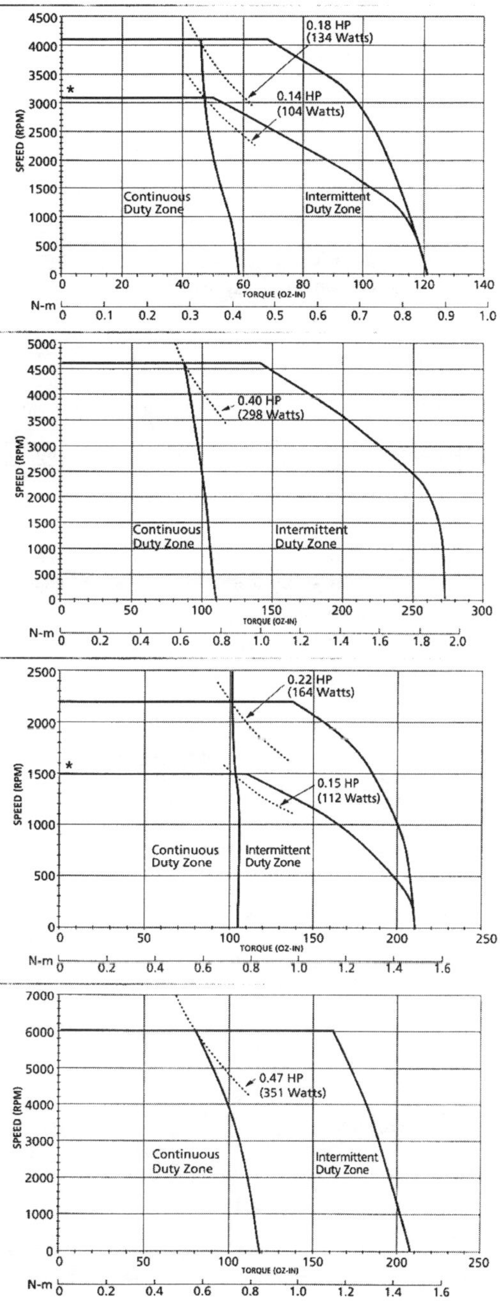

Figure 7.12 Torque-speed curves of high-performance permanent-magnet synchronous motors. (Courtesy of *Kollmorgen.*)

■ Motor: H-342-H ■ Amplifier: ROL-20012 ■ Bus Voltage: 150

Performance Specification	Symbol	Units	
Cont. Torque at stall	Tc	oz-in	163.0
		N-m	1.15
Peak Torque at stall	Tp	oz-in	290
		N-m	2.06
Cont. Power	HP rated	HP	0.40
	W rated	Watts	298
Max. Speed	N	RPM	6000
Motor Inertia x 10⁴	Jm	oz-in-s²	67.9
		kg-cm²	0.48
Motor Weight	Wt	lb	3.92
		kg	1.78

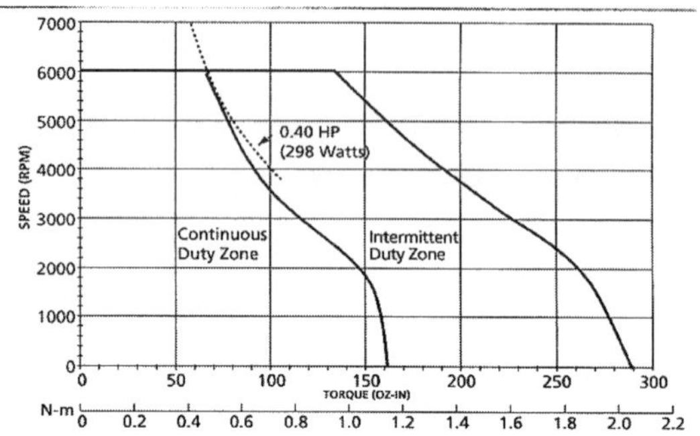

■ Motor: H-344-H ■ Amplifier: ROL-20012 ■ Bus Voltage: 150

Performance Specification	Symbol	Units	
Cont. Torque at stall	Tc	oz-in	297.0
		N-m	2.10
Peak Torque at stall	Tp	oz-in	532
		N-m	3.77
Cont. Power	HP rated	HP	0.94
	W rated	Watts	701
Max. Speed	N	RPM	4050
Motor Inertia x 10⁴	Jm	oz-in-s²	130.0
		kg-cm²	0.92
Motor Weight	Wt	lb	5.84
		kg	2.65

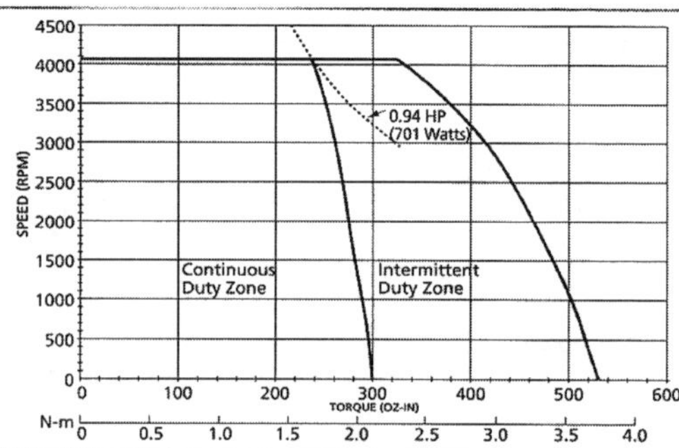

■ Motor: H-348-H ■ Amplifier: ROL-20012 ■ Bus Voltage: 150

Performance Specification	Symbol	Units	
Cont. Torque at stall	Tc	oz-in	786.0
		N-m	5.55
Peak Torque at stall	Tp	oz-in	1490
		N-m	10.56
Cont. Power	HP rated	HP	1.24
	W rated	Watts	925
Max. Speed	N	RPM	1850
Motor Inertia x 10⁴	Jm	oz-in-s²	377.2
		kg-cm²	2.66
Motor Weight	Wt	lb	13.9
		kg	6.32

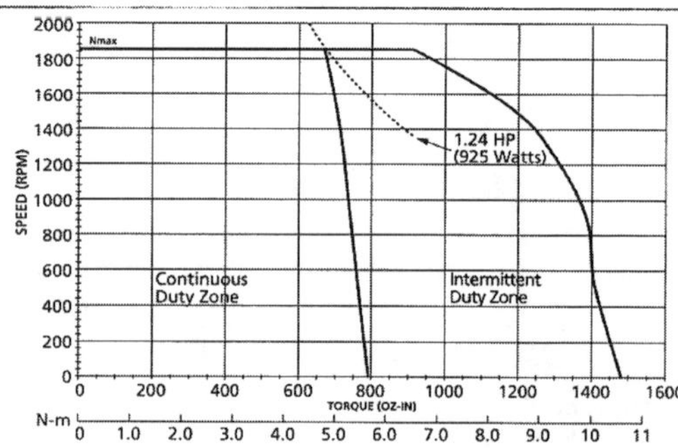

Figure 7.12 Continued.

Table 7.1 Characteristics and Parameters of Permanent-Magnet Synchronous Motors Courtesy of Kollmorgen

H-23x						
Motor Parameters Six Step Operation			Winding Data			
	Symbol	Units	H-232-G	H234-F	H234-G	H234-H
Continuous Power	HP rtd	HP	0.18	0.40	0.22	0.47
	Watts	W	134	298	164	351
** Speed	N rtd	RPM	4150	4600	2200	6000
Torque		oz-in	43.7	87.7	101	79
	T rtd	N-m	0.309	0.619	0.712	0.558
Maximum Speed	N max	RPM	4150	4600	2200	10000
* Continuous Torque (stall) @ 40°C ambient	TC	oz-in	58.8	111	106	118
		N-m	0.415	0.784	0.749	0.833
Cont. Line Current	lc	amps DC	6.85	3.89	7.18	12.6
* Peak Torque		oz-in	158	279	287	327
	TP	N-m	1.12	1.97	2.02	2.31
Peak Line Current	lp	amps DC	20.5	11.7	21.5	37.7
Theoretical Acceleration	acc	rad/sec^2	64900	85600	87900	100200
* Torque Sensitivity (Effective) +/– 10%		oz-in/ADC	8.59	28.5	14.8	9.39
	Kt	N-m/ADC	0.0607	0.201	0.104	0.0663
Back EMF (line to line) +/– 10%	Kb	$V_{O\text{-}PK}$/KRPM	7.21	23.9	12.4	7.89
Max Line to Line Volts	V max	$Volts_{O\text{-}PK}$	225	225	225	225
DC Res @ 25°C (line to line) +/– 10%	Rm	ohms	0.497	2.30	0.606	0.252
Inductance (line to line) +/– 30%	Lm	mh	0.941	4.63	1.27	0.52
Rotor Intertia		oz-in-sec^2	0.00244	0.00326	0.00326	0.00326
	Jm	kg-m^2	0.000017	0.000023	0.000023	0.000023
Weight including Encoder		lb	2.0	2.8	2.8	2.8
	Wt	kg (f)	0.91	1.27	1.27	1.27
Coulomb Friction		oz-in	4.05	6.2	6.2	6.2
	Tf	N-m	0.029	0.044	0.044	0.044
Thermal Time constant	TCT	minutes	8.1	12.5	12.5	12.5
Viscous Damping	Fi	oz-in/KRPM	0.174	0.41	0.41	0.41
Infinite Z Source		N-m/KRPM	0.0012	0.0029	0.0029	0.0029
Encoder Weight		lb	0.075	0.075	0.075	0.075

Table 7.1 Characteristics and Parameters of Permanent-Magnet Synchronous Motors Courtesy of Kollmorgen *(Continued)*

H-23x							
Motor Parameters Six Step Operation				Winding Data			
		Symbol	Units	H-232-G	H234-F	H234-G	H234-H
		Wt	kg (f)	0.0340	0.0340	0.0340	0.0340
Time Constant @ 25°C	Mech	Tm	msec	2.05	1.15	1.13	1.16
	Elec	Te	msec	1.89	2.01	2.09	2.08
Motor Constant @ 25°C		Km	oz-in / $\sqrt{\text{Watts}}$	13.8	21.3	21.5	21.2
Thermal Resistance		Rth	°C/Watt	3.4	2.27	2.54	1.99

H-34x					
Motor Parameters Six Step Operation			Winding Data		
	Symbol	Units	H-342-H	H344-H	H-348-H
Continuous Power	HP rtd	HP	0.40	0.94	1.24
	Watts	W	298	701	925
** Speed	N rtd	RPM	5500	4050	1850
Torque		oz-in	73.3	234	674
	T rtd	N-m	0.518	1.65	4.76
Maximum Speed	N max	RPM	8850	4050	1850
* Continuous Torque (stall) @ 40°C ambient	T	oz-in	163	297	786
		N-m	1.15	2.10	5.55
Cont. Line Current	lc	amps DC	11.0	11.7	12.1
* Peak Torque		oz-in	335	667	2323
	Tp	N-m	2.36	4.71	16.4
Peak Line Current	lp	amps DC	28.0	32.6	36.3

Theoretical Acceleration		acc	rad/sec^2	49300	51300	61600
* Torque Sensitivity (Effective) +/– 10%			oz-in/ADC	14.8	25.40	64.7
		Kt	N-m/ADC	0.105	0.179	0.457
Back EMF (line to line) +/– 10%		Kb	V_{O-PK}/KRPM	12.5	21.3	54.3
Max Line to Line Volts		V max	$Volts_{O-PK}$	225	225	225
DC Res @ 25°C (line to line) +/– 10%		Rm	0hms	0.318	0.360	0.582
Inductance (line to line) +/– 30%		Lm	mh	1.53	1.85	4.25
Rotor Inertia			oz-in-sec^2	0.00679	0.0130	0.0377
		Jm	kg-m^2	0.00048	0.000092	0.000266
Weight including Encoder			lb	3.92	5.8	13.9
		Wt	kg (f)	1.78	2.65	6.32
Coulomb Friction			oz-in	4.6	8.52	8.5
		Tf	N-m	0.032	0.060	0.060
Thermal Time Constant		TCT	minutes	11.4	16.2	33
Viscous Damping		Fi	oz-in/KRPM	0.304	0.951	4.73
Infinite Z Source			N-m/KRPM	0.00215	0.00672	0.0334
Encoder Weight			lb	0.075	0.075	0.075
		Wt	kg (f)	0.0340	0.0340	0.0340
Time Constant @ 25°C	Mech	Tm	msec	1.22	0.904	0.654
	Elec	Te	msec	4.81	5.14	7.3
Motor Constant @ 25°C		Km	oz-in / $\sqrt{\text{Watts}}$	29.9	48.1	96.3
Thermal Resistance		Rth	°C/Watt	2.07	1.61	0.92

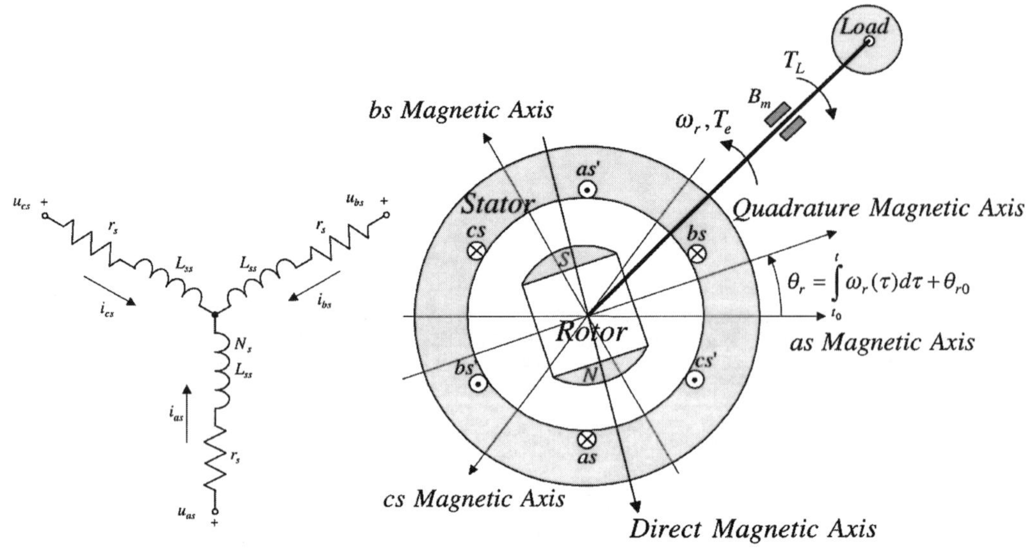

Figure 7.13 Two-pole permanent-magnet synchronous motor.

In matrix form, one has

$$\mathbf{u}_{abcs} = \mathbf{r}_s \mathbf{i}_{abcs} + \frac{d\boldsymbol{\psi}_{abcs}}{dt}, \quad \begin{bmatrix} u_{as} \\ u_{bs} \\ u_{cs} \end{bmatrix} = \begin{bmatrix} r_s & 0 & 0 \\ 0 & r_s & 0 \\ 0 & 0 & r_s \end{bmatrix} \begin{bmatrix} i_{as} \\ i_{bs} \\ i_{cs} \end{bmatrix} + \begin{bmatrix} \frac{d\psi_{as}}{dt} \\ \frac{d\psi_{bs}}{dt} \\ \frac{d\psi_{cs}}{dt} \end{bmatrix}$$

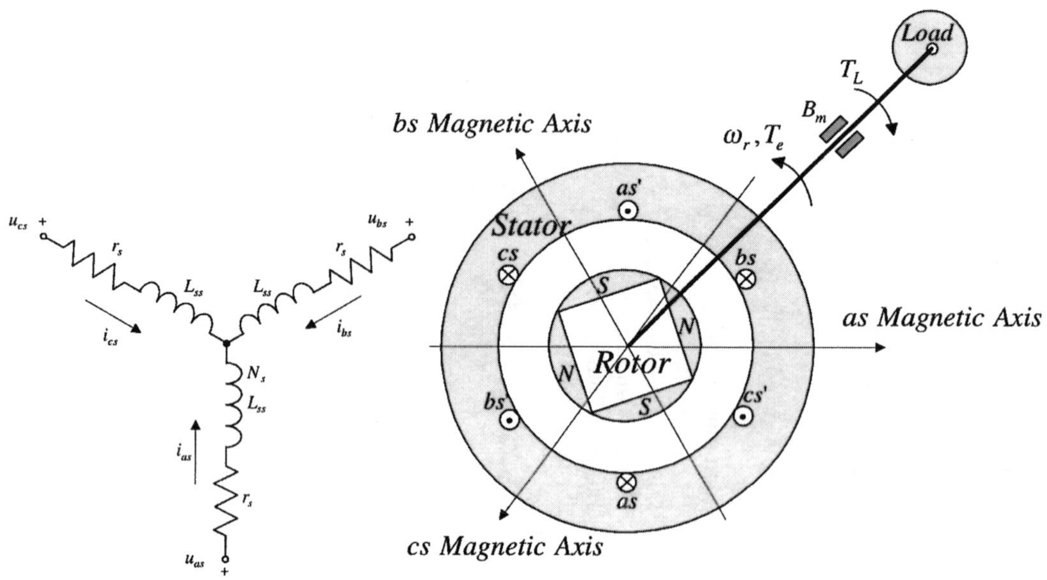

Figure 7.14 Four-pole permanent-magnet synchronous motor.

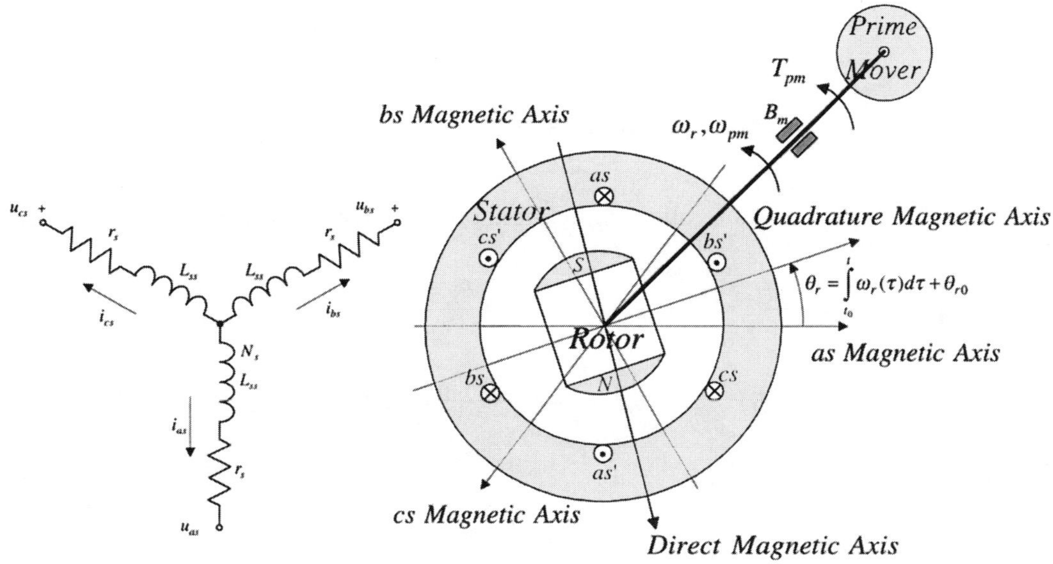

Figure 7.15 Three-phase wye-connected permanent-magnet synchronous generator.

The stator windings are displaced by 120°, and the flux linkages ψ_{asm}, ψ_{bsm}, and ψ_{csm}, established by the permanent magnet, which are periodic functions of θ_r, are assumed to obey the sine law. Using the magnitude of the flux linkages ψ_m established by the permanent magnet, we have

$$\psi_{asm} = \psi_m \sin\theta_r$$

$$\psi_{bsm} = \psi_m \sin\left(\theta_r - \frac{2}{3}\pi\right)$$

$$\psi_{csm} = \psi_m \sin\left(\theta_r + \frac{2}{3}\pi\right)$$

Self- and mutual inductances for three-phase permanent-magnet synchronous machines can be derived using the results given in Section 7.2 for three-phase synchronous reluctance motors (see expression (7.2)). Applying the expressions for inductances derived, we have the following equations for ψ_{as}, ψ_{bs}, and ψ_{cs}:

$$\psi_{as} = (L_{ls} + \bar{L}_m - L_{\Delta m}\cos 2\theta_r)i_{as} + \left(-\frac{1}{2}\bar{L}_m - L_{\Delta m}\cos 2\left(\theta_r - \frac{1}{3}\pi\right)\right)i_{bs}$$

$$+ \left(-\frac{1}{2}\bar{L}_m - L_{\Delta m}\cos 2\left(\theta_r + \frac{1}{3}\pi\right)\right)i_{cs} + \psi_m \sin\theta_r$$

$$\psi_{bs} = \left(-\frac{1}{2}\bar{L}_m - L_{\Delta m}\cos 2\left(\theta_r - \frac{1}{3}\pi\right)\right)i_{as} + \left(L_{ls} + \bar{L}_m - L_{\Delta m}\cos 2\left(\theta_r - \frac{2}{3}\pi\right)\right)i_{bs}$$

$$+ \left(-\frac{1}{2}\bar{L}_m - L_{\Delta m}\cos 2\theta_r\right)i_{cs} + \psi_m \sin\left(\theta_r - \frac{2}{3}\pi\right)$$

$$\psi_{cs} = \left(-\frac{1}{2}\bar{L}_m - L_{\Delta m}\cos 2\left(\theta_r + \frac{1}{3}\pi\right)\right)i_{as} + \left(-\frac{1}{2}\bar{L}_m - L_{\Delta m}\cos 2\theta_r\right)i_{bs}$$
$$+\left(L_{ls} + \bar{L}_m - L_{\Delta m}\cos 2\left(\theta_r + \frac{2}{3}\pi\right)\right)i_{cs} + \psi_m \sin\left(\theta_r + \frac{2}{3}\pi\right) \tag{7.12}$$

Using (7.12), one has the following expression in matrix form for the vector of flux linkages

$$\boldsymbol{\psi}_{abcs} = \mathbf{L}_s \mathbf{i}_{abcs} + \boldsymbol{\psi}_m$$
$$= \begin{bmatrix} L_{ls} + \bar{L}_m - L_{\Delta m}\cos 2\theta_r & -\frac{1}{2}\bar{L}_m - L_{\Delta m}\cos 2\left(\theta_r - \frac{1}{3}\pi\right) & -\frac{1}{2}\bar{L}_m - L_{\Delta m}\cos 2\left(\theta_r + \frac{1}{3}\pi\right) \\ -\frac{1}{2}\bar{L}_m - L_{\Delta m}\cos 2\left(\theta_r - \frac{1}{3}\pi\right) & L_{ls} + \bar{L}_m - L_{\Delta m}\cos 2\left(\theta_r - \frac{2}{3}\pi\right) & -\frac{1}{2}\bar{L}_m - L_{\Delta m}\cos 2\theta_r \\ -\frac{1}{2}\bar{L}_m - L_{\Delta m}\cos 2\left(\theta_r + \frac{1}{3}\pi\right) & -\frac{1}{2}\bar{L}_m - L_{\Delta m}\cos 2\theta_r & L_{ls} + \bar{L}_m - L_{\Delta m}\cos 2\left(\theta_r + \frac{2}{3}\pi\right) \end{bmatrix}$$
$$\times \begin{bmatrix} i_{as} \\ i_{bs} \\ i_{cs} \end{bmatrix} + \psi_m \begin{bmatrix} \sin\theta_r \\ \sin\left(\theta_r - \frac{2}{3}\pi\right) \\ \sin\left(\theta_r + \frac{2}{3}\pi\right) \end{bmatrix}$$

The matrix of inductances $\mathbf{L}_s$ is

$$\mathbf{L}_s = \begin{bmatrix} L_{ls} + \bar{L}_m - L_{\Delta m}\cos 2\theta_r & -\frac{1}{2}\bar{L}_m - L_{\Delta m}\cos 2\left(\theta_r - \frac{1}{3}\pi\right) & -\frac{1}{2}\bar{L}_m - L_{\Delta m}\cos 2\left(\theta_r + \frac{1}{3}\pi\right) \\ -\frac{1}{2}\bar{L}_m - L_{\Delta m}\cos 2\left(\theta_r - \frac{1}{3}\pi\right) & L_{ls} + \bar{L}_m - L_{\Delta m}\cos 2\left(\theta_r - \frac{2}{3}\pi\right) & -\frac{1}{2}\bar{L}_m - L_{\Delta m}\cos 2\theta_r \\ -\frac{1}{2}\bar{L}_m - L_{\Delta m}\cos 2\left(\theta_r + \frac{1}{3}\pi\right) & -\frac{1}{2}\bar{L}_m - L_{\Delta m}\cos 2\theta_r & L_{ls} + \bar{L}_m - L_{\Delta m}\cos 2\left(\theta_r + \frac{2}{3}\pi\right) \end{bmatrix}$$

It was shown that $\bar{L}_m$ and $L_{\Delta m}$ are found as

$$\bar{L}_m = \frac{1}{3}\left(\frac{N_s^2}{\Re_{mq}} + \frac{N_s^2}{\Re_{md}}\right) \text{ and } L_{\Delta m} = \frac{1}{3}\left(\frac{N_s^2}{\Re_{md}} - \frac{N_s^2}{\Re_{mq}}\right)$$

Permanent-magnet synchronous machines are round-rotor electrical machines. For round-rotor synchronous machines, the magnetic paths in the *quadrature* and *direct* magnetic axes are identical, and hence $\Re_{mq} = \Re_{md}$. Thus,

$$\bar{L}_m = \frac{2N_s^2}{3\Re_{mq}} = \frac{2N_s^2}{3\Re_{md}} \text{ and } L_{\Delta m} = 0$$

Hence, matrix $\mathbf{L}_s$ becomes

$$\mathbf{L}_s = \begin{bmatrix} L_{ls}+\bar{L}_m & -\frac{1}{2}\bar{L}_m & -\frac{1}{2}\bar{L}_m \\ -\frac{1}{2}\bar{L}_m & L_{ls}+\bar{L}_m & -\frac{1}{2}\bar{L}_m \\ -\frac{1}{2}\bar{L}_m & -\frac{1}{2}\bar{L}_m & L_{ls}+\bar{L}_m \end{bmatrix}$$

and the expressions for the flux linkages are found to be

$$\begin{aligned} \psi_{as} &= (L_{ls}+\bar{L}_m)i_{as} - \frac{1}{2}\bar{L}_m i_{bs} - \frac{1}{2}\bar{L}_m i_{cs} + \psi_m \sin\theta_r \\ \psi_{bs} &= -\frac{1}{2}\bar{L}_m i_{as} + (L_{ls}+\bar{L}_m)i_{bs} - \frac{1}{2}\bar{L}_m i_{cs} + \psi_m \sin\left(\theta_r - \frac{2}{3}\pi\right) \\ \psi_{cs} &= -\frac{1}{2}\bar{L}_m i_{as} - \frac{1}{2}\bar{L}_m i_{bs} + (L_{ls}+\bar{L}_m)i_{cs} + \psi_m \sin\left(\theta_r + \frac{2}{3}\pi\right) \end{aligned} \tag{7.13}$$

or

$$\boldsymbol{\psi}_{abcs} = \mathbf{L}_s\mathbf{i}_{abcs} + \boldsymbol{\psi}_m = \begin{bmatrix} L_{ls}+\bar{L}_m & -\frac{1}{2}\bar{L}_m & -\frac{1}{2}\bar{L}_m \\ -\frac{1}{2}\bar{L}_m & L_{ls}+\bar{L}_m & -\frac{1}{2}\bar{L}_m \\ -\frac{1}{2}\bar{L}_m & -\frac{1}{2}\bar{L}_m & L_{ls}+\bar{L}_m \end{bmatrix} \begin{bmatrix} i_{as} \\ i_{bs} \\ i_{cs} \end{bmatrix} + \psi_m \begin{bmatrix} \sin\theta_r \\ \sin\left(\theta_r - \frac{2}{3}\pi\right) \\ \sin\left(\theta_r + \frac{2}{3}\pi\right) \end{bmatrix}$$

Using (7.11) and (7.13), we have

$$\mathbf{u}_{abcs} = \mathbf{r}_s\mathbf{i}_{abcs} + \frac{d\boldsymbol{\psi}_{abcs}}{dt} = \mathbf{r}_s\mathbf{i}_{abcs} + \mathbf{L}_s\frac{d\mathbf{i}_{abcs}}{dt} + \frac{d\boldsymbol{\psi}_m}{dt}$$

where $\frac{d\boldsymbol{\psi}_m}{dt} = \psi_m \begin{bmatrix} \omega_r\cos\theta_r \\ \omega_r\cos\left(\theta_r - \frac{2}{3}\pi\right) \\ \omega_r\cos\left(\theta_r + \frac{2}{3}\pi\right) \end{bmatrix}$.

Hence, Cauchy's form can be found by making use of $\mathbf{L}_s^{-1}$. That is,

$$\frac{d\mathbf{i}_{abcs}}{dt} = -\mathbf{L}_s^{-1}\mathbf{r}_s\mathbf{i}_{abcs} - \mathbf{L}_s^{-1}\frac{d\boldsymbol{\psi}_m}{dt} + \mathbf{L}_s^{-1}\mathbf{u}_{abcs}$$

Denoting

$$L_{ss} = L_{ls} + \bar{L}_m$$

one finds the inverse matrix $\mathbf{L}_s^{-1}$ as

$$\mathbf{L}_s^{-1} = \frac{1}{L_{ss}^3 - \frac{3}{4}L_{ss}\bar{L}_m^2 - \frac{1}{4}\bar{L}_m^3}\begin{bmatrix} 4L_{ss}^2 - \bar{L}_m^2 & 2L_{ss}\bar{L}_m + \bar{L}_m^2 & 2L_{ss}\bar{L}_m + \bar{L}_m^2 \\ 2L_{ss}\bar{L}_m + \bar{L}_m^2 & 4L_{ss}^2 - \bar{L}_m^2 & 2L_{ss}\bar{L}_m + \bar{L}_m^2 \\ 2L_{ss}\bar{L}_m + \bar{L}_m^2 & 2L_{ss}\bar{L}_m + \bar{L}_m^2 & 4L_{ss}^2 - \bar{L}_m^2 \end{bmatrix}$$

Thus, the circuitry dynamics in Cauchy's form is given by

$$\begin{bmatrix} \frac{di_{as}}{dt} \\ \frac{di_{bs}}{dt} \\ \frac{di_{cs}}{dt} \end{bmatrix} = -\frac{1}{L_{ss}^3 - \frac{3}{4}L_{ss}\bar{L}_m^2 - \frac{1}{4}\bar{L}_m^3}\begin{bmatrix} 4L_{ss}^2 - \bar{L}_m^2 & 2L_{ss}\bar{L}_m + \bar{L}_m^2 & 2L_{ss}\bar{L}_m + \bar{L}_m^2 \\ 2L_{ss}\bar{L}_m + \bar{L}_m^2 & 4L_{ss}^2 - \bar{L}_m^2 & 2L_{ss}\bar{L}_m + \bar{L}_m^2 \\ 2L_{ss}\bar{L}_m + \bar{L}_m^2 & 2L_{ss}\bar{L}_m + \bar{L}_m^2 & 4L_{ss}^2 - \bar{L}_m^2 \end{bmatrix}\begin{bmatrix} r_s & 0 & 0 \\ 0 & r_s & 0 \\ 0 & 0 & r_s \end{bmatrix}\begin{bmatrix} i_{as} \\ i_{bs} \\ i_{cs} \end{bmatrix}$$

$$-\frac{\psi_m}{L_{ss}^3 - \frac{3}{4}L_{ss}\bar{L}_m^2 - \frac{1}{4}\bar{L}_m^3}\begin{bmatrix} 4L_{ss}^2 - \bar{L}_m^2 & 2L_{ss}\bar{L}_m + \bar{L}_m^2 & 2L_{ss}\bar{L}_m + \bar{L}_m^2 \\ 2L_{ss}\bar{L}_m + \bar{L}_m^2 & 4L_{ss}^2 - \bar{L}_m^2 & 2L_{ss}\bar{L}_m + \bar{L}_m^2 \\ 2L_{ss}\bar{L}_m + \bar{L}_m^2 & 2L_{ss}\bar{L}_m + \bar{L}_m^2 & 4L_{ss}^2 - \bar{L}_m^2 \end{bmatrix}\begin{bmatrix} \omega_r \cos\theta_r \\ \omega_r \cos\left(\theta_r - \frac{2}{3}\pi\right) \\ \omega_r \cos\left(\theta_r + \frac{2}{3}\pi\right) \end{bmatrix}$$

$$+\frac{1}{L_{ss}^3 - \frac{3}{4}L_{ss}\bar{L}_m^2 - \frac{1}{4}\bar{L}_m^3}\begin{bmatrix} 4L_{ss}^2 - \bar{L}_m^2 & 2L_{ss}\bar{L}_m + \bar{L}_m^2 & 2L_{ss}\bar{L}_m + \bar{L}_m^2 \\ 2L_{ss}\bar{L}_m + \bar{L}_m^2 & 4L_{ss}^2 - \bar{L}_m^2 & 2L_{ss}\bar{L}_m + \bar{L}_m^2 \\ 2L_{ss}\bar{L}_m + \bar{L}_m^2 & 2L_{ss}\bar{L}_m + \bar{L}_m^2 & 4L_{ss}^2 - \bar{L}_m^2 \end{bmatrix}\begin{bmatrix} u_{as} \\ u_{bs} \\ u_{cs} \end{bmatrix}$$

After simplification, we have

$$\begin{bmatrix} \frac{di_{as}}{dt} \\ \frac{di_{bs}}{dt} \\ \frac{di_{cs}}{dt} \end{bmatrix} = \begin{bmatrix} -\frac{r_s(2L_{ss} - \bar{L}_m)}{2L_{ss}^2 - L_{ss}\bar{L}_m - \bar{L}_m^2} & -\frac{r_s\bar{L}_m}{2L_{ss}^2 - L_{ss}\bar{L}_m - \bar{L}_m^2} & -\frac{r_s\bar{L}_m}{2L_{ss}^2 - L_{ss}\bar{L}_m - \bar{L}_m^2} \\ -\frac{r_s\bar{L}_m}{2L_{ss}^2 - L_{ss}\bar{L}_m - \bar{L}_m^2} & -\frac{r_s(2L_{ss} - \bar{L}_m)}{2L_{ss}^2 - L_{ss}\bar{L}_m - \bar{L}_m^2} & -\frac{r_s\bar{L}_m}{2L_{ss}^2 - L_{ss}\bar{L}_m - \bar{L}_m^2} \\ -\frac{r_s\bar{L}_m}{2L_{ss}^2 - L_{ss}\bar{L}_m - \bar{L}_m^2} & -\frac{r_s\bar{L}_m}{2L_{ss}^2 - L_{ss}\bar{L}_m - \bar{L}_m^2} & -\frac{r_s(2L_{ss} - \bar{L}_m)}{2L_{ss}^2 - L_{ss}\bar{L}_m - \bar{L}_m^2} \end{bmatrix}\begin{bmatrix} i_{as} \\ i_{bs} \\ i_{cs} \end{bmatrix}$$

$$+\begin{bmatrix} -\frac{\psi_m(2L_{ss} - \bar{L}_m)}{2L_{ss}^2 - L_{ss}\bar{L}_m - \bar{L}_m^2} & -\frac{\psi_m\bar{L}_m}{2L_{ss}^2 - L_{ss}\bar{L}_m - \bar{L}_m^2} & -\frac{\psi_m\bar{L}_m}{2L_{ss}^2 - L_{ss}\bar{L}_m - \bar{L}_m^2} \\ -\frac{\psi_m\bar{L}_m}{2L_{ss}^2 - L_{ss}\bar{L}_m - \bar{L}_m^2} & -\frac{\psi_m(2L_{ss} - \bar{L}_m)}{2L_{ss}^2 - L_{ss}\bar{L}_m - \bar{L}_m^2} & -\frac{\psi_m\bar{L}_m}{2L_{ss}^2 - L_{ss}\bar{L}_m - \bar{L}_m^2} \\ -\frac{\psi_m\bar{L}_m}{2L_{ss}^2 - L_{ss}\bar{L}_m - \bar{L}_m^2} & -\frac{\psi_m\bar{L}_m}{2L_{ss}^2 - L_{ss}\bar{L}_m - \bar{L}_m^2} & -\frac{\psi_m(2L_{ss} - \bar{L}_m)}{2L_{ss}^2 - L_{ss}\bar{L}_m - \bar{L}_m^2} \end{bmatrix}\begin{bmatrix} \omega_r \cos\theta_r \\ \omega_r \cos\left(\theta_r - \frac{2}{3}\pi\right) \\ \omega_r \cos\left(\theta_r + \frac{2}{3}\pi\right) \end{bmatrix}$$

$$+\begin{bmatrix} \frac{2L_{ss} - \bar{L}_m}{2L_{ss}^2 - L_{ss}\bar{L}_m - \bar{L}_m^2} & \frac{\bar{L}_m}{2L_{ss}^2 - L_{ss}\bar{L}_m - \bar{L}_m^2} & \frac{\bar{L}_m}{2L_{ss}^2 - L_{ss}\bar{L}_m - \bar{L}_m^2} \\ \frac{\bar{L}_m}{2L_{ss}^2 - L_{ss}\bar{L}_m - \bar{L}^2} & \frac{2L_{ss} - \bar{L}_m}{2L_{ss}^2 - L_{ss}\bar{L}_m - \bar{L}_m^2} & \frac{\bar{L}_m}{2L_{ss}^2 - L_{ss}\bar{L}_m - \bar{L}_m^2} \\ \frac{\bar{L}_m}{2L_{ss}^2 - L_{ss}\bar{L}_m - \bar{L}_m^2} & \frac{\bar{L}_m}{2L_{ss}^2 - L_{ss}\bar{L}_m - \bar{L}_m^2} & \frac{2L_{ss} - \bar{L}_m}{2L_{ss}^2 - L_{ss}\bar{L}_m - \bar{L}_m^2} \end{bmatrix}\begin{bmatrix} u_{as} \\ u_{bs} \\ u_{cs} \end{bmatrix}$$

In expanded form, one finds

$$
\begin{aligned}
\frac{di_{as}}{dt} = &-\frac{r_s(2L_{ss}-\bar{L}_m)}{2L_{ss}^2-L_{ss}\bar{L}_m-\bar{L}_m^2}i_{as}-\frac{r_s\bar{L}_m}{2L_{ss}^2-L_{ss}\bar{L}_m-\bar{L}_m^2}i_{bs}-\frac{r_s\bar{L}_m}{2L_{ss}^2-L_{ss}\bar{L}_m-\bar{L}_m^2}i_{cs}\\
&-\frac{\psi_m(2L_{ss}-\bar{L}_m)}{2L_{ss}^2-L_{ss}\bar{L}_m-\bar{L}_m^2}\omega_r\cos\theta_r-\frac{\psi_m\bar{L}_m}{2L_{ss}^2-L_{ss}\bar{L}_m-\bar{L}_m^2}\omega_r\cos\left(\theta_r-\frac{2}{3}\pi\right)-\frac{\psi_m\bar{L}_m}{2L_{ss}^2-L_{ss}\bar{L}_m-\bar{L}_m^2}\omega_r\cos\left(\theta_r+\frac{2}{3}\pi\right)\\
&+\frac{2L_{ss}-\bar{L}_m}{2L_{ss}^2-L_{ss}\bar{L}_m-\bar{L}_m^2}u_{as}+\frac{\bar{L}_m}{2L_{ss}^2-L_{ss}\bar{L}_m-\bar{L}_m^2}u_{bs}+\frac{\bar{L}_m}{2L_{ss}^2-L_{ss}\bar{L}_m-\bar{L}_m^2}u_{cs}\\
\frac{di_{bs}}{dt} = &-\frac{r_s\bar{L}_m}{2L_{ss}^2-L_{ss}\bar{L}_m-\bar{L}_m^2}i_{as}-\frac{r_s(2L_{ss}-\bar{L}_m)}{2L_{ss}^2-L_{ss}\bar{L}_m-\bar{L}_m^2}i_{bs}-\frac{r_s\bar{L}_m}{2L_{ss}^2-L_{ss}\bar{L}_m-\bar{L}_m^2}i_{cs}\\
&-\frac{\psi_m\bar{L}_m}{2L_{ss}^2-L_{ss}\bar{L}_m-\bar{L}_m^2}\omega_r\cos\theta_r-\frac{\psi_m(2L_{ss}-\bar{L}_m)}{2L_{ss}^2-L_{ss}\bar{L}_m-\bar{L}_m^2}\omega_r\cos\left(\theta_r-\frac{2}{3}\pi\right)-\frac{\psi_m\bar{L}_m}{2L_{ss}^2-L_{ss}\bar{L}_m-\bar{L}_m^2}\omega_r\cos\left(\theta_r+\frac{2}{3}\pi\right)\\
&+\frac{\bar{L}_m}{2L_{ss}^2-L_{ss}\bar{L}_m-\bar{L}_m^2}u_{as}+\frac{2L_{ss}-\bar{L}_m}{2L_{ss}^2-L_{ss}\bar{L}_m-\bar{L}_m^2}u_{bs}+\frac{\bar{L}_m}{2L_{ss}^2-L_{ss}\bar{L}_m-\bar{L}_m^2}u_{cs}\\
\frac{di_{cs}}{dt} = &-\frac{r_s\bar{L}_m}{2L_{ss}^2-L_{ss}\bar{L}_m-\bar{L}_m^2}i_{as}-\frac{r_s\bar{L}_m}{2L_{ss}^2-L_{ss}\bar{L}_m-\bar{L}_m^2}i_{bs}-\frac{r_s(2L_{ss}-\bar{L}_m)}{2L_{ss}^2-L_{ss}\bar{L}_m-\bar{L}_m^2}i_{cs}\\
&-\frac{\psi_m\bar{L}_m}{2L_{ss}^2-L_{ss}\bar{L}_m-\bar{L}_m^2}\omega_r\cos\theta_r-\frac{\psi_m\bar{L}_m}{2L_{ss}^2-L_{ss}\bar{L}_m-\bar{L}_m^2}\omega_r\cos\left(\theta_r-\frac{2}{3}\pi\right)-\frac{\psi_m(2L_{ss}-\bar{L}_m)}{2L_{ss}^2-L_{ss}\bar{L}_m-\bar{L}_m^2}\omega_r\cos\left(\theta_r+\frac{2}{3}\pi\right)\\
&+\frac{\bar{L}_m}{2L_{ss}^2-L_{ss}\bar{L}_m-\bar{L}_m^2}u_{as}+\frac{\bar{L}_m}{2L_{ss}^2-L_{ss}\bar{L}_m-\bar{L}_m^2}u_{bs}+\frac{2L_{ss}-\bar{L}_m}{2L_{ss}^2-L_{ss}\bar{L}_m-\bar{L}_m^2}u_{cs}
\end{aligned}
\tag{7.14}
$$

Having derived the differential equations to map the circuitry dynamics in (7.14), the transient behavior of the mechanical system must be incorporated. Otherwise, one cannot solve (7.14) where the electrical angular velocity ω_r and angular displacement θ_r are used as the state variables. Making use of Newton's second law $T_e - B_m\omega_{rm} - T_L = J\frac{d^2\theta_{rm}}{dt^2}$, we have

$$\frac{d\omega_{rm}}{dt} = \frac{1}{J}(T_e - B_m\omega_{rm} - T_L)$$

$$\frac{d\theta_{rm}}{dt} = \omega_{rm}$$

One concludes that the expression for the electromagnetic torque developed must be found. The coenergy is found to be

$$W_c = W_f = \frac{1}{2}\begin{bmatrix} i_{as} & i_{bs} & i_{cs}\end{bmatrix}\mathbf{L}_s\begin{bmatrix} i_{as}\\ i_{bs}\\ i_{cs}\end{bmatrix} + \begin{bmatrix} i_{as} & i_{bs} & i_{cs}\end{bmatrix}\begin{bmatrix}\psi_m\sin\theta_r\\ \psi_m\sin\left(\theta_r-\frac{2}{3}\pi\right)\\ \psi_m\sin\left(\theta_r+\frac{2}{3}\pi\right)\end{bmatrix} + W_{PM}$$

where W_{PM} is the energy stored in the permanent magnet.

Permanent-magnet synchronous machines are round-rotor machines. Hence, $\mathbf{L}_s$, which is given as $\mathbf{L}_s = \begin{bmatrix} L_{ls}+\bar{L}_m & -\frac{1}{2}\bar{L}_m & -\frac{1}{2}\bar{L}_m \\ -\frac{1}{2}\bar{L}_m & L_{ls}+\bar{L}_m & -\frac{1}{2}\bar{L}_m \\ -\frac{1}{2}\bar{L}_m & -\frac{1}{2}\bar{L}_m & L_{ls}+\bar{L}_m \end{bmatrix}$, as well as W_{PM}, are not functions of θ_r. Therefore, one obtains the following formula to calculate the electromagnetic torque for three-phase P-pole permanent-magnet synchronous motors

$$T_e = \frac{P}{2}\frac{\partial W_c}{\partial \theta_r} = \frac{P\psi_m}{2}\left(i_{as}\cos\theta_r + i_{bs}\cos\left(\theta_r - \frac{2}{3}\pi\right) + i_{cs}\cos\left(\theta_r + \frac{2}{3}\pi\right)\right)$$

Hence,

$$\frac{d\omega_{rm}}{dt} = \frac{P\psi_m}{2J}\left(i_{as}\cos\theta_r + i_{bs}\cos\left(\theta_r - \frac{2}{3}\pi\right) + i_{cs}\cos\left(\theta_r + \frac{2}{3}\pi\right)\right) - \frac{B_m}{J}\omega_{rm} - \frac{1}{J}T_L$$

$$\frac{d\theta_{rm}}{dt} = \omega_{rm}$$

Using the electrical angular velocity ω_r and displacement θ_r, related to the mechanical angular velocity and displacement as $\omega_{rm} = \frac{2}{P}\omega_r$ and $\theta_{rm} = \frac{2}{P}\theta_r$, the following differential equations to map the *torsional-mechanical* transient dynamics result

$$\frac{d\omega_r}{dt} = \frac{P^2\psi_m}{4J}\left(i_{as}\cos\theta_r + i_{bs}\cos\left(\theta_r - \frac{2}{3}\pi\right) + i_{cs}\cos\left(\theta_r + \frac{2}{3}\pi\right)\right) - \frac{B_m}{J}\omega_r - \frac{P}{2J}T_L$$
$$\frac{d\theta_r}{dt} = \omega_r \tag{7.15}$$

From (7.14) and (7.15), one obtains a nonlinear mathematical model in Cauchy's form, as given by a system of highly nonlinear differential equations

$$\begin{aligned}
\frac{di_{as}}{dt} = &-\frac{r_s(2L_{ss}-\bar{L}_m)}{2L_{ss}^2 - L_{ss}\bar{L}_m - \bar{L}_m^2}i_{as} - \frac{r_s\bar{L}_m}{2L_{ss}^2 - L_{ss}\bar{L}_m - \bar{L}_m^2}i_{bs} - \frac{r_s\bar{L}_m}{2L_{ss}^2 - L_{ss}\bar{L}_m - \bar{L}_m^2}i_{cs} \\
&- \frac{\psi_m(2L_{ss}-\bar{L}_m)}{2L_{ss}^2 - L_{ss}\bar{L}_m - \bar{L}_m^2}\omega_r\cos\theta_r - \frac{\psi_m\bar{L}_m}{2L_{ss}^2 - L_{ss}\bar{L}_m - \bar{L}_m^2}\omega_r\cos\left(\theta_r - \frac{2}{3}\pi\right) - \frac{\psi_m\bar{L}_m}{2L_{ss}^2 - L_{ss}\bar{L}_m - \bar{L}_m^2}\omega_r\cos\left(\theta_r + \frac{2}{3}\pi\right) \\
&+ \frac{2L_{ss}-\bar{L}_m}{2L_{ss}^2 - L_{ss}\bar{L}_m - \bar{L}_m^2}u_{as} + \frac{\bar{L}_m}{2L_{ss}^2 - L_{ss}\bar{L}_m - \bar{L}_m^2}u_{bs} + \frac{\bar{L}_m}{2L_{ss}^2 - L_{ss}\bar{L}_m - \bar{L}_m^2}u_{cs}
\end{aligned}$$

$$\begin{aligned}
\frac{di_{bs}}{dt} = &-\frac{r_s\bar{L}_m}{2L_{ss}^2 - L_{ss}\bar{L}_m - \bar{L}_m^2}i_{as} - \frac{r_s(2L_{ss}-\bar{L}_m)}{2L_{ss}^2 - L_{ss}\bar{L}_m - \bar{L}_m^2}i_{bs} - \frac{r_s\bar{L}_m}{2L_{ss}^2 - L_{ss}\bar{L}_m - \bar{L}_m^2}i_{cs} \\
&- \frac{\psi_m\bar{L}_m}{2L_{ss}^2 - L_{ss}\bar{L}_m - \bar{L}_m^2}\omega_r\cos\theta_r - \frac{\psi_m(2L_{ss}-\bar{L}_m)}{2L_{ss}^2 - L_{ss}\bar{L}_m - \bar{L}_m^2}\omega_r\cos\left(\theta_r - \frac{2}{3}\pi\right) - \frac{\psi_m\bar{L}_m}{2L_{ss}^2 - L_{ss}\bar{L}_m - \bar{L}_m^2}\omega_r\cos\left(\theta_r + \frac{2}{3}\pi\right)
\end{aligned}$$

$$
\begin{aligned}
&+\frac{\bar{L}_m}{2L_{ss}^2-L_{ss}\bar{L}_m-\bar{L}_m^2}u_{as}+\frac{2L_{ss}-\bar{L}_m}{2L_{ss}^2-L_{ss}\bar{L}_m-\bar{L}_m^2}u_{bs}+\frac{\bar{L}_m}{2L_{ss}^2-L_{ss}\bar{L}_m-\bar{L}_m^2}u_{cs}\\
&\frac{di_{cs}}{dt}=-\frac{r_s\bar{L}_m}{2L_{ss}^2-L_{ss}\bar{L}_m-\bar{L}_m^2}i_{as}-\frac{r_s\bar{L}_m}{2L_{ss}^2-L_{ss}\bar{L}_m-\bar{L}_m^2}i_{bs}-\frac{r_s(2L_{ss}-\bar{L}_m)}{2L_{ss}^2-L_{ss}\bar{L}_m-\bar{L}_m^2}i_{cs}\\
&-\frac{\psi_m\bar{L}_m}{2L_{ss}^2-L_{ss}\bar{L}_m-\bar{L}_m^2}\omega_r\cos\theta_r-\frac{\psi_m\bar{L}_m}{2L_{ss}^2-L_{ss}\bar{L}_m-\bar{L}_m^2}\omega_r\cos\left(\theta_r-\frac{2}{3}\pi\right)-\frac{\psi_m(2L_{ss}-\bar{L}_m)}{2L_{ss}^2-L_{ss}\bar{L}_m-\bar{L}_m^2}\omega_r\cos\left(\theta_r+\frac{2}{3}\pi\right)\\
&+\frac{\bar{L}_m}{2L_{ss}^2-L_{ss}\bar{L}_m-\bar{L}_m^2}u_{as}+\frac{\bar{L}_m}{2L_{ss}^2-L_{ss}\bar{L}_m-\bar{L}_m^2}u_{bs}+\frac{2L_{ss}-\bar{L}_m}{2L_{ss}^2-L_{ss}\bar{L}_m-\bar{L}_m^2}u_{cs}\\
&\frac{d\omega_r}{dt}=\frac{P^2\psi_m}{4J}\left(i_{as}\cos\theta_r+i_{bs}\cos\left(\theta_r-\frac{2}{3}\pi\right)+i_{cs}\cos\left(\theta_r+\frac{2}{3}\pi\right)\right)-\frac{B_m}{J}\omega_r-\frac{P}{2J}T_L\\
&\frac{d\theta_r}{dt}=\omega_r
\end{aligned}
\tag{7.16}
$$

In matrix for, we have

$$
\begin{bmatrix}\frac{di_{as}}{dt}\\ \frac{di_{bs}}{dt}\\ \frac{di_{cs}}{dt}\\ \frac{d\omega_r}{dt}\\ \frac{d\theta_r}{dt}\end{bmatrix}=
\begin{bmatrix}
-\frac{r_s(2L_{ss}-\bar{L}_m)}{2L_{ss}^2-L_{ss}\bar{L}_m-\bar{L}_m^2} & -\frac{r_s\bar{L}_m}{2L_{ss}^2-L_{ss}\bar{L}_m-\bar{L}_m^2} & -\frac{r_s\bar{L}_m}{2L_{ss}^2-L_{ss}\bar{L}_m-\bar{L}_m^2} & 0 & 0\\
-\frac{r_s\bar{L}_m}{2L_{ss}^2-L_{ss}\bar{L}_m-\bar{L}_m^2} & -\frac{r_s(2L_{ss}-\bar{L}_m)}{2L_{ss}^2-L_{ss}\bar{L}_m-\bar{L}_m^2} & -\frac{r_s\bar{L}_m}{2L_{ss}^2-L_{ss}\bar{L}_m-\bar{L}_m^2} & 0 & 0\\
-\frac{r_s\bar{L}_m}{2L_{ss}^2-L_{ss}\bar{L}_m-\bar{L}_m^2} & -\frac{r_s\bar{L}_m}{2L_{ss}^2-L_{ss}\bar{L}_m-\bar{L}_m^2} & -\frac{r_s(2L_{ss}-\bar{L}_m)}{2L_{ss}^2-L_{ss}\bar{L}_m-\bar{L}_m^2} & 0 & 0\\
0 & 0 & 0 & -\frac{B_m}{J} & 0\\
0 & 0 & 0 & 1 & 0
\end{bmatrix}
\begin{bmatrix}i_{as}\\ i_{bc}\\ i_{cs}\\ \omega_r\\ \theta_r\end{bmatrix}
$$

$$
+\begin{bmatrix}
-\frac{\psi_m(2L_{ss}-\bar{L}_m)}{2L_{ss}^2-L_{ss}\bar{L}_m-\bar{L}_m^2}\omega_r & -\frac{\psi_m\bar{L}_m}{2L_{ss}^2-L_{ss}\bar{L}_m-\bar{L}_m^2}\omega_r & -\frac{\psi_m\bar{L}_m}{2L_{ss}^2-L_{ss}\bar{L}_m-\bar{L}_m^2}\omega_r\\
-\frac{\psi_m\bar{L}_m}{2L_{ss}^2-L_{ss}\bar{L}_m-\bar{L}_m^2}\omega_r & -\frac{\psi_m(2L_{ss}-\bar{L}_m)}{2L_{ss}^2-L_{ss}\bar{L}_m-\bar{L}_m^2}\omega_r & -\frac{\psi_m\bar{L}_m}{2L_{ss}^2-L_{ss}\bar{L}_m-\bar{L}_m^2}\omega_r\\
-\frac{\psi_m\bar{L}_m}{2L_{ss}^2-L_{ss}\bar{L}_m-\bar{L}_m^2}\omega_r & -\frac{\psi_m\bar{L}_m}{2L_{ss}^2-L_{ss}\bar{L}_m-\bar{L}_m^2}\omega_r & -\frac{\psi_m(2L_{ss}-\bar{L}_m)}{2L_{ss}^2-L_{ss}\bar{L}_m-\bar{L}_m^2}\omega_r\\
\frac{P^2\psi_m}{4J}i_{as} & \frac{P^2\psi_m}{4J}i_{bs} & \frac{P^2\psi_m}{4J}i_{cs}\\
0 & 0 & 0
\end{bmatrix}
\begin{bmatrix}\cos\theta_r\\ \cos\left(\theta_r-\frac{2}{3}\pi\right)\\ \cos\left(\theta_r+\frac{2}{3}\pi\right)\end{bmatrix}
$$

$$
+\begin{bmatrix}
\frac{2L_{ss}-\bar{L}_m}{2L_{ss}^2-L_{ss}\bar{L}_m-\bar{L}_m^2} & \frac{\bar{L}_m}{2L_{ss}^2-L_{ss}\bar{L}_m-\bar{L}_m^2} & \frac{\bar{L}_m}{2L_{ss}^2-L_{ss}\bar{L}_m-\bar{L}_m^2}\\
\frac{\bar{L}_m}{2L_{ss}^2-L_{ss}\bar{L}_m-\bar{L}_m^2} & \frac{2L_{ss}-\bar{L}_m}{2L_{ss}^2-L_{ss}\bar{L}_m-\bar{L}_m^2} & \frac{\bar{L}_m}{2L_{ss}^2-L_{ss}\bar{L}_m-\bar{L}_m^2}\\
\frac{\bar{L}_m}{2L_{ss}^2-L_{ss}\bar{L}_m-\bar{L}_m^2} & \frac{\bar{L}_m}{2L_{ss}^2-L_{ss}\bar{L}_m-\bar{L}_m^2} & \frac{2L_{ss}-\bar{L}_m}{2L_{ss}^2-L_{ss}\bar{L}_m-\bar{L}_m^2}\\
0 & 0 & 0\\
0 & 0 & 0
\end{bmatrix}
\begin{bmatrix}u_{as}\\ u_{bs}\\ u_{cs}\end{bmatrix}-\begin{bmatrix}0\\0\\0\\ \frac{P}{2J}\\0\end{bmatrix}T_L
$$

To control the angular velocity, one regulates the currents fed or voltages applied to the stator windings. Neglecting the viscous friction coefficient, the analysis of Newton's second law,

$$T_e - T_L = J\frac{d\omega_{rm}}{dt}$$

indicates that the angular velocity increases if $T_e > T_L$, and ω_{rm} decreases if $T_e < T_L$. Furthermore, $\omega_{rm} = const$ if $T_e = T_L$. That is, to regulate the angular velocity, one changes the electromagnetic torque developed. The equation for the electromagnetic torque was found to be

$$T_e = \frac{P\psi_m}{2}\left(i_{as}\cos\theta_r + i_{bs}\cos\left(\theta_r - \frac{2}{3}\pi\right) + i_{cs}\cos\left(\theta_r + \frac{2}{3}\pi\right)\right)$$

If the motor is fed by a balanced three-phase current set

$$i_{as}(t) = \sqrt{2}i_M\cos(\omega_r t) = \sqrt{2}i_M\cos(\omega_e t) = \sqrt{2}i_M\cos\theta_r$$

$$i_{bs}(t) = \sqrt{2}i_M\cos\left(\omega_r t - \frac{2}{3}\pi\right) = \sqrt{2}i_M\cos\left(\omega_e t - \frac{2}{3}\pi\right) = \sqrt{2}i_M\cos\left(\theta_r - \frac{2}{3}\pi\right)$$

$$i_{cs}(t) = \sqrt{2}i_M\cos\left(\omega_r t + \frac{2}{3}\pi\right) = \sqrt{2}i_M\cos\left(\omega_e t + \frac{2}{3}\pi\right) = \sqrt{2}i_M\cos\left(\theta_r + \frac{2}{3}\pi\right)$$

one obtains

$$T_e = \frac{P\psi_m}{2}\sqrt{2}i_M\left(\cos^2\theta_r + \cos^2\left(\theta_r - \frac{2}{3}\pi\right) + \cos^2\left(\theta_r + \frac{2}{3}\pi\right)\right)$$

Taking note that $\cos^2\theta_r + \cos^2\left(\theta_r - \frac{2}{3}\pi\right) + \cos^2\left(\theta_r + \frac{2}{3}\pi\right) = \frac{3}{2}$, we have

$$T_e = \frac{3P\psi_m}{2\sqrt{2}}i_M$$

That is, the electromagnetic torque developed has been maximized, and the phase currents needed are

$$i_{as}(t) = \sqrt{2}i_M\cos\theta_r,\ i_{bs}(t) = \sqrt{2}i_M\cos\left(\theta_r - \frac{2}{3}\pi\right),\ \text{and}\ i_{cs}(t) = \sqrt{2}i_M\cos\left(\theta_r + \frac{2}{3}\pi\right)$$

For synchronous machines, the electrical and synchronous angular velocities are equal; that is, $\omega_r = \omega_e$. Furthermore, the electrical angular displacement is $\theta_r = \omega_r t = \omega_e t$. These features have been employed to find the balanced current set to be fed. One concludes that to regulate the angular velocity, the magnitude i_M must be changed. Furthermore, the phase currents $i_{as}(t)$, $i_{bs}(t)$, and $i_{cs}(t)$, which are shifted by $\frac{2}{3}\pi$, are the functions of the electrical angular displacement θ_r (measured using Hall sensors) to guarantee the maximum possible electromagnetic torque.

If voltage-fed power converters are used, one changes the magnitude of voltages $u_{as}(t)$, $u_{bs}(t)$, and $u_{cs}(t)$. The angular displacement θ_r needs to be measured or calculated in order to find a balanced three-phase voltage set. In particular, the *abc* voltages are given as

$$u_{as}(t) = \sqrt{2}u_M \cos(\theta_r + \varphi_u)$$

$$u_{bs}(t) = \sqrt{2}u_M \cos\left(\theta_r - \frac{2}{3}\pi + \varphi_u\right)$$

$$u_{cs}(t) = \sqrt{2}u_M \cos\left(\theta_r + \frac{2}{3}\pi + \varphi_u\right)$$

Neglecting the circuitry transients (assuming that inductances are negligible small), we have

$$u_{as}(t) = \sqrt{2}u_M \cos 0_r$$

$$u_{bs}(t) = \sqrt{2}u_M \cos\left(\theta_r - \frac{2}{3}\pi\right)$$

$$u_{cs}(t) = \sqrt{2}u_M \cos\left(\theta_r + \frac{2}{3}\pi\right)$$

Using a set of nonlinear differential equations (7.16) that map the transient dynamics of permanent-magnet synchronous motors, the corresponding block diagram is developed and documented in Figure 7.16. Here, $T_s = \dfrac{r_s(2L_{ss} - \bar{L}_m)}{2L_{ss}^2 - L_{ss}\bar{L}_m - \bar{L}_m^2}$.

Example 7.3

Using the block diagram of permanent-magnet synchronous motors, as illustrated in Figure 7.16, develop the SIMULINK diagram. Simulate a three-phase, two-pole permanent-magnet synchronous motor with the following parameters: r_s = 0.5 ohm, L_{ss} = 0.001 H, L_{ls} = 0.001 H, $\bar{L}_m$ = 0.0009 H, ψ_m = 0.069 V-s-rad^{-1} or ψ_m = 0.069 N-m-A^{-1}, B_m = 0.000015 N-m-s-rad^{-1}, and J = 0.000017 kg-m^2. Perform the transient analysis by supplying a balanced three-phase voltages set; u_M = 40 V.

Solution

To guarantee the balanced operating condition, the phase voltages to be applied are

$$u_{as}(t) = \sqrt{2}u_M \cos\theta_r,\ u_{bs}(t) = \sqrt{2}u_M \cos\left(\theta_r - \frac{2}{3}\pi\right),\ \text{and}\ u_{cs}(t) = \sqrt{2}u_M \cos\left(\theta_r + \frac{2}{3}\pi\right)$$

The motor accelerates from stall, and we apply the rated voltage; that is,

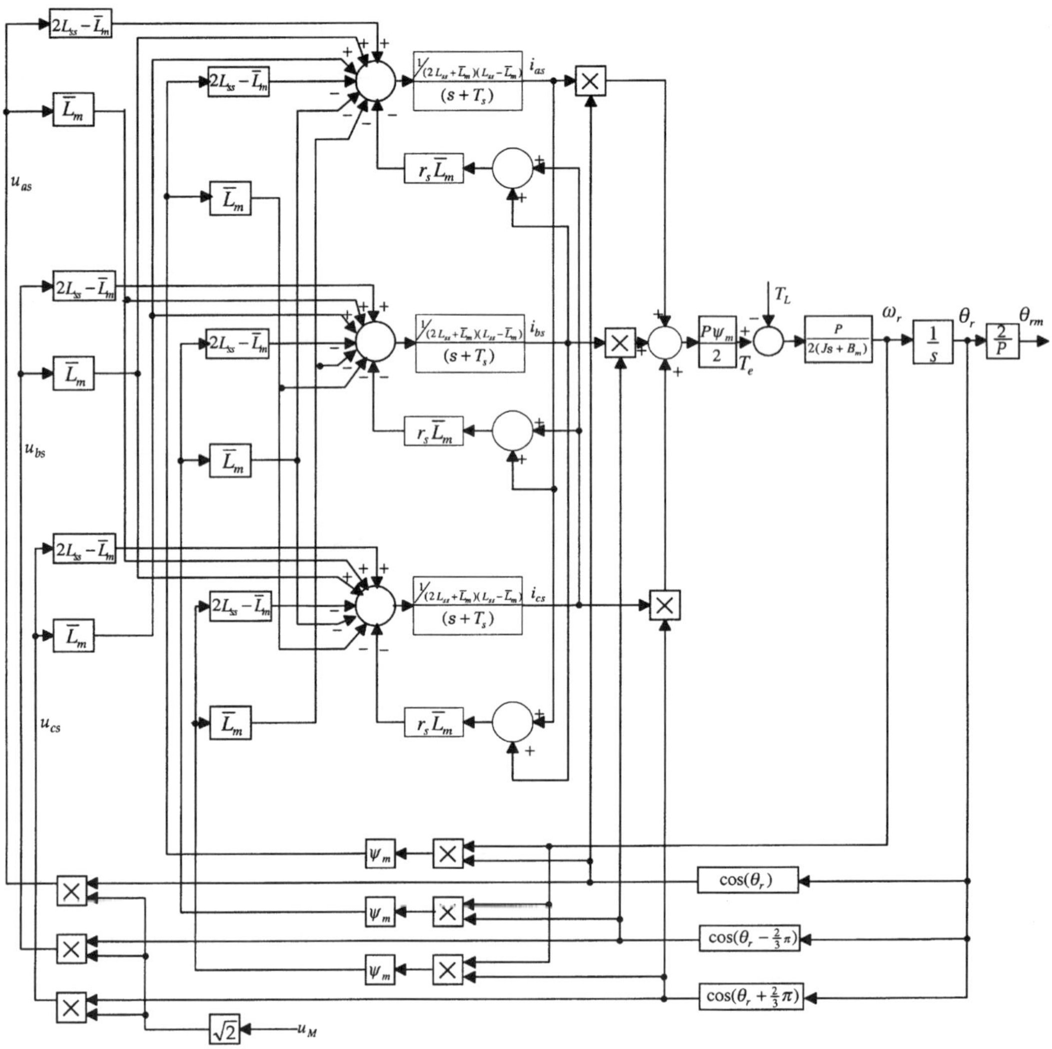

Figure 7.16 Block diagram of three-phase permanent-magnet synchronous motors controlled by supplying $u_{as}(t) = \sqrt{2}u_M \cos\theta_r$, $u_{bs}(t) = \sqrt{2}u_M \cos\left(\theta_r - \frac{2}{3}\pi\right)$, and $u_{cs}(t) = \sqrt{2}u_M \cos\left(\theta_r + \frac{2}{3}\pi\right)$.

$$u_{as}(t) = \sqrt{2}40\cos\theta_r,\ u_{bs}(t) = \sqrt{2}40\cos\left(\theta_r - \frac{2}{3}\pi\right),\ \text{and}\ u_{cs}(t) = \sqrt{2}40\cos\left(\theta_r + \frac{2}{3}\pi\right)$$

The SIMULINK diagram, which is documented in Figure 7.17, is developed using an *s*-domain diagram as shown in Figure 7.16. Nonlinear simulations were performed, and the transient dynamics of the stator currents in the *as*, *bs*, and *cs* windings $i_{as}(t)$, $i_{bs}(t)$, and $i_{cs}(t)$, as well as the rotor angular velocity $\omega_{rm}(t)$, are plotted in Figures 7.18 and 7.19. The motor accelerates from stall, and the load torques 0 and 0.3 N-m are applied at t = 0 s. The torque-speed curves $\omega_{rm} = \Omega_T(T_e)$ are also demonstrated. The acceleration capabilities of the permanent-magnet synchronous motor, documented in Figures 7.18 and 7.19, indicate that the motor reaches the steady-state angular velocity within 0.02 s. It is important to emphasize that permanent-magnet synchronous motors have superior

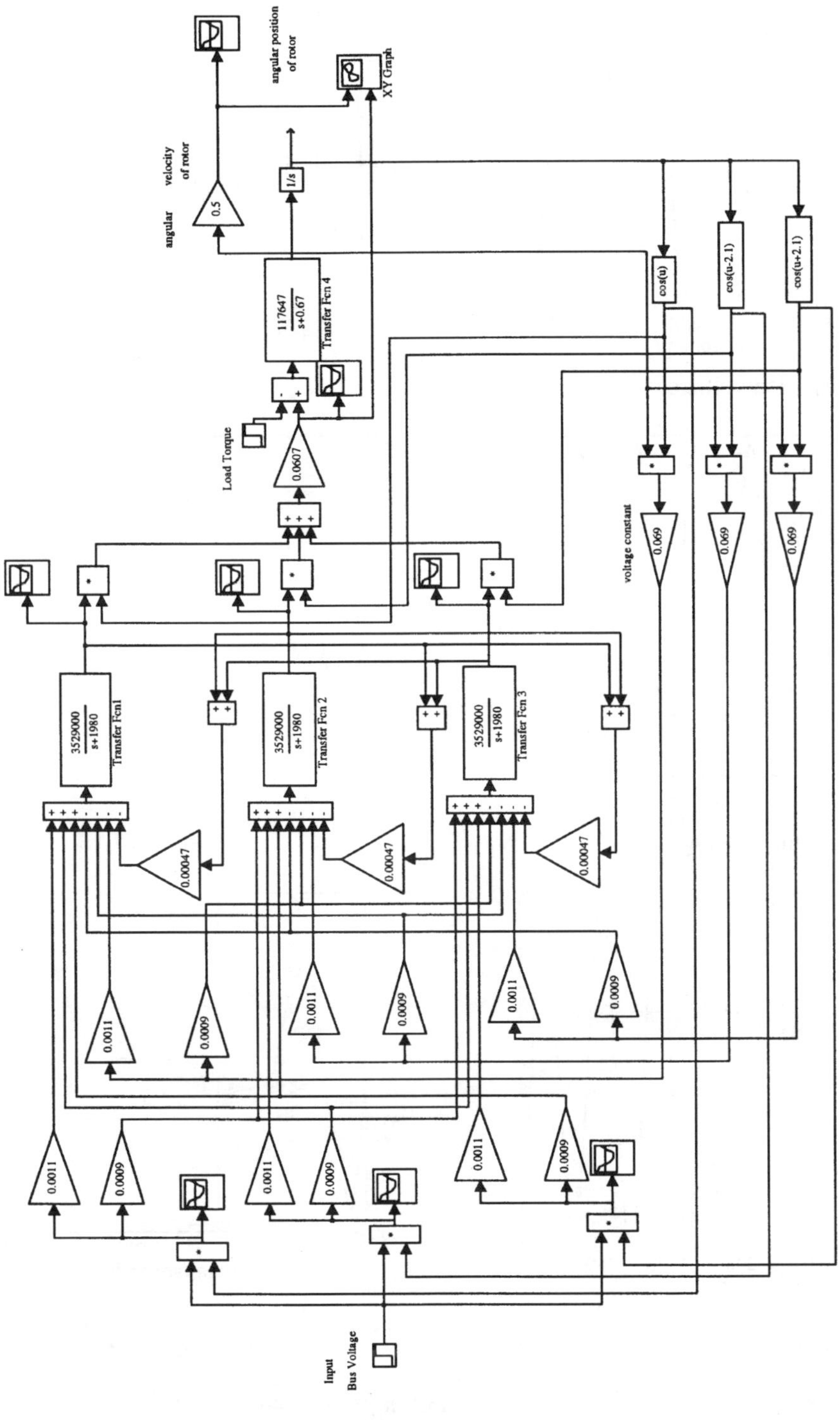

Figure 7.17 SIMULINK diagram used to model a permanent-magnet synchronous motor.

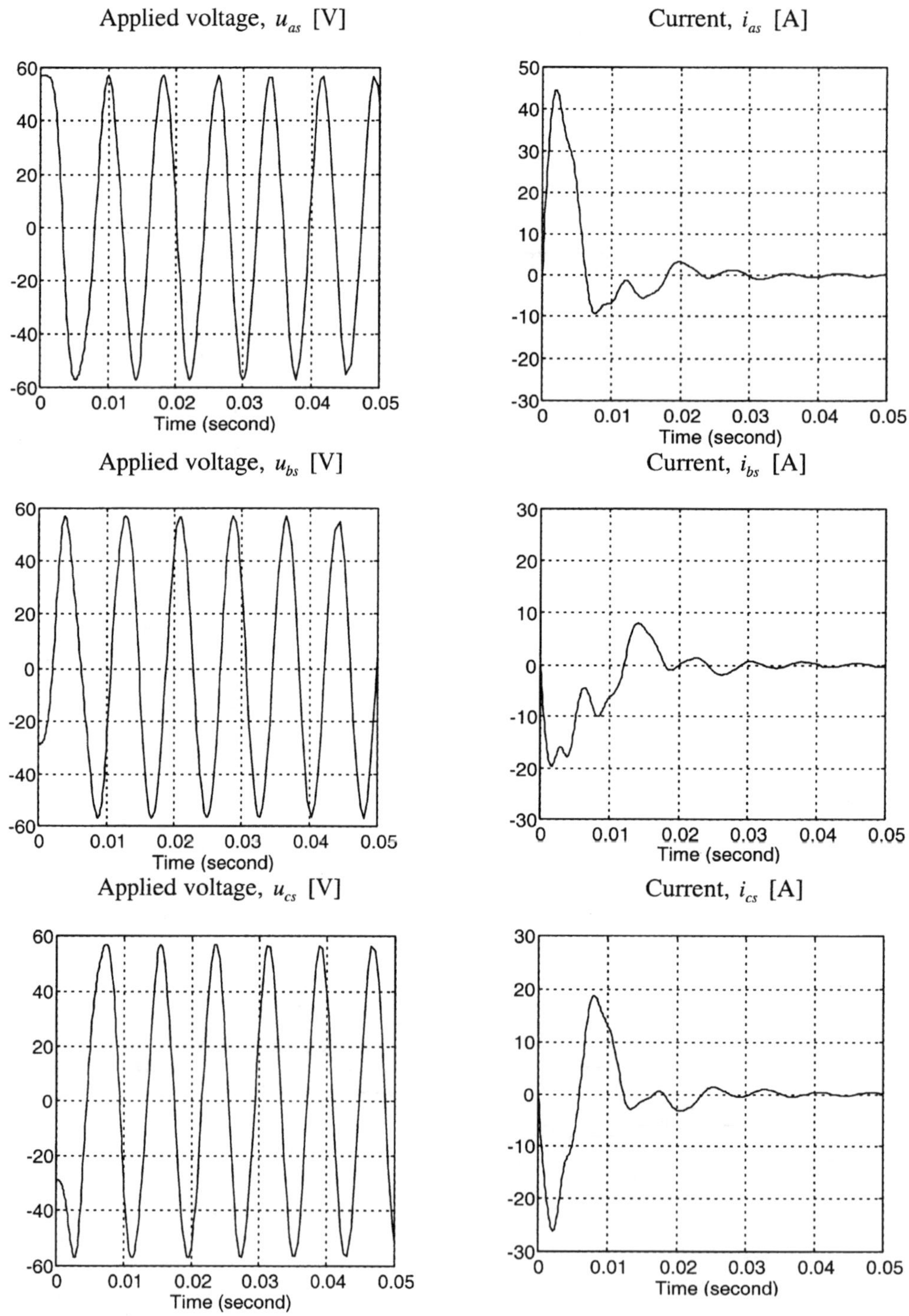

Figure 7.18 Acceleration of a three-phase permanent-magnet synchronous motor; $T_L = 0$ N-m, $u_{as}(t) = \sqrt{2}\,40\cos\theta_r$, $u_{bs}(t) = \sqrt{2}\,40\cos\left(\theta_r - \frac{2}{3}\pi\right)$, and $u_{cs}(t) = \sqrt{2}\,40\cos\left(\theta_r + \frac{2}{3}\pi\right)$.

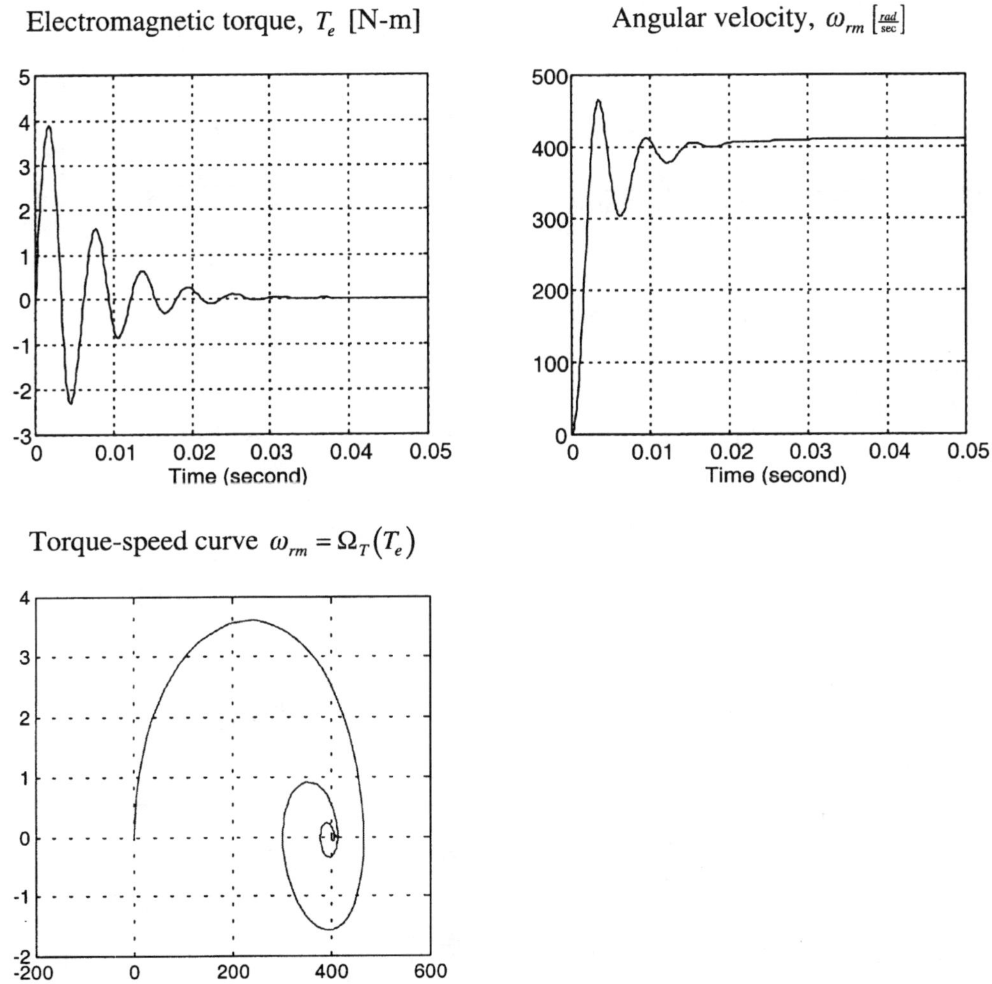

Figure 7.18 *Continued.*

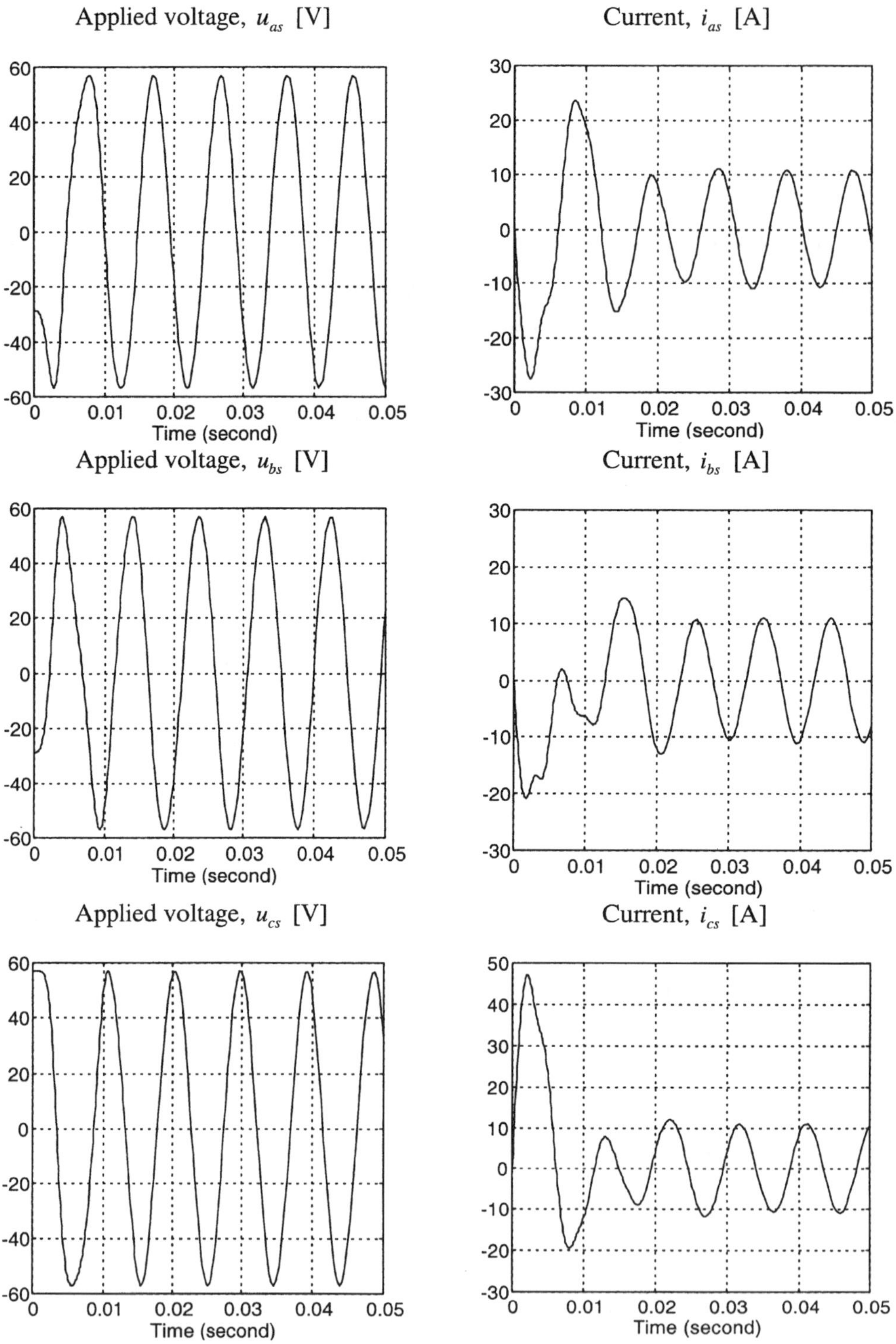

Figure 7.19 Acceleration of a three-phase permanent-magnet synchronous motor; $T_L = 0.3$ N-m, $u_{as}(t) = \sqrt{2}\,40\cos\theta_r$, $u_{bs}(t) = \sqrt{2}\,40\cos\left(\theta_r - \frac{2}{3}\pi\right)$, and $u_{cs}(t) = \sqrt{2}\,40\cos\left(\theta_r + \frac{2}{3}\pi\right)$.

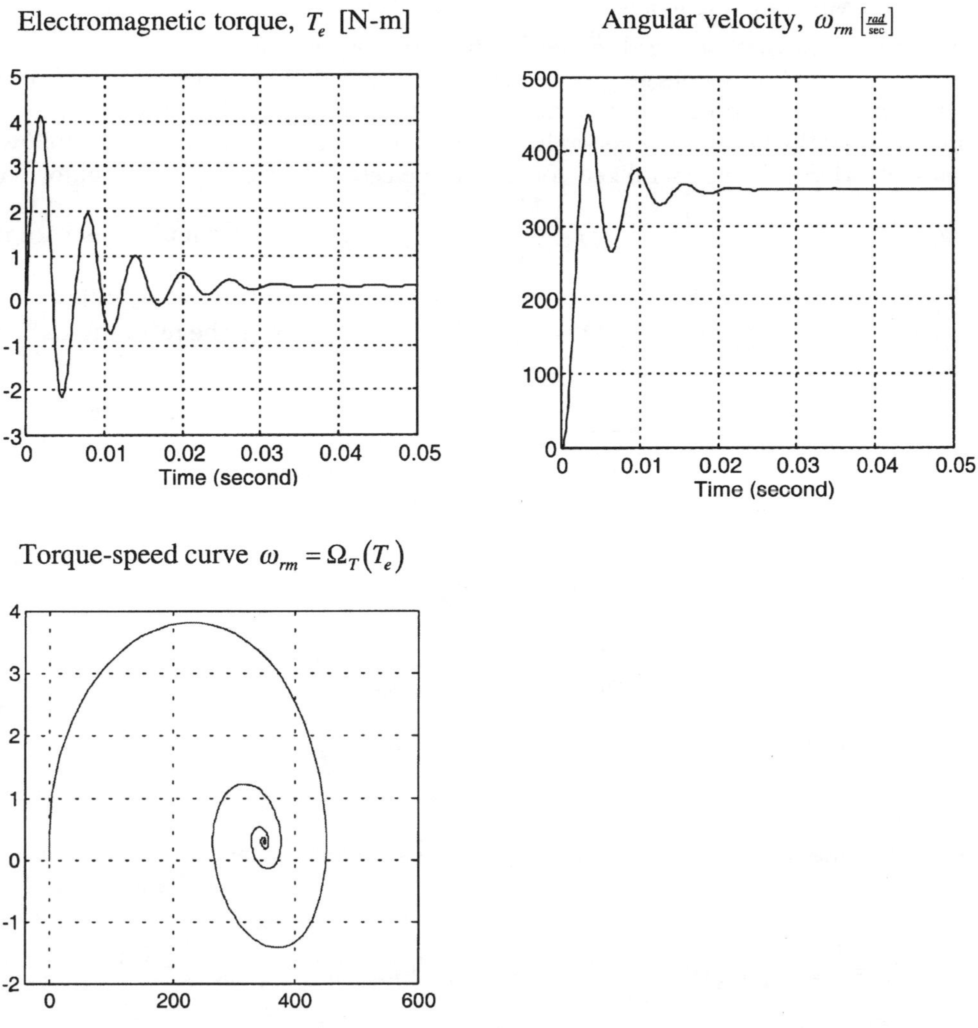

Figure 7.19 Continued.

performance compared with DC and induction motors. Due to high starting torque and excellent controllability capabilities in the full torque-speed operating envelope, permanent-magnet synchronous motors are widely used in high-performance electric drives and servo-systems.

□

The Lagrange equations of motion and dynamics of permanent-magnet synchronous motors

Having derived the mathematical model for three-phase permanent-magnet synchronous motors using Kirchhoff's voltage law, Newtonian mechanics, and the coenergy concept, one also can attack the problem of model development using the Lagrange concept. The generalized coordinates and the generalized forces must be chosen. We use the following variables as independent generalized coordinates: electric charges in the *abc* stator windings $q_1 = \frac{i_{as}}{s}$, $\dot{q}_1 = i_{as}$, $q_2 = \frac{i_{bs}}{s}$, $\dot{q}_2 = i_{bs}$, $q_3 = \frac{i_{cs}}{s}$, $\dot{q}_3 = i_{cs}$, and the angular displacement $q_4 = \theta_r$, $\dot{q}_4 = \omega_r$. The applied voltages to the *abc* windings and the load torque are the generalized forces; that is, $Q_1 = u_{as}$, $Q_2 = u_{bs}$, $Q_3 = u_{cs}$, and $Q_4 = -T_L$. The resulting Lagrange equations are

$$\frac{d}{dt}\left(\frac{\partial \Gamma}{\partial \dot{q}_1}\right) - \frac{\partial \Gamma}{\partial q_1} + \frac{\partial D}{\partial \dot{q}_1} + \frac{\partial \Pi}{\partial q_1} = Q_1$$

$$\frac{d}{dt}\left(\frac{\partial \Gamma}{\partial \dot{q}_2}\right) - \frac{\partial \Gamma}{\partial q_2} + \frac{\partial D}{\partial \dot{q}_2} + \frac{\partial \Pi}{\partial q_2} = Q_2$$

$$\frac{d}{dt}\left(\frac{\partial \Gamma}{\partial \dot{q}_3}\right) - \frac{\partial \Gamma}{\partial q_3} + \frac{\partial D}{\partial \dot{q}_3} + \frac{\partial \Pi}{\partial q_3} - Q_3$$

$$\frac{d}{dt}\left(\frac{\partial \Gamma}{\partial \dot{q}_4}\right) - \frac{\partial \Gamma}{\partial q_4} + \frac{\partial D}{\partial \dot{q}_4} + \frac{\partial \Pi}{\partial q_4} = Q_4$$

The total kinetic energy includes kinetic energies of electrical and mechanical systems; in particular,

$$\Gamma = \Gamma_E + \Gamma_M = \frac{1}{2}L_{asas}\dot{q}_1^2 + \frac{1}{2}(L_{asbs} + L_{bsas})\dot{q}_1\dot{q}_2 + \frac{1}{2}(L_{ascs} + L_{csas})\dot{q}_1\dot{q}_3 + \frac{1}{2}L_{bsbs}\dot{q}_2^2 + \frac{1}{2}(L_{bscs} + L_{csbs})\dot{q}_2\dot{q}_3$$
$$+ \frac{1}{2}L_{cscs}\dot{q}_3^2 + \psi_m\dot{q}_1\sin q_4 + \psi_m\dot{q}_2\sin\left(q_4 - \frac{2}{3}\pi\right) + \psi_m\dot{q}_3\sin\left(q_4 + \frac{2}{3}\pi\right) + \frac{1}{2}J\dot{q}_4^2$$

Here, the self- and mutual inductances are defined by their subscripts, and the flux established by the permanent magnet is denoted as ψ_m.

Then, we have

$$\frac{\partial \Gamma}{\partial q_1} = 0, \quad \frac{\partial \Gamma}{\partial \dot{q}_1} = L_{asas}\dot{q}_1 + \frac{1}{2}(L_{asbs} + L_{bsas})\dot{q}_2 + \frac{1}{2}(L_{ascs} + L_{csas})\dot{q}_3 + \psi_m \sin q_4$$

$$\frac{\partial \Gamma}{\partial q_2} = 0, \quad \frac{\partial \Gamma}{\partial \dot{q}_2} = \frac{1}{2}(L_{asbs} + L_{bsas})\dot{q}_1 + L_{bsbs}\dot{q}_2 + \frac{1}{2}(L_{bscs} + L_{csbs})\dot{q}_3 + \psi_m \sin\left(q_4 - \frac{2}{3}\pi\right)$$

$$\frac{\partial\Gamma}{\partial q_3}=0,\quad \frac{\partial\Gamma}{\partial \dot q_3}=\frac{1}{2}(L_{ascs}+L_{csas})\dot q_1+\frac{1}{2}(L_{bscs}+L_{csbs})\dot q_2+L_{cscs}\dot q_3+\psi_m\sin\left(q_4+\frac{2}{3}\pi\right)$$

$$\frac{\partial\Gamma}{\partial q_4}=\psi_m\dot q_1\cos q_4+\psi_m\dot q_2\cos\left(q_4-\frac{2}{3}\pi\right)+\psi_m\dot q_3\cos\left(q_4+\frac{2}{3}\pi\right),\quad \frac{\partial\Gamma}{\partial \dot q_4}=J\dot q_4$$

The total potential energy is zero; that is, $\Pi = 0$.

The total dissipated energy should be found as a sum of the heat energy dissipated by the electrical system and the heat energy dissipated by the mechanical system. That is,

$$D=\frac{1}{2}(r_s\dot q_1^2+r_s\dot q_2^2+r_s\dot q_3^2+B_m\dot q_4^2)$$

One obtains

$$\frac{\partial D}{\partial \dot q_1}=r_s\dot q_1,\quad \frac{\partial D}{\partial \dot q_2}=r_s\dot q_2,\quad \frac{\partial D}{\partial \dot q_3}=r_s\dot q_3,\quad and\quad \frac{\partial D}{\partial \dot q_4}=B_m\dot q_4$$

Then, taking note of $\dot q_1=i_{as}$, $\dot q_2=i_{bs}$, $\dot q_3=i_{cs}$, and $\dot q_4=\omega_r$, the Lagrange equations, which are expressed in terms of each independent coordinate, lead to four differential equations. In particular,

$$L_{asas}\frac{di_{as}}{dt}+\frac{1}{2}(L_{asbs}+L_{bsas})\frac{di_{bs}}{dt}+\frac{1}{2}(L_{ascs}+L_{csas})\frac{di_{cs}}{dt}+\psi_m\omega_r\cos\theta_r+r_si_{as}=u_{as}$$

$$\frac{1}{2}(L_{asbs}+L_{bsas})\frac{di_{as}}{dt}+L_{bsbs}\frac{di_{bs}}{dt}+\frac{1}{2}(L_{bscs}+L_{csbs})\frac{di_{cs}}{dt}+\psi_m\omega_r\cos\left(\theta_r-\frac{2}{3}\pi\right)+r_si_{bs}=u_{bs}$$

$$\frac{1}{2}(L_{ascs}+L_{csas})\frac{di_{as}}{dt}+\frac{1}{2}(L_{bscs}+L_{csbs})\frac{di_{bs}}{dt}+L_{cscs}\frac{di_{cs}}{dt}+\psi_m\omega_r\cos\left(\theta_r+\frac{2}{3}\pi\right)+r_si_{cs}=u_{cs}$$

$$J\frac{d^2\theta_r}{dt^2}-\psi_mi_{as}\cos\theta_r-\psi_mi_{bs}\cos\left(\theta_r-\frac{2}{3}\pi\right)-\psi_mi_{cs}\cos\left(\theta_r+\frac{2}{3}\pi\right)+B_m\frac{d\theta_r}{dt}=-T_L$$

For round-rotor permanent-magnet synchronous motors, using the expressions for inductances derived, one obtains

$$(L_{ls}+\bar L_m)\frac{di_{as}}{dt}-\frac{1}{2}\bar L_m\frac{di_{bs}}{dt}-\frac{1}{2}\bar L_m\frac{di_{cs}}{dt}+\psi_m\omega_r\cos\theta_r+r_si_{as}=u_{as}$$

$$-\frac{1}{2}\bar L_m\frac{di_{as}}{dt}+(L_{ls}+\bar L_m)\frac{di_{bs}}{dt}-\frac{1}{2}\bar L_m\frac{di_{cs}}{dt}+\psi_m\omega_r\cos\left(\theta_r-\frac{2}{3}\pi\right)+r_si_{bs}=u_{bs}$$

$$-\frac{1}{2}\bar L_m\frac{di_{as}}{dt}-\frac{1}{2}\bar L_m\frac{di_{bs}}{dt}+(L_{ls}+\bar L_m)\frac{di_{cs}}{dt}+\psi_m\omega_r\cos\left(\theta_r+\frac{2}{3}\pi\right)+r_si_{cs}=u_{cs}$$

$$J\frac{d\omega_r}{dt}+B_m\omega_r-\psi_m\left[i_{as}\cos\theta_r+i_{bs}\cos\left(\theta_r-\frac{2}{3}\pi\right)+i_{cs}\cos\left(\theta_r+\frac{2}{3}\pi\right)\right]=-T_L$$

$$\frac{d\theta_r}{dt}=\omega_r$$

From the fourth differential equation, one finds that the electromagnetic torque is

$$T_e = \psi_m \left[i_{as} \cos\theta_r + i_{bs} \cos\left(\theta_r - \frac{2}{3}\pi\right) + i_{cs} \cos\left(\theta_r + \frac{2}{3}\pi\right) \right]$$

Differential equations in Cauchy's form, as given by (7.16) for P-pole permanent-magnet synchronous motors, result. It was demonstrated that by applying the Lagrange concept, a complete mathematical model for permanent-magnet synchronous motors was straightforwardly developed.

Three-phase permanent-magnet synchronous generators

For permanent-magnet synchronous generators, as shown in Figure 7.15, the mathematical model can be developed using Kirchhoff's second law

$$\mathbf{u}_{abcs} = -\mathbf{r}_s \mathbf{i}_{abcs} + \frac{d\boldsymbol{\psi}_{abcs}}{dt}, \quad \begin{bmatrix} u_{as} \\ u_{bs} \\ u_{cs} \end{bmatrix} = -\begin{bmatrix} r_s & 0 & 0 \\ 0 & r_s & 0 \\ 0 & 0 & r_s \end{bmatrix} \begin{bmatrix} i_{as} \\ i_{bs} \\ i_{cs} \end{bmatrix} + \begin{bmatrix} \frac{d\psi_{as}}{dt} \\ \frac{d\psi_{bs}}{dt} \\ \frac{d\psi_{cs}}{dt} \end{bmatrix}$$

$$\boldsymbol{\psi}_{abcs} = -\mathbf{L}_s \mathbf{i}_{abcs} + \boldsymbol{\psi}_m =$$

$$-\begin{bmatrix} L_{ls} + \bar{L}_m - L_{\Delta m}\cos 2\theta_r & -\frac{1}{2}\bar{L}_m - L_{\Delta m}\cos 2\left(\theta_r - \frac{1}{3}\pi\right) & -\frac{1}{2}\bar{L}_m - L_{\Delta m}\cos 2\left(\theta_r + \frac{1}{3}\pi\right) \\ -\frac{1}{2}\bar{L}_m - L_{\Delta m}\cos 2\left(\theta_r - \frac{1}{3}\pi\right) & L_{ls} + \bar{L}_m - L_{\Delta m}\cos 2\left(\theta_r - \frac{2}{3}\pi\right) & -\frac{1}{2}\bar{L}_m - L_{\Delta m}\cos 2\theta_r \\ -\frac{1}{2}\bar{L}_m - L_{\Delta m}\cos 2\left(\theta_r + \frac{1}{3}\pi\right) & -\frac{1}{2}\bar{L}_m - L_{\Delta m}\cos 2\theta_r & L_{ls} + \bar{L}_m - L_{\Delta m}\cos\left(\theta_r + \frac{2}{3}\pi\right) \end{bmatrix}$$

$$\times \begin{bmatrix} i_{as} \\ i_{bs} \\ i_{cs} \end{bmatrix} + \psi_m \begin{bmatrix} \sin\theta_r \\ \sin\left(\theta_r - \frac{2}{3}\pi\right) \\ \sin\left(\theta_r + \frac{2}{3}\pi\right) \end{bmatrix}$$

and Newton's second law of motion, $-T_e - B_m\omega_{rm} + T_{pm} = J\frac{d^2\theta_{rm}}{dt^2}$, which gives

$$\frac{d\omega_{rm}}{dt} = \frac{1}{J}(-T_e - B_m\omega_{rm} + T_{pm})$$

$$\frac{d\theta_{rm}}{dt} = \omega_{rm}$$

The application of the results presented for the permanent-magnet synchronous motor results in the following set of differential equations

$$\begin{aligned}
\frac{di_{asc}}{dt} &= -\frac{r_s(2L_{ss}-\bar{L}_m)}{2L_{ss}^2-L_{ss}\bar{L}_m-\bar{L}_m^2}i_{as} - \frac{r_s\bar{L}_m}{2L_{ss}^2-L_{ss}\bar{L}_m-\bar{L}_m^2}i_{bs} - \frac{r_s\bar{L}_m}{2L_{ss}^2-L_{ss}\bar{L}_m-\bar{L}_m^2}i_{cs} \\
&+ \frac{\psi_m(2L_{ss}-\bar{L}_m)}{2L_{ss}^2-L_{ss}\bar{L}_m-\bar{L}_m^2}\omega_r\cos\theta_r + \frac{\psi_m\bar{L}_m}{2L_{ss}^2-L_{ss}\bar{L}_m-\bar{L}_m^2}\omega_r\cos\left(\theta_r-\frac{2}{3}\pi\right) + \frac{\psi_m\bar{L}_m}{2L_{ss}^2-L_{ss}\bar{L}_m-\bar{L}_m^2}\omega_r\cos\left(\theta_r+\frac{2}{3}\pi\right) \\
&- \frac{2L_{ss}-\bar{L}_m}{2L_{ss}^2-L_{ss}\bar{L}_m-\bar{L}_m^2}u_{as} - \frac{\bar{L}_m}{2L_{ss}^2-L_{ss}\bar{L}_m-\bar{L}_m^2}u_{bs} - \frac{\bar{L}_m}{2L_{ss}^2-L_{ss}\bar{L}_m-\bar{L}_m^2}u_{cs} \\
\frac{di_{bs}}{dt} &= -\frac{r_s\bar{L}_m}{2L_{ss}^2-L_{ss}\bar{L}_m-\bar{L}_m^2}i_{as} - \frac{r_s(2L_{ss}-\bar{L}_m)}{2L_{ss}^2-L_{ss}\bar{L}_m-\bar{L}_m^2}i_{bs} - \frac{r_s\bar{L}_m}{2L_{ss}^2-L_{ss}\bar{L}_m-\bar{L}_m^2}i_{cs} \\
&+ \frac{\psi_m\bar{L}_m}{2L_{ss}^2-L_{ss}\bar{L}_m-\bar{L}_m^2}\omega_r\cos\theta_r + \frac{\psi_m(2L_{ss}-\bar{L}_m)}{2L_{ss}^2-L_{ss}\bar{L}_m-\bar{L}_m^2}\omega_r\cos\left(\theta_r-\frac{2}{3}\pi\right) + \frac{\psi_m\bar{L}_m}{2L_{ss}^2-L_{ss}\bar{L}_m-\bar{L}_m^2}\omega_r\cos\left(\theta_r+\frac{2}{3}\pi\right) \\
&- \frac{\bar{L}_m}{2L_{ss}^2-L_{ss}\bar{L}_m-\bar{L}_m^2}u_{as} - \frac{2L_{ss}-\bar{L}_m}{2L_{ss}^2-L_{ss}\bar{L}_m-\bar{L}_m^2}u_{bs} - \frac{\bar{L}_m}{2L_{ss}^2-L_{ss}\bar{L}_m-\bar{L}_m^2}u_{cs} \\
\frac{di_{cs}}{dt} &= -\frac{r_s\bar{L}_m}{2L_{ss}^2-L_{ss}\bar{L}_m-\bar{L}_m^2}i_{as} - \frac{r_s\bar{L}_m}{2L_{ss}^2-L_{ss}\bar{L}_m-\bar{L}_m^2}i_{bs} - \frac{r_s(2L_{ss}-\bar{L}_m)}{2L_{ss}^2-L_{ss}\bar{L}_m-\bar{L}_m^2}i_{cs} \\
&+ \frac{\psi_m\bar{L}_m}{2L_{ss}^2-L_{ss}\bar{L}_m-\bar{L}_m^2}\omega_r\cos\theta_r + \frac{\psi_m\bar{L}_m}{2L_{ss}^2-L_{ss}\bar{L}_m-\bar{L}_m^2}\omega_r\cos\left(\theta_r-\frac{2}{3}\pi\right) + \frac{\psi_m(2L_{ss}-\bar{L}_m)}{2L_{ss}^2-L_{ss}\bar{L}_m-\bar{L}_m^2}\omega_r\cos\left(\theta_r+\frac{2}{3}\pi\right) \\
&- \frac{\bar{L}_m}{2L_{ss}^2-L_{ss}\bar{L}_m-\bar{L}_m^2}u_{as} - \frac{\bar{L}_m}{2L_{ss}^2-L_{ss}\bar{L}_m-\bar{L}_m^2}u_{bs} - \frac{2L_{ss}-\bar{L}_m}{2L_{ss}^2-L_{ss}\bar{L}_m-\bar{L}_m^2}u_{cs} \\
\frac{d\omega_r}{dt} &= -\frac{P^2\psi_m}{4J}\left(i_{as}\cos\theta_r + i_{bs}\cos\left(\theta_r-\frac{2}{3}\pi\right) + i_{cs}\cos\left(\theta_r+\frac{2}{3}\pi\right)\right) - \frac{B_m}{J}\omega_r + \frac{P}{2J}T_{pm} \\
\frac{d\theta_r}{dt} &= \omega_r
\end{aligned} \tag{7.17}$$

We have the following mathematical model of three-phase permanent-magnet synchronous generators in matrix form

$$\begin{bmatrix} \frac{di_{as}}{dt} \\ \frac{di_{bs}}{dt} \\ \frac{di_{cs}}{dt} \\ \frac{d\omega_r}{dt} \\ \frac{d\theta_r}{dt} \end{bmatrix} = \begin{bmatrix} -\frac{r_s(2L_{ss}-\bar{L}_m)}{2L_{ss}^2-L_{ss}\bar{L}_m-\bar{L}_m^2} & -\frac{r_s\bar{L}_m}{2L_{ss}^2-L_{ss}\bar{L}_m-\bar{L}_m^2} & -\frac{r_s\bar{L}_m}{2L_{ss}^2-L_{ss}\bar{L}_m-\bar{L}_m^2} & 0 & 0 \\ -\frac{r_s\bar{L}_m}{2L_{ss}^2-L_{ss}\bar{L}_m-\bar{L}_m^2} & -\frac{r_s(2L_{ss}-\bar{L}_m)}{2L_{ss}^2-L_{ss}\bar{L}_m-\bar{L}_m^2} & -\frac{r_s\bar{L}_m}{2L_{ss}^2-L_{ss}\bar{L}_m-\bar{L}_m^2} & 0 & 0 \\ -\frac{r_s\bar{L}_m}{2L_{ss}^2-L_{ss}\bar{L}_m-\bar{L}_m^2} & -\frac{r_s\bar{L}_m}{2L_{ss}^2-L_{ss}\bar{L}_m-\bar{L}_m^2} & -\frac{r_s(2L_{ss}-\bar{L}_m)}{2L_{ss}^2-L_{ss}\bar{L}_m-\bar{L}_m^2} & 0 & 0 \\ 0 & 0 & 0 & -\frac{B_m}{J} & 0 \\ 0 & 0 & 0 & 1 & 0 \end{bmatrix} \begin{bmatrix} i_{as} \\ i_{bs} \\ i_{cs} \\ \omega_r \\ \theta_r \end{bmatrix}$$

$$+\begin{bmatrix} \frac{\psi_m(2L_{ss}-\bar{L}_m)}{2L_{ss}^2-L_{ss}\bar{L}_m-\bar{L}_m^2}\omega_r & \frac{\psi_m\bar{L}_m}{2L_{ss}^2-L_{ss}\bar{L}_m-\bar{L}_m^2}\omega_r & \frac{\psi_m\bar{L}_m}{2L_{ss}^2-L_{ss}\bar{L}_m-\bar{L}_m^2}\omega_r \\ \frac{\psi_m\bar{L}_m}{2L_{ss}^2-L_{ss}\bar{L}_m-\bar{L}_m^2}\omega_r & \frac{\psi_m(2L_{ss}-\bar{L}_m)}{2L_{ss}^2-L_{ss}\bar{L}_m-\bar{L}_m^2}\omega_r & \frac{\psi_m\bar{L}_m}{2L_{ss}^2-L_{ss}\bar{L}_m-\bar{L}_m^2}\omega_r \\ \frac{\psi_m\bar{L}_m}{2L_{ss}^2-L_{ss}\bar{L}_m-\bar{L}_m^2}\omega_r & \frac{\psi_m\bar{L}_m}{2L_{ss}^2-L_{ss}\bar{L}_m-\bar{L}_m^2}\omega_r & \frac{\psi_m(2L_{ss}-\bar{L}_m)}{2L_{ss}^2-L_{ss}\bar{L}_m-\bar{L}_m^2}\omega_r \\ -\frac{P^2\psi_m}{4J}i_{as} & -\frac{P^2\psi_m}{4J}i_{bs} & -\frac{P^2\psi_m}{4J}i_{cs} \\ 0 & 0 & 0 \end{bmatrix} \begin{bmatrix} \cos\theta_r \\ \cos\left(\theta_r-\frac{2}{3}\pi\right) \\ \cos\left(\theta_r+\frac{2}{3}\pi\right) \end{bmatrix}$$

$$-\begin{bmatrix} \frac{2L_{ss}-\bar{L}_m}{2L_{ss}^2-L_{ss}\bar{L}_m-\bar{L}_m^2} & \frac{\bar{L}_m}{2L_{ss}^2-L_{ss}\bar{L}_m-\bar{L}_m^2} & \frac{\bar{L}_m}{2L_{ss}^2-L_{ss}\bar{L}_m-\bar{L}_m^2} \\ \frac{\bar{L}_m}{2L_{ss}^2-L_{ss}\bar{L}_m-\bar{L}_m^2} & \frac{2L_{ss}-\bar{L}_m}{2L_{ss}^2-L_{ss}\bar{L}_m-\bar{L}_m^2} & \frac{\bar{L}_m}{2L_{ss}^2-L_{ss}\bar{L}_m-\bar{L}_m^2} \\ \frac{\bar{L}_m}{2L_{ss}^2-L_{ss}\bar{L}_m-\bar{L}_m^2} & \frac{\bar{L}_m}{2L_{ss}^2-L_{ss}\bar{L}_m-\bar{L}_m^2} & \frac{2L_{ss}-\bar{L}_m}{2L_{ss}^2-L_{ss}\bar{L}_m-\bar{L}_m^2} \\ 0 & 0 & 0 \\ 0 & 0 & 0 \end{bmatrix} \begin{bmatrix} u_{as} \\ u_{bs} \\ u_{cs} \end{bmatrix} + \begin{bmatrix} 0 \\ 0 \\ 0 \\ \frac{P}{2J} \\ 0 \end{bmatrix} T_{pm}$$

7.3.2 *Development of mathematical models of permanent-magnet synchronous machines in the* arbitrary, *rotor, and synchronous reference frames*

Arbitrary reference frame

The *arbitrary* reference frame can be used, and this concept is an attractive avenue from analysis and control viewpoints. One can easily find mathematical models of permanent-magnet synchronous machines in the *arbitrary* reference frame. In particular, by fixing the reference frame with the rotor and making use of the *direct* Park transformations

$$\mathbf{u}_{qd0s} = \mathbf{K}_s \mathbf{u}_{abcs}, \ \mathbf{i}_{qd0s} = \mathbf{K}_s \mathbf{i}_{abcs}, \text{ and } \boldsymbol{\psi}_{qd0s} = \mathbf{K}_s \boldsymbol{\psi}_{abcs}$$

$$\mathbf{K}_s = \frac{2}{3}\begin{bmatrix} \cos\theta & \cos\left(\theta-\frac{2}{3}\pi\right) & \cos\left(\theta+\frac{2}{3}\pi\right) \\ \sin\theta & \sin\left(\theta-\frac{2}{3}\pi\right) & \sin\left(\theta+\frac{2}{3}\pi\right) \\ \frac{1}{2} & \frac{1}{2} & \frac{1}{2} \end{bmatrix}$$

equation (7.11)

$$\mathbf{u}_{abcs} = \mathbf{r}_s \mathbf{i}_{abcs} + \frac{d\boldsymbol{\psi}_{abcs}}{dt}$$

is rewritten in the *qd0* variables as

$$\mathbf{K}_s^{-1}\mathbf{u}_{qd0s} = \mathbf{r}_s\mathbf{K}_s^{-1}\mathbf{i}_{qd0s} + \frac{d(\mathbf{K}_s^{-1}\boldsymbol{\psi}_{qd0s})}{dt}, \quad \mathbf{K}_s^{-1} = \begin{bmatrix} \cos\theta & \sin\theta & 1 \\ \cos\left(\theta - \frac{2}{3}\pi\right) & \sin\left(\theta - \frac{2}{3}\pi\right) & 1 \\ \cos\left(\theta + \frac{2}{3}\pi\right) & \sin\left(\theta + \frac{2}{3}\pi\right) & 1 \end{bmatrix}$$

Multiplying left and right sides by $\mathbf{K}_s$, one obtains

$$\mathbf{K}_s\mathbf{K}_s^{-1}\mathbf{u}_{qd0s} = \mathbf{K}_s\mathbf{r}_s\mathbf{K}_s^{-1}\mathbf{i}_{qd0s} + \mathbf{K}_s\frac{d\mathbf{K}_s^{-1}}{dt}\boldsymbol{\psi}_{qd0s} + \mathbf{K}_s\mathbf{K}_s^{-1}\frac{d\boldsymbol{\psi}_{qd0s}}{dt}$$

The matrix $\mathbf{r}_s$ is diagonal, and thus

$$\mathbf{K}_s\mathbf{r}_s\mathbf{K}_s^{-1} = \mathbf{r}_s$$

Using

$$\frac{d\mathbf{K}_s^{-1}}{dt} = \omega\begin{bmatrix} -\sin\theta & \cos\theta & 0 \\ -\sin\left(\theta - \frac{2}{3}\pi\right) & \cos\left(\theta - \frac{2}{3}\pi\right) & 0 \\ -\sin\left(\theta + \frac{2}{3}\pi\right) & \cos\left(\theta + \frac{2}{3}\pi\right) & 0 \end{bmatrix}$$

we have

$$\mathbf{K}_s\frac{d\mathbf{K}_s^{-1}}{dt} = \omega\begin{bmatrix} 0 & 1 & 0 \\ -1 & 0 & 0 \\ 0 & 0 & 0 \end{bmatrix}$$

Therefore,

$$\mathbf{u}_{qd0s} = \mathbf{r}_s\mathbf{i}_{qd0s} + \omega\begin{bmatrix} \psi_{ds} \\ -\psi_{qs} \\ 0 \end{bmatrix} + \frac{d\boldsymbol{\psi}_{qd0s}}{dt} \tag{7.18}$$

The *quadrature-*, *direct-*, and *zero-*axis components of stator flux linkages are found using the Park transformation

$$\boldsymbol{\psi}_{qd0s} = \mathbf{K}_s\boldsymbol{\psi}_{abcs}$$

where $\boldsymbol{\psi}_{abcs} = \mathbf{L}_s \mathbf{i}_{abcs} + \boldsymbol{\psi}_m = \begin{bmatrix} L_{ls} + \bar{L}_m & -\frac{1}{2}\bar{L}_m & -\frac{1}{2}\bar{L}_m \\ -\frac{1}{2}\bar{L}_m & L_{ls} + \bar{L}_m & -\frac{1}{2}\bar{L}_m \\ -\frac{1}{2}\bar{L}_m & -\frac{1}{2}\bar{L}_m & L_{ls} + \bar{L}_m \end{bmatrix} \begin{bmatrix} i_{as} \\ i_{bs} \\ i_{cs} \end{bmatrix} + \psi_m \begin{bmatrix} \sin\theta_r \\ \sin\left(\theta_r - \frac{2}{3}\pi\right) \\ \sin\left(\theta_r + \frac{2}{3}\pi\right) \end{bmatrix}$

Therefore,

$$\boldsymbol{\psi}_{qd0s} = \mathbf{K}_s \mathbf{L}_s \mathbf{K}_s^{-1} \mathbf{i}_{qd0s} + \mathbf{K}_s \boldsymbol{\psi}_m \tag{7.19}$$

The matrix multiplication $\mathbf{K}_s \mathbf{L}_s \mathbf{K}_s^{-1}$ results in

$$\mathbf{K}_s \mathbf{L}_s \mathbf{K}_s^{-1} = \begin{bmatrix} L_{ls} + \frac{3}{2}\bar{L}_m & 0 & 0 \\ 0 & L_{ls} + \frac{3}{2}\bar{L}_m & 0 \\ 0 & 0 & L_{ls} \end{bmatrix}$$

and

$$\mathbf{K}_s \boldsymbol{\psi}_m = \frac{2}{3} \begin{bmatrix} \cos\theta & \cos\left(\theta - \frac{2}{3}\pi\right) & \cos\left(\theta + \frac{2}{3}\pi\right) \\ \sin\theta & \sin\left(\theta - \frac{2}{3}\pi\right) & \sin\left(\theta_r + \frac{2}{3}\pi\right) \\ \frac{1}{2} & \frac{1}{2} & \frac{1}{2} \end{bmatrix} \psi_m \begin{bmatrix} \sin\theta_r \\ \sin\left(\theta_r - \frac{2}{3}\pi\right) \\ \sin\left(\theta_r + \frac{2}{3}\pi\right) \end{bmatrix} = \psi_m \begin{bmatrix} -\sin(\theta - \theta_r) \\ \cos(\theta - \theta_r) \\ 0 \end{bmatrix}$$

From (7.19), one obtains

$$\boldsymbol{\psi}_{qd0s} = \begin{bmatrix} L_{ls} + \frac{3}{2}\bar{L}_m & 0 & 0 \\ 0 & L_{ls} + \frac{3}{2}\bar{L}_m & 0 \\ 0 & 0 & L_{ls} \end{bmatrix} \mathbf{i}_{qd0s} + \psi_m \begin{bmatrix} -\sin(\theta - \theta_r) \\ \cos(\theta - \theta_r) \\ 0 \end{bmatrix}$$

Using (7.18), we have

$$\mathbf{u}_{qd0s} = \mathbf{r}_s \mathbf{i}_{qd0s} + \omega \begin{bmatrix} \psi_{ds} \\ -\psi_{qs} \\ 0 \end{bmatrix} + \begin{bmatrix} L_{ls} + \frac{3}{2}\bar{L}_m & 0 & 0 \\ 0 & L_{ls} + \frac{3}{2}\bar{L}_m & 0 \\ 0 & 0 & L_{ls} \end{bmatrix} \frac{d\mathbf{i}_{qd0s}}{dt} + \psi_m \frac{d \begin{bmatrix} -\sin(\theta - \theta_r) \\ \cos(\theta - \theta_r) \\ 0 \end{bmatrix}}{dt}$$

Therefore, one finds at once

$$\mathbf{u}_{qd0s} = \mathbf{r}_s \mathbf{i}_{qd0s} + \omega \begin{bmatrix} \psi_{ds} \\ -\psi_{qs} \\ 0 \end{bmatrix} + \begin{bmatrix} L_{ls} + \frac{3}{2}\bar{L}_m & 0 & 0 \\ 0 & L_{ls} + \frac{3}{2}\bar{L}_m & 0 \\ 0 & 0 & L_{ls} \end{bmatrix} \frac{d\mathbf{i}_{qd0s}}{dt} + \psi_m(\omega - \omega_r) \begin{bmatrix} -\cos(\theta - \theta_r) \\ -\sin(\theta - \theta_r) \\ 0 \end{bmatrix}$$

That is, the resulting model for the permanent-magnet synchronous motors circuitry, as given by three differential equations, is found in the *arbitrary* reference frame. This model can be further specified as the angular velocity of the reference frame is specified. It has been shown that the stationary, rotor, and synchronous reference frames are commonly used.

Rotor reference frame

The electrical angular velocity is equal to the synchronous angular velocity, and the angular velocity of the reference frame is assigned as $\omega = \omega_r = \omega_e$. Then, using $\theta = \theta_r$, we have the following Park transformation matrix

$$\mathbf{K}_s^r = \frac{2}{3} \begin{bmatrix} \cos\theta_r & \cos\left(\theta_r - \frac{2}{3}\pi\right) & \cos\left(\theta_r + \frac{2}{3}\pi\right) \\ \sin\theta_r & \sin\left(\theta_r - \frac{2}{3}\pi\right) & \sin\left(\theta_r + \frac{2}{3}\pi\right) \\ \frac{1}{2} & \frac{1}{2} & \frac{1}{2} \end{bmatrix}$$

and one finds

$$\mathbf{K}_s^r \boldsymbol{\psi}_m = \frac{2}{3} \begin{bmatrix} \cos\theta_r & \cos\left(\theta_r - \frac{2}{3}\pi\right) & \cos\left(\theta_r + \frac{2}{3}\pi\right) \\ \sin\theta_r & \sin\left(\theta_r - \frac{2}{3}\pi\right) & \sin\left(\theta_r + \frac{2}{3}\pi\right) \\ \frac{1}{2} & \frac{1}{2} & \frac{1}{2} \end{bmatrix} \psi_m \begin{bmatrix} \sin\theta_r \\ \sin\left(\theta_r - \frac{2}{3}\pi\right) \\ \sin\left(\theta_r + \frac{2}{3}\pi\right) \end{bmatrix} = \begin{bmatrix} 0 \\ \psi_m \\ 0 \end{bmatrix}$$

Hence, equation (7.19) leads us to

$$\boldsymbol{\psi}_{qd0s}^r = \begin{bmatrix} L_{ls} + \frac{3}{2}\bar{L}_m & 0 & 0 \\ 0 & L_{ls} + \frac{3}{2}\bar{L}_m & 0 \\ 0 & 0 & L_{ls} \end{bmatrix} \mathbf{i}_{qd0s}^r + \begin{bmatrix} 0 \\ \psi_m \\ 0 \end{bmatrix}$$

In expanded form, we have

$$\psi_{qs}^r = \left(L_{ls} + \frac{3}{2}\bar{L}_m\right) i_{qs}^r$$

$$\psi_{ds}^r = \left(L_{ls} + \frac{3}{2}\bar{L}_m\right) i_{ds}^r + \psi_m$$

$$\psi_{0s}^r = L_{ls} i_{0s}^r$$

From (7.18), one finds

$$\frac{di_{qs}^r}{dt} = -\frac{r_s}{L_{ls}+\frac{3}{2}\bar{L}_m} i_{qs}^r - \frac{\psi_m}{L_{ls}+\frac{3}{2}\bar{L}_m}\omega_r - i_{ds}^r\omega_r + \frac{1}{L_{ls}+\frac{3}{2}\bar{L}_m} u_{qs}^r$$

$$\frac{di_{ds}^r}{dt} = -\frac{r_s}{L_{ls}+\frac{3}{2}\bar{L}_m} i_{ds}^r + i_{qs}^r\omega_r + \frac{1}{L_{ls}+\frac{3}{2}\bar{L}_m} u_{ds}^r \qquad (7.20)$$

$$\frac{di_{0s}^r}{dt} = -\frac{r_s}{L_{ls}} i_{0s}^r + \frac{1}{L_{ls}} u_{0s}^r$$

Applying the Park transformation, as specified for the rotor reference frame, the electromagnetic torque

$$T_e = \frac{P\psi_m}{2}\left[i_{as}\cos\theta_r + i_{bs}\cos\left(\theta_r - \frac{2}{3}\pi\right) + i_{cs}\cos\left(\theta_r + \frac{2}{3}\pi\right)\right]$$

should be found using the *qd0* currents.

Making use of

$$\begin{bmatrix} i_{as} \\ i_{bs} \\ i_{cs} \end{bmatrix} = \begin{bmatrix} \cos\theta_r & \sin\theta_r & 1 \\ \cos\left(\theta_r - \frac{2}{3}\pi\right) & \sin\left(\theta_r - \frac{2}{3}\pi\right) & 1 \\ \cos\left(\theta_r + \frac{2}{3}\pi\right) & \sin\left(\theta_r + \frac{2}{3}\pi\right) & 1 \end{bmatrix} \begin{bmatrix} i_{qs}^r \\ i_{ds}^r \\ i_{0s}^r \end{bmatrix}$$

and plugging

$$i_{as} = \cos\theta_r i_{qs}^r + \sin\theta_r i_{ds}^r + i_{0s}^r$$

$$i_{bs} = \cos\left(\theta_r - \frac{2}{3}\pi\right) i_{qs}^r + \sin\left(\theta_r - \frac{2}{3}\pi\right) i_{ds}^r + i_{0s}^r$$

$$i_{cs} = \cos\left(\theta_r + \frac{2}{3}\pi\right) i_{qs}^r + \sin\left(\theta_r + \frac{2}{3}\pi\right) i_{ds}^r + i_{0s}^r$$

into the expression for T_e using trigonometric identities, one finds at once that

$$T_e = \frac{3P\psi_m}{4} i_{qs}^r$$

Hence, for *P*-pole permanent-magnet synchronous motors, the *torsional-mechanical* dynamics are mapped by

$$\frac{d\omega_r}{dt} = \frac{3P^2\psi_m}{8J} i_{qs}^r - \frac{B_m}{J}\omega_r - \frac{P}{2J}T_L$$
$$\frac{d\theta_r}{dt} = \omega_r \tag{7.21}$$

Augmenting differential equations (7.20) and (7.21), we have

$$\frac{di_{qs}^r}{dt} = -\frac{r_s}{L_{ls}+\frac{3}{2}\bar{L}_m} i_{qs}^r - \frac{\psi_m}{L_{ls}+\frac{3}{2}\bar{L}_m}\omega_r - i_{ds}^r\omega_r + \frac{1}{L_{ls}+\frac{3}{2}\bar{L}_m}u_{qs}^r$$
$$\frac{di_{ds}^r}{dt} = -\frac{r_s}{L_{ls}+\frac{3}{2}\bar{L}_m} i_{ds}^r + i_{qs}^r\omega_r + \frac{1}{L_{ls}+\frac{3}{2}\bar{L}_m}u_{ds}^r$$
$$\frac{di_{0s}^r}{dt} = -\frac{r_s}{L_{ls}} i_{0s}^r + \frac{1}{L_{ls}}u_{0s}^r \tag{7.22}$$
$$\frac{d\omega_r}{dt} = \frac{3P^2\psi_m}{8J} i_{qs}^r - \frac{B_m}{J}\omega_r - \frac{P}{2J}T_L$$
$$\frac{d\theta_r}{dt} = \omega_r$$

In matrix form, the mathematical model of permanent-magnet synchronous motors in the rotor reference frame is given as

$$\begin{bmatrix} \frac{di_{qs}^r}{dt} \\ \frac{di_{ds}^r}{dt} \\ \frac{di_{0s}^r}{dt} \\ \frac{d\omega_r}{dt} \\ \frac{d\theta_r}{dt} \end{bmatrix} = \begin{bmatrix} -\frac{r_s}{L_{ls}+\frac{3}{2}\bar{L}_m} & 0 & 0 & -\frac{\psi_m}{L_{ls}+\frac{3}{2}\bar{L}_m} & 0 \\ 0 & -\frac{r_s}{L_{ls}+\frac{3}{2}\bar{L}_m} & 0 & 0 & 0 \\ 0 & 0 & -\frac{r_s}{L_{ls}} & 0 & 0 \\ \frac{3P^2\psi_m}{8J} & 0 & 0 & -\frac{B_m}{J} & 0 \\ 0 & 0 & 0 & 1 & 0 \end{bmatrix} \begin{bmatrix} i_{qs}^r \\ i_{ds}^r \\ i_{0s}^r \\ \omega_r \\ \theta_r \end{bmatrix} + \begin{bmatrix} -i_{ds}^r\omega_r \\ i_{qs}^r\omega_r \\ 0 \\ 0 \\ 0 \end{bmatrix}$$
$$+ \begin{bmatrix} \frac{1}{L_{ls}+\frac{3}{2}\bar{L}_m} & 0 & 0 \\ 0 & \frac{1}{L_{ls}+\frac{3}{2}\bar{L}_m} & 0 \\ 0 & 0 & \frac{1}{L_{ls}} \\ 0 & 0 & 0 \\ 0 & 0 & 0 \end{bmatrix} \begin{bmatrix} u_{qs}^r \\ u_{ds}^r \\ u_{0s}^r \end{bmatrix} - \begin{bmatrix} 0 \\ 0 \\ 0 \\ \frac{P}{2J} \\ 0 \end{bmatrix} T_L$$

A balanced three-phase current set, to be fed to the stator windings, is given by

$$i_{as}(t)=\sqrt{2}i_M\cos\theta_r,\ i_{bs}(t)=\sqrt{2}i_M\cos\left(\theta_r-\frac{2}{3}\pi\right),\ i_{cs}(t)=\sqrt{2}i_M\cos\left(\theta_r+\frac{2}{3}\pi\right)$$

Using the *direct* Park transformation

$$\begin{bmatrix} i_{qs}^r \\ i_{ds}^r \\ i_{0s}^r \end{bmatrix}=\frac{2}{3}\begin{bmatrix} \cos\theta_r & \cos\left(\theta_r-\frac{2}{3}\pi\right) & \cos\left(\theta_r+\frac{2}{3}\pi\right) \\ \sin\theta_r & \sin\left(\theta_r-\frac{2}{3}\pi\right) & \sin\left(\theta_r+\frac{2}{3}\pi\right) \\ \frac{1}{2} & \frac{1}{2} & \frac{1}{2} \end{bmatrix}\begin{bmatrix} i_{as} \\ i_{bs} \\ i_{cs} \end{bmatrix}$$

one obtains the *qd0* currents to regulate the angular velocity of permanent-magnet synchronous motors and guarantee balanced operating conditions. In particular, we have

$$\begin{bmatrix} i_{qs}^r \\ i_{ds}^r \\ i_{0s}^r \end{bmatrix}=\frac{2}{3}\begin{bmatrix} \cos\theta_r & \cos\left(\theta_r-\frac{2}{3}\pi\right) & \cos\left(\theta_r+\frac{2}{3}\pi\right) \\ \sin\theta_r & \sin\left(\theta_r-\frac{2}{3}\pi\right) & \sin\left(\theta_r+\frac{2}{3}\pi\right) \\ \frac{1}{2} & \frac{1}{2} & \frac{1}{2} \end{bmatrix}\begin{bmatrix} \sqrt{2}i_M\cos\theta_r \\ \sqrt{2}i_M\cos\left(\theta_r-\frac{2}{3}\pi\right) \\ \sqrt{2}i_M\cos\left(\theta_r+\frac{2}{3}\pi\right) \end{bmatrix}$$

and hence

$$i_{qs}^r(t)=\sqrt{2}i_M,\ i_{ds}^r(t)=0,\ i_{0s}^r(t)=0$$

Due to the self-inductances, the *abc* voltages should be supplied with advanced phase shifting. That is, one applies the following voltages to the stator windings

$$u_{as}(t)=\sqrt{2}u_M\cos(\theta_r+\varphi_u),\ u_{bs}(t)=\sqrt{2}u_M\cos\left(\theta_r-\frac{2}{3}\pi+\varphi_u\right),\ u_{cs}(t)=\sqrt{2}u_M\cos\left(\theta_r+\frac{2}{3}\pi+\varphi_u\right)$$

Taking note of

$$\begin{bmatrix} u_{qs}^r \\ u_{ds}^r \\ u_{0s}^r \end{bmatrix}=\frac{2}{3}\begin{bmatrix} \cos\theta_r & \cos\left(\theta_r-\frac{2}{3}\pi\right) & \cos\left(\theta_r+\frac{2}{3}\pi\right) \\ \sin\theta_r & \sin\left(\theta_r-\frac{2}{3}\pi\right) & \sin\left(\theta_r+\frac{2}{3}\pi\right) \\ \frac{1}{2} & \frac{1}{2} & \frac{1}{2} \end{bmatrix}\begin{bmatrix} u_{as} \\ u_{bs} \\ u_{cs} \end{bmatrix}$$

one finds

$$\begin{bmatrix} u_{qs}^r \\ u_{ds}^r \\ u_{0s}^r \end{bmatrix} = \frac{2}{3}\begin{bmatrix} \cos\theta_r & \cos\left(\theta_r - \frac{2}{3}\pi\right) & \cos\left(\theta_r + \frac{2}{3}\pi\right) \\ \sin\theta_r & \sin\left(\theta_r - \frac{2}{3}\pi\right) & \sin\left(\theta_r + \frac{2}{3}\pi\right) \\ \frac{1}{2} & \frac{1}{2} & \frac{1}{2} \end{bmatrix}\begin{bmatrix} \sqrt{2}u_M \cos(\theta_r + \varphi_u) \\ \sqrt{2}u_M \cos\left(\theta_r - \frac{2}{3}\pi + \varphi_u\right) \\ \sqrt{2}u_M \cos\left(\theta_r + \frac{2}{3}\pi + \varphi_u\right) \end{bmatrix}$$

Using the trigonometric identities, we have

$$u_{qs}^r(t) = \sqrt{2}u_M \cos\varphi_u, \quad u_{ds}^r(t) = -\sqrt{2}u_M \sin\varphi_u, \quad u_{0s}^r(t) = 0$$

Due to small inductances, $\varphi_u \approx 0$, the following phase voltages can be supplied

$$u_{as}(t) = \sqrt{2}u_M \cos\theta_r, \quad u_{bs}(t) = \sqrt{2}u_M\left(\theta_r - \frac{2}{3}\pi\right), \quad u_{cs}(t) = \sqrt{2}u_M \cos\left(\theta_r + \frac{2}{3}\pi\right)$$

For $\varphi_u = 0$, we have

$$u_{qs}^r(t) - \sqrt{2}u_M, \quad u_{ds}^r(l) = 0, \quad u_{0s}^r(t) = 0$$

To visualize the results, an *s*-domain block diagram in the *qd0* variables should be developed. From (7.22), the *s*-domain block diagram in Figure 7.20 results.

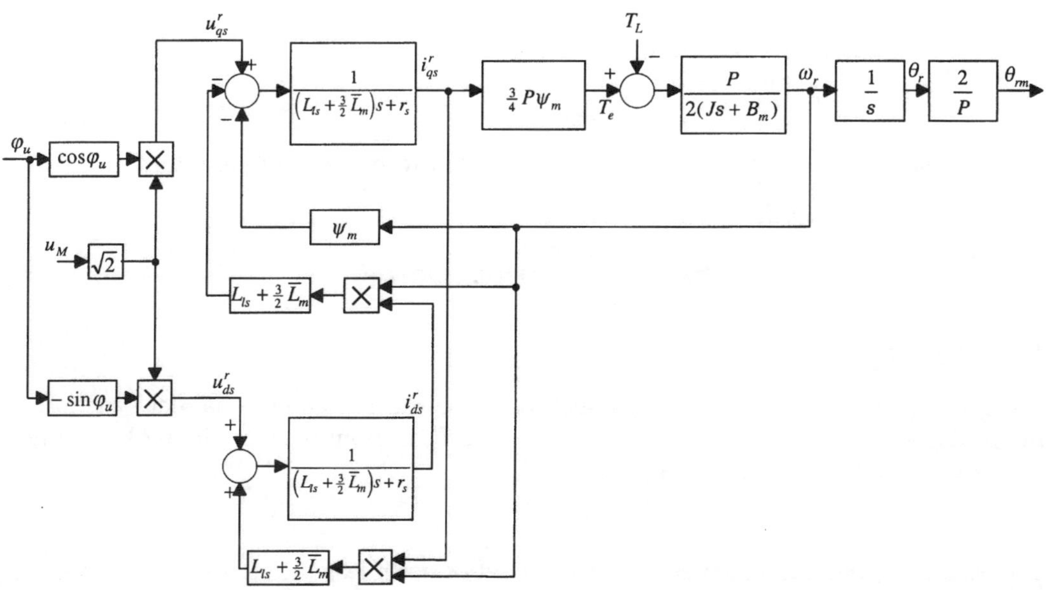

Figure 7.20 *s*-domain block diagram of permanent-magnet synchronous motors modeled in the rotor reference frame.

Synchronous reference frame

Analyzing permanent-magnet synchronous machines in the synchronous reference frame, one specifies the angular velocity of the reference frame to be $\omega = \omega_e$. Hence, $\theta = \theta_e$, and the Park transformation matrix is given as

$$\mathbf{K}_s^e = \frac{2}{3}\begin{bmatrix} \cos\theta_e & \cos\left(\theta_e - \frac{2}{3}\pi\right) & \cos\left(\theta_e + \frac{2}{3}\pi\right) \\ \sin\theta_e & \sin\left(\theta_e - \frac{2}{3}\pi\right) & \sin\left(\theta_e + \frac{2}{3}\pi\right) \\ \frac{1}{2} & \frac{1}{2} & \frac{1}{2} \end{bmatrix}$$

Taking note that $\omega_r = \omega_e$, from (7.22) we have the following system of differential equations that maps the permanent-magnet motor dynamics in the synchronous reference frame

$$\frac{di_{qs}^e}{dt} = -\frac{r_s}{L_{ls} + \frac{3}{2}\bar{L}_m} i_{qs}^e - \frac{\psi_m}{L_{ls} + \frac{3}{2}\bar{L}_m}\omega_r - i_{ds}^e\omega_r + \frac{1}{L_{ls} + \frac{3}{2}\bar{L}_m} u_{qs}^e$$

$$\frac{di_{ds}^e}{dt} = -\frac{r_s}{L_{ls} + \frac{3}{2}\bar{L}_m} i_{ds}^e + i_{qs}^e\omega_r + \frac{1}{L_{ls} + \frac{3}{2}\bar{L}_m} u_{ds}^e$$

$$\frac{di_{0s}^e}{dt} = -\frac{r_s}{L_{ls}} i_{0s}^e + \frac{1}{L_{ls}} u_{0s}^e$$

$$\frac{d\omega_r}{dt} = \frac{3P^2\psi_m}{8J} i_{qs}^e - \frac{B_m}{J}\omega_r - \frac{P}{2J}T_L$$

$$\frac{d\theta_r}{dt} = \omega_r$$

The *quadrature, direct,* and *zero* currents needed to be fed to guarantee the balanced operation are

$$i_{qs}^e(t) = \sqrt{2}i_M, \; i_{ds}^e(t) = 0, \text{ and } i_{0s}^e(t) = 0$$

Example 7.4

Using differential equations (7.22), which describe the permanent-magnet synchronous motor dynamics in the rotor reference frame, perform nonlinear modeling by applying the motor parameters as given in Example 7.3.

Solution

The SIMUNLINK diagram is developed using a set of nonlinear differential equations (7.22) and the *s*-domain block diagram illustrated in Figure 7.20. Letting $\varphi_u = 0$, the *quadrature, direct,* and *zero* voltages to be applied are

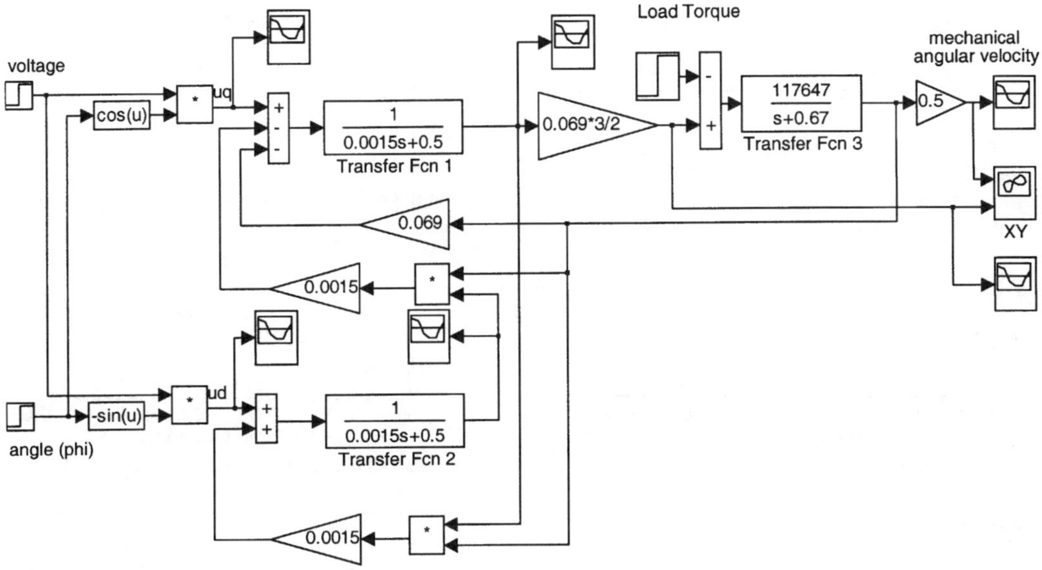

Figure 7.21 SIMULINK diagram to model a three-phase, two-pole permanent-magnet synchronous motor in the rotor reference frame.

$$u_{qs}^r(t)=\sqrt{2}u_M,\ \ u_{ds}^r(t)=0,\ \ u_{0s}^r(t)=0$$

The SIMULINK diagram is built (see Figure 7.21).
Supplying $u_M = 40$ V, we have

$$u_{qs}^r(t)=\sqrt{2}40,\ \ u_{ds}^r(t)=0,\ \ \text{and}\ \ u_{0s}^r(t)=0$$

The *quadrature* voltage, as well as the transient dynamics for the motor state variables, are illustrated in Figures 7.22 and 7.23 for different load conditions. The torque-speed characteristic curves are plotted for completeness of the modeling results. □

Modeling and analysis of three-phase permanent-magnet synchronous generators in the rotor and synchronous reference frames

In three-phase permanent-magnet synchronous generators, the phase voltages are induced in the *as*, *bs*, and *cs* windings, and the circuitry dynamics are mapped by the following differential equations

$$\mathbf{u}_{abcs}=-\mathbf{r}_s\mathbf{i}_{abcs}+\frac{d\boldsymbol{\psi}_{abcs}}{dt},\ \ \begin{bmatrix}u_{as}\\u_{bs}\\u_{cs}\end{bmatrix}=-\begin{bmatrix}r_s&0&0\\0&r_s&0\\0&0&r_s\end{bmatrix}\begin{bmatrix}i_{as}\\i_{bs}\\i_{cs}\end{bmatrix}+\begin{bmatrix}\frac{d\psi_{as}}{dt}\\\frac{d\psi_{bs}}{dt}\\\frac{d\psi_{cs}}{dt}\end{bmatrix}$$

where the flux linkages are expressed as

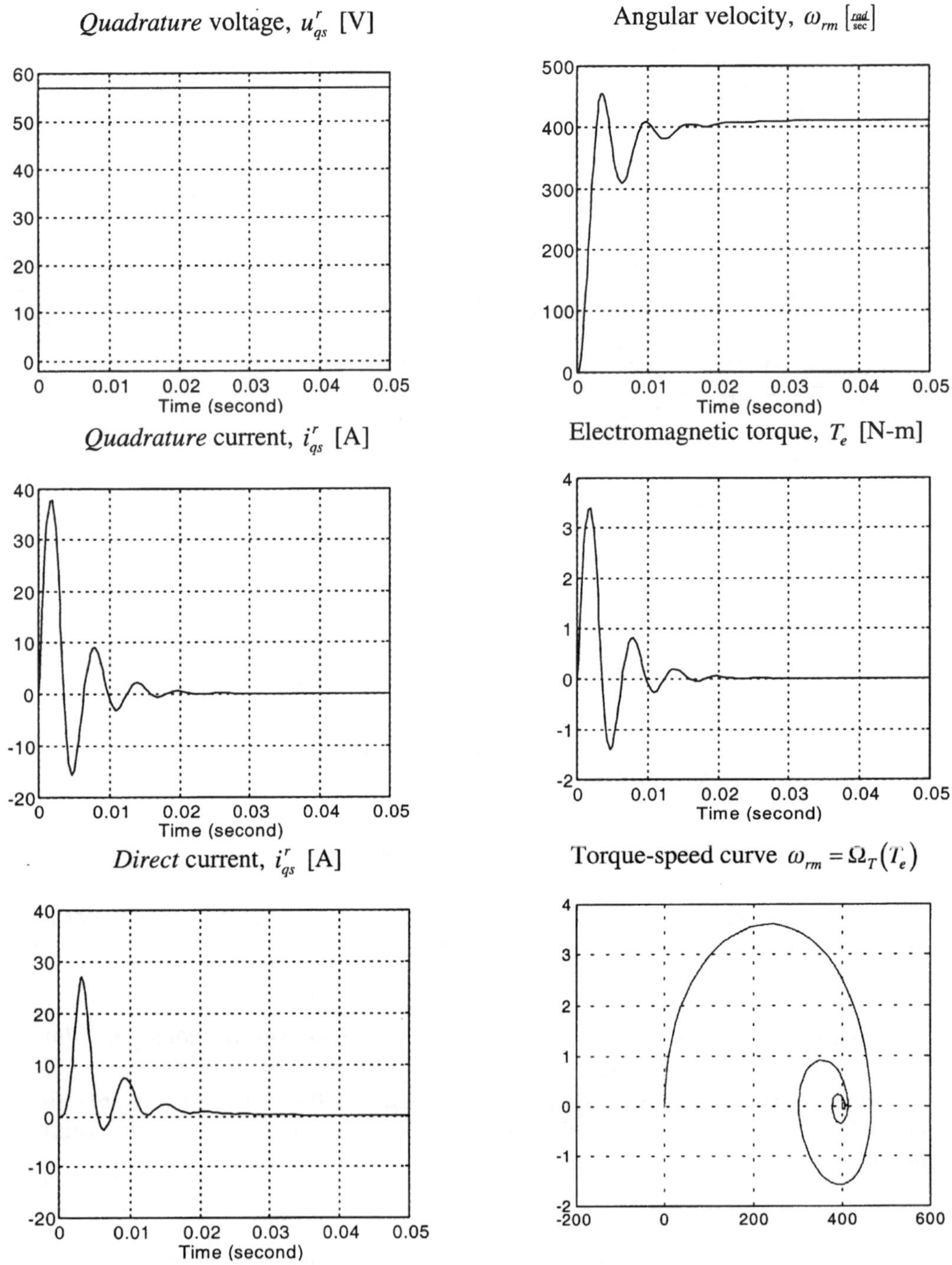

Figure 7.22 Acceleration of a three-phase permanent-magent synchronous motor; T_L = 0 N-m, $u^r_{qs}(t) = \sqrt{2}40$, $u^r_{ds}(t) = 0$, and $u^r_{0s}(t) = 0$.

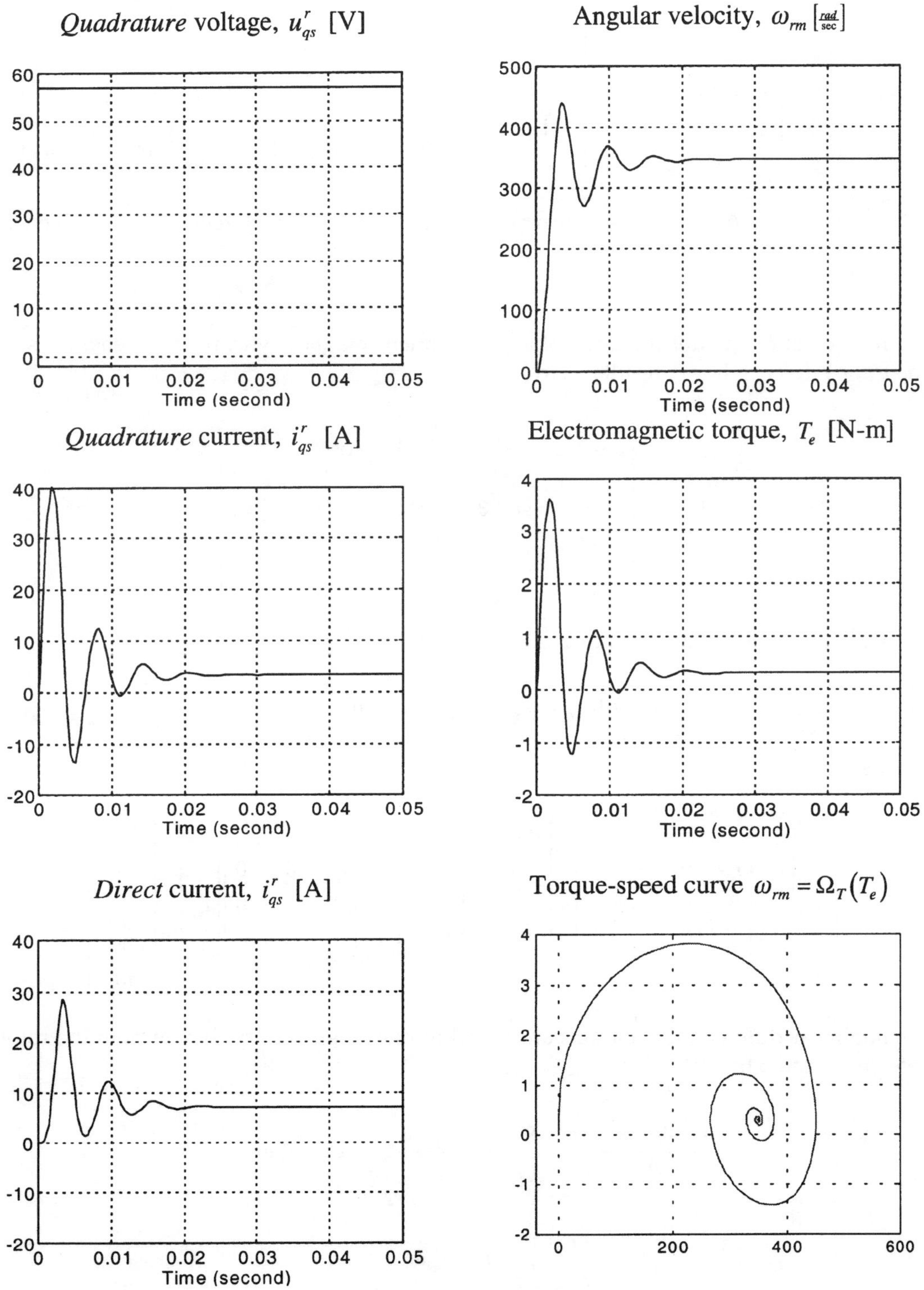

Figure 7.23 Acceleration of a three-phase permanent-magnet synchronous motor; T_L = 0.3 N-m, $u_{qs}^r(t) = \sqrt{2}40$, $u_{ds}^r(t) = 0$, and $u_{0s}^r(t) = 0$.

$$\boldsymbol{\psi}_{abcs} = \begin{bmatrix} \psi_{as} \\ \psi_{bs} \\ \psi_{cs} \end{bmatrix} = -\mathbf{L}_s \mathbf{i}_{abcs} + \boldsymbol{\psi}_m = -\begin{bmatrix} L_{ls} + \bar{L}_m & -\frac{1}{2}\bar{L}_m & -\frac{1}{2}\bar{L}_m \\ -\frac{1}{2}\bar{L}_m & L_{ls} + \bar{L}_m & -\frac{1}{2}\bar{L}_m \\ -\frac{1}{2}\bar{L}_m & -\frac{1}{2}\bar{L}_m & L_{ls} + \bar{L}_m \end{bmatrix} \begin{bmatrix} i_{as} \\ i_{bs} \\ i_{cs} \end{bmatrix} + \psi_m \begin{bmatrix} \sin\theta_r \\ \sin\left(\theta_r - \frac{2}{3}\pi\right) \\ \sin\left(\theta_r + \frac{2}{3}\pi\right) \end{bmatrix}$$

Applying the *direct* Park transformation, which in the rotor reference frame is given as

$$\mathbf{u}_{qd0s}^r = \mathbf{K}_s^r \mathbf{u}_{abcs}, \quad \mathbf{i}_{qd0s}^r = \mathbf{K}_s^r \mathbf{i}_{abcs}, \quad \text{and} \quad \boldsymbol{\psi}_{qd0s}^r = \mathbf{K}_s^r \boldsymbol{\psi}_{abcs}$$

the mathematical model for three-phase permanent-magnet synchronous generators in the rotor reference frame results. Here,

$$\mathbf{K}_s^r = \frac{2}{3} \begin{bmatrix} \cos\theta_r & \cos\left(\theta_r - \frac{2}{3}\pi\right) & \cos\left(\theta_r + \frac{2}{3}\pi\right) \\ \sin\theta_r & \sin\left(\theta_r - \frac{2}{3}\pi\right) & \sin\left(\theta_r + \frac{2}{3}\pi\right) \\ \frac{1}{2} & \frac{1}{2} & \frac{1}{2} \end{bmatrix}$$

For the resistive load, denoting the phase load resistances as R_L, we have

$$u_{as} = R_L i_{as}, \quad u_{bs} = R_L i_{bs}, \quad \text{and} \quad u_{cs} = R_L i_{cs}$$

Using

$$\begin{bmatrix} u_{as} \\ u_{bs} \\ u_{cs} \end{bmatrix} = -\begin{bmatrix} r_s & 0 & 0 \\ 0 & r_s & 0 \\ 0 & 0 & r_s \end{bmatrix} \begin{bmatrix} i_{as} \\ i_{bs} \\ i_{cs} \end{bmatrix} + \begin{bmatrix} \frac{d\psi_{as}}{dt} \\ \frac{d\psi_{bs}}{dt} \\ \frac{d\psi_{cs}}{dt} \end{bmatrix}, \quad \begin{bmatrix} R_L i_{as} \\ R_L i_{bs} \\ R_L i_{cs} \end{bmatrix} = -\begin{bmatrix} r_s & 0 & 0 \\ 0 & r_s & 0 \\ 0 & 0 & r_s \end{bmatrix} \begin{bmatrix} i_{as} \\ i_{bs} \\ i_{cs} \end{bmatrix} + \begin{bmatrix} \frac{d\psi_{as}}{dt} \\ \frac{d\psi_{bs}}{dt} \\ \frac{d\psi_{cs}}{dt} \end{bmatrix}$$

the nonlinear differential equations are found to map the circuitry dynamics in the rotor reference frame. In particular,

$$\frac{di_{qs}^r}{dt} = -\frac{r_s + R_L}{L_{ls} + \frac{3}{2}\bar{L}_m} i_{qs}^r + \frac{\psi_m}{L_{ls} + \frac{3}{2}\bar{L}_m} \omega_r - i_{ds}^r \omega_r$$

$$\frac{di_{ds}^r}{dt} = -\frac{r_s + R_L}{L_{ls} + \frac{3}{2}\bar{L}_m} i_{ds}^r + i_{qs}^r \omega_r$$

$$\frac{di_{0s}^r}{dt} = -\frac{r_s + R_L}{L_{ls}} i_{0s}^r$$

The *torsional-mechanical* dynamics are obtained by applying Newton's second law of motion using the mechanical angular velocity ω_{rm}, and

$$\frac{d\omega_{rm}}{dt} = -\frac{3P\psi_m}{4J} i_{qs}^r - \frac{B_m}{J}\omega_{rm} + \frac{1}{J}T_{p}$$

$$\frac{d\theta_{rm}}{dt} = \omega_{rm}$$

The *zero* current and the angular displacement can be omitted from analysis. As the state variable, the mechanical angular velocity ω_{rm} is used, and $\omega_{rm} = \frac{2}{P}\omega_r$. Augmenting the differential equations that describe the circuitry and *torsional-mechanical* dynamics, one obtains equations of motion for three-phase permanent-magnet synchronous generators with the resistive load in the rotor reference frame. In particular,

$$\frac{di_{qs}^r}{dt} = -\frac{r_s + R_L}{L_{ls} + \frac{3}{2}\bar{L}_m} i_{qs}^r + \frac{\psi_m}{L_{ls} + \frac{3}{2}\bar{L}_m}\frac{P}{2}\omega_{rm} - \frac{P}{2} i_{ds}^r \omega_{rm}$$

$$\frac{di_{ds}^r}{dt} = -\frac{r_s + R_L}{L_{ls} + \frac{3}{2}\bar{L}_m} i_{ds}^r + \frac{P}{2} i_{qs}^r \omega_{rm}$$

$$\frac{d\omega_{rm}}{dt} = -\frac{3P\psi_m}{4J} i_{qs}^r - \frac{B_m}{J}\omega_{rm} + \frac{1}{J}T_{pm}$$

In the synchronous reference frame, the mathematical model is expressed as

$$\frac{di_{qs}^e}{dt} = -\frac{r_s + R_L}{L_{ls} + \frac{3}{2}\bar{L}_m} i_{qs}^e + \frac{\psi_m}{L_{ls} + \frac{3}{2}\bar{L}_m}\frac{P}{2}\omega_{rm} - \frac{P}{2} i_{ds}^e \omega_{rm}$$

$$\frac{di_{ds}^e}{dt} = -\frac{r_s + R_L}{L_{ls} + \frac{3}{2}\bar{L}_m} i_{ds}^e + \frac{P}{2} i_{qs}^e \omega_{rm}$$

$$\frac{d\omega_{rm}}{dt} = -\frac{3P\psi_m}{4J} i_{qs}^e - \frac{B_m}{J}\omega_{rm} + \frac{1}{J}T_{pm}$$

Synchronous generators are driven by prime movers. Hence, the transient dynamics of generators must be integrated with the prime mover equations of motion. Unless the prime mover is not specified, one cannot find the augmented model needed. Even the expression for the T_{pm} is unknown. In fact, DC, induction, or synchronous motors can be used as prime movers, and the corresponding mathematical models needed are different. Consider the generation system when the permanent-magnet DC motor is used as a prime mover. The power generation system under consideration is illustrated in Figure 7.24.

An augmented mathematical model of this system can be easily derived as the basic foundations have been thoroughly studied. For the mechanical system permanent-magnet synchronous generator–prime mover, using Newton's law, we have

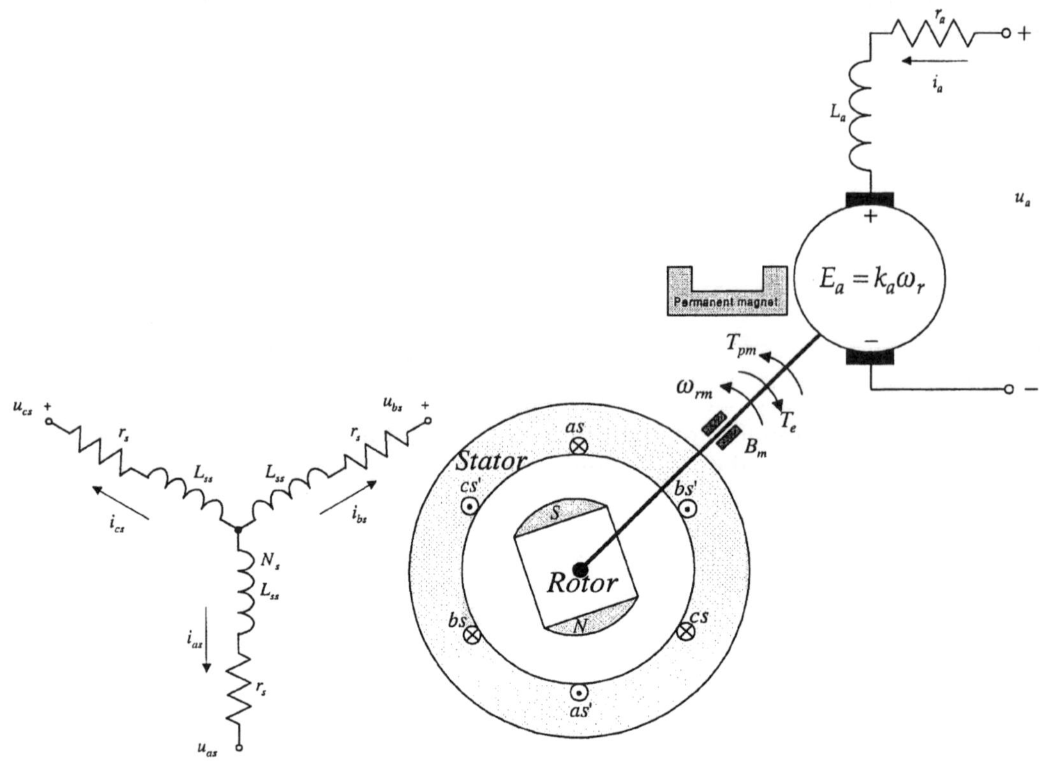

Figure 7.24 Three-phase wye-connected permanent-magnet synchronous generator driven by a permanent-magnet DC motor.

$$T_{pm} - T_e - B_m \omega_{rm} = J\frac{d\omega_{rm}}{dt}, \quad \frac{d\theta_{rm}}{dt} = \omega_{rm}$$

Here, the electromagnetic torque developed by the permanent-magnet DC motor is $T_{pm} = k_a i_a$; the load torque for the prime mover is the generator electromagnetic torque; and in the rotor reference frame, one finds $T_e = \frac{3P\psi_m}{4} i_{qs}^r$.

Therefore, the *torsional-mechanical* dynamics are mapped by

$$\frac{d\omega_{rm}}{dt} = \frac{k_a}{J} i_a - \frac{3P\psi_m}{4J} i_{qs}^r - \frac{B_m}{J}\omega_{rm}$$

and augmenting the circuitry dynamics of the permanent-magnet DC motor

$$\frac{di_a}{dt} = -\frac{r_a}{L_a} i_a - \frac{k_a}{L_a}\omega_r + \frac{1}{L_a} u_a$$

with the differential equations for the evolution of the *quadrature* and *direct* generator currents

$$\frac{di_{qs}^r}{dt} = -\frac{r_s + R_L}{L_{ls} + \frac{3}{2}\bar{L}_m} i_{qs}^r + \frac{\psi_m}{L_{ls} + \frac{3}{2}\bar{L}_m} \frac{P}{2}\omega_{rm} - \frac{P}{2} i_{ds}^r \omega_{rm}$$

$$\frac{di_{ds}^r}{dt} = -\frac{r_s + R_L}{L_{ls} + \frac{3}{2}\bar{L}_m} i_{ds}^r + \frac{P}{2} i_{qs}^r \omega_{rm}$$

the resulting set of differential equations for the generation system results. In particular, we have

$$\begin{aligned}
\frac{di_a}{dt} &= -\frac{r_a}{L_a} i_a - \frac{k_a}{L_a}\omega_r + \frac{1}{L_a} u_a \\
\frac{di_{qs}^r}{dt} &= -\frac{r_s + R_L}{L_{ls} + \frac{3}{2}\bar{L}_m} i_{qs}^r + \frac{\psi_m}{L_{ls} + \frac{3}{2}\bar{L}_m} \frac{P}{2}\omega_{rm} - \frac{P}{2} i_{ds}^r \omega_{rm} \\
\frac{di_{ds}^r}{dt} &= -\frac{r_s + R_L}{L_{ls} + \frac{3}{2}\bar{L}_m} i_{ds}^r + \frac{P}{2} i_{qs}^r \omega_{rm} \\
\frac{d\omega_{rm}}{dt} &= \frac{k_a}{J} i_a - \frac{3P\psi_m}{4J} i_{qs}^r - \frac{B_m}{J}\omega_{rm}
\end{aligned} \tag{7.23}$$

It should be observed that we have managed to find an augmented mathematical model using the *machine* variables for permanent-magnet DC motors (prime mover) and the *qd* variables for synchronous generators.

In the synchronous reference frame, one finds

$$\begin{aligned}
\frac{di_a}{dt} &= -\frac{r_a}{L_a} i_a - \frac{k_a}{L_a}\omega_r + \frac{1}{L_a} u_a \\
\frac{di_{qs}^e}{dt} &= -\frac{r_s + R_L}{L_{ls} + \frac{3}{2}\bar{L}_m} i_{qs}^e + \frac{\psi_m}{L_{ls} + \frac{3}{2}\bar{L}_m} \frac{P}{2}\omega_{rm} - \frac{P}{2} i_{ds}^e \omega \\
\frac{di_{ds}^e}{dt} &= -\frac{r_s + R_L}{L_{ls} + \frac{3}{2}\bar{L}_m} i_{ds}^e + \frac{P}{2} i_{qs}^e \omega_{rm} \\
\frac{d\omega_{rm}}{dt} &= \frac{k_a}{J} i_a - \frac{3P\psi_m}{4J} i_{qs}^e - \frac{B_m}{J}\omega_{rm}
\end{aligned}$$

Example 7.5

In the MATLAB environment, model a power generation system with the permanent-magnet synchronous generator and DC motor. A three-phase, four-pole, 135 W, 434 rad-s^{-1}, 40 V, 6.9 A, 23 NEMA size Kollmorgen permanent-magnet synchronous machine H-232-G is used. The parameters are: $r_s = 0.5$ ohm, $L_{ss} = L_{ls} + \frac{3}{2}\bar{L}_m = 0.01$ H, and $\psi_m = 0.054$ V-s-rad^{-1} or $\psi_m = 0.054$ N-m-A^{-1} (for the hot machine, the ψ_m decreases; and from the data given in Table 7.1 for the cold motor, we have $\psi_m = 0.0607$ N-m-A^{-1}). The parameters of the permanent-magnet DC motor are $r_a = 0.4$ ohm, $L_a = 0.015$ H, and $k_a = 0.25$ V-s-rad^{-1} or $k_a = 0.25$ N-m-A^{-1}. The viscous friction coefficient and equivalent moment of inertia are $B_m = 0.00006$ N-m-s-rad^{-1} and $J = 0.0075$ kg-m^2, respectively.

Solution

SIMULINK is used to perform the simulations; and by applying (7.23), the corresponding diagram is developed and documented in Figure 7.25.

The data m-file to download the parameters must be run first to simulate the transient dynamics using the SIMULINK diagram developed. We have

```
% Parameters of the permanent-magnet SG
P=4; rs=0.5; Lss=0.01; fm=0.054;
% Parameters of the permanent-magnet DC motor
ra=0.4; La=0.015; ka=0.25;
% Torsional-mechanical parameters
J=0.0075; Bm=0.00006;
```

In addition to the *quadrature* and *direct* currents (states) and voltages (outputs), the *abc* currents and voltages are of particular interest. The phase voltages induced by synchronous generators, as well as the currents in the *as*, *bs*, and *cs* windings, can be found using the *quadrature-*, *direct-*, and *zero*-axis components of voltages and currents. In particular, by applying the *inverse* Park transformation, we have

$$\mathbf{u}_{abcs} = \mathbf{K}_s^{r^{-1}} u_{qd0s}^r \text{ and } \mathbf{i}_{abcs} = \mathbf{K}_s^{r^{-1}} \mathbf{i}_{qd0s}^r$$

where $\mathbf{K}_s^{r^{-1}} = \begin{bmatrix} \cos\theta_r & \sin\theta_r & 1 \\ \cos\left(\theta_r - \frac{2}{3}\pi\right) & \sin\left(\theta_r - \frac{2}{3}\pi\right) & 1 \\ \cos\left(\theta_r + \frac{2}{3}\pi\right) & \sin\left(\theta_r + \frac{2}{3}\pi\right) & 1 \end{bmatrix}$.

That is,

$$\begin{bmatrix} u_{as} \\ u_{bs} \\ u_{cs} \end{bmatrix} = \begin{bmatrix} \cos\theta_r & \sin\theta_r & 1 \\ \cos\left(\theta_r - \frac{2}{3}\pi\right) & \sin\left(\theta_r - \frac{2}{3}\pi\right) & 1 \\ \cos\left(\theta_r + \frac{2}{3}\pi\right) & \sin\left(\theta_r + \frac{2}{3}\pi\right) & 1 \end{bmatrix} \begin{bmatrix} u_{qs}^r \\ u_{ds}^r \\ u_{0s}^r \end{bmatrix}$$

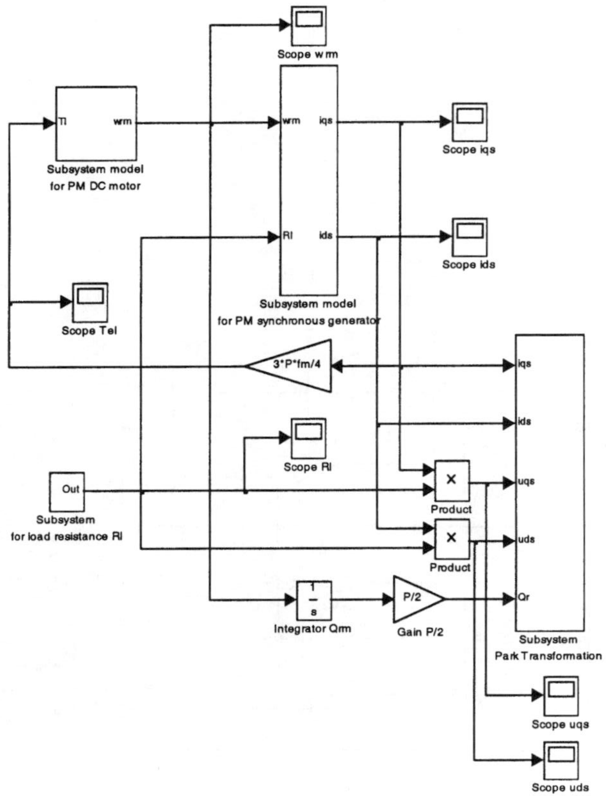

a

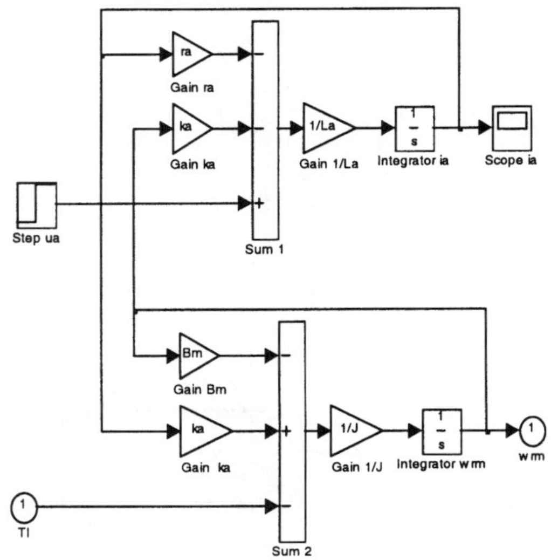

b

Figure 7.25 SIMULINK diagram to simulate generation systems: permanent-magnet synchronous generator/permanent-magnet DC motor: (a) augmented system; (b) SIMULINK subsystem diagram to model permanent-magnet DC motors; (c) SIMULINK subsystem diagram to model synchronous generators; (d) SIMULINK subsystem diagram to apply the resistive loads.

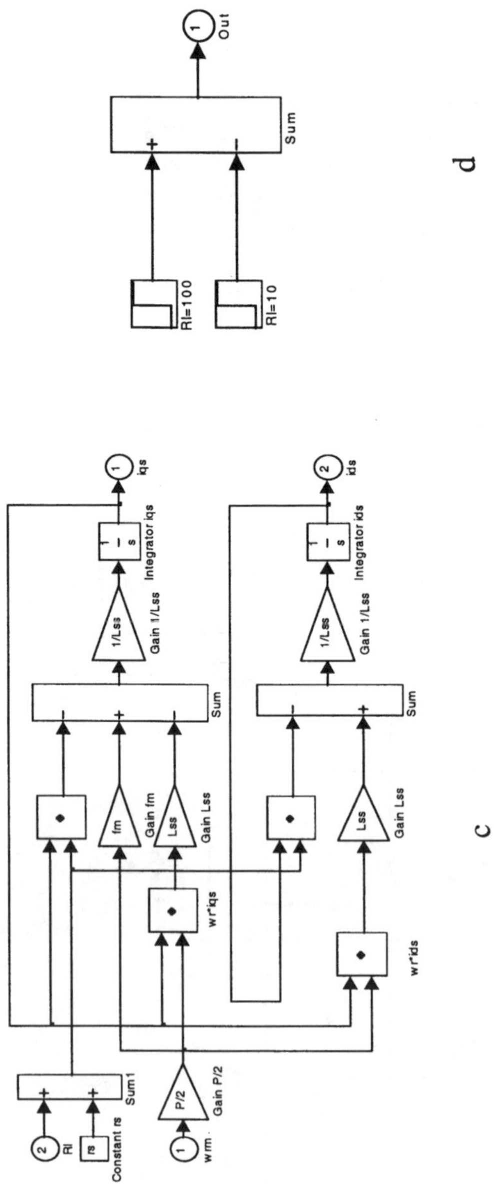

Figure 7.25 Continued.

$$u_{as} = \cos\theta_r u_{qs}^r + \sin\theta_r u_{ds}^r + u_{0s}^r$$

$$u_{bs} = \cos\left(\theta_r - \frac{2}{3}\pi\right)u_{qs}^r + \sin\left(\theta_r - \frac{2}{3}\pi\right)u_{ds}^r + u_{0s}^r$$

$$u_{cs} = \cos\left(\theta_r + \frac{2}{3}\pi\right)u_{qs}^r + \sin\left(\theta_r + \frac{2}{3}\pi\right)u_{ds}^r + u_{0s}^r$$

and

$$\begin{bmatrix} i_{as} \\ i_{bs} \\ i_{cs} \end{bmatrix} = \begin{bmatrix} \cos\theta_r & \sin\theta_r & 1 \\ \cos\left(\theta_r - \frac{2}{3}\pi\right) & \sin\left(\theta_r - \frac{2}{3}\pi\right) & 1 \\ \cos\left(\theta_r + \frac{2}{3}\pi\right) & \sin\left(\theta_r + \frac{2}{3}\pi\right) & 1 \end{bmatrix} \begin{bmatrix} i_{qs}^r \\ i_{ds}^r \\ i_{0s}^r \end{bmatrix}$$

$$i_{as} = \cos\theta_r i_{qs}^r + \sin\theta_r i_{ds}^r + i_{0s}^r$$

$$i_{bs} = \cos\left(\theta_r - \frac{2}{3}\pi\right)i_{qs}^r + \sin\left(\theta_r - \frac{2}{3}\pi\right)i_{ds}^r + i_{0s}^r$$

$$i_{cs} = \cos\left(\theta_r + \frac{2}{3}\pi\right)i_{qs}^r + \sin\left(\theta_r + \frac{2}{3}\pi\right)i_{ds}^r + i_{0s}^r$$

The *zero* voltage and current are equal to zero and, therefore,

$$u_{as} = \cos\theta_r u_{qs}^r + \sin\theta_r u_{ds}^r$$

$$u_{bs} = \cos\left(\theta_r - \frac{2}{3}\pi\right)u_{qs}^r + \sin\left(\theta_r - \frac{2}{3}\pi\right)u_{ds}^r$$

$$u_{cs} = \cos\left(\theta_r + \frac{2}{3}\pi\right)u_{qs}^r + \sin\left(\theta_r + \frac{2}{3}\pi\right)u_{ds}^r$$

and

$$i_{as} = \cos\theta_r i_{qs}^r + \sin\theta_r i_{ds}^r$$

$$i_{bs} = \cos\left(\theta_r - \frac{2}{3}\pi\right)i_{qs}^r + \sin\left(\theta_r - \frac{2}{3}\pi\right)i_{ds}^r$$

$$i_{cs} = \cos\left(\theta_r + \frac{2}{3}\pi\right)i_{qs}^r + \sin\left(\theta_r + \frac{2}{3}\pi\right)i_{ds}^r$$

Using these equations, the corresponding SIMULINK subsystem diagrams to perform the needed *inverse* Park transformation are illustrated in Figure 7.26. This subsystem, which allows one to find the phase currents i_{as}, i_{bs}, i_{cs} and voltages u_{as}, u_{bs}, u_{cs} through the

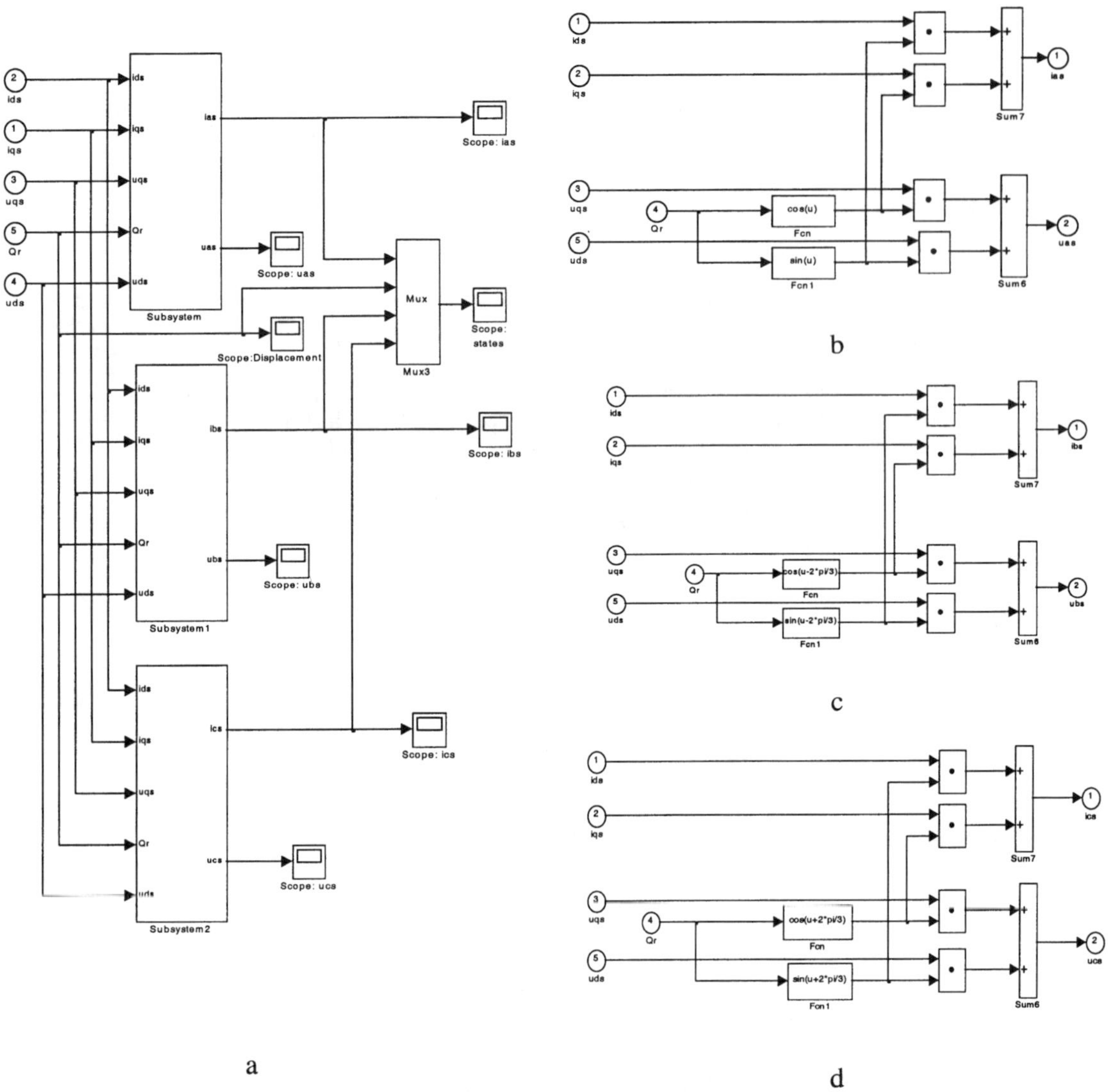

Figure 7.26 SIMULINK subsystem diagrams to perform the *inverse* Park transformations: (a) Park transformation diagram; (b) transformations of the *qd* variables to the *as* voltage and current; (c) transformations of the *qd* variables to the *bs* voltage and current; (d) transformations of the *qd* variables to the *cs* voltage and current.

Park transformations $\mathbf{u}_{abcs} = \mathbf{K}_s^{r^{-1}} \mathbf{u}_{qd0s}^r$ and $\mathbf{i}_{abcs} = \mathbf{K}_s^{r^{-1}} \mathbf{i}_{qd0s}^r$, was used in the SIMULINK diagram given in Figure 7.25a (see `Subsystem Park Transformation`).

Applying the constant voltage to the permanent-magnet DC motor $u_a = 100$ V and assigning the initial mechanical angular velocity to be 400 rad-s^{-1}, the phase load resistances R_L are inserted. In particular,

$$R_L = \begin{cases} 100 \text{ ohm}, \ \forall t \in [0 \quad 0.1) \text{ s} \\ 10 \text{ ohm}, \ \forall t \in [0 \quad 0.2) \text{ s} \end{cases}$$

The dynamics of the permanent-magnet synchronous generator and prime mover are documented in Figure 7.27. The state variables i_{qs}^r, i_{ds}^r, ω_{rm} are illustrated, and the *quadrature* and direct voltages induced u_{qs}^r, u_{ds}^r are the outputs. The waveforms for the phase voltages u_{as}, u_{bs}, u_{cs} and currents i_{as}, i_{bs}, i_{cs}, obtained using the *inverse* Park transformation are documented in Figure 7.27.

For the resistance R_L = 100 ohm, we have $u_{qs}^r = 42.6$ and $u_{ds}^r = 3.4$. The plotted electromagnetic torque of the synchronous generator T_e is the load torque applied to the prime mover, and T_e is a function of the phase resistances inserted. Using the *inverse* Park transformation, one finds that the lightly loaded permanent-magnet synchronous generator induces the ac voltages with frequency 127 Hz, and

$$u_{as} = 42.8\sin(798t)$$

$$u_{bs} = 42.8\sin\left(798t - \frac{2}{3}\pi\right)$$

$$u_{cs} = 42.8\sin\left(798t + \frac{2}{3}\pi\right)$$

If the phase resistances R_L = 10 ohm are inserted, we have the following steady-state values for the *quadrature* and *direct* voltages: $u_{qs}^r = 26$ and $u_{ds}^r = 19.6$. The *abc* terminal voltages induced have frequency 126 Hz. In particular, for the medium-loaded permanent-magnet synchronous generator, we have

$$u_{as} = 32.5\sin(791t)$$

$$u_{bs} = 32.5\sin\left(791t - \frac{2}{3}\pi\right)$$

$$u_{cs} = 32.5\sin\left(791t + \frac{2}{3}\pi\right)$$

7.4 *Conventional three-phase synchronous machines: dynamics in the **machine** variables, and in the rotor and synchronous reference frames*

In this section, mathematical models of three-phase synchronous machines in the *abc* and *qd0* variables will be developed. To find differential equations that map the transient dynamics, Kirchhoff's and Newton's second laws are used. Symmetrical synchronous motors and generators with sinusoidally distributed windings are illustrated in Figures 7.28 and 7.29, respectively.

7.4.1 *Dynamics of synchronous machines in the* machine *variables*

The application of Kirchhoff's voltage law for stator and rotor circuitry results in four differential equations for the stator (*as*, *bs*, and *cs*) and rotor (*fr*) windings. In particular, we have the following differential equations for the three-phase synchronous motor illustrated in Figure 7.28:

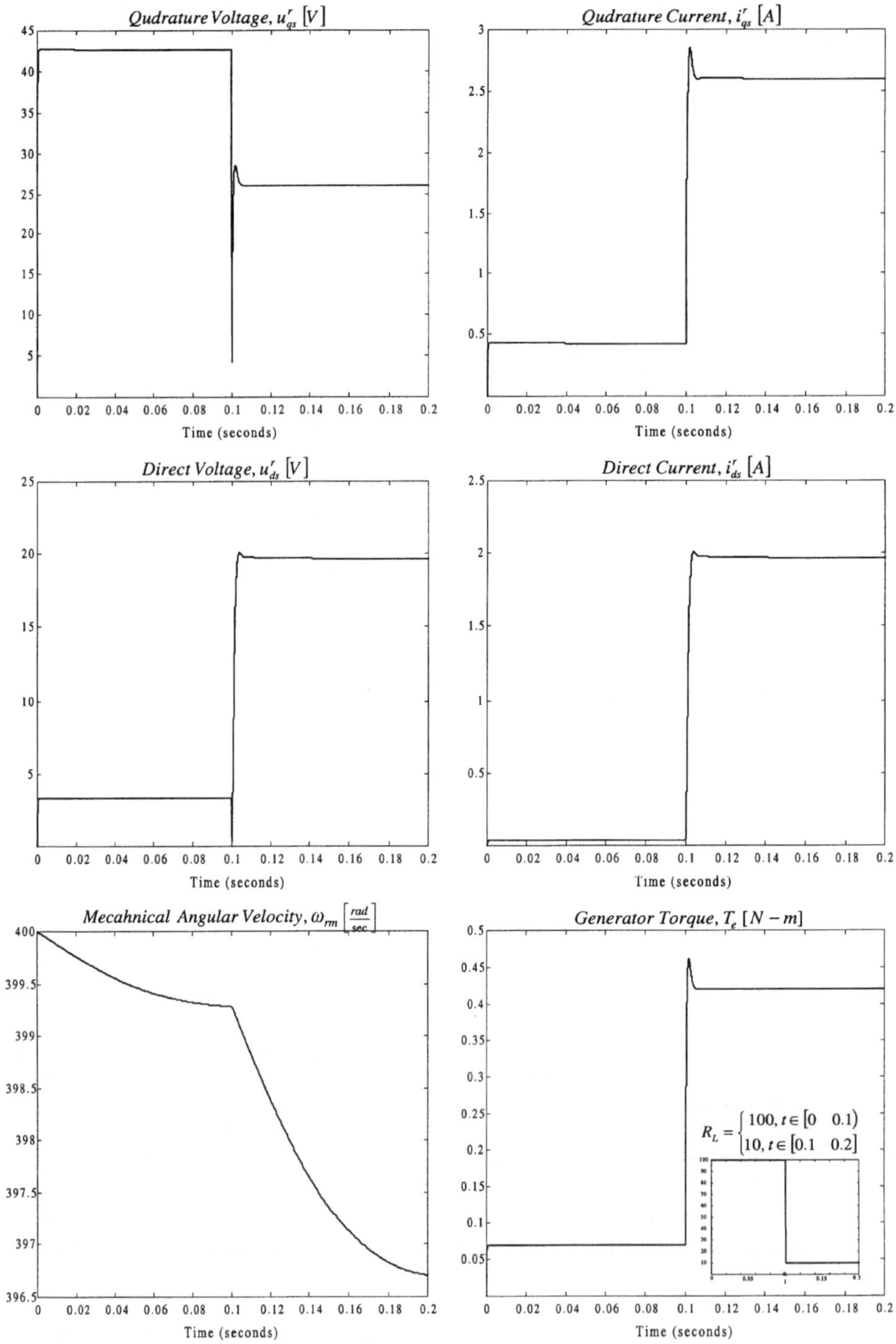

Figure 7.27 Transient dynamics for i_{qs}^r, i_{ds}^r, ω_{rm}, waveforms for induced voltages u_{qs}^r, u_{ds}^r, and u_{as}, u_{bs}, u_{cs}, and evolution of the phase currents i_{as}, i_{bs}, i_{cs}, $R_L = \begin{cases} 100 \text{ ohm}, \ \forall t \in [0 \quad 0.1) \text{ s} \\ 10 \text{ ohm}, \ \forall t \in [0 \quad 0.2) \text{ s} \end{cases}$.

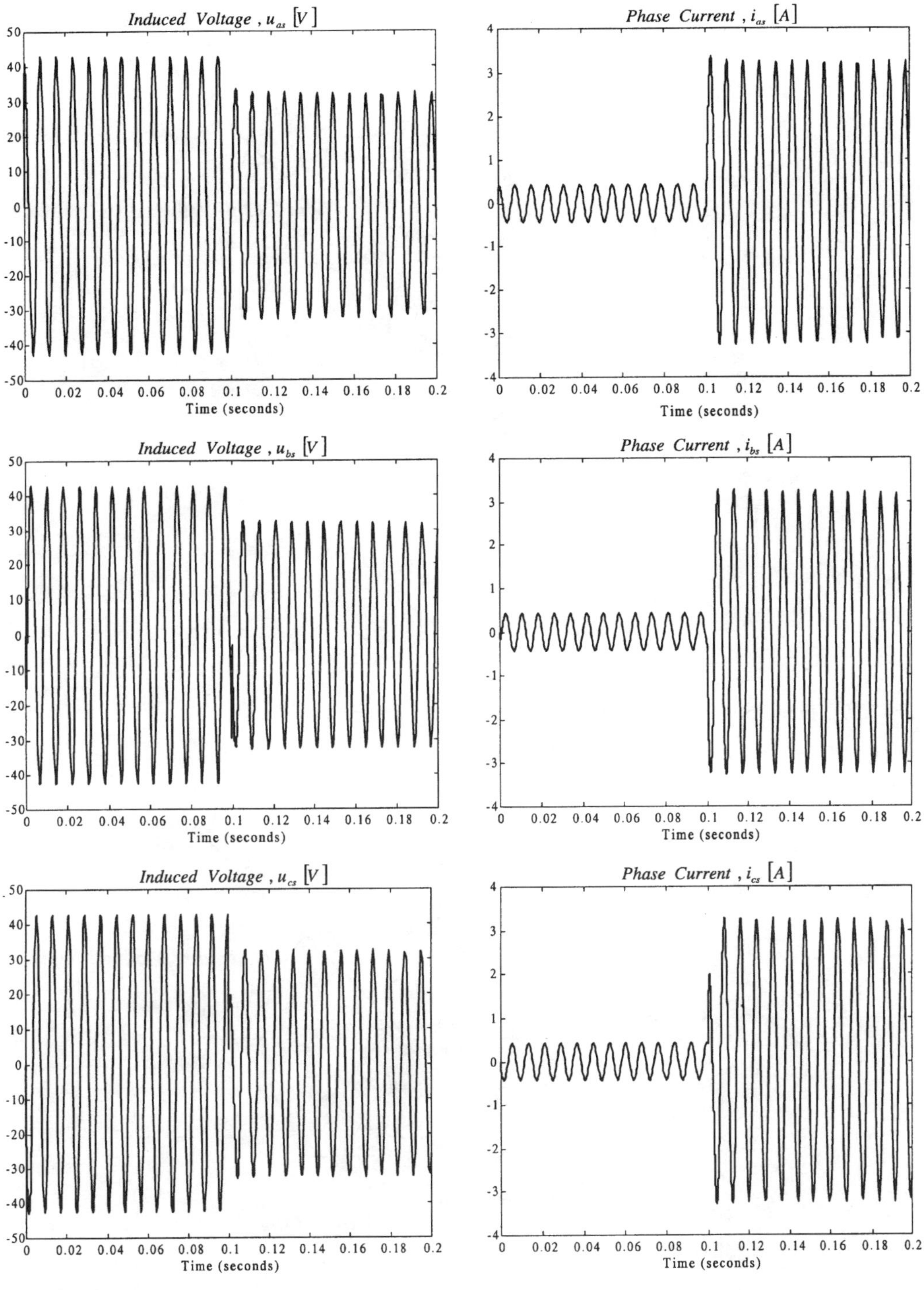

Figure 7.27 *Continued.*

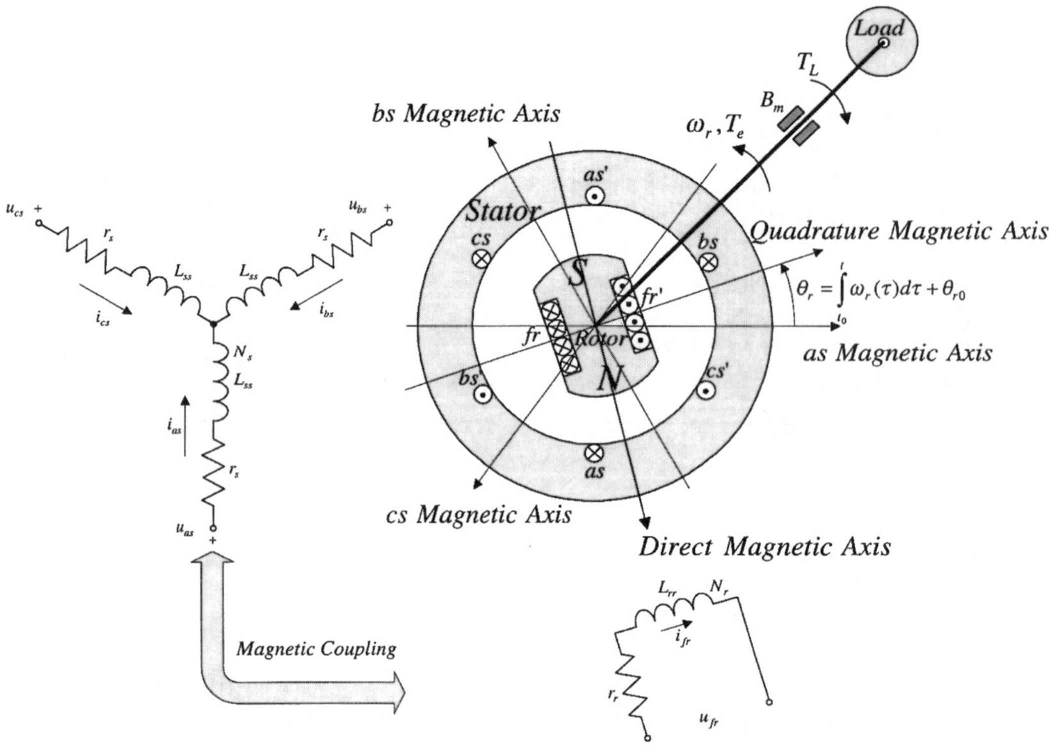

Figure 7.28 Three-phase wye-connected synchronous motor.

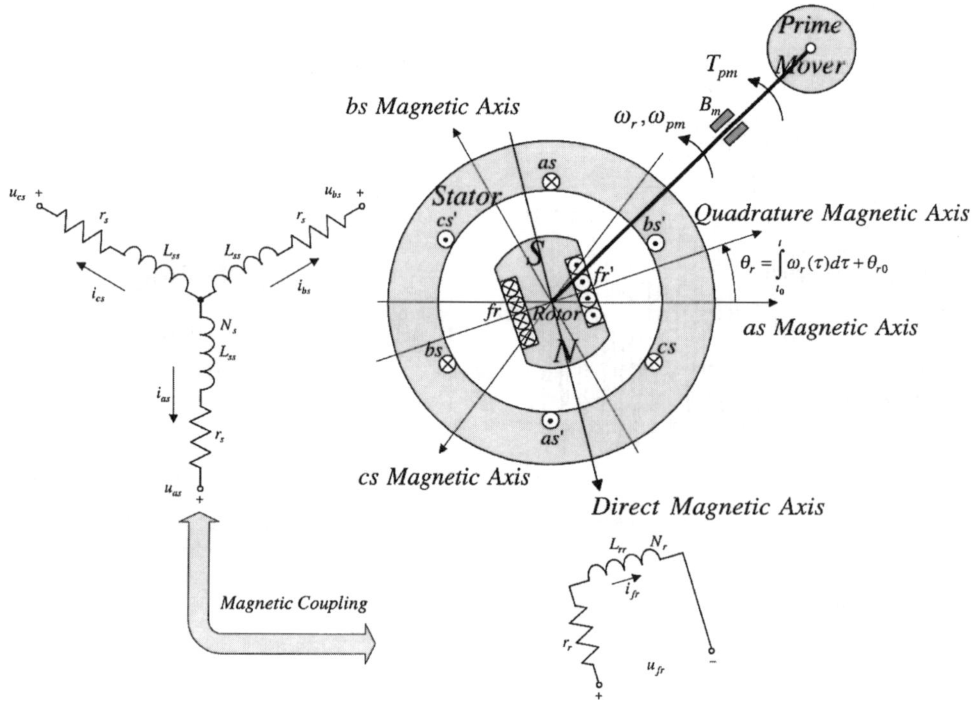

Figure 7.29 Three-phase wye-connected synchronous generator.

$$
\begin{aligned}
u_{as} &= r_s i_{as} + \frac{d\psi_{as}}{dt} \\
u_{bs} &= r_s i_{bs} + \frac{d\psi_{bs}}{dt} \\
u_{cs} &= r_s i_{cs} + \frac{d\psi_{cs}}{dt} \\
u_{fr} &= r_r i_{fr} + \frac{d\psi_{fr}}{dt}
\end{aligned}
\tag{7.24}
$$

where the flux linkages are given as

$$
\begin{aligned}
\psi_{as} &= L_{asas} i_{as} + L_{asbs} i_{bs} + L_{ascs} i_{cs} + L_{asfr} i_{fr} \\
\psi_{bs} &= L_{bsas} i_{as} + L_{bsbs} i_{bs} + L_{bscs} i_{cs} + L_{bsfr} i_{fr} \\
\psi_{cs} &= L_{csas} i_{as} + L_{csbs} i_{bs} + L_{cscs} i_{cs} + L_{csfr} i_{fr} \\
\psi_{fr} &= L_{fras} i_{as} + L_{frbs} i_{bs} + L_{frcs} i_{cs} + L_{frfr} i_{fr}
\end{aligned}
$$

From (7.24), in matrix form, one obtains

$$
\mathbf{u}_{abcs} = \mathbf{r}_s \mathbf{i}_{abcs} + \frac{d\boldsymbol{\psi}_{abcs}}{dt}
$$

$$
u_{fr} = r_f i_{fr} + \frac{d\psi_{fr}}{dt}
$$

or

$$
\begin{bmatrix} u_{as} \\ u_{bs} \\ u_{cs} \\ u_{fr} \end{bmatrix} = \begin{bmatrix} r_s & 0 & 0 & 0 \\ 0 & r_s & 0 & 0 \\ 0 & 0 & r_s & 0 \\ 0 & 0 & 0 & r_s \end{bmatrix} \begin{bmatrix} i_{as} \\ i_{bs} \\ i_{cs} \\ i_{fr} \end{bmatrix} + \begin{bmatrix} \dfrac{d\psi_{as}}{dt} \\ \dfrac{d\psi_{bs}}{dt} \\ \dfrac{d\psi_{cs}}{dt} \\ \dfrac{d\psi_{fr}}{dt} \end{bmatrix}
$$

For synchronous reluctance machines, the self- and mutual inductances between the stator windings were found, and the resulting expressions were given by equations (7.2). Making use of the equations for self- and mutual inductances (7.2), and performing the analysis of stator and rotor magnetic coupling, we have the following expressions for the stator and rotor flux linkages

$$\psi_{as} = (L_{ls} + \bar{L}_m - L_{\Delta m}\cos 2\theta_r)i_{as} + \left(-\frac{1}{2}\bar{L}_m - L_{\Delta m}\cos 2\left(\theta_r - \frac{1}{3}\pi\right)\right)i_{bs} + \left(-\frac{1}{2}\bar{L}_m - L_{\Delta m}\cos 2\left(\theta_r + \frac{1}{3}\pi\right)\right)i_{cs}$$
$$+ L_{md}\sin\theta_r i_{fr}$$

$$\psi_{bs} = \left(-\frac{1}{2}\bar{L}_m - L_{\Delta m}\cos 2\left(\theta_r - \frac{1}{3}\pi\right)\right)i_{as} + \left(L_{ls} + \bar{L}_m - L_{\Delta m}\cos 2\left(\theta_r - \frac{2}{3}\pi\right)\right)i_{bs} + \left(-\frac{1}{2}\bar{L}_m - L_{\Delta m}\cos 2\theta_r\right)i_{cs}$$
$$+ L_{md}\sin\left(\theta_r - \frac{2}{3}\pi\right)i_{fr}$$

$$\psi_{cs} = \left(-\frac{1}{2}\bar{L}_m - L_{\Delta m}\cos 2\left(\theta_r + \frac{1}{3}\pi\right)\right)i_{as} + \left(-\frac{1}{2}\bar{L}_m - L_{\Delta m}\cos 2\theta_r\right)i_{bs} + \left(L_{ls} + \bar{L}_m - L_{\Delta m}\cos 2\left(\theta_r + \frac{2}{3}\pi\right)\right)i_{cs}$$
$$+ L_{md}\sin\left(\theta_r + \frac{2}{3}\pi\right)i_{fr}$$

$$\psi_{fr} = L_{md}\sin\theta_r i_{as} + L_{md}\sin\left(\theta_r - \frac{2}{3}\pi\right)i_{bs} + L_{md}\sin\left(\theta_r + \frac{2}{3}\pi\right)i_{cs} + (L_{lf} + L_{mf})i_{fr} \tag{7.25}$$

where $L_{md} = \dfrac{N_r N_s}{\Re_{md}}$ and $L_{mf} = \dfrac{N_r^2}{\Re_{md}}$.

It should be emphasized that using the matrix of self- and mutual inductances $\mathbf{L}_{abcs/fr}$, as given by

$$\mathbf{L}_{abcs/fr} =$$
$$\begin{bmatrix} L_{ls} + \bar{L}_m - L_{\Delta m}\cos 2\theta_r & -\frac{1}{2}\bar{L}_m - L_{\Delta m}\cos 2\left(\theta_r - \frac{1}{3}\pi\right) & -\frac{1}{2}\bar{L}_m - L_{\Delta m}\cos 2\left(\theta_r + \frac{1}{3}\pi\right) & L_{md}\sin\theta_r \\ -\frac{1}{2}\bar{L}_m - L_{\Delta m}\cos 2\left(\theta_r - \frac{1}{3}\pi\right) & L_{ls} + \bar{L}_m - L_{\Delta m}\cos 2\left(\theta_r - \frac{2}{3}\pi\right) & -\frac{1}{2}\bar{L}_m - L_{\Delta m}\cos 2\theta_r & L_{md}\sin\left(\theta_r - \frac{2}{3}\pi\right) \\ -\frac{1}{2}\bar{L}_m - L_{\Delta m}\cos 2\left(\theta_r + \frac{1}{3}\pi\right) & -\frac{1}{2}\bar{L}_m - L_{\Delta m}\cos 2\theta_r & L_{ls} + \bar{L}_m - L_{\Delta m}\cos 2\left(\theta_r + \frac{2}{3}\pi\right) & L_{md}\sin\left(\theta_r + \frac{2}{3}\pi\right) \\ L_{md}\sin\theta_r & L_{md}\sin\left(\theta_r - \frac{2}{3}\pi\right) & L_{md}\sin\left(\theta_r + \frac{2}{3}\pi\right) & L_{lf} + L_{mf} \end{bmatrix}$$

one obtains

$$\boldsymbol{\psi}_{abcs/fr} = \mathbf{L}_{abcs/fr}\mathbf{i}_{abcs/fr}$$
$$= \begin{bmatrix} L_{ls} + \bar{L}_m - L_{\Delta m}\cos 2\theta_r & -\frac{1}{2}\bar{L}_m - L_{\Delta m}\cos 2\left(\theta_r - \frac{1}{3}\pi\right) & -\frac{1}{2}\bar{L}_m - L_{\Delta m}\cos 2\left(\theta_r + \frac{1}{3}\pi\right) & L_{md}\sin\theta_r \\ -\frac{1}{2}\bar{L}_m - L_{\Delta m}\cos 2\left(\theta_r - \frac{1}{3}\pi\right) & L_{ls} + \bar{L}_m - L_{\Delta m}\cos 2\left(\theta_r - \frac{2}{3}\pi\right) & -\frac{1}{2}\bar{L}_m - L_{\Delta m}\cos 2\theta_r & L_{md}\sin\left(\theta_r - \frac{2}{3}\pi\right) \\ -\frac{1}{2}\bar{L}_m - L_{\Delta m}\cos 2\left(\theta_r + \frac{1}{3}\pi\right) & -\frac{1}{2}\bar{L}_m - L_{\Delta m}\cos 2\theta_r & L_{ls} + \bar{L}_m - L_{\Delta m}\cos 2\left(\theta_r + \frac{2}{3}\pi\right) & L_{md}\sin\left(\theta_r + \frac{2}{3}\pi\right) \\ L_{md}\sin\theta_r & L_{md}\sin\left(\theta_r - \frac{2}{3}\pi\right) & L_{md}\sin\left(\theta_r + \frac{2}{3}\pi\right) & L_{lf} + L_{mf} \end{bmatrix}$$
$$\times \begin{bmatrix} i_{as} \\ i_{bs} \\ i_{cs} \\ i_{fr} \end{bmatrix}$$

From (7.24) and (7.25), we have the following set of four differential equations

$$\begin{aligned}
&\frac{d\left((L_{ls}+\bar{L}_m-L_{\Delta m}\cos 2\theta_r)i_{as}+\left(-\frac{1}{2}\bar{L}_m-L_{\Delta m}\cos 2\left(\theta_r-\frac{1}{3}\pi\right)\right)i_{bs}+\left(-\frac{1}{2}\bar{L}_m-L_{\Delta m}\cos 2\left(\theta_r+\frac{1}{3}\pi\right)\right)i_{cs}+L_{md}\sin\theta_r\, i_{fr}\right)}{dt}\\
&\quad=-r_s i_{as}+u_{as}\\
&\frac{d\left(\left(-\frac{1}{2}\bar{L}_m-L_{\Delta m}\cos 2\left(\theta_r-\frac{1}{3}\pi\right)\right)i_{as}+\left(L_{ls}+\bar{L}_m-L_{\Delta m}\cos 2\left(\theta_r-\frac{2}{3}\pi\right)\right)i_{bs}+\left(-\frac{1}{2}\bar{L}_m-L_{\Delta m}\cos 2\theta_r\right)i_{cs}+L_{md}\sin\left(\theta_r-\frac{2}{3}\pi\right)i_{fr}\right)}{dt}\\
&\quad=-r_s i_{bs}+u_{bs}\\
&\frac{d\left(\left(-\frac{1}{2}\bar{L}_m-L_{\Delta m}\cos 2\left(\theta_r+\frac{1}{3}\pi\right)\right)i_{as}+\left(-\frac{1}{2}\bar{L}_m-L_{\Delta m}\cos 2\theta_r\right)i_{bs}+\left(L_{ls}+\bar{L}_m-L_{\Delta m}\cos 2\left(\theta_r+\frac{2}{3}\pi\right)\right)i_{cs}+L_{md}\sin\left(\theta_r+\frac{2}{3}\pi\right)i_{fr}\right)}{dt}\\
&\quad=-r_s i_{cs}+u_{cs}\\
&\frac{d\left(L_{md}\sin\theta_r\, i_{as}+L_{md}\sin\left(\theta_r-\frac{2}{3}\pi\right)i_{bs}+L_{md}\sin\left(\theta_r+\frac{2}{3}\pi\right)i_{cs}+(L_{lf}+L_{mf})i_{fr}\right)}{dt}=-r_r i_{fr}+u_{fr}
\end{aligned}\tag{7.26}$$

Using (7.26), Cauchy's form of the differential equations to map the circuitry dynamics of the conventional three-phase synchronous motors is derived by applying Symbolic Math Toolbox, and an m-file is developed. In this Matlab file, the following notations are used `S1` $=\sin\theta_r$, `S2` $=\sin\left(\theta_r-\frac{2}{3}\pi\right)$, `S3` $=\sin\left(\theta_r+\frac{2}{3}\pi\right)$, `C1` $=\cos\theta_r$, `C2` $=\cos\left(\theta_r-\frac{2}{3}\pi\right)$, `C3` $=\cos\left(\theta_r+\frac{2}{3}\pi\right)$, `Lls` $=L_{ls}$, `Lmb` $=\bar{L}_m$, `Lmd` $=L_{md}$, `Lmf` $=L_{mf}$, and `Llf` $=L_{lf}$.

Matlab script

```
L=sym('[Lls+Lmb,-Lmb/2,-Lmb/2,Lmd*S1;
     -Lmb/2,Lls+Lmb,-Lmb/2,Lmd*S2;
     -Lmb/2,-Lmb/2,Lls+Lmb,Lmd*S3;
     Lmd*S1,Lmd*S2,Lmd*S3,Llf+Lmf]');
R=sym('[-rs 0 0 0;
        0 -rs 0 0;
        0 0 -rs 0;
        0 0 0 -rr]');
I=sym('[ias; ibs; ics; ifr]');
V=sym('[uas;ubs;ucs;ufr]');
K=sym('[Lmd*C1*wr*ifr;
     Lmd*C2*wr*ifr;
     Lmd*C3*wr*ifr;
     Lmd*wr*(ias*C1+ibs*C2+ics*C3)]');
L1=inv(L); L2=simplify(L1)
FS1=L2*R*I;
FS2=simplify(FS1)
FS3=L2*V;
FS4=simplify(FS3)
FS5=L2*K
FS6=simplify(FS5)
FS7=FS2+FS4-FS6
FS=simplify(FS7)
```

Taking note of the results derived in the symbolic form and trigonometric identities, the following nonlinear differential equations are obtained

$$\frac{di_{as}}{dt}=\frac{1}{(2L_{ls}+3\bar{L}_m)(3L_{md}^2L_{ls}-(2L_{ls}^2+3L_{ls}\bar{L}_m)L_{ff})}[(3L_{md}^2\bar{L}_m+4L_{ls}L_{md}^2-(3\bar{L}_m^2+4L_{ls}^2+8L_{ls}\bar{L}_m)L_{ff}$$
$$+2L_{md}^2L_{ls}\cos 2\theta_r)(-r_si_{as}+u_{as})+(3L_{md}^2\bar{L}_m+L_{ls}L_{md}^2-(3\bar{L}_m^2+2L_{ls}\bar{L}_m)L_{ff}+2L_{md}^2L_{ls}\cos 2\left(\theta_r-\frac{1}{3}\pi\right))(-r_si_{bs}+u_{bs})$$
$$+\left(3L_{md}^2\bar{L}_m+L_{ls}L_{md}^2-(3\bar{L}_m^2+2L_{ls}\bar{L}_m)L_{ff}+2L_{md}^2L_{ls}\cos 2\left(\theta_r+\frac{1}{3}\pi\right)\right)(-r_si_{cs}+u_{cs})+(6L_{ls}L_{md}\bar{L}_m+4L_{ls}^2L_{md})\sin\theta_r(-r_ri_{fr}+u_{fr})$$
$$-(6L_{ls}L_{md}^2\bar{L}_m+4L_{ls}^2L_{md}^2)\left(i_{as}\cos\theta_r+i_{bs}\cos\left(\theta_r-\frac{2}{3}\pi\right)+i_{cs}\cos\left(\theta_r+\frac{2}{3}\pi\right)\right)\omega_r\sin\theta_r$$
$$+((6L_{md}L_{ls}\bar{L}_m+4L_{md}L_{ls}^2)L_{ff}-6L_{md}^3L_{ls})i_{fr}\omega_r\cos\theta_r]$$

$$\frac{di_{bs}}{dt}=\frac{1}{(2L_{ls}+3\bar{L}_m)(3L_{md}^2L_{ls}-(2L_{ls}^2+3L_{ls}\bar{L}_m)L_{ff})}[(3L_{md}^2\bar{L}_m+L_{ls}L_{md}^2-(3\bar{L}_m^2+2L_{ls}\bar{L}_m)L_{ff}+2L_{md}^2L_{ls}\cos 2\left(\theta_r-\frac{1}{3}\pi\right))(-r_si_{as}+u_{as})$$
$$+\left(3L_{md}^2\bar{L}_m+4L_{ls}L_{md}^2-(3\bar{L}_m^2+4L_{ls}^2+8L_{ls}\bar{L}_m)L_{ff}+2L_{md}^2L_{ls}\cos 2\left(\theta_r-\frac{2}{3}\pi\right)\right)(-r_si_{bs}+u_{bs})$$
$$+(3L_{md}^2\bar{L}_m+L_{ls}L_{md}^2-(3\bar{L}_m^2+2L_{ls}\bar{L}_m)L_{ff}+2L_{md}^2L_{ls}\cos 2\theta_r)(-r_si_{cs}+u_{cs})+(6L_{ls}L_{md}\bar{L}_m+4L_{ls}^2L_{md})\sin\left(\theta_r-\frac{2}{3}\pi\right)(-r_ri_{fr}+u_{fr})$$
$$-(6L_{ls}L_{md}^2\bar{L}_m+4L_{ls}^2L_{md}^2)\left(i_{as}\cos\theta_r+i_{bs}\cos\left(\theta_r-\frac{2}{3}\pi\right)+i_{cs}\cos\left(\theta_r+\frac{2}{3}\pi\right)\right)\omega_r\sin\left(\theta_r-\frac{2}{3}\pi\right)$$
$$+((6L_{md}L_{ls}\bar{L}_m+4L_{md}L_{ls}^2)L_{ff}-6L_{md}^3L_{ls})i_{fr}\omega_r\cos\left(\theta_r-\frac{2}{3}\pi\right)]$$

$$\frac{di_{cs}}{dt}=\frac{1}{(2L_{ls}+3\bar{L}_m)(3L_{md}^2L_{ls}-(2L_{ls}^2+3L_{ls}\bar{L}_m)L_{ff})}\left[\left(3L_{md}^2\bar{L}_m+L_{ls}L_{md}^2-(3\bar{L}_m^2+2L_{ls}\bar{L}_m)L_{ff}+2L_{md}^2L_{ls}\cos 2\left(\theta_r+\frac{1}{3}\pi\right)\right)(-r_si_{as}+u_{as})\right.$$
$$+(3L_{md}^2\bar{L}_m+L_{ls}L_{md}^2-(3\bar{L}_m^2+2L_{ls}\bar{L}_m)L_{ff}+2L_{md}^2L_{ls}\cos 2\theta_r)(-r_si_{bs}+u_{bs})$$
$$+\left(3L_{md}^2\bar{L}_m+4L_{ls}L_{md}^2-(3\bar{L}_m^2+4L_{ls}^2+8L_{ls}\bar{L}_m)L_{ff}+2L_{md}^2L_{ls}\cos 2\left(\theta_r+\frac{2}{3}\pi\right)\right)(-r_si_{cs}+u_{cs})+(6L_{ls}L_{md}\bar{L}_m+4L_{ls}^2L_{md})\sin\left(\theta_r+\frac{2}{3}\pi\right)(-r_ri_{fr}+u_{fr})$$
$$\left.+((6L_{md}L_{ls}\bar{L}_m+4L_{md}L_{ls}^2)L_{ff}-6L_{md}^3L_{ls})i_{fr}\omega_r\cos\left(\theta_r+\frac{2}{3}\pi\right)\right]$$

$$\frac{di_{fr}}{dt}=\frac{1}{3L_{md}^2L_{ls}-(2L_{ls}^2+3L_{ls}\bar{L}_m)L_{ff}}\left[2L_{md}L_{ls}\left(\sin\theta_r(-r_si_{as}+u_{as})+\sin\left(\theta_r-\frac{2}{3}\pi\right)(-r_si_{bs}+u_{bs})+\sin\left(\theta_r+\frac{2}{3}\pi\right)(-r_si_{cs}+u_{cs})\right)\right.$$
$$\left.-(2L_{ls}^2+3\bar{L}_mL_{ls})(-r_ri_{fr}+u_{fr})+(3L_{ls}L_{md}\bar{L}_m+2L_{ls}^2L_{md})\left(i_{as}\omega_r\cos\theta_r+i_{bs}\omega_r\cos\left(\theta_r-\frac{2}{3}\pi\right)+i_{cs}\omega_r\cos\left(\theta_r+\frac{2}{3}\pi\right)\right)\right]$$

(7.27)

Here, $L_{ff}=L_{lf}+L_{mf}$. Hence, in matrix form, we have

$$\begin{bmatrix}\frac{di_{as}}{dt}\\ \frac{di_{bs}}{dt}\\ \frac{di_{cs}}{dt}\\ \frac{di_{fr}}{dt}\end{bmatrix}=\begin{bmatrix}-\frac{r_sL_{Ds}}{L_{\Sigma s}} & -\frac{r_sL_{Ms}}{L_{\Sigma s}} & -\frac{r_sL_{Ms}}{L_{\Sigma s}} & 0\\ -\frac{r_sL_{Ms}}{L_{\Sigma s}} & -\frac{r_sL_{Ds}}{L_{\Sigma s}} & -\frac{r_sL_{Ms}}{L_{\Sigma s}} & 0\\ -\frac{r_sL_{Ms}}{L_{\Sigma s}} & -\frac{r_sL_{Ms}}{L_{\Sigma s}} & -\frac{r_sL_{Ds}}{L_{\Sigma s}} & 0\\ 0 & 0 & 0 & \frac{r_r(2L_{ls}^2+3\bar{L}_mL_{ls})}{L_{\Sigma f}}\end{bmatrix}\begin{bmatrix}i_{as}\\ i_{bs}\\ i_{cs}\\ i_{fr}\end{bmatrix}$$

$$
+\begin{bmatrix}
-\frac{2r_sL_{md}^2L_{ls}}{L_{\Sigma s}}\cos 2\theta_r & -\frac{2r_sL_{md}^2L_{ls}}{L_{\Sigma s}}\cos 2\left(\theta_r-\frac{1}{3}\pi\right) & -\frac{2r_sL_{md}^2L_{ls}}{L_{\Sigma s}}\cos 2\left(\theta_r+\frac{1}{3}\pi\right) & -\frac{r_rL_{Mf}}{L_{\Sigma s}}\sin\theta_r \\
-\frac{2r_sL_{md}^2L_{ls}}{L_{\Sigma s}}\cos 2\left(\theta_r-\frac{1}{3}\pi\right) & -\frac{2r_sL_{md}^2L_{ls}}{L_{\Sigma s}}\cos 2\left(\theta_r-\frac{2}{3}\pi\right) & -\frac{2r_sL_{md}^2L_{ls}}{L_{\Sigma s}}\cos 2\theta_r & -\frac{r_rL_{Mf}}{L_{\Sigma s}}\sin\left(\theta_r-\frac{2}{3}\pi\right) \\
-\frac{2r_sL_{md}^2L_{ls}}{L_{\Sigma s}}\cos 2\left(\theta_r+\frac{1}{3}\pi\right) & -\frac{2r_sL_{md}^2L_{ls}}{L_{\Sigma s}}\cos 2\theta_r & -\frac{2r_sL_{md}^2L_{ls}}{L_{\Sigma s}}\cos 2\left(\theta_r+\frac{2}{3}\pi\right) & -\frac{r_sL_{Mf}}{L_{\Sigma s}}\sin\left(\theta_r+\frac{2}{3}\pi\right) \\
-\frac{2r_rL_{md}L_{ls}}{L_{\Sigma f}}\sin\theta_r & -\frac{2r_rL_{md}L_{ls}}{L_{\Sigma f}}\sin\left(\theta_r-\frac{2}{3}\pi\right) & -\frac{2r_rL_{md}L_{ls}}{L_{\Sigma f}}\sin\left(\theta_r+\frac{2}{3}\pi\right) & 0
\end{bmatrix}
\begin{bmatrix} i_{as} \\ i_{bs} \\ i_{cs} \\ i_{fr} \end{bmatrix}
$$

$$
+\begin{bmatrix}
-\frac{6L_{ls}L_{md}^2\bar{L}_m+4L_{ls}^2L_{md}^2}{L_{\Sigma s}}\sin\theta_r & -\frac{6L_{ls}L_{md}^2\bar{L}_m+4L_{ls}^2L_{md}^2}{L_{\Sigma s}}\sin\theta_r & -\frac{6L_{ls}L_{md}^2\bar{L}_m+4L_{ls}^2L_{md}^2}{L_{\Sigma s}}\sin\theta_r \\
-\frac{6L_{ls}L_{md}^2\bar{L}_m+4L_{ls}^2L_{md}^2}{L_{\Sigma s}}\sin\left(\theta_r-\frac{2}{3}\pi\right) & -\frac{6L_{ls}L_{md}^2\bar{L}_m+4L_{ls}^2L_{md}^2}{L_{\Sigma s}}\sin\left(\theta_r-\frac{2}{3}\pi\right) & -\frac{6L_{ls}L_{md}^2\bar{L}_m+4L_{ls}^2L_{md}^2}{L_{\Sigma s}}\sin\left(\theta_r-\frac{2}{3}\pi\right) \\
-\frac{6L_{ls}L_{md}^2\bar{L}_m+4L_{ls}^2L_{md}^2}{L_{\Sigma s}}\sin\left(\theta_r+\frac{2}{3}\pi\right) & -\frac{6L_{ls}L_{md}^2\bar{L}_m+4L_{ls}^2L_{md}^2}{L_{\Sigma s}}\sin\left(\theta_r+\frac{2}{3}\pi\right) & -\frac{6L_{ls}L_{md}^2\bar{L}_m+4L_{ls}^2L_{md}^2}{L_{\Sigma s}}\sin\left(\theta_r+\frac{2}{3}\pi\right) \\
-\frac{3L_{ls}L_{md}\bar{L}_m+2L_{ls}^2L_{md}}{L_{\Sigma f}} & -\frac{3L_{ls}L_{md}\bar{L}_m+2L_{ls}^2L_{md}}{L_{\Sigma f}} & -\frac{3L_{ls}L_{md}\bar{L}_m+2L_{ls}^2L_{md}}{L_{\Sigma f}}
\end{bmatrix}
\begin{bmatrix} i_{as}\omega_r\cos\theta_r \\ i_{bs}\omega_r\cos\left(\theta_r-\frac{2}{3}\pi\right) \\ i_{cs}\omega_r\cos\left(\theta_r+\frac{2}{3}\pi\right) \end{bmatrix}
$$

$$
+\begin{bmatrix}
\frac{(6L_{md}L_{ls}\bar{L}_m+4L_{md}L_{ls}^2)L_{ff}-6L_{md}^3L_{ls}}{L_{\Sigma s}}i_{fr}\omega_r\cos\theta_r \\
\frac{(6L_{md}L_{ls}\bar{L}_m+4L_{md}L_{ls}^2)L_{ff}-6L_{md}^3L_{ls}}{L_{\Sigma s}}i_{fr}\omega_r\cos\left(\theta_r-\frac{2}{3}\pi\right) \\
\frac{(6L_{md}L_{ls}\bar{L}_m+4L_{md}L_{ls}^2)L_{ff}-6L_{md}^3L_{ls}}{L_{\Sigma s}}i_{fr}\omega_r\cos\left(\theta_r+\frac{2}{3}\pi\right) \\
0
\end{bmatrix}
$$

$$
+\begin{bmatrix}
\frac{L_{Ds}+2L_{md}^2L_{ls}\cos 2\theta_r}{L_{\Sigma s}} & \frac{L_{Ms}+2L_{md}^2L_{ls}\cos 2\left(\theta_r-\frac{1}{3}\pi\right)}{L_{\Sigma s}} & \frac{L_{Ms}+2L_{md}^2L_{ls}\cos 2\left(\theta_r+\frac{1}{3}\pi\right)}{L_{\Sigma s}} & \frac{L_{Mf}}{L_{\Sigma s}}\sin\theta_r \\
\frac{L_{Ms}+2L_{md}^2L_{ls}\cos 2\left(\theta_r-\frac{1}{3}\pi\right)}{L_{\Sigma s}} & \frac{L_{Ds}+2L_{md}^2L_{ls}\cos 2\left(\theta_r-\frac{2}{3}\pi\right)}{L_{\Sigma s}} & \frac{L_{Ms}+2L_{md}^2L_{ls}\cos 2\theta_r}{L_{\Sigma s}} & \frac{L_{Mf}}{L_{\Sigma s}}\sin\left(\theta_r-\frac{2}{3}\pi\right) \\
\frac{L_{Ms}+2L_{md}^2L_{ls}\cos 2\left(\theta_r+\frac{1}{3}\pi\right)}{L_{\Sigma s}} & \frac{L_{Ms}+2L_{md}^2L_{ls}\cos 2\theta_r}{L_{\Sigma s}} & \frac{L_{Ds}+2L_{md}^2L_{ls}\cos 2\left(\theta_r+\frac{2}{3}\pi\right)}{L_{\Sigma s}} & \frac{L_{Mf}}{L_{\Sigma s}}\sin\left(\theta_r+\frac{2}{3}\pi\right) \\
\frac{2L_{md}L_{ls}\sin\theta_r}{L_{\Sigma f}} & \frac{2L_{md}L_{ls}\sin\left(\theta_r-\frac{2}{3}\pi\right)}{L_{\Sigma f}} & \frac{2L_{md}L_{ls}\sin\left(\theta_r+\frac{2}{3}\pi\right)}{L_{\Sigma f}} & -\frac{2L_{ls}^2+3\bar{L}_mL_{ls}}{L_{\Sigma f}}
\end{bmatrix}
\begin{bmatrix} u_{as} \\ u_{bs} \\ u_{cs} \\ u_{fr} \end{bmatrix}
$$

where the following notations are used:

$$L_{Ds}=3L_{md}^2\bar{L}_m+4L_{ls}L_{md}^2-(3\bar{L}_m^2+4L_{ls}^2+8L_{ls}\bar{L}_m)(L_{lf}+L_{mf})$$

$$L_{Ms}=3L_{md}^2\bar{L}_m+L_{ls}L_{md}^2-(3\bar{L}_m^2+2L_{ls}\bar{L}_m)(L_{lf}+L_{mf})$$

$$L_{\Sigma s}=(2L_{ls}+3\bar{L}_m)[3L_{md}^2L_{ls}-(2L_{ls}^2+3L_{ls}\bar{L}_m)(L_{lf}+L_{mf})]$$

$$L_{\Sigma f}=3L_{md}^2L_{ls}-(2L_{ls}^2+3L_{ls}\bar{L}_m)(L_{lf}+L_{mf})$$

$$L_{Mf}=6L_{ls}L_{md}\bar{L}_m+4L_{ls}^2L_{md}$$

The *torsional-mechanical* equations are obtained by taking note of Newton's second law

$$T_e-B_m\omega_{rm}-T_L=J\frac{d^2\theta_{rm}}{dt^2}$$

Hence,

$$\frac{d\omega_{rm}}{dt} = \frac{1}{J}(T_e - B_m\omega_{rm} - T_L)$$
$$\frac{d\theta_{rm}}{dt} = \omega_{rm} \tag{7.28}$$

The mechanical angular velocity ω_{rm} and displacement θ_{rm} are related to the electrical angular velocity ω_r and displacement θ_r. In particular,

$$\omega_{rm} = \frac{2}{P}\omega_r$$
$$\theta_{rm} = \frac{2}{P}\theta_r \tag{7.29}$$

Hence, from (7.28) and (7.29), one obtains

$$\frac{d\omega_r}{dt} = \frac{P}{2J}T_e - \frac{B_m}{J}\omega_r - \frac{P}{2J}T_L$$
$$\frac{d\theta_r}{dt} = \omega_r \tag{7.30}$$

To find the expression for the electromagnetic torque, the coenergy concept is used. The electromagnetic torque developed by three-phase, P-pole synchronous motors, as given by

$$T_e = \frac{P}{2}\frac{\partial W_c(i_{as}, i_{bs}, i_{cs}, i_{fr}, \theta_r)}{\partial \theta_r} = \frac{P}{2}\frac{\partial W_c\left(\frac{1}{2}[i_{as} \quad i_{bs} \quad i_{cs} \quad i_{fr}]\mathbf{L}_{abcs/fr}\begin{bmatrix} i_{as} \\ i_{bs} \\ i_{cs} \\ i_{fr} \end{bmatrix}\right)}{\partial \theta_r}$$

is found to be

$$T_e = \frac{P}{2}\left[L_{\Delta m}\left(i_{as}^2 \sin 2\theta_r + 2i_{as}i_{bs}\sin 2\left(\theta_r - \frac{1}{3}\pi\right) + 2i_{as}i_{cs}\sin 2\left(\theta_r + \frac{1}{3}\pi\right) + i_{bs}^2 \sin 2\left(\theta_r - \frac{2}{3}\pi\right)\right.\right.$$
$$\left.\left. + 2i_{bs}i_{cs}\sin 2\theta_r + i_{cs}^2\sin 2\left(\theta_r + \frac{2}{3}\pi\right)\right) + L_{md}i_{fr}\left(i_{as}\cos\theta_r + i_{bs}\cos\left(\theta_r - \frac{2}{3}\pi\right) + i_{cs}\cos\left(\theta_r + \frac{2}{3}\pi\right)\right)\right] \tag{7.31}$$

For round-rotor synchronous machines, $L_{\Delta m} = 0$ and, hence,

$$T_e = \frac{PL_{md}}{2} i_{fr} \left[i_{as} \cos\theta_r + i_{bs} \cos\left(\theta_r - \frac{2}{3}\pi\right) + i_{cs} \cos\left(\theta_r + \frac{2}{3}\pi\right) \right]$$

Remark. Using trigonometric identities, from (7.31), an alternative form is found as

$$T_e = \frac{P}{2}\left[\frac{L_{md} - L_{mq}}{3} \left(\left(i_{as}^2 - \frac{1}{2} i_{bs}^2 - \frac{1}{2} i_{cs}^2 - i_{as} i_{bs} - i_{as} i_{cs} + 2 i_{bs} i_{cs} \right) \sin 2\theta_r + \frac{\sqrt{3}}{2} (i_{bs}^2 - i_{cs}^2 - 2 i_{as} i_{bs} + 2 i_{as} i_{cs}) \cos 2\theta_r \right) \right.$$
$$\left. + L_{md} i_{fr} \left(\left(i_{as} - \frac{1}{2} i_{bs} - \frac{1}{2} i_{cs} \right) \cos\theta_r + \frac{\sqrt{3}}{2} (i_{bs} - i_{cs}) \sin\theta_r \right) \right]$$

□

From (7.31) and using the *torsional-mechanical* differential equations of motion (7.30), one concludes that the mechanical dynamics are mapped by two differential equations. In particular,

$$\frac{d\omega_r}{dt} = \frac{P^2}{4J}\left[L_{\Delta m} \left(i_{as}^2 \sin 2\theta_r + 2 i_{as} i_{bs} \sin 2\left(\theta_r - \frac{1}{3}\pi\right) + 2 i_{as} i_{cs} \sin 2\left(\theta_r + \frac{1}{3}\pi\right) + i_{bs}^2 \sin 2\left(\theta_r - \frac{2}{3}\pi\right) \right.\right.$$
$$\left.\left. + 2 i_{bs} i_{cs} \sin 2\theta_r + i_{cs}^2 \sin 2\left(\theta_r + \frac{2}{3}\pi\right) \right) + L_{md} i_{fr} \left(i_{as} \cos\theta_r + i_{bs} \cos\left(\theta_r - \frac{2}{3}\pi\right) + i_{cs} \cos\left(\theta_r + \frac{2}{3}\pi\right) \right) \right]$$
$$- \frac{B_m}{J} \omega_r - \frac{P}{2J} T_L$$
$$\frac{d\theta_r}{dt} = \omega_r$$

(7.32)

Augmenting (7.27) and (7.32), the following highly nonlinear differential equations are found to map the transient dynamics of three-phase synchronous motors

$$\frac{di_{as}}{dt} = \frac{1}{(2L_{ls} + 3\bar{L}_m)(3L_{md}^2 L_{ls} - (2L_{ls}^2 + 3L_{ls}\bar{L}_m)L_{ff})} \left[(3L_{md}^2 \bar{L}_m + 4L_{ls} L_{md}^2 - (3\bar{L}_m^2 + 4L_{ls}^2 + 8L_{ls}\bar{L}_m)L_{ff} + 2L_{md}^2 L_{ls} \cos 2\theta_r)(-r_s i_{as} + u_{as}) \right.$$
$$+ \left(3L_{md}^2 \bar{L}_m + L_{ls} L_{md}^2 - (3\bar{L}_m^2 + 2L_{ls}\bar{L}_m)L_{ff} + 2L_{md}^2 L_{ls} \cos 2\left(\theta_r - \frac{1}{3}\pi\right) \right)(-r_s i_{bs} + u_{bs})$$
$$+ \left(3L_{md}^2 \bar{L}_m + L_{ls} L_{md}^2 - (3\bar{L}_m^2 + 2L_{ls}\bar{L}_m)L_{ff} + 2L_{md}^2 L_{ls} \cos 2\left(\theta_r + \frac{1}{3}\pi\right) \right)(-r_s i_{cs} + u_{cs}) + (6L_{ls} L_{md} \bar{L}_m + 4L_{ls}^2 L_{md}) \sin\theta_r (-r_r i_{fr} + u_{fr})$$
$$- (6L_{ls} L_{md}^2 \bar{L}_m + 4L_{ls}^2 L_{md}^2) \left(i_{as} \cos\theta_r + i_{bs} \cos\left(\theta_r - \frac{2}{3}\pi\right) + i_{cs} \cos\left(\theta_r + \frac{2}{3}\pi\right) \right) \omega_r \sin\theta_r$$
$$\left. + ((6L_{md} L_{ls} \bar{L}_m + 4L_{md} L_{ls}^2) L_{ff} - 6L_{md}^3 L_{ls}) i_{fr} \omega_r \cos\theta_r \right]$$

$$\begin{aligned}
\frac{di_{bs}}{dt} &= \frac{1}{(2L_{ls}+3\bar{L}_m)(3L_{md}^2L_{ls}-(2L_{ls}^2+3L_{ls}\bar{L}_m)L_{ff})}\left[\left(3L_{md}^2\bar{L}_m+L_{ls}L_{md}^2-(3\bar{L}_m^2+2L_{ls}\bar{L}_m)L_{ff}+2L_{md}^2L_{ls}\cos 2\left(\theta_r-\frac{1}{3}\pi\right)\right)(-r_si_{as}+u_{as})\right.\\
&+\left(3L_{md}^2\bar{L}_m+4L_{ls}L_{md}^2-(3\bar{L}_m^2+4L_{ls}^2+8L_{ls}\bar{L}_m)L_{ff}+2L_{md}^2L_{ls}\cos 2\left(\theta_r-\frac{2}{3}\pi\right)\right)(-r_si_{bs}+u_{bs})\\
&+(3L_{md}^2\bar{L}_m+L_{ls}L_{md}^2-(3\bar{L}_m^2+2L_{ls}\bar{L}_m)L_{ff}+2L_{md}^2L_{ls}\cos 2\theta_r)(-r_si_{cs}+u_{cs})+(6L_{ls}L_{md}\bar{L}_m+4L_{ls}^2L_{md})\sin\left(\theta_r-\frac{2}{3}\pi\right)(-r_si_{fr}+u_{fr})\\
&-(6L_{ls}L_{md}^2\bar{L}_m+4L_{ls}^2L_{md}^2)\left(i_{as}\cos\theta_r+i_{bs}\cos\left(\theta_r-\frac{2}{3}\pi\right)+i_{cs}\cos\left(\theta_r+\frac{2}{3}\pi\right)\right)\omega_r\sin\left(\theta_r-\frac{2}{3}\pi\right)\\
&\left.+((6L_{md}L_{ls}\bar{L}_m+4L_{md}L_{ls}^2)L_{ff}-6L_{md}^3L_{ls})i_{fr}\omega_r\cos\left(\theta_r-\frac{2}{3}\pi\right)\right]\\
\frac{di_{cs}}{dt} &= \frac{1}{(2L_{ls}+3\bar{L}_m)(3L_{md}^2L_{ls}-(2L_{ls}^2+3L_{ls}\bar{L}_m)L_{ff})}\left[\left(3L_{md}^2\bar{L}_m+L_{ls}L_{md}^2-(3\bar{L}_m^2+2L_{ls}\bar{L}_m)L_{ff}+2L_{md}^2L_{ls}\cos 2\left(\theta_r+\frac{1}{3}\pi\right)\right)(-r_si_{as}+u_{as})\right.\\
&+(3L_{md}^2\bar{L}_m+L_{ls}L_{md}^2-(3\bar{L}_m^2+2L_{ls}\bar{L}_m)L_{ff}+2L_{md}^2L_{ls}\cos 2\theta_r)(-r_si_{bs}+u_{bs})\\
&+\left(3L_{md}^2\bar{L}_m+4L_{ls}L_{md}^2-(3\bar{L}_m^2+4L_{ls}^2+8L_{ls}\bar{L}_m)L_{ff}+2L_{md}^2L_{ls}\cos 2\left(\theta_r+\frac{2}{3}\pi\right)\right)(-r_si_{cs}+u_{cs})+(6L_{ls}L_{md}\bar{L}_m+4L_{ls}^2L_{md})\sin\left(\theta_r+\frac{2}{3}\pi\right)(-r_ri_{fr}+u_{fr})\\
&-(6L_{ls}L_{md}^2\bar{L}_m+4L_{ls}^2L_{md}^2)(i_{as}\cos\theta_r+i_{bs}\cos(\theta_r-\frac{2}{3}\pi)+i_{cs}\cos(\theta_r+\frac{2}{3}\pi))\omega_r\sin(\theta_r+\frac{2}{3}\pi)\\
&\left.+((6L_{md}L_{ls}\bar{L}_m+4L_{md}L_{ls}^2)L_{ff}-6L_{md}^3L_{ls})i_{fr}\omega_r\cos\left(\theta_r+\frac{2}{3}\pi\right)\right]\\
\frac{di_{fr}}{dt} &= \frac{1}{3L_{md}^2L_{ls}-(2L_{ls}^2+3L_{ls}\bar{L}_m)L_{ff}}\left[2L_{md}L_{ls}\left(\sin\theta_r(-r_si_{as}+u_{as})+\sin\left(\theta_r-\frac{2}{3}\pi\right)(-r_si_{bs}+u_{bs})+\sin\left(\theta_r+\frac{2}{3}\pi\right)(-r_si_{cs}+u_{cs})\right)\right.\\
&\left.-(2L_{ls}^2+3\bar{L}_mL_{ls})(-r_ri_{fr}+u_{fr})+(3L_{ls}L_{md}\bar{L}_m+2L_{ls}^2L_{md})\left(i_{as}\omega_r\cos\theta_r+i_{bs}\omega_r\cos\left(\theta_r-\frac{2}{3}\pi\right)+i_{cs}\omega_r\cos\left(\theta_r+\frac{2}{3}\pi\right)\right)\right]\\
\frac{d\omega_r}{dt} &= \frac{P^2}{4J}\left[L_{\Delta m}\left(i_{as}^2\sin 2\theta_r+2i_{as}i_{bs}\sin 2\left(\theta_r-\frac{1}{3}\pi\right)+2i_{as}i_{cs}\sin 2\left(\theta_r+\frac{1}{3}\pi\right)+i_{bs}^2\sin 2\left(\theta_r-\frac{2}{3}\pi\right)\right.\right.\\
&\left.\left.+2i_{bs}i_{cs}\sin 2\theta_r+i_{cs}^2\sin 2\left(\theta_r+\frac{2}{3}\pi\right)\right)+L_{md}i_{fr}\left(i_{as}\cos\theta_r+i_{bs}\cos\left(\theta_r-\frac{2}{3}\pi\right)+i_{cs}\cos\left(\theta_r+\frac{2}{3}\pi\right)\right)\right]-\frac{B_m}{J}\omega_r-\frac{P}{2J}T_L\\
\frac{d\theta_r}{dt} &= \omega_r
\end{aligned}\tag{7.33}$$

The angular velocity of synchronous motors is locked with the frequency of the applied phase voltages to the stator windings, and the torque-speed characteristics can be represented as a family of horizontal lines (see Figure 7.30). The electrical angular velocity is equal to the synchronous angular velocity, and $\omega_e = \frac{4\pi f}{P}$. Thus, the rotor angular velocity can be regulated by changing the frequency of the phase voltages applied. It should be emphasized that the designer has to study the maximum load conditions $T_{L\max}$ to satisfy the requirement $T_{pullout} > T_{L\max}$. If the $T_{pullout} > T_{L\max}$ criterion is not met, as the $T_{L\max}$ is applied, the rotor magnetic field is no longer locked to the stator magnetic field. In particular, the rotor magnetic field slips behind the stator field. Due to the loss of synchronization, the electromagnetic torque developed by the motor reverses (surges) with each pass. Therefore, the condition $T_{pullout} > T_{L\max}$ must be guaranteed. For round-rotor synchronous machines, $L_{\Delta m} = 0$. Hence, from (7.31), as a balanced three-phase sinusoidal set of $i_{as}(t)$, $i_{bs}(t)$, and $i_{cs}(t)$ is fed, we have

$$T_e = \frac{PL_{md}}{2}i_{fr}\sqrt{2}i_M\left(\cos^2\theta_r+\cos^2\left(\theta_r-\frac{2}{3}\pi\right)+\cos^2\left(\theta_r+\frac{2}{3}\pi\right)\right)=\frac{3PL_{md}}{2\sqrt{2}}i_{fr}i_M$$

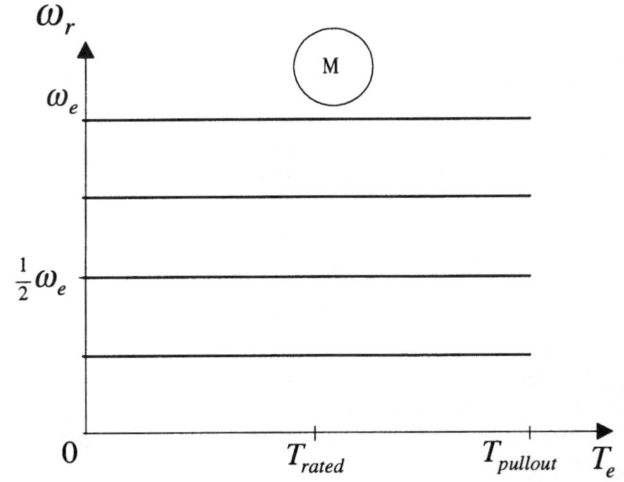

Figure 7.30 Typical torque-speed characteristic curves of synchronous motors.

Example 7.6

Model a three-phase, two-pole synchronous motor with the following parameters: r_s = 0.25 ohm, r_r = 0.47 ohm, L_{ls} = 0.0001 H, L_{mq} = 0.00095 H, L_{md} = 0.00095 H, L_{lf} = 0.00022 H, L_{mf} = 0.0001 H, L_{md} = 0.00035 H, J = 0.003 kg-m^2, and B_m = 0.00072 N-m-s-rad^{-1}. Apply the following phase voltages to the stator windings

$$u_{as}(t)=\sqrt{2}150\cos(377t),\ \ u_{bs}(t)=\sqrt{2}150\cos\left(377t-\frac{2}{3}\pi\right),\ \text{and}\ \ u_{cs}(t)=\sqrt{2}150\cos\left(377t+\frac{2}{3}\pi\right)$$

The field voltage is u_{fr} = 5 V.
The load conditions are specified as

$$T_L=\begin{cases}0\ \text{N}-\text{m}, \forall t\in[0\quad 0.2)\text{s}\\ 1\ \text{N}-\text{m}, \forall t\in[0.2\quad 0.5)\text{s}\end{cases}$$

Solution

The derived differential equations in Cauchy's form are used to model the conventional synchronous motors, and MATLAB files are developed and used (`crm_01.m` and `crm_02.m`).

MATLAB script `crm_01.m`

```
tspan = [0 0.5]; y0=[0 0 0 0 0 0]';
options = odeset('RelTol',5e-3,'AbsTol',[1e-4 1e-4 1e-4 1e-4 1e-4 1e-4]);
[t,y] = ODE45('crm_02',tspan,y0,options);
subplot(3,2,1); plot(t,y(:,1));
ylabel('ias'); axis([0,0.5,-500,500]); grid;
subplot(3,2,3); plot(t,y(:,2));
ylabel('ibs'); axis([0,0.5,-500,500]); grid;
subplot(3,2,5); plot(t,y(:,3));
```

```
ylabel('ics'); axis([0,0.5,-500,500]); grid;
subplot(3,2,2); plot(t,y(:,4));
ylabel('ifr'); axis([0,0.5,-200,200]); grid;
subplot(3,2,4); plot(t,y(:,5));
ylabel('wr'); axis([0,0.5,0,400]); grid;
subplot(3,2,6); plot(t,y(:,6));
ylabel('Q'); axis([0,0.5,0,200]); grid;
```

MATLAB script 2 `crm_02.m`

```
function yprime=difer(t,y);
% Stator leakage and magnetizing inductances
Lls=0.0001; Lmqs=0.00095; Lmds=0.00095;
% Rotor leakage and magnetizing inductances
Llf=0.00022; Lmf=0.0001;
% Mutual inductance
Lmd=0.00035;
% Average value of the magnetizing inductance
Lmb=(Lmqs+Lmds)/3;
% Resistances of the stator and rotor windings
rs=0.25; rr=0.47;
% Equivalent moment of inertia and viscous friction coefficient
J=0.003; Bm=0.00072;
% Number of poles
P=2;
% Magnitude of the supplied phase voltages and angular frequency
um=sqrt(2)*150; w=377;
% Voltage applied to the rotor field winding
ufr=5;
% Load torque, applied at time tTl sec
if t<=0.2
    Tl=0;
else
    Tl=1;
end
% Applied phase voltages to the abc windings
uas=um*cos(w*t);
ubs=um*cos(w*t-2*pi/3);
ucs=um*cos(w*t+2*pi/3);
% Numerical expressions used
Ld=2*Lls*Lmd^2; Ldr=2*Lls*Lmd;
Lss=(-3*Lmb^2-4*Lls^2-8*Lls*Lmb)*(Lmf+Llf)+3*Lmd^2*Lmb+4*Lls*Lmd^2;
Ldens=(2*Lls+3*Lmb)*((-2*Lls^2-3*Lmb*Lls)*(Llf+Lmf)+3*Lmd^2*Lls);
Ldenr=(-2*Lls^2-3*Lmb*Lls)*(Llf+Lmf)+3*Lmd^2*Lls;
Lms=(-3*Lmb^2-2*Lls*Lmb)*(Lmf+Llf)+3*Lmd^2*Lmb+Lls*Lmd^2;
Lrr=6*Lls*Lmd*Lmb+4*Lls^2*Lmd;
Llr=-(2*Lls^2+3*Lmb*Lls);
% Variables used
S1=sin(y(6,:)); S2=sin(y(6,:)-2*pi/3); S3=sin(y(6,:)+2*pi/3);
IUas=-rs*y(1,:)+uas; IUbs=-rs*y(2,:)+ubs; IUcs=-rs*y(3,:)+ucs;
IUfr=-rr*y(4,:)+ufr;
```

```
C1=cos(y(6,:)); C2=cos(y(6,:)-2*pi/3); C3=cos(y(6,:)+2*pi/3);
C12=cos(2*y(6,:)); C22=cos(2*y(6,:)-2*pi/3); C32=cos(2*y(6,:)
 +2*pi/3);
Nsts=(-6*Lmd^2*Lls*Lmb-
 4*Lmd^2*Lls^2)*y(5,:).*(C1.*y(1,:)+C2.*y(2,:)+C3.*y(3,:));
Nstr=(3*Lmd*Lmb*Lls+2*Lmd*Lls^2)*y(5,:).*(C1.*y(1,:)+C2.*y(2,:)
 +C3.*y(3,:));
Nct=((6*Lmd*Lls*Lmb+4*Lmd*Lls^2)*(Llf+Lmf)-6*Lmd^3*Lls)*
 y(5,:).*y(4,:);
Te=P*(Lmd*y(4,:).*(y(1,:).*C1+y(2,:).*C2+y(3,:).*C3))/2;
% Differential equations
yprime=[((Lss+Ld*C12).*IUas+(Lms+Ld*C22).*IUbs+(Lms+Ld*C32).*
 IUcs+Lrr*S1.*IUfr+Nsts.*S1+Nct.*C1)/Ldens;...
((Lms+Ld*C22).*IUas+(Lss+Ld*C32).*IUbs+(Lms+Ld*C12).*IUcs+Lrr*S2.*
 IUfr+Nsts.*S2+Nct.*C2)/Ldens;...
((Lms+Ld*C32).*IUas+(Lms+Ld*C12).*IUbs+(Lss+Ld*C22).*IUcs+Lrr*S3.*
 IUfr+Nsts.*S3+Nct.*C3)/Ldens;...
(Ldr*S1.*IUas+Ldr*S2.*IUbs+Ldr*S3.*IUcs+Llr*IUfr+Nstr)/Ldenr;...
(P*Te-2*Bm*y(5,:)-P*Tl)/(2*J);...
y(5,:)];
```

The transient dynamics, found through the numerical solution of a set of highly nonlinear differential equations (7.33), is illustrated in Figure 7.31. In particular, a two-pole motor starts from stall, and it is easy to see that the motor rotates with the synchronous angular velocity, which is 377 rad-s^{-1}. The load torque is applied, and the following load conditions were used in simulations $T_L = \begin{cases} 0 \text{ N-m}, & \forall t \in [0 \quad 0.2)\text{s} \\ 1 \text{ N-m}, & \forall t \in [\,0.2 \quad 0.5)\text{s} \end{cases}$ The results show that the angular velocity of the motor is locked with the synchronous angular velocity.

Dynamics of three-phase synchronous generators

Three-phase synchronous generators are driven by the prime mover, and Figure 7.29 illustrates the *generic* configuration of generation system. Using Kirchhoff's second law, we have four equations for stator and rotor circuitry. That is,

$$\begin{aligned} u_{as} &= -r_s i_{as} + \frac{d\psi_{as}}{dt} \\ u_{bs} &= -r_s i_{bs} + \frac{d\psi_{bs}}{dt} \\ u_{cs} &= -r_s i_{cs} + \frac{\psi_{cs}}{dt} \\ u_{fr} &= r_r i_{fr} + \frac{d\psi_{fr}}{dt} \end{aligned} \tag{7.34}$$

Hence, one writes

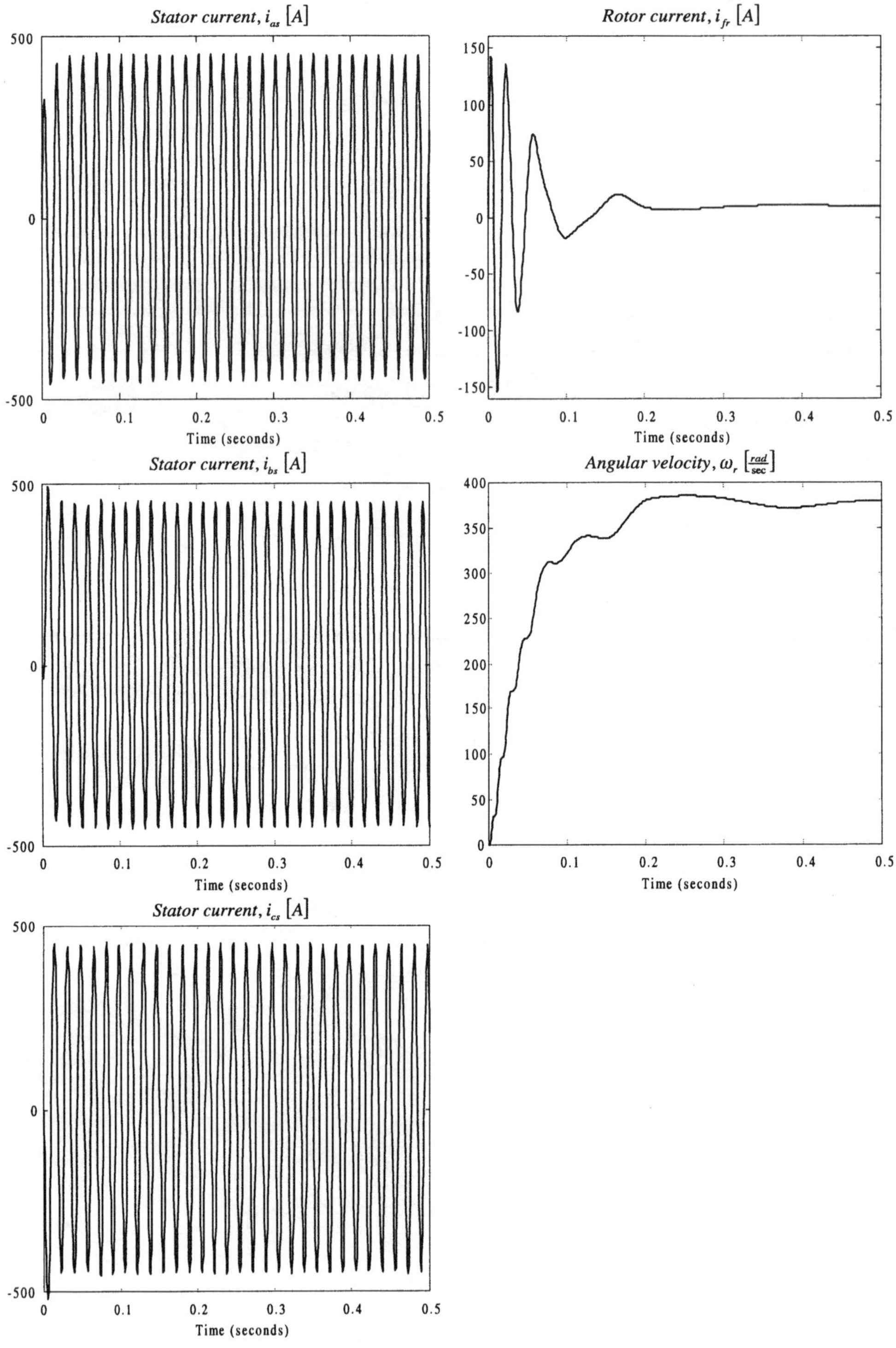

Figure 7.31 Transient dynamics of the conventional three-phase synchronous motor; $T_L = \begin{cases} 0 \text{ N-m}, \forall t \in [0 \quad 0.2)\text{s} \\ 1 \text{ N-m}, \forall t \in [0.2 \quad 0.5)\text{s} \end{cases}$.

$$\begin{bmatrix} u_{as} \\ u_{bs} \\ u_{cs} \\ u_{fr} \end{bmatrix} = \begin{bmatrix} r_s & 0 & 0 & 0 \\ 0 & r_s & 0 & 0 \\ 0 & 0 & r_s & 0 \\ 0 & 0 & 0 & r_r \end{bmatrix} \begin{bmatrix} -i_{as} \\ -i_{bs} \\ -i_{cs} \\ i_{fr} \end{bmatrix} + \begin{bmatrix} \frac{d\psi_{as}}{dt} \\ \frac{d\psi_{bs}}{dt} \\ \frac{d\psi_{cs}}{dt} \\ \frac{d\psi_{fr}}{dt} \end{bmatrix}$$

In (7.34), the flux linkages must be expressed as functions of the currents and the electrical angular displacement. We have

$$\psi_{as} = -(L_{ls} + \bar{L}_m - L_{\Delta m}\cos 2\theta_r)i_{as} - \left(-\frac{1}{2}\bar{L}_m - L_{\Delta m}\cos 2\left(\theta_r - \frac{1}{3}\pi\right)\right)i_{bs} - \left(-\frac{1}{2}\bar{L}_m - L_{\Delta m}\cos 2\left(\theta_r + \frac{1}{3}\pi\right)\right)i_{cs}$$
$$+ L_{md}\sin\theta_r i_{fr}$$

$$\psi_{bs} = -\left(-\frac{1}{2}\bar{L}_m - L_{\Delta m}\cos 2\left(\theta_r - \frac{1}{3}\pi\right)\right)i_{as} - \left(L_{ls} + \bar{L}_m - L_{\Delta m}\cos 2\left(\theta_r - \frac{2}{3}\pi\right)\right)i_{bs} - \left(-\frac{1}{2}\bar{L}_m - L_{\Delta m}\cos 2\theta_r\right)i_{cs}$$
$$+ L_{md}\sin\left(\theta_r - \frac{2}{3}\pi\right)i_{fr}$$

$$\psi_{cs} = \left(-\frac{1}{2}\bar{L}_m - L_{\Delta m}\cos 2\left(\theta_r + \frac{1}{3}\pi\right)\right)i_{as} + \left(-\frac{1}{2}\bar{L}_m - L_{\Delta m}\cos 2\theta_r\right)i_{bs} + \left(L_{ls} + \bar{L}_m - L_{\Delta m}\cos 2\left(\theta_r + \frac{2}{3}\pi\right)\right)i_{cs}$$
$$+ L_{md}\sin\left(\theta_r + \frac{2}{3}\pi\right)i_{fr}$$

$$\psi_{fr} = -L_{md}\sin\theta_r i_{as} - L_{md}\sin\left(\theta_r - \frac{2}{3}\pi\right)i_{bs} - L_{md}\sin\left(\theta_r + \frac{2}{3}\pi\right)i_{cs} + (L_{lf} + L_{mf})i_{fr}$$

(7.35)

Thus, for three-phase synchronous generators, one finds

$$\boldsymbol{\psi}_{abcs/fr} = \mathbf{L}_{abcs/fr}\mathbf{i}_{abcs/fr} =$$

$$\begin{bmatrix} L_{ls} + \bar{L}_m - L_{\Delta m}\cos 2\theta_r & -\frac{1}{2}\bar{L}_m - L_{\Delta m}\cos 2\left(\theta_r - \frac{1}{3}\pi\right) & -\frac{1}{2}\bar{L}_m - L_{\Delta m}\cos 2\left(\theta_r + \frac{1}{3}\pi\right) & L_{md}\sin\theta_r \\ -\frac{1}{2}\bar{L}_m - L_{\Delta m}\cos 2\left(\theta_r - \frac{1}{3}\pi\right) & L_{ls} + \bar{L}_m - L_{\Delta m}\cos 2\left(\theta_r - \frac{2}{3}\pi\right) & -\frac{1}{2}\bar{L}_m - L_{\Delta m}\cos 2\theta_r & L_{md}\sin\left(\theta_r - \frac{2}{3}\pi\right) \\ -\frac{1}{2}\bar{L}_m - L_{\Delta m}\cos 2\left(\theta_r + \frac{1}{3}\pi\right) & -\frac{1}{2}\bar{L}_m - L_{\Delta m}\cos 2\theta_r & L_{ls} + \bar{L}_m - L_{\Delta m}\cos 2\left(\theta_r + \frac{2}{3}\pi\right) & L_{md}\sin\left(\theta_r + \frac{2}{3}\pi\right) \\ L_{md}\sin\theta_r & L_{md}\sin\left(\theta_r - \frac{2}{3}\pi\right) & L_{md}\sin\left(\theta_r + \frac{2}{3}\pi\right) & L_{lf} + L_{mf} \end{bmatrix}$$

$$\times \begin{bmatrix} -i_{as} \\ -i_{bs} \\ -i_{cs} \\ i_{fr} \end{bmatrix}$$

From (7.34) and (7.35), taking note that for round-rotor synchronous machines $L_{\Delta m} = 0$, we have

$$\frac{di_{as}}{dt}=\frac{1}{L_{ls}+\bar{L}_m}\left(-r_s i_{as}+\frac{d\left(\frac{1}{2}\bar{L}_m i_{bs}+\frac{1}{2}\bar{L}_m i_{cs}+L_{md}\sin\left(\frac{P}{2}\theta_{rm}\right)i_{fr}\right)}{dt}-u_{as}\right)$$

$$\frac{di_{bs}}{dt}=\frac{1}{L_{ls}+\bar{L}_m}\left(-r_s i_{bs}+\frac{d\left(\frac{1}{2}\bar{L}_m i_{as}+\frac{1}{2}\bar{L}_m i_{cs}+L_{md}\sin\left(\frac{P}{2}\theta_{rm}-\frac{2}{3}\pi\right)i_{fr}\right)}{dt}-u_{bs}\right)$$

$$\frac{di_{cs}}{dt}=\frac{1}{L_{ls}+\bar{L}_m}\left(-r_s i_{cs}+\frac{d\left(\frac{1}{2}\bar{L}_m i_{as}+\frac{1}{2}\bar{L}_m i_{bs}+L_{md}\sin\left(\frac{P}{2}\theta_{rm}+\frac{2}{3}\pi\right)i_{fr}\right)}{dt}-u_{cs}\right)$$

$$\frac{di_{fr}}{dt}=\frac{1}{L_{lf}+L_{mf}}\left(-r_s i_{fr}+\frac{d\left(L_{md}\sin\left(\frac{P}{2}\theta_{rm}\right)i_{as}+L_{md}\sin\left(\frac{P}{2}\theta_{rm}-\frac{2}{3}\pi\right)i_{bs}+L_{md}\sin\left(\frac{P}{2}\theta_{rm}+\frac{2}{3}\pi\right)i_{cs}\right)}{dt}+u_{fr}\right)$$

Cauchy's form of the circuitry dynamics for conventional synchronous generators is found to be

$$\begin{aligned}\frac{di_{as}}{dt}=&\frac{1}{(2L_{ls}+3\bar{L}_m)(3L_{md}^2L_{ls}-(2L_{ls}^2+3L_{ls}\bar{L}_m)L_{ff})}[(3L_{md}^2\bar{L}_m+4L_{ls}L_{md}^2-(3\bar{L}_m^2+4L_{ls}^2+8L_{ls}\bar{L}_m)L_{ff}+2L_{md}^2L_{ls}\cos\theta_r)(-r_s i_{as}-u_{as})\\
&+\left(3L_{md}^2\bar{L}_m+L_{ls}L_{md}^2-(3\bar{L}_m^2+2L_{ls}\bar{L}_m)L_{ff}+2L_{md}^2L_{ls}\cos 2\left(\theta_r-\frac{1}{3}\pi\right)\right)(-r_s i_{bs}-u_{bs})\\
&+\left(3L_{md}^2\bar{L}_m+L_{ls}L_{md}^2-(3\bar{L}_m^2+2L_{ls}\bar{L}_m)L_{ff}+2L_{md}^2L_{ls}\cos 2\left(\theta_r+\frac{1}{3}\pi\right)\right)(-r_s i_{cs}-u_{cs})-(6L_{ls}L_{md}\bar{L}_m+4L_{ls}^2L_{md})\sin\theta_r(-r_r i_{fr}+u_{fr})\\
&-(6L_{ls}L_{md}^2\bar{L}_m+4L_{ls}^2L_{md}^2)\left(i_{as}\cos\theta_r+i_{bs}\cos\left(\theta_r-\frac{2}{3}\pi\right)+i_{cs}\cos\left(\theta_r+\frac{2}{3}\pi\right)\right)\omega_r\sin\theta_r\\
&-((6L_{md}L_{ls}\bar{L}_m+4L_{md}L_{ls}^2)L_{ff}-6L_{md}^3L_{ls})i_{fr}\omega_r\cos\theta_r]\end{aligned}$$

$$\begin{aligned}\frac{di_{bs}}{dt}=&\frac{1}{(2L_{ls}+3\bar{L}_m)(3L_{md}^2L_{ls}-(2L_{ls}^2+3L_{ls}\bar{L}_m)L_{ff})}\left[\left(3L_{md}^2\bar{L}_m+L_{ls}L_{md}^2-(3\bar{L}_m^2+2L_{ls}\bar{L}_m)L_{ff}+2L_{md}^2L_{ls}\cos 2\left(\theta_r-\frac{1}{3}\pi\right)\right)(-r_s i_{as}-u_{as})\right.\\
&+\left(3L_{md}^2\bar{L}_m+4L_{ls}L_{md}^2-(3\bar{L}_m^2+4L_{ls}^2+8L_{ls}\bar{L}_m)L_{ff}+2L_{md}^2L_{ls}\cos 2\left(\theta_r-\frac{2}{3}\pi\right)\right)(-r_s i_{bs}-u_{bs})\\
&+(3L_{md}^2\bar{L}_m+L_{ls}L_{md}^2-(3\bar{L}_m^2+2L_{ls}\bar{L}_m)L_{ff}+2L_{md}^2L_{ls}\cos 2\theta_r)(-r_s i_{cs}-u_{cs})-(6L_{ls}L_{md}\bar{L}_m+4L_{ls}^2L_{md})\sin\left(\theta_r-\frac{2}{3}\pi\right)(-r_r i_{fr}+u_{fr})\\
&-(6L_{ls}L_{md}^2\bar{L}_m+4L_{ls}^2L_{md}^2)\left(i_{as}\cos\theta_r+i_{bs}\cos\left(\theta_r-\frac{2}{3}\pi\right)+i_{cs}\cos\left(\theta_r+\frac{2}{3}\right)\right)\omega_r\sin\left(\theta_r-\frac{2}{3}\pi\right)\\
&\left.-((6L_{md}L_{ls}\bar{L}_m+4L_{md}L_{ls}^2)L_{ff}-6L_{md}^3L_{ls})i_{fr}\omega_r\cos\left(\theta_r-\frac{2}{3}\pi\right)\right]\end{aligned}$$

$$\begin{aligned}\frac{di_{cs}}{dt}=&\frac{1}{(2L_{ls}+3\bar{L}_m)(3L_{md}^2L_{ls}-(2L_{ls}^2+3L_{ls}\bar{L}_m)L_{ff})}\left[\left(3L_{md}^2\bar{L}_m+L_{ls}L_{md}^2-(3\bar{L}_m^2+2L_{ls}\bar{L}_m)L_{ff}+2L_{md}^2L_{ls}\cos 2\left(\theta_r+\frac{1}{3}\pi\right)\right)(-r_s i_{as}-u_{as})\right.\\
&+(3L_{md}^2\bar{L}_m+L_sL_{md}^2-(3\bar{L}_m^2+2L_{ls}\bar{L}_m)L_{ff}+2L_{md}^2L_{ls}\cos 2\theta_r)(-r_s i_{bs}-u_{bs})\\
&+\left(3L_{md}^2\bar{L}_m+4L_{ls}L_{md}^2-(3\bar{L}_m^2+4L_{ls}^2+8L_{ls}\bar{L}_m)L_{ff}+2L_{md}^2L_{ls}\cos 2\left(\theta_r+\frac{2}{3}\pi\right)\right)(-r_s i_{cs}-u_{cs})-(6L_{ls}L_{md}\bar{L}_m+4L_{ls}^2L_{md})\sin\left(\theta_r+\frac{2}{3}\pi\right)(-r_r i_{fr}+u_{fr})\end{aligned}$$

$$-(6L_{ls}L_{md}^2\bar{L}_m+4L_{ls}^2L_{md}^2)\left(i_{as}\cos\theta_r+i_{bs}\cos\left(\theta_r-\frac{2}{3}\pi\right)+i_{cs}\cos\left(\theta_r+\frac{2}{3}\pi\right)\right)\omega_r\sin\left(\theta_r+\frac{2}{3}\pi\right)$$

$$-((6L_{md}L_{ls}\bar{L}_m+4L_{md}L_{ls}^2)L_{ff}-6L_{md}^3L_{ls})i_{fr}\omega_r\cos\left(\theta_r+\frac{2}{3}\pi\right)\Bigg]$$

$$\frac{di_{fr}}{dt}=\frac{1}{3L_{md}^2L_{ls}-(2L_{ls}^2+3L_{ls}\bar{L}_m)L_{ff}}\Bigg[2L_{md}L_{ls}\left(\sin\theta_r(r_si_{as}+u_{as})+\sin\left(\theta_r-\frac{2}{3}\pi\right)(r_si_{bs}+u_{bs})+\sin\left(\theta_r+\frac{2}{3}\pi\right)(r_si_{cs}+u_{cs})\right)$$

$$-(2L_{ls}^2+3\bar{L}_mL_{ls})(-r_ri_{fr}+u_{fr})-(3L_{ls}L_{md}\bar{L}_m+2L_{ls}^2L_{md})\left(i_{as}\omega_r\cos\theta_r+i_{bs}\omega_r\cos\left(\theta_r-\frac{2}{3}\pi\right)+i_{cs}\omega_r\cos\left(\theta_r+\frac{2}{3}\pi\right)\right)\Bigg]$$

Hence, the following matrix equation maps the circuitry dynamics of conventional synchronous generators

$$\begin{bmatrix}\frac{di_{as}}{dt}\\ \frac{di_{bs}}{dt}\\ \frac{di_{cs}}{dt}\\ \frac{di_{fr}}{dt}\end{bmatrix}=\begin{bmatrix}-\frac{r_sL_{Ds}}{L_{\Sigma s}} & -\frac{r_sL_{Ms}}{L_{\Sigma s}} & -\frac{r_sL_{Ms}}{L_{\Sigma s}} & 0\\ -\frac{r_sL_{Ms}}{L_{\Sigma s}} & -\frac{r_sL_{Ds}}{L_{\Sigma s}} & -\frac{r_sL_{Ms}}{L_{\Sigma s}} & 0\\ -\frac{r_sL_{Ms}}{L_{\Sigma s}} & -\frac{r_sL_{Ms}}{L_{\Sigma s}} & -\frac{r_sL_{Ds}}{L_{\Sigma s}} & 0\\ 0 & 0 & 0 & \frac{r_r(2L_{ls}^2+3\bar{L}_mL_{ls})}{L_{\Sigma f}}\end{bmatrix}\begin{bmatrix}i_{as}\\ i_{bs}\\ i_{cs}\\ i_{fr}\end{bmatrix}$$

$$+\begin{bmatrix}-\frac{2r_sL_{md}^2L_{ls}}{L_{\Sigma s}}\cos 2\theta_r & -\frac{2r_sL_{md}^2L_{ls}}{L_{\Sigma s}}\cos 2\left(\theta_r-\frac{1}{3}\pi\right) & -\frac{2r_sL_{md}^2L_{ls}}{L_{\Sigma s}}\cos 2\left(\theta_r+\frac{1}{3}\pi\right) & \frac{r_rL_{Mf}}{L_{\Sigma s}}\sin\theta_r\\ -\frac{2r_sL_{md}^2L_{ls}}{L_{\Sigma s}}\cos 2\left(\theta_r-\frac{1}{3}\pi\right) & -\frac{2r_sL_{md}^2L_{ls}}{L_{\Sigma s}}\cos 2\left(\theta_r-\frac{2}{3}\pi\right) & -\frac{2r_sL_{md}^2L_{ls}}{L_{\Sigma s}}\cos 2\theta_r & \frac{r_rL_{Mf}}{L_{\Sigma s}}\sin\left(\theta_r-\frac{2}{3}\pi\right)\\ -\frac{2r_sL_{md}^2L_{ls}}{L_{\Sigma s}}\cos 2\left(\theta_r+\frac{1}{3}\pi\right) & -\frac{2r_sL_{md}^2L_{ls}}{L_{\Sigma s}}\cos 2\theta_r & -\frac{2r_sL_{md}^2L_{ls}}{L_{\Sigma s}}\cos 2\left(\theta_r+\frac{2}{3}\pi\right) & \frac{r_rL_{Mf}}{L_{\Sigma s}}\sin\left(\theta_r+\frac{2}{3}\pi\right)\\ \frac{2r_rL_{md}L_{ls}}{L_{\Sigma f}}\sin\theta_r & \frac{2r_sL_{md}L_{ls}}{L_{\Sigma f}}\sin\left(\theta_r-\frac{2}{3}\pi\right) & \frac{2r_rL_{md}L_{ls}}{L_{\Sigma f}}\sin\left(\theta_r+\frac{2}{3}\pi\right) & 0\end{bmatrix}\begin{bmatrix}i_{as}\\ i_{bs}\\ i_{cs}\\ i_{fr}\end{bmatrix}$$

$$+\begin{bmatrix}-\frac{6L_{ls}L_{md}^2\bar{L}_m+4L_{ls}^2L_{md}^2}{L_{\Sigma s}}\sin\theta_r & -\frac{6L_{ls}L_{md}^2\bar{L}_m+4L_{ls}^2L_{md}^2}{L_{\Sigma s}}\sin\theta_r & -\frac{6L_{ls}L_{md}^2\bar{L}_m+4L_{ls}^2L_{md}^2}{L_{\Sigma s}}\sin\theta_r\\ -\frac{6L_{ls}L_{md}^2\bar{L}_m+4L_{ls}^2L_{md}^2}{L_{\Sigma s}}\sin\left(\theta_r-\frac{2}{3}\pi\right) & -\frac{6L_{ls}L_{md}^2\bar{L}_m+4L_{ls}^2L_{md}^2}{L_{\Sigma s}}\sin\left(\theta_r-\frac{2}{3}\pi\right) & -\frac{6L_{ls}L_{md}^2\bar{L}_m+4L_{ls}^2L_{md}^2}{L_{\Sigma s}}\sin\left(\theta_r-\frac{2}{3}\pi\right)\\ -\frac{6L_{ls}L_{md}^2\bar{L}_m+4L_{ls}^2L_{md}^2}{L_{\Sigma s}}\sin\left(\theta_r+\frac{2}{3}\pi\right) & -\frac{6L_{ls}L_{md}^2\bar{L}_m+4L_{ls}^2L_{md}^2}{L_{\Sigma s}}\sin\left(\theta_r+\frac{2}{3}\pi\right) & -\frac{6L_{ls}L_{md}^2\bar{L}_m+4L_{ls}^2L_{md}^2}{L_{\Sigma s}}\sin\left(\theta_r+\frac{2}{3}\pi\right)\\ -\frac{3L_{ls}L_{md}\bar{L}_m+2L_{ls}^2L_{md}}{L_{\Sigma f}} & -\frac{3L_{ls}L_{md}\bar{L}_m+2L_{ls}^2L_{md}}{L_{\Sigma f}} & -\frac{3L_{ls}L_{md}\bar{L}_m+2L_{ls}^2L_{md}}{L_{\Sigma f}}\end{bmatrix}$$

$$\times\begin{bmatrix}i_{as}\omega_r\cos\theta_r\\ i_{bs}\omega_r\cos\left(\theta_r-\frac{2}{3}\pi\right)\\ i_{cs}\omega_r\cos\left(\theta_r+\frac{2}{3}\pi\right)\end{bmatrix}-\begin{bmatrix}\frac{(6L_{md}L_{ls}\bar{L}_m+4L_{md}L_{ls}^2)L_{ff}-6L_{md}^3L_{ls}}{L_{\Sigma s}}i_{fr}\omega_r\cos\theta_r\\ \frac{(6L_{md}L_{ls}\bar{L}_m+4L_{md}L_{ls}^2)L_{ff}-6L_{md}^3L_{ls}}{L_{\Sigma s}}i_{fr}\omega_r\cos\left(\theta_r-\frac{2}{3}\pi\right)\\ \frac{(6L_{md}L_{ls}\bar{L}_m+4L_{md}L_{ls}^2)L_{ff}-6L_{md}^3L_{ls}}{L_{\Sigma s}}i_{fr}\omega_r\cos\left(\theta_r+\frac{2}{3}\pi\right)\\ 0\end{bmatrix}$$

$$+\begin{bmatrix}-\frac{L_{Ds}+2L_{md}^2L_{ls}\cos 2\theta_r}{L_{\Sigma s}} & -\frac{L_{Ms}+2L_{md}^2L_{ls}\cos 2\left(\theta_r-\frac{1}{3}\pi\right)}{L_{\Sigma s}} & -\frac{L_{Ms}+2L_{md}^2L_{ls}\cos 2\left(\theta_r+\frac{1}{3}\pi\right)}{L_{\Sigma s}} & -\frac{L_{Mf}}{L_{\Sigma s}}\sin\theta_r\\ -\frac{L_{Ms}+2L_{md}^2L_{ls}\cos 2\left(\theta_r-\frac{1}{3}\pi\right)}{L_{\Sigma s}} & -\frac{L_{Ds}+2L_{md}^2L_{ls}\cos 2\left(\theta_r-\frac{2}{3}\pi\right)}{L_{\Sigma s}} & -\frac{L_{Ms}+2L_{md}^2L_{ls}\cos 2\theta_r}{L_{\Sigma s}} & -\frac{L_{Mf}}{L_{\Sigma s}}\sin\left(\theta_r-\frac{2}{3}\pi\right)\\ -\frac{L_{Ms}+2L_{md}^2L_{ls}\cos 2\left(\theta_r+\frac{1}{3}\pi\right)}{L_{\Sigma s}} & -\frac{L_{Ms}+2L_{md}^2L_{ls}\cos 2\theta_r}{L_{\Sigma s}} & -\frac{L_{Ds}+2L_{md}^2L_{ls}\cos 2\left(\theta_r+\frac{2}{3}\pi\right)}{L_{\Sigma s}} & -\frac{L_{Mf}}{L_{\Sigma s}}\sin\left(\theta_r+\frac{2}{3}\pi\right)\\ \frac{2L_{md}L_{ls}\sin\theta_r}{L_{\Sigma f}} & \frac{2L_{md}L_{ls}\sin\left(\theta_r-\frac{2}{3}\pi\right)}{L_{\Sigma f}} & \frac{2L_{md}L_{ls}\sin\left(\theta_r+\frac{2}{3}\pi\right)}{L_{\Sigma f}} & -\frac{2L_{ls}^2+3L_mL_{ls}}{L_{\Sigma f}}\end{bmatrix}\begin{bmatrix}u_{as}\\ u_{bs}\\ u_{cs}\\ u_{fr}\end{bmatrix}$$

The *torsional-mechanical dynamics* are expressed as

$$\frac{d\omega_r}{dt} = -\frac{P^2 L_{md}}{4J} i_{fr}\left[i_{as}\cos\theta_r + i_{bs}\cos\left(\theta_r - \frac{2}{3}\pi\right) + i_{cs}\cos\left(\theta_r + \frac{2}{3}\pi\right)\right] - \frac{B_m}{J}\omega_r + \frac{P}{2J}T_{pm}$$

$$\frac{d\theta_r}{dt} = \omega_r$$

Therefore, the resulting mathematical model for three-phase synchronous generators is

$$\frac{di_{as}}{dt} = \frac{1}{(2L_{ls} + 3\bar{L}_m)(3L_{md}^2 L_{ls} - (2L_{ls}^2 + 3L_{ls}\bar{L}_m)L_{ff})}[(3L_{md}^2\bar{L}_m + 4L_{ls}L_{md}^2 - (3\bar{L}_m^2 + 4L_{ls}^2 + 8L_{ls}\bar{L}_m)L_{ff} + 2L_{md}^2 L_{ls}\cos 2\theta_r)(-r_s i_{as} - u_{as})$$
$$+\left(3L_{md}^2\bar{L}_m + L_{ls}L_{md}^2 - (3\bar{L}_m^2 + 2L_{ls}\bar{L}_m)L_{ff} + 2L_{md}^2 L_{ls}\cos 2\left(\theta_r - \frac{1}{3}\pi\right)\right)(-r_s i_{bs} - u_{bs})$$
$$+\left(3L_{md}^2\bar{L}_m + L_{ls}L_{md}^2 - (3\bar{L}_m^2 + 2L_{ls}\bar{L}_m)L_{ff} + 2L_{md}^2 L_{ls}\cos 2\left(\theta_r + \frac{1}{3}\pi\right)\right)(-r_s i_{cs} - u_{cs}) - (6L_{ls}L_{md}\bar{L}_m + 4L_{ls}^2 L_{md})\sin\theta_r(-r_r i_{fr} + u_{fr})$$
$$-(6L_{ls}L_{md}^2\bar{L}_m + 4L_{ls}^2 L_{md}^2)\left(i_{as}\cos\theta_r + i_{bs}\cos\left(\theta_r - \frac{2}{3}\pi\right) + i_{cs}\cos\left(\theta_r + \frac{2}{3}\pi\right)\right)\omega_r\sin\theta_r$$
$$-((6L_{md}L_{ls}\bar{L}_m + 4L_{md}L_{ls}^2)L_{ff} - 6L_{md}^3 L_{ls})i_{fr}\omega_r\cos\theta_r]$$

$$\frac{di_{bs}}{dt} = \frac{1}{(2L_{ls} + 3\bar{L}_m)(3L_{md}^2 L_{ls} - (2L_{ls}^2 + 3L_{ls}\bar{L}_m)L_{ff})}\left[\left(3L_{md}^2\bar{L}_m + L_{ls}L_{md}^2 - (3\bar{L}_m^2 + 2L_{ls}\bar{L}_m)L_{ff} + 2L_{md}^2 L_{ls}\cos 2\left(\theta_r - \frac{1}{3}\pi\right)\right)(-r_s i_{as} - u_{as})\right.$$
$$+\left(3L_{md}^2\bar{L}_m + 4L_{ls}L_{md}^2 - (3\bar{L}_m^2 + 4L_{ls}^2 + 8L_{ls}\bar{L}_m)L_{ff} + 2L_{md}^2 L_{ls}\cos 2\left(\theta_r - \frac{2}{3}\pi\right)\right)(-r_s i_{bs} - u_{bs})$$
$$+(3L_{md}^2\bar{L}_m + L_{ls}L_{md}^2 - (3\bar{L}_m^2 + 2L_{ls}\bar{L}_m)L_{ff} + 2L_{md}^2 L_{ls}\cos 2\theta_r)(-r_s i_{cs} - u_{cs}) - (6L_{ls}L_{md}\bar{L}_m + 4L_{ls}^2 L_{md})\sin\left(\theta_r - \frac{2}{3}\pi\right)(-r_r i_{fr} + u_{fr})$$
$$-(6L_{ls}L_{md}^2\bar{L}_m + 4L_{ls}^2 L_{md}^2)\left(i_{as}\cos\theta_r + i_{bs}\cos\left(\theta_r - \frac{2}{3}\pi\right) + i_{cs}\cos\left(\theta_r + \frac{2}{3}\pi\right)\right)\omega_r\sin\left(\theta_r - \frac{2}{3}\pi\right)$$
$$\left.-((6L_{md}L_{ls}\bar{L}_m + 4L_{md}L_{ls}^2)L_{ff} - 6L_{md}^3 L_{ls})i_{fr}\omega_r\cos\left(\theta_r - \frac{2}{3}\pi\right)\right]$$

$$\frac{di_{cs}}{dt} = \frac{1}{(2L_{ls} + 3\bar{L}_m)(3L_{md}^2 L_{ls} - (2L_{ls}^2 + 3L_{ls}\bar{L}_m)L_{ff})}\left[\left(3L_{md}^2\bar{L}_m + L_{ls}L_{md}^2 - (3\bar{L}_m^2 + 2L_{ls}\bar{L}_m)L_{ff} + 2L_{md}^2 L_{ls}\cos 2\left(\theta_r + \frac{1}{3}\pi\right)\right)(-r_s i_{as} - u_{as})\right.$$
$$+(3L_{md}^2\bar{L}_m + L_{ls}L_{md}^2 - (3\bar{L}_m^2 + 2L_{ls}\bar{L}_m)L_{ff} + 2L_{md}^2 L_{ls}\cos 2\theta_r)(-r_s i_{bs} - u_{bs})$$
$$+\left(3L_{md}^2\bar{L}_m + 4L_{ls}L_{md}^2 - (3\bar{L}_m^2 + 4L_{ls}^2 + 8L_{ls}\bar{L}_m)L_{ff} + 2L_{md}^2 L_{ls}\cos 2\left(\theta_r + \frac{2}{3}\pi\right)\right)(-r_s i_{cs} - u_{cs}) - (6L_{ls}L_{md}\bar{L}_m + 4L_{ls}^2 L_{md})\sin\left(\theta_r + \frac{2}{3}\pi\right)(-r_r i_{fr} + u_{fr})$$
$$-(6L_{ls}L_{md}^2\bar{L}_m + 4L_{ls}^2 L_{md}^2)\left(i_{as}\cos\theta_r + i_{bs}\cos\left(\theta_r - \frac{2}{3}\pi\right) + i_{cs}\cos\left(\theta_r + \frac{2}{3}\right)\right)\omega_r\sin\left(\theta_r + \frac{2}{3}\pi\right)$$
$$\left.-((6L_{md}L_{ls}\bar{L}_m + 4L_{md}L_{ls}^2)L_{ff} - 6L_{md}^3 L_{ls})i_{fr}\omega_r\cos\left(\theta_r + \frac{2}{3}\pi\right)\right]$$

$$\frac{di_{fr}}{dt} = \frac{1}{3L_{md}^2 L_{ls} - (2L_{ls}^2 + 3L_{ls}\bar{L}_m)L_{ff}}\left[2L_{md}L_{ls}\left(\sin\theta_r(r_s i_{as} + u_{as}) + \sin\left(\theta_r - \frac{2}{3}\pi\right)(r_s i_{bs} + u_{bs}) + \sin\left(\theta_r + \frac{2}{3}\pi\right)(r_s i_{cs} + u_{cs})\right)\right.$$
$$\left.-(2L_{ls}^2 + 3\bar{L}_m L_{ls})(-r_r i_{fr} + u_{fr}) - (3L_{ls}L_{md}\bar{L}_m + 2L_{ls}^2 L_{md})\left(i_{as}\omega_r\cos\theta_r + i_{bs}\omega_r\cos\left(\theta_r - \frac{2}{3}\pi\right) + i_{cs}\omega_r\cos\left(\theta_r + \frac{2}{3}\pi\right)\right)\right]$$

$$\frac{d\omega_r}{dt} = -\frac{P^2 L_{md}}{4J} i_{fr}\left[i_{as}\cos\theta_r + i_{bs}\cos\left(\theta_r - \frac{2}{3}\pi\right) + i_{cs}\cos\left(\theta_r + \frac{2}{3}\pi\right)\right] - \frac{B_m}{J}\omega_r + \frac{P}{2J}T_{pm}$$

$$\frac{d\theta_r}{dt} = \omega_r$$

Specifying the prime mover, one augments the transient dynamics of the electric motor with the derived differential equations for conventional synchronous generators.

Example 7.7

Model a generation system if a three-phase synchronous generator is driven by a permanent-magnet DC motor. Use the following parameters for the generator circuitry: $r_s = 0.25$ ohm, $r_r = 0.47$ ohm, $L_{ls} = 0.0001$ H, $L_{mq} = 0.00095$ H, $L_{md} = 0.00095$ H, $L_{lf} = 0.00022$ H, $L_{mf} = 0.0001$ H, and $L_{md} = 0.00035$ H. The equivalent moment of inertia and the viscous friction coefficient are $J = 0.003$ kg-m^2 and $B_m = 0.00072$ N-m-s-rad^{-1}. The parameters of the prime mover are $r_{apm} = 0.4$ ohm, $L_{apm} = 0.015$ H, $k_{apm} = 0.25$ V-s-rad^{-1}, and $k_{apm} = 0.25$ N-m-A^{-1}. The armature voltage applied to the prime mover is $u_{apm} = 100$ V. The voltage supplied to the generator field winding is $u_{fr} = 200$ V.

Solution

The differential equations for permanent-magnet DC motors are

$$\frac{di_{apm}}{dt} = -\frac{r_{apm}}{L_{apm}} i_{apm} - \frac{k_{apm}}{L_{apm}} \omega_{rm} + \frac{1}{L_{apm}} u_{apm}$$

$$\frac{d\omega_{rm}}{dt} = \frac{k_a}{J} i_{apm} - \frac{B_m}{J} \omega_{rm} - T_L$$

Therefore, the following system of differential equations to map the generation system dynamics results

$$\frac{di_{apm}}{dt} = -\frac{r_{apm}}{L_{apm}} i_{apm} - \frac{k_{apm}}{L_{apm}} \omega_{rm} + \frac{1}{L_{apm}} u_{ap}$$

$$\begin{aligned}
\frac{di_{as}}{dt} = {} & \frac{1}{(2L_{ls}+3\bar{L}_m)(3L_{md}^2L_{ls}-(2L_{ls}^2+3L_{ls}\bar{L}_m)L_{ff})}[(3L_{md}^2\bar{L}_m+4L_{ls}L_{md}^2-(3\bar{L}_m^2+4L_{ls}^2+8L_{ls}\bar{L}_m)L_{ff}+2L_{md}^2L_{ls}\cos 2\theta_r)(-r_si_{as}-u_{as}) \\
& +\left(3L_{md}^2\bar{L}_m+L_{ls}L_{md}^2-(3\bar{L}_m^2+2L_{ls}\bar{L}_m)L_{ff}+2L_{md}^2L_{ls}\cos 2\left(\theta_r-\frac{1}{3}\pi\right)\right)(-r_si_{bs}-u_{bs}) \\
& +\left(3L_{md}^2\bar{L}_m+L_{ls}L_{md}^2-(3\bar{L}_m^2+2L_{ls}\bar{L}_m)L_{ff}+2L_{md}^2L_{ls}\cos 2\left(\theta_r+\frac{1}{3}\pi\right)\right)(-r_si_{cs}-u_{cs})-(6L_{ls}L_{md}\bar{L}_m+4L_{ls}^2L_{md})\sin\theta_r(-r_ri_{fr}+u_{fr}) \\
& -(6L_{ls}L_{md}^2\bar{L}_m+4L_{ls}^2L_{md}^2)\left(i_{as}\cos\theta_r+i_{bs}\cos\left(\theta_r-\frac{2}{3}\pi\right)+i_{cs}\cos\left(\theta_r+\frac{2}{3}\pi\right)\right)\omega_r\sin\theta_r \\
& -((6L_{md}L_{ls}\bar{L}_m+4L_{md}L_{ls}^2)L_{ff}-6L_{md}^3L_{ls})i_{fr}\omega_r\cos\theta_r]
\end{aligned}$$

$$\begin{aligned}
\frac{di_{bs}}{dt} = {} & \frac{1}{(2L_{ls}+3\bar{L}_m)(3L_{md}^2L_{ls}-(2L_{ls}^2+3L_{ls}\bar{L}_m)L_{ff})}\left[\left(3L_{md}^2\bar{L}_m+L_{ls}L_{md}^2-(3\bar{L}_m^2+2L_{ls}\bar{L}_m)L_{ff}+2L_{md}^2L_{ls}\cos 2\left(\theta_r-\frac{1}{3}\pi\right)\right)(-r_si_{as}-u_{as})\right. \\
& +\left(3L_{md}^2\bar{L}_m+4L_{ls}L_{md}^2-(3\bar{L}_m^2+4L_{ls}^2+8L_{ls}\bar{L}_m)L_{ff}+2L_{md}^2L_{ls}\cos 2\left(\theta_r-\frac{2}{3}\pi\right)\right)(-r_si_{bs}-u_{bs}) \\
& +(3L_{md}^2\bar{L}_m+L_{ls}L_{md}^2-(3\bar{L}_m^2+2L_{ls}\bar{L}_m)L_{ff}+2L_{md}^2L_{ls}\cos 2\theta_r)(-r_si_{cs}-u_{cs})-(6L_{ls}L_{md}\bar{L}_m+4L_{ls}^2L_{md})\sin\left(\theta_r-\frac{2}{3}\pi\right)(-r_ri_{fr}+u_{fr}) \\
& -(6L_{ls}L_{md}^2\bar{L}_m+4L_{ls}^2L_{md}^2)\left(i_{as}\cos\theta_r+i_{bs}\cos\left(\theta_r-\frac{2}{3}\pi\right)+i_{cs}\cos\left(\theta_r+\frac{2}{3}\pi\right)\right)\omega_r\sin\left(\theta_r-\frac{2}{3}\pi\right) \\
& \left.-((6L_{md}L_{ls}\bar{L}_m+4L_{md}L_{ls}^2)L_{ff}-6L_{md}^3L_{ls})i_{fr}\omega_r\cos\left(\theta_r-\frac{2}{3}\pi\right)\right]
\end{aligned}$$

$$\frac{di_{cs}}{dt}=\frac{1}{(2L_{ls}+3\bar{L}_m)(3L_{md}^2L_{ls}-(2L_{ls}^2+3L_{ls}\bar{L}_m)L_{ff})}\Bigg[\Bigg(3L_{md}^2\bar{L}_m+L_{ls}L_{md}^2-(3\bar{L}_m^2+2L_{ls}\bar{L}_m)L_{ff}+2L_{md}^2L_{ls}\cos 2\Big(\theta_r+\frac{1}{3}\pi\Big)\Bigg)(-r_si_{as}-u_{as})$$

$$+(3L_{md}^2\bar{L}_m+L_{ls}L_{md}^2-(3\bar{L}_m^2+2L_{ls}\bar{L}_m)L_{ff}+2L_{md}^2L_{ls}\cos 2\theta_r)(-r_si_{bs}-u_{bs})$$

$$+\Bigg(3L_{md}^2\bar{L}_m+4L_{ls}L_{md}^2-(3\bar{L}_m^2+4L_{ls}^2+8L_{ls}\bar{L}_m)L_{ff}+2L_{md}^2L_{ls}\cos 2\Big(\theta_r+\frac{2}{3}\pi\Big)\Bigg)(-r_si_{cs}-u_{cs})-(6L_{ls}L_{md}\bar{L}_m+4L_{ls}^2L_{md})\sin\Big(\theta_r+\frac{2}{3}\pi\Big)(-r_ri_{fr}+u_{fr})$$

$$-(6L_{ls}L_{md}^2\bar{L}_m+4L_{ls}^2L_{md}^2)\Bigg(i_{as}\cos\theta_r+i_{bs}\cos\Big(\theta_r-\frac{2}{3}\pi\Big)+i_{cs}\cos\Big(\theta_r+\frac{2}{3}\pi\Big)\Bigg)\omega_r\sin\Big(\theta_r+\frac{2}{3}\Big)$$

$$-((6L_{md}L_{ls}\bar{L}_m+4L_{md}L_{ls}^2)L_{ff}-6L_{md}^3L_{ls})i_{fr}\omega_r\cos\Big(\theta_r+\frac{2}{3}\pi\Big)\Bigg]$$

$$\frac{di_{fr}}{dt}=\frac{1}{3L_{md}^2L_{ls}-(2L_{ls}^2+3L_{ls}\bar{L}_m)L_{ff}}\Bigg[2L_{md}L_{ls}\Bigg(\sin\theta_r(r_si_{as}+u_{as})+\sin\Big(\theta_r-\frac{2}{3}\pi\Big)(r_si_{bs}+u_{bs})+\sin\Big(\theta_r+\frac{2}{3}\pi\Big)(r_si_{cs}+u_{cs})\Bigg)$$

$$-(2L_{ls}^2+3\bar{L}_mL_{ls})(-r_ri_{fr}+u_{fr})-(3L_{ls}L_{md}\bar{L}_m+2L_{ls}^2L_{md})\Bigg(i_{as}\omega_r\cos\theta_r+i_{bs}\omega_r\cos\Big(\theta_r-\frac{2}{3}\pi\Big)+i_{cs}\omega_r\cos\Big(\theta_r+\frac{2}{3}\pi\Big)\Bigg)\Bigg]$$

$$\frac{d\omega_{rm}}{dt}=\frac{k_{apm}}{J}i_{apm}-\frac{PL_{md}}{2J}i_{fr}\Bigg[i_{as}\cos\Big(\frac{P}{2}\theta_{rm}\Big)+i_{bs}\cos\Big(\frac{P}{2}\theta_{rm}-\frac{2}{3}\pi\Big)+i_{cs}\cos\Big(\frac{P}{2}\theta_{rm}+\frac{2}{3}\pi\Big)\Bigg]-\frac{B_m}{J}\omega_{rm}$$

$$\frac{d\theta_{rm}}{dt}=\omega_{rm}$$

In the equations for synchronous generator circuitry, the electrical angular velocity and displacement are used; that is, $\omega_r=\frac{1}{2}P\omega_{rm}$ and $\theta_r=\frac{1}{2}P\theta_{rm}$.

To model the generator dynamics in the MATLAB environment, two m-files are developed (csg_01.m and csg_02.m).

MATLAB script csg_01.m

```
tspan = [0 .7]; y0=[0 0 0 0 0 380 0]';
options=odeset('RelTol',5e-2,'AbsTol',[1e-41e-41e-41e-41e-41e-41e-4]);
[t,y] = ODE45('csg_02',tspan,y0,options);
subplot(3,2,4);plot(t,y(:,1));
ylabel('iapm');axis([0,0.7,0,30]);grid;
subplot(3,2,1);plot(t,y(:,2));
ylabel('ias');axis([0,0.7,-25,25]);grid;
subplot(3,2,3);plot(t,y(:,3));
ylabel('ibs');axis([0,0.7,-25,25]);grid;
subplot(3,2,5);plot(t,y(:,4));
ylabel('ics');axis([0,0.7,-25,25]);grid;
subplot(3,2,2);plot(t,y(:,5));
ylabel('ifr');axis([0,0.7,0,450]);grid;
subplot(3,2,6);plot(t,y(:,6));
ylabel('wr');axis([0,0.7,330,400]);grid;pause
```

MATLAB script csg_02.m

```
function yprime=difer(t,y);
% Voltage applied to the prime mover
uapm=100;
% Parameters of the permanent-magnet DC motor
```

```
rapm=0.4; Lapm=0.015; kapm=0.25;
% Stator leakage and magnetizing inductances
Lls=0.0001; Lmqs=0.00095; Lmds=0.00095;
% Rotor leakage and magnetizing inductances
Llf=0.00022; Lmf=0.0001;
% Mutual inductance
Lmd=0.00035;
% Average value of the magnetizing inductance
Lmb=(Lmqs+Lmds)/3;
% Resistances of the stator and rotor windings
rs=0.25;rr=0.47;
% Equivalent moment of inertia and viscous friction coefficient
J=0.003; Bm=0.00072;
% Number of poles
P=2;
% Voltage applied to the rotor field winding
ufr=200;
% Load resistance
if t<=0.35
    Rl=2;
else
    Rl=5;
end
wr=2*y(6,:)/P;
% Numerical expressions used
Ld=2*Lls*Lmd^2;
Ldr=2*Lls*Lmd;
Lss=(-3*Lmb^2-4*Lls^2-8*Lls*Lmb)*(Lmf+Llf)+3*Lmd^2*Lmb+4*Lls*Lmd^2;
Ldens=(2*Lls+3*Lmb)*((-2*Lls^2-3*Lmb*Lls)*(Llf+Lmf)+3*Lmd^2*Lls);
Ldenr=(-2*Lls^2-3*Lmb*Lls)*(Llf+Lmf)+3*Lmd^2*Lls;
Lms=(-3*Lmb^2-2*Lls*Lmb)*(Lmf+Llf)+3*Lmd^2*Lmb+Lls*Lmd^2;
Lrr=6*Lls*Lmd*Lmb+4*Lls^2*Lmd;
Llr=-(2*Lls^2+3*Lmb*Lls);
% Variables used
S1=sin(y(7,:));S2=sin(y(7,:)-2*pi/3);S3=sin(y(7,:)+2*pi/3);
Ias=-(rs+Rl)*y(2,:);Ibs=-(rs+Rl)*y(3,:);Ics=-(rs+Rl)*y(4,:);
Ifr=-rr*y(5,:)+ufr;
C1=cos(y(7,:));C2=cos(y(7,:)-2*pi/3);C3=cos(y(7,:)+2*pi/3);
C12=cos(2*y(7,:));C22=cos(2*y(7,:)-2*pi/3);C32=cos(2*y(7,:)+2*pi/3);
Nsts=(-6*Lmd^2*Lls*Lmb-
 4*Lmd^2*Lls^2)*wr.*(C1.*y(2,:)+C2.*y(3,:)+C3.*y(4,:));
Nstr=(3*Lmd*Lmb*Lls+2*Lmd*Lls^2)*wr.*(C1.*y(2,:)+C2.*y(3,:)+C3.*y(4,:));
Nct=((6*Lmd*Lls*Lmb+4*Lmd*Lls^2)*(Llf+Lmf)-6*Lmd^3*Lls)*wr.*y(5,:);
Te=P*(Lmd*y(5,:).*(y(2,:).*C1+y(3,:).*C2+y(4,:).*C3))/2;
% Differential equations
yprime=[(-rapm*y(1,:)-kapm*y(6,:)+uapm)/Lapm;...
((Lss+Ld*C12).*Ias+(Lms+Ld*C22).*Ibs+(Lms+Ld*C32).*Ics-
 Lrr*S1.*Ifr+Nsts.*S1-Nct.*C1)/Ldens;...
  ((Lms+Ld*C22).*Ias+(Lss+Ld*C32).*Ibs+(Lms+Ld*C12).*Ics-
 Lrr*S2.*Ifr+Nsts.*S2-Nct.*C2)/Ldens;...
  ((Lms+Ld*C32).*Ias+(Lms+Ld*C12).*Ibs+(Lss+Ld*C22).*Ics-
 Lrr*S3.*Ifr+Nsts.*S3-Nct.*C3)/Ldens;...
  (-Ldr*S1.*Ias-Ldr*S2.*Ibs-Ldr*S3.*Ics+Llr*Ifr-Nstr)/Ldenr;...
```

```
(-Te-P*Bm*y(6,:)+kapm*y(1,:))/J;...
wr];
```

The transient dynamics, as modeled using a set of differential equations derived, are documented in Figure 7.32. Figure 7.33 illustrates the resulting waveforms for the phase voltages induced by the generator u_{as}, u_{bs}, and u_{cs} under different values of the phase resistance, $R_L = \begin{cases} 3\ \text{ohm}, \ \forall t \in [0 \quad 0.35)\text{s} \\ 5\ \text{ohm}, \ \forall t \in [0.35 \quad 0.7)\text{s} \end{cases}$. In the steady-state, the frequency of the induced voltage is proportional to the angular velocity, and the magnitude of u_{as}, u_{bs}, and u_{cs} depends on the resistance inserted.

7.4.2 *Mathematical models of synchronous machines in the rotor and synchronous reference frames*

It was emphasized that the mathematical models of synchronous machines (motors and generators) are identical when one specifies that the angular velocity of the reference frame is equal to the electrical or synchronous angular velocity. This is due the fact that $\omega_r = \omega_e$, and hence, the *direct* and *inverse* Park transformations are identical: $\mathbf{K}_s^r = \mathbf{K}_s^e$, and $\mathbf{K}_s^{r^{-1}} = \mathbf{K}_s^{e^{-1}}$.

Making use of the Park transformation, which for the rotor reference frame is given as

$$\mathbf{u}_{qd0s}^r = \mathbf{K}_s^r u_{absc}, \quad \mathbf{i}_{qd0s}^r = \mathbf{K}_s^r \mathbf{i}_{abcs}, \quad \text{and} \quad \boldsymbol{\psi}_{qd0s}^r = \mathbf{K}_s^r \boldsymbol{\psi}_{abcs}$$

$$\mathbf{K}_s^r = \frac{2}{3}\begin{bmatrix} \cos\theta_r & \cos\left(\theta_r - \frac{2}{3}\pi\right) & \cos\left(\theta_r + \frac{2}{3}\pi\right) \\ \sin\theta_r & \sin\left(\theta_r - \frac{2}{3}\pi\right) & \sin\left(\theta_r + \frac{2}{3}\pi\right) \\ \frac{1}{2} & \frac{1}{2} & \frac{1}{2} \end{bmatrix}$$

one transforms the *abc* variables used to map the stator circuitry dynamics to the *qd0* quantities. It was shown that synchronous machines circuitry dynamics are described by matrix differential equations, which are given for motors as

$$\mathbf{u}_{abcs} = \mathbf{r}_s \mathbf{i}_{abcs} + \frac{d\boldsymbol{\psi}_{abcs}}{dt}$$

in contrast, for generators, one has

$$\mathbf{u}_{abcs} = -\mathbf{r}_s \mathbf{i}_{abcs} + \frac{d\boldsymbol{\psi}_{abcs}}{dt}$$

The application of the Park transformations, which was reported in Section 7.2.2 in detail, results in four differential equations for synchronous motors

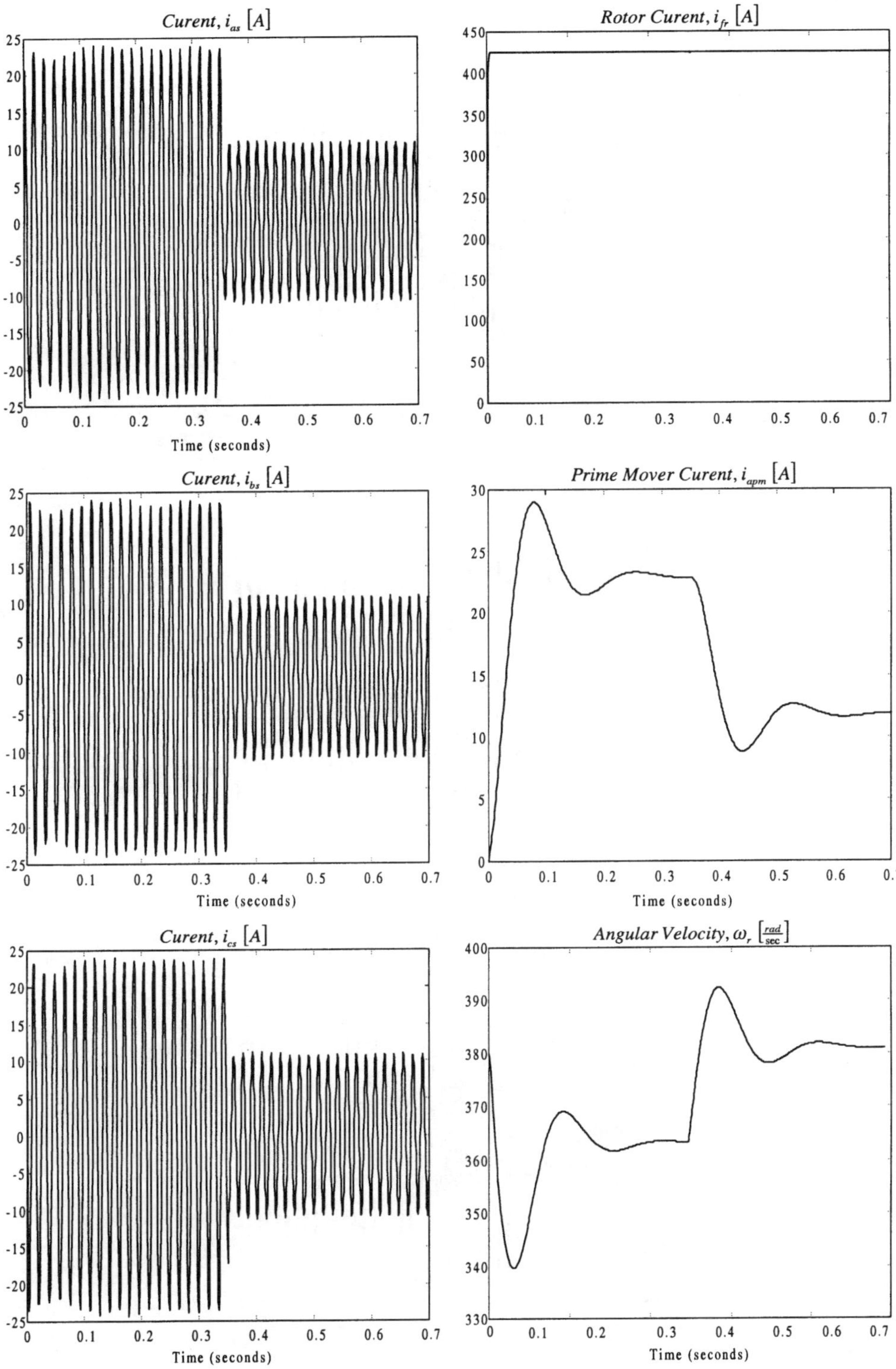

Figure 7.32 Transient dynamics of the synchronous generator and prime mover; $R_L = \begin{cases} 3\ \text{ohm}, & \forall t \in [0 \quad 0.35)\text{s} \\ 5\ \text{ohm}, & \forall t \in [0.35 \quad 0.7)\text{s} \end{cases}$.

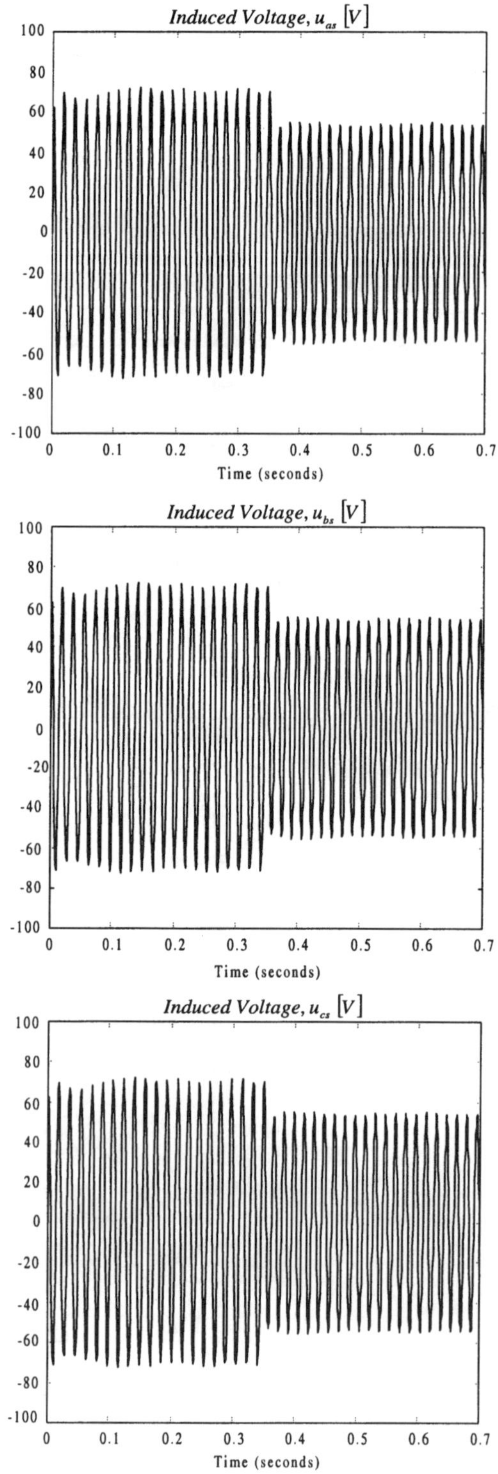

Figure 7.33 Voltages induced by a three-phase generator; $R_L = \begin{cases} 3 \text{ ohm}, & \forall t \in [0 \quad 0.35)\text{s} \\ 5 \text{ ohm}, & \forall t \in [0.35 \quad 0.7)\text{s} \end{cases}$.

$$u_{qs}^r = r_s i_{qs}^r + \omega_r \psi_{ds}^r + \frac{d\psi_{qs}^r}{dt}$$

$$u_{qs}^r = r_s i_{ds}^r - \omega_r \psi_{qs}^r + \frac{d\psi_{ds}^r}{dt}$$

$$u_{0s}^r = r_s i_{0s}^r + \frac{d\psi_{0s}^r}{dt}$$

$$u_{fr} = r_r i_{fr} + \frac{d\psi_{fr}}{dt}$$

For synchronous generators, we have

$$u_{qs}^r = -r_s i_{qs}^r + \omega_r \psi_{ds}^r + \frac{d\psi_{qs}^r}{dt}$$

$$u_{qs}^r = -r_s i_{ds}^r - \omega_r \psi_{qs}^r + \frac{d\psi_{ds}^r}{dt}$$

$$u_{0s}^r = -r_s i_{0s}^r + \frac{d\psi_{0s}^r}{dt}$$

$$u_{fr} = r_r i_{fr} + \frac{d\psi_{fr}}{dt}$$

Having obtained the corresponding equations for the flux linkages for synchronous motors, see the results reported for synchronous reluctance and permanent-magnet synchronous machines,

$$\psi_{qs}^r = (L_{ls} + L_{mq}) i_{qs}^r$$

$$\psi_{ds}^r = (L_{ls} + L_{md}) i_{ds}^r + L_{md} i_{fr}$$

$$\psi_{0s}^r = L_{ls} i_{0s}^r$$

$$\psi_{fr} = L_{md} i_{ds}^r + (L_{lf} + L_{md}) i_{fr}$$

the differential equations in the *qd0* variables result. In particular, from a set of differential equations

$$u_{qs}^r = r_s i_{qs}^r + (L_{ls} + L_{mq}) \frac{di_{qs}^r}{dt} + (L_{ls} + L_{md}) i_{ds}^r \omega_r + L_{md} i_{fr} \omega_r$$

$$u_{ds}^r = r_s i_{ds}^r + (L_{ls} + L_{md}) \frac{di_{ds}^r}{dt} - (L_{ls} + L_{mq}) i_{qs}^r \omega_r + L_{md} \frac{di_{fr}}{dt}$$

$$u_{0s}^r = r_s i_{0s}^r + L_{ls} \frac{di_{0s}^r}{dt}$$

$$u_{fr} = r_r i_{fr} + L_{md} \frac{di_{ds}^r}{dt} + (L_{lf} + L_{md}) \frac{di_{fr}}{dt}$$

Cauchy's form can be easily derived.

In the synchronous reference frame, one has

$$u_{qs}^e = r_s i_{qs}^e + (L_{ls} + L_{mq})\frac{di_{qs}^e}{dt} + (L_{ls} + L_{md})i_{ds}^e\omega_r + L_{md}i_{fr}\omega_r$$

$$u_{ds}^e = r_s i_{ds}^e + (L_{ls} + L_{md})\frac{di_{ds}^e}{dt} - (L_{ls} + L_{mq})i_{qs}^e\omega_r + L_{md}\frac{di_{fr}}{dt}$$

$$u_{0s}^e = r_s i_{0s}^e + Ll_s\frac{di_{0s}^e}{dt}$$

$$u_{fr} = r_r i_{fr} + L_{md}\frac{di_{ds}^e}{dt} + (L_{lf} + L_{md})\frac{di_{fr}}{dt}$$

We must find an explicit expression for the electromagnetic torque, and using the Park transformation and (7.31) for the rotor and synchronous reference frames, we have

$$T_e = \frac{3P}{4}[L_{mq}i_{qs}^r i_{ds}^r - L_{md}i_{qs}^r(i_{ds}^r - i_{fr})]$$

and

$$T_e = \frac{3P}{4}[L_{mq}i_{qs}^e i_{ds}^e - L_{md}i_{qs}^e(i_{ds}^e - i_{fr})]$$

Hence, the *torsional-mechanical* dynamics for conventional synchronous motors in the rotor and synchronous reference frames are

$$\frac{d\omega_r}{dt} = \frac{3P^2}{8J}[L_{mq}i_{qs}^r i_{ds}^r - L_{md}i_{qs}^r(i_{ds}^r - i_{fr})] - \frac{B_m}{J}\omega_r - \frac{P}{2J}T_L, \quad \frac{d\theta_r}{dt} = \omega_r$$

and

$$\frac{d\omega_r}{dt} = \frac{3P^2}{8J}[L_{mq}i_{qs}^e i_{ds}^e - L_{md}i_{qs}^e(i_{ds}^e - i_{fr})] - \frac{B_m}{J}\omega_r - \frac{P}{2J}T_L, \quad \frac{d\theta_r}{dt} = \omega_r$$

Therefore, augmenting the circuitry and *torsional-mechanical* dynamics, one finds the models that need to be used. For example, in the rotor reference frame, we obtain

$$u_{qs}^r = r_s i_{qs}^r + (L_{ls} + L_{mq})\frac{di_{qs}^r}{dt} + (L_{ls} + L_{md})i_{ds}^r\omega_r + L_{md}i_{fr}\omega_r$$

$$u_{ds}^r = r_s i_{ds}^r + (L_{ls} + L_{md})\frac{di_{ds}^r}{dt} - (L_{ls} + L_{mq})i_{qs}^r\omega_r + L_{md}\frac{di_{fr}}{dt}$$

$$u_{0s}^r = r_s i_{0s}^r + L_{ls}\frac{di_{0s}^r}{dt}$$

$$u_{fr} = r_r i_{fr} + L_{md}\frac{di^r_{ds}}{dt} + (L_{lf} + L_{md})\frac{di_{fr}}{dt}$$

$$\frac{d\omega_r}{dt} = \frac{3P^2}{8J}[L_{mq}i^r_{qs}i^r_{ds} - L_{md}i^r_{qs}(i^r_{ds} - i_{fr})] - \frac{B_m}{J}\omega_r - \frac{P}{2J}T_L$$

For synchronous generators, plugging the expressions for the flux linkages

$$\psi^r_{qs} = -(L_{ls} + L_{mq})i^r_{qs}$$

$$\psi^r_{ds} = -(L_{ls} + L_{md})i^r_{ds} + L_{md}i_{fr}$$

$$\psi^r_{0s} = -L_{ls}i^r_{0s}$$

$$\psi_{fr} = -L_{md}i^r_{ds} + (L_{lf} + L_{md})i_{fr}$$

into Kirchhoff's differential equations, one obtains, without difficulty, the resulting set of four differential equations to map the circuitry transient behavior. We have

$$u^r_{qs} = -r_s i^r_{qs} - (L_{ls} + L_{mq})\frac{di^r_{qs}}{dt} - (L_{ls} + L_{md})i^r_{ds}\omega_r + L_{md}i_{fr}\omega_r$$

$$u^r_{ds} = -r_s i^r_{ds} - (L_{ls} + L_{md})\frac{di^r_{ds}}{dt} + (L_{ls} + L_{mq})i^r_{qs}\omega_r + L_{md}\frac{di_{fr}}{dt}$$

$$u^r_{0s} = -r_s i^r_{0s} - L_{ls}\frac{di^r_{0s}}{dt}$$

$$u_{fr} = r_r i_{fr} - L_{md}\frac{di^r_{ds}}{dt} + (L_{lf} + L_{md})\frac{di_{fr}}{dt}$$

The *torsional-mechanical* dynamics for synchronous generators under consideration are

$$\frac{d\omega_r}{dt} = -\frac{3P^2}{8J}[L_{mq}i^r_{qs}i^r_{ds} - L_{md}i^r_{qs}(i^r_{ds} - i_{fr})] - \frac{B_m}{J}\omega_r + \frac{P}{2J}T_{pm}, \quad \frac{d\theta_r}{dt} = \omega_r$$

Therefore,

$$u^r_{qs} = -r_s i^r_{qs} - (L_{ls} + L_{mq})\frac{di^r_{qs}}{dt} - (L_{ls} + L_{md})i^r_{ds}\omega_r + L_{md}i_{fr}\omega_r$$

$$u^r_{ds} = -r_s i^r_{ds} - (L_{ls} + L_{md})\frac{di^r_{ds}}{dt} + (L_{ls} + L_{mq})i^r_{qs}\omega_r + L_{md}\frac{di_{fr}}{dt}$$

$$u^r_{0s} = -r_s i^r_{0s} - L_{ls}\frac{di^r_{0s}}{dt}$$

$$u_{fr} = r_r i_{fr} - L_{md}\frac{di^r_{qs}}{dt} + (L_{lf} + L_{md})\frac{di_{fr}}{dt}$$

$$\frac{d\omega_r}{dt} = -\frac{3P^2}{8J}[L_{mq}i^r_{qs}i^r_{ds} - L_{md}i^r_{qs}(i^r_{ds} - i_{fr})] - \frac{B_m}{J}\omega_r + \frac{P}{2J}T_{pm}$$

In the synchronous reference frame, conventional synchronous generators are mapped by

$$u_{qs}^e = -r_s i_{qs}^e - (L_{ls} + L_{mq})\frac{di_{qs}^e}{dt} - (L_{ls} + L_{md})i_{ds}^e \omega_r + L_{md} i_{fr} \omega_r$$

$$u_{ds}^e = -r_s i_{ds}^e - (L_{ls} + L_{md})\frac{di_{ds}^e}{dt} + (L_{ls} + L_{mq})i_{qs}^e \omega_r + L_{md}\frac{di_{fr}}{dt}$$

$$u_{0s}^e = -r_s i_{0s}^e - L_{ls}\frac{di_{0s}^e}{dt}$$

$$u_{fr} = r_r i_{fr} - L_{md}\frac{di_{ds}^e}{dt} + (L_{lf} + L_{md})\frac{di_{fr}}{dt}$$

$$\frac{d\omega_r}{dt} = -\frac{3P^2}{8J}[L_{mq} i_{qs}^e i_{ds}^e - L_{md} i_{qs}^e (i_{ds}^e - i_{fr})] - \frac{B_m}{J}\omega_r + \frac{P}{2J}T_{pm}$$

The derived highly nonlinear differential equations must be used to analyze and model conventional synchronous machines in the rotor and synchronous reference frames

7.5 Stepper motors

Two basic classes of stepper motors are the permanent-magnet and the variable-reluctance motors. Severe requirements on weight and size, efficiency and torque density, steady-state and dynamic characteristics, ruggedness and reliability, survivability, and maintenance, imposed on high-performance electromechanical systems, facilitate the need for analysis, design, and deployment of advanced actuator technology. Actuators must guarantee the desired performance in the full operating envelope, develop the required torque, ensure the specified accuracy in spite of disturbances, and satisfy the specified integrity and packaging. In high-performance servo-systems, stepper motors are perfectly suited as actuators due to their efficiency and reliability, low volume-, weight-, and size-to-torque ratios, high power and torque densities, low cost and maintenance, simplicity, ruggedness, etc. It should be emphasized that electromechanical actuators driven by permanent-magnet synchronous and stepper motors have displaced hydraulic technology (hydraulic actuators are larger in size and weight than synchronous motors; additionally, leaks and reliability have been persistent challenges in the application of hydraulic technology). There is a critical need to guarantee the specified controllability and stability requirements, reduce weight and size, and increase efficiency and reliability using advanced high-performance stepper motors that are directly attached to mechanisms. This solution enhances the reliability by eliminating geartrains and ensures compact, highly integrated assembling features compared with existing solutions. Recent advances in electric machinery and power electronics allow one to ensure significant improvements in design and manufacturing of advanced, highly integrated, state-of-the-art electromechanical servo-systems actuated by stepper motors. The application of permanent-magnet stepper motors guarantees the desired spectrum of the specified requirements, and these motors can be easily integrated into servos to directly actuate mechanisms. This direct connection of motors without matching mechanical coupling or gears allows one to achieve a remarkable level of efficiency, reliability, and performance. It should be emphasized that stepper motors are used in rotating (pointing) systems, as well as to actuate the liner positioning systems.

Stepper motors must be controlled to ensure stability and precision tracking, disturbance rejection and robustness to parameter changes, desired steady-state and dynamic performance, zero steady-state error, etc. To approach the motion control problem, complete nonlinear mathematical models of stepper motors must be found, and basic features should be thoroughly discussed.

By energizing the stator windings in the proper sequence, the rotor rotates in the *counterclockwise* or *clockwise* direction due to the electromagnetic torque developed. In particular, the rotor displaces by full or half step (for conventional stepper motors, one step is equal to 0.9 or 1.8°, depending on the number of the rotor teeth). Hence, by energizing windings, one achieves the angular increment equal to a full or half step. The angular velocity is regulated, and stepper motors rotate at the angular velocity determined by the frequency of the phase currents fed or voltages supplied to the phase windings. This frequency is rated by electromagnetic and mechanical dynamic capabilities. Recall that other synchronous motors are controlled in a similar, but slightly different way using θ_r, and the phase currents fed or voltages supplied to the stator windings are found as functions of the angular displacement to maximize the electromagnetic torque. Stepper motors are widely used in positioning servo-systems; and simple regulating features (through energizing the stator windings) allow one to operate stepper motors in an *open-loop* mode. That is, *open-loop* stepper motor servos have few components, do not have an optical encoder and, therefore, are not expensive. Usually, voltage-fed power amplifiers are used, and the phase voltages are applied (pulse train voltage inputs) to attain the rate from 100 to 1000 steps per second (usually, one step corresponds to 0.9° or 1.8° of the rotor displacement). However, *open-loop* stepper motor servos do not ensure the required accuracy and dynamic performances. Furthermore, due to disturbances (loads) and parameter variations (inertia changes), servo motors can miss steps; as a result, the real displacement will be different from the specified one. Therefore, the angular velocity is low because stepper motors operate in the so-called slow-stepping mode to avoid missing steps. The electromagnetic torque produced by stepper motors can be maximized only in the closed-loop configuration by solving the torque maximization problem using the measured angular displacement.

The assigned tracking accuracy, desired stability margins, microstepping capabilities, and disturbance attenuation are ensured by closed-loop control configurations, and mathematical models must be obtained to solve complex analysis, design, and optimization problems to improve the performance of servo-systems. There are two basic classes of stepper motors: permanent-magnet stepper motors and variable reluctance stepper motors.

Permanent-magnet stepper motors

Permanent-magnet stepper motors, which is equipped with optical encoder, is shown in Figure 7.34. Using the results and basic principles in analysis of operation of permanent-magnet synchronous motors, stepper motors will be studied. The stator windings are wound over the full axial length, and the permanent-magnet rotor rotates through a specific angle in response to each pulse supplied to the phase windings. These pulses are applied sequentially to the stator windings, and the angular rotor displacement depends on the number of rotor teeth. The *direct* magnetic axis of a cylindrical permanent-magnet rotor tends to align with the corresponding magnetic axis of the stator windings, depending on which winding is excited. A stepper motor with 50 rotor teeth is illustrated in Figure 7.35.

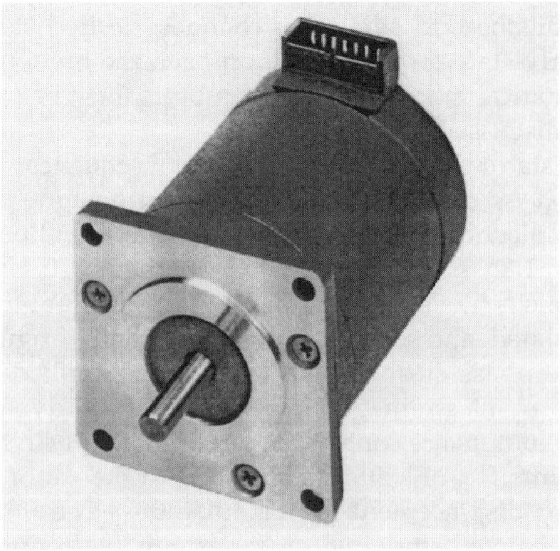

Figure 7.34 Permanent-magnet stepper motor. (Courtesy of Warner Electric.)

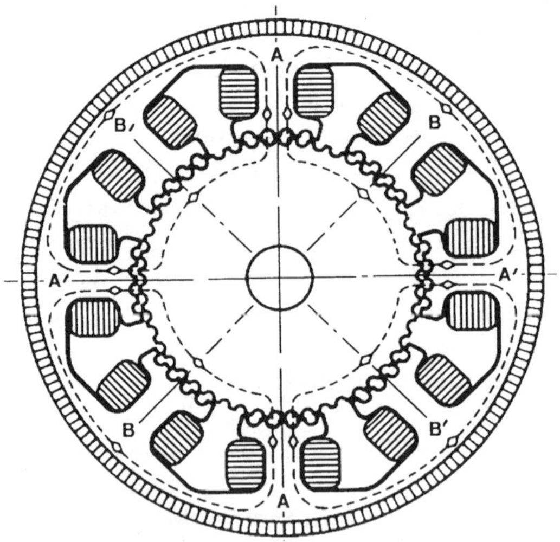

Figure 7.35 View of permanent magnet stepper motor. (Courtesy of Warner Electric.)

The catalog data for permanent-magnet 23 and 42 NEMA size stepper motors is given in Table 7.2; the torque-speed characteristic curves, which allow one to analyze the operating envelopes, are given in Figure 7.36.

To visualize the operation principles of two-pole, four-phase permanent-magnet stepper motors, we should discuss how one must energize the stator windings (see Figure 7.37).

If positive and negative voltages are applied (or currents are fed) to stator phases *as* and *cs*, the resulting magnetic polarities are indicated in Figure 7.37a, and the rotor has a vertical position (step 1 in Figure 7.38). By deenergizing the *as* and *cs* windings, and energizing phases *bs* and *ds*, the magnetic polarities are changed, and an alignment electromagnetic torque is produced. The rotor turns by 90° and has a horizontal position (see Figure 7.37b and step 2 in Figure 7.38). When one energizes phases *as* and *cs* by applying negative and positive voltages, the rotor turns in the *counterclockwise* direction by 90° (step 3 in Figure 7.38). The direction of rotation can be reversed to *clockwise* by changing the polarity of the voltages applied (or the direction of the currents fed) to the stator windings.

Table 7.2 Permanent-Magnet Stepper Motors.

60MM (NEMA 23) RATINGS and SPECIFICATIONS

MOTOR TYPE 3% ACCURACY	CONNECTIONS		TYPICAL TIME FOR SINGLE STEP (mS) (1)	UNIPOLAR CONFIGURATION: NOMINAL DC VOLTS (3)	RATED AMPERES PER WINDING	NOMINAL RESISTANCE PER WINDING (25 C) OHMS (2)	NOMINAL INDUCTANCE PER PHASE (MILLI-HENRYS) (2) (4)	MINIMUM HOLDING TORQUE OZ-IN (Ncm)		BIPOLAR CONNECTIONS: SERIES CONNECTION				MINIMUM HOLDING TORQUE OZ-IN (Ncm)		PARALLEL CONNECTION				MINIMUM HOLDING TORQUE OZ-IN (Ncm)	
	NUMBER	TYPE						2Ø ON	1Ø ON	VOLTS	AMPERES	R	L	2Ø ON	1Ø ON	VOLTS	AMPERES	R	L	2Ø ON	1Ø ON
M061-CE02	8	CONN.	2.5	5	1	5	9.57	60 (42)	36 (25)	7	0.7	10	38.3	75 (53)	45 (32)	3.54	1.4	2.5	9.57	75 (53)	45 (32)
M061-LE02	8	LEADS	2.5	5	1	5	9.57	60 (42)	36 (25)	7	0.7	10	38.3	75 (53)	45 (32)	3.54	1.4	2.5	9.57	75 (53)	45 (32)
M061-CS02	6	CONN.	2.5	5	1	5	9.57	60 (42)	36 (25)	7	0.7	10	38.3	75 (53)	45 (32)	—	—	—	—	—	—
M061-LS02	6	LEADS	2.5	5	1	5	9.57	60 (42)	36 (25)	7	0.7	10	38.3	75 (53)	45 (32)	—	—	—	—	—	—
M061-CE08	8	CONN.	2	1.25	3.8	0.33	0.635	60 (42)	36 (25)	1.8	2.7	0.66	2.54	75 (53)	45 (32)	0.89	5.4	0.165	0.635	75 (53)	45 (32)
M061-LE08	8	LEADS	2	1.25	3.8	0.33	0.635	60 (42)	36 (25)	1.8	2.7	0.66	2.54	75 (53)	45 (32)	0.89	5.4	0.165	0.635	75 (53)	45 (32)
M061-CS08	6	CONN.	2	1.25	3.8	0.33	0.635	60 (42)	36 (25)	1.8	2.7	0.66	2.54	75 (53)	45 (32)	—	—	—	—	—	—
M061-LS08	6	LEADS	2	1.25	3.8	0.33	0.635	60 (42)	36 (25)	1.8	2.7	0.66	2.54	75 (53)	45 (32)	—	—	—	—	—	—
M061-FF206	4	LEADS	—	—	—	—	—	—	—	6.3	1	6.3	24.8	75 (53)	—	—	—	—	—	—	—
M061-CS-301*	6	CONN.	7.5	11	0.44	22.5	38.4	53 (37)	32 (23)	15.6	0.3	45	154	65 (46)	39 (28)	—	—	—	—	—	—
M061-LS-301*	6	LEADS	7.5	11	0.44	22.5	38.4	53 (37)	32 (23)	15.6	0.3	45	154	65 (46)	39 (28)	—	—	—	—	—	—
M061-CS-302*	6	CONN.	4.2	5.5	0.88	6.2	9.57	53 (37)	32 (23)	7.7	0.6	12.4	38.3	65 (46)	39 (28)	—	—	—	—	—	—
M061-LS-302*	6	LEADS	4.2	5.5	0.88	6.2	9.57	53 (37)	32 (23)	7.7	0.6	12.4	38.3	65 (46)	39 (28)	—	—	—	—	—	—
M061-CS-311*	6	CONN.	12	20	0.22	91	126	47 (33)	28 (20)	28.3	0.16	182	504	58 (41)	35 (25)	—	—	—	—	—	—
M061-LS-311*	6	LEADS	12	20	0.22	91	126	47 (33)	28 (20)	28.3	0.16	182	504	58 (41)	35 (25)	—	—	—	—	—	—
M061-CF-408	4	CONN.	—	—	—	—	—	—	—	8	0.5	15.9	61	75 (53)	45 (32)	—	—	—	—	—	—
M061-LR-408	4	LEADS	—	—	—	—	—	—	—	8	0.5	15.9	61	75 (53)	45 (32)	—	—	—	—	—	—
M062-CS03	6	CONN.	3.5	5.3	1.6	3.3	8.28	100 (71)	60 (42)	7.5	1.13	6.6	33.12	75 (53)	75 (53)	—	—	—	—	—	—
M062-LS03	6	LEADS	3.5	5.3	1.6	3.3	8.28	100 (71)	60 (42)	7.5	1.13	6.6	33.12	125 (88)	75 (53)	—	—	—	—	—	—
M062-CS04	6	CONN.	2.8	4.2	1.9	2.2	5.89	100 (71)	60 (42)	5.9	1.34	4.4	23.56	125 (88)	75 (53)	—	—	—	—	—	—
M062-LS04	6	LEADS	2.8	4.2	1.9	2.2	5.89	100 (71)	60 (42)	5.9	1.34	4.4	23.56	125 (88)	75 (53)	—	—	—	—	—	—
M062-CE04	8	CONN.	2.8	4.2	1.9	2.2	5.89	100 (71)	60 (42)	5.9	1.34	4.4	23.56	125 (88)	75 (53)	3	2.7	1.1	5.89	125 (88)	75 (53)
M062-LE04	8	LEADS	2.8	4.2	1.9	2.2	5.89	100 (71)	60 (42)	5.9	1.34	4.4	23.56	125 (88)	75 (53)	3	2.7	1.1	5.89	125 (88)	75 (53)
M062-CS06	6	CONN.	—	2.6	3.1	0.88	2	100 (71)	60 (42)	3.9	2.2	1.76	8	125 (88)	75 (53)	—	—	—	—	—	—
M062-LS06	6	LEADS	—	2.6	3.1	0.88	2	100 (71)	60 (42)	3.9	2.2	1.76	8	125 (88)	75 (53)	—	—	—	—	—	—
M062-CE06	8	CONN.	—	2.6	3.1	0.88	2	100 (71)	60 (42)	3.9	2.2	1.76	8	125 (88)	75 (53)	1.9	4.4	0.44	2	125 (88)	75 (53)
M062-LE06	8	LEADS	—	2.6	3.1	0.88	2	100 (71)	60 (42)	3.9	2.2	1.76	8	125 (88)	75 (53)	1.9	4.4	0.44	2	125 (88)	75 (53)
M062-CS09	6	CONN.	2.2	1.65	4.7	0.35	0.8	100 (71)	60 (42)	2.3	3.3	0.7	3.2	125 (88)	75 (53)	—	—	—	—	—	—
M062-LS09	6	LEADS	2.2	1.65	4.7	0.35	0.8	100 (71)	60 (42)	2.3	3.3	0.7	3.2	125 (88)	75 (53)	—	—	—	—	—	—
M062-CE09	8	CONN.	2.2	1.65	4.7	0.35	0.8	100 (71)	60 (42)	2.3	3.3	0.7	3.2	125 (88)	75 (53)	1.2	6.65	0.175	0.8	125 (88)	75 (53)
M062-LE09	8	LEADS	2.2	1.65	4.7	0.35	0.8	100 (71)	60 (42)	2.3	3.3	0.7	3.2	125 (88)	75 (53)	1.2	6.65	0.175	0.8	125 (88)	75 (53)
M062-FF206	4	LEADS	—	—	—	—	—	—	—	4.2	1.5	2.8	12.5	125 (88)	—	—	—	—	—	—	—
M062-CF-401	4	CONN.	—	—	—	—	—	—	—	6.6	1	6.6	33	125 (88)	75 (53)	—	—	—	—	—	—
M062-LF-401	4	LEADS	—	—	—	—	—	—	—	6.6	1	6.6	33	125 (88)	75 (53)	—	—	—	—	—	—
M063-CE06	8	CONN.	2.6	3.36	2.9	1.16	2.85	150 (106)	90 (64)	4.9	2.1	2.32	11.4	190 (134)	115 (81)	2.4	4.1	0.58	2.85	190 (134)	115 (81)
M063-LE06	8	LEADS	2.6	3.36	2.9	1.16	2.85	150 (106)	90 (64)	4.9	2.1	2.32	11.4	190 (134)	115 (81)	2.4	4.1	0.58	2.85	190 (134)	115 (81)
M063-CS06	6	CONN.	2.6	3.36	2.9	1.16	2.85	150 (106)	90 (64)	4.9	2.1	2.32	11.4	190 (134)	115 (81)	—	—	—	—	—	—
M063-LS06	6	LEAD	2.6	3.36	2.9	1.16	2.85	150 (106)	90 (64)	4.9	2.1	2.32	11.4	190 (134)	115 (81)	—	—	—	—	—	—
M063-CE09	8	CONN.	2.4	2.25	4.6	0.49	1.15	150 (106)	90 (64)	3.2	3.25	0.98	4.6	190 (134)	115 (81)	1.6	6.5	0.245	1.15	190 (134)	115 (81)
M063-LE09	8	LEADS	2.4	2.25	4.6	0.49	1.15	150 (106)	90 (64)	3.2	3.25	0.98	4.6	190 (134)	115 (81)	1.6	6.5	0.245	1.15	190 (134)	115 (81)
M063-CS09	6	CONN.	2/4	2.25	4.6	0.49	1.15	150 (106)	90 (64)	3.2	3.25	0.98	4.6	190 (134)	115 (81)	—	—	—	—	—	—
M063-LS09	6	LEADS	2/4	2.25	4.6	0.49	1.15	150 (106)	90 (64)	3.2	3.25	0.98	4.6	190 (134)	115 (81)	—	—	—	—	—	—
M063-FF206	4	LEADS	—	—	—	—	—	—	—	5.4	1.5	3.6	14.5	190 (134)	—	—	—	—	—	—	—
M063-CF-401	4	CONN.	—	—	—	—	—	—	—	5.66	1	5.66	29	160 (113)	96 (68)	—	—	—	—	—	—
M063-LF-401	4	LEADS	—	—	—	—	—	—	—	5.66	1	5.66	29	160 (113)	96 (68)	—	—	—	—	—	—

110mm (NEMA Size 42) & 165mm (NEMA Size 66) RATINGS and SPECIFICATIONS

MOTOR TYPE 5% ACCURACY	CONNECTIONS		TYPICAL TIME FOR SINGLE STEP (mS) (1)	UNIPOLAR CONFIGURATION						BIPOLAR CONNECTIONS											
										SERIES CONNECTION						PARALLEL CONNECTION					
				NOMINAL DC VOLTS (3)	RATED AMPERES PER WINDING	NOMINAL RESISTANCE PER WINDING (25 C) OHMS (2)	NOMINAL INDUCTANCE PER PHASE (MILLI-HENRYS) (2) (4)	MINIMUM HOLDING TORQUE OZ-IN (Ncm)						MINIMUM HOLDING TORQUE OZ-IN (Ncm)						MINIMUM HOLDING TORQUE OZ-IN (Ncm)	
	NUMBER	TYPE						2Ø ON	1Ø ON	VOLTS	AMPERES	R	L	2Ø ON	1Ø ON	VOLTS	AMPERES	R	L	2Ø ON	1Ø ON
M111-FD12	6	TERM.	4.4	2.26	6.1	0.37	2.3	625 (441)	375 (265)	3.2	4.3	0.74	0.92	850 (600)	500 (353)	—	—	—	—	—	—
M111-FD16*	6	TERM.	4.4	1.7	8	0.21	1.1	625 (441)	375 (265)	2.4	5.7	0.42	4.4	850 (600)	500 (353)	—	—	—	—	—	—
M111-FF206	4	TERM.	—	—	—	—	—	—	—	3.5	5	0.7	9.2	760 (537)	—	—	—	—	—	—	—
M111-FD-327*	6	TERM.	7	4.1	3.5	1.17	7.2	625 (441)	375 (265)	5.85	2.5	2.34	28.8	850 (600)	500 (353)	—	—	—	—	—	—
M111-FF-401	4	TERM.	—	—	—	—	—	—	—	4	3.4	1.14	17.7	850 (600)	500 (353)	—	—	—	—	—	—
MX111-FF-401	4	TERM.	—	—	—	—	—	—	—	4	3.4	1.14	17.7	850 (600)	500 (353)	—	—	—	—	—	—
M111-FD-8003*	8	TERM.	—	6.93	1.55	4.47	26.2	625 (441)	375 (265)	9.8	1.1	8.94	105	850 (600)	500 (353)	4.9	2.2	2.24	26.2	850 (600)	500 (353)
M111-FD-8007*	8	TERM.	7	4.1	3.5	1.17	7.2	625 (441)	375 (265)	5.85	2.5	2.34	28.8	850 (600)	500 (353)	2.93	5	0.585	7.2	850 (600)	500 (353)
M111-FD-8012	8	TERM.	4.4	2.26	6.1	0.37	2.3	625 (441)	375 (265)	3.2	4.3	0.74	9.2	850 (600)	500 (353)	1.6	8.63	0.185	2.3	850 (600)	500 (353)
M111-FD-8016*	8	TERM.	4.4	1.7	8	0.21	1.1	625 (441)	375 (265)	2.4	5.7	0.42	4.4	850 (600)	500 (353)	1.2	11.3	0.105	1.1	850 (600)	500 (353)
M112-FD08*	6	TERM.	7	5.8	3.8	1.53	14	1125 (794)	675 (477)	8.25	2.7	3.06	56	1390(981)	830 (586)	—	—	—	—	—	—
M112-FD12	6	TERM.	5.5	3.66	6.1	0.6	5.3	1125 (794)	675 (477)	5.2	4.3	1.2	21.2	1390(981)	830 (586)	—	—	—	—	—	—
M112-FJ12	6	TERM.	5.5	3.66	6.1	0.6	5.3	1125 (794)	675 (477)	5.2	4.3	1.2	21.2	1300(981)	830 (586)	—	—	—	—	—	—
M111-FF206	4	TERM.	—	—	—	—	—	—	—	2.95	6	0.49	8.8	1390(981)	—	—	—	—	—	—	—
M112-FF-401	4	TERM.	—	—	—	—	—	—	—	1.95	4	0.49	8.8	950 (671)	675 (477)	—	—	—	—	—	—
MX112-FF401U	4	TERM.	—	—	—	—	—	—	—	1.95	4	0.49	8.8	950 (671)	675 (477)	—	—	—	—	—	—
M112-FJ-326*	6	TERM.	6	1.52	15.2	0.1	0.88	1125 (794)	675 (477)	2.15	10.75	0.2	3.52	1390 (981)	830 (586)	—	—	—	—	—	—
M112-FJ-327*	6	TERM.	6	2.26	9.2	0.246	2.2	1125 (794)	675 (477)	3.2	6.5	0.492	8.8	1390 (981)	830 (586)	—	—	—	—	—	—
M112-FJ-335(5)*	6	TERM.	6	2.26	9.2	0.246	2.2	1125 (794)	675 (477)	3.2	6.5	0.492	8.8	1390 (981)	830 (586)	—	—	—	—	—	—
M112-FJ-344(5)*	6	TERM.	6	1.52	15.2	0.1	0.88	1125 (794)	675 (477)	2.215	10.75	0.2	3.52	1390 (981)	830 (586)	—	—	—	—	—	—
M112-FJ-8008*	8	TERM.	7	5.8	3.8	1.53	14	1125 (794)	675 (477)	8.25	2.7	3.06	56	1390 (981)	830 (586)	4.1	5.4	0.765	14	1390 (981)	830 (586)
M112-FD-8012*	8	TERM.	5.5	3.66	6.1	0.6	5.3	1125 (794)	675 (477)	5.2	4.3	1.2	21.2	1390 (981)	830 (586)	2.6	8.6	0.3	5.3	1390 (981)	830 (586)
M112-FJ-8012	8	TERM.	5.5	3.66	6.1	0.6	5.3	1125 (794)	675 (477)	5.2	4.3	1.2	21.2	1390 (981)	830 (586)	2.6	8.6	0.3	5.3	1390 (981)	830 (586)
M112-FJ-8018*	8	TERM.	6	2.1	9.2	0.242	2.1	1125 (794)	675 (477)	3	6.5	0.483	8.4	1390 (981)	830 (586)	1.5	13	0.12	2.1	1390 (981)	830 (586)
M112-FJ-8025*	8	TERM.	6	1.75	12.7	0.137	1	1125 (794)	675 (477)	2.5	9	0.274	4	1390 (981)	830 (586)	1.25	18	0.069	1	1390 (981)	830 (586)
M112-FJ-8030*	8	TERM.	6	1.52	15.2	0.1	0.88	1125 (794)	675 (477)	2.15	10.75	0.2	3.52	1390 (981)	830 (586)	1.1	21.5	0.05	0.88	1390 (981)	830 (586)
M113-FF-401	4	TERM.	—	—	—	—	—	—	—	4.5	6	0.75	17	2150(1518)	1290 (911)	—	—	—	—	—	—
M172-FD-306*	6	TERM.	24	2.35	15	0.15	1.98	2700 (1906)	1600 (1130)	3.3	10.6	0.3	7.92	3300(2330)	1980(1398)	—	—	—	—	—	—
M172-FD-308*	6	TERM.	24	1.45	20	0.075	1.06	2700 (1906)	1600 (1130)	2.1	14.1	0.15	4.24	3300(2330)	1980(1398)	—	—	—	—	—	—
M172-FF-401	4	TERM.	—	—	—	—	—	—	—	2.6	4	0.65	14.6	2000(1412)	1200 (847)	—	—	—	—	—	—
M172-FD-8030	8	TERM.	24	2.35	15	0.15	2.50	2700 (1906)	1600 (1130)	3.3	10.6	0.3	10.0	3300(2330)	1980(1398)	1.7	21.5	0.075	2.50	3300(2330)	1980 (1398)
M172-FD-8040*	8	TERM.	24	1.45	20	0.075	1.06	2700 (1906)	1600 (1130)	2.1	14.1	0.15	4.24	3300(2330)	1980(1398)	1	28.3	0.0375	1.06	3300(2330)	1980 (1398)

(Courtesy of Warner Electric.)

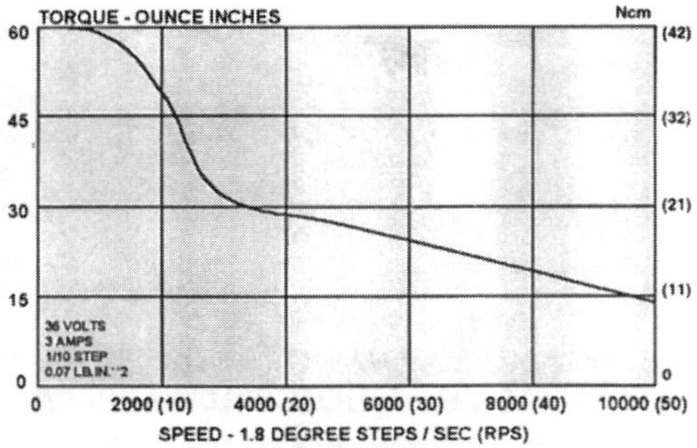

M061-C308 OR M061-LE08
MOTOR WITH SS2000MD4 3.0 AMP, 36 VDC
SLO-SYN MOTION CONTROL

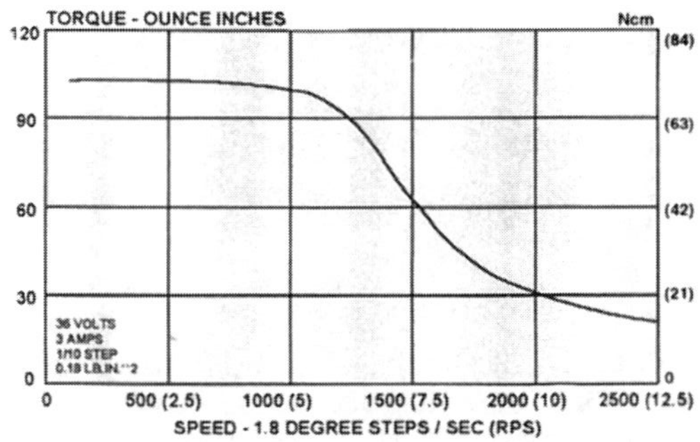

M062-CE09 OR M062-LE09
MOTOR WITH SS2000MD4 3.0 AMP, 36 VDC
SLO-SYN MOTION CONTROL

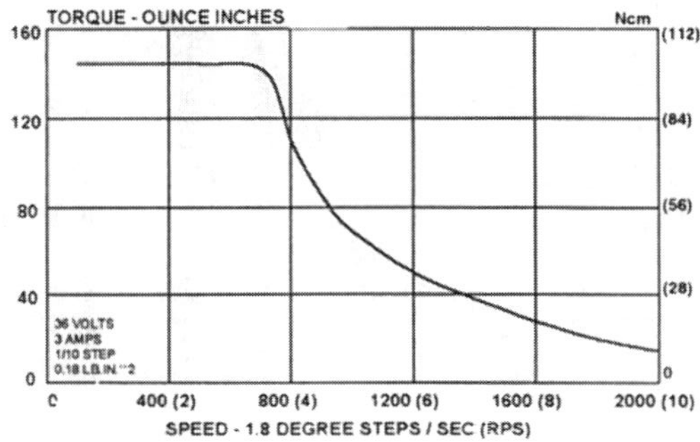

M063-CE09 OR M063-LE09 MOTOR WITH SS2000MD4
3.0 AMP, 36 VDC SLO-SYN MOTION CONTROL

Figure 7.36 Torque-speed characteristics. (Courtesy of Warner Electric.)

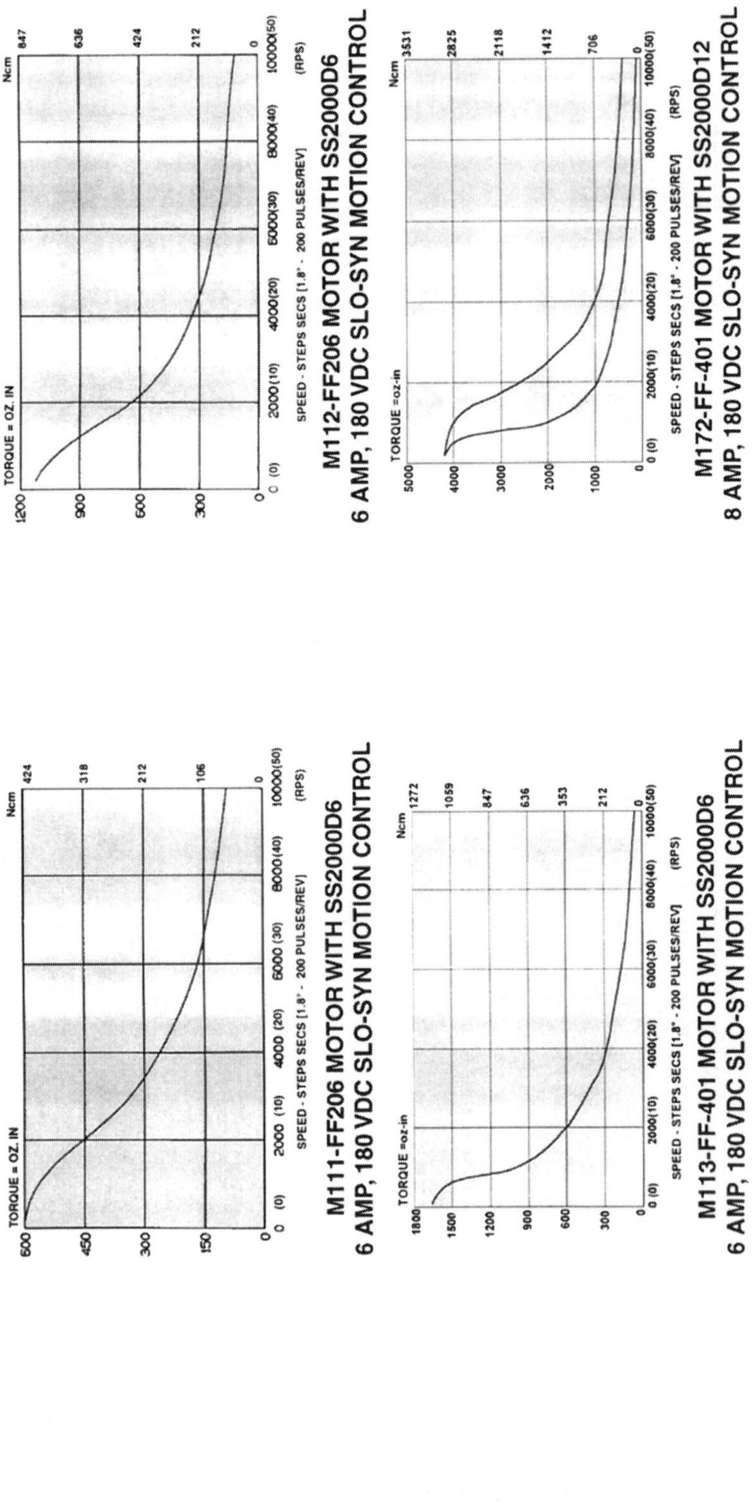

Figure 7.36 Continued.

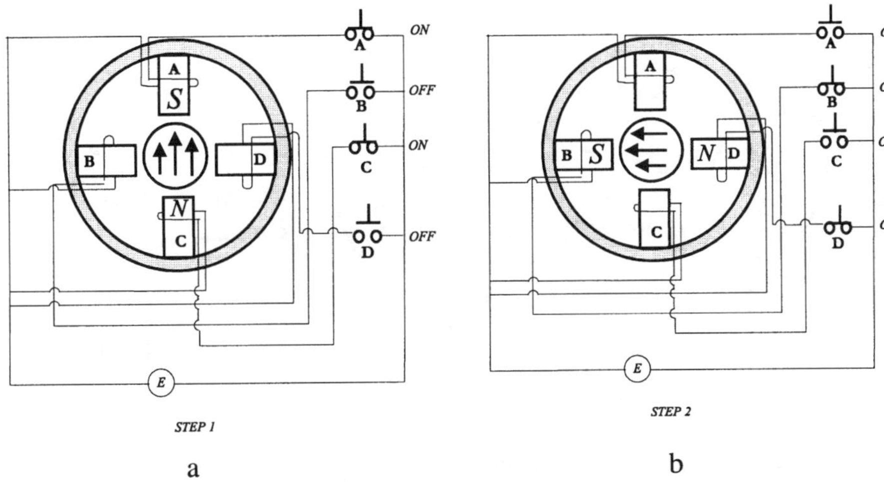

Figure 7.37 Operation of a two-pole, four-phase permanent-magnet stepper motor: (a) phases *as* and *cs* are excited; (b) phases *bs* and *ds* are excited.

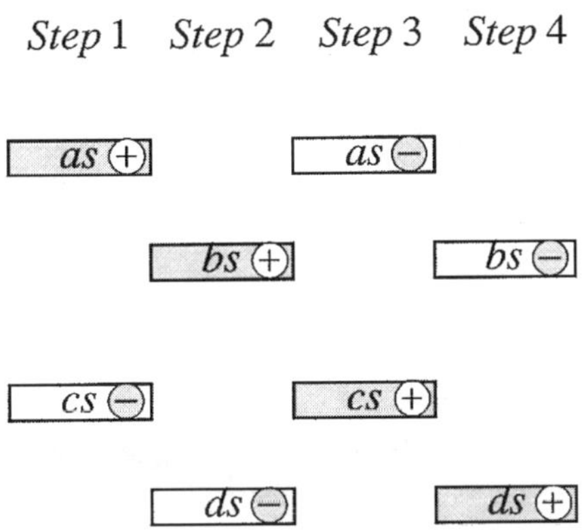

Figure 7.38 Stepping operation of permanent-magnet stepper motors: *counterclockwise* rotation.

Conventional permanent-magnet stepper motors have unifilar or bifilar winding configurations. For an unifilar winding configuration, as shown in Figure 7.39a, one has two phases and four lead wires. Hence, bidirectional phase voltages u_{as} and u_{bs} should be supplied to the *as* and *bs* windings. That is, to drive a stepper motor, positive and negative voltages must be applied to the phase windings. This increases the complexity of power converters (the so-called *servo-drivers* or power *amplifiers*); the bifilar winding configuration is a preferable choice in many applications. Here, each phase winding has a pair of windings, and four-phase stepper motors have six or eight lead wires connected to the voltage- or current-fed converters (see Figures 7.39b and 7.39c). Unipolar and bipolar *servo-drivers* are predominantly used. Permanent-magnet stepper motors with bifilar windings can be considered as four-phase devices. However, magnetically, they are two-phase devices.

Variable-reluctance stepper motors

The operating principles of variable-reluctance stepper motors are similar to those of synchronous reluctance machines, and the configuration of variable-reluctance motors is illustrated in Figure 7.40.

The stator windings are sequentially fed, and if the *as* winding is energized, the rotor aligns with the stator phase energized. Where phase *as* is turned off, and phase *bs* is

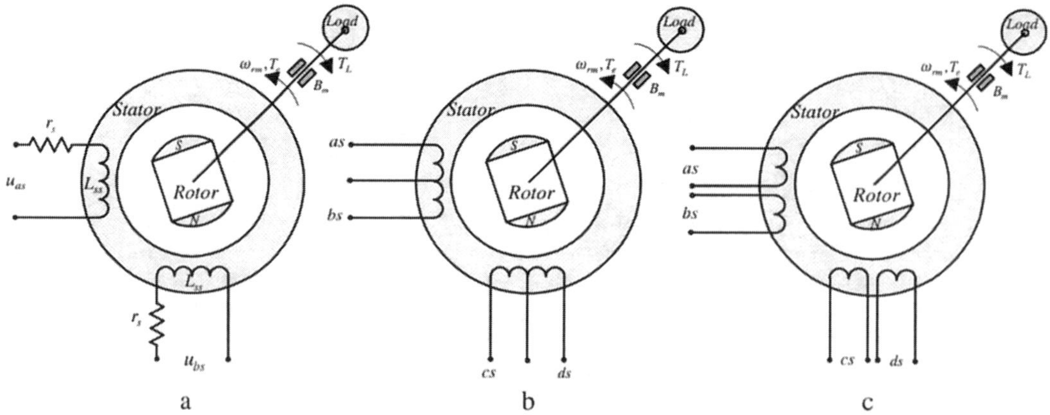

Figure 7.39 Permanent-magnet stepper motor winding configurations: (a) unifilar winding configuration; (b) six-lead bifilar winding configuration; (c) eight-lead bifilar winding configuration.

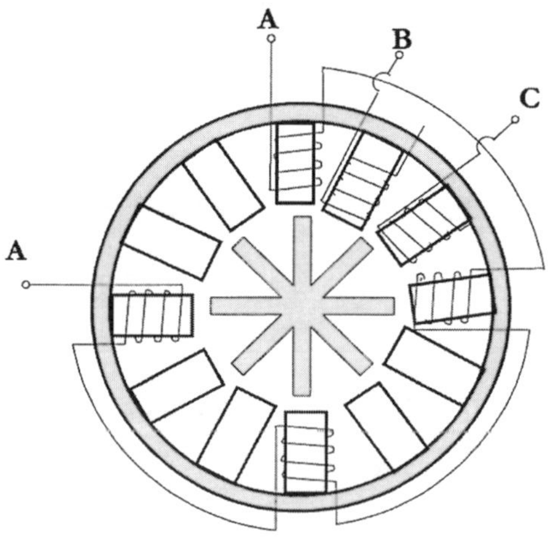

Figure 7.40 Configuration of variable-reluctance stepper motors.

energized, the rotor rotates 15° *counterclockwise* due to magnetic realignment. This rotation continues as one feeds the phase *cs*. The *clockwise* rotation occurs if the sequence is reversed.

7.5.1 *Mathematical model of permanent-magnet stepper motors in the* machine *variables*

Using Kirchhoff's voltage law for two-phase permanent-magnet stepper motors, we have

$$u_{as} = r_s i_{as} + \frac{d\psi_{as}}{dt}$$
$$u_{bs} = r_s i_{bs} + \frac{d\psi_{bs}}{dt} \tag{7.36}$$

where the flux linkages are expressed as

$$\begin{aligned}\psi_{as} &= L_{asas}i_{as} + L_{asbs}i_{bs} + \psi_{asm}\\ \psi_{bs} &= L_{bsas}i_{as} + L_{bsbs}i_{bs} + \psi_{bsm}\end{aligned} \tag{7.37}$$

For stepper motors, the flux linkages are a function of the number of the rotor teeth *RT*; using the results obtained for permanent-magnet synchronous motors, we have

$$\begin{aligned}\psi_{asm} &= \psi_m \cos(RT\theta_{rm})\\ \psi_{bsm} &= \psi_m \sin(RT\theta_{rm})\end{aligned} \tag{7.38}$$

The self-inductance of the stator windings is found to be

$$L_{ss} = L_{asas} = L_{bsbs} = L_{ls} + \bar{L}_m \tag{7.39}$$

The stator windings are displaced by 90 electrical degrees; hence, the mutual inductances between the stator windings are $L_{asbs} = L_{bsas} = 0$. Thus, using (7.37), (7.38), and (7.39), we have

$$\begin{aligned}\psi_{as} &= L_{ss}i_{as} + \psi_m \cos(RT\theta_{rm})\\ \psi_{bs} &= L_{ss}i_{bs} + \psi_m \sin(RT\theta_{rm})\end{aligned} \tag{7.40}$$

From (7.36) and (7.40), one finds

$$u_{as} = r_s i_{as} + \frac{d(L_{ss}i_{as} + \psi_m \cos(RT\theta_{rm}))}{dt} = r_s i_{as} + L_{ss}\frac{di_{as}}{dt} - RT\psi_m\omega_{rm}\sin(RT\theta_{rm})$$

$$u_{bs} = r_s i_{bs} + \frac{d(L_{ss}i_{bs} + \psi_m \sin(RT\theta_{rm}))}{dt} = r_s i_{bs} + L_{ss}\frac{di_{bs}}{dt} + RT\psi_m\omega_{rm}\cos(RT\theta_{rm})$$

That is,

$$\begin{aligned}\frac{di_{as}}{dt} &= -\frac{r_s}{L_{ss}}i_{as} + \frac{RT\psi_m}{L_{ss}}\omega_{rm}\sin(RT\theta_{rm}) + \frac{1}{L_{ss}}u_{as}\\ \frac{di_{bs}}{dt} &= -\frac{r_s}{L_{ss}}i_{bs} - \frac{RT\psi_m}{L_{ss}}\omega_{rm}\cos(RT\theta_{rm}) + \frac{1}{L_{ss}}u_{bs}\end{aligned} \tag{7.41}$$

Using Newton's second law $T_e - B_m\omega_{rm} - T_L = J\dfrac{d^2\theta_{rm}}{dt^2}$, we have

$$\frac{d\omega_{rm}}{dt} = \frac{1}{J}(T_e - B_m\omega_{rm} - T_L)$$

$$\frac{d\theta_{rm}}{dt} = \omega_{rm}$$

The expression for the electromagnetic torque developed by permanent-magnet stepper motors can be obtained using the coenergy

$$W_c = \frac{1}{2}(L_{ss}i_{as}^2 + L_{ss}i_{bs}^2) + \psi_m i_{as}\cos(RT\theta_{rm}) + \psi_m i_{bs}\sin(RT\theta_{rm}) + W_{PM}$$

Then, one obtains

$$T_e = \frac{\partial W_c}{\partial \theta_{rm}} = -RT\psi_m[i_{as}\sin(RT\theta_{rm}) - i_{bs}\cos(RT\theta_{rm})]$$

That is, the following differential equations for the rotor angular velocity ω_{rm} and displacement θ_{rm} result

$$\begin{aligned}\frac{d\omega_{rm}}{dt} &= -\frac{RT\psi_m}{J}[i_{as}\sin(RT\theta_{rm}) - i_{bs}\cos(RT\theta_{rm})] - \frac{B_m}{J}\omega_{rm} - \frac{1}{J}T_L\\ \frac{d\theta_{rm}}{dt} &= \omega_{rm}\end{aligned} \tag{7.42}$$

It should be emphasized that the electrical angular velocity and displacement can be found using the number of rotor teeth. In particular,

$$\omega_r = RT\omega_{rm} \text{ and } \theta_r = RT\theta_{rm}$$

From equations (7.41) and (7.42), one finds

$$\begin{aligned}\frac{di_{as}}{dt} &= -\frac{r_s}{L_{ss}}i_{as} + \frac{RT\psi_m}{L_{ss}}\omega_{rm}\sin(RT\theta_{rm}) + \frac{1}{L_{ss}}u_{as}\\ \frac{di_{bs}}{dt} &= -\frac{r_s}{L_{ss}}i_{bs} - \frac{RT\psi_m}{L_{ss}}\omega_{rm}\cos(RT\theta_{rm}) + \frac{1}{L_{ss}}u_{bs}\\ \frac{d\omega_{rm}}{dt} &= -\frac{RT\psi_m}{J}[i_{as}\sin(RT\theta_{rm}) - i_{bs}\cos(RT\theta_{rm})] - \frac{B_m}{J}\omega_{rm} - \frac{1}{J}T_L\\ \frac{d\theta_{rm}}{dt} &= \omega_{rm}\end{aligned} \tag{7.43}$$

From (7.43), we have

$$\begin{bmatrix} \frac{di_{as}}{dt} \\ \frac{di_{bs}}{dt} \\ \frac{d\omega_{rm}}{dt} \\ \frac{d\theta_{rm}}{dt} \end{bmatrix} = \begin{bmatrix} -\frac{r_s}{L_{ss}} & 0 & 0 & 0 \\ 0 & -\frac{r_s}{L_{ss}} & 0 & 0 \\ 0 & 0 & -\frac{B_m}{J} & 0 \\ 0 & 0 & 1 & 0 \end{bmatrix} \begin{bmatrix} i_{as} \\ i_{bs} \\ \omega_{rm} \\ \theta_{rm} \end{bmatrix} + \begin{bmatrix} \frac{RT\psi_m}{L_{ss}} \omega_{rm} \sin(RT\theta_{rm}) \\ -\frac{RT\psi_m}{L_{ss}} \omega_{rm} \cos(RT\theta_{rm}) \\ -\frac{RT\psi_m}{J} [i_{as} \sin(RT\theta_{rm}) - i_{bs} \cos(RT\theta_{rm})] \\ 0 \end{bmatrix}$$

$$+ \begin{bmatrix} \frac{1}{L_{ss}} & 0 \\ 0 & \frac{1}{L_{ss}} \\ 0 & 0 \\ 0 & 0 \end{bmatrix} \begin{bmatrix} u_{as} \\ u_{bs} \end{bmatrix} - \begin{bmatrix} 0 \\ 0 \\ \frac{1}{J} \\ 0 \end{bmatrix} T_L$$

Using (7.43), an s-domain block diagram can be developed. The resulting block diagram is illustrated in Figure 7.41.

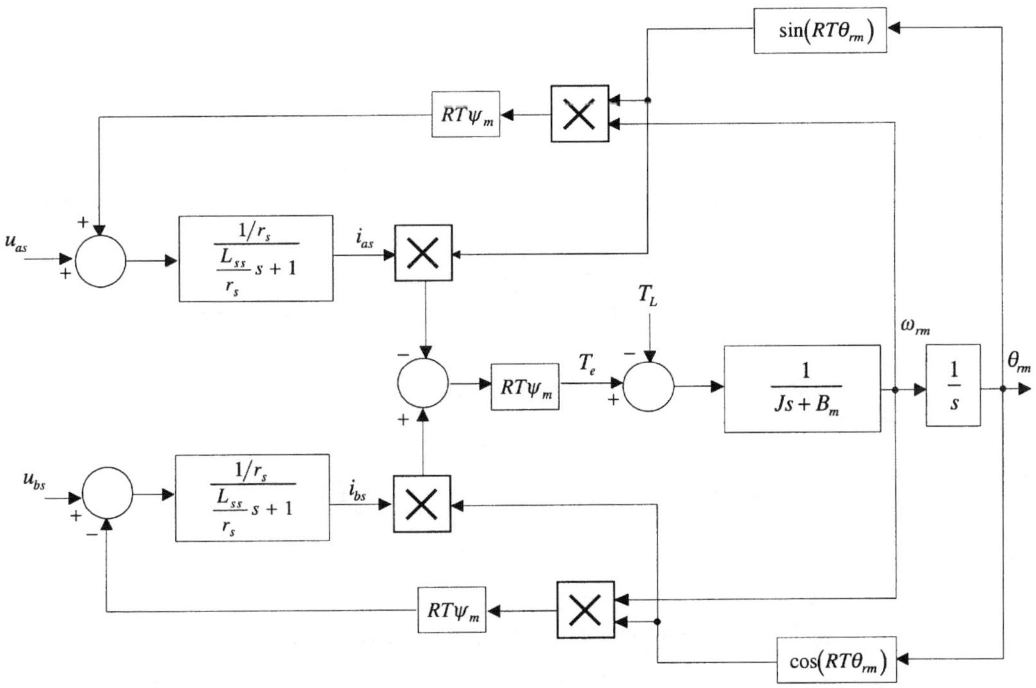

Figure 7.41 Block diagram of permanent-magnet stepper motors.

To rotate stepper motors, the pulse waveforms of the phase voltages can be applied to the stator windings, or currents can be fed to the *as* and *bs* windings. However, an analysis of the torque equation

$$T_e = -RT\psi_m[i_{as} \sin(RT\theta_{rm}) - i_{bs} \cos(RT\theta_{rm})]$$

guides one to the conclusion that the expressions for a balanced two-phase current sinusoidal set are

$$i_{as} = -\sqrt{2}i_M \sin(RT\theta_{rm})$$
$$i_{bs} = \sqrt{2}i_M \cos(RT\theta_{rm}) \tag{7.44}$$

If these phase currents are fed, the electromagnetic torque developed is given as

$$T_e = \sqrt{2}RT\psi_m i_M$$

Hence, the electromagnetic torque is maximized by feeding $i_{as} = -\sqrt{2}i_M \sin(RT\theta_{rm})$ and $i_{bs} = \sqrt{2}i_M \cos(RT\theta_{rm})$, which are obviously the functions of the rotor angular displacement. It is evident that the magnitude i_M should be changed to control permanent-magnet stepper motors.

Assuming that the inductances are negligiblly small, we have the following phase voltages needed to be supplied:

$$u_{as} = -\sqrt{2}u_M \sin(RT\theta_{rm})$$
$$u_{bs} = \sqrt{2}u_M \cos(RT\theta_{rm}) \tag{7.45}$$

One concludes that, to maximize the electromagnetic torque, the rotor angular displacement should be measured and fed to the controller. When permanent-magnet stepper motors are operated using the balanced current or voltage sets, they are similar to permanent-magnet synchronous motors. An *s*-domain block diagram of permanent-magnet stepper motors if voltages $u_{as} = -\sqrt{2}u_M \sin(RT\theta_{rm})$ and $u_{bs} = \sqrt{2}u_M \cos(RT\theta_{rm})$ are applied is shown in Figure 7.42.

As mentioned, other possible phase current and voltage waveforms are pulse trains, and the corresponding two-phase sets are given as

$$i_{as} = -i_M \operatorname{sgn}(\sin(RT\theta_{rm})),\ i_{bs} = i_M \operatorname{sgn}(\cos(RT\theta_{rm}))$$

and

$$u_{as} = -u_M \operatorname{sgn}(\sin(RT\theta_{rm})),\ u_{bs} = u_M \operatorname{sgn}(\cos(RT\theta_{rm}))$$

An *s*-domain block diagram is documented in Figure 7.43.

Example 7.8

Simulate the closed-loop servo actuated by a Pacific Scientific NEMA 23 size, two-phase 1.8° full-step 5.4 V (*rms*) and 1.4 N-m permanent-magnet stepper motor P22NSXA. Use the following motor parameters: r_s = 1.68 ohm, L_{ss} = 0.0057 H, ψ_m = 0.0064 V-s-rad^{-1}, ψ_m = 0.0064 N-m-A^{-1}, B_m = 0.000074 N-n-s-rad^{-1}, J = 0.000024 kg-m^2, and RT = 50. Study the motor performance by applying the following two-phase voltages to the stator windings

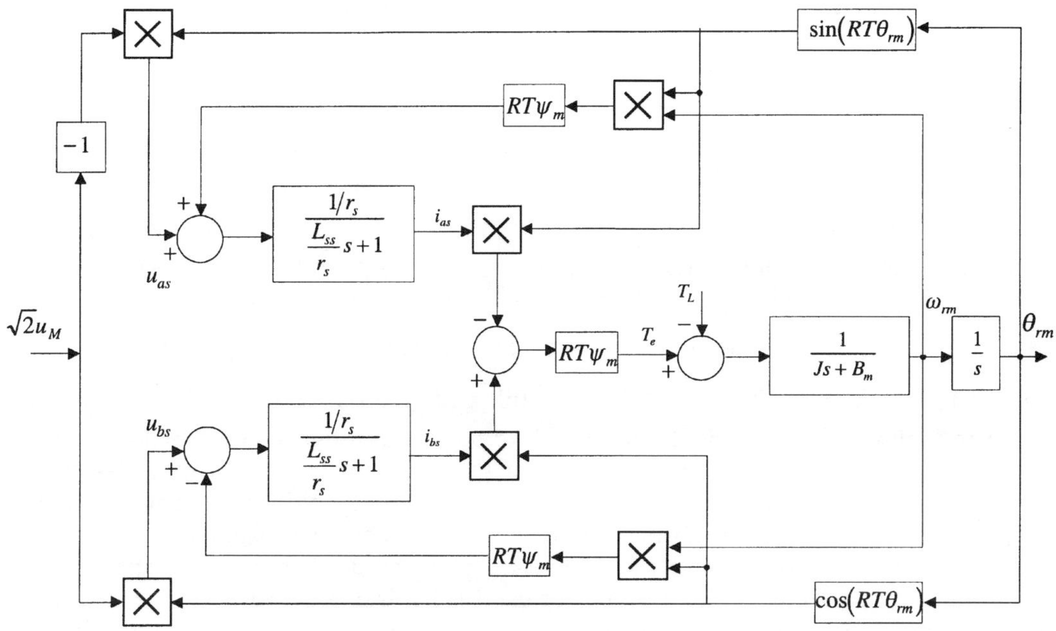

Figure 7.42 Block diagram of permanent-magnet stepper motors; $u_{as} = -\sqrt{2}u_M \sin(RT\theta_{rm})$ and $u_{bs} = \sqrt{2}u_M \cos(RT\theta_{rm})$.

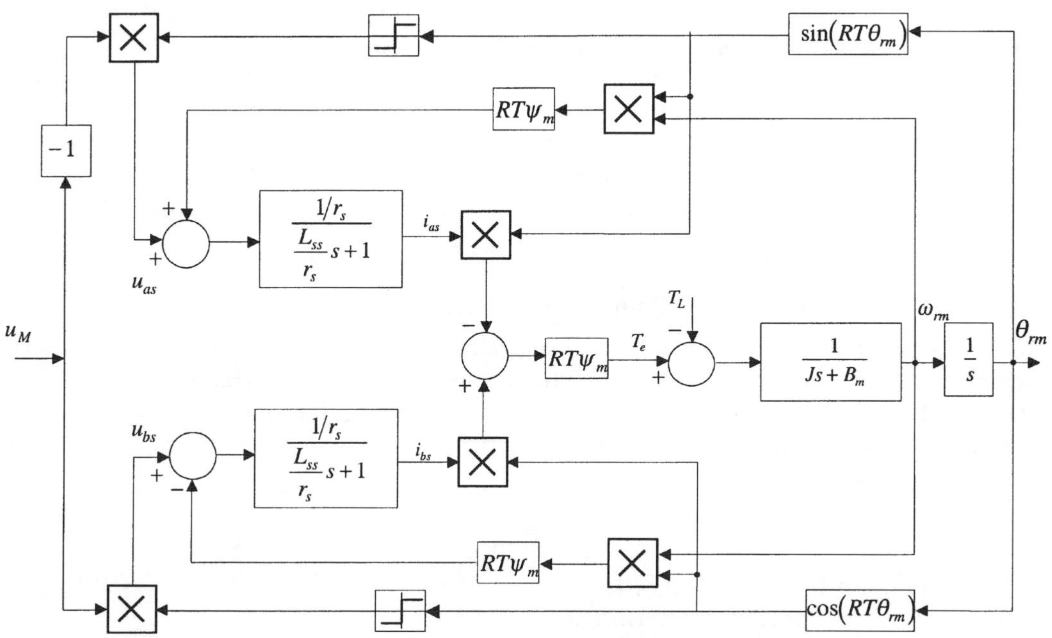

Figure 7.43 Block diagram of permanent-magnet stepper motors; $u_{as} = -u_M\ \mathrm{sgn}(\sin(RT\theta_{rm}))$ and $u_{bs} = u_M\ \mathrm{sgn}(\cos(RT\theta_{rm}))$.

$$u_{as} = -u_M \operatorname{sgn}(\sin(RT\theta_{rm})) \text{ and } u_{bs} = u_M \operatorname{sgn}(\cos(RT\theta_{rm}))$$

and

$$u_{as} = -\sqrt{2}u_M \sin(RT\theta_{rm}),\ u_{bs} = \sqrt{2}u_M \cos(RT\theta_{rm})$$

Solution

The state-space model of the studied stepper motor is mapped by equations (7.43). To find the voltages to be applied to the *as* and *bs* windings, the rotor angular displacement $\theta_{rm}(t)$ is used. Taking note of the motor parameters, the simulation diagram for $u_{as} = -\sqrt{2}u_M \sin(RT\theta_{rm})$ and $u_{bs} = \sqrt{2}u_M \cos(RT\theta_{rm})$ is built and given in Figure 7.44.

The analysis of the transient dynamics is of particular interest, and the motor behavior is illustrated in Figure 7.45. The analysis of the transient responses indicates that the settling time is 0.02 s. Due to the excellent acceleration capabilities, permanent-magnet stepper motors are used in high-performance electromechanical systems. The steady-state mechanical angular velocity of permanent-magnet stepper motors is relatively low. It should be emphasized that the number of teeth is 50. Using $\omega_r = RT\omega_{rm}$, the steady-state value of the electrical angular velocity is found to be 975 rad-s^{-1}. It should also be noted that stepper motors develop high electromagnetic torque because the number of rotor teeth is 50 or higher. Therefore, stepper motors are commonly used in direct-drive applications.

For the permanent-magnet stepper motor studied, the phase voltages applied can be expressed as $u_{as} = -u_M \operatorname{sgn}(\sin(RT\theta_{rm}))$ and $u_{bs} = u_M \operatorname{sgn}(\cos(RT\theta_{rm}))$. The corresponding SIMULINK diagram is developed and documented in Figure 7.46.

Figure 7.47 depicts the dynamics of a stepper motor if one applies the rated voltage; that is, $u_M = 5.4$ V. The applied phase voltages to the motor windings $u_{as}(t)$, $u_{bs}(t)$, which are given as $u_{as} = -5.4 \operatorname{sgn}(\sin(RT\theta_{rm}))$ and $u_{bs} = 5.4 \operatorname{sgn}(\cos(RT\theta_{rm}))$, as well as the stator currents $i_{as}(t)$, $i_{bs}(t)$, mechanical angular velocity $\omega_{rm}(t)$, and rotor displacement $\theta_{rm}(t)$, are illustrated. One concludes that the torque production is degraded compared with the results obtained as the sinusoidal voltages to the stator windings $u_{as} = -\sqrt{2}u_M \sin(RT\theta_{rm})$ and $u_{bs} = \sqrt{2}u_M \cos(RT\theta_{rm})$ were supplied. The reader can easily perform the analysis of the waveforms for the currents $i_{as}(t)$, $i_{bs}(t)$, as well as the electromagnetic torque developed (see Figures 7.45 and 7.47). Due to the chattering of the phase currents and the electromagnetic torque, motor overheating and ripple of the angular velocity result. However, from software and hardware standpoints, the pulse train algorithms can be easily implemented.

□

7.5.2 *Mathematical models of permanent-magnet stepper motors in the rotor and synchronous reference frames*

It was shown that using the *machine* variables, Kirchhoff's voltage law results in two nonlinear differential equations (see equations (7.41)):

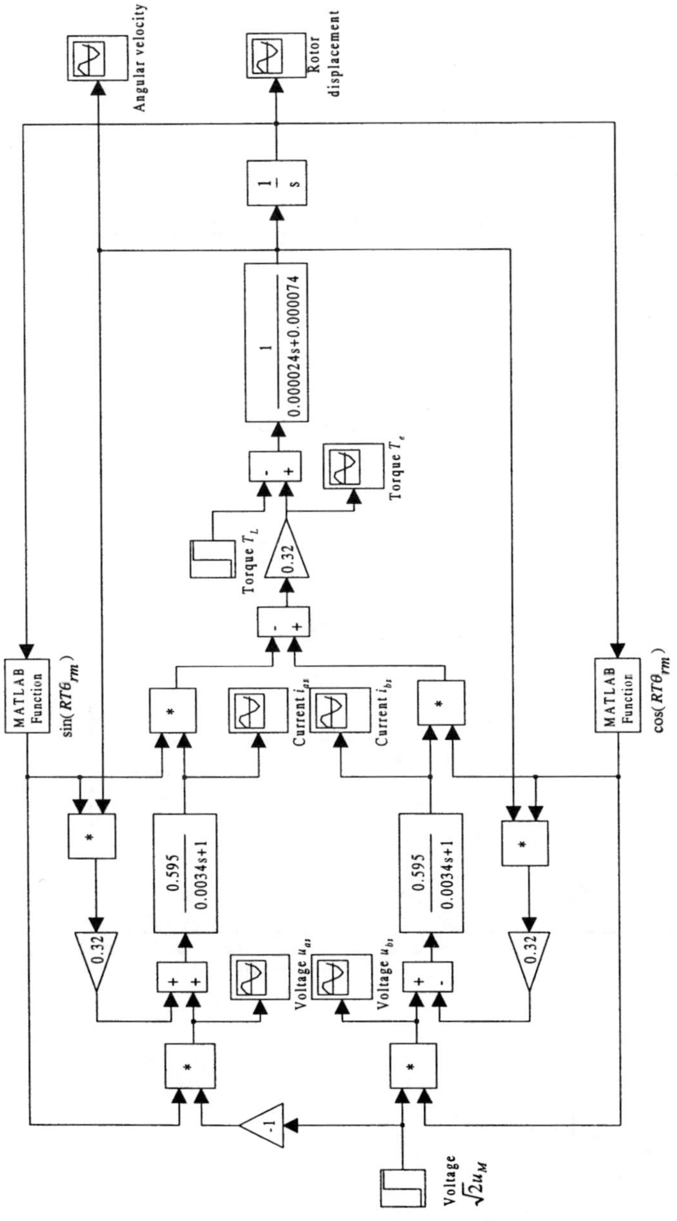

Figure 7.44 SIMULINK diagram for simulation of a stepper motor if the phase voltages are $u_{as} = -\sqrt{2}u_M \sin(RT\theta_{rm})$ and $u_{bs} = \sqrt{2}u_M \cos(RT\theta_{rm})$.

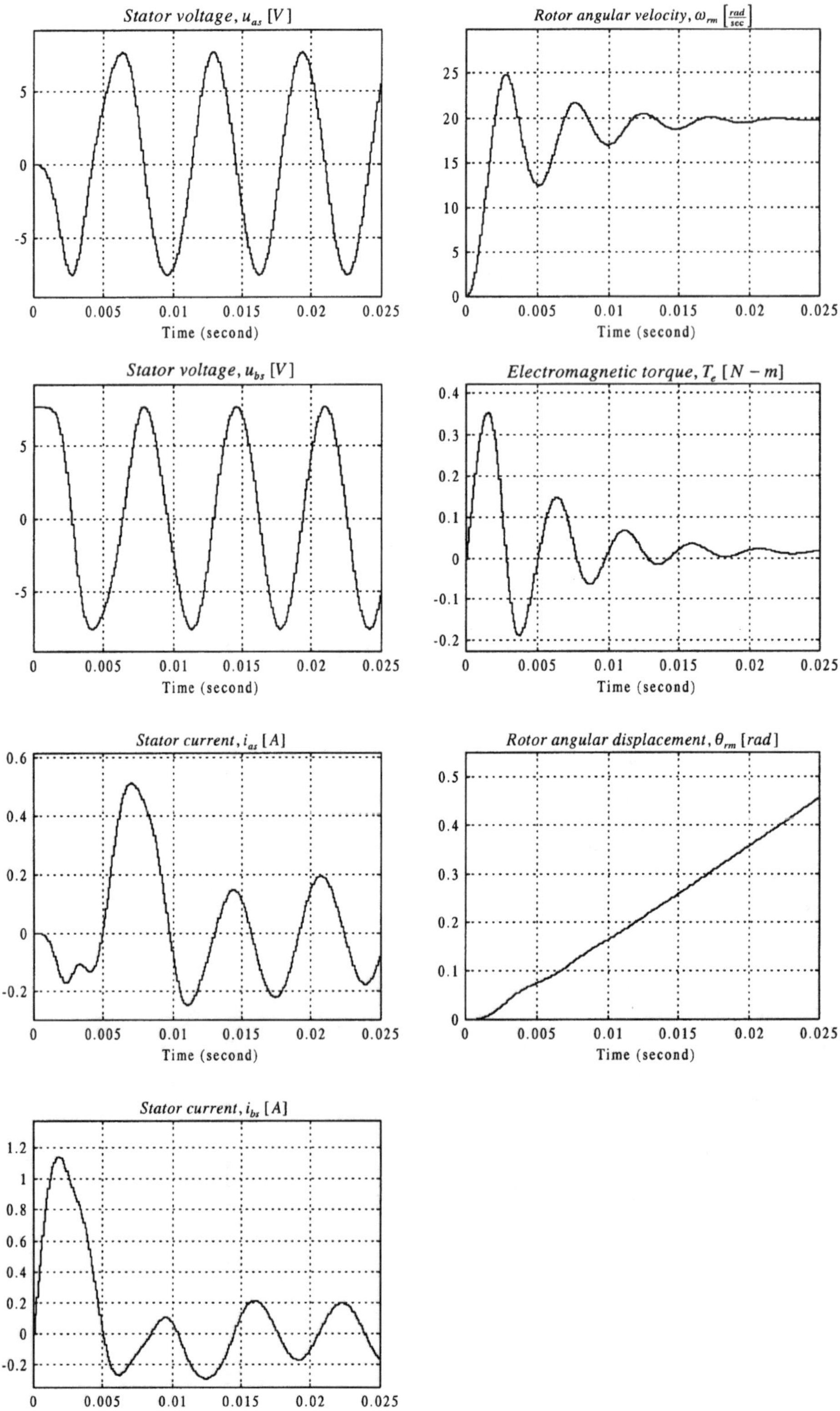

Figure 7.45 Dynamics of a permament-magnet stepper motor;

$$u_{as} = -\sqrt{2}5.4\sin(RT\theta_{rm}) \text{ and } u_{bs} = \sqrt{2}5.4\cos(RT\theta_{rm}).$$

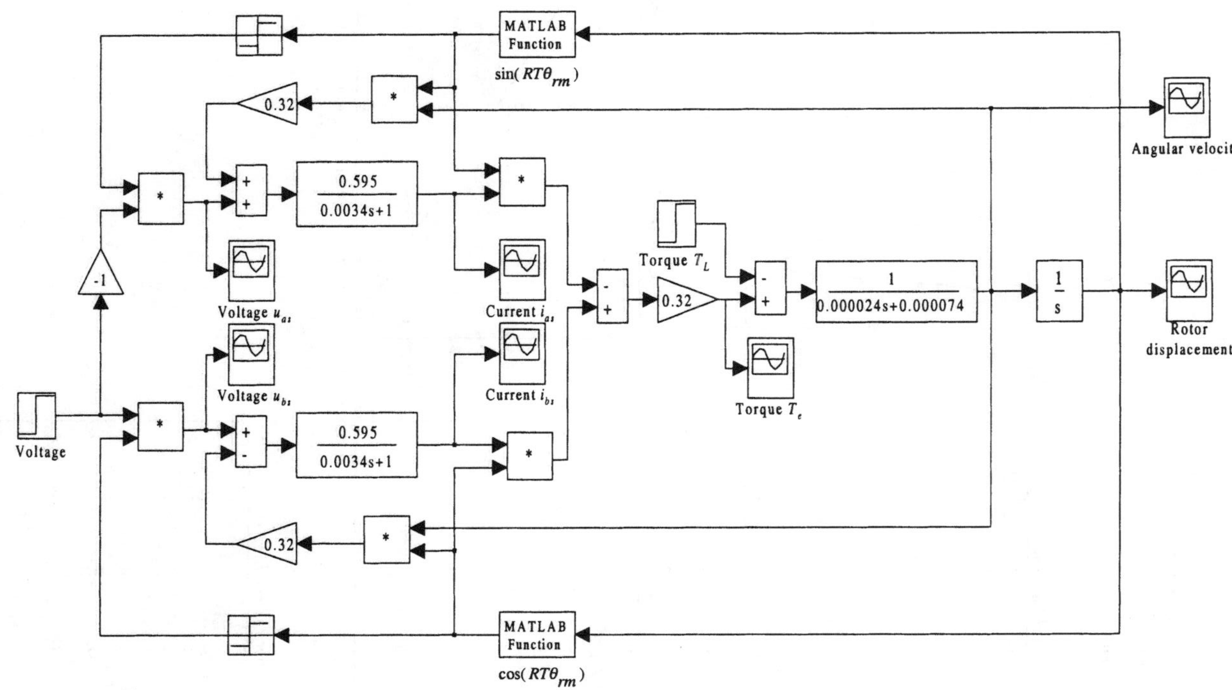

Figure 7.46 SIMULINK diagram for simulation of a stepper motor if $u_{as} = -u_M \operatorname{sgn}(\sin(RT\theta_{rm}))$ and $u_{bs} = u_M \operatorname{sgn}(\cos(RT\theta_{rm}))$.

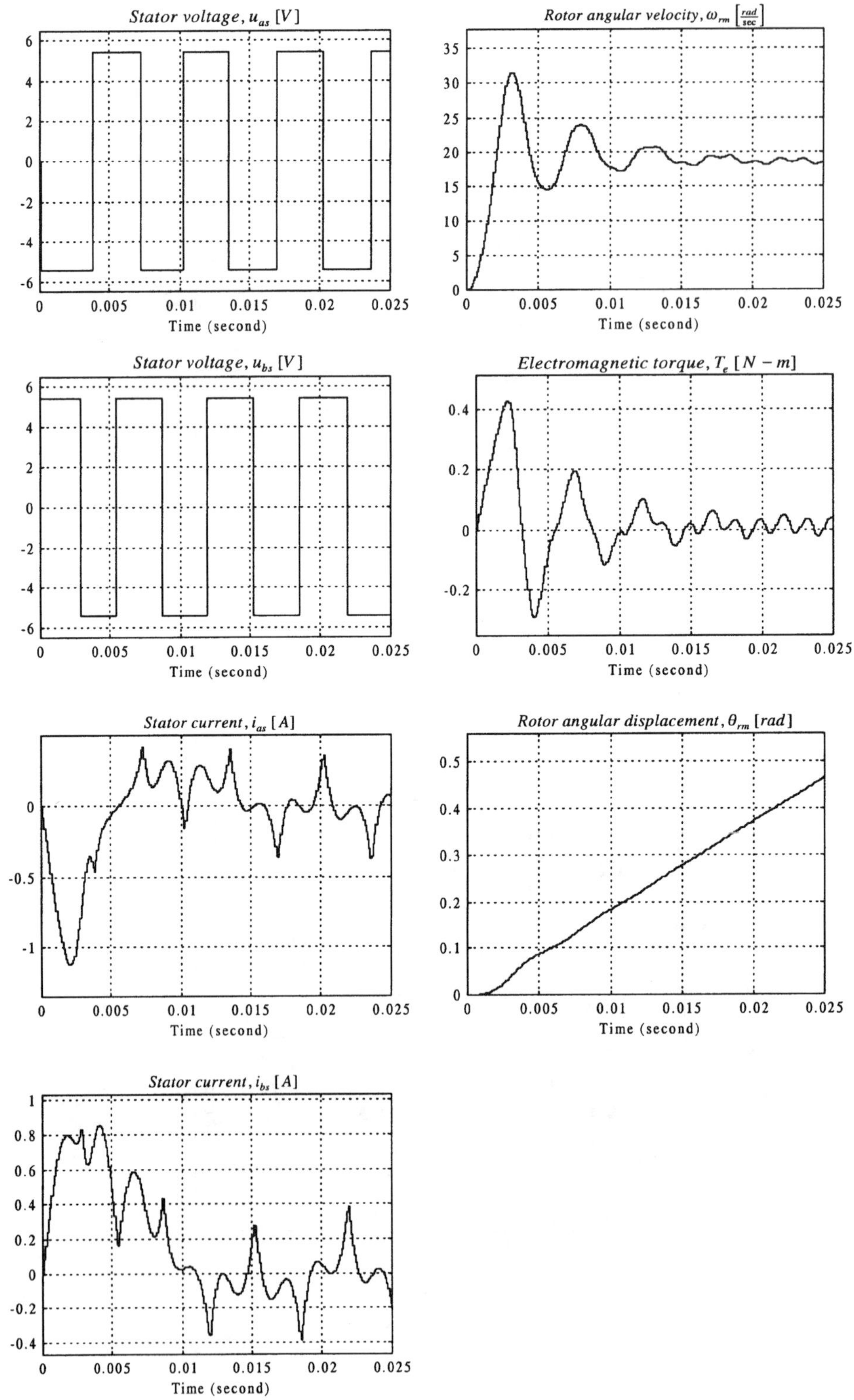

Figure 7.47 Transient dynamics of a permanent-magnet stepper motor; $u_{as} = -5.4\ \mathrm{sgn}(\sin(RT\theta_{rm}))$ and $u_{bs} = 5.4\ \mathrm{sgn}(\cos(RT\theta_{rm}))$.

$$u_{as} = r_s i_{as} + L_{ss}\frac{di_{as}}{dt} - RT\psi_m \omega_{rm}\sin(RT\theta_{rm})$$

$$u_{bs} = r_s i_{bs} + L_{ss}\frac{di_{bs}}{dt} + RT\psi_m \omega_{rm}\cos(RT\theta_{rm})$$

Applying the *direct* Park formation, which in the rotor reference frame is given as

$$\begin{bmatrix} u_{qs}^r \\ u_{ds}^r \end{bmatrix} = \begin{bmatrix} -\sin(RT\theta_{rm}) & \cos(RT\theta_{rm}) \\ \cos(RT\theta_{rm}) & \sin(RT\theta_{rm}) \end{bmatrix} \begin{bmatrix} u_{as} \\ u_{bs} \end{bmatrix}, \quad \begin{bmatrix} i_{qs}^r \\ i_{ds}^r \end{bmatrix} = \begin{bmatrix} -\sin(RT\theta_{rm}) & \cos(RT\theta_{rm}) \\ \cos(RT\theta_{rm}) & \sin(RT\theta_{rm}) \end{bmatrix} \begin{bmatrix} i_{as} \\ i_{bs} \end{bmatrix}$$

it is easy to find the corresponding differential equations in the *qd* quantities. In particular,

$$u_{qs}^r = r_s i_{qs}^r + L_{ss}\frac{di_{qs}^r}{dt} + RT\psi_m \omega_{rm} + RTL_{ss} i_{ds}^r \omega_{rm}$$

$$u_{ds}^r = r_s i_{ds}^r + L_{ss}\frac{di_{ds}^r}{dt} - RTL_{ss} i_{qs}^r \omega_{rm}$$

Hence, one obtains

$$\begin{aligned} \frac{di_{qs}^r}{dt} &= -\frac{r_s}{L_{ss}} i_{qs}^r - \frac{RT\psi_m}{L_{ss}}\omega_{rm} - RTi_{ds}^r\omega_{rm} + \frac{1}{L_{ss}}u_{qs}^r \\ \frac{di_{ds}^r}{dt} &= -\frac{r_s}{L_{ss}} i_{ds}^r + RTi_{qs}^r\omega_{rm} + \frac{1}{L_{ss}}u_{ds}^r \end{aligned} \tag{7.46}$$

The torque equation in the *machine* variables was found as

$$T_e = -RT\psi_m[i_{as}\sin(RT\theta_{rm}) - i_{bs}\cos(RT\theta_{rm})]$$

The expression for the electromagnetic torque in the rotor reference frame is obtained by expressing i_{as} and i_{bs} in terms of i_{qs}^r and i_{ds}^r. Using the *inverse* Park transformation, one obtains

$$\begin{bmatrix} i_{as} \\ i_{bs} \end{bmatrix} = \begin{bmatrix} -\sin(RT\theta_{rm}) & \cos(RT\theta_{rm}) \\ \cos(RT\theta_{rm}) & \sin(RT\theta_{rm}) \end{bmatrix} \begin{bmatrix} i_{qs}^r \\ i_{ds}^r \end{bmatrix}$$

which gives

$$i_{as} = -i_{qs}^r\sin(RT\theta_{rm}) + i_{ds}^r\cos(RT\theta_{rm}) \text{ and } i_{bs} = i_{qs}^r\cos(RT\theta_{rm}) + i_{ds}^r\sin(RT\theta_{rm})$$

and we have

$$T_e = RT\psi_m i_{qs}^r$$

From Newton's second law of motion $T_e - B_m\omega_{rm} - T_L = J\dfrac{d^2\theta_{rm}}{dt^2}$ one obtains

$$\frac{d\omega_{rm}}{dt} = \frac{RT\psi_m}{J} i_{qs}^r - \frac{B_m}{J}\omega_{rm} - \frac{1}{J}T_L$$
$$\frac{d\theta_{rm}}{dt} = \omega_{rm} \tag{7.47}$$

Augmenting differential equations (7.46) and (7.47), the following mathematical model of permanent-magnet synchronous motors in the rotor reference frame results

$$\frac{di_{qs}^r}{dt} = -\frac{r_s}{L_{ss}} i_{qs}^r - \frac{RT\psi_m}{L_{ss}}\omega_{rm} - RTi_{ds}^r\omega_{rm} + \frac{1}{L_{ss}}u_{qs}^r$$
$$\frac{di_{ds}^r}{dt} = -\frac{r_s}{L_{ss}} i_{ds}^r + RTi_{qs}^r\omega_{rm} + \frac{1}{L_{ss}}u_{ds}^r$$
$$\frac{d\omega_{rm}}{dt} = \frac{RT\psi_m}{J} i_{qs}^r - \frac{B_m}{J}\omega_{rm} - \frac{1}{J}T_L \tag{7.48}$$
$$\frac{d\theta_{rm}}{dt} = \omega_{rm}$$

In matrix form, we have

$$\begin{bmatrix} \frac{di_{qs}^r}{dt} \\ \frac{di_{ds}^r}{dt} \\ \frac{d\omega_{rm}}{dt} \\ \frac{d\theta_{rm}}{dt} \end{bmatrix} = \begin{bmatrix} -\frac{r_s}{L_{ss}} & 0 & -\frac{RT\psi_m}{L_{ss}} & 0 \\ 0 & -\frac{r_s}{L_{ss}} & 0 & 0 \\ \frac{RT\psi_m}{J} & 0 & -\frac{B_m}{J} & 0 \\ 0 & 0 & 1 & 0 \end{bmatrix} \begin{bmatrix} i_{qs}^r \\ i_{ds}^r \\ \omega_{rm} \\ \theta_{rm} \end{bmatrix} + \begin{bmatrix} -RTi_{ds}^r\omega_{rm} \\ RTi_{qs}^r\omega_{rm} \\ 0 \\ 0 \end{bmatrix} + \begin{bmatrix} \frac{1}{L_{ss}} & 0 \\ 0 & \frac{1}{L_{ss}} \\ 0 & 0 \\ 0 & 0 \end{bmatrix} \begin{bmatrix} u_{qs}^r \\ u_{ds}^r \end{bmatrix} - \begin{bmatrix} 0 \\ 0 \\ \frac{1}{J} \\ 0 \end{bmatrix} T_L$$

It is important to find the *quadrature* and *direct* currents and voltages needed to guarantee balanced operating conditions. The *as* and *bs* currents and voltages have been derived. From equations (7.44) and (7.45), we have the following expressions for the currents and voltages to maximize the electromagnetic torque

$$i_{as} = -\sqrt{2}i_M \sin(RT\theta_{rm}),\; i_{bs} = \sqrt{2}i_M \cos(RT\theta_{rm})$$

and

$$u_{as} = -\sqrt{2}u_M \sin(RT\theta_{rm}),\; u_{bs} = \sqrt{2}u_M \cos(RT\theta_{rm})$$

Using

$$\begin{bmatrix} i_{qs}^r \\ i_{ds}^r \end{bmatrix} = \begin{bmatrix} -\sin(RT\theta_{rm}) & \cos(RT\theta_{rm}) \\ \cos(RT\theta_{rm}) & \sin(RT\theta_{rm}) \end{bmatrix} \begin{bmatrix} i_{as} \\ i_{bs} \end{bmatrix}$$

we have

$$i_{qs}^r = -i_{as}\sin(RT\theta_{rm}) + i_{bs}\cos(RT\theta_{rm}),\ i_{ds}^r = i_{as}\cos(RT\theta_{rm}) + i_{bs}\sin(RT\theta_{rm})$$

Substituting $i_{as} = -\sqrt{2}i_M \sin(RT\theta_{rm})$ and $i_{bs} = \sqrt{2}i_M \cos(RT\theta_{rm})$ in the equations for the *quadrature* and *direct* currents, we have

$$i_{qs}^r = \sqrt{2}i_M \sin^2(RT\theta_{rm}) + \sqrt{2}i_M \cos^2(RT\theta_{rm}) = \sqrt{2}i_M$$

and

$$i_{ds}^r = -\sqrt{2}i_M \sin(RT\theta_{rm})\cos(RT\theta_{rm}) + \sqrt{2}i_M \sin(RT\theta_{rm})\cos(RT\theta_{rm}) = 0$$

That is, the *quadrature* and *direct* currents are found to be

$$i_{qs}^r = \sqrt{2}i_M \text{ and } i_{ds}^r = 0$$

For the *quadrature* and *direct* voltages, which are related to the *as* and *bs* voltages as

$$\begin{bmatrix} u_{qs}^r \\ u_{ds}^r \end{bmatrix} = \begin{bmatrix} -\sin(RT\theta_{rm}) & \cos(RT\theta_{rm}) \\ \cos(RT\theta_{rm}) & \sin(RT\theta_{rm}) \end{bmatrix} \begin{bmatrix} u_{as} \\ u_{bs} \end{bmatrix}$$

or

$$u_{qs}^r = -u_{as}\sin(RT\theta_{rm}) + u_{bs}\cos(RT\theta_{rm})$$

$$u_{ds}^r = u_{as}\cos(RT\theta_{rm}) + u_{bs}\sin(RT\theta_{rm})$$

we have

$$u_{qs}^r = \sqrt{2}u_M \text{ and } u_{ds}^r = 0$$

It has been shown that if the inductances cannot be neglected, the advanced shifting should be used. One finds the *quadrature* and *direct* voltages as

$$u_{qs}^r = \sqrt{2}u_M \cos\varphi_u \text{ and } u_{ds}^r = -\sqrt{2}u_M \sin\varphi_u$$

Using the resulting differential equations (7.48), the block diagram of permanent-magnet stepper motors modeled in the rotor reference frame is developed and illustrated in Figure 7.48.

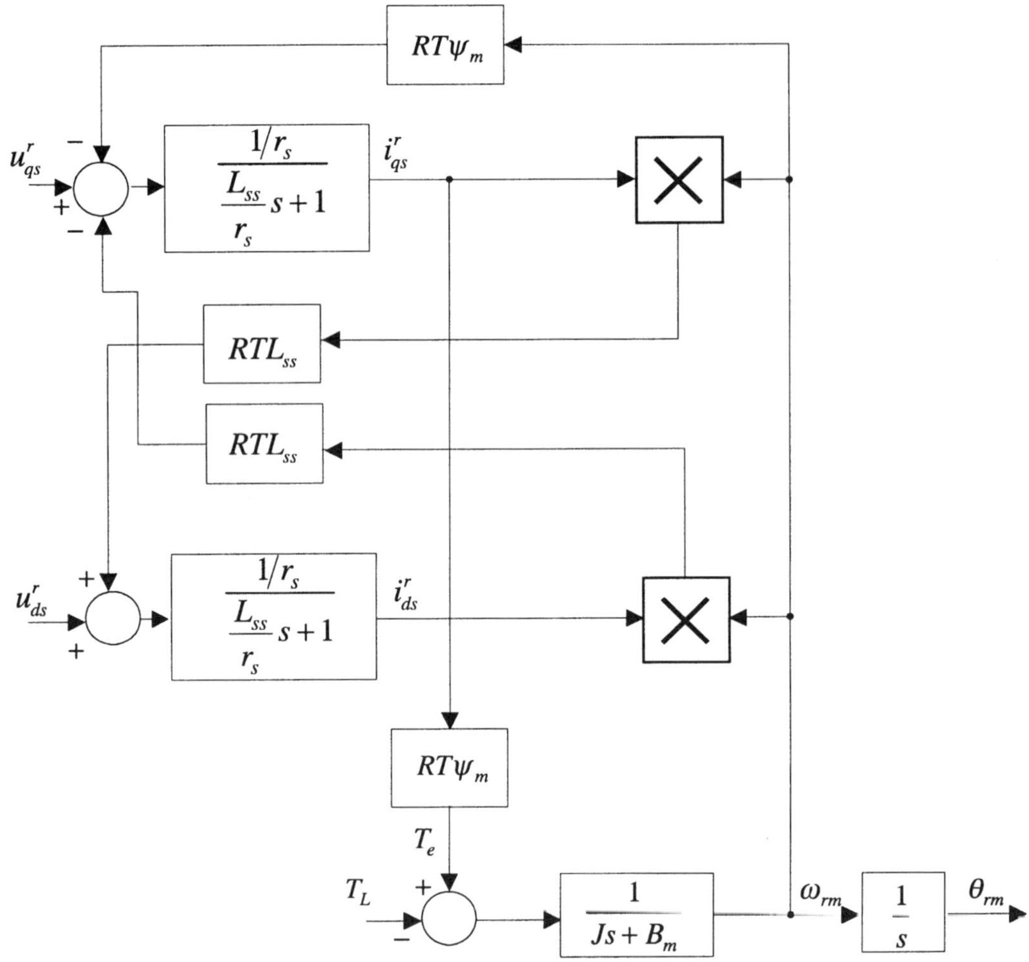

Figure 7.48 Block diagram of permanent-magnet stepper motors modeled in the rotor reference frame.

Using the synchronous reference frame, the resulting system of differential equations is found, and we have

$$\frac{di_{qs}^e}{dt} = -\frac{r_s}{L_{ss}} i_{qs}^e - \frac{RT\psi_m}{L_{ss}} \omega_{rm} - RTi_{ds}^e \omega_{rm} + \frac{1}{L_{ss}} u_{qs}^e$$

$$\frac{di_{ds}^e}{dt} = -\frac{r_s}{L_{ss}} i_{ds}^e + RTi_{qs}^e \omega_{rm} + \frac{1}{L_{ss}} u_{ds}^e$$

$$\frac{d\omega_{rm}}{dt} = \frac{RT\psi_m}{J} i_{qs}^e - \frac{B_m}{J} \omega_{rm} - \frac{1}{J} T_L$$

$$\frac{d\theta_{rm}}{dt} = \omega_{rm}$$

Hence, in matrix form one has

$$\begin{bmatrix} \frac{di_{qs}^e}{dt} \\ \frac{di_{ds}^e}{dt} \\ \frac{d\omega_{rm}}{dt} \\ \frac{d\theta_{rm}}{dt} \end{bmatrix} = \begin{bmatrix} -\frac{r_s}{L_{ss}} & 0 & -\frac{RT\psi_m}{L_{ss}} & 0 \\ 0 & -\frac{r_s}{L_{ss}} & 0 & 0 \\ \frac{RT\psi_m}{L_{ss}} & 0 & -\frac{B_m}{J} & 0 \\ 0 & 0 & 1 & 0 \end{bmatrix} \begin{bmatrix} i_{qs}^e \\ i_{ds}^e \\ \omega_{rm} \\ \theta_{rm} \end{bmatrix} + \begin{bmatrix} -RTi_{ds}^e\omega_{rm} \\ RTi_{qs}^e\omega_{rm} \\ 0 \\ 0 \end{bmatrix} + \begin{bmatrix} \frac{1}{L_{ss}} & 0 \\ 0 & \frac{1}{L_{ss}} \\ 0 & 0 \\ 0 & 0 \end{bmatrix} \begin{bmatrix} u_{qs}^e \\ u_{ds}^e \end{bmatrix} - \begin{bmatrix} 0 \\ 0 \\ \frac{1}{J} \\ 0 \end{bmatrix} T_L$$

It is evident that the models of permanent-magnet stepper motors in the rotor and synchronous reference frames are identical because $\omega_r = \omega_e$. In conclusion, it should be emphasized that the models derived cannot be linearized, and there do not exist any difficulties in performing modeling and analysis of permanent-magnet stepper motors.

Example 7.9

Develop the SIMULINK diagram and perform modeling of a Pacific Scientific permanent-magnet stepper motor P22NSXA in the rotor reference frame. The stepper motor parameters are given in Example 7.8.

Solution

The dynamics of permanent-magnet stepper motors in the rotor reference frame are mapped by a set of nonlinear differential equations (7.48), and the *quadrature* and *direct* voltages, which guarantee balanced operation, are $u_{qs}^r = \sqrt{2}u_M \cos\varphi_u$ and $u_{ds}^r = -\sqrt{2}u_M \sin\varphi_u$. Making use of an *s*-domain block diagram, as illustrated in Figure 7.48, the SIMULINK diagram is built (see Figure 7.49).

For the studied motor, the self-inductances are negligibly small, and $\varphi_u \approx 0$. Therefore, we supply $u_{qs}^r = \sqrt{2}u_M$ and $u_{ds}^r = 0$.

The transient dynamics have been studied by assigning different load conditions. In particular, the load torques were assigned as 0 and 0.05 N-m. Figures 7.50 and 7.51 illustrate the motor dynamics if the motor starts from stall and $u_M = 5.4$ V.

One might conclude that there is no need to use the angular displacement to find the *quadrature* and *direct* voltages. However, the designer must understand that to control the stepper motor, one applies the phase voltages u_{as} and u_{bs} — not the dc voltages u_{qs}^r and u_{ds}^r. That is, using state-of-the-art software and hardware, the *inverse* Park transformation

$$\begin{bmatrix} u_{as} \\ u_{bs} \end{bmatrix} = \begin{bmatrix} -\sin(RT\theta_{rm}) & \cos(RT\theta_{rm}) \\ \cos(RT\theta_{rm}) & \sin(RT\theta_{rm}) \end{bmatrix} \begin{bmatrix} u_{qs}^r \\ u_{ds}^r \end{bmatrix}$$

must be implemented in real-time in order to calculate u_{as} and u_{bs} that need to be applied. Therefore, there is a critical need to measure the rotor angular displacement θ_{rm} in order to calculate the *as* and *bs* voltages that need to be supplied to the phase windings.

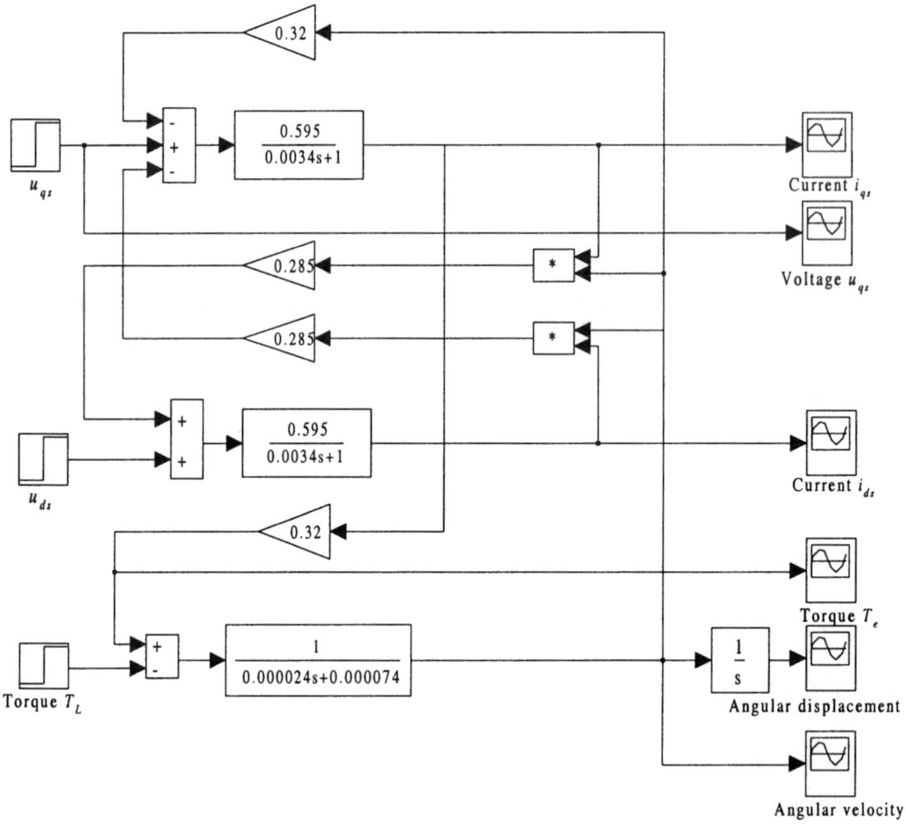

Figure 7.49 SIMULINK diagram for simulation of a stepper motor in the rotor reference frame; $u^r_{qs} = \sqrt{2}u_M \cos\varphi_u$ and $u^r_{ds} = -\sqrt{2}u_M \sin\varphi_u$.

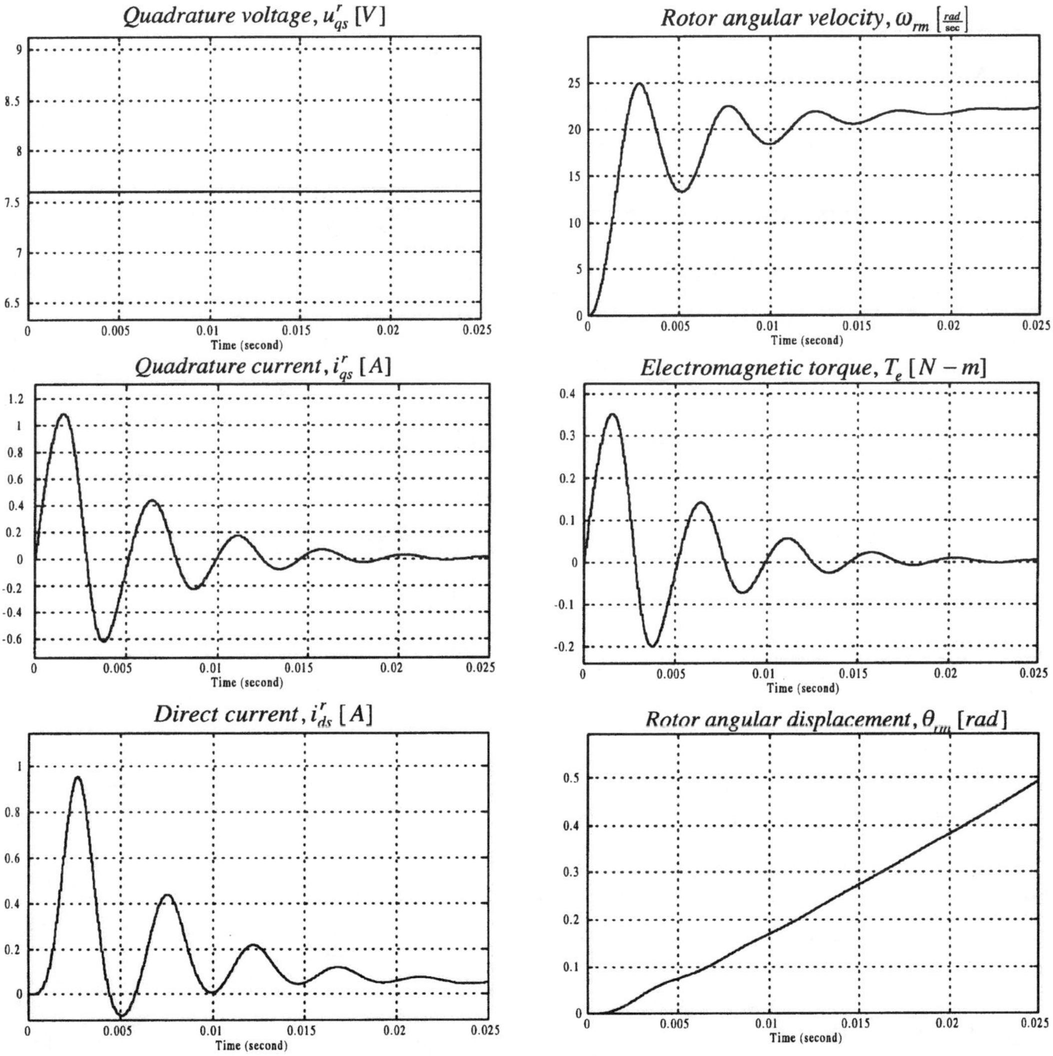

Figure 7.50 Transient dynamics of a permanent-magnet stepper motor; $T_L = 0$ N-m, $u_{qs}^r = \sqrt{2}5.4$ V, and $u_{ds}^r = 0$.

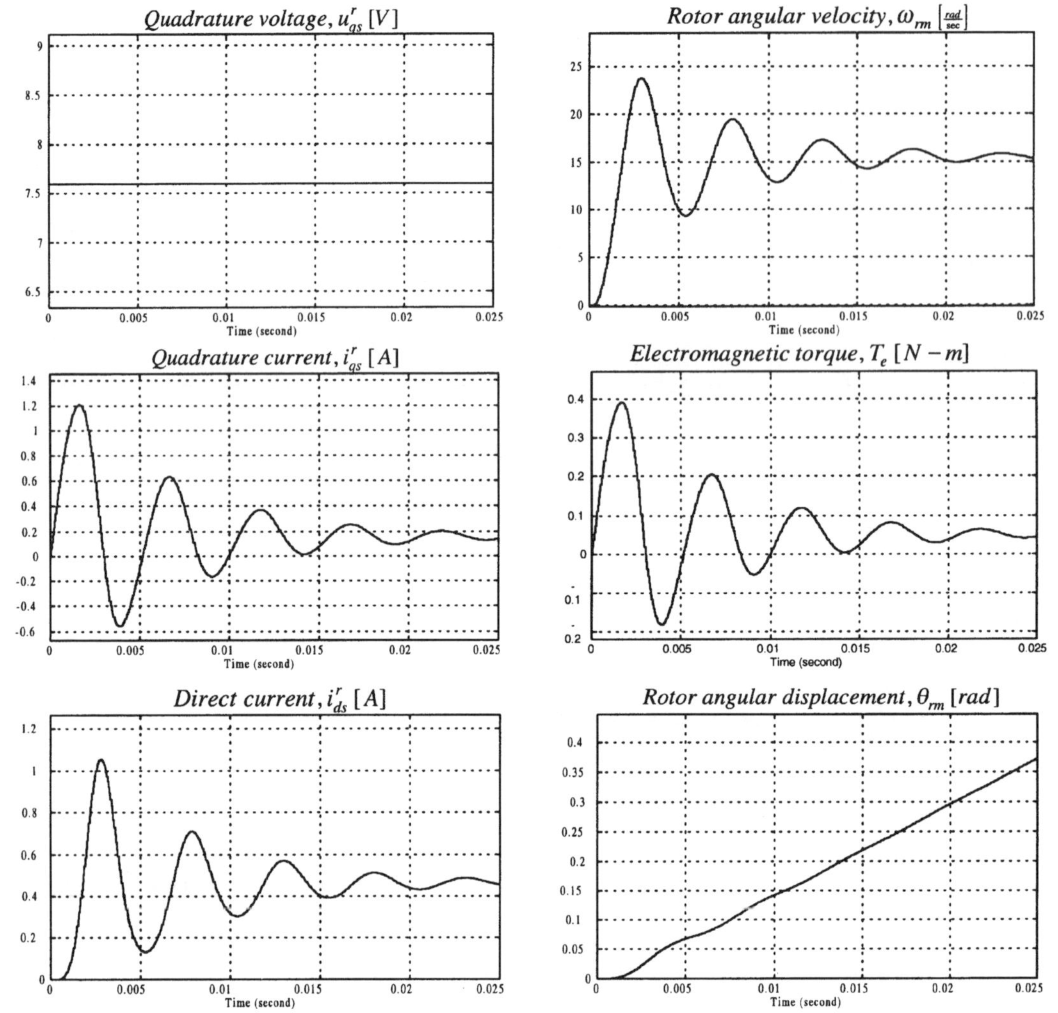

Figure 7.51 Transient dynamics of a permanent-magnet stepper motor; T_L = 0.05 N-m, $u_{qs}^r = \sqrt{2}5.4$ V, and $u_{ds}^r = 0$.

References

Krause, P. C. and Wasynczuk, O., *Electromechanical Motion Devices*, McGraw-Hill, New York, 1989.

chapter eight

Mechatronic systems

8.1 Introduction

The current trends in the development and deployment of advanced electromechanical systems have facilitated unified activities in the analysis and design of state-of-the-art motion devices, electric motors, power electronics, and digital controllers. In this chapter, a variety of problems in motion control are thoroughly studied and solved for mechatronic systems, including electric motors, power electronics, and motion controllers. Using high-performance electric machines and switching converters, this chapter demonstrates how to control electromechanical systems to ensure a spectrum of performance objectives imposed on advanced mechatronic systems. Electric machines are operated and controlled in the rated torque-speed envelope with maximum angular velocity and electromagnetic torque specified; the maximum allowable magnitude of the applied voltages or currents fed to the motor windings are limited; the currents in power converters are bounded; there exist lower and upper limits of the duty ratio of converters; and many other phenomena contribute to the complexity of problems to be solved in mechatronic systems. To approach design trade-offs and analyze performance (accuracy, settling time, overshoot, stability margins, and other quantities), the imposed constraints, model nonlinearities, and parameter variations must be studied and attenuated. The goal is to ensure the specified characteristics of mechatronic systems, avoiding deficiencies associated with a one-dimensional problem consideration based on pure machinery, electronics, or controller formulations. To approach and solve motion control problems, mechatronic philosophy is used to ensure performance improvements and operating envelope extension, stability and robust tracking, specified accuracy, and disturbance rejection. To illustrate the reported concept, high-performance electromechanical systems are thoroughly covered in this chapter using state-of-the-art control architectures to regulate electric machines and power electronics. Analytical and experimental results, advanced power electronics and microcontrollers to interface and control electric motors are presented, and catalog data, given in Appendices, should be very useful in engineering practice. Comprehensive analyses with experimental verification of control algorithms are reported.

The problem of analyzing, designing, and controlling mechatronic systems is a very important one in many industrial applications. Nonlinear analysis, optimization, and control are central issues to be approached and solved, and advanced hardware and software must be deployed to improve the steady-state and dynamic characteristics of electric drives and servos used in a wide range of industrial devices. The intensive research in analysis and motion control, and extensive studies in interfacing and augmentation of electromechanical systems, have been performed. However, the assumptions in analysis and design, along with inadequate hardware configurations, limit operating envelopes and mechatronic systems capabilities. The practical use and applicability of high-performance electric machines, power converters, and microcontrollers are documented in this chapter.

8.2 Mechatronic systems with permanent-magnet direct-current motors

It is shown that augmented nonlinear motor–converter dynamics cannot be linearized, and hard bounds imposed cannot be neglected. Specific requirements are assigned, and the *absolute* limit of motor and converter performances can be placed in the scope of practical design. The general line of attack is associated with nonlinear analysis and synthesis. Recognizing the potential difficulties associated with ignoring nonlinearities and constraints, the studies must focus on nonlinear analysis, constrained control, and advanced hardware. A number of important issues in the design of PID-type control algorithms were addressed and solved in Chapter 3. Despite the manageable results in control of permanent-magnet DC motors, covered in Sections 3.2 and 3.4, fully integrated mechatronic systems must be researched to guarantee the validity and applicability of analytical and numerical results that must be implemented. In particular, power converters should be incorporated in analysis and design. The trade-off between model accuracy and model simplicity is a central problem. It is recognized that the steady-state and dynamic characteristics, stability margins, and performance of mechatronic systems depend on the validity of the models used. Nonlinear models of electromechanical systems result if one integrates electric machines with the corresponding power converters (in Chapters 4 through 7, differential equations are found using Kirchhoff and Newton laws, as well as by applying the Lagrange equations). New features in nonlinear analysis and motion control of mechatronic systems arise, and difficulties associated with the classical formulation, which is based on transfer function concept or linear state-space theory, limit the validity and usefulness of analytical results as well as the implementation capabilities. We study mechatronic systems; and it was shown that only permanent-magnet DC machines are mapped by linear differential equations. However, transient dynamics of power converters (needed to be used to control electric motors) are nonlinear. Therefore, nonlinear models with a minimum level of simplifications and assumptions must be used. The bounded control law, introduced in Chapter 3, as given by

$$u(t) = \mathrm{sat}_{u_{\min}}^{u_{\max}}\left(\sum_{j=0}^{\varsigma} k_{pj} e^{\frac{2j+1}{2\beta+1}} + \sum_{j=0}^{\sigma} k_{ij} \frac{1}{s} e^{\frac{2j+1}{2\mu+1}} + \sum_{j=0}^{\eta} k_{dj} s e^{\frac{2j+1}{2\gamma+1}}\right), \quad u_{\min} \le u \le u_{\max} \tag{8.1}$$

$$e(t) = \omega_{reference}(t) - \omega_{rm}(t), \quad e(t) = \theta_{reference}(t) - \theta_{rm}(t)$$

is the most general form of PID-type controller to be designed and deployed. The application of this controller leads one to the solution of tracking, disturbance rejection, and stabilization problems. To solve the motion control problem for mechatronic systems, this chapter section studies a practical, experimentally verified nonlinear PID-type controller.

The mathematical model of permanent-magnet DC motors was developed in Section 5.7. In particular, the transient dynamics of electric drives with permanent-magnet DC motors (see Figure 8.1), are described by the following linear differential equations

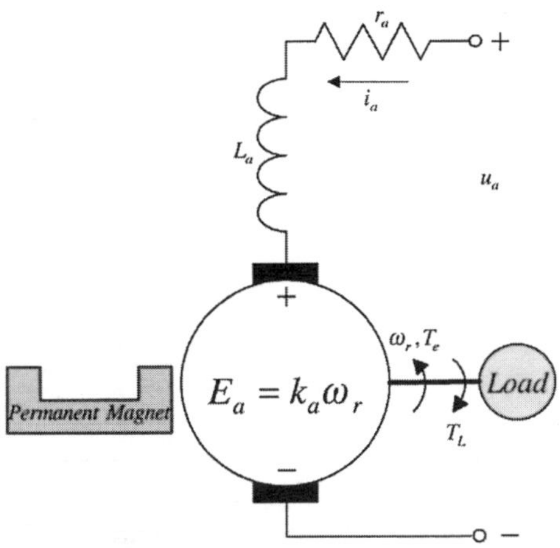

Figure 8.1 Permanent-magnet DC motor.

$$\frac{di_a}{dt} = -\frac{r_a}{L_a} i_a - \frac{k_a}{L_a} \omega_r + \frac{1}{L_a} u_a$$

$$\frac{d\omega_r}{dt} = \frac{k_a}{J} i_a - \frac{B_m}{J} \omega_r - \frac{1}{J} T_L$$

For servo-systems, the rotor angular displacement must be considered, and one has

$$\frac{di_a}{dt} = -\frac{r_a}{L_a} i_a - \frac{k_a}{L_a} \omega_r + \frac{1}{L_a} u_a$$

$$\frac{d\omega_r}{dt} = \frac{k_a}{J} i_a - \frac{B_m}{J} \omega_r - \frac{1}{J} T_L \qquad (8.2)$$

$$\frac{d\theta_r}{dt} = \omega_r$$

Permanent-magnet DC motors, which are actuators in mechatronic systems, need to be controlled using power amplifiers. Hence, the armature voltage u_a cannot be considered as a control, as was assumed in Sections 3.2 and 3.4. In fact, u_a is the converter output, and u_a is a state or is related to the state variables. For example, using the results of Section 4.3.1, for the *buck* converter, we have $u_a = u_C + r_c i_L - r_c i_a$. Hence, power converters must be integrated into analysis and design. Control of the motor angular velocity ω_r or the rotor angular displacement θ_r is accomplished by changing the armature voltage applied to the armature winding using PWM dc-dc high-frequency switching converters.

Permanent-magnet direct-current motors controlled by step-down converters

Figure 8.2 shows the permanent-magnet motor with a high-frequency *step-down* switching converter.

The voltage applied to the motor winding u_a is regulated by controlling the switching *on* and *off* durations, t_{on} and t_{off}. That is, one changes the duty ratio d_D, which is given as

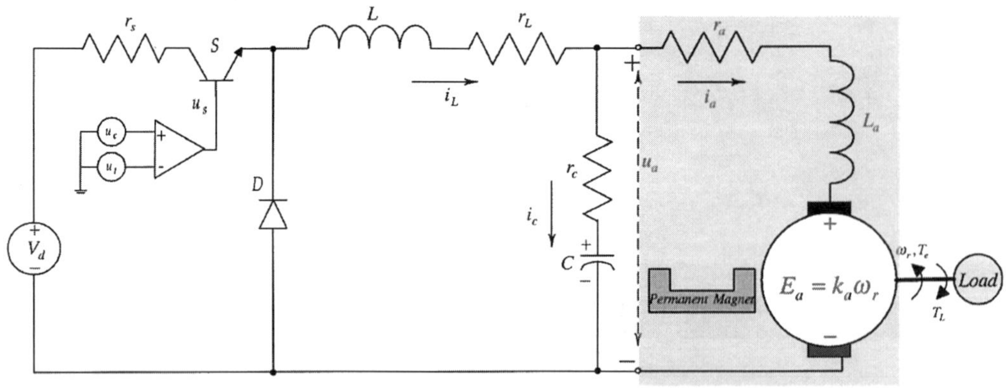

Figure 8.2 Permanent-magnet DC motor with *step-down* switching converter.

$d_D = \dfrac{t_{on}}{t_{on} + t_{off}}$. From equations (8.2), making use of the differential equations for the *buck* converter as developed in Section 4.3.1 for the *R-L* load in the following form

$$\frac{du_C}{dt} = \frac{1}{C}(i_L - i_a)$$

$$\frac{di_L}{dt} = \frac{1}{L}(-u_C - (r_L + r_c)i_L + r_c i_a - r_s i_L d_D + V_d d_D)$$

$$\frac{di_a}{dt} = \frac{1}{L_a}(u_C + r_c i_L - (r_a + r_c)i_a - E_a)$$

we have a set of five first-order differential equations that map the augmented circuitry and *torsional-mechanical* dynamics. In particular, the mathematical model is given as

$$\frac{du_C}{dt} = \frac{1}{C}(i_L - i_a)$$

$$\frac{di_L}{dt} = \frac{1}{L}(-u_C - (r_L + r_c)i_L + r_c i_a - r_s i_L d_D + V_d d_D)$$

$$\frac{di_a}{dt} = \frac{1}{L_a}(u_C + r_c i_L - (r_a + r_c)i_a - k_a\omega_r) \tag{8.3}$$

$$\frac{d\omega_r}{dt} = \frac{k_a}{J}i_a - \frac{B_m}{J}\omega_r - \frac{1}{J}T_L$$

$$\frac{d\theta_r}{dt} = \omega_r$$

The duty ratio is regulated by changing the signal-level control voltage u_c, which cannot be greater than $u_{t\,\max}$ or less than $u_{t\,\min}$; that is, $u_{t\,\min} \le u_c \le u_{t\,\max}$. Hence, u_c is bounded, and if $u_{t\,\min} = 0$, we have

$$d_D = \frac{u_c}{u_{t\max}} \in [0 \quad 1],\ u_c \in [0 \quad u_{c\max}],\ u_{c\max} = u_{t\max}$$

One concludes that the signal-level control voltage u_c should be considered as a control, and from (8.3), the resulting nonlinear model is found as

$$\frac{du_C}{dt} = \frac{1}{C}(i_L - i_a)$$

$$\frac{di_L}{dt} = \frac{1}{L}\left(-u_C - (r_L + r_c)i_L + r_c i_a - \frac{r_s}{u_{t\max}} i_L u_c + \frac{V_d}{u_{t\max}} u_c\right)$$

$$\frac{di_a}{dt} = \frac{1}{L_a}(u_C + r_c i_L - (r_a + r_c)i_a - k_a \omega_r) \tag{8.4}$$

$$\frac{d\omega_r}{dt} = \frac{k_a}{J} i_a - \frac{B_m}{J}\omega_r - \frac{1}{J}T_L$$

$$\frac{d\theta_r}{dt} = \omega_r$$

The mathematical model of the mechatronic system studied (permanent-magnet DC motors augmented with *buck* converter) is nonlinear due to a nonlinear term $\frac{r_s}{Lu_{t\max}} i_L u_c$ and hard control bounds imposed $u_c \in [0 \quad u_{c\max}]$.

Permanent-magnet direct-current motors controlled by boost converters

The armature voltage applied to the armature and field motor windings can be regulated using one-quadrant *boost* (*step-up*) dc-dc switching converters. Figure 8.3 illustrates the electromechanical system when the *boost* converter is used to control the applied armature voltage to the rotor winding.

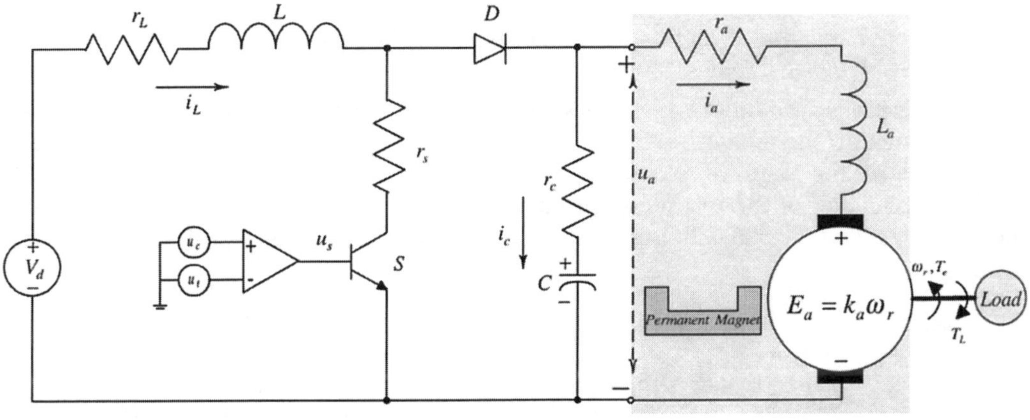

Figure 8.3 Permanent-magnet DC motor controlled by *boost* converter.

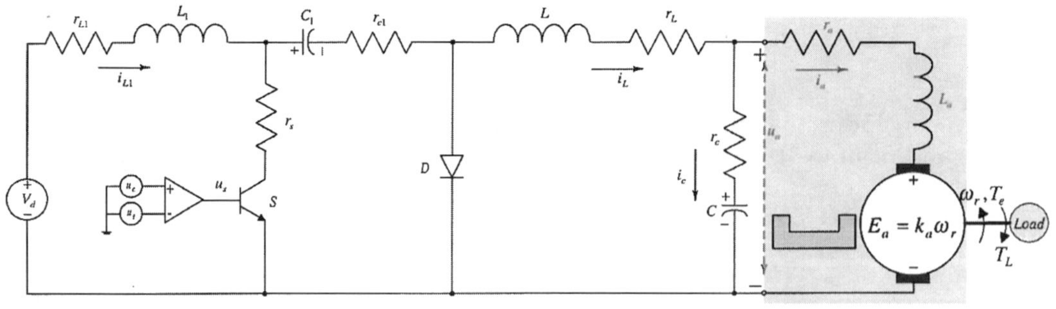

Figure 8.4 Cuk high-frequency switching converter.

Using the mathematical model of the *boost* converter developed in Section 4.3.1, and taking note of the expression for the *back emf* $E_a = k_a\omega_r$, we have

$$\frac{du_C}{dt} = \frac{1}{C}\left(i_L - i_a - \frac{1}{u_{t\max}} i_L u_c\right)$$

$$\frac{di_L}{dt} = \frac{1}{L}\left(-u_C - (r_L + r_c)i_L + r_c i_a + \frac{1}{u_{t\max}} u_C u_c + \frac{(r_c - r_s)}{u_{t\max}} i_L u_c - \frac{r_c}{u_{t\max}} i_a u_c + V_d\right)$$

$$\frac{di_a}{dt} = \frac{1}{L_a}\left(u_C + r_c i_L - (r_a + r_c)i_a - \frac{r_c}{u_{t\max}} i_L u_c - k_a\omega_r\right)$$

$$\frac{d\omega_r}{dt} = \frac{k_a}{J} i_a - \frac{B_m}{J}\omega_r - \frac{1}{J} T_L$$

$$\frac{d\theta_r}{dt} = \omega_r$$

The last two differential equations map the *torsional-mechanical* transients.

Permanent-magnet direct-current motors controlled by the Cuk converter

Consider permanent-magnet DC motors regulated by the Cuk dc-dc converter (see Figure 8.4). In particular, the armature voltage supplied to the rotor winding u_a is regulated in order to control the angular velocity ω_r.

The application of the results reported in Section 4.3.1, where the Cuk converter was studied, as well as the *torsional-mechanical* motor dynamics, give the resulting mathematical model as

$$\frac{du_{C1}}{dt} = \frac{1}{C_1}\left(i_{L1} - \frac{1}{u_{t\max}} i_{L1} u_c + \frac{1}{u_{t\max}} i_L u_c\right)$$

$$\frac{du_C}{dt} = \frac{1}{C}(i_L - i_a)$$

$$\frac{di_{L1}}{dt}=\frac{1}{L_1}\left(-u_{C1}-(r_{L1}+r_{c1})i_{L1}+\frac{1}{u_{t\max}}u_{C1}u_c+\frac{r_{c1}-r_s}{u_{t\max}}i_{L1}u_c+\frac{r_s}{u_{t\max}}i_Lu_c+V_d\right)$$

$$\frac{di_L}{dt}=\frac{1}{L}\left(-u_C-(r_L+r_c)i_L+r_ci_a-\frac{1}{u_{t\max}}u_{C1}u_c+\frac{r_s}{u_{t\max}}i_{L1}u_c-\frac{r_{c1}+r_s}{u_{t\max}}i_Lu_c\right)$$

$$\frac{di_a}{dt}=\frac{1}{L_a}(u_C+r_ci_L-(r_c+r_a)i_a-k_a\omega_r)$$

$$\frac{d\omega_r}{dt}=\frac{k_a}{J}i_a-\frac{B_m}{J}\omega_r-\frac{1}{J}T_L$$

$$\frac{d\theta_r}{dt}=\omega_r$$

Having discussed the application of the *step-down, step-up,* and Cuk converters to control permanent-magnet DC motors, it should be emphasized that other converters, thoroughly studied in Section 4.3, can be used to regulate the armature and field voltages. The corresponding mathematical models can be straightforwardly developed because in Section 4.3, the *R-L* load with voltage source E_a was used. That is, taking note that $E_a = k_a\omega_r$ and using *torsional-mechanical* motor dynamics

$$\frac{d\omega_r}{dt}=\frac{k_a}{J}i_a-\frac{B_m}{J}\omega_r-\frac{1}{J}T_L$$

$$\frac{d\theta_r}{dt}=\omega_r$$

the resulting mathematical models occur.

The dc-dc switching converters are used to regulate the field voltage supplied to the field windings. That is, for example, for separately excited DC motors, two converters must be used. One-quadrant converters can be used to excite conventional synchronous motors and generators (dc field voltage u_{fr} needs to be applied to the rotor *fr* winding). Mathematical models of separately excited, shunt-, series-, and compound-connected DC motors, as developed and thoroughly covered in Chapter 5, can be easily augmented with the high-frequency switching dc-dc converters studied. The resulting mathematical models must be used in analytical and numerical analysis of mechatronic systems.

8.3 Analysis and design of an electric drive with experimental verification

Consider an electric drive driven by the permanent-magnet DC motor JDH-2250 that is integrated with a *step-down* converter (see Figure 8.2). A low-pass filter ensures the specified 5% voltage ripple. The parameters are: r_s = 0.035 ohm, r_L = 0.02 ohm, r_c = 0.15 ohm, r_a = 3.2 ohm, C = 0.003 F, L = 0.0007 H, L_a = 0.004 H, k_a = 0.105 V-s-rad^{-1}, k_a = 0.105 N-m-A^{-1}, J = 0.00005 kg-m^2, and B_m = 0.00001 N-m-s-rad^{-1}.

Using the MATLAB environment, one can now model the open-loop electric drive.

Taking note of differential equations (8.3) for the electric drive, we have

$$\frac{du_C}{dt} = \frac{1}{C}(i_L - i_a)$$

$$\frac{di_L}{dt} = \frac{1}{L}(-u_C - (r_L + r_c)i_L + r_c i_a - r_s i_L d_D + V_d d_D)$$

$$\frac{di_a}{dt} = \frac{1}{L_a}(u_C + r_c i_L - (r_a + r_c)i_a - k_a \omega_r) \tag{8.5}$$

$$\frac{d\omega_r}{dt} = \frac{k_a}{J} i_a - \frac{B_m}{J}\omega_r - \frac{1}{J}T_L$$

and the following m-files are written to perform the simulation assuming that the motor is unloaded, the supplied dc voltage is 30 V, and the duty ratio is 0.5; that is, $T_L = 0$ N-m, $V_d = 30$ V, and $d_D = 0.5$.

MATLAB script (ch_8_01.m)

```
t0=0; tfinal=0.06; tol=1e-6; trace=1e-6; y0=[0 0 0 0]';
[t,y]=ode45('ch_8_02',t0,tfinal,y0,tol,trace);
subplot(2,2,1); plot(t,y(:,1),'-');
xlabel('Time (seconds)'); title('Voltage uc, [V]');
subplot(2,2,2); plot(t,y(:,2),'-');
xlabel('Time (seconds)'); title('Current il, [A]');
subplot(2,2,3); plot(t,y(:,3),'-');
xlabel('Time (seconds)'); title('Current ia, [A]');
subplot(2,2,4); plot(t,y(:,4),'-');
xlabel('Time (seconds)'); title('Angular velocity wr, [rad/sec]');
disp('End');
```

MATLAB script (ch_8_02.m)

```
% Dynamics of the permanent-magnet DC motor with buck converter
function yprime=difer(t,y);
% parameters
Vd=30; rs=0.035; rl=0.02; rc=0.15; ra=3.2;
C=0.003; L=0.0007; La=0.004; ka=0.105; Bm=0.00001; J=0.00005;
Tl=0; D=0.5;
% differential equations for PMDC motors with buck converters
yprime=[(y(2,:)-y(3,:))/C;...
(-y(1,:)-(rl+rc)*y(2,:)+rc*y(3,:)-rs*y(2,:)*D+Vd*D)/L;...
(y(1,:)+rc*y(2,:)-(rc+ra)*y(3,:)-ka*y(4,:))/La;...
(ka*y(3,:)-Bm*y(4,:)-Tl)/J];
```

The state variables for the electric drive with *step-down* converter are $u_C(t)$, $i_L(t)$, $i_a(t)$, $\omega_r(t)$. The angular velocity of the motor $\omega_r(t)$ is considered as the output. The transient dynamics, found by numerically solving differential equations (8.5), are illustrated in Figure 8.5, and the settling time is 0.06 s.

After the nonlinear analysis problem is solved, the closed-loop system should be designed and tested. The control u_c is bounded $u_c \in [0 \quad u_{c\,max}]$; and from (8.4), we have

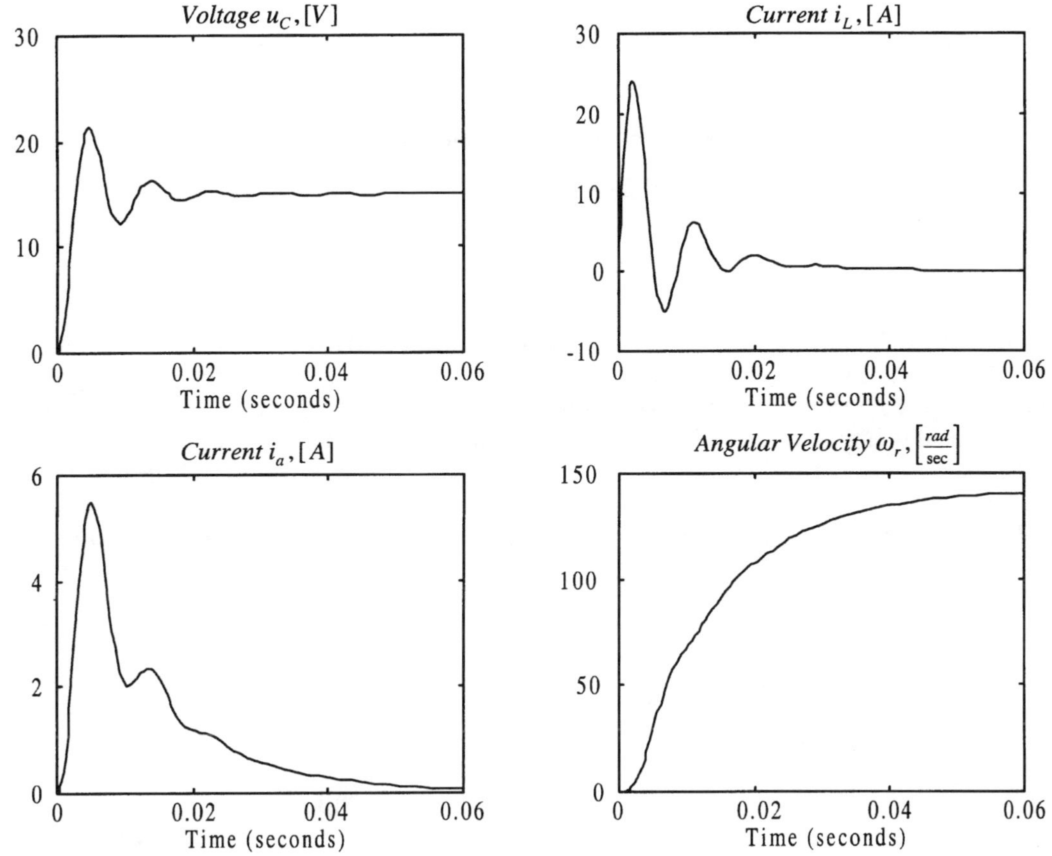

Figure 8.5 Transient dynamics of the permanent-magnet DC motor with *buck* converter.

$$\frac{du_C}{dt} = \frac{1}{C}(i_L - i_a)$$

$$\frac{di_L}{dt} = \frac{1}{L}\left(-u_C - (r_L + r_c)i_L + r_c i_a - \frac{r_s}{u_{t\max}} i_L u_c + \frac{V_d}{u_{t\max}} u_c\right)$$

$$\frac{di_a}{dt} = \frac{1}{L_a}(u_C + r_c i_L - (r_a + r_c)i_a - k_a \omega_r)$$

$$\frac{d\omega_r}{dt} = \frac{k_a}{J} i_a - \frac{B_m}{J}\omega_r - \frac{1}{J}T_L$$

The proportional-integral control law is designed using the error between the reference and rotor (actual) angular velocities. That is, the tracking error is found to be

$$e(t) = \omega_{reference}(t) - \omega_r(t)$$

In designing a nonlinear PI-type controller, one uses (8.1) and assigns $\varsigma = \sigma = 1$ and $\beta = \mu = 0$. That is, one has the following nonlinear algorithm

$$u_c(t) = \text{sat}_{u_{\min}}^{u_{\max}}\left(\sum_{j=0}^{1} k_{pj}e^{2j+1} + \sum_{j=0}^{1} k_{ij}\frac{1}{s}e^{2j+1}\right) = \text{sat}_{u_{\min}}^{u_{\max}}\left(k_{p0}e + k_{p1}e^3 + \frac{1}{s}k_{i0}e + \frac{1}{s}k_{i1}e^3\right), \quad 0 \le u_c \le 10 \text{ V}$$

The proportional and integral feedback coefficients are found, and

$$u_c = \text{sat}_0^{10}\left(1.8e + 0.25e^3 + 7.3\int edt + 0.92\int e^3 dt\right) \tag{8.6}$$

The experimental validation of stability, tracking, and disturbance attenuation is needed. The efficacy of the controller (8.6) is verified through comprehensive experiments. Different operating conditions and scenarios were studied to analyze the dynamic performance of an electric drive. The worst dynamics are observed in cases where the motor temperature is high. For T = 140°C (the maximum operating temperature), the resistance of the armature winding reaches the maximum value $r_a = r_{a\,\max}$, and the *torque* constant is minimum $k_a = k_{a\,\min}$. Therefore, the minimum electromagnetic torque, which is expressed as $T_e = k_a i_a$, results. One examines the tracking error $e(t)$, which represents the difference

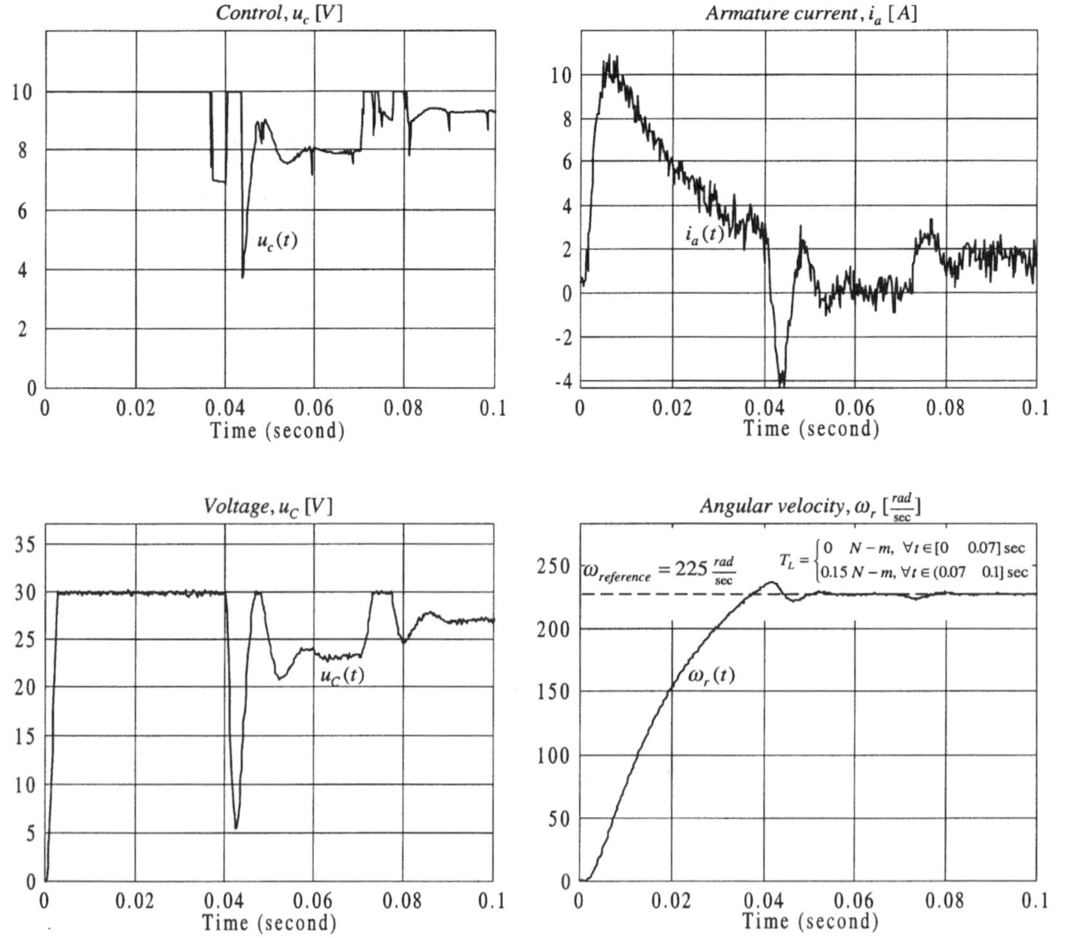

Figure 8.6 Transient dynamics of the resulting closed-loop electromechanical system.

between the actual rotor angular velocity $\omega_r(t)$ and its desired (assigned) reference value $\omega_{reference}(t)$. The disturbance attenuation features are also of interest because in real industrial systems it is required that the angular velocity remains equal to the reference value if the load torque T_L is applied. Figure 8.6 depicts the measured transient dynamics for control $u_c(t)$ and states $u_C(t)$, $i_a(t)$, $\omega_r(t)$ of the closed-loop drive when $\omega_{reference}$ = 225 rad-s^{-1}and T = 140°C. A motor reaches the desired (reference) angular velocity within 0.05 s with overshoot 4.5%, and the steady-state error of $e(t)$, $e(t) = \omega_{reference}(t) - \omega_r(t)$ is zero. Observe that a motor starts from stall, and the states $u_C(t)$, $i_a(t)$, $\omega_r(t)$ have been measured and prefiltered to digitally implement the controller (8.6) with a sampling time of 0.0001 s. The analysis of the experimental results indicates that the tracking error $e(t) = \omega_{reference}(t) - \omega_r(t)$ converges to zero. Furthermore, the disturbance rejection has been studied. In particular, a load torque of 0.15 N-m is applied at t = 0.07 s. It is evident that the disturbance rejection is achieved (see the evolution of the angular velocity illustrated in Figure 8.6). By analyzing the motor angular velocity $\omega_r(t)$ as the load is applied, one concludes that the settling time is 0.01 s with 2.5% deflection from $\omega_{reference}(t)$, and the steady-state error is zero. From the experimental data, it follows that the desired performance has been achieved, and the angular velocity precisely follows the given reference speed.

It is important to compare the nonlinear controller (8.6) with the conventional linear proportional-integral control law. In particular, a high-frequency (36 kHz with 2.5 kHz banwidth) state-of-the-art PWM servo-amplifier 25A8 (Advanced Motion Controls) is used. This servo-amplifier integrates motion control features, and the 25A8 model is illustrated in Figure 8.7.

The linear proportional-integral feedback was used to vary the amplitude of the applied armature voltage. This linear proportional-integral control algorithm, as given by

$$u = 1.8e + 7.3\int edt \tag{8.7}$$

was implemented.

The experimental results are illustrated in Figure 8.8 (motor starts from stall, and the load torque is applied at t = 0.08 s). It is easy to see that the settling time is 0.075 s with overshoot 5.2%. Furthermore, as the load torque of 0.15 N-m is applied at t = 0.08 s, a steady-state error of 1.9% results, and the settling time is 0.01 s with 3.4% maximum deflection from the assigned $\omega_{reference}(t)$. The analysis of two experimentally examined control algorithms shows that the acceleration rate remains the same. In particular, the motor reaches the maximum angular velocity at 0.041 s and 0.038 s when the nonlinear proportional-integral controller (8.6) and the linear proportional-integral control law (8.7), respectively, are used. This can be easily justified because the same rated voltage of 30 V is applied to the armature winding as the motor starts. However, the documented experimental results illustrate that the nonlinear controller (8.6) significantly improves the closed-loop dynamics and steady-state performance due to the use of nonlinear proportional and integral error feedback. Hence, compared with conventional proportional-integral controllers, one concludes that dynamic characteristics can be improved due to the use of nonlinear feedback. The reported analytical and experimental results illustrate that nonlinear controllers guarantee better dynamic responses, precise tracking and disturbance attenuation, as well as superior robustness and stability.

This section has addressed and solved the problem of nonlinear analysis and design of mechatronic systems by augmenting permanent-magnet DC motors with power converters. The presented analytical and experimental studies for a completely featured electric drive (controller–switching converter–permanent-magnet motor) demonstrate that

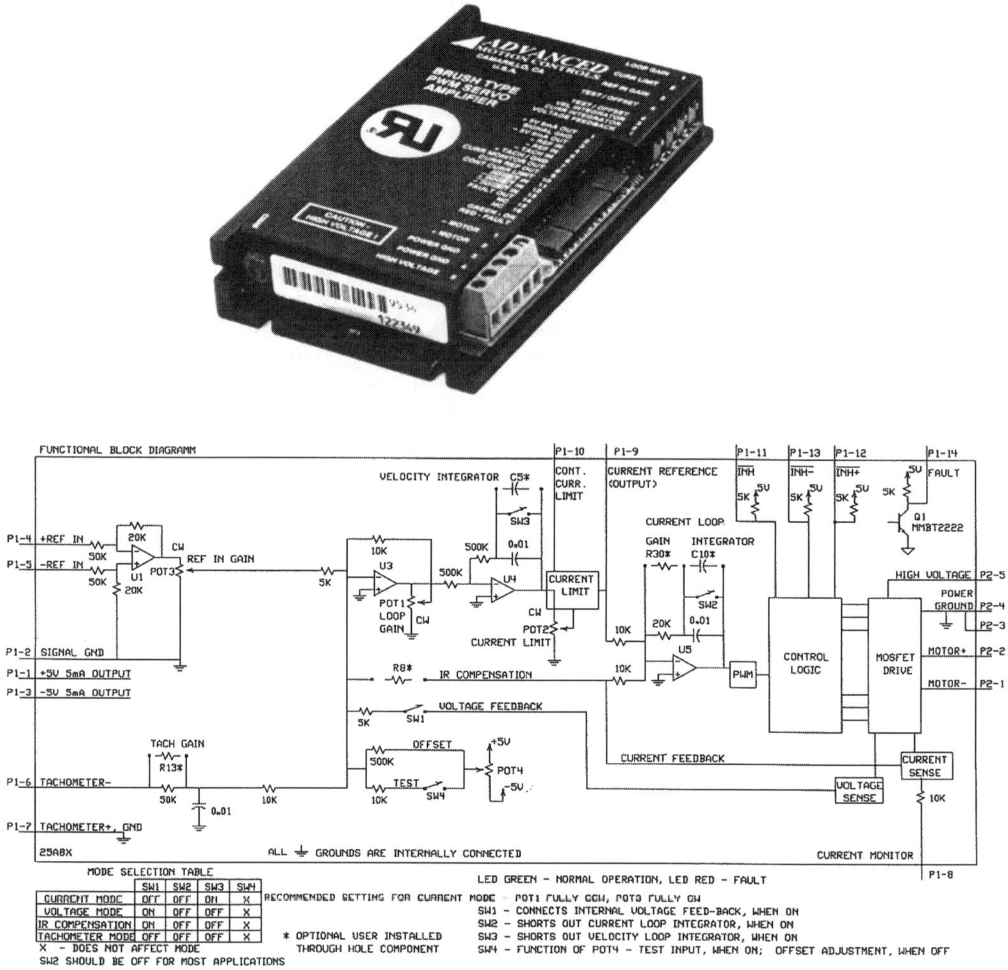

Figure 8.7 25A8 PWM amplifier. (Courtesy of Advanced Motion Controls.)

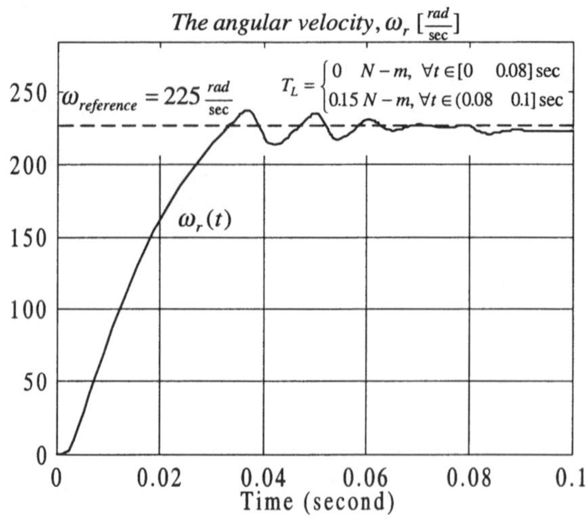

Figure 8.8 Dynamics of the permanent-magnet DC motor with the 25A8 servo-amplifier.

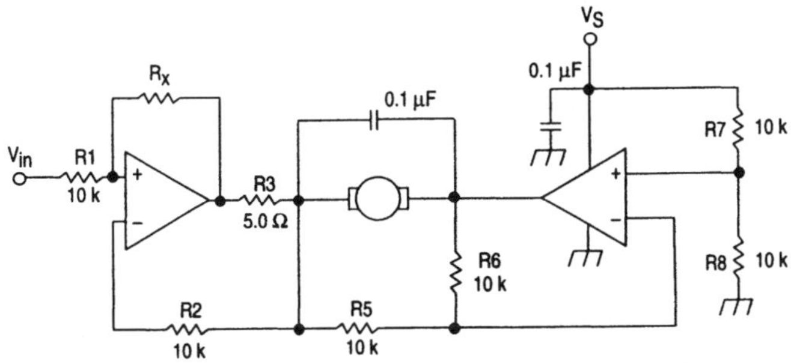

Figure 8.9 DC motor driven by the TCA0372 monolithic power amplifier. (Copyright of Motorola, used with permission.)

nonlinear proportional-integral control laws provide superior performance and ensure requirements imposed on industrial drives (stability, robustness, fast dynamics, precise tracking, disturbance rejection, etc.).

Two-quadrant power operational amplifiers can be used in electric drives and servo-systems when small DC motors are used. In particular, the bidirectional control of the angular velocity of DC motors can be achieved using the configuration shown in Figure 8.9 with the TCA0372 Motorola monolithic power operational amplifier. The rated voltage and current are 40 V and 1 A, respectively, and the bandwidth is 1.1 MHz. Figure 8.10 illustrates the packages, pin connections, and block diagram. The frequency-domain characteristics (Bode plots) and transient responses of the TCA0372 monolithic power amplifier are documented in Figure 8.11.

The analysis of the dynamics documented in Figure 8.11 indicates that due to high switching frequency and monolithic design, which eliminates parasitic inductances and capacitances, the transient behavior of the TCA0372 power amplifier can be neglected. For the rated control voltage, the settling time for the output voltage is 1×10^{-5} s.

It should be emphasized that the control algorithms are implemented to control DC motors. Small motors can be controlled using monolithic servo-amplifiers, and the MC33030 monolithic servo-motor driver integrates on-chip operational amplifier and comparator, driving and braking logic, PWM four-quadrant converter, etc. The rated output voltage and current are 36 V and 1 A, respectively, and these capabilities allow one to implement this driver to control small DC motors or use the MC33030 to regulate the field voltage. This monolithic servo-motor driver contains 119 active transistors, and the data is given in Appendix 8.1. As in conventional configurations, the difference between the reference and actual angular velocity or displacement, linear velocity or position, is compared by the *error amplifier,* and two comparators are used. A *pnp* differential output power stage ensures driving and braking capabilities, and the four-quadrant H-configured power stage guarantees high performance and efficiency. For a complete description of the MC33030 motor driver, the reader should refer to Appendix 8.1.

8.4 Mechatronic systems with direct-current motors

Single-, two-, and four-quadrant dc-dc power converters are widely used. In high-performance mechatronic systems, electric machines must be controlled. In particular, motors should be regulated in the first and third quadrants, and generators are controlled in the second and fourth quadrants. Hence, to reverse the angular velocity and to attain the generator capabilities, four-quadrant operation needs to be achieved using power converters. The first and third quadrants of the torque-speed envelope map the motor oper-

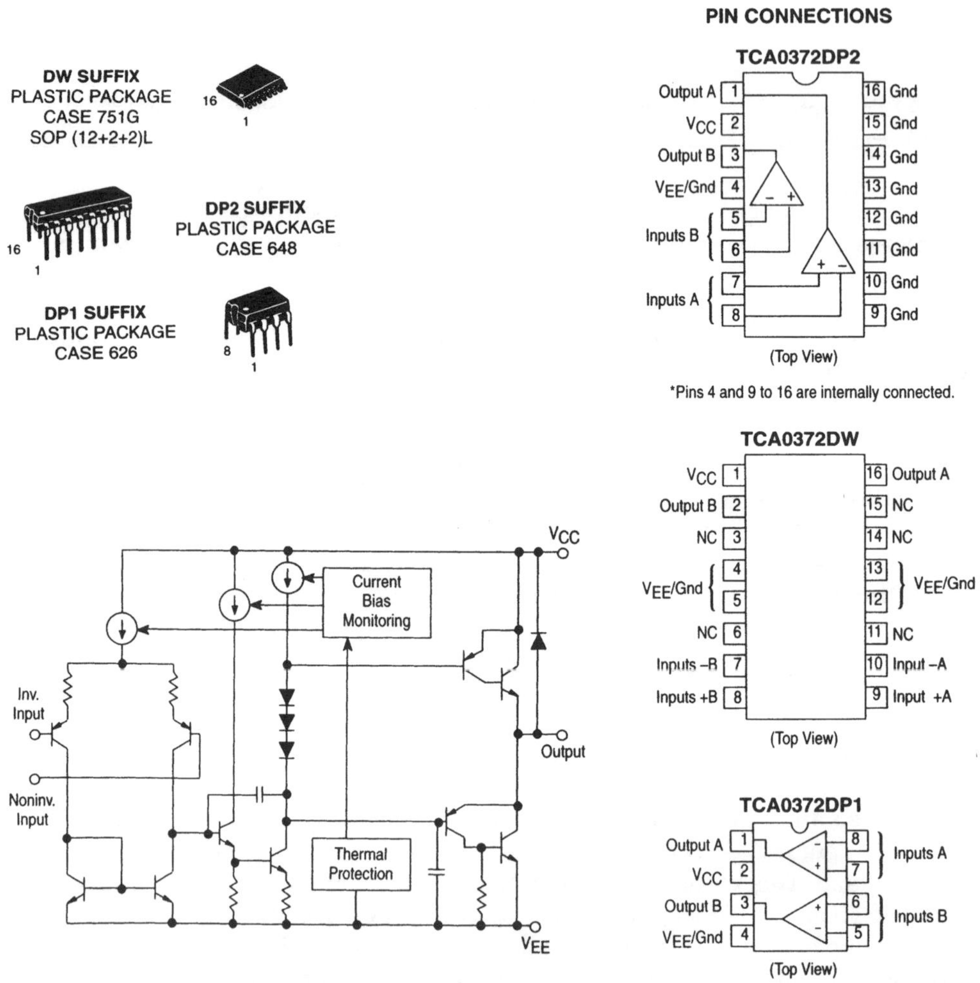

Figure 8.10 Packages, pin connections, and block diagram of the TCA0372 monolithic power amplifier. (Copyright of Motorola, used with permission.)

Figure Voltage Gain and Phase versus Frequency

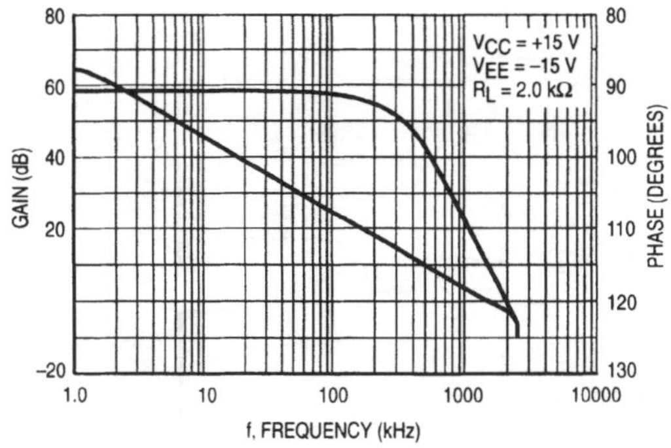

Figure Phase Margin versus Output Load Capacitance

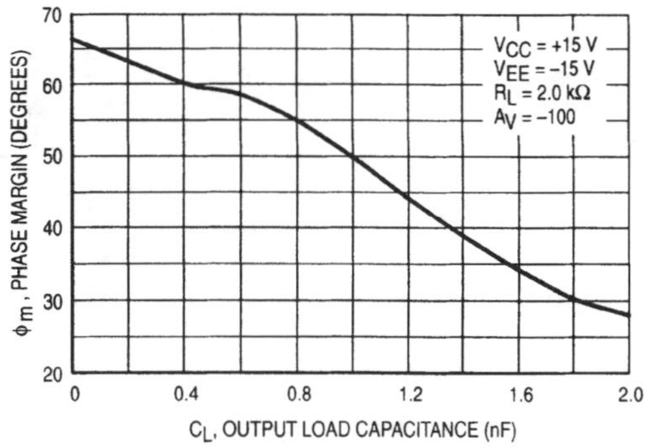

Figure Small Signal Transient Response

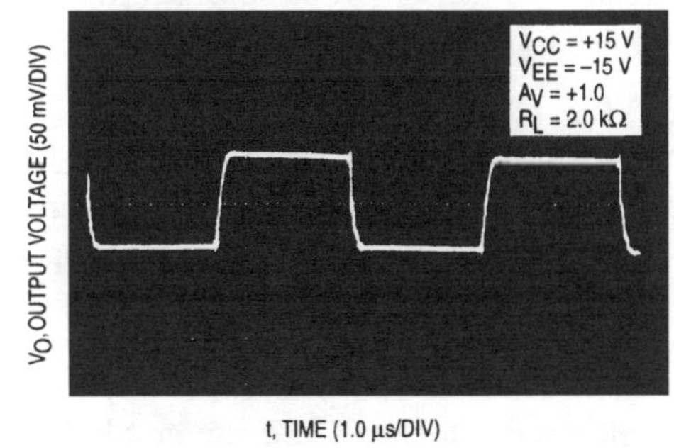

Figure Large Signal Transient Response

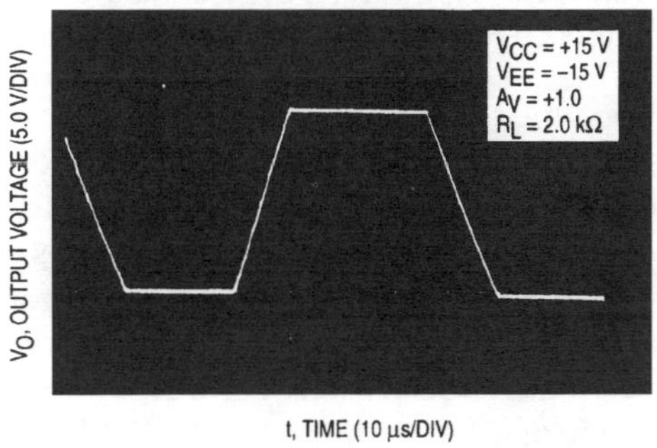

Figure Sine Wave Reponse

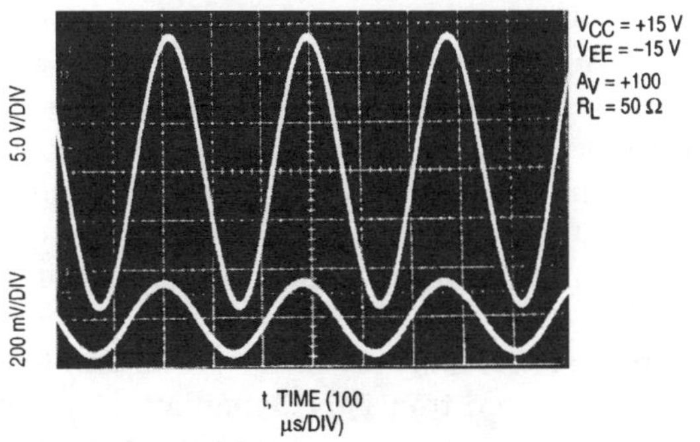

Figure 8.11 Operating characteristic and transient dynamics of the TCA0372 amplifier. (Copyright of Motorola, used with permission.)

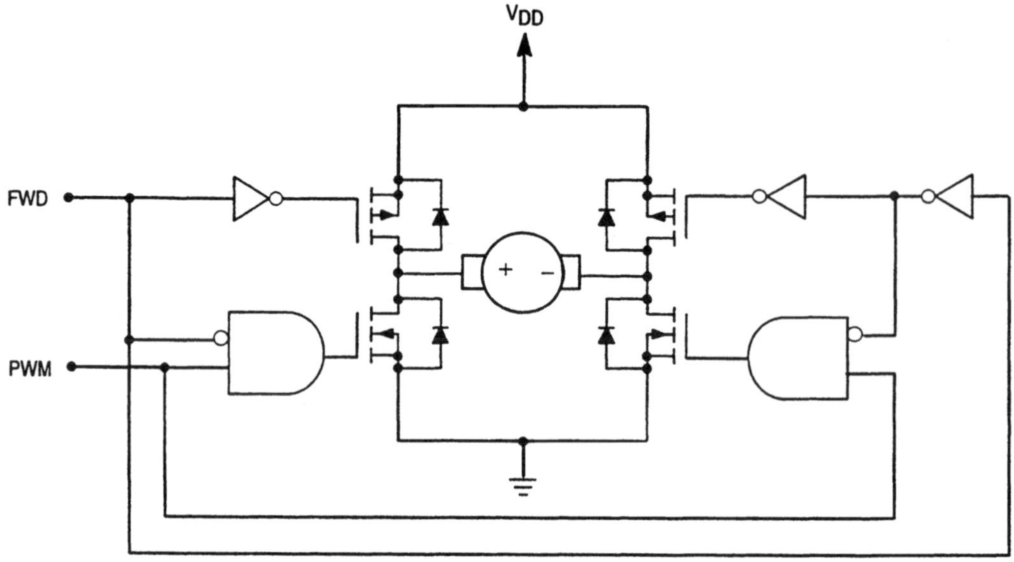

Figure 8.12 Application of two-quadrant converter to control the armature voltage. (Copyright of Motorola, used with permission.)

ation when electric motors develop positive or negative electromagnetic torque and rotate *counterclockwise* or *clockwise*. The reverse and forward braking of electric machines are accomplished in the second and fourth quadrants, which correspond to the generator operation. The H-bridge dc-dc high-switching converters allow one to reverse applied voltage and armature current. Therefore, the H-bridge topology is used in high-performance mechatronic systems. Considering only the armature motor winding, the simplified two-quadrant converter configuration with four MOSFETs and clamp diodes is illustrated in Figure 8.12. A signal-level PWM signal drives either the left or right lower MOSFET, while the FWD signal assigns the direction of rotation. Due to limited current capabilities, the maximum armature current and electromagnetic torque are rated. Figure 8.13a illustrates the studied two-quadrant converter configuration with current limit. Assume that the lower transistor is *on*. If the voltage across the sense resistor exceeds the reference PWM signal, the comparator output sets the Reset-Set latch high. The lower transistor is turned *off*. When the signal-level PWM signal becomes greater than the inserted current limit, the Reset-Set latch switches low, and the low transistor is turned *on*. A two-quadrant 5 A (rated current) power converter with two pairs of *p*- and *n*-channel MTB23P06E and MTB36N06E MOSFETs, driven by the MC33151D high-speed MOSFET drivers, is documented in Figure 8.13b. The armature or field motor windings are connected to the pins +OUTPUT and –OUTPUT using the connector CON3.

The voltage and current waveforms in the MTB36N06E transistor are illustrated in Figure 8.14. The analysis of the experimentally measured transistor dynamics documents that the settling time is 0.7 μs. Due to fast transients of operational amplifiers, MOSFET drivers, and high-frequency power transistors, these unmodeled dynamics can be omitted from analysis. It should be emphasized that the power converter studied can be used to regulate armature and field voltages applied to the rotor and stator windings of DC electric motors. To excite separately excited DC generators as well as conventional three-phase synchronous generators, this converter configuration can be easily implemented.

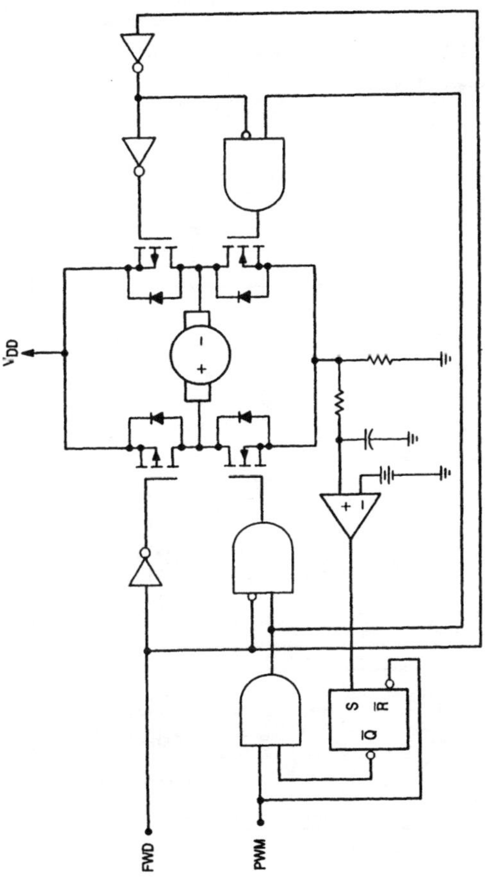

a

Figure 8.13 Two-quadrant power converter: (a) basic configuration with the current limit; (b) circuit schematics: MTB23P06E and MTB36N06E MOSFETs driven by the MC33151D high-speed drivers. (Copyright of Motorola, used with permission.)

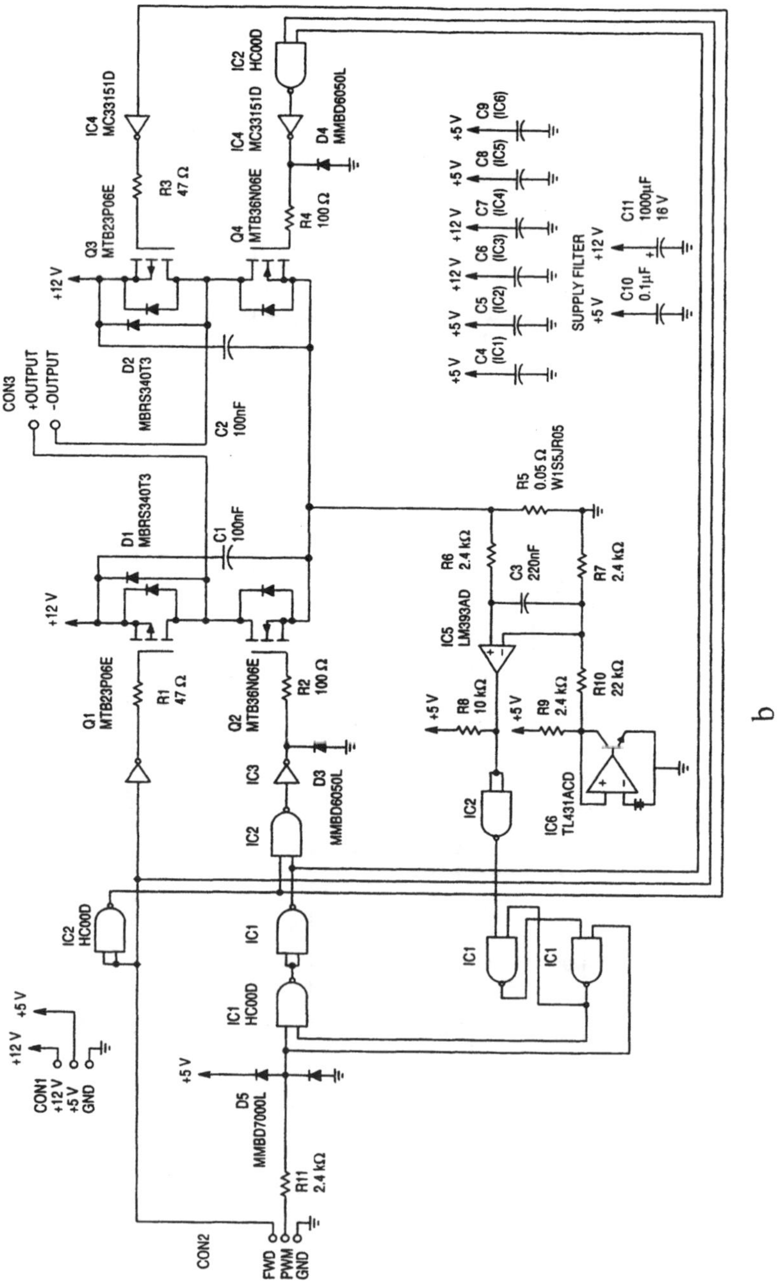

Figure 8.13 Continued.

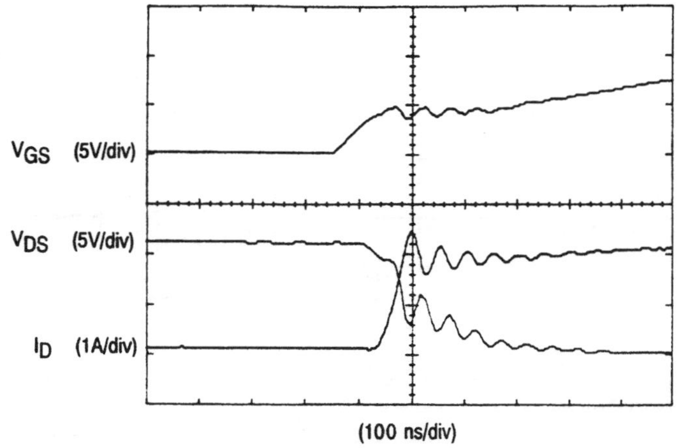

Figure 8.14 Voltage and current waveforms in the MTB36N06E. (Copyright of Motorola, used with permission.)

8.5 Mechatronic systems with induction motors

One needs to control rectifiers and inverters to regulate the angular velocity of induction motors. The majority of induction motors used in industry are small and medium motors from 0.25 to 5 kW. To implement the voltage-frequency control, the frequency is changed and the magnitude of the phase voltages is regulated. For example, in general-purpose 10 kW induction machines, frequency and phase voltages are usually regulated from 1 to 120 Hz and 10 to 480 V, respectively.

Consider the ac-ac power converter with the Motorola integrated power stages (see the schematic diagram shown in Figure 8.15). The input terminals are denoted as R, S, and T, and the *as*, *bs*, and *cs* stator windings are connected to the U, V, and W inverter outputs.

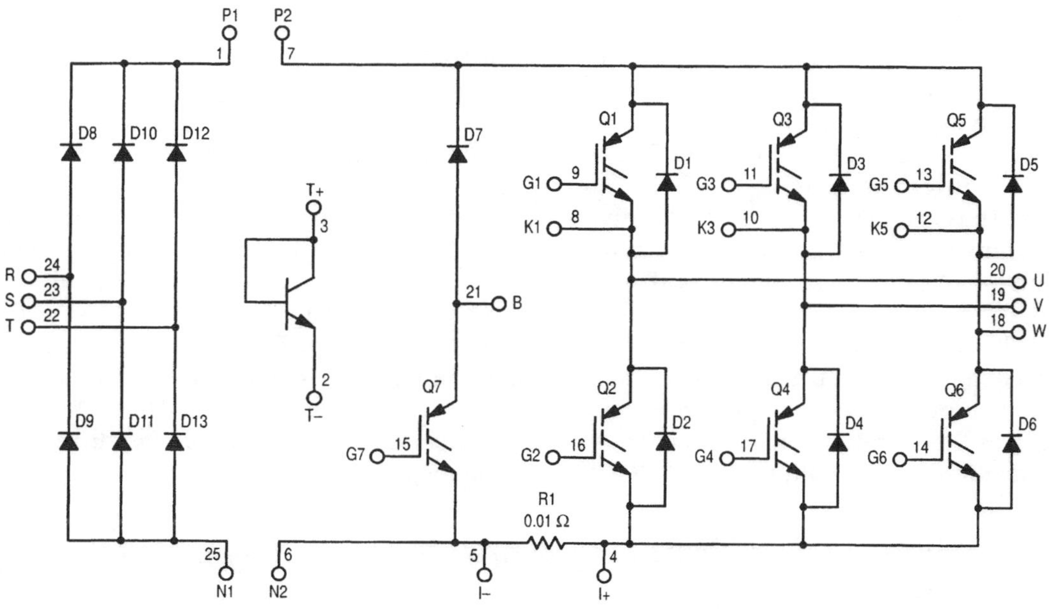

Figure 8.15 Schematic diagram of a power converter. (Copyright of Motorola, used with permission.)

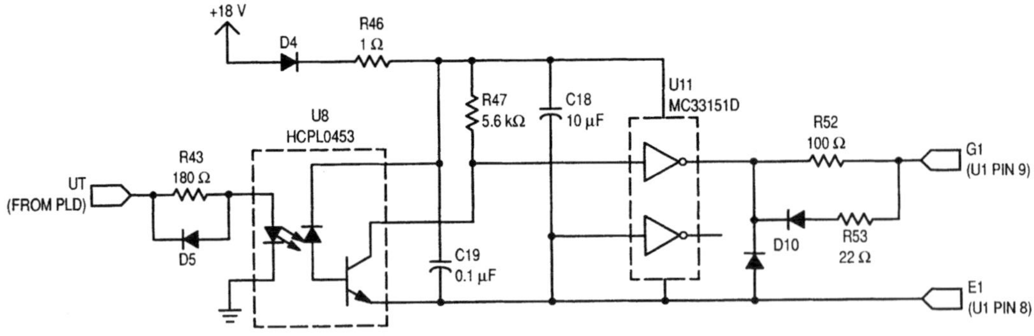

Figure 8.16 High-speed HCPL0453 optocoupler and MC33151D IGBTs gate driver. (Copyright of Motorola, used with permission.)

Two thermistors are inserted between the rectifier output P1-N1 and inverter input P2-N2 to limit the inrush current and protect the rectifier diodes from the surge current.

Hybrid power stage 600 or 1200 V, 8, 15, 20, or 30 A modules MHPM7A15A60A, MHPM7A20A60A, MHPM7A30A60A, MHPM7A8A120A, MHPM7A12A120A, and MHPM7A16A120A with six IGBTs are manufactured by Motorola to drive 0.5 to 2.5 kW three-phase induction motors. To drive IGBTs in these hybrid power stage modules, the MC33151D gate drivers with 2 A capability are used, and the basic configuration is illustrated in Figure 8.16. To limit the turn-on voltage acceleration rate to the specified level, the turn-on resistor is used. It has been emphasized that the pulse-width-modulation concept is used to control the angular velocity of induction machines by changing the voltage. Microprocessors are used to generate the PWM signal, which is the input to the HCPL0453 optocoupler–MC33151D driver configuration (this input is denoted as UT in Figure 8.16). The UT is the output of the programmable logic device (PLD). It should be emphasized that the HCPL0453 high-speed optocoupler is used to interface the PLD and MC33151D. We consider the HCPL0453 optocoupler and MC33151D driver to drive the IGBT in the first lag of the inverter illustrated in Figure 8.15. See the pins G1-9 and K1-8 in Figure 8.15, as well as the pins G1 (U1 PIN 9) and E1 (U1 PIN 8) in Figure 8.16. The positive voltage (+18 V) is supplied to the bootstrap circuit, which includes a diode D4, resistor R46, and capacitor C18. When the lower transistor Q2 is turned *on*, the capacitor C18 charges through diode D4 and resistor R46. The lower transistor is turned *off* and the upper transistor Q1 is turned *on*. The charge stored in the capacitor C18 provides the gate drive current needed.

The current sense resistor R1 with resistance 0.01 ohm is inserted between the rectifier and inverter (see Figure 8.15). The sensed current is used to protect power converters and motors. A possible schematic of the current sense circuitry is documented in Figure 8.17.

The programmable logic device (PLD), which can be easily implemented using the 68HC family of Motorola microprocessors or DSPs, generates the PWM signal-level signals needed to drive IGBTs. In practice, as was shown in Figure 8.16, the signal from the PLD is the input to the high-speed optocoupler which drives the IGBT driver. This gate drive drives the high-frequency transistor. A complete microprocessor-based system to control induction motors with the MC68HC708MP16 microcontroller is covered in Section 8.7.

8.6 *Mechatronic systems with permanent-magnet synchronous motors*

The basic components of mechatronic systems with permanent-magnet synchronous motors are synchronous motors with attached mechanisms, power converters, angular

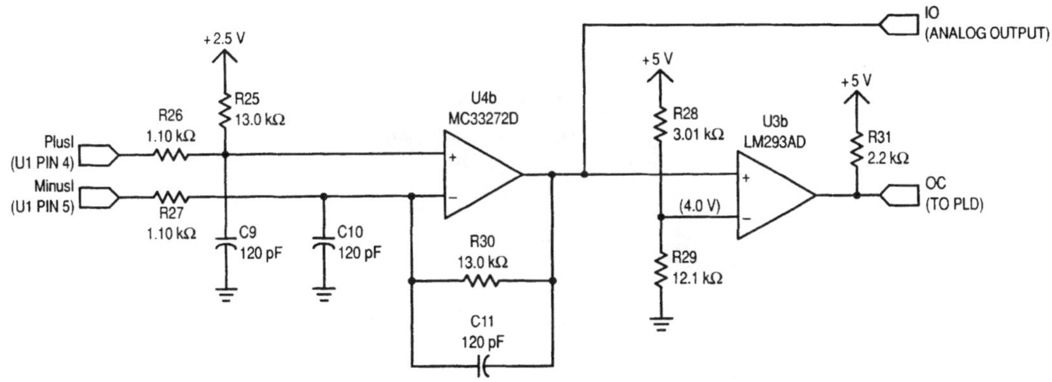

Figure 8.17 Current sense circuitry configuration. (Copyright of Motorola, used by permission.)

displacement (Hall), and currents sensors, as well as analog or digital controllers. Figures 8.18 and 8.19 illustrate the B15A8 and BE15A8 PWM amplifier with 2.5 kHz bandwidth, manufactured by Advanced Motion Controls, to control the phase voltages applied to the stator windings of permanent-magnet synchronous motors. The angular velocity of permanent-magnet synchronous motors is regulated in the torque-speed operating envelope, as was studied in Section 7.3. The analog proportional-integral controllers are integrated into these PWM amplifiers. Stability, tracking accuracy, disturbance attenuation, and other specified steady-state and dynamic characteristics can be achieved by changing the proportional and integral feedback gains. The angular velocity can be measured by tachogenerators or optical encoders. The voltage induced by the tachogenerator, which is directly attached to the rotor, is proportional to the angular velocity. This generated voltage is applied to the input pins P1-6 and P1-7 of the B15A8 *amplifier* (see Figure 8.18). If optical encoders are used, the output is connected to the D/A card (see the input pins P1-6 and P1-7 for the BE15A8 amplifier in Figure 8.19). The reference (desired) angular velocity is assigned by applying the signal-level input to pins P1-4 and P1-5. The tracking error signal, which is the difference between the assigned and actual angular velocities, is used to generate the signal-level control input that through the operational amplifier U4, drives the high-switching transistors. The analog proportional and proportional-integral control algorithms are implemented using U2 and U3, and the reader is referred to Table 4.1, which gives the corresponding transfer functions of the inverting operational amplifiers. The PWM B15A8 and BE15A8 amplifiers have maximum current protection, and the so-called *current feedback* is integrated. The phase voltages are the PWM amplifier outputs, and pins P2-1, P2-2, and P2-3 must be connected to the permanent-magnet synchronous motor stator windings *as*, *bs*, and *cs*.

Control of permanent-magnet synchronous motors is straightforward, and the basic operation principles are covered in Section 7.3. To maximize the electromagnetic torque, one applies a balanced three-phase voltage sinusoidal set to the stator windings. In particular, it was shown that

$$u_{as}(t)=\sqrt{2}u_M\cos\theta_r,\quad u_{bs}(t)=\sqrt{2}u_M\cos\left(\theta_r-\frac{2}{3}\pi\right),\quad u_{cs}(t)=\sqrt{2}u_M\cos\left(\theta_r+\frac{2}{3}\pi\right)$$

The electrical angular displacement θ_r must be measured and used to calculate the phase voltages. The angular displacement is measured using Hall sensors. In studying permanent-magnet stepper motors in Section 7.5, it was emphasized that to avoid some

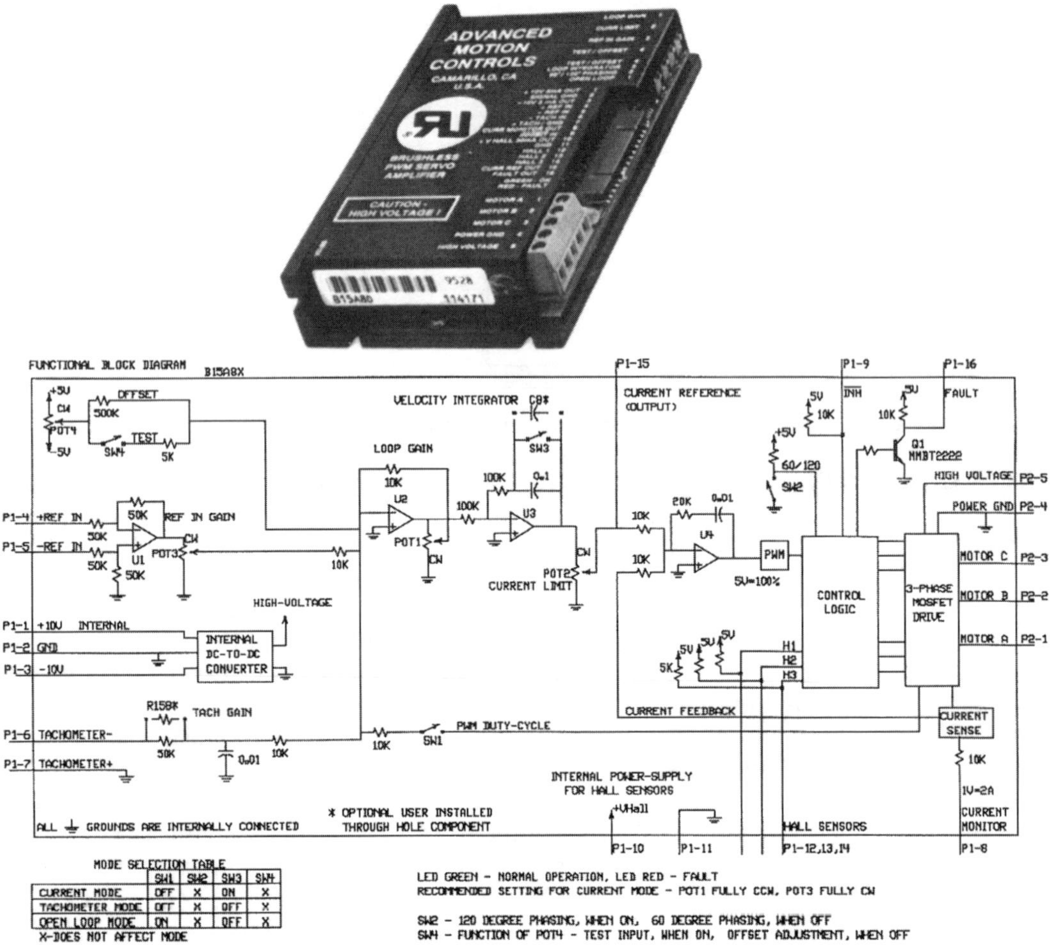

Figure 8.18 33 kHz, 7.5 A B15A8 PWM amplifier. (Courtesy of Advanced Motion Controls.)

difficulties associated with the need to use the cosine look-up tables, one can simplify the expressions for the phase voltages needed to be applied. In particular,

$$u_{as}(t) = u_M \operatorname{sgn}(\cos\theta_r),\quad u_{bs}(t) = u_M \operatorname{sgn}\left(\cos\left(\theta_r - \frac{2}{3}\pi\right)\right),\quad u_{cs}(t) = u_M \operatorname{sgn}\left(\cos\left(\theta_r + \frac{2}{3}\pi\right)\right)$$

This degrades the torque production features, results in angular velocity ripple, leads to currents and torque chattering, and therefore leads to additional losses. However, in many applications, to avoid sophisticated digital control hardware needed to optimize the torque production, these pulse train phase voltages can be applied.

One concludes that the magnitude u_M of the phase voltages u_{as}, u_{bs}, and u_{cs} is regulated, based on the tracking error signal e, which is proportional to the difference between the reference and actual angular velocities, $e \equiv \omega_{reference} - \omega_{rm}$. The measured (by Hall sensors) angular displacement is used to generate the phase voltages u_{as}, u_{bs}, and u_{cs}. In particular, the sensed angular displacement is supplied to the control logic, and pins P1-12, P1-13, and P1-14 are used (see Figures 8.18 and 8.19). It should be emphasized that the B15A8

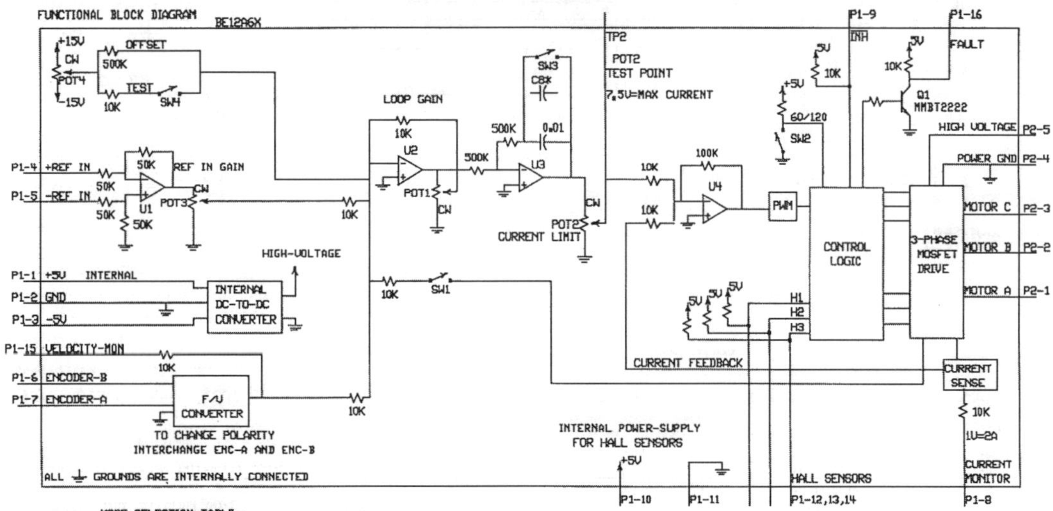

MODE SELECTION TABLE

	SW1	SW2	SW3	SW4	ENCODER
CURRENT MODE	OFF	X	ON	X	NOT CONNECTED
ENCODER MODE	OFF	X	OFF	X	CONNECTED
OPEN LOOP MODE	ON	X	OFF	X	NOT CONNECTED

X-DOES NOT AFFECT MODE

Figure 8.19 33 kHz, 7.5 A BE15A8 PWM amplifier. (Courtesy of Advanced Motion Controls.)

and BE15A8 amplifiers can be integrated with microcontrollers or DSPs, and the control algorithms can be implemented digitally.

Integrated circuits and microcontrollers are widely used to control permanent-magnet synchronous motors. It was shown that the applied phase voltages to the *abc* windings, which are shifted by 120 electrical degrees, are functions of the angular displacement. Figure 8.20 illustrates the power amplifier schematics designed to control 12 V, 2 A permanent-magnet synchronous motors. This configuration is based on using the MC33035DW motor controller, which is given in Appendix 8.2, a two-quadrant H-bridge converter with six MOSFETs, and the MC33039D angular velocity estimator that calculates the rotor angular velocity based on the sensed (by Hall sensors) rotor angular displacement. Appendix 8.3 documents the MC33039D angular velocity estimator along with important characteristics and data needed for practical engineers.

Two-quadrant H-bridge converter topology is commonly used, and six MOSFETs are driven by the MC33035DW integrated circuit driver. In this converter, *p*- and *n*-channel transistors are used, and complementary packaged half-bridges (e.g., Motorola MMDF2C05E module) are available. The transistors are turned *on* and *off*, depending on

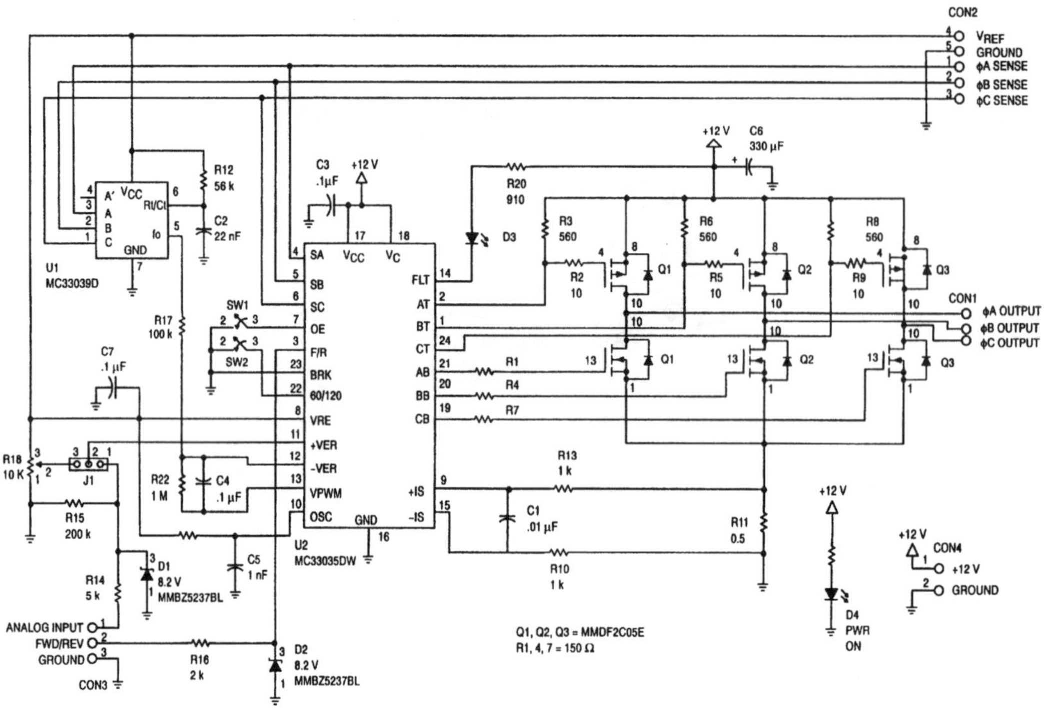

Figure 8.20 Control of permanent-magnet synchronous motors using the MC33035DW controller, MC33039D angular velocity estimator, and H-bridge power converter. (Copyright of Motorola, used by permission.)

the rotor angular displacement; and the signal-level voltages from the Hall-effect sensors are fed to the MC33039D integrated circuit. The reference and actual angular velocities are compared, and the error signal is used to regulate the magnitude of the phase voltages applied. Based on the angular displacement and the error signal, the *abc* stator windings are commutated. Current protection is needed. The current is proportional to the voltage across the current sensor resistor R11, and the low-pass filter, which is formed by resistors R10, R13, and capacitor C1, guarantees noise immunity. To control the angular velocity, the reference and actual speeds are compared. The desired angular velocity is assigned by the voltage supplied to the pin 1 (analog input) (see Figure 8.20). The *counterclockwise* and *clockwise* directions are assigned by applying the voltage to pin 2 (FWD/REV). It was emphasized that the rotor angular velocity is calculated by the MC33039D using the measured (by Hall sensors) rotor angular displacement, and the filter (resistances R17, R22, and capacitor C4) attenuates the noise. The error amplifier provides the signal-level control voltage to regulate the motor angular velocity by changing the magnitude u_M of the phase voltages u_{as}, u_{bs}, and u_{cs}. A complete description of the MC33035 permanent-magnet synchronous motor driver is given in Appendix 8.4.

Mechatronic system with MC68HC705MC4 microcontroller

Another simple and effective solution is to design and implement high-performance mechatronic systems with PWM amplifiers controlled by microcontrollers. This chapter section describes the design and analysis of microprocessor-controlled mechatronic systems with permanent-magnet synchronous motors. The MC68HC705MC4 microcontroller and PWM amplifier are used. A highly integrated and flexible solution can be achieved using low-cost available hardware and software, and the Motorola HC05 microcontroller family is uniquely suitable for use in the control of mechatronic systems. Hall sensors measure the rotor angular displacement. For three-phase permanent-magnet synchronous motors, we found that the *as, bs,* and *cs* phase currents fed and voltages to be applied are given, respectively, as

$$i_{as}(t)=\sqrt{2}i_M\cos\theta_r,\ i_{bs}(t)=\sqrt{2}i_M\cos\left(\theta_r-\frac{2}{3}\pi\right),\ i_{cs}(t)=\sqrt{2}i_M\cos\left(\theta_r+\frac{2}{3}\pi\right)$$

and

$$u_{as}(t)=\sqrt{2}u_M\cos(\theta_r+\varphi_u),\ u_{bs}(t)=\sqrt{2}u_M\cos\left(\theta_r-\frac{2}{3}\pi+\varphi_u\right),\ u_{cs}(t)=\sqrt{2}u_M\cos\left(\theta_r+\frac{2}{3}\pi+\varphi_u\right)$$

It was illustrated that the electromagnetic torque is developed if

$$i_{as}(t)=i_M\,\text{sgn}(\cos\theta_r),\ i_{bs}(t)=i_M\,\text{sgn}\left(\cos\left(\theta_r-\frac{2}{3}\pi\right)\right),\ i_{cs}(t)=i_M\,\text{sgn}\left(\cos\left(\theta_r+\frac{2}{3}\pi\right)\right)$$

and

$$u_{as}(t)=u_M\,\text{sgn}(\cos(\theta_r+\varphi_u)),\ u_{bs}(t)=u_M\,\text{sgn}\left(\cos\left(\theta_r-\frac{2}{3}\pi+\varphi_u\right)\right),$$

$$u_{cs}(t)=u_M\,\text{sgn}\left(\cos\left(\theta_r+\frac{2}{3}\pi+\varphi_u\right)\right)$$

Other commonly used commutation sequences, the so-called six-step sequence, to ensure the *counterclockwise* and *clockwise* rotation are presented in Table 8.1.

The angular velocity is controlled by changing the duty ratio of the high-frequency switching transistors (IGBTs or MOSFETs); and the maximum angular velocity is limited by the rated current or voltage (electrical limit) as well as by the maximum mechanical angular velocity (mechanical limit). The ability of the controllers and drivers to drive IGBTs or MOSFETs at the desired frequency is a critical issue. The pulse-width-modulation concept ensures a high-performance solution. Permanent-magnet synchronous motors are driven by inverters, consisting of complementary pairs of high-frequency switches (typically one pair of IGBTs or MOSFETs per phase), which are controlled by the PWM signals generated by microcontrollers. The average magnitude u_M of the supplied phase voltages is controlled by changing the duty ratio of the PWM signals. The Hall sensor supplies the angular displacement to the microcontroller, and Figure 8.21 documents the permanent-magnet synchronous motor with the MC68HC705MC4 microcontroller. It was shown that the balanced three-phase voltage set must be supplied (see the corresponding balanced sets, as well as six-step commutation sequences for *counterclockwise* and *clockwise* rotations documented in Table 8.1). The angular velocity of synchronous motors is controlled by changing u_M; and to regulate the magnitude of the phase voltages, the duty ratio of the transistors is controlled using the PWM principle. Digital PID algorithms can be easily implemented to attain the desired system performance. The closed-loop mechatronic system is illustrated in Figure 8.21. This system includes the Motorola ITC127 motion

Table 8.1 Commutation Sequences for *Counterclockwise* and *Clockwise* Rotation

Rotor Angle θ_r,°	Hall Sensor *as*	Hall Sensor *bs*	Hall Sensor *cs*	Phase *as*	Phase *bs*	Phase *cs*
$\theta_r \in -[0/360 \quad 180]$	1	0	0	$-\sqrt{2}i_M$ or $-\sqrt{2}u_M$	$\sqrt{2}i_M$ or $\sqrt{2}u_M$	0
$\theta_r \in -[30 \quad 210]$	1	0	1	0	$\sqrt{2}i_M$ or $\sqrt{2}u_M$	$-\sqrt{2}i_M$ or $-\sqrt{2}u_M$
$\theta_r \in -[60 \quad 240]$	0	0	1	$\sqrt{2}i_M$ or $\sqrt{2}u_M$	0	$-\sqrt{2}i_M$ or $-\sqrt{2}u_M$
$\theta_r \in -[90 \quad 270]$	0	1	1	$\sqrt{2}i_M$ or $\sqrt{2}u_M$	$-\sqrt{2}i_M$ or $-\sqrt{2}u_M$	0
$\theta_r \in -[120 \quad 300]$	0	1	0	0	$-\sqrt{2}i_M$ or $-\sqrt{2}u_M$	$\sqrt{2}i_M$ or $\sqrt{2}u_M$
$\theta_r \in -[150 \quad 330]$	1	1	0	$-\sqrt{2}i_M$ or $-\sqrt{2}u_M$	0	$\sqrt{2}i_M$ or $\sqrt{2}u_M$
$\theta_r \in [0/360 \quad 180]$	1	0	0	$\sqrt{2}i_M$ or $\sqrt{2}u_M$	$-\sqrt{2}i_M$ or $-\sqrt{2}u_M$	0
$\theta_r \in [30 \quad 210]$	1	1	0	$\sqrt{2}i_M$ or $\sqrt{2}u_M$	0	$-\sqrt{2}i_M$ or $-\sqrt{2}u_M$
$\theta_r \in [60 \quad 240]$	0	1	0	0	$\sqrt{2}i_M$ or $\sqrt{2}u_M$	$-\sqrt{2}i_M$ or $-\sqrt{2}u_M$
$\theta_r \in [90 \quad 270]$	0	1	1	$-\sqrt{2}i_M$ or $-\sqrt{2}u_M$	$\sqrt{2}i_M$ or $\sqrt{2}u_M$	0
$\theta_r \in [120 \quad 300]$	0	0	1	$-\sqrt{2}i_M$ or $-\sqrt{2}u_M$	0	$\sqrt{2}i_M$ or $\sqrt{2}u_M$
$\theta_r \in [150 \quad 330]$	1	0	1	0	$-\sqrt{2}i_M$ or $-\sqrt{2}u_M$	$\sqrt{2}i_M$ or $\sqrt{2}u_M$

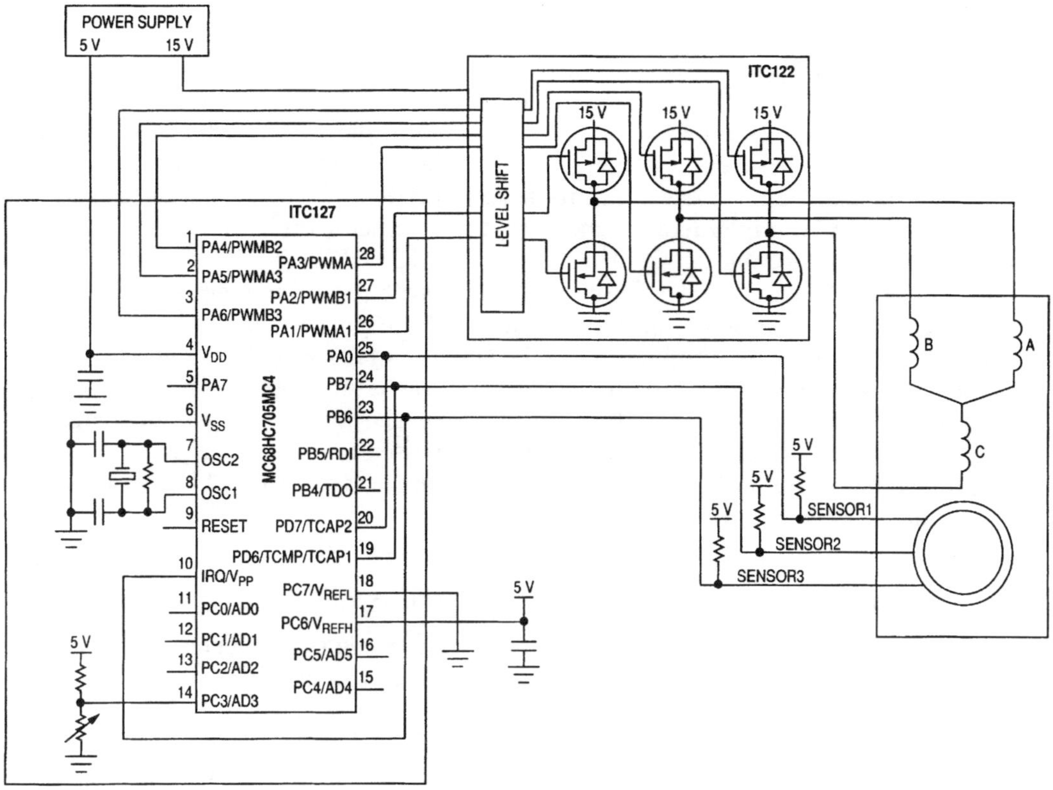

Figure 8.21 Permanent-magnet synchronous motor controlled by the ITC122 inverter and MC68HC705MC4 microcontroller. (Copyright of Motorola, used with permission.)

control development board with MC68HC705MC4, ITC122 inverter, and three-phase permanent-magnet synchronous motor. The MC68HC705MC4 integrates a high-speed (up to 23.4 kHz), two-channel, eight-bit double-buffered PWM module with a commutation multiplexer (three pins per PWM channel), 3-MHz bus with 333-ns instruction cycle, and an interlock mechanism for coherent updates. A six-channel, eight-bit A/D converter can be used for calculating the angular velocity, based on the rotor angular displacement measured by the Hall sensors.

A 6-MHz signal generator across the OSC1 and OSC2 pins allows one to ensure an eight-bit resolution, 23.4 kHz output PWM signal, which is the input to the motion control development board. The outputs of Hall sensors are connected to the MC68HC705MC4 microcontroller. The angular velocity can be measured using optical encoders or tachogenerators. However, one can calculate the angular velocity using the sensed rotor displacement. Taking note that $\frac{d\theta_{rm}(t)}{dt} = \omega_{rm}(t)$ and using the Euler formula, the rotor angular velocity can be easily estimated. In particular, the signals from the A and B Hall sensors are fed to the MC68HC705MC4 using pins PD6/TCAP1 and PD7/TCAP2 to calculate the angular velocity. In particular, by reading the MC68HC705MC4 16-bit timer at two different rotor angular positions sensed by the Hall sensors, which are fed by pins PD6/TCAP1 and PD7/TCAP2, one finds ω_{rm}.

To generate the balanced three-phase voltage set, the rotor angular displacement is fed to the MC68HC705MC4 through pins PA0, PB6, and PB7, and the signal-level voltages are generated to drive the IGBTs drivers. Hence, the transistor drivers are controlled by the MC68HC705MC4, and pins PA1 to PA6 are the microcontroller outputs. The ITC122 inverter is designed to drive three-phase, fractional horsepower permanent-magnet synchronous motors, and the rated voltage is 40 V. The MC68HC705MC4 assembly code is relatively simple due to the use of a commutation multiplexer on each PWM channel in combination with an interlock mechanism that allows one to perform a coherent update in less than 100 instruction cycles. Electronic commutation, PID control, diagnostics, and other features can be implemented, and the Motorola Semiconductor Application Note AN1702 "Brushless DC Motor Control Using the MC68HC705MC4" indicates that the motor control software kernel consumes 298 bytes of program memory. The angular velocity is controlled by changing the magnitude of the phase voltages applied, and the desired angular velocity is compared with the actual motor speed. This reference angular velocity is set by applying the signal-level voltage, which is the input of the A/D converter (see the external potentiometer and pin PC3/AD3). The motor control and diagnostics consume 298 bytes of user ROM and 10 bytes of user RAM. Hence, due to memory availability, other important routines such as communications protocols, diagnostics, and A/D calculations can be embedded and performed.

8.7 Digital control of mechatronic systems

With a focus on system performance, efficient mechatronic systems should be designed and deployed. Efficient PWM amplifiers and high-speed microcontrollers are available, and the MC68HC708MP16 microcontroller offers the cost-effective solution needed (see the Motorola Semiconductor Application Notes AN1624 "ITC137 68HC708MP16 Motion Control Development Board" and AN1712 "Get Your Motor Running with the MC68HC708MP16." In addition to effective digital implementation of comprehensive control algorithms, diagnostics and energy management can be performed. The Motorola HC05 family of microcontrollers offers rich enhancements to the instruction set and maintains object code compatibility. A 16-bit index register and stack pointer, five-cycle multiply instruction, and high-level language support make these microcontrollers ideally suited for motion control applications. Another important feature of the MC68HC708MP16 is the PWM generator module, which can be configured and partitioned in a variety of ways for AC and DC motors. Functional block diagrams of the MC68HC708MP16 microcontroller and the PWM module are illustrated in Figure 8.22.

The MC68HC708MP16 PWM module generates six independent PWM signals that are synchronized by the same counter/timer. These signal-level signals drive six IGBT or MOSFET drivers. The MC68HC708MP16 microcontroller can be used to control three-phase conventional synchronous motors. One recalls that, in addition to a balanced three-phase voltage set needed (six PWM signals are needed to drive six high-frequency power transistors), the excitation voltage must be controlled. The additional PWM signal, obtained from the timer modules, is used to regulate the voltage applied to the rotor field winding. The MC68HC708MP16 PWM module uniquely suits drive DC and AC motors, and the ITC137 motion control development board can be used (see Figure 8.23). This highly augmented, flexible solution allows one to design high-performance mechatronic systems to satisfy a broad spectrum of objectives and requirements.

The ITC137 motion control development board is designed to be used with the M68HC08MP16 microcontroller and its emulator. The outputs of the board are connected to the ITC122 or ITC132 power stages, and six ITC137 outputs will sink 20 mA can drive optocouplers. High-frequency transistors, driven by the IGBTs or MOSFETs drivers, allow one to implement the PWM concept to control three-phase induction or synchronous motors. Direct-current motors can be controlled as well, and one changes the armature

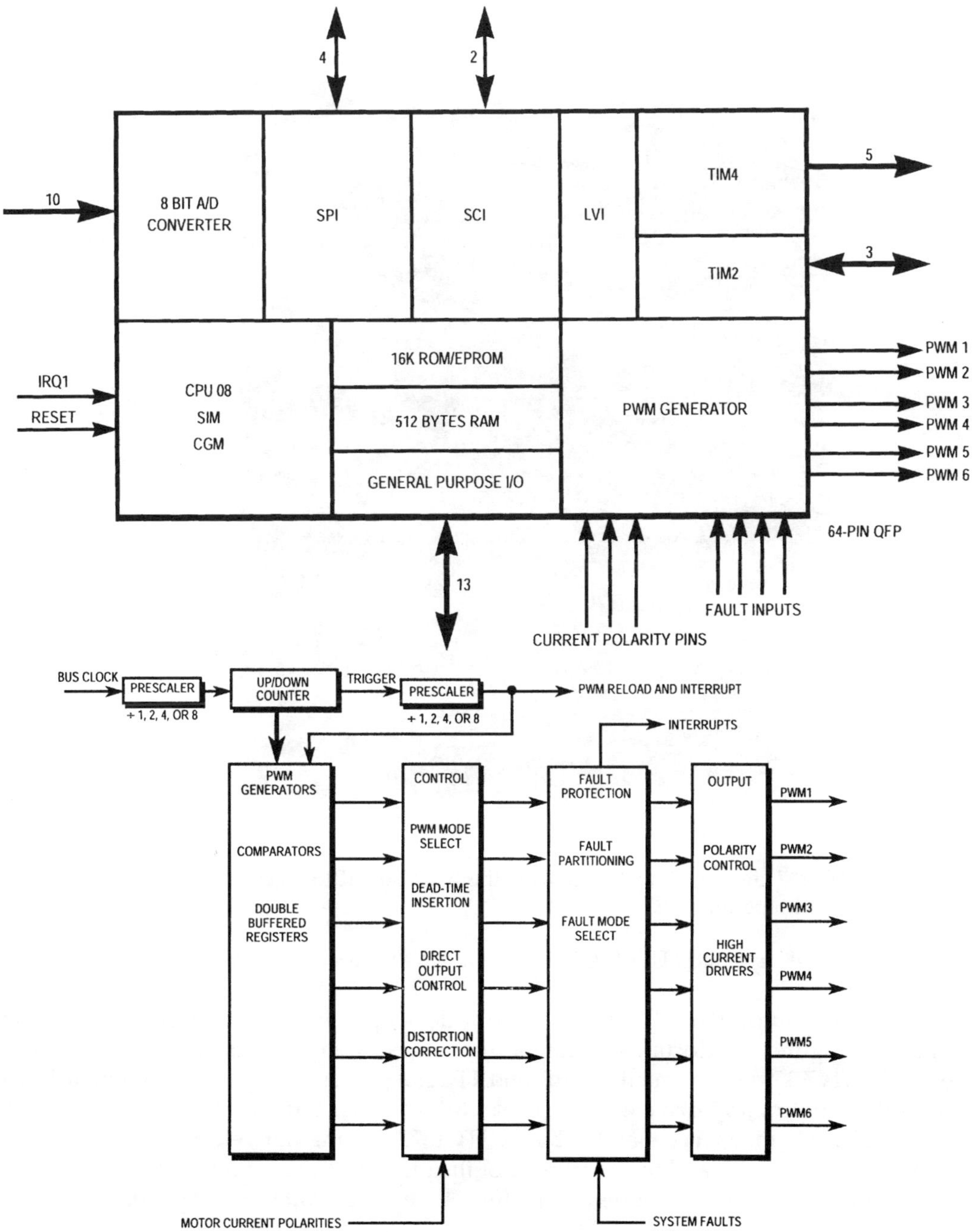

Figure 8.22 Functional block diagrams of the MC68HC708MP16 with PWM module. (Copyright of Motorola, used with permission.)

Figure 8.23 ITC137 development board with the 68HC708MP16 microcontroller. (Copyright of Motorola, used with permission.)

and field voltages applied. The ITC137 schematics are documented in Figure 8.24, along with a Parts List.

Inputs and outputs of the motion control development board are grouped into six connectors. Figure 8.25 documents the mechatronic system that integrates the induction motor, the ITC137 development board, and ITC132 power stage. These two boards are designed to control and drive electromechanical systems. A three phase, center-aligned, sine-wave PWM voltage is generated by the ITC137. The angular velocity is controlled by changing the frequency, and the amplitude of the phase voltages is controlled as well. The direction of the rotation is changed by shifting the phases voltages supplied.

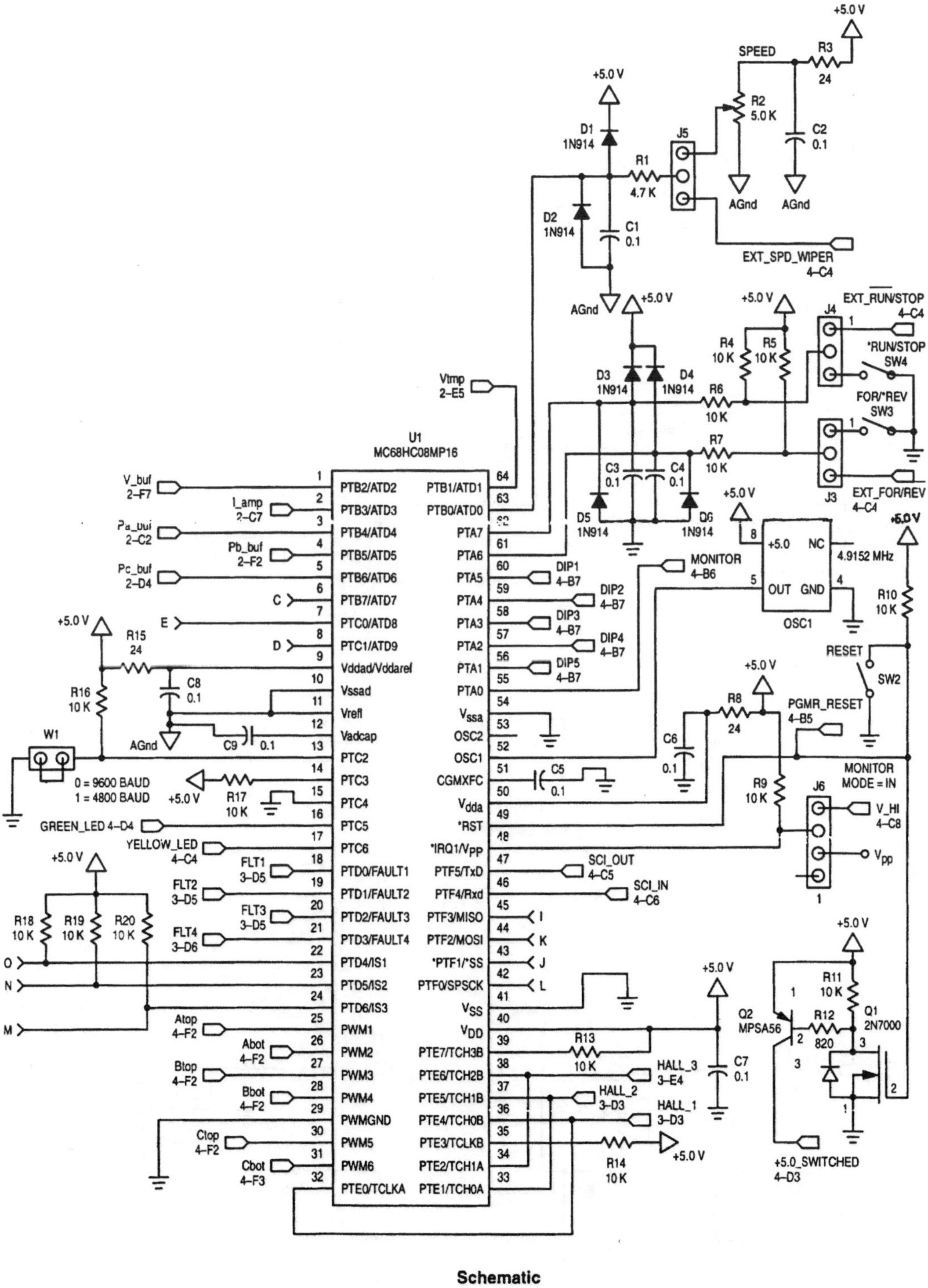

Figure 8.24 The ITC137 schematics and parts list. (Copyright of Motorola, used with permission.)

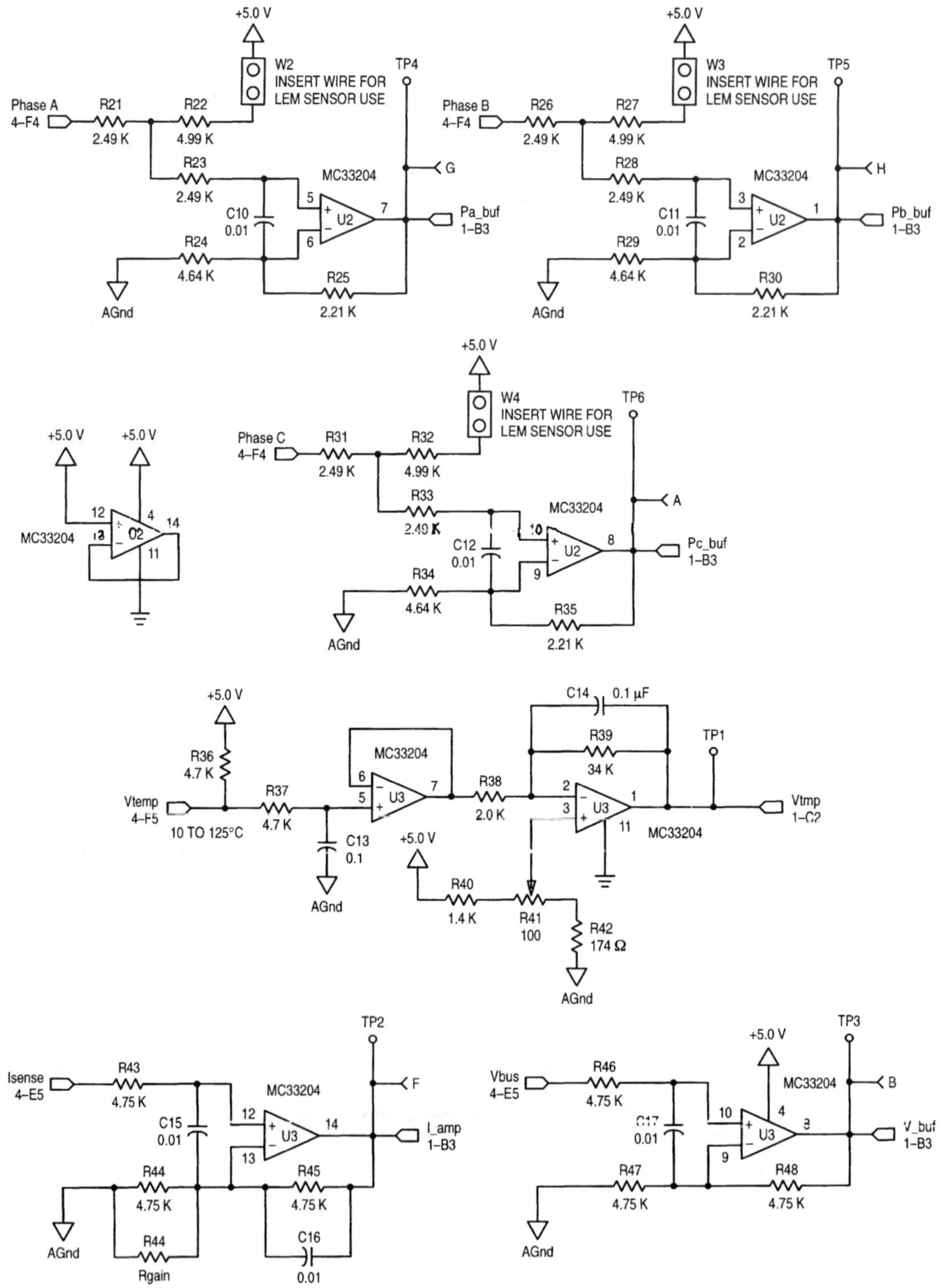

Schematic

Figure 8.24 *Continued.*

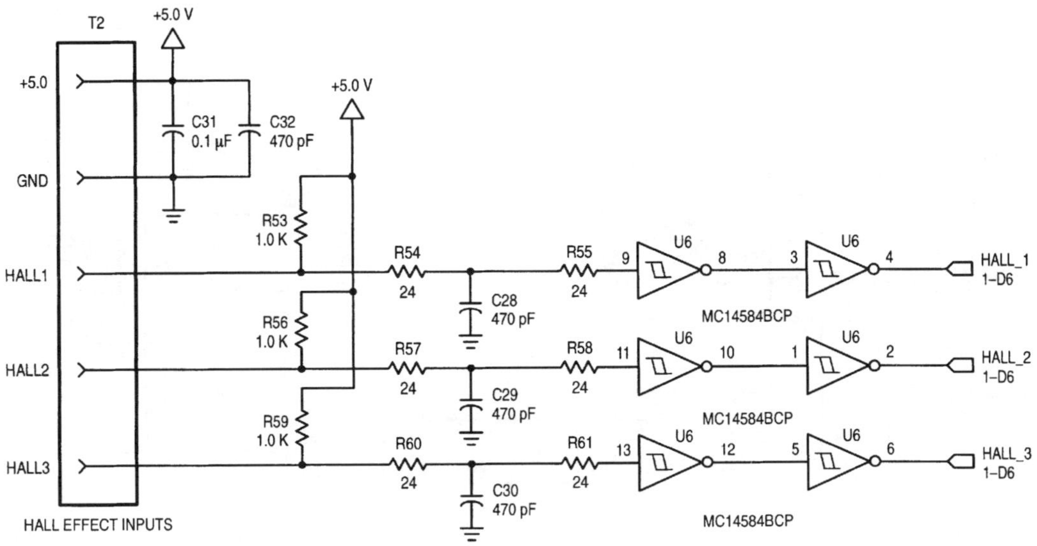

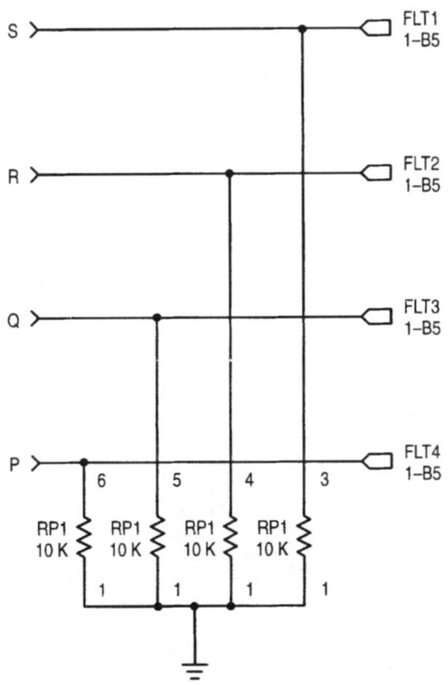

Schematic

Figure 8.24 Continued.

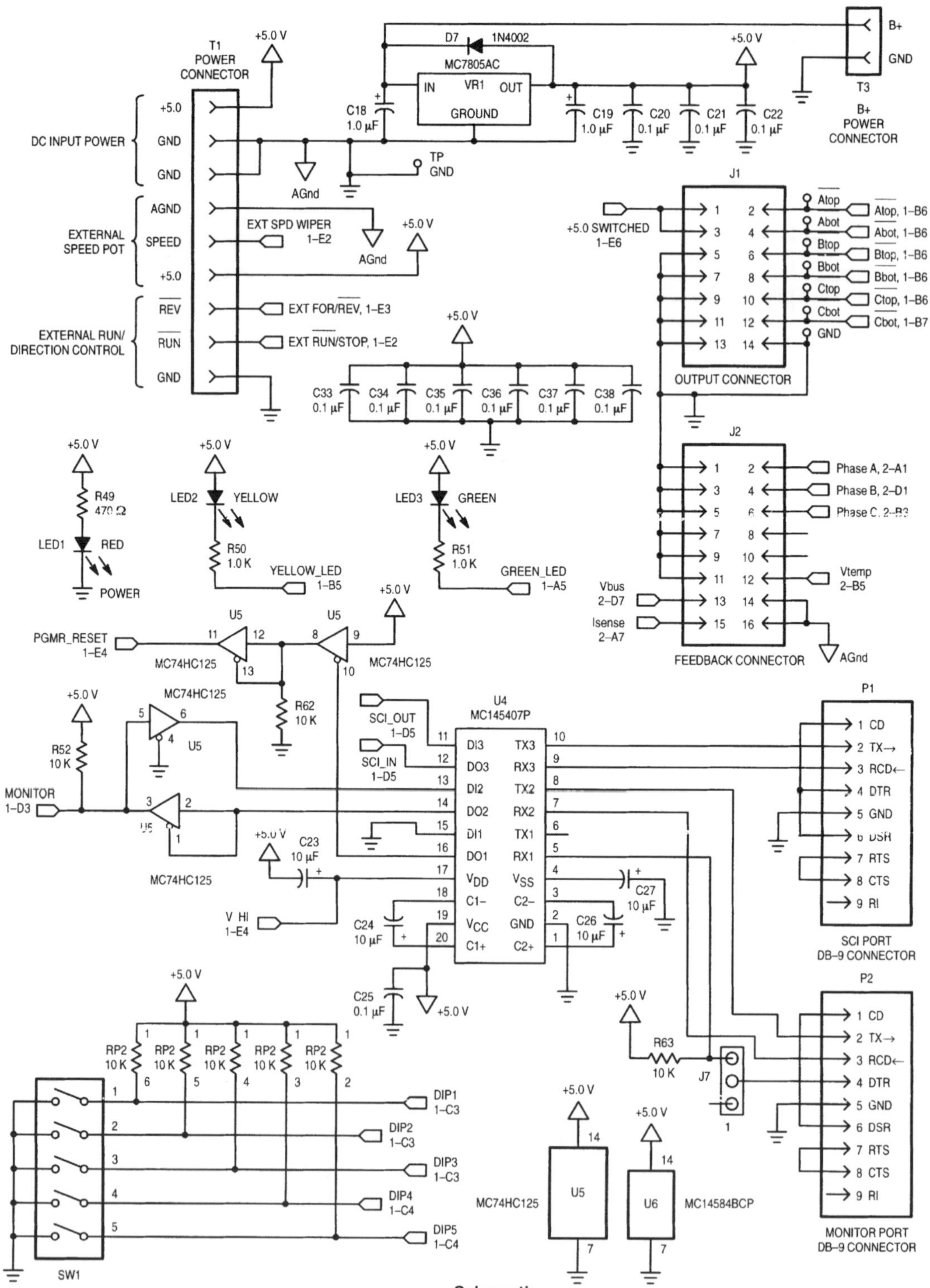

Schematic

Figure 8.24 Continued.

Parts List

Designators	Qty	Description	Manufacturer	Part Number
C1–C9,C13,C14,C20–C22 C25,C31,C33–C38	22	0.1 ;µf Capacitor	Sprague	1C105Z5U104M050B
C10–12,C15–C17	6	0.01 ;µf Capacitor	Sprague	1C105Z5U103M050B
C18,C19	2	1 ;µF Electrolytic cap	Mepco–Centralab	CN15A220K
C23,C24,C26,C27	4	10 ;µF Electrolytic cap	Digi–Key	P5272
C28–C30,C32	4	470 pf Capacitor	Sprague	1C105Z5U471M050B
D1–6	6	Small signal Diode		1N914
D7	1	General–purpose diode	Motorola	1N4002
J1	1	2 × 7 .1o.c. Jumper block note: 2	Digi–Key	S2011–36–ND
J2	1	2 × 8 .1o.c. Jumper block note: 2	Digi–Key	S2011–36–ND
J3,J4,J5,J7	4	1 × 3 .1o.c. Jumper block note: 3	Digi–Key	S1011–36–ND
J6	1	1 × 4 .1o.c. Jumper block note: 3	digi–Key	S1011–36–ND
LED1	1	Red LED General Insruments	MV5774C	
LED2	1	Yellow LED	General Insruments	MV5374C
LED3	1	Green LED	General Insruments	MV5474C
P1,P2	2	DB–9 Connector (Female	Digi–Key	A2100–ND
Q1	1	Small signal FET transistor	Motorola	2N7000
Q2	1	Small signal *pnp* transistor	Motorola	MPSA56
PR1,RP2	2	10k ohm 6 pin SIP resistor pack	Digi–Key	770–61R10K–ND
R2	1	5k ohm Variable resistor	Clarostat	392JB–5k0–S
R4–R7,R9,R10,R11,R13,R14 R16–R20,R52,R62,R63	17	10k ohm Resistor	Yageo	
R12	1	820 ohm Resistor	Yageo	
R21,R23,R26,R28,R31,R33	6	2.49K ohm Resistor 1%	Yageo	
R22,R27,R32	3	4.99K ohm Resistor 1%	Yageo	
R24,R29,R34	3	4.64K ohm Resistor 1%	Yageo	
R25,R30,R35	3	2.21K ohm Resistor 1%	Yageo	
R1,R36,R37	3	4.7k ohm Resistor	Yageo	
R38	1	2.00K ohm Resistor 1%	Yageo	
R39	1	34.0K ohm Resistor 1%	Yageo	
R40	1	1.40K ohm Resistor 1%	Yageo	
R41	1	100 ohm Trim–pot	Digi–Key	3386P–101–ND
R42	1	174 ohm Resistor 1%	Yageo	
R43–48	6	4.75K ohm Resistor 1%	Yageo	
R49	1	470 ohm Resistor	Yageo	
R50,R51,R53,R56,R59	5	1K ohm Resistor	Yageo	
R3,R8,R15,R54,R55 R57,R58,R60,R61	9	24 ohm Resistor	Yageo	
SW1	1	5–Pos. dip switch	CTS	CT2065–ND
SW2	1	SPST pushbutton switch	NKK	AB15AP–FA
SW3,SW4	2	SPST toggle switch	NKK	A12AH
T1	1	8 Screw terminal connector Phoenix Contact	MKDSN 1,5/8–5,08	
T2	1	5 Screw terminal connector	Phoenix Contact	MKDSN 1,5/5–5,08
T3	1	2–Screw terminal connector	Phoenix Contact	MKDSN 1,5/2–5,08
U1 Socketed	1	Microprocessor	Motorola	MC68HC708MP16
U1X U1 Socket	1	QFP 64 pin socket	Prine Distributors	FPQ–64–0.8–10A
U2,U3	1	Quad Op–Amp	Motorola	MC33204P
U4	1	RS–232 Driver/Receiver	Motorola	MC145407P
U5	1	Quad bus driver	Motorola	MC74HC125P
U6	1	Hex Schmitt trigger	Motorola	MC14585BCP
VR1	1	Voltage regulator	Motorola	MC7805ACT
OSC1	1	4.9152 MHz Oscillator	Digi–Key/CTS	CTS156–ND
GND,GND,AGND	3	Test point black	Components Corp.	TP–104–01–00
Atop,Btop,Ctop,TP1,TP5,TP6	6	Test point red	Components Corp.	TP–104–01–02
Abot,Bbot,Cbot,TP2,TP3,TP4,Vpp	7	Test point yellow	components Corp.	TP–104–01–04
No designator	1	4–40 × 1/4″ Screw for VR1		
No designator	1	4–40 Nuts for VR1		
No designator	5	Shorting jumpers for J3–J7	Digi–Key	929955–06–ND
No designator	6	Self–stick rubber feet		
ITC137	1	PC board		

Figure 8.24 *Continued*

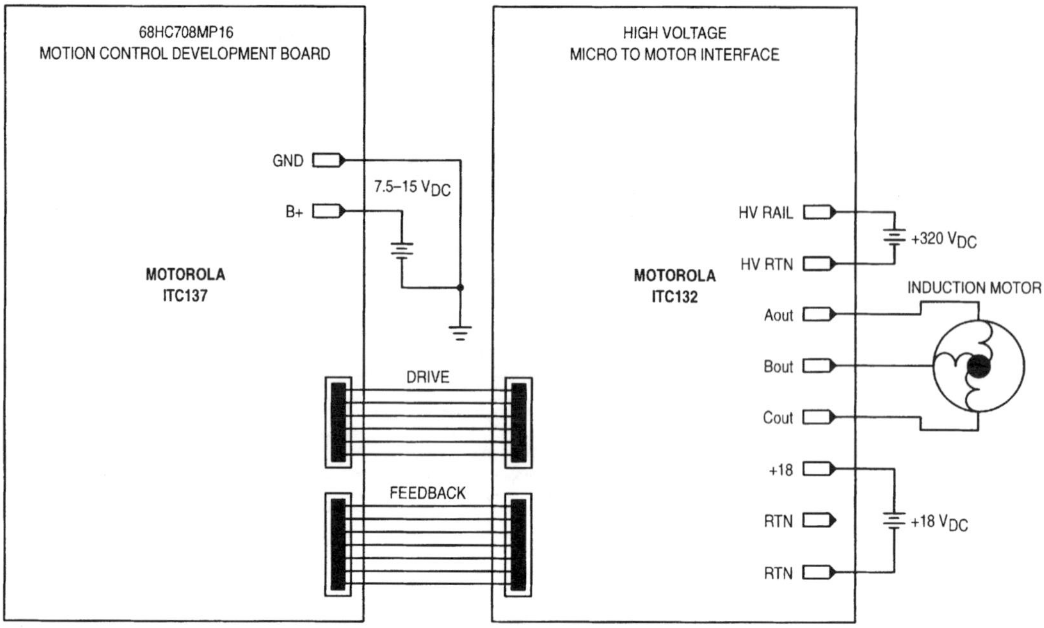

Figure 8.25 Induction motor controlled by the ITC137 motion control development board. (Copyright of Motorola, used with permission.)

References

Berringer, K., AC motor drive using integrated power stage, *Motorola Semiconductor Application Note AN1524*, Motorola, AZ, 1995.

Deatherage, J. and Hunsinger, J., Brushless DC motor control using the MC68HC705MC4, *Motorola Semiconductor Application Note AN1321*, Motorola, AZ, 1997.

Chavez, W., Brushless DC motor drive incorporates small integrated circuit package MOSFETs, *Motorola Semiconductor Application Note AN1321*, Motorola, AZ, 1993.

Gray, J., Lucas B., and Schultz, ITC137 68HC708MP16 Motion Control Development Board, *Motorola Semiconductor Application Note AN1624*, Motorola, AZ, 1997.

Krause, P. C., Wasynczuk, O., and Sudhoff, S. D., *Analysis of Electric Machinery*, IEEE Press, New York, 1995.

Krause, P. C. and Wasynczuk, O., *Electromechanical Motion Devices*, McGraw-Hill, New York, 1989.

Wilson, D., Get your motor running with the MC68HC708MP16, *Motorola Semiconductor Application Note AN1712*, Motorola, AZ, 1997.

Appendix 8.1.1 Direct-Current Servo Motor Driver.
(Copyright of Motorola, used with permission.)

MOTOROLA

MC33030

DC Servo Motor Controller/Driver

DC SERVO MOTOR CONTROLLER/DRIVER

SEMICONDUCTOR TECHNICAL DATA

The MC33030 is a monolithic DC servo motor controller providing all active functions necessary for a complete closed loop system. This device consists of an on–chip op amp and window comparator with wide input common–mode range, drive and brake logic with direction memory, Power H–Switch driver capable of 1.0 A, independently programmable over–current monitor and shutdown delay, and over–voltage monitor. This part is ideally suited for almost any servo positioning application that requires sensing of temperature, pressure, light, magnetic flux, or any other means that can be converted to a voltage.

Although this device is primarily intended for servo applications, it can be used as a switchmode motor controller.

- On–Chip Error Amp for Feedback Monitoring
- Window Detector with Deadband and Self Centering Reference Input
- Drive/Brake Logic with Direction Memory
- 1.0 A Power H–Switch
- Programmable Over–Current Detector
- Programmable Over–Current Shutdown Delay
- Over–Voltage Shutdown

P SUFFIX
PLASTIC PACKAGE
CASE 648C
(DIP–16)

DW SUFFIX
PLASTIC PACKAGE
CASE 751G
(SOP–16L)

Representative Block Diagram

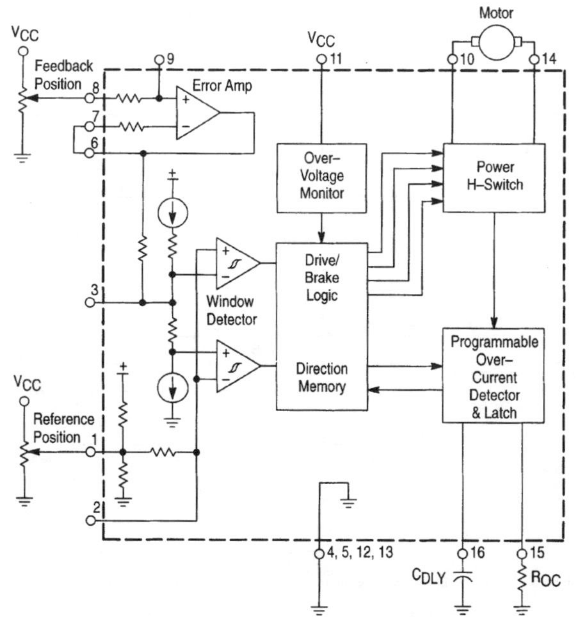

This device contains 119 active transistors.

PIN CONNECTIONS

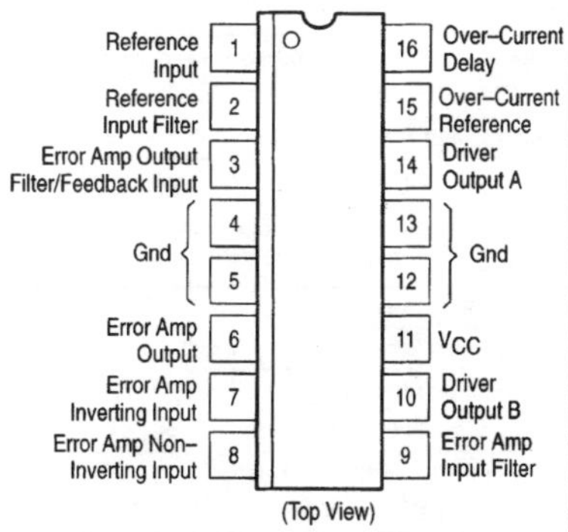

Pins 4, 5, 12 and 13 are electrical ground and heat sink pins for IC.

ORDERING INFORMATION

Device	Operating Temperature Range	Package
MC33030DW	$T_A = -40°$ to $+85°C$	SOP–16L
MC33030P		DIP–16

Appendix 8.1.2

MC33030

MAXIMUM RATINGS

Rating	Symbol	Value	Unit
Power Supply Voltage	V_{CC}	36	V
Input Voltage Range Op Amp, Comparator, Current Limit (Pins 1, 2, 3, 6, 7, 8, 9, 15)	V_{IR}	−0.3 to V_{CC}	V
Input Differential Voltage Range Op Amp, Comparator (Pins 1, 2, 3, 6, 7, 8, 9)	V_{IDR}	−0.3 to V_{CC}	V
Delay Pin Sink Current (Pin 16)	$I_{DLY(sink)}$	20	mA
Output Source Current (Op Amp)	I_{source}	10	mA
Drive Output Voltage Range (Note 1)	V_{DRV}	−0.3 to (V_{CC} + V_F)	V
Drive Output Source Current (Note 2)	$I_{DRV(source)}$	1.0	A
Drive Output Sink Current (Note 2)	$I_{DRV(sink)}$	1.0	A
Brake Diode Forward Current (Note 2)	I_F	1.0	A
Power Dissipation and Thermal Characteristics P Suffix, Dual In Line Case 648C Thermal Resistance, Junction–to–Air Thermal Resistance, Junction–to–Case (Pins 4, 5, 12, 13) DW Suffix, Dual In Line Case 751G Thermal Resistance, Junction–to–Air Thermal Resistance, Junction–to–Case (Pins 4, 5, 12, 13)	 $R_{\theta JA}$ $R_{\theta JC}$ $R_{\theta JA}$ $R_{\theta JC}$	 80 15 94 18	°C/W
Operating Junction Temperature	T_J	+150	°C
Operating Ambient Temperature Range	T_A	−40 to +85	°C
Storage Temperature Range	T_{stg}	−65 to +150	°C

NOTES: 1. The upper voltage level is clamped by the forward drop, V_F, of the brake diode.
2. These values are for continuous DC current. Maximum package power dissipation limits must be observed.

ELECTRICAL CHARACTERISTICS (V_{CC} = 14 V, T_A = 25°C, unless otherwise noted.)

Characteristic	Symbol	Min	Typ	Max	Unit
ERROR AMP					
Input Offset Voltage (− 40°C ≤ T_A ≤ 85°C) $V_{Pin\ 6}$ = 7.0 V, R_L = 100 k	V_{IO}	–	1.5	10	mV
Input Offset Current ($V_{Pin\ 6}$ = 1.0 V, R_L = 100 k)	I_{IO}	–	0.7	–	nA
Input Bias Current ($V_{Pin\ 6}$ = 7.0 V, R_L = 100 k)	I_{IB}	–	7.0	–	nA
Input Common–Mode Voltage Range ΔV_{IO} = 20 mV, R_L = 100 k	V_{ICR}	–	0 to (V_{CC} − 1.2)	–	V
Slew Rate, Open Loop (V_{ID} = 0.5 V, C_L = 15 pF)	SR	–	0.40	–	V/μs
Unity–Gain Crossover Frequency	f_c	–	550	–	kHz
Unity–Gain Phase Margin	ϕm	–	63	–	deg.
Common–Mode Rejection Ratio ($V_{Pin\ 6}$ = 7.0 V, R_L = 100 k)	CMRR	50	82	–	dB
Power Supply Rejection Ratio V_{CC} = 9.0 to 16 V, $V_{Pin\ 6}$ = 7.0 V, R_L = 100 k	PSRR	–	89	–	dB
Output Source Current ($V_{Pin\ 6}$ = 12 V)	I_{O+}	–	1.8	–	mA
Output Sink Current ($V_{Pin\ 6}$ = 1.0 V)	I_{O-}	–	250	–	μA
Output Voltage Swing (R_L = 17 k to Ground)	V_{OH} V_{OL}	12.5 –	13.1 0.02	– –	V V

NOTES: 3. The upper or lower hysteresis will be lost when operating the Input, Pin 3, close to the respective rail. Refer to Figure 4.
4. Low duty cycle pulse techniques are used during test to maintain junction temperature as close to ambient temperature as possible.

Appendix 8.1.3

MC33030

ELECTRICAL CHARACTERISTICS (continued) (V_{CC} = 14 V, T_A = 25°C, unless otherwise noted.)

Characteristic	Symbol	Min	Typ	Max	Unit
WINDOW DETECTOR					
Input Hysteresis Voltage ($V_1 - V_4$, $V_2 - V_3$, Figure 18)	V_H	25	35	45	mV
Input Dead Zone Range ($V_2 - V_4$, Figure 18)	V_{IDZ}	166	210	254	mV
Input Offset Voltage ($\|[V_2 - V_{Pin\ 2}] - [V_{Pin\ 2} - V_4]\|$ Figure 18)	V_{IO}	–	25	–	mV
Input Functional Common–Mode Range (Note 3) Upper Threshold Lower Threshold	 V_{IH} V_{IL}	 – –	 (V_{CC} – 1.05) 0.24	 – –	V
Reference Input Self Centering Voltage Pins 1 and 2 Open	V_{RSC}	–	(1/2 V_{CC})	–	V
Window Detector Propagation Delay Comparator Input, Pin 3, to Drive Outputs V_{ID} = 0.5 V, $R_{L(DRV)}$ = 390 Ω	$t_{p(IN/DRV)}$	–	2.0	–	µs
OVER–CURRENT MONITOR					
Over–Current Reference Resistor Voltage (Pin 15)	R_{OC}	3.9	4.3	4.7	V
Delay Pin Source Current V_{DLY} = 0 V, R_{OC} = 27 k, I_{DRV} = 0 mA	$I_{DLY(source)}$	–	5.5	6.9	µA
Delay Pin Sink Current (R_{OC} = 27 k, I_{DRV} = 0 mA) V_{DLY} = 5.0 V V_{DLY} = 8.3 V V_{DLY} = 14 V	$I_{DLY(sink)}$	 – – –	 0.1 0.7 16.5	 – – –	mA
Delay Pin Voltage, Low State (I_{DLY} = 0 mA)	$V_{OL(DLY)}$	–	0.3	0.4	V
Over–Current Shutdown Threshold V_{CC} = 14 V V_{CC} = 8.0 V	$V_{th(OC)}$	 6.8 5.5	 7.5 6.0	 8.2 6.5	V
Over–Current Shutdown Propagation Delay Delay Capacitor Input, Pin 16, to Drive Outputs, V_{ID} = 0.5 V	$t_{p(DLY/DRV)}$	–	1.8	–	µs
POWER H–SWITCH					
Drive–Output Saturation (– 40°C ≤ T_A ≤ + 85°C, Note 4) High–State (I_{source} = 100 mA) Low–State (I_{sink} = 100 mA)	 $V_{OH(DRV)}$ $V_{OL(DRV)}$	 (V_{CC} – 2) –	 (V_{CC} – 0.85) 0.12	 – 1.0	V
Drive–Output Voltage Switching Time (C_L = 15 pF) Rise Time Fall Time	 t_r t_f	 – –	 200 200	 – –	ns
Brake Diode Forward Voltage Drop (I_F = 200 mA, Note 4)	V_F	–	1.04	2.5	V
TOTAL DEVICE					
Standby Supply Current	I_{CC}	–	14	25	mA
Over–Voltage Shutdown Threshold (– 40°C ≤ T_A ≤ + 85°C)	$V_{th(OV)}$	16.5	18	20.5	V
Over–Voltage Shutdown Hysteresis (Device "off" to "on")	$V_{H(OV)}$	0.3	0.6	1.0	V
Operating Voltage Lower Threshold (– 40°C ≤ T_A ≤ + 85°C)	V_{CC}	–	7.5	8.0	V

NOTES: 3. The upper or lower hysteresis will be lost when operating the Input, Pin 3, close to the respective rail. Refer to Figure 4.
4. Low duty cycle pulse techniques are used during test to maintain junction temperature as close to ambient temperature as possible.

Appendix 8.1.4

MC33030

Figure 1. Error Amp Input Common–Mode Voltage Range versus Temperature

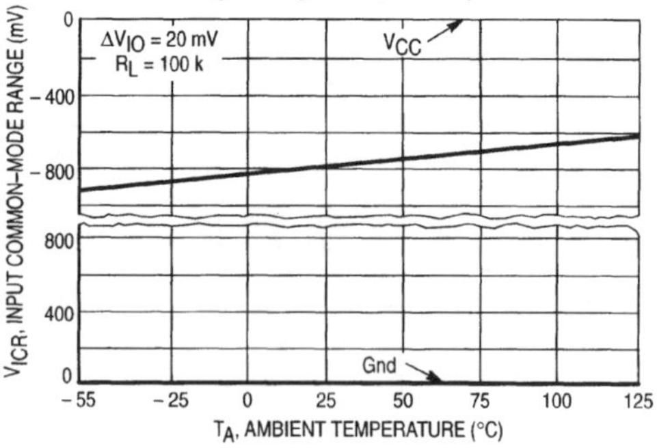

Figure 2. Error Amp Output Saturation versus Load Current

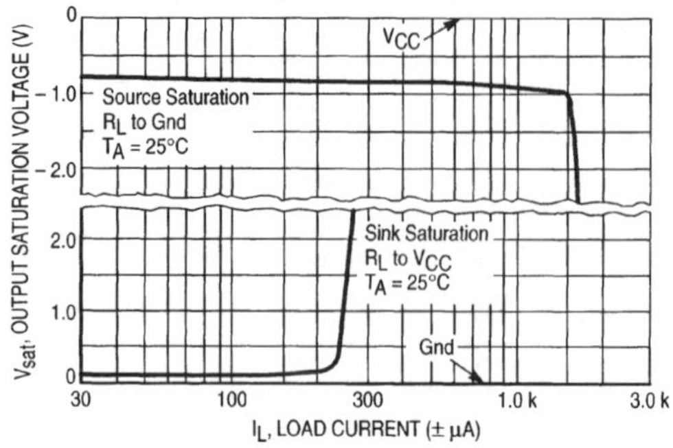

Figure 3. Open Loop Voltage Gain and Phase versus Frequency

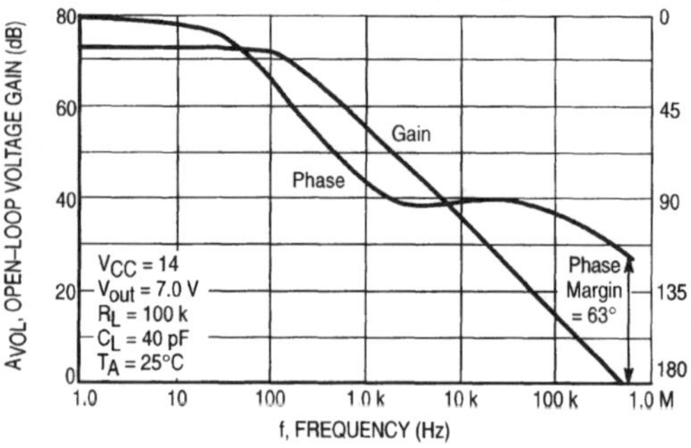

Figure 4. Window Detector Reference–Input Common–Mode Voltage Range versus Temperature

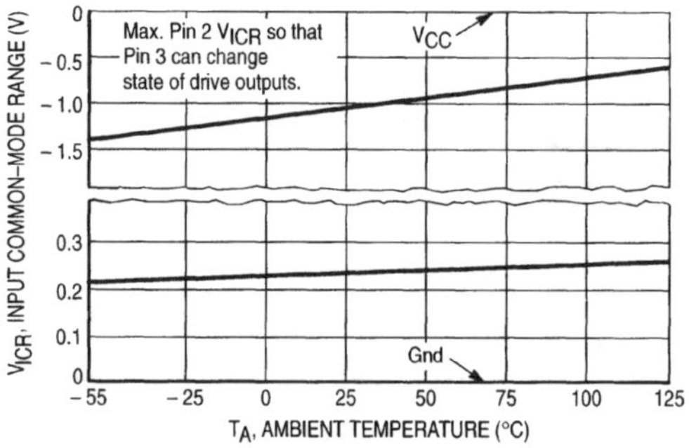

Figure 5. Window Detector Feedback–Input Thresholds versus Temperature

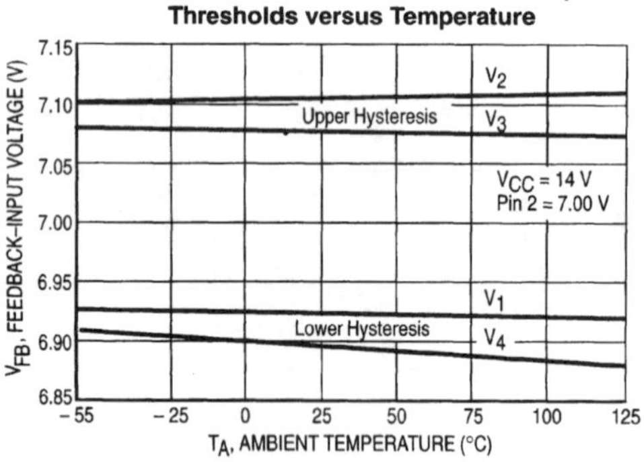

Figure 6. Output Driver Saturation versus Load Current

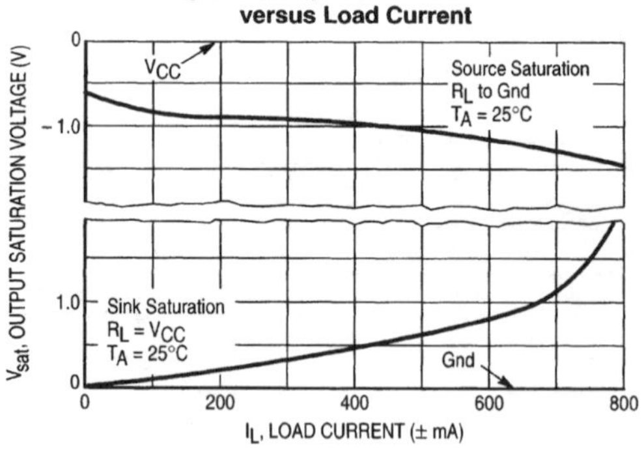

Appendix 8.1.5

MC33030

Figure 7. Brake Diode Forward Current versus Forward Voltage

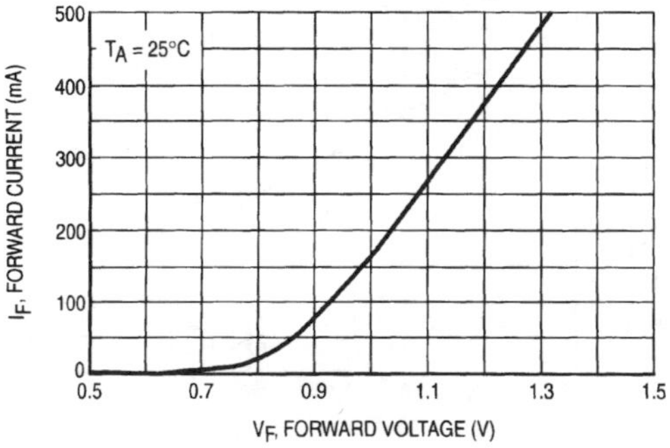

Figure 8. Output Source Current–Limit versus Over–Current Reference Resistance

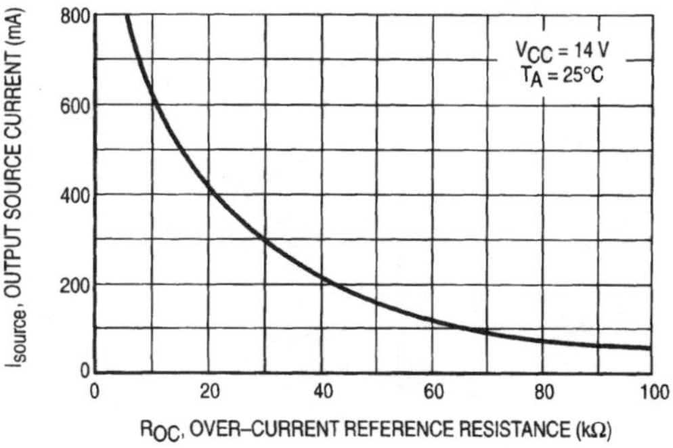

Figure 9. Output Source Current–Limit versus Temperature

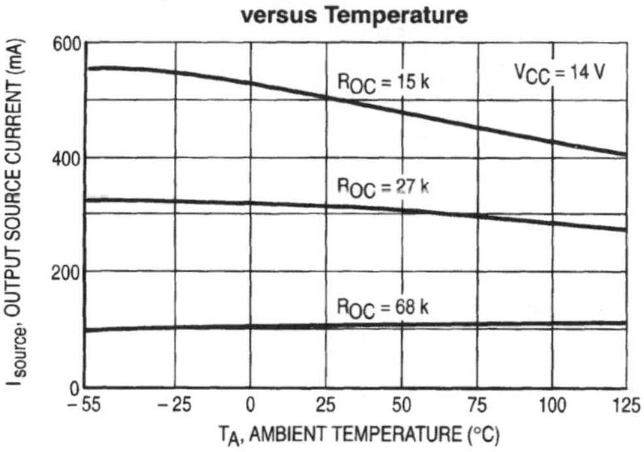

Figure 10. Normalized Delay Pin Source Current versus Temperature

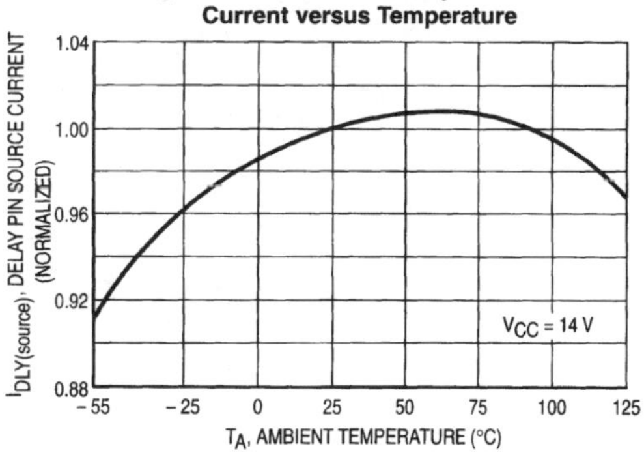

Figure 11. Normalized Over–Current Delay Threshold Voltage versus Temperature

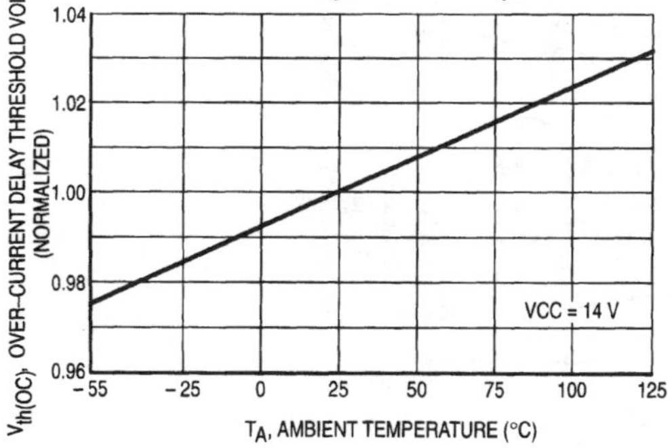

Figure 12. Supply Current versus Supply Voltage

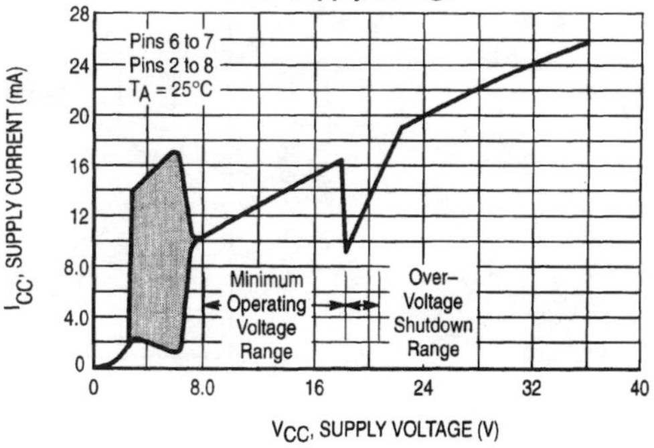

Appendix 8.1.6

MC33030

Figure 13. Normalized Over–Voltage Shutdown Threshold versus Temperature

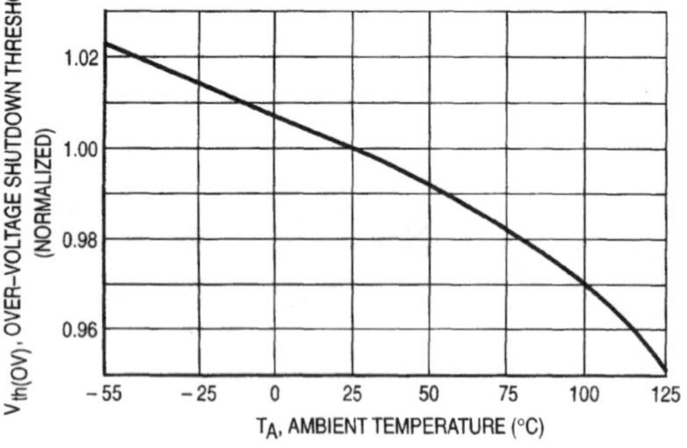

Figure 14. Normalized Over–Voltage Shutdown Hysteresis versus Temperature

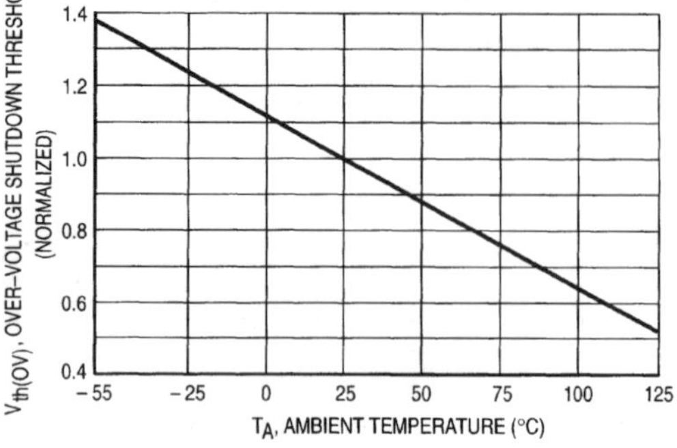

Figure 15. P Suffix (DIP–16) Thermal Resistance and Maximum Power Dissipation versus P.C.B. Copper Length

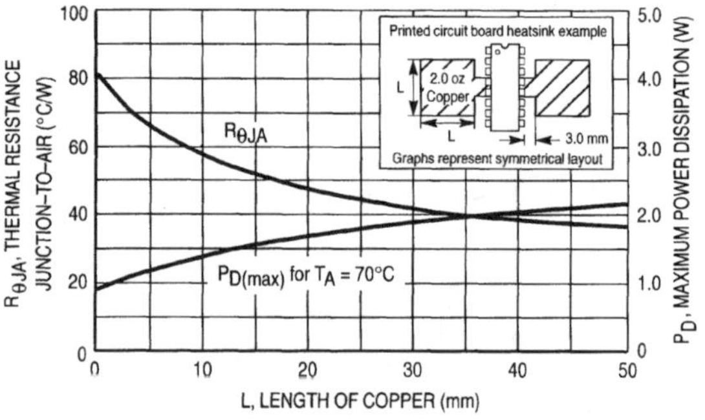

Figure 16. DW Suffix (SOP–16L) Thermal Resistance and Maximum Power Dissipation versus P.C.B. Copper Length

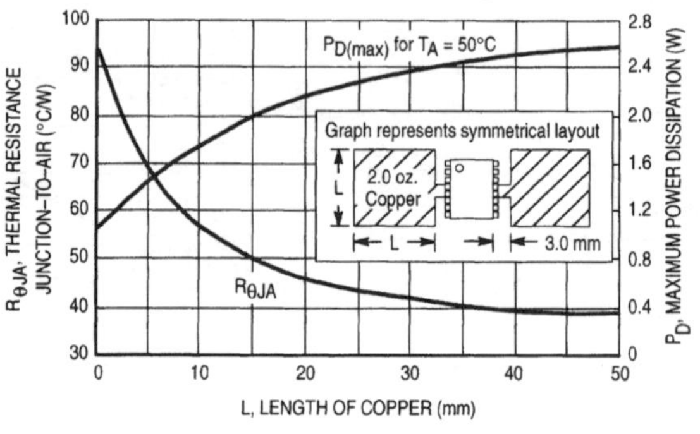

Appendix 8.1.7

MC33030

Figure 17. Representative Block Diagram and Typical Servo Application

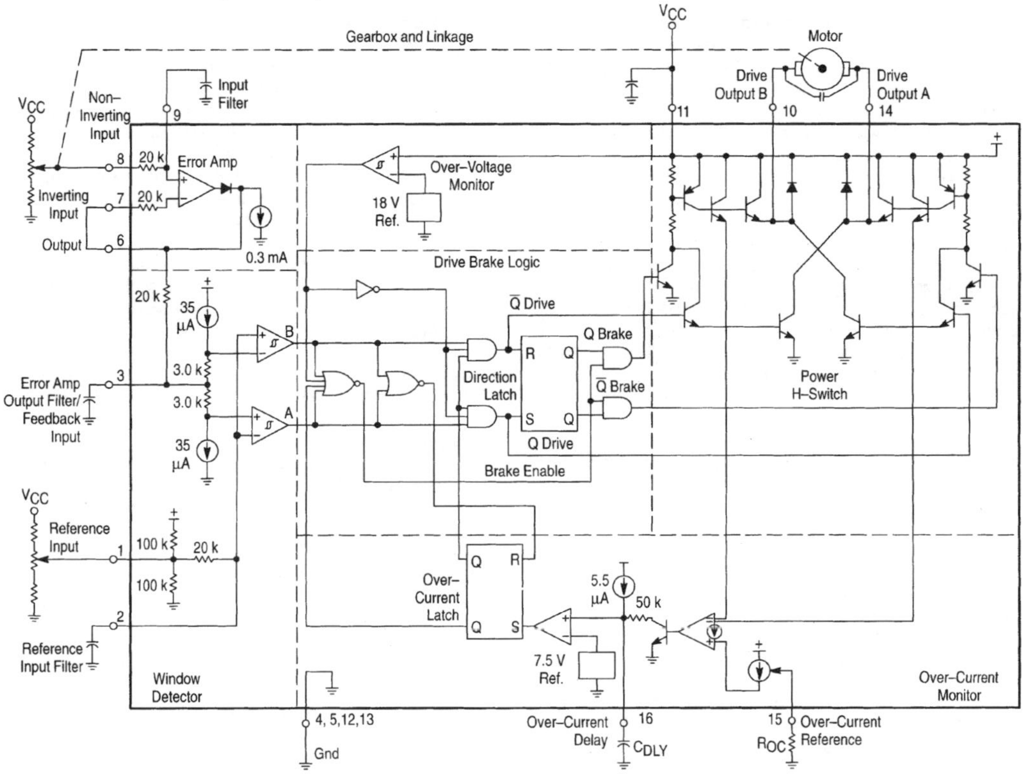

If $V_{Pin\ 3}$ should continue to rise and become greater than V_2, the actuator will have over shot the dead zone range and cause the motor to run in Direction A until $V_{Pin\ 3}$ is equal to V_3. The Drive/Brake behavior for Direction A is identical to that of B. Overshooting the dead zone range in both directions can cause the servo system to continuously hunt or oscillate. Notice that the last motor run–direction is stored in the direction latch. This information is needed to determine whether Q or $\overline{Q}$ Brake is to be enabled when $V_{Pin\ 3}$ enters the dead zone range. The dashed lines in [8,9] indicate the resulting waveforms of an over–current condition that has exceeded the programmed time delay. Notice that both Drive Outputs go into a high impedance state until $V_{Pin\ 2}$ is readjusted so that $V_{Pin\ 3}$ enters or crosses through the dead zone [7, 4].

The inputs of the Error Amp and Window Detector can be susceptible to the noise created by the brushes of the DC motor and cause the servo to hunt. Therefore, each of these inputs are provided with an internal series resistor and are pinned out for an external bypass capacitor. It has been found that placing a capacitor with *short leads* directly across the brushes will significantly reduce noise problems. Good quality RF bypass capacitors in the range of 0.001 to 0.1 μF may be required. Many of the more economical motors will generate significant levels of RF energy over a spectrum that extends from DC to beyond 200 MHz. The capacitance value and method of noise filtering must be determined on a system by system basis.

Thus far, the operating description has been limited to servo systems in which the motor mechanically drives a potentiometer for position sensing. Figures 19, 20, 27, and 31 show examples that use light, magnetic flux, temperature, and pressure as a means to drive the feedback element. Figures 21, 22 and 23 are examples of two position, open loop servo systems. In these systems, the motor runs the actuator to each end of its travel limit where the Over–Current Monitor detects a locked rotor condition and shuts down the drive. Figures 32 and 33 show two possible methods of using the MC33030 as a switching motor controller. In each example a fixed reference voltage is applied to Pin 2. This causes $V_{Pin\ 3}$ to be less than V_4 and Drive Output A, Pin 14, to be in a low state saturating the TIP42 transistor. In Figure 32, the motor drives a tachometer that generates an ac voltage proportional to RPM. This voltage is rectified, filtered, divided down by the speed set potentiometer, and applied to Pin. 8. The motor will accelerate until $V_{Pin\ 3}$ is equal to V_1 at which time Pin 14 will go to a high state and terminate the motor drive. The motor will now coast until $V_{Pin\ 3}$ is less than V_4 where upon drive is then reapplied. The system operation of Figure 31 is identical to that of 32 except the signal at Pin 3 is an amplified average of the motors drive and back EMF voltages. Both systems exhibit excellent control of RPM with variations of V_{CC}; however, Figure 32 has somewhat better torque characteristics at low RPM.

Appendix 8.1.8

MC33030

OPERATING DESCRIPTION

The MC33030 was designed to drive fractional horsepower DC motors and sense actuator position by voltage feedback. A typical servo application and representative internal block diagram are shown in Figure 17. The system operates by setting a voltage on the reference input of the Window Dectector (Pin 1) which appears on (Pin 2). A DC motor then drives a position sensor, usually a potentiometer driven by a gear box, in a corrective fashion so that a voltage proportional to position is present at Pin 3. The servo motor will continue to run until the voltage at Pin 3 falls within the dead zone, which is centered about the reference voltage.

The Window Detector is composed of two comparators, A and B, each containing hysteresis. The reference input, common to both comparators, is pre–biased at 1/2 V_{CC} for simple two position servo systems and can easily be overriden by an external voltage divider. The feedback voltage present at Pin 3 is connected to the center of two resistors that are driven by an equal magnitude current source and sink. This generates an offset voltage at the input of each comparator which is centered about Pin 3 that can float virtually from V_{CC} to ground. The sum of the upper and lower offset voltages is defined as the window detector input dead zone range.

To increase system flexibility, an on–chip Error Amp is provided. It can be used to buffer and/or gain–up the actuator position voltage which has the effect of narrowing the dead zone range. A PNP differential input stage is provided so that the input common–mode voltage range will include ground. The main design goal of the error amp output stage was to be able to drive the window detector input. It typically can source 1.8 mA and sink 250 μA. Special design considerations must be made if it is to be used for other applications.

The Power H–Switch provides a direct means for motor drive and braking with a maximum source, sink, and brake current of 1.0 A continuous. Maximum package power dissipation limits must be observed. Refer to Figure 15 for thermal information. For greater drive current requirements, a method for buffering that maintains all the system features is shown in Figure 30.

The Over–Current Monitor is designed to distinguish between motor start–up or locked rotor conditions that can occur when the actuator has reached its travel limit. A fraction of the Power H–Switch source current is internally fed into one of the two inverting inputs of the current comparator, while the non–inverting input is driven by a programmable current reference. This reference level is controlled by the resistance value selected for R_{OC}, and must be greater than the required motor run–current with its mechanical load over temperature; refer to Figure 8. During an over–current condition, the comparator will turn off and allow the current source to charge the delay capacitor, C_{DLY}. When C_{DLY} charges to a level of 7.5 V, the set input of the over–current latch will go high, disabling the drive and brake functions of the Power H–Switch. The programmable time delay is determined by the capacitance value–selected for C_{DLY}.

$$t_{DLY} = \frac{V_{ref}\,C_{DLY}}{I_{DLY(source)}} = \frac{7.5\,C_{DLY}}{5.5\ \mu A} = 1.36\,C_{DLY} \text{ in } \mu F$$

This system allows the Power H–Switch to supply motor start–up current for a predetermined amount of time. If the rotor is locked, the system will time–out and shut–down. This feature eliminates the need for servo end–of–travel or limit switches. Care must be taken so as not to select too large of a capacitance value for C_{DLY}. An over–current condition for an excessively long time–out period can cause the integrated circuit to overheat and eventually fail. Again, the maximum package power dissipation limits must be observed. The over–current latch is reset upon power–up or by readjusting $V_{Pin\ 2}$ as to cause $V_{Pin\ 3}$ to enter or pass through the dead zone. This can be achieved by requesting the motor to reverse direction.

An Over–Voltage Monitor circuit provides protection for the integrated circuit and motor by disabling the Power H–Switch functions if V_{CC} should exceed 18 V. Resumption of normal operation will commence when V_{CC} falls below 17.4 V.

A timing diagram that depicts the operation of the Drive/Brake Logic section is shown in Figure 18. The waveforms grouped in [1] show a reference voltage that was preset, appearing on Pin 2, which corresponds to the desired actuator position. The true actuator position is represented by the voltage on Pin 3. The points V_1 through V_4 represent the input voltage thresholds of comparators A and B that cause a change in their respective output state. They are defined as follows:

V_1 = Comparator B turn–off threshold
V_2 = Comparator A turn–on threshold
V_3 = Comparator A turn–off threshold
V_4 = Comparator B turn–on threshold
V_1–V_4 = Comparator B input hysteresis voltage
V_2–V_3 = Comparator A input hysteresis voltage
V_2–V_4 = Window detector input dead zone range
$|(V_2 - V_{Pin2}) - (V_{Pin2} - V_4)|$ = Window detector input voltage

It must be remembered that points V_1 through V_4 always try to follow and center about the reference voltage setting if it is within the input common–mode voltage range of Pin 3; Figures 4 and 5. Initially consider that the feedback input voltage level is somewhere on the dashed line between V_2 and V_4 in [1]. This is within the dead zone range as defined above and the motor will be off. Now if the reference voltage is raised so that $V_{Pin\ 3}$ is less than V_4, comparator B will turn–on [3] enabling $\overline{Q}$ Drive, causing Drive Output A to sink and B to source motor current [8]. The actuator will move in Direction B until $V_{Pin\ 3}$ becomes greater than V_1. Comparator B will turn–off, activating the brake enable [4] and $\overline{Q}$ Brake [6] causing Drive Output A to go high and B to go into a high impedance state. The inertia of the mechanical system will drive the motor as a generator creating a positive voltage on Pin 10 with respect to Pin 14. The servo system can be stopped quickly, so as not to over–shoot through the dead zone range, by braking. This is accomplished by shorting the motor/generator terminals together. Brake current will flow into the diode at Drive Output B, through the internal V_{CC} rail, and out the emitter of the sourcing transistor at Drive Output A. The end of the solid line and beginning of the dashed for $V_{Pin\ 3}$ [1] indicates the possible resting position of the actuator after braking.

Appendix 8.1.9

MC33030

Figure 18. Timing Diagram

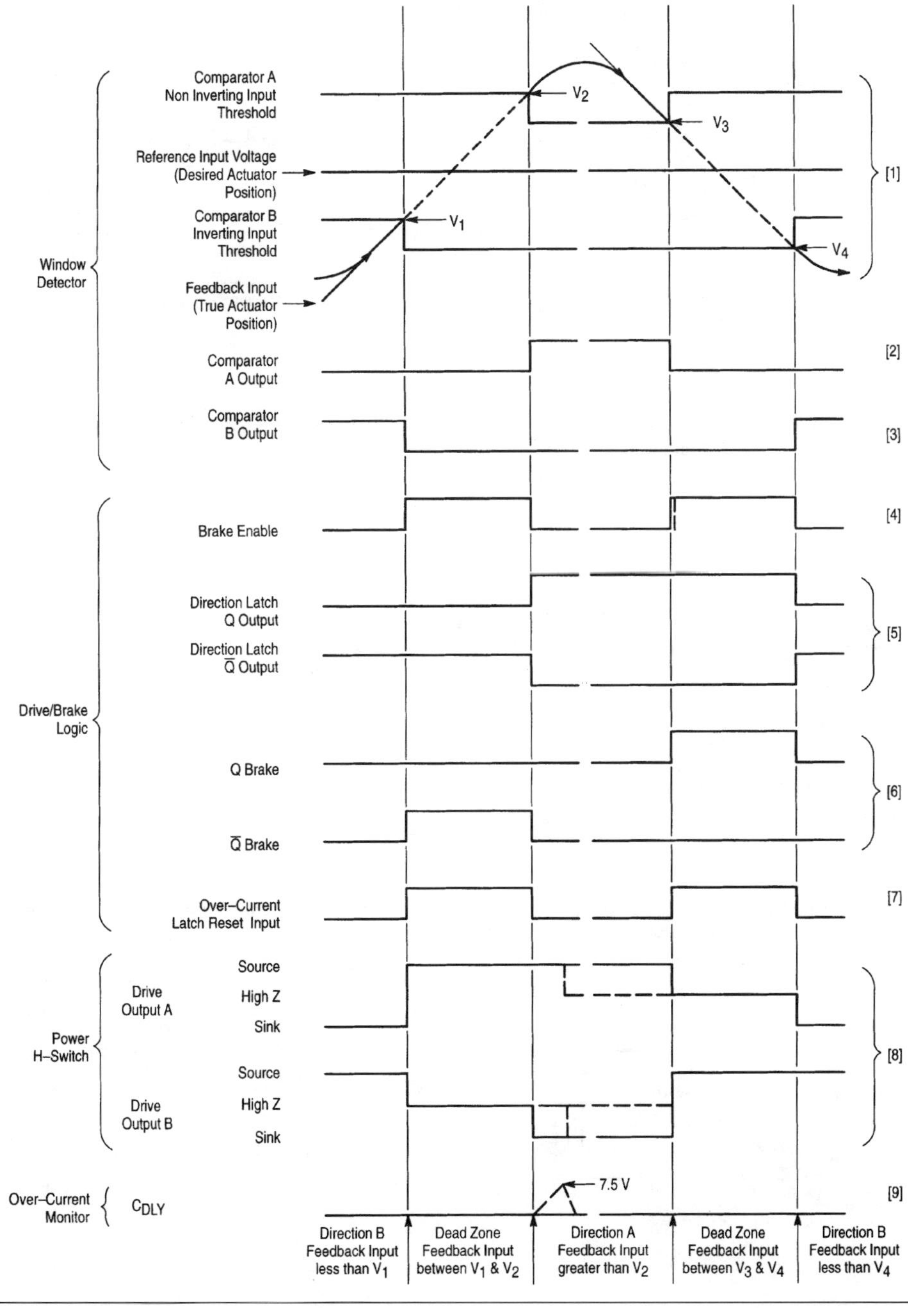

Appendix 8.1.10

MC33030

Figure 19. Solar Tracking Servo System

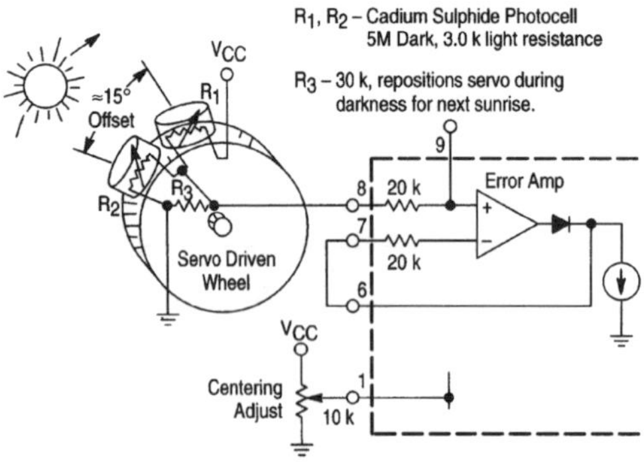

Figure 20. Magnetic Sensing Servo System

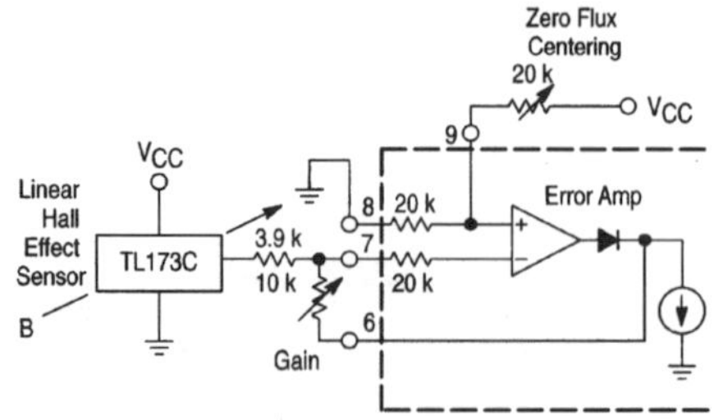

Typical sensitivity with gain set at 3.9 k is 1.5 mV/gauss.
Servo motor controls magnetic field about sensor.

Figure 21. Infrared Latched Two Position Servo System

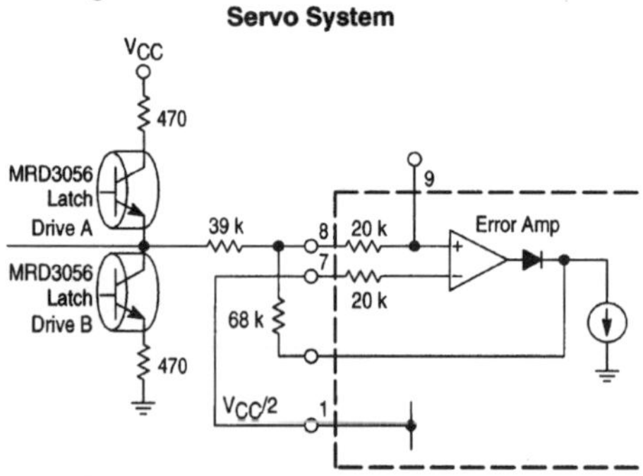

Over–current monitor (not shown) shuts down servo when end stop is reached.

Figure 22. Digital Two Position Servo System

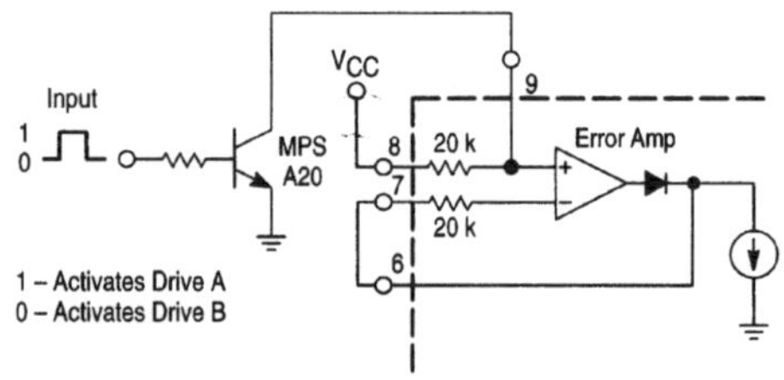

Over–current monitor (not shown) shuts down servo when end stop is reached.

Figure 23. 0.25 Hz Square–Wave Servo Agitator

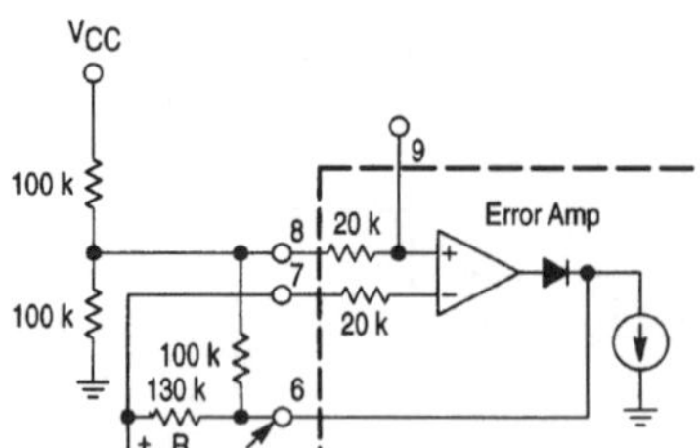

$$f \approx \frac{0.72}{RC}$$

$$R \geq 20\text{ k}$$

Figure 24. Second Order Low–Pass Active Filter

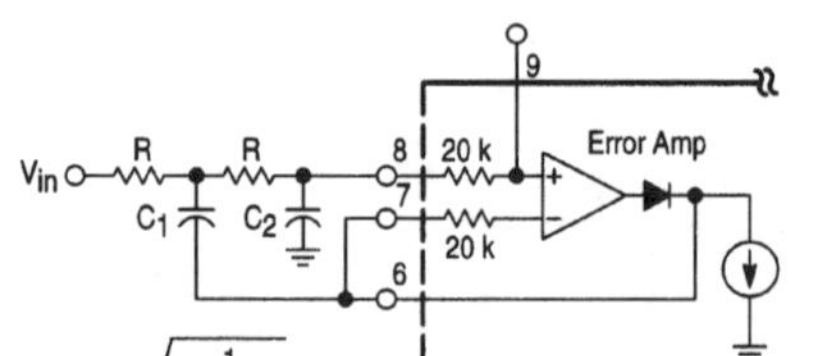

$$f_o = \frac{\sqrt{\frac{1}{R^2 C_1 C_2}}}{2\pi}$$

$$Q = \frac{\sqrt{\frac{C_1}{C_2}}}{2}$$

R = 1.0 M
C_1 = 1000 pF
C_2 = 100 pF

Appendix 8.1.11

MC33030

Figure 25. Notch Filter

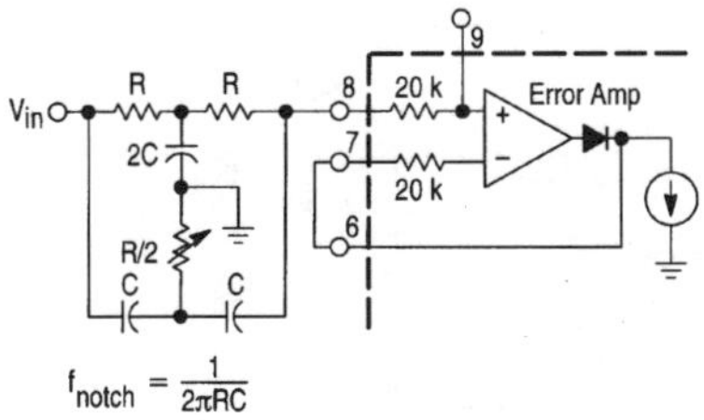

$$f_{notch} = \frac{1}{2\pi RC}$$

For 60 Hz R = 53.6 k, C = 0.05

Figure 26. Differential Input Amplifier

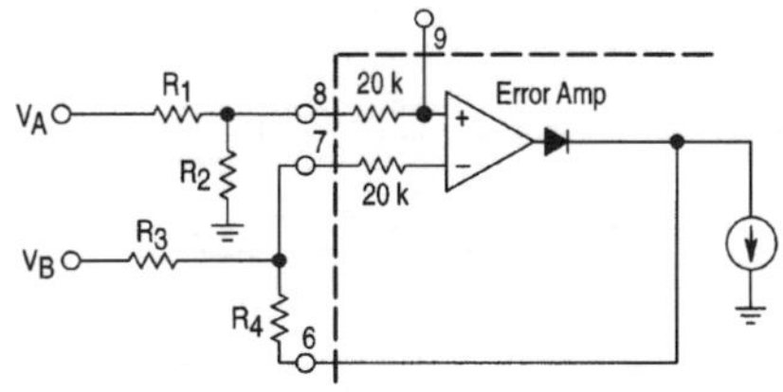

$$V_{Pin\ 6} = V_A\left(\frac{R_3 + R_4}{R_1 + R_2}\right)\frac{R_2}{R_3} - \left(\frac{R_4}{R_3}V_B\right)$$

Figure 27. Temperature Sensing Servo System

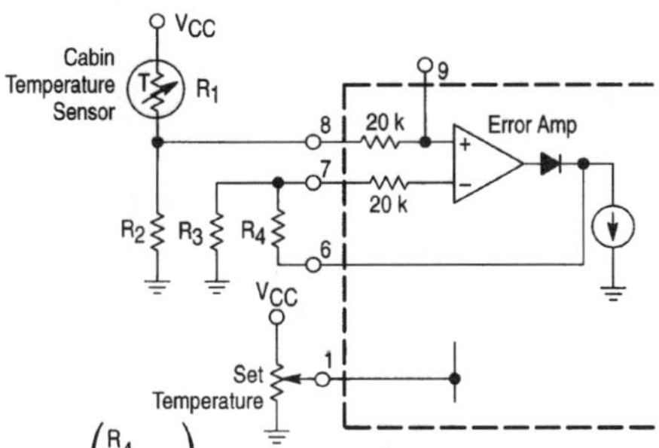

$$V_{Pin\ 6} = \frac{V_{CC}\left(\frac{R_4}{R_3} + 1\right)}{\left(\frac{R_1}{R_2} + 1\right)}$$

In this application the servo motor drives the heat/air conditioner modulator door in a duct system.

Figure 28. Bridge Amplifier

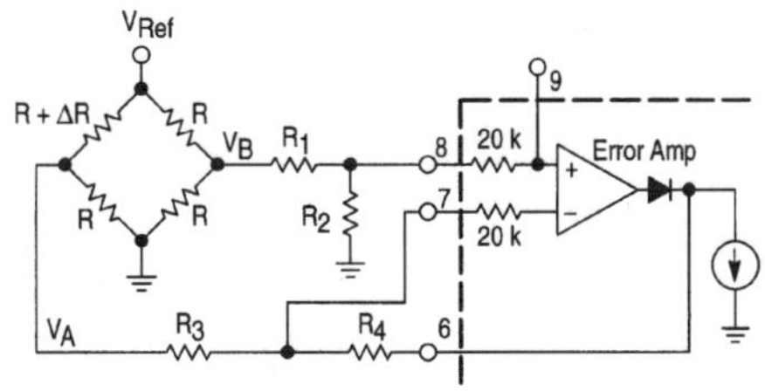

$$V_A - V_B = V_{Ref}\left(\frac{\Delta R}{4R + 2\Delta R}\right)$$

$$R_1 = R_3,\ R_2 = R_4,\ R_1 >> R$$

$$V_{Pin\ 6} = \frac{R_4}{R_3}(V_A - V_B)$$

Figure 29. Remote Latched Shutdown

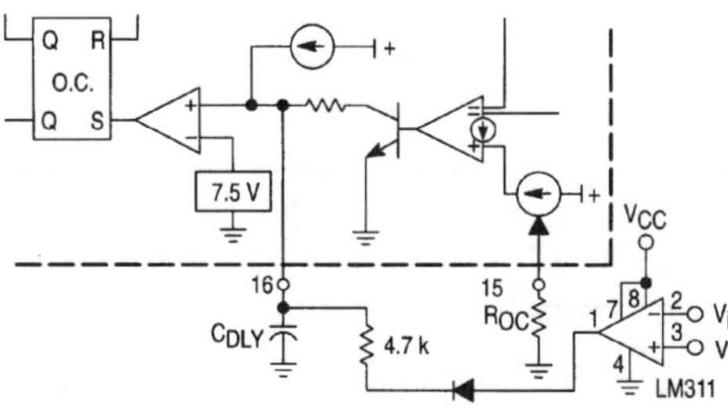

A direction change signal is required at Pins 2 or 3 to reset the over-current latch.

Figure 30. Power H-Switch Buffer

$$R_E \approx \frac{V_{F(D_1)} + V_{F(D_2)} - V_{BE(ON)}}{I_{MOTOR} - I_{DRV(max)}}$$

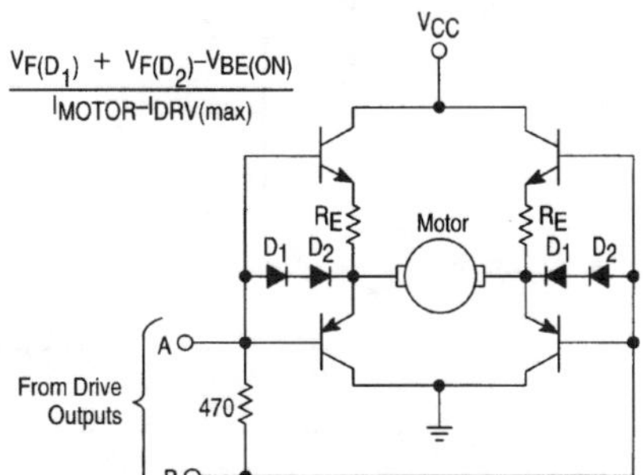

This circuit maintains the brake and over-current features of the MC33030. Set R_{OC} to 15 k for $I_{DRV(max)} \approx 0.5$ A.

Appendix 8.1.12

MC33030

Figure 31. Adjustable Pressure Differential Regulator

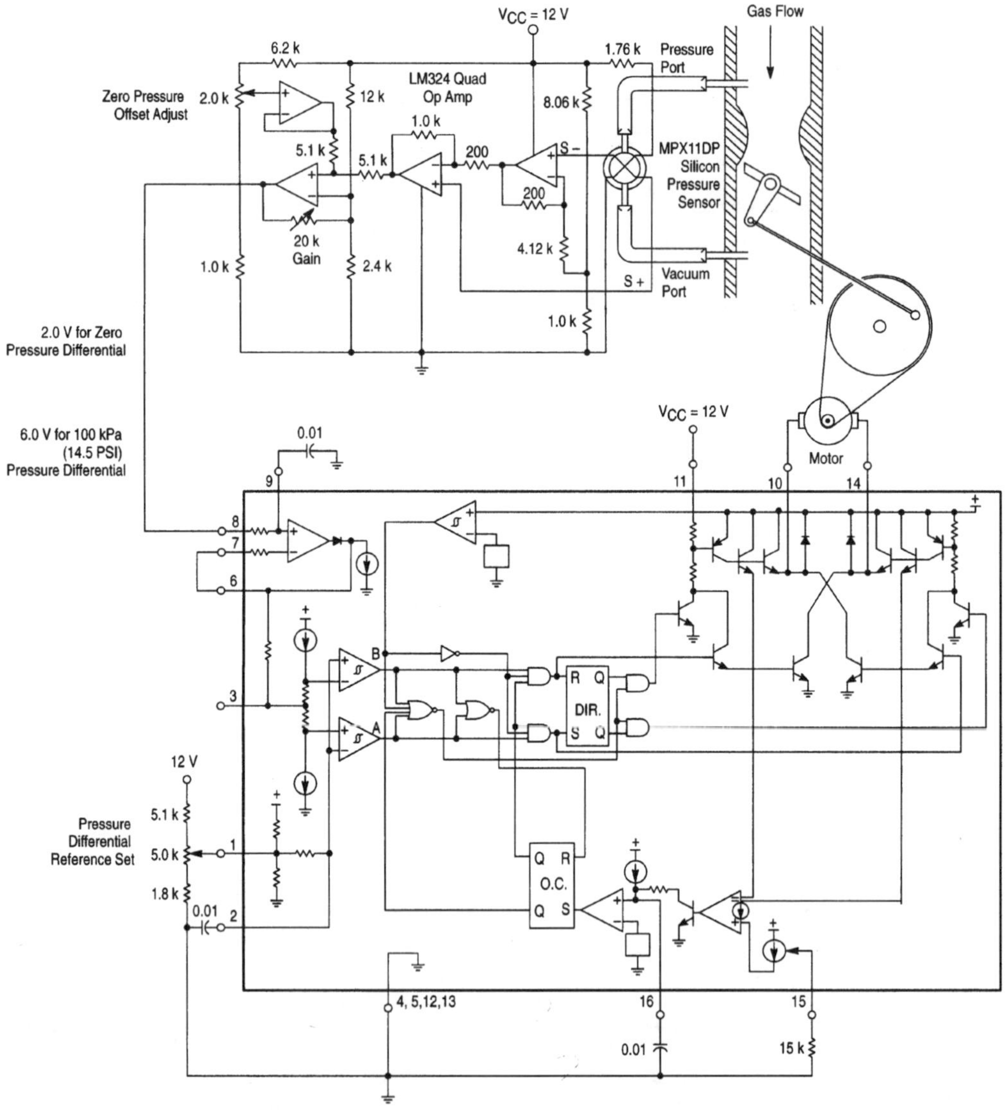

Appendix 8.1.13

MC33030

Figure 32. Switching Motor Controller With Buffered Output and Tach Feedback

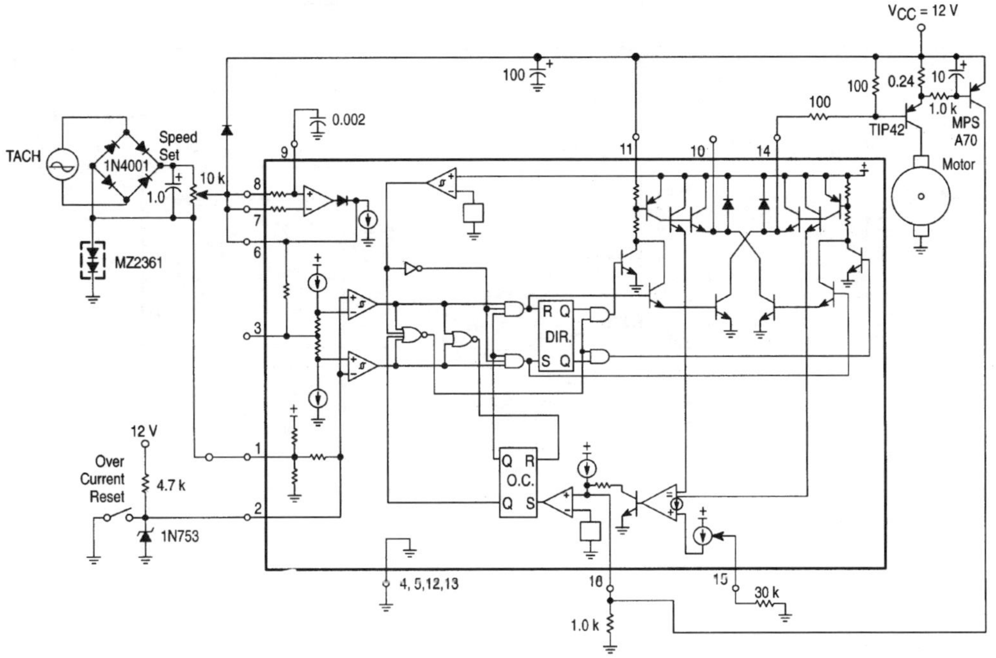

Appendix 8.1.14

MC33030

Figure 33. Switching Motor Controller With Buffered Output and Back EMF Sensing

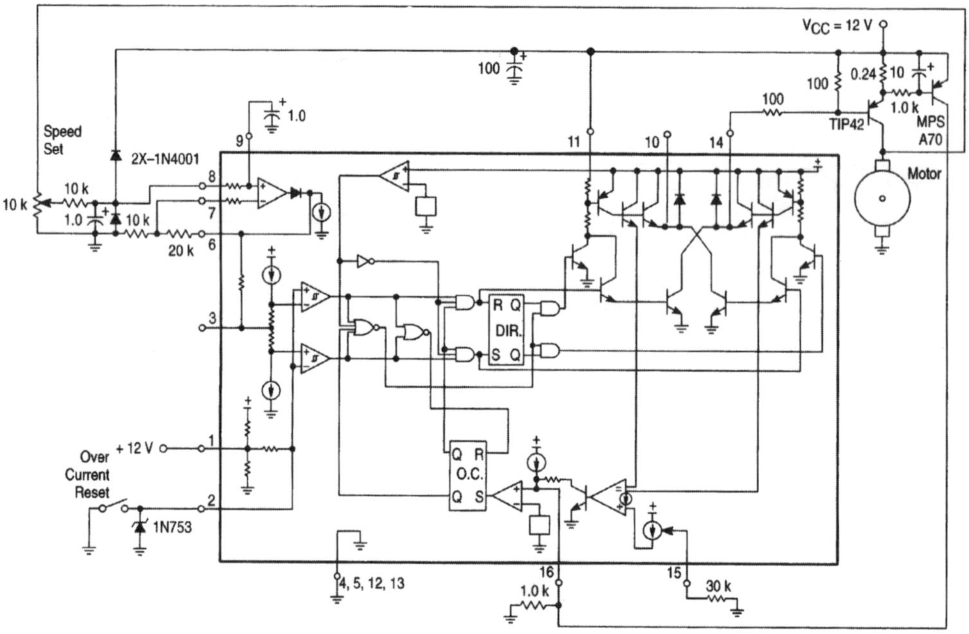

Appendix 8.2.1 Permanent-Magnet Synchronous Motor Driver.
(Copyright of Motorola, used with permission.)

MOTOROLA

MC3479

Stepper Motor Driver

STEPPER MOTOR DRIVER

SEMICONDUCTOR TECHNICAL DATA

The MC3479 is designed to drive a two–phase stepper motor in the bipolar mode. The circuit consists of four input sections, a logic decoding/sequencing section, two driver–stages for the motor coils, and an output to indicate the $\overline{\text{Phase A}}$ drive state.

- Single Supply Operation: 7.2 to 16.5 V
- 350 mA/Coil Drive Capability
- Clamp Diodes Provided for Back–EMF Suppression
- Selectable $\overline{\text{CW}}$/CCW and $\overline{\text{Full}}$/Half Step Operation
- Selectable High/Low Output Impedance (Half Step Mode)
- TTL/CMOS Compatible Inputs
- Input Hysteresis: 400 mV Minimum
- Phase Logic Can Be Initialized to $\overline{\text{Phase A}}$
- $\overline{\text{Phase A}}$ Output Drive State Indication (Open–Collector)
- Available in Standard DIP and Surface Mount

P SUFFIX
PLASTIC PACKAGE
CASE 648C

Figure 1. Representative Block Diagram

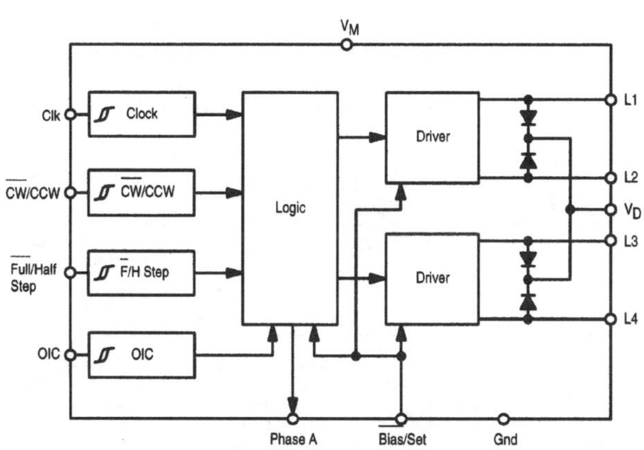

PIN CONNECTIONS

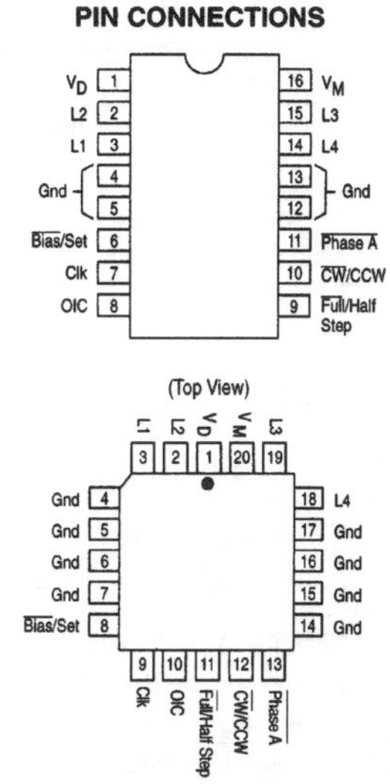

ORDERING INFORMATION

Device	Operating Temperature Range	Package
MC3479P	T_A = 0° to +70°C	Plastic

INPUT TRUTH TABLE

	Input Low	Input High
$\overline{\text{CW}}$/CCW	CW	CCW
$\overline{\text{Full}}$/Half Step	Full Step	Half Step
OIC	Hi Z	Low Z
Clk	Positive Edge Triggered	

Appendix 8.2.2

MC3479

MAXIMUM RATINGS

Rating	Symbol	Value	Unit
Supply Voltage	V_M	+ 18	Vdc
Clamp Diode Cathode Voltage (Pin 1)	V_D	V_M + 5.0	Vdc
Driver Output Voltage	V_{OD}	V_M + 6.0	Vdc
Drive Output Current/Coil	I_{OD}	± 500	mA
Input Voltage (Logic Controls)	V_{in}	– 0.5 to + 7.0	Vdc
$\overline{\text{Bias}}$/Set Current	I_{BS}	– 10	mA
$\overline{\text{Phase A}}$ Output Voltage	V_{OA}	+ 18	Vdc
$\overline{\text{Phase A}}$ Sink Current	I_{OA}	20	mA
Junction Temperature	T_J	+ 150	°C
Storage Temperature Range	T_{stg}	– 65 to + 150	°C

RECOMMENDED OPERATING CONDITIONS

Characteristic	Symbol	Min	Max	Unit
Supply Voltage	V_M	+ 7.2	+ 16.5	Vdc
Clamp Diode Cathode Voltage	V_D	V_M	V_M + 4.5	Vdc
Driver Output Current (Per Coil) (Note 1)	I_{OD}	—	350	mA
Input Voltage (Logic Controls)	V_{in}	0	+ 5.5	Vdc
$\overline{\text{Bias}}$/Set Current (Outputs Active)	I_{BS}	– 300	– 75	μA
$\overline{\text{Phase A}}$ Output Voltage	V_{OA}	—	V_M	Vdc
$\overline{\text{Phase A}}$ Sink Current	I_{OA}	0	8.0	mA
Operating Ambient Temperature	T_A	0	+ 70	°C

NOTE: 1. See section on Power Dissipation in Application Information.

DC ELECTRICAL CHARACTERISTICS (Specifications apply over the recommended supply voltage and temperature range, [Notes 2, 3] unless otherwise noted.)

Characteristic	Pins	Symbol	Min	Typ	Max	Unit
INPUT LOGIC LEVELS						
Threshold Voltage (Low–to–High)	7, 8, 9, 10	V_{TLH}	—	—	2.0	Vdc
Threshold Voltage (High–to–Low)		V_{THL}	0.8	—	—	Vdc
Hysteresis		V_{HYS}	0.4	—	—	Vdc
Current: (V_I = 0.4 V) (V_I = 5.5 V) (V_I = 2.7 V)		I_{IL}	–100 — —	— — —	— +100 +20	μA
DRIVER OUTPUT LEVELS						
Output High Voltage (I_{BS} = – 300 μA): (I_{OD} = – 350 mA) (I_{OD} = – 0.1 mA)	2, 3, 14, 15	V_{OHD}	V_M – 2.0 V_M – 1.2	— —	— —	Vdc
Output Low Voltage (I_{BS} = – 300 μA, I_{OD} = 350 mA)		V_{OLD}	—	—	0.8	Vdc
Differential Mode Output Voltage Difference (Note 4) (I_{BS} = – 300 μA, I_{OD} = 350 mA)		DV_{OD}	—	—	0.15	Vdc
Common Mode Output Voltage Difference (Note 5) (I_{BS} = – 300 μA, I_{OD} = – 0.1 mA)		CV_{OD}	—	—	0.15	Vdc
Output Leakage, Hi Z State ($0 \leq V_{OD} \leq V_M$, I_{BS} = – 5.0 μA) ($0 \leq V_{OD} \leq V_M$, I_{BS} = – 300 μA, F/H = 2.0 V, OIC = 0.8 V)		 I_{OZ1} I_{OZ2}	 – 100 – 100	 — —	 + 100 + 100	μA

NOTES: 2. Algebraic convention rather than absolute values is used to designate limit values.
3. Current into a pin is designated as positive. Current out of a pin is designated as negative.
4. $DV_{OD} = |V_{OD1,2} - V_{OD3,4}|$ where: $V_{OD1,2} = (V_{OHD1} - V_{OLD2})$ or $(V_{OHD2} - V_{OLD1})$, and $V_{OD3,4} = (V_{OHD3} - V_{OLD4})$ or $(V_{OHD4} - V_{OLD3})$.
5. $CV_{OD} = |V_{OHD1} - V_{OHD2}|$ or $|V_{OHD3} - V_{OHD4}|$.

Appendix 8.2.3

MC3479

DC ELECTRICAL CHARACTERISTICS (Specifications apply over the recommended supply voltage and temperature range, [Notes 2, 3] unless otherwise noted.)

Characteristic	Pins	Symbol	Min	Typ	Max	Unit
CLAMP DIODES						
Forward Voltage (I_D = 350 mA)	1, 2, 3, 14, 15	V_{DF}	—	2.5	3.0	Vdc
Leakage Current (Per Diode) (Pin 1 = 21 V; Outputs = 0 V; I_{BS} = 0 μA)		I_{DR}	—	—	100	μA
$\overline{\text{PHASE A}}$ OUTPUT						
Output Low Voltage (I_{OA} = 8.0 mA)	11	V_{OLA}	—	—	0.4	Vdc
Off State Leakage Current (V_{OHA} = 16.5 V)		I_{OHA}	—	—	100	μA
POWER SUPPLY						
Power Supply Current (I_{OD} = 0 μA, I_{BS} = – 300 μA)	16					mA
(L1 = V_{OHD}, L2 = V_{OLD}, L3 = V_{OHD}, L4 = V_{OLD})		I_{MW}	—	—	70	
(L1 = V_{OHD}, L2 = V_{OLD}, L3 = Hi Z, L4 = Hi Z)		I_{MZ}	—	—	40	
(L1 = V_{OHD}, L2 = V_{OLD}, L3 = V_{OHD}, L4 = V_{OHD})		I_{MN}	—	—	75	
$\overline{\text{BIAS}}$/SET CURRENT						
To Set $\overline{\text{Phase A}}$	6	I_{BS}	– 5.0	—	—	μA

PACKAGE THERMAL CHARACTERISTICS

Characteristic	Symbol	Min	Typ	Max	Unit
Thermal Resistance, Junction–to–Ambient (No Heatsink)	$R_{\theta JA}$	—	45	—	°C/W

AC SWITCHING CHARACTERISTICS (T_A = + 25°C, V_M = 12 V) (See Figures 2, 3, 4)

Characteristic	Pins	Symbol	Min	Typ	Max	Unit
Clock Frequency	7	f_{CK}	0	—	50	kHz
Clock Pulse Width (High)	7	PW_{CKH}	10	—	—	μs
Clock Pulse Width (Low)	7	PW_{CKL}	10	—	—	μs
$\overline{\text{Bias}}$/Set Pulse Width	6	PW_{BS}	10	—	—	μs
Setup Time ($\overline{\text{CW}}$/CCW and $\overline{\text{F}}$/HS)	10–7 9–7	t_{su}	5.0	—	—	μs
Hold Time ($\overline{\text{CW}}$/CCW and $\overline{\text{F}}$/HS)	10–7 9–7	t_h	10	—	—	μs
Propagation Delay (Clk–to–Driver Output)		t_{PCD}	—	8.0	—	μs
Propagation Delay ($\overline{\text{Bias}}$/Set–to–Driver Output)		t_{PBSD}	—	1.0	—	μs
Propagation Delay (Clk–to–$\overline{\text{Phase A}}$ Low)	7–11	t_{PHLA}	—	12	—	μs
Propagation Delay (Clk–to–$\overline{\text{Phase A}}$ High)	7–11	t_{PLHA}	—	5.0	—	μs

NOTES: 2. Algebraic convention rather than absolute values is used to designate limit values.
3. Current into a pin is designated as positive. Current out of a pin is designated as negative.

Appendix 8.2.4

MC3479

Figure 2. AC Test Circuit

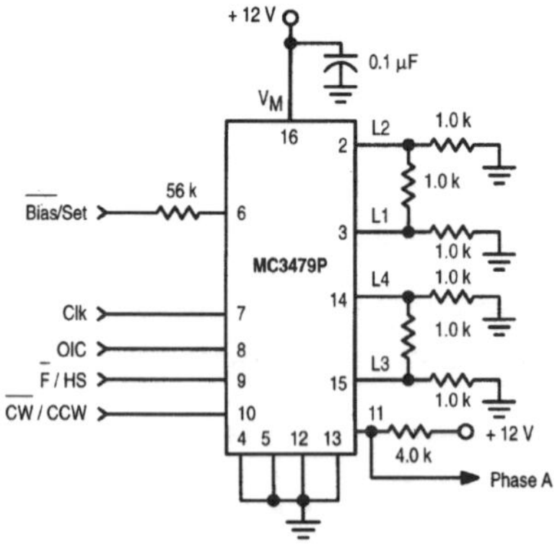

Figure 3. Bias/Set Timing (Refer to Figure 2)

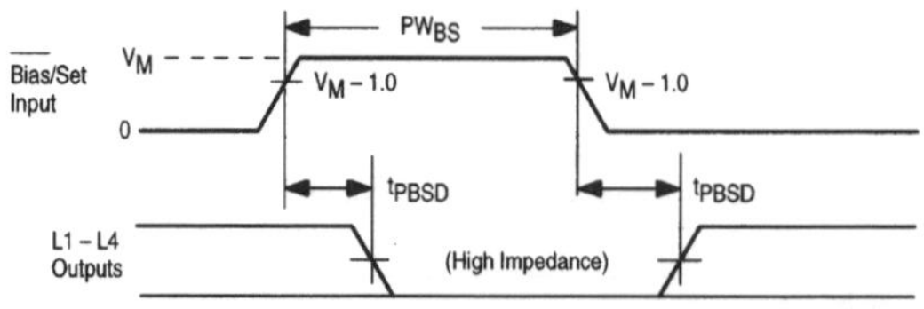

Note: t_r, t_f (10% to 90%) for input signals are ≤ 25 ns.

PIN FUNCTION DESCRIPTION

Pin No.		Function	Symbol	Description
20–Pin	16–Pin			
20	16	Power Supply	V_M	Power supply pin for both the logic circuit and the motor coil current. Voltage range is + 7.2 to + 16.5 volts.
4, 5, 6, 7, 14, 15, 16, 17	4, 5, 12, 13	Ground	Gnd	Ground pins for the logic circuit and the motor coil current. The physical configuration of the pins aids in dissipating heat from within the IC package.
1	1	Clamp Diode Voltage	V_D	This pin is used to protect the outputs where large voltage spikes may occur as the motor coils are switched. Typically a diode is connected between this pin and Pin 16. See Figure 11.
2, 3, 18, 19	2, 3, 14, 15	Driver Outputs	L1, L2 L3, L4	High current outputs for the motor coils. L1 and L2 are connected to one coil, and L3 and L4 to the other coil.
8	6	$\overline{\text{Bias}}$/Set	$\overline{\text{B}}$/S	This pin is typically 0.7 volts below V_M. The current out of this pin (through a resistor to ground) determines the maximum output sink current. If the pin is opened (I_{BS} < 5.0 μA) the outputs assume a high impedance condition, while the internal logic presets to a $\overline{\text{Phase A}}$ condition.
9	7	Clock	Clk	The positive edge of the clock input switches the outputs to the next position. This input has no effect if Pin 6 is open.
11	9	$\overline{\text{Full}}$/Half Step	$\overline{\text{F}}$/HS	When low (Logic "0"), each clock input pulse will cause the motor to rotate one full step. When high, each clock pulse will cause the motor to rotate one–half step. See Figure 7 for sequence.
12	10	$\overline{\text{Clockwise}}$/ Counterclockwise	$\overline{\text{CW}}$/CCW	This input allows reversing the rotation of the motor. See Figure 7 for sequence.
10	8	Output Impedance Control	OIC	This input is relevant only in the half step mode (Pin 9 > 2.0 V). When low (Logic "0"), the two driver outputs of the non–energized coil will be in a high impedance condition. When high the same driver outputs will be at a low impedance referenced to V_M. See Figure 7.
13	11	Phase A	Ph A	This open–collector output indicates (when low) that the driver outputs are in the $\overline{\text{Phase A}}$ condition (L1 = L3 = V_{OHD}, L2 = L4 = V_{OLD}).

APPLICATION INFORMATION

General

The MC3479 integrated circuit is designed to drive a stepper positioning motor in applications such as disk drives and robotics. The outputs can provide up to 350 mA to each of two coils of a two–phase motor. The outputs change state with each low–to–high transition of the clock input, with the new output state depending on the previous state, as well as the input conditions at the logic controls.

Outputs

The outputs (L1–L4) are high current outputs (see Figure 5), which when connected to a two–phase motor, provide two full–bridge configurations (L3 and L4 are not shown in Figure 5). The polarities applied to the motor coils depend on which transistor (Q_H or Q_L) of each output is on, which in turn depends on the inputs and the decoding circuitry.

Appendix 8.2.5

MC3479

Figure 4. Clock Timing
(Refer to Figure 2)

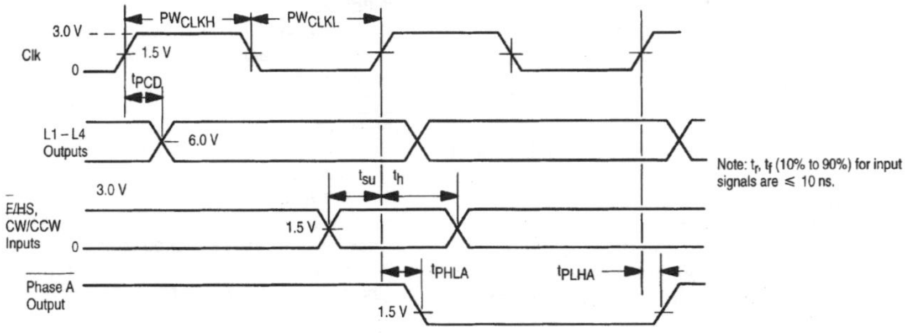

Figure 5. Output Stages

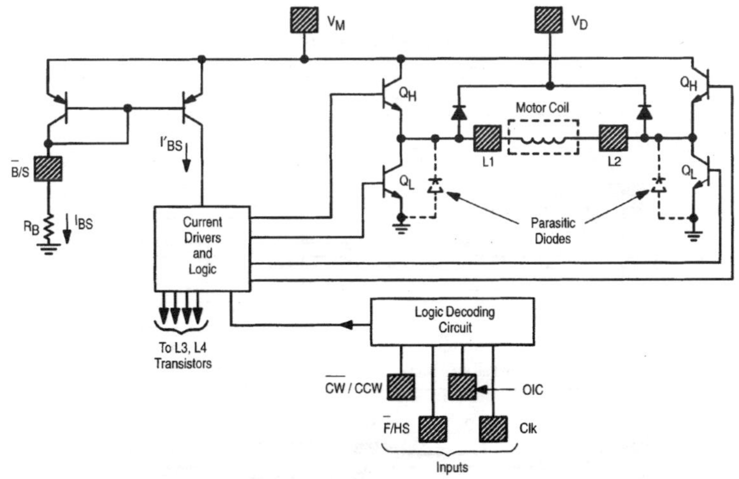

The maximum sink current available at the outputs is a function of the resistor connected between Pin 6 and ground (see section on $\overline{\text{Bias}}$/Set operation). Whenever the outputs are to be in a high impedance state, both transistors (Q_H and Q_L of Figure 5) of each output are off.

V_D

This pin allows for provision of a current path for the motor coil current during switching, in order to suppress back–EMF voltage spikes. V_D is normally connected to V_M (Pin 16) through a diode (zener or regular), a resistor, or directly. The peaks instantaneous voltage at the outputs must not exceed V_M by more than 6.0 V. The voltage drop across the internal clamping diodes must be included in this portion of the design (see Figure 6). Note the parasitic diodes (Figure 5) across each Q_L of each output provide for a complete circuit path for the switched current.

Figure 6. Clamp Diode Characteristics

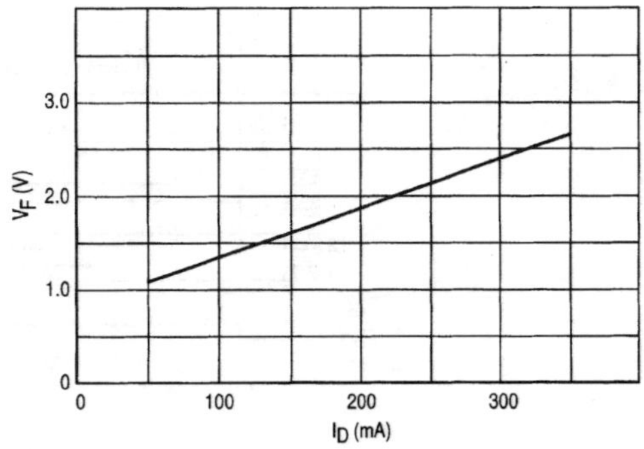

Appendix 8.2.6

MC3479

$\overline{\text{Full}}$/Half Step

When this input is at a Logic "0" (<0.8 V), the outputs change a full step with each clock cycle, with the sequence direction depending on the $\overline{\text{CW}}$/CCW input. There are four steps ($\overline{\text{Phase A}}$, $\overline{\text{B}}$, $\overline{\text{C}}$, $\overline{\text{D}}$) for each complete cycle of the sequencing logic. Current flows through both motor coils during each step, as shown in Figure 7.

When taken to a Logic "1" (>2.0 V), the outputs change a half step with each clock cycle, with the sequence direction depending on the $\overline{\text{CW}}$/CCW input. Eight steps ($\overline{\text{Phase A}}$ to $\overline{\text{H}}$) result for each complete cycle of the sequencing logic. $\overline{\text{Phase A}}$, $\overline{\text{C}}$, $\overline{\text{E}}$ and $\overline{\text{G}}$ correspond (in polarity) to $\overline{\text{Phase A}}$, $\overline{\text{B}}$, $\overline{\text{C}}$, and $\overline{\text{D}}$, respectively, of the full step sequence. $\overline{\text{Phase B}}$, $\overline{\text{D}}$, $\overline{\text{F}}$ and $\overline{\text{H}}$ provide current to one motor coil, while de–energizing the other coil. The condition of the outputs of the de–energized coil depends on the OIC input, see Figure 7 timing diagram.

OIC

The output impedance control input determines the output impedance to the de–energized coil when operating in the half–step mode. When the outputs are in $\overline{\text{Phase B}}$, $\overline{\text{D}}$, $\overline{\text{F}}$ or $\overline{\text{H}}$ (Figure 7) and this input is at a Logic "0" (<0.8 V), the two outputs to the de–energized coil are in a high impedance condition — Q_L and Q_H of both outputs (Figure 5) are off. When this input is at a Logic "1" (>2.0 V), a low impedance output is provided to the de–energized coil as both outputs have Q_H on (Q_L off). To complete the low impedance path requires connecting V_D to V_M as described elsewhere in this data sheet.

$\overline{\text{Bias}}$/Set

This pin can be used for three functions: a) determining the maximum output sink current; b) setting the internal logic to a known state; and c) reducing power consumption.

a) The maximum output sink current is determined by the base drive current supplied to the lower transistors (Q_Ls of Figure 5) of each output, which in turn, is a function of I_{BS}. The appropriate value of I_{BS} is determined by:

$$I_{BS} = I_{OD} \times 0.86$$

where I_{BS} is in microamps, and I_{OD} is the motor current/coil in milliamps.

Figure 7. Output Sequence

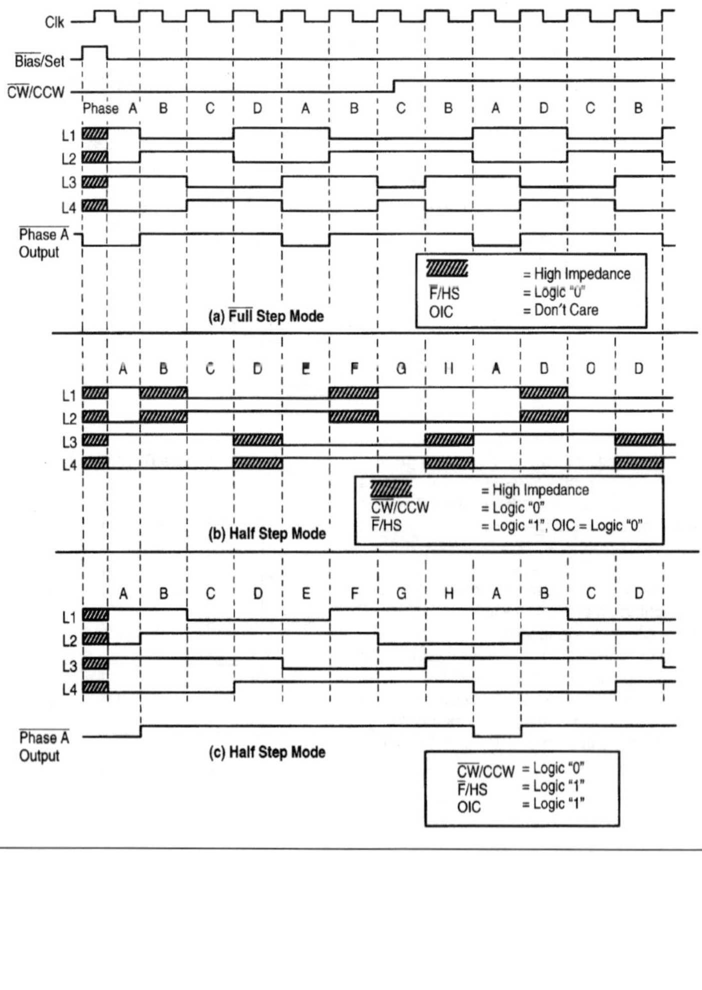

Appendix 8.2.7

MC3479

The value of R_B (between this pin and ground) is then determined by:

$$R_B = \frac{V_M - 0.7\ V}{I_{BS}}$$

b) When this pin is opened (raised to V_M) such that I_{BS} is <5.0 μA, the internal logic is set to the Phase A condition, and the four driver outputs are put into a high impedance state. The Phase A output (Pin 11) goes active (low), and input signals at the controls are ignored during this time. Upon re–establishing I_{BS}, the driver outputs become active, and will be in the Phase A position (L1 = L3 = V_{OHD}, L2 = L4 = V_{OLD}). The circuit will then respond to the inputs at the controls.

The Set function (opening this pin) can be used as a power–up reset while supply voltages are settling. A CMOS logic gate (powered by V_M) can be used to control this pin as shown in Figure 11.

c) Whenever the motor is not being stepped, power dissipation in the IC and in the motor may be lowered by reducing I_{BS}, so as to reduce the output (motor) current. Setting I_{BS} to 75 μA will reduce the motor current, but will not reset the internal logic as described above. See Figure 12 for a suggested circuit.

Power Dissipation

The power dissipated by the MC3479 must be such that the junction temperature (T_J) does not exceed 150°C. The power dissipated can be expressed as:

$$P = (V_M \times I_M) + (2 \times I_{OD})\ [(V_M - V_{OHD}) + V_{OLD}]$$

where V_M = Supply voltage;
I_M = Supply current other than I_{OD};
I_{OD} = Output current to each motor coil;
V_{OHD} = Driver output high voltage;
V_{OLD} = Driver output low voltage.

The power supply current (I_M) is obtained from Figure 8. After the power dissipation is calculated, the junction temperature can be calculated using:

$$T_J = (P \times R_{\theta JA}) + T_A$$

where $R_{\theta JA}$ = Junction–to–ambient thermal resistance (52°C/W for the DIP, 72°C/W for the FN Package);
T_A = Ambient Temperature.

Figure 8. Power Supply Current

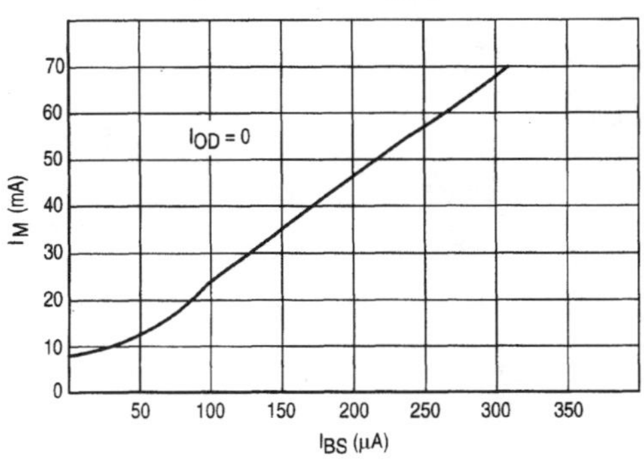

For example, assume an application where V_M = 12 V, the motor requires 200 mA/coil, operating at room temperature with no heatsink on the IC. I_{BS} is calculated:

$I_{BS} = 200 \times 0.86$
$I_{BS} = 172\ \mu A$

R_B is calculated:

$R_B = (12 - 0.7)\ V/172\ \mu A$
$R_B = 65.7\ k\Omega$

From Figure 8, I_M (max) is determined to be 40 mA. From Figure 9, V_{OLD} is 0.46 volts, and from Figure 10, ($V_M - V_{OHD}$) is 1.4 volts.

$P = (12 \times 0.040) + (2 \times 0.2)\ (1.4 + 0.46)$
$P = 1.22\ W$
$T_J = (1.22\ W \times 52°C/W) + 25°C$
$T_J = 88°C$

This temperature is well below the maximum limit. If the calculated T_J had been higher than 150°C, a heatsink such as the Staver Co. V–7 Series, Aavid #5802, or Thermalloy #6012 could be used to reduce $R_{\theta JA}$. In extreme cases, forced air cooling should be considered.

The above calculation, and $R_{\theta JA}$, assumes that a ground plane is provided under the MC3479 (either or both sides of the PC board) to aid in the heat dissipation. Single nominal width traces leading from the four ground pins should be avoided as this will increase T_J, as well as provide potentially disruptive ground noise and I_R drops when switching the motor current.

Figure 9. Maximum Saturation Voltage — Driver Output Low

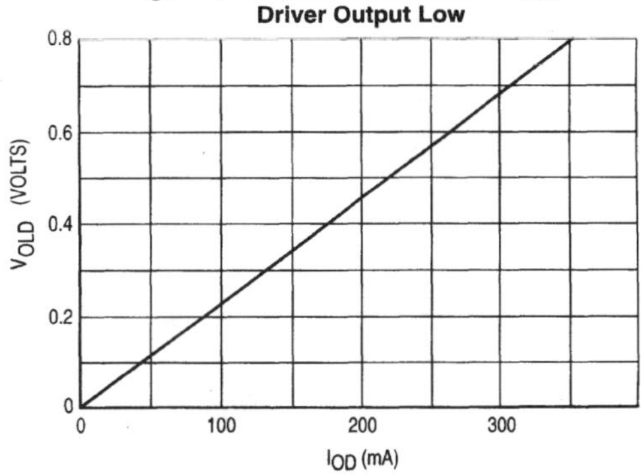

Figure 10. Maximum Saturation Voltage — Driver Output High

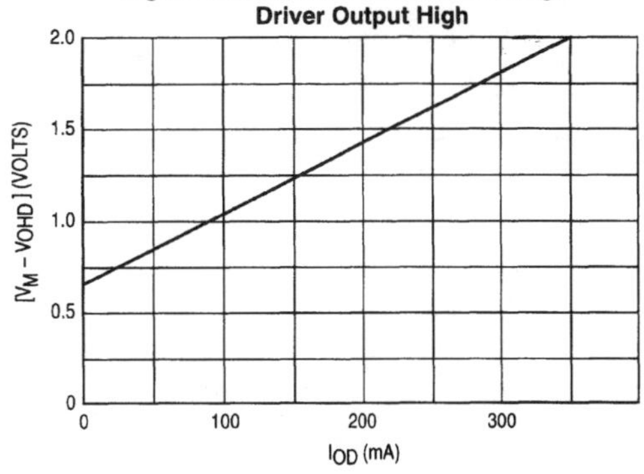

Appendix 8.2.8

MC3479

Figure 11. Typical Applications Circuit

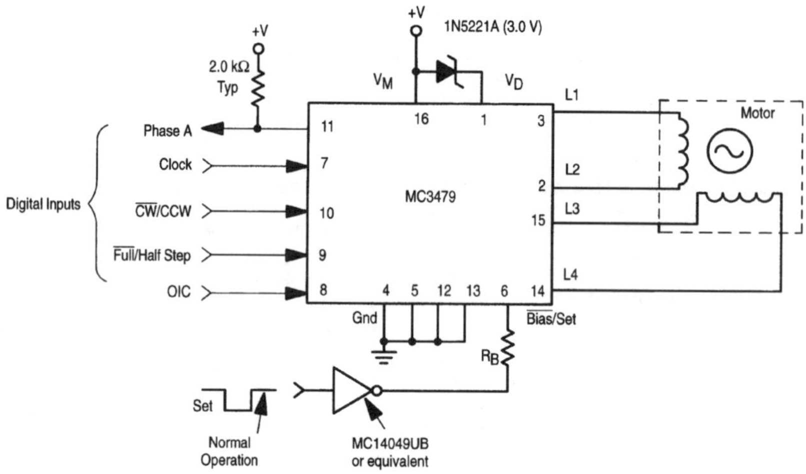

Figure 12. Power Reduction

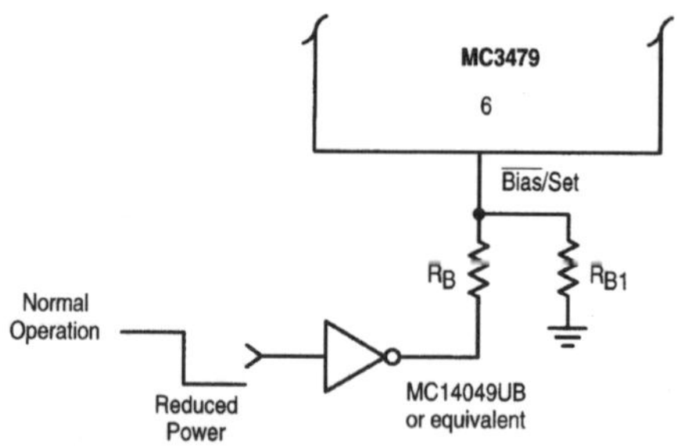

— Suggested value for R_{B1} (V_M = 12 V) is 150 kΩ.
— R_B calculation (see text) must take into account the current through R_{B1}.

Appendix 8.3.1 Angular Velocity Estimator. (Copyright of Motorola, used with permission.)

MOTOROLA

MC33039

Closed Loop Brushless Motor Adapter

CLOSED LOOP BRUSHLESS MOTOR ADAPTER

SEMICONDUCTOR TECHNICAL DATA

The MC33039 is a high performance closed–loop speed control adapter specifically designed for use in brushless DC motor control systems. Implementation will allow precise speed regulation without the need for a magnetic or optical tachometer. This device contains three input buffers each with hysteresis for noise immunity, three digital edge detectors, a programmable monostable, and an internal shunt regulator. Also included is an inverter output for use in systems that require conversion of sensor phasing. Although this device is primarily intended for use with the MC33035 brushless motor controller, it can be used cost effectively in many other closed–loop speed control applications.

- Digital Detection of Each Input Transition for Improved Low Speed Motor Operation
- TTL Compatible Inputs With Hysteresis
- Operation Down to 5.5 V for Direct Powering from MC33035 Reference
- Internal Shunt Regulator Allows Operation from a Non–Regulated Voltage Source
- Inverter Output for Easy Conversion between 60°/300° and 120°/240° Sensor Phasing Conventions

P SUFFIX
PLASTIC PACKAGE
CASE 626

D SUFFIX
PLASTIC PACKAGE
CASE 751
(SO–8)

Representative Block Diagram

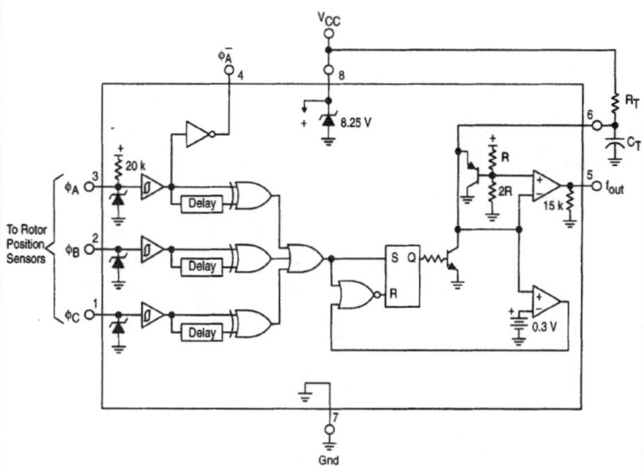

PIN CONNECTIONS

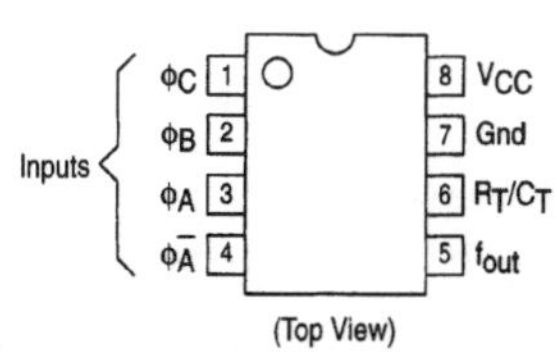

ORDERING INFORMATION

Device	Operating Temperature Range	Package
MC33039D	$T_A = -40°$ to $+85°C$	SO–8
MC33039P		Plastic DIP

Appendix 8.3.2

MC33039

MAXIMUM RATINGS

Rating	Symbol	Value	Unit
V_{CC} Zener Current	$I_{Z(V_{CC})}$	30	mA
Logic Input Current (Pins 1, 2, 3)	I_{IH}	5.0	mA
Output Current (Pins 4, 5), Sink or Source	I_{DRV}	20	mA
Power Dissipation and Thermal Characteristics Maximum Power Dissipation @ T_A = + 85°C Thermal Resistance, Junction–to–Air	 P_D $R_{\theta JA}$	 650 100	 mW °C/W
Operating Junction Temperature	T_J	+ 150	°C
Operating Ambient Temperature Range	T_A	– 40 to + 85	°C
Storage Temperature Range	T_{stg}	– 65 to + 150	°C

ELECTRICAL CHARACTERISTICS (V_{CC} = 6.25 V, R_T = 10 k, C_T = 22 nF, T_A = 25°C, unless otherwise noted)

Characteristic	Symbol	Min	Typ	Max	Unit
LOGIC INPUTS					
Input Threshold Voltage High State Low State Hysteresis	 V_{IH} V_{IL} V_H	 2.4 — 0.4	 2.1 1.4 0.7	 — 1.0 0.9	V
Input Current High State (V_{IH} = 5.0 V) ϕA ϕB, ϕC Low State (V_{IL} = 0 V) ϕA ϕB, ϕC	 I_{IH} I_{IL}	 – 40 — – 190 —	 – 60 – 0.3 – 300 – 0.3	 – 80 – 5.0 – 380 – 5.0	µA
MONOSTABLE AND OUTPUT SECTIONS					
Output Voltage High State f_{out} (I_{source} = 5.0 mA) ϕA (I_{source} = 2.0 mA) Low State f_{out} (I_{sink} = 10 mA) ϕA (I_{sink} = 10 mA)	 V_{OH} V_{OL}	 3.60 4.20 — —	 3.95 4.75 0.25 0.25	 4.20 — 0.50 0.50	V
Capacitor C_T Discharge Current	I_{dischg}	20	35	60	mA
Output Pulse Width (Pin 5)	t_{PW}	205	225	245	µs
POWER SUPPLY SECTION					
Power Supply Operating Voltage Range (T_A = – 40° to + 85°C)	V_{CC}	5.5	—	V_Z	V
Power Supply Current	I_{CC}	1.8	3.9	5.0	mA
Zener Voltage (I_Z = 10 mA)	V_Z	7.5	8.25	9.0	V
Zener Dynamic Impedance (ΔI_Z = 10 mA to 20 mA, f ≤ 1.0 kHz)	$\lvert Z_{ka} \rvert$	—	2.0	5.0	Ω

Appendix 8.3.3

MC33039

Figure 1. Typical Three Phase, Six Step Motor Application

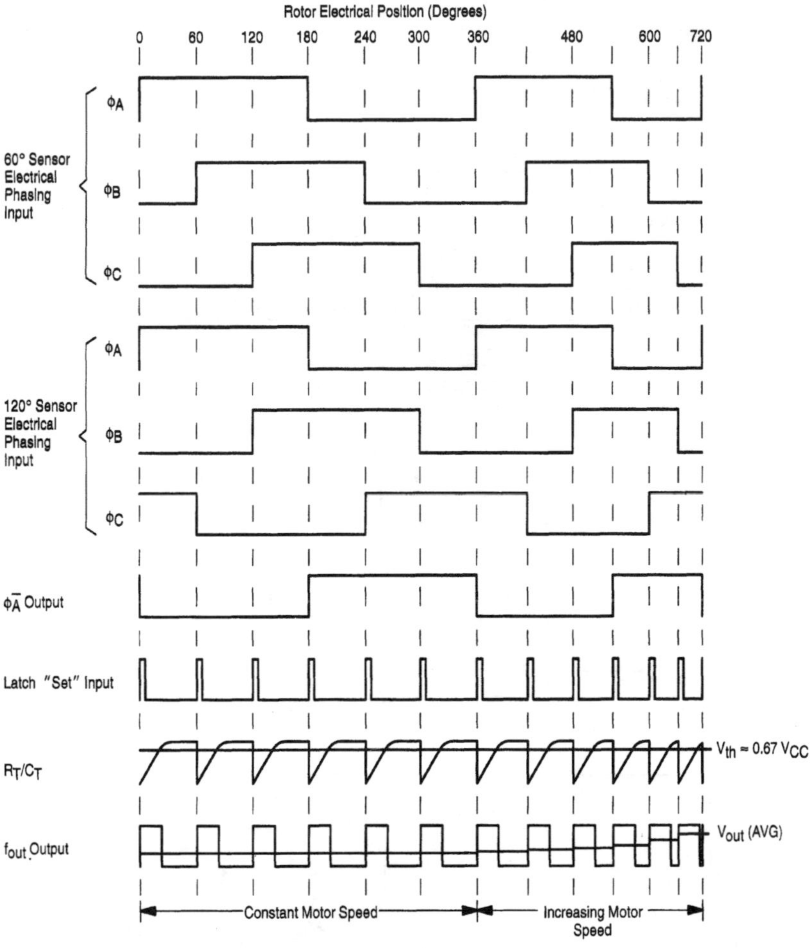

OPERATING DESCRIPTION

The MC33039 provides an economical method of implementing closed–loop speed control of brushless DC motors by eliminating the need for a magnetic or optical tachometer. Shown in the timing diagram of Figure 1, the three inputs (Pins 1, 2, 3) monitor the brushless motor rotor position sensors. Each sensor signal transition is digitally detected, OR'ed at the Latch 'Set' Input, and causes C_T to discharge. A corresponding output pulse is generated at f_{out} (Pin 5) of a defined amplitude, and programmable width determined by the values selected for R_T and C_T (Pin 6). The average voltage of the output pulse train increases with motor speed. When fed through a low pass filter or integrator, a DC voltage proportional to speed is generated. Figure 2 shows the proper connections for a typical closed loop application using the MC33035 brushless motor controller. Constant speed operation down to 100 RPM is possible with economical three phase four pole motors.

The $\phi_{\bar{A}}$ inverter output (Pin 4) is used in systems where the controller and motor sensor phasing conventions are not compatible. A method of converting from either convention to the other is shown in Figure 3. For a more detailed explanation of this subject, refer to the text above Figure 39 on the MC33035 data sheet.

The output pulse amplitude V_{OH} is constant with temperature and controlled by the supply voltage on V_{CC} (Pin 8). Operation down to 5.5 V is guaranteed over temperature. For systems without a regulated power supply, an internal 8.25 V shunt regulator is provided.

Appendix 8.3.4

MC33039

Figure 2. Typical Closed Loop Speed Control Application

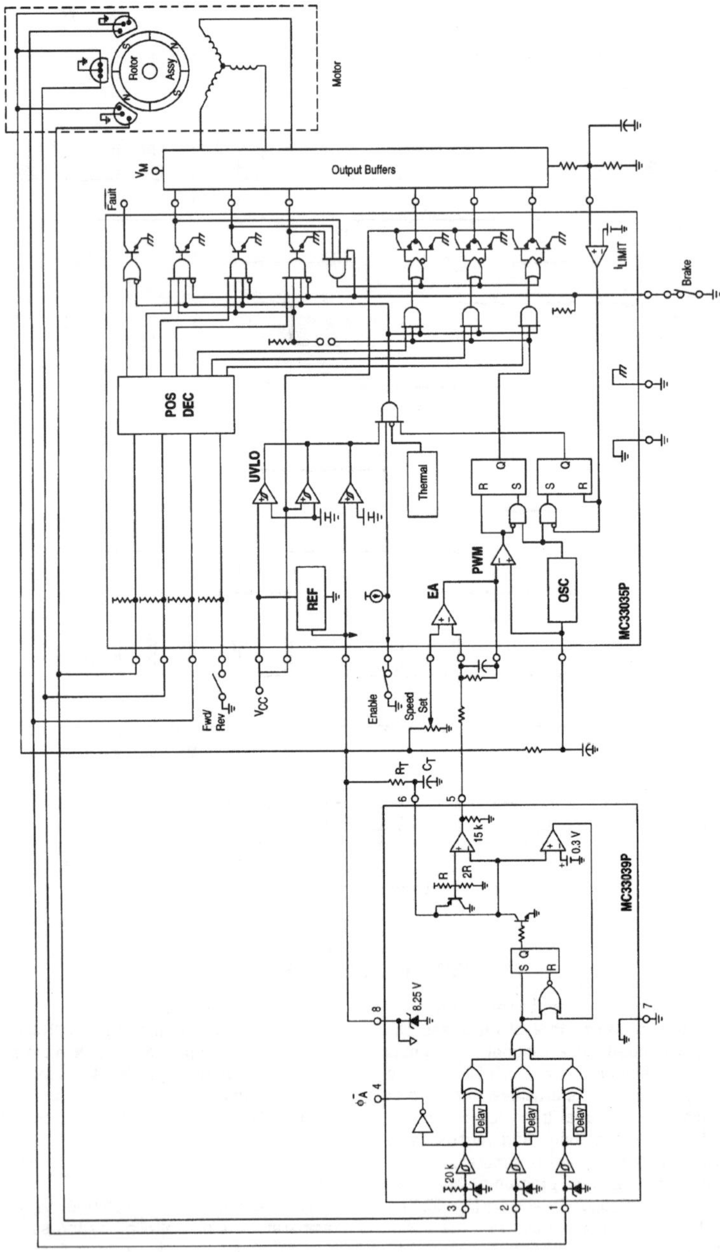

Appendix 8.3.5

MC33039

Figure 3. f_{out}, Pulse Width versus Timing Resistor

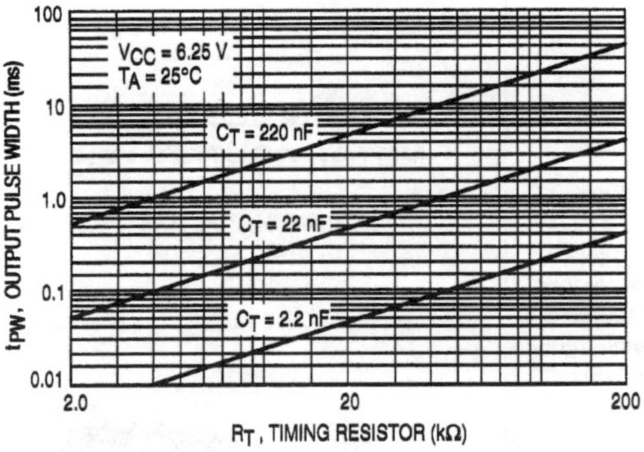

Figure 4. f_{out}, Pulse Width Change versus Temperature

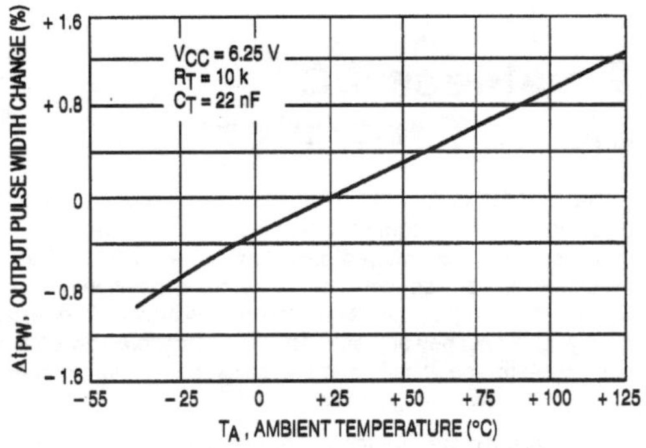

Figure 5. f_{out}, Pulse Width Change versus Supply Voltage

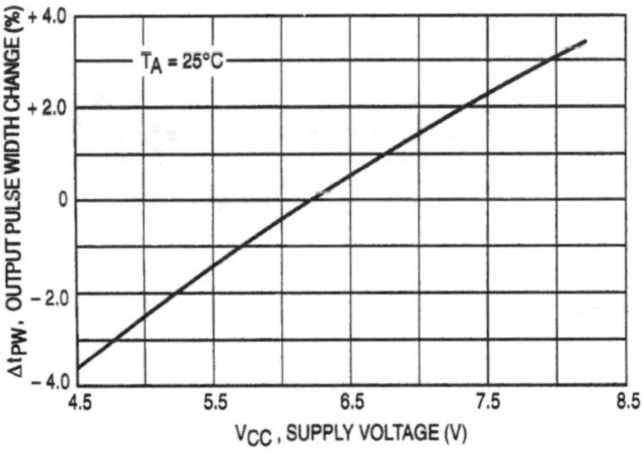

Figure 6. Supply Current versus Supply Voltage

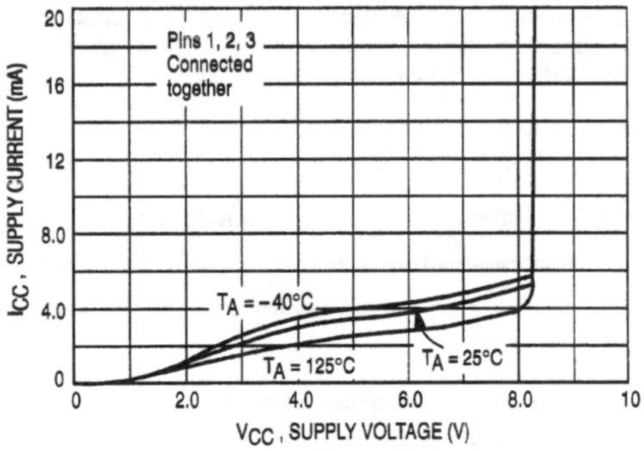

Figure 7. f_{out}, Saturation versus Load Current

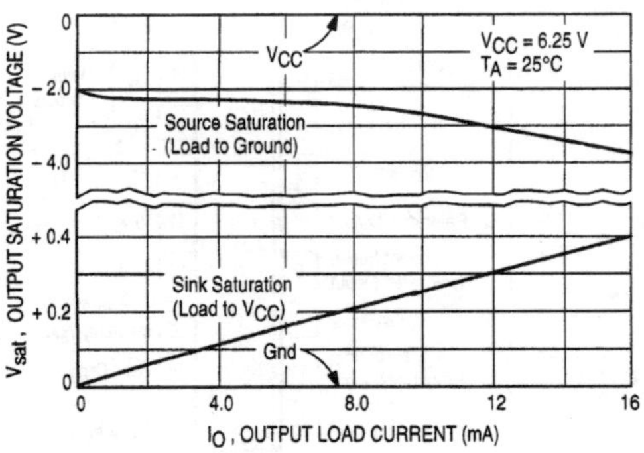

Figure 8. f_{out}, Saturation Change versus Temperature

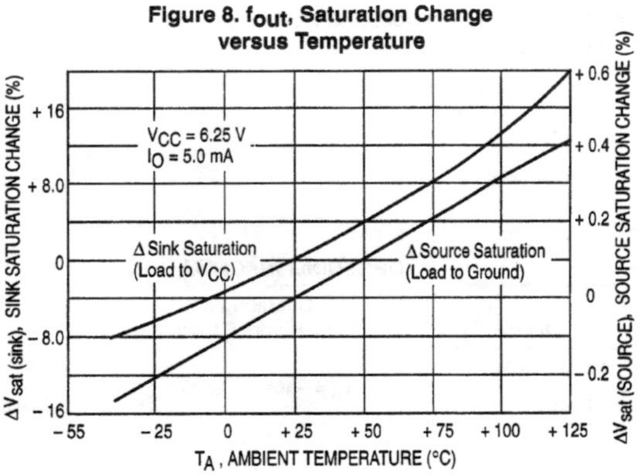

Appendix 8.4.1 Stepper Motor Driver Appendix. (Copyright of Motorola, used with permission.)

MOTOROLA

MC33035

Brushless DC Motor Controller

BRUSHLESS DC MOTOR CONTROLLER

SEMICONDUCTOR TECHNICAL DATA

The MC33035 is a high performance second generation monolithic brushless DC motor controller containing all of the active functions required to implement a full featured open loop, three or four phase motor control system. This device consists of a rotor position decoder for proper commutation sequencing, temperature compensated reference capable of supplying sensor power, frequency programmable sawtooth oscillator, three open collector top drivers, and three high current totem pole bottom drivers ideally suited for driving power MOSFETs.

Also included are protective features consisting of undervoltage lockout, cycle–by–cycle current limiting with a selectable time delayed latched shutdown mode, internal thermal shutdown, and a unique fault output that can be interfaced into microprocessor controlled systems.

Typical motor control functions include open loop speed, forward or reverse direction, run enable, and dynamic braking. The MC33035 is designed to operate with electrical sensor phasings of 60°/300° or 120°/240°, and can also efficiently control brush DC motors.

- 10 to 30 V Operation
- Undervoltage Lockout
- 6.25 V Reference Capable of Supplying Sensor Power
- Fully Accessible Error Amplifier for Closed Loop Servo Applications
- High Current Drivers Can Control External 3–Phase MOSFET Bridge
- Cycle–By–Cycle Current Limiting
- Pinned–Out Current Sense Reference
- Internal Thermal Shutdown
- Selectable 60°/300° or 120°/240° Sensor Phasings
- Can Efficiently Control Brush DC Motors with External MOSFET H–Bridge

P SUFFIX
PLASTIC PACKAGE
CASE 724

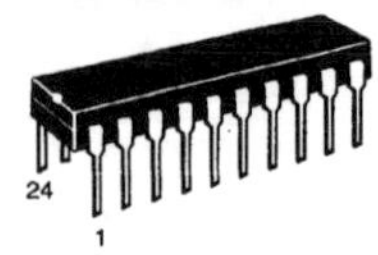

DW SUFFIX
PLASTIC PACKAGE
CASE 751E
(SO–24L)

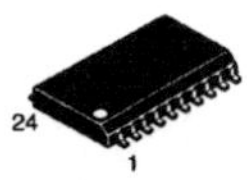

PIN CONNECTIONS

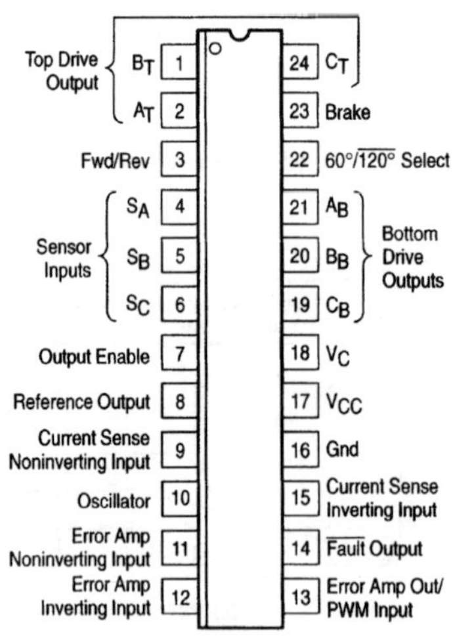

(Top View)

ORDERING INFORMATION

Device	Operating Temperature Range	Package
MC33035DW	$T_A = -40°$ to $+85°C$	SO–24L
MC33035P		Plastic DIP

Appendix 8.4.2

MC33035

Representative Schematic Diagram

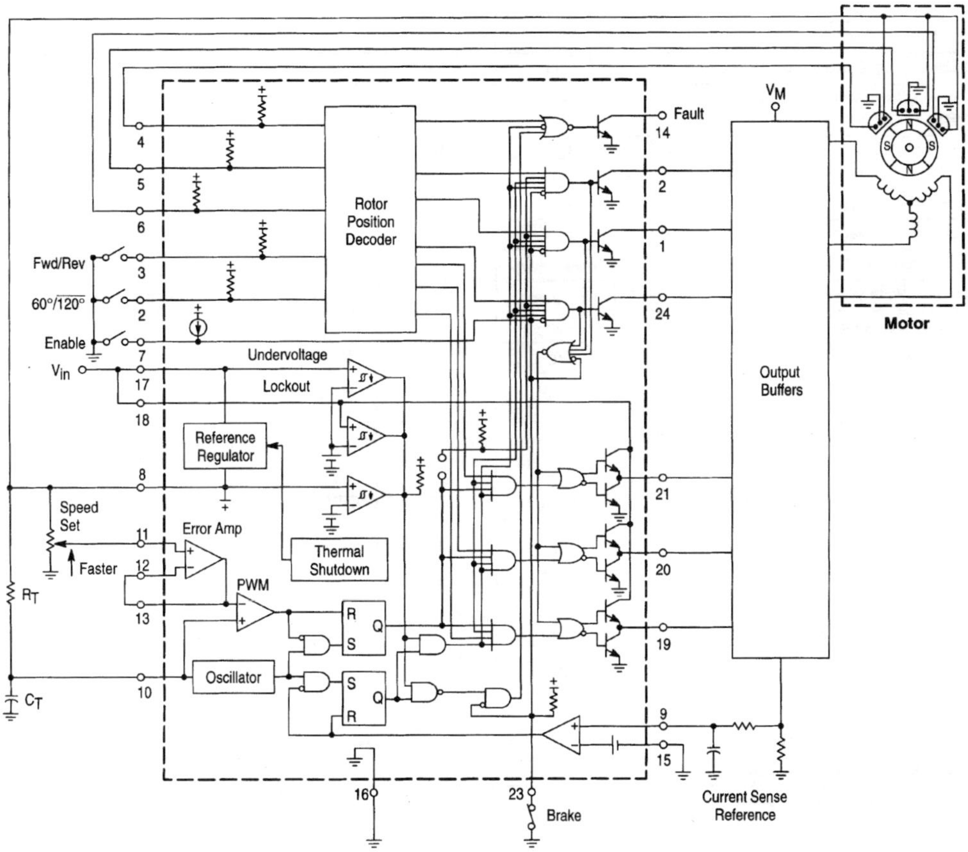

This device contains 285 active transistors.

Appendix 8.4.3

MC33035

MAXIMUM RATINGS

Rating	Symbol	Value	Unit
Power Supply Voltage	V_{CC}	40	V
Digital Inputs (Pins 3, 4, 5, 6, 22, 23)	–	V_{ref}	V
Oscillator Input Current (Source or Sink)	I_{OSC}	30	mA
Error Amp Input Voltage Range (Pins 11, 12, Note 1)	V_{IR}	−0.3 to V_{ref}	V
Error Amp Output Current (Source or Sink, Note 2)	I_{Out}	10	mA
Current Sense Input Voltage Range (Pins 9, 15)	V_{Sense}	−0.3 to 5.0	V
$\overline{\text{Fault}}$ Output Voltage	$V_{CE(\overline{Fault})}$	20	V
$\overline{\text{Fault}}$ Output Sink Current	$I_{Sink(\overline{Fault})}$	20	mA
Top Drive Voltage (Pins 1, 2, 24)	$V_{CE(top)}$	40	V
Top Drive Sink Current (Pins 1, 2, 24)	$I_{Sink(top)}$	50	mA
Bottom Drive Supply Voltage (Pin 18)	V_C	30	V
Bottom Drive Output Current (Source or Sink, Pins 19, 20, 21)	I_{DRV}	100	mA
Power Dissipation and Thermal Characteristics			
P Suffix, Dual In Line, Case 724			
Maximum Power Dissipation @ T_A = 85°C	P_D	867	mW
Thermal Resistance, Junction–to–Air	$R_{\theta JA}$	75	°C/W
DW Suffix, Surface Mount, Case 751E			
Maximum Power Dissipation @ T_A = 85°C	P_D	650	mW
Thermal Resistance, Junction–to–Air	$R_{\theta JA}$	100	°C/W
Operating Junction Temperature	T_J	150	°C
Operating Ambient Temperature Range	T_A	−40 to +85	°C
Storage Temperature Range	T_{stg}	−65 to +150	°C

ELECTRICAL CHARACTERISTICS ($V_{CC} = V_C = 20$ V, $R_T = 4.7$ k, $C_T = 10$ nF, $T_A = 25$°C, unless otherwise noted.)

Characteristic	Symbol	Min	Typ	Max	Unit
REFERENCE SECTION					
Reference Output Voltage (I_{ref} = 1.0 mA)	V_{ref}				V
T_A = 25°C		5.9	6.24	6.5	
T_A = −40° to +85°C		5.82	–	6.57	
Line Regulation (V_{CC} = 10 to 30 V, I_{ref} = 1.0 mA)	Reg_{line}	–	1.5	30	mV
Load Regulation (I_{ref} = 1.0 to 20 mA)	Reg_{load}	–	16	30	mV
Output Short Circuit Current (Note 3)	I_{SC}	40	75	–	mA
Reference Under Voltage Lockout Threshold	V_{th}	4.0	4.5	5.0	V
ERROR AMPLIFIER					
Input Offset Voltage (T_A = −40° to +85°C)	V_{IO}	–	0.4	10	mV
Input Offset Current (T_A = −40° to +85°C)	I_{IO}	–	8.0	500	nA
Input Bias Current (T_A = −40° to +85°C)	I_{IB}	–	−46	−1000	nA
Input Common Mode Voltage Range	V_{ICR}		(0 V to V_{ref})		V
Open Loop Voltage Gain (V_O = 3.0 V, R_L = 15 k)	A_{VOL}	70	80	–	dB
Input Common Mode Rejection Ratio	CMRR	55	86	–	dB
Power Supply Rejection Ratio ($V_{CC} = V_C$ = 10 to 30 V)	PSRR	65	105	–	dB

NOTES: 1. The input common mode voltage or input signal voltage should not be allowed to go negative by more than 0.3 V.
2. The compliance voltage must not exceed the range of −0.3 to V_{ref}.
3. Maximum package power dissipation limits must be observed.

Appendix 8.4.4

MC33035

ELECTRICAL CHARACTERISTICS (continued) (V_{CC} = V_C = 20 V, R_T = 4.7 k, C_T = 10 nF, T_A = 25°C, unless otherwise noted.)

Characteristic	Symbol	Min	Typ	Max	Unit
ERROR AMPLIFIER					
Output Voltage Swing					V
High State (R_L = 15 k to Gnd)	V_{OH}	4.6	5.3	–	
Low State (R_L = 15 k to V_{ref})	V_{OL}	–	0.5	1.0	
OSCILLATOR SECTION					
Oscillator Frequency	f_{OSC}	22	25	28	kHz
Frequency Change with Voltage (V_{CC} = 10 to 30 V)	$\Delta f_{OSC}/\Delta V$	–	0.01	5.0	%
Sawtooth Peak Voltage	$V_{OSC(P)}$	–	4.1	4.5	V
Sawtooth Valley Voltage	$V_{OSC(V)}$	1.2	1.5	–	V
LOGIC INPUTS					
Input Threshold Voltage (Pins 3, 4, 5, 6, 7, 22, 23)					V
High State	V_{IH}	3.0	2.2	–	
Low State	V_{IL}	–	1.7	0.8	
Sensor Inputs (Pins 4, 5, 6)					μA
High State Input Current (V_{IH} = 5.0 V)	I_{IH}	–150	–70	–20	
Low State Input Current (V_{IL} = 0 V)	I_{IL}	–600	–337	–150	
Forward/Reverse, 60°/$\overline{120°}$ Select (Pins 3, 22, 23)					μA
High State Input Current (V_{IH} = 5.0 V)	I_{IH}	–75	–36	–10	
Low State Input Current (V_{IL} = 0 V)	I_{IL}	–300	–175	–75	
Output Enable					μA
High State Input Current (V_{IH} = 5.0 V)	I_{IH}	–60	–29	–10	
Low State Input Current (V_{IL} = 0 V)	I_{IL}	–60	–29	–10	
CURRENT–LIMIT COMPARATOR					
Threshold Voltage	V_{th}	85	101	115	mV
Input Common Mode Voltage Range	V_{ICR}	–	3.0	–	V
Input Bias Current	I_{IB}	–	–0.9	–5.0	μA
OUTPUTS AND POWER SECTIONS					
Top Drive Output Sink Saturation (I_{sink} = 25 mA)	$V_{CE(sat)}$	–	0.5	1.5	V
Top Drive Output Off–State Leakage (V_{CE} = 30 V)	$I_{DRV(leak)}$	–	0.06	100	μA
Top Drive Output Switching Time (C_L = 47 pF, R_L = 1.0 k)					ns
Rise Time	t_r	–	107	300	
Fall Time	t_f	–	26	300	
Bottom Drive Output Voltage					V
High State (V_{CC} = 20 V, V_C = 30 V, I_{source} = 50 mA)	V_{OH}	(V_{CC} –2.0)	(V_{CC} –1.1)	–	
Low State (V_{CC} = 20 V, V_C = 30 V, I_{sink} = 50 mA)	V_{OL}	–	1.5	2.0	
Bottom Drive Output Switching Time (C_L = 1000 pF)					ns
Rise Time	t_r	–	38	200	
Fall Time	t_f	–	30	200	
$\overline{\text{Fault}}$ Output Sink Saturation (I_{sink} = 16 mA)	$V_{CE(sat)}$	–	225	500	mV
$\overline{\text{Fault}}$ Output Off–State Leakage (V_{CE} = 20 V)	$I_{FLT(leak)}$	–	1.0	100	μA
Under Voltage Lockout					V
Drive Output Enabled (V_{CC} or V_C Increasing)	$V_{th(on)}$	8.2	8.9	10	
Hysteresis	V_H	0.1	0.2	0.3	
Power Supply Current					mA
Pin 17 (V_{CC} = V_C = 20 V)	I_{CC}	–	12	16	
Pin 17 (V_{CC} = 20 V, V_C = 30 V)		–	14	20	
Pin 18 (V_{CC} = V_C = 20 V)	I_C	–	3.5	6.0	
Pin 18 (V_{CC} = 20 V, V_C = 30 V)		–	5.0	10	

Appendix 8.4.5

MC33035

Figure 1. Oscillator Frequency versus Timing Resistor

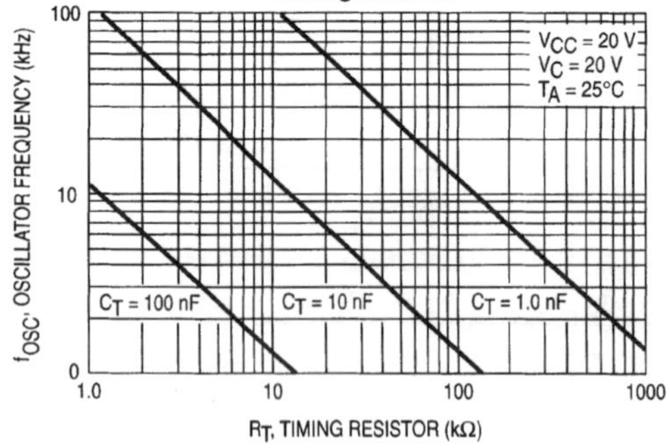

Figure 2. Oscillator Frequency Change versus Temperature

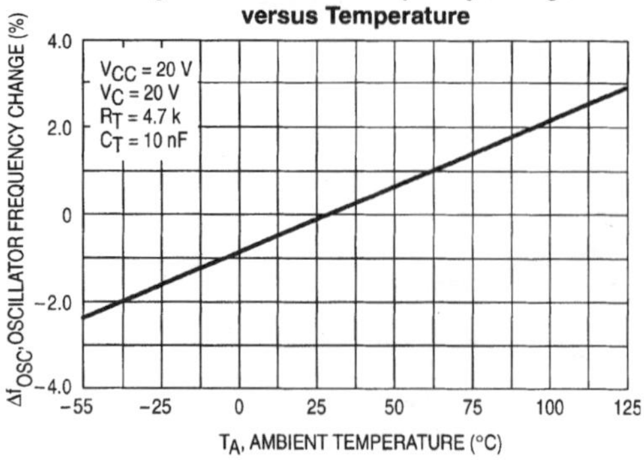

Figure 3. Error Amp Open Loop Gain and Phase versus Frequency

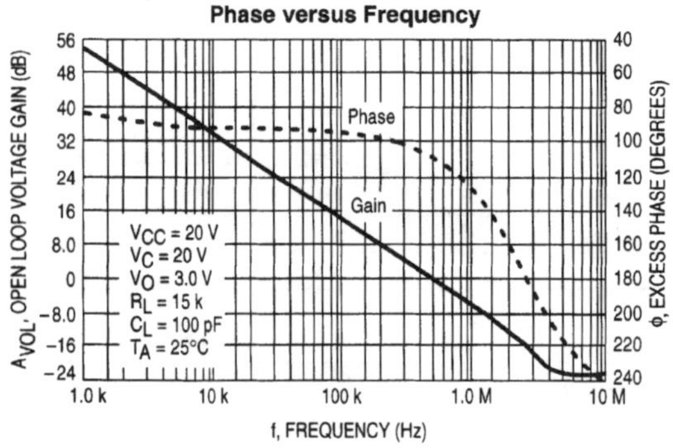

Figure 4. Error Amp Output Saturation Voltage versus Load Current

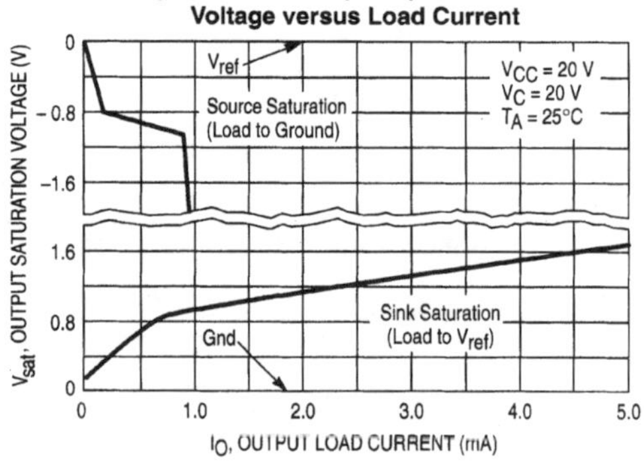

Figure 5. Error Amp Small-Signal Transient Response

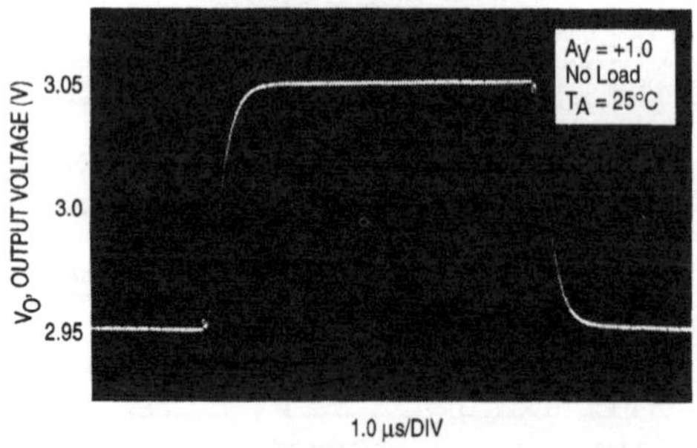

Figure 6. Error Amp Large-Signal Transient Response

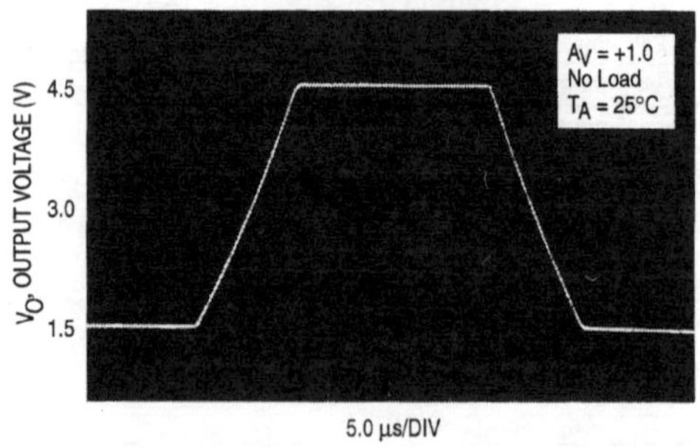

Appendix 8.4.6

MC33035

Figure 7. Reference Output Voltage Change versus Output Source Current

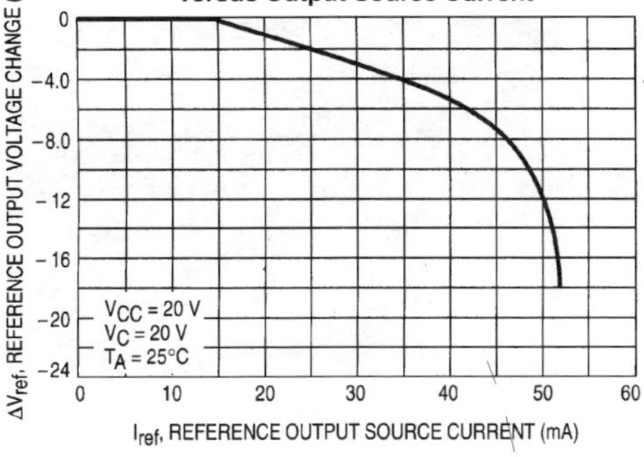

Figure 8. Reference Output Voltage versus Supply Voltage

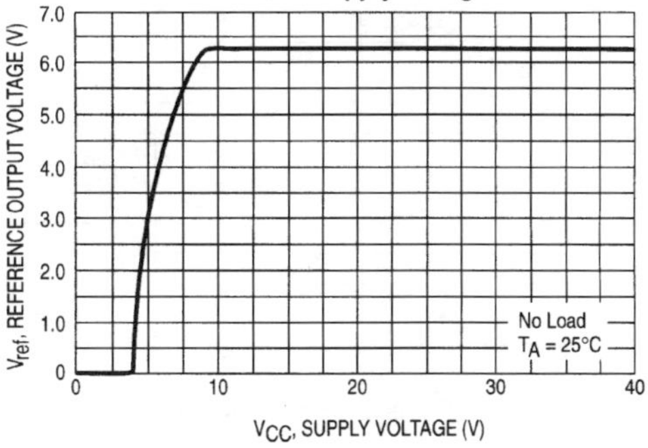

Figure 9. Reference Output Voltage versus Temperature

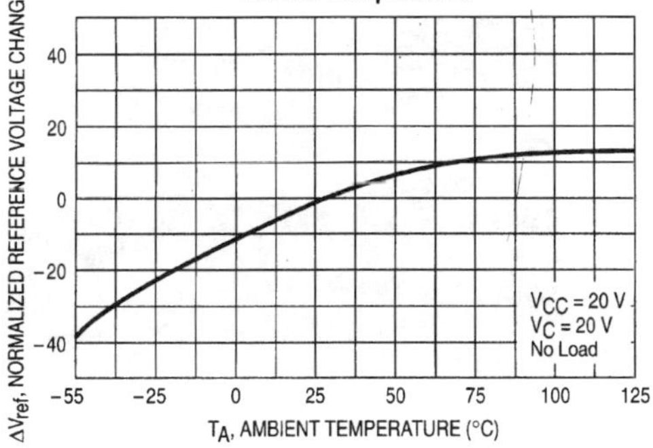

Figure 10. Output Duty Cycle versus PWM Input Voltage

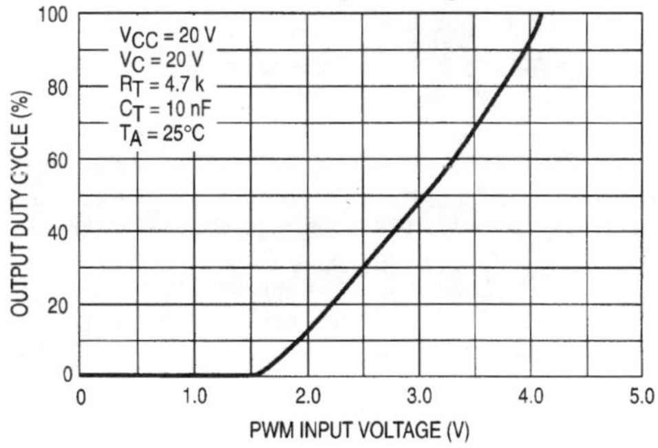

Figure 11. Bottom Drive Response Time versus Current Sense Input Voltage

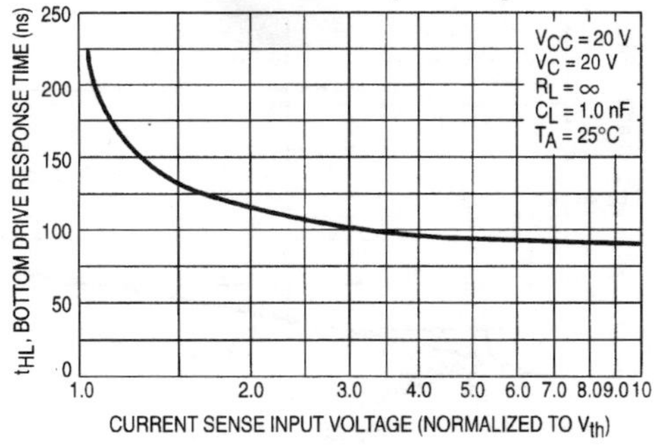

Figure 12. $\overline{\text{Fault}}$ Output Saturation versus Sink Current

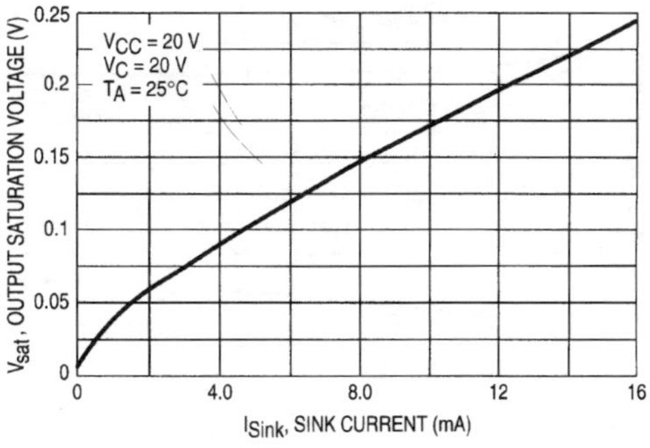

Appendix 8.4.7

MC33035

Figure 13. Top Drive Output Saturation Voltage versus Sink Current

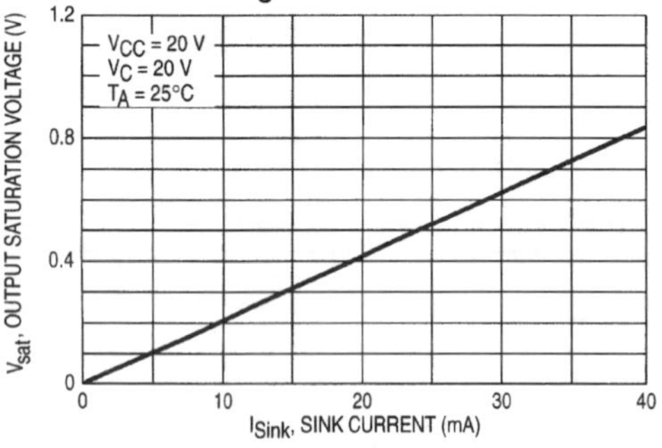

Figure 14. Top Drive Output Waveform

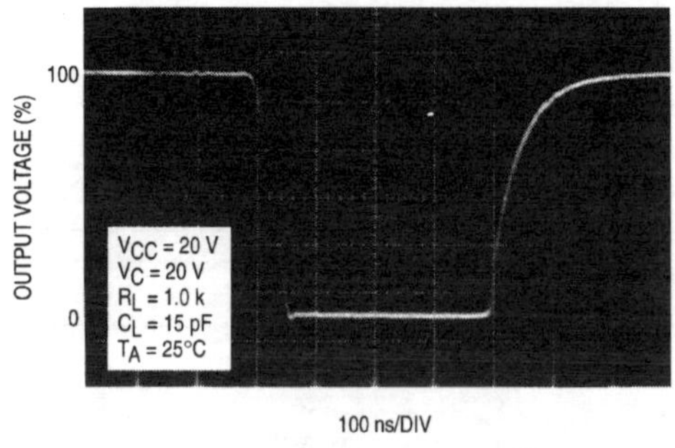

Figure 15. Bottom Drive Output Waveform

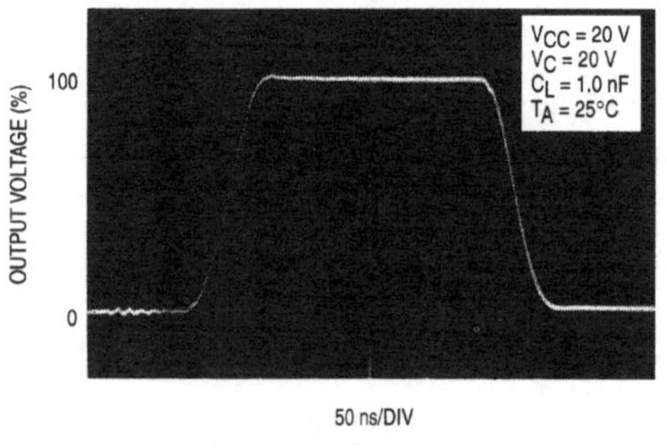

Figure 16. Bottom Drive Output Waveform

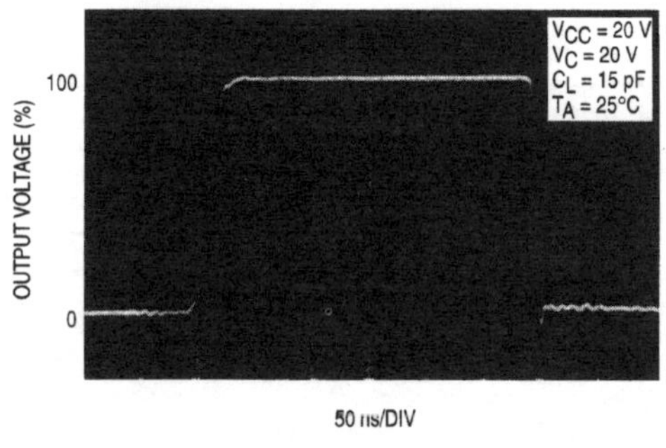

Figure 17. Bottom Drive Output Saturation Voltage versus Load Current

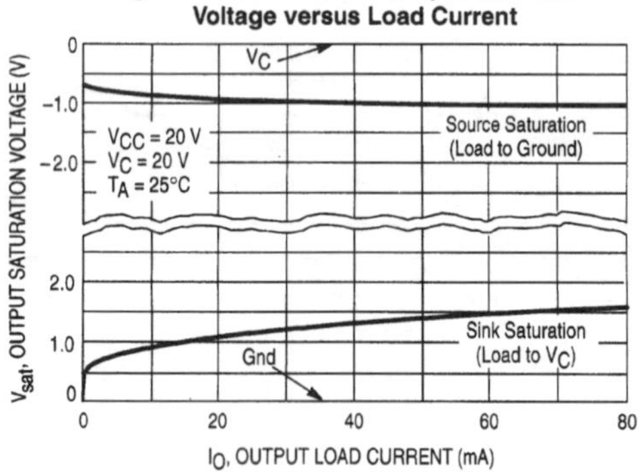

Figure 18. Power and Bottom Drive Supply Current versus Supply Voltage

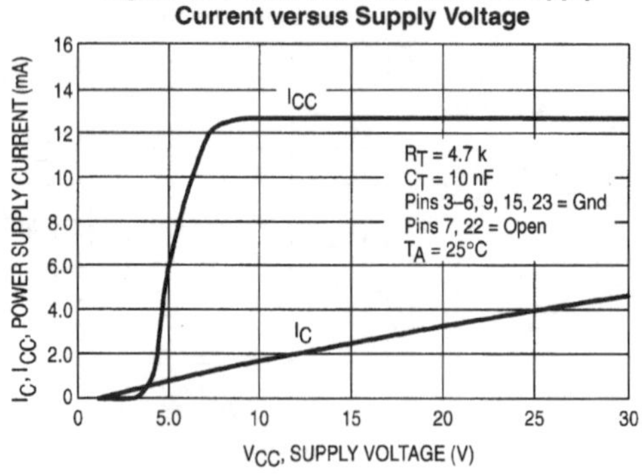

Appendix 8.4.8

MC33035

PIN FUNCTION DESCRIPTION

Pin	Symbol	Description
1, 2, 24	B_T, A_T, C_T	These three open collector Top Drive outputs are designed to drive the external upper power switch transistors.
3	Fwd/Rev	The Forward/Reverse Input is used to change the direction of motor rotation.
4, 5, 6	S_A, S_B, S_C	These three Sensor Inputs control the commutation sequence.
7	Output Enable	A logic high at this input causes the motor to run, while a low causes it to coast.
8	Reference Output	This output provides charging current for the oscillator timing capacitor C_T and a reference for the error amplifier. It may also serve to furnish sensor power.
9	Current Sense Noninverting Input	A 100 mV signal, with respect to Pin 15, at this input terminates output switch conduction during a given oscillator cycle. This pin normally connects to the top side of the current sense resistor.
10	Oscillator	The Oscillator frequency is programmed by the values selected for the timing components, R_T and C_T.
11	Error Amp Noninverting Input	This input is normally connected to the speed set potentiometer.
12	Error Amp Inverting Input	This input is normally connected to the Error Amp Output in open loop applications.
13	Error Amp Out/PWM Input	This pin is available for compensation in closed loop applications.
14	$\overline{\text{Fault}}$ Output	This open collector output is active low during one or more of the following conditions: Invalid Sensor Input code, Enable Input at logic 0, Current Sense Input greater than 100 mV (Pin 9 with respect to Pin 15), Undervoltage Lockout activation, and Thermal Shutdown.
15	Current Sense Inverting Input	Reference pin for internal 100 mV threshold. This pin is normally connected to the bottom side of the current sense resistor.
16	Gnd	This pin supplies a ground for the control circuit and should be referenced back to the power source ground.
17	V_{CC}	This pin is the positive supply of the control IC. The controller is functional over a minimum V_{CC} range of 10 to 30 V.
18	V_C	The high state (V_{OH}) of the Bottom Drive Outputs is set by the voltage applied to this pin. The controller is operational over a minimum V_C range of 10 to 30 V.
19, 20, 21	C_B, B_B, A_B	These three totem pole Bottom Drive Outputs are designed for direct drive of the external bottom power switch transistors.
22	60°/$\overline{120°}$ Select	The electrical state of this pin configures the control circuit operation for either 60° (high state) or 120° (low state) sensor electrical phasing inputs.
23	Brake	A logic low state at this input allows the motor to run, while a high state does not allow motor operation and if operating causes rapid deceleration.

INTRODUCTION

The MC33035 is one of a series of high performance monolithic DC brushless motor controllers produced by Motorola. It contains all of the functions required to implement a full–featured, open loop, three or four phase motor control system. In addition, the controller can be made to operate DC brush motors. Constructed with Bipolar Analog technology, it offers a high degree of performance and ruggedness in hostile industrial environments. The MC33035 contains a rotor position decoder for proper commutation sequencing, a temperature compensated reference capable of supplying a sensor power, a frequency programmable sawtooth oscillator, a fully accessible error amplifier, a pulse width modulator comparator, three open collector top drive outputs, and three high current totem pole bottom driver outputs ideally suited for driving power MOSFETs.

Included in the MC33035 are protective features consisting of undervoltage lockout, cycle–by–cycle current limiting with a selectable time delayed latched shutdown mode, internal thermal shutdown, and a unique fault output that can easily be interfaced to a microprocessor controller.

Typical motor control functions include open loop speed control, forward or reverse rotation, run enable, and dynamic braking. In addition, the MC33035 has a 60°/$\overline{120°}$ select pin which configures the rotor position decoder for either 60° or 120° sensor electrical phasing inputs.

FUNCTIONAL DESCRIPTION

A representative internal block diagram is shown in Figure 19 with various applications shown in Figures 36, 38, 39, 43, 45, and 46. A discussion of the features and function of each of the internal blocks given below is referenced to Figures 19 and 36.

Rotor Position Decoder

An internal rotor position decoder monitors the three sensor inputs (Pins 4, 5, 6) to provide the proper sequencing of the top and bottom drive outputs. The sensor inputs are designed to interface directly with open collector type Hall Effect switches or opto slotted couplers. Internal pull–up resistors are included to minimize the required number of external components. The inputs are TTL compatible, with their thresholds typically at 2.2 V. The MC33035 series is designed to control three phase motors and operate with four of the most common conventions of sensor phasing. A 60°/$\overline{120°}$ Select (Pin 22) is conveniently provided and affords the MC33035 to configure itself to control motors having either 60°, 120°, 240° or 300° electrical sensor phasing. With three sensor inputs there are eight possible input code combinations, six of which are valid rotor positions. The remaining two codes are invalid and are usually caused by an open or shorted sensor line. With six valid input codes, the

Appendix 8.4.9

MC33035

decoder can resolve the motor rotor position to within a window of 60 electrical degrees.

The Forward/Reverse input (Pin 3) is used to change the direction of motor rotation by reversing the voltage across the stator winding. When the input changes state, from high to low with a given sensor input code (for example 100), the enabled top and bottom drive outputs with the same alpha designation are exchanged (A_T to A_B, B_T to B_B, C_T to C_B). In effect, the commutation sequence is reversed and the motor changes directional rotation.

Motor on/off control is accomplished by the Output Enable (Pin 7). When left disconnected, an internal 25 μA current source enables sequencing of the top and bottom drive outputs. When grounded, the top drive outputs turn off and the bottom drives are forced low, causing the motor to coast and the $\overline{\text{Fault}}$ output to activate.

Dynamic motor braking allows an additional margin of safety to be designed into the final product. Braking is accomplished by placing the Brake Input (Pin 23) in a high state. This causes the top drive outputs to turn off and the bottom drives to turn on, shorting the motor–generated back EMF. The brake input has unconditional priority over all other inputs. The internal 40 kΩ pull–up resistor simplifies interfacing with the system safety–switch by insuring brake activation if opened or disconnected. The commutation logic truth table is shown in Figure 20. A four input NOR gate is used to monitor the brake input and the inputs to the three top drive output transistors. Its purpose is to disable braking until the top drive outputs attain a high state. This helps to prevent simultaneous conduction of the the top and bottom power switches. In half wave motor drive applications, the top drive outputs are not required and are normally left disconnected. Under these conditions braking will still be accomplished since the NOR gate senses the base voltage to the top drive output transistors.

Error Amplifier

A high performance, fully compensated error amplifier with access to both inputs and output (Pins 11, 12, 13) is provided to facilitate the implementation of closed loop motor speed control. The amplifier features a typical DC voltage gain of 80 dB, 0.6 MHz gain bandwidth, and a wide input common mode voltage range that extends from ground to V_{ref}. In most open loop speed control applications, the amplifier is configured as a unity gain voltage follower with the noninverting input connected to the speed set voltage source. Additional configurations are shown in Figures 31 through 35.

Oscillator

The frequency of the internal ramp oscillator is programmed by the values selected for timing components R_T and C_T. Capacitor C_T is charged from the Reference Output (Pin 8) through resistor R_T and discharged by an internal discharge transistor. The ramp peak and valley voltages are typically 4.1 V and 1.5 V respectively. To provide a good compromise between audible noise and output switching efficiency, an oscillator frequency in the range of 20 to 30 kHz is recommended. Refer to Figure 1 for component selection.

Figure 19. Representative Block Diagram

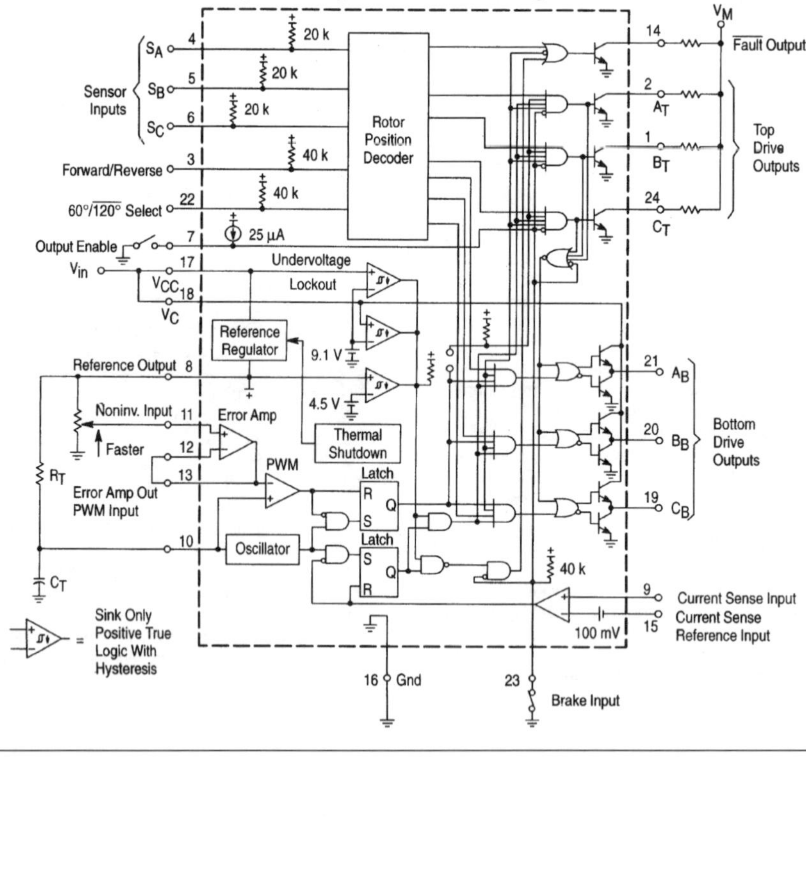

Appendix 8.4.10

MC33035

Figure 20. Three Phase, Six Step Commutation Truth Table (Note 1)

Inputs (Note 2)										Outputs (Note 3)							
Sensor Electrical Phasing (Note 4)										Top Drives			Bottom Drives				
60°			120°														
S_A	S_B	S_C	S_A	S_B	S_C	F/R	Enable	Brake	Current Sense	A_T	B_T	C_T	A_B	B_B	C_B	$\overline{\text{Fault}}$	
1	0	0	1	0	0	1	1	0	0	0	1	1	0	0	1	1	**(Note 5)** F/R = 1
1	1	0	1	1	0	1	1	0	0	1	0	1	0	0	1	1	
1	1	1	0	1	0	1	1	0	0	1	0	1	1	0	0	1	
0	1	1	0	1	1	1	1	0	0	1	1	0	1	0	0	1	
0	0	1	0	0	1	1	1	0	0	1	1	0	0	1	0	1	
0	0	0	1	0	1	1	1	0	0	0	1	1	0	1	0	1	
1	0	0	1	0	0	0	1	0	0	1	1	0	1	0	0	1	**(Note 5)** F/R = 0
1	1	0	1	1	0	0	1	0	0	1	1	0	0	1	0	1	
1	1	1	0	1	0	0	1	0	0	0	1	1	0	1	0	1	
0	1	1	0	1	1	0	1	0	0	0	1	1	0	0	1	1	
0	0	1	0	0	1	0	1	0	0	1	0	1	0	0	1	1	
0	0	0	1	0	1	0	1	0	0	1	0	1	1	0	0	1	
1	0	1	1	1	1	X	X	0	X	1	1	1	0	0	0	0	**(Note 6)** Brake = 0
0	1	0	0	0	0	X	X	0	X	1	1	1	0	0	0	0	
1	0	1	1	1	1	X	X	1	X	1	1	1	1	1	1	0	**(Note 7)** Brake = 1
0	1	0	0	0	0	X	X	1	X	1	1	1	1	1	1	0	
V	V	V	V	V	V	X	1	1	X	1	1	1	1	1	1	1	**(Note 8)**
V	V	V	V	V	V	X	0	1	X	1	1	1	1	1	1	0	**(Note 9)**
V	V	V	V	V	V	X	0	0	X	1	1	1	0	0	0	0	**(Note 10)**
V	V	V	V	V	V	X	1	0	1	1	1	1	0	0	0	0	**(Note 11)**

NOTES:
1. V = Any one of six valid sensor or drive combinations X = Don't care.
2. The digital inputs (Pins 3, 4, 5, 6, 7, 22, 23) are all TTL compatible. The current sense input (Pin 9) has a 100 mV threshold with respect to Pin 15. A logic 0 for this input is defined as < 85 mV, and a logic 1 is > 115 mV.
3. The fault and top drive outputs are open collector design and active in the low (0) state.
4. With 60°/120° select (Pin 22) in the high (1) state, configuration is for 60° sensor electrical phasing inputs. With Pin 22 in low (0) state, configuration is for 120° sensor electrical phasing inputs.
5. Valid 60° or 120° sensor combinations for corresponding valid top and bottom drive outputs.
6. Invalid sensor inputs with brake = 0; All top and bottom drives off, Fault low.
7. Invalid sensor inputs with brake = 1; All top drives off, all bottom drives on, Fault low.
8. Valid 60° or 120° sensor inputs with brake = 1; All top drives off, all bottom drives on, Fault high.
9. Valid sensor inputs with brake = 1 and enable = 0; All top drives off, all bottom drives on, Fault low.
10. Valid sensor inputs with brake = 0 and enable = 0, All top and bottom drives off, Fault low.
11. All bottom drives off, Fault low.

Pulse Width Modulator

The use of pulse width modulation provides an energy efficient method of controlling the motor speed by varying the average voltage applied to each stator winding during the commutation sequence. As C_T discharges, the oscillator sets both latches, allowing conduction of the top and bottom drive outputs. The PWM comparator resets the upper latch, terminating the bottom drive output conduction when the positive–going ramp of C_T becomes greater than the error amplifier output. The pulse width modulator timing diagram is shown in Figure 21. Pulse width modulation for speed control appears only at the bottom drive outputs.

Current Limit

Continuous operation of a motor that is severely over–loaded results in overheating and eventual failure. This destructive condition can best be prevented with the use of cycle–by–cycle current limiting. That is, each on–cycle is treated as a separate event. Cycle–by–cycle current limiting is accomplished by monitoring the stator current build–up each time an output switch conducts, and upon sensing an over current condition, immediately turning off the switch and holding it off for the remaining duration of oscillator ramp–up period. The stator current is converted to a voltage by inserting a ground–referenced sense resistor R_S (Figure 36) in series with the three bottom switch transistors (Q_4, Q_5, Q_6). The voltage developed across the sense resistor is monitored by the Current Sense Input (Pins 9 and 15), and compared to the internal 100 mV reference. The current sense comparator inputs have an input common mode range of approximately 3.0 V. If the 100 mV current sense threshold is exceeded, the comparator resets the lower sense latch and terminates output switch conduction. The value for the current sense resistor is:

$$R_S = \frac{0.1}{I_{stator(max)}}$$

The $\overline{\text{Fault}}$ output activates during an over current condition. The dual–latch PWM configuration ensures that only one single output conduction pulse occurs during any given oscillator cycle, whether terminated by the output of the error amp or the current limit comparator.

Figure 21. Pulse Width Modulator Timing Diagram

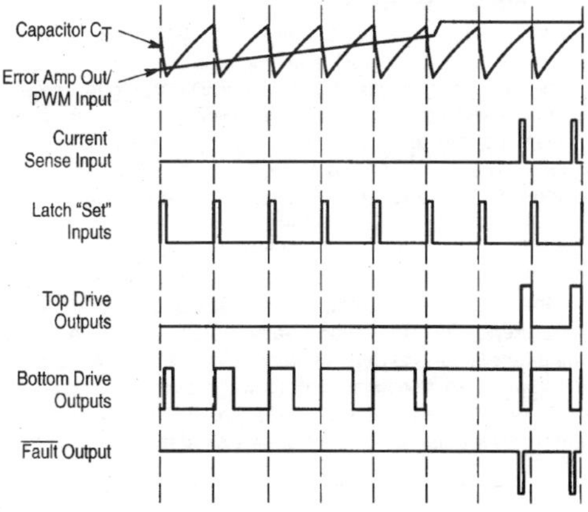

Appendix 8.4.11

MC33035

Reference

The on–chip 6.25 V regulator (Pin 8) provides charging current for the oscillator timing capacitor, a reference for the error amplifier, and can supply 20 mA of current suitable for directly powering sensors in low voltage applications. In higher voltage applications, it may become necessary to transfer the power dissipated by the regulator off the IC. This is easily accomplished with the addition of an external pass transistor as shown in Figure 22. A 6.25 V reference level was chosen to allow implementation of the simpler NPN circuit, where $V_{ref} - V_{BE}$ exceeds the minimum voltage required by Hall Effect sensors over temperature. With proper transistor selection and adequate heatsinking, up to one amp of load current can be obtained.

Figure 22. Reference Output Buffers

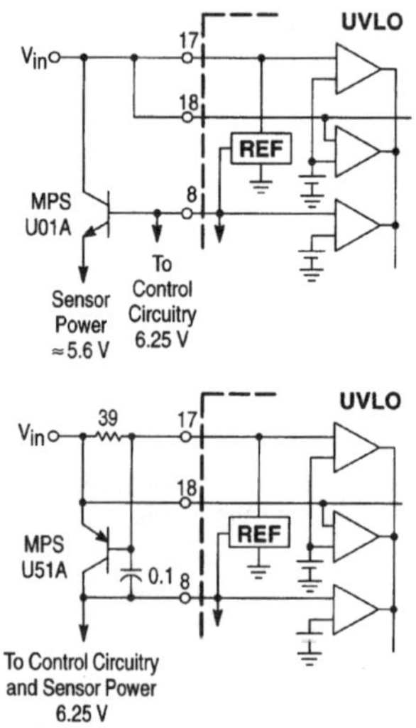

The NPN circuit is recommended for powering Hall or opto sensors, where the output voltage temperature coefficient is not critical. The PNP circuit is slightly more complex, but is also more accurate over temperature. Neither circuit has current limiting.

Undervoltage Lockout

A triple Undervoltage Lockout has been incorporated to prevent damage to the IC and the external power switch transistors. Under low power supply conditions, it guarantees that the IC and sensors are fully functional, and that there is sufficient bottom drive output voltage. The positive power supplies to the IC (V_{CC}) and the bottom drives (V_C) are each monitored by separate comparators that have their thresholds at 9.1 V. This level ensures sufficient gate drive necessary to attain low $R_{DS(on)}$ when driving standard power MOSFET devices. When directly powering the Hall sensors from the reference, improper sensor operation can result if the reference output voltage falls below 4.5 V. A third comparator is used to detect this condition. If one or more of the comparators detects an undervoltage condition, the $\overline{\text{Fault}}$ Output is activated, the top drives are turned off and the bottom drive outputs are held in a low state. Each of the comparators contain hysteresis to prevent oscillations when crossing their respective thresholds.

$\overline{\text{Fault}}$ Output

The open collector $\overline{\text{Fault}}$ Output (Pin 14) was designed to provide diagnostic information in the event of a system malfunction. It has a sink current capability of 16 mA and can directly drive a light emitting diode for visual indication. Additionally, it is easily interfaced with TTL/CMOS logic for use in a microprocessor controlled system. The $\overline{\text{Fault}}$ Output is active low when one or more of the following conditions occur:

1) Invalid Sensor Input code
2) Output Enable at logic [0]
3) Current Sense Input greater than 100 mV
4) Undervoltage Lockout, activation of one or more of the comparators
5) Thermal Shutdown, maximum junction temperature being exceeded

This unique output can also be used to distinguish between motor start–up or sustained operation in an overloaded condition. With the addition of an RC network between the $\overline{\text{Fault}}$ Output and the enable input, it is possible to create a time–delayed latched shutdown for overcurrent. The added circuitry shown in Figure 23 makes easy starting of motor systems which have high inertial loads by providing additional starting torque, while still preserving overcurrent protection. This task is accomplished by setting the current limit to a higher than nominal value for a predetermined time. During an excessively long overcurrent condition, capacitor C_{DLY} will charge, causing the enable input to cross its threshold to a low state. A latch is then formed by the positive feedback loop from the $\overline{\text{Fault}}$ Output to the Output Enable. Once set, by the Current Sense Input, it can only be reset by shorting C_{DLY} or cycling the power supplies.

Drive Outputs

The three top drive outputs (Pins 1, 2, 24) are open collector NPN transistors capable of sinking 50 mA with a minimum breakdown of 30 V. Interfacing into higher voltage applications is easily accomplished with the circuits shown in Figures 24 and 25.

The three totem pole bottom drive outputs (Pins 19, 20, 21) are particularly suited for direct drive of N–Channel MOSFETs or NPN bipolar transistors (Figures 26, 27, 28 and 29). Each output is capable of sourcing and sinking up to 100 mA. Power for the bottom drives is supplied from V_C (Pin 18). This separate supply input allows the designer added flexibility in tailoring the drive voltage, independent of V_{CC}. A zener clamp should be connected to this input when driving power MOSFETs in systems where V_{CC} is greater than 20 V so as to prevent rupture of the MOSFET gates.

The control circuitry ground (Pin 16) and current sense inverting input (Pin 15) must return on separate paths to the central input source ground.

Thermal Shutdown

Internal thermal shutdown circuitry is provided to protect the IC in the event the maximum junction temperature is exceeded. When activated, typically at 170°C, the IC acts as though the Output Enable was grounded.

Appendix 8.4.12

MC33035

Figure 23. Timed Delayed Latched Over Current Shutdown

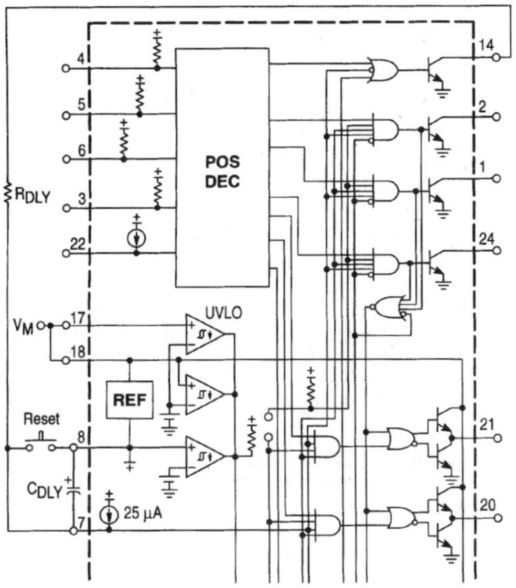

$$t_{DLY} \approx R_{DLY}\, C_{DLY} \ln\left(\frac{V_{ref} - (I_{IL}\ \text{enable}\ R_{DLY})}{V_{th}\ \text{enable} - (I_{IL}\ \text{enable}\ R_{DLY})}\right)$$

$$\approx R_{DLY}\, C_{DLY} \ln\left(\frac{6.25 - (20 \times 10^{-6}\, R_{DLY})}{1.4 - (20 \times 10^{-6}\, R_{DLY})}\right)$$

Figure 24. High Voltage Interface with NPN Power Transistors

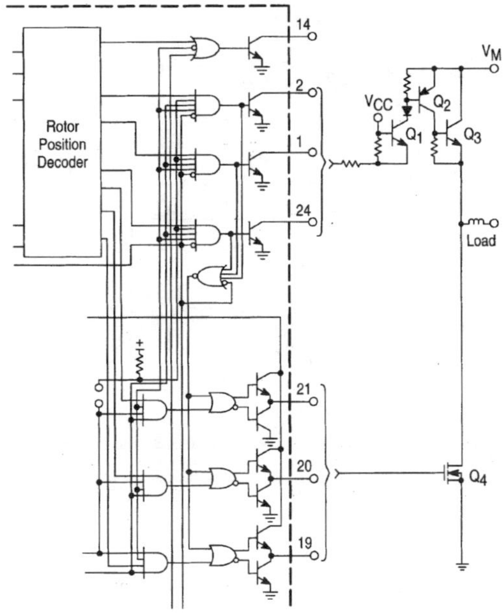

Transistor Q_1 is a common base stage used to level shift from V_{CC} to the high motor voltage, V_M. The collector diode is required if V_{CC} is present while V_M is low.

Figure 25. High Voltage Interface with N-Channel Power MOSFETs

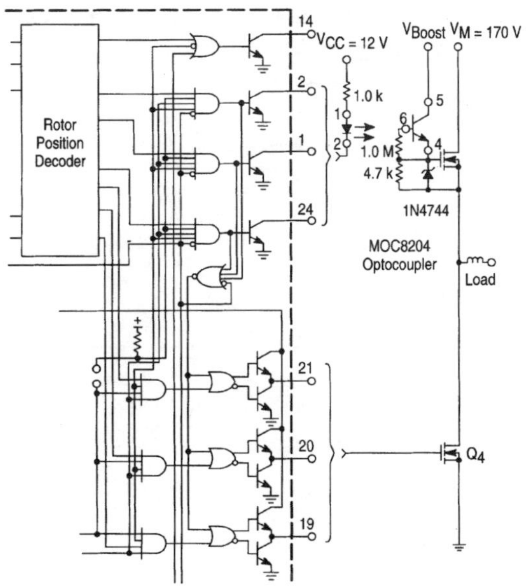

Figure 26. Current Waveform Spike Suppression

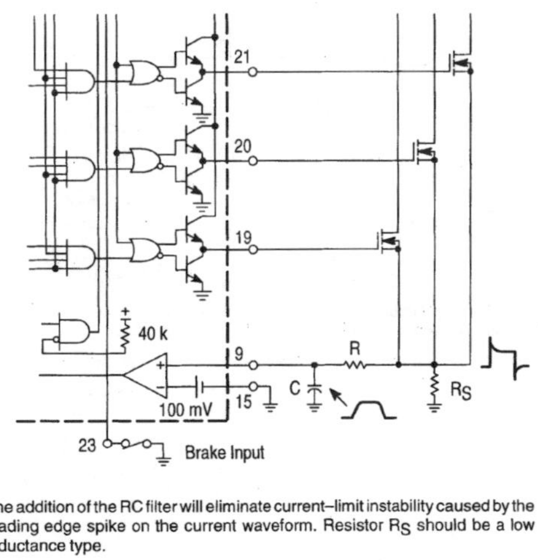

The addition of the RC filter will eliminate current-limit instability caused by the leading edge spike on the current waveform. Resistor R_S should be a low inductance type.

Appendix 8.4.13

MC33035

Figure 27. MOSFET Drive Precautions

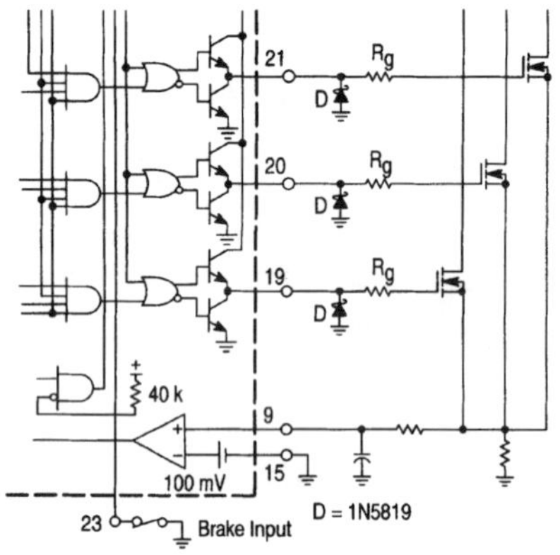

Series gate resistor R_g will dampen any high frequency oscillations caused by the MOSFET input capacitance and any series wiring induction in the gate–source circuit. Diode D is required if the negative current into the Bottom Drive Outputs exceeds 50 mA.

Figure 28. Bipolar Transistor Drive

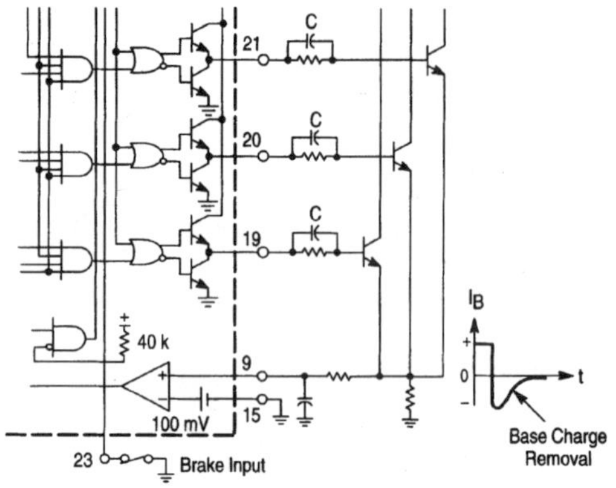

The totem–pole output can furnish negative base current for enhanced transistor turn–off, with the addition of capacitor C.

Figure 29. Current Sensing Power MOSFETs

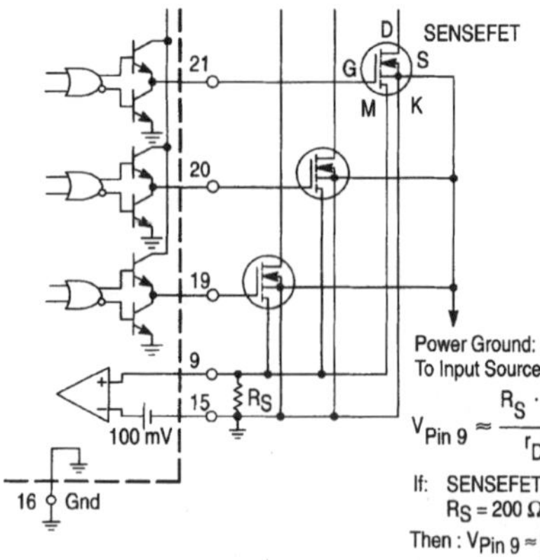

$$V_{Pin\ 9} \approx \frac{R_S \cdot I_{pk} \cdot R_{DS(on)}}{r_{DM(on)} + R_S}$$

If: SENSEFET = MPT10N10M
$R_S = 200\ \Omega$, 1/4 W

Then : $V_{Pin\ 9} \approx 0.75\ I_{pk}$

Control Circuitry Ground (Pin 16) and Current Sense Inverting Input (Pin 15) must return on separate paths to the Central Input Source Ground.

Virtually lossless current sensing can be achieved with the implementation of SENSEFET power switches.

Figure 30. High Voltage Boost Supply

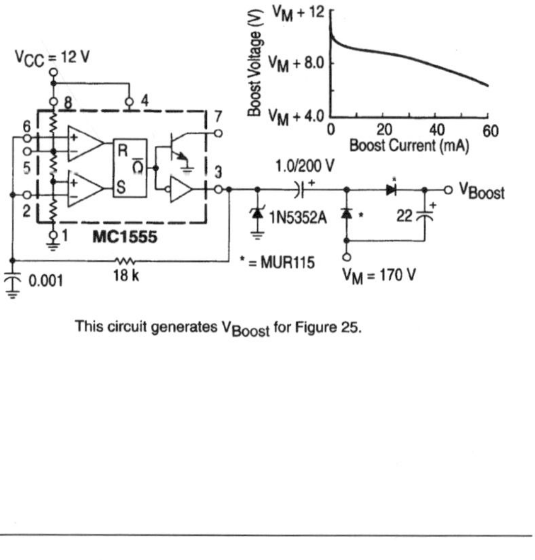

This circuit generates V_{Boost} for Figure 25.

Appendix 8.4.14

MC33035

Figure 31. Differential Input Speed Controller

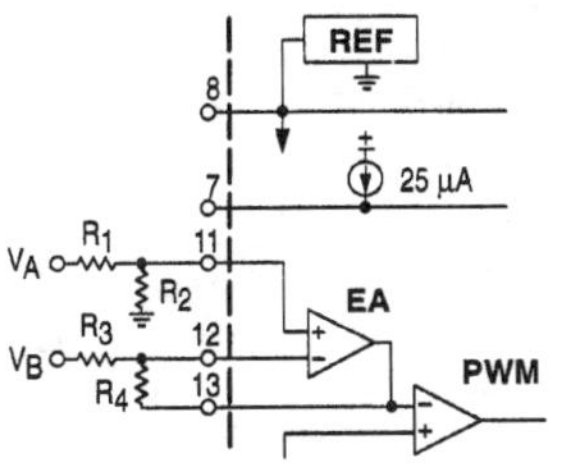

$$V_{Pin\ 13} = V_A \left(\frac{R_3 + R_4}{R_1 + R_2}\right) \frac{R_2}{R_3} - \left(\frac{R_4}{R_3} V_B\right)$$

Figure 32. Controlled Acceleration/Deceleration

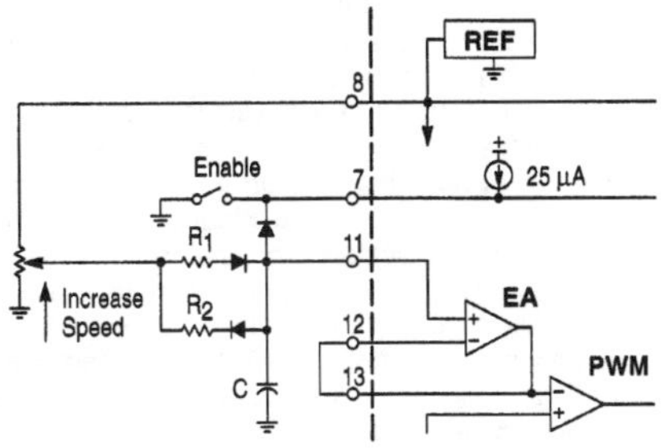

Resistor R_1 with capacitor C sets the acceleration time constant while R_2 controls the deceleration. The values of R_1 and R_2 should be at least ten times greater than the speed set potentiometer to minimize time constant variations with different speed settings.

Figure 33. Digital Speed Controller

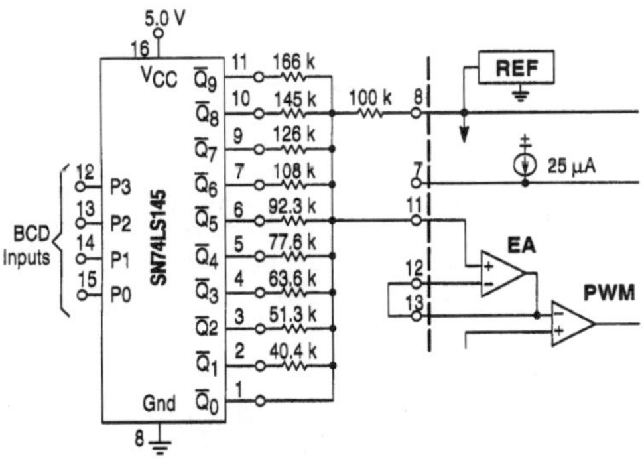

The SN74LS145 is an open collector BCD to One of Ten decoder. When connected as shown, input codes 0000 through 1001 steps the PWM in increments of approximately 10% from 0 to 90% on–time. Input codes 1010 through 1111 will produce 100% on–time or full motor speed.

Figure 34. Closed Loop Speed Control

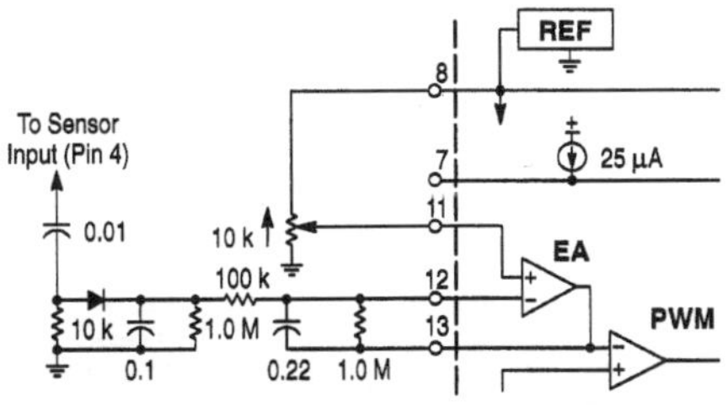

The rotor position sensors can be used as a tachometer. By differentiating the positive–going edges and then integrating them over time, a voltage proportional to speed can be generated. The error amp compares this voltage to that of the speed set to control the PWM.

Figure 35. Closed Loop Temperature Control

$$V_{Pin\ 3} = V_{ref} \left(\frac{R_3 + R_4}{R_1 + R_2}\right) \frac{R_2}{R_3} - \left(\frac{R_4}{R_3} V_B\right)$$

$$V_B = \frac{V_{ref}}{\left(\frac{R_5}{R_6} + 1\right)}$$

$$R_3 >> R_5 \parallel R_6$$

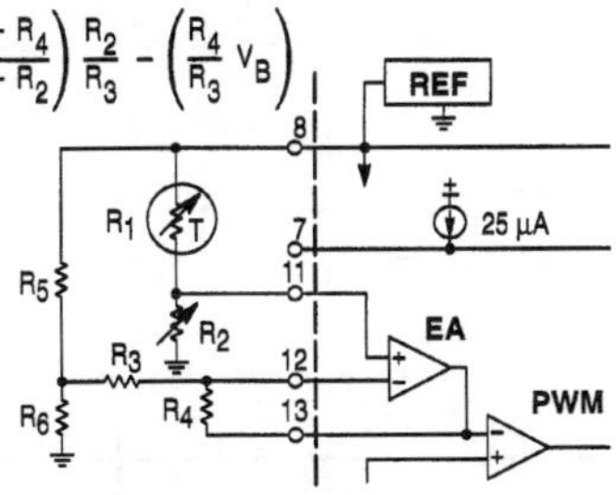

This circuit can control the speed of a cooling fan proportional to the difference between the sensor and set temperatures. The control loop is closed as the forced air cools the NTC thermistor. For controlled heating applications, exchange the positions of R_1 and R_2.

Appendix 8.4.15

MC33035

SYSTEM APPLICATIONS

Three Phase Motor Commutation

The three phase application shown in Figure 36 is a full–featured open loop motor controller with full wave, six step drive. The upper power switch transistors are Darlingtons while the lower devices are power MOSFETs. Each of these devices contains an internal parasitic catch diode that is used to return the stator inductive energy back to the power supply. The outputs are capable of driving a delta or wye connected stator, and a grounded neutral wye if split supplies are used. At any given rotor position, only one top and one bottom power switch (of different totem poles) is enabled. This configuration switches both ends of the stator winding from supply to ground which causes the current flow to be bidirectional or full wave. A leading edge spike is usually present on the current waveform and can cause a current–limit instability. The spike can be eliminated by adding an RC filter in series with the Current Sense Input. Using a low inductance type resistor for R_S will also aid in spike reduction. Care must be taken in the selection of the bottom power switch transistors so that the current during braking does not exceed the device rating. During braking, the peak current generated is limited only by the series resistance of the conducting bottom switch and winding.

$$I_{peak} = \frac{V_M + EMF}{R_{switch} + R_{winding}}$$

If the motor is running at maximum speed with no load, the generated back EMF can be as high as the supply voltage, and at the onset of braking, the peak current may approach twice the motor stall current. Figure 37 shows the commutation waveforms over two electrical cycles. The first cycle (0° to 360°) depicts motor operation at full speed while the second cycle (360° to 720°) shows a reduced speed with about 50% pulse width modulation. The current waveforms reflect a constant torque load and are shown synchronous to the commutation frequency for clarity.

Figure 36. Three Phase, Six Step, Full Wave Motor Controller

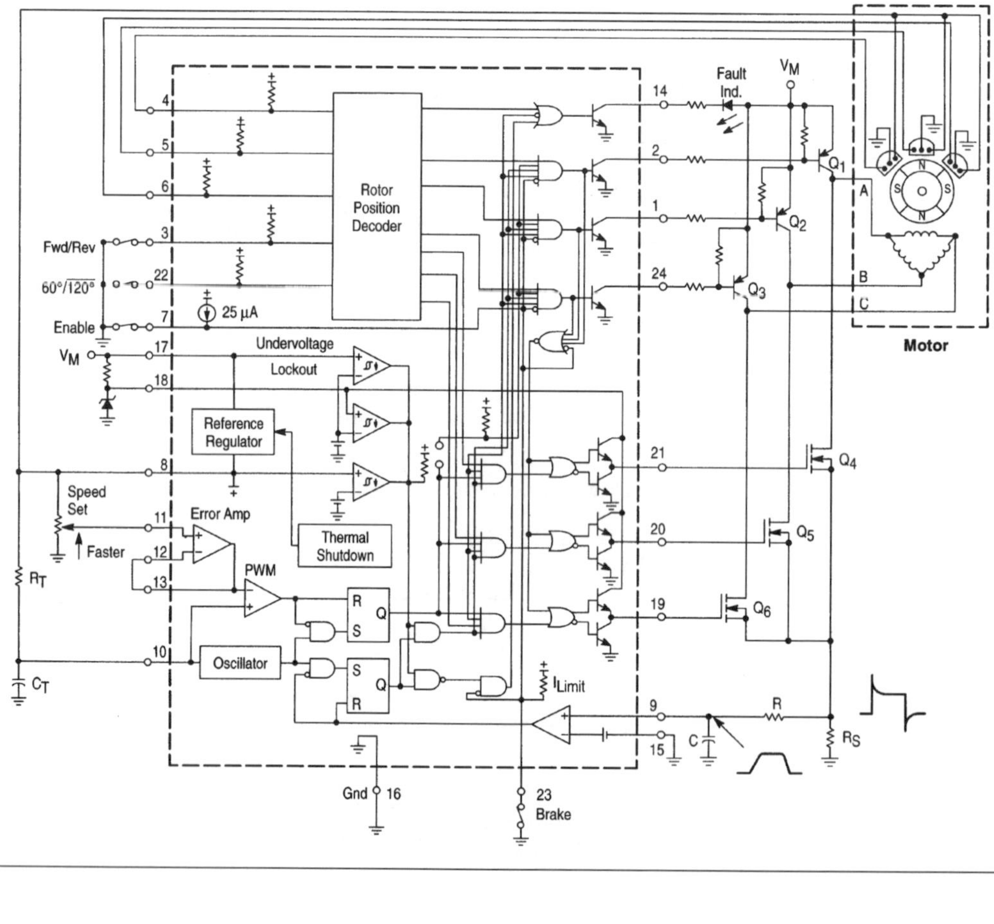

Appendix 8.4.16

MC33035

Figure 37. Three Phase, Six Step, Full Wave Commutation Waveforms

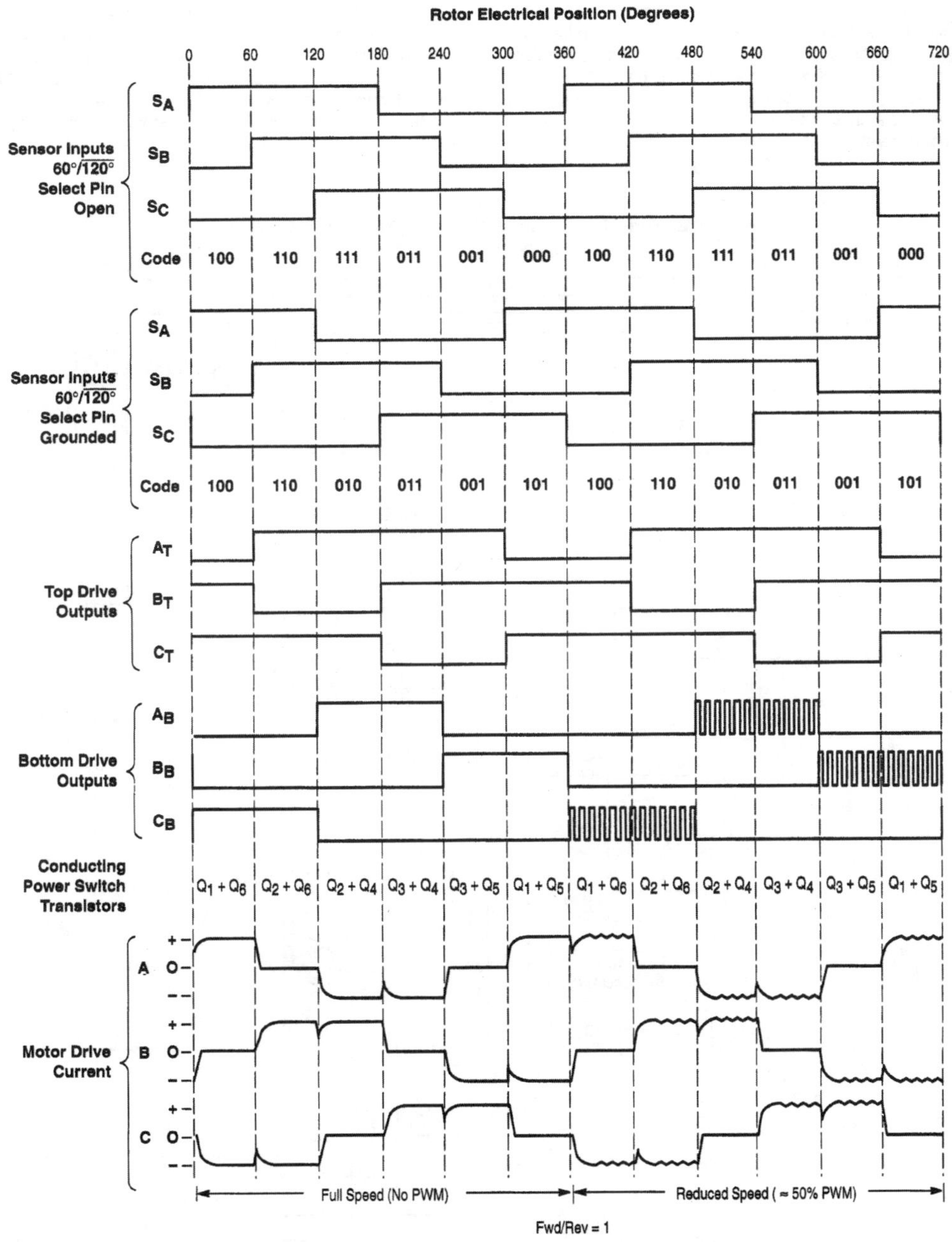

Appendix 8.4.17

MC33035

Figure 38 shows a three phase, three step, half wave motor controller. This configuration is ideally suited for automotive and other low voltage applications since there is only one power switch voltage drop in series with a given stator winding. Current flow is unidirectional or half wave because only one end of each winding is switched. Continuous braking with the typical half wave arrangement presents a motor overheating problem since stator current is limited only by the winding resistance. This is due to the lack of upper power switch transistors, as in the full wave circuit, used to disconnect the windings from the supply voltage V_M. A unique solution is to provide braking until the motor stops and then turn off the bottom drives. This can be accomplished by using the Fault Output in conjunction with the Output Enable as an over current timer. Components R_{DLY} and C_{DLY} are selected to give the motor sufficient time to stop before latching the Output Enable and the top drive AND gates low. When enabling the motor, the brake switch is closed and the PNP transistor (along with resistors R_1 and R_{DLY}) are used to reset the latch by discharging C_{DLY}. The stator flyback voltage is clamped by a single zener and three diodes.

Figure 38. Three Phase, Three Step, Half Wave Motor Controller

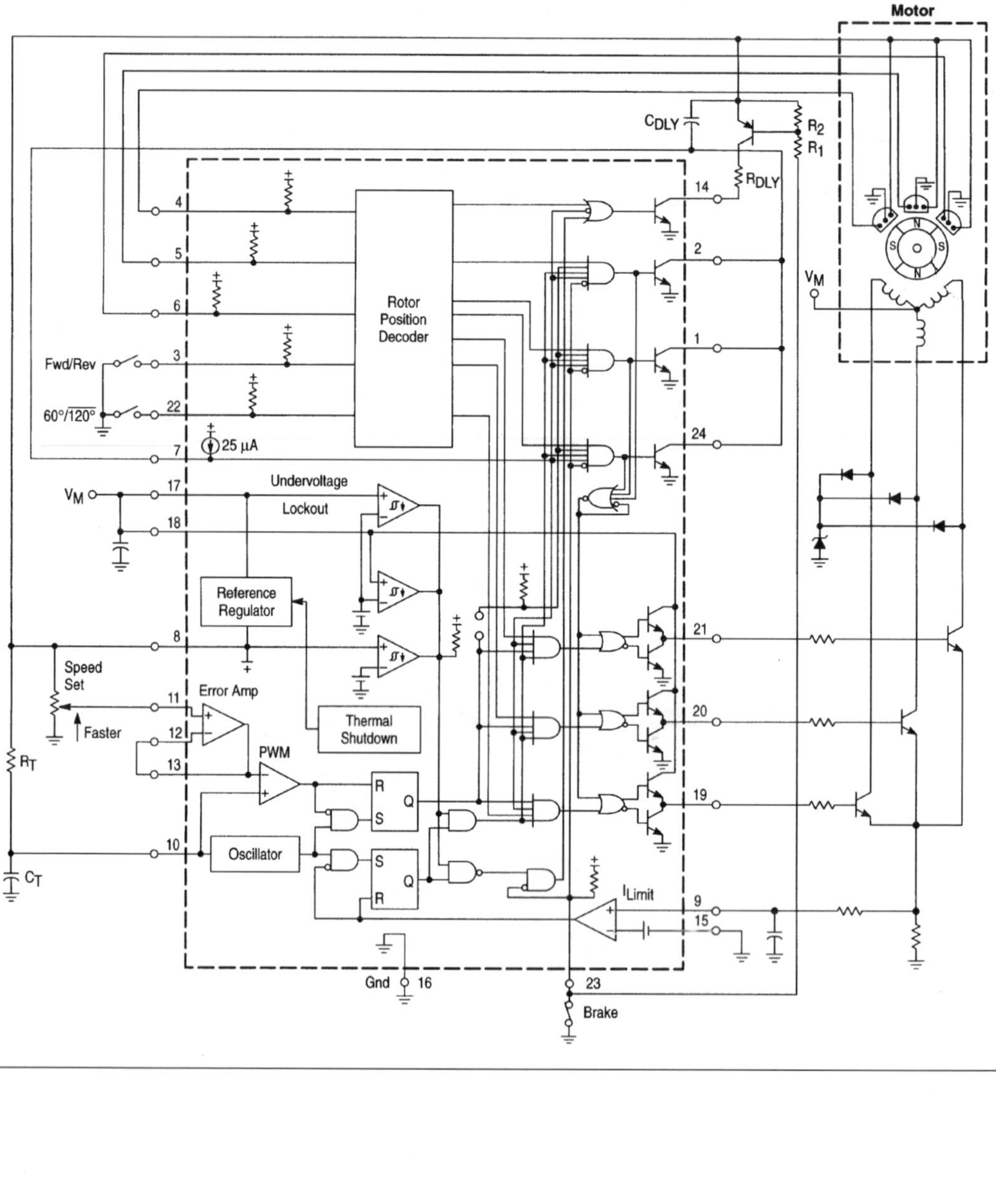

Appendix 8.4.18

MC33035

Three Phase Closed Loop Controller

The MC33035, by itself, is only capable of open loop motor speed control. For closed loop motor speed control, the MC33035 requires an input voltage proportional to the motor speed. Traditionally, this has been accomplished by means of a tachometer to generate the motor speed feedback voltage. Figure 39 shows an application whereby an MC33039, powered from the 6.25 V reference (Pin 8) of the MC33035, is used to generate the required feedback voltage without the need of a costly tachometer. The same Hall sensor signals used by the MC33035 for rotor position decoding are utilized by the MC33039. Every positive or negative going transition of the Hall sensor signals on any of the sensor lines causes the MC33039 to produce an output pulse of defined amplitude and time duration, as determined by the external resistor R_1 and capacitor C_1. The output train of pulses at Pin 5 of the MC33039 are integrated by the error amplifier of the MC33035 configured as an integrator to produce a DC voltage level which is proportional to the motor speed. This speed proportional voltage establishes the PWM reference level at Pin 13 of the MC33035 motor controller and closes the feedback loop. The MC33035 outputs drive a TMOS power MOSFET 3–phase bridge. High currents can be expected during conditions of start–up, breaking, and change of direction of the motor.

The system shown in Figure 39 is designed for a motor having 120/240 degrees Hall sensor electrical phasing. The system can easily be modified to accommodate 60/300 degree Hall sensor electrical phasing by removing the jumper (J_2) at Pin 22 of the MC33035.

Figure 39. Closed Loop Brushless DC Motor Control Using The MC33035 and MC33039

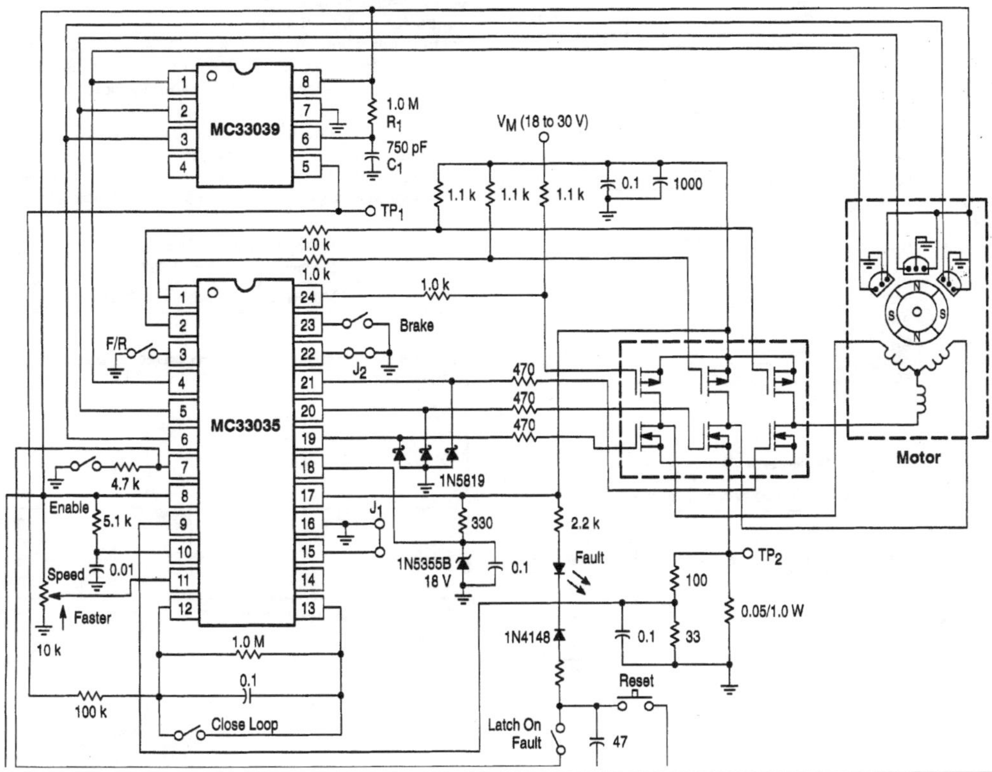

Appendix 8.4.19

MC33035

Sensor Phasing Comparison

There are four conventions used to establish the relative phasing of the sensor signals in three phase motors. With six step drive, an input signal change must occur every 60 electrical degrees; however, the relative signal phasing is dependent upon the mechanical sensor placement. A comparison of the conventions in electrical degrees is shown in Figure 40. From the sensor phasing table in Figure 41, note that the order of input codes for 60° phasing is the reverse of 300°. This means the MC33035, when configured for 60° sensor electrical phasing, will operate a motor with either 60° or 300° sensor electrical phasing, but resulting in opposite directions of rotation. The same is true for the part when it is configured for 120° sensor electrical phasing; the motor will operate equally, but will result in opposite directions of rotation for 120° for 240° conventions.

Figure 40. Sensor Phasing Comparison

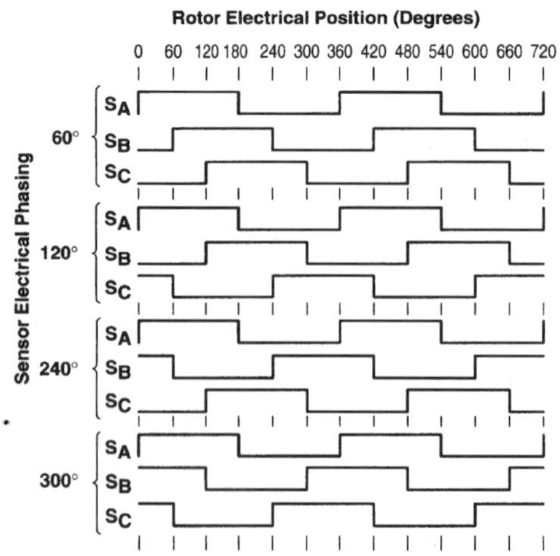

Figure 41. Sensor Phasing Table

Sensor Electrical Phasing (Degrees)											
60°			120°			240°			300°		
S_A	S_B	S_C	S_A	S_B	S_C	S_A	S_B	S_C	S_A	S_B	S_C
1	0	0	1	0	1	1	1	0	1	1	1
1	1	0	1	0	0	1	0	0	1	1	0
1	1	1	1	1	0	1	0	1	1	0	0
0	1	1	0	1	0	0	0	1	0	0	0
0	0	1	0	1	1	0	1	1	0	0	1
0	0	0	0	0	1	0	1	0	0	1	1

In this data sheet, the rotor position is always given in electrical degrees since the mechanical position is a function of the number of rotating magnetic poles. The relationship between the electrical and mechanical position is:

$$\text{Electrical Degrees} = \text{Mechanical Degrees}\left(\frac{\#\text{Rotor Poles}}{2}\right)$$

An increase in the number of magnetic poles causes more electrical revolutions for a given mechanical revolution. General purpose three phase motors typically contain a four pole rotor which yields two electrical revolutions for one mechanical.

Two and Four Phase Motor Commutation

The MC33035 is also capable of providing a four step output that can be used to drive two or four phase motors. The truth table in Figure 42 shows that by connecting sensor inputs S_B and S_C together, it is possible to truncate the number of drive output states from six to four. The output power switches are connected to B_T, C_T, B_B, and C_B. Figure 43 shows a four phase, four step, full wave motor control application. Power switch transistors Q_1 through Q_8 are Darlington type, each with an internal parasitic catch diode. With four step drive, only two rotor position sensors spaced at 90 electrical degrees are required. The commutation waveforms are shown in Figure 44.

Figure 45 shows a four phase, four step, half wave motor controller. It has the same features as the circuit in Figure 38, except for the deletion of speed control and braking.

Figure 42. Two and Four Phase, Four Step, Commutation Truth Table

MC33035 (60°/$\overline{120°}$ Select Pin Open)						
Inputs			Outputs			
Sensor Electrical Spacing* = 90°			Top Drives		Bottom Drives	
S_A	S_B	F/R	B_T	C_T	B_B	C_B
1	0	1	1	1	0	1
1	1	1	0	1	0	0
0	1	1	1	0	0	0
0	0	1	1	1	1	0
1	0	0	1	0	0	0
1	1	0	1	1	1	0
0	1	0	1	1	0	1
0	0	0	0	1	0	0

*With MC33035 sensor input S_B connected to S_C.

Appendix 8.4.20

MC33035

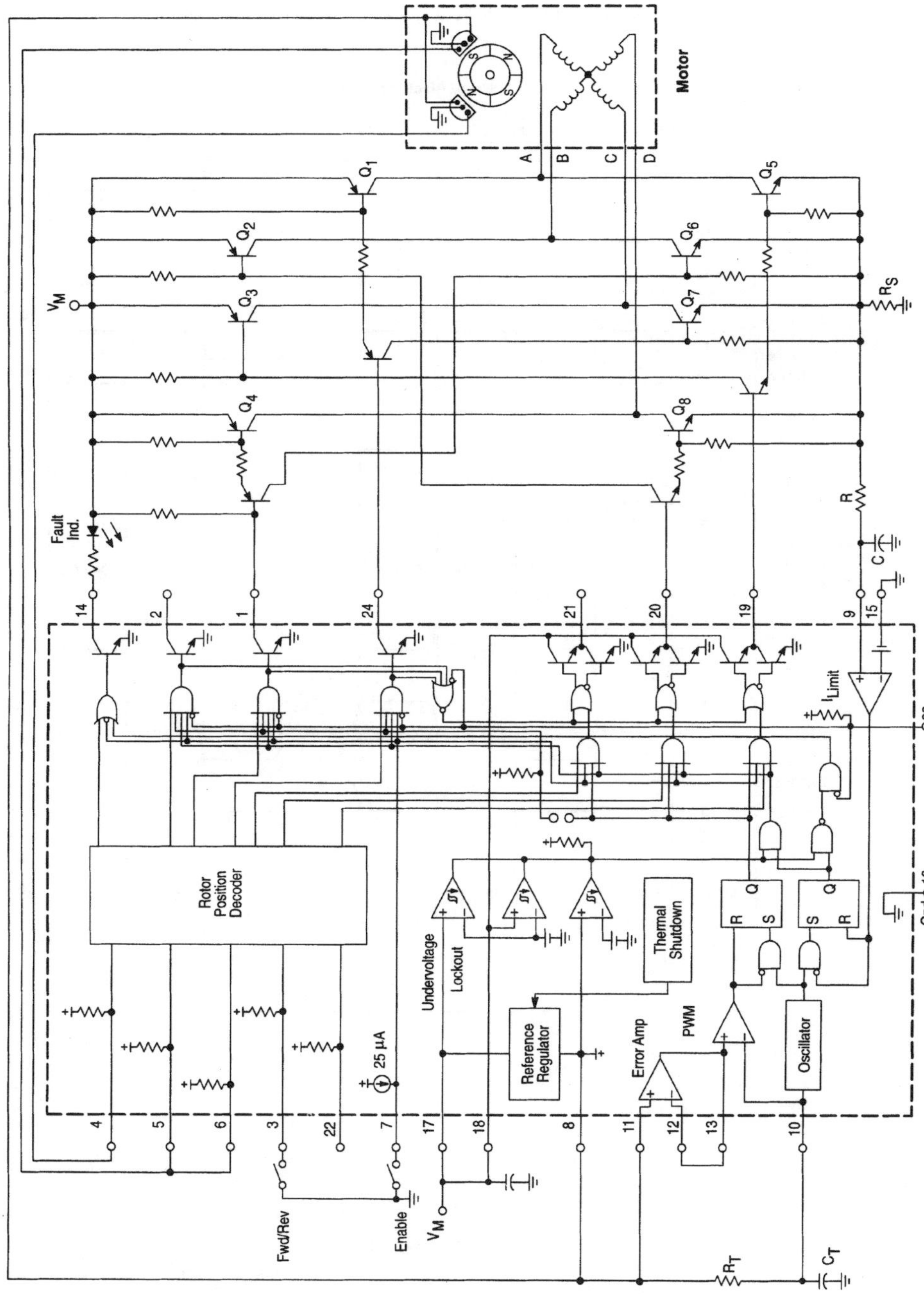

Figure 43. Four Phase, Four Step, Full Wave Motor Controller

Appendix 8.4.21

MC33035

Figure 44. Four Phase, Four Step, Full Wave Motor Controller

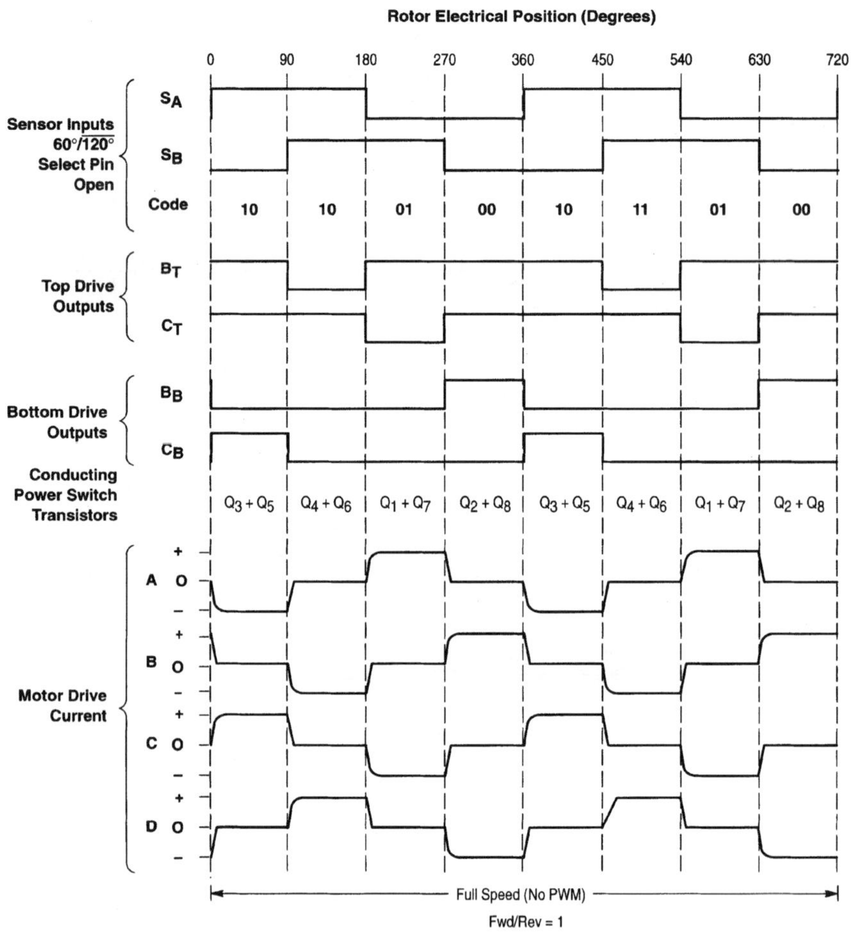

Appendix 8.4.22

MC33035

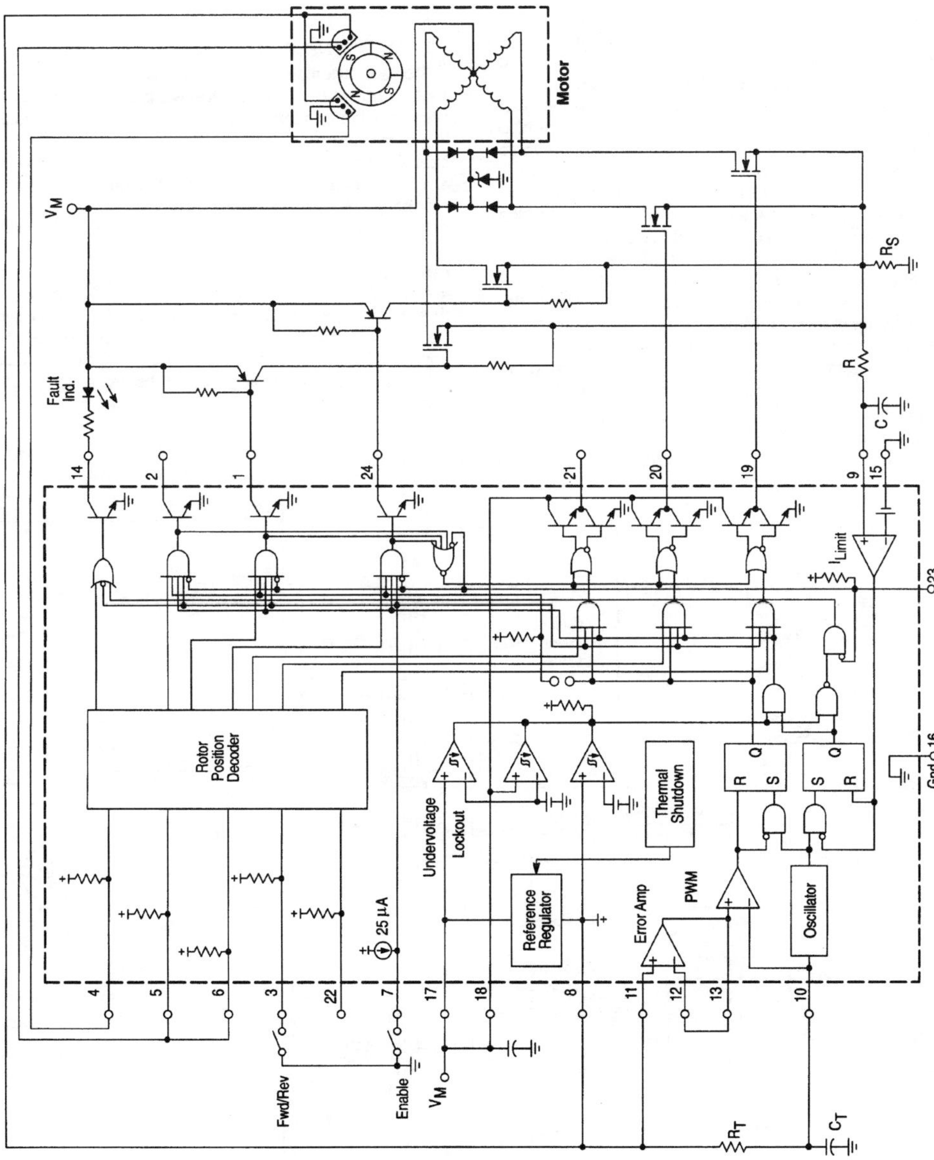

Figure 45. Four Phase, Four Step, Half Wave Motor Controller

Appendix 8.4.23

MC33035

Brush Motor Control

Though the MC33035 was designed to control brushless DC motors, it may also be used to control DC brush type motors. Figure 46 shows an application of the MC33035 driving a MOSFET H–bridge affording minimal parts count to operate a brush–type motor. Key to the operation is the input sensor code [100] which produces a top–left (Q_1) and a bottom–right (Q_3) drive when the controller's forward/reverse pin is at logic [1]; top–right (Q_4), bottom–left (Q_2) drive is realized when the Forward/Reverse pin is at logic [0]. This code supports the requirements necessary for H–bridge drive accomplishing both direction and speed control.

The controller functions in a normal manner with a pulse width modulated frequency of approximately 25 kHz. Motor speed is controlled by adjusting the voltage presented to the noninverting input of the error amplifier establishing the PWM's slice or reference level. Cycle–by–cycle current limiting of the motor current is accomplished by sensing the voltage (100 mV) across the R_S resistor to ground of the H–bridge motor current. The over current sense circuit makes it possible to reverse the direction of the motor, using the normal forward/reverse switch, on the fly and not have to completely stop before reversing.

LAYOUT CONSIDERATIONS

Do not attempt to construct any of the brushless motor control circuits on wire–wrap or plug–in prototype boards. High frequency printed circuit layout techniques are imperative to prevent pulse jitter. This is usually caused by excessive noise pick–up imposed on the current sense or error amp inputs. The printed circuit layout should contain a ground plane with low current signal and high drive and output buffer grounds returning on separate paths back to the power supply input filter capacitor V_M. Ceramic bypass capacitors (0.1 μF) connected
close to the integrated circuit at V_{CC}, V_C, V_{ref} and the error amp noninverting input may be required depending upon circuit layout. This provides a low impedance path for filtering any high frequency noise. All high current loops should be kept as short as possible using heavy copper runs to minimize radiated EMI.

Figure 46. H–Bridge Brush–Type Controller

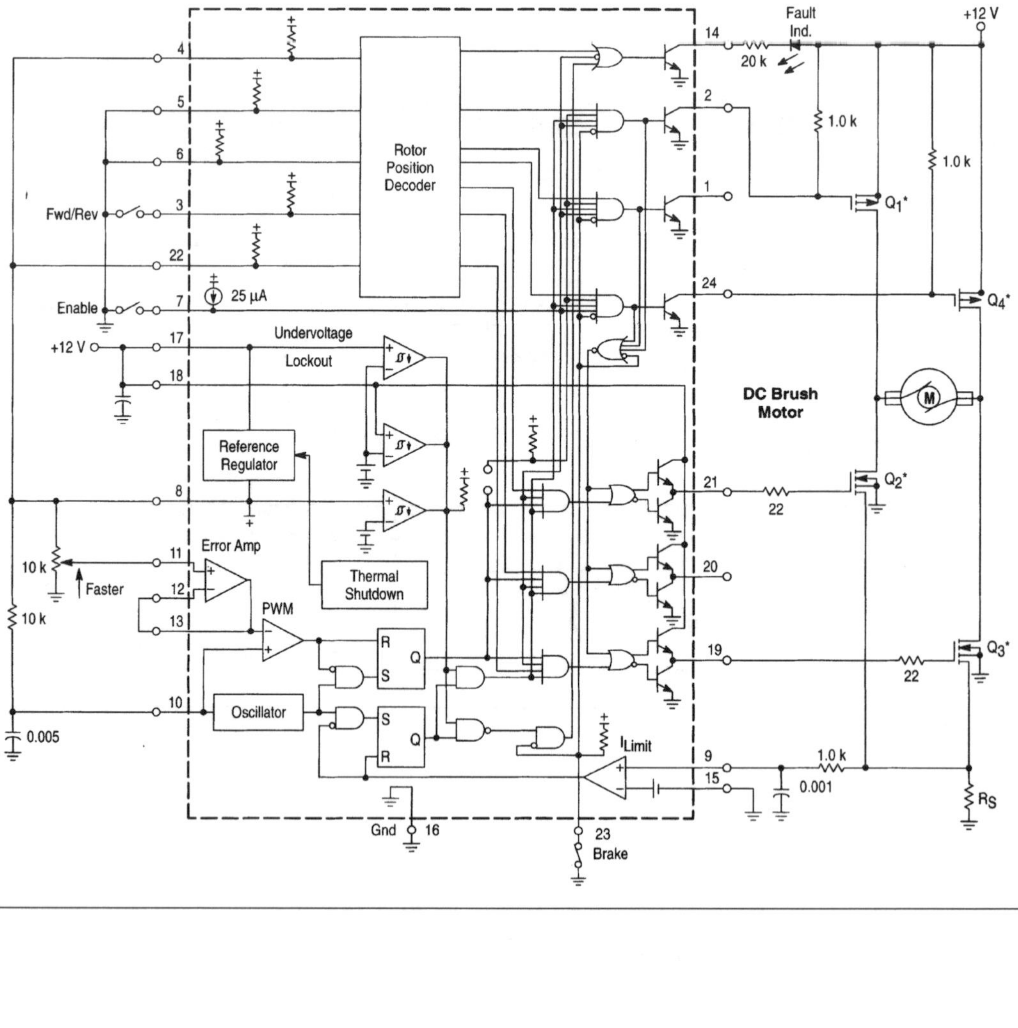

Index

A

B

C

D

E

F

G

H

I

L

M

N

O

P

Q

R

S

T

W

Z